Categorical, Combinatorial and Geometric Representation Theory and Related Topics

Proceedings of Symposia in

PURE MATHEMATICS

Volume 108

Categorical, Combinatorial and Geometric Representation Theory and Related Topics

2015-2021 Southeastern Lie Theory Workshop Series

Algebraic and Combinatorial Representation Theory
October 9–11, 2015, North Carolina State University,
Raleigh, North Carolina

Algebraic Groups, Quantum Groups and Geometry
May 23–27, 2016, University of Virginia, Charlottesville, Virginia

Representation Theory of Algebraic Groups and Lie Superalgebras
June 10–12, 2018, University of Georgia, Athens, Georgia

Geometric and Categorical Representation Theory
May 13–15, 2019, Louisiana State University, Baton Rouge, Louisiana

Lie Algebras, Lie Groups and Applications
October 15–17, 2021, College of Charleston,
Charleston, South Carolina

Pramod N. Achar
Kailash C. Misra
Daniel K. Nakano
Editors

AMERICAN MATHEMATICAL SOCIETY
Providence, Rhode Island

2020 *Mathematics Subject Classification.* Primary 17B05, 17B10, 17B37, 17B55, 17B65, 18G10, 18M20, 20C11, 20G05, 20G43.

Library of Congress Cataloging-in-Publication Data

Names: Achar, Pramod N., 1976– editor. | Misra, Kailash C., 1954– editor. | Nakano, Daniel K. (Daniel Ken), 1964– editor.

Title: Categorical, combinatorial and geometric representation theory and related topics / Pramod N. Achar, Kailash C. Misra, Daniel K. Nakano, editors.

Description: Providence, Rhode Island : American Mathematical Society, [2024] | Series: Proceedings of symposia in pure mathematics, 0082-0717 ; volume 108 | "Southeastern Lie Theory Workshop Series. Algebraic and Combinatorial Representation Theory, October 9–11, 2015, North Carolina State University; Algebraic Groups, Quantum Groups and Geometry, May 23–27, 2016, University of Virginia; Representation Theory of Algebraic Groups and Lie Superalegbras, June 10–12, 2018, University of Georgia; Geometric and Categorical Representation Theory, May 13–15, 2019, Louisiana State University; Lie Algebras, Lie Groups and Applications, October 15–17, 2021, College of Charleston." | Includes bibliographical references.

Identifiers: LCCN 2024001826 | ISBN 9781470471170 (paperback) | ISBN 9781470477295 (ebook)

Subjects: LCSH: Group theory–Congresses. | Lie algebras–Congresses. | AMS: Nonassociative rings and algebras – Lie algebras and Lie superalgebras – Structure theory. | Nonassociative rings and algebras – Lie algebras and Lie superalgebras – Representations, algebraic theory (weights). | Nonassociative rings and algebras – Lie algebras and Lie superalgebras – Quantum groups (quantized enveloping algebras) and related deformations. | Nonassociative rings and algebras – Lie algebras and Lie superalgebras – Homological methods in Lie (super)algebras. | Nonassociative rings and algebras – Lie algebras and Lie superalgebras – Infinite-dimensional Lie (super)algebras. | Category theory; homological algebra – Homological algebra – Resolutions; derived functors. | Group theory and generalizations – Representation theory of groups – p-adic representations of finite groups. | Group theory and generalizations – Linear algebraic groups and related topics – Representation theory. | Group theory and generalizations – Linear algebraic groups and related topics – Schur and q-Schur algebras.

Classification: LCC QA176 .C38 2024 | DDC 512/.2–dc23/eng/20240310

LC record available at https://lccn.loc.gov/2024001826

Color graphic policy. Any graphics created in color will be rendered in grayscale for the printed version unless color printing is authorized by the Publisher. In general, color graphics will appear in color in the online version.

∞ The paper used in this book is acid-free and falls within the guidelines
established to ensure permanence and durability.
Visit the AMS home page at `https://www.ams.org/`

10 9 8 7 6 5 4 3 2 1   29 28 27 26 25 24

To the memory of
Ben L. Cox and Brian J. Parshall

Contents

Preface ix

Quantum generalized Kac–Moody algebras via Hall algebras of complexes
JONATHAN D. AXTELL and KYU-HWAN LEE 1

Cohomology of algebraic groups, Lie algebras, and finite groups of Lie type
CHRISTOPHER P. BENDEL 35

Idempotent modules, locus of compactness and local supports
JON F. CARLSON 59

Real and symmetric quasi-maps
TSAO-HSIEN CHEN and DAVID NADLER 77

Support varieties for Lie superalgebras in characteristic 2
CHRISTOPHER M. DRUPIESKI and JONATHAN R. KUJAWA 99

Canonical bases for the modified quantum $\mathfrak{gl}_{m|n}$
JIE DU, HAIXIA GU, and ZHONGGUO ZHOU 127

An exact category approach to Hecke endomorphism algebras
JIE DU, BRIAN PARSHALL, and LEONARD SCOTT,
WITH AN APPENDIX BY VERONICA SHALOTENKO 159

On two constructions of Gelfand-Tsetlin modules with arbitrary characters
VYACHESLAV FUTORNY, DIMITAR GRANTCHAROV, LUIS ENRIQUE RAMIREZ,
and PABLO ZADUNAISKY 293

Lie algebra cohomology of the positive part of twisted affine Lie algebras
JIUZU HONG and SHRAWAN KUMAR 301

Realizing rings of regular functions via the cohomology of quantum groups
ZONGZHU LIN and DANIEL K. NAKANO 337

$\imath$Hall algebras and $\imath$quantum groups
MING LU and WEIQIANG WANG 371

Unipotent blocks and weighted affine Weyl groups
G. LUSZTIG 435

From conjugacy classes in the Weyl group to representations
GEORGE LUSZTIG and ZHIWEI YUN 453

Geometric crystal on unipotent variety
TOSHIKI NAKASHIMA 465

Difference operators for wreath Macdonald polynomials
DANIEL ORR and MARK SHIMOZONO 483

Universal central extensions of Krichever-Novikov algebras and orthogonal polynomials
FELIPE ALBINO DOS SANTOS, VYACHESLAV FUTORNY, and KAIMING ZHAO 493

Notes on Fock space
PETER TINGLEY 507

Preface

Representation theory of Lie algebras, quantum groups, and related algebraic structures has become a central research area in mathematics, with numerous applications in mathematics and theoretical physics. Research topics include quantized enveloping algebras, quantum function algebras, Kac–Moody Lie algebras, Hecke algebras, canonical bases and crystal bases, vertex operator algebras, Hall algebras, A_∞-algebras, quivers, cluster algebras, Hopf algebras, and Khovanov–Lauda–Rouquier algebras. In particular, quantized Kac–Moody Lie algebras and related categorical and geometric constructions have taken on a leading role not only within Lie theory but also in other fields including combinatorics, group theory, number theory, integrable systems, low-dimensional topology and conformal field theory.

In 1985, the interaction of semisimple and Kac–Moody Lie algebras with integrable systems led Drinfeld and Jimbo to introduce a new class of algebraic objects known as quantized universal enveloping algebras (also known as quantum groups) associated with symmetrizable Kac–Moody Lie algebras. The abstract theory of integrable representations of quantum groups given by Lusztig illustrates the similarity between quantum groups and Kac–Moody Lie algebras.

Much of the theory of quantum groups, vertex algebras and Kac–Moody Lie algebras is based on results developed for algebraic groups and Lie groups. Much of the classical theory deals with algebraic groups over algebraically closed fields or local fields, but in recent decades, there has been an increasing focus on Kac–Moody groups. Recent developments include the introduction of a building-like structure for Kac–Moody groups, with applications to Kac–Moody symmetric varieties. The full impact of these results on algebraic groups, quantum groups and Kac–Moody Lie groups is yet to be realized.

The theory of canonical basis for quantum groups (developed independently by Lusztig and Kashiwara around 1990) has had a huge impact on the field and has provided important insights to representation theory at large. In particular, the canonical base at $q = 0$ (i.e., crystal base) provides a beautiful combinatorial tool to study the representations of quantum groups. Explicit realizations of crystal bases for finite-dimensional simple Lie algebras and affine Lie algebras led to interesting applications in combinatorics and mathematical physics. Furthermore, various geometric realizations of crystals have revealed natural connections with quiver varieties and affine Grassmannians. The theory of geometric crystals introduced by Berenstein and Kazhdan in 2000 has also brought new ideas to the area.

The canonical basis theory of Lusztig–Kashiwara has motivated the Khovanov–Lauda–Rouquier (KLR) categorification, whose striking applications include Broué's conjecture for symmetric groups, $\mathbb{Z}$-gradings on blocks of symmetric groups, quiver varieties, rational Cherednik algebras, and Reshetikhin-Turaev invariants.

The $\imath$canonical bases yielded geometric interpretations via flag varieties of classical type and they afforded the positivity property, as shown by Y. Li and collaborators. A general theory of quantum symmetric pairs has been developed by G. Letzter and others. The canonical bases can be generalized to more general quantum symmetric pairs. In the affine Kac–Moody case, the $\imath$canonical basis should be closely related to modular representation of classical type algebraic groups and quantum groups at roots of unity. The canonical basis of quantum symmetric pairs is also leading to a new KLR-type categorification and new diagrammatic algebras.

For quantum groups, Arkhipov, Bezrukavnikov, and Ginzburg (ABG) used the geometry of the nilpotent cone $\mathcal{N}$, whose coordinate algebra identifies with the cohomology ring, as a starting point for their seminal work. They established derived equivalences between the principal block for the quantum group, (equivariant) coherent sheaves on the Springer resolution of $\mathcal{N}$, and perverse sheaves on the loop Grassmannian. These equivalences of triangulated categories yielded a proof of Lusztig's character formula for quantum groups when $l > h$ (where l is the order of the root of unity). The ABG approach illustrated that to calculate characters of simple modules at the representation theoretic level one needs to examine the structure of the underlying tensor triangulated category at the derived level where the geometry becomes more transparent. An important piece of their work involved the lifting of the support variety theory for quantum groups to obtain the equivalence between the principal block and coherent sheaves on the Springer resolution.

In 2009, Kailash Misra, Daniel Nakano, and Brian Parshall established a network of Lie theorists in the southeastern region and proposed an annual regional workshop series of 3 to 4 days in Lie theory. The aim of these workshops was to bring together senior and junior researchers as well as graduate students to build and foster cohesive research groups in the region. Since that time, there have been thirteen successful workshops at North Carolina State University (2009, 2012, 2015 and 2023), the University of Georgia (2010, 2014 and 2018), the University of Virginia (2011 and 2016), the College of Charleston (2012 and 2021) and Louisiana State University (2013 and 2019). Each of these workshops included expository talks by senior researchers, invited talks on recent developments by mid-career and junior researchers, contributed talks by post-docs and senior graduate students along with afternoon discussion sessions, intended as an educational opportunity for graduate students and junior researchers interested in Lie theory. This book is a collection of papers contributed by some of the invited speakers at the workshops held during 2015–2021. The previous two volumes PSPUM 86 and PSPUM 92 in this series edited by Misra, Nakano, and Parshall covered the workshops during 2009–2011 and 2012–2014 respectively. We dearly miss Ben Cox, a member of the scientific committee and Brian Parshall, a founding member of the southeastern Lie theory network due to their sudden passing in September 2019 and January 2022 respectively and dedicate this volume to their memory.

The editors take this opportunity to thank the conference participants, the contributors, the anonymous referees and the editorial offices of the American Mathematical Society for making this volume possible.

Pramod N. Achar
Kailash C. Misra
Daniel K. Nakano

Proceedings of Symposia in Pure Mathematics
Volume **108**, 2024
https://doi.org/10.1090/pspum/108/01965

Quantum generalized Kac–Moody algebras via Hall algebras of complexes

Jonathan D. Axtell and Kyu-Hwan Lee

Abstract. We establish an embedding of the quantum enveloping algebra of a symmetric generalized Kac–Moody algebra into a localized Hall algebra of $\mathbb{Z}_2$-graded complexes of representations of a quiver with (possible) loops. To overcome difficulties resulting from the existence of infinite dimensional projective objects, we consider the category of finitely-presented representations and the category of $\mathbb{Z}_2$-graded complexes of projectives with finite homology.

1. Introduction

Let $\mathcal{A}$ be an abelian category such that the sets $\mathrm{Hom}(A, B)$ and $\mathrm{Ext}^1(A, B)$ are both finite for all $A, B \in \mathcal{A}$. The *Hall algebra* of $\mathcal{A}$ is defined to be the $\mathbb{C}$-vector space with basis elements indexed by isomorphism classes in $\mathcal{A}$ and with associative multiplication which encodes information about extensions of objects. Typical examples of such abelian categories arise as the category $\mathrm{rep}_{\Bbbk}(\mathcal{Q})$ of finite-dimensional representations of an acyclic quiver $\mathcal{Q}$ over a finite field $\Bbbk := \mathbb{F}_q$. This category became a focal point of intensive research when C. Ringel [**20**] realized one half of a quantum group via a twisted Hall algebra of the category. This twisted Hall algebra is usually called the *Ringel–Hall* algebra. The construction was further generalized by J. A. Green [**9**] to one half of the quantum group of an arbitrary Kac–Moody algebra.

Even though there is a construction, called *Drinfeld double*, which glues together two copies of one-half quantum group to obtain the whole quantum group, it is desirable to have an explicit realization of the whole quantum group in terms of a Hall algebra. Among various attempts, the idea of using a category of $\mathbb{Z}_2$-graded complexes was suggested by the works of M. Kapranov [**15**], L. Peng and J. Xiao [**17, 18**]. In his seminal work [**4**], T. Bridgeland successfully utilized this idea to achieve a Hall algebra realization of the whole quantum group. More precisely, given a Kac–Moody algebra $\mathfrak{g}$, he took the category $\mathrm{rep}_{\Bbbk}(\mathcal{Q})$ of finite dimensional representations of an acyclic quiver $\mathcal{Q}$ associated with $\mathfrak{g}$, and considered the full subcategory $\mathcal{P}$ of projective objects in $\mathrm{rep}_{\Bbbk}(\mathcal{Q})$. By studying the category $\mathcal{C}(\mathcal{P})$ of $\mathbb{Z}_2$-graded complexes in $\mathcal{P}$, he showed that the whole quantum group is embedded into the reduced localization of a twisted Hall algebra of $\mathcal{C}(\mathcal{P})$.

The first author was supported by the National Research Foundation of Korea (NRF) funded by the Ministry of Science (NRF-2017R1C1B5018384). The second author was partially supported by a grant from the Simons Foundation (#712100).

The purpose of this paper is to extend Bridgeland's construction to *generalized Kac–Moody algebras.* These algebras were introduced by R. Borcherds [**2**] around 1988. He used a generalized Kac–Moody algebra, called the *Monster Lie algebra*, to prove the celebrated Moonshine Conjecture [**3**]. Since then, many of the constructions in the theory of Kac–Moody algebras have been extended to generalized Kac–Moody algebras. In particular, the quantum group of a generalized Kac–Moody algebra was defined by Kang [**12**], and one half of the quantum group was realized via a Hall algebra by Kang and Schiffmann [**13**], following Ringel–Green's construction.

The main difference from the usual Kac–Moody case is that the quiver $\mathcal{Q}$ may have loops in order to account for imaginary simple roots. A natural question arises:

> *Is it possible to realize the whole quantum group of a generalized Kac–Moody algebra in terms of a Hall algebra of $\mathbb{Z}_2$-graded complexes?*

A straightforward approach would run into an obstacle. Namely, when there is a loop, a projective object may well be infinite dimensional, and the Hall product would not be defined.

In this paper, we show that this difficulty can be overcome by considering the category $\mathcal{R}$ of finitely-presented representations of a locally finite quiver $\mathcal{Q}$, possibly with loops, and the category $\mathcal{C}_{\mathsf{fin}}(\mathcal{P})$ of $\mathbb{Z}_2$-graded complexes in the category of projectives $\mathcal{P} \subset \mathcal{R}$ with finite homology. In contrast to Bridgeland's construction, however, it is not clear that the corresponding product in the Hall algebra $\mathcal{H}(\mathcal{C}_{\mathsf{fin}}(\mathcal{P}))$ is associative. Therefore the proof of associativity for the (localized) Hall algebra is one of the main results of this paper.

Following the approach of [**4**], we work in the more general setting of a category $\mathcal{R}$ satisfying certain natural assumptions listed in the next subsection (Section 1.1) which we keep throughout Sections 2–4. The main theorem (Theorem 4.6) for this general setting states that a certain localization $\mathcal{DH}(\mathcal{R})$ of the Hall algebra $\mathcal{H}(\mathcal{C}_{\mathsf{fin}}(\mathcal{P}))$ is isomorphic to the Drinfeld double of the (extended) Hall algebra of $\mathcal{R}$, generalizing a result of Yanagida [**23**]. As a corollary (Corollary 4.7), the localized Hall algebra is shown to be an associative algebra.

In Section 5, we show that the category $\mathcal{R}$ of finitely-presented representations of a locally finite quiver satisfies all the assumptions in Section 1.1 under some minor restrictions on the quiver. As a consequence of the results of Section 4 in the general setting, we obtain the main result for the quantum group (Theorem 5.12) which establishes an embedding

$$\Xi\colon \mathbf{U}_v \hookrightarrow \mathcal{DH}_{\mathsf{red}}(\mathcal{R})$$

of the whole quantum group $\mathbf{U}_v = \mathbf{U}_v(\mathfrak{g})$ of a generalized Kac–Moody algebra $\mathfrak{g}$ into a reduced version of the localized Hall algebra $\mathcal{DH}(\mathcal{R})$.

1.1. Assumptions. Given an abelian category $\mathcal{C}$, let $K(\mathcal{C})$ denote its Grothendieck group and write $\boldsymbol{k}_{\mathcal{C}}(X) \in K(\mathcal{C})$ to denote the class of an object $X \in \mathcal{C}$.

Throughout this paper $\mathcal{R}$ is an abelian category. Let $\mathcal{P} \subset \mathcal{R}$ denote the full subcategory of projectives and $\mathcal{A} \subset \mathcal{R}$ the full subcategory of objects $A \in \mathcal{R}$ such that $\mathrm{Hom}_{\mathcal{R}}(M, A)$ is a finite set for any $M \in \mathcal{R}$. There are several conditions that we will impose on the triple $(\mathcal{R}, \mathcal{P}, \mathcal{A})$. Precisely, we shall always assume that

(a) $\mathcal{R}$ is essentially small and linear over $\Bbbk = \mathbb{F}_q$,

(b) $\mathcal{R}$ is hereditary, that is of global dimension at most 1, and has enough projectives,
(c) for any objects $P, Q, M \in \mathcal{P}$, the relation $M \oplus P \cong M \oplus Q$ implies $P \cong Q$,
(d) every element in $K(\mathcal{R})$ is a $\mathbb{Q}$-linear combination of elements in $\{\boldsymbol{k}_{\mathcal{R}}(A) \mid A \in \mathcal{A}\}$,
(e) the identity $\boldsymbol{k}_{\mathcal{R}}(A) = \boldsymbol{k}_{\mathcal{R}}(B)$ implies $|\operatorname{Hom}(P, A)| = |\operatorname{Hom}(P, B)|$ for all $P \in \mathcal{P}$, $A, B \in \mathcal{A}$.

It is clear that the category $\mathcal{A}$ is Hom-finite, that is $\operatorname{Hom}_{\mathcal{R}}(A, B)$ is a finite set for all $A, B \in \mathcal{A}$. Since $\mathcal{R}$ has enough projectives by (b), it follows that the subcategory $\mathcal{A}$ is abelian and hence a Krull–Schmidt category by [**16**]. It is also easy to check that $\mathcal{A}$ is closed under extensions in $\mathcal{R}$. We note, however, that the category $\mathcal{R}$ is not necessarily Krull-Schmidt in general.

The condition (c) is required in the proof of Proposition 3.18 to show that the localized Hall algebra $\mathcal{DH}(\mathcal{R})$ is a free module over the group algebra $\mathbb{C}[K(\mathcal{R}) \times K(\mathcal{R})]$. Conditions (d) and (e) are needed in Section 2.3 to ensure that the Euler form on $\mathcal{A}$ can be lifted to a bilinear form on the whole category $\mathcal{R}$, albeit with values in $\mathbb{Q}$.

The class of categories $\mathcal{R}$ satisfying the above conditions (a)-(e) generalizes the class of categories $\mathcal{A}$ satisfying Bridgeland's conditions, also denoted (a)-(e) in [**4**], although our conditions do not correspond precisely. In particular, if the category $\mathcal{R}$ is Hom-finite then $\mathcal{A} = \mathcal{R}$ so that condition (d) is superfluous, and it is also clear that (c) holds since $\mathcal{R}$ is Krull-Schmidt in this case. One may check using Proposition 2.4 that the remaining conditions (a), (b), (e) hold in case $\mathcal{A} = \mathcal{R}$ if and only if the conditions (a)-(e) in [**4**] hold for $\mathcal{A}$.

1.2. Notation. Assume throughout that $\Bbbk = \mathbb{F}_q$ is a finite field ($q \geq 2$) with q elements. Let $v \in \mathbb{R}_{>0}$ be such that $v^2 = q$. We denote by $v^{1/n}$ the positive real $2n$-th root of q for $n \in \mathbb{Z}_{\geq 1}$, and write $q^{1/n} = v^{2/n}$. We also write $\mathbb{Z}_2 = \mathbb{Z}/2\mathbb{Z}$.

2. Hall algebras

We assume that the triple $(\mathcal{R}, \mathcal{P}, \mathcal{A})$ satisfies axioms (a)-(e) in Section 1.1. The presentation at the begininning of this section follows [**4**] up to a suitable generalization, while Sections 2.3–2.5 contain new material.

2.1. Hall algebras. Given a small category $\mathcal{C}$, denote by $\operatorname{Iso}(\mathcal{C})$ the set of isomorphism classes in $\mathcal{C}$. Suppose that $\mathcal{C}$ is abelian. Given objects $X, Y, Z \in \mathcal{C}$, define $\operatorname{Ext}^1_{\mathcal{C}}(X, Y)_Z$ to be the set of (equivalence classes of) extensions with middle term isomorphic to Z in $\operatorname{Ext}^1_{\mathcal{C}}(X, Y)$.

Since $\mathcal{R}$ has enough projectives, it follows from the definition of $\mathcal{A}$ that the set $\operatorname{Ext}^1_{\mathcal{A}}(A, B)$ is finite for all $A, B \in \mathcal{A}$. The *Hall algebra* $\mathcal{H}(\mathcal{A})$ is defined to be the $\mathbb{C}$-vector space with basis indexed by elements $A \in \operatorname{Iso}(\mathcal{A})$, and with associative multiplication defined by

$$[A] \diamond [B] = \sum_{C \in \operatorname{Iso}(\mathcal{A})} \frac{|\operatorname{Ext}^1_{\mathcal{A}}(A, B)_C|}{|\operatorname{Hom}_{\mathcal{A}}(A, B)|} [C]. \tag{2.1}$$

The unit is given by $[0]$, where 0 is the zero object in $\mathcal{A}$.

Recall from [**4**] that the multiplication (2.1) is a variant of the usual Hall product (see e.g. [**20**]) defined as follows. Given objects $A, B, C \in \mathcal{A}$, define the number

$$g_{A,B}^{C} = \left|\{B' \subset C : B' \cong B,\ C/B' \cong A\}\right|. \tag{2.2}$$

Writing $a_A = |\operatorname{Aut}_{\mathcal{A}}(A)|$ to denote the cardinality of the automorphism group of an object A, recall that

$$g_{A,B}^{C} = \frac{|\operatorname{Ext}^1_{\mathcal{A}}(A,B)_C|}{|\operatorname{Hom}_{\mathcal{A}}(A,B)|} \cdot \frac{a_C}{a_A\, a_B}.$$

Hence using $[[A]] = [A] \cdot a_A^{-1}$ as alternative generators, the product takes the form

$$[[A]] \diamond [[B]] = \sum_{C \in \operatorname{Iso}(\mathcal{A})} g_{A,B}^{C} \cdot [[C]].$$

The associativity of multiplication in $\mathcal{H}(\mathcal{A})$ then reduces to the equality

$$\sum_{C_1} g_{A,B}^{C_1}\, g_{C_1,C}^{D} = \sum_{D_1} g_{A,D_1}^{D}\, g_{B,C}^{D_1} \tag{2.3}$$

which holds for any $A, B, C, D \in \operatorname{Iso}(\mathcal{A})$.

2.2. Twisted and extended Hall algebras. The Euler form on $\mathcal{A}$ is a bilinear mapping

$$\langle -,- \rangle : \mathcal{A} \times \mathcal{A} \to \mathbb{Z}$$

defined by

$$\langle A, B \rangle := \dim_{\Bbbk} \operatorname{Hom}(A,B) - \dim_{\Bbbk} \operatorname{Ext}^1(A,B) \tag{2.4}$$

for all $A, B \in \mathcal{A}$. This form factors through the Grothendieck group $K(\mathcal{A})$ (see Lemma 2.1). We also introduce the associated symmetric form $(A,B) = \langle A,B\rangle + \langle B,A\rangle$.

We define the *twisted Hall algebra* $\mathcal{H}_v(\mathcal{A})$ to be the same as the Hall algebra of $\mathcal{A}$ with a new multiplication given by

$$[A] * [B] = v^{\langle \boldsymbol{k}_{\mathcal{A}}(A),\, \boldsymbol{k}_{\mathcal{A}}(B)\rangle} [A] \diamond [B]. \tag{2.5}$$

We extend $\mathcal{H}_v(\mathcal{A})$ by adjoining new generators K_α for all $\alpha \in K(\mathcal{A})$ with the relations

$$K_\alpha * K_\beta = K_{\alpha+\beta}, \qquad K_\alpha * [A] * K_{-\alpha} = v^{(\alpha,\, \boldsymbol{k}_{\mathcal{A}}(A))}[A], \tag{2.6}$$

where we use the symmetric form $(\ ,\ )$. The resulting algebra will be called the *extended Hall algebra* and denoted by $\widetilde{\mathcal{H}_v}(\mathcal{A})$.

2.3. Generalized Euler form. Since the category $\mathcal{R}$ has enough projectives, it follows from the definition of $\mathcal{A}$ that $\operatorname{Ext}^1_{\mathcal{R}}(M,A)$ is a finite set for any $M \in \mathcal{R}$ and $A \in \mathcal{A}$. Define an Euler form $\langle -,- \rangle : \mathcal{R} \times \mathcal{A} \to \mathbb{Z}$ by setting

$$\langle M, A \rangle := \dim_{\Bbbk} \operatorname{Hom}_{\mathcal{R}}(M,A) - \dim_{\Bbbk} \operatorname{Ext}^1_{\mathcal{R}}(M,A)$$

for all $M \in \mathcal{R}, A \in \mathcal{A}$. The following lemma shows that the Euler form induces a bilinear map

$$\langle -,- \rangle : K(\mathcal{R}) \times K(\mathcal{A}) \to \mathbb{Z}.$$

LEMMA 2.1. *The Euler form $\langle M, A\rangle$ depends only on the classes of objects $M \in \mathcal{R}$ and $A \in \mathcal{A}$ in the Grothendieck groups $K(\mathcal{R})$ and $K(\mathcal{A})$, respectively.*

PROOF. Suppose $0 \to M' \to M \to M'' \to 0$ is an exact sequence in $\mathcal{R}$. Then there is a long exact sequence

$$0 \to \mathrm{Hom}(M'', A) \to \mathrm{Hom}(M, A) \to \mathrm{Hom}(M', A)$$
$$\to \mathrm{Ext}(M'', A) \to \mathrm{Ext}(M, A) \to \mathrm{Ext}(M', A) \to 0$$

which shows that

$$\langle M'', A\rangle - \langle M, A\rangle + \langle M', A\rangle = 0.$$

So the Euler form is well-defined on the class of M. The proof for the class of A is similar. □

In the remainder, let us write $\hat{X} = \boldsymbol{k}_{\mathcal{R}}(X)$ to denote the class of an object X in $K(\mathcal{R})$, and continue to write $\boldsymbol{k}_{\mathcal{A}}(-)$ for classes in $K(\mathcal{A})$. The inclusion $\mathcal{A} \subset \mathcal{R}$ induces a canonical map

$$K(\mathcal{A}) \to K(\mathcal{R}), \quad \boldsymbol{k}_{\mathcal{A}}(A) \mapsto \hat{A}. \tag{2.7}$$

Write $\bar{K}(\mathcal{A}) \subset K(\mathcal{R})$ to denote the image of $K(\mathcal{A})$ under this map.

Suppose that $\{P_i\}_{i\in I}$ is a complete list of isomorphism classes of indecomposable projective objects in $\mathcal{P}$, for some indexing set I. Then define the *dimension vector* of an object $A \in \mathcal{A}$ to be

$$\mathbf{dim}\, A := (\dim_{\Bbbk} \mathrm{Hom}_{\mathcal{R}}(P_i, A))_{i\in I} \in \mathbb{Z}^{\oplus I}.$$

From condition (e), we have

$$\mathbf{dim}\, A = \mathbf{dim}\, B$$

for any objects $A, B \in \mathcal{A}$, whenever the classes $\hat{A} = \hat{B}$ are equal in the Grothendieck group $K(\mathcal{R})$.

LEMMA 2.2. *Suppose that $A, A' \in \mathcal{A}$. Then $\langle M, A\rangle = \langle M, A'\rangle$ for all $M \in \mathcal{R}$ if and only if*

$$\dim_{\Bbbk} \mathrm{Hom}(P, A) = \dim_{\Bbbk} \mathrm{Hom}(P, A')$$

for all $P \in \mathcal{P}$.

PROOF. The only-if-part follows from the fact that $\langle P, A\rangle = \dim_{\Bbbk} \mathrm{Hom}(P, A)$ for all $P \in \mathcal{P}$ and $A \in \mathcal{A}$. For the if-part, let $M \in \mathcal{R}$, $A \in \mathcal{A}$, and suppose $0 \to P \to Q \to M \to 0$ is a projective resolution. Then there is a long exact sequence

$$0 \longrightarrow \mathrm{Hom}_{\mathcal{R}}(M, A) \longrightarrow \mathrm{Hom}_{\mathcal{R}}(Q, A) \longrightarrow \mathrm{Hom}_{\mathcal{R}}(P, A) \longrightarrow \mathrm{Ext}_{\mathcal{R}}(M, A) \longrightarrow 0,$$

which shows that

$$\langle M, A\rangle = \dim_{\Bbbk} \mathrm{Hom}(Q, A) - \dim_{\Bbbk} \mathrm{Hom}(P, A).$$

The converse statement now follows. □

The above lemma now has the following consequence.

COROLLARY 2.3. *The Euler form on $\mathcal{R} \times \mathcal{A}$ factors through a bilinear form*

$$\langle -, -\rangle : K(\mathcal{R}) \times \bar{K}(\mathcal{A}) \to \mathbb{Z}.$$

Now suppose $X, Y \in \mathcal{R}$. Then it follows from condition (d) that there exist objects $A_1, \ldots, A_r \in \mathcal{A}$ such that

$$\hat{Y} = c_1 \hat{A}_1 + \cdots + c_r \hat{A}_r$$

for some coefficients $c_i \in \mathbb{Q}$. Define a bilinear form

$$K(\mathcal{R}) \times K(\mathcal{R}) \to \mathbb{Q}$$

by setting

$$\langle \hat{X}, \hat{Y} \rangle = \sum_i c_i \cdot \langle \hat{X}, \hat{A}_i \rangle. \tag{2.8}$$

and extend it through linearity. We again define a symmetric version by setting $(\hat{X}, \hat{Y}) = \langle \hat{X}, \hat{Y} \rangle + \langle \hat{Y}, \hat{X} \rangle$ for any $X, Y \in \mathcal{R}$.

PROPOSITION 2.4. *Let A be a nonzero object in $\mathcal{A}$. Then the classes $\boldsymbol{k}_{\mathcal{A}}(A) \in K(\mathcal{A})$ and $\hat{A} \in K(\mathcal{R})$ are both nonzero.*

PROOF. Consider a nonzero object $A \in \mathcal{A}$. Since $\mathcal{R}$ has enough projectives, there exists a surjection $P \to A \to 0$ for some projective $P \in \mathcal{P}$. It follows that $\operatorname{Hom}(P, A)$ is a nonzero set. We obtain

$$\langle P, A \rangle = \dim_{\Bbbk} \operatorname{Hom}(P, A) - \dim_{\Bbbk} \operatorname{Ext}(P, A) = \dim_{\Bbbk} \operatorname{Hom}(P, A) \geq 1.$$

By Lemma 2.2 and Corollary 2.3, we obtain

$$\langle P, A \rangle = \langle \hat{P}, \boldsymbol{k}_{\mathcal{A}}(A) \rangle = \langle \hat{P}, \hat{A} \rangle \geq 1.$$

Thus neither $\hat{A}$ nor $\boldsymbol{k}_{\mathcal{A}}(A)$ is zero. □

Denote by $K_{\geq 0}(\mathcal{A}) \subset K(\mathcal{A})$ the positive cone in the Grothendieck group generated by the classes $\boldsymbol{k}_{\mathcal{A}}(A)$ for $A \in \mathcal{A}$. Define

$$\alpha \leq \beta \iff \beta - \alpha \in K_{\geq 0}(\mathcal{A}). \tag{2.9}$$

Then it follows from Proposition 2.4 that $\leq$ is a partial order on $K(\mathcal{A})$.

2.4. The extended Hall algebra $\widetilde{\mathcal{H}_v}(\mathcal{R})$. The algebra $\mathcal{H}_v(\mathcal{A})$ is naturally graded by the Grothendieck group $K(\mathcal{R})$:

$$\mathcal{H}_v(\mathcal{A}) = \bigoplus_{\alpha \in K(\mathcal{R})} \mathcal{H}_v(\mathcal{A})_{(\alpha)}, \qquad \mathcal{H}_v(\mathcal{A})_{(\alpha)} := \bigoplus_{\hat{A} = \alpha} \mathbb{C}[A].$$

We define a slight modification of the extended Hall algebra $\widetilde{\mathcal{H}_v}(\mathcal{A})$ from Section 2.2. Starting again from $\mathcal{H}_v(\mathcal{A})$ with multiplication (2.5), the *extended Hall algebra* $\widetilde{\mathcal{H}_v}(\mathcal{R})$ is defined by adjoining generators K_α for all $\alpha \in K(\mathcal{R})$ with the relations

$$K_\alpha * K_\beta = K_{\alpha+\beta}, \qquad K_\alpha * [A] * K_{-\alpha} = v^{(\alpha, \hat{A})}[A]. \tag{2.10}$$

The algebra $\widetilde{\mathcal{H}_v}(\mathcal{R})$ is also $K(\mathcal{R})$-graded, with the degree of each K_α equal to zero. Note that the multiplication map

$$\mathcal{H}_v(\mathcal{A}) \otimes_{\mathbb{C}} \mathbb{C}[K(\mathcal{R})] \to \widetilde{\mathcal{H}_v}(\mathcal{R})$$

is an isomorphism of vector spaces.

Following Green [**9**] and Xiao [**22**], we define a coalgebra structure on $\widetilde{\mathcal{H}_v}(\mathcal{R})$. We refer to [**23**] for the definition of a *topological* coalgebra, which involves a completed tensor product.

DEFINITION 2.5 ([**9**, **22**]). In the extended Hall algebra $\mathcal{H} = \widetilde{\mathcal{H}_v}(\mathcal{R})$ define $\Delta : \mathcal{H} \to \mathcal{H} \widehat{\otimes}_{\mathbb{C}} \mathcal{H}$, and $\epsilon : \mathcal{H} \to \mathbb{C}$, by setting

$$\Delta([A]K_\alpha) := \sum_{B,C \in \mathrm{Iso}(\mathcal{A})} v^{\langle B,C\rangle} g_{B,C}^{A} \cdot ([B]K_{\hat{C}+\alpha}) \otimes ([C]K_\alpha), \qquad \epsilon([A]K_\alpha) := \delta_{A,0}$$

for all $A \in \mathrm{Iso}(\mathcal{A})$, $\alpha \in K(\mathcal{R})$, where the numbers $g_{B,C}^{A}$ are defined in (2.2) and $\mathcal{H} \widehat{\otimes}_{\mathbb{C}} \mathcal{H}$ is the completed tensor product. This gives $\widetilde{\mathcal{H}_v}(\mathcal{R})$ the structure of a topological coassociative coalgebra.

As noted in [**21**, Remark 1.6], the coproduct Δ on $\mathcal{H} = \widetilde{\mathcal{H}_v}(\mathcal{R})$ takes values in $\mathcal{H} \otimes \mathcal{H}$, instead of the completion $\mathcal{H} \widehat{\otimes} \mathcal{H}$, if and only if the following condition holds:

(2.11) *Any fixed object $A \in \mathcal{A} \subset \mathcal{R}$ has only finitely many subobjects $B \subset A$.*

This condition is satisfied if $\mathcal{R}$ is the category of quiver representations considered in Section 5.

Now we have an algebra structure and a coalgebra structure on $\widetilde{\mathcal{H}_v}(\mathcal{R})$. It follows from [**9**, **22**] that these structures are compatible to give a (topological) *bialgebra* structure. Below, we simply denote by $\widetilde{\mathcal{H}_v}(\mathcal{R})$ the bialgebra $(\widetilde{\mathcal{H}_v}(\mathcal{R}), *, [0], \Delta, \epsilon)$. The bialgebra $\widetilde{\mathcal{H}_v}(\mathcal{R})$ admits a natural bilinear form compatible with the bialgebra structure called a *Hopf pairing.*

DEFINITION 2.6 ([**9**, **22**, **23**]). Define a bilinear form $(\cdot,\cdot)_H$ on $\widetilde{\mathcal{H}_v}(\mathcal{R})$ by setting

$$([A]K_\alpha, [B]K_\beta)_H := v^{(\alpha,\beta)} a_A\, \delta_{A,B}$$

for $\alpha, \beta \in K(\mathcal{R})$, where $a_A = |\operatorname{Aut}_{\mathcal{A}}(A)|$ as before.

It is clear that the restriction of this bilinear form to the subalgebra $\mathcal{H}_v(\mathcal{A}) \subset \widetilde{\mathcal{H}_v}(\mathcal{R})$ is nondegenerate. The following result was stated for $\widetilde{\mathcal{H}_v}(\mathcal{A})$ in [**22**]. It is easy to check that it holds for $\widetilde{\mathcal{H}_v}(\mathcal{R})$ as well.

PROPOSITION 2.7 ([**21**, **22**]). *The bilinear form $(\cdot,\cdot)_H$ is a Hopf pairing on the bialgebra $\widetilde{\mathcal{H}_v}(\mathcal{R})$, that is, for any $x, y, z \in \widetilde{\mathcal{H}_v}(\mathcal{R})$, one has*

$$(1, x)_H = \epsilon(x), \qquad (x * y, z)_H = (x \otimes y, \Delta(z))_H$$

where we use the usual pairing on the tensor product space:

$$(x \otimes y, z \otimes w)_H = (x, z)_H\, (y, w)_H.$$

2.5. The Drinfeld double. We briefly recall the Drinfeld double construction for Hall bialgebras. A complete treatment of Drinfeld doubles is given in [**10**, §3.2] and [**21**, §5.2].

In [**22**], Xiao showed that the extended Hall algebra $\widetilde{\mathcal{H}_v}(\mathcal{A})$ is a Hopf algebra and gave an explicit formula for both the antipode σ and its inverse σ^{-1}, provided that $\mathcal{A}$ is a category of quiver representations. It can be shown that Xiao's formulas hold more generally provided that $\mathcal{A} \subset \mathcal{R}$ satisfies (2.11).

Athough the formula for σ is no longer well-defined in the case where $\mathcal{A}$ does not satisfy (2.11), there is a more general condition that ensures the formula for σ^{-1} is still defined. Recall from [**8**] that an *anti-equivalence* between two objects $A, B \in \mathcal{A}$ is a pair of strict filtrations

$$0 = L_{n+1} \subsetneq L_n \subsetneq \dots L_1 \subsetneq L_0 = A, \qquad 0 = L'_{n+1} \subsetneq L'_n \subsetneq \dots L'_1 \subsetneq L'_0 = B$$

such that $L'_i/L'_{i+1} \cong L_{n-i}/L_{n-i+1}$ for all i. Two objects A and B in $\mathcal{A}$ are called *anti-equivalent* if there exists at least one anti-equivalence between them.

It follows from results in [**5**] and [**8**] that Xiao's formula for the map σ^{-1} is still well-defined provided that the following condition holds for every pair of objects $A, B \in \mathcal{A}$:

(2.12) *There are finitely many anti-equivalences (if any) between A and B.*

Since the map σ^{-1} generally takes values in a certain completion of $\widetilde{\mathcal{H}_v}(\mathcal{R})$, an extra condition is needed to ensure that corresponding relations in the bialgebra, such as

$$m \circ (\sigma^{-1} \otimes \mathrm{id}) \circ \Delta^{\mathrm{op}} = i \circ \epsilon,$$

are still well-defined. The required condition can be stated as follows:

(2.13)
Given any $A, B \in \mathcal{A}$, there are finitely many pairs (A', B') of anti-equivalent objects such that $A' \hookrightarrow A$ and $B \twoheadrightarrow B'$.

Up to minor modifications, the formulas for (and corresponding properties of) σ and σ^{-1} continue to hold, respectively, in $\widetilde{\mathcal{H}_v}(\mathcal{R})$, whenever they are defined in $\widetilde{\mathcal{H}_v}(\mathcal{A})$.

If the conditions (2.12) and (2.13) are both satisfied by the category $\mathcal{A} \subset \mathcal{R}$, then the *Drinfeld double* of $\mathcal{H} = \widetilde{\mathcal{H}_v}(\mathcal{R})$ is the vector space $\mathcal{H} \otimes \mathcal{H}$ equipped with the multiplication $\circ$ uniquely determined by the following conditions:

(D1) The maps

$$\mathcal{H} \longrightarrow \mathcal{H} \otimes_{\mathbb{C}} \mathcal{H}, \quad a \longmapsto a \otimes 1$$

and

$$\mathcal{H} \longrightarrow \mathcal{H} \otimes_{\mathbb{C}} \mathcal{H}, \quad a \longmapsto 1 \otimes a$$

are injective homomorphisms of $\mathbb{C}$-algebras;

(D2) For all elements $a, b \in \mathcal{H}$, one has

$$(a \otimes 1) \circ (1 \otimes b) = a \otimes b;$$

(D3) For all elements $a, b \in \mathcal{H}$, one has

$$(1 \otimes b) \circ (a \otimes 1) = \sum (b_{(1)}, a_{(3)})_H (\sigma^{-1}(b_{(3)}), a_{(1)})_H \cdot a_{(2)} \otimes b_{(2)}$$

where $\Delta^2(a) = \sum a_{(1)} \otimes a_{(2)} \otimes a_{(3)}$ and $\Delta^2(b) = \sum b_{(1)} \otimes b_{(2)} \otimes b_{(3)}$.

The last identity is equivalent to

$$\sum (a_{(2)}, b_{(1)})_H \cdot a_{(1)} \otimes b_{(2)} = \sum (a_{(1)}, b_{(2)})_H \cdot (1 \otimes b_{(1)}) \circ (a_{(2)} \otimes 1) \tag{2.14}$$

for all $a, b \in \mathcal{H}$, where $\Delta(a) = \sum a_{(1)} \otimes a_{(2)}$ and $\Delta(b) = \sum b_{(1)} \otimes b_{(2)}$.

An argument similar to the proof given in [**10**, Lemma 3.2.2] shows that the multiplication $\circ$ is associative.

3. Hall algebras of complexes

The material at the beginning of this section is based on [**4**] with a suitable generalization. The main new material is contained in Lemma 3.3, Proposition 3.6 and Sections 3.6 and 3.7.

Assume that $\mathcal{R}$ is an abelian category for which the triple $(\mathcal{R}, \mathcal{P}, \mathcal{A})$ satisfies axioms (a)-(e) of Section 1.1. We now introduce certain categories of complexes over $\mathcal{R}$ and define corresponding Hall algebras and their localizations. The associativity of multiplication in the localized Hall algebras will be established later in Section 4.

3.1. Categories of complexes. Define a $\mathbb{Z}_2$-graded chain complex in $\mathcal{R}$ to be a diagram

$$M_1 \underset{d_0}{\overset{d_1}{\rightleftarrows}} M_0$$

such that $d_i \circ d_{i+1} = 0$ for all $i \in \mathbb{Z}_2$.

A morphism $s_\bullet \colon M_\bullet \to \tilde{M}_\bullet$ consists of a diagram

$$\begin{array}{ccc} M_1 & \underset{d_0}{\overset{d_1}{\rightleftarrows}} & M_0 \\ {\scriptstyle s_1}\downarrow & & \downarrow{\scriptstyle s_0} \\ \tilde{M}_1 & \underset{\tilde{d}_0}{\overset{\tilde{d}_1}{\rightleftarrows}} & \tilde{M}_0 \end{array}$$

with $s_{i+1} \circ d_i = \tilde{d}_i \circ s_i$.

Let $\mathcal{C}_{\mathbb{Z}_2}(\mathcal{R})$ denote the category of all $\mathbb{Z}_2$-graded chain complexes in $\mathcal{R}$ with morphisms defined above. Two morphisms $s_\bullet, t_\bullet \colon M_\bullet \to \tilde{M}_\bullet$ are *homotopic* if there are morphisms $h_i \colon M_i \to \tilde{M}_{i+1}$ such that

$$t_i - s_i = \tilde{d}_{i+1} \circ h_i + h_{i+1} \circ d_i.$$

We write $\mathrm{Ho}_{\mathbb{Z}_2}(\mathcal{R})$ for the category obtained from $\mathcal{C}_{\mathbb{Z}_2}(\mathcal{R})$ by identifying homotopic morphisms.

The *shift functor* defines an involution

$$\mathcal{C}_{\mathbb{Z}_2}(\mathcal{R}) \qquad \overset{*}{\longleftrightarrow} \qquad \mathcal{C}_{\mathbb{Z}_2}(\mathcal{R})$$

which shifts the grading and changes the sign of the differential

$$M_1 \underset{d_0}{\overset{d_1}{\rightleftarrows}} M_0 \qquad \overset{*}{\longleftrightarrow} \qquad M_0 \underset{-d_1}{\overset{-d_0}{\rightleftarrows}} M_1.$$

The image of $M_\bullet$ under the shift functor will be denoted by $M_\bullet^*$. Every complex $M_\bullet \in \mathcal{C}_{\mathbb{Z}_2}(\mathcal{R})$ defines a class $\hat{M}_\bullet := \hat{M}_0 - \hat{M}_1 \in K(\mathcal{R})$ in the Grothendieck group of $\mathcal{R}$.

We are mostly concerned with the full subcategories

$$\mathcal{C}_{\mathbb{Z}_2}(\mathcal{P}),\ \mathcal{C}_{\mathbb{Z}_2}(\mathcal{A}) \subset \mathcal{C}_{\mathbb{Z}_2}(\mathcal{R}) \quad \text{and} \quad \mathrm{Ho}_{\mathbb{Z}_2}(\mathcal{P}) \subset \mathrm{Ho}_{\mathbb{Z}_2}(\mathcal{R})$$

consisting of complexes of objects in $\mathcal{P}$ and $\mathcal{A}$, respectively.

3.2. Root category. Let $\mathrm{D}^b(\mathcal{R})$ denote the ($\mathbb{Z}$-graded) bounded derived category of $\mathcal{R}$, with its shift functor [1]. Let $\mathrm{Rt}(\mathcal{R}) = \mathrm{D}^b(\mathcal{R})/[2]$ be the *orbit category*, also known as the *root category* of $\mathcal{R}$. This has the same objects as $\mathrm{D}^b(\mathcal{R})$, but the morphisms are given by

$$\mathrm{Hom}_{\mathrm{Rt}(\mathcal{R})}(X, Y) := \bigoplus_{i \in \mathbb{Z}} \mathrm{Hom}_{\mathrm{D}^b(\mathcal{R})}(X, Y[2i]).$$

Since $\mathcal{R}$ is an abelian category of finite global dimension (≤ 1) with enough projectives, the category $\mathrm{D}^b(\mathcal{R})$ is equivalent to the ($\mathbb{Z}$-graded) bounded homotopy category $\mathrm{Ho}^b(\mathcal{P})$. Thus we can equally well define $\mathrm{Rt}(\mathcal{R})$ as the orbit category of $\mathrm{Ho}^b(\mathcal{P})$.

LEMMA 3.1 ([**4**]). *There is a fully faithful functor*

$$D\colon \mathrm{Rt}(\mathcal{R}) \longrightarrow \mathrm{Ho}_{\mathbb{Z}_2}(\mathcal{P})$$

sending a $\mathbb{Z}$-graded complex of projectives $(P_i)_{i\in\mathbb{Z}}$ to the $\mathbb{Z}_2$-graded complex

$$\bigoplus\nolimits_{i\in\mathbb{Z}} P_{2i+1} \underset{0}{\overset{0}{\rightleftarrows}} \bigoplus\nolimits_{i\in\mathbb{Z}} P_{2i}.$$

3.3. Decompositions. From now on, we omit $\mathbb{Z}_2$ in the notations for categories and write

$$\begin{array}{ccc} \mathcal{C}(\mathcal{P}) = \mathcal{C}_{\mathbb{Z}_2}(\mathcal{P}), & \mathcal{C}(\mathcal{A}) = \mathcal{C}_{\mathbb{Z}_2}(\mathcal{A}), & \mathcal{C}(\mathcal{R}) = \mathcal{C}_{\mathbb{Z}_2}(\mathcal{R}), \\ \mathrm{Ho}(\mathcal{P}) = \mathrm{Ho}_{\mathbb{Z}_2}(\mathcal{P}), & \mathrm{Ho}(\mathcal{R}) = \mathrm{Ho}_{\mathbb{Z}_2}(\mathcal{R}). & \end{array}$$

The homology of a complex $M_\bullet \in \mathcal{C}(\mathcal{R})$ will be denoted

$$H_\bullet(M_\bullet) := (H_1(M_\bullet) \underset{0}{\overset{0}{\rightleftarrows}} H_0(M_\bullet)) \in \mathcal{C}(\mathcal{R}).$$

To each morphism $f : P \to Q$ in the category $\mathcal{P}$, we associate the following complexes

$$C_f := (P \underset{0}{\overset{f}{\rightleftarrows}} Q), \qquad C_f^* := (Q \underset{-f}{\overset{0}{\rightleftarrows}} P) \tag{3.1}$$

in $\mathcal{C}(\mathcal{P})$.

LEMMA 3.2. *Every complex of projectives $M_\bullet \in \mathcal{C}(\mathcal{P})$ can be decomposed uniquely, up to isomorphism, as a direct sum of complexes of the form*

$$M_\bullet = C_f \oplus C_g^*$$

for some injective morphisms f, g in $\mathcal{P}$ such that $H_0(M_\bullet) = \mathrm{coker}(f)$ and $H_1(M_\bullet) = \mathrm{coker}(g)$.

PROOF. Consider the short exact sequences

$$0 \longrightarrow \ker(d_1) \xrightarrow{\ i\ } M_1 \xrightarrow{\ p\ } \mathrm{im}(d_1) \longrightarrow 0,$$

$$0 \longrightarrow \ker(d_0) \xrightarrow{\ j\ } M_0 \xrightarrow{\ q\ } \mathrm{im}(d_0) \longrightarrow 0.$$

Since the the category $\mathcal{R}$ is hereditary by assumption, all the objects appearing in these sequences are projective. Thus the sequences split, and we can find morphisms

$$r\colon M_1 \to \ker(d_1), \quad k\colon \mathrm{im}(d_0) \to M_0, \quad l\colon \mathrm{im}(d_1) \to M_1, \quad s\colon M_0 \to \ker(d_0)$$

such that $r \circ i = \mathrm{id}$, $q \circ k = \mathrm{id}$, $p \circ l = \mathrm{id}$, and $s \circ j = \mathrm{id}$. This yields the following split exact sequence of morphisms of complexes

$$\begin{array}{ccc}
\ker(d_1) & \underset{m}{\overset{0}{\rightleftarrows}} & \mathrm{im}(d_0) \\
{\scriptstyle i}\downarrow & & \downarrow{\scriptstyle k} \\
M_1 & \underset{d_0}{\overset{d_1}{\rightleftarrows}} & M_0 \\
{\scriptstyle p}\downarrow & & \downarrow{\scriptstyle s} \\
\mathrm{im}(d_1) & \underset{0}{\overset{m'}{\rightleftarrows}} & \ker(d_0)
\end{array}$$

where m, m' denote the obvious inclusions. (Note that $d_0 = i \circ m \circ q$ and $d_1 = j \circ m' \circ p$.) The desired decomposition of $M_\bullet$ is thus given by setting: $f = m \circ q$, $g = -m' \circ l$.

Now suppose there is an isomorphism $M_\bullet \cong C_{f'} \oplus C^*_{g'}$ for some other pair f', g' of injective morphisms in $\mathcal{P}$. Then one can easily define corresponding isomorphisms of complexes $C_{f'} \cong C_f$ and $C_{g'} \cong C_g$, showing uniqueness. □

Given $M_\bullet \in \mathcal{C}(\mathcal{P})$, it will be convenient to write the decomposition in Lemma 3.2 as

$$M_\bullet = M^+_\bullet \oplus M^-_\bullet$$

where $M^+_\bullet = C_f$ and $M^-_\bullet = C^*_g$. Let the sign map

$$\varepsilon : \mathbb{Z}_2 \to \{+, -\}$$

be defined by $\varepsilon(0) = +$, $\varepsilon(1) = -$.

LEMMA 3.3. *Let $M_\bullet, N_\bullet \in \mathcal{C}(\mathcal{P})$. Then there is an isomorphism*

$$\mathrm{Hom}_{\mathcal{C}(\mathcal{R})}(M_\bullet, N_\bullet) \cong \mathrm{Hom}_{\mathcal{C}(\mathcal{R})}(H_\bullet(M_\bullet), H_\bullet(N_\bullet)) \oplus \Big\{ \bigoplus_{i,j \in \mathbb{Z}_2} \mathrm{Hom}_{\mathcal{R}}(M_i^{\varepsilon(j)}, N_{j+1}^{\varepsilon(i)}) \Big\}$$

of $\Bbbk$-vector spaces.

PROOF. First suppose that $P \xrightarrow{f} Q$, $P' \xrightarrow{g} Q'$ is a pair of injective morphisms in $\mathcal{P}$. We note that there is a short exact sequence

$$0 \longrightarrow \mathrm{Hom}_{\mathcal{R}}(Q, P') \longrightarrow \mathrm{Hom}_{\mathcal{C}(\mathcal{R})}(C_f, C_g) \longrightarrow \mathrm{Hom}_{\mathcal{R}}(X, Y) \longrightarrow 0 \tag{3.2}$$

for $X = \mathrm{coker}\, f$, $Y = \mathrm{coker}\, g$. One may also check directly that

$$\mathrm{Hom}_{\mathcal{C}(\mathcal{R})}(C_f, C^*_g) \cong \mathrm{Hom}_{\mathcal{R}}(P, Q'). \tag{3.3}$$

The decomposition of $\mathrm{Hom}_{\mathcal{C}(\mathcal{R})}(M_\bullet, N_\bullet)$ now follows easily by applying Lemma 3.2 to both $M_\bullet$ and $N_\bullet$, respectively, and by using the involutive shift functor $*$ together with (3.2) and (3.3). □

3.4. Acyclic complexes. Given a projective object $P \in \mathcal{P}$, there are associated acyclic complexes

$$K_P := (P \underset{0}{\overset{\mathrm{id}}{\rightleftarrows}} P), \qquad K^*_P := (P \underset{-\mathrm{id}}{\overset{0}{\rightleftarrows}} P). \tag{3.4}$$

Notice that $M_\bullet \in \mathcal{C}(\mathcal{P})$ is acyclic precisely if $M_\bullet \cong 0$ in $\mathrm{Ho}(\mathcal{P})$.

LEMMA 3.4. *If $M_\bullet \in \mathcal{C}(\mathcal{P})$ is an acyclic complexes of projectives, then there exist objects $P, Q \in \mathcal{P}$, unique up to isomorphism, such that $M_\bullet \cong K_P \oplus K^*_Q$.*

PROOF. If $M_\bullet$ is acyclic, then by Lemma 3.2 we have $M_\bullet = C_f \oplus C_g^*$ for some isomorphisms $f : P \cong P'$ and $g : Q \cong Q'$ of projectives. It follows that $M_\bullet \cong K_P \oplus K_Q^*$. Since the complexes K_P and K_Q^* are unique up to isomorphism, the objects P, Q are unique up to isomorphism as well. □

LEMMA 3.5. *Suppose that* $P \xrightarrow{f} Q$ *and* $P' \xrightarrow{f'} Q'$ *are injective morphisms in* $\mathcal{P}$. *Then* $\operatorname{coker} f \cong \operatorname{coker} f'$ *in* $\mathcal{R}$, *if and only if there is an isomorphism*

$$C_f \oplus K_{L'} \cong K_L \oplus C_{f'}$$

of complexes in $\mathcal{C}(\mathcal{P})$, *for some objects* $L, L' \in \mathcal{P}$.

PROOF. This is a reformulation of Schanuel's lemma. We refer to [**7**, Theorem 0.5.3] for the proof. □

PROPOSITION 3.6. *Suppose* $M_\bullet, M'_\bullet \in \mathcal{C}(\mathcal{P})$. *Then there exists an isomorphism*

$$M_\bullet \oplus K'_\bullet \cong K_\bullet \oplus M'_\bullet$$

for some acyclic complexes $K_\bullet, K'_\bullet \in \mathcal{C}(\mathcal{P})$, *if and only if* $H_\bullet(M_\bullet) \cong H_\bullet(M'_\bullet)$ *in* $\mathcal{C}(\mathcal{R})$.

PROOF. This follows directly from Lemmas 3.2 and 3.5, and by applying $*$ to the latter. □

3.5. Extensions of complexes. Given any morphism $s_\bullet : M_\bullet \to N_\bullet$ of complexes in $\mathcal{C}(\mathcal{P})$, we can form a corresponding exact sequence

$$0 \longrightarrow N_\bullet^* \longrightarrow \operatorname{Cone}(s_\bullet) \longrightarrow M_\bullet \longrightarrow 0$$

of complexes in $\mathcal{C}(\mathcal{P})$, where the middle term is defined by

$$\operatorname{Cone}(s_\bullet) = (N_0 \oplus M_1 \underset{d_0}{\overset{d_1}{\rightleftarrows}} N_1 \oplus M_0)$$

with

$$d_0 := \begin{bmatrix} -d_1^N & s_0 \\ 0 & d_0^M \end{bmatrix}, \qquad d_1 := \begin{bmatrix} -d_0^N & s_1 \\ 0 & d_1^M \end{bmatrix}.$$

This leads to the following result.

LEMMA 3.7 ([**4**]). *Let* $M_\bullet, N_\bullet \in \mathcal{C}(\mathcal{P})$. *The mapping* $s_\bullet \mapsto \operatorname{Cone}(s_\bullet)$ *defines an isomorphism*

$$\operatorname{Hom}_{\mathrm{Ho}(\mathcal{R})}(M_\bullet, N_\bullet^*) \cong \operatorname{Ext}^1_{\mathcal{C}(\mathcal{R})}(M_\bullet, N_\bullet).$$

We also have the following.

LEMMA 3.8. *Suppose* $P \xrightarrow{f} Q$, $P' \xrightarrow{g} Q'$ *are injective morphisms in the category* $\mathcal{P}$. *Let* $X, Y \in \mathcal{R}$ *denote the cokernels:* $X = \operatorname{coker} f$, $Y = \operatorname{coker} g$. *Then the following hold.*

(i) $\operatorname{Hom}_{\mathrm{Ho}(\mathcal{R})}(C_f, C_g) \cong \operatorname{Hom}_{\mathcal{R}}(X, Y)$;
(ii) $\operatorname{Hom}_{\mathrm{Ho}(\mathcal{R})}(C_f, C_g^*) \cong \operatorname{Ext}^1_{\mathcal{R}}(X, Y)$.

PROOF. The category $\mathcal{R}$ can be identified as a full subcategory of $\mathrm{D}^b(\mathcal{R})$ by considering any object in $\mathcal{R}$ as a complex concentrated in degree 0. It follows that

$$\operatorname{Hom}_{\mathcal{R}}(X, Y) \cong \operatorname{Hom}_{\mathrm{D}^b(\mathcal{R})}(X, Y) \cong \operatorname{Hom}_{\mathrm{Rt}(\mathcal{R})}(X, Y).$$

The objects X, Y have the projective resolutions

$$0 \to P \xrightarrow{f} Q \twoheadrightarrow X \to 0, \qquad 0 \to P' \xrightarrow{g} Q' \twoheadrightarrow Y \to 0.$$

So the complexes C_f, C_g are quasi-isomorphic to X, Y respectively, and isomorphisms (i), (ii) thus follow by Lemmas 3.1. □

Note that the isomorphism in Lemma 3.8 (i) may be given explicitly by $s_\bullet \mapsto s$, where $s : A \to B$ is the unique morphism making the diagram

$$\begin{array}{ccccccccc} 0 & \longrightarrow & P & \longrightarrow & Q & \longrightarrow & X & \longrightarrow & 0 \\ & & \downarrow{\scriptstyle s_1} & & \downarrow{\scriptstyle s_0} & & \downarrow{\scriptstyle s} & & \\ 0 & \longrightarrow & P' & \longrightarrow & Q' & \longrightarrow & Y & \longrightarrow & 0 \end{array} \tag{3.5}$$

commutative.

3.6. Hall algebras of complexes. We denote by $\mathcal{C}_{\mathsf{fin}}(\mathcal{P})$ the full subcategory of $\mathcal{C}(\mathcal{P})$ consisting of all complexes with finite homology, i.e. complexes $M_\bullet$ such that

$$H_\bullet(M_\bullet) = (H_1(M_\bullet) \underset{0}{\overset{0}{\rightleftarrows}} H_0(M_\bullet)) \in \mathcal{C}(\mathcal{A}).$$

The following result will be crucial for our definition of the Hall algebra of $\mathcal{C}_{\mathsf{fin}}(\mathcal{P})$.

LEMMA 3.9. *The set* $\operatorname{Ext}^1_{\mathcal{C}(\mathcal{R})}(M_\bullet, N_\bullet)$ *is finite for all* $M_\bullet, N_\bullet \in \mathcal{C}_{\mathsf{fin}}(\mathcal{P})$.

PROOF. This follows by using the involution $*$ together with Lemma 3.8 and combining with Lemmas 3.2 and 3.7. □

Since the category $\mathcal{C}_{\mathsf{fin}}(\mathcal{P})$ is not necessarily Hom-finite, we must consider a generalization of the coefficients appearing in the definition (2.1) of the Hall product. First define a bilinear map, $\mu : \mathcal{C}(\mathcal{P}) \times \mathcal{C}(\mathcal{P}) \to \mathbb{Q}$, given by

$$\begin{aligned} \mu(M_\bullet, N_\bullet) &:= \langle \hat{M}_0^+, \hat{N}_1^+ \rangle + \langle \hat{M}_1^+, \hat{N}_1^- \rangle + \langle \hat{M}_0^-, \hat{N}_0^+ \rangle + \langle \hat{M}_1^-, \hat{N}_0^- \rangle \\ &= \sum_{i,j} \langle \hat{M}_i^{\varepsilon(j)}, \hat{N}_{j+1}^{\varepsilon(i)} \rangle \end{aligned} \tag{3.6}$$

where $\langle\ ,\ \rangle : K(\mathcal{R}) \times K(\mathcal{R}) \to \mathbb{Q}$ denotes the generalized Euler form (Section 2.3).

Then for any $M_\bullet, N_\bullet, P_\bullet \in \mathcal{C}_{\mathsf{fin}}(\mathcal{P})$, we define

$$\mathrm{h}(M_\bullet, N_\bullet) := q^{\mu(M_\bullet, N_\bullet)} \,|\operatorname{Hom}_{\mathcal{C}(\mathcal{R})}(H_\bullet(M_\bullet), H_\bullet(N_\bullet))| \tag{3.7}$$

where q is the cardinality of $\Bbbk$, and we also write

$$\mathrm{e}(M_\bullet, N_\bullet)_{P_\bullet} := |\operatorname{Ext}^1_{\mathcal{C}(\mathcal{R})}(M_\bullet, N_\bullet)_{P_\bullet}|$$

which is well-defined by Lemma 3.9.

In the remainder, let us write $\mathcal{X} = \operatorname{Iso}(\mathcal{C}_{\mathsf{fin}}(\mathcal{P}))$ for the set of isomorphism classes in $\mathcal{C}_{\mathsf{fin}}(\mathcal{P})$.

DEFINITION 3.10. The Hall algebra $\mathcal{H}(\mathcal{C}_{\mathsf{fin}}(\mathcal{P}))$ is defined to be the $\mathbb{C}$-vector space with basis elements $[M_\bullet]$ indexed by isoclasses $M_\bullet \in \mathcal{X}$, and with multiplication defined by

$$[M_\bullet] \circledast [N_\bullet] := v^{\langle \hat{M}_0, \hat{N}_0 \rangle + \langle \hat{M}_1, \hat{N}_1 \rangle} \sum_{P_\bullet \in \mathcal{X}} \frac{\mathrm{e}(M_\bullet, N_\bullet)_{P_\bullet}}{\mathrm{h}(M_\bullet, N_\bullet)} [P_\bullet],$$

for all $M_\bullet, N_\bullet \in \mathcal{C}_{\mathsf{fin}}(\mathcal{P})$.

REMARK 3.11. Suppose there are complexes $M_\bullet, N_\bullet \in \mathcal{C}_{\mathsf{fin}}(\mathcal{P})$ such that $\mathrm{Hom}_{\mathcal{C}(\mathcal{R})}(M_\bullet, N_\bullet)$ is a finite set. Then it can be checked using Lemma 3.3 and the definition of Euler form that

$$|\,\mathrm{Hom}_{\mathcal{C}(\mathcal{R})}(M_\bullet, N_\bullet)| = \mathrm{h}(M_\bullet, N_\bullet).$$

The above definition thus generalizes the (twisted) Hall algebras of complexes of projectives defined by Bridgeland in [**4**].

3.7. Localization. As before, let us write $\hat{M}_\bullet = \hat{M}_0 - \hat{M}_1 \in K(\mathcal{P})$, for each $M_\bullet \in \mathcal{C}(\mathcal{P})$. The following result shows that the acyclic complexes K_P introduced in Section 3.4 define elements of $\mathcal{H}(\mathcal{C}_{\mathsf{fin}}(\mathcal{P}))$ with particularly simple properties.

LEMMA 3.12. *For any projective object $P \in \mathcal{P}$ and any complex $M_\bullet \in \mathcal{C}_{\mathsf{fin}}(\mathcal{P})$ the following identities hold in $\mathcal{H}(\mathcal{C}_{\mathsf{fin}}(\mathcal{P}))$:*

$$[K_P] \circledast [M_\bullet] = v^{\langle \hat{P}, \hat{M}_\bullet \rangle} \cdot [K_P \oplus M_\bullet], \tag{3.8}$$

$$[M_\bullet] \circledast [K_P] = v^{-\langle \hat{M}_\bullet, \hat{P} \rangle} \cdot [K_P \oplus M_\bullet]. \tag{3.9}$$

PROOF. It is easy to check directly from (3.6) that

$$\mu(K_P, M_\bullet) = \langle \hat{P}, \hat{M}_1^+ \rangle + \langle \hat{P}, \hat{M}_1^- \rangle, \quad \mu(M_\bullet, K_P) = \langle \hat{M}_0^+, \hat{P} \rangle + \langle \hat{M}_0^-, \hat{P} \rangle$$

so that

$$\mathrm{h}(K_P, M_\bullet) = q^{\langle \hat{P}, \hat{M}_1 \rangle}, \quad \mathrm{h}(M_\bullet, K_P) = q^{\langle \hat{M}_0, \hat{P} \rangle}.$$

The complexes K_P are homotopy equivalent to the zero complex, so Lemma 3.7 shows that the extension group in the definition of the Hall product vanishes. Taking into account Definition 3.10 gives the result. □

LEMMA 3.13. *For any projective object $P \in \mathcal{P}$ and any complex $M_\bullet \in \mathcal{C}_{\mathsf{fin}}(\mathcal{P})$ the following identities are true in $\mathcal{H}(\mathcal{C}_{\mathsf{fin}}(\mathcal{P}))$:*

$$[K_P] \circledast [M_\bullet] = v^{(\hat{P}, \hat{M}_\bullet)} [M_\bullet] \circledast [K_P], \tag{3.10}$$

$$[K_P^*] \circledast [M_\bullet] = v^{-(\hat{P}, \hat{M}_\bullet)} [M_\bullet] \circledast [K_P^*]. \tag{3.11}$$

PROOF. Equation (3.10) is immediate from Lemma 3.12. Equation (3.11) follows by applying the involution $*$. □

In particular, since $\hat{K}_P = 0 \in K(\mathcal{R})$, we have for $P, Q \in \mathcal{P}$,

$$[K_P] \circledast [K_Q] = [K_P \oplus K_Q], \quad [K_P] \circledast [K_Q^*] = [K_P \oplus K_Q^*], \tag{3.12}$$

$$[[K_P], [K_Q]] = [[K_P], [K_Q^*]] = [[K_P^*], [K_Q^*]] = 0, \tag{3.13}$$

where $[x, y] := x \circledast y - y \circledast x$. Note that any element of the form $[K_P] \circledast [K_P^*]$ is central.

Let us write $\mathcal{C}_0 \subset \mathcal{C}_{\mathsf{fin}}(\mathcal{P})$ to denote the full subcategory of all acyclic complexes of projectives. It then follows from (3.12) and (3.13) that the subspace $\mathcal{H}(\mathcal{C}_0) \subset \mathcal{H}(\mathcal{C}_{\mathsf{fin}}(\mathcal{P}))$ spanned by the isoclasses of objects in $\mathcal{C}_0$ is closed under the multiplication $\circledast$ and has the structure of a commutative associative algebra.

The following result is also clear.

LEMMA 3.14. *The left and right actions of $\mathcal{H}(\mathcal{C}_0)$ on $\mathcal{H}(\mathcal{C}_{\mathsf{fin}}(\mathcal{P}))$ given by restricting multiplication make the Hall algebra $\mathcal{H}(\mathcal{C}_{\mathsf{fin}}(\mathcal{P}))$ into an $\mathcal{H}(\mathcal{C}_0)$-bimodule.*

Notice that the basis $\mathcal{Z} := \{[M_\bullet] \in \mathcal{H}(\mathcal{C}_0)\}$ is a multiplicative subset in $\mathcal{H}(\mathcal{C}_0)$. Let us write $\mathcal{DH}_0(\mathcal{R}) = \mathcal{H}(\mathcal{C}_0)_{\mathcal{Z}}$ to denote the localization of $\mathcal{H}(\mathcal{C}_0)$ at $\mathcal{Z}$. More explicitly, we have

$$\mathcal{DH}_0(\mathcal{R}) = \mathcal{H}(\mathcal{C}_0)\big[[M_\bullet]^{-1} : M_\bullet \in \mathcal{C}_0\big].$$

The assignment $P \mapsto K_P$ extends to a group homomorphism

$$K\colon K(\mathcal{R}) \longrightarrow \mathcal{DH}_0(\mathcal{R})^\times.$$

This map is given explicitly by writing an element $\alpha \in K(\mathcal{R})$ in the form $\alpha = \hat{P} - \hat{Q}$ for objects $P, Q \in \mathcal{P}$ and then setting $K_\alpha := K_{\hat{P}} \circledast K_{\hat{Q}}^{-1}$. Composing with the involution $*$ gives another map

$$K^*\colon K(\mathcal{R}) \longrightarrow \mathcal{DH}_0(\mathcal{R})^\times.$$

Taking these maps together and extending linearly defines a $\mathbb{C}$-linear map from the group algebra

$$\mathbb{C}[K(\mathcal{R}) \times K(\mathcal{R})] \xrightarrow{\sim} \mathcal{DH}_0(\mathcal{R}),$$

which is an isomorphism by Lemma 3.4. It follows that the set $\{K_\alpha \circledast K_\beta^* \mid \alpha, \beta \in K(\mathcal{R})\}$ gives a $\mathbb{C}$-basis of $\mathcal{DH}_0(\mathcal{R})$

DEFINITION 3.15. The *localized Hall algebra*, $\mathcal{DH}(\mathcal{R})$, is the right $\mathcal{DH}_0(\mathcal{R})$-module obtained from $\mathcal{H}(\mathcal{C}_{\mathsf{fin}}(\mathcal{P}))$ by extending scalars,

$$\mathcal{DH}(\mathcal{R}) := \mathcal{H}(\mathcal{C}_{\mathsf{fin}}(\mathcal{P})) \otimes_{\mathcal{H}(\mathcal{C}_0)} \mathcal{DH}_0(\mathcal{R}).$$

That is, $\mathcal{DH}(\mathcal{R})$ is the localization $\mathcal{H}(\mathcal{C}_{\mathsf{fin}}(\mathcal{P}))_{\mathcal{Z}}$. We also consider $\mathcal{DH}(\mathcal{R})$ as a $\mathcal{DH}_0(\mathcal{R})$-bimodule by setting

$$(K_\alpha \circledast K_\beta^*) \circledast [M_\bullet] := v^{(\alpha-\beta,\hat{M}_\bullet)} \cdot [M_\bullet] \circledast (K_\alpha \circledast K_\beta^*) \tag{3.14}$$

for all $M_\bullet \in \mathcal{C}_{\mathsf{fin}}(\mathcal{P})$ and $\alpha, \beta \in K(\mathcal{R})$. We thus have a well-defined binary operation

$$- \circledast - : \mathcal{DH}(\mathcal{R}) \times \mathcal{DH}(\mathcal{R}) \to \mathcal{DH}(\mathcal{R})$$

which agrees with the map induced from the multiplication in Definition 3.10 by restricting along the canonical map $\mathcal{H}(\mathcal{C}_{\mathsf{fin}}(\mathcal{P})) \to \mathcal{DH}(\mathcal{R})$.

Given a complex $M_\bullet \in \mathcal{C}_{\mathsf{fin}}(\mathcal{P})$, define a corresponding element $E_{M_\bullet}$ in $\mathcal{DH}(\mathcal{R})$ given by

$$E_{M_\bullet} := v^{\langle \hat{M}_1^+ - \hat{M}_0^-, \hat{M}_\bullet\rangle} K_{-\hat{M}_1^+} \circledast K^*_{-\hat{M}_0^-} \circledast [M_\bullet].$$

Then we claim that

$$E_{M_\bullet \oplus K_\bullet} = E_{M_\bullet} \tag{3.15}$$

for any acyclic complex of projectives $K_\bullet \in \mathcal{C}_0$. Indeed, suppose that $K_\bullet = K_P \oplus K_Q^*$ for some $P, Q \in \mathcal{P}$. Then clearly $M_\bullet \hat{\oplus} K_\bullet = \hat{M}_\bullet$, and it follows by Lemma 3.12 that

$$\begin{aligned} E_{M_\bullet \oplus K_\bullet} &= v^{\langle M_1^+ \hat{\oplus} P - M_0^- \hat{\oplus} Q,\, M_\bullet \hat{\oplus} K_\bullet\rangle} \cdot K_{-M_1^+ \hat{\oplus} P} \circledast K^*_{-M_0^- \hat{\oplus} Q} \circledast [K_P \oplus K_Q^* \oplus M_\bullet] \\ &= v^{\langle \hat{M}_1^+ - \hat{M}_0^-,\, \hat{M}_\bullet\rangle} \cdot K_{-\hat{M}_1^+} \circledast K^*_{-\hat{M}_0^-} \circledast [M_\bullet], \end{aligned}$$

so we get the same element $E_{M_\bullet}$.

We note that a minimal projective resolution of $A \in \mathcal{A}$ need not be unique because the category $\mathcal{C}(\mathcal{P})$ is not Krull–Schmidt in general. However, it will be convenient to fix a (not necessarily minimal) resolution for each object $A \in A$.

DEFINITION 3.16.

(i) For each object $A \in \mathcal{A}$, fix a projective resolution

$$0 \longrightarrow P_A \xrightarrow{f_A} Q_A \longrightarrow A \longrightarrow 0 \tag{3.16}$$

and the complex C_A is defined to be $C_{f_A} \in \mathcal{C}(\mathcal{P})$.

(ii) Given objects $A, B \in \mathcal{A}$, write $E_{A,B}$ to denote the element $E_{C_A \oplus C_B^*}$ in $\mathcal{DH}(\mathcal{R})$.

The next lemma shows that the definition of $E_{A,B}$ is independent of the choice of resolutions defining C_A and C_B.

LEMMA 3.17. *Suppose $A, B \in \mathcal{A}$, and let $M_\bullet \in \mathcal{C}_{\mathsf{fin}}(\mathcal{P})$ be any complex such that $A \cong H_0(M_\bullet)$ and $B \cong H_1(M_\bullet)$. Then $E_{M_\bullet} = E_{A,B}$.*

PROOF. Let $M_\bullet$ be such a complex. By Proposition 3.6 there exist acyclic complexes $K_\bullet, K'_\bullet$ in $\mathcal{C}_0(\mathcal{P})$ such that $[M_\bullet \oplus K_\bullet] \cong [C_A \oplus C_B^* \oplus K'_\bullet]$, and the result follows from (3.15). □

The following result provides an explicit basis for the localized Hall algebra.

PROPOSITION 3.18. *The algebra $\mathcal{DH}(\mathcal{R})$ is free as a right $\mathcal{DH}_0(\mathcal{R})$-module, with basis consisting of elements $E_{A,B}$ indexed by all pairs of objects $A, B \in \mathrm{Iso}(\mathcal{A})$.*

PROOF. Suppose $M_\bullet \in \mathcal{X}$, and set $A = H_0(M_\bullet)$, $B = H_1(M_\bullet)$. Then $E_{M_\bullet} = E_{A,B}$ by Lemma 3.17, and one may check using (3.14) that

$$[M_\bullet] = v^{\langle \hat{Q} - \hat{P}, \hat{M}_\bullet \rangle} E_{A,B} \circledast [K_P \oplus K_Q^*],$$

for $P = M_1^+$ and $Q = M_0^-$. This shows that the elements $E_{A,B}$ span $\mathcal{DH}(\mathcal{R})$ as a $\mathcal{DH}_0(\mathcal{R})$-module.

It remains to check that the elements $E_{A,B}$ are $\mathcal{DH}_0(\mathcal{R})$–linearly independent. Notice that the Hall algebra $\mathcal{H}(\mathcal{C}_{\mathsf{fin}}(\mathcal{P}))$ is naturally graded as a $\mathbb{C}$-vector space by the set $\mathrm{Iso}(\mathcal{A}) \times \mathrm{Iso}(\mathcal{A})$:

$$\mathcal{H}(\mathcal{C}_{\mathsf{fin}}(\mathcal{P})) = \bigoplus_{(A,B) \in \mathrm{Iso}(\mathcal{A})^2} \mathcal{H}_{(A,B)}, \qquad \mathcal{H}_{(A,B)} := \bigoplus_{H_0(M_\bullet) \simeq A,\, H_1(M_\bullet) \simeq B} \mathbb{C}[M_\bullet].$$

Since the action of $\mathcal{H}(\mathcal{C}_0)$ on $\mathcal{H}(\mathcal{C}_{\mathsf{fin}}(\mathcal{P}))$ is $\mathrm{Iso}(\mathcal{A})^2$-homogeneous, it follows that $\mathcal{DH}(\mathcal{R})$ also has an $\mathrm{Iso}(\mathcal{A})^2$-grading. It is thus clear that the elements $E_{A,B}$ span distinct graded components of $\mathcal{DH}(\mathcal{R})$. To see that each component is a free $\mathcal{DH}_0(\mathcal{R})$-module of rank one, it remains to check that for each $y \in \mathcal{DH}(\mathcal{R})$, we have $E_{A,B} \circledast y = 0$ implies $y = 0$.

Let us write $M_\bullet = C_A \oplus C_B^*$. Then it will suffice to show that for any $x \in \mathcal{H}(\mathcal{C}_0)$, the element

$$[M_\bullet] \circledast x \in \mathcal{H}(\mathcal{C}_{\mathsf{fin}}(\mathcal{P}))$$

is a $\mathcal{Z}$-torsion element only if $x = 0$. Suppose that

$$x = c_1 z_1 + \cdots + c_n z_n$$

for some constants $c_1, \ldots, c_n \in \mathbb{C}$ and distinct elements $z_1, \ldots, z_n \in \mathcal{Z}$. One may check using Lemma 3.4 and condition (c) in Section 1.1, that for any $z \in \mathcal{Z}$ the elements $z_1 \circledast z, \ldots, z_n \circledast z$ are also distinct. Next suppose that $[M_\bullet] \circledast (x \circledast z) = 0$ in $\mathcal{H}(\mathcal{C}_{\mathsf{fin}}(\mathcal{P}))$. This gives an equation

$$c_1 \cdot [M_\bullet] \circledast z_1 \circledast z + \cdots + c_n \cdot [M_\bullet] \circledast z_n \circledast z = 0.$$

One may again use condition (c) together with Lemmas 3.2 and 3.4 to check that the terms appearing in this dependence relation are unit multiples of distinct basis elements in $\mathcal{X}$. So the relation must be trivial: $c_1 = \cdots = c_n = 0$, which gives $x = 0$. This completes the proof. □

4. Associativity via the Drinfeld double

In this section we prove that $\mathcal{DH}(\mathcal{R})$ is the Drinfeld double of the bialgebra $\widetilde{\mathcal{H}_v}(\mathcal{R})$ under suitable finiteness conditions. As a corollary, we show that $\mathcal{DH}(\mathcal{R})$ is an associative algebra with respect to the multiplication described in the previous section.

4.1. Multiplication formulas. Suppose $A, B \in \mathcal{A}$ and recall the element $E_{A,B}$ in $\mathcal{DH}(\mathcal{R})$ defined in Definition 3.16. Notice that the image under the involution $*$ is given by $E^*_{A,B} = E_{B,A}$. Let us write

$$E_A := E_{A,0}, \qquad F_B := E_{0,B}$$

so that $F_A = E^*_A$.

LEMMA 4.1. *Suppose $A, B \in \mathcal{A}$. The following equality holds in $\mathcal{DH}(\mathcal{R})$.*

$$E_A \circledast E_B = v^{\langle A,B\rangle} \sum_{C\in \mathrm{Iso}(\mathcal{A})} \frac{|\operatorname{Ext}^1_{\mathcal{R}}(A,B)_C|}{|\operatorname{Hom}_{\mathcal{R}}(A,B)|} E_C$$

PROOF. Let C_A, C_B be the complexes associated to A, B in Definition 3.16. Then using the formula

$$\mathrm{h}(C_A, C_B) = q^{\langle \hat{Q}_A, \hat{P}_B\rangle} \cdot |\operatorname{Hom}_{\mathcal{R}}(A,B)|$$

together with the relations $\hat{A} = \hat{Q}_A - \hat{P}_A$ and $\hat{B} = \hat{Q}_B - \hat{P}_B$ in the Grothendieck group $K(\mathcal{R})$, we have

$$[C_A] \circledast [C_B] = v^{\langle \hat{P}_A,\hat{P}_B\rangle + \langle \hat{Q}_A,\hat{Q}_B\rangle} \sum_{M_\bullet\in\mathcal{X}} \frac{\mathrm{e}(C_A,C_B)_{M_\bullet}}{\mathrm{h}(C_A,C_B)} [M_\bullet]$$

$$= v^{\langle A,B\rangle - \langle \hat{A},\hat{P}_B\rangle + \langle \hat{P}_A,\hat{B}\rangle} \sum_{M_\bullet\in\mathcal{X}} \frac{|\operatorname{Ext}^1_{\mathcal{C}(\mathcal{R})}(C_A,C_B)_{M_\bullet}|}{|\operatorname{Hom}_{\mathcal{R}}(A,B)|} [M_\bullet].$$

It follows by (3.14) that

$$E_A \circledast E_B = v^{\langle \hat{P}_A,\hat{A}\rangle + \langle \hat{P}_B,\hat{B}\rangle + (\hat{P}_B,\hat{A})} \cdot K_{-P_A \hat{\oplus} P_B} \circledast [C_A] \circledast [C_B]$$

$$= v^{\langle A,B\rangle + \langle \hat{P}_A+\hat{P}_B,\, \hat{A}+\hat{B}\rangle} \sum_{M_\bullet\in\mathcal{X}} \frac{|\operatorname{Ext}^1_{\mathcal{C}(\mathcal{R})}(C_A,C_B)_{M_\bullet}|}{|\operatorname{Hom}_{\mathcal{R}}(A,B)|} K_{-P_A\hat{\oplus}P_B} \circledast [M_\bullet]. \tag{4.1}$$

Consider an extension

$$0 \longrightarrow C_B \longrightarrow M_\bullet \longrightarrow C_A \longrightarrow 0. \tag{4.2}$$

By Lemma 3.7, we may assume $M_\bullet = \mathrm{Cone}(s_\bullet)$ for some morphism $s_\bullet : C_A \to C^*_B$, so that

$$M_\bullet = P_B \oplus P_A \underset{d_0}{\overset{d_1}{\rightleftarrows}} Q_B \oplus Q_A,$$

where

$$d_1 = \begin{pmatrix} f_B & s_1 \\ 0 & f_A \end{pmatrix}, \qquad d_0 = \begin{pmatrix} 0 & s_0 \\ 0 & 0 \end{pmatrix}.$$

Since f_A, f_B are monomorphisms, so is d_1. Thus $d_1 \circ d_0 = 0$ implies that $s_0 = 0$. Setting $C = H_0(M_\bullet)$, it follows that (4.2) induces an extension

$$0 \longrightarrow B \longrightarrow C \longrightarrow A \longrightarrow 0.$$

One may check that this extension agrees with the corresponding image of (4.2) under the isomorphism

$$\operatorname{Ext}^1_{\mathcal{C}(\mathcal{R})}(C_A, C_B) \cong \operatorname{Ext}^1_{\mathcal{R}}(A, B)$$

given by Lemma 3.7 and Lemma 3.8 (ii). It follows that

$$\sum_{H_0(M_\bullet)=C} |\operatorname{Ext}^1_{\mathcal{C}(\mathcal{R})}(C_A, C_B)_{M_\bullet}| \;=\; |\operatorname{Ext}^1_{\mathcal{R}}(A, B)_C|.$$

Finally, notice that $\hat{C} = \hat{A} + \hat{B}$ for any extension C of A by B. Putting everything together shows that equation (4.1) becomes

$$E_A \circledast E_B = v^{\langle A,B\rangle} \sum_{M_\bullet \in \mathcal{X}} v^{\langle \hat{P}_A + \hat{P}_B,\, \hat{H}_0(M_\bullet)\rangle} \cdot \frac{|\operatorname{Ext}^1_{\mathcal{C}(\mathcal{R})}(C_A, C_B)_{M_\bullet}|}{|\operatorname{Hom}_{\mathcal{R}}(A, B)|} K_{-P_A \hat{\oplus} P_B} \circledast [M_\bullet]$$

$$= v^{\langle A,B\rangle} \sum_{C \in \operatorname{Iso}(\mathcal{A})} \frac{|\operatorname{Ext}^1_{\mathcal{R}}(A, B)_C|}{|\operatorname{Hom}_{\mathcal{R}}(A, B)|} E_C$$

which completes the proof. □

LEMMA 4.2. *Let $A, B \in \mathcal{A}$. The following equations hold in $\mathcal{DH}(\mathcal{R})$,*

(i) $E_A \circledast F_B = \displaystyle\sum_{A_1, B_1, B_2} v^{\langle \hat{B} - \hat{B}_1,\, \hat{A} - \hat{B}\rangle} g^B_{B_1, B_2} g^A_{B_2, A_1} a_{B_2} \cdot K^*_{\hat{B} - \hat{B}_1} \circledast E_{A_1, B_1},$

(ii) $F_B \circledast E_A = \displaystyle\sum_{A_1, A_2, B_1} v^{\langle \hat{A} - \hat{A}_1,\, \hat{B} - \hat{A}\rangle} g^A_{A_1, A_2} g^B_{A_2, B_1} a_{A_2} \cdot K_{\hat{A} - \hat{A}_1} \circledast E_{A_1, B_1},$

where each sum runs over classes of objects in $\operatorname{Iso}(\mathcal{A})$.

PROOF.

(i) Again let C_A, C_B be complexes associated to A, B as in Definition 3.16. By definition, the product $[C_A] \circledast [C^*_B]$ is equal to

$$v^{\langle \hat{Q}_A, \hat{P}_B\rangle + \langle \hat{P}_A, \hat{Q}_B\rangle} \sum_{M_\bullet \in \mathcal{X}} \frac{\mathrm{e}(C_A, C^*_B)_{M_\bullet}}{\mathrm{h}(C_A, C^*_B)} [M_\bullet].$$

Then using $\mu(C_A, C^*_B) = \langle \hat{P}_A, \hat{Q}_B\rangle$, it is easy to check that

$$\mathrm{h}(C_A, C^*_B) = q^{\langle \hat{P}_A, \hat{Q}_B\rangle}.$$

This gives

$$[C_A] \circledast [C^*_B] = v^{\langle \hat{A}, \hat{P}_B\rangle - \langle \hat{P}_A, \hat{B}\rangle} \sum_{M_\bullet \in \mathcal{X}} \mathrm{e}(C_A, C^*_B)_{M_\bullet} \cdot [M_\bullet],$$

where we have used the equalities $\hat{Q}_A = \hat{P}_A + \hat{A}$ and $\hat{Q}_B = \hat{P}_B + \hat{B}$ in $K(\mathcal{R})$. It thus follows by (3.14) that

$$
\begin{aligned}
E_A \circledast F_B &= v^{\langle \hat{P}_A, \hat{A}\rangle + \langle \hat{P}_B, \hat{B}\rangle - (\hat{P}_B, \hat{A})} \cdot K_{-\hat{P}_A} \circledast K^*_{-\hat{P}_B} \circledast [C_A] \circledast [C_B^*] \\
&= v^{\langle \hat{P}_A - \hat{P}_B, \hat{A} - \hat{B}\rangle} \cdot K_{-\hat{P}_A} \circledast K^*_{-\hat{P}_B} \circledast \sum_{M_\bullet \in \mathcal{X}} \mathrm{e}(C_A, C_B^*)_{M_\bullet} \cdot [M_\bullet]. \qquad (4.3)
\end{aligned}
$$

Now suppose that $M_\bullet$ is an extension of C_A by C_B^*. By Lemma 3.7, we may assume that $M_\bullet = \mathrm{Cone}(s_\bullet)$ for some $s_\bullet \in \mathrm{Hom}_{\mathrm{Ho}(\mathcal{R})}(C_A, C_B)$. The extension thus takes the form

$$
\begin{array}{ccc}
Q_B & \overset{0}{\underset{-f_B}{\rightleftarrows}} & P_B \\
{\scriptstyle i_1}\downarrow & & \downarrow{\scriptstyle i_0} \\
Q_B \oplus P_A & \overset{f_1}{\underset{f_0}{\rightleftarrows}} & P_B \oplus Q_A \\
{\scriptstyle p_1}\downarrow & & \downarrow{\scriptstyle p_0} \\
P_A & \overset{f_A}{\underset{0}{\rightleftarrows}} & Q_A
\end{array}
$$

where $f_0 = \begin{pmatrix} -f_B & s_0 \\ 0 & 0 \end{pmatrix}$, $f_1 = \begin{pmatrix} 0 & s_1 \\ 0 & f_A \end{pmatrix}$, and f_A, f_B are defined in (3.16). This extension induces an exact commutative diagram

$$
\begin{array}{ccccccc}
 & & 0 & \longrightarrow & H_0(M_\bullet) & \longrightarrow & H_0(C_A) & & \\
 & & \downarrow & & \downarrow & & \downarrow & & \\
 & & P_B & \xrightarrow{i_0} & M_0/\operatorname{im} f_1 & \xrightarrow{p_0} & Q_A/\operatorname{im} f_A & \longrightarrow & 0 \\
 & & {\scriptstyle -f_B}\downarrow & & \downarrow{\scriptstyle f_0} & & \downarrow & & \\
0 & \longrightarrow & Q_B & \xrightarrow{i_1} & \ker f_1 & \xrightarrow{p_1} & 0 & & \\
 & & \downarrow & & \downarrow & & \downarrow & & \\
 & & H_1(C_B^*) & \longrightarrow & H_1(M_\bullet) & \longrightarrow & 0, & &
\end{array} \qquad (4.4)
$$

where "i_0" denotes the map induced by i_0, etc.

Since the map induced by i_1 in (4.4) is an isomorphism, it follows that the direct summands in the decomposition $M_\bullet = M_\bullet^+ \oplus M_\bullet^-$ of Lemma 3.2 have the form

$$
M_\bullet^+ = (P_A \overset{f_1^+}{\underset{0}{\rightleftarrows}} M_0^+), \quad M_\bullet^- = (Q_B \overset{0}{\underset{f_0^-}{\rightleftarrows}} M_0^-)
$$

where the maps f_0^-, f_1^+ are obtained from f_0, f_1 by restriction.

The objects $A_1 = H_0(M_\bullet)$ and $B_1 = H_1(M_\bullet)$ thus have projective resolutions

$$
P_A \xrightarrow{f_1^+} M_0^+ \longrightarrow A_1 \longrightarrow 0, \qquad M_0^- \xrightarrow{f_0^-} Q_B \longrightarrow B_1 \longrightarrow 0
$$

respectively. This gives relations

$$
\hat{M}_1^+ = \hat{P}_A, \qquad \hat{M}_0^- = \hat{Q}_B - \hat{B}_1 = \hat{P}_B + \hat{B} - \hat{B}_1
$$

in $K(\mathcal{R})$. Substituting in (4.3), we have

$$
\begin{aligned}
&E_A \circledast F_B \\
&= \sum_{M_\bullet \in \mathcal{X}} v^{\langle \hat{M}_1^+ - \hat{M}_0^- + \hat{B} - \hat{B}_1, \hat{A} - \hat{B}\rangle} \cdot \mathrm{e}(C_A, C_B^*)_{M_\bullet} \cdot K_{-\hat{M}_0^+} \circledast K^*_{-\hat{M}_0^- + \hat{B} - \hat{B}_1} \circledast [M_\bullet] \\
&= \sum_{M_\bullet \in \mathcal{X}} v^{\langle \hat{B} - \hat{B}_1, \hat{A} - \hat{B}\rangle} \cdot \mathrm{e}(C_A, C_B^*)_{M_\bullet} \cdot K^*_{\hat{B} - \hat{B}_1} \circledast E_{M_\bullet}.
\end{aligned} \tag{4.5}
$$

One may check directly using (4.4) that the map $s : A \to B$ induced by (3.5) coincides with the connecting homomorphism $H_0(C_A) \xrightarrow{\delta} H_1(C_B)$ in the long exact sequence of cohomology. In particular, note that $H_0(M_\bullet) \simeq \ker s$, and $H_1(M_\bullet) \simeq \operatorname{coker} s$.

Hence, we may conclude that

$$
\begin{aligned}
\sum_{\substack{P_\bullet \in \mathcal{X} \\ H_0(P_\bullet) \simeq A_1,\, H_1(P_\bullet) \simeq B_1}} & \mathrm{e}(C_A, C_B^*)_{P_\bullet} \cdot E_{P_\bullet} \\
&= |\{h \in \mathrm{Hom}_{\mathcal{R}}(A, B) \mid \ker h \simeq A_1, \operatorname{coker} h \simeq B_1\}| \cdot E_{A_1, B_1}.
\end{aligned} \tag{4.6}
$$

By the equality on [**23**, p.984], the preceding equation may be rewritten as

$$
\sum_{\substack{H_0(P_\bullet) \simeq A_1, \\ H_1(P_\bullet) \simeq B_1}} \mathrm{e}(C_A, C_B^*)_{P_\bullet} \cdot E_{P_\bullet} = \sum_{B_2 \in \mathrm{Iso}(\mathcal{A})} g^B_{B_1, B_2} g^A_{B_2, A_1} a_{B_2} \cdot E_{A_1, B_1}. \tag{4.7}
$$

The equality in part (i) is now obtained by combining (4.3), (4.7) and (2.3).

(ii) This follows by interchanging A and B in (i) and taking $*$ on both sides. □

4.2. Embedding $\widetilde{\mathcal{H}}_v(\mathcal{R})$ in $\mathcal{DH}(\mathcal{R})$. In this subsection we make some more precise statements about the relationships between the various Hall algebras we have been considering.

Consider the injective linear map $I_+ \colon \widetilde{\mathcal{H}}_v(\mathcal{R}) \hookrightarrow \mathcal{DH}(\mathcal{R})$ defined by

$$[A] * K_\alpha \mapsto E_A \circledast K_\alpha,$$

and let $\mathcal{DH}^+(\mathcal{R}) \subset \mathcal{DH}(\mathcal{R})$ denote the image of this map.

PROPOSITION 4.3. *The restriction of multiplication in $\mathcal{DH}(\mathcal{R})$ makes the subspace $\mathcal{DH}^+(\mathcal{R})$ into an associative algebra, and the embedding $I_+ \colon \widetilde{\mathcal{H}}_v(\mathcal{R}) \hookrightarrow \mathcal{DH}(\mathcal{R})$ restricts to an isomorphism $\widetilde{\mathcal{H}}_v(\mathcal{R}) \cong \mathcal{DH}^+(\mathcal{R})$ of (unital) associative algebras.*

PROOF. The result follows from Lemma 4.1, together with a comparison of the relations (2.6) defining the extended Hall algebra with the relation (3.14) in the localized Hall algebra. □

Composing with the involution $*$ gives another embedding

$$I_- \colon \widetilde{\mathcal{H}}_v(\mathcal{R}) \hookrightarrow \mathcal{DH}(\mathcal{R}),$$

defined by $[B] * K_\beta \mapsto F_B \circledast K_\beta^*$, whose image $\mathcal{DH}^-(\mathcal{R})$ is again an associative algebra such that I_- restricts to an algebra isomorphism $\widetilde{\mathcal{H}}_v(\mathcal{R}) \cong \mathcal{DH}^-(\mathcal{R})$.

4.3. Drinfeld double of $\widetilde{\mathcal{H}_v}(\mathcal{R})$.

LEMMA 4.4. *The multiplication map* $\nabla\colon a \otimes b \mapsto I_+(a) \circledast I_-(b)$ *defines an isomorphism of vector spaces*

$$\nabla\colon \widetilde{\mathcal{H}_v}(\mathcal{R}) \otimes_{\mathbb{C}} \widetilde{\mathcal{H}_v}(\mathcal{R}) \longrightarrow \mathcal{DH}(\mathcal{R}).$$

PROOF. It follows from Proposition 3.18 that the algebra $\mathcal{DH}(\mathcal{R})$ has a $\mathbb{C}$-basis consisting of elements

$$E_{A,B} \circledast K_\alpha \circledast K^*_\beta, \quad A, B \in \operatorname{Iso}(\mathcal{A}), \quad \alpha, \beta \in K(\mathcal{R}).$$

Recall the partial order on $K(\mathcal{A})$ defined in (2.9) and define $\mathcal{DH}_{\leq\gamma}$ for $\gamma \in K(\mathcal{A})$ to be the subspace of $\mathcal{DH}(\mathcal{R})$ spanned by elements from this basis for which $\boldsymbol{k}_{\mathcal{A}}(A) + \boldsymbol{k}_{\mathcal{A}}(B) \leq \gamma$. We claim that

$$\mathcal{DH}_{\leq\gamma} \circledast \mathcal{DH}_{\leq\delta} \subset \mathcal{DH}_{\leq\gamma+\delta}, \qquad \gamma, \delta \in K(\mathcal{A}), \tag{4.8}$$

so that this defines a filtration on $\mathcal{DH}(\mathcal{R})$.

Suppose that $M_\bullet, N_\bullet \in \mathcal{C}_{\mathsf{fin}}(\mathcal{P})$ and let

$$\gamma = \boldsymbol{k}_{\mathcal{A}}(H_0(M_\bullet)) + \boldsymbol{k}_{\mathcal{A}}(H_1(M_\bullet)), \quad \delta = \boldsymbol{k}_{\mathcal{A}}(H_0(N_\bullet)) + \boldsymbol{k}_{\mathcal{A}}(H_1(N_\bullet)).$$

Then consider an extension of complexes

$$0 \longrightarrow M_\bullet \longrightarrow P_\bullet \longrightarrow N_\bullet \longrightarrow 0.$$

The long exact sequence in homology can be split to give two long exact sequences

$$0 \longrightarrow K \longrightarrow H_0(M_\bullet) \longrightarrow H_0(P_\bullet) \longrightarrow H_0(N_\bullet) \longrightarrow L \longrightarrow 0,$$
$$0 \longrightarrow L \longrightarrow H_1(M_\bullet) \longrightarrow H_1(P_\bullet) \longrightarrow H_1(N_\bullet) \longrightarrow K \longrightarrow 0$$

for some objects $K, L \in \mathcal{A}$. It follows that there is a relation in $K(\mathcal{A})$,

$$\gamma + \delta = \boldsymbol{k}_{\mathcal{A}}(H_0(P_\bullet)) + \boldsymbol{k}_{\mathcal{A}}(H_1(P_\bullet)) + 2(\boldsymbol{k}_{\mathcal{A}}(K) + \boldsymbol{k}_{\mathcal{A}}(L))$$

which proves (4.8).

Suppose now that $N_\bullet = C_A$ and $M_\bullet = C^*_B$ for some objects $A, B \in \mathcal{A}$. Then $K = 0$, and by Lemmas 3.1, 3.7, and 3.8

$$\operatorname{Ext}^1_{\mathcal{C}(\mathcal{R})}(N_\bullet, M_\bullet) = \operatorname{Hom}_{\mathcal{R}}(A, B),$$

and the extension class is completely determined by the connecting morphism $H_0(N_\bullet) \to H_1(M_\bullet)$. By Proposition 2.4, we therefore know that $\boldsymbol{k}_{\mathcal{A}}(L) = 0$ exactly when the extension is trivial. It follows that in the graded algebra associated to the filtered algebra $\mathcal{DH}(\mathcal{R})$, one has a relation

$$\nabla([A] * K_\alpha \otimes [B] * K_\beta) = v^{-(\alpha,\hat{B})} \cdot E_{A,B} \circledast K_\alpha \circledast K^*_\beta.$$

It follows that ∇ takes a basis to a basis and is hence an isomorphism. □

As a corollary, we have

COROLLARY 4.5. *The algebra* $\mathcal{DH}(\mathcal{R})$ *has a linear basis consisting of elements*

$$E_A \circledast K_\alpha \circledast F_B \circledast K^*_\beta, \quad A, B \in \operatorname{Iso}(\mathcal{A}),\ \alpha, \beta \in K(\mathcal{R}).$$

Now we state the main result of this section.

THEOREM 4.6. *Suppose that* $\mathcal{A} \subset \mathcal{R}$ *satisfies conditions* (2.12) *and* (2.13). *Then the algebra* $\mathcal{DH}(\mathcal{R})$ *is isomorphic to the Drinfeld double of the bialgebra* $\widetilde{\mathcal{H}_v}(\mathcal{R})$.

PROOF. Because of the description of the basis of $\mathcal{DH}(\mathcal{R})$ (Corollary 4.5) and the definition of Drinfeld double, the proof of the theorem is reduced to check equation (2.14) for the elements consisting of the basis of $\widetilde{\mathcal{H}_v}(\mathcal{R})$.

Let us write equation (2.14) in the present situation:

$$\sum(a_2,b_1)_H\cdot I_+(a_1)\circledast I_-(b_2)\stackrel{?}{=}\sum(a_1,b_2)_H\cdot I_-(b_1)\circledast I_+(a_2). \tag{4.9}$$

Now let $A,B\in\mathcal{A}$ and $\alpha,\beta\in K(\mathcal{R})$. Let us write

$$\Delta([A]K_\alpha)=\sum_{A_1,A_2}v^{\langle A_1,A_2\rangle}g^A_{A_1,A_2}\cdot([A_1]K_{\hat{A}_2+\alpha})\otimes([A_2]K_\alpha),$$
$$\Delta([B]K_\beta)=\sum_{B_2,B_1}v^{\langle B_2,B_1\rangle}g^B_{B_2,B_1}\cdot([B_2]K_{\hat{B}_1+\beta})\otimes([B_1]K_\beta).$$

By the Hopf pairing (Definition 2.6 and Proposition 2.7) and (3.14), the left hand side of (4.9) becomes

$$\begin{aligned}\text{LHS of (4.9)}=&\sum_{A_1,A_2,B_1,B_2}v^{\langle A_1,A_2\rangle}g^A_{A_1,A_2}v^{\langle B_2,B_1\rangle}g^B_{B_2,B_1}([A_2]K_\alpha,[B_2]K_{\hat{B}_1+\beta})_H\\&\qquad\cdot E_{A_1}\circledast K_{\hat{A}_2+\alpha}\circledast F_{B_1}\circledast K^*_\beta\\=&v^{(\alpha,\beta)}\sum_{A_1,A_2,B_1,B_2}v^{\langle A_1,A_2\rangle+\langle B_2,B_1\rangle}g^A_{A_1,A_2}g^B_{B_2,B_1}([A_2],[B_2])_H\\&\qquad\cdot E_{A_1}\circledast K_{\hat{A}_2}\circledast F_{B_1}\circledast K_\alpha\circledast K^*_\beta.\end{aligned}$$

Similarly, the right hand side becomes

RHS of (4.9)

$$=v^{(\alpha,\beta)}\sum_{\substack{A_1,A_2,\\B_1,B_2}}v^{\langle A_1,A_2\rangle+\langle B_2,B_1\rangle}g^A_{A_1,A_2}g^B_{B_2,B_1}([A_1],[B_1])_H\cdot F_{B_2}\circledast K^*_{\hat{B}_1}\circledast E_{A_2}\circledast K_\alpha\circledast K^*_\beta.$$

After removing the term $v^{(\alpha,\beta)}\cdot K_\alpha\circledast K^*_\beta$ from both sides, equation (4.9) reduces to

(4.10)

$$\begin{aligned}&\sum_{A_1,A_2,B_1,B_2}v^{\langle A_1,A_2\rangle+\langle B_2,B_1\rangle}g^A_{A_1,A_2}g^B_{B_2,B_1}([A_2],[B_2])_H\cdot E_{A_1}\circledast K_{\hat{A}_2}\circledast F_{B_1}\\&\stackrel{?}{=}\sum_{A_1,A_2,B_1,B_2}v^{\langle A_1,A_2\rangle+\langle B_2,B_1\rangle}g^A_{A_1,A_2}g^B_{B_2,B_1}([A_1],[B_1])_H\cdot F_{B_2}\circledast K^*_{\hat{B}_1}\circledast E_{A_2}.\end{aligned}$$

By Definition 2.6 and (3.14), the left hand side of (4.10) becomes

$$\begin{aligned}&\text{LHS of (4.10)}\\&=\sum_{A_1,A_2,B_1,B_2}v^{\langle A_1,A_2\rangle+\langle B_2,B_1\rangle}g^A_{A_1,A_2}g^B_{B_2,B_1}a_{A_2}\delta_{A_2,B_2}\cdot E_{A_1}\circledast K_{\hat{A}_2}\circledast F_{B_1}\\&=\sum_{A_1,A_2,B_1}v^{\langle A_1,A_2\rangle+\langle A_2,B_1\rangle}g^A_{A_1,A_2}g^B_{A_2,B_1}a_{A_2}\cdot E_{A_1}\circledast K_{\hat{A}_2}\circledast F_{B_1}\\&=\sum_{A_1,A_2,B_1}v^{\langle\hat{A}_2,\hat{B}_1\rangle-\langle\hat{A}_2,\hat{A}_1\rangle}g^A_{A_1,A_2}g^B_{A_2,B_1}a_{A_2}\cdot K_{\hat{A}_2}\circledast E_{A_1}\circledast F_{B_1}.\end{aligned}$$

Thus by Lemma 4.2 (i) we have

$$\text{LHS of (4.10)} = \sum_{A_1,A_2,A_3,B_1,B_2,B_3} v^{\langle \hat{B}_2-\hat{A}_2,\, \hat{A}-\hat{B}\rangle} g^{A_1}_{B_2,A_3} g^{A}_{A_1,A_2} g^{B}_{A_2,B_1} g^{B_1}_{B_3,B_2} \cdot a_{A_2} a_{B_2} \cdot K_{\hat{A}_2} \circledast K^*_{\hat{B}_2} \circledast E_{A_3,B_3}.$$

Similar computations using Lemma 4.2 (ii) show that the right hand side of (4.10) becomes

$$\text{RHS of (4.10)} = \sum_{A_1,A_2,A_3,B_1,B_2,B_3} v^{\langle \hat{A}_1-\hat{B}_1,\, \hat{B}-\hat{A}\rangle} g^{A_2}_{A_3,A_1} g^{B_2}_{A_1,B_3} g^{B}_{B_2,B_1} g^{A}_{B_1,A_2} \cdot a_{B_1} a_{A_1} \cdot K_{\hat{A}_1} \circledast K^*_{\hat{B}_1} \circledast E_{A_3,B_3}.$$

It follows by associativity (2.3) that this can be rewritten as

$$\text{RHS of (4.10)} = \sum_{A_1,A_3,B_1,B_3} v^{\langle \hat{A}_1-\hat{B}_1,\, \hat{B}-\hat{A}\rangle} \sum_{A'_2,B'_2} g^{A'_2}_{B_1,A_3} g^{A}_{A'_2,A_1} g^{B}_{A_1,B'_2} g^{B'_2}_{B_3,B_1} \cdot a_{B_1} a_{A_1} \cdot K_{\hat{A}_1} \circledast K^*_{\hat{B}_1} \circledast E_{A_3,B_3}.$$

One may check that this expression agrees with the LHS of (4.10), which completes the proof. □

It is now possible to verify that the multiplication in $\mathcal{DH}(\mathcal{R})$ is associative.

COROLLARY 4.7. *If the category $\mathcal{R}$ is Hom-finite (so that $\mathcal{A} = \mathcal{R}$) or if the subcategory $\mathcal{A} \subset \mathcal{R}$ satisfies conditions* (2.12) *and* (2.13), *then the algebra $\mathcal{DH}(\mathcal{R})$ is associative.*

PROOF. If $\mathcal{R} = \mathcal{A}$, then it follows by Remark 3.11 that the algebra $\mathcal{DH}(\mathcal{R})$ is isomorphic to the localized Hall algebra $\mathcal{DH}(\mathcal{A})$ defined in [**4**]. It follows by results in [**4**] that the category $\mathcal{C}(\mathcal{A})$ is Hom-finite, so that $\mathcal{H}(\mathcal{C}(\mathcal{P}))$ and $\mathcal{DH}(\mathcal{A})$ are both associative in this case.

The remaining statement is a consequence of Theorem 4.6 since the multiplication in the Drinfeld double is associative. □

REMARK 4.8. From the above result we can only conclude that the algebra $\mathcal{H}(\mathcal{C}_{\mathsf{fin}}(\mathcal{P}))$ is "locally associative" in general, in the sense that given $x, y, z \in \mathcal{H}(\mathcal{C}_{\mathsf{fin}}(\mathcal{P}))$ we have

$$u \circledast (x \circledast (y \circledast z) - (x \circledast y) \circledast z) = 0$$

for some element $u \in \mathcal{Z}$.

4.4. Reduction. Define the reduced localized Hall algebra by setting $[M_\bullet] = 1$ in $\mathcal{DH}(\mathcal{R})$ whenever $M_\bullet$ is an acyclic complex, invariant under the shift functor. More formally, we set

$$\mathcal{DH}_{\mathsf{red}}(\mathcal{R}) = \mathcal{DH}(\mathcal{R})/\big([M_\bullet] - 1 : H_\bullet(M_\bullet) = 0,\ M_\bullet \cong M^*_\bullet\big).$$

By Lemma 3.4 this is the same as setting

$$[K_P] \circledast [K^*_P] = 1$$

for all $P \in \mathcal{P}$. One can check that the shift functor $*$ defines involutions of $\mathcal{DH}_{\mathsf{red}}(\mathcal{R})$.

We have the following triangular decomposition.

PROPOSITION 4.9. *The multiplication map* $[A] \otimes \alpha \otimes [B] \mapsto E_A \circledast K_\alpha \circledast F_B$ *defines an isomorphism of vector spaces*

$$\mathcal{H}(\mathcal{A}) \otimes_{\mathbb{C}} \mathbb{C}[K(\mathcal{R})] \otimes_{\mathbb{C}} \mathcal{H}(\mathcal{A}) \longrightarrow \mathcal{DH}_{\mathsf{red}}(\mathcal{R}).$$

PROOF. The same argument given for Lemma 4.4 also applies here. □

4.5. Commutation relations. In this subsection, we prove commutation relations among generators that are important to understand $\mathcal{DH}_{\mathsf{red}}(\mathcal{R})$.

LEMMA 4.10. *Suppose* $A_1, A_2 \in \mathcal{A}$ *satisfy*

$$\mathrm{Hom}_{\mathcal{R}}(A_1, A_2) = 0 = \mathrm{Hom}_{\mathcal{R}}(A_2, A_1).$$

Then $[E_{A_1}, F_{A_2}] = 0$.

PROOF. It follows from (4.5) and (4.6) that $E_{A_1} \circledast F_{A_2} = E_{A_1,A_2}$. Exchanging A_1 and A_2 in this equation and taking $*$ on both sides gives $F_{A_2} \circledast E_{A_1} = E_{A_1,A_2}$ as well, and the result follows. □

LEMMA 4.11. *Suppose* $A \in \mathcal{A}$ *satisfies* $\mathrm{End}_{\mathcal{A}}(A) = \Bbbk$. *Then*

$$[E_A, F_A] = (q-1) \cdot (K^*_{\hat{A}} - K_{\hat{A}}).$$

PROOF. Using formulas (4.5) and (4.6) again, we have $E_A \circledast F_A = E_{A,A} + (q-1) \cdot K^*_{\hat{A}}$. Taking $*$ on both sides gives the equation $F_A \circledast E_A = E_{A,A} + (q-1) \cdot K_A$, since $E_{A,A}$ is $*$-invariant. The result follows by subtracting these equations. □

5. Realization of quantum groups

5.1. Quivers. Let $\mathcal{Q}$ be a locally finite quiver with vertex set I and (oriented) edge set Ω. For $\sigma \in \Omega$ we denote by $h(\sigma)$ and $t(\sigma)$ the head and tail, respectively, and sometimes use the notation $t(\sigma) \xrightarrow{\sigma} h(\sigma)$. We will denote by c_i the number of loops at $i \in I$ (i.e., the number of edges σ with $h(\sigma) = t(\sigma) = i$). A (finite) *path* in $\mathcal{Q}$ is a sequence $\sigma_m \cdots \sigma_1$ of edges which satisfies $h(\sigma_i) = t(\sigma_{i+1})$ for $1 \le i < m$. For each $i \in I$, we let e_i denote the trivial path. We again let $h(x)$ and $t(x)$ denote the head and tail vertices of a path x.

Consider the sub-quiver $\bar{\mathcal{Q}} \subset \mathcal{Q}$ with vertex set $\bar{I} = I$ and edge set $\bar{\Omega} = \Omega \backslash \{\sigma | h(\sigma) = t(\sigma)\}$. We make the following assumptions throughout:

(A) There are no infinite paths of the form $i_0 \to i_1 \to i_2 \to \cdots$ in $\bar{\mathcal{Q}}$. In particular, $\bar{\mathcal{Q}}$ is acyclic and $\mathcal{Q}$ has no oriented cycles other than loops.

(B) Each vertex of the quiver $\mathcal{Q}$ has either zero loops or more than one loop, i.e. $c_i \neq 1$ for all $i \in I$.

It follows from (A) that I is partially ordered, with $i \preceq j$ if there exists a path x such that $t(x) = j$ and $h(x) = i$, and the set $(I, \preceq)$ satisfies the descending chain condition.

From now on, we assume that the quiver $\mathcal{Q}$ satisfies the conditions (A) and (B).

EXAMPLE 5.1. Write $\mathbb{L}_n$ to denote the quiver consisting of a single vertex $I = \{1\}$ and n loops. If $n \geq 2$, then the quiver $\mathcal{Q} = \mathbb{L}_n$ trivially satisfies the assumptions (A) and (B). Below is a diagram for $\mathbb{L}_4$.

$\mathbb{L}_4$:

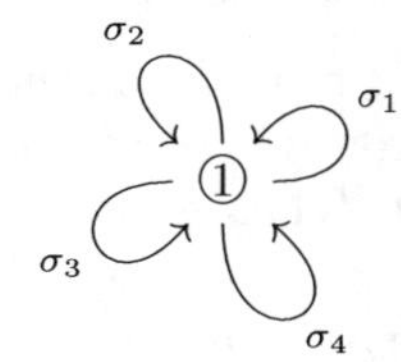

Let $R = \Bbbk\mathcal{Q}$ denote the path algebra, with basis given by the set of paths and multiplication defined via concatenation. The elements e_i are pairwise orthogonal idempotents. It follows from (A) that the subring e_iRe_i is isomorphic to a free associative $\Bbbk$-algebra on c_i generators. (If $c_i = 0$ then $e_iRe_i \cong \Bbbk$.) Hence each left ideal $P_i := Re_i$ is an indecomposable projective R-module. It can then be checked that R splits as a direct sum

$$R = \bigoplus_{i\in I} P_i \tag{5.1}$$

of pairwise non-isomorphic projective left R-submodules.

A representation of $\mathcal{Q}$ over $\Bbbk$ is a collection $(V_i, x_\sigma)_{i\in I,\sigma\in\Omega}$, where V_i is a (possibly infinite dimensional) $\Bbbk$-vector space and $x_\sigma \in \mathrm{Hom}_\Bbbk(V_{t(\sigma)}, V_{h(\sigma)})$. We let $\mathrm{Rep}_\Bbbk(\mathcal{Q})$ denote the abelian category consisting of the representations of $\mathcal{Q}$ over $\Bbbk$ which are of *finite support*, i.e. such that $V_i = 0$ for almost all i. A representation $(V_i, x_\sigma) \in \mathrm{Rep}_\Bbbk(\mathcal{Q})$ is called *finite-dimensional* if each V_i is finite-dimensional. For such a representation, set $\mathbf{dim}(V_i, x_\sigma) = \mathbf{dim}(V_i) = (\dim_\Bbbk V_i) \in \mathbb{N}^{\oplus I}$. We denote by $\mathrm{rep}_\Bbbk(\mathcal{Q})$ the full subcategory of $\mathcal{R}$ consisting of the finite dimensional representations of $\mathcal{Q}$.

Any representation of $\mathcal{Q}$ is naturally an R-module for the path algebra R. Whenever it is convenient, particularly in the next subsection, we will consider representations of $\mathcal{Q}$ as R-modules. It follows from the decomposition (5.1) that any left R-module M has a decomposition into $\Bbbk$-subspaces, $M = \bigoplus_{i\in I} M(i)$, with $M(i) = e_iM$. Then we have $\mathbf{dim}(M) = (\dim_\Bbbk M(i)) \in \mathbb{N}^{\oplus I}$ for a finite dimensional R-module M, which is equal to the dimension vector as a representation. We say that an R-module is of finite support if the associated representation is.

Recall that any $M \in \mathrm{Rep}_\Bbbk(\mathcal{Q})$ has the standard presentation of the form

$$0 \longrightarrow \bigoplus_{\sigma\in\Omega} P_{h(\sigma)} \otimes_\Bbbk e_{t(\sigma)}M \xrightarrow{f} \bigoplus_{i\in I} P_i \otimes_\Bbbk e_iM \xrightarrow{g} M \longrightarrow 0. \tag{5.2}$$

Let $\mathrm{Proj}_\Bbbk(\mathcal{Q})$ denote the full subcategory of $\mathrm{Rep}_\Bbbk(\mathcal{Q})$ whose objects are finitely-generated, projective R-modules. Let $\mathcal{R}$ be the full subcategory of $\mathrm{Rep}_\Bbbk(\mathcal{Q})$ whose objects are finitely presented representations of $\mathcal{Q}$, i.e. the full subcategory consisting of all objects M for which there exists a presentation

$$P \to Q \to M \to 0$$

for some $P, Q \in \mathrm{Proj}_\Bbbk(\mathcal{Q})$.

As in Section 1.1, define $\mathcal{P} \subset \mathcal{R}$ to be the full subcategory of projectives in $\mathcal{R}$ and $\mathcal{A} \subset \mathcal{R}$ the full subcategory of objects $A \in \mathcal{R}$ such that $\mathrm{Hom}_\mathcal{R}(M, A)$ is a finite set for any $M \in \mathcal{R}$. It is easy to see that $\mathcal{P} = \mathrm{Proj}_\Bbbk(\mathcal{Q})$. For the category $\mathcal{A}$, we have the following characterization.

LEMMA 5.2. *The category $\mathcal{A}$ is equal to the full subcategory of $\mathcal{R}$ consisting of all finite-dimensional representations, i.e.* $\mathcal{A} = \mathrm{rep}_{\Bbbk}(\mathcal{Q})$.

PROOF. Assume that $A \in \mathrm{rep}_{\Bbbk}(\mathcal{Q})$. From the standard resolution (5.2), we see that $A \in \mathcal{R}$. Clearly, A is finite as a set. Since any $M \in \mathcal{R}$ is finitely generated, the set $\mathrm{Hom}_{\mathcal{R}}(M, A)$ is also finite. Thus A is an object of $\mathcal{A}$. For the converse, assume that $M \in \mathrm{Rep}_{\Bbbk}(\mathcal{Q})$ is infinite dimensional. Then there is a vertex $i \in I$ such that $e_i M$ is infinite dimensional. For each $a \in e_i M$, we have a homomorphism $P_i \to M$ given by $e_i \mapsto a$. Thus $\mathrm{Hom}_{\mathcal{R}}(P_i, M)$ is an infinite set and M does not belong to $\mathcal{A}$. □

5.2. Krull–Schmidt property for $\mathrm{Proj}_{\Bbbk}(\mathcal{Q})$**.** Since the endomorphism ring $\mathrm{End}_R(P_i)$ is not local in general, the usual Krull–Schmidt theorem does not hold in the category $\mathrm{Proj}_{\Bbbk}(\mathcal{Q})$. In this subsection, we describe a suitable analogue.

First note the following.

LEMMA 5.3. *Suppose $i, j \in I$ are distinct vertices such that $(Re_iR) \cap (Re_jR) \neq 0$. Then either $i \prec j$ or $j \prec i$.*

PROOF. It follows from the stated condition that there are paths x, x', y, y' such that xe_iy and $x'e_jy'$ are both nonzero and

$$xy = xe_iy = x'e_jy' = x'y'.$$

We then have $xy = x'y' = \sigma_1 \cdots \sigma_n$, for some $\sigma_1, \ldots, \sigma_n \in \Omega$. Let $1 \leq l, l' \leq n$ be such that $x = \sigma_1 \cdots \sigma_l$, $y = \sigma_{l+1} \cdots \sigma_n$, $x' = \sigma_1 \cdots \sigma_{l'}$, and $y' = \sigma_{l'+1} \cdots \sigma_n$. Suppose without loss of generality that $l < l'$. Then $z = \sigma_{l+1} \cdots \sigma_{l'}$ is a path such that $h(z) = i$ and $t(z) = j$. Thus $i \prec j$. □

Let R-Mod denote the abelian category of all left R-modules of finite support. We identify R-Mod with $\mathrm{Rep}_{\Bbbk}(\mathcal{Q})$, and consider $\mathcal{P} = \mathrm{Proj}_{\Bbbk}(\mathcal{Q})$ as a full subcategory of R-Mod. We write $\mathrm{supp}(M) := \{i \mid M(i) \neq 0\}$ for $M \in R$-Mod. Recall that for any idempotent $e \in R$ there is an exact functor from the category R-Mod to (eRe)-Mod given by $M \mapsto eM$. Now suppose $M \in \mathcal{P}$. Then $M(i)$ is a finitely-generated, projective (e_iRe_i)-module. Since e_iRe_i is a free associative $\Bbbk$-algebra, $M(i)$ is a free (e_iRe_i)-module of finite rank, say $r_i(M)$. (See, for example, [**7**].) Write $\mathbf{rk}_{\mathcal{P}}(M) = (r_i(M))_{i \in I} \in \mathbf{N}^{\oplus I}$ to denote the vector formed by these ranks. Notice that $r_i(P_i) = 1$ and $r_j(P_i) = 0$ unless $j \preceq i$. It follows that the vectors, $\mathbf{rk}_{\mathcal{P}}(P_i)$, form a basis for $\mathbb{Z}^{\oplus I}$.

Given a subset $J \subseteq I$, consider the set $J_{\preceq} = \{i \in I \mid i \preceq j \text{ for some } j \in J\}$. If J is finite then so is $J_{\preceq}$ by (A), and there are corresponding idempotents

$$e_J = \sum_{j \in J} e_j \qquad \text{and} \qquad e_{\preceq J} = \sum_{i \in J_{\preceq}} e_i.$$

LEMMA 5.4. *Suppose $J \subseteq I$ is a finite subset and write $e = e_{\preceq J}$. Then there is an equivalence between (eRe)-Mod and the full subcategory of R-Mod consisting of modules M such that $\mathrm{supp}(M) \subseteq J$.*

PROOF. Consider the functor $R\text{-Mod} \to (eRe)\text{-Mod} : M \mapsto eM$. Then $eRe = Re$, and an inverse functor is given by extending the action of Re on $M \in (eRe)$-Mod to all of R by letting $\bigoplus_{i \notin J_{\preceq}} P_i$ act by zero. □

LEMMA 5.5. *Suppose $\mathfrak{a}$ is a finitely generated left ideal of R. For $i, j \in I$, the following hold.*

(i) *The ideal* $\mathfrak{a}$ *is a projective left* R*-module.*
(ii) *Any nonzero* R*-module homomorphism,* $\phi : P_i \to \mathfrak{a}$*, is injective.*
(iii) *Suppose* $\phi_i : P_i \to \mathfrak{a}$ *and* $\phi_j : P_j \to \mathfrak{a}$ *are homomorphisms such that* $\mathrm{im}(\phi_i) \cap \mathrm{im}(\phi_j) \neq 0$*. Then either* $\mathrm{im}(\phi_i) \subseteq \mathrm{im}(\phi_j)$ *or* $\mathrm{im}(\phi_j) \subseteq \mathrm{im}(\phi_i)$*.*
(iv) *As a left* R*-module,* $\mathfrak{a}$ *is isomorphic to a finite direct sum of copies of the modules* $\{P_i\}_{i\in I}$*.*

PROOF.

(i) Suppose $\mathfrak{a} \subseteq R$ is a left ideal with $S \subseteq \mathfrak{a}$ a finite set of generators. Then $S \subseteq Re_J$ for some finite set J and it follows that $\mathrm{supp}(\mathfrak{a}) \subseteq J_{\preceq}$. If we set $e = e_{\preceq J}$, then $\mathfrak{a} \subseteq eRe$, which shows that $\mathfrak{a}$ is a left ideal of a hereditary ring and thus projective as an eRe-module. It follows that $\mathfrak{a}$ is a projective R-module by Lemma 5.4.

(ii) The image $\mathrm{im}(\phi) \subseteq \mathfrak{a}$ is a projective left R-module by (i). So the exact sequence

$$0 \to \ker(\phi) \to P_i \to \mathrm{im}(\phi) \to 0$$

splits. Since P_i is indecomposable, it follows that ϕ is injective.

(iii) Letting $\phi_i(e_i) = v$ and $\phi_j(e_j) = w$, we have $\mathrm{im}(\phi_i) = Rv$ and $\mathrm{im}(\phi_j) = Rw$. We thus have $(Re_i v) \cap (Re_j w) \neq 0$. It follows by Lemma 5.3 that $i \preceq j$ or $j \preceq i$. Assume without loss of generality that $j \preceq i$. It then follows from the proof of Lemma 5.3, that there exists a path x such that $w = e_j w = e_j x e_i v = xv$. It follows that $\mathrm{im}(\phi_j) = Rw = R(xv) \subseteq Rv = \mathrm{im}(\phi_i)$.

(iv) First set $J^1 := \mathrm{supp}(\mathfrak{a})$, and choose a maximal vertex $j_1 \in J^1$. Then $\mathfrak{a}_{j_1}$ is an $e_{j_1}Re_{j_1}$-module with a finite set, say $\mathcal{S}^1$, of free generators of size $n_1 := r_{j_1}(\mathfrak{a}) \neq 0$. It follows from (ii) that for each generator $v \in \mathcal{S}^1$ the mapping, $P_{j_1} \to \mathfrak{a} : e_{j_1} \mapsto v$, defines an injective R-module homomorphism. By (iii), we thus have a corresponding isomorphism

$$(P_{j_1})^{\oplus n_1} \xrightarrow{\sim} R\mathcal{S}^1 \subseteq \mathfrak{a}.$$

Next choose a maximal vertex j_2 belonging to the subset

$$J^2 := \{j \in J^1 \mid r_j(\mathfrak{a}) - n_1 \cdot r_j(P_{j_1}) > 0\}.$$

It follows that $\mathfrak{a}_{j_2} = e_{j_2}\mathfrak{a}$ is a free $e_{j_2}Re_{j_2}$-module of rank $n_1 \cdot r_j(P_{j_1}) + n_2$, for some $n_2 > 0$. Let $\mathcal{S}^2 = \mathcal{S}^2_1 \sqcup \mathcal{S}^2_2$ be a set of free $e_{j_2}Re_{j_2}$-generators such that $\mathcal{S}^2_1$ generates $e_{j_2}(R\mathcal{S}^1)$. Then $\mathcal{S}^2_2$ has size n_2, and it follows as in the previous paragraph that we have an embedding

$$(P_{j_2})^{\oplus n_2} \xrightarrow{\sim} R\mathcal{S}^2_2 \subseteq \mathfrak{a}.$$

By the maximality of j_1 we have $j_1 \not\preceq j_2$. It is also clear that $R\mathcal{S}^2_2 \not\subseteq R\mathcal{S}^1$. It follows by (iii) that $(R\mathcal{S}^1) \cap (R\mathcal{S}^2_2) = 0$. We thus obtain an embedding

$$(P_{j_1})^{\oplus n_1} \oplus (P_{j_2})^{\oplus n_2} \xrightarrow{\sim} R\mathcal{S}^1 + R\mathcal{S}^2_2 \subseteq \mathfrak{a}.$$

Continuing in this way the process eventually terminates, since $\mathrm{supp}(\mathfrak{a})$ is finite, yielding the desired decomposition. □

The following is an analogue of the Krull–Schmidt theorem for the category of finitely-generated projective R-modules.

PROPOSITION 5.6. *Given any finitely-generated, projective left R-module $M \in \mathrm{Proj}_{\Bbbk}(\mathcal{Q})$, there is an R-module isomorphism*

$$M \cong \bigoplus_{i \in I} P_i^{\oplus n_i}$$

for some nonnegative integers n_i, only finitely many of which are nonzero. Moreover, given another such decomposition, $M \cong \bigoplus_{i\in I} P_i^{\oplus m_i}$, we must have $m_i = n_i$ for all i.

PROOF. Since M is finitely generated, $J = \mathrm{supp}(M)$ is finite. Letting $e = e_{\preceq J}$, we see that M is a projective module of the hereditary ring eRe. It follows from [**6**, Theorem 5.3] or [**14**] that M is isomorphic to a finitely generated left ideal of eRe. Hence Lemma 5.5 yields a decomposition

$$M \cong \bigoplus_{i \in I} P_i^{\oplus n_i}.$$

It follows that

$$\mathbf{rk}_{\mathcal{P}}(M) = \sum_{i \in I} n_i \cdot \mathbf{rk}_{\mathcal{P}}(P_i).$$

Since the vectors $\{\mathbf{rk}_{\mathcal{P}}(P_i)\}_{i\in I}$ form a basis for $\mathbb{Z}^{\oplus I}$, the decomposition must be unique. $\square$

5.3. Assumptions (a)-(e). Let S_i be a simple module supported only at $i \in I$. Then we obtain from (5.2) the standard resolution

$$0 \longrightarrow P_i' \longrightarrow P_i \longrightarrow S_i \longrightarrow 0,$$

where $P_i' = \bigoplus_{j\in I} P_j^{\oplus n_j}$ for some integers n_j. Then clearly $n_j = 0$ unless $j \preceq i$. In particular, if i is a minimal vertex then $P_i' = P_i^{\oplus c_i}$ and hence

$$(1 - c_i)\,\hat{P}_i = \hat{S}_i \tag{5.3}$$

which is non-zero by assumption (B). Now if $i \in I$ is not minimal, then by assumption (A) the set $\{j \mid j \preceq i\}$ is finite. We may thus use (5.2) and (5.3) inductively to write $\hat{P}_i$ as a linear combination

$$\hat{P}_i = \frac{1}{1 - c_i}\hat{S}_i + \sum_{j \prec i} r_{ij}\hat{S}_j, \qquad r_{ij} \in \mathbb{Q}. \tag{5.4}$$

The following proposition makes it possible to apply the results in the general setting of the previous sections to the category $\mathcal{R}$ of finitely-presented quiver representations.

PROPOSITION 5.7. *The triple $(\mathcal{R}, \mathcal{P}, \mathcal{A})$ satisfies the assumptions (a)-(e) in Section* 1.1.

PROOF.

(a) It is clear.

(b) Since the path algebra R is hereditary, the category $\mathcal{R}$ is hereditary as well. Furthermore, $\mathcal{R}$ has enough projectives by definition.

(c) It is clear that $\mathcal{P} = \mathrm{Proj}_{\Bbbk}(\mathcal{Q})$, so this condition follows easily from Proposition 5.6.

(d) It follows from the expression (5.4).

(e) If $\boldsymbol{k}_{\mathcal{R}}(A) = \boldsymbol{k}_{\mathcal{R}}(B)$ for $A, B \in \mathcal{A}$, the standard resolution (5.2) tells us that $e_i A$ and $e_i B$ have the same number of elements for each $i \in I$. Then $|\operatorname{Hom}(P_i, A)| = |\operatorname{Hom}(P_i, B)|$ for each $i \in I$, and thus $|\operatorname{Hom}(P, A)| = |\operatorname{Hom}(P, B)|$ for $P \in \mathcal{P}$ by Proposition 5.6. □

It is also clear that the subcategory $\mathcal{A} \subset \mathcal{R}$ satisfies the finite subobjects condition (2.11), since each object $A \in \mathcal{A}$ is a finite dimensional vector space by Lemma 5.2. Thus $\mathcal{A}$ also satisfies conditions (2.12) and (2.13).

REMARK 5.8. The quiver $\mathbb{L}_1$ is called the *Jordan quiver*, and its path algebra is isomorphic to the polynomial algebra $\Bbbk[x]$. For each $\lambda \in \Bbbk$, there is the one-dimensional simple modules S_λ over $\Bbbk[x]$ where x acts as λ. Considering the standard resolution (5.2), we see $\widehat{S}_\lambda = 0$, and the assumptions (d) and (e) are not satisfied. Nonetheless, the Jordan quiver is related to classical examples of Hall algebras. One can find details, for example, in [**21**].

5.4. Quantum generalized Kac-Moody algebras. In this subsection we recall the basic definitions concerning quantum generalized Kac–Moody algebras. We keep the assumptions on the choice of v as in Section 1.2.

Let I be a countable index set, and fix a symmetric *Borcherds–Cartan* matrix $A = (a_{ij})_{i,j\in I}$ whose entries a_{ij}, by definition, satisfy (i) $a_{ii} \in \{2, 0, -2, -4, \ldots\}$ and (ii) $a_{ij} = a_{ji} \in \mathbb{Z}_{\leq 0}$ for all i, j. Put $I^{re} = \{i \in I \mid a_{ii} = 2\}$ and $I^{im} = I \backslash I^{re}$, and assume that we are given a collection of positive integers $\mathbf{m} = (m_i)_{i\in I}$, called the *charge* of A, with $m_i = 1$ whenever $i \in I^{re}$. We put

$$[n] = \frac{v^n - v^{-n}}{v - v^{-1}}, \qquad [n]! = [1][2]\cdots[n], \qquad \begin{bmatrix} n \\ k \end{bmatrix} = \frac{[n]!}{[n-k]![k]!}.$$

The *quantum generalized Kac–Moody algebra* associated with $(A, \mathbf{m})$ is defined to be the (unital) $\mathbb{C}$-algebra $\mathbf{U}_v$ generated by the elements $K_i, K_i^{-1}, E_{ik}, F_{ik}$ for $i \in I$, $k = 1, \ldots, m_i$, subject to the following set of relations: for $i, j \in I$, $k = 1, \ldots, m_i$ and $l = 1, \ldots, m_j$,

$$K_i K_i^{-1} = K_i^{-1} K_i = 1, \quad K_i K_j = K_j K_i, \tag{5.5}$$

$$K_i E_{jl} K_i^{-1} = v^{a_{ij}} E_{jl}, \qquad K_i F_{jl} K_i^{-1} = v^{-a_{ij}} F_{jl}, \tag{5.6}$$

$$E_{ik} F_{jl} - F_{jl} E_{ik} = \delta_{lk} \delta_{ij} \frac{K_i - K_i^{-1}}{v - v^{-1}}, \tag{5.7}$$

$$E_{ik} E_{jl} - E_{jl} E_{ik} = F_{ik} F_{jl} - F_{jl} F_{ik} = 0 \quad \text{if } a_{ij} = 0, \quad \text{and} \tag{5.8}$$

$$\sum_{n=0}^{1-a_{ij}} (-1)^n \begin{bmatrix} 1 - a_{ij} \\ n \end{bmatrix} E_{ik}^{1-a_{ij}-n} E_{jl} E_{ik}^n$$
$$= \sum_{n=0}^{1-a_{ij}} (-1)^n \begin{bmatrix} 1 - a_{ij} \\ n \end{bmatrix} F_{ik}^{1-a_{ij}-n} F_{jl} F_{ik}^n = 0 \qquad \text{if } i \in I^{re} \text{ and } i \neq j.$$

The algebra $\mathbf{U}_v$ is equipped with a Hopf algebra structure as follows (see [**1, 12**]):

$$\Delta(K_i) = K_i \otimes K_i,$$
$$\Delta(E_{ik}) = E_{ik} \otimes K_i^{-1} + 1 \otimes E_{ik}, \qquad \Delta(F_{ik}) = F_{ik} \otimes 1 + K_i \otimes F_{ik},$$
$$\epsilon(K_i) = 1, \quad \epsilon(E_{ik}) = \epsilon(F_{ik}) = 0,$$
$$S(K_i) = K_i^{-1}, \quad S(E_{ik}) = -E_{ik} K_i, \quad S(F_{ik}) = -K_i^{-1} F_{ik}.$$

We have an involution $\tau : \mathbf{U}_v \longrightarrow \mathbf{U}_v$ defined by $E_{ik} \mapsto F_{ik}$, $F_{ik} \mapsto E_{ik}$ and $K_i \mapsto K_i^{-1}$ for $i \in I$ and $k = 1, \ldots, m_i$. We denote by $\mathbf{U}_v^0$ the subalgebra generated by $K_i^{\pm 1}$, and by $\mathbf{U}_v^+$ (resp. $\mathbf{U}_v^-$) the subalgebra generated by E_{ik} (resp. F_{ik}). Similarly, we define $\mathbf{U}_v^{\geq 0}$ (resp. $\mathbf{U}_v^{\leq 0}$) to be the subalgebra generated by $K_i^{\pm 1}$ and E_{ik} (resp. $K_i^{\pm 1}$ and F_{ik}) for $i \in I$ and $k = 1, \ldots, m_i$. Then

$$\mathbf{U}_v^0 \cong \mathbb{C}[K_i^{\pm 1}]_{i \in I},$$

and the involution τ identifies $\mathbf{U}_v^+$ with $\mathbf{U}_v^-$.

The following result provides a triangular decomposition for $\mathbf{U}_v$.

PROPOSITION 5.9 ([**1**]). *The multiplication maps*

$$\mathbf{U}_v^0 \otimes_{\mathbb{C}} \mathbf{U}_v^+ \longrightarrow \mathbf{U}_v^{\geq 0}, \qquad \mathbf{U}_v^0 \otimes_{\mathbb{C}} \mathbf{U}_v^- \longrightarrow \mathbf{U}_v^{\leq 0},$$

and

$$\mathbf{U}_v^+ \otimes_{\mathbb{C}} \mathbf{U}_v^0 \otimes_{\mathbb{C}} \mathbf{U}_v^- \longrightarrow \mathbf{U}_v$$

are isomorphisms of vector spaces.

5.5. Embedding of $\mathbf{U}_v^{\geq 0}$ into a Hall algebra. Let $A = (a_{ij})_{i,j \in I}$ be a symmetric Borcherds–Cartan matrix such that each row has only finitely many nonzero entries and $a_{ii} \neq 0$ for any $i \in I$. Fix a locally finite quiver $\mathcal{Q}$ associated to A satisfying conditions (A) and (B): each vertex i has $1 - a_{ii}/2$ loops, and two distinct vertices i and j are connected with $-a_{ij}$ arrows for $i \neq j$. Then, since $a_{ii} \neq 0$ for any $i \in I$, the condition (B) is satisfied, and we can always choose an orientation for $\mathcal{Q}$ so that (A) is satisfied.

If $i \in I^{re}$, then there exists a unique simple object $S_i \in \mathrm{Rep}_{\Bbbk}(\mathcal{Q})$ supported at i. On the other hand, if $i \in I^{im}$ then the set of simple objects supported at i is in bijection with $\Bbbk^{c_i}$: if $\sigma_1, \ldots, \sigma_{c_i}$ denote the simple loops at i then to $\underline{\lambda} = (\lambda_1, \ldots, \lambda_{c_i}) \in \Bbbk^{c_i}$ corresponds the simple module $S_{i,\underline{\lambda}} = (V_j, x_\sigma)_{j \in I, \sigma \in \Omega}$ with $\dim_{\Bbbk} V_j = \delta_{ij}$ and $x_{\sigma_k} = \lambda_k \cdot \mathrm{id}$ for $k = 1, \ldots, c_i$.

Let us now assume that the charge $\mathbf{m} = (m_i)_{i \in I}$ satisfies

$$m_i \leq |\Bbbk^{c_i}| = q^{c_i} \qquad \text{for each } i \in I.$$

We choose $\underline{\lambda}^{(l)} \in \Bbbk^{c_i}$ for $l = 1, \ldots, m_i$ in such a way that $\underline{\lambda}^{(l)} \neq \underline{\lambda}^{(l')}$ for $l \neq l'$. Then we set $S_{il} = S_{i,\underline{\lambda}^{(l)}}$ for $i \in I^{im}$ and $l = 1, \ldots, m_i$, and simply set $S_{i1} = S_i$ for $i \in I^{re}$. Since S_{il} have the same projectives in the standard resolution (5.2) for all $l = 1, \ldots, m_i$, they define a unique class in $K(\mathcal{R})$. We will denote this unique class by $\hat{S}_i$ for any $i \in I$.

The following lemma will be used in the proof of Theorem 5.11.

LEMMA 5.10. *Let $\bar{K}(\mathcal{A})$ denote the image of the map $K(\mathcal{A}) \to K(\mathcal{R})$ defined in* (2.7)*. Then the map* $\mathbf{dim} : \bar{K}(\mathcal{A}) \to \mathbb{Z}^{\oplus I}$ *is well-defined and is an isomorphism. That is,*

$$\mathbb{Z}^{\oplus I} \simeq \bar{K}(\mathcal{A}) \hookrightarrow K(\mathcal{R}).$$

PROOF. The map $\mathbf{dim} : \bar{K}(\mathcal{A}) \to \mathbb{Z}^{\oplus I}$ is well-defined by condition (e) which is verified in Proposition 5.7. Clearly, the map is surjective. As noted already, there is a unique class $\hat{S}_i$ which represents all simple modules S_{il} in $\mathcal{A}$ for each $i \in I$. Thus the map is injective. □

The following theorem, due to Kang and Schiffmann, is an extension of well-known results of Ringel [**20**] and Green [**9**] from the case of a finite quiver without loops to a locally finite quiver with loops:

THEOREM 5.11 ([**13**]). *Suppose* $\Bbbk = \mathbb{F}_q$ *is such that* $|\Bbbk^{c_i}| \geq m_i$ *for all* $i \in I$. *Then there are injective homomorphisms of algebras*

$$\mathbf{U}_v^+ \hookrightarrow \mathcal{H}_v(\mathcal{A}) \qquad \textit{and} \qquad \mathbf{U}_v^{\geq 0} \hookrightarrow \widetilde{\mathcal{H}_v}(\mathcal{R})$$

defined on generators by $K_i^{\pm 1} \mapsto K_{\hat{S}_i}^{\pm 1}$ *for* $i \in I$,

$$E_{il} \mapsto [S_{il}] \cdot (q-1)^{-1} \qquad \textit{for } i \in I,$$

where S_{il} *and* $\hat{S}_i$ *are defined right before Lemma* 5.10.

PROOF. Let $\widetilde{\mathcal{H}_{v,\Bbbk}}(\mathcal{Q})$ be the extended, twisted Hall algebra defined in [**13**]. It is shown in [**13**] that

$$\mathbf{U}_v^+ \hookrightarrow \mathcal{H}_v(\mathcal{A}) \qquad \text{and} \qquad \mathbf{U}_v^{\geq 0} \hookrightarrow \widetilde{\mathcal{H}_{v,\Bbbk}}(\mathcal{Q}).$$

Thus we have only to check that there exists an injective algebra homomorphism $\widetilde{\mathcal{H}_{v,\Bbbk}}(\mathcal{Q}) \hookrightarrow \widetilde{\mathcal{H}_v}(\mathcal{R})$. The only difference between the two algebras is that, while $\widetilde{\mathcal{H}_v}(\mathcal{R})$ is extended by $K(\mathcal{R})$, the algebra $\widetilde{\mathcal{H}_{v,\Bbbk}}(\mathcal{Q})$ is extended by $\mathbb{Z}^{\oplus I}$. Thus the embedding of $\widetilde{\mathcal{H}_{v,\Bbbk}}(\mathcal{Q})$ into $\widetilde{\mathcal{H}_v}(\mathcal{R})$ follows from Lemma 5.10. □

5.6. Embedding of $\mathbf{U}_v$ into $\mathcal{DH}_{\mathsf{red}}(\mathcal{R})$. We keep the notations in the previous subsection. In particular, the matrix A is a Borcherds–Cartan Matrix and $\mathcal{Q}$ is a fixed quiver corresponding to A. Suppose that $\mathbf{U}_v$ is the quantum group of the generalized Kac-Moody algebra associated with A.

Now we state and prove the main result of this paper.

THEOREM 5.12. *There is an injective homomorphism of algebra*

$$\Xi\colon \mathbf{U}_v \hookrightarrow \mathcal{DH}_{\mathsf{red}}(\mathcal{R}),$$

defined on generators by

$$\Xi(E_{il}) = (q-1)^{-1} \cdot E_{S_{il}}, \qquad \Xi(F_{il}) = (-v) \cdot (q-1)^{-1} \cdot F_{S_{il}},$$
$$\Xi(K_i) = K_{\hat{S}_i}, \qquad \Xi(K_i^{-1}) = K_{\hat{S}_i}^*,$$

where $l = 1, 2, \ldots, m_i$ *and* $\hat{S}_i$ *is the unique class representing* S_{il} *in* $K(\mathcal{R})$ *for each* $i \in I$.

PROOF. We have a commutative diagram of linear maps

$$\begin{array}{ccc} \mathbf{U}_v^+ \otimes \mathbf{U}_v^0 \otimes \mathbf{U}_v^- & \xrightarrow{\quad\Theta\quad} & \mathcal{H}_v(\mathcal{A}) \otimes \mathbb{C}[K(\mathcal{R})] \otimes \mathcal{H}_v(\mathcal{A}) \\ \downarrow & & \downarrow \\ \mathbf{U}_v & \xrightarrow{\quad\Xi\quad} & \mathcal{DH}_{\mathsf{red}}(\mathcal{R}) \end{array}$$

where the vertical arrows are the isomorphisms described in Propositions 4.9 and 5.9, respectively, and the homomorphism Θ is constructed out of the homomorphisms of Theorem 5.11. The map Ξ is a well-defined algebra homomorphism by Theorem 5.11 and by Lemmas 4.10, 4.11 and Corollary 4.7, which show that the generators satisfy the defining relations of the quantum group. It is clear from this diagram that Ξ is injective. □

References

[1] Georgia Benkart, Seok-Jin Kang, and Duncan Melville, *Quantized enveloping algebras for Borcherds superalgebras*, Trans. Amer. Math. Soc. **350** (1998), no. 8, 3297–3319, DOI 10.1090/S0002-9947-98-02058-3. MR1451594

[2] Richard Borcherds, *Generalized Kac-Moody algebras*, J. Algebra **115** (1988), no. 2, 501–512, DOI 10.1016/0021-8693(88)90275-X. MR943273

[3] Richard E. Borcherds, *Monstrous moonshine and monstrous Lie superalgebras*, Invent. Math. **109** (1992), no. 2, 405–444, DOI 10.1007/BF01232032. MR1172696

[4] Tom Bridgeland, *Quantum groups via Hall algebras of complexes*, Ann. of Math. (2) **177** (2013), no. 2, 739–759, DOI 10.4007/annals.2013.177.2.9. MR3010811

[5] Igor Burban and Olivier Schiffmann, *On the Hall algebra of an elliptic curve, I*, Duke Math. J. **161** (2012), no. 7, 1171–1231, DOI 10.1215/00127094-1593263. MR2922373

[6] Henri Cartan and Samuel Eilenberg, *Homological algebra*, Princeton University Press, Princeton, NJ, 1956. MR77480

[7] P. M. Cohn, *Free ideal rings and localization in general rings*, New Mathematical Monographs, vol. 3, Cambridge University Press, Cambridge, 2006, DOI 10.1017/CBO9780511542794. MR2246388

[8] Tim Cramer, *Double Hall algebras and derived equivalences*, Adv. Math. **224** (2010), no. 3, 1097–1120, DOI 10.1016/j.aim.2009.12.021. MR2628805

[9] James A. Green, *Hall algebras, hereditary algebras and quantum groups*, Invent. Math. **120** (1995), no. 2, 361–377, DOI 10.1007/BF01241133. MR1329046

[10] Anthony Joseph, *Quantum groups and their primitive ideals*, Ergebnisse der Mathematik und ihrer Grenzgebiete (3) [Results in Mathematics and Related Areas (3)], vol. 29, Springer-Verlag, Berlin, 1995, DOI 10.1007/978-3-642-78400-2. MR1315966

[11] Victor G. Kac, *Infinite-dimensional Lie algebras*, 3rd ed., Cambridge University Press, Cambridge, 1990, DOI 10.1017/CBO9780511626234. MR1104219

[12] Seok-Jin Kang, *Quantum deformations of generalized Kac-Moody algebras and their modules*, J. Algebra **175** (1995), no. 3, 1041–1066, DOI 10.1006/jabr.1995.1226. MR1341758

[13] Seok-Jin Kang and Olivier Schiffmann, *Canonical bases for quantum generalized Kac-Moody algebras*, Adv. Math. **200** (2006), no. 2, 455–478, DOI 10.1016/j.aim.2005.01.002. MR2200853

[14] Irving Kaplansky, *Modules over Dedekind rings and valuation rings*, Trans. Amer. Math. Soc. **72** (1952), 327–340, DOI 10.2307/1990759. MR46349

[15] M. Kapranov, *Heisenberg doubles and derived categories*, J. Algebra **202** (1998), no. 2, 712–744, DOI 10.1006/jabr.1997.7323. MR1617651

[16] Henning Krause, *Krull-Schmidt categories and projective covers*, Expo. Math. **33** (2015), no. 4, 535–549, DOI 10.1016/j.exmath.2015.10.001. MR3431480

[17] Liangang Peng and Jie Xiao, *Root categories and simple Lie algebras*, J. Algebra **198** (1997), no. 1, 19–56, DOI 10.1006/jabr.1997.7152. MR1482975

[18] Liangang Peng and Jie Xiao, *Triangulated categories and Kac-Moody algebras*, Invent. Math. **140** (2000), no. 3, 563–603, DOI 10.1007/s002220000062. MR1760751

[19] I. Reiten and M. Van den Bergh, *Noetherian hereditary abelian categories satisfying Serre duality*, J. Amer. Math. Soc. **15** (2002), no. 2, 295–366, DOI 10.1090/S0894-0347-02-00387-9. MR1887637

[20] Claus Michael Ringel, *Hall algebras and quantum groups*, Invent. Math. **101** (1990), no. 3, 583–591, DOI 10.1007/BF01231516. MR1062796

[21] Olivier Schiffmann, *Lectures on Hall algebras* (English, with English and French summaries), Geometric methods in representation theory. II, Sémin. Congr., vol. 24, Soc. Math. France, Paris, 2012, pp. 1–141. MR3202707

[22] Jie Xiao, *Drinfeld double and Ringel-Green theory of Hall algebras*, J. Algebra **190** (1997), no. 1, 100–144, DOI 10.1006/jabr.1996.6887. MR1442148

[23] Shintarou Yanagida, *A note on Bridgeland's Hall algebra of two-periodic complexes*, Math. Z. **282** (2016), no. 3-4, 973–991, DOI 10.1007/s00209-015-1573-x. MR3473652

University College, Sungkyunkwan University, 2066 Seobu-ro, Jangan-gu, Suwon-si, Gyeonggi-do 16419, Republic of Korea
Email address: jaxtell@skku.edu

Department of Mathematics, University of Connecticut, 341 Mansfield Road U1009, Storrs, Connecticut 06269-1009
Email address: khlee@math.uconn.edu

Proceedings of Symposia in Pure Mathematics
Volume **108**, 2024
https://doi.org/10.1090/pspum/108/01957

Cohomology of algebraic groups, Lie algebras, and finite groups of Lie type

Christopher P. Bendel

In memory of Georgia M. Benkart and Brian J. Parshall

ABSTRACT. Let G be a reductive algebraic group over a field of prime characteristic. One can associate to G (or subgroups thereof) its Lie algebra, its Frobenius kernels, and the finite Chevalley group of points over a finite field. The representation theories of these structures are highly interconnected. This expository article will focus specifically on the cohomology theories of these structures and the relationships between them with the aim of highlighting a few key developments over the past 20 years and related open questions.

1. Introduction

1.1. Let k be an algebraically closed field of characteristic $p > 0$ and G be a connected, reductive algebraic group scheme over k. For example, G may be the general linear group GL_n or the (semi-simple) special linear group SL_n or one of the other classical matrix groups. Associated to G are several standard subgroup schemes: a maximal torus T, a Borel subgroup B containing T, and the unipotent radical U of B. For example, in the case of GL_n, T consists of the diagonal matrices, B might be taken to be the (non-strictly) lower triangular matrices, and then U is the strictly lower triangular matrices.

There are a number of other algebraic structures related to G: Frobenius kernels, Lie algebras, and finite groups of Lie type. Let $F : G \to G$ denote the Frobenius map (given in the matrix example by raising each matrix entry to the pth power). The first Frobenius kernel of G is the scheme-theoretic kernel of F, denoted G_1. More generally, for an integer $r \geq 1$, the rth Frobenius kernel G_r is the kernel of the rth iterate F^r of F. One similarly has Frobenius kernels for the aforementioned subgroup schemes B and U.

The first Frobenius kernel G_1 is closely related to the Lie algebra $\mathfrak{g}$ of G. In this modular setting, $\mathfrak{g}$ admits the additional structure of a p-restricted Lie algebra. In the matrix example, the defining $[p]$-power mapping is given by pth power of matrices. Similarly, one has (p-restricted) Lie algebras associated to B and U. The representations of G_1 are equivalent to the restricted representations of $\mathfrak{g}$. Formally, there is an isomorphism of Hopf algebras $k[G_1]^* \cong u(\mathfrak{g})$, where $k[G_1]^*$

2020 *Mathematics Subject Classification.* Primary 17B56, 20G10, 20J06.

Research of the author was supported in part by Simons Foundation Collaboration Grant 317062.

denotes the k-linear dual of the coordinate algebra of G_1 and $u(\mathfrak{g})$ denotes the restricted enveloping algebra of $\mathfrak{g}$. Geometrically, one may think of the Lie algebra $\mathfrak{g}$ as a "tangent space" approximation to G. Translating that to Frobenius kernels, as r increases, the representation theory of G_r provides a better approximation to that of G. See Section 1.3 for an example of this behavior.

Let $\mathbb{F}_p$ denote the finite field of p elements and, more generally, $\mathbb{F}_q$ the field with $q := p^r$ elements. The fixed points of the Frobenius map (or its iterates) give rise to finite groups of Lie type: $G(\mathbb{F}_p) := G^F$ and $G(\mathbb{F}_q) := G^{F^r}$. In our matrix example, these are the finite matrix groups with entries in the given finite field. One similarly has finite groups associated to B and U. In addition, one has "twisted" groups of Lie type that arise by composing the Frobenius map with an automorphism of G induced from a non-trivial automorphism of the associated Dynkin diagram. See for example [**Hum**, 1.5]. Many of the results presented in this paper hold also for these twisted groups. For convenience, the discussion here will be restricted to the "untwisted" case.

Any rational G-module may be restricted to a module for any of the above objects, and the representation theories of these structures are well-known to be highly interconnected. For a thorough discussion, the reader is directed to the excellent books by Jantzen [**Jan03**] and Humphreys [**Hum**], noting that the notation used here generally follows that of [**Jan03**]. The goal of this paper is not to reproduce the material found therein, rather it is an incomplete attempt to highlight the current status of a few fundamental cohomological questions where progress has been made since those books were published. It is also hoped that this work complements and updates the article of Nakano [**Nak**]. As done in [**Nak**], this work aims to include examples demonstrating how the cohomology of these various structures are interrelated.

Sections 2 and 3 consider the cohomology of Frobenius kernels, beginning with trivial coefficients in Section 2. Computations of of $\mathrm{H}^i(G_r,k)$, $\mathrm{H}^i(B_r,k)$, and $\mathrm{H}^i(U_r,k)$ in particular degrees will be considered as well as the global ring structure of $\mathrm{H}^\bullet(G_r,k) := \bigoplus_{i\geq 0}\mathrm{H}^i(G_r,k)$ (or for B or U). In Section 3, non-trivial coefficients are considered, with the primary focus being on the computation of the groups $\mathrm{H}^i(G_r,H^0(\lambda))$ for a standard induced module $H^0(\lambda)$, along with the related computation of $\mathrm{H}^i(B_r,\lambda)$ for the one-dimensional B-module with weight λ. See Section 1.3 for details on the module notation. This also leads to a discussion of the ordinary Lie algebra cohomology of $\mathfrak{u}$, i.e., $\mathrm{H}^i(\mathfrak{u},k)$.

The discussion for Frobenius kernels leads to algebraic group computations, including a discussion of B-cohomology $\mathrm{H}^i(B,\lambda)$ in Section 4 and of G-cohomology in Section 5 with a new look at generic cohomology and the connections between G-cohomology and $G(\mathbb{F}_q)$-cohomology. Cline, Parshall, Scott, and van der Kallen [**CPSvdK**] showed that the cohomology of the finite group $G(\mathbb{F}_q)$ stabilizes as r increases and may be identified with certain G-cohomology. Section 5 will discuss recent improvements to that result and further connections between $G(\mathbb{F}_q)$- and G-cohomology for simple modules. These recent results are highly dependent on the use of the induction functor $\mathrm{ind}_{G(\mathbb{F}_q)}^G(-)$, a tool that was introduced into the theory by Bendel, Nakano, and Pillen and has proved to serve as a way to conceptually unify a number of finite group cohomology questions, as will be seen in applications in Sections 5, 6, 7, and 8.

Section 6 will continue to discuss $G(\mathbb{F}_q)$-cohomology, bringing in ideas from Sections 3 and 5. Dating to work of Quillen [**Qui**], it has been observed that $\mathrm{H}^i(G(\mathbb{F}_q), k)$ is zero in "small" positive degrees. This section will discuss results giving the least positive degree with non-trivial cohomology. Low degree $G(\mathbb{F}_q)$-cohomology of simple modules is considered in Section 7. Lastly, Section 8 will consider, in multiple contexts, the problem of determining the maximum dimension of a cohomology group. Ideas from previous sections will again play a role in the results presented here.

This paper will primarily consider cohomology, which can of course be viewed from the perspective of extension groups. E.g., for a rational G-module M, $\mathrm{H}^i(G, M) \cong \mathrm{Ext}^i_G(k, M)$. In some cases, explicit mention of related Ext-questions will be made, with other generalizations left to the interested reader. With this commentary on extensions, the reader may note that absent from the above list of topics is the Lusztig Conjecture on the character of a simple G-module; a topic with strong extension/cohomological ties. There have been many exciting developments on that question in recent years, and a thorough discussion thereof is beyond the scope of this article. The interested reader is directed to the recent work of Riche and Williamson [**RW**] (and references therein).

1.2. Notation. From now on, except as otherwise noted, G will denote a simple, simply connected algebraic group which is defined and split over $\mathbb{F}_p$. As above, T denotes a maximal torus. Associated to G is an irreducible root system Φ with positive roots Φ^+ and simple roots S. B will be a Borel subgroup containing T corresponding to the negative roots, with U the unipotent radical so that $B = T \ltimes U$.

Associated to the root system Φ, the following notation will be used:

- α_0: the maximal short root.
- $\mathbb{E}$: the Euclidean space spanned by Φ with inner product $\langle \, , \, \rangle$ normalized so that $\langle \alpha, \alpha \rangle = 2$ for $\alpha \in \Phi$ any short root.
- $\alpha^\vee = 2\alpha/\langle \alpha, \alpha \rangle$: the coroot of $\alpha \in \Phi$.
- ρ: the Weyl weight defined by $\rho = \frac{1}{2} \sum_{\alpha \in \Phi^+} \alpha$.
- h: the Coxeter number of Φ, given by $h = \langle \rho, \alpha_0^\vee \rangle + 1$. Explicit values are given here:

Φ	A_n	B_n	C_n	D_n	E_6	E_7	E_8	F_4	G_2
h	$n+1$	$2n$	$2n$	$2n-2$	12	18	30	12	6

- W: the Weyl group of Φ.
- $\ell : W \to \mathbb{N}$: the usual length function on W.
- $X = \mathbb{Z}\varpi_1 \oplus \cdots \oplus \mathbb{Z}\varpi_n$: the weight lattice, where the fundamental dominant weights $\varpi_i \in \mathbb{E}$ are defined by $\langle \varpi_i, \alpha_j^\vee \rangle = \delta_{ij}$, $1 \le i, j \le n$.
- $X^+ = \mathbb{N}\varpi_1 + \cdots + \mathbb{N}\varpi_n$: the cone of dominant weights.
- $X_r = \{\lambda \in X^+ : 0 \le \langle \lambda, \alpha^\vee \rangle < p^r, \quad \forall \alpha \in S\}$: the set of p^r-restricted dominant weights.
- $\lambda^\star := -w_0\lambda$: where w_0 is the longest word in the Weyl group W and $\lambda \in X$.
- $\le$: the usual partial ordering on X, where $\lambda \le \mu$ iff $\lambda - \mu$ is a non-negative sum of simple roots.
- The dot action $\cdot$ on X: for $w \in W$ and $\lambda \in X$, $w \cdot \lambda = w(\lambda + \rho) - \rho$.

1.3. Modules. Let H be an arbitrary affine algebraic group scheme over k and M be a rational H-module. M^* will denote the ordinary k-linear dual of M. Let $M^{(r)}$ denote the "twisted" representation obtained by composing the underlying representation of H with F^r (the rth iterate of the Frobenius morphism on H). If one has an isomorphism $M \cong N^{(r)}$ of rational H-modules, then one sometimes writes $M^{(-r)}$ for the module N.

Continuing with our notation as above, consider the Borel subgroup B and a weight $\lambda \in X$. By definition, such a weight λ defines a one-dimensional T-module. Letting U act trivially gives rise to a one-dimensional B-module that will also (abusively) be denoted λ. Induction gives rise to one of our standard G-modules: set $H^0(\lambda) := \operatorname{ind}_B^G \lambda$. It is well-known that $H^0(\lambda) \neq 0$ iff $\lambda \in X^+$. When λ is dominant, $H^0(\lambda)$ has highest weight λ and simple socle, denoted $L(\lambda)$. The set of such $L(\lambda)$ over X^+ gives a complete set of finite-dimensional simple G-modules. The induced module $H^0(\lambda)$ is dual to a Weyl module. Let $V(\lambda)$ denote the Weyl module with highest weight λ. One has $V(\lambda) \cong H^0(\lambda^*)^*$ and the character of $V(\lambda)$ or $H^0(\lambda)$ is given by Weyl's character formula (cf. [**Jan03**, II.5.10,5.11]).

Given a rational G-module M, one may restrict to get a module over B, U, G_r, B_r, U_r, and $G(\mathbb{F}_q)$ (as well as subgroups thereof). In particular, the finite-dimensional simple modules for G_r and $G(\mathbb{F}_q)$ arise in such a way. Given $\lambda \in X_r$, the restriction of $L(\lambda)$ to G_r or $G(\mathbb{F}_q)$ remains simple and the set of all such $L(\lambda)$ gives a complete set of finite-dimensional simple modules. This is perhaps the most fundamental example of the connections between the representation theories of G_r and $G(\mathbb{F}_q)$, and of both groups to G. This also illustrates the conceptual idea noted in Section 1.1 that G_r-representation theory approximates that of G: as r increases the set of G_r-simples approaches the set of G-simples.

1.4. Spectral sequences. Various spectral sequences have proven to be key tools in the study of cohomology. Most fundamental is the Lyndon-Hochschild-Serre spectral sequence associated to a normal subgroup of a group. See for example [**Jan03**, Prop. I.6.6].

A second spectral sequence gives an immediate connection between the (ordinary) cohomology of Lie algebras and the cohomology of a first Frobenius kernel (cf. [**Jan03**, Rem. I.9.20]). Let H be an arbitrary affine algebraic group scheme over k and assume $p \neq 2$, with $\mathfrak{h}$ denoting the Lie algebra of H. Let M be an H_1-module, then there exists a first-quadrant spectral sequence with E_2-term as follows

$$E_2^{2i,j} = S^i(\mathfrak{h}^*)^{(1)} \otimes \mathrm{H}^j(\mathfrak{h}, k) \Rightarrow \mathrm{H}^{2i+j}(H_1, M). \tag{1.4.1}$$

A third relevant spectral sequence relates G-cohomology to B-cohomology via the induction functor. For a B-module M such that $R^m \operatorname{ind}_B^G(M) = 0$ for all $m > 0$, there exists a first quadrant spectral sequence (cf. [**Jan03**, II.12.2])

$$E_2^{i,j} = R^i \operatorname{ind}_B^G \left(\mathrm{H}^j(B_r, M)^{(-r)}\right) \Rightarrow \mathrm{H}^{i+j}\left(G_r, \operatorname{ind}_B^G(M)\right)^{(-r)}. \tag{1.4.2}$$

This may be applied for example to $M = \lambda$ for a dominant weight λ.

2. Cohomology of Frobenius kernels with trivial coefficients

2.1. Consider the Frobenius kernel G_r. Of interest here is not only the cohomology group $\mathrm{H}^i(G_r, k)$ for a given degree but also the cohomology ring

$$\mathrm{H}^\bullet(G_r, k) := \bigoplus_{i=0}^{\infty} \mathrm{H}^i(G_r, k).$$

This is a graded commutative ring and, by work of Friedlander and Suslin [**FS**], known to be finitely-generated. While these rings have been described geometrically in a sense through the use of cohomological varieties[1] (cf. [**SFB1**, **SFB2**]), explicit ring computations remain minimal.

2.2. In the $r = 1$ case, for sufficiently large p, identification of $\mathrm{H}^\bullet(G_1, k)$ preceded the general finite-generation result of Friedlander-Suslin. It was shown by Friedlander and Parshall [**FP86a**] and also by Andersen and Jantzen [**AJ**] that, for $p > h$, there is an algebra isomorphism

$$\mathrm{H}^\bullet(G_1, k)^{(-1)} \cong k[\mathcal{N}],$$

where $\mathcal{N} \subseteq \mathfrak{g}$ is the set of nilpotent elements in $\mathfrak{g}$. A key idea in the proof was the use of induction from B to G.

2.3. The induction question. Recall the induction spectral sequence relating G_r-cohomology to B_r-cohomology given in (1.4.2). Optimally, one would have an affirmative answer to the following key question (for a B-module M with $R^i \operatorname{ind}_B^G(M) = 0$ for $i > 0$):

QUESTION 2.3.1. Is $R^i \operatorname{ind}_B^G \left(\mathrm{H}^j(B_r, M)^{(-r)}\right) = 0$ for $i > 0$?

If yes, the spectral sequence would collapse to give

$$\mathrm{H}^j \left(G_r, \operatorname{ind}_B^G(M)\right)^{(-r)} \cong \operatorname{ind}_B^G \left(\mathrm{H}^j(B_r, M)^{(-r)}\right).$$

For both answering the question and then making a final computation, it is often useful to use the fact that $\mathrm{H}^i(B_r, M) \cong \mathrm{H}^i(U_r, M)^{T_r}$.

Indeed, for $M = k$, this strategy works to give the above computation of G_1-cohomology. The condition $p > h$ arises when considering T_1-fixed points. See Section 2.5 below for more on this. This approach will be applied at multiple points below, and Question 2.3.1 remains a highly relevant question in its own right, in addition to its potential applications.

2.4. Small primes. For small primes, examples in [**AJ**] show that the answer is necessarily different, with the nilpotent cone (or nullcone) $\mathcal{N}$ needing to be replaced by a smaller variety. Observe that $\mathcal{N}$ may be identified with $G \cdot \mathfrak{u}$, the G-orbit of the Lie algebra of U. When $2 < p \leq h$ (and some further mild restrictions on p), work of Bendel, Nakano, Parshall, and Pillen [**BNPP**][2] provides an answer in some cases and a conjectural answer more generally in terms of certain G-orbits.

Assume that $p \geq 3$ and that p is a *good* prime for the root system. More precisely, the latter condition requires that $p \geq 5$ in types E_6, E_7, F_4, and G_2 and

[1] Another topic with an extensive literature that will not be discussed here.

[2] The primary result of that work was an identification of the complex cohomology ring of what is often known as Lusztig's "small" quantum group for $\mathfrak{g}$ at a root of unity.

$p \geq 7$ in type E_8. In type A_n, assume also that p does not divide $n+1$ (p is said to be *very good*). Consider the subset

$$\Phi_0 := \{\alpha \in \Phi \mid \langle \rho, \alpha \rangle \equiv 0 \mod p\} \subseteq \Phi.$$

When $p \geq h$, Φ_0 is empty, whereas, when $p < h$, Φ_0 is a closed nonempty subsystem of Φ. There exists a subset $J \subseteq S$ of simple roots and an element $w \in W$ such that $w(\Phi_0) = \Phi_J$ (the root subsystem generated by J) and $w(\Phi_0^+) = \Phi_J^+$. Associated to J is a standard parabolic subgroup $B \subseteq P_J = L_J \ltimes U_J \subseteq G$ with Levi subgroup L_J and unipotent radical U_J. Let $\mathfrak{u}_J$ denote the Lie algebra of U_J. Note that, if J was the empty set, we would have $P_J = B$, $L_J = T$, and $U_J = U$.

One can replace the spectral sequence of (1.4.2) with one involving the parabolic subgroup P_J and the question of computing T_1 fixed points with understanding the following homomorphism group:

$$\mathrm{Hom}_{(L_J)_1}\left(k, \operatorname{ind}_B^{P_J} w \cdot 0 \otimes H^j(\mathfrak{u}_J, k)\right).$$

This can be shown to be zero unless $j = \ell(w)$, in which case it is the trivial module k. One then needs to know the vanishing of certain higher right derived induction functors:

ASSUMPTION 1. $R^i \operatorname{ind}_{P_J}^G S^\bullet(\mathfrak{u}_J^*) = 0$ for $i > 0$.

If true, one may conclude that

- $\mathrm{H}^{2\bullet+1}(G_1, k) = 0$
- $\mathrm{H}^{2\bullet}(G_1, k)^{(-1)} \cong \operatorname{ind}_{P_j}^G S^\bullet(\mathfrak{u}_J^*)$.

One would like to identify $\operatorname{ind}_{P_j}^G S^\bullet(\mathfrak{u}_J^*)$ with $k[G \cdot \mathfrak{u}_J]$. That can be shown to hold under a second geometric assumption:

ASSUMPTION 2. The Richardson orbit closure $G \cdot \mathfrak{u}_J$ is a normal subvariety of $\mathcal{N}$.

Note that Assumptions 1 and 2 trivially hold in the case $p = h$, but that occurs only in type A_n where $p = n+1$, which is precisely one of our excluded cases. This will be discussed more below. The assumptions are also known to hold in the case $p = h - 1$, where $G \cdot \mathfrak{u}_J$ is the closure of the subregular orbit. More generally, with the assumptions on p as above, cohomology will arise only in even degrees and one may identify $\mathrm{H}^\bullet(G_1, k)^{(-1)}$ with $k[G \cdot \mathfrak{u}_J]$ under the following assumptions:

Type	A_n, C_n, D_n	B_n	E_6, G_2	F_4	E_7	E_8
Condition	$p > h/2$	$p \geq h/2$	$p \geq 5$	$p \geq 7$	$p \geq 11$	$p \geq 17$

Again, the reader is referred to [**BNPP**] for more details and explicit information on the subset J (and element $w \in W$). There one will also find a discussion of the type A_n case when p divides $n+1$, where the answer has a conjecturally different format (as already seen in [**AJ**]).

2.5. ***B_1*-cohomology.** As part of the computation of $\mathrm{H}^\bullet(G_1, k)$, when $p > h$, one finds that $\mathrm{H}^\bullet(B_1, k) \cong S^\bullet(\mathfrak{u}^*)^{(1)}$ (with the generators in degree 2); a result that fails when $p \leq h$. By the nature of the small prime argument just discussed, no new information is gained on the B_1-cohomology and this remains an open question in general. There are some known results for special linear groups as discussed in the next subsection and more generally in low degrees as discussed in Section 3.3.

2.6. Higher Frobenius kernels. Much less is known about $\mathrm{H}^\bullet(G_r, k)$ for $r > 1$. For $G = SL_2$, $\mathrm{H}^\bullet(U_r, k)$ was identified in [**AJ**], from which some observations on B_r-cohomology were made for $r = 1, 2$. Ngo [**Ngo13**] expanded on this, providing complete calculations of the B_r- and G_r-cohomology. This is done in the more general context (discussed further in Section 3) of computations of $\mathrm{H}^i(B_r, \lambda)$ and $\mathrm{H}^i(G_r, H^0(\lambda))$ for $\lambda \in X^+$. In this case, one has an affirmative answer to Question 2.3.1. For the rings $\mathrm{H}^\bullet(B_r, k)$ and $\mathrm{H}^\bullet(G_r, k)$, Ngo also identified the ring structure of the reduced rings.

More recent work of Ngo [**Ngo19**] for arbitrary G shows that B_r-cohomology is related to B_1-cohomology. More precisely, it is shown that, for $i < \frac{p}{c}$,

$$\mathrm{H}^i(B_r, k) \cong \mathrm{H}^i(B_1, k)^{(r-1)} \cong \begin{cases} S^{\frac{i}{2}}(\mathfrak{u}^*)^{(r)} & \text{if } i \text{ is even,} \\ 0 & \text{if } i \text{ is odd,} \end{cases}$$

where c depends on the root system and is given by the following table:

Type	A_n	B_n, C_n, D_n	E_6	E_7	E_8	F_4	G_2
c	1	2	3	4	6	4	3

Using the ideas of Section 2.3 one may conclude that $\mathrm{H}^i(G_r, k)^{(-r)} \cong \mathrm{ind}_B^G\left(\mathrm{H}^i(B_r, k)^{(-r)}\right)$ for $i < \frac{p}{c}$.

Such a result had been foreshadowed by earlier observations in type A_n made by Friedlander and Parshall [**FP86b**] and Kaneda, Shimada, Tezuka, and Yagita [**KSTY**] where such a relationship was shown to hold in degrees less than $2p - 1$ for $p > h$. Kaneda et al. explicitly computed $\mathrm{H}^{2p-1}(B_r, k)$ for $r > 1$, finding it to be non-zero, thus demonstrating that the $2p - 1$ bound was strict. Note also that $\mathrm{H}^{2p-1}(G_r, k)^{(-r)} \cong H^0(\alpha_0)$. The work of Kaneda et al. also gave a complete description of $\mathrm{H}^\bullet(B_2, k)$ for SL_3 for $p \geq 3$.

More recently, Friedlander [**Fr19**] has investigated U_r-cohomology, specifically considering the case that U is the unipotent radical of GL_3. The cohomology ring $\mathrm{H}^\bullet(U_r, k)$ is approximately identified in a sense. It is shown that there is an algebra given by known generators and relations that embeds into $\mathrm{H}^\bullet(U_r, k)$ with image containing all pth powers. This arises as a special case of a more general theory developed for the unipotent radical U_J of a standard parabolic in the general linear group GL_n and considering the quotient by a term in its descending central series.

3. Cohomology of Frobenius kernels with non-trivial coefficients

3.1. More generally, one may consider cohomology groups $\mathrm{H}^i(G_r, M)$, $\mathrm{H}^i(B_r, M)$, or $\mathrm{H}^i(U_r, M)$ for a G_r-module (or B_r-module or U_r-module) M. Over G_r, natural modules of interest are a simple module $L(\lambda)$ or a standard induced module $H^0(\lambda)$. Thanks to relationships (seen above) between B_r- and G_r-cohomology, much more is known in the latter case.

3.2. Induced modules. The seminal result in this context (for $r = 1$) is due to Andersen and Jantzen [**AJ**], with certain exceptional cases dealt with later by

Kumar, Lauritzen, and Thomsen [**KLT**]. Assume $p > h$ and $\lambda \in X^+$. Then

$$\begin{aligned} \mathrm{H}^i(G_1, H^0(\lambda))^{(-1)} &\cong \mathrm{ind}_B^G \left(\mathrm{H}^i(B_1,\lambda)^{(-1)}\right) \\ &\cong \begin{cases} \mathrm{ind}_B^G \left(S^{\frac{i-\ell(w)}{2}}(\mathfrak{u}^*) \otimes \mu\right) & \text{if } \lambda = w\cdot 0 + p\mu, \\ 0 & \text{otherwise,} \end{cases} \end{aligned}$$

where $w \in W$ and $\mu \in X^+$. Implicit in the statement is not only a computation of $\mathrm{H}^i(G_1, H^0(\lambda))$ but also the computation of $\mathrm{H}^i(B_1,\lambda)$ generalizing that of Section 2.5 in the case $\lambda = 0$. Again, as noted in [**AJ**], this formula breaks down for $p \leq h$.

For small primes and higher r, there is only one general result, and that is when $G = SL_2$, found in the aforementioned work of Ngo [**Ngo13**]. Indeed, the results mentioned in Section 2.6 are special cases of this work, where explicit computations of $\mathrm{H}^i(U_r,\lambda)$, $\mathrm{H}^i(B_r,\lambda)$, and $\mathrm{H}^i(G_r, H^0(\lambda))$ are given for all i, r, and $\lambda \in X^+$.

3.3. Small degrees. In degree 1, Jantzen [**Jan91**] computed all $\mathrm{H}^1(B_1,\lambda)$ and $\mathrm{H}^1(G_1, H^0(\lambda))$. Those computations were extended inductively to B_r and G_r by Bendel, Nakano, and Pillen [**BNP04c**] and then to degree 2 for $p \geq 3$ [**BNP07**]. The degree 2 and $p = 2$ case was done by Wright [**Wri**]. Bendel, Nakano, and Pillen [**BNP16**] extended such computations to degree 3 as long as the prime is not too small: $p \geq 5$ in Types A_n ($p \neq 5$ in type A_4 and $p \neq 7$ in type A_6), C_n, D_n or E_n; and $p \geq 7$ in types B_n ($n \geq 3$), F_4, or G_2.

The general approach follows the ideas of Section 2 for trivial coefficients, but becomes more intricate as the degree increases. Following Section 2.3 the goal is to conclude that

$$\mathrm{H}^i(G_r, H^0(\lambda))^{(-1)} \cong \mathrm{ind}_B^G \left(\mathrm{H}^i(B_r,\lambda)^{(-1)}\right),$$

which requires not simply knowledge of $\mathrm{H}^i(B_r,\lambda)^{(-1)}$ per se, but the specific nature of those groups to be able to conclude that higher right derived induction functors vanish. B_r-cohomology is related to B_1-cohomology through the Lyndon-Hochschild-Serre spectral sequence. Then B_1-cohomology is related to U_1-cohomology:

$$\mathrm{H}^i(B_1,\lambda) \cong \left(\mathrm{H}^i(U_1,\lambda)^{T_1}\right) \cong \left(\mathrm{H}^i(U_1,k)\otimes\lambda\right)^{T_1}.$$

Lastly, one uses (1.4.1) to relate U_1-cohomology to $\mathfrak{u}$-cohomology.[3]

In practice, the computational process is more intertwined than the above overview might suggest. For example, to make computations of $\mathfrak{u}$-cohomology one sometimes has to partially reverse the above process: via the connections between $\mathfrak{u}$, U_1, and B_1, using the Lyndon-Hochschild-Serre spectral sequence for B_1 in B, one may use information about B-cohomology to get $\mathfrak{u}$-cohomology information. In degree 3, one also begins to see that not only is information on $\mathrm{H}^i(\mathfrak{u},k)$ necessary, but also Lie algebra cohomology for non-trivial coefficients, for example, $\mathrm{H}^i(\mathfrak{u},\mathfrak{u}^*)$. In higher degrees, this approach would potentially need $\mathrm{H}^i(\mathfrak{u}, S^j(\mathfrak{u}^*))$.

An interesting feature of this spiraling/inductive strategy is that while the initial aim might be understanding $\mathrm{H}^i(G_r, H^0(\lambda))$, one obtains new cohomology results for B_r, B, and $\mathfrak{u}$ that are of interest in their own right. The latter two cases will be discussed in Sections 4 and 3.4 respectively.

[3]For $p = 2$, the spectral sequence is not available. There is a complex that may be used to compute U_1-cohomology (cf. [**Jan03**, Lemma I.9.15]).

If the underlying root system Φ is of type C_2 or F_4 with $p = 2$ or G_2 with $p = 3$, G admits a purely inseparable isogeny whose square is the Frobenius morphism. This gives rise to so-called half Frobenius kernels $G_{r/2}$ for an odd positive integer r.[4] Similar to the B_r- and G_r-computations above, Radu [**Rad**] has computed $\mathrm{H}^1(B_{r/2}, \lambda)$ and $\mathrm{H}^1(G_{r/2}, H^0(\lambda))$ in these cases.

3.4. Lie algebra cohomology. For a Lie algebra $\mathfrak{h}$ the ordinary Lie algebra cohomology is equivalently that of its universal enveloping algebra. For an $\mathfrak{h}$-module M, $\mathrm{H}^i(\mathfrak{h}, M)$ may be computed from a complex of exterior powers: $M \otimes \Lambda^\bullet(\mathfrak{h}^*)$. See for example [**Jan03**, I.9.17] and note that $\mathrm{H}^i(\mathfrak{h}, M) = 0$ for $i > \dim \mathfrak{h}$. The primary case of interest here is for the Lie algebra $\mathfrak{u}$ with trivial coefficients.

A classical result of Kostant [**Kos**][5] is

$$\mathrm{H}^i(\mathfrak{u}, k) \cong \bigoplus_{\ell(w)=i, w\in W} -w \cdot 0.$$

This was extended to prime characteristic by Friedlander and Parshall [**FP86b**] for $p > h$ and improved to $p \geq h - 1$ by Polo and Tilouine [**PT**] and, more recently, by the University of Georgia Vigre Algebra group [**UGA09**]. There are also more general versions of this result for $\mathfrak{u}_J$ associated to a parabolic subgroup for a set of simple roots J.

For $p < h - 1$, one will necessarily have cohomology classes with weights as above, but there may be more. Indeed, the University of Georgia VIGRE Algebra Group showed that there is always more cohomology in a global character sense:

$$\mathrm{ch}\, \mathrm{H}^\bullet(\mathfrak{u}, k) \neq \mathrm{ch}\, \mathrm{H}^\bullet(\mathfrak{u}, \mathbb{C}).$$

As a result, there is necessarily some degree i where $\dim \mathrm{H}^i(\mathfrak{u}, k) > \dim \mathrm{H}^i(\mathfrak{u}, \mathbb{C})$, but that does not mean that the dimensions differ in all degrees. As part of the University of Georgia VIGRE Algebra Group work, there is a MAGMA program written by Brian Boe to compute $\mathrm{H}^i(\mathfrak{u}, k)$. The author also had undergraduate students at the University of Wisconsin-Stout making computer computations. The dimensions of all $\mathrm{H}^i(\mathfrak{u}, k)$ are known at least for all rank 2 and 3 groups, as well as in type A_4 and A_5. Partial results are known in some higher ranks, with practical computational power being the limiting factor.

For small primes, as implied in Section 3.3, complete degree 1 computations were made by Jantzen [**Jan91**] and degree 2 computations by Bendel, Nakano, and Pillen [**BNP04c**] and Wright [**Wri**]. One finds that Kostant's Theorem holds in all types if $p > 3$. In degree 3, Bendel, Nakano, and Pillen [**BNP16**] showed that Kostant's Theorem holds with stronger conditions on p as given in the following chart:

Root System	A_n, $n \leq 3$; B_2	A_n, $n \geq 4$; B_3; C_n, $n \geq 3$; D_n; E_n; G_2	B_n, $n \geq 4$; F_4
$p \geq$	3	5	7

[4]These isogenies are also involved in the construction of the finite Suzuki-Ree groups.

[5]Kostant's result is actually for more general coefficients in a simple module, as are subsequent prime characteristic results that are mentioned below.

3.5. Simple modules. Returning to the question of G_1- or G_r-cohomology, a more challenging problem, and one where much less is known, is to compute $\mathrm{H}^i(G_1, L(\lambda))$ for $\lambda \in X_1$ when $L(\lambda) \neq H^0(\lambda)$. One may try to use the long exact sequence in cohomology associated to the short exact sequence

$$0 \to L(\lambda) \to H^0(\lambda) \to H^0(\lambda)/L(\lambda) \to 0.$$

Some results in that direction were provided in degree 1 by Jantzen [**Jan91**] for G_1 and by Bendel, Nakano, and Pillen [**BNP04c**] for $r > 1$. This is a special case of determining $\mathrm{Ext}^1_{G_1}(L(\lambda), L(\mu))$ for $\lambda, \mu \in X_1$, where some results are known in small rank: for example, Yehia [**Yeh**] for type A_2, Sin [**Sin**] and Dowd and Sin [**DS**] in characteristic 2 for ranks at most 4, Sin [**Sin**] for type G_2 and $p = 3$, and Lin [**Lin**] for type G_2 and $p \geq 3(h-1)$. Radu [**Rad**] has recently made computations of $\mathrm{Ext}^1_{G_r}(L(\lambda), L(\mu))$ for $\lambda, \mu \in X_r$ for general r in types C_2 or F_4 with $p = 2$ and G_2 with $p = 3$, as well as for the half Frobenius kernel $G_{r/2}$ mentioned in Section 3.3.

A finite-dimensional rational G-module M is said to be a *tilting* module if it admits both a good and Weyl filtration (cf. [**Jan03**, II.E.1, II.4.16, II.4.19]). That is, a filtration with factors of the form $H^0(\lambda)$ (good filtration) or $V(\lambda)$ (Weyl filtration) for $\lambda \in X^+$. A fundamental question for Ext-groups is the following.

QUESTION 3.5.1. Let $\lambda, \mu \in X_1$. Is $\mathrm{Ext}^1_{G_1}(L(\lambda), L(\mu))^{(-1)}$ a tilting module?

This is known to hold for large primes, where it follows from a stronger statement. For $p \geq 3h-3$, Andersen [**And**, Prop. 5.5] showed that $\mathrm{Ext}^1_{G_1}(L(\lambda), L(\mu))^{(-1)}$ is semi-simple with each composition factor $L(\sigma)$ having highest weight σ lying in the closure of the fundamental alcove. For such a σ, one has $L(\sigma) = H^0(\sigma) = V(\sigma)$, from which the existence of good and Weyl filtrations are immediate. The bound on the prime for this stronger statement was lowered to $p \geq 2h-2$ by Bendel, Nakano, and Pillen [**BNP04c**, Cor. 5.5B] and then to $p \geq 2h-4$ by Bendel, Nakano, Pillen, and Sobaje [**BNPS24b**, Thm. 4.3.1].

For small primes, there exist negative answers to Question 3.5.1. For example, when $p = 2$ and Φ is of type B_n $(n \geq 3)$ or G_2, Jantzen [**Jan91**, Prop. 6.9] showed that $\mathrm{H}^1(G_1, L(\varpi_2))^{(-1)} \cong \mathrm{Ext}^1_{G_1}(k, L(\varpi_2))^{(-1)} \cong H^0(\varpi_1)$ (following the Bourbaki ordering of roots). But $H^0(\varpi_1)$ is not simple (and hence not tilting). For $p = 3$ and Φ being of type C_3, from [**Jan91**, Prop. 4.1, §4.2, Prop. 4.5], one can show that $\mathrm{Ext}^1_{G_1}(k, L(\varpi_1+\varpi_2+\varpi_3))^{(-1)}$ is not tilting (cf. [**BNPS20**, §4.5.3]). Recent work of Bendel, Nakano, Pillen, and Sobaje [**BNPS20, BNPS24a, BNPS24b**] has shown that there is close connection between Question 3.5.1 and a conjecture of Donkin on tilting modules. Let $\lambda \in X_r$ and $Q_r(\lambda)$ denote the injective hull (equivalently, projective cover) of $L(\lambda)$ as a G_r-module. A yet unresolved question dating to work of Humphreys and Verma [**HV**] is whether $Q_r(\lambda)$ admits the structure of a rational G-module.[6] Donkin [**Don**] conjectured that this lift should be a specific indecomposable tilting module: $T(2(p-1)\rho + w_0\lambda)$. The negative answers to Question 3.5.1 noted above have led to the first known counterexamples to Donkin's conjecture (cf. [**BNPS20, BNPS24a**]).

[6]It is known to hold for $p \geq 2h-4$, as well as for smaller primes for some small rank groups.

4. B-cohomology

4.1. Leaving the domain of Frobenius kernels and returning to the full algebraic group setting, consider the Borel subgroup B. Its cohomology with trivial coefficients agrees with that of G and is well-known:

$$\mathrm{H}^i(G,k) \cong \mathrm{H}^i(B,k) = \begin{cases} k & \text{if } i = 0, \\ 0 & \text{if } i > 0. \end{cases}$$

Indeed, for a rational G-module M, one has (cf. [**Jan03**, Cor. II.4.7]) $\mathrm{H}^i(G,M) \cong \mathrm{H}^i(B,M)$ and one might attempt to use B-information to translate to G.

However, one is often interested in $\mathrm{H}^i(B,M)$ where M is a B-module that may not be a G-module. In general B-cohomology can be identified with an inverse limit over r of B_r-cohomology (cf. [**Jan03**, Cor. II.4.12]):

$$\mathrm{H}^i(B,M) \cong \varprojlim \mathrm{H}^i(B_r,M); \tag{4.1.1}$$

a statement that also hods for G or any parabolic subgroup thereof. One also has some general knowledge on the cohomological degrees in which $\mathrm{H}^i(B,M)$ can be non-zero based on weights of the module M (cf. [**Jan03**, Prop. II.4.10]), but specific computations are limited. One fundamental case of interest (suggested by earlier discussions) is a one-dimensional B-module.

4.2. One-dimensional B-modules. In characteristic zero, it is well-known that, for $\lambda \in X$, $\mathrm{H}^i(B,\lambda)$ is zero unless $\lambda = w \cdot 0$. For such a λ, non-vanishing occurs only in degree $i = \ell(w)$, where one has $\mathrm{H}^{\ell(w)}(B,\lambda) = k$. This follows from the Borel-Bott-Weil Theorem that describes $R^i \operatorname{ind}_B^G \lambda$ for all λ and i, along with the fact that G-cohomology vanishes in positive degree.

In prime characteristic, much less is known about $\mathrm{H}^i(B,\lambda)$. In degree 1, Andersen [**And**] showed that the only non-zero cohomology is the following:

$$\mathrm{H}^1(B, -p^m\alpha) = k$$

for $m \geq 0$, $\alpha \in S$.

The answer is also known in degree 2 due to work of Andersen [**And**], O'Halloran [**OHal**], Bendel, Nakano, and Pillen [**BNP07**], and Wright [**Wri**]. The cohomology is always at most one-dimensional. For $p > 3$, $\mathrm{H}^2(B,\lambda)$ is non-zero in the following cases:

$$\lambda = \begin{cases} p^i w \cdot 0, & \text{for } i \geq 0, \ell(w) = 2, \\ -p^i\alpha, & \text{for } i > 0, \alpha \in S, \\ -p^i\alpha - p^j\beta, & \text{for } 0 \leq i < j, \alpha, \beta \in S. \end{cases}$$

For $p = 3$, additional cases occur in type G_2, and, for $p = 2$, additional cases occur in all non-simply laced root systems (i.e., types B_n, C_n, F_4, and G_2).

There are many more cases where $\mathrm{H}^3(B,\lambda) \neq 0$, as well as cases where the cohomology is 2-dimensional (even for large p).[7] For $p > h$, complete computations of $\mathrm{H}^3(B,\lambda)$ were given by Andersen and Rian [**AR**] using the Lyndon-Hochschild-Serre spectral sequence for B_1 in B, along with known computations for B_1-cohomology (as discussed in Section 3.3). Bendel, Nakano, and Pillen [**BNP16**] recovered the [**AR**] results and extended them to smaller primes (with restrictions

[7] For example, one such case is if $\lambda = -p^i\alpha - p^j\beta$ for $i > j \geq 1$ and $\alpha, \beta \in S$.

on p as given in the table in Section 3.4 by using (4.1.1) and computations of $\mathrm{H}^i(B_r,\lambda)$.

4.3. U-cohomology. While the computation of $\mathrm{H}^i(B,k)$ is trivial, the closely related question of determining $\mathrm{H}^i(U,k)$ is more complex and seems to have received minimal study. Fairly recently, Friedlander [**Fr19**] initiated an attempt to better understand $\mathrm{H}^i(U,k)$. In particular, he observed that $\mathrm{H}^i(U,k)$ can be identified with the inverse limit over r of the $\mathrm{H}^i(U_r,k)$, analogous to the long-known result (4.1.1). If computations could be made for U_r, analogous to the approached just mentioned for B-cohomology, this would provide an avenue for computing U-cohomology.

5. Rational and generic cohomology

5.1. One of the most fundamental connections between cohomology groups is the seminal work of Cline, Parshall, Scott, and van der Kallen [**CPSvdK**] relating G-cohomology and $G(\mathbb{F}_q)$-cohomology. Specifically, they showed that, for a finite-dimensional rational G-module M and sufficiently large r and s (depending on i), the restriction map

$$\mathrm{H}^i(G,M^{(s)}) \to \mathrm{H}^i(G(\mathbb{F}_q),M^{(s)})$$

is an isomorphism. Since the Frobenius morphism on G is an automorphism on the subgroup $G(\mathbb{F}_q)$ (for any r), $\mathrm{H}^i(G(\mathbb{F}_q),M^{(s)}) \cong \mathrm{H}^i(G(\mathbb{F}_q),M)$ as a $G(\mathbb{F}_q)$-module for any $s \geq 1$. In other words, the above isomorphism becomes

$$\mathrm{H}^i(G,M^{(s)}) \cong \mathrm{H}^i(G(\mathbb{F}_q),M), \tag{5.1.1}$$

noting that the right-hand side is independent of s.

From the perspective of $G(\mathbb{F}_q)$-cohomology, (5.1.1) says that one can equate $\mathrm{H}^i(G(\mathbb{F}_q),M)$ with G-cohomology as long as one applies a sufficiently high twist to M. Recent developments in regards to removing (in a sense) the need for the twist are discussed in Section 5.5 below. The isomorphism (5.1.1) leads to two stability consequences:

Rational Stability: The cohomology $\mathrm{H}^i(G,M^{(s)})$ eventually stabilizes as s increases. That is, for a given i, there exists an s such that

$$\mathrm{H}^i(G,M^{(s)}) \cong \mathrm{H}^i(G,M^{(s+1)}). \tag{5.1.2}$$

Generic Cohomology: The cohomology $\mathrm{H}^i(G(\mathbb{F}_q),M)$ eventually stabilizes as r increases. This stable value is known as the *generic* cohomology $\mathrm{H}^i_{\mathrm{gen}}(G,M)$.

A new perspective on these ideas was taken in work of Bendel, Nakano, and Pillen [**BNP14**]. This will be discussed in Sections 5.3 and 5.4 which depends heavily on certain induced modules that are considered in the next subsection.

5.2. Inducing from $G(\mathbb{F}_q)$ to G. A direct relationship between $G(\mathbb{F}_q)$-cohomology and G-cohomology is given by generalized Frobenius reciprocity (or Shapiro's lemma). For a $G(\mathbb{F}_q)$-module M, one has

$$\mathrm{H}^i(G(\mathbb{F}_q),M) \simeq \mathrm{H}^i(G,\mathrm{ind}_{G(\mathbb{F}_q)}^G(M)). \tag{5.2.1}$$

For a rational G-module M, by the tensor identity, one has $\mathrm{ind}_{G(\mathbb{F}_q)}^G(M) \cong M \otimes \mathrm{ind}_{G(\mathbb{F}_q)}^G(k)$ and hence

$$\mathrm{H}^i(G(\mathbb{F}_q),M) \simeq \mathrm{H}^i(G,M \otimes \mathrm{ind}_{G(\mathbb{F}_q)}^G(k)).$$

Unfortunately, the module $\operatorname{ind}_{G(\mathbb{F}_q)}^G(k)$ is infinite-dimensional.

As observed by Bendel, Nakano, and Pillen [**BNP11**, Prop. 2.4] (cf. also [**PSS, BNPPSS**]), $\operatorname{ind}_{G(\mathbb{F}_q)}^G(k)$ admits a filtration with factors of the form $H^0(\lambda) \otimes H^0(\lambda^*)^{(r)}$, where there is precisely one such factor for each $\lambda \in X^+$. That is, there exists a filtration

$$0 \subseteq F_0 \subseteq F_1 \subseteq F_2 \subseteq \cdots \subseteq \operatorname{ind}_{G(\mathbb{F}_q)}^G(k),$$

where $F_i/F_{i-1} \cong H^0(\lambda_i) \otimes H^0(\lambda_i^*)^{(r)}$ for $\lambda_i \in X^+$, $\cup_i F_i = \operatorname{ind}_{G(\mathbb{F}_q)}^G(k)$, and each $\lambda \in X^+$ arises precisely once as a λ_i. Moreover, this filtration behaves nicely with respect to the size of weights. Given a positive integer m, $\operatorname{ind}_{G(\mathbb{F}_q)}^G(k)$ admits a finite-dimensional submodule I_m that has a (necessarily finite) filtration as above, where each λ_i appearing in a filtration factor satisfies $\langle \lambda_i, \alpha_0^\vee \rangle \leq m$.

Such a filtration had been foreshadowed in earlier work of Bendel, Nakano, and Pillen [**BNP01, BNP02, BNP04a, BNP04b, BNP06**], where truncations of $\operatorname{ind}_{G(\mathbb{F}_q)}^G(k)$ (to certain bounded weight categories) were considered and used to directly relate $G(\mathbb{F}_q)$- and G-cohomology (and extensions more generally). The reader is referred to [**Nak**, §5] for an overview of those results.

5.3. Generic cohomology revisited. The filtration of Section 5.2 was used in [**BNP14**] to give a new proof of (5.1.1). The idea is to to consider a short exact sequence

$$0 \to k \to \operatorname{ind}_{G(\mathbb{F}_q)}^G(k) \to N \to 0.$$

Tensoring with $M^{(s)}$ gives another exact sequence:

$$0 \to M^{(s)} \to M^{(s)} \otimes \operatorname{ind}_{G(\mathbb{F}_q)}^G(k) \to M^{(s)} \otimes N \to 0. \tag{5.3.1}$$

One can then take the long exact sequence in G-cohomology associated to (5.3.1). Using the identification $\mathrm{H}^i(G, M^{(s)} \otimes \operatorname{ind}_{G(\mathbb{F}_q)}^G(k)) \cong \mathrm{H}^i(G, \operatorname{ind}_{G(\mathbb{F}_q)}^G(M^{(s)})) \cong \mathrm{H}^i(G(\mathbb{F}_q), M^{(s)})$, this becomes

$$\begin{aligned}
0 \longrightarrow \operatorname{Hom}_G(k, M^{(s)}) &\xrightarrow{\text{res}} \operatorname{Hom}_{G(\mathbb{F}_q)}(k, M^{(s)}) \longrightarrow \operatorname{Hom}_G(k, M^{(s)} \otimes N) \\
\longrightarrow \mathrm{H}^1(G, M^{(s)}) &\xrightarrow{\text{res}} \mathrm{H}^1(G(\mathbb{F}_q), M^{(s)}) \longrightarrow \mathrm{H}^1(G, M^{(s)} \otimes N) \\
\longrightarrow \mathrm{H}^2(G, M^{(s)}) &\xrightarrow{\text{res}} \mathrm{H}^2(G(\mathbb{F}_q), M^{(s)}) \longrightarrow \mathrm{H}^2(G, M^{(s)} \otimes N) \\
\longrightarrow \mathrm{H}^3(G, M^{(s)}) &\xrightarrow{\text{res}} \mathrm{H}^3(G(\mathbb{F}_q), M^{(s)}) \longrightarrow \mathrm{H}^3(G, M^{(s)} \otimes N) \longrightarrow \cdots.
\end{aligned}$$

One sees that the groups in the third column are the obstruction to the restriction map being an isomorphism. To show that $\mathrm{H}^i(G, M^{(s)} \otimes N)$ vanishes, one shows the vanishing of all $\mathrm{H}^i(G, M^{(s)} \otimes H^0(\lambda_j) \otimes H^0(\lambda_j^*)^{(r)})$ for the $H^0(\lambda_j) \otimes H^0(\lambda_j^*)^{(r)}$ appearing as filtration factors in N.

While these generic cohomology statements are existence in nature, both the original work [**CPSvdK**] and the newer work [**BNP14**] give explicit values for r and s where (5.1.1) is guaranteed to hold. One distinction between the methods is that the values in [**CPSvdK**] depend on the root system (in addition to the degree i and prime p), whereas those in [**BNP14**] depend only on the degree and the prime. In [**BNP14**, §8] a comparison of bounds is made, showing a general improvement (decrease) in the size of r and s needed.

5.4. Rational stability revisited. As noted above, a consequence of (5.1.1) is the rational stability result 5.1.2. In [**BNP14**], the authors gave an independent proof of this using new results on B-cohomology of one-dimensional B-modules as discussed in Section 4.2.

5.5. Shifted cohomology. A particular case of interest is when the rational G-module M is a simple module $L(\lambda)$ with $\lambda \in X_r$, so that $L(\lambda)$ remains simple upon restriction to $G(\mathbb{F}_q)$. Then one has $\mathrm{H}^i(G(\mathbb{F}_q), L(\lambda)) \cong \mathrm{H}^i(G, L(\lambda)^{(s)})$. It turns out that one can remove the s-twist in a qualified sense. The potential for such a result was seen for example in [**BNP06**, Thm. 5.5], where it was shown that, for sufficiently large r (depending on the prime p and Coxeter number h), either $\mathrm{H}^1(G(\mathbb{F}_q), L(\lambda)) \cong \mathrm{H}^1(G, L(\lambda))$ or $\mathrm{H}^1(G(\mathbb{F}_q), L(\lambda)) \cong \mathrm{H}^1(G, L(\tilde{\lambda}))$ for a *different* p^r-restricted weight $\tilde{\lambda}$ (obtained by a certain shifting algorithm applied to the weight λ). A version for extensions between two simple modules was also shown to hold, where the highest weights of either or both simple modules might need to be shifted.

Later work of Stewart [**Ste12b**] for degree 2 cohomology with G being SL_3 suggested that such a result might hold in higher degrees. Parshall, Scott, and Stewart [**PSS**] showed that one may choose r sufficiently large (depending on the root system and cohomological degree m) so that, for $i \leq m$ and $\lambda \in X_r$, $\mathrm{H}^i(G(\mathbb{F}_q), L(\lambda)) \cong \mathrm{H}^i(G, L(\lambda'))$ for some potentially shifted weight $\lambda' \in X_r$. A version for Ext-groups between simples was also given.

6. Vanishing ranges

6.1. Dating back to work of Quillen [**Qui**], it has been observed that $\mathrm{H}^i(G(\mathbb{F}_q), k)$ is zero in "low" positive degrees. That is

$$\mathrm{H}^i(G(\mathbb{F}_q), k) = 0 \text{ for } 0 < i < m$$

for some m depending on G and r. Ideas discussed in the previous section have been used to provide new results on this problem, discussion of which may also be found in [**Nak**, §6].

6.2. The general linear group. Quillen [**Qui**] showed that $\mathrm{H}^i(GL_n(\mathbb{F}_q), k) = 0$ for $0 < i < r(p-1)$, while making no claim about what happened in degree $r(p-1)$. The question of interest then is to identify precisely the least positive degree with non-zero cohomology. Friedlander and Parshall [**FP83**] (for $p > 2$) expanded Quillen's original vanishing range to $0 < i < r(2p-3)$ by showing such vanishing for the Borel subgroup $B(\mathbb{F}_q)$. Further, they showed the bound was sharp for $B(\mathbb{F}_q)$. That is, $\mathrm{H}^{r(2p-3)}(B(\mathbb{F}_q), k) \neq 0$. But they were not able to show non-vanishing of $\mathrm{H}^{r(2p-3)}(GL_n(\mathbb{F}_q), k)$. Barbu [**Bar**] showed that $\mathrm{H}^{r(2p-2)}(GL_n(\mathbb{F}_q), k) \neq 0$ and conjectured that it should be related (via a Bockstein map) to a non-zero class in degree $r(2p-3)$. Non-vanishing in degree $r(2p-3)$ was shown by Bendel, Nakano, and Pillen [**BNP11**] under the condition that $p \geq n+2$. The bound on the prime was lowered to $p \geq n$ by Sprehn [**Spr**] who took a more constructive approach. More recent work of Lahtinen and Sprehn [**LS**] constructed non-zero cohomology elements when $p < n$. See more discussion in Section 6.5 below.

6.3. Other groups. Quillen [**Qui**] commented (without any specifics) that there should be a similar vanishing range for other classical groups. Vanishing ranges (but not the least degree of non-vanishing) were found by Friedlander [**Fr76**] for special orthogonal and symplectic groups and by Hiller [**Hil**] for simply connected G in all types.

The first precise bounds were given in work of Bendel, Nakano, and Pillen [**BNP11, BNP12**]. As with earlier arguments for Frobenius kernels, one may focus on the $r = 1$ case (i.e., $G(\mathbb{F}_p)$) and then attempt to use induction to obtain results for higher r. One has the isomorphism

$$\mathrm{H}^i(G(\mathbb{F}_p), k) \simeq \mathrm{H}^i(G, \mathrm{ind}_{G(\mathbb{F}_p)}^G(k))$$

of (5.2.1) along with the filtration of $\mathrm{ind}_{G(\mathbb{F}_p)}^G(k)$ discussed in Section 5.2. One then wishes to show the vanishing of

$$\mathrm{H}^i(G, H^0(\lambda) \otimes H^0(\lambda^*)^{(1)}) \cong \mathrm{Ext}_G^i(V(\lambda)^{(1)}, H^0(\lambda))$$

in small degrees. To that end, one may use the Lyndon-Hochschild-Serre spectral sequence for $G_1 \trianglelefteq G$:

$$E_2^{i,j} = \mathrm{Ext}_{G/G_1}^i(V(\lambda)^{(1)}, \mathrm{H}^j(G_1, H^0(\lambda))) \Rightarrow \mathrm{Ext}_G^{i+j}(V(\lambda)^{(1)}, H^0(\lambda)).$$

One may now apply the result of Andersen-Jantzen from Section 3.2 on $\mathrm{H}^j(G_1, H^0(\lambda))$ (where the assumption $p > h$ is needed). This leads to combinatorial questions based on weight spaces in $S^\bullet(\mathfrak{u}^*)$ and Kostant's partition function.

For G simple, simply connected as in our standing assumption, for $p > h$, precise bounds are obtained for $G(\mathbb{F}_p)$ in all types except F_4 and G_2. Generically, $\mathrm{H}^i(G(\mathbb{F}_p), k) = 0$ for $0 < i < 2p - 3$ and $\mathrm{H}^{2p-3}(G(\mathbb{F}_p), k) \neq 0$. However, that is not universally the case with exceptions in type C_n, "small" rank, and in type A_n when $p = n + 2 = h + 1$. A complete summary of these bounds is given in the following table, where the sharp bound D means that $\mathrm{H}^D(G(\mathbb{F}_p), k) \neq 0$ and $\mathrm{H}^i(G(\mathbb{F}_p), k) = 0$ for $0 < i < D$.

Root System	p	Sharp Vanishing Bound D
$A_n, n \geq 4$	$p > n+2$	$2p-3$
$A_n, n \geq 3$	$p = n+2$	$p-2$
A_3	$p > 5$	$2p-6$
A_2	$p = 3k+1,\ k \geq 2$	$2p-6$
A_2	$p = 3k+2,\ k \geq 1$	$2p-3$
$B_n, n \geq 7$	$p > 2n$	$2p-3$
$B_n, n \in \{5,6\}$	$p > 13$	$2p-3$
$B_n, n \in \{5,6\}$	$p = 13$	$2p-5$
B_5	$p = 11$	$2p-7=3$
$B_n,\ n \in \{3,4\}$	$p > 2n$	$2p-8$
$C_n, n \geq 1$	$p > 2n$	$p-2$
$D_n, n \geq 4$	$p > 2n-2$	$2p-2n$
E_6	$p > 13$	$2p-3$
E_6	$p = 13$	$2p-10=16$
E_7	$p > 23$	$2p-3$
E_7	$p = 23$	$2p-7=39$
E_7	$p = 19$	$2p-9=27$
E_8	$p \geq 31$	$2p-3$
F_4	$p \geq 13$	$2p-9 \leq D \leq 2p-3$
G_2	$p \geq 7$	$2p-8 \leq D \leq 2p-3$

For $r > 1$, one can usually obtain bounds inductively, with the sharp bound given by multiplying D as above by r. E.g., $r(2p-3)$ or $r(2p-2n)$. However, this is not possible in all cases, particularly types B_n and E_6, where additional subtleties arise. The reader is referred to [**BNP11**, **BNP12**] for precise details on the known bounds.

The aforementioned work of Sprehn [**Spr**] is purely constructive in nature, showing the existence of cohomology classes in certain degrees (but not the vanishing of any cohomology). Sprehn considers an arbitrary connected reductive group G and formally works over the finite field $\mathbb{F}_p$ assuming that $p \geq h$. In that work, it is shown that $\mathrm{H}^\bullet(G(\mathbb{F}_q), \mathbb{F}_p)$ contains (as a graded module over the Steenrod algebra) $\mathrm{H}^\bullet(GL_2(\mathbb{F}_q), \mathbb{F}_p)$, from which it follows that $\mathrm{H}^{r(2p-3)}(G(\mathbb{F}_q), \mathbb{F}_p) \neq 0$. By similar means, the lower degree non-vanishing is also observed for the symplectic group (type C_n): $\mathrm{H}^{r(p-2)}(Sp_{2n}(\mathbb{F}_q), \mathbb{F}_p) \neq 0$ (for $p \geq 2n$).

6.4. Adjoint type. Suppose for this subsection that G is simple of *adjoint* type rather than simply connected, so that the root lattice coincides with the weight lattice. For example, we are considering $PGL_n(\mathbb{F}_q)$ rather than $SL_n(\mathbb{F}_q)$ in type A. For the simply-laced root systems A_n, D_n, and E_n, still with $p > h$, there is a uniform sharp vanishing bound of $r(2p-3)$ (cf. [**BNP12**]).

6.5. Small primes. In addition to the remaining questions for large primes, the task of identifying precise vanishing ranges for $p < h$ remains much more open. Using a more direct approach to showing the vanishing of $\mathrm{Ext}^j_G(V(\lambda)^{(1)}, H^0(\lambda))$, Bendel, Nakano, and Pillen [**BNP14**] obtained smaller vanishing ranges for arbitrary primes:

- $p = 2$: $\mathrm{H}^i(G(\mathbb{F}_q), k) = 0$ for $0 < i < r$,

- $p > 2$: $\mathrm{H}^i(G(\mathbb{F}_q), k) = 0$ for $0 < i < r(p-2)$.

For an arbitrary (simple, simply connected) G, those are the best possible ranges as seen by long-known computations of Carlson [**Car**] for SL_2: for $p = 2$, $\mathrm{H}^r(SL_2(\mathbb{F}_q), k) \neq 0$, and, for $p > 2$, $\mathrm{H}^{r(p-2)}(SL_2(\mathbb{F}_q), k) \neq 0$. These ranges are of course not necessarily strict in general.

Following the characteristic class approach of [**Spr**], Lahtinen and Sprehn [**LS**] gave a generalization in a sense of the aforementioned generic $r(2p-3)$ non-vanishing result. For $n \geq 2$, let s be such that $p^{s-1} < n \leq p^s$. They showed that $\mathrm{H}^d(GL_n(\mathbb{F}_q), \mathbb{F}_p) \neq 0$ for $d = r(2p^s - 2p^{s-1} - 1)$.[8] Note that, when $s = 1$, $d = r(2p-3)$. For $r = 1$, some stronger results are given on the existence of non-nilpotent elements. For example, with $p = 2$, suppose $2 \leq n \leq 2^s$. If the sum of the binary digits of d is at least s, then there is a non-nilpotent element in $\mathrm{H}^d(GL_n(\mathbb{F}_p), \mathbb{F}_p)$. For $p > 2$, the result involves the p-ary expansion of the degree.

7. Finite group cohomology of simple modules

7.1. Similar to the situation with Frobenius kernels, computations of $G(\mathbb{F}_q)$-cohomology with non-trivial coefficients are limited. See [**Hum**, §12 & §14, particularly §12.11 & §14.2] for some early results. We consider here the basic case of $\mathrm{H}^i(G(\mathbb{F}_q), L(\lambda))$ for $\lambda \in X_r$. Jones [**Jon**], Cline, Parshall, and Scott [**CPS75**], and Jones and Parshall [**JP**] computed $\mathrm{H}^1(G(\mathbb{F}_q), L(\lambda))$ for dominant weights λ that are *minimal* in the usual partial ordering of weights.

7.2. Degree 1. More recent work of the University of Georgia VIGRE Algebra Group [**UGA13**] took a new approach to these degree 1 computations and expanded on the known results. The approach of that work (related to the ideas in Section 5) was to relate $G(\mathbb{F}_q)$-cohomology to G-cohomology. Specifically they showed that for $q > 3$ and p given in the following table

Root System	A_n, D_n	B_n, C_n, E_6, E_7, F_4, G_2	E_8
$p >$	2	3	5

,

if $\lambda \leq \varpi_j$ for some fundamental dominant weight ϖ_j, then $\mathrm{H}^i(G(\mathbb{F}_q), L(\lambda)) \cong \mathrm{H}^1(G, L(\lambda))$. This was done by relating $G(\mathbb{F}_q)$-cohomology to U_1-cohomology. With this result and a slightly stronger condition on the prime ($p > 7$ in types E_7 and E_8), the cohomology groups were computed and usually seen to be zero.[9] In particular, under the assumptions on q and p as in the above table, complete computations of $\mathrm{H}^1(G(\mathbb{F}_q), L(\varpi_j))$ are given.

7.3. Degree 2. Using the induction functor $\mathrm{ind}_{G(\mathbb{F}_q)}^G(-)$ and the ideas of Section 5.3 and (again) connections with U_r-extensions, in later work, the University of Georgia VIGRE Algebra group [**UGA12**] investigated degree 2 cohomology and identified conditions under which $\mathrm{H}^2(G(\mathbb{F}_q), L(\lambda)) \cong \mathrm{H}^2(G, L(\lambda))$. Again, the focus is on the case that $\lambda \leq \varpi_j$. Such an isomorphism is given under the following assumptions on p and q[10]

[8] They also gave the existence of non-nilpotent elements (in larger degree).

[9] With exceptions in type C_n, F_4, E_7, and E_8, where the result is the trivial module k.

[10] † In type C_n, if λ lies in the root lattice, one requires $q > 5$.

Root System	A_n, B_n, $C_n^{\dagger}$, D_n, E_6	E_7, F_4	G_2
$p >$	3	3	5
$q >$	3	5	5

,

with some exceptions on the weight λ. Explicit computations are made in almost all cases, as well as computations for most of the exceptional cases. These latter cases give some examples where $\mathrm{H}^2(G, L(\lambda)) = 0$ but $\mathrm{H}^2(G(\mathbb{F}_q), L(\lambda)) \neq 0$.

8. Bounding dimensions

8.1. For a finite simple group, a longstanding question has been to bound the size of the first cohomology group of a faithful, absolutely irreducible module. In our context, the specific question is to understand the dimension of $\mathrm{H}^1(G(\mathbb{F}_q), L(\sigma))$ for $\sigma \in X_1$. Even in the broader finite simple group context, known examples (such as those in Section 7) suggested that the size should be small, potentially with a universal bound. That remained the consensus until a workshop held at the American Institute of Mathematics in 2012, where discussions at the meeting and soon thereafter led to the discovery of extremely large-dimensional cohomology groups. For a discussion of some of the history and development of the problem, the reader is referred to [**AIM**] (cf. also [**BNPPSS**]).

The existence of these large-dimensional cohomology groups fundamentally changed the perspective on the problem, in particular, increasing the importance of earlier work of Cline, Parshall, and Scott [**CPS09**] who had shown that there is a bound

$$\dim \mathrm{H}^1(G(\mathbb{F}_q), L(\lambda)) \leq C(\Phi),$$

where $C(\Phi)$ depends only on the root system, as well as similar work of Guralnick and Tiep [**GT**], where the *cross-characteristic* case was considered. That is, where one is considering $G(\mathbb{F}_q)$ with p (the characteristic of the base field k) *not* dividing q; a situation generally not under consideration in this paper.

8.2. ***G*-cohomology.** The Cline-Parshall-Scott result along with further developments below provide another example of the key interplay between cohomology groups over different structures. The bound on $\mathrm{H}^1(G(\mathbb{F}_q), L(\lambda))$ was achieved by showing that such a bound held for G-cohomology of simple modules, and then using a result of Bendel, Nakano, and Pillen [**BNP06**] (as referenced in Section 5.5) that related $G(\mathbb{F}_q)$-cohomology with G-cohomology.

Later work of Parshall and Scott [**PS**] extended these bounds on G-cohomology to higher degree cohomology groups, showing that there exist bounds on $\dim \mathrm{H}^i(G, L(\lambda))$ depending only on Φ and the degree i. More generally, it was shown that such a bound exists for $\dim \mathrm{Ext}^i_G(L(\lambda), L(\mu))$ for a pair of simple modules.

8.3. Finite groups continued. Parker and Stewart [**PaSt**] improved upon [**CPS09**] by finding explicit bounds on the dimension of $\mathrm{H}^1(G(\mathbb{F}_q), L(\lambda))$ based on the Coxeter number h. That work also makes use of the relationship with G-cohomology but takes a more direct approach to bounding the G-cohomology through the use of the Jantzen sum formula [**Jan03**, II.8.19] for a Weyl module and bounding the number of composition factors in a Weyl module. The explicit nature of their bound also allows one to see how the bound grows as the rank grows.

8.4. Finite group cohomology in higher degrees. Using the bounds on G-cohomology in higher degrees from [**PS**], Bendel, Nakano, Parshall, Pillen, Scott, and Stewart [**BNPPSS**] showed that there exist bounds

$$\dim \mathrm{H}^i(G(\mathbb{F}_q), L(\lambda)) \leq C(\Phi, i),$$

where $C(\Phi, i)$ depends only on Φ and the degree i. More generally, such a bound was given for extensions between simple $G(\mathbb{F}_q)$-modules. This made use of a modification of ideas discussed in Sections 5 and 6.

For a simple G-module $L(\sigma)$, as in Section 5.2, we know

$$\mathrm{H}^i(G(\mathbb{F}_q), L(\sigma)) \cong \mathrm{H}^i(G, \mathrm{ind}_{G(\mathbb{F}_q)}^G(L(\sigma)) \cong \mathrm{H}^i(G, L(\sigma) \otimes \mathrm{ind}_{G(\mathbb{F}_q)}^G(k))$$

and $\mathrm{ind}_{G(\mathbb{F}_q)}^G(k)$ admits a filtration with factors of the form $H^0(\lambda) \otimes H^0(\lambda^*)^{(r)}$, one for each $\lambda \in X^+$. For a given i, there exists a positive integer b such that $\mathrm{H}^j(G, L(\sigma) \otimes H^0(\lambda) \otimes H^0(\lambda^*)^{(r)}) = 0$ for $j \leq i$ and $\langle \lambda, \alpha_0^\vee \rangle > b$. This allows one to truncate $\mathrm{ind}_{G(\mathbb{F}_q)}^G(k)$ giving a finite-dimensional module I_b with

$$\mathrm{H}^i(G(\mathbb{F}_q), L(\sigma)) \cong \mathrm{H}^i(G, L(\sigma) \otimes I_b),$$

where I_b admits a filtration with factors $H^0(\lambda) \otimes H^0(\lambda^*)^{(r)}$ where $\langle \lambda, \alpha_0^\vee \rangle \leq b$. This then allows one to transfer G-bounds to $G(\mathbb{F}_q)$.

REMARK 8.4.1. In the [**BNPPSS**] work, it was similarly shown that bounds exist on $\mathrm{H}^i(G_r, L(\sigma))$ and, more generally, for Ext-groups between simple modules over Frobenius kernels.

REMARK 8.4.2. Given the potentially large dimensions of cohomology groups, it is of interest to determine just how large they can be. Stewart [**Ste12a**] demonstrated that the dimensions could be exponential in the cohomology degree by explicitly constructing a simple SL_2-module L (depending on n) with $\dim \mathrm{H}^{2n}(SL_2, L) \geq 2^{n-1}$.

8.5. Bounds based on the dimension of the coefficient module. In a different direction from finding absolute bounds on dimensions, there have been attempts to bound cohomology in terms of the dimension of the coefficient module. That is, for a group H and an H-module M, one may try to show that there is a constant C (likely depending on the degree i) such that

$$\dim \mathrm{H}^i(H, M) \leq C \cdot \dim M.$$

For example, let H be a finite group and M be an irreducible kH-module on which H acts faithfully, then Guralnick and Hoffman [**GH**] showed that

$$\dim \mathrm{H}^1(H, M) \leq \frac{1}{2} \cdot \dim M.$$

In degree 2, for a quasi-simple group H and any finite-dimensional kH-module M, Guralnick, Kantor, Kassabov, and Lubotsky [**GKKL**] showed that

$$\dim \mathrm{H}^2(H, M) \leq 17.5 \cdot \dim M.$$

For an arbitrary finite group and faithful, irreducible module M, they found a bound of $C = 18.5$.

In our setting, this question was investigated with some progress at the aforementioned 2012 AIM meeting by Bendel, Boe, Drupieski, Nakano, Parshall, Pillen, and Wright [**AIM**]. For the algebraic group G and M an arbitrary finite-dimensional rational G-module, the following bounds were found:

- $\dim \mathrm{H}^1(G, M) \leq \frac{1}{h} \cdot \dim M \leq \frac{1}{2} \cdot \dim M$,
- $\dim \mathrm{H}^2(G, M) \leq \dim M$,
- for $p > h$, $\dim \mathrm{H}^3(G, M) \leq 2 \cdot \dim M$.

These results use the fact that $\mathrm{H}^i(G, M) \cong \mathrm{H}^i(B, M)$ for a rational G-module M (as noted in Section 4.1) and results from Section 4.2 on $\mathrm{H}^i(B, \lambda)$. In the degree 3 case, the condition $p > h$ can be weakened to those given in Section 3.4 by using the $\mathrm{H}^3(B, \lambda)$ computations of [**BNP16**]. Using these G-cohomology bounds, one may show that, by choosing r sufficiently large, the same bounds may be given for $G(\mathbb{F}_q)$ (and any $kG(\mathbb{F}_q)$-module M).

REMARK 8.5.1. Outside the context of this paper, in [**AIM**], bounds were also obtained on the first cohomology of a symmetric group with coefficients in a simple module.

Acknowledgments

I would like to thank Daniel Nakano and Cornelius Pillen for their many years of friendship and mathematical collaboration that led to a number of the results discussed in this paper. I also thank the organizers for the opportunity to speak at the Southeastern Lie Theory Workshop X held in June 2018 at the University of Georgia. This article stemmed from the material given in a series of of lectures presented at that meeting.

The year 2022 saw the tragic passing of two stalwarts in the field: Georgia Benkart and Brian Parshall. Both were in attendance at that SE Lie Theory Workshop, and Georgia was my "partner" in giving a second minicourse at the event. I would like to acknowledge their impact on my mathematical career. I received my first speaking invitation from Georgia, and we interacted regularly over the years, particularly as she tried to promote representation theory in Wisconsin and the surrounding region. I also met Brian early in my career, when he visited Northwestern University. I eventually had the pleasure of collaborating with him on multiple projects. I remain ever grateful for a wonderful semester-long sabbatical I spent at the University of Virginia; time that bore mathematical fruit for many years thereafter.

References

[And] Henning Haahr Andersen, *Extensions of modules for algebraic groups*, Amer. J. Math. **106** (1984), no. 2, 489–504, DOI 10.2307/2374311. MR737781

[AJ] Henning Haahr Andersen and Jens Carsten Jantzen, *Cohomology of induced representations for algebraic groups*, Math. Ann. **269** (1984), no. 4, 487–525, DOI 10.1007/BF01450762. MR766011

[AR] Henning Haahr Andersen and Tarik Rian, *B-cohomology*, J. Pure Appl. Algebra **209** (2007), no. 2, 537–549, DOI 10.1016/j.jpaa.2006.07.009. MR2293326

[Bar] Adrian Barbu, *On the range of non-vanishing p-torsion cohomology for* $\mathrm{GL}_n(\mathbb{F}_p)$, J. Algebra **278** (2004), no. 2, 456–472, DOI 10.1016/j.jalgebra.2004.02.016. MR2071647

[AIM] Christopher P. Bendel, Brian D. Boe, Christopher M. Drupieski, Daniel K. Nakano, Brian J. Parshall, Cornelius Pillen, and Caroline B. Wright, *Bounding the dimensions of rational cohomology groups*, Developments and retrospectives in Lie theory, Dev. Math., vol. 38, Springer, Cham, 2014, pp. 51–69, DOI 10.1007/978-3-319-09804-3_2. MR3308777

[BNPP] Christopher P. Bendel, Daniel K. Nakano, Brian J. Parshall, and Cornelius Pillen, *Cohomology for quantum groups via the geometry of the nullcone*, Mem. Amer. Math. Soc. **229** (2014), no. 1077, x+93. MR3204911

[BNPPSS] Christopher P. Bendel, Daniel K. Nakano, Brian J. Parshall, Cornelius Pillen, Leonard L. Scott, and David Stewart, *Bounding cohomology for finite groups and Frobenius kernels*, Algebr. Represent. Theory **18** (2015), no. 3, 739–760, DOI 10.1007/s10468-014-9514-7. MR3357946

[BNP01] Christopher P. Bendel, Daniel K. Nakano, and Cornelius Pillen, *On comparing the cohomology of algebraic groups, finite Chevalley groups and Frobenius kernels*, J. Pure Appl. Algebra **163** (2001), no. 2, 119–146, DOI 10.1016/S0022-4049(01)00024-X. MR1846657

[BNP02] Christopher P. Bendel, Daniel K. Nakano, and Cornelius Pillen, *Extensions for finite Chevalley groups. II*, Trans. Amer. Math. Soc. **354** (2002), no. 11, 4421–4454, DOI 10.1090/S0002-9947-02-03073-8. MR1926882

[BNP04a] Christopher P. Bendel, Daniel K. Nakano, and Cornelius Pillen, *Extensions for finite Chevalley groups. I*, Adv. Math. **183** (2004), no. 2, 380–408, DOI 10.1016/S0001-8708(03)00091-4. MR2041903

[BNP04b] Christopher P. Bendel, Daniel K. Nakano, and Cornelius Pillen, *Extensions for finite groups of Lie type: twisted groups*, Finite groups 2003, Walter de Gruyter, Berlin, 2004, pp. 29–46. MR2125064

[BNP04c] Christopher P. Bendel, Daniel K. Nakano, and Cornelius Pillen, *Extensions for Frobenius kernels*, J. Algebra **272** (2004), no. 2, 476–511, DOI 10.1016/j.jalgebra.2003.04.003. MR2028069

[BNP06] Christopher P. Bendel, Daniel K. Nakano, and Cornelius Pillen, *Extensions for finite groups of Lie type. II. Filtering the truncated induction functor*, Representations of algebraic groups, quantum groups, and Lie algebras, Contemp. Math., vol. 413, Amer. Math. Soc., Providence, RI, 2006, pp. 1–23, DOI 10.1090/conm/413/07837. MR2262362

[BNP07] Christopher P. Bendel, Daniel K. Nakano, and Cornelius Pillen, *Second cohomology groups for Frobenius kernels and related structures*, Adv. Math. **209** (2007), no. 1, 162–197, DOI 10.1016/j.aim.2006.05.001. MR2294220

[BNP11] Christopher P. Bendel, Daniel K. Nakano, and Cornelius Pillen, *On the vanishing ranges for the cohomology of finite groups of Lie type*, Int. Math. Res. Not. IMRN **12** (2012), 2817–2866, DOI 10.1093/imrn/rnr130. MR2942711

[BNP12] Christopher P. Bendel, Daniel K. Nakano, and Cornelius Pillen, *On the vanishing ranges for the cohomology of finite groups of Lie type II*, Recent developments in Lie algebras, groups and representation theory, Proc. Sympos. Pure Math., vol. 86, Amer. Math. Soc., Providence, RI, 2012, pp. 25–73, DOI 10.1090/pspum/086/1410. MR2976996

[BNP14] Christopher P. Bendel, Daniel K. Nakano, and Cornelius Pillen, *Extensions for finite Chevalley groups III: Rational and generic cohomology*, Adv. Math. **262** (2014), 484–519, DOI 10.1016/j.aim.2014.04.014. MR3228435

[BNP16] Christopher P. Bendel, Daniel K. Nakano, and Cornelius Pillen, *Third cohomology for Frobenius kernels and related structures*, Lie algebras, Lie superalgebras, vertex algebras and related topics, Proc. Sympos. Pure Math., vol. 92, Amer. Math. Soc., Providence, RI, 2016, pp. 81–118, DOI 10.1090/pspum/092/01579. MR3644227

[BNPS20] Christopher P. Bendel, Daniel K. Nakano, Cornelius Pillen, and Paul Sobaje, *Counterexamples to the tilting and (p,r)-filtration conjectures*, J. Reine Angew. Math. **767** (2020), 193–202, DOI 10.1515/crelle-2019-0036. MR4160306

[BNPS24a] Christopher P. Bendel, Daniel K. Nakano, Cornelius Pillen, and Paul Sobaje, *On Donkin's tilting module conjecture II: counterexamples*, Compos. Math. **160** (2024), no. 6, 1167–1193, DOI 10.1112/S0010437X24007115. MR4740592

[BNPS24b] Christopher P. Bendel, Daniel K. Nakano, Cornelius Pillen, and Paul Sobaje, *On Donkin's Tilting Module Conjecture III: New generic lower bounds*, Journal of Algebra, DOI 10.1016/j.jalgebra.2023.07.001. https://www.sciencedirect.com/science/article/pii/S0021869323003198

[Car] Jon F. Carlson, *The cohomology of irreducible modules over* $\mathrm{SL}(2,p^n)$, Proc. London Math. Soc. (3) **47** (1983), no. 3, 480–492, DOI 10.1112/plms/s3-47.3.480. MR716799

[CPS75] Edward Cline, Brian Parshall, and Leonard Scott, *Cohomology of finite groups of Lie type. I*, Inst. Hautes Études Sci. Publ. Math. **45** (1975), 169–191. MR399283

[CPS09] Edward T. Cline, Brian J. Parshall, and Leonard L. Scott, *Reduced standard modules and cohomology*, Trans. Amer. Math. Soc. **361** (2009), no. 10, 5223–5261, DOI 10.1090/S0002-9947-09-04633-9. MR2515810

[CPSvdK] E. Cline, B. Parshall, L. Scott, and Wilberd van der Kallen, *Rational and generic cohomology*, Invent. Math. **39** (1977), no. 2, 143–163, DOI 10.1007/BF01390106. MR439856

[Don] Stephen Donkin, *On tilting modules for algebraic groups*, Math. Z. **212** (1993), no. 1, 39–60, DOI 10.1007/BF02571640. MR1200163

[DS] Michael F. Dowd and Peter Sin, *On representations of algebraic groups in characteristic two*, Comm. Algebra **24** (1996), no. 8, 2597–2686, DOI 10.1080/00927879608542648. MR1393277

[Fr76] Eric M. Friedlander, *Computations of K-theories of finite fields*, Topology **15** (1976), no. 1, 87–109, DOI 10.1016/0040-9383(76)90054-9. MR394660

[Fr19] Eric M. Friedlander, *Cohomology of unipotent group schemes*, Algebr. Represent. Theory **22** (2019), no. 6, 1427–1455, DOI 10.1007/s10468-018-9827-z. MR4034790

[FP83] E. M. Friedlander and B. J. Parshall, *On the cohomology of algebraic and related finite groups*, Invent. Math. **74** (1983), no. 1, 85–117, DOI 10.1007/BF01388532. MR722727

[FP86a] Eric M. Friedlander and Brian J. Parshall, *Cohomology of Lie algebras and algebraic groups*, Amer. J. Math. **108** (1986), no. 1, 235–253 (1986), DOI 10.2307/2374473. MR821318

[FP86b] Eric M. Friedlander and Brian J. Parshall, *Cohomology of infinitesimal and discrete groups*, Math. Ann. **273** (1986), no. 3, 353–374, DOI 10.1007/BF01450727. MR824427

[FS] Eric M. Friedlander and Andrei Suslin, *Cohomology of finite group schemes over a field*, Invent. Math. **127** (1997), no. 2, 209–270, DOI 10.1007/s002220050119. MR1427618

[GH] Robert M. Guralnick and Corneliu Hoffman, *The first cohomology group and generation of simple groups*, Groups and geometries (Siena, 1996), Trends Math., Birkhäuser, Basel, 1998, pp. 81–89, DOI 10.1007/bf01214004. MR1644977

[GKKL] Robert Guralnick, William M. Kantor, Martin Kassabov, and Alexander Lubotzky, *Presentations of finite simple groups: profinite and cohomological approaches*, Groups Geom. Dyn. **1** (2007), no. 4, 469–523, DOI 10.4171/GGD/22. MR2357481

[GT] Robert M. Guralnick and Pham Huu Tiep, *First cohomology groups of Chevalley groups in cross characteristic*, Ann. of Math. (2) **174** (2011), no. 1, 543–559, DOI 10.4007/annals.2011.174.1.16. MR2811608

[Hil] Howard L. Hiller, *Cohomology of Chevalley groups over finite fields*, J. Pure Appl. Algebra **16** (1980), no. 3, 259–263, DOI 10.1016/0022-4049(80)90031-6. MR558490

[Hum] James E. Humphreys, *Modular representations of finite groups of Lie type*, London Mathematical Society Lecture Note Series, vol. 326, Cambridge University Press, Cambridge, 2006. MR2199819

[HV] J. E. Humphreys and D. N. Verma, *Projective modules for finite Chevalley groups*, Bull. Amer. Math. Soc. **79** (1973), 467–468, DOI 10.1090/S0002-9904-1973-13220-3. MR320176

[Jan91] Jens C. Jantzen, *First cohomology groups for classical Lie algebras*, Representation theory of finite groups and finite-dimensional algebras (Bielefeld, 1991), Progr. Math., vol. 95, Birkhäuser, Basel, 1991, pp. 289–315. MR1112165

[Jan03] Jens Carsten Jantzen, *Representations of algebraic groups*, 2nd ed., Mathematical Surveys and Monographs, vol. 107, American Mathematical Society, Providence, RI, 2003. MR2015057

[Jon] Wayne Russell Jones, *COHOMOLOGY OF FINITE GROUPS OF LIE TYPE*, ProQuest LLC, Ann Arbor, MI, 1975. Thesis (Ph.D.)–University of Minnesota. MR2625445

[JP] Wayne Jones and Brian Parshall, *On the 1-cohomology of finite groups of Lie type*, Proceedings of the Conference on Finite Groups (Univ. Utah, Park City, Utah, 1975), Academic Press, New York-London, 1976, pp. 313–328. MR404470

[KSTY] Masaharu Kaneda, Nobuo Shimada, Michishige Tezuka, and Nobuaki Yagita, *Cohomology of infinitesimal algebraic groups*, Math. Z. **205** (1990), no. 1, 61–95, DOI 10.1007/BF02571225. MR1069485

[Kos] Bertram Kostant, *Lie algebra cohomology and the generalized Borel-Weil theorem*, Ann. of Math. (2) **74** (1961), 329–387, DOI 10.2307/1970237. MR142696

[KLT] Shrawan Kumar, Niels Lauritzen, and Jesper Funch Thomsen, *Frobenius splitting of cotangent bundles of flag varieties*, Invent. Math. **136** (1999), no. 3, 603–621, DOI 10.1007/s002220050320. MR1695207

[LS] Anssi Lahtinen and David Sprehn, *Modular characteristic classes for representations over finite fields*, Adv. Math. **323** (2018), 1–37, DOI 10.1016/j.aim.2017.10.029. MR3725872

[Lin] Zong Zhu Lin, *Extensions between simple modules for Frobenius kernels*, Math. Z. **207** (1991), no. 4, 485–499, DOI 10.1007/BF02571404. MR1119953

[Nak] Daniel K. Nakano, *Cohomology of algebraic groups, finite groups, and Lie algebras: interactions and connections*, Lie theory and representation theory, Surv. Mod. Math., vol. 2, Int. Press, Somerville, MA, 2012, pp. 151–176. MR3076141

[Ngo13] Nham V. Ngo, *Cohomology for Frobenius kernels of SL_2*, J. Algebra **396** (2013), 39–60, DOI 10.1016/j.jalgebra.2013.07.033. MR3108071

[Ngo19] Nham V. Ngo, *Low degree cohomology of Frobenius kernels*, Advances in algebra, Springer Proc. Math. Stat., vol. 277, Springer, Cham, 2019, pp. 255–259, DOI 10.1007/978-3-030-11521-0_13. MR3922649

[OHal] Joyce O'Halloran, *A vanishing theorem for the cohomology of Borel subgroups*, Comm. Algebra **11** (1983), no. 14, 1603–1606, DOI 10.1080/00927878308822921. MR702726

[PaSt] Alison E. Parker and David I. Stewart, *First cohomology groups for finite groups of Lie type in defining characteristic*, Bull. Lond. Math. Soc. **46** (2014), no. 2, 227–238, DOI 10.1112/blms/bdt087. MR3194742

[PS] Brian J. Parshall and Leonard L. Scott, *Bounding Ext for modules for algebraic groups, finite groups and quantum groups*, Adv. Math. **226** (2011), no. 3, 2065–2088, DOI 10.1016/j.aim.2010.09.021. MR2739773

[PSS] Brian J. Parshall, Leonard L. Scott, and David I. Stewart, *Shifted generic cohomology*, Compos. Math. **149** (2013), no. 10, 1765–1788, DOI 10.1112/S0010437X13007331. MR3123309

[PT] Patrick Polo and Jacques Tilouine, *Bernstein-Gelfand-Gelfand complexes and cohomology of nilpotent groups over $\mathbb{Z}_{(p)}$ for representations with p-small weights* (English, with English and French summaries), Astérisque **280** (2002), 97–135. Cohomology of Siegel varieties. MR1944175

[Rad] Aura-Cristiana Radu, *On the cohomology of the Ree groups and kernels of exceptional isogenies*, J. Algebra **612** (2022), 1–54, DOI 10.1016/j.jalgebra.2022.09.001. MR4484818

[RW] Simon Riche and Geordie Williamson, *A simple character formula* (English, with English and French summaries), Ann. H. Lebesgue **4** (2021), 503–535, DOI 10.5802/ahl.79. MR4275245

[Qui] Daniel Quillen, *On the cohomology and K-theory of the general linear groups over a finite field*, Ann. of Math. (2) **96** (1972), 552–586, DOI 10.2307/1970825. MR315016

[Sin] Peter Sin, *Extensions of simple modules for special algebraic groups*, J. Algebra **170** (1994), no. 3, 1011–1034, DOI 10.1006/jabr.1994.1373. MR1305273

[Spr] David Sprehn, *Nonvanishing cohomology classes on finite groups of Lie type with Coxeter number at most p*, J. Pure Appl. Algebra **219** (2015), no. 6, 2396–2404, DOI 10.1016/j.jpaa.2014.09.006. MR3299737

[Ste12a] David I. Stewart, *Unbounding Ext*, J. Algebra **365** (2012), 1–11, DOI 10.1016/j.jalgebra.2012.04.026. MR2928449

[Ste12b] David I. Stewart, *The second cohomology of simple SL_3-modules*, Comm. Algebra **40** (2012), no. 12, 4702–4716, DOI 10.1080/00927872.2011.618858. MR2989676

[SFB1] Andrei Suslin, Eric M. Friedlander, and Christopher P. Bendel, *Infinitesimal 1-parameter subgroups and cohomology*, J. Amer. Math. Soc. **10** (1997), no. 3, 693–728, DOI 10.1090/S0894-0347-97-00240-3. MR1443546

[SFB2] Andrei Suslin, Eric M. Friedlander, and Christopher P. Bendel, *Support varieties for infinitesimal group schemes*, J. Amer. Math. Soc. **10** (1997), no. 3, 729–759, DOI 10.1090/S0894-0347-97-00239-7. MR1443547

[UGA09] University of Georgia VIGRE Algebra Group, *On Kostant's theorem for Lie algebra cohomology*, Representation theory, Contemp. Math., vol. 478, Amer. Math. Soc.,

Providence, RI, 2009, pp. 39–60, DOI 10.1090/conm/478/09318. University of Georgia VIGRE Algebra Group: Irfan Bagci, Brian D. Boe, Leonard Chastkofsky, Benjamin Connell, Bobbe J. Cooper, Mee Seong Im, Tyler Kelly, Jonathan R. Kujawa, Wenjing Li, Daniel K. Nakano, Kenyon J. Platt, Emilie Wiesner, Caroline B. Wright and Benjamin Wyser. MR2513265

[UGA12] University of Georgia VIGRE Algebra Group, *Second cohomology for finite groups of Lie type*, J. Algebra **360** (2012), 21–52, DOI 10.1016/j.jalgebra.2012.02.028. University of Georgia VIGRE Algebra Group: Brian D. Boe, Brian Bonsignore, Theresa Brons, Jon F. Carlson, Leonard Chastkofsky, Christopher M. Drupieski, Niles Johnson, Wenjing Li, Phong Thanh Luu, Tiago Macedo, Daniel K. Nakano, Nham Vo Ngo, Brandon L. Samples, Andrew J. Talian, Lisa Townsley and Benjamin J. Wyser. MR2914632

[UGA13] University of Georgia VIGRE Algebra Group, *First cohomology for finite groups of Lie type: simple modules with small dominant weights*, Trans. Amer. Math. Soc. **365** (2013), no. 2, 1025–1050, DOI 10.1090/S0002-9947-2012-05664-9. University of Georgia VIGRE Algebra Group: Brian D. Boe, Adrian M. Brunyate, Jon F. Carlson, Leonard Chastkofsky, Christopher M. Drupieski, Niles Johnson, Benjamin F. Jones, Wenjing Li, Daniel K. Nakano, Nham Vo Ngo, Duc Duy Nguyen, Brandon L. Samples, Andrew J. Talian, Lisa Townsley and Benjamin J. Wyser. MR2995382

[Wri] Caroline B. Wright, *Second cohomology groups for algebraic groups and their Frobenius kernels*, J. Algebra **330** (2011), 60–75, DOI 10.1016/j.jalgebra.2011.01.013. MR2774617

[Yeh] Samy El Badawy Yehia, *Extensions of simple modules for the universal Chevalley group and its parabolic subgroups*, Ph.D. thesis, University of Warwick, 1982.

Department of Mathematics, Statistics and Computer Science, University of Wisconsin-Stout, Menomonie, Wisconsin 54751

Email address: bendelc@uwstout.edu

Proceedings of Symposia in Pure Mathematics
Volume **108**, 2024
https://doi.org/10.1090/pspum/108/01958

Idempotent modules, locus of compactness and local supports

Jon F. Carlson

To the memory of Brian Parshall and Georgia Benkart, good friends and wonderful colleagues

ABSTRACT. Let kG be the group algebra of a finite group scheme defined over a field k of characteristic $p > 0$. Associated to any closed subset V of the projectivized prime ideal spectrum $\operatorname{Proj} \mathrm{H}^*(G, k)$ is a thick tensor ideal subcategory of the stable category of finitely generated kG-module, whose closure under arbitrary direct sums is a localizing tensor ideal in the stable category of all kG-modules. The colocalizing functor from the big stable category to this localizing subcategory is given by tensoring with an idempotent module $\mathcal{E}$. A property of the idempotent module is that its restriction along any flat map $\alpha : k[t]/(t^p) \to kG$ is a compact object. For a kG-module M, we define its locus of compactness in terms of such restrictions. With some added hypothesis, in the case that V is a closed point, for a kG-module M, we show that in the stable category $\operatorname{Hom}_{kG}(\mathcal{E}, M)$ is finitely generated over the endomorphism ring of $\mathcal{E}$, provided the restriction along an associated flat map is a compact object. This leads to a notion of local supports. We prove some of its properties, give a realization theorem and display the images of the L_ζ modules under the colocalizing functor.

1. Introduction and notation

Suppose that k is a field of characteristic $p > 0$ and that G is a finite group scheme defined over k. Let kG denote its group algebra, which is a finite dimensional cocommutative Hopf algebra. Examples of such algebras include the group algebras of finite groups, restricted enveloping algebras of restricted p-Lie algebras and infinitesimal subgroups of algebraic groups defined over k. For any such algebra the stable categories $\mathbf{stmod}(kG)$ of finitely generated kG-modules and $\mathbf{StMod}(kG)$ of all kG-modules are tensor triangulated categories. There is a notion of support varieties, and the thick tensor ideals in $\mathbf{stmod}(kG)$ have been classified using this construction and the spectrum $V_G(k) = \operatorname{Proj} \mathrm{H}^*(G, k)$ of the cohomology ring [**5**, **12**].

Associated to any thick tensor ideal in $\mathbf{stmod}(kG)$ is a triangle consisting of the trivial module and a pair of idempotent modules $\mathcal{E}$ and $\mathcal{F}$. Tensoring with these modules induce colocalizing and localizing functors on the stable category. Except

2020 *Mathematics Subject Classification.* Primary 20C20; Secondary 20J06, 18G80.

Key words and phrases. Finite group schemes, representations, idempotent modules, thick subcategories, support varieties.

Research partially supported by Simons Foundation grant 054813-01.

in trivial cases, these module are infinitely generated. None the less, they have the unusual property that their restrictions along any flat map (called a π-point) $A = k[t]/(t^p) \to kG$ is a compact object in $\mathbf{StMod}(A)$. So it would seem that the property of having compact restrictions is related to the structure of the category $\mathbf{StMod}(kG)$. In this paper we begin an investigation of this issue.

We introduce a locus of compactness for a kG-module M. It is the collection of all points in $V_G(k)$ such that the restriction of M along any associated π-point is compact. The locus has some properties similar to support varieties, and in particular, it identifies some thick subcategories of $\mathbf{StMod}(kG)$. In the case that kG is the restricted enveloping algebra of a Lie algebra, the properties are particularly nice, and the subcategories closed under tensor products. In other cases there is some dependence on the Hopf structure on kG. Given a module M, the inclusion of a point of $V_G(k)$ in the locus of M depends on the image of M in the colocalized category generated by the idempotent $\mathcal{E}$ module.

The idempotent $\mathcal{E}$ module acts by tensor product as the identity on the colocalized subcategory that it generates. As a consequence, the group of homomorphisms between two modules in the subcategory is module over the endomorphism ring of the idempotent module. Using earlier work [**9**], we know precisely the structure of the endomorphism ring of the $\mathcal{E}$ module in the case that kG is the group algebra of an elementary abelian p-group and the variety of the $\mathcal{E}$ module is a single point V in $V_G(k)$. While it is true in this case that the endomorphism ring is infinitely generated, modules in the subcategory can still be distinguished by the annihilator of their endomorphisms and homomorphisms. We show, still assuming that V is a closed point, that given any noninveritble element ζ of the stable endomorphism ring $\underline{\mathrm{End}}_{kG}(\mathcal{E})$, there is a kG-module M_ζ such that the annihilator of $\underline{\mathrm{Hom}}_{kG}(\mathcal{E}, M)$ in $\underline{\mathrm{End}}_{kG}(\mathcal{E})$ is close to being the ideal generated by ζ. That is, the radicals of the ideals coincide. In addition, if the point V is in the locus of compactness of a module M in the subcategory, then the homomorphism group in the stable category $\underline{\mathrm{Hom}}_{kG}(\mathcal{E}, M)$ is finitely generated over the endomorphism ring $\underline{\mathrm{End}}_{kG}(\mathcal{E})$.

After an early version of this paper was posted, it came to light that the group consisting of Dave Benson, Srikanth Iyengar, Henning Krause and Julia Pevtsova have results that overlap significantly with those in this paper. The BIKP collaboration approaches the subject from an abstract categorical perspective that contrasts sharply with the more elementary hands-on development presented here. In several cases, they prove stronger theorems, but without as much of the detailed information on the modules and maps. I am grateful to the group for letting me see their unfinished manuscript [**8**] and letting me borrow at least one of their ideas. I want to also thank the Hausdorff Research Institute for Mathematics at the University of Bonn for support and for the stimulating program during which some of work on this paper was completed.

For references on group representation and cohomology see the texts [**3**] or [**10**]. For triangulated categories, see [**17**] or the early sections of [**2**].

2. Background

Throughout the paper kG is the group algebra of a finite group scheme G that is defined over the field k of characteristic $p > 0$. Let $\mathbf{mod}(kG)$ denote the category of finitely generated kG-modules and kG-module homomorphisms, and let $\mathbf{Mod}(kG)$ be the category of all kG-modules and homomorphisms. The stable category **stmod**(kG) has the same objects as **mod**(kG), but in the stable category the set of morphisms from object M to object N is given by

$$\underline{\mathrm{Hom}}_{kG}(M,N) = \frac{\mathrm{Hom}_{kG}(M,N)}{\mathrm{PHom}_{kG}(M,N)}$$

where $\mathrm{PHom}_{kG}(M,N)$ is the set of all homomorphisms that factor through a projective module. The category **StMod**(kG) of all kG-modules is constructed the same way, by factoring out any map that factors through a projective module. Note that **stmod**(kG) is the subcategory of compact objects in **StMod**(kG).

For a kG-module M, the modules $\Omega(M)$ and $\Omega^{-1}(M)$ are defined to be the kernel of a projective cover $P \twoheadrightarrow M$ and the cokernel of an injective hull $M \hookrightarrow I$, respectively. Recall that kG is a self-injective ring, so that the injective module I is projective and the projective module P is injective. Consequently, $\Omega(\Omega^{-1}(M)) \cong \Omega^{-1}(\Omega(M)$ is the largest direct summand of M that has no projective summands.

The stable categories **stmod**(kG) and **StMod**(kG) are tensor triangulated categories. The tensor product of modules is over the base field k with the action of kG defined using the coalgebra structure on kG. The triangles correspond roughly to short exact sequences in the module category. That is, a triangle has the form

$$\longrightarrow X \xrightarrow{\alpha} Y \xrightarrow{\beta} Z \xrightarrow{\gamma} \Omega^{-1}(X) \longrightarrow$$

where for some projective module P and maps α', β' representing the classes α and β, there is an exact sequence

$$0 \longrightarrow X \xrightarrow{\alpha'} Y \oplus P \xrightarrow{\beta'} Z \longrightarrow 0 \quad .$$

The operator Ω^{-1} is the translation or shift functor on the stable categories.

A π-point [**12**] is a flat map $\alpha : K[t]/(t^p) \to KG_K$, where K is some extension of k, and the map factors by flat maps through the group algebra KE of some unipotent abelian subgroup scheme E of G_K. Two π-points, $\alpha_K : K[t]/(t^p) \to KG_K$ and $\beta_L : L[t]/(t^p) \to LG_L$, are equivalent if for any finitely generated kG-module M the restriction of $K \otimes_k M$ to $K[t]/(t^p)$, denoted $\alpha_K^*(K \otimes M)$, and $\beta_L^*(L \otimes M)$ are either both projective or both not projective.

In the case that G is a finite group, a unipotent subgroup scheme would be a subalgebra kE where $E = \langle g_1, \ldots, g_r \rangle$ is an elementary abelian subgroup of order p^r for some r. Then a π-point that is defined over k would have the form $\alpha : k[t]/(t^p) \to kE \subseteq kG$ given by $\alpha(1) = \sum_{i=1}^r \alpha_i(g_i - 1) + w$ where w is some element in $\mathrm{Rad}^2(kE)$ and for some i, $\alpha_i \neq 0$. The equivalence class of α depends only on the element $[\alpha_1| \ldots |\alpha_r] \in \mathcal{P}_k^{r-1}$. Let $V_G^r(k)$ denote the collection of all equivalence classes of π-points.

The cohomology ring $\mathrm{H}^*(G,k)$ is a finitely generated graded-commutative k-algebra. This was proved by Evens, Golod and Venkov for finite groups and by Friedlander and Suslin [**14**] for general finite group schemes. The projectivized prime ideal spectrum $V_G(k) = \mathrm{Proj}\,\mathrm{H}^*(G,k)$ is a finite dimensional projective variety. For M a finitely generated kG-module, let $V_G(M) \subseteq V_G(k)$ be the collection of all homogeneous prime ideals that contain the annihilator of $\mathrm{Ext}_{kG}^*(M,M)$.

Given a closed subvariety $V \subseteq V_G(k)$, the collection $\mathcal{M}_V$ of all finitely generated kG-modules M with $V_G(M) \subseteq V$ is a thick tensor ideal in **stmod**(kG), meaning that it is a triangulated subcategory that is both closed under taking direct summands and closed under tensor products with arbitrary objects in **stmod**(kG).

For any such V and $\mathcal{M}_V$, there is a distinguished triangle [**18**] in $\mathbf{StMod}(kG)$ having the form

$$\mathcal{S}_V: \qquad \longrightarrow \mathcal{E}_V \xrightarrow{\sigma_V} k \xrightarrow{\tau_V} \mathcal{F}_V \longrightarrow \Omega(\mathcal{E}_V) \longrightarrow$$

where $\mathcal{E}_V$ and $\mathcal{F}_V$ are idempotent modules in that $\mathcal{E}_V \otimes \mathcal{E}_V \cong \mathcal{E}_V$, and $\mathcal{F}_V \otimes \mathcal{F}_V \cong \mathcal{F}_V$ in the stable category. In addition, $\mathcal{E}_V \otimes \mathcal{F}_V \cong 0$, meaning that $\mathcal{E}_V \otimes \mathcal{F}_V$ is a projective module. The triangle has a couple of universal properties. Let M be a finitely generated kG-module. The first property is that any map $\gamma : N \to M$, with N in $\mathcal{M}_V$, factors through $\sigma_V \otimes \mathrm{Id}_M : \mathcal{E}_V \otimes M \to k \otimes M \cong M$. The other property is that any map $\gamma : M \to N$, such that the third object in the triangle of γ is in $\mathcal{M}_V$, factors through $\tau_V \otimes \mathrm{Id}_M : M \cong k \otimes M \to \mathcal{F}_V \otimes M$.

As to support varieties, for any kG-module M, $\mathcal{V}_G(M)$ is the collection of all points in $V_G(k)$ with the property that the pull back along a corresponding π-point $\alpha_K : K[t]/(t^p) \to KG_K$ of the extended module $K \otimes M$ is not projective. In the case of the idempotent modules, with V closed, $\mathcal{V}_G(\mathcal{E}_V)$ is the set of all points in V, while $\mathcal{V}_G(\mathcal{F}_V) = \mathcal{V}_G(k) \setminus \mathcal{V}_G(\mathcal{E}_V)$, its complement.

3. The locus of compactness

First we make a series of observations that lead to the definition of many thick subcategories of $\mathbf{StMod}(kG)$ associated to subsets the projectived prime ideal spectrum of the cohomology ring $\mathrm{H}^*(G,k) \cong \mathrm{Ext}^*_{kG}(k,k)$. We show that under certain circumstance, there exist some different sorts of support varieties based on cardinalities of dimensions of restrictions of the objects along π-points.

By an idempotent module, we always mean a kG-module M such that $M \otimes M \cong M$ in the stable category. Or said another way, M is an idempotent module if $M \otimes M \cong M \oplus P$ where P is a projective module. We recall that the only finitely generated idempotent module is the trivial module k. We refer to the modules $\mathcal{E}_V$ and $\mathcal{F}_V$, for V closed in $V_G(k)$ as the Rickard idempotent modules. In addition, there are many other idempotent modules, including those defined for more general collections of points in $V_G(k)$. For example, a countable direct sum of copies of the trivial module is an idempotent module. If we let $M = \sum_{n \geq 0} \Omega^n(k)$, then a countable direct sum of copies of M is an idempotent module.

For all of this, the Rickard idempotent modules have a very special property. To be precise we have the following.

LEMMA 3.1. *Suppose that M is one of $\mathcal{E}_V$ or $\mathcal{F}_V$ for V a closed subvariety of $V_G(k)$. Then for any π-point $\alpha_K : K[t]/(t^p) \to KG_K$ we have that the restriction $\alpha_K^*(K \otimes M)$ is either the zero module or is isomorphic to K in $\mathbf{StMod}(K[t]/(t^p))$. The same is true of the tensor $\mathcal{E}_V \otimes \mathcal{F}_{V'}$ for V and V' closed subvarieties.*

PROOF. Consider the extended triangle $K \otimes_k \mathcal{S}_V$. If the π-point α_K corresponds to a point in V, then it is does not correspond to any point in $\mathcal{V}_k(\mathcal{F}_V)$. Consequently, $\alpha_K^*(K \otimes \mathcal{F}_V)$ is a projective module, and by the triangle, $\alpha_K^*(K \otimes \mathcal{E}_V) \cong K$ in the stable category. On the other hand, if α_K does not correspond to a point in V, then $\alpha_K^*(K \otimes \mathcal{E}_V)$ is projective and $\alpha_K^*(K \otimes \mathcal{F}_V) \cong K$. □

The above lemma implies that all of the Rickard idempotent modules are contained in the full subcategory $\mathcal{SF}(kG)$ of $\mathbf{StMod}(kG)$ consisting of all kG-modules M with the property that $\alpha_K^*(K \otimes M)$ is finite dimensional for all π-points

$$\alpha_K : K[t]/(t^p) \to KG_K.$$

In other words, the restriction along any π-point of a module in $\mathcal{SF}(kG)$ is a compact object in the stable category of $K[t]/(t^p)$. The category $\mathcal{SF}(kG)$ is triangulated because the restriction maps preserve triangles and if two objects in a triangle are compact then so it the third. It is easy to see that it is also closed under finite direct sums and taking direct summands. In addition, by an arguments that is very similar to that of the proof of Lemma 3.1, it contains the tensor product of any finite dimensional module with a Rickard idempotent module.

An observation of Benson, Iyengar, Krause and Pevtsova helps us to make sense of this.

PROPOSITION 3.2 ([**8**]). *Suppose that $\alpha, \beta : k[t]/(t^p) \to kG$ are equivalent π-points that are defined over k. Let M be a kG-module such that $\alpha^*(M)$ is a compact object in* $\mathbf{StMod}(k[t]/(t^p))$*. Then $\beta^*(M)$ is also compact.*

PROOF. Let $M^* = \operatorname{Hom}_k(M, k)$ be the k-dual of M. There is a natural map $M \to M^{**}$ from M to its double dual. This is a kG-homomorphism. That is, while the action of kG on M^* is defined using the antipode of the Hopf structure, the action on M^{**} involves applying the antipode twice, which is the identity. Assume $\alpha^*(M)$ is compact. Then the map to the double dual is an isomorphism in the stable category. In particular, the cokernel U of the natural map $M \to M^{**}$, which is an injective map, is a projective module on restriction along α. But then, $\beta^*(U)$ is also projective since β is equivalent to α. It follows, that $\beta^*(M)$ is a compact object. □

Suppose that $\mathcal{V}$ is any subset of points in $V_G(k)$. Let $\mathcal{N}_{\mathcal{V}}$ be the full subcategory of all kG-modules M with the property that for any π-point $\alpha_K : K[t]/(t^p) \to KG_K$ whose equivalence class is in the collection $\mathcal{V}$, the restriction $\alpha_K^*(K \otimes M)$ is a compact object, meaning isomorphic to a finite dimensional object in the stable category. Then $\mathcal{N}_{\mathcal{V}}$ is a thick triangulated subcategory of $\mathbf{StMod}(kG)$. All of this suggests a variant support variety. We thank Paul Balmer for suggesting the name.

DEFINITION 3.3. For a kG-module M, let $\mathcal{U}_G(M)$ be the collection of points in $V_G(k)$ with the property that for any π-point associated to this point,

$$\alpha_K : A = K[t]/(t^p) \longrightarrow KG_K,$$

the restriction $\alpha_K^*(K \otimes M)$ is isomorphic to a finite dimensional module in the stable category $\mathbf{StMod}(A)$. We call $\mathcal{U}_G(M)$ the locus of compactness or compact locus of the module M.

Note that $\mathcal{U}_G(M)$ is not a closed subvariety of $V_G(k)$, but rather only a subset of points. Indeed, we see from item (2) in the next proposition, that $\mathcal{U}_G(M)$ is likely to contain an open set of $V_G(M)$. We list some of the properties of $\mathcal{U}_G$.

PROPOSITION 3.4. *Suppose that M and N are kG-modules.*

(1) *If M is a compact object or if M is in $\mathcal{SF}(kG)$, then $\mathcal{U}_G(M) = V_G(k)$.*
(2) *If V is a point that is not in $\mathcal{V}_G(M)$, then $V \in \mathcal{U}_G(M)$.*
(3) *As noted above, any subset $\mathcal{V} \subset V_G(k)$ defines a thick subcategory $\mathcal{N}_{\mathcal{V}}$ in* $\mathbf{StMod}(kG)$.
(4) *Suppose that $L \to M \to N$ is a triangle in* $\mathbf{StMod}(kG)$*. Then*

$$\mathcal{U}_G(L) \cap \mathcal{U}_G(N) \subset \mathcal{U}_G(M).$$

(5) *Suppose that V is a closed subvariety of $V_G(k)$. Then*

$$\mathcal{U}_G(M \otimes \mathcal{E}_V) = (V \cap \mathcal{U}_G(M)) \cup c(V) \quad \textit{and}$$
$$\mathcal{U}_G(M \otimes \mathcal{F}_V) = V \cap (\mathcal{U}_G(M) \cap c(V)).$$

where $c(V)$ is the complement of V in $V_G(k)$. In particular, if M is a compact object, then $\mathcal{U}_G(M \otimes \mathcal{E}_V) = V_G(k) = \mathcal{U}_G(M \otimes \mathcal{F}_V)$.

PROOF. All but the last of the items are obvious. For the last item, suppose that $\alpha_K : K[t]/(t^p) \to KG_K$ is a π-point for some extension K of k. If the equivalence class of α_K is in V, then $\alpha_K^*(K \otimes \mathcal{F}_V)$ is projective. It follows that in this case,

$$\alpha_K^*(K \otimes (\mathcal{E}_V \otimes M)) = \alpha_K^*(K \otimes M).$$

Hence, $V \cap \mathcal{U}_G(M) = V \cap \mathcal{U}_G(\mathcal{E}_V \otimes M)$.

On the other hand, suppose that the class of α_K is not in V. Then $\alpha_K^*(K \otimes \mathcal{E}_V)$ is projective and so is $\alpha_K^*(K \otimes (\mathcal{E}_V \otimes M))$. Hence, the class of α_K is in $\mathcal{U}_G(M)$. This proves the first statement. The proof of the decomposition of $\mathcal{U}_G(M \otimes \mathcal{F}_V)$ is very similar. The last statement follows from the previous ones, noting that $\mathcal{U}_G(M) = V_G(k)$. □

We remark that statement (4) of the proposition turns the usual condition for support varieties on its head. That is, for $L \to M \to N$ a triangle in **StMod**(kG) we have that

$$\mathcal{V}_G(M) \quad \subseteq \quad \mathcal{V}_G(L) \cup \mathcal{V}_G(N)$$

which is something like a reverse of the condition in (4).

There is one case where we can go even further with known results.

EXAMPLE 3.5. Suppose that kG is the restricted enveloping algebra of a restricted p-Lie algebra. In this case, $V_G(k)$ is identified with the restricted null cone $\mathcal{N}_1$ (see [**13, 15**]). That is, we have an embedding of the Lie algebra into its restricted enveloping algebra, which is the free tensor algebra on the Lie algebra modulo the relations given by the Lie bracket and the pth-power operation. The restricted null cone $\mathcal{N}$ is the set of all elements x of the Lie algebra such that $x^{[p]} = 0$, where $x \mapsto x^{[p]}$ is the pth-power operation. In this case, every element of $V_G(k) \cong \mathcal{N}$ has a unique distinguished π-point that is a map of Hopf algebras. Consequently, we can define $\mathcal{U}_G(M)$ to be the collection of all points in $V_G(k)$ such that the restriction of M along the distinguished π-point associated to this point is compact. That is, instead of letting $\mathcal{U}_G(M)$, for a module M, be a set of equivalence classes of π-points with the finite dimensionality property for *all* elements of the equivalence class, we measure $\mathcal{U}_G(M)$ only on the distinguished π-point in each class.

The main fact is that if $\alpha_K : K[t]/(t^p) \to KG_K$ is such a distinguished π-point, then α_K is a Hopf algebra map. Most importantly, restriction along α_K commutes with the tensor product operation. This means that there is a tensor product theorem, which says for modules M and N that

$$\mathcal{U}_G(M \otimes N) = (\mathcal{U}_G(M) \cap \mathcal{U}_G(N)) \cup c(\mathcal{V}_G(M)) \cup c(\mathcal{V}_G(N)),$$

where $c(\mathcal{V}_G(M))$ is the complement of $\mathcal{V}_G(M)$ in $\mathcal{V}_G(k)$. That is, for a distinguished π-point α_K, $\alpha_K^*(K \otimes (M \otimes N))$ is finite dimensional in the stable category if the restrictions of both M and N are finite dimensional or if one of the two is projective.

In particular, if V is closed subvariety of $V_G(k)$ we have that $\mathcal{N}_V$ is closed under tensor products in $\mathbf{StMod}(kG)$.

4. Dependence on Hopf structure

We notice that in some cases a given algebra over k may have many possible Hopf algebra structures. In particular, if B is the truncated polynomial ring $B = k[t_1, \dots, t_r]/(t_1^p, \dots, t_r^p)$, then we can make B into a Hopf algebra by choosing any collection $X_1, \dots, X_r \in \mathrm{Rad}(kG)$ that generate the radical of B. This means that the elements should be linearly independent modulo $\mathrm{Rad}^2(B)$. Then the elements $1+X_1, \dots, 1+X_r$ generate an elementary abelian subgroup H order p^r in the group of units of B, and B can be taken to be the group algebra kH. We can then impose the Hopf algebra structure of kH on B. Similarly we can take the subspace of B generated by $X_1, \dots, X_r$ to be a commutative restricted Lie algebra L with trivial p^{th}-power operation. Then B is isomorphic to the restricted enveloping algebra of L, and it can be equipt with the induced Hopf structure.

One implication of the above is the following, a result that we find useful.

LEMMA 4.1. *Let $kG = k[t_1, \dots, t_r]/(t_1^p, \dots, t_r^p)$. Assume that $A = k[t]/(t^p)$ is made into a Hopf algebra by regarding it either as a group algebra (the group being generated by $g = 1 + t$) or as a restricted enveloping algebra of the one dimensional Lie algebra $< t >$. Then for any π-point defined over k, $\alpha : A \to kG$, there is a Hopf algebra structure on kG such that α is a homomorphism of Hopf algebras.*

The ability to choose the Hopf algebra structure is important when considering restrictions on the categories. In particular, if we are given a subalgebra A of kG that is a flat embedding, then the restriction map to the subalgebra induces a functor on stable categories. However, it does not preserve the tensor structure unless A is a Hopf subalgebra.

On the other hand, there are some cases where it does not matter. The following is one such case.

PROPOSITION 4.2. *Suppose that kG is a finite group scheme and $\alpha : A = k[t]/(t^p) \to kG$ is a π-point defined over k. Let V be the closed point of $V_G(k)$ associated to α and let $\mathcal{E}_V$ be the associated idempotent module. For any kG-module M, the restriction $\alpha^*(M)$ is isomorphic in $\mathbf{StMod}(A)$ to a finite dimensional module if and only if the restriction $\alpha^*(\mathcal{E}_V \otimes M)$ is also isomorphic to a finite dimensional module.*

PROOF. The point is that the subcategory $\mathcal{M}_V$ of all kG-modules whose variety is contained in V is the same regardless of the tensor product. Likewise, the triangle

$$\mathcal{S}_V : \qquad \longrightarrow M \otimes \mathcal{E}_V \xrightarrow{1 \otimes \sigma_V} M \otimes k \xrightarrow{1 \otimes \tau_V} M \otimes \mathcal{F}_V \longrightarrow$$

is distinguished by universal properties that don't depend on the tensor product (see Theorem 2.6 of [**2**]). In particular, $\alpha^*(M \otimes \mathcal{F}_V) = \{0\}$ in the stable category $\mathbf{StMod}(A)$, since the variety of $\mathcal{F}_V$ does not contain V. Hence, we have an isomorphism $\alpha^*(M \otimes \mathcal{E}_V) \cong \alpha^*(M)$. □

5. Endomorphism rings

In this section we investigate the endomorphism rings of the module $\mathcal{E}_V$ and associated modules. Earlier work by Daugulis [**11**], showed that under reasonable hypotheses on V, $\underline{\mathrm{End}}_{kG}(\mathcal{E}_V)$ is local and locally nilpotent. Krause [**16**] derived some properties of $\mathcal{E}_V$ using the observation that $\mathcal{E}_V$ is endofinite. He also showed that $\underline{\mathrm{End}}_{kG}(\mathcal{E}_V)$ is commutative, since it is the center of the localizing subcategory of all kG-modules with variety contained in V. That is, it is the algebra of natural transformations on the identity functor on that category. In this section, we give an explicit calculation of $\underline{\mathrm{End}}_{kG}(\mathcal{E}_V)$ in a specific case.

We assume throughout that $kG = k[t_1, \dots, t_r]/(t_1^p, \dots, t_r^p)$ is the group algebra of an elementary abelian p-group or the restricted enveloping algebra of a commutative Lie algebra with vanishing p^{th}-power operation. We assume a coalgebra structure, but in the end, it does not matter what that structure is.

Assume that we have a π-point $\alpha : k[t]/(t^p) \to kG$ representing a closed point V in $V_G(k)$. The structure of the idempotent module $\mathcal{E}_V$ is presented in [**9**]. It is described as follows. For notation, let kC be the image of α and let $Z = \alpha(t)$. Choose elements $X_1, \dots, X_{r-1}$ in $\mathrm{Rad}(kG)$ such $X_1, \dots, X_{r-1}, Z$ generate $\mathrm{Rad}(kG)$. Let kH be the subalgebra generated by 1 and $X_1, \dots, X_{r-1}$, so that $kG \cong kH \otimes kC$.

THEOREM 5.1 ([**9**, Proposition 5.4]). *Suppose that*

$$(P_*, \varepsilon) : \qquad \longrightarrow P_2 \xrightarrow{\partial} P_1 \xrightarrow{\partial} P_0 \xrightarrow{\varepsilon} k \longrightarrow 0$$

is a minimal projective kH-resolution of kG. Let U be the inflation to kG of the indecomposable kC-module of dimension $p-1$, so that $U_{\downarrow C} \cong \Omega_{kC}(k)$, has basis 1, $Z, \dots, Z^{p-2}$ and has trivial action by H. Note that if $p = 2$, then $U \cong k$. Then the restriction of $\mathcal{E}_V$ to kH is the direct sum

$$(\mathcal{E}_V)_{\downarrow H} \cong P_0 \oplus (P_1 \otimes U) \oplus P_2 \oplus (P_3 \otimes U) \oplus \dots.$$

The action of Z on $\mathcal{E}_V$ is given by the following formula. Assume that $x \in P_n$. For n odd let

$$Z(x \otimes Z^j) = \begin{cases} x \otimes Z^{j+1} \in P_n \otimes U & \text{if } 0 \leq j < p-2, \\ \partial(x) \in P_{n-1} & \text{if } j = p-2, \end{cases}$$

while for n even

$$Z(x) = \begin{cases} \partial(x) \otimes 1 \in P_{n-1} \otimes U & \text{if } n > 0, \\ 0 & \text{if } n = 0. \end{cases}$$

Next we consider the endomorphism ring of $\mathcal{E}_V$. The fact that (P_*, ε) is a minimal resolution implies that $\partial(P_i) \subseteq \mathrm{Rad}_{kH}(P_{i-1})$ for all $i \geq 1$. As a consequence, $\underline{\mathrm{Hom}}_{kG}(\mathcal{E}_V, k) \cong \prod_{i=0}^{\infty} \mathrm{Hom}_{kH}(P_i, k)$. That is, the minimality of the resolution also means that $\mathrm{Hom}_{kH}(P_i, k) \cong \mathrm{H}^i(H, k)$.

In addition, applying $\underline{\mathrm{Hom}}_{kG}(\mathcal{E}_V, -)$ to the distinguished triangle $\mathcal{S}_V$, we get that

$$\underline{\mathrm{Hom}}_{kG}(\mathcal{E}_V, \mathcal{E}_V) \xrightarrow{(\sigma_V)_*} \underline{\mathrm{Hom}}_{kG}(\mathcal{E}_V, k)$$

is an isomorphism. Putting these facts together, we have established the following.

PROPOSITION 5.2. *The endomorphism ring $\underline{\mathrm{End}}_{kG}(\mathcal{E}_V) \cong \prod_{i=0}^{\infty} \mathrm{H}^i(H, k)$ as an additive group.*

Any element $\gamma \in \mathrm{H}^d(H, k)$ is represented by a unique cocycle $\gamma : P_d \to k$, which in turn, extends to a chain map $\{\gamma_i\} : P_* \to P_*$ that is unique up to homotopy. Here γ_i maps P_{d+i} to P_i. The product on the cohomology ring $\mathrm{H}^*(H, k)$ can be taken to be the homotopy class of the composition of these chain maps. Moreover, as in the proof of [**9**, Theorem 7.1] such a chain map induces an endomorphism $\hat{\gamma} : \mathcal{E}_V \to \mathcal{E}_V$.

In the case that the degree d of γ is even or that $p = 2$, the map $\hat{\gamma}$ is given on the kH-summands as follows. For n odd, and $x \in P_{n+d}$,

$$\hat{\gamma}_n(x \otimes Z^j) = \gamma_n(x) \otimes Z^j,$$

while for n even, $\hat{\gamma}_n : P_{n+d} \to P_n$ by

$$\hat{\gamma}_n(x) = \gamma_n(x).$$

The formula is somewhat more complicated when d is an odd integer and $p > 2$. In that case, for n odd, $\hat{\gamma}_n : P_{n+d} \to P_n \otimes U$ is given by $\hat{\gamma}_n(x) = \gamma_n(x) \otimes Z^{p-2}$. For n even, define $\hat{\gamma}_n : P_{n+d} \to P_n \oplus (P_{n-1} \otimes U)$ (assuming that $P_{-1} = \{0\}$) by

$$\hat{\gamma}_n(x \otimes Z^j) = \begin{cases} (\gamma_n(x), 0) & \text{if } j = 0 \\ (0, \partial\gamma_n(x) \otimes Z^{j-1}) & \text{if } j > 0 \end{cases}$$

Notice that in all cases, $\hat{\gamma}$ is a kH-homomorphism since $\mathcal{E}_V$ is free as a kH-module. To prove that it is a kG-homomorphism, it is only necessary to show that $\hat{\gamma}$ commutes with the action of Z. We leave this as an exercise for the reader.

Taking compositions, we get the following.

LEMMA 5.3. *Suppose that the characteristic p is odd. Let* $\hat{}\ : \mathrm{H}^*(H,k) \to \underline{\mathrm{End}}_{kG}(\mathcal{E}_V)$ *be the map that sends the cohomology element to the endomorphism as described above. Then we have for $\gamma \in \mathrm{H}^n(G,k)$ and $\mu \in \mathrm{H}^m(G,k)$, the product is given as follows.*

(1) *If either n or m is even, the $\hat{\gamma}\hat{\mu} = \widehat{\gamma\mu}$.*
(2) *if both n and m are odd, then $\hat{\gamma}\hat{\mu} = 0$.*

PROOF. The proof of the first item is a matter of choosing chain maps representing the two elements and using the formula for the composition as above. For the second, we note that the image of the composition of the two chain maps is contained in $Z^{p-2}\mathcal{E}_V$. But this is in the kernel of the map $\sigma_V : \mathcal{E}_V \to k$. Because $\sigma_V^* : \underline{\mathrm{Hom}}_{kG}(\mathcal{E}_V, \mathcal{E}_V) \to \underline{\mathrm{Hom}}_{kG}(\mathcal{E}_V, k)$ is an isomorphism, we conclude that $\hat{\gamma}\hat{\mu} = 0$. □

In this way we can characterize $\underline{\mathrm{End}}_{kG}(\mathcal{E}_V)$. We must first define a variant of the cohomology ring. Assume that $p > 2$. Recall that

$$\mathrm{H}^*(H,k) \cong k[\zeta_1, \dots, \zeta_{r-1}] \otimes \Lambda(\eta_1, \dots, \eta_{r-1})$$

where Λ is the exterior algebra generated by the degree one elements η_i and the elements ζ_i have degree 2. Let $R \subset \Lambda$ be the subalgebra spanned by the elements of even degree. Let $\Theta = R[\eta_1', \dots, \eta_{r-1}']/J$ where J is the ideal generated by all products $\eta_i'\eta_j'$ and all monomials $\eta_i'x$ such that $x \in R$ and $\eta_i x = 0$ in Λ. Notice that Θ is a graded k-algebra and as k-vector spaces, $\Theta \cong \Lambda$. Let $\Gamma = k[\zeta_1, \dots, \zeta_{r-1}] \otimes \Theta$. Note that Γ is a commutative k-algebra, and that we have an isomorphism of k-vector spaces $\varphi : \mathrm{H}^*(G,k) \to \Gamma$ that is the identity on $k[\zeta_1, \dots, \zeta_{r-1}]$ and on R and sends η_i to η_i'.

In the case that $p = 2$, let $\Gamma = \mathrm{H}^*(G,k)$ and let φ be the identity. Then we have the following.

PROPOSITION 5.4. *If* $\alpha = (\alpha_1, \alpha_2, \dots)$ *and* $\beta = (\beta_1, \beta_2, \dots)$ *are elements of* $\underline{\mathrm{End}}_{kG}(\mathcal{E}_V)$ *as in Proposition 5.2 with* $\alpha_i, \beta_i \in \mathrm{H}^i(H,k)$, *then the product* $\alpha\beta = \gamma = (\gamma_1, \gamma_2, \dots)$ *is given by the relation*

$$\gamma_n = \sum_{i=0}^{n} \varphi^{-1}(\varphi(\alpha_i)\varphi(\beta_{n-i})).$$

That is, $\underline{\mathrm{End}}_{kG}(\mathcal{E}_V) \cong \prod_{i=0}^{\infty} \Gamma_i$.

REMARK 5.5. One of the features that makes the construction of $\mathcal{E}_V$ and its endomorphism ring (as above) possible is that kC acts trivially on the trivial module k. A similar thing can be done for any kG module M that is inflated from a kH-module, so that Z annihilates M. That is, we can construct $\mathcal{E}_V(M) \cong \mathcal{E}_V \otimes M$ from a minimal projective kH-resolution of $M_{\downarrow H}$ with the action of Z given by formulas as above. The final result is that $\underline{\mathrm{End}}_{kG}(\mathcal{E}_V(M))$ is, as a k-vector space, the direct product $\prod_{i=0}^{\infty} \mathrm{Ext}^i_{kH}(M,M)$. The computation of the product poses similar problems as above, though products of even degree elements coincide with the cohomology products.

6. Finite generation

Assume that kG is a unipotent commutative group scheme, and V is a cloed point in $V_G(k)$. In this section, we show that if the restriction of a module M along a π-point is a compact object in the stable category, then $\underline{\mathrm{Hom}}_{kG}(\mathcal{E}_V, M)$ is finitely generated as a module over the endomorphism ring of the unit object $\mathcal{E}_V$ in the colocalized category. This leads to a definition of a local support variety for such a module. In the proof, we assume that the π-point is a map of Hopf algebras. This can be done as in Lemma 4.1. However, in the end, the consequences of the analysis do not depend on the Hopf algebra structure. Hence the consequences hold without the assumption. Note that this is not an issue in the case that kG is the restricted enveloping algebra of a commutative Lie algebra with trivial p^{th}-power operation, as we saw in Example 3.5. The proof also uses the detailed knowledge of the structure of the idempotent module $\mathcal{E}_V$.

We first note a general principle.

LEMMA 6.1. *Suppose that* V *is a closed subvariety of* $V_G(k)$. *For any* kG-*module* M, *we have that* $\underline{\mathrm{Hom}}_{kG}(\mathcal{E}_V, M) \cong \underline{\mathrm{Hom}}_{kG}(\mathcal{E}_V, \mathcal{E}_V \otimes M)$ *as modules over* $\underline{\mathrm{End}}_{kG}(\mathcal{E}_V)$.

PROOF. This follows from the distinguished triangle for M associated to the tensor ideal $\mathcal{M}_V$:

$$\dots \longrightarrow \Omega(\mathcal{F}_V) \otimes M \longrightarrow \mathcal{E}_V \otimes M \longrightarrow k \otimes M \longrightarrow \mathcal{F}_V \otimes M \longrightarrow \dots$$

in that $\underline{\mathrm{Hom}}_{kG}(\mathcal{E}_V, \Omega(\mathcal{F}_V) \otimes M) = 0 = \underline{\mathrm{Hom}}_{kG}(\mathcal{E}_V, \mathcal{F}_V \otimes M)$. □

Let $kG = k[t_1, \dots, t_r]/(t_1^p, \dots, t_r^p)$. We assume the notation of Theorem 5.1. In particular, we let $\alpha : A = k[t]/(t^p) \to kG$ be a π-point that is a map of Hopf algebras. Let kC be the image of α with $Z = \alpha(t)$. We choose a complementary subalgebra kH so that $kG \cong kH \otimes kC$. Let V be the variety consisting of the closed point represented by α.

The module $\mathcal{E}_V$ has a sequence of finitely generated submodules

$$\mathcal{E}_0 \subseteq \mathcal{E}_1 \subseteq \mathcal{E}_2 \subseteq \dots$$

where

$$(\mathcal{E}_n)_{\downarrow H} = P_0 \oplus (P_1 \otimes U) \oplus P_2 \cdots \oplus (P_n \otimes W)$$

for $W = U$ if n is odd and $W = k$ if n is even, exactly as in Theorem 5.1. Let $\iota_n : \mathcal{E}_n \to \mathcal{E}_V$ be the inclusion. Suppose that M is a kG-module such that $\alpha^*(M)$ is compact in the stable category $\mathbf{StMod}(A)$.

DEFINITION 6.2. For an element $\zeta \in \underline{\mathrm{Hom}}_{kG}(\mathcal{E}_V, M)$, the degree of ζ is the largest integer d such that there exist a representative $\zeta' \in \mathrm{Hom}_{kG}(\mathcal{E}_V, M)$ of the class of ζ such that $\zeta'\iota_{d-1} = 0$. The leading term of ζ is the class in $\underline{\mathrm{Hom}}_{kG}(\mathcal{E}_d/\mathcal{E}_{d-1}, M)$ of the map induced by $\zeta'\iota_d$, where ζ' is a representative of the degree-d element ζ with $\zeta'\iota_{d-1} = 0$.

LEMMA 6.3. *Suppose that $\zeta : \mathcal{E}_V \to M$ is a kG-homomorphism such that, for some n, $\zeta(\mathcal{E}_n) = \{0\}$ and the induced map $\hat{\zeta} : \mathcal{E}_{n+1}/\mathcal{E}_n \to M$ factors through a kG-projective module. Then there exist $\gamma : \mathcal{E}_V \to M$ such that $(\zeta - \gamma)(\mathcal{E}_{n+1}) = \{0\}$ and γ factors through a projective kG-module. In particular, the degree of ζ is at least $n + 2$.*

PROOF. First notice that since $\hat{\zeta}$ factors through a projective module, then so does $\zeta\iota_{n+1}$. So, we have a diagram

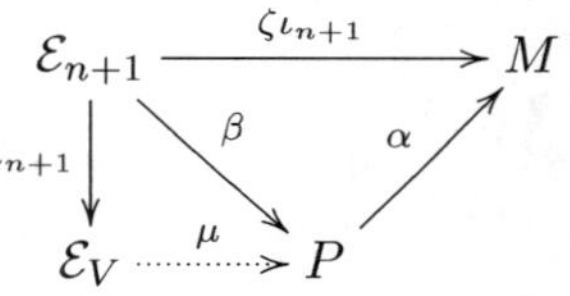

where P is projective and $\alpha\beta = \zeta\iota_{n+1}$. The point is that P is also injective, so that there exists μ with $\mu\iota_{n+1} = \beta$. Then let $\gamma = \alpha\mu$. □

Using the last lemma it is an easy exercise to prove the following. We leave this exercise to the reader.

PROPOSITION 6.4. *Every element ζ in $\underline{\mathrm{Hom}}_{kG}(\mathcal{E}_V, M)$ has a finite degree and well defined leading term. In particular, if $\zeta(\mathcal{E}_{d-1}) = \{0\}$, then the degree of ζ is d if and only if $\zeta\iota_d$ does not factor through a projective module.*

PROPOSITION 6.5. *Suppose that $\gamma : \mathcal{E}_n/\mathcal{E}_{n-1} \to M$ is a kG-homomorphism. Then γ factors through a projective module if and only if either n is odd and $\gamma(\mathcal{E}_n/\mathcal{E}_{n-1}) \subset ZM$ or n is even and $\gamma(\mathcal{E}_n/\mathcal{E}_{n-1}) \subset Z^{p-1}M$.*

PROOF. We prove the case for n odd. The case for n even is even easier. The point is that for n odd, $\mathcal{E}_n/\mathcal{E}_{n-1} \cong P_n \otimes U$. Because P_n is a free kH-module, it is a direct sum of copies of kH and Z acts trivially on it. Thus, $\mathcal{E}_n/\mathcal{E}_{n-1}$ is a direct sum of copies of $kH \otimes U$. If a map from $kH \otimes U$ factors through a projective module, then it factors through the injective hull $\psi : kH \otimes U \to kG$, where $\psi(1 \otimes 1) = Z$. Thus, if γ factors through a projective, then its image is in ZM. On the other hand, if we have, for some summand of $\mathcal{E}_n/\mathcal{E}_{n-1}$ that is isomorphic to $kH \otimes U$, that $\gamma(1 \otimes 1) = Zm$, for $m \in M$, then we can define $\theta : kG \to M$ by $\theta(1) = m$, and $\theta\psi$ coincides with γ on that summand. □

We recall that $\underline{\mathrm{End}}_{kG}(\mathcal{E}_V) \cong \prod_{n \geq 0} \mathrm{H}^*(H, k)$. The action of a nonzero element $\gamma \in \mathrm{H}^n(G, k) \subseteq \underline{\mathrm{End}}_{kG}(\mathcal{E}_V)$, on $\mathcal{E}_V$ is determined by the chain map $\{\gamma_*\} : P_* \to P_*$ of degree n. Hence, its degree as an element of $\underline{\mathrm{End}}_{kG}(\mathcal{E}_V)$ is n and its leading term is the map $\nu: \mathcal{E}_n/\mathcal{E}_{n-1} \to P_0 \cong \mathcal{E}_0 \subseteq \mathcal{E}_V$, defined by γ_0. (See the discussion following Proposition 5.2.)

For any $i \geq 0$, there is a shift $\Omega^i(\nu)$ which is the induced map

$$\Omega^i(\nu) = \mathcal{E}_{n+i}/\mathcal{E}_{n+i-1} \longrightarrow \mathcal{E}_i/\mathcal{E}_{i-1}.$$

Note that, $\gamma(\mathcal{E}_{n+i-1}) \subseteq \mathcal{E}_{i-1}$, so there is such a map.

PROPOSITION 6.6. *Suppose that an element $\zeta \in \underline{\mathrm{Hom}}_{kG}(\mathcal{E}_V, M)$ has degree d and leading term μ. Let $\gamma \in \underline{\mathrm{Hom}}_{kG}(\mathcal{E}_V, \mathcal{E}_V)$ be an element with degree n and leading term ν. If $\mu\Omega^d(\nu) \neq 0$ (meaning a representative does not factor through a projective module), then $\gamma\zeta$ has degree $d+n$ and $\mu\Omega^d(\nu)$ is its leading term. Otherwise, it has degree greater than $d+n$.*

PROOF. We have that γ has a representative (call it γ) in $\mathrm{Hom}_{kG}(\mathcal{E}_V, \mathcal{E}_V)$ such that $\gamma(\mathcal{E}_{n-1}) = 0$. Likewise, ζ has a representative such that $\zeta(\mathcal{E}_{d-1}) = 0$. As a result, $\gamma\zeta(\mathcal{E}_{n+d-1}) = 0$ and the composition $\gamma\zeta$ induces a map that is a composition

$$\mathcal{E}_{n+d}/\mathcal{E}_{n+d-1} \longrightarrow \mathcal{E}_n/\mathcal{E}_{n-1} \longrightarrow M$$

that is easily seen to be $\mu\Omega^d(\nu)$. □

To explain the products of leading elements, we should note the following. For notation, let $\Gamma(M) = M^C/Z^{p-1}M$, where M^C is the submodule of kC-fixed points. With the assumption that $\alpha^*(M)$ is finite dimensional, so also is $\Gamma(M)$.

LEMMA 6.7. *For any n we have that*

$$\underline{\mathrm{Hom}}_{kG}(\mathcal{E}_n/\mathcal{E}_{n-1}, M) \cong \begin{cases} \Gamma(M) \otimes \mathrm{H}^n(H,k) & \text{if } n \text{ is even}, \\ \Gamma(M \otimes U) \otimes \mathrm{H}^n(H,k) & \text{if } n \text{ is odd}. \end{cases}$$

PROOF. We note that, if n is even, then Z annihilates $\mathcal{E}_n/\mathcal{E}_{n-1}$. Consequently, the image of any homomorphism $\vartheta : \mathcal{E}_n/\mathcal{E}_{n-1} \to M$ lies in M^C. On the other hand, if this image lies in $Z^{p-1}M$, then ϑ factors through a projective module. This proves the first case. For the case that n is odd, we notice that

$$\underline{\mathrm{Hom}}_{kG}(\mathcal{E}_n/\mathcal{E}_{n-1}, M) = \underline{\mathrm{Hom}}_{kG}(P_n \otimes U, M) \cong \underline{\mathrm{Hom}}_{kG}(P_n, M \otimes U)$$

since U is self-dual. □

Putting the last two results together we easily get the following.

LEMMA 6.8. *In the notation of Proposition 6.6, if ζ has even degree n and leading term $\mu = \sum_{i=1}^t m_i \otimes \vartheta_i$ for $m_i \in \Gamma(M)$ and $\vartheta_i \in \mathrm{H}^n(H,k)$ as in Lemma 6.7, then $\mu\Omega^d(\nu) = \sum_{i=1}^t m_i \otimes \vartheta_i\nu$ where by $\vartheta_i\nu$ we mean the product in $\mathrm{H}^*(H,k)$.*

We can now prove the main theorem of the section.

THEOREM 6.9. *Let $kG = k[t_1, \dots, t_r]/(t_1^p, \dots, t_r^p)$. We let $\alpha : A = k[t]/(t^p) \to kG$ be a π-point defined over k. Let V be the variety consisting of the closed point represented by α. If M is a kG-module whose support variety consists of the single point V, and if the restriction $\alpha^*(M)$ is a compact object in* **StMod**(A)*, then $\underline{\mathrm{Hom}}_{kG}(\mathcal{E}_V, M)$ is finitely generated as a module over $\underline{\mathrm{End}}_{kG}(\mathcal{E}_V)$.*

PROOF. As before, we write $kG \cong kH \otimes kC$ where kC is the image of α and kH is a complementary subalgebra. By Lemma 4.1, there is a Hopf algebra structure on kG such that α is a map of Hopf algebras. We assume this structure so that the previous results of this section hold. Let $\mathrm{H}^*_{ev} = \sum_{n\geq 0} \mathrm{H}^{2n}(H,k)$. Notice that $\mathrm{H}^*(H,k)$ is finitely generated over H^*_{ev}.

Let $\mathcal{L}_n$ be the collection of all leading terms of elements in $\underline{\mathrm{Hom}}_{kG}(\mathcal{E}_V, M)$ having degree at most n. Thus,

$$\mathcal{L}_n \subseteq \sum_{j=0}^{n} (\Gamma(M) \oplus \Gamma(M \otimes U)) \otimes \mathrm{H}^j(H,k).$$

Let $\mathcal{H} = \sum_{j\geq 0}(\Gamma(M) \oplus \Gamma(M \otimes U)) \otimes \mathrm{H}^j(H,k)$. Let $I_n \subset \mathcal{H}$ be the H^*_{ev}-submodule of $\mathcal{H}$ generated by $\mathcal{L}_n$. Then we have an increasing sequence of submodules $I_0 \subseteq I_1 \subseteq I_2 \ldots$. Because, $\mathcal{H}$ is finitely generated over H^*_{ev}, which is noetherian, the sequence terminates. That is, $I = \cup_n I_n$ is a finitely generated H^*_{ev}-submodule of $\mathcal{H}$ and $I = I_N$ for some N.

For each $j = 0, \ldots, N$, choose a finite set of elements $U_j \subseteq \underline{\mathrm{Hom}}_{kG}(\mathcal{E}_V, M)$ such that the set of leading terms of the elements of U_j generated I_j/I_{j-1}. Let $U = \{u_1, u_2, \ldots, u_t\} = \cup_{j=0}^N U_j$. We claim that U is a set of generators for $\underline{\mathrm{Hom}}_{kG}(\mathcal{E}_V, M)$ as a module over H^*_{ev}.

For each j, let v_j be the leading term of u_j. Suppose that $\gamma \in \underline{\mathrm{Hom}}_{kG}(\mathcal{E}_V, M)$. Let s be the degree of γ and let μ be its leading term. Then we can write $\mu = \sum_{i=1}^t u_i\zeta_i$ where each ζ_i is an element of $\mathrm{H}^*(H,k)$. Let ζ_i denote also an element in $\underline{\mathrm{End}}_{kG}(\mathcal{E}_V) = \prod_i \mathrm{H}^i(H,k)$, that has leading term ζ_i. Let $\beta_1 = \sum_{i=0}^t v_i\zeta_i$. Note that β_1 has degree s and has the same leading term as $\gamma_1 = \gamma$. Then $\gamma_2 = \gamma_1 - \beta_1$ has larger degree than s, the degree of γ. Now repeat this process with γ_2 in place of γ. We get β_2 with the same degree and leading term as γ_2, so that $\gamma_3 = \gamma_2 - \beta_2$ has higher degree than γ_2.

Taking a limit of this process we obtain a sequence $\beta_1, \beta_2, \ldots$ such that $\gamma = \beta_1 + \beta_2 + \ldots$. Note that we can add the elements of this infinite sequence because they have different degrees. For each j, we have that $\beta_j = \sum_{i=1}^t v_i\zeta_{i,j}$ for some elements $\zeta_{i,j} \in \underline{\mathrm{End}}_{kG}(\mathcal{E}_V)$. Thus,

$$\alpha = \sum_{j\geq 1} \beta_j = \sum_{i=0}^{t} v_i (\sum_{j\geq 1} \zeta_{i,j})$$

It follows that the $\{v_i\}_{1\leq i\leq t}$ is a set of generators for $\underline{\mathrm{Hom}}_{kG}(\mathcal{E}_V, M)$ as asserted.

Finally, we point out that the conclusion of the theorem is independent of the Hopf algebra structure on kG. Consequently, the theorem is true even without the assumption on the Hopf nature of the π-point at the beginning of the proof. □

REMARK 6.10. In [**8**] it is proved that Theorem 6.9 has a strong converse. That is, the authors of [**8**] show that the condition on the module M in the hypothesis is equivalent to the condition that the module by dualizable, which in turn is equivalent to the finite generation condition on $\underline{\mathrm{Hom}}_{kG}(\mathcal{E}_V, M)$

7. (Co)Local supports

If a module is finitely generated over its base ring, then its annihilator in that ring is an ideal, and we can define the support variety of the module. Rings constructed as $\underline{\mathrm{End}}_{kG}(\mathcal{E}_V)$, as in Section 5, are not often finitely generated as algebras over the base field k. We expect that the prime ideal spectrum of such a ring is chaotic. None the less, the annihilator of a module such as $\underline{\mathrm{Hom}}_{kG}(\mathcal{E}_V, M)$ for a M a kG-module, is an invariant and can be used to distinguish modules. In this section we take a brief look at some of the possibilities. In particular, we prove a form of a realization theorem.

To this end, let $\mathfrak{A}_V(M) = \mathrm{Ann}_{\mathcal{C}}(\underline{\mathrm{Hom}}_{kG}(\mathcal{E}_V, M))$ where $\mathcal{C} = \underline{\mathrm{End}}_{kG}(\mathcal{E}_V)$ and V is a closed subvariety of $V_G(k)$. From Lemma 6.1, we see that if M is a finite dimensional kG-module, then $\mathfrak{A}_V(M) = \mathfrak{A}_V(M \otimes \mathcal{E}_V)$.

We begin with a couple of easy result.

PROPOSITION 7.1. *Suppose that the variety of M is in V for V as above, a closed subvariety of $V_G(k)$. Then, $\underline{\mathrm{Hom}}_{kG}(\mathcal{E}_V, M) \cong \underline{\mathrm{Hom}}_{kG}(k, M)$.*

PROOF. In the stable category $M \otimes \mathcal{F}_V = 0$. Hence, $\underline{\mathrm{Hom}}_{kG}(\mathcal{E}_V, M \otimes \mathcal{F}_V) = 0$. Thus the result follows from the distinguished triangle $\mathcal{S}_V$ involving $\mathcal{E}_V$. □

The next result follows directly from Remark 5.5.

PROPOSITION 7.2. *Suppose that $kG \cong kH \otimes kC$ as in Theorem 5.1. That is, $kC \cong k[t]/(t^p)$ is the image of a π-point α whose class V in $V_G(k)$ is a closed point, and kH is the group algebra of an elementary abelian p-group. Suppose that M is the inflation of a finitely generated kH-module. Then, $\mathfrak{A}_V(M)$ consists of all tuples $(\gamma_1, \gamma_2, \dots)$ such that γ_i annilates $\mathrm{H}^*(G, M)$ for all i.*

Before further exploration, we need a technical result.

LEMMA 7.3. *Let V be a closed subvariety of $V_G(k)$. Suppose that*

$$\longrightarrow L \xrightarrow{\ \alpha\ } M \xrightarrow{\ \beta\ } N \longrightarrow \Omega^{-1}(L)$$

is a triangle of kG-modules and that $\zeta \in \mathfrak{A}_V(N)$, $\gamma \in \mathfrak{A}_V(L)$. Then $\zeta\gamma \in \mathfrak{A}_V(M)$.

PROOF. For $\mu \in \underline{\mathrm{Hom}}_{kG}(\mathcal{E}_V, M)$, we have that $\beta(\mu\zeta) = (\beta\mu)\zeta = 0$ in the stable category. Consequently, there exists $\nu \in \underline{\mathrm{Hom}}_{kG}(\mathcal{E}_V, L)$ such that $\mu\zeta = \alpha\nu$. Thus, $\mu\zeta\gamma = \alpha(\nu\gamma) = 0$. □

Let $\ell = \ell(kG)$ denote the Loewy length of kG, i.e. the least integer n such that $\mathrm{Rad}^n(kG) = \{0\}$. Recall that for any kG-module M, $\mathrm{Rad}^{\ell}(M) = \mathrm{Rad}^{\ell}(kG)M = \{0\}$.

PROPOSITION 7.4. *Let V be a closed subvariety of $V_G(k)$ and M a kG-module. Suppose that $\zeta \in \mathfrak{A}_V(M)$. Then, for any finite dimensional kG-module N, $\zeta^{\ell(kG)} \in \mathfrak{A}_V(M \otimes N)$.*

PROOF. First notice that it is sufficient to prove the proposition in the case that G is a p-group. This is because whenever a kG-map has the property that its restriction to the Sylow p-subgroup P of G factors through a kP-projective module, then the map factors through a kG-projective module.

Let $R_j = \mathrm{Rad}^j(N)$, so that we have a radical filtration

$$\{0\} = R_n \subset R_{n-1} \subset \cdots \subset R_1 \subset R_0 = N, \qquad n \leq \ell,$$

where every one of the quotients R_i/R_{i+1} is a direct sum of copies of the trivial module k. Let $M_i = M \otimes R_i$. Thus, for $0 \leq i < n$ we have a triangle

$$\longrightarrow \underline{\mathrm{Hom}}_{kG}(\mathcal{E}_V, M_{i+1}) \longrightarrow \underline{\mathrm{Hom}}_{kG}(\mathcal{E}_V, M_i) \longrightarrow \underline{\mathrm{Hom}}_{kG}(\mathcal{E}_V, M_i/M_{i+1}) \longrightarrow .$$

Fix the index i. The quotient

$$M_i/M_{i+1} \cong M \otimes (R_i/R_{i+1}) \cong \oplus_{j\in J} M \otimes k \cong \oplus_{j \in J} M$$

for some finite indexing set J.

Let $\hat{\zeta} : \mathcal{E}_V \to \mathcal{E}_V$ be a representative of the class of ζ. Choose any homomorphism $\mu : \mathcal{E}_V \to M \otimes (R_i/R_{i+1})$. The composition $\mu\hat{\zeta}$ factors through a projective module since R_i/R_{i+1} has finite dimension. Thus, we have shown that ζ annihilates $\underline{\mathrm{Hom}}_{kG}(\mathcal{E}_V, (M \otimes R_i)/(M \otimes R_{i+1}))$.

Using this fact we prove the proposition by applying Lemma 7.3 repeatedly, beginning with $i = n-1$. □

There is something of a realization theorem. First we need the following.

LEMMA 7.5. *Suppose that V is a closed subvariety of $V_G(k)$. Suppose that $\zeta \in \underline{\mathrm{End}}_{kG}(\mathcal{E}_V)$ and that $\gamma \in \underline{\mathrm{Hom}}_{kG}(\mathcal{E}_V, \Omega(\mathcal{E}_V))$. Then $\gamma\zeta = \Omega(\zeta)\gamma$ in* $\mathbf{StMod}(kG)$.

PROOF. The lemma is a consequence of general principles. That is, the kG-module $\mathcal{E}_V$ is the unit object in the subcategory $\mathcal{M}_V$ and its graded endomorphism ring $\mathrm{End}^*_{\mathcal{M}}(\mathcal{E}_V) = \sum_{n\in\mathbb{Z}} \underline{\mathrm{Hom}}_{kG}(\mathcal{E}_V, \Omega^{-n}(\mathcal{E}_V))$ is graded commutative. □

THEOREM 7.6. *Suppose that V is a closed subvariety of $V_G(k)$. Choose an element $\zeta \in \underline{\mathrm{End}}_{kG}(\mathcal{E}_V)$ and let M_ζ be the third object in the triangle of ζ:*

$$\longrightarrow \Omega(\mathcal{E}_V) \xrightarrow{\ \gamma\ } M_\zeta \xrightarrow{\ \beta\ } \mathcal{E}_V \xrightarrow{\ \zeta\ } \mathcal{E}_V \longrightarrow \Omega^{-1}(M_\zeta) \longrightarrow$$

Then the radical of the ideal $\mathfrak{A}_V(M_\zeta)$ is the same as that of the ideal generated by ζ.

PROOF. We use the diagram below to show that if $\mu \in \mathfrak{A}_V(M_\zeta)$, then μ^ℓ is a multiple of ζ. To start, notice that $\Omega^{-1}(M_\zeta) \cong \Omega^{-1}(k) \otimes M_\zeta$ in the stable category and by Proposition 7.4, μ^ℓ is in $\mathfrak{A}_V(\Omega^{-1}(M_\zeta))$. In the diagram the vertical maps are right multiplication by μ^ℓ.

$$\begin{array}{ccccccc}
\longrightarrow \underline{\mathrm{Hom}}_{kG}(\mathcal{E}_V, \mathcal{E}_V) & \xrightarrow{\ \zeta\ } & \underline{\mathrm{Hom}}_{kG}(\mathcal{E}_V, \mathcal{E}_V) & \xrightarrow{\ \beta\ } & \underline{\mathrm{Hom}}_{kG}(\mathcal{E}_V, \Omega^{-1}(M_\zeta)) \longrightarrow \\
\downarrow \mu^\ell_* & \overset{\sigma}{\swarrow} & \downarrow \mu^\ell_* & & \downarrow \mu^\ell_* \\
\longrightarrow \underline{\mathrm{Hom}}_{kG}(\mathcal{E}_V, \mathcal{E}_V) & \xrightarrow{\ \zeta\ } & \underline{\mathrm{Hom}}_{kG}(\mathcal{E}_V, \mathcal{E}_V) & \xrightarrow{\ \beta\ } & \underline{\mathrm{Hom}}_{kG}(\mathcal{E}_V, \Omega^{-1}(M_\zeta)) \longrightarrow
\end{array}$$

Because the composition $\beta\mu^\ell_* = \mu^\ell_*\beta$ is zero on the middle term $\underline{\mathrm{Hom}}_{kG}(\mathcal{E}_V, \mathcal{E}_V)$, we have that there is a map $\sigma : \underline{\mathrm{Hom}}_{kG}(\mathcal{E}_V, \mathcal{E}_V) \to \underline{\mathrm{Hom}}_{kG}(\mathcal{E}_V, \mathcal{E}_V)$ such that $\zeta\sigma = \mu^\ell_*$. Thus we have that $\sigma(\mathrm{Id}_M) \circ \zeta = \mu^\ell_*(\mathrm{Id}_M) = \mathrm{Id}_M \circ \mu^\ell = \mu^\ell$.

Next we show that $\zeta^2 \in \mathfrak{A}_V(M_\zeta)$. First, we see from the exact sequence

$$\longrightarrow \underline{\mathrm{Hom}}_{kG}(\mathcal{E}_V, M_\zeta) \xrightarrow{\ \beta_*\ } \underline{\mathrm{Hom}}_{kG}(\mathcal{E}_V, \mathcal{E}_V) \xrightarrow{\ \zeta\ } \underline{\mathrm{Hom}}_{kG}(\mathcal{E}_V, \mathcal{E}_V) \longrightarrow$$

that $\zeta\beta = 0$ where here ζ means left composition with ζ. Hence, in the diagram

$$\begin{array}{ccccccc}
\longrightarrow \underline{\mathrm{Hom}}_{kG}(\mathcal{E}_V, \Omega(\mathcal{E}_V)) & \xrightarrow{\ \gamma_*\ } & \underline{\mathrm{Hom}}_{kG}(\mathcal{E}_V, M_\zeta) & \xrightarrow{\ \beta_*\ } & \underline{\mathrm{Hom}}_{kG}(\mathcal{E}_V, \mathcal{E}_V) \longrightarrow \\
\downarrow \zeta^* & \overset{\tau}{\swarrow} & \downarrow \zeta^* & & \downarrow \zeta^* \\
\longrightarrow \underline{\mathrm{Hom}}_{kG}(\mathcal{E}_V, \Omega(\mathcal{E}_V)) & \xrightarrow{\ \gamma_*\ } & \underline{\mathrm{Hom}}_{kG}(\mathcal{E}_V, M_\zeta) & \xrightarrow{\ \beta_*\ } & \underline{\mathrm{Hom}}_{kG}(\mathcal{E}_V, \mathcal{E}_V) \longrightarrow
\end{array}$$

the composition $\beta\zeta$ is also zero. Note here, the down arrows marked ζ^* denote right multiplication by ζ. However, there is no problem since $\underline{\mathrm{Hom}}_{kG}(\mathcal{E}_V, \mathcal{E}_V)$ is commutative. As a result, the map τ exists with $\gamma\tau = \zeta$.

Finally, we consider the diagram

$$\begin{array}{ccccccc} \longrightarrow \underline{\mathrm{Hom}}_{kG}(\mathcal{E}_V, \Omega(\mathcal{E}_V)) & \xrightarrow{\Omega(\zeta)_*} & \underline{\mathrm{Hom}}_{kG}(\mathcal{E}_V, \Omega(\mathcal{E}_V)) & \xrightarrow{\gamma_*} & \underline{\mathrm{Hom}}_{kG}(\mathcal{E}_V, M_\zeta) \longrightarrow \\ \downarrow \zeta^* & & \downarrow \zeta^* & & \downarrow \zeta^* \\ \longrightarrow \underline{\mathrm{Hom}}_{kG}(\mathcal{E}_V, \Omega(\mathcal{E}_V)) & \xrightarrow{\Omega(\zeta)_*} & \underline{\mathrm{Hom}}_{kG}(\mathcal{E}_V, \Omega(\mathcal{E}_V)) & \xrightarrow{\gamma_*} & \underline{\mathrm{Hom}}_{kG}(\mathcal{E}_V, M_\zeta) \longrightarrow . \end{array}$$

For $\mu \in \underline{\mathrm{Hom}}_{kG}(\mathcal{E}_V, M_\zeta)$ we have that $\mu\zeta = \gamma \circ (\tau(\mu))$. Here we regard $\tau(\mu)$ as an element of the object in the middle of the upper row of the diagram. Recall that γ_* is left composition with the map $\gamma : \Omega(\mathcal{E}_V) \to M_\zeta$ in the triangle that contains ζ. Thus we have that

$$\mu\zeta^2 = (\gamma(\tau(\mu))\zeta = (\gamma \circ \tau(\mu)) \circ \zeta = \gamma \circ (\Omega(\zeta) \circ \tau(\mu)) = \gamma(\Omega(\zeta)(\tau(\mu))) = 0,$$

by Lemma 7.5 and the triangle of γ. □

Suppose that $0 \neq \zeta \in \mathrm{H}^n(G, k)$. Assume that n is even if p is odd. Associated to ζ is a module that generates the thick tensor ideal in $\mathbf{stmod}(kG)$ of all modules whose support variety is contained in the closed set of all prime ideals that contain ζ. The module L_ζ is defined to be the kernel the map $\hat{\zeta}$

$$0 \longrightarrow L_\zeta \longrightarrow \Omega^n(k) \xrightarrow{\hat{\zeta}} k \longrightarrow 0$$

where $\hat{\zeta}$ is a cocyle representing ζ. Suppose that V is a single point as in the last theorem. Tensoring the above sequence with $\mathcal{E}_V$ yields an exact sequence

$$0 \longrightarrow L_\zeta \otimes \mathcal{E}_V \longrightarrow \Omega^n(k) \otimes \mathcal{E}_V \xrightarrow{1\otimes\hat{\zeta}} \mathcal{E}_V \longrightarrow 0.$$

But recall that $\mathcal{E}_V$ is periodic of period 1 if $p = 2$ and period 2 otherwise. That is, for example, if $p > 2$, then the tensor of $\mathcal{E}_V$ with the sequence

$$0 \longrightarrow k \longrightarrow k(G/H) \xrightarrow{Z} k(G/H) \longrightarrow k \longrightarrow 0$$

where $k(G/H)$ is the permutation module on which kH-acts trivially and the middle map is multiplication by Z, yields a sequence with $\mathcal{E}_V$ at the ends and middle terms projective. Thus we have that $\Omega^n(k) \otimes \mathcal{E}_V \cong \mathcal{E}_V$ in the stable category. The consequence of this is that $L_\zeta \otimes \mathcal{E}_V$ is isomorphic to some M_γ. So what is γ? We sketch a proof of an answer to the question.

For notation, we recall that $\mathrm{H}^*(C, k) \cong k[\mu]$ if $p = 2$ and, otherwise, $\mathrm{H}^*(C, k) \cong k[\mu, \nu]/(\mu^2)$ where μ and ν are in degrees one and two. Thus for all n, $\mathrm{H}^n(C, k)$ has dimension one and has a basis element which we demote μ_n. So in the case $p > 2$, $\mu_{2n+1} = \mu\nu^n$ while $\mu_{2n} = \nu^n$. Recall also that $\mathrm{H}^*(G, k) = \mathrm{H}^*(H, k) \otimes \mathrm{H}^*(C, k)$.

PROPOSITION 7.7. *Assume the hypothesis of the Theorem 7.6. Let ζ be an element in $\mathrm{H}^n(G, k)$. Write $\zeta = \sum_{i=0}^n \mu_i\gamma_{n-i}$ for $\gamma_{n-i} \in \mathrm{H}^{n-i}(H, k)$. Then in the stable category, $L_\zeta \otimes \mathcal{E}_V \cong M_\gamma$ where*

$$\gamma = (\gamma_0, \dots, \gamma_n, 0, 0, 0, 0, \dots) \in \prod_{i\geq 0} \mathrm{H}^i(H, k) \cong \underline{\mathrm{End}}_{kG}(\mathcal{E}_V).$$

PROOF. The secret to the proof is finding a splitting of the middle vertical map in the diagram:

$$\begin{array}{ccccccccc} 0 & \longrightarrow & L_\zeta \otimes \mathcal{E}_V & \longrightarrow & \Omega^n(k) \otimes \mathcal{E}_V & \xrightarrow{1\otimes\hat{\zeta}} & \mathcal{E}_V & \longrightarrow & 0 \\ & & \downarrow & & \downarrow & & \| & & \\ 0 & \longrightarrow & M_\zeta & \longrightarrow & \mathcal{E}_V & \longrightarrow & \mathcal{E}_V & \longrightarrow & 0. \end{array}$$

The map itself can be taken to be $(\mu_n \otimes 1) \otimes 1$, since the kernel L_{μ_n} of the cocycle $\mu_n \otimes 1 : \Omega^n(k) \to k$ is free on restriction to kC, and $L_{\mu_n} \otimes \mathcal{E}_V$ is a projective module.

We construct the splitting θ as follows. Let $k\hat{G} = kC \otimes kH \otimes kH$ and construct the idempotent $\mathcal{E}$-module $\hat{\mathcal{E}}$ as in Theorem 5.1. If $p = 2$, then on restriction to $kH \otimes kH$, $\hat{\mathcal{E}}$ is a sum of the terms of the projective resolution $((P \otimes P)_*, \varepsilon \otimes \varepsilon)$ where (P_*, ε) is the minimal projective kH-resolution of k. For p odd, the restriction is a sum of terms of the form $(P \otimes P)_n \otimes X$ where X is either U or k depending on the parity of n. In either case we have a chain map $\hat{\psi} : P_* \to (P \otimes P)_*$ that defines cup product on cohomology and splits both $1 \otimes \varepsilon$ and $\varepsilon \otimes 1$. This gives us a map $\psi : \mathcal{E}_V \to \hat{\mathcal{E}}$. At the same time, $\hat{\mathcal{E}}$ when viewed as a kG-module by restriction (taking the diagonal embedding of kH into $kH \otimes kH$) is very naturally isomorphic to $\mathcal{E}_V \otimes \mathcal{E}_V$.

Next we notice that the module V whose restriction to kH is the direct sum

$$V_{\downarrow kH} = P_0 \oplus (P_1 \otimes U) \oplus P_2 \oplus (P_3 \otimes U) \oplus \cdots \oplus (P_{n-1} \otimes U) \oplus P_n/\partial(P_{n+1})$$

where $U = k$, if $p = 2$. Here, Z acts as in the formula in Theorem 5.1, where defined, and by the induce map $Z : P_n \to P_{n+1}$ on the last summand. But now, we see from the formula for minimal projective kG-resolution of k as the tensor product of kC and kH resolutions, that $V \cong \Omega^n(k)$. Letting $\sigma : \mathcal{E}_V \to V$ be the obvious quotient, we get the splitting

$$\mathcal{E}_V \xrightarrow{\psi} \hat{\mathcal{E}} \cong \mathcal{E}_V \otimes \mathcal{E}_V \xrightarrow{\sigma\otimes 1} \Omega^n(k) \otimes \mathcal{E}_V.$$

Now the theorem follows by composition of maps, using the splittings. □

In the case that $\gamma_0 \neq 0$, then the γ is a unit in $\underline{\mathrm{End}}_{kG}(\mathcal{E}_V)$, implying that $L_\zeta \otimes \mathcal{E}_V$ is the zero module. The situation is consistent with what we know in that $\gamma_0 \neq 0$ implies that $\mathrm{res}_{G,C}(\zeta) \neq 0$, and hence L_ζ is projective on restriction to kC and $L_\zeta \otimes \mathcal{E}_V$ is a projective module.

We end by pointing out another curosity. Again, suppose that G, H, C, V are as in Proposition 7.2. Let ζ be an element $\underline{\mathrm{End}}_{kG}(\mathcal{E}_V)$ with ζ not invertible, Let $\mathcal{N}_\zeta$ be the subcategory of all kG-modules M whose variety is in V and such that such that ζ annihilates $\underline{\mathrm{Hom}}_{kG}(\mathcal{E}_V, M)$. Then $\mathcal{N}_\zeta$ is a thick subcategory of **StMod**(kG). It is not however a localizing subcategory. We know this because the localizing tensor ideal have been classified [**6, 7**]. But also, we can see by Theorem 7.6, that the direct sum $M = \sum_{n=1}^{\infty} M_{\zeta^n}$ is not in $\mathcal{N}_\zeta$. Moreover, under some mild assumptions, the module $\mathcal{E}_V$ can be recovered as the third object in a triangle involving an endomorphism of the module M, above. This is a sort of homotopy colimit construction and will be explored in subsequent work.

References

[1] Paul Balmer, *The spectrum of prime ideals in tensor triangulated categories*, J. Reine Angew. Math. **588** (2005), 149–168, DOI 10.1515/crll.2005.2005.588.149. MR2196732

[2] Paul Balmer and Giordano Favi, *Generalized tensor idempotents and the telescope conjecture*, Proc. Lond. Math. Soc. (3) **102** (2011), no. 6, 1161–1185, DOI 10.1112/plms/pdq050. MR2806103

[3] D. J. Benson, *Representations and cohomology. II*, Cambridge Studies in Advanced Mathematics, vol. 31, Cambridge University Press, Cambridge, 1991. Cohomology of groups and modules. MR1156302

[4] D. J. Benson, Jon F. Carlson, and J. Rickard, *Complexity and varieties for infinitely generated modules. II*, Math. Proc. Cambridge Philos. Soc. **120** (1996), no. 4, 597–615, DOI 10.1017/S0305004100001584. MR1401950

[5] D. J. Benson, Jon F. Carlson, and Jeremy Rickard, *Thick subcategories of the stable module category*, Fund. Math. **153** (1997), no. 1, 59–80, DOI 10.4064/fm-153-1-59-80. MR1450996

[6] David J. Benson, Srikanth B. Iyengar, and Henning Krause, *Stratifying modular representations of finite groups*, Ann. of Math. (2) **174** (2011), no. 3, 1643–1684, DOI 10.4007/annals.2011.174.3.6. MR2846489

[7] Dave Benson, Srikanth B. Iyengar, Henning Krause, and Julia Pevtsova, *Stratification for module categories of finite group schemes*, J. Amer. Math. Soc. **31** (2018), no. 1, 265–302, DOI 10.1090/jams/887. MR3718455

[8] D. J. Benson, S. B. Iyengar, H. Krause, and J. Pevtsova, *Local dualisable objects in local algebra*, `arXiv:2302.08562`, 2023.

[9] Jon F. Carlson, *Negative cohomology and the endomorphism ring of the trivial module*, J. Pure Appl. Algebra **226** (2022), no. 9, Paper No. 107046, 13, DOI 10.1016/j.jpaa.2022.107046. MR4381675

[10] Jon F. Carlson, Lisa Townsley, Luis Valeri-Elizondo, and Mucheng Zhang, *Cohomology rings of finite groups*, Algebra and Applications, vol. 3, Kluwer Academic Publishers, Dordrecht, 2003. With an appendix: Calculations of cohomology rings of groups of order dividing 64 by Carlson, Valeri-Elizondo and Zhang, DOI 10.1007/978-94-017-0215-7. MR2028960

[11] Peteris Daugulis, *Stable endomorphism rings of idempotent Epsilon-modules*, ProQuest LLC, Ann Arbor, MI, 1998. Thesis (Ph.D.)–University of Georgia. MR2698267

[12] Eric M. Friedlander and Julia Pevtsova, *Π-supports for modules for finite group schemes*, Duke Math. J. **139** (2007), no. 2, 317–368, DOI 10.1215/S0012-7094-07-13923-1. MR2352134

[13] Eric M. Friedlander and Brian J. Parshall, *Support varieties for restricted Lie algebras*, Invent. Math. **86** (1986), no. 3, 553–562, DOI 10.1007/BF01389268. MR860682

[14] Eric M. Friedlander and Andrei Suslin, *Cohomology of finite group schemes over a field*, Invent. Math. **127** (1997), no. 2, 209–270, DOI 10.1007/s002220050119. MR1427618

[15] J. C. Jantzen, *Kohomologie von p-Lie-Algebren und nilpotente Elemente* (German), Abh. Math. Sem. Univ. Hamburg **56** (1986), 191–219, DOI 10.1007/BF02941516. MR882415

[16] Henning Krause, *Decomposing thick subcategories of the stable module category*, Math. Ann. **313** (1999), no. 1, 95–108, DOI 10.1007/s002080050252. MR1666825

[17] Amnon Neeman, *Triangulated categories*, Annals of Mathematics Studies, vol. 148, Princeton University Press, Princeton, NJ, 2001, DOI 10.1515/9781400837212. MR1812507

[18] Jeremy Rickard, *Idempotent modules in the stable category*, J. London Math. Soc. (2) **56** (1997), no. 1, 149–170, DOI 10.1112/S0024610797005309. MR1462832

Department of Mathematics, University of Georgia, Athens, Georgia 30602
Email address: `jfc@math.uga.edu`

Proceedings of Symposia in Pure Mathematics
Volume **108**, 2024
https://doi.org/10.1090/pspum/108/01959

Real and symmetric quasi-maps

Tsao-Hsien Chen and David Nadler

Abstract. Let $G_{\mathbb{R}}$ be a real reductive group and let X be the corresponding complex symmetric variety under the Cartan bijection. We construct a stratified homeomorphism between the based polynomial arc group of $G_{\mathbb{R}}$ and the based polynomial arc space of X. We also prove a multi-point version where we replace arcs by moduli spaces of quasi-maps from the projective line $\mathbb{P}^1$ to $G_{\mathbb{R}}$ and X. The key ingredients in the proof include: (i) a multi-point generalization of the "Gram-Schmidt" factorization of loop groups, and (ii) a nodal degeneration of moduli spaces of quasi-maps. As an application, we show that for the closures of real spherical orbits in the real affine Grassmannian, their singularities near the base point are locally homeomorphic to complex algebraic varieties.

Contents

1. Introduction
2. Group data
3. Factorization
4. Quasi-maps
5. Nodal degeneration
6. Stratified homeomorphisms
Acknowledgments
References

1. Introduction

1.1. Overview. Let G be a connected complex reductive Lie group with real form $G_{\mathbb{R}}$, with maximal compact $K_c \subset G_{\mathbb{R}}$, with complexification $K \subset G$, and corresponding complex symmetric variety $X = K\backslash G$. (See Section 2 for a collection of standard Lie theory constructions downstream of this starting point.)

This paper gives a direct geometric explanation for why much of the local spherical geometry of the real affine Grassmannian $\mathrm{Gr}_{\mathbb{R}} = G_{\mathbb{R}}((t))/G_{\mathbb{R}}[[t]]$ behaves like complex geometry (for example, satisfying the semisimplicity of the Decomposition Theorem as shown in [**N**]). For a spherical orbit $S_{\mathbb{R}}^{\lambda} = G_{\mathbb{R}}[[t]] \cdot \lambda \subset \mathrm{Gr}_{\mathbb{R}}$, we show the intersection of its closure $\overline{S_{\mathbb{R}}^{\lambda}} \subset \mathrm{Gr}_{\mathbb{R}}$ with the open cospherical orbit

The research of the first author was supported by NSF grant DMS-2001257 and DMS-2143722, and that of the second author by NSF grant DMS-2101466.

$T^0_\mathbb{R} = G_\mathbb{R}[t^{-1}] \cdot 1 \subset \mathrm{Gr}_\mathbb{R}$ is stratified homeomorphic to a complex algebraic variety. More precisely, we view the open cospherical orbit as the based polynomial arc group

$$T^0_\mathbb{R} \simeq G_\mathbb{R}[t^{-1}]_1 := \{g : \mathbb{P}^1_\mathbb{R} \setminus \{0\} \to G_\mathbb{R} \mid g(\infty) = 1\} \tag{1.1}$$

and construct a stratified homeomorphism

$$G_\mathbb{R}[t^{-1}]_1 \simeq X[t^{-1}]_1 \tag{1.2}$$

to the based polynomial arc space of the symmetric variety

$$X[t^{-1}]_1 := \{x : \mathbb{P}^1 \setminus \{0\} \to X \mid x(\infty) = 1\} \tag{1.3}$$

(In the definitions of (1.1) and (1.3), the subscript 1 conveys the condition $g(\infty) = 1$.)

Going further, our main theorem provides a generalization of the stratified homeomorphism (1.2) to spaces of maps where we allow poles at multiple points. Write $\mathcal{G}_\mathbb{R} \to \mathbb{R}^m$ for the group ind-scheme whose fiber over $(z_1, \ldots, z_m) \in \mathbb{R}^m$ is the group of maps $\gamma : \mathbb{P}^1_\mathbb{R} \setminus \{z_1, \ldots, z_m\} \to G_\mathbb{R}$ such that $\gamma(\infty) = e$. Similarly, write $\mathcal{X} \to \mathbb{R}^m$ for the ind-scheme whose fiber over $(z_1, \ldots, z_m) \in \mathbb{R}^m$ is the space of maps $\gamma : \mathbb{P}^1 \setminus \{z_1, \ldots, z_m\} \to X$ such that $\gamma(\infty) = e$.

THEOREM 1.1 (See Theorem 6.9). *There is a K_c-equivariant stratified homeomorphism*

$$\mathcal{G}_\mathbb{R} \simeq \mathcal{X} \tag{1.4}$$

over $\mathbb{R}^m$ that restricts to real analytic isomorphisms on spherical strata.

The construction of the stratified homeomorphism (1.4) involves two ingredients of independent interest.

(1) First, we establish a multi-point generalization of the "Gram-Schmidt" factorization of the polynomial loop group

$$G[t, t^{-1}] \simeq \Omega G_c \cdot G[t] \tag{1.5}$$

given by multiplication of the factors on the right. Here $\Omega G_c \subset G[t, t^{-1}]$ is the polynomial based loop group of the maximal compact $G_c \subset G$, i.e., the subgroup of maps that take the unit circle $S^1 \in \mathbb{A}^1_\mathbb{C}$ to $G_c \subset G$ and $1 \in S^1$ to $1 \in G$.

The following states one version of our multi-point generalization; the classical case (1.5) results from taking $S = \emptyset$, $S^+ = \{0\}$, $S^- = \{\infty\}$, and $s_0 = 1$.

THEOREM 1.2 (See Theorem 3.3). *Let $S^1 \subset \mathbb{A}^1_\mathbb{C}$ be the unit circle, $\mathbb{D}^+ \subset \mathbb{A}^1_\mathbb{C}$ the open unit disk, and $\mathbb{D}^- \subset \mathbb{P}^1_\mathbb{C}$ the complementary open disk. Let $\kappa(z) = \bar{z}^{-1}$ be the conjugation of $\mathbb{P}^1_\mathbb{C}$ with real form $S^1 \subset \mathbb{A}^1_\mathbb{C}$.*

Suppose given a finite set of points $S \subset S^1$, a finite set of points $S^+ \subset \mathbb{D}^+$, with conjugates $S^- = \kappa(S^+) \subset \mathbb{D}^-$. Set $\mathbb{S} = S \cup S^+ \cup S^-$ to be their union.

Let $G[\mathbb{P}^1 \setminus \mathbb{S}]$ be the group of polynomial maps $\mathbb{P}^1 \setminus \mathbb{S} \to G$, and $G[\mathbb{P}^1 \setminus \mathbb{S}]_c \subset G[\mathbb{P}^1 \setminus \mathbb{S}]$ the subgroup that takes $S^1 \setminus S$ to G_c.

Fix a point $s_0 \in S^1 \setminus S$, and let $G[\mathbb{P}^1 \setminus \mathbb{S}]_{c,s_0} \subset G[\mathbb{P}^1 \setminus \mathbb{S}]_c$ be the further subgroup that takes s_0 to 1.

Then multiplication provides a homeomorphism

$$G[\mathbb{P}^1 \setminus \mathbb{S}]_{c,s_0} \times G[\mathbb{P}^1 \setminus \{S^- \cup S\}] \xrightarrow{\sim} G[\mathbb{P}^1 \setminus \mathbb{S}] \tag{1.6}$$

After passing to the quotient by $G[t]$, one can interpret (1.5) as a homeomorphism $\mathrm{Gr} = G[t,t^{-1}]/G[t] \simeq \Omega G_c$ of the affine Grassmannian with the based loop group. More generally, one can similarly use (1.6) to obtain a homeomorphism of a Belinson-Drinfeld Grassmannian with the quotient of a mapping space. For our application, this is useful in it gives one access to the global mapping space via local data.

(2) Second, in Section 5, we study how maps, and more generally quasi-maps, behave under collisions of marked points (where the maps or quasi-maps are allowed to have poles) and degenerations of the domain curve itself.

Specifically, for our application, we study two such families: the bubbling of the domain curve $\mathbb{P}^1$ to the nodal curve $\mathbb{P}^1 \vee \mathbb{P}^1$ given in local coordinates by $xy = a^2$, for a parameter $a \in \mathbb{A}^1$; and the collision of distinct but Galois-conjugate marked points $x \neq \bar{x} \in \mathbb{P}^1$ of the fixed domain curve to a single Galois fixed-point $x_0 = \bar{x}_0 \in \mathbb{P}^1$.

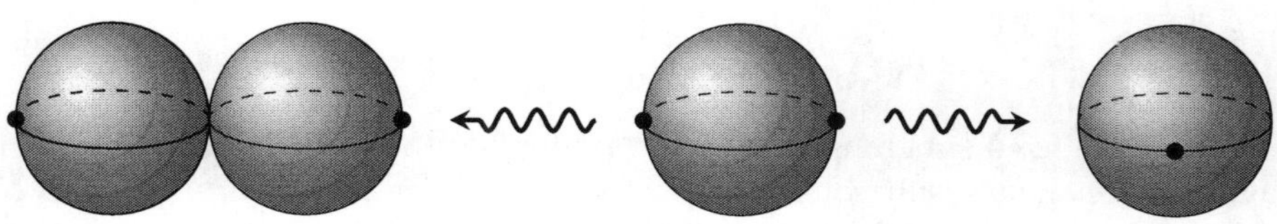

FIGURE 1. Degenerations of curve $\mathbb{P}^1$ and marked points S. Nodal degeneration of curve $\mathbb{P}^1$ to the left, collision of marked points S to the right.

A detailed study of quasi-maps under the collision of distinct but Galois-conjugate marked points can be found in [**CN1**]. Here let us focus on the key Galois-theoretic property of the bubbling degeneration that we exploit. We start at the parameter $a = 1$ with the conjugation of $\mathbb{P}^1$ given by $x \mapsto \bar{x}^{-1}$ in the local coordinate x. So its real form is the unit circle $S^1 \subset \mathbb{P}^1_{\mathbb{C}}$ and it exchanges the points 0 and ∞. Then we extend this over $a \neq 0$ by taking $x \mapsto \bar{a}^2\bar{x}^{-1}$. When we degenerate to $a = 0$, we find the conjugation $(x, y) \mapsto (\bar{y}, \bar{x})$ exchanging the two components of $\mathbb{P}^1 \vee \mathbb{P}^1$. Thus the Galois-action on the special fiber is almost free with the node the only real point. This allows us to describe Galois-equivariant maps, and more generally quasi-maps, on $\mathbb{P}^1 \vee \mathbb{P}^1$ in almost completely complex algebraic terms: the real structure constrains their values at the node alone.

Finally, within these families, we single out stratified subfamilies, defined over real parameters $a \in \mathbb{A}^1_{\mathbb{R}}$, that are locally constant with respect to the parameters. This provides homeomorphisms from their general fibers, which involve evident real structures, to their special fibers, where the real structures "disappear" due to the almost free nature of the Galois-action. The proof of Theorem 1.1 is an application of this idea: the general fiber is the group $\mathcal{G}_{\mathbb{R}}$ of real maps and the special fiber $\mathcal{X}$ is also a space of "real" maps but has a completely complex algebraic interpretation.

2. Group data

We collect here notation and standard constructions used throughout the rest of the paper (for further discussion, see for example [**CN1**, **N**]).

2.1. Real forms. Let G be a connected complex reductive Lie group with Lie algebra $\mathfrak{g}$.

Let $G_{\mathbb{R}} \subset G$ be a real form, defined by a conjugation $\eta : G \to G$, with Lie algebra $\mathfrak{g}_{\mathbb{R}} \subset \mathfrak{g}$.

Choose a Cartan conjugation $\delta : G \to G$ that commutes with η, and let $G_c \subset G$ be the corresponding maximal compact subgroup with Lie algebra $\mathfrak{g}_c \subset \mathfrak{g}$.

Introduce the involution $\theta = \delta \circ \eta : G \to G$, and let $K \subset G$ be the fixed subgroup of θ.

One can organize the above groups into the diagram:

(2.1)

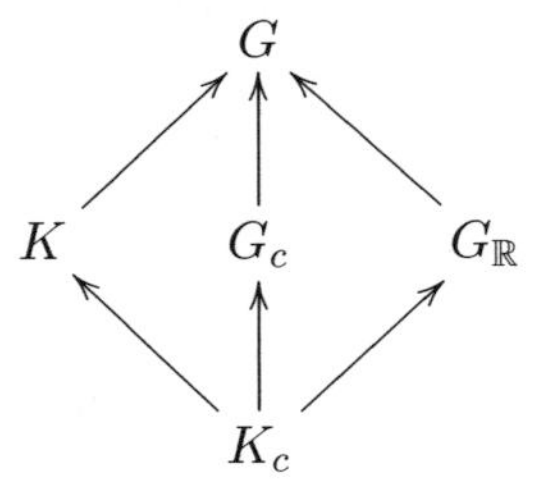

Here K_c is the fixed subgroup of θ, δ, and η together (or any two of the three) and the maximal compact subgroup of $G_{\mathbb{R}}$ with complexification K.

Fix a maximal δ-stable split torus $A_{\mathbb{R}} \subset G_{\mathbb{R}}$ with Lie algebra $\mathfrak{a}_{\mathbb{R}} \subset \mathfrak{g}_{\mathbb{R}}$, and complexification $A \subset G$ with Lie algebra $\mathfrak{a} \subset \mathfrak{g}$. Fix a maximal δ-stable torus $T_{\mathbb{R}} \subset G_{\mathbb{R}}$ containing $A_{\mathbb{R}}$ with Lie algebra $\mathfrak{t}_{\mathbb{R}} \subset \mathfrak{g}_{\mathbb{R}}$, and complexification $T \subset G$ with Lie algebra $\mathfrak{t} \subset \mathfrak{g}$.

Fix a Borel subgroup $B \subset G$ containing T with Lie algebra $\mathfrak{b} \subset \mathfrak{g}$, and unipotent radical $U \subset B$ with Lie algebra $\mathfrak{u} \subset \mathfrak{b}$. Let $H = B/U$ be the universal Cartan with Lie algebra $\mathfrak{h} = \mathfrak{b}/\mathfrak{u}$. Note the composition $T \to B \to H$ is an isomorphism.

Let $W = N_G(\mathfrak{t})/Z_G(\mathfrak{t}) = N_G(T)/T$ denote the Weyl group, and $W_0 = N_K(\mathfrak{a})/Z_K(\mathfrak{a})$ the "baby Weyl group".

Let $\Lambda_T = \mathrm{Hom}(\mathbb{G}_m, T)$ denote the coweight lattice of T with dominant coweights $\Lambda_T^+ \subset \Lambda_T$. Let $R_G \subset \Lambda_T$ denote the coroot lattice generated by the simple coroots $\Delta_G \subset R_G$ and let $R_G^+ = R_G \cap \Lambda_T^+$. Let $\pi_1(G)$ denote the fundamental group based at the identity $e \in G$, and recall the natural isomorphism $\Lambda_T/R_G \xrightarrow{\sim} \pi_1(G)$.

Similarly, let $\Lambda_A = \mathrm{Hom}(\mathbb{G}_m, A)$ denote the coweight lattice of A with dominant coweights $\Lambda_A^+ = \Lambda_A \cap \Lambda_T^+$.

Introduce the symmetric variety $X = K\backslash G$. The map $\pi : G \to G$, $\pi(g) = \theta(g)^{-1}g$ factors through X, and descends to an isomorphism $X \xrightarrow{\sim} G^0_{sym}$, where $G_{sym} = \{g \in G \mid \theta(g^{-1}) = g\}$, and $G^0_{sym} \subset G_{sym}$ denotes its neutral component. Note the inclusion $A \subset G^0_{sym}$ induces a map $\Lambda_A \to \pi_1(G_{sym}) \simeq \pi_1(X)$.

Consider the induced map $\pi_* : \pi_1(G) \to \pi_1(X)$. Define $\mathcal{L} \subset \Lambda_A$ to be the inverse image of $\pi_*(\pi_1(G)) \subset \pi_1(X)$ under the natural map $\Lambda_A \to \pi_1(X)$, and set $\mathcal{L}^+ = \mathcal{L} \cap \Lambda_T^+$. Note that $\mathcal{L} = \Lambda_A$ if and only if K is connected.

EXAMPLE 2.1 (Complex groups). An important special case is when the real form is itself a complex group. To avoid potential confusion in this case, we will use the alternative notation $\mathrm{sw} : G\times G \to G\times G$, $\mathrm{sw}(g, h) = (h, g)$ for the "swap" Cartan involution with fixed-point subgroup the diagonal $G \subset G \times G$. The corresponding conjugation is the composition $\mathrm{sw}_\delta = \mathrm{sw} \circ (\delta \times \delta)$, and the compact conjugation is the product $\delta \times \delta$. We identify the corresponding symmetric variety with the group

$G \simeq G\backslash(G \times G)$ via inclusion to the left factor, which coincides with the subspace $G \simeq \{(g,h) \in G \times G \mid h = g^{-1}\}$.

Introduce the affine quotient

$$\mathfrak{c} = \mathfrak{t}/\!\!/W = \mathrm{Spec}(\mathcal{O}(\mathfrak{t})^W) \simeq \mathfrak{g}/\!\!/G = \mathrm{Spec}(\mathcal{O}(\mathfrak{g})^G) \tag{2.2}$$

The conjugation η descends to a real structure on $\mathfrak{c}$, and we denote its real points by $\mathfrak{c}_\mathbb{R}$. We also have the real characteristic polynomial map $\mathfrak{g}_\mathbb{R} \to \mathfrak{c}_\mathbb{R}$ from real matrices to their unordered eigenvalues.

2.2. Loop groups. Let $\mathcal{K} = \mathbb{C}((z))$ denote the field of Laurent series, $\mathcal{O} = \mathbb{C}[[z]]$ the ring of power series, and $\mathcal{O}^- = \mathbb{C}[z^{-1}]$ the ring of Laurent poles. Let $\mathrm{Gr} = G(\mathcal{K})/G(\mathcal{O})$ be the affine Grassmannian of G.[1] For any $g \in G(\mathcal{K})$ we denote by $[g] \in \mathrm{Gr}$ the corresponding coset.

Via the natural inclusion $\Lambda_T \to T(\mathcal{K}) \subset G(\mathcal{K})$, any coweight $\lambda \in \Lambda_T$ defines a point $[\lambda] \in \mathrm{Gr}$. For a dominant coweight $\lambda \in \Lambda_T^+$, introduce the $G(\mathcal{O})$-orbit $S^\lambda = G(\mathcal{O}) \cdot [\lambda] \subset \mathrm{Gr}$ (spherical stratum) and $G(\mathcal{O}^-)$-orbit $T^\lambda = G(\mathcal{O}^-) \cdot [\lambda] \subset \mathrm{Gr}$ (cospherical stratum). Recall the disjoint union decompositions

$$\mathrm{Gr} = \coprod_{\lambda \in \Lambda_T^+} S^\lambda \qquad \mathrm{Gr} = \coprod_{\lambda \in \Lambda_T^+} T^\lambda \tag{2.3}$$

We can similarly repeat the above constructions over the real numbers. Let $\mathcal{K}_\mathbb{R} = \mathbb{R}((z))$ denote the field of real Laurent series, $\mathcal{O}_\mathbb{R} = \mathbb{R}[[z]]$ the ring of real power series, and $\mathcal{O}^- = \mathbb{R}[z^{-1}]$ the ring of real Laurent poles. Let $\mathrm{Gr}_\mathbb{R} = G_\mathbb{R}(\mathcal{K}_\mathbb{R})/G_\mathbb{R}(\mathcal{O}_\mathbb{R})$ be the real affine Grassmannian of the real form $G_\mathbb{R}$.

For a dominant coweight $\lambda \in \Lambda_A^+ = \Lambda_A \cap \Lambda_T^+$, set $S_\mathbb{R}^\lambda = G_\mathbb{R}(\mathcal{O}_\mathbb{R}) \cdot [\lambda] = S^\lambda \cap \mathrm{Gr}_\mathbb{R}$ (real spherical stratum) and $T_\mathbb{R}^\lambda = G_\mathbb{R}(\mathcal{O}_\mathbb{R}^-) \cdot [\lambda] = T^\lambda \cap \mathrm{Gr}_\mathbb{R}$ (real cospherical stratum). We have the disjoint union decompositions

$$\mathrm{Gr}_\mathbb{R} = \coprod_{\lambda \in \Lambda_A^+} S_\mathbb{R}^\lambda \qquad \mathrm{Gr}_\mathbb{R} = \coprod_{\lambda \in \Lambda_A^+} T_\mathbb{R}^\lambda \tag{2.4}$$

Note that all of above constructions result from the natural conjugation $\eta_\mathcal{K}$: $G(\mathcal{K}) \to G(\mathcal{K})$, $\eta_\mathcal{K}(g(z)) = \eta(g(\bar{z}))$ with real form $G_\mathbb{R}(\mathcal{K}_\mathbb{R}) \subset G(\mathcal{K})$.

Now let us recall some parallel constructions where we work with the global curve $\mathbb{G}_m = \mathrm{Spec}(\mathbb{C}[z, z^{-1}]) = \mathbb{P}^1 \setminus \{0, \infty\}$ in place of the punctured disk $D^\times = \mathrm{Spec}(\mathcal{K})$, and similarly $\mathbb{A}^1 = \mathrm{Spec}(\mathbb{C}[z]) = \mathbb{P}^1 \setminus \{\infty\}$ in place of the disk $D = \mathrm{Spec}(\mathcal{O})$.

Introduce the polynomial loop group $LG = G(\mathbb{C}[z, z^{-1}]) \subset G(\mathcal{K})$ of maps $\mathbb{G}_m \to G$, and similarly, the polynomial arc group $L_+G = G(\mathbb{C}[z]) \subset G(\mathcal{O})$ of maps $\mathbb{A}^1 \to G$. Recall the natural map is an isomorphism $LG/L_+G \xrightarrow{\sim} G(\mathcal{K})/G(\mathcal{O}) = \mathrm{Gr}$.

We have an additional compact conjugation $\kappa : \mathbb{G}_m \to \mathbb{G}_m$, $\kappa(z) = \bar{z}^{-1}$ with real points the unit circle $S^1 \subset \mathbb{C}^\times$.[2] Note that κ does not preserve the punctured disk $D^\times \subset \mathbb{G}_m$.

We extend the conjugations $\eta, \delta : G \to G$ to conjugations $\eta, \delta : LG \to LG$ by the formulas $\eta(g)(z) = \eta(g(\kappa(z))), \delta(g)(z) = \delta(g(\kappa(z)))$. The corresponding real forms $LG_\mathbb{R}, LG_c \subset LG$ consist of maps $g : \mathbb{G}_m \to G$ that take $S^1 \subset \mathbb{G}_m$

[1] Our concerns in this paper will be exclusively topological, and we will ignore any non-reduced structure throughout.

[2] There is another conjugation $z \mapsto -\bar{z}^{-1}$ of the curve $\mathbb{G}_m$ with empty real points, but it will not appear in the developments of this paper.

respectively to $G_{\mathbb{R}}, G_c \subset G$. We denote by $\Omega G_{\mathbb{R}} \subset LG_{\mathbb{R}}$, $\Omega G_c \subset LG_c$ the based subgroups of maps that take $1 \in S^1$ to the identity $e \in G$. Note that multiplication gives isomorphisms $\Omega G_{\mathbb{R}} \times G_{\mathbb{R}} \xrightarrow{\sim} LG_{\mathbb{R}}$, $\Omega G_c \times G_c \xrightarrow{\sim} LG_c$.

3. Factorization

In this section, we record a useful extension of a well-known loop group factorization.

Recall the "Gram-Schmidt" factorization

$$\Omega G_c \times L_+G \xrightarrow{\sim} LG \tag{3.1}$$

of the polynomial loop group $LG = G(\mathbb{C}[z, z^{-1}])$, its arc subgroup $L_+G = G(\mathbb{C}[z])$, and the based loop group $\Omega G_c \subset LG$. Note that (3.1) is equivalent to the fact that the ΩG_c-action on the base point $[e] \in \mathrm{Gr} \simeq LG/L_+G$ induces a homeomorphism

$$\Omega G_c \xrightarrow{\sim} \mathrm{Gr} \tag{3.2}$$

Consider the projective line $\mathbb{P}^1 = \mathrm{Proj}(\mathbb{C}[z_0, z_1])$. Using the local coordinate $z = z_1/z_0$, we will regard the complex points of $\mathbb{P}^1$ as the Riemann sphere $\mathbb{P}^1(\mathbb{C}) = \mathbb{C} \cup \{\infty\}$.

Given any finite set of points $\mathbb{S} \subset \mathbb{P}^1$, we will denote by $G[\mathbb{P}^1 \setminus \mathbb{S}]$ the group ind-scheme of maps $\mathbb{P}^1 \setminus \mathbb{S} \to G$. We will only be interested in the complex points of $G[\mathbb{P}^1 \setminus \mathbb{S}]$ equipped with the classical topology, and thus ignore any non-reduced structure.

For example, if $\mathbb{S} = \emptyset$, we have $G[\mathbb{P}^1] = G$; if $\mathbb{S} = \{\infty\}$, we have $G[\mathbb{P}^1 \setminus \{\infty\}] = L_+G$; and if $\mathbb{S} = \{0, \infty\}$, we have $G[\mathbb{P}^1 \setminus \{0, \infty\}] = LG$.

For a pair $s = (s_+, s_-)$ of points $s_+, s_- \in \mathbb{P}^1$, we will write $L_sG = G[\mathbb{P}^1 \setminus \{s_+, s_-\}]$. If we choose an isomorphism $\varphi : \mathbb{P}^1 \xrightarrow{\sim} \mathbb{P}^1$, $\varphi(0) = s_+$, $\varphi(\infty) = s_-$, then we obtain an isomorphism

$$\varphi^* : L_sG \xrightarrow{\sim} LG \qquad \varphi^*(g) = g \circ \varphi \tag{3.3}$$

To uniquely prescribe φ, we can further require $\varphi(p) = q$, for some $p, q \in \mathbb{P}^1$ with $p \neq 0, \infty$, $q \neq s_+, s_-$.

LEMMA 3.1. *Suppose $S^+, S^- \subset \mathbb{P}^1$ are disjoint, finite sets of points with S^- non-empty.*

For $\mathbb{S} = S^+ \cup S^-$, the natural map is an isomorphism

$$G[\mathbb{P}^1 \setminus \mathbb{S}]/G[\mathbb{P}^1 \setminus S^-] \xrightarrow{\sim} \prod_{s \in S^+} \mathrm{Gr}_s \tag{3.4}$$

where $\mathrm{Gr}_s := G(\mathcal{K}_s)/G(\mathcal{O}_s)$ denotes the affine Grassmannian at s (which is isomorphic to the affine Grassmannian Gr by the choice of a local coordinate).

PROOF. Set $U^\pm = \mathbb{P}^1 \setminus S^\pm$. Since S^- is non-empty, the restriction $\mathcal{E}|_{U^-}$ of any G-bundle $\mathcal{E}$ on $\mathbb{P}^1$ may be trivialized. Thus the left hand side classifies a trivial G-bundle $\mathcal{E}_0^+$ on U^+, a G-bundle $\mathcal{E}^-$ on U^-, and an isomorphism between them over $U^+ \cap U^-$. This is equivalent to a G-bundle $\mathcal{E}$ on $\mathbb{P}^1$ with a trivialization over U^+. It is standard that this factorizes to give the right hand side. □

Fix a non-zero real number a and consider the conjugation $\kappa_a : \mathbb{P}^1 \to \mathbb{P}^1$, $\kappa(z) = \frac{1}{a^2 \bar{z}}$ with real points being the circle $S^1_{a^{-1}} = \{|z| = |a^{-1}|\}$. Let $\mathbb{D}_a^+ = \{|z| < |a^{-1}|\}$ denote the open disk, and $\mathbb{D}_a^- = \{|z| > |a^{-1}|\}$ the complementary open disk.

If a finite set of points $\mathbb{S} \subset \mathbb{P}^1$ is invariant under κ_a, then the conjugation $\delta : G \to G$ induces a conjugation $\delta : G[\mathbb{P}^1 \setminus \mathbb{S}] \to G[\mathbb{P}^1 \setminus \mathbb{S}]$, $\delta(g)(z) = \delta(g(\kappa_a(z))$. The corresponding real points, denoted by $G[\mathbb{P}^1 \setminus \mathbb{S}]_c$, consist of maps $g : \mathbb{P}^1 \setminus \mathbb{S} \to G$ that take $S^1 \setminus (S^1 \cap \mathbb{S})$ to the maximal compact $G_c \subset G$. For example, if $\mathbb{S} = \emptyset$, we have $G[\mathbb{P}^1]_c = G_c$; if $\mathbb{S} = \{\infty\}$, we have $G[\mathbb{P}^1 \setminus \{\infty\}]_c = G_c$.

For a pair $s = (s_+, s_-)$ of points $s_+ \in \mathbb{D}_a^+$, $s_- = \kappa_a(s_+) \in \mathbb{D}_a^-$, we write $L_s G_c = G[\mathbb{P}^1 \setminus \{s_+, s_-\}]_c$. Observe that the map $\varphi : \mathbb{P}^1 \xrightarrow{\sim} \mathbb{P}^1$, $\varphi(0) = s_+$, $\varphi(\infty) = s_-$, $\varphi(a^{-1}) = a^{-1}$ commutes with the conjugation κ_a. Thus the isomorphism (3.3) restricts to an isomorphism on real points

$$\varphi^* : L_s G_c \xrightarrow{\sim} LG_c \tag{3.5}$$

For any $s_0 \in S^1$, we will write $\Omega_{s,s_0} G_c \subset L_s G_c$ for the based subgroup of those maps with $g(s_0) = e$.

For finite subsets $\mathbb{S} \subset S^1$, we have the following:

Lemma 3.2. *For any finite subset $\mathbb{S} \subset S^1$, any element of $G[\mathbb{P}^1 \setminus \mathbb{S}]_c$ is constant, and hence $G[\mathbb{P}^1 \setminus \mathbb{S}]_c = G_c$.*

Proof. Choose an embedding $G \subset \mathrm{GL}(N)$ so that $G_c \subset U(N)$. For any $g \in G[\mathbb{P}^1 \setminus \mathbb{S}]_c$, consider a matrix entry $g_{ij} : \mathbb{P}^1 \setminus \mathbb{S} \to \mathbb{A}^1$. The restriction $g|_{S^1 \setminus \mathbb{S}}$ lands in $G_c \subset U(N)$, hence the restricted matrix entry $g_{ij}|_{S^1 \setminus \mathbb{S}}$ is bounded. Thus g_{ij} is bounded, and hence constant. □

Now we are ready to state the main result of this section. The special case when $S = \emptyset$ and $S^+ = \{0\}$, so that $S^- = \{\infty\}$, recovers the "Gram-Schmidt" factorization (3.1).

Theorem 3.3. *Suppose we are given a finite set of points $S \subset S^1$, and a finite set of points $S^+ \subset \mathbb{D}^+$, with conjugates $S^- = \kappa(S^+) \subset \mathbb{D}^-$. Set $\mathbb{S} = S \cup S^+ \cup S^-$ to be their union.*

Fix a point $s_0 \in S^1 \setminus S$, and consider the kernel of the evaluation

$$G[\mathbb{P}^1 \setminus \mathbb{S}]_{c,s_0} := \ker(\mathrm{ev}_{s_0} : G[\mathbb{P}^1 \setminus \mathbb{S}]_c \to G_c) \tag{3.6}$$

Then we have:

1) *Multiplication provides a homeomorphism*

$$G[\mathbb{P}^1 \setminus \mathbb{S}]_{c,s_0} \times G[\mathbb{P}^1 \setminus \{S^- \cup S\}] \xrightarrow{\sim} G[\mathbb{P}^1 \setminus \mathbb{S}] \tag{3.7}$$

2) *The natural action map provides a homeomorphism*

$$G[\mathbb{P}^1 \setminus \mathbb{S}]_{c,s_0} \xrightarrow{\sim} G[\mathbb{P}^1 \setminus \mathbb{S}]/G[\mathbb{P}^1 \setminus \{S^- \cup S\}] \simeq \prod_{s \in S^+} \mathrm{Gr}_s \tag{3.8}$$

3) *For any ordering $S^+ = \{s_1, \ldots, s_k\}$, multiplication provides a homeomorphism*

$$\Omega_{s_1,s_0} G_c \times \cdots \times \Omega_{s_k,s_0} G_c \xrightarrow{\sim} G[\mathbb{P}^1 \setminus \mathbb{S}]_{c,s_0} \tag{3.9}$$

Proof. Clearly 1) and 2) are equivalent. We will prove 2) and 3) simultaneously by considering the diagram

$$\Omega_{s_1,s_0} G_c \times \cdots \times \Omega_{s_k,s_0} G_c \to G[\mathbb{P}^1 \setminus \mathbb{S}]_{c,s_0} \to G[\mathbb{P}^1 \setminus \mathbb{S}]/G[\mathbb{P}^1 \setminus \{S^- \cup S\}] \simeq \prod_{s \in S^+} \mathrm{Gr}_s$$

Clearly the composite map $\Omega_{s_1,s_0}G_c \times \cdots \times \Omega_{s_k,s_0}G_c \to \prod_{s\in S^+} \mathrm{Gr}_s$ is a homeomorphism by (3.2). Thus it suffices to show the map $G[\mathbb{P}^1 \setminus \mathbb{S}]_{c,s_0} \to \prod_{s\in S^+} \mathrm{Gr}_s$ is injective, i.e., the $G[\mathbb{P}^1 \setminus \mathbb{S}]_{c,s_0}$-action on the base-point $[e] \in \prod_{s\in S^+} \mathrm{Gr}_s$ is free. Any element in the stabilizer must extend across S^+, hence also across S^-, and so lie in $G[\mathbb{P}^1 \setminus S]_{c,s_0}$. Thus by Lemma 3.2, any element in the stabilizer must be constant. Since its evaluation at s_0 is trivial, the element must in fact be trivial. □

4. Quasi-maps

4.1. Definitions. Let B be a smooth complex base-scheme.

Let $\pi : \mathcal{Z} \to B$ be a projective family of curves, with fibers denoted $\mathcal{Z}_b = \pi^{-1}(b)$. We do not assume the total space $\mathcal{Z}$ or fibers $\mathcal{Z}_b$ are smooth. See the next section for the specific setting used in this paper.

Let $\mathrm{Bun}_G(\mathcal{Z}/B)$ denote the moduli stack of a point $b \in B$ and a G-bundle $\mathcal{E}$ on the fiber $\mathcal{Z}_b$. More precisely, an S-point consists of an S-point $a : \operatorname{Spec} S \to B$ and a G-bundle $\mathcal{E}$ on the corresponding fiber product $\mathcal{Z} \times_B \operatorname{Spec} S$. Denote by $p : \mathrm{Bun}_G(\mathcal{Z}/B) \to B$ the evident projection with fibers $p^{-1}(b) = \mathrm{Bun}_G(\mathcal{Z}_b)$.

Let $\sigma = (\sigma_1, \ldots, \sigma_n) : B \to \mathcal{Z}^n$ be an ordered n-tuple of sections of π. We allow the sections to intersect or coincide, but require $\sigma(b) \in \mathcal{Z}_b^n$ to be a smooth point, for all $b \in B$.

For an affine G-variety X, let $QM_{G,X}(\mathcal{Z}/B, \sigma)$ denote the ind-stack of quasi-maps classifying a point $b \in B$, a G-bundle $\mathcal{E}$ on the fiber $\mathcal{Z}_b$, and a section

$$s : \mathcal{Z}_b \setminus \{\sigma_1(b), \ldots, \sigma_n(b)\} \longrightarrow X_{\mathcal{E}} \tag{4.1}$$

to the associated X-bundle over the complement of the points $\sigma_1(b), \ldots, \sigma_n(b) \in \mathcal{Z}_b$. We have the evident forgetful maps

$$q : QM_{G,X}(\mathcal{Z}/B, \sigma) \longrightarrow \mathrm{Bun}_G(\mathcal{Z}/B) \longrightarrow B \tag{4.2}$$

with fibers $q^{-1}(b) = QM_{G,X}(\mathcal{Z}_b, \sigma(b))$.

Let $\xi : B \to \mathcal{Z}$ be another section of π such that $\xi(b) \neq \sigma_i(b)$, for all $b \in B$, and $i = 1, \ldots, n$.

Let $QM_{G,X}(\mathcal{Z}/B, \sigma, \xi)$ denote the ind-stack of rigidified quasimaps classifying quadruples $(b, \mathcal{E}, s, \iota)$ where $(b, \mathcal{E}, s)$ is a quasi-map as above and $\iota : \mathcal{E}_K|_{\xi(b)} \simeq K$ is a trivialization, where $\mathcal{E}_K$ is the K-reduction of $\mathcal{E}$ on $\mathcal{Z}_b \setminus \{\sigma_1(b), \ldots, \sigma_n(b)\}$ given by the section s. We have the evident forgetful maps

$$r : QM_{G,X}(\mathcal{Z}/B, \sigma, \xi) \longrightarrow \mathrm{Bun}_G(\mathcal{Z}/B) \longrightarrow B \tag{4.3}$$

with fibers $r^{-1}(b) = QM_{G,X}(\mathcal{Z}_b, \sigma(b), \xi(b))$.

Suppose given conjugations $c_{\mathcal{Z}} : \mathcal{Z} \to \mathcal{Z}$ and $c_B : B \to B$ such that $\pi \circ c_{\mathcal{Z}} = c_B \circ \pi$.

The conjugation η of G induces a conjugation $\eta : \mathrm{Bun}_G(\mathcal{Z}/B) \to \mathrm{Bun}_G(\mathcal{Z}/B)$ given by $\eta(b, \mathcal{E}) = (c_B(b), c_{\mathcal{Z}}^* \mathcal{E}_\eta)$ where we write $c_{\mathcal{Z}}^* \mathcal{E}_\eta$ for the bundle $c_{\mathcal{Z}}^* \mathcal{E}$ with its η-twisted G-action. We denote by $\mathrm{Bun}_G(\mathcal{Z}/B)_{\mathbb{R}}$ the corresponding real points.

Suppose $n = 2m$ so that we have $\sigma = (\sigma^+, \sigma^-) : B \to \mathcal{Z}^n$ with components $\sigma^\pm = (\sigma_1^\pm, \ldots, \sigma_m^\pm) : B \to \mathcal{Z}^m$. Suppose further that $c_{\mathcal{Z}} \circ \sigma_i^\pm \circ c_B = \sigma_i^\mp$, for $i = 1, \ldots, m$. Then the conjugation η of G induces a conjugation $\eta : QM_{G,X}(\mathcal{Z}/B, \sigma) \to QM_{G,X}(\mathcal{Z}/B, \sigma)$. We denote by $QM_{G,X}(\mathcal{Z}/B, \sigma)_{\mathbb{R}}$ the corresponding real points.

Suppose further that $c_{\mathcal{Z}} \circ \xi = \xi \circ c_B$. Then the conjugation η of G similarly induces a conjugation $\eta : QM_{G,X}(\mathcal{Z}/B, \sigma, \xi) \to QM_{G,X}(\mathcal{Z}/B, \sigma, \xi)$. We denote by $QM_{G,X}(\mathcal{Z}/B, \sigma, \xi)_{\mathbb{R}}$ the corresponding real points.

4.2. Uniformizations. Let $\mathrm{Gr}_{G,\mathcal{Z},\sigma_i}$ (resp. $\mathrm{Gr}_{G,\mathcal{Z},\sigma}$) denote the Beilinson-Drinfeld Grassmannian of a point $b \in B$, a G-bundle $\mathcal{E}$ on $\mathcal{Z}_b$, and a section $s : \mathcal{Z}_b \setminus \sigma_i(b) \to \mathcal{E}$ (resp. $s : \mathcal{Z}_b \setminus \{\sigma_1(b), \dots, \sigma_n(b)\} \to \mathcal{E}$).

Let $G[\mathcal{Z}, \hat{\sigma}_i]$ denote the group scheme of a point $b \in B$ and a section $\hat{D}_{\sigma_i(b)} \to G$, where $\hat{D}_{\sigma_i(b)}$ is the formal disk around $\sigma_i(b)$.

Let $G[\mathcal{Z}, \sigma_i]$ (resp. $G[\mathcal{Z}, \sigma]$) denote the group ind-scheme of a point $b \in B$ and a section $s : \mathcal{Z}_b \setminus \{\sigma_i(b)\} \to G$ (resp. $s : \mathcal{Z}_b \setminus \{\sigma_1(b), \dots, \sigma_n(b)\} \to G$).

Let $G[\mathcal{Z}, \sigma_i, \xi]$ (resp. $G[\mathcal{Z}, \sigma, \xi]$) denote the subgroup ind-scheme of $G[\mathcal{Z}, \sigma_i]$ (resp. $G[\mathcal{Z}, \sigma]$) consisting of $(b, s) \in G[\mathcal{Z}, \sigma_i]$ (resp. $(b, s) \in G[\mathcal{Z}, \sigma]$) such that $s(\xi(b)) = e$.

For any $b \in B$, we write $\mathrm{Gr}_{G,\mathcal{Z},\sigma,b}$, $G[\mathcal{Z}, \sigma, b]$, etc., for the respective fibers over b.

The conjugations $c_{\mathcal{Z}}$, c_B, η induce conjugations on $\mathrm{Gr}_{G,\mathcal{Z},\sigma}$, $G[\mathcal{Z}, \sigma]$, etc., and we denote by $\mathrm{Gr}_{G,\mathcal{Z},\sigma,\mathbb{R}}$, $G[\mathcal{Z}, \sigma]_{\mathbb{R}}$, etc., the respective real points.

For any $b \in B(\mathbb{R})$, we write $\mathrm{Gr}_{G,\mathcal{Z},\sigma,b,\mathbb{R}}$, $G[\mathcal{Z}, \sigma, b]_{\mathbb{R}}$, etc., for the respective fibers over b.

The group ind-scheme $G[\mathcal{Z}, \sigma]$ (resp. $G[\mathcal{Z}, \sigma_i]$) naturally acts on $\mathrm{Gr}_{G,\mathcal{Z},\sigma}$ (resp. $\mathrm{Gr}_{G,\mathcal{Z},\sigma_i}$) and we have uniformizations morphisms

$$G[\mathcal{Z}, \sigma_i]\backslash \mathrm{Gr}_{G,\mathcal{Z},\sigma_i} \longrightarrow \mathrm{Bun}_G(\mathcal{Z}/B) \qquad G[\mathcal{Z}, \sigma]\backslash \mathrm{Gr}_{G,\mathcal{Z},\sigma} \longrightarrow \mathrm{Bun}_G(\mathcal{Z}/B) \tag{4.4}$$

$$K[\mathcal{Z}, \sigma]\backslash \mathrm{Gr}_{G,\mathcal{Z},\sigma} \longrightarrow QM_{G,X}(\mathcal{Z}/B, \sigma) \tag{4.5}$$

$$K[\mathcal{Z}, \sigma, \xi]\backslash \mathrm{Gr}_{G,\mathcal{Z},\sigma} \longrightarrow QM_{G,X}(\mathcal{Z}/B, \sigma, \xi) \tag{4.6}$$

The uniformizations are compatible with the given conjugations, hence induce uniformizations on real points:

(4.7)
$$G[\mathcal{Z}, \sigma_i]_{\mathbb{R}}\backslash \mathrm{Gr}_{G,\mathcal{Z},\sigma_i,\mathbb{R}} \longrightarrow \mathrm{Bun}_G(\mathcal{Z}/B)_{\mathbb{R}} \qquad G[\mathcal{Z}, \sigma]_{\mathbb{R}}\backslash \mathrm{Gr}_{G,\mathcal{Z},\sigma,\mathbb{R}} \longrightarrow \mathrm{Bun}_G(\mathcal{Z}/B)_{\mathbb{R}}$$

$$K[\mathcal{Z}, \sigma]_{\mathbb{R}}\backslash \mathrm{Gr}_{G,\mathcal{Z},\sigma,\mathbb{R}} \longrightarrow QM_{G,X}(\mathcal{Z}/B, \sigma)_{\mathbb{R}} \tag{4.8}$$

$$K[\mathcal{Z}, \sigma, \xi]_{\mathbb{R}}\backslash \mathrm{Gr}_{G,\mathcal{Z},\sigma,\mathbb{R}} \longrightarrow QM_{G,X}(\mathcal{Z}/B, \sigma, \xi)_{\mathbb{R}} \tag{4.9}$$

4.3. Morphisms. Let G_1 and G_2 be two reductive groups with complex conjugations η_1 and η_2 and Cartan involutions θ_1 and θ_2 respectively. Then the constructions of quasi-maps, rigidified quasi-maps, uniformization morphisms, and real forms of those are functorial with respect to homomorphisms $f : G_1 \to G_2$ that intertwine η_1, η_2 and θ_1, θ_2.

5. Nodal degeneration

We invoke here the preceding constructions in a situation to be studied in the remainder of the paper.

5.1. Universal family. Let $\mathbb{A}^1 = \mathrm{Spec}(\mathbb{C}[a])$ be the affine line with coordinate a. Consider the product $\mathbb{P}^1_x \times \mathbb{P}^1_y$ with respective homogeneous coordinates $[x_0, x_1], [y_0, y_1]$, local coordinates $x = x_1/x_0, y = y_1/y_0$, and projections $p_x, p_y : \mathbb{P}^1_x \times \mathbb{P}^1_y \to \mathbb{P}^1$. For convenience, we will also set $t^+ = x^{-1}, t^- = y^{-1}$.

Introduce the surface $Z \subset \mathbb{P}^1_x \times \mathbb{P}^1_y \times \mathbb{A}^1$ cut out by $x_1 y_1 = a^2 x_0 y_0$. We will regard Z as a family of curves via the evident projection $p : Z \to \mathbb{A}^1$. We denote the fibers by $Z_a = p^{-1}(a)$, for $a \in \mathbb{A}^1$. When $a \neq 0$, projection along p_x or p_y provides an isomorphism $Z_a \simeq \mathbb{P}^1$. When $a = 0$, the image of the inclusion $Z_0 \subset \mathbb{P}^1 \times \mathbb{P}^1$ is the nodal curve

$$\mathbb{P}^1_x \vee \mathbb{P}^1_y := (\mathbb{P}^1_x \times \{0\}) \cup (\{0\} \times \mathbb{P}^1_y) \tag{5.1}$$

Equip $\mathbb{A}^1$ with the usual conjugation $c(a) = \bar{a}$ with real points $\mathbb{A}^1(\mathbb{R}) \simeq \mathbb{R}$. Equip Z with the twisted conjugation $c_Z(x, y, a) = (\bar{y}, \bar{x}, \bar{a})$. When $a \neq 0 \in \mathbb{A}^1(\mathbb{R})$, under the identification $p_x : Z \xrightarrow{\sim} \mathbb{P}^1_x$, we have $c_Z(x) = a^2/\bar{x}$, and thus $p_x : Z_a(\mathbb{R}) \xrightarrow{\sim} \mathbb{RP}^1$. When $a = 0 \in \mathbb{A}^1(\mathbb{R})$, the components of Z_0 are exchanged by c_Z, and $Z_0(\mathbb{R})$ is the single point $x = 0, y = 0$.

Note that in terms of the coordinate t^+ of $\mathbb{P}^1$ the complex conjugation is given by $c_{\mathcal{Z}}(t^+) = \frac{1}{a^2 \bar{t}^+}$. Let $\mathbb{A}^1_+ = \mathrm{Spec}(\mathbb{C}[t^+])$, $\mathbb{A}^1_- = \mathrm{Spec}(\mathbb{C}[t^-])$ be the affine lines with respective coordinates $t^+ = x^{-1}, t^- = y^{-1}$. Thus we have natural open embeddings $\mathbb{A}^1_+ \subset \mathbb{P}^1_x$, $\mathbb{A}^1_- \subset \mathbb{P}^1_y$.

Fix $n = 2m$. Consider the base scheme

$$B = \mathbb{A}^1_a \times B^+ \times B^- \qquad B^\pm = (\mathbb{A}^1_\pm)^m \tag{5.2}$$

with coordinates $(a, \mathbf{t}^+, \mathbf{t}^-)$ where $\mathbf{t}^\pm = (t^\pm_1, \dots, t^\pm_m)$. Equip B with the twisted conjugation $c_A(a, \mathbf{t}^+, \mathbf{t}^-) = (\bar{a}, \bar{\mathbf{t}}^-, \bar{\mathbf{t}}^+)$ where $\bar{\mathbf{t}}^\pm = (\bar{t}^\pm_1, \dots, \bar{t}^\pm_m)$. Projection provides an identification of real points

$$B(\mathbb{R}) \simeq \mathbb{A}^1(\mathbb{R}) \times B^+(\mathbb{C}) \simeq \mathbb{R} \times \mathbb{C}^m \tag{5.3}$$

Consider the family of curves

$$p : \mathcal{Z} = Z \times_{\mathbb{A}^1} B \longrightarrow B \tag{5.4}$$

with the tautological sections

$$\sigma = (\sigma^+, \sigma^-) : B \longrightarrow \mathcal{Z}^{2m} \tag{5.5}$$

$$\sigma^+_i(a, \mathbf{t}^+, \mathbf{t}^-) = (a, [t^+_i, 1], [1, t^+_i a^2], \mathbf{t}^+, \mathbf{t}^-) \qquad i = 1, \dots, m \tag{5.6}$$

$$\sigma^-_i(a, \mathbf{t}^+, \mathbf{t}^-) = (a, [1, t^-_i a^2], [t^-_i, 1], \mathbf{t}^+, \mathbf{t}^-) \qquad i = 1, \dots, m \tag{5.7}$$

Equip $\mathcal{Z}$ with the twisted conjugation $c_{\mathcal{Z}}(a, x, y, \mathbf{t}^+, \mathbf{t}^-) = (\bar{a}, \bar{y}, \bar{x}, \bar{\mathbf{t}}^-, \bar{\mathbf{t}}^+)$. Projection provides an identification of real points

$$\mathcal{Z}(\mathbb{R}) \simeq Z(\mathbb{R}) \times B^+(\mathbb{C}) \simeq Z(\mathbb{R}) \times \mathbb{C}^m \tag{5.8}$$

Introduce the canonical section

$$\xi : B \longrightarrow \mathcal{Z} \qquad \xi(a, \mathbf{t}^+, \mathbf{t}^-) = (a, [1, a], [1, a], \mathbf{t}^+, \mathbf{t}^-) \tag{5.9}$$

We denote by B' the open subset of B consisting of (a, t^+, t^-) with $a \neq 0$ and we define $\mathcal{Z}' := \mathcal{Z} \times_B B'$.

Note that $\xi(a, \mathbf{t}^+, \mathbf{t}^-) = \sigma^\pm_i(a, \mathbf{t}^+, \mathbf{t}^-)$ if and only if $t^\pm_i = a^{-1}$. Denote by $B_\circ \subset B$ the open complement of such coincidences. Introduce the base-change

$$p : \mathcal{Z}_\circ = \mathcal{Z} \times_B B_\circ \longrightarrow B_\circ \tag{5.10}$$

and note, by construction, the tautological sections

$$\sigma = (\sigma^+, \sigma^-) : B_\circ \longrightarrow \mathcal{Z}^{2m}_\circ \tag{5.11}$$

do not intersect the canonical section

$$\xi : B_\circ \longrightarrow \mathcal{Z}_\circ \tag{5.12}$$

With the above choices fixed, we will study the real points of the ind-stacks of quasimaps with their natural projections

$$q : QM_{G,X}(\mathcal{Z}/B, \sigma)_{\mathbb{R}} \longrightarrow B(\mathbb{R}) \tag{5.13}$$

$$r : QM_{G,X}(\mathcal{Z}_\circ/B_\circ, \sigma, \xi)_{\mathbb{R}} \longrightarrow B_\circ(\mathbb{R}) \tag{5.14}$$

5.2. Complex groups. We specialize here our prior constructions to the distinguished case of complex groups.

Recall $\delta = \theta \circ \eta = \eta \circ \theta$ denotes the Cartan conjugation of G with compact real form G_c. Equip $G \times G$ with the swap involution $\mathrm{sw}(g,h) = (h,g)$ and the conjugation $\mathrm{sw}_\delta(g,h) = (\delta(h), \delta(g))$. The fixed-point subgroup of sw is the diagonal $G \subset G \times G$, and the corresponding symmetric space is isomorphic to the group $G\backslash(G \times G) \simeq G$.

LEMMA 5.1. *The uniformization morphisms* (4.8), (4.9) *are isomorphisms:*

$$G[\mathcal{Z}', \sigma]_{\mathbb{R}} \backslash \mathrm{Gr}_{G\times G, \mathcal{Z}', \sigma, \mathbb{R}} \times_{B'} B'(\mathbb{R}) \xrightarrow{\sim} QM_{G\times G, G}(\mathcal{Z}'/B', \sigma)_{\mathbb{R}} \tag{5.15}$$

$$G[\mathcal{Z}'_\circ, \sigma, \xi]_{\mathbb{R}} \backslash \mathrm{Gr}_{G\times G, \mathcal{Z}', \sigma, \mathbb{R}} \times_{B'} B'_\circ(\mathbb{R}) \xrightarrow{\sim} QM_{G\times G, G}(\mathcal{Z}'_\circ/B'_\circ, \sigma, \xi)_{\mathbb{R}} \tag{5.16}$$

PROOF. Note that the above uniformization morphisms over a base point $b \in B'$ (resp. $b \in B'_\circ$) are the (multipoint version) of the real or complex uniformization morphisms in [**CN1**, Section 6.3]. Since the fixed-point subgroup $G \subset G \times G$ is connected and $H^1(\mathrm{Gal}(\mathbb{C}/\mathbb{R}), G \times G)$ is trivial for the Galois-action given by sw_δ, it follows from [**CN1**, Remark 5.8 and Lemma 6.1] that any real bundles on the curve $\mathcal{Z}_b(\mathbb{R}) \simeq \mathbb{RP}^1$ (associated to the real form sw_δ of $G \times G$) admit either real or complex uniformizations. The lemma follows by standard arguments. □

Next, the projection maps $\mathrm{pr}_1, \mathrm{pr}_2 : G \times G \to G$, $\mathrm{pr}_i(g_1, g_2) = g_i$, provide an isomorphism

$$\mathrm{Gr}_{G\times G, \mathcal{Z}, \sigma} \xrightarrow{\sim} \mathrm{Gr}_{G, \mathcal{Z}, \sigma} \times_B \mathrm{Gr}_{G, \mathcal{Z}, \sigma} \tag{5.17}$$

The conjugations $c_{\mathcal{Z}}$ of $\mathcal{Z}$ and sw_δ of $G \times G$ together induce a conjugation of $\mathrm{Gr}_{G\times G, \mathcal{Z}, \sigma}$ which, under the isomorphism (5.17), is given by the map on pairs of bundles with sections

$$(\mathcal{E}, s, \mathcal{E}', s') \longmapsto (c^*_{\mathcal{Z}} \mathcal{E}'_\delta, c^*_{\mathcal{Z}}(s'), c^*_{\mathcal{Z}} \mathcal{E}_\delta, c^*_{\mathcal{Z}}(s)) \tag{5.18}$$

Thus the isomorphism (5.17) followed by pr_1 provides an isomorphism

$$\mathrm{Gr}_{G\times G, \mathcal{Z}, \sigma, \mathbb{R}} \xrightarrow{\sim} \mathrm{Gr}_{G, \mathcal{Z}, \sigma} \times_B B(\mathbb{R}) \tag{5.19}$$

of real analytic spaces over $B(\mathbb{R}) \simeq \mathbb{P}^1(\mathbb{R}) \times_{\mathbb{P}^1(\mathbb{C})} B^+(\mathbb{C})$.

Thus the preceding lemma has the following consequence.

COROLLARY 5.2. *There are natural isomorphisms*

$$G[\mathcal{Z}', \sigma]_{\mathbb{R}} \backslash (\mathrm{Gr}_{G, \mathcal{Z}', \sigma} \times_{B'} B'(\mathbb{R})) \xrightarrow{\sim} QM_{G\times G, G}(\mathcal{Z}'/B', \sigma)_{\mathbb{R}} \tag{5.20}$$

$$G[\mathcal{Z}'_\circ, \sigma, \xi]_{\mathbb{R}} \backslash (\mathrm{Gr}_{G, \mathcal{Z}'_\circ, \sigma} \times_{B'_\circ} B'_\circ(\mathbb{R})) \xrightarrow{\sim} QM_{G\times G, G}(\mathcal{Z}'_\circ/B'_\circ, \sigma, \xi)_{\mathbb{R}} \tag{5.21}$$

Next, consider the natural map

$$\mathrm{Gr}_{G, \mathcal{Z}, \sigma^+} \longrightarrow \mathrm{Gr}_{G, \mathcal{Z}, \sigma} \tag{5.22}$$

given by restricting a trivialization of a G-bundle defined away from the section σ^+ to the complement of both sections σ^+ and σ^-.

We will find open loci where the map (5.22) induces an isomorphism.

First, consider the restriction of (5.22) to the generic family

$$\mathrm{Gr}_{G,\mathcal{Z}',\sigma^+} \longrightarrow \mathrm{Gr}_{G,\mathcal{Z}',\sigma} \tag{5.23}$$

PROPOSITION 5.3. *The map* (5.23) *induces isomorphisms*

$$G_c\backslash \mathrm{Gr}_{G,\mathcal{Z}',\sigma^+} \times_{B'} B'(\mathbb{R}) \xrightarrow{\sim} G[\mathcal{Z}',\sigma]_{\mathbb{R}}\backslash(\mathrm{Gr}_{G,\mathcal{Z}',\sigma} \times_{B'} B'(\mathbb{R})) \tag{5.24}$$

$$\mathrm{Gr}_{G,\mathcal{Z}'_\circ,\sigma^+} \times_{B'_\circ} B'_\circ(\mathbb{R}) \xrightarrow{\sim} G[\mathcal{Z}'_\circ,\sigma,\xi]_{\mathbb{R}}\backslash(\mathrm{Gr}_{G,\mathcal{Z}'_\circ,\sigma} \times_{B'_\circ} B'_\circ(\mathbb{R})) \tag{5.25}$$

PROOF. It suffices to establish the second with its natural G_c-equivariance then glue to obtain the first. Let $b = (a, t_1^+, \dots, t_n^+) \in B'_\circ(\mathbb{R})$ and let $z_1, \dots, z_k \in \mathbb{C}$ be k-distinct points such that $\{z_1, .., z_k\} = \{t_1^+, \dots, t_n^+\}$. It follows from the standard factorization property of Beilinson-Drinfeld Grassmannian and Theorem 3.3 that, over the based point b, the second map above can be identified with the map

$$\prod_{i=1}^k \mathrm{Gr}_{z_i} \to \prod_{i=1}^k (\mathrm{Gr}_{z_i} \times \mathrm{Gr}_{\frac{1}{a^2\bar{z}_i}}) \to \prod_{i=1}^k \Omega_{z_i,a^{-1}} G_c\backslash(\mathrm{Gr}_{z_i} \times \mathrm{Gr}_{\frac{1}{a^2\bar{z}_i}}) \tag{5.26}$$

where the first map is the left copy embedding sending γ to (γ, e), where e is the base point of $\mathrm{Gr}_{\frac{1}{a^2\bar{z}_i}}$, and the second map is the natural quotient map. (Note that in terms of the local coordinate z_i the complex conjugation on $\mathcal{Z}_b \simeq \mathbb{P}^1_{z_i}$ is given by $z_i \to \frac{1}{a^2\bar{z}_i}$.) Thus the assertion follows from the fact that the based loop group $\Omega_{z_i,a^{-1}} G_c$ acts freely on Gr_{z_i} and $\mathrm{Gr}_{\frac{1}{a^2\bar{z}_i}}$. □

COROLLARY 5.4. *There are natural isomorphisms*

$$G_c\backslash \mathrm{Gr}_{G,\mathcal{Z}',\sigma^+} \times_{B'} B'(\mathbb{R}) \xrightarrow{\sim} QM_{G\times G,G}(\mathcal{Z}'/B',\sigma)_{\mathbb{R}} \tag{5.27}$$

$$\mathrm{Gr}_{G,\mathcal{Z}'_\circ,\sigma^+} \times_{B'_\circ} B'_\circ(\mathbb{R}) \xrightarrow{\sim} QM_{G\times G,G}(\mathcal{Z}'_\circ/B'_\circ,\sigma,\xi)_{\mathbb{R}} \tag{5.28}$$

Next, let $\mathrm{Bun}^0_G(\mathcal{Z}/B) \subset \mathrm{Bun}_G(\mathcal{Z}/B)$ denote the open sub-stack of a point $b \in B$ and a trivializable G-bundle on $\mathcal{Z}_b$. Denote by

$$QM^0_{G\times G,G}(\mathcal{Z}/B,\sigma)_{\mathbb{R}} \simeq G_c\backslash \mathrm{Gr}^0_{G,\mathcal{Z},\sigma^+} \qquad QM^0_{G\times G,G}(\mathcal{Z}_\circ/B_\circ,\sigma,\xi)_{\mathbb{R}} \simeq \mathrm{Gr}^0_{G,\mathcal{Z}^+_\circ,\sigma^+} \tag{5.29}$$

the base-changes to $\mathrm{Bun}^0_G(\mathcal{Z}/B) \subset \mathrm{Bun}_G(\mathcal{Z}/B)$. Consider the restriction of (5.22) to the trivial bundle locus

$$\mathrm{Gr}^0_{G,\mathcal{Z},\sigma^+} \longrightarrow \mathrm{Gr}^0_{G,\mathcal{Z},\sigma} \tag{5.30}$$

PROPOSITION 5.5. *The map* (5.30) *induces isomorphisms*

$$G_c\backslash \mathrm{Gr}^0_{G,\mathcal{Z},\sigma^+} \times_B B(\mathbb{R}) \xrightarrow{\sim} G[\mathcal{Z},\sigma]_{\mathbb{R}}\backslash(\mathrm{Gr}^0_{G,\mathcal{Z},\sigma} \times_B B(\mathbb{R})) \tag{5.31}$$

$$\mathrm{Gr}^0_{G,\mathcal{Z}_\circ,\sigma^+} \times_{B_\circ} B_\circ(\mathbb{R}) \xrightarrow{\sim} G[\mathcal{Z}_\circ,\sigma,\xi]_{\mathbb{R}}\backslash(\mathrm{Gr}^0_{G,\mathcal{Z}_\circ,\sigma} \times_{B_\circ} B_\circ(\mathbb{R})) \tag{5.32}$$

PROOF. By Proposition 5.3, it suffices to establish the maps are isomorphisms at the special fiber $b = (a, t_1^+, \dots, t_n^+) \in B(\mathbb{R})$ with $a = 0$. Moreover, it suffices to establish the second with its natural G_c-equivariance then glue to obtain the first.

Now it is elementary to see we have a natural isomorphism at the special fiber

$$\mathrm{Gr}_{G,\mathcal{Z}_\circ,\sigma} \times_{B_\circ} B_\circ(\mathbb{R})|_b \xrightarrow{\sim} \mathrm{Gr}_{G,\mathcal{Z}_\circ,\sigma^+,b} \times \mathrm{Gr}_{G,\mathcal{Z}_\circ,\sigma^-,b} \tag{5.33}$$

And similarly, we have a natural isomorphism of groups at the special fiber

$$G[\mathcal{Z}_\circ, \sigma, \xi]_{\mathbb{R}}|_b \xrightarrow{\sim} \Delta_{\mathbb{R}} \subset G[\mathbb{P}^1, \sigma^+, \xi, b] \times G[\mathbb{P}^1, \sigma^-, \xi, b] \tag{5.34}$$

where $\Delta_{\mathbb{R}}$ denotes the conjugate diagonal. Thus the assertion follows from the facts that the isomorphism (5.33) restricts to an isomorphism

$$\mathrm{Gr}^0_{G,\mathcal{Z}_\circ,\sigma} \times_{B_\circ} B_\circ(\mathbb{R})|_b \xrightarrow{\sim} \mathrm{Gr}^0_{G,\mathcal{Z}_\circ,\sigma^+,b} \times \mathrm{Gr}^0_{G,\mathcal{Z}_\circ,\sigma^-,b} \tag{5.35}$$

and the action of $G[\mathbb{P}^1, \sigma^+, \xi, b]$ on $\mathrm{Gr}^0_{G,\mathcal{Z}_\circ,\sigma^+,b}$ is free and transitive. □

COROLLARY 5.6. *There are natural isomorphisms*

$$G_c \backslash \mathrm{Gr}^0_{G,\mathcal{Z},\sigma^+} \times_B B(\mathbb{R}) \xrightarrow{\sim} QM^0_{G\times G,G}(\mathcal{Z}/B, \sigma)_{\mathbb{R}} \tag{5.36}$$

$$\mathrm{Gr}^0_{G,\mathcal{Z}_\circ,\sigma^+} \times_{B_\circ} B_\circ(\mathbb{R}) \xrightarrow{\sim} QM^0_{G\times G,G}(\mathcal{Z}_\circ/B_\circ, \sigma, \xi)_{\mathbb{R}} \tag{5.37}$$

REMARK 5.7. The isomorphisms of Cor. 5.4 and 5.37 coincide on the intersections of their domains.

5.3. Stratification. Recall the Beilinson-Drinfeld Grassmannian $\mathrm{Gr}^{(m)} \to (\mathbb{P}^1)^m$ of a G-bundle $\mathcal{E}$ on $\mathbb{P}^1$, a point $(z_1, \dots, z_m) \in (\mathbb{P}^1)^m$, and a section $s : \mathbb{P}^1 \setminus \{z_1, \dots, z_m\} \to \mathcal{E}$. We denote by $\mathrm{Gr}^{(m),0} \subset \mathrm{Gr}^{(m)}$ the open subset consisting of $(\mathcal{E}, z_1, \dots, z_m, s)$ such that $\mathcal{E}$ is trivializable.

First, let us stratify the base $(\mathbb{P}^1)^m$ by coincidences among points. For any partition $\mathfrak{p}$ of the set $\{1, \dots, m\}$, denote by $(\mathbb{P}^1)^{\mathfrak{p}} \subset (\mathbb{P}^1)^m$ the locus where $z_i = z_j$ if and only if i and j are in the same part of $\mathfrak{p}$. This provides a Whitney stratification of $(\mathbb{P}^1)^m$.

Next, for any partition $\mathfrak{p}$ of the set $\{1, \dots, m\}$, and any map $\lambda_{\mathfrak{p}} : \mathfrak{p} \to \Lambda_T^+$, denote by $\mathcal{S}^{\lambda_{\mathfrak{p}}} \subset \mathrm{Gr}^{(m)}$ (resp. $\mathcal{S}^{\lambda_{\mathfrak{p}},0} \subset \mathrm{Gr}^{(m),0}$) the spherical stratum of a G-bundle $\mathcal{E}$ (resp. a trivializable G-bundle $\mathcal{E}$), a point $(z_1, \dots, z_m) \in (\mathbb{P}^1)^{\mathfrak{p}}$, and a section $s : \mathbb{P}^1 \setminus \{z_1, \dots, z_m\} \to \mathcal{E}$ of modification type $\lambda_{\mathfrak{p}}$. This provides a Whitney stratification of $\mathrm{Gr}^{(m)}$ (resp. $\mathrm{Gr}^{(m),0}$) compatible with that of $(\mathbb{P}^1)^m$.

Note that the coordinate x^{-1} provides an isomorphism

$$(\mathbb{R} \times \mathrm{Gr}^{(m)}) \times_{\mathbb{R}\times(\mathbb{P}^1)^m} B(\mathbb{R}) \simeq \mathrm{Gr}_{G,\mathcal{Z},\sigma^+} \times_B B(\mathbb{R}) \tag{5.38}$$

here we identify $B(\mathbb{R}) \overset{(5.3)}{\simeq} \mathbb{R} \times \mathbb{C}^m \subset \mathbb{R} \times (\mathbb{P}^1)^m$. The isomorphism above restricts to an isomorphism

$$(\mathbb{R} \times \mathrm{Gr}^{(m),0}) \times_{\mathbb{R}\times(\mathbb{P}^1)^m} B(\mathbb{R}) \simeq \mathrm{Gr}^0_{G,\mathcal{Z},\sigma^+} \times_B B(\mathbb{R}) \tag{5.39}$$

between the corresponding open loci and, by Corollary 5.4 and 5.6, we obtain:

$$(\mathbb{R} \times \mathrm{Gr}^{(m)}) \times_{\mathbb{R}\times(\mathbb{P}^1)^m} B'_\circ(\mathbb{R}) \simeq QM_{G\times G,G}(\mathcal{Z}'_\circ/B'_\circ, \sigma, \xi)_{\mathbb{R}}, \tag{5.40}$$

$$(\mathbb{R} \times \mathrm{Gr}^{(m),0}) \times_{\mathbb{R}\times(\mathbb{P}^1)^m} B_\circ(\mathbb{R}) \simeq QM^0_{G\times G,G}(\mathcal{Z}_\circ/B_\circ, \sigma, \xi)_{\mathbb{R}}. \tag{5.41}$$

Let us equip $\mathbb{R} \times \mathrm{Gr}^{(m)}$ (resp. $\mathbb{R} \times \mathrm{Gr}^{(m),0}$) with the product stratification $\{\mathbb{R}\times\mathcal{S}^{\lambda_{\mathfrak{p}}}\}$ (resp. $\{\mathbb{R}\times\mathcal{S}^{\lambda_{\mathfrak{p}},0}\}$) and transport the spherical strata $\mathbb{R}\times\mathcal{S}^{\lambda_{\mathfrak{p}}} \subset \mathbb{R}\times\mathrm{Gr}^{(m)}$ (resp. $\mathbb{R}\times\mathcal{S}^{\lambda_{\mathfrak{p}},0} \subset \mathbb{R}\times\mathrm{Gr}^{(m),0}$) across the isomorphisms (5.40) and (5.41) and denote the resulting strata by

$$\mathcal{S}^{\lambda_{\mathfrak{p}}}_{\mathcal{Z},\mathbb{R}} \subset QM_{G\times G,G}(\mathcal{Z}'_\circ/B'_\circ, \sigma, \xi)_{\mathbb{R}} \qquad (\text{resp. } \mathcal{S}^{\lambda_{\mathfrak{p}},0}_{\mathcal{Z},\mathbb{R}} \subset QM^0_{G\times G,G}(\mathcal{Z}_\circ/B_\circ, \sigma, \xi)_{\mathbb{R}}). \tag{5.42}$$

Note that $\mathcal{S}_{\mathcal{Z},\mathbb{R}}^{\lambda_\mathfrak{p},0}$ and $\mathcal{S}_{\mathcal{Z},\mathbb{R}}^{\lambda_\mathfrak{p},0}$ are non-empty if and only if the total coweight $|\lambda_\mathfrak{p}| \in \Lambda_T^+$, given by summing the values of $\lambda_\mathfrak{p}$ over the parts of $\mathfrak{p}$, in fact lies in $R_G^+ \subset \Lambda_T^+$.

5.4. Symmetry. We continue here with the distinguished case of complex groups. We will describe an involution of quasi-maps.

DEFINITION 5.8 (Swap involution $\delta_\mathcal{Z}$). The conjugation $\delta\times\delta$ of $G\times G$ commutes with the conjugation sw_δ. Hence together with the conjugation $c_\mathcal{Z}$ of $\mathcal{Z}$ it induces a fiberwise involution of $QM_{G\times G,G}(\mathcal{Z}/B,\sigma)_\mathbb{R}$ and $QM_{G\times G,G}(\mathcal{Z}_\circ/B_\circ,\sigma,\xi)_\mathbb{R}$ denoted by $\delta_\mathcal{Z}$.

In the remainder of this section, we will give concrete descriptions of how the involution $\delta_\mathcal{Z}$ act at the generic and special fibers in terms of our prior uniformizations.

5.4.1. *Generic fiber.* Let $b = (a \neq 0, t_1^+, \ldots, t_m^+) \in B_\circ'(\mathbb{R})$. Let $z_1, \ldots, z_k \in \mathbb{C}$ be the distinct k points such that there is an equality of sets $\{z_1, \ldots, z_k\} = \{t_1^+, \ldots, t_m^+\}$ and consider the isomorphism

$$(5.43)\quad v_b : \mathrm{Gr}_{z_1}\times\cdots\times\mathrm{Gr}_{z_k} \simeq (\mathbb{R}\times\mathrm{Gr}^{(m)})|_b \overset{(5.38)}{\simeq} \mathrm{Gr}_{G,\mathcal{Z},\sigma^+,b} \overset{(5.28)}{\simeq} QM_{G\times G,G}(\mathcal{Z}_\circ/B_\circ,\sigma,\xi,b)_\mathbb{R}.$$

Here the first map is the factorization isomorphism. For $i = 1, \ldots, k$, we define

$$v_i : \mathrm{Gr}_{z_i} \hookrightarrow \mathrm{Gr}_{z_1}\times\cdots\times\mathrm{Gr}_{z_k} \overset{v_b}{\simeq} QM_{G\times G,G}(\mathcal{Z}_\circ/B_\circ,\sigma,\xi,b)_\mathbb{R}$$

PROPOSITION 5.9. *For $i = 1, \ldots, k$, the involution $\delta_\mathcal{Z}$ satisfies:*

(1) *When $z_i \in S^1_{a^{-1}} = \{|z| = |a^{-1}|\} \subset \mathbb{P}^1$, we have*

$$(5.44)\qquad \delta_\mathcal{Z}\circ v_i = v_i \circ \delta$$

where δ is the conjugation on Gr_{z_i} induced by the conjugations $c(z) = \frac{1}{a^2\bar{z}}$ on $\mathbb{P}^1$ and δ on G.

(2) *When $z_i \in \mathbb{C}\setminus S^1_{a^{-1}}$, we have*

$$(5.45)\qquad \delta_\mathcal{Z}\circ v_i = v_i \circ \mathrm{inv}$$

where $\mathrm{inv}(g)(z) = g(z)^{-1}$ is the group-inverse on $\Omega_{z_i,a^{-1}}G_c \simeq \mathrm{Gr}_{z_i}$.

PROOF. We first note that the isomorphism

$$\prod_{i=1}^k \Omega_{z_i,a^{-1}}G_c\backslash(\mathrm{Gr}_{z_i}\times\mathrm{Gr}_{c(z_i)}) \simeq G[\mathcal{Z}_\circ',\sigma,\xi,b]_\mathbb{R}\backslash\mathrm{Gr}_{G,\mathcal{Z}_\circ',\sigma,b} \simeq QM_{G\times G,G}(\mathcal{Z}_\circ/B_\circ,\sigma,\xi,b)_\mathbb{R}$$

intertwines the conjugation $\delta_\mathcal{Z}$ with the one $\delta_\mathcal{Z}'$ on the product

$$\prod_{i=1}^k \Omega_{z_i,a^{-1}}G_c\backslash(\mathrm{Gr}_{z_i}\times\mathrm{Gr}_{c(z_i)})$$

given by

$$\delta_\mathcal{Z}'(\gamma_{i,1},\gamma_{i,2}) = (\gamma_{i,1}',\gamma_{i,2}'),\quad i = 1,\ldots,k$$

where $\gamma_{i,j}' = \delta(\gamma_{i,j})$, $j = 1, 2$, if $z_i = c(z_i) \in S^1_{a^{-1}}$, otherwise $\gamma_{i,1}' = \delta(\gamma_{i,2})$ and $\gamma_{i,2}' = \delta(\gamma_{i,1})$ if $z_i \neq c(z_i)$. Here δ is the map $\delta : \mathrm{Gr}_{z_i} \to \mathrm{Gr}_{c(z_i)}$ induced by the complex conjugation $c(z) = \frac{1}{a^2\bar{z}}$ on $\mathbb{P}^1$ and δ on G.

On the other hand, a direct computation shows that the isomorphism

$$\prod_{i=1}^{k} \mathrm{Gr}_{z_i} \simeq \prod_{i=1}^{k} \Omega_{z_i,a^{-1}} G_c \backslash (\mathrm{Gr}_{z_i} \times \mathrm{Gr}_{c(z_i)})$$

induced by the left copy embedding $\gamma \to (\gamma, e)$ intertwines the map $\delta'_{\mathcal{Z}}$ above with the one $\delta''_{\mathcal{Z}}$ on Gr_{z_i} given by $\delta''_{\mathcal{Z}}(\gamma) = \delta(\gamma)$ when $z_i = c(z_i)$ and $\delta''_{\mathcal{Z}}(\gamma) = \gamma^{-1}$ is the group-inverse on $\Omega_{z_i,a^{-1}} G_c \simeq \mathrm{Gr}_{z_i}$ when $z_i \neq c(z_i)$.

To deduce the proposition, we note that the map v_b in (5.43) is equal to the composition of the above two isomorphisms. $\square$

5.4.2. *Special fiber.* Let $b = (0, t_1^+, \ldots, t_m^+) \in B_\circ(\mathbb{R})$. Consider the map

(5.46)

$$v_b : G[\mathbb{P}^1 \setminus \{t_1^+, \ldots, t_m^+\}]_{\infty \to e} \simeq (\mathbb{R} \times \mathrm{Gr}^{(m),0})|_b \overset{(5.39)}{\simeq} \mathrm{Gr}^0_{G,\mathcal{Z},\sigma^+,b} \overset{(5.37)}{\simeq} QM^0_{G\times G,G}(\mathcal{Z}_\circ/B_\circ, \sigma, \xi, b)_{\mathbb{R}}.$$

Here and in what follows, the subscript $\infty \to e$ conveys the condition $g(\infty) = e$ on a mapping.

PROPOSITION 5.10. *The involution $\delta_{\mathcal{Z}}$ satisfies*

$$\delta_{\mathcal{Z}} \circ v_b = v_b \circ \mathrm{inv}$$

here $\mathrm{inv}(g(z)) = g(z)^{-1}$ *is the group inverse on* $G[\mathbb{P}^1 \setminus \{t_1^+, \ldots, t_m^+\}]_{\infty \to e}$.

PROOF. We first note that the isomorphism

$$G[\mathbb{P}^1 \setminus \{t_1^+, \ldots, t_m^+\}]_{\infty \to e} \backslash (\mathrm{Gr}^0_{G,\mathcal{Z},\sigma^+,b} \times \mathrm{Gr}^0_{G,\mathcal{Z},\sigma^-,b}) \simeq G[\mathcal{Z}'_\circ, \sigma, \xi, b]_{\mathbb{R}} \backslash \mathrm{Gr}^0_{G,\mathcal{Z}'_\circ,\sigma,b} \simeq QM^0_{G\times G,G}(\mathcal{Z}_\circ/B_\circ, \sigma, \xi, b)_{\mathbb{R}}$$

intertwines the involution $\delta_{\mathcal{Z}}$ with the one $\delta'_{\mathcal{Z}}$ on

$$G[\mathbb{P}^1 \setminus \{t_1^+, \ldots, t_m^+\}]_{\infty \to e} \backslash (\mathrm{Gr}^0_{G,\mathcal{Z},\sigma^+,b} \times \mathrm{Gr}^0_{G,\mathcal{Z},\sigma^-,b})$$

given by $\delta'_{\mathcal{Z}}(\gamma_1, \gamma_2) = (\delta(\gamma_2), \delta(\gamma_1))$ (note that the group $G[\mathbb{P}^1 \setminus \{t_1^+, \ldots, t_m^+\}]_{\infty \to e}$ acts on the product via the conjugate diagonal embedding $\gamma \to (\gamma, \delta(\gamma))$). On the other hand, the isomorphism

$$G[\mathbb{P}^1 \setminus \{t_1^+, \ldots, t_m^+\}]_{\infty \to e} \simeq \mathrm{Gr}^0_{G,\mathcal{Z},\sigma^+,b} \simeq G[\mathbb{P}^1 \setminus \{t_1^+, \ldots, t_m^+\}]_{\infty \to e} \backslash (\mathrm{Gr}^0_{G,\mathcal{Z},\sigma^+,b} \times \mathrm{Gr}^0_{G,\mathcal{Z},\sigma^-,b})$$

where the fist map is the action map and the second map is induced by the left copy embedding intertwines the involution $\delta'_{\mathcal{Z}}$ with the group inverse on $G[\mathbb{P}^1 \setminus \{t_1^+, \ldots, t_m^+\}]_{\infty \to e}$. To deduce the proposition, we observe that the map v_b is equal to the composition of the above two isomorphisms. $\square$

5.5. Compatibility with stratifications. We have the following compatibility of the involution $\delta_{\mathcal{Z}}$ with stratifications. Let $w_0 \in W$ denote the longest element of the Weyl group. For any partition $\mathfrak{p}$ of the set $\{1, \ldots, m\}$, and map $\lambda_{\mathfrak{p}} : \mathfrak{p} \to \Lambda_T^+$, we set $-w_0(\lambda_{\mathfrak{p}}) = -w_0 \circ \lambda_{\mathfrak{p}}$.

LEMMA 5.11. (1) *The involution $\delta_{\mathcal{Z}}$ preserves the open subspace*

(5.47) $$QM^0_{G\times G,G}(\mathcal{Z}_\circ/B_\circ, \sigma, \xi)_{\mathbb{R}} \subset QM_{G\times G,G}(\mathcal{Z}_\circ/B_\circ, \sigma, \xi)_{\mathbb{R}}$$

(2) *The involution $\delta_{\mathcal{Z}}$ restricts to a map on spherical strata*

$$\mathcal{S}^{\lambda_{\mathfrak{p}}}_{\mathcal{Z},\mathbb{R}} \longrightarrow \mathcal{S}^{-w_0(\lambda_{\mathfrak{p}})}_{\mathcal{Z},\mathbb{R}} \tag{5.48}$$

within $QM_{G\times G,G}(\mathcal{Z}'_{\circ}/B'_{\circ},\sigma,\xi)_{\mathbb{R}}$, and also the spherical strata

$$\mathcal{S}^{\lambda_{\mathfrak{p}},0}_{\mathcal{Z},\mathbb{R}} \longrightarrow \mathcal{S}^{-w_0(\lambda_{\mathfrak{p}}),0}_{\mathcal{Z},\mathbb{R}} \tag{5.49}$$

within $QM^0_{G\times G,G}(\mathcal{Z}_{\circ}/B_{\circ},\sigma,\xi)_{\mathbb{R}}$.

PROOF. The claim follows from the facts that the conjugation δ of Gr maps S^{λ} (resp. Gr^0) to $S^{-w_0(\lambda)}$ (resp. Gr^0), and the involution inv of ΩG_c maps S^{λ} to $S^{-w_0(\lambda)}$. □

5.6. The involution $\eta_{\mathcal{Z}}$. Recall θ denotes the Cartan involution of G with fixed-point subgroup K, and $\delta=\theta\circ\eta=\eta\circ\theta$ the Cartan conjugation of G with compact real form G_c. Since η and θ commute with δ, the conjugation $\eta\times\eta$ and involution $\theta\times\theta$ of $G\times G$ define involutions of the real moduli $QM_{G\times G,G}(\mathcal{Z}_{\circ}/B_{\circ},\sigma,\xi)_{\mathbb{R}}$ of rigidified quasi-maps which we denote respectively by $\eta_{\mathcal{Z}}$ and $\theta_{\mathcal{Z}}$.

Propositions 5.9 and 5.10, and Lemma 5.11 immediately imply:

PROPOSITION 5.12. *The involution $\eta_{\mathcal{Z}}$ on $QM_{G\times G,G}(\mathcal{Z}_{\circ}/B_{\circ},\sigma,\xi)_{\mathbb{R}}$ satisfies:*

(1) *The involution commutes with the natural G_c-action and preserves the open subfamilies $QM_{G\times G,G}(\mathcal{Z}'_{\circ}/B'_{\circ},\sigma,\xi)_{\mathbb{R}}$ and $QM^0_{G\times G,G}(\mathcal{Z}_{\circ}/B_{\circ},\sigma,\xi)_{\mathbb{R}}$ and their spherical stratifications $\{\mathcal{S}^{\lambda_{\mathfrak{p}}}_{\mathcal{Z},\mathbb{R}}\}$ and $\{\mathcal{S}^{\lambda_{\mathfrak{p}},0}_{\mathcal{Z},\mathbb{R}}\}$.*

(2) *At $b=(a\neq 0,t_1^+,\dots,t_m^+)\in B_{\circ}(\mathbb{R})$, the isomorphism*

$$v_b:\mathrm{Gr}_{z_1}\times\cdots\times\mathrm{Gr}_{z_k}\xrightarrow{\ \sim\ } QM_{G\times G,G}(\mathcal{Z}_{\circ}/B_{\circ},\sigma,\xi,b)_{\mathbb{R}}$$

of (5.43) intertwines the involution $\delta_{\mathcal{Z}}$ with the involution on $\mathrm{Gr}_{z_1}\times\cdots\times\mathrm{Gr}_{z_k}$ given by $(\gamma_1,\dots,\gamma_k)\to(\gamma'_1,\dots,\gamma'_k)$, where $\gamma'_i=\eta(\gamma_i)$ if $z_i\in S^1_{a^{-1}}$ and η is the conjugation on Gr_{z_i} induced by $c(z)=\frac{1}{a^2\bar z}$ on $\mathbb{P}^1$ and η on G, otherwise $\gamma'_i=\mathrm{inv}\circ\theta(\gamma_i)$ if $z_i\in\mathbb{C}\setminus S^1_{a^{-1}}$ and $\mathrm{inv}\circ\theta$ is the involution on $\Omega_{z_i,a^{-1}}G_c\simeq\mathrm{Gr}_{z_i}$.

(3) *At $b=(0,t_1^+,,,.t_m^+)\in B_{\circ}(\mathbb{R})$, the isomorphism*

$$v_b: G[\mathbb{P}^1\setminus\{t_1^+,,,.t_m^+\}]_{\infty\to e}\xrightarrow{\ \sim\ } QM^0_{G\times G,G}(\mathcal{Z}_{\circ}/B_{\circ},\sigma,\xi,b)_{\mathbb{R}}$$

of (5.46) intertwines the involution $\eta_{\mathcal{Z}}$ with the involution $\mathrm{inv}\circ\theta(\gamma)=\theta(\gamma)^{-1}$ on $G[\mathbb{P}^1\setminus\{t_1^+,,,.t_m^+\}]_{\infty\to e}$.

6. Stratified homeomorphisms

6.1. Trivializations. Set $\mathbb{B}=[0,1]\times\mathbb{R}^m$ and $\mathbb{B}'=(0,1]\times\mathbb{R}^m$.

DEFINITION 6.1. An embedding

$$\zeta:\mathbb{B}=[0,1]\times\mathbb{R}^m\longrightarrow B_{\circ}(\mathbb{R})\subset\mathbb{R}\times\mathbb{C}^m$$

is called *admissible* if it is given by $\zeta(a,z_1,\dots,z_m)=(a,f_a(z_1),\dots,f_a(z_m))$ where $f:[0,1]\to\mathrm{Aut}(\mathbb{P}^1)$ is a real analytic map satisfying $f_0(z)=z$ and $f_1(z)=\frac{z-i}{z+i}$.

REMARK 6.2. Each admissible ζ defines a one-parameter family of embeddings $\zeta_a:=\zeta|_{a\times\mathbb{R}^m}:\mathbb{R}^m\to\mathbb{C}^m$, $a\in[0,1]$ satisfying $\zeta_0(\mathbb{R}^m)=\mathbb{R}^m$ and $\zeta_1(\mathbb{R}^m)=(S^1\setminus\{1\})^m$.

EXAMPLE 6.3. Consider the map $f : [0,1] \to \mathrm{Aut}(\mathbb{P}^1)$ given by $f_a(z) = \frac{z-ai}{az+ai+(1-a)}$. A direct computation shows that $f_a(z) \neq a^{-1}$ for all $a \in \mathbb{R}$, and $f_0(z) = z$, $f_1(z) = \frac{z-i}{z+i}$, and hence the corresponding embedding $\zeta : \mathbb{B} \to B_\circ(\mathbb{R})$ is admissible.

Let $\zeta : \mathbb{B} \to B_\circ(\mathbb{R})$ be an admissible embedding. Consider the following base changes

$$QM_{G\times G,G}(\mathcal{Z}/\mathbb{B},\sigma,\xi)_{\mathbb{R}} := QM_{G\times G,G}(\mathcal{Z}_\circ/B_\circ,\sigma,\xi)_{\mathbb{R}} \times_{B_\circ(\mathbb{R})} \mathbb{B} \longrightarrow \mathbb{B}$$

$$QM^0_{G\times G,G}(\mathcal{Z}/\mathbb{B},\sigma,\xi)_{\mathbb{R}} := QM^0_{G\times G,G}(\mathcal{Z}_\circ/B_\circ,\sigma,\xi)_{\mathbb{R}} \times_{B_\circ(\mathbb{R})} \mathbb{B} \longrightarrow \mathbb{B}$$

$$QM_{G\times G,G}(\mathcal{Z}'/\mathbb{B}',\sigma,\xi)_{\mathbb{R}} := QM_{G\times G,G}(\mathcal{Z}'_\circ/B'_\circ,\sigma,\xi)_{\mathbb{R}} \times_{B'_\circ(\mathbb{R})} \mathbb{B}' \longrightarrow \mathbb{B}'$$

Then (5.40) and (5.41) restrict to isomorphisms

$$(\mathbb{R}\times \mathrm{Gr}^{(m)}) \times_{\mathbb{R}\times(\mathbb{P}^1)^m} \mathbb{B}' \xrightarrow{\sim} QM_{G\times G,G}(\mathcal{Z}'/\mathbb{B}',\sigma,\xi)_{\mathbb{R}}, \tag{6.1}$$

$$(\mathbb{R}\times \mathrm{Gr}^{(m),0}) \times_{\mathbb{R}\times(\mathbb{P}^1)^m} \mathbb{B} \xrightarrow{\sim} QM^0_{G\times G,G}(\mathcal{Z}/\mathbb{B},\sigma,\xi)_{\mathbb{R}}. \tag{6.2}$$

Note that there is an isomorphism over $[0,1]$

$$[0,1]\times(\mathrm{Gr}^{(m)} \times_{(\mathbb{P}^1)^m}\mathbb{R}^m) \xrightarrow{\sim} (\mathbb{R}\times\mathrm{Gr}^{(m)}) \times_{\mathbb{R}\times(\mathbb{P}^1)^m} \mathbb{B}$$

given by

$$(a,(\mathcal{E},z_1,\dots,z_m,s)) = (a,(f_a)_*\mathcal{E},f_a(z_1),\dots,f_a(z_m),(f_a)_*s)$$

and in view of (6.1), (6.2), we obtain

PROPOSITION 6.4. *Each admissible embedding $\zeta : \mathbb{B} \to B_\circ(\mathbb{R})$ induces isomorphisms:*

$$(0,1]\times(\mathrm{Gr}^{(m)} \times_{(\mathbb{P}^1)^m}\mathbb{R}^m) \xrightarrow{\sim} QM_{G\times G,G}(\mathcal{Z}'/\mathbb{B}',\sigma,\xi)_{\mathbb{R}}, \tag{6.3}$$

$$[0,1]\times(\mathrm{Gr}^{(m),0} \times_{(\mathbb{P}^1)^m}\mathbb{R}^m) \xrightarrow{\sim} QM^0_{G\times G,G}(\mathcal{Z}/\mathbb{B},\sigma,\xi)_{\mathbb{R}}. \tag{6.4}$$

Note that the isomorphisms above coincide on the intersections of their domains.

6.2. Families of involutions. Let $p : \mathcal{G} \to \mathbb{R}^m$ be the group ind-scheme over $\mathbb{R}^m$ whose fiber over $(z_1,\dots,z_m) \in \mathbb{R}^m$ is the group ind-scheme $G[\mathbb{P}^1 \setminus \{z_1,\dots,z_m\}]_{\infty\to e}$ of maps $\gamma : \mathbb{P}^1 \setminus \{z_1,\dots,z_m\} \to G$ such that $\gamma(\infty) = e$. The acton of $\mathcal{G}$ on the base point of $\mathrm{Gr}^{(m)} \times_{(\mathbb{P}^1)^m}\mathbb{R}^m$ defines an isomorphism

$$\mathcal{G} \xrightarrow{\sim} \mathrm{Gr}^{(m),0} \times_{(\mathbb{P}^1)^m}\mathbb{R}^m \tag{6.5}$$

Consider the transported strata $\mathcal{G}^{\lambda_{\mathfrak{p}}} := \mathcal{S}^{\lambda_{\mathfrak{p}},0} \cap \mathcal{G}$ of $\mathcal{G}$ and equip $[0,1]\times\mathcal{G}$ with the product stratification $\{[0,1]\times\mathcal{G}^{\lambda_{\mathfrak{p}}}\}$. By Proposition 6.4, we obtain a stratified isomorphism

$$p_\zeta : [0,1]\times\mathcal{G} \xrightarrow{\sim} QM^0_{G\times G,G}(\mathcal{Z}/\mathbb{B},\sigma,\xi)_{\mathbb{R}}. \tag{6.6}$$

Recall the involution $\eta_{\mathcal{Z}}$ on $QM^0_{G\times G,G}(\mathcal{Z}_\circ/B_\circ,\sigma,\xi)_{\mathbb{R}}$ of Section 5.6. Via the isomorphism p_ζ, the involution $\eta_{\mathcal{Z}}$ gives rise to a family of involutions

$$\alpha_a : \mathcal{G} \longrightarrow \mathcal{G}, \quad a \in [0,1], \tag{6.7}$$

where $\alpha_a(\gamma) := p_\zeta^{-1} \circ \eta_{\mathcal{Z}} \circ p_\zeta(\{a\} \times \gamma)$. The following proposition follows from Proposition 5.12:

THEOREM 6.5. *The family of involutions $\alpha_a : \mathcal{G} \longrightarrow \mathcal{G}$, $a \in [0,1]$ satisfy the following:*

(1) *We have $q \circ \alpha_a = q : \mathcal{G} \to \mathbb{R}^m$.*
(2) *α_a is G_c-equivariant and preserves the stratification $\{\mathcal{G}^{\lambda_{\mathfrak{p}}}\}$.*
(3) *At $a = 0$, we have $\alpha_0(\gamma(z)) = \theta(\gamma(z))^{-1}$.*
(4) *At $a = 1$, we have $\alpha_1(\gamma(z)) = \eta(\gamma(\bar{z}))$.*

PROOF. Part (1), (2), (3) is clear. Part (4) follows from the fact that the automorphism $f_1(z) = \frac{z-i}{z+i}$ satisfies $f_1(\bar{z}) = \frac{\bar{z}-i}{\bar{z}+i} = (\overline{f_1(z)})^{-1}$. □

6.3. Trivializations of fixed-points. Our aim is to trivialize the fixed-point of the family involution α_a. To that end, we will invoke the following lemma:

LEMMA 6.6. *Let $I \subset \mathbb{R}$ be an interval. Let $M \to I$ and $N \to I$ be two stratified real analytic submersions of real analytic Whitney stratified ind-varieties M and N (where I is equipped with the trivial stratification). Let $f : M \to N$ be a Thom map.*

(1) *Assume there is a compact group $H \times \mathbb{Z}/2$ acting real analytically on M such that the action preserves the stratifications and f is $H\times\mathbb{Z}/2$-invariant (where $H\times\mathbb{Z}/2$ acts trivially on N). Then the $\mathbb{Z}/2$-fixed-point ind-variety $M^{\mathbb{Z}/2}$ is Whitney stratified by the fixed-points of the strata and the induced map $f^{\mathbb{Z}/2} : M^{\mathbb{Z}/2} \to N$ is an H-equivariant Thom map.*
(2) *Assume further that f is ind-proper and there is an H-equivariant stratified trivialization of $f : M \to N$ over I, that is, there are stratified preserving homeomorphisms h_M and h_N fitting into a commutative diagram*

$$\begin{array}{ccc} M & \xrightarrow{h_M} & I \times M_0 \\ \downarrow{\scriptstyle f} & & \downarrow{\scriptstyle \mathrm{id} \times f_0} \\ N & \xrightarrow{h_N} & I \times N_0 \end{array}$$

that are real analytic on each stratum. Then there is an H-equivariant stratified trivialization of $f^{\mathbb{Z}/2} : M^{\mathbb{Z}/2} \to N$ that is real analytic on each stratum.

PROOF. Part (1) is proved in [**N**, Lemma 4.5.1]. For part (2), the H-equivariant stratified trivialization of $f : M \to N$ provides a horizontal lift of the coordinate vector field ∂_t on I to a continuous H-invariant vector field v on M that is tangent to and real analytic along each stratum. Let w be the average of v with respect to the $\mathbb{Z}/2\mathbb{Z}$-action. As f is ind-proper and the $\mathbb{Z}/2\mathbb{Z}$-action is real analytic, the vector field w is complete and the integral curves of w define an H-equivariant stratified trivialization of $f^{\mathbb{Z}/2} : M^{\mathbb{Z}/2} \to N$ over $[0,1]$ that is real analytic along each stratum. □

Now let us apply the above lemma to the map

$$QM_{G\times G,G}(\mathcal{Z}'/\mathbb{B}', \sigma, \xi)_{\mathbb{R}} \longrightarrow \mathbb{B}'$$

$$(\text{resp.}\quad QM^0_{G\times G,G}(\mathcal{Z}/\mathbb{B}, \sigma, \xi)_{\mathbb{R}} \longrightarrow \mathbb{B})$$

with the stratifications $\{\mathcal{S}^{\lambda_\mathfrak{p}}_{\mathcal{Z},\mathbb{R},\mathbb{B}'}\}$ and $\{(\mathbb{B}')^\mathfrak{p} = (0,1] \times \mathbb{R}^\mathfrak{p}\}$ (resp. $\{\mathcal{S}^{\lambda_\mathfrak{p},0}_{\mathcal{Z},\mathbb{R},\mathbb{B}}\}$ and $\{(\mathbb{B}^\mathfrak{p} = [0,1] \times \mathbb{R}^\mathfrak{p}\})$. We will consider the $H \times \mathbb{Z}/2$-action given by $K_c \times \langle \eta_\mathcal{Z} \rangle$.

PROPOSITION 6.7.

(1) *There is a K_c-equivariant topological trivialization of the fixed-points of $\eta_\mathcal{Z}$ of the map*

$$QM_{G\times G,G}(\mathcal{Z}'/\mathbb{B}', \sigma, \xi)_\mathbb{R} \longrightarrow \mathbb{B}'$$

over $I = (0,1]$.

(2) *There is a K_c-equivariant topological trivialization of the fixed-points of $\eta_\mathcal{Z}$ of the map*

$$QM^0_{G\times G,G}(\mathcal{Z}/\mathbb{B}, \sigma, \xi)_\mathbb{R} \longrightarrow \mathbb{B}$$

over $I = [0,1]$.

PROOF. For part (1), by Proposition 6.4, there is a K_c-equivariant stratified trivialization of

$$QM_{G\times G,G}(\mathcal{Z}'/\mathbb{B}', \sigma, \xi)_\mathbb{R} \longrightarrow \mathbb{B}' = (0,1] \times \mathbb{R}^m \tag{6.8}$$

Applying Lemma 6.6 with $H \times \mathbb{Z}/2 = K_c \times \langle \eta_\mathcal{Z} \rangle$, we obtain part (1).

For part (2), by Proposition 6.4, there is a K_c-equivariant stratified trivialization of

$$QM^0_{G\times G,G}(\mathcal{Z}/\mathbb{B}, \sigma, \xi)_\mathbb{R} \longrightarrow \mathbb{B} = [0,1] \times \mathbb{R}^m \tag{6.9}$$

Following the proof of Lemma 6.6, consider the averaged vector field w with respect to the $\mathbb{Z}/2\mathbb{Z}$-action given by $\langle \eta_\mathcal{Z} \rangle$. We claim that w is complete, hence the integral curves of w provide the desired trivialization.

To prove the claim, observe that, over the open locus $\mathbb{B}' \subset \mathbb{B}$ in the base, the K_c-equivariant trivialization in (6.9) extends to a K_c-equivariant trivialization of the ind-proper family in (6.8).

Thus for any $b = (a, z_1, \ldots, z_m) \in \mathbb{B}'$, any integral curve $p(t)$ for w with initial point $p(a) \in QM^0_{G\times G,G}(\mathcal{Z}'/\mathbb{B}', \sigma, \xi, b)_\mathbb{R}$ exists for $t \geq a$. Together with the local existence of integral curves with initial point in the special fiber $QM^0_{G\times G,G}(\mathcal{Z}'/\mathbb{B}', \sigma, \xi, b)_\mathbb{R}$, $b = (0, z_1, \ldots, z_m)$, this implies $p(t)$ exists for all $t \in [0,1]$. Hence w is complete and we have proved the claim. □

6.4. Real and symmetric spherical strata. Let us summarize here the results obtained by the preceding considerations.

Recall we write $\mathcal{G} \simeq \mathrm{Gr}^{(m),0} \times_{(\mathbb{P}^1)^m} \mathbb{R}^m \to \mathbb{R}^m$ for the group ind-scheme over $\mathbb{R}^m$ whose fiber over $(z_1, \ldots, z_m) \in \mathbb{R}^m$ is the group of maps $\gamma : \mathbb{P}^1 \setminus \{z_1, \ldots, z_m\} \to G$ such that $\gamma(\infty) = e$. Recall the transported spherical strata $\mathcal{G}^{\lambda_\mathfrak{p}} = \mathcal{G} \cap \mathcal{S}^{\lambda_\mathfrak{p}}$.

The isomorphism $p_\zeta : [0,1] \times \mathcal{G} \simeq QM^0_{G\times G,G}(\mathcal{Z}/\mathbb{B}, \sigma, \xi)_\mathbb{R}$ in (6.6) together with Theorem 6.5 and Proposition 6.7 immediately imply:

THEOREM 6.8. *There is a K_c-equivariant stratified homeomorphism between the fixed-points of η and* $\mathrm{inv} \circ \theta$ *on $\mathcal{G}$ compatibile with projections to $\mathbb{R}^m$. The homeomorphism restricts to a K_c-equivariant real analytic isomorphism between the fixed-points of η and* $\mathrm{inv} \circ \theta$ *on $\mathcal{G}^{\lambda_\mathfrak{p}}$.*

Observe that the fixed-points $\mathcal{G}^\eta$ coincide with the group ind-scheme $\mathcal{G}_\mathbb{R} \to \mathbb{R}^m$ of a point $(z_1, \ldots, z_m) \in \mathbb{R}^m$ and a map $\gamma : \mathbb{P}^1 \setminus \{z_1, \ldots, z_m\} \to G$, such that $\gamma(\mathbb{P}^1(\mathbb{R}) \setminus \{z_1, \ldots, z_m\}) \subset G_\mathbb{R}$ and $\gamma(\infty) = e$. Denote by $\mathcal{G}^{\lambda_\mathfrak{p}}_\mathbb{R} = \mathcal{G}^{\lambda_\mathfrak{p}} \cap \mathcal{G}_\mathbb{R}$ its spherical strata. Similarly, observe that the fixed-points $(\mathcal{G})^{\mathrm{inv}\circ\theta}$ coincides with the space $\mathcal{X} \to \mathbb{R}^m$ of a point $(z_1, \dots, z_m) \in \mathbb{R}^m$ and a map $\gamma : \mathbb{P}^1 \setminus \{z_1, \dots, z_m\} \to X \subset G$ such that $\gamma(\infty) = e$. Denote by $\mathcal{X}^{\lambda_{\mathfrak{p}}} = \mathcal{G}^{\lambda_{\mathfrak{p}}} \cap \mathcal{X}$ its spherical strata.

We can restate the above theorem in the form:

THEOREM 6.9. *There is a K_c-equivariant stratified homeomorphism*

$$\mathcal{G}_{\mathbb{R}} \xrightarrow{\sim} \mathcal{X} \tag{6.10}$$

fitting into the diagram

$$\begin{array}{ccc} \mathcal{G}_{\mathbb{R}} & \xrightarrow{\sim} & \mathcal{X} \\ \downarrow & & \downarrow \\ \mathbb{R}^m & \xrightarrow{\mathrm{id}} & \mathbb{R}^m \end{array}$$

that restricts to real analytic isomorphisms on strata

$$\mathcal{G}_{\mathbb{R}}^{\lambda_{\mathfrak{p}}} \xrightarrow{\sim} \mathcal{X}^{\lambda_{\mathfrak{p}}} \tag{6.11}$$

where $\lambda_{\mathfrak{p}} : \mathfrak{p} \to \mathcal{L}^+$ (for the subset $\mathcal{L}^+ \subset \Lambda_A^+$ defined in Section 2.1) with $|\lambda_{\mathfrak{p}}| \in R_G^+$.

REMARK 6.10. Since the stratum $\mathcal{G}_{\mathbb{R}}^{\lambda_{\mathfrak{p}}} = S_{\mathbb{R}}^{\lambda_{\mathfrak{p}}} \cap \mathcal{G}_{\mathbb{R}}$ is the intersection of the real spherical stratum $S_{\mathbb{R}}^{\lambda_{\mathfrak{p}}}$ in the real Beilinson-Drinfeld grassmannian $\mathrm{Gr}_{\mathbb{R}}^{(m)}$ with the open cospherical stratum $\mathcal{G}_{\mathbb{R}}$, the above theorem implies that the singularities of the closures of the real spherical strata $\overline{S_{\mathbb{R}}^{\lambda_{\mathfrak{p}}}}$ are *locally* homeomorphic to complex algebraic varieties. However, one should note that the closures $\overline{S_{\mathbb{R}}^{\lambda_{\mathfrak{p}}}}$ are not in general *globally* homeomorphic to complex projective varieties. For example, for the rank one group $G_{\mathbb{R}} = \mathrm{SL}_2(\mathbb{H})$, and the generator $\lambda \in \Lambda_A^+$ (i.e., $m = 1$ and $\lambda = \lambda^{\mathfrak{p}} : \mathfrak{p} = \{1\} \to \Lambda_A^+$), one finds that $\overline{S_{\mathbb{R}}^{\lambda}} \subset \mathrm{Gr}_{\mathbb{R}}$ is the one-point compactification of the cotangent bundle $T^*\mathbb{HP}^1$ of the quaternionic projective line. Thus its intersection cohomology Poincaré polynomial is $1+t^4+t^8$ so does not satisfy the Hard Lefschetz Theorem.

6.5. Kostant-Sekiguchi homeomorphism for GL_n. Here we explain how the prior construction recover Theorem 1.1 and Theorem 1.2 in [**CN2**] but by a completely different argument.[3]

Consider the case when $G = \mathrm{GL}_n(\mathbb{C})$ with real form $G_{\mathbb{R}} = \mathrm{GL}_n(\mathbb{R})$. We have $K = \mathrm{O}_n(\mathbb{C})$ and $K_c = \mathrm{O}_n(\mathbb{R})$ the complex and real orthogonal groups respectively. The conjugation η and involution θ on $\mathrm{GL}_n(\mathbb{C})$ are given by $\eta(M) = \overline{M}$ and $\theta(M) = (M^t)^{-1}$. Let $\mathfrak{g}_n = \mathfrak{gl}_n(\mathbb{C})$ be the Lie algebra of $G = \mathrm{GL}_n(\mathbb{C})$, $\mathfrak{t}_n \subset \mathfrak{g}_n$ the subspace of diagonal matrices, and $\mathfrak{c}_n = \mathfrak{t}_n /\!/ \mathrm{S}_n$ the quotient of $\mathfrak{t}_n$ by the symmetric group S_n on n letters. Let $\chi : \mathfrak{g}_n \to \mathfrak{g}_n /\!/ \mathrm{GL}_n(\mathbb{C}) \simeq \mathfrak{c}_n$ be the Chevalley map.

As observed by G. Lusztig and B.C. Ngo, for any $(M, z_1, \dots, z_n) \in \mathfrak{g}_n \times_{\mathfrak{c}_n} \mathfrak{t}_n(\mathbb{R})$, the formula $\gamma := \mathrm{id} - z^{-1}M$ defines a map from $\gamma : \mathbb{P}^1 \setminus \{z_1, .., z_n\} \to \mathrm{GL}_n(\mathbb{C})$ satisfying $\gamma(\infty) = \mathrm{id}$. Indeed, the collection $\{z_1, .., z_n\}$ is the set of eigenvalues of M and hence the matrix $\mathrm{id} - z^{-1}M$ is invertible for $z \in \mathbb{P}^1 \setminus \{z_1, .., z_n\}$. Thus we have an embedding

$$\mathfrak{g}_n \times_{\mathfrak{c}_n} \mathfrak{t}_n(\mathbb{R}) \longrightarrow \mathcal{G} \simeq \mathrm{Gr}^{(n),0} \times_{\mathbb{P}^1} \mathbb{R}^n$$

[3] The argument in [**CN2**] used quiver varieties and hyperkähler rotations.

given by

$$(M, z_1, \ldots, z_n) \to (1 - z^{-1}M, z_1, \ldots, z_n)$$

compatible with the natural projection maps to $\mathfrak{t}_n(\mathbb{R}) \simeq \mathbb{R}^n$. Moreover, by [**Ngo**, Lemma 2.3.1], the image of $\mathfrak{g}_n \times_{\mathfrak{c}_n} \mathfrak{t}_n(\mathbb{R})$ in $\mathcal{G}$ is a union of strata and the restriction of those strata to the fibers of the projection $\mathfrak{g}_n \times_{\mathfrak{c}_n} \mathfrak{t}_n(\mathbb{R}) \to \mathfrak{t}_n(\mathbb{R})$ are unions of orbits under the natural adjoint action of $\mathrm{GL}_n(\mathbb{C})$ on $\mathfrak{g}_n \times_{\mathfrak{c}_n} \mathfrak{t}_n(\mathbb{R})$. Thus the family of involutions in Theorem 6.5 restricts to a family of involutions

$$\alpha_a : \mathfrak{g}_n \times_{\mathfrak{c}_n} \mathfrak{t}_n(\mathbb{R}) \to \mathfrak{g}_n \times_{\mathfrak{c}_n} \mathfrak{t}_n(\mathbb{R}), \quad a \in [0,1] \tag{6.12}$$

satisfying the following properties:

THEOREM 6.11. *The family of involutions* $\alpha_a : \mathfrak{g}_n \times_{\mathfrak{c}_n} \mathfrak{t}_n(\mathbb{R}) \to \mathfrak{g}_n \times_{\mathfrak{c}_n} \mathfrak{t}_n(\mathbb{R})$, $a \in [0,1]$ *satisfy the following:*

(1) *We have* $\mathrm{pr} \circ \alpha_a = \mathrm{pr} : \mathfrak{g}_n \times_{\mathfrak{c}_n} \mathfrak{t}_n(\mathbb{R}) \to \mathfrak{t}_n(\mathbb{R})$, *for all* $a \in [0,1]$.
(2) α_a *is* $\mathrm{O}_n(\mathbb{R})$*-equivariant and takes a* $\mathrm{GL}_n(\mathbb{C})$*-orbit real analytically to a* $\mathrm{GL}_n(\mathbb{C})$*-orbit.*
(3) *At* $a = 0$, *we have* $\alpha_0(M, z_1, \ldots, z_n) = (\overline{M}, z_1, \ldots, z_n)$.
(4) *At* $a = 1$, *we have* $\alpha_\infty(M, z_1, \ldots, z_n) = (M^t, z_1, \ldots, z_n)$.

PROOF. Only part (3) and (4) require proof, and they follow from the following identities: for $\gamma(z) = \mathrm{id} - z^{-1}M$ we have $\alpha_0(\gamma(z)) = \overline{(\gamma(\bar{z}))} = \overline{(\mathrm{id} - \bar{z}^{-1}M)} = \mathrm{id} - z^{-1}\overline{M}$ and $\alpha_1(\gamma(z)) = \theta(\gamma(z))^{-1} = (\mathrm{id} - z^{-1}M)^t = \mathrm{id} - z^{-1}M^t$. □

Let $\mathfrak{g}_n(\mathbb{R})$ be the space of real $n \times n$-matrices and let $\mathfrak{g}_n^{\mathrm{sym}}$ be the space of $n \times n$ complex symmetric matrices. Theorem 6.9 implies the following result which can be viewed as a lift of the well-known Kostant-Sekiguchi bijection between real and symmetric nilpotent orbits to a stratified homeomorphism:

THEOREM 6.12. *There is an* $\mathrm{O}_n(\mathbb{R})$*-equivariant homeomorphism*

$$\mathfrak{g}_n(\mathbb{R}) \times_{\mathfrak{c}_n} \mathfrak{t}_n(\mathbb{R}) \xrightarrow{\sim} \mathfrak{g}_n^{\mathrm{sym}} \times_{\mathfrak{c}_n} \mathfrak{t}_n(\mathbb{R}) \tag{6.13}$$

fitting into the diagram

$$\begin{array}{ccc} \mathfrak{g}_n(\mathbb{R}) \times_{\mathfrak{c}_n} \mathfrak{t}_n(\mathbb{R}) & \xrightarrow{\sim} & \mathfrak{g}_n^{\mathrm{sym}} \times_{\mathfrak{c}_n} \mathfrak{t}_n(\mathbb{R}) \\ \downarrow & & \downarrow \\ \mathfrak{t}_n(\mathbb{R}) & \xrightarrow{\mathrm{id}} & \mathfrak{t}_n(\mathbb{R}) \end{array}$$

that restricts to real analytic isomorphisms between $\mathrm{GL}_n(\mathbb{R})$ *and* $\mathrm{O}_n(\mathbb{C})$ *adjoint orbits on* $\mathfrak{g}_n(\mathbb{R}) \times_{\mathfrak{c}_n} \mathfrak{t}_n(\mathbb{R})$ *and* $\mathfrak{g}_n^{\mathrm{sym}} \times_{\mathfrak{c}_n} \mathfrak{t}_n(\mathbb{R})$.

REMARK 6.13. It follows from the construction that the family of involutions in (6.7) are in fact equivariant under the natural S_n-action on $\mathcal{G} \simeq \mathrm{Gr}^{(m),0} \times_{(\mathbb{P}^1)^m} \mathbb{R}^m$. Thus the involutions in (6.12) are also equivariant under the natural S_n-action on $\mathfrak{g}_n \times_{\mathfrak{c}_n} \mathfrak{t}_n(\mathbb{R})$. Hence Theorem 6.11 and Theorem 6.12 have a direct analogy for the quotient $\mathfrak{g}_n \times_{\mathfrak{c}_n} \mathfrak{t}_n(\mathbb{R}) /\!/ \mathrm{S}_n$.

On the other hand, the quotient $\mathfrak{g}_n \times_{\mathfrak{c}_n} \mathfrak{t}_n(\mathbb{R}) /\!/ \mathrm{S}_n$ is isomorphic to the subset $\mathfrak{g}_n' \subset \mathfrak{g}_n$ consisting of matrices with real eigenvalues. Thus in this way we recover Theorem 1.1 and Theorem 1.2 in [**CN2**], but by a completely different argument.

Acknowledgments

The authors would like to thank the generous organizers of the 2019 SE Lie Theory workshop at LSU where these results were presented.

T.-H. Chen would also like to thank the Institute of Mathematics Academia Sinica in Taipei for support, hospitality, and a nice research environment.

References

[CN1] T. H. Chen and D. Nadler, *Affine Matsuki correspondence for sheaves*, Preprint, arXiv:1805.06514.

[CN2] Tsao-Hsien Chen and David Nadler, *Real and symmetric matrices*, Duke Math. J. **172** (2023), no. 9, 1623–1672, DOI 10.1215/00127094-2022-0076. MR4608328

[N] David Nadler, *Perverse sheaves on real loop Grassmannians*, Invent. Math. **159** (2005), no. 1, 1–73, DOI 10.1007/s00222-004-0382-3. MR2142332

[Ngo] Báo Châu Ngô, *Faisceaux pervers, homomorphisme de changement de base et lemme fondamental de Jacquet et Ye* (French, with English and French summaries), Ann. Sci. École Norm. Sup. (4) **32** (1999), no. 5, 619–679, DOI 10.1016/S0012-9593(01)80002-1. MR1710755

School of Mathematics, University of Minnesota, Minneapolis, Vincent Hall, Minnesota, 55455

Email address: chenth@umn.edu

Department of Mathematics, UC Berkeley, Evans Hall, Berkeley, California 94720

Email address: nadler@math.berkeley.edu

Proceedings of Symposia in Pure Mathematics
Volume **108**, 2024
https://doi.org/10.1090/pspum/108/01960

Support varieties for Lie superalgebras in characteristic 2

Christopher M. Drupieski and Jonathan R. Kujawa

Dedicated to Brian Parshall with gratitude.

Abstract. This paper investigates cohomology and support varieties for Lie superalgebras and restricted Lie superalgebras over a field of characteristic 2. The existence of an underlying ordinary Lie algebra allows us to obtain results that are still open in odd characteristic, and also to establish results that have no non-super analogues in characteristic 2.

1. Introduction

1.1. Overview. In the mid-1980s, Friedlander and Parshall [**18, 19**] introduced support varieties to the representation theory of restricted Lie algebras. Through their work and the work of those who followed, support varieties for restricted Lie algebras are comparatively well-understood. Almost four decades later, support varieties and their generalizations remain a subject of active investigation in non-semisimple representation theory, and the work of Friedlander and Parshall continues to serve as a model for those hoping to bring their methods to other settings. Relevant to this paper, one such setting is that of Lie superalgebras.

Roughly, a 'super' version of a mathematical structure incorporates $\mathbb{Z}_2$-gradings on objects, together with $\pm$ signs that are inserted into the defining equations whenever elements of odd superdegree move past one another. Lie superalgebras arise in algebraic topology, spin representations of the symmetric group, mathematical physics, and elsewhere. Work by Boe, Kujawa, and Nakano [**7, 8**], the authors [**14–16**], Benson, Iyengar, Krause, and Pevtsova [**4, 5**], and others demonstrates that support varieties generalize in interesting and nontrivial ways to the super setting.

The $\pm$ signs that show up in super formulas are one of the defining features of the theory. However, when the ground field is of characteristic 2 the $\pm$ signs effectively disappear, and by forgetting the $\mathbb{Z}_2$-gradings one seems to obtain just a classical, non-super version of the object in question. Indeed, it is sometimes assumed that super-mathematics over fields of characteristic 2 is uninteresting. Nevertheless, Lie superalgebras over fields of characteristic 2 have appeared in the

---

2020 *Mathematics Subject Classification.* Primary 17B56; Secondary 20G10, 17B50.

Key words and phrases. Lie superalgebra, restricted Lie superalgebra, support variety.

The first author was supported in part by Simons Collaboration Grant for Mathematicians No. 426905, and by a DePaul University Research Council paid faculty leave in Spring 2022.

The second author was supported in part by Simons Collaboration Grant for Mathematicians No. 525043.

literature in various forms at least as far back as the 1980s (see for example the 'adjusted Lie algebras' defined by Sjödin [**26**]), and there is currently an extensive literature on the classification of Lie superalgebras over fields of characteristic 2 (see for example [**10**] and the references therein). The current paper hopefully adds additional evidence against the assumption that this topic is uninteresting.

The goal of this paper is to study support varieties for Lie superalgebras (both restricted and un-restricted) over fields of characteristic 2. As we will see, the overlap between the super and non-super theories in characteristic 2 is a feature, not a bug. One can use this overlap as leverage to prove results for the super-theory that are merely conjectural in odd characteristics, as well as results that have no non-super analogues in characteristic 2. This paper thus exists at the intersection of topics that are well-understood (support varieties for restricted Lie algebras) and topics that are not (support varieties for graded objects).

1.2. Main results. We begin in Section 2 by recalling the definition of a Lie superalgebra over a field of arbitrary characteristic. Roughly, a Lie superalgebra in characteristic 2 is a $\mathbb{Z}_2$-graded ordinary Lie algebra $L = L_{\overline{0}} \oplus L_{\overline{1}}$, together with a distinguished quadratic operator $q : L_{\overline{1}} \to L_{\overline{0}}$. The quadratic operator effectively serves as a replacement for the map given in other characteristics by $x \mapsto \frac{1}{2}[x,x]$; see Remark 2.2. A representation of a Lie superalgebra in characteristic 2 is then a $\mathbb{Z}_2$-graded representation of the underlying ordinary Lie algebra $\overline{L}$ that is 'partially restricted' to ensure compatibility with the quadratic operator.

A restricted Lie superalgebra is a Lie superalgebra L equipped with a p-operation $x \mapsto x^{[p]}$ on $L_{\overline{0}}$ that makes $L_{\overline{0}}$ into an ordinary restricted Lie algebra, and that makes $L_{\overline{1}}$ into a restricted $L_{\overline{0}}$-module. In Lemma 2.14 we observe that, given a restricted Lie superalgebra over a field of characteristic $p = 2$, the 2-operation $(-)^{[2]} : L_{\overline{0}} \to L_{\overline{0}}$ and the quadratic operator $q : L_{\overline{1}} \to L_{\overline{0}}$ can be combined to define a 2-operation $z \mapsto z^{\{2\}}$ on all of L,

$$z = z_{\overline{0}} + z_{\overline{1}} \quad \mapsto \quad z^{\{2\}} := (z_{\overline{0}})^{[2]} + q(z_{\overline{1}}) + [z_{\overline{1}}, z_{\overline{0}}],$$

which then makes the underlying ordinary Lie algebra $\overline{L}$ of L into a restricted Lie algebra.[1] Moreover, this restricted structure on $\overline{L}$ then induces a canonical identification between $V(L)$, the restricted enveloping superalgebra of the restricted Lie superalgebra L, and $V(\overline{L})$, the restricted enveloping algebra of the ordinary restricted Lie algebra $\overline{L}$.

1.2.1. *Restricted Lie superalgebras.* In Section 3 we explore in detail the implications for support varieties of the canonical identification $V(L) = V(\overline{L})$, for L a finite-dimensional restricted Lie superalgebra over a (for simplicity, algebraically closed) field k of characteristic 2. The cohomology ring $\mathrm{H}^\bullet(V(L), k)$ is a finitely-generated commutative k-algebra that is naturally graded by $\mathbb{Z} \times \mathbb{Z}_2$; the first component of the grading is the 'external' cohomological grading and the second component is the 'internal' grading induced by the superalgebra structure of $V(L)$. Write $|V(L)|$ for the maximal ideal spectrum of the k-algebra $\mathrm{H}^\bullet(V(L), k)$. We

[1]This observation has doubtlessly been made before and is presumably well-known to characteristic 2 experts.

show in (3.9) that $|V(L)|$ identifies with the restricted nullcone of $\overline{L}$:

$$|V(L)| \simeq \mathcal{N}_1(\overline{L}) = \{z \in \overline{L} : z^{\{2\}} = 0\}$$
$$= \{z_{\overline{0}} + z_{\overline{1}} \in L : (z_{\overline{0}})^{[2]} + q(z_{\overline{1}}) = 0 \text{ and } [z_{\overline{1}}, z_{\overline{0}}] = 0\}.$$

This extends to characteristic $p = 2$ a description previously obtained by the authors for restricted Lie superalgebras in characteristics $p \geq 3$; cf. [**14**, Proposition 5.4.4].

Given a finite-dimensional $V(L)$-supermodule M, let $I(M)$ be the annihilator ideal for the cup product action of $\mathrm{H}^\bullet(V(L), k)$ on $\mathrm{Ext}^\bullet_{V(L)}(M, M)$. The support variety of M, denoted $|V(L)|_M$, is the Zariski closed subset of $|V(L)|$ defined by the ideal $I(M)$. Each nonzero element $z \in \mathcal{N}_1(\overline{L})$ generates a k-subalgebra of $V(L) = V(\overline{L})$ of the form $k[z]/(z^2)$. Appealing to support variety theory for ordinary restricted Lie algebras, the support variety $|V(L)|_M$ then admits the following "rank variety" description:

$$|V(L)|_M \simeq \{z \in \mathcal{N}_1(\overline{L}) : M \text{ is not free as a } k[z]/(z^2)\text{-module}\} \cup \{0\}. \tag{1.1}$$

This immediately implies that support varieties for $V(L)$-supermodules satisfy the Tensor Product Property (Proposition 3.2). In Proposition 3.4, we reformulate the condition that M is not free as a $k[z]/(z^2)$-module in terms of the projective dimension of a pullback of M to a hypersurface ring. As discussed in Remark 3.5, this alternate characterization of $|V(L)|_M$ can be interpreted as confirming, for characteristic 2, a conjecture we previously made for the support varieties of restricted Lie superalgebras (and more generally, for arbitrary infinitesimal supergroup schemes) over fields of odd characteristic.

Now let $\mathsf{st\text{-}smod}_{V(L)}$ be the stable category of finite-dimensional $V(L)$-supermodules. The Hopf algebra structure on $V(L)$ makes $\mathsf{st\text{-}smod}_{V(L)}$ into a tensor triangulated category. Following Balmer, one can use the tensor triangulated structure to define the spectrum $\mathrm{Spc}(\mathsf{st\text{-}smod}_{V(L)})$ of $\mathsf{st\text{-}smod}_{V(L)}$ and the support function

$$\mathrm{supp}' : \mathsf{st\text{-}smod}_{V(L)} \to 2^{\mathrm{Spc}(\mathsf{st\text{-}smod}_{V(L)})}.$$

Here $2^{\mathrm{Spc}(\mathsf{st\text{-}smod}_{V(L)})}$ denotes the power set of $\mathrm{Spc}(\mathsf{st\text{-}smod}_{V(L)})$. Given its abstract definition, it is not surprising that it is desirable to have a more concrete description of the Balmer spectrum and support function.

Let $X_s = \mathrm{Proj}_s(A)$ be the set of all bi-homogeneous ideals in $A = \mathrm{H}^\bullet(V(L), k)$ (i.e., ideals that are homogeneous with respect to the $\mathbb{Z} \times \mathbb{Z}_2$-grading) that are prime among the bi-homogeneous ideals, and which are properly contained in the maximal ideal $\mathrm{H}^{>0}(V(L), k)$ of elements of positive cohomological degree. We view X_s as a topological space via the Zariski topology. We note that X_s is a coarser space than the usual projective spectrum $\mathrm{Proj}(A)$ (which consists of prime ideals in $\mathrm{H}^\bullet(V(L), k)$ which are homogeneous with respect to just the $\mathbb{Z}$-component of the grading).

In Section 3.4 we show that X_s is a Zariski space and that there is a support function

$$\mathrm{supp}_s : \mathsf{st\text{-}smod}_{V(L)} \to 2^{X_s}.$$

Applying work of Dell'Ambrogio [**12**] and of Benson, Iyengar, Krause, and Pevtsova [**3**], we show in Theorem 3.10 that there is a canonical homeomorphism

$$\mathrm{Proj}_s(A) \simeq \mathrm{Spc}(\mathsf{st\text{-}smod}_{V(L)})$$

that identifies the two support functions supp' and supp_s. This gives a non-categorical realization of the Balmer spectrum and support function. We further deduce that the pair $(\mathrm{Proj}_s(A), \mathrm{supp}_s)$ provides a classification of the thick tensor ideal subcategories of $\mathsf{st\text{-}smod}_{V(L)}$. Concretely, this implies that if M and N are two finite-dimensional $V(L)$-supermodules, then M can be obtained from N by a finite sequence of the standard operations available in $\mathsf{st\text{-}smod}_{V(L)}$ (direct sums and summands, tensor products, extensions, syzygies) if and only if $|V(L)|_M \subseteq |V(L)|_N$. Thus the question of whether one supermodule can be built from another can be reformulated into a statement about their support varieties, which in turn is reduced to a linear-algebraic problem by (1.1).

1.2.2. *Lie superalgebras.* In Section 4 we turn to the case of an arbitrary finite-dimensional Lie superalgebra L over k. Again, the cohomology ring $\mathrm{H}^\bullet(L,k) = \mathrm{H}^\bullet(U(L),k)$ is a finitely-generated commutative k-algebra graded by $\mathbb{Z} \times \mathbb{Z}_2$, and one can define support varieties for finite-dimensional L-supermodules. Writing $|U(L)|$ for the maximal ideal spectrum of $\mathrm{H}^\bullet(L,k)$, we show in Theorem 4.5 that $|U(L)|$ identifies with the odd nullcone of L:[2]

$$|U(L)| \simeq \mathcal{N}_{\mathrm{odd}}(L) := \{x \in L_{\overline{1}} : q(x) = 0\}.$$

Likewise, Theorem 4.6 shows that the support variety of a finite-dimensional L-supermodule M admits a rank variety description:

$$|U(L)|_M \simeq \{x \in \mathcal{N}_{\mathrm{odd}}(L) : M \text{ is not free as a } k[x]/(x^2)\text{-supermodule}\} \cup \{0\}.$$

This description implies that the support varieties of finite-dimensional L-supermodules satisfy the Tensor Product Property (Corollary 4.7). Then Theorem 4.8 shows that $|U(L)|_M = \{0\}$ if and only if M has finite projective dimension as a $U(L)$-supermodule. In particular, $\mathcal{N}_{\mathrm{odd}}(L) = \{0\}$ if and only if $U(L)$ is of finite global dimension. This provides for $p = 2$ a new proof of a theorem of Bøgvad [**9**] on graded Lie algebras of finite global dimension.

1.3. Future directions. Since Premet's proof of the Kac–Weisfieler conjecture [**25**], p-divisibility of a module's dimension has been known to be closely related to the dimension of the module's support variety. Some divisibility results along these lines are known in super-representation theory (see, e.g., [**6**, **30**]), but much remains unknown, however, and it would be interesting to improve upon these results.

It would also be interesting to compute, in all positive characteristics, the Balmer spectrum for the stable module category of finite-dimensional $V(L)$-supermodules, and for the singularity category of finite-dimensional $U(L)$-supermodules. More generally, there are conjectures and questions for Lie superalgebras and infinitesimal supergroup schemes that the present paper addresses, but which remain open in odd characteristics. For $V(L)$ we were able to take advantage of coincidences that happen only in characteristic 2. In odd characteristic new ideas will be required for both superalgebras.

Finally, interesting alternate notions of super-mathematics in characteristic 2 were proposed in [**23**, **29**]. It would be interesting to consider the questions covered in this paper within that framework.

---

[2]There is also a Frobenius twist involved in the identification, which we ignore for the sake of the introduction.

1.4. Conventions. Throughout, k will be a field of positive characteristic p (usually $p = 2$). All vector spaces will be k-vector spaces, and all unadorned tensor products will denote tensor products over k. Let $\mathbb{N} = \{0, 1, 2, 3, \dots\}$ be the set of non-negative integers.

Set $\mathbb{Z}_2 = \mathbb{Z}/2\mathbb{Z} = \{\overline{0}, \overline{1}\}$. Following the literature, we use the prefix 'super' to indicate that an object is $\mathbb{Z}_2$-graded. We denote the decomposition of a vector superspace into its $\mathbb{Z}_2$-homogeneous components by $V = V_{\overline{0}} \oplus V_{\overline{1}}$, calling $V_{\overline{0}}$ and $V_{\overline{1}}$ the even and odd subspaces of V, respectively, and writing $\overline{v} \in \mathbb{Z}_2$ to denote the superdegree of a homogeneous element $v \in V_{\overline{0}} \cup V_{\overline{1}}$. When written without additional adornment, we consider the field k to be a superspace concentrated in even superdegree. Whenever we state a formula in which homogeneous degrees of elements are specified, we mean that the formula is true as written for homogeneous elements, and that it extends linearly to non-homogeneous elements.

2. Definitions

2.1. Lie superalgebras.

DEFINITION 2.1. Let k be a field (of any characteristic). A Lie superalgebra over k is a $\mathbb{Z}_2$-graded k-vector space $L = L_{\overline{0}} \oplus L_{\overline{1}}$, together with an even (i.e., $\mathbb{Z}_2$-degree preserving) bilinear map $[\,,\,] : L \otimes L \to L$ and a quadratic operator $q : L_{\overline{1}} \to L_{\overline{0}}$ (i.e., a function such that $q(\alpha \cdot y) = \alpha^2 \cdot q(y)$ for all $\alpha \in k$ and $y \in L_{\overline{1}}$), such that the following axioms hold:

(L1) $[x, y] = -(-1)^{\overline{x} \cdot \overline{y}}[y, x]$ for all homogeneous $x, y \in L$.
(L2) $[x, [y, z]] = [[x, y], z] + (-1)^{\overline{x} \cdot \overline{y}}[y, [x, z]]$ for all homogeneous $x, y, z \in L$ (Jacobi identity).
(L3) $[x, x] = 0$ for all $x \in L_{\overline{0}}$.
(L4) $[y, [y, y]] = 0$ for all $y \in L_{\overline{1}}$.
(L5) $[x, y] = q(x + y) - q(x) - q(y)$ for all $x, y \in L_{\overline{1}}$.
(L6) $[y, [y, z]] = [q(y), z]$ for all $y \in L_{\overline{1}}$ and $z \in L$.

REMARK 2.2. Let L be a Lie superalgebra and $y \in L_{\overline{1}}$. Then (L5) asserts that

$$[y, y] = q(y+y) - q(y) - q(y) = q(2 \cdot y) - 2 \cdot q(y) = 4 \cdot q(y) - 2 \cdot q(y) = 2 \cdot q(y). \tag{2.1}$$

For $p \neq 2$, this shows that $q(y) = \frac{1}{2}[y, y]$, while for $p = 2$, it shows that $[y, y] = 0$ for all $y \in L_{\overline{1}}$. Thus for $p = 2$, if we forget the quadratic operator $q : L_{\overline{1}} \to L_{\overline{0}}$, the Lie bracket $[\,,\,] : L \otimes L \to L$ makes L into an ordinary Lie algebra, which we may denote $\overline{L}$.

REMARK 2.3. Depending on the characteristic p of the field k, some of the axioms of a Lie superalgebra are redundant:

(1) If $p \neq 2$, then (L3) is a consequence of (L1).
(2) If $p \neq 3$, then (L4) is a consequence of (L1) and (L2).
(3) Suppose $p \neq 2$. Then (L5) and (L6) follow from (L1), (L2), and the formula $q(y) = \frac{1}{2}[y, y]$. Conversely, if L is a k-superspace equipped with an even bilinear map $[\,,\,] : L \otimes L \to L$ satisfying (L1)–(L4), then $q : L_{\overline{1}} \to L_{\overline{0}}$ defined by $q(y) = \frac{1}{2}[y, y]$ satisfies (L5) and (L6).

It is necessary to assume (L3) when $p = 2$, and to assume (L4) when $p = 3$, to ensure that L can be embedded into its universal enveloping algebra $U(L)$; see [**2**, 1.1.10 and 3.2.3].

REMARK 2.4. For those who (like the authors of this paper) have spent most of their lives only considering Lie superalgebras over fields of characteristic $p \neq 2$, the inclusion of the quadratic operator $q : L_{\overline{1}} \to L_{\overline{0}}$ in Definition 2.1 may require explanation. Remarks 2.2 and 2.3 provide some motivation for why it is natural to include the quadratic operator in the definition of a Lie superalgebra. Another source of motivation is the fact that it permits a uniform, characteristic-free description of the Koszul resolution (Theorem 4.1). Here is yet another motivating observation:

Let k be a field (of any characteristic), let A be a k-superalgebra, and let $D \in \mathrm{Der}_k(A, A)_{\overline{1}}$ be an odd superderivation of A. Thus D is an odd linear map such that for all homogeneous elements $a, b \in A$, one has $D(ab) = D(a) \cdot b + (-1)^{\overline{a}} a \cdot D(b)$. Then

$$\begin{aligned} D^2(ab) &= D\left(D(a) \cdot b + (-1)^{\overline{a}} a \cdot D(b)\right) \\ &= D^2(a) \cdot b + (-1)^{\overline{a}+\overline{1}} D(a) \cdot D(b) + (-1)^{\overline{a}} D(a) \cdot D(b) + a \cdot D^2(b) \\ &= D^2(a) \cdot b + a \cdot D^2(b), \end{aligned}$$

so $D^2 \in \mathrm{Der}_k(A, A)_{\overline{0}}$. Now, one of the ways to describe the Lie superalgebra $\mathrm{Lie}(G)$ of an affine algebraic supergroup G is as the space of left-invariant superderivations of its coordinate superalgebra $k[G]$.[3] The preceding calculation then implies that, in arbitrary characteristic, the squaring map $D \mapsto D^2$ defines a quadratic operator $q : \mathrm{Lie}(G)_{\overline{1}} \to \mathrm{Lie}(G)_{\overline{0}}$.

EXAMPLE 2.5. Let k be a field (of any characteristic), and let A be a k-superalgebra. Define an even bilinear map $[\,,\,] : A \otimes A \to A$ and a quadratic operator $q : A_{\overline{1}} \to A_{\overline{0}}$ by

- $[a, b] = ab - (-1)^{\overline{a} \cdot \overline{b}} ba$ for $a, b \in A$ homogeneous, and
- $q(a) = a^2$ for $a \in A_{\overline{1}}$.

These operations define a Lie superalgebra structure on the underlying k-superspace of A, which we may denote by $\mathrm{Lie}(A)$.

DEFINITION 2.6. Let k be a field (of any characteristic), let L be a Lie superalgebra over k, and let $T(L) = \bigoplus_{i \geq 0} L^{\otimes i}$ be the tensor (super)algebra on L. The universal enveloping algebra of L is the k-superalgebra $U(L) := T(L)/I$, where I is the ideal in $T(L)$ generated by the following elements:

- $x \otimes y - (-1)^{\overline{x} \cdot \overline{y}} y \otimes x - [x, y]$ for $x, y \in L$ homogeneous, and
- $y \otimes y - q(y)$ for $y \in L_{\overline{1}}$.

The proof of the following theorem is standard.

THEOREM 2.7 (Poincaré–Birkhoff–Witt Theorem for $U(L)$). *Let L be a Lie superalgebra over a field k (of any characteristic), and let $X = \{x_\lambda : \lambda \in \Omega\}$ be an ordered basis for L, with the ordering on basis elements denoted $\leq$. Then $U(L)$ admits a k-basis consisting of all monomials of the form $x_1 x_2 \cdots x_n$, where $n \in \mathbb{N}$, $x_i \in X$, $x_i \leq x_{i+1}$, and $x_i \neq x_{i+1}$ if $x_i \in L_{\overline{1}}$. In particular, the inclusion $L = L^{\otimes 1} \hookrightarrow T(L)$ induces an injective Lie superalgebra homomorphism $\iota : L \hookrightarrow \mathrm{Lie}(U(L))$.*

[3] Recall that, writing $\Delta : k[G] \to k[G] \otimes k[G]$ for the coproduct on $k[G]$, a superderivation $D : k[G] \to k[G]$ is left-invariant if $\Delta \circ D = (\mathrm{id} \otimes D) \circ \Delta$.

REMARK 2.8. Let L be a Lie superalgebra over a field k of characteristic p.

(1) The algebra $U(L)$ is a cocommutative Hopf superalgebra, with the subspace L consisting of primitive elements. If $p = 2$, then $U(L)$ is a Hopf algebra in the ordinary sense.

(2) Suppose $p = 2$, and let $\overline{L}$ be the ordinary ($\mathbb{Z}_2$-graded) Lie algebra obtained by forgetting the quadratic operator $q : L_{\overline{1}} \to L_{\overline{0}}$. Then

$$U(L) = U(\overline{L})/\langle y^2 - q(y) : y \in L_{\overline{1}}\rangle,$$

and the canonical quotient map $U(\overline{L}) \twoheadrightarrow U(L)$ is a Hopf (super)algebra homomorphism.

(3) Suppose $p = 2$. More-or-less by definition, a (left) L-supermodule is a (left) supermodule over the universal enveloping superalgebra $U(L)$. Because $y^2 - q(y) = 0$ in $U(L)$ for each $y \in L_{\overline{1}}$, but $y^2 - q(y) \neq 0$ in $U(\overline{L})$, this shows that the category of L-supermodules is not simply the category of ($\mathbb{Z}_2$-graded) $\overline{L}$-modules.

2.2. Restricted Lie superalgebras.

DEFINITION 2.9. Let k be a field of characteristic $p \geq 2$, and let L be a Lie superalgebra over k. We say that L is a restricted Lie superalgebra if there exists a map $x \mapsto x^{[p]}$ on $L_{\overline{0}}$ (called a *p-map* or a *p-operation*) such that the following are satisfied for all $x, y \in L_{\overline{0}}$:

(R1) $(\alpha \cdot x)^{[p]} = \alpha^p \cdot x^{[p]}$ for all $\alpha \in k$,

(R2) $\mathrm{ad}(x^{[p]}) = \mathrm{ad}(x)^p$

(R3) $(x+y)^{[p]} = x^{[p]} + y^{[p]} + \sum_{i=1}^{p-1} s_i(x,y)$, where $i \cdot s_i(x,y)$ is the coefficient of t^{i-1} in the formal expression $\mathrm{ad}(t \cdot x + y)^{p-1}(x)$.

In other words, L is a restricted Lie superalgebra if $L_{\overline{0}}$ is a restricted Lie algebra in the usual sense, and $L_{\overline{1}}$ is a restricted $L_{\overline{0}}$-module.

REMARK 2.10. For $p = 2$, the condition (R3) becomes

$$(x+y)^{[2]} = x^{[2]} + y^{[2]} + [y, x]. \tag{2.2}$$

EXAMPLE 2.11. Let k be a field of characteristic $p \geq 2$, and let A be a k-superalgebra. Then the map $a \mapsto a^p$ defines a p-operation on $A_{\overline{0}}$ that makes $\mathrm{Lie}(A)$ into a restricted Lie superalgebra.

DEFINITION 2.12. Let k be a field of characteristic $p \geq 2$, and let L be a restricted Lie superalgebra over k. The restricted enveloping algebra of L is the k-superalgebra

$$V(L) := U(L)/\langle x^p - x^{[p]} : x \in L_{\overline{0}}\rangle.$$

The proof of the following theorem is also standard.

THEOREM 2.13 (Poincaré–Birkhoff–Witt Theorem for $V(L)$). *Let L be a restricted Lie superalgebra over a field k of characteristic $p \geq 2$, and let $X = \{x_\lambda : \lambda \in \Omega\}$ be an ordered basis for L, with the ordering on basis elements denoted $\leq$. Then $V(L)$ admits a k-basis consisting of all monomials of the form $x_1^{\lambda_1} x_2^{\lambda_2} \cdots x_n^{\lambda_n}$, where $n \in \mathbb{N}$, $x_i \in X$, $x_1 < x_2 < \cdots < x_n$, $0 \leq \lambda_i < p$ if $x_i \in L_{\overline{0}}$, and $0 \leq \lambda_i < 2$ if $x_i \in L_{\overline{1}}$. The natural inclusion $L \hookrightarrow T(L)$ induces an injective homomorphism of restricted Lie superalgebras $\iota : L \hookrightarrow \mathrm{Lie}(V(L))$.*

Let G be an abelian group, written additively. Recall that a *G-graded Lie algebra* is a (ordinary) Lie algebra L together with a family of subspaces $(L_g)_{g\in G}$ of L such that $L = \bigoplus_{g\in G} L_g$ and $[L_g, L_h] \subseteq L_{g+h}$ for all $g, h \in G$. If L is a (ordinary) restricted Lie algebra with p-map $x \mapsto x^{[p]}$, then L is a *restricted G-graded Lie algebra* if the p-map satisfies $(L_g)^{[p]} \subseteq L_{pg}$ for all $g \in G$.

LEMMA 2.14. *Let L be a restricted Lie superalgebra over a field k of characteristic $p = 2$, with p-operation $x \mapsto x^{[2]}$ on $L_{\overline{0}}$, and let $\overline{L}$ be the ordinary Lie algebra over k obtained by forgetting the quadratic operator $q : L_{\overline{1}} \to L_{\overline{0}}$. Given $z \in L$, let $z = z_{\overline{0}} + z_{\overline{1}}$ be the decomposition of z into its even and odd parts. Then the map*

$$z \mapsto z^{\{2\}} := (z_{\overline{0}})^{[2]} + q(z_{\overline{1}}) + [z_{\overline{1}}, z_{\overline{0}}] \tag{2.3}$$

defines a p-operation on $\overline{L}$ that makes $\overline{L}$ into a restricted ($\mathbb{Z}_2$-graded) Lie algebra over k. The canonical quotient map $U(\overline{L}) \twoheadrightarrow U(L)$ induces a Hopf (super)algebra isomorphism $V(\overline{L}) \cong V(L)$.

PROOF. Let $x, y \in \overline{L}$, and let $x = x_{\overline{0}}+x_{\overline{1}}$ and $y = y_{\overline{0}}+y_{\overline{1}}$ be the decompositions of x and y into their $\mathbb{Z}_2$-homogeneous components. First, for $\alpha \in k$, one has

$$\begin{aligned}(\alpha \cdot x)^{\{2\}} &= (\alpha \cdot x_{\overline{0}} + \alpha \cdot x_{\overline{1}})^{\{2\}} \\ &= (\alpha \cdot x_{\overline{0}})^{[2]} + q(\alpha \cdot x_{\overline{1}}) + [\alpha \cdot x_{\overline{1}}, \alpha \cdot x_{\overline{0}}] \\ &= \alpha^2 \cdot (x_{\overline{0}})^{[2]} + \alpha^2 \cdot q(x_{\overline{1}}) + \alpha^2 \cdot [x_{\overline{1}}, x_{\overline{0}}] \\ &= \alpha^2 \cdot x^{\{2\}},\end{aligned}$$

so (R1) is satisfied by the map $z \mapsto z^{\{2\}}$. Second,

$$\begin{aligned}&\mathrm{ad}(x^{\{2\}})(y) \\ &= [x^{\{2\}}, y] \\ &= [(x_{\overline{0}})^{[2]} + q(x_{\overline{1}}) + [x_{\overline{1}}, x_{\overline{0}}], y] \\ &= [(x_{\overline{0}})^{[2]}, y] + [q(x_{\overline{1}}), y] + [[x_{\overline{1}}, x_{\overline{0}}], y] \\ &= [x_{\overline{0}}, [x_{\overline{0}}, y]] + [q(x_{\overline{1}}), y] + [[x_{\overline{1}}, x_{\overline{0}}], y] && \text{by (R2)} \\ &= [x_{\overline{0}}[x_{\overline{0}}, y]] + [x_{\overline{1}}, [x_{\overline{1}}, y]] + [[x_{\overline{1}}, x_{\overline{0}}], y] && \text{by (L6)} \\ &= [x_{\overline{0}}, [x_{\overline{0}}, y]] + [x_{\overline{1}}, [x_{\overline{1}}, y]] + [x_{\overline{1}}, [x_{\overline{0}}, y]] + [x_{\overline{0}}, [x_{\overline{1}}, y]] && \text{by (L2) and (L1)} \\ &= [x_{\overline{0}} + x_{\overline{1}}, [x_{\overline{0}} + x_{\overline{1}}, y]] \\ &= \mathrm{ad}(x_{\overline{0}} + x_{\overline{1}})^2(y),\end{aligned}$$

so (R2) is satisfied by the map $z \mapsto z^{\{2\}}$. Third,

$$
\begin{aligned}
&(x+y)^{\{2\}} \\
&\quad = ((x_{\bar 0}+y_{\bar 0})+(x_{\bar 1}+y_{\bar 1}))^{\{2\}} \\
&\quad = (x_{\bar 0}+y_{\bar 0})^{[2]}+q(x_{\bar 1}+y_{\bar 1})+[x_{\bar 1}+y_{\bar 1},x_{\bar 0}+y_{\bar 0}] \\
&\quad = (x_{\bar 0})^{[2]}+(y_{\bar 0})^{[2]}+[y_{\bar 0},x_{\bar 0}]+q(x_{\bar 1}+y_{\bar 1})+[x_{\bar 1}+y_{\bar 1},x_{\bar 0}+y_{\bar 0}] && \text{by (2.2)} \\
&\quad = (x_{\bar 0})^{[2]}+(y_{\bar 0})^{[2]}+[y_{\bar 0},x_{\bar 0}]+[x_{\bar 1},y_{\bar 1}]+q(x_{\bar 1})+q(y_{\bar 1}) \\
&\qquad\qquad +[x_{\bar 1}+y_{\bar 1},x_{\bar 0}+y_{\bar 0}] && \text{by (L5)} \\
&\quad = (x_{\bar 0})^{[2]}+(y_{\bar 0})^{[2]}+[y_{\bar 0},x_{\bar 0}]+[x_{\bar 1},y_{\bar 1}]+q(x_{\bar 1})+q(y_{\bar 1}) \\
&\qquad\qquad +[x_{\bar 1},x_{\bar 0}]+[x_{\bar 1},y_{\bar 0}]+[y_{\bar 1},x_{\bar 0}]+[y_{\bar 1},y_{\bar 0}] \\
&\quad = x^{\{2\}}+y^{\{2\}}+[y_{\bar 0},x_{\bar 0}]+[y_{\bar 1},x_{\bar 1}]+[y_{\bar 0},x_{\bar 1}]+[y_{\bar 1},x_{\bar 0}] \\
&\quad = x^{\{2\}}+y^{\{2\}}+[y_{\bar 0}+y_{\bar 1},x_{\bar 0}+x_{\bar 1}] \\
&\quad = x^{\{2\}}+y^{\{2\}}+[y,x]
\end{aligned}
$$

so (R3) is satisfied by the map $z \mapsto z^{\{2\}}$. Finally, let $j : U(\overline{L}) \twoheadrightarrow U(L)$ and $\pi : U(L) \twoheadrightarrow V(L)$ be the canonical quotient maps. For $z = z_{\bar 0} + z_{\bar 1} \in L$, in the algebra $U(\overline{L})$ one has

$$
\begin{aligned}
z^2 - z^{\{2\}} &= (z_{\bar 0}+z_{\bar 1})^2 - \big((z_{\bar 0})^{[2]} + q(z_1) + [z_{\bar 1}, z_{\bar 0}]\big) \\
&= \big((z_{\bar 0})^2 + z_{\bar 0}z_{\bar 1} + z_{\bar 1}z_{\bar 0} + (z_{\bar 1})^2\big) - \big((z_{\bar 0})^{[2]} + q(z_1) + [z_{\bar 1}, z_{\bar 0}]\big) \\
&= \big((z_{\bar 0})^2 - (z_{\bar 0})^{[2]}\big) + \big((z_{\bar 1})^2 - q(z_1)\big).
\end{aligned}
$$

This is a sum of a defining relation in $V(L)$ and a defining relation in $U(L)$, so the composite algebra homomorphism $\pi \circ j : U(\overline{L}) \twoheadrightarrow V(L)$ factors through a surjective homomorphism $V(\overline{L}) \twoheadrightarrow V(L)$, which can then be seen to be an injection (and hence an isomorphism) by Theorem 2.13. $\square$

REMARK 2.15. The converse of Lemma 2.14 is true. That is, let $\overline{L}$ be a restricted $\mathbb{Z}_2$-graded Lie algebra over a field of characteristic $p = 2$, with 2-map $z \mapsto z^{\{2\}}$. For $z \in \overline{L}_{\bar 1}$, set $q(z) = z^{\{2\}} \in \overline{L}_{2\cdot\bar 1} = \overline{L}_{\bar 0}$, and for $z \in \overline{L}_{\bar 0}$ set $z^{[2]} = z^{\{2\}} \in \overline{L}_{2\cdot\bar 0} = \overline{L}_{\bar 0}$. Then $q : \overline{L}_{\bar 1} \to \overline{L}_{\bar 0}$ defines a quadratic operator that makes the underlying $\mathbb{Z}_2$-graded Lie algebra of $\overline{L}$ into a Lie superalgebra, which we may denote L, and the map $z \mapsto z^{[2]}$ defines a p-operation on $L_{\bar 0}$ that makes L into a restricted Lie superalgebra.

3. Support for restricted Lie superalgebras in characteristic 2

3.1. Cohomology. Let L be a finite-dimensional restricted Lie superalgebra over a field k of characteristic $p = 2$, and let $V(L)$ be its restricted enveloping algebra. Let $\mathsf{sMod} = \mathsf{sMod}_{V(L)}$ be the category of left $V(L)$-supermodules, and let $\mathsf{smod} = \mathsf{smod}_{V(L)}$ be the subcategory of finite-dimensional $V(L)$-supermodules. An object in $\mathsf{sMod}_{V(L)}$ is a k-superspace M equipped with an even bilinear map $V(L) \otimes M \to M$ that makes M into a left $V(L)$-module. Given $V(L)$-supermodules M and N, one has

$$\mathrm{Hom}_{\mathsf{sMod}}(M, N) = \mathrm{Hom}_{\mathsf{sMod}}(M, N)_{\bar 0} \oplus \mathrm{Hom}_{\mathsf{sMod}}(M, N)_{\bar 1}$$

where $\mathrm{Hom}_{\mathsf{sMod}}(M,N)_{\overline{0}}$ (resp. $\mathrm{Hom}_{\mathsf{sMod}}(M,N)_{\overline{1}}$) is the subspace of all even (resp. odd) k-linear maps $f : M \to N$ such that $f(u.m) = u.f(m)$ for all $u \in V(L)$ and $m \in M$.[4] The category $\mathsf{sMod}_{V(L)}$ is not an abelian category (because the kernel or image of a non-homogeneous homomorphism need not be the direct sum of its even and odd subspaces), but the underlying even subcategory $(\mathsf{sMod}_{V(L)})_{\mathrm{ev}}$ of $\mathsf{sMod}_{V(L)}$, having the same objects as $\mathsf{sMod}_{V(L)}$ but only the even homomorphisms as morphisms, is an abelian category.

REMARK 3.1. For $p \neq 2$, the category $(\mathsf{sMod}_{V(L)})_{\mathrm{ev}}$ identifies with the category of left modules for the smash product algebra $V(L)\#k\mathbb{Z}_2$. Here $V(L)$ is considered as a $k\mathbb{Z}_2$-module algebra, with the nontrivial element of $\mathbb{Z}_2$ acting on $V(L)$ via the sign automorphism $u \mapsto (-1)^{\overline{u}}u$. This identification of categories no longer holds for $p = 2$, because then $k\mathbb{Z}_2$-modules no longer decompose into ± 1 eigenspaces, and hence you can no longer use the action of $k\mathbb{Z}_2$ to recover the superspace structure of modules, and you cannot use $k\mathbb{Z}_2$-equivariance of maps to ensure that you are only dealing with even $V(L)$-supermodule homomorphisms.

Each $V(L)$-supermodule is the image, via an even homomorphism, of a direct sum of copies of $V(L)$ and its parity shift $\Pi(V(L))$.[5] Then the category $(\mathsf{sMod}_{V(L)})_{\mathrm{ev}}$ has enough projectives, so given $V(L)$-supermodules M and N, we can define $\mathrm{Ext}^n_{\mathsf{sMod}}(M,N)$ to be the n-th right derived functor of the contravariant functor $\mathrm{Hom}_{\mathsf{sMod}}(-,N) : (\mathsf{sMod}_{V(L)})_{\mathrm{ev}} \to \mathsf{svec}_k$, i.e.,

$$\mathrm{Ext}^n_{\mathsf{sMod}_{V(L)}}(M,N) = \big(R^n \mathrm{Hom}_{\mathsf{sMod}_{V(L)}}(-,N)\big)(M).$$

The category $(\mathsf{sMod}_{V(L)})_{\mathrm{ev}}$ also has enough injectives: Given a $V(L)$-supermodule N, define $\varphi : N \to \mathrm{Hom}_k(V(L),N)$ by $\varphi(n) = \varphi_n$, where $\varphi_n(u) = (-1)^{\overline{u}\cdot\overline{n}}u.n$. Then φ is an even $V(L)$-supermodule homomorphism, where the left $V(L)$-module structure on $\mathrm{Hom}_k(V(L),N)$ comes from the right action of $V(L)$ on itself, and the Hom-tensor adjunction

$$\mathrm{Hom}_{V(L)}(M, \mathrm{Hom}_k(V(L),N)) \cong \mathrm{Hom}_k(V(L) \otimes_{V(L)} M, N) \cong \mathrm{Hom}_k(M,N)$$

implies that $\mathrm{Hom}_k(V(L),N)$ is injective in $(\mathsf{sMod}_{V(L)})_{\mathrm{ev}}$. In fact, $V(L)$ is itself an injective object in $(\mathsf{sMod}_{V(L)})_{\mathrm{ev}}$. This can be seen from (3.1) below, using the fact that $V(L) = V(\overline{L})$ is self-injective in the category of arbitrary $V(\overline{L})$-modules because $V(\overline{L})$ is a finite-dimensional Hopf algebra; cf. [**14**, Lemma 2.3.2] for $p \neq 2$. From this it follows that $(\mathsf{sMod}_{V(L)})_{\mathrm{ev}}$ is a Frobenius category.

Let $\mathsf{Mod} = \mathsf{Mod}_{V(L)}$ be the category of arbitrary (i.e., not necessarily $\mathbb{Z}_2$-graded) $V(L)$-modules. The forgetful functor, Forget : $\mathsf{sMod}_{V(L)} \to \mathsf{Mod}_{V(L)}$, which simply ignores the $\mathbb{Z}_2$-gradings on objects, defines a canonical identification[6]

$$\mathrm{Hom}_{\mathsf{sMod}}(M,N) = \mathrm{Hom}_{\mathsf{Mod}}(M,N) \quad \text{for all } M,N \in \mathsf{sMod}_{V(L)}. \tag{3.1}$$

[4]If $p \neq 2$, then $\mathrm{Hom}_{\mathsf{sMod}}(M,N)_{\overline{1}}$ consists of all odd k-linear maps $f : M \to N$ such that $f(u.m) = (-1)^{\overline{u}}u.f(m)$ for all homogeneous $u \in V(L)$ and all $m \in M$.

[5]For $p = 2$, the parity shift $\Pi(M)$ of a $V(L)$-supermodule M is the $V(L)$-supermodule that has the same underlying k-vector space as M, and the same action of $V(L)$ on that k-vector space, but with $\Pi(M)_{\overline{0}} = M_{\overline{1}}$ and $\Pi(M)_{\overline{1}} = M_{\overline{0}}$. For $p \neq 2$, there is more than one choice of parity change functor; see [**15**, §2.3].

[6]In general, the forgetful functor acts linearly on morphisms, with Forget$(f) = f$ if f is even, and Forget$(f) = \widetilde{f}$ if f is odd. Here $\widetilde{f} : M \to N$ is the k-linear map defined by $\widetilde{f}(m) = (-1)^{\overline{m}} \cdot f(m)$. For $p = 2$, this means that Forget$(f) = f$ for all $f \in \mathrm{Hom}_{\mathsf{sMod}}(M,N)$.

More generally, since each object in $\mathsf{sMod}_{V(L)}$ can be resolved in $(\mathsf{sMod}_{V(L)})_{\mathrm{ev}}$ by free $V(L)$-modules, it follows that the forgetful functor induces for each $n \in \mathbb{N}$ a canonical identification

$$\mathrm{Ext}^n_{\mathsf{sMod}}(M,N) = \mathrm{Ext}^n_{\mathsf{Mod}}(M,N) \quad \text{for all } M,N \in \mathsf{sMod}_{V(L)}. \tag{3.2}$$

From now on, we will simply write $\mathrm{Hom}_{V(L)}(M,N)$ and $\mathrm{Ext}^n_{V(L)}(M,N)$ to denote the common values indicated by (3.1) and (3.2).

The identification in (3.2) is also compatible with cup products. This can be seen using the description of cup products given, for example, in [**31**, §9.3]. The key point to check to ensure that this description makes sense for $\mathsf{sMod}_{V(L)}$ is that if P and Q are projective modules in $\mathsf{sMod}_{V(L)}$, then the tensor product $P \otimes Q$ with the diagonal $V(L)$-action is also projective in $\mathsf{sMod}_{V(L)}$. It suffices to assume that P is a free $V(L)$-supermodule, and then the proof of [**31**, Lemma 9.2.9] produces an isomorphism in $(\mathsf{sMod}_{V(L)})_{\mathrm{ev}}$ between $P \otimes Q$ and the free $V(L)$-supermodule $P \otimes Q_{triv}$. Here Q_{triv} denotes the $V(L)$-supermodule having the same underlying superspace as Q, but with the action of $V(L)$ defined by the augmentation map $\varepsilon : V(L) \to k$.

By Lemma 2.14, $V(L)$ identifies with $V(\overline{L})$, the restricted enveloping algebra of the ordinary restricted Lie algebra $\overline{L}$, whose underlying Lie algebra is obtained by forgetting the quadratic operator $q : L_{\overline{1}} \to L_{\overline{0}}$, and whose p-operation $z \mapsto z^{\{2\}}$ is defined by (2.3). Then the module categories $\mathsf{Mod}_{V(L)}$ and $\mathsf{Mod}_{V(\overline{L})}$ are the same, so by (3.2) one gets a canonical identification

$$\mathrm{Ext}^\bullet_{V(L)}(M,N) = \mathrm{Ext}^\bullet_{V(\overline{L})}(M,N) \quad \text{for all } M,N \in \mathsf{sMod}_{V(L)}, \tag{3.3}$$

which is again evidently compatible with cup products. To emphasize, the left-hand side of (3.3) represents the value calculated in the category of $V(L)$-supermodules, while the right-hand side represents the value calculated without regard for the $\mathbb{Z}_2$-gradings.

3.2. Support. Given a Hopf superalgebra A over k, the cohomology ring $\mathrm{H}^\bullet(A,k) = \mathrm{Ext}^\bullet_A(k,k)$ is a graded-commutative superalgebra [**14**, Corollary 2.3.6]. For $p = 2$, this means that $\mathrm{H}^\bullet(A,k)$ is an ordinary commutative ring. We define the cohomological spectrum of A, denoted $|A|$, to be

$$|A| = \mathrm{Spec}\,\mathrm{H}^\bullet(A,k), \tag{3.4}$$

the prime ideal spectrum of $\mathrm{H}^\bullet(A,k)$, considered as a topological space via the Zariski topology. If the field k is algebraically closed, one can take the maximal ideal spectrum instead, and then $|A|$ will be an affine algebraic variety (assuming that $\mathrm{H}^\bullet(A,k)$ is a finitely-generated k-algebra, which it is for the examples we consider) rather than an affine scheme. We will adopt this latter, more classical perspective later in Section 4 when discussing support varieties for $U(L)$.

Given A-supermodules M and N, let $I_A(M,N)$ be the annihilator ideal for the cup product action of $\mathrm{H}^\bullet(A,k)$ on $\mathrm{Ext}^\bullet_A(M,N)$. The ideal $I_A(M,N)$ is bihomogeneous (i.e., homogeneous with respect to both the 'external' cohomological degree and the 'internal' super degree) because the cup product action is compatible with both the cohomological and super degrees. The relative support variety of the pair (M,N) is then defined to be

$$|A|_{(M,N)} = \mathrm{Spec}\left(\mathrm{H}^\bullet(A,k)/I_A(M,N)\right), \tag{3.5}$$

a Zariski closed subset of $|A|$, and the support variety of M is defined to be

$$|A|_M = |A|_{(M,M)} = \operatorname{Spec}\left(\mathrm{H}^\bullet(A,k)/I_A(M,M)\right). \tag{3.6}$$

For more details on basic notions concerning support varieties for Hopf (super)-algebras, see for example [**14**, §2.3] or [**31**, §8.3].

Taking $M = N = k$ in (3.3), the identification of cohomology rings

$$\mathrm{H}^\bullet(V(L),k) = \mathrm{H}^\bullet(V(\overline{L}),k)$$

implies the equality of cohomological spectra:

$$|V(L)| = |V(\overline{L})|.$$

By [**27**, Lemma 1.6] and [**28**, Theorem 5.2], there is a finite universal homeomorphism between $|V(\overline{L})|$ and the restricted nullcone of $\overline{L}$:[7]

$$|V(\overline{L})| \simeq \mathcal{N}_1(\overline{L}) := \{z \in \overline{L} : z^{\{2\}} = 0\}. \tag{3.7}$$

Writing $z = z_{\overline{0}} + z_{\overline{1}}$ for the decomposition of z into its even and odd parts, one has $z^{\{2\}} = 0$ if and only if $(z_{\overline{0}})^{[2]} + q(z_{\overline{1}}) + [z_{\overline{1}}, z_{\overline{0}}] = 0$. Since $(z_{\overline{0}})^{[2]} + q(z_{\overline{1}})$ is of even superdegree and $[z_{\overline{1}}, z_{\overline{0}}]$ is of odd superdegree, this implies that

$$z^{\{2\}} = 0 \quad \text{if and only if} \quad (z_{\overline{0}})^{[2]} + q(z_{\overline{1}}) = 0 \text{ and } [z_{\overline{1}}, z_{\overline{0}}] = 0. \tag{3.8}$$

Thus, for the spectrum of $V(L)$ one gets

$$|V(L)| \simeq \{z = z_{\overline{0}} + z_{\overline{1}} \in L : (z_{\overline{0}})^{[2]} + q(z_{\overline{1}}) = 0 \text{ and } [z_{\overline{1}}, z_{\overline{0}}] = 0\}. \tag{3.9}$$

This extends to $p = 2$ the description of $|V(L)|$ we previously obtained for characteristic $p \geq 3$ in [**14**, Proposition 5.4.4]. In fact, the result here for $p = 2$ is stronger than the result in [**14**], since the identification there is only proved up to a finite morphism of varieties.

Next, the compatibility of (3.3) with cup products implies that

$$|V(L)|_M = |V(\overline{L})|_M \quad \text{for all } M \in \mathsf{sMod}_{V(L)}. \tag{3.10}$$

By [**28**, Corollary 6.3.1], the homeomorphism (3.7) restricts for M finite-dimensional to[8]

$$|V(\overline{L})|_M \simeq \{z \in \overline{L} : z^{\{2\}} = 0 \text{ and } M|_{\langle z\rangle} \text{ is not free}\} \cup \{0\}. \tag{3.11}$$

Here $M|_{\langle z\rangle}$ denotes the restriction of M to the k-subalgebra of $V(\overline{L})$ generated by z. For $z \neq 0$ this subalgebra has the form $k[z]/(z^2)$, while for $z = 0$ the subalgebra is just the field k. (This subalgebra is not a sub-*superalgebra* of $V(\overline{L})$ unless z is homogeneous.) An immediate consequence of (3.10), (3.11), and the representation theory of $k[z]/(z^2)$ is the following (cf. [**28**, Theorem 7.2]):

---

[7] The results in [**27**, **28**] are stated in the language of finite group schemes, so the reader must make the standard translation from the category of $V(\overline{L})$-modules to the equivalent category of rational modules for the finite group scheme G whose coordinate algebra $k[G]$ is $V(\overline{L})^*$, the Hopf algebra dual to $V(\overline{L})$.

[8] This identification is true as written if the field k is algebraically closed, and if one defines $|V(\overline{L})|$ and support varieties in terms of the maximal ideal spectrum. But if one is working at the level of schemes, the precise statement of the identification involves certain field extensions; see [**28**, Proposition 6.1]. For simplicity of exposition, we have suppressed the extra details.

PROPOSITION 3.2 (Tensor Product Property). *Let L be a finite-dimensional restricted Lie superalgebra over an algebraically closed field of characteristic $p = 2$, and let M and N be finite-dimensional $V(L)$-supermodules. Then*

$$|V(L)|_{M\otimes N} = |V(L)|_M \cap |V(L)|_N.$$

Next we give an alternative characterization of the condition that $M|_{\langle z\rangle}$ is not free.

LEMMA 3.3. *Let M be a finite-dimensional $V(\overline{L})$-module, and let $z = z_{\overline{0}} + z_{\overline{1}} \in \overline{L}$ with $z^{\{2\}} = 0$. Set $P = k[u,v]/(u^2+v^2)$, and let $\sigma_z : P \to V(\overline{L})$ be the homomorphism defined by $\sigma_z(u) = z_{\overline{0}}$ and $\sigma_z(v) = z_{\overline{1}}$. Write $M{\downarrow_{\sigma_z}}$ for the P-module obtained by pulling back M along σ_z. Then for $z \neq 0$,*

$$M|_{\langle z\rangle} \text{ is not free} \qquad \text{if and only if} \qquad \operatorname{projdim}_P(M{\downarrow_{\sigma_z}}) = \infty.$$

For $z = 0$, one has that $M|_{\langle z\rangle}$ is free (i.e., free over k) and $\operatorname{projdim}_P(M{\downarrow_{\sigma_z}}) = \infty$.

PROOF. Suppose $z \neq 0$, and let Q be the k-subalgebra of P generated by $u+v$. Then $Q \cong k[z]/(z^2)$, so $M|_{\langle z\rangle}$ is not free if and only if $\operatorname{projdim}_Q(M{\downarrow_{\sigma_z}}) = \infty$. Next, $P = Q[u]$ is a polynomial extension of the ring Q, so $\operatorname{flatdim}_P(M{\downarrow_{\sigma_z}}) = \infty$ if and only if $\operatorname{flatdim}_Q(M{\downarrow_{\sigma_z}}) = \infty$ by [**1**, §5.3]. Finally, $\operatorname{projdim}_P(M{\downarrow_{\sigma_z}}) = \operatorname{flatdim}_P(M{\downarrow_{\sigma_z}})$ and $\operatorname{projdim}_Q(M{\downarrow_{\sigma_z}}) = \operatorname{flatdim}_Q(M{\downarrow_{\sigma_z}})$ because P and Q are commutative Noetherian rings over which M is finitely generated. If $z = 0$, then σ_z is the augmentation map, and $\operatorname{projdim}_P(M{\downarrow_{\sigma_z}}) = \operatorname{projdim}_P(k) = \infty$. □

PROPOSITION 3.4. *Let L be a finite-dimensional restricted Lie superalgebra over an algebraically closed field of characteristic $p = 2$, and let M be a finite-dimensional $V(L)$-supermodule. Let P and $\sigma_z : P \to V(\overline{L})$ be as defined in Lemma 3.3. Then there exists a homeomorphism*

$$(3.12)\quad |V(L)|_M \simeq \{z \in L : (z_{\overline{0}})^{[2]} + q(z_{\overline{1}}) = 0 \text{ and } [z_{\overline{1}}, z_{\overline{0}}] = 0 \text{ and } \operatorname{projdim}_P(M{\downarrow_{\sigma_z}}) = \infty\}.$$

A $V(L)$-supermodule M is projective if and only if $|V(L)|_M = \{0\}$, or equivalently, if and only if $\operatorname{projdim}_P(M{\downarrow_{\sigma_z}}) < \infty$ *for all nonzero $z \in L$ such that $z^{\{2\}} = 0$.*

PROOF. The asserted homeomorphism follows immediately from (3.8), (3.10), (3.11), and the lemma. Next, the fact that the forgetful functor

$$\text{Forget} : \mathsf{sMod}_{V(L)} \to \mathsf{Mod}_{V(L)} = \mathsf{Mod}_{V(\overline{L})}$$

is surjective on morphisms implies that if a $V(L)$-supermodule is projective as an object in $\mathsf{Mod}_{V(\overline{L})}$, then it is also projective as an object in (the underlying even subcategory of) $\mathsf{sMod}_{V(L)}$. Then the last sentence of of the proposition is true by the well-known fact that a finite-dimensional $V(\overline{L})$-module is projective if and and only if $|V(\overline{L})|_M = \{0\}$; cf. [**19**, Proposition 1.5]. □

REMARK 3.5. In [**16**, Lemma 3.1.3], we interpreted our previous work from [**15**] to show that an identification analogous to (3.12) holds for the support varieties of arbitrary finite-dimensional *p-nilpotent* restricted Lie superalgebras over (algebraically closed) fields of characteristic $p \geq 3$. In [**15**, §1.2], we conjectured that a similar identification should hold for all finite-dimensional restricted Lie superalgebras (and more generally, for all infinitesimal supergroup schemes) in characteristic $p \geq 3$. The primary obstruction to proving such a generalization in characteristic

$p \geq 3$ is the lack of a characterization of projectivity similar to that stated in the last sentence of Proposition 3.4.

Recall that a closed set $W \subseteq |A|$ is *conical* if it is defined by a homogeneous (with respect to the cohomological degree) ideal in $\mathrm{H}^\bullet(A,k)$. Given a Hopf superalgebra A, let us say that a closed set $W \subseteq |A|$ is *super-conical* if it is defined by a bi-homogeneous ideal in $\mathrm{H}^\bullet(A,k)$, i.e., an ideal that is homogeneous with respect to both the 'external' cohomological degree and the 'internal' super degree. Given a $V(L)$-supermodule M, the support variety $|V(L)|_M$ is super-conical because its defining ideal $I_{V(L)}(M,M)$ is automatically bi-homogeneous. The next proposition states that every super-conical subset of $|V(L)|$ is of this form.

PROPOSITION 3.6 (Realizability). *Let L be a finite-dimensional restricted Lie superalgebra over an algebraically closed field of characteristic 2. Then $W \subseteq |V(L)|$ is the support variety $|V(L)|_M$ of a finite-dimensional $V(L)$-supermodule if and only if W is a closed, super-conical subset of $|V(L)|$.*

PROOF. We have already explained the 'only if' portion of the statement, so we just need to prove the 'if' portion. Let $W \subseteq |V(L)|$ be a closed, super-conical subset. Then W is defined by a bi-homogeneous ideal in $\mathrm{H}^\bullet(V(L),k)$, which in turn is generated by a finite set of bi-homogeneous elements. By the Tensor Product Property (Proposition 3.2), it suffices to show that if $W = Z(\zeta)$ is the closed set defined by a single bi-homogeneous element $\zeta \in \mathrm{H}^n(V(L),k)$, then $W = |V(L)|_M$ for some finite-dimensional $V(L)$-supermodule M.

Choose a projective resolution $(P_\bullet, d)$ in $(\mathsf{sMod}_{V(L)})_{\mathrm{ev}}$ of the trivial $V(L)$-supermodule k, and a homogeneous (with respect to the super degree) cocycle representative $g : P_n \to k$ for ζ. Let M be the kernel of the induced map $\widehat{g} : P_n/d(P_{n+1}) \to k$. (As an ungraded module in $\mathsf{Mod}_{V(\overline{L})}$, and perhaps modulo a projective summand, $P_n/d(P_{n+1})$ is the n-th syzygy module $\Omega^n(k)$ for the trivial $V(\overline{L})$-module k.) Then M is a $V(L)$-supermodule because $d : P_{n+1} \to P_n$ and $g : P_n \to k$ are both homogeneous $V(L)$-supermodule homomorphisms. Now the proof of [**28**, Theorem 7.5] (cf. also [**19**, §4]) shows that $|V(\overline{L})|_M = Z(\zeta)$, hence $|V(L)|_M = Z(\zeta)$ by (3.10). □

3.3. $\mathrm{Proj}(A)$ **and** $\mathrm{Proj}_s(A)$**.** In anticipation of Section 3.4, we now discuss a modification of the projective spectrum that is useful for graded(-commutative) superalgebras in characteristic 2. If we were in characteristic other than 2, then $\mathbb{Z}_2$-gradings could be encoded by an action of the cyclic group C_2 and the following would essentially be the theory of C_2-$\mathrm{Proj}(A)$ as described in the literature; see for example [**24**]. Since we are in characteristic 2, and for the sake of completeness, we provide the relevant details.

Let A be a commutative k-algebra that is graded by $\mathbb{N} \times \mathbb{Z}_2$. Let $A = \bigoplus_{i \in \mathbb{N}} A^i$ be the decomposition of A into its $\mathbb{N}$-graded components, and assume that $A^0 = k$. Let $\mathrm{Proj}(A)$ be the subset of $\mathrm{Spec}(A)$ consisting of the prime ideals that are homogeneous with respect to the $\mathbb{N}$-grading on A, and which are properly contained in the maximal ideal $A^+ = \bigoplus_{i \geq 1} A^i$. A subset W of $\mathrm{Proj}(A)$ is closed in the Zariski topology if there exists an $\mathbb{N}$-homogeneous ideal I of A such that

$$W = Z(I) := \{\mathfrak{p} \in \mathrm{Proj}(A) : I \subseteq \mathfrak{p}\}.$$

We say that an ideal I of A is *bi-homogeneous* if it is homogeneous with respect to the full $\mathbb{N} \times \mathbb{Z}_2$-grading on A. For example, the ideal A^+ is bi-homogeneous.

Given an ideal I of A, let $I_s \subseteq I$ be the largest bi-homogeneous subideal of I. Equivalently, I_s is the ideal in A generated by the bi-homogeneous elements of I. Of course, $I = I_s$ if I is bi-homogeneous. If J is a bi-homogeneous ideal and I is an arbitrary ideal, then $J \subseteq I$ if and only if $J \subseteq I_s$. If $\{I_\lambda : \lambda \in \Omega\}$ is any family of ideals in A, then $(\bigcap_{\lambda\in\Omega} I_\lambda)_s = \bigcap_{\lambda\in\Omega}(I_\lambda)_s$.

Given an ideal I of A, let $\sqrt{I}$ be the usual radical of I, and set $\sqrt[s]{I} = (\sqrt{I})_s$. Then $I_s \subseteq \sqrt[s]{I}$, and $\sqrt[s]{I}$ has the property that if $a \in A$ is a bi-homogeneous element with $a^n \in I_s$ for some $n \geq 1$, then $a \in \sqrt[s]{I}$. We say that a bi-homogeneous ideal J is an *s-radical ideal* if $J = \sqrt[s]{J}$. We say that a bi-homogeneous ideal P is *s-prime* if whenever I and J are bi-homogeneous ideals with $IJ \subseteq P$, then either $I \subseteq P$ or $J \subseteq P$. Equivalently, P is s-prime if whenever a and b are bi-homogeneous elements with $ab \in P$, either $a \in P$ or $b \in P$. If P is a prime ideal in A, then P_s is an s-prime ideal. Every s-prime ideal in A is an s-radical ideal.

LEMMA 3.7. *Let P be an s-prime ideal of A. Then $\sqrt{P}$ is an $\mathbb{N}$-homogeneous prime ideal of A.*

PROOF. Ignoring the $\mathbb{Z}_2$-gradings, P is an $\mathbb{N}$-homogeneous ideal of A, so its radical is also an $\mathbb{N}$-homogeneous ideal. To show that $\sqrt{P}$ is prime, let $x, y \in A$, and suppose $xy \in \sqrt{P}$, say with $x^n y^n = (xy)^n \in P$. Replacing x and y with x^n and y^n, we may assume that $xy \in P$. Then also $x^2y^2 = (xy)^2 \in P$. Write $x = \sum x_i$ with $x_i \in A^i$, and write $x_i = x_{i,\overline{0}} + x_{i,\overline{1}}$ with $x_{i,\overline{0}} \in (A^i)_{\overline{0}}$ and $x_{i,\overline{1}} \in (A^i)_{\overline{1}}$. Since A is commutative, the binomial theorem modulo 2 implies that $x^2 = \sum(x_i)^2 = \sum[(x_{i,\overline{0}})^2 + (x_{i,\overline{1}})^2]$. Then x^2 is a sum of terms of even superdegree, and we can rewrite x^2 as $x^2 = \sum x'_{2i}$, with $x'_{2i} \in (A^{2i})_{\overline{0}}$. Similarly, $y^2 = \sum y'_{2i}$ with $y'_{2i} \in (A^{2i})_{\overline{0}}$.

Now suppose that $x^2 \notin P$ and $y^2 \notin P$. Then there exist minimal indices s and t such that $x'_{2s} \notin P$ and $y'_{2t} \notin P$. Since x^2 and y^2 are each of even superdegree, the $\mathbb{N}$-graded components of x^2y^2 are each bi-homogeneous. Since P is bi-homogeneous, and since $x^2y^2 \in P$, this means in particular that the component of $\mathbb{N}$-degree $s+t$ in x^2y^2 is an element of P. This component is equal to $\sum_{\ell\in\mathbb{Z}} x'_{2s-\ell}y'_{2t+\ell}$. By the minimality of s and t, each summand $x'_{2s-\ell}y'_{2t+\ell}$ with $\ell \neq 0$ is an element of P, because at least one of the factors is an element of P by assumption. Then $x'_{2s}y'_{2t}$ must also be an element of P. The factors x'_{2s} and y'_{2t} are each bi-homogeneous, so the fact that P is an s-prime ideal implies that either $x'_{2s} \in P$ or $y'_{2t} \in P$, in contradiction to the assumptions on s and t. Thus, it must be the case that either $x^2 \in P$ or $y^2 \in P$, and hence $x \in \sqrt{P}$ or $y \in \sqrt{P}$. □

DEFINITION 3.8. Define $\mathrm{Proj}_s(A)$ to be the set of all (bi-homogeneous) s-prime ideals in A that are properly contained in the (bi-homogeneous, s-prime) maximal ideal A^+. We say that a subset W of $\mathrm{Proj}_s(A)$ is *closed* if there exists a bi-homogeneous ideal I of A such that

$$W = Z_s(I) := \{P \in \mathrm{Proj}_s(A) : I \subseteq P\}.$$

The properties listed below are now straightforward to verify. Unless stated otherwise, I and J denote bi-homogeneous ideals in A.

(P1) $Z_s(I) \cup Z_s(J) = Z_s(IJ)$.
(P2) $Z_s(I) \cap Z_s(J) = Z_s(I + J)$.
(P3) If $\mathfrak{p} \in \mathrm{Proj}(A)$, then $\mathfrak{p}_s \in \mathrm{Proj}_s(A)$.

(P4) If $\varphi : \mathrm{Proj}(A) \to \mathrm{Proj}_s(A)$ is the function $\varphi(\mathfrak{p}) = \mathfrak{p}_s$, then $\varphi^{-1}(Z_s(I)) = Z(I)$.

(P5) If $P \in \mathrm{Proj}_s(A)$, then $\sqrt{P} \in \mathrm{Proj}(A)$.

(P6) If $\psi : \mathrm{Proj}_s(A) \to \mathrm{Proj}(A)$ is the function $\psi(P) = \sqrt{P}$, then $(\varphi \circ \psi)(P) = \sqrt[s]{P} = P$.

The first two properties imply that finite unions and arbitrary intersections of closed sets are again closed. Thus the closed subsets of $\mathrm{Proj}_s(A)$ define a topology, which we call the Zariski topology on $\mathrm{Proj}_s(A)$. Property (P4) then implies that $\varphi : \mathrm{Proj}(A) \to \mathrm{Proj}_s(A)$ is a continuous function.

If K is an arbitrary $\mathbb{N}$-homogeneous ideal in A, one has $\varphi(Z(K)) \subseteq Z_s(K_s)$. On the other hand, if I is a bi-homogeneous ideal, and if $P \in Z_s(I)$, then $I \subseteq P \subseteq \sqrt{P}$. Then property (P6) implies:

(P7) If I is bi-homogeneous, then $\varphi(Z(I)) = Z_s(I)$.

Next observe that if $Z_s(I) \subseteq Z_s(J)$, then

$$Z(I) = \varphi^{-1}(Z_s(I)) \subseteq \varphi^{-1}(Z_s(J)) = Z(J).$$

But $Z(I) \subseteq Z(J)$ implies that $\sqrt{I} \supseteq \sqrt{J}$, and hence $\sqrt[s]{I} \supseteq \sqrt[s]{J}$. Now one can check that:

(P8) $Z_s(I) = Z_s(\sqrt[s]{I})$.

(P9) If $I \subseteq J$, then $Z_s(I) \supseteq Z_s(J)$.

(P10) If $Z_s(I) \subseteq Z_s(J)$, then $\sqrt[s]{I} \supseteq \sqrt[s]{J}$.

(P11) If A is a Noetherian ring, then $\mathrm{Proj}_s(A)$ is a Noetherian topological space, i.e., if $Z_s(I_1) \supseteq Z_s(I_2) \supseteq Z_s(I_3) \supseteq \cdots$ is a descending chain of closed sets, then there exists an integer m such that $Z_s(I_j) = Z_s(I_m)$ for all $j \geq m$.

LEMMA 3.9. *If $W = Z_s(I)$ is an irreducible closed set in* $\mathrm{Proj}_s(A)$, *then $\sqrt[s]{I}$ is an s-prime ideal. In particular, if W is nonempty, then W is the closure of the point $\sqrt[s]{I} \in$* $\mathrm{Proj}_s(A)$, *and $\sqrt[s]{I}$ is the unique point in* $\mathrm{Proj}_s(A)$ *with this property.*

PROOF. To show that $\sqrt[s]{I}$ is an s-prime ideal, one can apply the usual argument for $\mathrm{Spec}(A)$; see for example the proof of [**11**, Proposition 4.5.3]. Then $W = Z_s(I) = Z_s(\sqrt[s]{I})$ by property (P8) above, while property (P10) implies that if $W = \overline{\{P\}} = Z_s(P)$ for some $P \in \mathrm{Proj}_s(A)$, then $P = \sqrt[s]{I}$. □

Combining property (P11) and Lemma 3.9, we get:

(P12) If A is a Noetherian ring, then $\mathrm{Proj}_s(A)$ is a Zariski space, i.e., it is a Noetherian topological space, and every nonempty closed irreducible subset is the closure of a unique point.

3.4. Tensor triangular geometry. Now take $A = \mathrm{H}^\bullet(V(L), k)$. In this section we discuss how the geometry of $\mathrm{Proj}_s(A) = \mathrm{Proj}_s(\mathrm{H}^\bullet(V(L), k))$ classifies certain categories of $V(L)$-supermodules. First we recall the corresponding results for the category of ungraded $V(\overline{L})$-modules.

Let $\mathbf{K}$ be a symmetric tensor triangulated category. Recall that a triangulated subcategory of $\mathbf{K}$ is *thick* if it is closed under taking direct summands, and is a *tensor ideal* (or $\otimes$-*ideal*) if it is closed under $M \otimes -$ for any object M in $\mathbf{K}$. A $\otimes$-ideal $\mathcal{J}$ is *radical* if whenever $M^{\otimes n} \in \mathcal{J}$ for some $n \geq 1$, then also $M \in \mathcal{J}$. A thick $\otimes$-ideal $\mathcal{P}$ is *prime* if whenever $M, N \in \mathbf{K}$ and $M \otimes N \in \mathcal{P}$, then either $M \in \mathcal{P}$ or $N \in \mathcal{P}$. The *Balmer spectrum* of $\mathbf{K}$, denoted $\mathrm{Spc}(\mathbf{K})$, is the set of all proper prime thick $\otimes$-ideals of $\mathbf{K}$, considered as a topological space via the Zariski topology. Write $2^{\mathrm{Spc}(\mathbf{K})}$ for the power set of $\mathrm{Spc}(\mathbf{K})$. Then the support function

$$\mathrm{supp}' : \mathbf{K} \to 2^{\mathrm{Spc}(\mathbf{K})}$$

is defined by $\mathrm{supp}'(M) = \{\mathcal{P} \in \mathrm{Spc}(\mathbf{K}) : M \notin \mathcal{P}\}$ for each object $M \in \mathbf{K}$.

Let $\mathsf{stmod}_{V(\overline{L})}$ (resp., $\mathsf{StMod}_{V(\overline{L})}$) be the stable module category of finite-dimensional (resp., all) $V(\overline{L})$-modules. These are the categories obtained from $\mathsf{mod}_{V(\overline{L})}$ and $\mathsf{Mod}_{V(\overline{L})}$, respectively, by quotienting out the morphisms that factor through a projective. The objects of the stable category are the same as in the corresponding ordinary module category, but with projective objects now isomorphic to zero. The categories $\mathsf{stmod}_{V(\overline{L})}$ and $\mathsf{StMod}_{V(\overline{L})}$ are tensor triangulated categories, with suspension Σ equal to $\Omega^{-1}(-)$, the inverse of the syzygy functor. The category $\mathsf{StMod}_{V(\overline{L})}$ is compactly generated, with subcategory of compact objects $(\mathsf{StMod}_{V(\overline{L})})^c = \mathsf{stmod}_{V(\overline{L})}$.

Now set $X = \mathrm{Proj}(\mathrm{H}^\bullet(V(\overline{L}), k))$. In **[3]**, Benson, Iyengar, Krause, and Pevtsova use localization functors to define a support function

$$\mathrm{supp} : \mathsf{StMod}_{V(\overline{L})} \to 2^X \tag{3.13}$$

that satisfies the properties listed below.[9] Except when indicated, M, M_i, and N denote arbitrary objects in $\mathsf{StMod}_{V(\overline{L})}$.

(S0) $\mathrm{supp}(0) = \emptyset$
(S1) $\mathrm{supp}(k) = X$
(S2) $\mathrm{supp}(M \oplus N) = \mathrm{supp}(M) \cup \mathrm{supp}(N)$
(S3) $\mathrm{supp}(\Sigma M) = \mathrm{supp}(M)$
(S4) For each distinguished triangle $M_1 \to M_2 \to M_3 \to \Sigma M_1$ in $\mathsf{StMod}_{V(\overline{L})}$,

$$\mathrm{supp}(M_2) \subseteq \mathrm{supp}(M_1) \cup \mathrm{supp}(M_3).$$

(S5) Tensor Product Property: For all compact M and arbitrary N,[10]

$$\mathrm{supp}(M \otimes N) = \mathrm{supp}(M) \cap \mathrm{supp}(N).$$

(S6) $\mathrm{supp}(\bigoplus_{i\in I} M_i) = \bigcup_{i\in I} \mathrm{supp}(M_i)$ for any set I.
(S7) For all compact M, $\mathrm{supp}(M)$ is Zariski closed with quasi-compact complement.[11]
(S8) For every Zariski closed subset $W \subseteq X$ with quasi-compact complement, there exists a compact object M such that $\mathrm{supp}(M) = W$.
(S9) If $\mathrm{supp}(M) = \emptyset$, then $M \cong 0$.
(S10) Given a finite-dimensional $V(\overline{L})$-module M, let $M^* = \mathrm{Hom}_k(M, k)$ be its linear dual. Then $\mathrm{supp}(M^*) = \mathrm{supp}(M)$.

We note that the 'quasi-compact complement' condition in (S7) and (S8) is automatically satisfied by any Zariski closed subset of X, as a consequence of the fact that $\mathrm{H}^\bullet(V(\overline{L}), k)$ is a finitely-generated commutative k-algebra and hence a Noetherian ring. Elaborating further on (S7), if M is a compact object in

[9] Again, the results in **[3]** are stated in the language of finite group schemes, so the reader must translate to the category of rational G-modules, where $k[G] = V(\mathfrak{g})^*$.

[10] Actually, this property holds without the assumption that M is compact, but we will find it convenient to use this more restricted formulation of the Tensor Product Property.

[11] As used in algebraic geometry, the term *quasi-compact* means that every open cover has a finite subcover.

$\mathsf{StMod}_{V(\overline{L})}$ (equivalently, if M is isomorphic, modulo projective summands, to a finite-dimensional $V(\overline{L})$-module), then $\operatorname{supp}(M)$ is the Zariski closed subset of X defined by $I(M)$, the annihilator ideal for the cup product action of $\mathrm{H}^\bullet(V(\overline{L}),k)$ on $\operatorname{Ext}^\bullet_{V(\overline{L})}(M,M)$.

Properties (S0)–(S9) imply by Theorem 1.5 of [**12**] that the restriction of (3.13) to $\mathsf{stmod}_{V(\overline{L})}$ is a classifying support datum in the sense of [**12**, Theorem 2.19]; cf. also Theorems 3.4.1 and 3.5.1 of [**8**]. Recall that a subset of a topological space is *specialization closed* if it is a union of arbitrarily many closed sets. Then there is a canonical homeomorphism from $X = \operatorname{Proj}(\mathrm{H}^\bullet(V(\overline{L}),k))$ to the Balmer spectrum of the category $\mathsf{stmod}_{V(\overline{L})}$,

$$\operatorname{Proj}(\mathrm{H}^\bullet(V(\overline{L}),k)) \simeq \operatorname{Spc}(\mathsf{stmod}_{V(\overline{L})}), \tag{3.14}$$

which identifies the support functions supp and supp′. Furthermore, there are inverse bijections

$$\{\text{specialization closed subsets } \mathcal{V} \text{ of } \operatorname{Proj}(\mathrm{H}^\bullet(V(\overline{L}),k))\} \underset{\Theta}{\overset{\Gamma}{\rightleftarrows}} \{\text{thick } \otimes\text{-ideal subcategories } \mathcal{J} \text{ of } \mathsf{stmod}_{V(\overline{L})}\}, \tag{3.15}$$

defined by $\Gamma(\mathcal{V}) = \{M \in \mathsf{stmod}_{V(\overline{L})} : \operatorname{supp}(M) \subseteq \mathcal{V}\}$ and $\Theta(\mathcal{J}) = \bigcup_{M\in\mathcal{J}} \operatorname{supp}(M)$. We note that [**12**, Theorem 2.19] is stated in terms of *radical* thick $\otimes$-ideals, but using the fact that each finite-dimensional $V(\overline{L})$-module M is a direct summand of $M \otimes M^* \otimes M$, one can see that every thick $\otimes$-ideal of $\mathsf{stmod}_{V(\overline{L})}$ is automatically radical.

Now we consider the super setting. Recall from Section 3.1 that $(\mathsf{sMod}_{V(L)})_{\mathrm{ev}}$ is a Frobenius category. Then we can consider the corresponding stable category (see, e.g., [**22**, I.2]), which we denote $\mathsf{St\text{-}sMod}_{V(L)}$. It is a compactly generated tensor-triangulated category, with subcategory of compact objects

$$(\mathsf{St\text{-}sMod}_{V(L)})^c = \mathsf{st\text{-}smod}_{V(L)},$$

the stable category of $(\mathsf{smod}_{V(L)})_{\mathrm{ev}}$. The forgetful functor $\mathsf{sMod}_{V(L)} \to \mathsf{Mod}_{V(L)}$ induces a (exact) triangulated functor

$$\text{Forget} : \mathsf{St\text{-}sMod}_{V(L)} \to \mathsf{StMod}_{V(L)} = \mathsf{StMod}_{V(\overline{L})}.$$

In particular, the forgetful functor maps projective objects in $\mathsf{sMod}_{V(L)}$ to projective objects in $\mathsf{Mod}_{V(L)}$. Via this functor, we will consider objects from $\mathsf{St\text{-}sMod}_{V(L)}$ as objects in $\mathsf{StMod}_{V(L)}$.

Our goal now is to establish analogues of (3.14) and (3.15) for $\mathsf{st\text{-}smod}_{V(L)}$. Set

$$X_s = \operatorname{Proj}_s(\mathrm{H}^\bullet(V(L),k)),$$

and let

$$\varphi : \operatorname{Proj}(\mathrm{H}^\bullet(V(L),k)) \to \operatorname{Proj}_s(\mathrm{H}^\bullet(V(L),k))$$

be the function $\mathfrak{p} \mapsto \mathfrak{p}_s$ as defined in (P4) above. Using the fact that $\mathrm{H}^\bullet(V(L),k)$ is a Noetherian ring and the properties listed in Section 3.3, one can again deduce that every Zariski closed set in X_s has quasi-compact complement. Define

$$\operatorname{supp}_s : \mathsf{St\text{-}sMod}_{V(L)} \to 2^{X_s}$$

by

$$\operatorname{supp}_s(M) = \varphi\left(\operatorname{supp}(\text{Forget}(M))\right).$$

THEOREM 3.10. *Let L be a finite-dimensional restricted Lie superalgebra over a field k of characteristic $p = 2$. There is a canonical homeomorphism from $X_s = \mathrm{Proj}_s(\mathrm{H}^\bullet(V(L), k))$ to the Balmer spectrum of the category* $\mathsf{st}\text{-}\mathsf{smod}_{V(L)}$,

$$\mathrm{Proj}_s(\mathrm{H}^\bullet(V(\overline{L}), k)) \simeq \mathrm{Spc}(\mathsf{stmod}_{V(L)}), \tag{3.16}$$

which identifies supp_s *with* supp'. *Furthermore, there are inverse bijections*

$$\{\textit{specializataion closed subsets } \mathcal{V} \textit{ of } \mathrm{Proj}_s(\mathrm{H}^\bullet(V(L), k))\} \underset{\Theta}{\overset{\Gamma}{\rightleftarrows}} \{\textit{thick } \otimes\textit{-ideal subcategories } \mathcal{J} \textit{ of } \mathsf{st}\text{-}\mathsf{smod}_{V(L)}\}, \tag{3.17}$$

defined by

$$\Gamma(\mathcal{V}) = \{M \in \mathsf{st}\text{-}\mathsf{smod}_{V(L)} : \mathrm{supp}_s(M) \subseteq \mathcal{V}\}, \quad \textit{and}$$
$$\Theta(\mathcal{J}) = \textstyle\bigcup_{M \in \mathcal{J}} \mathrm{supp}_s(M).$$

PROOF. The theorem follows directly from [**12**, Theorem 1.5], once we verify that the analogues of (S0)–(S9) hold for supp_s. Many of the properties are immediate consequences of the fact that the forgetful functor Forget : $\mathsf{St}\text{-}\mathsf{sMod}_{V(L)} \to \mathsf{StMod}_{V(L)}$ is a triangulated functor, and the fact that the function $\varphi : X \to X_s$ is a continuous surjection. We will make a few comments about those properties that are not entirely immediate.

(S7): Let M be a finite-dimensional $V(L)$-supermodule. Then $\mathrm{supp}(M) = Z(I(M))$, the Zariski closed subset of $\mathrm{Proj}\,\mathrm{H}^\bullet(V(L), k)$ defined by $I(M)$, the anihilator ideal for the cup product action of $\mathrm{H}^\bullet(V(L), k)$ on $\mathrm{Ext}^\bullet_{V(L)}(M, M)$. The ideal $I(M)$ is bi-homogeneous in $\mathrm{H}^\bullet(V(L), k)$, so then

$$\mathrm{supp}_s(M) = \varphi\,(Z(I(M))) = Z_s(I(M))$$

by property (P7). So $\mathrm{supp}_s(M)$ is Zariski closed (with quasi-compact complement) in X_s.

(S5): Let M be a finite-dimensional $V(L)$-supermodule, and let N be an arbitrary $V(L)$-supermodule. By property (S5) for supp, one has $\mathrm{supp}(M \otimes N) = \mathrm{supp}(M) \cap \mathrm{supp}(N)$. Then

$$\mathrm{supp}_s(M \otimes N) = \varphi\,(\mathrm{supp}(M) \cap \mathrm{supp}(N)) \subseteq \mathrm{supp}_s(M) \cap \mathrm{supp}_s(N).$$

For the reverse set inclusion, let $P \in \mathrm{supp}_s(M) \cap \mathrm{supp}_s(N)$. Then there exist $\mathfrak{p} \in \mathrm{supp}(M)$ and $\mathfrak{q} \in \mathrm{supp}(N)$ such that $\varphi(\mathfrak{p}) = P = \varphi(\mathfrak{q})$, i.e., $\mathfrak{p}_s = P = \mathfrak{q}_s$. Since $\mathrm{supp}(M) = Z(I(M))$ and $I(M)$ is bi-homogeneous, this implies that $I(M) \subseteq \mathfrak{p}_s = \mathfrak{q}_s \subseteq \mathfrak{q}$. Then $\mathfrak{q} \in \mathrm{supp}(M) \cap \mathrm{supp}(N) = \mathrm{supp}(M \otimes N)$, and $P = \mathfrak{q}_s = \varphi(\mathfrak{q}) \in \mathrm{supp}_s(M \otimes N)$.

(S8): Let $W \subseteq X_s$ be a Zarsiki closed set (with quasi-compact complement). Then $W = Z_s(I)$ for some bi-homogeneous ideal I in $\mathrm{H}^\bullet(V(L), k)$. By Proposition 3.6, there exists a finite-dimensional $V(L)$-supermodule M such that $\mathrm{supp}(M) = Z(I)$. Then $\mathrm{supp}_s(M) = Z_s(I) = W$ by (P7).

(S9): If $M \cong 0$ in $\mathsf{St}\text{-}\mathsf{sMod}_{V(L)}$, then $\mathrm{supp}_s(M) = \emptyset$ by property (S0) for supp and the fact that the forgetful functor maps the zero object of $\mathsf{St}\text{-}\mathsf{sMod}_{V(L)}$ to the zero object of $\mathsf{StMod}_{V(L)}$. Conversely, if $\mathrm{supp}_s(M) = \emptyset$, then $\mathrm{supp}(M) = \emptyset$, and hence $M \cong 0$ in $\mathsf{StMod}_{V(L)}$ by property (S9) for supp. Then M is projective as an object in $\mathsf{Mod}_{V(L)}$. Since the forgetful functor $\mathsf{sMod}_{V(L)} \to \mathsf{Mod}_{V(L)}$ is exact and fully faithful, this implies that M is projective in $(\mathsf{sMod}_{V(L)})_{\mathrm{ev}}$, and hence $M \cong 0$ in $\mathsf{St}\text{-}\mathsf{sMod}_{V(L)}$. Thus, $\mathrm{supp}_s(M) = \emptyset$ if and only if $M \cong 0$ in $\mathsf{St}\text{-}\mathsf{sMod}_{V(L)}$. □

4. Support for non-restricted Lie superalgebras in characteristic 2

4.1. Lie superalgebra cohomology. Given a k-vector space V, let $S(V)$, $\Lambda(V)$, and $\Gamma(V)$ be the symmetric, exterior,[12] and divided power algebras on V, respectively. We denote a typical monomial in $S(V)$ by $v_1^{a_1} \cdots v_t^{a_t}$, and a typical monomial in $\Lambda(V)$ by $\langle v_1 \cdots v_n \rangle$. The divided power algebra $\Gamma(V)$ is generated by the symbols $\gamma_r(v)$ for $r \in \mathbb{N}$ and $v \in V$. These generators satisfy the relations $\gamma_0(v) = 1$, $\gamma_r(0) = 0$ if $r \geq 1$, $\gamma_r(v)\gamma_s(v) = \binom{r+s}{r}\gamma_{r+s}(v)$, and $\gamma_r(v + v') = \sum_{i=0}^{r} \gamma_i(v)\gamma_{r-i}(v')$. Since $p = \operatorname{char}(k) = 2$, the algebras $S(V)$, $\Lambda(V)$, and $\Gamma(V)$ are each naturally Hopf algebras: the subspace V consists of primitive elements in each of $S(V)$ and $\Lambda(V)$, and the coproduct on $\Gamma(V)$ satisfies $\Delta(\gamma_r(v)) = \sum_{i+j=r} \gamma_i(v) \otimes \gamma_j(v)$.[13] The algebras $S(V)$, $\Lambda(V)$, and $\Gamma(V)$ are each naturally $\mathbb{N}$-graded, and these $\mathbb{N}$-gradings are compatible with the Hopf algebra structures.

Now let $L = L_{\overline{0}} \oplus L_{\overline{1}}$ be Lie superalgebra over k, and let $\overline{Y}(L) = \Lambda(L_{\overline{0}}) \otimes \Gamma(L_{\overline{1}})$ be the tensor product of algebras.[14] Then $\overline{Y}(L)$ inherits the structure of an $\mathbb{N}$-graded Hopf algebra, with $\overline{Y}_\ell(L) = \bigoplus_{i+j=\ell} \Lambda^i(L_{\overline{0}}) \otimes \Gamma^j(L_{\overline{1}})$. In particular, $\overline{Y}_1(L) \cong L$, and we write $s : L \to \overline{Y}_1(L)$ for the natural identification. The right adjoint action of L on itself extends to a right action of L on $\overline{Y}(L)$ such that $(z_1 z_2).u = z_1(z_2.u) + (-1)^{\overline{u}\cdot\overline{z}_2}(z_1.u)z_2$ for all $z_1, z_2 \in \overline{Y}(L)$ and $u \in L$ (i.e., u acts by right superderivations), and such that $\langle x \rangle.u = s([x, u])$ and $\gamma_r(y).u = \gamma_{r-1}(y)s([y, u])$ for all $x \in L_{\overline{0}}$, $y \in L_{\overline{1}}$, and $r \in \mathbb{N}$. The right action of L on $\overline{Y}(L)$ then extends to a right action of the universal enveloping superalgebra $U(L)$ on $\overline{Y}(L)$. Now define $Y(L)$ to be the corresponding smash product algebra $U(L)\#\overline{Y}(L)$. The coproducts on $U(L)$ and $\overline{Y}(L)$ induce a bialgebra structure on $Y(L)$. Then $Y(L)$ is an $\mathbb{N}$-graded bialgebra with $U(L)$ concentrated in $\mathbb{N}$-degree 0. As a vector space and as a left $U(L)$-module, $Y(L) = U(L) \otimes \overline{Y}(L)$. From now on we denote the product in $Y(L)$ by juxtaposition. Then $Y(L)$ is spanned by the monomials

$$u\langle x_1 \cdots x_s \rangle \gamma_{a_1}(y_1) \cdots \gamma_{a_t}(y_t), \tag{4.1}$$

with $u \in U(L)$, $x_i \in L_{\overline{0}}$, $y_j \in L_{\overline{1}}$, and $a_j \in \mathbb{N}$.

THEOREM 4.1. *Let L be a finite-dimensional Lie superalgebra over the field k. There exists a differential $d : Y(L) \to Y(L)$ (which we call the Koszul differential) making $Y(L)$ into a differential graded bialgebra and into a left $U(L)$-free resolution of the trivial module k (which we call the Koszul resolution of k). Given $u \in U(L)$, $x \in L_{\overline{0}}$, and $y \in L_{\overline{1}}$, one has*

$$\begin{aligned} d(u) &= 0, \\ d(\langle x \rangle) &= x, \quad \textit{and} \\ d(\gamma_r(y)) &= y\gamma_{r-1}(y) - \langle q(y) \rangle \gamma_{r-2}(y). \end{aligned}$$

PROOF. Apply the same reasoning as in the proof of [**13**, Theorem 3.1.1]. □

Let M and N be left $U(L)$-supermodules. Given $n \in \mathbb{N}$, set

$$C^n(L, M, N) = \operatorname{Hom}_{U(L)}(Y_n(L) \otimes M, N).$$

---

[12]For $p = 2$, one has $\Lambda(V) \cong S(V)/\langle x^2 : x \in V \rangle$; i.e., $\Lambda(V)$ is isomorphic to a truncated polynomial ring.

[13]If $p \neq 2$, then $\Lambda(V)$ is not an ordinary Hopf algebra but is instead a Hopf superalgebra.

[14]If $p \neq 2$, take the graded tensor product of superalgebras, in which $(a \otimes b)(c \otimes d) = (-1)^{\overline{b}\cdot\overline{c}+\deg(b)\cdot\deg(c)} ac \otimes bd$.

Here we consider $Y(L) \otimes M$ as a $U(L)$-module via the coproduct, and as a chain complex with differential induced by the Koszul differential d on $Y(L)$. In particular, set $C^\bullet(L,k) = C^\bullet(L,k,k)$. The Koszul differential induces a differential ∂ on $C^\bullet(L,M,N)$, making $C^\bullet(L,M,N)$ into a cochain complex. Then the Lie superalgebra cohomology group $\operatorname{Ext}_L^\bullet(M,N) = \operatorname{Ext}_{U(L)}^\bullet(M,N)$ can be computed as the n-th cohomology group of the complex $C^\bullet(L,M,N)$. The coalgebra structure of $Y(L)$ induces the structure of a differential graded algebra on $C^\bullet(L,k)$, and the structure of a differential graded $C^\bullet(L,k)$-bimodule on $C^\bullet(L,M,N)$. Specifically,

$$C^\bullet(L,k) \cong (\overline{Y}_\bullet(L))^* \cong \Lambda(L_{\overline{0}}^*) \otimes S(L_{\overline{1}}^*) \tag{4.2}$$

as an $\mathbb{N}$-graded algebra.[15] Passing to cohomology, one gets the cup product structure on

$$\mathrm{H}^\bullet(L,k) = \operatorname{Ext}_L^\bullet(k,k) = \operatorname{Ext}_{U(L)}^\bullet(k,k),$$

and the $\mathrm{H}^\bullet(L,k)$-bimodule structure on

$$\operatorname{Ext}_L^\bullet(M,N) = \operatorname{Ext}_{U(L)}^\bullet(M,N).$$

As in Section 3, let $\mathsf{sMod}_{U(L)}$ be the category of left $U(L)$-supermodules, and let $\mathsf{Mod}_{U(L)}$ be the category of arbitrary (not necessarily $\mathbb{Z}_2$-graded) left $U(L)$-modules. For each $M \in \mathsf{sMod}_{U(L)}$, the tensor product $Y(L) \otimes M$ provides a resolution of M in $(\mathsf{sMod}_{U(L)})_{\mathrm{ev}}$ by free $U(L)$-supermodules, so by the same reasoning as for (3.2), one gets

$$\operatorname{Ext}^\bullet_{\mathsf{sMod}_{U(L)}}(M,N) = \operatorname{Ext}^\bullet_{\mathsf{Mod}_{U(L)}}(M,N) \quad \text{for all } M,N \in \mathsf{sMod}_{U(L)}. \tag{4.3}$$

We will simply write $\operatorname{Ext}_{U(L)}^\bullet(M,N)$ for the common value in (4.3).

Let $\varphi : k \to k$ be the Frobenius map, $\lambda \mapsto \lambda^2$, and given a k-vector space V, let $V^{(1)} = k \otimes_\varphi V$ denotes its Frobenius twist. Let $L_{\overline{1}}^*[2]$ be the superspace $L_{\overline{1}}^*$ concentrated in $\mathbb{N}$-degree 2. The squaring operation on $S(L_{\overline{1}}^*)$ induces a homomorphism of graded algebras $S(L_{\overline{1}}^*[2])^{(1)} \to S(L_{\overline{1}}^*)$, $\lambda \otimes_\varphi z \mapsto \lambda \cdot z^2$. This is not a homomorphism of graded superalgebras, because the odd superdegree generators in $S(L_{\overline{1}}^*[2])^{(1)}$ are mapped to even superdegree elements in $S(L_{\overline{1}}^*)$. Composing with the inclusion $S(L_{\overline{1}}^*) \hookrightarrow \Lambda(L_{\overline{0}}^*) \otimes S(L_{\overline{1}}^*) \cong C^\bullet(L,k)$, one gets a homomorphism of graded algebras $\widehat{\varphi} : S(L_{\overline{1}}^*[2])^{(1)} \to C^\bullet(L,k)$. Since ∂ acts by derivations on the ring $C^\bullet(L,k)$, it follows that the image of $\widehat{\varphi}$ consists of cocycles in $C^\bullet(L,k)$, and hence $\widehat{\varphi}$ induces a graded algebra map

$$\varphi : S(L_{\overline{1}}^*[2])^{(1)} \to \mathrm{H}^\bullet(L,k), \tag{4.4}$$

which by abuse of notation we also denote φ. The map φ is natural with respect to L, i.e., with respect to even homomorphisms $L \to L'$ of Lie superalgebras.

THEOREM 4.2. *Let L be a finite-dimensional Lie superalgebra over k, and let M and N be finite-dimensional $U(L)$-supermodules. Then $\mathrm{H}^\bullet(L,k)$ is finite over the image of φ, and $\operatorname{Ext}_L^\bullet(M,N) \cong \operatorname{Ext}_L^\bullet(k, \operatorname{Hom}_k(M,N))$ is finite as either a left or right $\mathrm{H}^\bullet(L,k)$-module.*

PROOF. Same as for $p \geq 3$; see [**13**]. □

In fact, a more general version of Theorem 4.2 is true:

[15] If $p \neq 2$, this is the graded tensor product of algebras.

THEOREM 4.3. *Let L be a finite-dimensional Lie superalgebra over k. Let M and N be arbitrary (i.e., not necessarily $\mathbb{Z}_2$-graded) $U(L)$-modules, with M finite-dimensional and N finitely-generated over $U(L)$. Then $\operatorname{Ext}^\bullet_{U(L)}(M,N)$ is finite under the cup product action of $\mathrm{H}^\bullet(L,k)$.*

PROOF. This can be shown in precisely the same manner as in [**16**, §§2.2–2.3]. For the arguments presented there, it is immaterial whether or not the modules are $\mathbb{Z}_2$-graded. The exterior algebra $\Lambda(\widetilde{\mathfrak{g}}_1)$ is a genuine cocommutative Hopf algebra in characteristic 2, so in the argument at the end of the proof of [**16**, Lemma 2.3.1], one can appeal to the results of [**20**] rather than [**13**, Theorem 3.2.4] to argue that the E_2-page of [**16**, (2.3.5)] is a finite module over the cohomology ring $\mathrm{H}^\bullet(\Lambda(\widetilde{\mathfrak{g}}_1),k)$. □

4.2. Support varieties. *For the rest of this section, assume that the field k is algebraically closed.* Taking $A = U(L)$ in (3.4), (3.5), and (3.6), and using the maximal ideal spectrum rather than the prime ideal spectrum, we get the definition of the cohomological spectrum $|U(L)|$, and the definition of (relative) support varieties for $U(L)$-supermodules. Given $M, N \in \mathsf{sMod}_{U(L)}$, let $J_{U(L)}(M,N) = \varphi^{-1}(I_{U(L)}(M,N))$, the inverse image of $I_{U(L)}(M,N)$ under the map (4.4). Then $J_{U(L)}(M,N)$ is homogeneous with respect to the $\mathbb{N}$-grading on $S(L_{\overline{1}}^*[2])^{(1)}$. Set

$$\mathcal{X}_L(M,N) = \operatorname{MaxSpec}\left(S(L_{\overline{1}}^*[2])^{(1)}/J_L(M,N)\right).$$

and set $\mathcal{X}_L(M) = \mathcal{X}_L(M,M)$.

LEMMA 4.4. *The graded algebra homomorphism $\varphi : S(L_{\overline{1}}^*[2])^{(1)} \to \mathrm{H}^\bullet(L,k)$ induces for each pair of finite-dimensional $U(L)$-supermodules M and N a homeomorphism*

$$\varphi^*_{M,N} : |U(L)|_{(M,N)} \simeq \mathcal{X}_L(M,N).$$

PROOF. Same as for $p \geq 3$; see [**14**, §4.2]. □

If X is an affine variety with coordinate algebra $k[X]$, we write $X^{(1)}$ for the affine variety with coordinate algebra $k[X^{(1)}] = k[X]^{(1)}$. If $Y = X^{(1)}$, then we may write $X = Y^{(-1)}$.

THEOREM 4.5. *Let L be a finite-dimensional Lie superalgebra over an algebraically closed field k of characteristic $p = 2$. Then*

$$\mathcal{X}_L(k)^{(-1)} = \mathcal{N}_{\mathrm{odd}}(L) := \{x \in L_{\overline{1}} : q(x) = 0\}.$$

PROOF. Essentially the same as for $p \geq 3$; see the proof of [**14**, Theorem 4.2.4]. □

We call $\mathcal{N}_{\mathrm{odd}}(L)$ the *odd nullcone* of L. For $p \neq 2$, one has $\mathcal{N}_{odd}(L) = \{x \in L_{\overline{1}} : [x,x] = 0\}$.

THEOREM 4.6. *Let L be a finite-dimensional Lie superalgebra over an algebraically closed field k of characteristic $p = 2$, and let M be a finite-dimensional L-supermodule. Then*

$$\mathcal{X}_L(M)^{(-1)} = \{x \in \mathcal{N}_{\mathrm{odd}}(L) : M|_{\langle x\rangle} \textit{ is not free}\} \cup \{0\}. \tag{4.5}$$

In (4.5), $M|_{\langle x\rangle}$ denotes the restriction of M to the k-subalgebra of $U(L)$ generated by x. If $x \neq 0$, this subalgebra has the form $k[x]/(x^2)$, while for $x = 0$ the subalgebra is just the field k. The set on the right-hand side of (4.5) corresponds to the *associated variety* originally considered in characteristic zero by Duflo and Serganova; see [**17, 21**].

PROOF OF THEOREM 4.6. Let

$$\mathcal{X}'_L(M)^{(-1)} = \{x \in \mathcal{N}_{\mathrm{odd}}(L) : M|_{\langle x\rangle} \text{ is not free}\} \cup \{0\}$$

be the 'rank variety' defined by the right-hand side of (4.5). For convenience, we will henceforth omit the (-1) superscript. First, using naturality with respect to L, one can argue as in the proof of [**14**, Proposition 4.3.1] to show that $\mathcal{X}'_L(M) \subseteq \mathcal{X}_L(M)$. To establish the reverse inclusion, we relate support varieties for $U(L)$ to support varieties for $V(L)$ by way of the quotient map $\pi : U(L) \twoheadrightarrow V(L)$. The following argument replaces, for $p = 2$, the arguments given in [**16**, §§3.2–3.3] for $p \geq 3$.

Consider the increasing monomial-length filtrations on $U(L)$ and $V(L)$, and the corresponding associated graded algebras

$$\begin{gathered} \operatorname{gr} U(L) \cong S(L_{\overline{0}}) \otimes \Lambda(L_{\overline{1}}) \quad \text{and} \\ \operatorname{gr} V(L) \cong S(L_{\overline{0}})/(x^2 : x \in L_{\overline{0}}) \otimes \Lambda(L_{\overline{1}}) \cong \Lambda(L_{\overline{0}}) \otimes \Lambda(L_{\overline{1}}) \cong \Lambda(L). \end{gathered}$$

The algebras $\operatorname{gr} U(L)$ and $\operatorname{gr} V(L)$ are $\mathbb{N}$-graded, with the degree-0 component equal to the field k, and the degree-1 component identifying with the superspace L. Now in the same manner described in [**13**, §3.4] (cf. also [**16**, §2.3]), the monomial-length filtrations on $U(L)$ and $V(L)$ induce decreasing filtrations on the corresponding bar complexes for each algebra, and the decreasing filtrations on the bar complexes then give rise to spectral sequences of algebras

$$\text{(4.6)} \qquad {}^V E_1^{i,j} = \mathrm{H}^{i+j}(\operatorname{gr} V(L), k)_{-i} \Rightarrow \mathrm{H}^\bullet(V(L), k), \quad \text{and}$$

$$\text{(4.7)} \qquad {}^U E_1^{i,j} = \mathrm{H}^{i+j}(\operatorname{gr} U(L), k)_{-i} \Rightarrow \mathrm{H}^\bullet(U(L), k).$$

The non-negative gradings on $\operatorname{gr} U(L)$ and $\operatorname{gr} V(L)$ induce non-positive gradings on their cohomology rings, which are indicated by the subscript $-i$ in (4.7) and (4.6).

By the descriptions of $\operatorname{gr} U(L)$ and $\operatorname{gr} V(L)$ given above, one has

$$\begin{aligned} &\mathrm{H}^\bullet(\operatorname{gr} V(L), k) \cong S(L^*), \quad \text{and} \\ &\mathrm{H}^\bullet(\operatorname{gr} U(L), k) \cong \Lambda(L_{\overline{0}}^*) \otimes S(L_{\overline{1}}^*) \cong S(L^*)/(x^2 : x \in L_{\overline{0}}^*). \end{aligned}$$

In both cases, the superspace $L^* = L_{\overline{0}}^* \oplus L_{\overline{1}}^*$ is concentrated in the $E_1^{1,0}$ term of the spectral sequence. By multiplicativity, this implies that both spectral sequences are concentrated in the row $j = 0$, and consequently the abutment is equal to the cohomology of the E_1-page. Since the differential of each spectral sequence acts by algebra derivations, it follows that the square of any element in $S(L^*)$ is a cocycle in ${}^V E_1^{\bullet,0}$. As a complex, ${}^U E_1^{\bullet,0}$ is simply the Koszul cochain complex described above in (4.2). The quotient map $\pi : U(L) \twoheadrightarrow V(L)$ induces a morphism of spectral sequences ${}^V E \to {}^U E$. On the abutment this is just the inflation map induced by π, while on the E_1-page this is the quotient homomorphism $S(L^*) \twoheadrightarrow S(L^*)/(x^2 : x \in L_{\overline{0}}^*)$.

From the observations of the preceding paragraph, we get the commutative diagram

$$(4.8)\qquad \begin{array}{ccc} S(L^*[2])^{(1)} & \xrightarrow{\ \phi\ } & \mathrm{H}^\bullet(V(L),k) \\ \downarrow{\scriptstyle \mathrm{res}} & & \downarrow{\scriptstyle \pi^*} \\ S(L_{\overline{1}}^*[2])^{(1)} & \xrightarrow{\ \varphi\ } & \mathrm{H}^\bullet(U(L),k). \end{array}$$

Here φ is the homomorphism of (4.4), and ϕ is the analogous map for $\mathrm{H}^\bullet(V(L),k)$ that arises from the graded k-algebra homomorphism $S(L^*[2])^{(1)} \to S(L^*) \cong {}^V E_1^{\bullet,0}$ induced by the squaring operation on $S(L^*)$. Commutativity of (4.8) then implies commutativity of the diagram

$$(4.9)\qquad \begin{array}{ccccccc} L^{(1)} & & \xleftarrow{\quad \phi^* \quad} & & |V(L)| & \hookleftarrow & |V(L)|_M \\ \uparrow{\scriptstyle \mathrm{res}^*} & & & & \uparrow{\scriptstyle \pi_*} & & \uparrow{\scriptstyle \pi_*} \\ L_{\overline{1}}^{(1)} & \hookleftarrow & \mathcal{X}_L(k) & \xleftarrow{\ \simeq\ } & |U(L)| & \hookleftarrow & |U(\widetilde{\mathfrak{g}})|_M. \end{array}$$

By the classical theory (cf. [**18**]),

$$\phi^*(|V(L)|_M) = \{z \in L : z^{\{2\}} = 0 \text{ and } M|_{\langle z\rangle} \text{ is not free}\} \cup \{0\}.$$

Then commutativity of the diagram implies that

$$\mathcal{X}_L(M) \subseteq \{z \in \mathcal{N}_{\mathrm{odd}}(L) : M|_{\langle z\rangle} \text{ is not free}\} \cup \{0\} = \mathcal{X}_L'(M).$$

Then $\mathcal{X}_L'(M) \subseteq \mathcal{X}_L(M)$ and $\mathcal{X}_L(M) \subseteq \mathcal{X}_L'(M)$, so $\mathcal{X}_L'(M) = \mathcal{X}_L(M)$. □

COROLLARY 4.7 (Tensor Product Property). *Let L be a finite-dimensional Lie superalgebra over an algebraically closed field k of characteristic $p = 2$, and let M and N be finite-dimensional L-supermodules. Then*

$$|U(L)|_{M\otimes N} = |U(L)|_M \cap |U(L)|_N.$$

PROOF. Follows immediately from the theorem, in the same manner as Proposition 3.2. □

THEOREM 4.8. *Let L be a finite-dimensional Lie superalgebra over an algebraically closed field k of characteristic $p = 2$. Let M be a finite-dimensional L-supermodule. Then*

$$\{x \in \mathcal{N}_{\mathrm{odd}}(L) : M|_{\langle x\rangle} \textit{ is not free}\} = \emptyset \quad \textit{if and only if} \quad \mathrm{projdim}_{U(L)}(M) < \infty.$$

PROOF. The 'if' portion of the statement is an immediate consequence of the identification (4.5). For the 'only if' portion of statement, the line of reasoning is essentially the same as that for the proof of [**16**, Theorem 3.4.2]. First, one argues as in the proof of [**16**, Lemma 2.4.1] that $U(L)$ is free of finite rank over a central (purely even) polynomial subalgebra $\mathcal{O}$, and hence $U(L)$ is a Noether $\mathcal{O}$-algebra in the terminology of [**16**, Definition A.1.1]. Next, one uses Theorems 4.6 and 4.3 to deduce for each finitely-generated $U(L)$-module N that $\mathrm{Ext}^i_{U(L)}(M,N) = 0$ for all $i \gg 0$. By [**16**, Theorem A.1.2], M has finite projective dimension in $\mathsf{Mod}_{U(L)}$. By (4.3), this implies that M has finite projective dimension in $\mathsf{sMod}_{U(L)}$. □

REMARK 4.9. An immediate consequence of Theorem 4.8 is that

$$\mathcal{N}_{\mathrm{odd}}(L) = \{0\} \quad \text{if and only if} \quad U(L) \text{ is of finite global dimension.}$$

On the one hand, if the global dimension of $U(L)$ is finite, then $\mathrm{H}^\bullet(U(L), k)$ is a finite-dimensional k-algebra, and hence $\mathcal{N}_{\mathrm{odd}}(L) \simeq |U(L)| = \{0\}$. On the other hand, if $\mathcal{N}_{\mathrm{odd}}(L) = \{0\}$, then $\mathrm{projdim}_{U(L)}(k) < \infty$. Since $\mathrm{projdim}_{U(L)}(M) \leq \mathrm{projdim}_{U(L)}(k)$ for any $U(L)$-supermodule M, this implies that $U(L)$ has finite global dimension. This gives in characteristic 2 a new 'geometric' proof of a theorem of Bøgvad [**9**] characterizing graded Lie algebras of finite global dimension.

In any positive characteristic it would be interesting to find a tensor triangulated category whose thick tensor ideals are governed by $\mathrm{Proj}_s(\mathrm{H}^\bullet(U(L), k))$ in a manner analogous to (3.16) and (3.17). Instead of quotienting out the projective supermodules, as one does for the stable category, one would evidently need to factor out the supermodules of finite projective dimension, since these are the modules on which the support theory for $U(L)$ vanishes. This suggests a version of the singularity category will be the appropriate setting to consider these questions.

Acknowledgments

The authors thank Luchezar Avramov for prodding us to address what happens in characteristic two.

References

[1] Luchezar L. Avramov and Srikanth B. Iyengar, *Restricting homology to hypersurfaces*, Geometric and topological aspects of the representation theory of finite groups, Springer Proc. Math. Stat., vol. 242, Springer, Cham, 2018, pp. 1–23, DOI 10.1007/978-3-319-94033-5_1. MR3901154

[2] Yuri A. Bahturin, Alexander A. Mikhalev, Viktor M. Petrogradsky, and Mikhail V. Zaicev, *Infinite-dimensional Lie superalgebras*, De Gruyter Expositions in Mathematics, vol. 7, Walter de Gruyter & Co., Berlin, 1992, DOI 10.1515/9783110851205. MR1192546

[3] Dave Benson, Srikanth B. Iyengar, Henning Krause, and Julia Pevtsova, *Stratification for module categories of finite group schemes*, J. Amer. Math. Soc. **31** (2018), no. 1, 265–302, DOI 10.1090/jams/887. MR3718455

[4] Dave Benson, Srikanth B. Iyengar, Henning Krause, and Julia Pevtsova, *Rank varieties and π-points for elementary supergroup schemes*, Trans. Amer. Math. Soc. Ser. B **8** (2021), 971–998, DOI 10.1090/btran/74. MR4340831

[5] Dave Benson, Srikanth B. Iyengar, Henning Krause, and Julia Pevtsova, *Stratification and duality for unipotent finite supergroup schemes*, Equivariant topology and derived algebra, London Math. Soc. Lecture Note Ser., vol. 474, Cambridge Univ. Press, Cambridge, 2022, pp. 241–275. MR4327102

[6] Brian D. Boe, Jonathan R. Kujawa, and Daniel K. Nakano, *Cohomology and support varieties for Lie superalgebras. II*, Proc. Lond. Math. Soc. (3) **98** (2009), no. 1, 19–44, DOI 10.1112/plms/pdn019. MR2472160

[7] Brian D. Boe, Jonathan R. Kujawa, and Daniel K. Nakano, *Cohomology and support varieties for Lie superalgebras*, Trans. Amer. Math. Soc. **362** (2010), no. 12, 6551–6590, DOI 10.1090/S0002-9947-2010-05096-2. MR2678986

[8] Brian D. Boe, Jonathan R. Kujawa, and Daniel K. Nakano, *Tensor triangular geometry for classical Lie superalgebras*, Adv. Math. **314** (2017), 228–277, DOI 10.1016/j.aim.2017.04.022. MR3658717

[9] Rikard Bøgvad, *Some elementary results on the cohomology of graded Lie algebras*, Algebraic homotopy and local algebra (Luminy, 1982), Astérisque, vol. 113, Soc. Math. France, Paris, 1984, pp. 156–166. MR749051

[10] Sofiane Bouarroudj, Pavel Grozman, Alexei Lebedev, Dimitry Leites, and Irina Shchepochkina, *Simple vectorial Lie algebras in characteristic 2 and their superizations*, SIGMA Symmetry Integrability Geom. Methods Appl. **16** (2020), Paper No. 089, 101, DOI 10.3842/SIGMA.2020.089. MR4153159

[11] David A. Cox, John Little, and Donal O'Shea, *Ideals, varieties, and algorithms*, 4th ed., Undergraduate Texts in Mathematics, Springer, Cham, 2015. An introduction to computational algebraic geometry and commutative algebra, DOI 10.1007/978-3-319-16721-3. MR3330490

[12] Ivo Dell'Ambrogio, *Tensor triangular geometry and KK-theory*, J. Homotopy Relat. Struct. **5** (2010), no. 1, 319–358. MR2812924

[13] Christopher M. Drupieski, *Cohomological finite generation for restricted Lie superalgebras and finite supergroup schemes*, Represent. Theory **17** (2013), 469–507, DOI 10.1090/S1088-4165-2013-00440-5. MR3096330

[14] Christopher M. Drupieski and Jonathan R. Kujawa, *On support varieties for Lie superalgebras and finite supergroup schemes*, J. Algebra **525** (2019), 64–110, DOI 10.1016/j.jalgebra.2018.12.033. MR3909818

[15] Christopher M. Drupieski and Jonathan R. Kujawa, *Support schemes for infinitesimal unipotent supergroups*, Adv. Math. **384** (2021), Paper No. 107754, 61, DOI 10.1016/j.aim.2021.107754. MR4246972

[16] Christopher M. Drupieski and Jonathan R. Kujawa, *Support varieties and modules of finite projective dimension for modular Lie superalgebras*, Algebra Number Theory **15** (2021), no. 5, 1157–1180, DOI 10.2140/ant.2021.15.1157. With an appendix by Luchezar L. Avramov and Srikanth B. Iyengar. MR4283100

[17] M. Duflo and V. Serganova, *On associated variety for Lie superalgebras*, unpublished, 2005, arXiv:0507198.

[18] Eric M. Friedlander and Brian J. Parshall, *Support varieties for restricted Lie algebras*, Invent. Math. **86** (1986), no. 3, 553–562, DOI 10.1007/BF01389268. MR860682

[19] Eric M. Friedlander and Brian J. Parshall, *Geometry of p-unipotent Lie algebras*, J. Algebra **109** (1987), no. 1, 25–45, DOI 10.1016/0021-8693(87)90161-X. MR898334

[20] Eric M. Friedlander and Andrei Suslin, *Cohomology of finite group schemes over a field*, Invent. Math. **127** (1997), no. 2, 209–270, DOI 10.1007/s002220050119. MR1427618

[21] M. Gorelik, C. Hoyt, V. Serganova, and A. Sherman, *The Duflo–Serganova functor, vingt ans après*, J. Indian Inst. Sci. **102** (2022), no. 3, 961–1000. https://doi.org/10.1007/s41745-022-00334-9

[22] Dieter Happel, *Triangulated categories in the representation theory of finite-dimensional algebras*, London Mathematical Society Lecture Note Series, vol. 119, Cambridge University Press, Cambridge, 1988, DOI 10.1017/CBO9780511629228. MR935124

[23] A. Kaufer, *Superalgebra in characteristic 2*, preprint, 2018, arXiv:1804.00824.

[24] Martin Lorenz, *Algebraic group actions on noncommutative spectra*, Transform. Groups **14** (2009), no. 3, 649–675, DOI 10.1007/s00031-009-9059-8. MR2534802

[25] Alexander Premet, *Irreducible representations of Lie algebras of reductive groups and the Kac-Weisfeiler conjecture*, Invent. Math. **121** (1995), no. 1, 79–117, DOI 10.1007/BF01884291. MR1345285

[26] Gunnar Sjödin, *Hopf algebras and derivations*, J. Algebra **64** (1980), no. 1, 218–229, DOI 10.1016/0021-8693(80)90143-X. MR575792

[27] Andrei Suslin, Eric M. Friedlander, and Christopher P. Bendel, *Infinitesimal 1-parameter subgroups and cohomology*, J. Amer. Math. Soc. **10** (1997), no. 3, 693–728, DOI 10.1090/S0894-0347-97-00240-3. MR1443546

[28] Andrei Suslin, Eric M. Friedlander, and Christopher P. Bendel, *Support varieties for infinitesimal group schemes*, J. Amer. Math. Soc. **10** (1997), no. 3, 729–759, DOI 10.1090/S0894-0347-97-00239-7. MR1443547

[29] Siddharth Venkatesh, *Hilbert basis theorem and finite generation of invariants in symmetric tensor categories in positive characteristic*, Int. Math. Res. Not. IMRN **16** (2016), 5106–5133, DOI 10.1093/imrn/rnv305. MR3556434

[30] Weiqiang Wang and Lei Zhao, *Representations of Lie superalgebras in prime characteristic. I*, Proc. Lond. Math. Soc. (3) **99** (2009), no. 1, 145–167, DOI 10.1112/plms/pdn057. MR2520353

[31] Sarah J. Witherspoon, *Hochschild cohomology for algebras*, Graduate Studies in Mathematics, vol. 204, American Mathematical Society, Providence, RI, [2019] ©2019, DOI 10.1090/gsm/204. MR3971234

Department of Mathematical Sciences, DePaul University, Chicago, Illinois 60614
Email address: c.drupieski@depaul.edu

Department of Mathematics, University of Oklahoma, Norman, Oklahoma 73019
Current address: Department of Mathematics, Oregon State University, Corvallis, Oregon 97331
Email address: kujawaj@oregonstate.edu

Proceedings of Symposia in Pure Mathematics
Volume **108**, 2024
https://doi.org/10.1090/pspum/108/01961

Canonical bases for the modified quantum $\mathfrak{gl}_{m|n}$

Jie Du, Haixia Gu, and Zhongguo Zhou

In Memory of Professor Brian J. Parshall (1945–2022)

Abstract. We extend the construction of the regular (matrix) representation for the quantum linear supergroups $\mathbf{U} = \mathbf{U}(\mathfrak{gl}_{m|n})$ developed in [Proc. Amer. Math. Soc., 148 (2020), 111–124] to the modified quantum linear supergroups $\dot{\mathbf{U}}$. We then obtain a new basis for $\dot{\mathbf{U}}$ that is similar to the basis for the algebra $\mathbf{K}$ in [Duke Math J., 61 (1990), 655–677] whose $\mathbb{Z}[v, v^{-1}]$-span $\dot{U}_{\mathbb{Z}}$ becomes an integral form for $\dot{\mathbf{U}}$. We also develop a canonical basis theory for $\dot{U}_{\mathbb{Z}}$ and display its connection to the canonical bases for v-Schur superalgebras.

Contents

1. Introduction
2. The $\mathbf{U}(\mathfrak{gl}_{m|n})$-supermodules $\mathcal{S}^k_{0|1}$ and $\mathcal{S}^l_{1|0}$
3. The $\mathbf{U}(\mathfrak{gl}_{m|n})$-supermodule $\mathcal{K}^{m|n}$
4. Triangular actions on $\mathcal{K}^{m|n}$ by monomial basis elements
5. The modified quantum supergroup $\dot{\mathbf{U}}$ of $\mathbf{U} = \mathbf{U}(\mathfrak{gl}_{m|n})$
6. The regular (matrix) representation of $\dot{\mathbf{U}}$
7. A canonical basis theory for $\dot{\mathbf{U}}$
8. Canonical bases for v-Schur superalgebras
Appendix A. A construction of $\mathbf{U}(\mathfrak{gl}_{m|n})$ from $\dot{\mathbf{U}}$
Appendix B. v-Schur superalgebras and some multiplication formulas
References

1. Introduction

The first example of the modified quantum group was constructed for quantum $\mathfrak{gl}_n$ by Beilinson–Lusztig–MacPherson (BLM) in [**BLM**]. Using the geometry of partial flag varieties and their associated convolution algebras (or q-Schur algebras), certain multiplication (or convolution product) formulas are computed, and their stabilisation properties are derived to define a new algebra $\mathbf{K}$. This algebra is

2020 *Mathematics Subject Classification.* Primary 20B30, 20G43, 17A70. 20C08.

Key words and phrases. Quantum linear supergroup, modified quantum group, regular representation, canonical basis, v-Schur superalgebra.

The first author was partially supported by the Science FRG.

The second author was partially supported by NSFC 12071129. The second author was the corresponding author.

isomorphic to the modified quantum group $\dot{\mathbf{U}}$, generally defined in [**Lubk**, Ch. 23], in the type A case. The BLM construction has soon been generalised to affine type A case; see [**GV**, **Lu**] for quantum affine $\mathfrak{sl}_n$ and [**DF2**] for quantum affine $\mathfrak{gl}_n$. For types other than A, the convolution algebras via the geometry of type B/C have recently been developed to similarly construct the modified $\imath$-quantum groups in [**BKLW**].

For $\mathbf{U} := \mathbf{U}(\mathfrak{gl}_{m|n})$, similar multiplication formulas in υ-Schur superalgebras have been established in [**DG1**, Props. 4.4&4.5]. However, the stabilisation property cannot be seen from them.[1] Thus, a BLM type realization for $\dot{\mathbf{U}}$ in this super case is not known.

In [**DZ**], the first and third authors develop a simple approach to present quantum linear supergroups $\mathbf{U}$ via their regular representations in matrix form. By using quantum differential operators, we directly introduce a supermodule structure on a certain polynomial superalgebra and extend this structure to the associated formal power series algebra. The latter contains a submodule that is naturally a supermodule. We then prove that this supermodule is isomorphic to the regular representations of the quantum linear supergroup. In this way, we present $\mathbf{U}$ by a basis and its associated matrix representations of generators. Note that this differential operator approach has already been generalized to quantum queer supergroups in [**DLZ2**].

In this paper, we reveal a surprising by-product of the theory. This time, we extend the $\mathbf{U}$-supermodule structure on the polynomial superalgebra $\mathcal{S}^{m|n}$ to a (partially) Laurent polynomial superalgebra $\mathcal{K}^{m|n}$. This supermodule structure induces naturally a unital $\dot{\mathbf{U}}$-supermodule $\mathcal{K}^{m|n}$ (in the sense of [**Lubk**, 23.1.4]) for the modified quantum supergroup $\dot{\mathbf{U}}$ of $\mathbf{U}$. We then prove that this $\dot{\mathbf{U}}$-supermodule $\mathcal{K}^{m|n}$ is isomorphic to the regular (unital) module ${}_{\dot{\mathbf{U}}}\dot{\mathbf{U}}$. In this way, we obtain a presentation of $\dot{\mathbf{U}}$ by a basis that arises from the monomial basis for $\mathcal{K}^{m|n}$ and their multiplication formulas by generators that arise from the action formulas on $\mathcal{K}^{m|n}$ by generators. This new basis together with the monomial basis allows us to establish a canonical basis theory for $\dot{\mathbf{U}}$.

We organize the paper as follows. We start with extending the $\mathbf{U}$-modules $\mathcal{S}_{0|1}$ and $\mathcal{S}_{1|0}$ in [**DZ**] to $\mathcal{S}^k_{0|1}$ and $\mathcal{S}^l_{1|0}$ in §2, where a superscript s indicates that the s-th even variable X_s is replaced by a Laurent variable $X_s^{\pm 1}$. The $\mathbf{U}$-module $\mathcal{K}^{m|n}$ is a tensor product of various $\mathcal{S}^k_{0|1}$ and $\mathcal{S}^l_{1|0}$. Explicit action formulas by generators K_i, E_h, F_h are derived in §3. In §4, we discuss the actions on $\mathcal{K}^{m|n}$ by monomial basis vectors with a certain triangular property. The modified quantum supergroup $\dot{\mathbf{U}}$ is introduced and various properties are discussed in §5. In §6, the $\mathbf{U}$-supermodule $\mathcal{K}^{m|n}$ is turned into a unital $\dot{\mathbf{U}}$-supermodule which is proved to be isomorphic to the regular supermodule ${}_{\dot{\mathbf{U}}}\dot{\mathbf{U}}$. Thus, we obtain a new basis for $\dot{\mathbf{U}}$ and a presentation via the basis together with some explicit multiplication formulas by the generators for $\dot{\mathbf{U}}$ (Theorem 6.4). With this new basis, a canonical basis theory is developed in §7 (Theorem 7.5), and finally, the relationship between $\dot{\mathbf{U}}$ and υ-Schur superalgebras and between their canonical bases are discussed in the last section.

In Appendix A, we give an application of the BLM basis for $\dot{\mathbf{U}}$ in (6.4.1). We follow the approach in [**BLM**] to construct a completion superalgebra $\widehat{\dot{\mathbf{U}}}$ from $\dot{\mathbf{U}}$

[1] Motivated from the current work, we will discuss possible new multiplication formulas in Remark 6.6(1) and Appendix B from which the stabilisation property may be established.

and to prove that a subsuperalgebra $\mathfrak{U}(m|n)$ of $\widehat{\mathbf{U}}$ is isomorphic to $\mathbf{U}(\mathfrak{gl}_{m|n})$. Instead of repeating the steps in [**BLM**, §5], we directly show that $\mathfrak{U}(m|n)$ is isomorphic to the subsuperalgebra $\mathfrak{A}(m|n)$ of the direct product of v-Schur superalgebras (see [**DG1**, Ths. 8.1&8.4]). Appendix B gives a further discussion about the new basis mentioned in Remark 6.6(1). Multiplication formulas relative to the new basis for v-Schur superalgebras are given. We compare them with those in Theorem 6.4.

Throughout the paper, we always regard $\mathbb{Z}_2 = \{0, 1\}$ as a subset of $\mathbb{Z}$ unless it is used as the grading set for a superspace. If the latter occurs, we regard $\mathbb{Z}_2$ as the abelian group of order 2 and denote its elements by $\bar{0}, \bar{1}$.

Let $\mathcal{Z} = \mathbb{Z}[v, v^{-1}]$ be the Laurent polynomial ring over $\mathbb{Z}$, and let $\mathbb{Q}(v)$ be the fraction field of $\mathcal{Z}$. For an integers s and $a \geq 1$, let

$$[s] = \frac{v^s - v^{-s}}{v - v^{-1}}, [a]^! = [1][2] \cdots [a] \text{ and } \begin{bmatrix} s \\ a \end{bmatrix} = \prod_{i=1}^{a} \frac{v^{s-i+1} - v^{-s+i-1}}{v^i - v^{-i}}.$$

Note that $[-i] = -[i]$, $[0] = 0$, $[0]^! = 1$, and $\left[\begin{smallmatrix} s \\ 0 \end{smallmatrix}\right] = 1$. Note also that if $\left[\begin{smallmatrix} s \\ a \end{smallmatrix}\right]_q$ denotes the value at q for any q in a ring, then $\left[\begin{smallmatrix} s \\ a \end{smallmatrix}\right]_{v^{-1}} = \left[\begin{smallmatrix} s \\ a \end{smallmatrix}\right]$.

2. The $\mathbf{U}(\mathfrak{gl}_{m|n})$-supermodules $\mathcal{S}^k_{0|1}$ and $\mathcal{S}^l_{1|0}$

Let $m, n \geq 0$ (not both 0) be integers. Let $[1, m+n] := \{1, \cdots, m+n\}$, and $[1, m+n) = [1, m+n] \backslash \{m+n\}$. Let $\{\mathbf{e}_1, \mathbf{e}_2, \cdots, \mathbf{e}_{m+n}\}$ be the standard basis for $\mathbb{Z}^{m+n}$, and define the "super dot product" on $\mathbb{Z}^{m+n}$ by

(2.0.1) $$\mathbf{e}_i \centerdot_s \mathbf{e}_j = (\mathbf{e}_i, \mathbf{e}_j)_s = (-1)^{\wp(i)} \delta_{ij},$$

where the parity function $\wp : [1, m+n] \to \mathbb{Z}_2 = \{\bar{0}, \bar{1}\}, i \mapsto \wp(i)$ by

$$\wp(i) = \begin{cases} \bar{0}, & \text{if } 1 \leq i \leq m; \\ \bar{1}, & \text{if } m+1 \leq i \leq m+n. \end{cases}$$

For $i \in [1, m+n)$, if we set $\alpha_i = \mathbf{e}_i - \mathbf{e}_{i+1}$, then

(2.0.2) $$\alpha_1, \ldots, \alpha_{m-1}, \alpha_m, \alpha_{m+1}, \ldots, \alpha_{m+n-1}$$

form the simple root system for $\mathbf{U}$. The parity of α_i is $\wp(i) + \wp(i+1)$. Thus, α_i is odd if and only if $i = m$.

Recall from [**Zh**], [**EK**], and [**DG1**] that the quantum supergroup $\mathbf{U} := \mathbf{U}(\mathfrak{gl}_{m|n})$ over $\mathbb{Q}(v)$ is generated by

$$\begin{cases} \text{even generators:} & E_h, F_h, K_j^{\pm 1}, \quad 1 \leq h, j \leq m+n, h \neq m, m+n; \\ \text{odd generators:} & E_m, F_m. \end{cases}$$

which satisfy the relations (QG1)–(QG6) below (see e.g., [**DZ**, Def. 2.1]):

(QG1) $K_i K_j = K_j K_i$, $K_j K_j^{-1} = K_j^{-1} K_j = 1$;

(QG2) $K_j E_h = v^{\mathbf{e}_j \centerdot_s \alpha_h} E_h K_j$, $K_j F_h = v^{-\mathbf{e}_j \centerdot_s \alpha_h} F_h K_j$;

(QG3) $[E_a, F_b] = \delta_{a,b} \frac{K_a K_{a+1}^{-1} - K_a^{-1} K_{a+1}}{v_a - v_a^{-1}}$;

(QG4) $E_a E_b = E_b E_a$, $F_a F_b = F_b F_a$, if $|a - b| > 1$;

(QG5) For $a \neq m$ and $|a - b| = 1$,

$$E_a^2 E_b - (v_a + v_a^{-1}) E_a E_b E_a + E_b E_a^2 = 0,$$
$$F_a^2 F_b - (v_a + v_a^{-1}) F_a F_b F_a + F_b F_a^2 = 0;$$

(QG6) $E_m^2 = F_m^2 = [E_m, E_{m-1,m+2}] = [F_m, E_{m+2,m-1}] = 0$, where

$E_{m-1,m+2} = E_{m-1}E_mE_{m+1} - vE_{m-1}E_{m+1}E_m - v^{-1}E_mE_{m+1}E_{m-1} + E_{m+1}E_mE_{m-1}$,

$E_{m+2,m-1} = F_{m+1}F_mF_{m-1} - v^{-1}F_mF_{m+1}F_{m-1} - vF_{m-1}F_{m+1}F_m + F_{m-1}F_mF_{m+1}$.

Note that if we set $v_h = v^{(-1)^{\wp(h)}}$ for all $h \in [1, m+n]$, then $v^{\mathbf{e}_a \bullet_s \alpha_b} = v_a^{\mathbf{e}_a \bullet \alpha_b}$ with the usual dot product.

Note also that $\mathbf{U}$ is a Hopf superalgebra with the comultiplication $\Delta : \mathbf{U} \to \mathbf{U} \otimes \mathbf{U}$ defined by

$$\begin{aligned} \Delta(K_i) &= K_i \otimes K_i, \\ \Delta(E_h) &= E_h \otimes \widetilde{K}_h + 1 \otimes E_h, \quad \Delta(F_h) = F_h \otimes 1 + \widetilde{K}_h^{-1} \otimes F_h, \end{aligned} \tag{2.0.3}$$

where $\widetilde{K}_h = K_h K_{h+1}^{-1}$. (We omit the counit and antipode maps as they are not used.)

For later use, we also record a $\mathbb{Q}(v)$-algebra anti-automorphisms of order 2:

$$\tau : \mathbf{U} \longrightarrow \mathbf{U}, \quad E_h \mapsto F_h, F_h \mapsto E_h, K_j \mapsto K_j, \tag{2.0.4}$$

and a ring automorphism of order 2

$$\bar{\ } : \mathbf{U} \longrightarrow \mathbf{U} \text{ with } \bar{v} = v^{-1}, \overline{E}_h = E_h, \overline{F}_h = F_h, \overline{K}_i = K_i^{-1}. \tag{2.0.5}$$

The latter is known as the "bar involution" which will be used in the construction of canonical bases.

We first extend the $\mathbf{U}$-supermodule structure on the polynomial superalgebras

$$\begin{aligned} \mathcal{S}_{0|1} &:= \mathbb{Q}(v)[X_1, X_2, \cdots, X_{m+n}] \text{ with } \wp(X_i) = \bar{0}, \wp(X_{m+j}) = \bar{1}, \\ \mathcal{S}_{1|0} &:= \mathbb{Q}(v)[X_1, X_2, \cdots, X_{m+n}] \text{ with } \wp(X_i) = \bar{1}, \wp(X_{m+j}) = \bar{0}, \end{aligned} \tag{2.0.6}$$

for all $i \in [1, m], j \in [1, n]$ (see [**DZ**, (3.0.1),(3.0.2)]), to their partial Laurent polynomial superalgebras

(2.0.7)
$$\begin{aligned} \mathcal{S}_{0|1}^k &:= \mathbb{Q}(v)[X_1, \ldots, X_{k-1}, X_k, X_k^{-1}, X_{k+1} \cdots, X_{m+n}] \quad (k \in [1, m]), \\ \mathcal{S}_{1|0}^l &:= \mathbb{Q}(v)[X_1, \ldots, X_{l-1}, X_l, X_l^{-1}, X_{l+1}, \ldots, X_{m+n}] \quad (l \in [m+1, m+n]). \end{aligned}$$

Recall the basis

$$X^{(\mathbf{a})} = X_1^{(a_1)} X_2^{(a_2)} \cdots X_{m+n}^{(a_{m+n})}, \tag{2.0.8}$$

where $X_i^{(a_i)} = \frac{X_i^{a_i}}{[a_i]!}$ and $\mathbf{a} = (a_1, \cdots, a_{m+n}) \in \mathbb{N}^m \times \mathbb{Z}_2^n$ for $\mathcal{S}_{0|1}$, respectively, $\mathbf{a} = (a_1, \cdots, a_{m+n}) \in \mathbb{Z}_2^m \times \mathbb{N}^n$ for $\mathcal{S}_{1|0}$. Note that $X^{(\mathbf{0})} = 1$ is the identity element of $\mathcal{S}_{0|1}$ or $\mathcal{S}_{1|0}$.

The $\mathbf{U}$-supermodule $\mathcal{S}_{0|1}$ (resp., $\mathcal{S}_{1|0}$) are defined as follows (see [**DZ**, (3.0.2)]): for $1 \le h, i \le m+n$, $h \ne m+n$, and $\mathbf{a} = (a_1, \cdots, a_{m+n}) \in \mathbb{N}^m \times \mathbb{Z}_2^n$ (resp., $\mathbb{Z}_2^m \times \mathbb{N}^n$),

$$\begin{aligned} K_i.X^{(\mathbf{a})} &= v_i^{a_i} X^{(\mathbf{a})}, \\ E_h.X^{(\mathbf{a})} &= \begin{cases} [a_h+1]X^{(\mathbf{a}+\alpha_h)}, & \text{if } a_{h+1} > 0; \\ 0, & \text{otherwise.} \end{cases} \\ F_h.X^{(\mathbf{a})} &= \begin{cases} [a_{h+1}+1]X^{(\mathbf{a}-\alpha_h)}, & \text{if } a_h > 0; \\ 0, & \text{otherwise.} \end{cases} \end{aligned} \tag{2.0.9}$$

Here $\alpha_h = \mathbf{e}_h - \mathbf{e}_{h+1}$ are simple roots.

For $k \in [1,m], l = m+k' \in [m+1, m+n]$, the superalgebras $\mathcal{S}^k_{0|1}, \mathcal{S}^l_{1|0}$ as $\mathbb{Q}(v)$-vector spaces have bases, respectively,

$$\{X^{(\mathbf{a})} \mid \mathbf{a} = (a_1, \cdots, a_{m+n})^t \in \mathbb{N}^m(k,\mathbb{Z}) \times \mathbb{Z}_2^n\}; \tag{2.0.10}$$
$$\{X^{(\mathbf{a})} \mid \mathbf{a} = (a_1, \cdots, a_{m+n})^t \in \mathbb{Z}_2^m \times \mathbb{N}^n(k',\mathbb{Z})\},$$

where $\mathbb{N}^s(t,\mathbb{Z}) := \mathbb{N}^{t-1} \times \mathbb{Z} \times \mathbb{N}^{s-t}$ (i.e., replacing the tth factor $\mathbb{N}$ by $\mathbb{Z}$) for all $t \in [1,s]$. Here, for later use, $\mathbf{a}$ is a column vector. Note that if $a \in \mathbb{Z}$ is negative, then $X_i^{(a)} = \frac{X_i^a}{[a]!} = (-1)^{-a}\frac{X_i^a}{[-a]!}$ for $i = k$ or l is well defined. Thus, $X^{(\mathbf{a})}$ is similarly defined as in (2.0.8).

The parity, induced from the parity on $\mathcal{S}_{0|1}$ and $\mathcal{S}_{1|0}$ defined in (2.0.6), of each basis element $X^{(\mathbf{a})}$ is defined by

$$\wp(X^{(\mathbf{a})}) = \begin{cases} \sum_{j=1}^n a_{m+j} \pmod 2, & \text{if } \mathbf{a} \in \mathbb{N}^m(k,\mathbb{Z}) \times \mathbb{Z}_2^n \\ \sum_{j=1}^m a_j \pmod 2, & \text{if } \mathbf{a} \in \mathbb{Z}_2^m \times \mathbb{N}^n(k',\mathbb{Z}). \end{cases} \tag{2.0.11}$$

Thus, $\mathcal{S}^k_{0|1}, \mathcal{S}^l_{1|0}$ are superspaces. We now extend the $\mathbf{U}$-supermodule structure on $\mathcal{S}_{0|1}, \mathcal{S}_{1|0}$ to $\mathcal{S}^k_{0|1}, \mathcal{S}^l_{1|0}$. We also define, for each basis element $X^{(\mathbf{a})}$, its degree

$$\deg(X^{(\mathbf{a})}) = \sum_{i=1}^{m+n} a_i.$$

LEMMA 2.1. *For all $k \in [1,m], l = m+k' \in [m+1, m+n]$, $\mathcal{S}^k_{0|1}$ (resp. $\mathcal{S}^l_{1|0}$) is a $\mathbf{U}$-supermodule under the following actions: for all $i, h \in [1, m+n]$ $(h \neq m+n)$ and $\mathbf{a} \in \mathbb{N}^m(k,\mathbb{Z}) \times \mathbb{Z}_2^n$ (resp., $\mathbf{a} \in \mathbb{Z}_2^m \times \mathbb{N}^n(k',\mathbb{Z})$),*

$$K_i.X^{(\mathbf{a})} = v_i^{a_i} X^{(\mathbf{a})}, \tag{2.1.1}$$
$$E_h.X^{(\mathbf{a})} = \begin{cases} [a_h+1]X^{(\mathbf{a}+\alpha_h)}, & \text{if } a_{h+1} > 0 \text{ and } h+1 \neq k \text{ (resp., } l); \\ [a_h+1]X^{(\mathbf{a}+\alpha_h)}, & \text{if } h+1 = k \text{ (resp., } l); \\ 0, & \text{otherwise}, \end{cases}$$
$$F_h.X^{(\mathbf{a})} = \begin{cases} [a_{h+1}+1]X^{(\mathbf{a}-\alpha_h)}, & \text{if } a_h > 0 \text{ and } h \neq k \text{ (resp., } l); \\ [a_{h+1}+1]X^{(\mathbf{a}-\alpha_h)}, & \text{if } h = k \text{ (resp., } l); \\ 0, & \text{otherwise.} \end{cases}$$

Moreover,

$$\mathcal{S}^k_{0|1} = \bigoplus_{r\in\mathbb{Z}} \mathcal{S}^k_{0|1}(r), \quad \mathcal{S}^l_{1|0} = \bigoplus_{r\in\mathbb{Z}} \mathcal{S}^l_{1|0}(r),$$

where $\mathcal{S}^k_{0|1}(r)$, $\mathcal{S}^l_{1|0}(r)$ $(r \in \mathbb{Z})$ are the homogeneous components of degree r and are all subsupermodules.

PROOF. We need to show that the actions defined in (2.1.1) satisfy the defining relations for $\mathbf{U} = \mathbf{U}(\mathfrak{gl}_{m|n})$. We just prove the $\mathcal{S}^k_{0|1}$ case; the proof for $\mathcal{S}^l_{1|0}$ is similar.

Let R be one of the defining relation for $\mathbf{U}$. We want to prove that, for every $\mathbf{a} \in \mathbb{N}^m(k,\mathbb{Z}) \times \mathbb{Z}_2^n$, R.$X^{(\mathbf{a})} = 0$. We now use the idea in the proof of the stabilisation property in [**BLM**] to prove this.

For $p \in \mathbb{Z}$, let $X_p^{(\mathbf{a})} = X^{(\mathbf{a}+p\mathbf{e}_k)}$, $1 \le k \le m$. Then, for sufficiently large $p \gg 0$, $a_k + p > 0$. Thus, we have $X_p^{(\mathbf{a})} \in \mathcal{S}_{0|1}$. Hence, for any $g \in \mathfrak{G}$, where

$$\mathfrak{G} = \{E_h, F_h, K_i^{\pm 1} \mid h, i \in [1, m+n], h < m+n\},$$

the action formula for $g.X_p^{(\mathbf{a})}$ is given in (2.0.9). More precisely, only for $g \in \{K_k, E_k, F_{k-1}\}$, the coefficients in the action formulas involve p and have the form

$$v_k^{a_k+p}, \text{ or } [a_k+p+1] = \frac{v^{a_k+1}v^p - v^{-a_k-1}v^{-p}}{v - v^{-1}}.$$

For other g's, the action formula for $g.X_p^{(\mathbf{a})}$ agrees with that for $g.X^{(\mathbf{a})}$. Thus, if we put

$$f_{K_k,\mathbf{a}}(v') = v_k^{a_k}v' \text{ and } f_{E_k,\mathbf{a}}(v') = \frac{v^{a_k+1}v' - v^{-a_k-1}v'^{-1}}{v - v^{-1}} = f_{F_{k-1},\mathbf{a}}(v'),$$

and, for other g's, let $f_{g,\mathbf{a}}(v')$ be the coefficient (including 0) in $g.X^{(\mathbf{a})}$, then $f_{g,\mathbf{a}}(v') \in \mathbb{Q}(v)[v', v'^{-1}]$. By definition, we have, for all $g \in \mathfrak{G}$ and $p \gg 0$,

(2.1.2) (a) $g.X_p^{(\mathbf{a})} = f_{g,\mathbf{a}}(v^p)X_p^{(\mathbf{b})}$, as given in (2.0.9), and

(b) $g.X^{(\mathbf{a})} = f_{g,\mathbf{a}}(1)X^{(\mathbf{b})}$, as given in (2.1.1).

Thus, for the action on $X_p^{(\mathbf{a})}$ by a relation R, there exist elements $f_{\mathrm{R},\mathbf{a}}^{\mathbf{b}}(v') \in \mathbb{Q}(v)[v', v'^{-1}]$ such that we have correspondingly,

(a) in $\mathcal{S}_{0|1}$, $\quad 0 = \mathrm{R}.X_p^{(\mathbf{a})} = \sum_{\mathbf{b}} f_{\mathrm{R},\mathbf{a}}^{\mathbf{b}}(v^p)X_p^{(\mathbf{b})}$, for all $p \gg 0$,

(b) in $\mathcal{S}_{0|1}^k$, $\quad \mathrm{R}.X^{(\mathbf{a})} = \sum_{\mathbf{b}} f_{\mathrm{R},\mathbf{a}}^{\mathbf{b}}(1)X^{(\mathbf{b})}$.

Hence $f_{\mathrm{R},\mathbf{a}}^{\mathbf{b}}(v^p) = 0$ for all $p \gg 0$ and $\mathbf{b}$. This implies all $f_{\mathrm{R},\mathbf{a}}^{\mathbf{b}}(v') \equiv 0$. Therefore, all $f_{\mathrm{R},\mathbf{a}}^{\mathbf{b}}(1) = 0$ and consequently, $\mathrm{R}.X^{(\mathbf{a})} = 0$. This completes the proof. □

We also may prove this lemma directly by verifying all defining relations for the actions of generators on $X^{(\mathbf{a})} \in \mathcal{S}_{0|1}^k$ or in $\mathcal{S}_{1|0}^l$ similarly as done in [**DZ**, Lemma 3.1]. This is possible because all relevant equations involving Gauss binomial coefficients still hold when some positive integers are replaced by negative integers. For example:

$$[a_m] + [a_m+2] - (v+v^{-1})[a_m+1] = 0$$

is true for all $a_m \in \mathbb{Z}$ (see the proof of [**DZ**, Lemma 3.1]).

We illustrate the action with the following example. For $m = n = 2$, the **U**-action on $\mathcal{S}_{0|1}^2$ takes the form (setting $[\mathbf{a}] := X^{(\mathbf{a})}$ for simplicity):

$$E_1.\begin{bmatrix}1\\-2\\0\\1\end{bmatrix} = [2]\begin{bmatrix}2\\-3\\0\\1\end{bmatrix},\ F_2.\begin{bmatrix}1\\-2\\0\\1\end{bmatrix} = [1]\begin{bmatrix}1\\-3\\1\\1\end{bmatrix},\ F_3.\begin{bmatrix}1\\-2\\1\\1\end{bmatrix} = [2]\begin{bmatrix}1\\-2\\0\\2\end{bmatrix} = 0.$$

$$E_1.\begin{bmatrix}1\\0\\0\\1\end{bmatrix} = [2]\begin{bmatrix}2\\-1\\0\\1\end{bmatrix},\ F_1.\begin{bmatrix}2\\-1\\0\\1\end{bmatrix} = [-1+1]\begin{bmatrix}1\\0\\1\\1\end{bmatrix} = 0.$$

REMARKS 2.2.

(1) We used the idea of the stabilisation property discussed in [**BLM**, §5] to derive the new action formulas.

(2) Unlike the actions on $\mathcal{S}_{0|1}$ and $\mathcal{S}_{1|0}$, the new actions cannot be interpreted as quantum differential operators. For example, $F_1.X^{(\mathbf{0})} = X^{(-\mathbf{e}_1+\mathbf{e}_2)}$ is not 0 in $\mathcal{S}^1_{0|1}$.

(3) The new action formulas in (2.1.1) agree with those in (2.0.9) when the kth (or lth)-component of $\mathbf{a}$ satisfies $a_k > 0$ (or $a_l > 0$).

(4) The actions may result in some odd component of $\mathbf{b}$ being 2, if $X^{(\mathbf{b})}$ appears in $E_h.X^{(\mathbf{a})}$ or $F_h.X^{(\mathbf{a})}$. When this occurs, it means that the term is 0. (Recall from [**DZ**, Ex. 2.2(2)] a generator d for the exterior algebra satisfies $d^2 = 0$.) For simplity we set $X^{(\mathbf{a})} = 0$ if any odd component $a_i \geq 2$.

3. The $\mathbf{U}(\mathfrak{gl}_{m|n})$-supermodule $\mathcal{K}^{m|n}$

We now form a certain tensor product $\mathcal{K}^{m|n}$ of $\mathcal{S}^k_{0|1}, \mathcal{S}^l_{1|0}$ and derive the action formulas for the $\mathbf{U}$-supermodule $\mathcal{K}^{m|n}$. The following constructions and results in the rest of this section are paralleled to those in [**DZ**].

Recall the tensor product $\mathcal{S}^{m|n} = (\mathcal{S}_{0|1})^{\otimes m} \otimes (\mathcal{S}_{1|0})^{\otimes n}$ of multiple $\mathcal{S}_{0|1}$ and $\mathcal{S}_{1|0}$ from [**DZ**, (3.2.1)]. We now define its extension $\mathcal{K}^{m|n}$:

(3.0.1) $$\mathcal{K}^{m|n} := \mathcal{S}^1_{0|1} \otimes \cdots \otimes \mathcal{S}^m_{0|1} \otimes \mathcal{S}^{m+1}_{1|0} \otimes \cdots \otimes \mathcal{S}^{m+n}_{1|0}.$$

Since each tensor factor is a $\mathbf{U}$-module, the comultiplication in (2.0.3) makes $\mathcal{K}^{m|n}$ into a $\mathbf{U}$-module.

Like the polynomial superalgebra $\mathcal{S}^{m|n}$, $\mathcal{K}^{m|n}$ may be regarded as a *diagonally Laurent* polynomial superalgebra:

$$\mathcal{K}^{m|n} \cong \mathbb{Q}(v)[X_{ij}, X_{i,i}, X^{-1}_{i,i} \mid i, j \in [1, m+n], i \neq j].$$

We first describe a monomial basis for $\mathcal{K}^{m|n}$. Let

(3.0.2) $$\begin{aligned} M(m|n) &= \left\{ \begin{pmatrix} X & Q \\ Q' & Y \end{pmatrix} \,\middle|\, \begin{matrix} X \in M_{m\times m}(\mathbb{N}), Q \in M_{m\times n}(\mathbb{Z}_2) \\ Q' \in M_{n\times m}(\mathbb{Z}_2), Y \in M_{n\times n}(\mathbb{N}) \end{matrix} \right\} \\ M(m|n)^{\pm} &= \{A^{\pm} := A^- + A^+ \mid A \in M(m|n)\} \\ \widetilde{M}(m|n) &= \{A + aI \mid A \in M(m|n), a \in \mathbb{Z}\} \\ \widetilde{M}(m|n)^{\pm} &= M(m|n)^{\pm}, \end{aligned}$$

Here, for such a square matrix A, we naturally decompose A into the sum of its upper/lower triangular part A^+/A^- and its diagonal part

$$A^0 = \mathrm{diag}(a_{1,1}, \cdots, a_{m+n,m+n}) \,:\, A = A^- + A^0 + A^+.$$

Note that the set $M(m|n)$ is used to define the monomial basis for $\mathcal{S}^{m|n}$ in [**DZ**].

For $A = (a_{i,j}) \in \widetilde{M}(m|n)$, let $|A| = \sum_{i,j} a_{i,j}$ and define the parity of A, induced from (2.0.11), by

(3.0.3) $$\wp(A) := \sum_{\wp(i)+\wp(j)=\bar{1}} a_{i,j} \pmod 2.$$

We also define the row-/column-sum vectors

$$\text{ro}(A) = (\sum_j a_{1,j}, \ldots, \sum_j a_{m+n,j}), \text{co}(A) = (\sum_i a_{i,1}, \ldots, \sum_i a_{i,m+n}). \tag{3.0.4}$$

Following [**DZ**, (5.5.1)], let

$$s(A) = \sum_{\substack{1\le i,\, k\le m\\ m<j<l\le m+n}} a_{i,j}a_{k,l} = \sum_{m<j<l\le m+n} \Big(\sum_{i=1}^{m} a_{i,j}\Big)\Big(\sum_{k=1}^{m} a_{k,l}\Big). \tag{3.0.5}$$

Write $A = (\mathbf{c}_1|\mathbf{c}_2|\cdots|\mathbf{c}_{m+n})$ and define

$$X^{[A]} = X^{(\mathbf{c}_1)} \otimes \cdots \otimes X^{(\mathbf{c}_{m+n})}, \qquad {}'X^{[A]} := (-1)^{s(A)} X^{[A]}, \tag{3.0.6}$$

where $X^{(\mathbf{c}_k)} \in \mathcal{S}^k_{0|1}$ (resp., $X^{(\mathbf{c}_l)} \in \mathcal{S}^l_{1|0}$) if $k \in [1, m]$ (resp., $l \in [m+1, m+n]$).

REMARK 3.1. Note that the basis element $X^{[A]}$, for $A \in \widetilde{M}(m|n)$, is the natural extension of the basis element $X^{[A]}$, for $A \in M(m|n)$, defined in [**DZ**, p.116]. Their signed version ${}'X^{[A]}$ will be convenient for the construction of the canonical basis below.

Recall also the sign rule: for supermodules V_1, V_2 over a superalgebra U, if $u_1, u_2 \in U, v_i \in V_i$ with u_2, v_1 homogeneous, then

$$(u_1 \otimes u_2).(v_1 \otimes v_2) = (-1)^{\wp(u_2)\wp(v_1)} u_1v_1 \otimes u_2v_2.$$

We now extend the $\mathbf{U}$-action formulas on $\mathcal{S}^{m|n}$ given in [**DZ**, Lem. 3.3] to $\mathcal{K}^{m|n}$. For $A \in M(m|n)$, $i \in [1, m+n]$, the following are defined in [**DZ**, (3.2.3),(3.2.4)]:

$$\sigma(i, A) = \begin{cases} \sum\limits_{s>m,t<i} a_{s,t}, & \text{if } 1 \le i \le m; \\ \sum\limits_{s>m,t\le m} a_{s,t} + \sum\limits_{s\le m,\, m<t<i} a_{s,t}, & \text{if } m+1 \le i \le m+n, \end{cases} \tag{3.1.1}$$

$$f(i, A) = f_h(i, A) = \sum_{j>i} a_{h,j} - (-1)^{\delta_{h,m}} \sum_{j>i} a_{h+1,j},$$

$$g(i, A) = g_h(i, A) = \sum_{j<i} a_{h+1,j} - (-1)^{\delta_{h,m}} \sum_{j<i} a_{h,j}.$$

With the adjustment of the sign (3.0.5), we need the following:

$$\tilde{\sigma}(i, A) = \begin{cases} \sum\limits_{s>m,t<i} a_{s,t}, & \text{if } 1 \le i \le m; \\ \sum\limits_{s>m,t\le m} a_{s,t} + \sum\limits_{s\le m,\, t>i} a_{s,t}, & \text{if } m+1 \le i \le m+n, \end{cases} \tag{3.1.2}$$

Let $\tilde{\sigma}_{h,m}(i, A) = \delta_{h,m}\tilde{\sigma}(i, A)$. The following action formulas generalise [**DZ**, Lemma 3.3].

PROPOSITION 3.2. *The set* $\{'X^{[A]} \mid A \in \widetilde{M}(m|n)\}$ *forms a* $\mathbb{Q}(v)$*-basis for the* $\mathbf{U}$*-supermodule* $\mathcal{K}^{m|n}$ *which has the following action formulas:*

(3.2.1)

$$
\begin{aligned}
(1)\quad & K_i.'X^{[A]} = v_i^{\sum_{1\le j\le m+n} a_{i,j}}\,'X^{[A]} = v_i^{\mathrm{ro}(A)_i}\,'X^{[A]},\\
(2)\quad & E_h.'X^{[A]} = \sum_{\substack{1\le i\le m+n, i\ne h+1\\ a_{h+1,i}\ge 1}} (-1)^{\tilde\sigma_{h,m}(i,A)} v_h^{f(i,A)}[a_{h,i}+1]'X^{[A+E_{h,i}-E_{h+1,i}]}\\
& \qquad + (-1)^{\tilde\sigma_{h,m}(h+1,A)} v_h^{f(h+1,A)}[a_{h,h+1}+1]'X^{[A+E_{h,h+1}-E_{h+1,h+1}]},\\
(3)\quad & F_h.'X^{[A]} = \sum_{\substack{1\le i\le m+n, i\ne h\\ a_{h,i}\ge 1}} (-1)^{\tilde\sigma_{h,m}(i,A)} v_{h+1}^{g(i,A)}[a_{h+1,i}+1]'X^{[A-E_{h,i}+E_{h+1,i}]}\\
& \qquad + (-1)^{\tilde\sigma_{h,m}(h,A)} v_{h+1}^{g(h,A)}[a_{h+1,h}+1]'X^{[A-E_{h,h}+E_{h+1,h}]}.
\end{aligned}
$$

The differences between these action formulas with those in [**DZ**, Lemma 3.3] are the conditions $a_{h+1,h+1} \ge 1$ in (2) and $a_{h,h} \ge 1$ in (3) which are dropped here.

PROOF. By applying the comultiplication $\Delta^{(m+n-1)} : \mathbf{U} \to \mathbf{U}^{\otimes(m+n)}$, as given in [**DZ**, (3.3.2)], to K_i, E_h, F_h, the action formulas in (2.1.1) induce action formulas for $K_i.X^{[A]}, E_h.X^{[A]}$, and $F_h.X^{[A]}$ whose right hand sides are obtained from (3.2.1) with all $\tilde\sigma_{h,m}(i,A)$ replaced by $\sigma_{h,m}(i,A)$. Applying the formulas stated in [**DZ**, Lem. 5.6] yields

$$
\begin{aligned}
s(A)+\sigma_{h,m}(i,A) &= s(A+E_{h,i}-E_{h+1,i}) + \delta_{h,m}\Big(\sum_{\substack{s>m\\ t\le\min\{i-1,m\}}} a_{s,t} - \delta^{>}_{i,m}\sum_{\substack{s\le m\\ t>i}} a_{s,t}\Big)\\
&= s(A+E_{h,i}-E_{h+1,i}) + \tilde\sigma_{h,m}(i,A) - 2\delta^{>}_{i,m}\sum_{s\le m, t>i} a_{s,t}.
\end{aligned}
$$

Here, $\delta^{>}_{k,m}=1$ if $k>m$ and 0 otherwise. Substituting gives the desired (3.2.1). □

A $\mathbf{U}$-supermodule M is called a weight supermodule, if M has a weight space decomposition $M = \bigoplus_{\lambda\in\mathbb{Z}^{m+n}} M_\lambda$ where

$$M_\lambda = \{m \in M \mid K_i.m = v_i^{\lambda_i} m,\ \forall 1\le i\le m+n\}.$$

Clearly, we have $E_i M_\lambda \subseteq M_{\lambda+\alpha_i}$ and $F_i M_\lambda \subseteq M_{\lambda-\alpha_i}$.

By Proposition 3.2, the $\mathbf{U}$-supermodule $\mathcal{K}^{m|n}$ is a weight $\mathbf{U}$-supermodule:

$$\mathcal{K}^{m|n} = \bigoplus_{\lambda\in\mathbb{Z}^{m+n}} \mathcal{K}^{m|n}_{\lambda,\bullet}, \tag{3.2.2}$$

where $\mathcal{K}^{m|n}_{\lambda,\bullet} = \mathrm{span}\{'X^{[A]} \mid A \in \widetilde{M}(m|n), \lambda = \mathrm{ro}(A)\}$.

For $\lambda, \mu \in \mathbb{Z}^{m+n}$, let

$$
\begin{aligned}
&{}_\lambda\widetilde{M}(m|n)_\mu = \{A \in \widetilde{M}(m|n) \mid \lambda = \mathrm{ro}(A), \mu = \mathrm{co}(A)\} \qquad (3.2.3)\\
&\mathcal{K}^{m|n}_{\lambda,\mu} = \begin{cases} \mathrm{span}\{'X^{[A]} \mid A \in {}_\lambda\widetilde{M}(m|n)_\mu\}, & \text{if } {}_\lambda\widetilde{M}(m|n)_\mu \ne \emptyset;\\ 0, & \text{if } {}_\lambda\widetilde{M}(m|n)_\mu = \emptyset.\end{cases}
\end{aligned}
$$

Then, we have $\mathcal{K}^{m|n}_{\lambda,\bullet} = \bigoplus_{\mu\in\mathbb{Z}^{m+n}} \mathcal{K}^{m|n}_{\lambda,\mu}$.

COROLLARY 3.3. *For any fixed* $\mu \in \mathbb{Z}^{m+n}$, *the subspace* $\mathcal{K}^{m|n}_{\bullet,\mu} := \bigoplus_{\lambda\in\mathbb{Z}^{m+n}} \mathcal{K}^{m|n}_{\lambda,\mu}$ *is a weighted* $\mathbf{U}$-*subsupermodule of* $\mathcal{K}^{m|n}$, *and* $\mathcal{K}^{m|n} = \bigoplus_{\mu\in\mathbb{Z}^{m+n}} \mathcal{K}^{m|n}_{\bullet,\mu}$.

This is seen easily from Proposition 3.2, since $\mathrm{co}(A) = \mathrm{co}(A \pm E_{h,i} \mp E_{h+1,i})$, for all $i \in [1, m+n]$ and $h \in [1, m+n)$.

4. Triangular actions on $\mathcal{K}^{m|n}$ by monomial basis elements

We now have a closer look at the action formulas by certain monomials in K_j and divided powers $E_h^{(s)} := E_h^s/[s]^!$ and $F_h^{(s)} := F_h^s/[s]^!$.

For $A = (a_{i,j}) \in M(m|n)^{\pm}$, let

$$A^{\llcorner}_{s,t} = \sum_{i\le s, j\ge t} a_{i,j} \text{ if } s<t, \quad A^{\urcorner}_{s,t} = \sum_{i\ge s, j\le t} a_{i,j} \text{ if } s>t.$$

Following [**BLM**, §3.5] or [**DG1**, (8.0.1)], define a partial order relation on $M(m|n)^{\pm}$:

$$A \preceq B \iff \begin{cases} A^{\llcorner}_{s,t} \le B^{\llcorner}_{s,t}, & \text{for all } s<t \ (\iff A^+ \preceq B^+); \\ A^{\urcorner}_{s,t} \le B^{\urcorner}_{s,t}, & \text{for all } s>t \ (\iff A^- \preceq B^-). \end{cases} \tag{4.0.1}$$

Note that this is a partial order relation on $M(m|n)^{\pm}$. Extend this partial order to $\widetilde{M}(m|n)$ by setting

$$A \preceq B \iff A^{\pm} \preceq B^{\pm}, \quad \text{for } A, B \in \widetilde{M}(m|n). \tag{4.0.2}$$

The $\mathbf{U}$-action formulas in Proposition 3.2 satisfy certain "triangular relations" relative to $\preceq$. The "lower terms" below means a linear combination of $'X^{[B]}$ with $B \prec$ the leading matrix.

For $1 \le k \le m+n$, let

$$\Xi_k^+ = \{A \in \widetilde{M}(m|n) \mid a_{s,t} = 0 \text{ for } s>t \text{ or } s<t, t>k\};$$

$$\Xi_k^- = \{A \in \widetilde{M}(m|n) \mid a_{s,t} = 0 \text{ for } s>t,\ t<k\};$$

LEMMA 4.1. (1) *If* $A \in \Xi_k^+$ *then, for all* $h<k$, $\tilde{\sigma}_{h,m}(k,A) = 0$, $f_h(k,A) = 0$, *and*

$$E_h.'X^{[A]} \in \mathrm{span}\{'X^{[A]} \mid A \in \Xi_k^+\}.$$

If, in addition, $a_{h,k} = 0$ *then, for* $a > 0$ *(and* $a \le a_{h+1,k}$ *when* $h+1 \ne k$*), we have*

$$E_h^{(a)}.'X^{[A]} = {}'X^{[A-aE_{h+1,k}+aE_{h,k}]} + \text{(lower terms)}.$$

(2) *If* $A \in \Xi_k^-$ *then, for* $h > k-1$, $\tilde{\sigma}_{h,m}(k,A) = 0$, $g_h(k,A) = 0$, *and*

$$F_h.'X^{[A]} \in \mathrm{span}\{'X^{[A]} \mid A \in \Xi_k^-\}.$$

If, in addition, $a_{h+1,k} = 0$, *then, for* $a>0$ *(and* $a \le a_{h,k}$ *when* $h \ne k$*), we have*

$$F_h^{(a)}.'X^{[A]} = {}'X^{[A-aE_{h,k}+aE_{h+1,k}]} + \text{(lower terms)}.$$

PROOF. (1) Inspect the action formula in Proposition 3.2(2). Since $A \in \Xi_k^+$, i.e., $a_{s,t} = 0$ for $s<t$, $t>k$, then $h<k$ implies

$$f_h(k,A) = \sum_{j>k} a_{h,j} - (-1)^{\delta_{h,m}} \sum_{j>k} a_{h+1,j} = 0.$$

If $h = m$ then $m < k$, so

$$\tilde{\sigma}_{h,m}(k, A) = \tilde{\sigma}(k, A) = \sum_{s>m,t\leq m} a_{s,t} + \sum_{s\leq m,\, t>k} a_{s,t} = 0.$$

Then,

$$\begin{aligned}E_h.'X^{[A]} &= [a_{h,k}+1]'X^{[A+E_{h,k}-E_{h+1,k}]}\\ &+ \sum_{1\leq i<k, i\neq h+1} (-1)^{\tilde{\sigma}_{h,m}(i,A)} v_h^{f(i,A)} [a_{h,i}+1]'X^{[A+E_{h,i}-E_{h+1,i}]}\\ &+ \delta^{<}_{h+1,k}(-1)^{\tilde{\sigma}_{h,m}(h+1,A)} v_h^{f(h+1,A)} [a_{h,h+1}+1]'X^{[A+E_{h,h+1}-E_{h+1,h+1}]}\end{aligned}$$

is in the span. If, in addition, $a_{h,k} = 0$ if $h < k$ and $a > 0$, then, for any $a > 0$ and assuming $a_{h+1,k} \geq a$ if $h+1 \neq k$, repeatedly applying the above action formula gives $E_h^a.'X^{[A]} = [a]^!{}'X^{[A-aE_{h+1,k}+aE_{h,k}]} +$ (lower terms). Here the "lower terms" follows from the order relations: $A - aE_{h+1,i} + aE_{h,i} \prec A - aE_{h+1,k} + aE_{h,k}$ if $i < k$ and $h < i$ (or $h > i$). This proves (1).

Assertion (2) can be proved similarly. □

For any $A = (a_{i,j}) \in \widetilde{M}(m|n)^{\pm}, \mathbf{j} \in \mathbb{Z}^{m+n}$, recall the following definition in [**DZ**, Theorem 5.3]

(4.1.1)

$$\begin{aligned}\mathfrak{m}_j^+ &= \mathfrak{m}_j^+(A) = E_{j-1}^{(a_{j-1,j})}(E_{j-2}^{(a_{j-2,j})}E_{j-1}^{(a_{j-2,j})})\cdots(E_1^{(a_{1,j})}E_2^{(a_{1,j})}\cdots E_{j-1}^{(c_{1,j})}),\\ \mathfrak{m}_k^- &= \mathfrak{m}_k^-(A) = F_k^{(a_{k+1,k})}(F_{k+1}^{(a_{k+2,k})}F_k^{(a_{k+2,k})})\cdots(F_{m+n-1}^{(a_{m+n,k})}\cdots F_{k+1}^{(a_{m+n,k})}F_k^{(a_{m+n,k})}),\\ \mathfrak{m}^{A,\mathbf{j}} &= K^{\mathbf{j}}\cdot \mathfrak{m}_1^-\mathfrak{m}_2^-\cdots\mathfrak{m}_{m+n-1}^-\cdot\mathfrak{m}_{m+n}^+\cdots\mathfrak{m}_3^+\mathfrak{m}_2^+.\end{aligned}$$

For example, if $m = 2$, $n = 2$, $A \in \widetilde{M}(2|2)$, then

$$\begin{aligned}\mathfrak{m}^{A^{\pm},\mathrm{diag}(A)} = {}& K_1^{a_{11}}K_2^{a_{22}}K_3^{a_{33}}K_4^{a_{44}}\\ & (F_1^{(a_{21})}F_2^{(a_{31})}F_1^{(a_{31})}F_3^{(a_{41})}F_2^{(a_{41})}F_1^{(a_{41})})(F_2^{(a_{32})}F_3^{(a_{42})}F_2^{(a_{42})})F_3^{(a_{43})}\\ & (E_3^{(a_{34})}E_2^{(a_{24})}E_3^{(a_{24})}E_1^{(a_{14})}E_2^{(a_{14})}E_3^{(a_{14})})(E_2^{(a_{23})}E_1^{(a_{13})}E_2^{(a_{13})})E_1^{(a_{12})}.\end{aligned}$$

Repeatedly applying Proposition 3.2 or see [**DZ**, Theorem 5.3] for a similar argument, we obtain

PROPOSITION 4.2. *For any $A \in \widetilde{M}(m|n)^{\pm}$, we have*[2]

$$\mathfrak{m}^{A,\mathbf{0}}.'X^{[\lambda]} = 'X^{[A+\lambda-\mathrm{co}(A)]} + \textit{(lower terms)}_{\prec}.$$

In particular, for any $A \in \widetilde{M}(m|n)$, $\mathfrak{m}^{A^{\pm},\mathbf{0}}.'X^{[\mathrm{co}(A)]} = 'X^{[A]} +$ (lower terms)$_{\prec}$. Moreover, for any $'X^{[B]}$ appearing in lower terms, we have $\mathrm{ro}(A) = \mathrm{ro}(B)$ *and* $\mathrm{co}(A) = \mathrm{co}(B)$.

PROOF. Since $\mathfrak{m}^{A,\mathbf{0}} = \mathfrak{m}_1^-\mathfrak{m}_2^-\cdots\mathfrak{m}_{m+n-1}^-\mathfrak{m}_{m+n}^+\cdots\mathfrak{m}_3^+\mathfrak{m}_2^+$, the action of $\mathfrak{m}^{A,\mathbf{0}}$ is a sequence of actions of $\mathfrak{m}_j^+$ for $j = 2, 3, \ldots, m+n$, followed by actions of $\mathfrak{m}_k^-$ for $k = m+n-1, \ldots, 2, 1$. By repeatedly applying Lemma 4.1 and tracking the leading term A_j^+ after applying $\mathfrak{m}_2^+, \ldots, \mathfrak{m}_j^+$, we see that A_2^+ is the matrix $\lambda + a_{1,2}E_{1,2} - a_{1,2}E_{2,2}$, A_3^+ is the matrix $A_2^+ + a_{1,3}E_{1,3} + a_{2,3}E_{2,3} - (a_{1,3} + a_{2,3})E_{3,3}$,

[2]Note that those $\lambda \in \mathbb{Z}^{m+n}$ appearing in $X^{[\lambda]}$ of $X^{[A+\lambda]}$ should be understood as the diagonal matrix $\mathrm{diag}(\lambda)$.

etc., and $A^+_{m+n} = \lambda + A^+ - \mathrm{co}(A^+)$. This completes the filling of the upper triangular part. Now, tracking the leading terms A^-_{N-k} ($N = m+n$) after applying $\mathfrak{m}^-_{N-1}, \ldots, \mathfrak{m}^-_{N-k}$ for $k = N-1, N-2, \ldots, 2, 1$, we see that $A^-_{N-1} = A^+_N + a_{N,N-1}E_{N,N-1} - a_{N,N-1}E_{N-1,N-1}$, etc., and $A^-_1 = \lambda + A^\pm - \mathrm{co}(A^\pm)$. In each action, the actions on lower terms result in lower terms by the same lemma.

The last assertion follows from Proposition 3.2. □

5. The modified quantum supergroup $\dot{\mathbf{U}}$ of $\mathbf{U} = \mathbf{U}(\mathfrak{gl}_{m|n})$

We first follow [**Lubk**, Ch. 23] to give a candidate of the modified quantum general linear supergroup $\dot{\mathbf{U}}$. Recall that there is a triangular decomposition $\mathbf{U} = \mathbf{U}^-\mathbf{U}^0\mathbf{U}^+$; see [**EK**, (29)] and [**Zh**, Prop. 1]. Here $\mathbf{U}^0$ (resp., $\mathbf{U}^+$, $\mathbf{U}^-$) is the subalgebra generated by all $K_j^{\pm 1}$ (resp., E_i, F_i).

Consider the root lattice R spanned by the simple roots displayed in (2.0.2). The root lattice is used as a grading group for a quantum group $\mathbf{U}$ in the non-super case. This grading $\mathbf{U} = \oplus_{\nu\in\mathsf{R}}\mathbf{U}(\nu)$ is crucial for the definition for $\dot{\mathbf{U}}$, especially for the multiplication for $\dot{\mathbf{U}}$.

However, if $\mathbf{U} = \mathbf{U}(\mathfrak{gl}_{m|n})$ is the quantum linear supergroup, a similar grading structure $\mathbf{U} = \oplus_{\nu\in\mathsf{R}}\mathbf{U}(\nu)$ with $E_i \in \mathbf{U}(\alpha_i)$, $F_i \in \mathbf{U}(-\alpha_i)$ and $K_i \in \mathbf{U}(0)$ may be defined for this $\mathbf{U}$. Here

$$\mathbf{U}(\nu) = \{u \in \mathbf{U} \mid K^{\mathbf{j}}u = \upsilon^{\mathbf{j}\cdot_s\nu}uK^{\mathbf{j}},\ \forall \mathbf{j} \in \mathbb{Z}^{m+n}\}. \tag{5.0.1}$$

Thus, by the definition, $\mathbf{U}(\nu)\mathbf{U}(\nu') \subseteq \mathbf{U}(\nu+\nu')$ and $K^{\mathbf{j}}\mathbf{U}(\nu) = \upsilon^{\mathbf{j}\cdot_s\nu}\mathbf{U}(\nu)K^{\mathbf{j}}$.

Note that the sizes of $\mathbf{U}(\nu)$'s are dramatically reduced. For example, since $E_m^2 = 0$, the grading space $\mathbf{U}(2\alpha_m) = 0$; while $\mathbf{U}(\alpha_{m-1} + 2\alpha_m)$ has dimension 1 instead of 3.

LEMMA 5.1. *For $A \in M(m|n)^\pm$, let $\mathfrak{m}^{A,\mathbf{j}}$ be the monomial basis element defined in* (4.1.1). *Then $\mathfrak{m}^{A,\mathbf{0}} \in \mathbf{U}(\nu)$ with*

$$\nu = \mathrm{ro}(A) - \mathrm{co}(A) = \sum_{j=2}^{m+n}\sum_{i=1}^{j-1}(a_{i,j}\alpha_{i,j} - a_{j,i}\alpha_{i,j}),$$

where $\alpha_{i,j} = \mathbf{e}_i - \mathbf{e}_j$ for all $i < j$.

PROOF. By the relation (QG2), we have $K^{\mathbf{j}}E_i = \upsilon^{\mathbf{j}\cdot_s\alpha_i}E_iK^{\mathbf{j}}$ and $K^{\mathbf{j}}F_i = \upsilon^{-\mathbf{j}\cdot_s\alpha_i}F_iK^{\mathbf{j}}$ for all $i \in [1, m+n)$. Then, by (4.1.1), $K^{\mathbf{j}}\mathfrak{m}_j^+ = \upsilon^{\mathbf{j}\cdot_s\nu^+(j)}\mathfrak{m}_j^+K^{\mathbf{j}}$ for all $j \in [2, m+n]$, where

$$\begin{aligned}\nu^+(j) &= a_{j-1,j}\alpha_{j-1} + a_{j-2,j}(\alpha_{j-2} + \alpha_{j-1}) + \cdots + a_{1,j}(\alpha_1 + \cdots + \alpha_{j-1})\\ &= a_{j-1,j}\alpha_{j-1,j} + a_{j-2,j}\alpha_{j-2,j} + \cdots + a_{1,j}\alpha_{1,j} = \sum_{i=1}^{j-1} a_{i,j}(\mathbf{e}_i - \mathbf{e}_j).\end{aligned}$$

Thus, $K^{\mathbf{j}}(\mathfrak{m}^+_{m+n}\cdots\mathfrak{m}^+_3\mathfrak{m}^+_2) = \upsilon^{\mathbf{j}\cdot_s\nu^+}(\mathfrak{m}^+_{m+n}\cdots\mathfrak{m}^+_3\mathfrak{m}^+_2)K^{\mathbf{j}}$, where

$$\nu^+ = \sum_{j=2}^{m+n}\nu^+(j) = \sum_{j=2}^{m+n}\sum_{i=1}^{j-1}a_{i,j}\mathbf{e}_i - \sum_{j=2}^{m+n}\sum_{i=1}^{j-1}a_{i,j}\mathbf{e}_j = \mathrm{ro}(A^+) - \mathrm{co}(A^+).$$

Similarly, we have $K^{\mathbf{j}}(\mathfrak{m}^-_1\mathfrak{m}^-_2\cdots\mathfrak{m}^-_{m+n-1}) = \upsilon^{\mathbf{j}\cdot_s\nu^-}(\mathfrak{m}^-_1\mathfrak{m}^-_2\cdots\mathfrak{m}^-_{m+n-1})K^{\mathbf{j}}$ with $\nu^- = \mathrm{ro}(A^-) - \mathrm{co}(A^-)$. Hence, the assertion follows. □

Define the proper subset of the root lattice R:

(5.1.1) $\mathsf{R}^{\mathsf{s}} := \{\mathrm{ro}(A) - \mathrm{co}(A) \mid A \in M(m|n)^{\pm}\} = \{\mathrm{ro}(A) - \mathrm{co}(A) \mid A \in \widetilde{M}(m|n)\}$.

Note $\mathrm{ro}(A) - \mathrm{co}(A) = \mathrm{ro}(A + \lambda I) - \mathrm{co}(A + \lambda I)$, and there exist $A, B \in M(m|n)^{\pm}$ with $A \neq B$, $\mathrm{ro}(A) - \mathrm{co}(A) = \mathrm{ro}(B) - \mathrm{co}(B)$.

By the definition in (3.2.3), we have ${}_\lambda\widetilde{M}(m|n)_\mu \neq \emptyset$ if and only if $\lambda - \mu \in \mathsf{R}^{\mathsf{s}}$.

COROLLARY 5.2. *The grading subspace* $\mathbf{U}(\nu) = 0$ *for all* $\nu \in \mathsf{R}\backslash\mathsf{R}^{\mathsf{s}}$, *and* $\mathbf{U}(\nu)\mathbf{U}(\nu') = 0$ *for all* $\nu, \nu' \in \mathsf{R}$ *with* $\nu + \nu' \notin \mathsf{R}^{\mathsf{s}}$. *Moreover, for* $\nu \in \mathsf{R}$ *and* $\lambda, \mu \in \mathbb{Z}^{m+n}$ *with* $\nu = \lambda - \mu$, *we have*

$$\mathbf{U}(\nu) = \begin{cases} \mathrm{span}\{\mathfrak{m}^{A^{\pm},\mathbf{j}} \mid A \in {}_\lambda\widetilde{M}(m|n)_\mu, \mathbf{j} \in \mathbb{Z}^{m+n}\}, & \text{if } {}_\lambda\widetilde{M}(m|n)_\mu \neq \emptyset; \\ 0, & \text{if } {}_\lambda\widetilde{M}(m|n)_\mu = \emptyset. \end{cases}$$

Note that R^{s} is not an abelian subgroup of R, since, for some ν, ν', it may occur that $\nu + \nu' \notin \mathsf{R}^{\mathsf{s}}$, therefore, we won't use the refined R^{s} to grade $\mathbf{U}$.

For $\lambda, \mu \in \mathbb{Z}^{m+n}$, let ${}_\lambda\mathbf{U}_\mu = \mathbf{U}/{}_\lambda\mathbf{K}_\mu$, where[3]

$$\begin{aligned} {}_\lambda\mathbf{K}_\mu &:= \sum_{\mathbf{j} \in \mathbb{Z}^{m+n}} (K^{\mathbf{j}} - \upsilon^{\lambda \bullet_s \mathbf{j}})\mathbf{U} + \sum_{\mathbf{j} \in \mathbb{Z}^{m+n}} \mathbf{U}(K^{\mathbf{j}} - \upsilon^{\mu \bullet_s \mathbf{j}}) \\ &= \sum_{i=1}^{m+n} (K_i - \upsilon_i^{\lambda_i})\mathbf{U} + \sum_{i=1}^{m+n} \mathbf{U}(K_i - \upsilon_i^{\mu_i}). \end{aligned} \tag{5.2.1}$$

Let $\pi_{\lambda,\mu} : \mathbf{U} \to {}_\lambda\mathbf{U}_\mu$ be the canonical projection. In order to define an algebra $\dot{\mathbf{U}}$ without 1, we first prove the following. Note that part (2) is stated in [**Lubk**, 23.1.1]. For completeness, we include a proof.

LEMMA 5.3.

(1) *For* $\lambda, \mu \in \mathbb{Z}^{m+n}$, *and* $\nu \in \mathsf{R}^{\mathsf{s}}$, *if* $\lambda - \mu \neq \nu$, *then* $\pi_{\lambda,\mu}(\mathbf{U}(\nu)) = 0$. *In particular, we have* ${}_\lambda\widetilde{M}(m|n)_\mu = \emptyset \implies {}_\lambda\mathbf{U}_\mu = 0$.

(2) *For any* $\lambda \in \mathbb{Z}^{m+n}$, ${}_\lambda\mathbf{U}_\lambda$ *is an algebra with multiplication induced from that of the subalgebra* $\mathbf{U}(0)$ *and* ${}_\lambda\mathbf{U}_\lambda \neq 0$.

PROOF. For $u \in \mathbf{U}(\nu)$, since $K^{\mathbf{j}}u = \upsilon^{\nu \bullet_s \mathbf{j}}uK^{\mathbf{j}}$ for all $\mathbf{j} \in \mathbb{Z}^{m+n}$, it follows that

$$\upsilon^{\lambda \bullet_s \mathbf{j}}\pi_{\lambda,\mu}(u) = \pi_{\lambda,\mu}(K^{\mathbf{j}}u) = \upsilon^{\nu \bullet_s \mathbf{j}}\pi_{\lambda,\mu}(uK^{\mathbf{j}}) = \upsilon^{\nu \bullet_s \mathbf{j} + \mu \bullet_s \mathbf{j}}\pi_{\lambda,\mu}(u).$$

Now, $\lambda - \mu \neq \nu$ implies that there exists $\mathbf{j}$ such that $(\lambda - \mu) \bullet_s \mathbf{j} \neq \nu \bullet_s \mathbf{j}$, forcing $\pi_{\lambda,\mu}(u) = 0$.[4]

To see the last assertion of (1), we observe that the above shows $\nu \neq \lambda - \mu \implies \pi_{\lambda,\mu}(\mathbf{U}(\nu)) = 0 \implies \mathbf{U}(\nu) \subseteq {}_\lambda\mathbf{K}_\mu$, while the hypothesis ${}_\lambda\widetilde{M}(m|n)_\mu = \emptyset$ implies $\nu = \lambda - \mu$ cannot occur. Thus, $\mathbf{U} = \bigoplus_{\nu \in \mathsf{R}} \mathbf{U}(\nu) \subseteq {}_\lambda\mathbf{K}_\mu$. Hence, ${}_\lambda\mathbf{U}_\mu = 0$, proving (1).

We now prove (2). By (1), we have $\pi_{\lambda,\lambda}(\mathbf{U}(\nu)) = 0$, for all $\nu \neq 0$. Thus, $\sum_{\nu \neq 0} \mathbf{U}(\nu) \subseteq {}_\lambda\mathbf{K}_\lambda$. So, the abelian group

$${}_\lambda\mathbf{U}_\lambda = (\mathbf{U}(0) + {}_\lambda\mathbf{K}_\lambda)/{}_\lambda\mathbf{K}_\lambda \cong \mathbf{U}(0)/(\mathbf{U}(0) \cap {}_\lambda\mathbf{K}_\lambda).$$

---

[3]Putting $\mathbf{j}_i = j_i\mathbf{e}_i$, the inclusion "$\subseteq$" follows by writing $K^{\mathbf{j}} - \upsilon^{\lambda \bullet_s \mathbf{j}}$ as $(K^{\mathbf{j}_1} - \upsilon^{\lambda \bullet_s \mathbf{j}_1})K^{\mathbf{j}-\mathbf{j}_1} + \upsilon^{\lambda \bullet_s \mathbf{j}_1}(K^{\mathbf{j}_2} - \upsilon^{\lambda \bullet_s \mathbf{j}_2})K^{\mathbf{j}-\mathbf{j}_1-\mathbf{j}_2} + \upsilon^{\lambda \bullet_s (\mathbf{j}_1 - \mathbf{j}_2)}(K^{\mathbf{j}_3} - \upsilon^{\lambda \bullet_s \mathbf{j}_3})K^{\mathbf{j}-\mathbf{j}_1-\mathbf{j}_2-\mathbf{j}_3} + \cdots$.

[4]This proof works in the general context of [**Lubk**, 4.1.1]. Thus, the direct-sum decomposition of ${}_{\lambda'}\dot{\mathbf{U}}_{\lambda''}$ given in [**Lubk**, page 186, line 4] collapses to a single summand: ${}_{\lambda'}\dot{\mathbf{U}}_{\lambda''} = {}_{\lambda'}\dot{\mathbf{U}}_{\lambda''}(\lambda' - \lambda'') := \pi_{\lambda',\lambda''}(\mathbf{U}(\lambda' - \lambda''))$.

Since $K^{\mathbf{j}}$ commutes with every element in $\mathbf{U}(0)$, it follows that

$$\mathbf{U}(0) \cap {}_\lambda\mathbf{K}_\lambda = \sum_{\mathbf{j}\in\mathbb{Z}^{m+n}} (K^{\mathbf{j}} - v^{\lambda\cdot_s\mathbf{j}})\mathbf{U}(0) = \sum_{\mathbf{j}\in\mathbb{Z}^{m+n}} \mathbf{U}(0)(K^{\mathbf{j}} - v^{\lambda\cdot_s\mathbf{j}})$$

is an ideal of $\mathbf{U}(0)$. Hence, ${}_\lambda\mathbf{U}_\lambda$ is an algebra.

Finally, by (5.2.1), the last assertion in (2) is clear since the ideal of $\mathbf{U}^0$ generated by $K_i - v_i^{\lambda_i}$ for all $i \in [1, m+n]$ is a proper ideal. Hence, the ideal $\mathbf{U}(0) \cap {}_\lambda\mathbf{K}_\lambda$ does not contain the identity element 1. □

We will see in Corollary 6.3 below that if ${}_\lambda\widetilde{M}(m|n)_\mu \neq \emptyset$ then ${}_\lambda\mathbf{U}_\mu \neq 0$.

We now follow [**Lubk**, 23.1.1(b)] to define an associative superalgebra $\dot{\mathbf{U}}$ over $\mathbb{Q}(v)$ without identity. The following result is largely stated in [**Lubk**, §23.1.1]. For completeness, we include a proof.

PROPOSITION 5.4. *The* $\mathbb{Q}(v)$*-space*

$$\dot{\mathbf{U}} := \bigoplus_{\lambda,\mu\in\mathbb{Z}^{m+n}} {}_\lambda\mathbf{U}_\mu.$$

together with the multiplication defined in [**Lubk**, 23.1.1] *becomes an associative superalgebra (without identity).*

PROOF. First, we observe that $\dot{\mathbf{U}}$ is a superspace inherited from the superspace structure on $\mathbf{U} = \mathbf{U}_{\bar{0}} \oplus \mathbf{U}_{\bar{1}}$. This is seen as follows. If $t \in \mathbf{U}_{\bar{j}}$ ($\bar{j} \in \mathbb{Z}_2$) is homogeneous and $\pi_{\lambda,\mu}(t) \neq 0$, then we define the parity of $\pi_{\lambda,\mu}(t)$ by setting $\wp(\pi_{\lambda,\mu}(t)) := \wp(t) = \bar{j}$. We need to verify that this is well defined. In fact, if $\pi_{\lambda,\mu}(t) = \pi_{\lambda,\mu}(t')$ with t' homogeneous of degree $\wp(t') = \bar{j}'$, then $t - t' \in {}_\lambda\mathbf{K}_\mu$. By writing $t - t' = (t_{\bar{0}} - t'_{\bar{0}}) + (t_{\bar{1}} - t'_{\bar{1}})$, we have $t_{\bar{i}} - t'_{\bar{i}} \in ({}_\lambda\mathbf{K}_\mu)_{\bar{i}} := \sum_{\mathbf{j}\in\mathbb{Z}^{m+n}} (K^{\mathbf{j}} - v^{\lambda\cdot_s\mathbf{j}})\mathbf{U}_{\bar{i}} + \sum_{\mathbf{j}\in\mathbb{Z}^{m+n}} \mathbf{U}_{\bar{i}}(K^{\mathbf{j}} - v^{\mu\cdot_s\mathbf{j}})$ for $\bar{i} = \bar{0}, \bar{1}$. This implies $\wp(\pi_{\lambda,\mu}(t'))$ is the same as $\wp(\pi_{\lambda,\mu}(t))$. For example, if $\bar{j} = \bar{0}$, then $t_{\bar{1}} = 0$ and so $t' \neq t'_{\bar{1}}$ since $\pi_{\lambda,\mu}(t'_{\bar{1}}) = 0$. Hence, $t' = t'_{\bar{0}}$.

Second, we now define a multiplication on $\dot{\mathbf{U}}$ which extends the multiplications on all ${}_\lambda\mathbf{U}_\lambda$ discussed in Lemma 5.3(2). Following [**Lubk**, 23.1.1], for $\lambda, \mu, \lambda', \mu' \in \mathbb{Z}^{m+n}$, $t \in \mathbf{U}(\lambda - \mu)$, and $s \in \mathbf{U}(\lambda' - \mu')$, define

$$\pi_{\lambda,\mu}(t) \cdot \pi_{\lambda',\mu'}(s) = \delta_{\mu,\lambda'}\pi_{\lambda,\mu'}(ts).$$

We claim that this multiplication is well defined. In other words, for all $t_1, t_2 \in \mathbf{U}(\lambda - \mu)$ and $s_1, s_2 \in \mathbf{U}(\lambda' - \mu')$, we have

$$\pi_{\lambda,\mu}(t_1) = \pi_{\lambda,\mu}(t_2), \pi_{\lambda',\mu'}(s_1) = \pi_{\lambda',\mu'}(s_2)$$
$$\implies \pi_{\lambda,\mu}(t_1)\pi_{\lambda',\mu'}(s_1) = \pi_{\lambda,\mu}(t_2)\pi_{\lambda',\mu'}(s_2).$$

This follows from the following fact: for $t \in \mathbf{U}(\lambda - \mu)$ and $s \in \mathbf{U}(\mu - \mu')$,

$$\pi_{\lambda,\mu}(t) = 0 \text{ (resp., } \pi_{\mu,\mu'}(s) = 0) \implies \pi_{\lambda,\mu'}(ts) = 0. \tag{5.4.1}$$

We now prove the first case of (5.4.1). Suppose $t \in \mathbf{U}(\lambda-\mu)$ and $\pi_{\lambda,\mu}(t) = 0$. Then t is an element of

$$\begin{aligned}
&\sum_{\mathbf{j}\in\mathbb{Z}^{m+n}} (K^{\mathbf{j}} - v^{\lambda\cdot_s\mathbf{j}})\mathbf{U}(\lambda-\mu) + \sum_{\mathbf{j}\in\mathbb{Z}^{m+n}} \mathbf{U}(\lambda-\mu)(K^{\mathbf{j}} - v^{\mu\cdot_s\mathbf{j}}) \\
&= \sum_{\mathbf{j}\in\mathbb{Z}^{m+n}} (K^{\mathbf{j}} - v^{\lambda\cdot_s\mathbf{j}})\mathbf{U}(\lambda-\mu) + \sum_{\mathbf{j}\in\mathbb{Z}^{m+n}} (v^{(\mu-\lambda)\cdot_s\mathbf{j}}K^{\mathbf{j}} - v^{\mu\cdot_s\mathbf{j}})\mathbf{U}(\lambda-\mu) \\
&= \sum_{\mathbf{j}\in\mathbb{Z}^{m+n}} (K^{\mathbf{j}} - v^{\lambda\cdot_s\mathbf{j}})\mathbf{U}(\lambda-\mu).
\end{aligned}$$

Thus, $ts \in {}_\lambda\mathbf{K}_{\mu'}$. Hence $\pi_{\lambda,\mu'}(ts) = 0$. The other case is proved similarly.

Finally, the multiplication rule defined above implies that $\dot{\mathbf{U}}_{\bar{i}}\dot{\mathbf{U}}_{\bar{j}} \subseteq \dot{\mathbf{U}}_{\bar{i}+\bar{j}}$. Hence, $\dot{\mathbf{U}}$ becomes an associative superalgebra. $\square$

The superalgebra $\dot{\mathbf{U}}$ has infinitely many orthogonal idempotents $1_\lambda := \pi_{\lambda,\lambda}(1)$, for all $\lambda \in \mathbb{Z}^{m+n}$; see Lemma 5.3(2). We call this superalgebra (without 1) the *modified quantum linear supergroup.*

REMARKS 5.5.

(1) The elements of $\dot{\mathbf{U}}$ can be expressed as finite sum

$$\bar{u} = \sum_{i=1}^{n} \pi_{\lambda^{(i)},\mu^{(i)}}(t_i), \tag{5.5.1}$$

where $\lambda^{(i)}, \mu^{(i)} \in \mathbb{Z}^{m+n}$ and $t_i \in \mathbf{U}(\lambda^{(i)} - \mu^{(i)})$. Moreover, such a $\bar{u} = 0 \iff \pi_{\lambda^{(i)},\mu^{(i)}}(t_i) = 0$, for all i. Notice that by setting $\bar{u}_\mu = \sum_{i,\mu^{(i)}=\mu} \pi_{\lambda^{(i)},\mu^{(i)}}(t_i)$, we have $\bar{u} = \sum_\mu \bar{u}_\mu \mathbf{l}_\mu$.

(2) By Lemma 5.3(1), each summand ${}_\lambda\mathbf{U}_\mu = \pi_{\lambda,\mu}(\mathbf{U}(\lambda-\mu))$. Hence,

$$\dot{\mathbf{U}} = \bigoplus_{\lambda,\mu\in\mathbb{Z}^{m+n}} \pi_{\lambda,\mu}(\mathbf{U}(\lambda-\mu)).$$

(3) Recall from (2.0.4) the anti-involution τ on $\mathbf{U}$. Since τ fixes every $K_i - v^a$, it follows that τ induces a linear isomorphism $\dot{\tau} : {}_\lambda\mathbf{U}_\mu \to {}_\mu\mathbf{U}_\lambda$. Hence, τ induces an anti-involution

$$\dot{\tau} : \dot{\mathbf{U}} \longrightarrow \dot{\mathbf{U}}. \tag{5.5.2}$$

Like the non-super case, there is a $\mathbf{U}$-bimodule action on $\dot{\mathbf{U}}$ defined as follows: for $\lambda, \mu \in \mathbb{Z}^{m+n}, t_1 \in \mathbf{U}(\nu), t_2 \in \mathbf{U}(\nu'), s \in \mathbf{U}$, the $\mathbf{U}$-bimodule actions on $\dot{\mathbf{U}}$ given by

$$t_1\pi_{\lambda,\mu}(s)t_2 = \pi_{\lambda+\nu,\,\mu-\nu'}(t_1st_2).$$

Since $s = \sum_\alpha s_\alpha$ with $\alpha \in \mathbf{U}(\alpha)$ and $\pi_{\lambda,\mu}(s_\alpha) = 0 = \pi_{\lambda+\nu,\mu-\nu'}(t_1s_\alpha t_2)$ unless $\alpha = \lambda - \mu$, we may assume in the definition above that $s \in \mathbf{U}(\lambda-\mu)$. Putting $\lambda' = \lambda + \nu$ and $\mu' = \mu - \nu$, we have $t_1 \in \mathbf{U}(\lambda' - \lambda)$ and $t_2 \in \mathbf{U}(\mu - \mu')$ and so,

$$t_1\pi_{\lambda,\mu}(s)t_2 = \pi_{\lambda+\nu,\,\mu-\nu'}(t_1st_2) = \pi_{\lambda',\lambda}(t_1)\pi_{\lambda,\mu}(s)\pi_{\mu,\mu'}(t_2).$$

Hence, by linear extension, $\dot{\mathbf{U}}$ becomes a $\mathbf{U}$-$\mathbf{U}$-*superbimodule* which further induces the regular $\dot{\mathbf{U}}$-$\dot{\mathbf{U}}$-superbimodule ${}_{\dot{\mathbf{U}}}\dot{\mathbf{U}}_{\dot{\mathbf{U}}}$.

By the bimodule actions, for $t \in \mathbf{U}(\lambda-\mu)$ we have

$$\pi_{\lambda,\mu}(t) = \pi_{\lambda,\mu}(t1) = t\pi_{\mu,\mu}(1) = t1_\mu = 1_\lambda t = 1_\lambda t1_\mu.$$

Hence, for a given $\mu \in \mathbb{Z}^{m+n}$,

$$\dot{\mathbf{U}}_\mu := \bigoplus_{\lambda \in \mathbb{Z}^{m+n}} {}_\lambda\mathbf{U}_\mu = \mathbf{U}1_\mu = \dot{\mathbf{U}}1_\mu \tag{5.5.3}$$

is a $\dot{\mathbf{U}}$-supermodule.

Following [**Lubk**, 23.1.4], we make the following definition.

DEFINITION 5.6. A $\dot{\mathbf{U}}$-supermodule M is said to be *unital* if

(a) for any given $x \in M$, $1_\lambda x = 0$, for all but finitely many $\lambda \in \mathbb{Z}^{m+n}$;
(b) for any given $x \in M$, $\sum_{\lambda \in \mathbb{Z}^{m+n}} 1_\lambda x = x$.

Note that the orthogonality of the idempotents $1_\lambda, \lambda \in \mathbb{Z}^{m+n}$, shows that the regular $\dot{\mathbf{U}}$-supermodule ${}_{\dot{\mathbf{U}}}\dot{\mathbf{U}}$ is a unital supermodule. The $\dot{\mathbf{U}}$-supermodule $\dot{\mathbf{U}}_\mu$ above is also a unital subsupermodule of ${}_{\dot{\mathbf{U}}}\dot{\mathbf{U}}$ and $\dot{\mathbf{U}} = \bigoplus_{\mu \in \mathbb{Z}^{m+n}} \dot{\mathbf{U}}_\mu$.

6. The regular (matrix) representation of $\dot{\mathbf{U}}$

We now bring the $\mathbf{U}$-supermodule $\mathcal{K}^{m|n}$ defined in (3.0.1) to the game. Recall from Proposition 3.2 the signed basis for $\mathcal{K}^{m|n}$:

$$\{'X^{[A]} \mid A \in \widetilde{M}(m|n)\}$$

and $\mathbf{U}$ acts on $\mathcal{K}^{m|n}$ via the action formulas in Proposition 3.2.

LEMMA 6.1. *The $\mathbf{U}$-supermodule structure on $\mathcal{K}^{m|n}$ induces naturally a $\dot{\mathbf{U}}$-supermodule structure on $\mathcal{K}^{m|n}$ which is given by:*

$$\pi_{\lambda,\mu}(t).'X^{[A]} = \delta_{\lambda-\mu,\nu}\delta_{\mu,\mathrm{ro}(A)}t.'X^{[A]}, \tag{6.1.1}$$

for all $A \in \widetilde{M}(m|n), \lambda, \mu \in \mathbb{Z}^{m+n}, \nu \in \mathsf{R}$, and $t \in \mathbf{U}(\nu)$.

PROOF. Recall the decompositions $\mathbf{U} = \bigoplus_{\nu \in \mathsf{R}} \mathbf{U}(\nu)$ and

$$\mathcal{K}^{m|n} = \bigoplus_{\alpha,\beta \in \mathbb{Z}^{m+n}} \mathcal{K}^{m|n}_{\alpha,\beta} = \bigoplus_{\alpha \in \mathbb{Z}^{m+n}} \mathcal{K}^{m|n}_{\alpha,\bullet},$$

where $\mathcal{K}^{m|n}_{\alpha,\bullet} = \bigoplus_{\beta \in \mathbb{Z}^{m+n}} \mathcal{K}^{m|n}_{\alpha,\beta}$. Let $\lambda, \mu \in \mathbb{Z}^{m+n}$ be arbitrarily given.

The "or" case. If $\lambda - \mu \neq \nu$ or $\mu \neq \alpha$, then the action $\pi_{\lambda,\mu}(t)$ with $t \in \mathbf{U}(\nu)$ on $\mathcal{K}^{m|n}_{\alpha,\bullet}$ is trivial, since $\pi_{\lambda,\mu}(t).'X^{[A]} = 0$ for all $t \in \mathbf{U}(\nu)$ or A with $\mu \neq \mathrm{ro}(A)$.

The "and" case. If $\lambda - \mu = \nu$ and $\mu = \mathrm{ro}(A)$, then

$$(K^{\mathbf{j}} - v^{\lambda \cdot_s \mathbf{j}})\mathbf{U}(\nu) = v^{\nu \cdot_s \mathbf{j}}\mathbf{U}(\nu)(K^{\mathbf{j}} - v^{\mu \cdot_s \mathbf{j}}).$$

Since $(K^{\mathbf{j}} - v^{\mu \cdot_s \mathbf{j}}).'X^{[A]} = 0$ by Proposition 3.2(1), this together with the "or" case implies that ${}_\lambda\mathbf{K}_\mu.'X^{[A]} = 0$. Thus, it induces a well-defined action of ${}_\lambda\mathbf{U}_\mu$ on $\mathcal{K}^{m|n}$. Hence, we obtain a well-defined action of $\dot{\mathbf{U}}$ on $\mathcal{K}^{m|n}$. □

In particular, the idempotents $1_\lambda, \lambda \in \mathbb{Z}^{m+n}$ have the following action formulas:

$$1_\lambda.'X^{[A]} = \delta_{\lambda,\mathrm{ro}(A)}{'X^{[A]}}. \tag{6.1.2}$$

Hence, $\mathcal{K}^{m|n}$ is a *unital* $\dot{\mathbf{U}}$*-supermodule* in the sense of Definition 5.6. Recall the $\dot{\mathbf{U}}$-subsupermodule $\mathcal{K}^{m|n}_{\bullet,\mu}$ and $\dot{\mathbf{U}}_\mu$ defined in Corollary 3.3 and (5.5.3).

THEOREM 6.2. *For any $\mu \in \mathbb{Z}^{m+n}$, there is a (unital) supermodule isomorphism*

$$\gamma_\mu : \dot{\mathbf{U}}_\mu \longrightarrow \mathcal{K}^{m|n}_{\bullet,\mu}, \quad u \longmapsto u.'X^{[\mu]}.$$

In particular, we obtain a $\dot{\mathbf{U}}$-supermodule isomorphism $\gamma = \oplus_\mu \gamma_\mu : \dot{\mathbf{U}} \to \mathcal{K}^{m|n}$.

PROOF. Clearly, by definition, γ_μ is a module homomorphism. For any $\lambda, \mu \in \mathbb{Z}^{m+n}$ and homogeneous $t \in \mathbf{U}(\lambda - \mu)$ of degree $\bar{i}$, the vector $\delta_{\lambda-\mu,\nu} t.'X^{[\mu]} = \pi_{\lambda,\mu}(t).'X^{[\mu]}$ is homogeneous of the same degree and $\gamma_\mu(\pi_{\lambda,\mu}(t)) = \pi_{\lambda,\mu}(t).'X^{[\mu]}$. Hence, γ_μ is a supermodule homomorphism. It remains to prove that γ_μ is a bijection.

For any $A \in \widetilde{M}(m|n)$ with $\lambda = \mathrm{ro}(A)$ and $\mu = \mathrm{co}(A)$, by Lemma 5.1 and Proposition 4.2,

$$\gamma_\mu(\pi_{\lambda,\mu}(\mathfrak{m}^{A^\pm,\mathbf{0}})) = \mathfrak{m}^{A^\pm,\mathbf{0}}.'X^{[\mu]} = 'X^{[A]} + (\text{lower terms})_{\prec}. \tag{6.2.1}$$

Since the ideal $(-\infty, A^\pm]$ of the poset $(M(m|n)^\pm, \preceq)$ is finite, by induction on $\preceq$, we see that every $'X^{[A]}$ is in the image of γ_μ. Hence, the map γ_μ is surjective.

Suppose now that $\bar{u} = \sum_{i=1}^n \pi_{\lambda^{(i)},\mu^{(i)}}(t_i) \in \ker(\gamma_\mu)$, where $\lambda^{(i)}, \mu^{(i)} \in \mathbb{Z}^{m+n}$ with $\lambda^{(i)} - \mu^{(i)} = \nu^{(i)}$ and $t_i \in \mathbf{U}(\nu^{(i)})$. Then $t_i.'X^{[\mu^{(i)}]} = \pi_{\lambda^{(i)},\mu^{(i)}}(t_i)'X^{[\mu^{(i)}]} = 0$ for all i. By Corollary 5.2 and noting $\mathfrak{m}^{A^\pm,\mathbf{j}} = v^a \mathfrak{m}^{A^\pm,\mathbf{0}} K^{\mathbf{j}}$, we may write $t_i = \sum_{A,\mathbf{j}} f_{A,\mathbf{j}} \mathfrak{m}^{A^\pm,\mathbf{0}} K^{\mathbf{j}}$, for finitely many $A \in {}_{\lambda^{(i)}}\widetilde{M}(m|n)_{\mu^{(i)}}, \mathbf{j} \in \mathbb{Z}^{m+n}$ with $f_{A,\mathbf{j}} \in \mathbb{Q}(v)$. Thus,

$$0 = t_i.'X^{[\mu^{(i)}]} = \sum_{A,\mathbf{j}} f_{A,\mathbf{j}} \mathfrak{m}^{A^\pm,\mathbf{0}} K^{\mathbf{j}}.'X^{[\mu^{(i)}]} = \sum_{A,\mathbf{j}} f_{A,\mathbf{j}} v^{\mu^{(i)} \centerdot_s \mathbf{j}} \mathfrak{m}^{A^\pm,\mathbf{0}}.'X^{[\mu^{(i)}]}.$$

So, for a fixed A, we have $\sum_{\mathbf{j}} v^{\mu^{(i)} \centerdot_s \mathbf{j}} f_{A,\mathbf{j}} = 0$. Hence,

(6.2.2)

$$\begin{aligned} t_i &= \sum_A \sum_{\mathbf{j}} f_{A,\mathbf{j}} \mathfrak{m}^{A^\pm,\mathbf{0}} K^{\mathbf{j}} = \sum_A \sum_{\mathbf{j}} f_{A,\mathbf{j}} \mathfrak{m}^{A^\pm,\mathbf{0}} K^{\mathbf{j}} - \sum_A \sum_{\mathbf{j}} v^{\mu^{(i)} \centerdot_s \mathbf{j}} f_{A,\mathbf{j}} \mathfrak{m}^{A^\pm,\mathbf{0}} \\ &= \sum_A \sum_{\mathbf{j}} f_{A,\mathbf{j}} \mathfrak{m}^{A^\pm,\mathbf{0}} (K^{\mathbf{j}} - v^{\mu^{(i)} \centerdot_s \mathbf{j}}) \in {}_{\lambda^{(i)}}\mathbf{K}_{\mu^{(i)}}. \end{aligned}$$

This implies $\pi_{\lambda^{(i)},\mu^{(i)}}(t_i) = 0$, for all i. Hence, $\bar{u} = 0$, and so the map γ_μ is injective. □

COROLLARY 6.3. *For all $\lambda, \mu \in \mathbb{Z}^{m+n}$, we have vector space isomorphism ${}_\lambda\mathbf{U}_\mu \cong \mathcal{K}^{m|n}_{\lambda,\mu}$.*

PROOF. This is clear since $\gamma_\mu({}_\lambda\mathbf{U}_\mu) = \gamma_\mu(1_\lambda \dot{\mathbf{U}}_\mu) = 1_\lambda \mathcal{K}^{m|n}_{\bullet,\mu} = \mathcal{K}^{m|n}_{\lambda,\mu}$, by Lemma 6.1. □

We now use the isomorphism $\gamma : \dot{\mathbf{U}} \to \mathcal{K}^{m|n}, u \mapsto u.'X^{[\mu]}$ $(u \in \dot{\mathbf{U}}_\mu)$ to derive certain multiplication formulas in $\dot{\mathbf{U}}$ from the action formulas of $\dot{\mathbf{U}}$ on $\mathcal{K}^{m|n}$ (cf. [**DZ**, Lem. 5.1]).

First, the basis $\{'X^{[A]} \mid A \in \widetilde{M}(m|n)\}$ gives rise to a basis $\{[A] \mid A \in \widetilde{M}(m|n)\}$ for $\dot{\mathbf{U}}$, where

$$[A] = \gamma^{-1}('X^{[A]}) \in \dot{\mathbf{U}}.$$

Since $[A] = \gamma_\mu^{-1}('X^{[A]})$ if $\mu = \mathrm{co}(A)$, we have

$$'X^{[A]} = [A].'X^{[\mu]} \text{ and } 1_\lambda = [\lambda](:= [\mathrm{diag}(\lambda)]). \tag{6.3.1}$$

Moreover, (6.1.1) implies that

(6.3.2) $[A][B]=0$ unless $\mathrm{co}(A)=\mathrm{ro}(B)$, $[A]1_\lambda=\delta_{\mathrm{co}(A),\lambda}[A]$, $1_\lambda[B]=\delta_{\lambda,\mathrm{ro}(B)}[B]$.

Note that ${}_\lambda\mathbf{U}_\mu=\mathrm{span}\{[A]\mid A\in{}_\lambda\widetilde{M}(m|n)_\mu\}$.

Now, Propositions 4.2 and 3.2 imply the following presentation for $\dot{\mathbf{U}}$.

THEOREM 6.4. *For $\mathbf{U}=U_\upsilon(\mathfrak{gl}_{m|n})$, the modified quantum linear supergroup $\dot{\mathbf{U}}$ has a basis*

$$\{[A]\mid A\in\widetilde{M}(m|n)\}\tag{6.4.1}$$

and is generated by $1_\lambda,[E_{h,h+1}+\lambda],[E_{h+1,h}+\lambda]$ for all $\lambda\in\mathbb{Z}^{m+n}$, $h\in[1,m+n)$. Moreover, the following multiplication formulas hold in $\dot{\mathbf{U}}$:

(1) $1_\lambda\cdot[A]=\delta_{\lambda,\mathrm{ro}(A)}[A]$ *and* $[A]\cdot 1_\lambda=\delta_{\mathrm{co}(A),\lambda}[A]$;

(2)
$$\begin{aligned}&[E_{h,h+1}+\mathrm{ro}(A)-\mathbf{e}_{h+1}]\cdot[A]\\&=\sum_{\substack{1\le i\le m+n,i\neq h+1\\a_{h+1,i}\ge1}}(-1)^{\tilde{\sigma}_{h,m}(i,A)}\upsilon_h^{f(i,A)}[a_{h,i}+1][A+E_{h,i}-E_{h+1,i}],\\&\quad+(-1)^{\tilde{\sigma}_{h,m}(h+1,A)}\upsilon_h^{f(h+1,A)}[a_{h,h+1}+1][A+E_{h,h+1}-E_{h+1,h+1}],\end{aligned}$$

(3)
$$\begin{aligned}&[E_{h+1,h}+\mathrm{ro}(A)-\mathbf{e}_h]\cdot[A]\\&=\sum_{\substack{1\le i\le m+n,i\neq h\\a_{h,i}\ge1}}(-1)^{\tilde{\sigma}_{h,m}(i,A)}\upsilon_{h+1}^{g(i,A)}[a_{h+1,i}+1][A-E_{h,i}+E_{h+1,i}].\\&\quad+(-1)^{\tilde{\sigma}_{h,m}(h,A)}\upsilon_{h+1}^{g(h,A)}[a_{h+1,h}+1][A-E_{h,h}+E_{h+1,h}].\end{aligned}$$

PROOF. The basis assertion follows from the isomorphism γ. To see the generating set, we observe from Proposition 4.2 that, for $A\in\widetilde{M}(m|n)$, we have

$$\mathfrak{m}^{A^\pm,\mathbf{0}}{}'X^{[\mathrm{co}(A)]}={}'X^{[A]}+\sum_{B\in\widetilde{M}(m|n),\,B\prec A}g_{B,A}{}'X^{[B]}\quad(g_{B,A}\in\mathbb{Q}(\upsilon)).$$

Thus, (6.3.1) implies that

$$\Big(\mathfrak{m}^{A^\pm,\mathbf{0}}1_{\mathrm{co}(A)}-[A]-\sum_{B\in\widetilde{M}(m|n),B\prec A}g_{B,A}[B]\Big)\cdot{}'X^{[\mathrm{co}(A)]}=0.$$

Hence,

$$\mathfrak{m}^{A^\pm,\mathbf{0}}1_{\mathrm{co}(A)}=[A]+\sum_{B\in\widetilde{M}(m|n),\,B\prec A}g_{B,A}[B]\tag{6.4.2}$$

and

$$\{\mathfrak{m}^{A^\pm,\mathbf{0}}1_{\mathrm{co}(A)}\mid A\in\widetilde{M}(m|n)\}\tag{6.4.3}$$

forms a new basis for $\dot{\mathbf{U}}$. Since, by Proposition 3.2, $E_h\cdot{}'X^{[\lambda]}={}'X^{[E_{h,h+1}-\mathbf{e}_{h+1}+\lambda]}$ and $F_h\cdot{}'X^{[\lambda]}={}'X^{[E_{h+1,h}-\mathbf{e}_h+\lambda]}$, it follows that

$$E_h1_\lambda=[E_{h,h+1}-\mathbf{e}_{h+1}+\lambda]\text{ and }F_h1_\lambda=[E_{h+1,h}-\mathbf{e}_h+\lambda],$$

and every element $\mathfrak{m}^{A^\pm,\mathbf{0}}1_{\mathrm{co}(A)}$ is a product of $[E_{h,h+1}+\lambda]$ or $[E_{h+1,h}+\lambda]$, proving the generating-set assertion. Finally, the multiplication formulas follow from Proposition 3.2 and a similar argument as above. □

The bases (6.4.1) and (6.4.3) are called, respectively, the *BLM basis* and *monomial basis* for $\dot{\mathbf{U}}$.

The BLM basis for $\dot{\mathbf{U}}$ allows us to decompose $\dot{\mathbf{U}}$ into a direct sum of two-sided ideals labeled by its degree.

For each $r \in \mathbb{Z}$, let

$$\widetilde{M}(m|n)_r := \{A \in \widetilde{M}(m|n) \mid r = |A|\}, \tag{6.4.4}$$

and define $M(m|n)_r$, $(\mathbb{Z}^{m+n})_r$, and $(\mathbb{N}^{m+n})_r$ similarly. Let

$$\dot{\mathbf{U}}(r) := \text{span}\{[A] \mid A \in \widetilde{M}(m|n)_r\}.$$

COROLLARY 6.5. *Each $\dot{\mathbf{U}}(r)$ is an ideal of $\dot{\mathbf{U}}$. The algebra $\dot{\mathbf{U}}$ has a decomposition:*

$$\dot{\mathbf{U}} = \oplus_{r \in \mathbb{Z}} \dot{\mathbf{U}}(r).$$

PROOF. The decomposition is clear. For $r, r' \in \mathbb{Z}$, if $r \neq r'$, then $\dot{\mathbf{U}}(r)\dot{\mathbf{U}}(r') = 0$ by (6.3.2). On the other hand, the proof above shows that $\dot{\mathbf{U}}(r)$ is the subalgebra generated by $1_\alpha, [E_{h,h+1}+\mu], [E_{h+1,h}+\mu]$, for all $\alpha \in (\mathbb{Z}^{m+n})_r$ and $\mu \in (\mathbb{Z}^{m+n})_{r-1}$. □

REMARK 6.6.

(1) The existence of the BLM type construction for $\dot{\mathbf{U}}$ in the theorem above raises naturally the following question: is there a stablisation property approach to the construction of $\dot{\mathbf{U}}$ as developed in [**BLM**, §4]? We now can give an answer to this question. As we pointed out in the introduction, the stabilisation property cannot be derived from the multiplication formulas given in [**DG1**, Props. 4.4&4.5]. However, if we inspect the definition in [**DG1**, (5.1.1)] and introduce a new basis $\{[A]^\sharp \mid A \in M(m|n)_r\}$ for the v-Schur algebra $\mathcal{S}_{\mathcal{Z}}(m|n, r)$, where $[A]^\sharp = (-1)^{\overline{A}}\xi_A$ with $\overline{A}$ being defined in [**DG1**, (5.0.1)],[5] then the stabilisation property may be derived from the multiplication formulas associated with this new basis, similar to [**DG1**, Props. 4.4&4.5], and a construction for $\dot{\mathbf{U}}$, similar to the algebra $\mathbf{K}$ in [**BLM**, §4], can be completed. For more details regarding v-Schur superalgebras and the multiplication formulas associated with this new basis, see Appendix B.

(2) As remarked in [**DZ**, Remark 5.5], the quantum differential operator approach to the regular representation of $\mathbf{U}(\mathfrak{gl}_{m|n})$ provides a new method to the realisation problem of $\mathbf{U}(\mathfrak{gl}_m)$ when $n = 0$. Likewise, the construction developed in this paper gives also a new approach to the regular representation $\mathcal{K}^{m|0}$ of $\dot{\mathbf{U}}(\mathfrak{gl}_m)$ in the classical or nonsuper case. However, as mentioned in the introduction of [**DZ**], using the quantum differential operator to construct certain representations of quantum $\mathfrak{gl}_m$ is not new; see [**DZ**] and the references therein.

7. A canonical basis theory for $\dot{\mathbf{U}}$

We first construct a $\mathcal{Z}$-form $\dot{U}_{\mathcal{Z}}$ for $\dot{\mathbf{U}}$. We then define a bar involution on $\dot{U}_{\mathcal{Z}}$, which is induced from a bar involution on $\mathbf{U}$, and, finally, construct a canonical basis for $\dot{U}_{\mathcal{Z}}$.

[5]The elements ξ_A are endomorphisms of the tensor space; see [**DG2**, Lem. 2.3] for an interpretation as homomorphisms between v-permutation modules.

PROPOSITION 7.1. *The free $\mathcal{Z}$-module $\dot{U}_{\mathcal{Z}} := \sum_A \mathcal{Z}[A]$ is a $\mathcal{Z}$-subalgebra of $\dot{\mathbf{U}}$. Moreover, every monomial basis element $\mathfrak{m}^{A^\pm,\mathbf{0}}1_{\mathrm{co}(A)} \in \dot{U}_{\mathcal{Z}}$ and the set*

$$\{\mathfrak{m}^{A^\pm,\mathbf{0}}1_{\mathrm{co}(A)} \mid A \in \widetilde{M}(m|n)\}$$

forms a $\mathcal{Z}$-basis for $\dot{U}_{\mathcal{Z}}$. Moreover, $\dot{U}_{\mathcal{Z}}$ can be presented by generators $1_\lambda, [sE_{h,h+1}+\lambda], [sE_{h+1,h}+\lambda]$, for all $s \geq 0$, $\lambda \in \mathbb{Z}^{m+n}$ (with $s = 1$ if $h = m$).

PROOF. Let $\mathcal{K}_{\mathcal{Z}}^{m|n} := \sum_{A\in\widetilde{M}(m|n)} \mathcal{Z}'X^{[A]}$. Then $\dot{U}_{\mathcal{Z}} = \gamma^{-1}(\mathcal{K}_{\mathcal{Z}}^{m|n})$. To prove that $\dot{U}_{\mathcal{Z}}$ is a subalgebra, it is equivalent to prove that $\mathcal{K}_{\mathcal{Z}}^{m|n}$ is a $U_{\mathcal{Z}}$-module, where $U_{\mathcal{Z}}$ is the Lusztig type $\mathcal{Z}$-form of $\mathbf{U}$, generated by divided powers $E_h^{(s)} := \frac{E_h^s}{[s]^!}$, $F_h^{(s)} := \frac{F_h^s}{[s]^!}$, K_i, and $\begin{bmatrix} K_i;0 \\ t \end{bmatrix} := \prod_{a=1}^t \frac{K_i v_i^{-a+1} - K_i^{-1} v_i^{a-1}}{v_i^a - v_i^{-a}}$. In other words, we need to show that, for any $A \in \widetilde{M}(m|n)$,

$$K_i.'X^{[A]}, \quad \begin{bmatrix} K_i;0 \\ t \end{bmatrix}.'X^{[A]}, \quad E_h^{(s)}.'X^{[A]}, \quad F_h^{(s)}.'X^{[A]} \in \mathcal{K}_{\mathcal{Z}}^{m|n}.$$

For the first two elements, by Proposition 3.2, we have

$$K_i.'X^{[A]} = v_i^{\lambda_i}{}'X^{[A]} \quad \text{and} \quad \begin{bmatrix} K_i;0 \\ t \end{bmatrix}.'X^{[A]} = \begin{bmatrix} \lambda_i \\ t \end{bmatrix}'X^{[A]} \quad (\lambda = \mathrm{ro}(A)).$$

We now check that the last two elements are in $\mathcal{K}_{\mathcal{Z}}^{m|n}$.

By (2.1.1), for $0 < s \in \mathbb{N}$, we have the following action formula on $\mathcal{S}_{0|1}^k$ resp., $\mathcal{S}_{1|0}^l$:

(7.1.1)

$$E_h^{(s)}.X^{(\mathbf{a})} = \begin{cases} \left[\begin{smallmatrix} a_h+s \\ s \end{smallmatrix}\right] X^{(\mathbf{a}+s\alpha_h)}, & \text{if } a_{h+1} \geq s \text{ or } h+1 \neq k, \text{ resp., } h+1 \neq l; \\ \left[\begin{smallmatrix} a_h+s \\ s \end{smallmatrix}\right] X^{(\mathbf{a}+s\alpha_h)}, & \text{if } h+1 = k, \text{ resp., } h+1 = l; \\ 0, & \text{otherwise,} \end{cases}$$

A similar action formula holds for $F_h^{(s)}.X^{(\mathbf{a})}$. On the other hand, using the commutation formula $(E_h \otimes \widetilde{K}_h)(1 \otimes E_h) = v_h^2(1 \otimes E_h)(E_h \otimes \widetilde{K}_h)$ and the quantum binomial theorem [**DDPW**, (0.4.6)], we have, for $h \neq m$,

$$\Delta(E_h^{(s)}) = \sum_{i=0}^s v_h^{i(s-i)} E_h^{(i)} \otimes E_h^{(s-i)} \widetilde{K}_h^i. \tag{7.1.2}$$

Combining (7.1.2) with (7.1.1) proves $E_h^{(s)}.'X^{[A]} \in \mathcal{K}_{\mathcal{Z}}^{m|n}$. The proof for $F_h^{(s)}.'X^{[A]} \in \mathcal{K}_{\mathcal{Z}}^{m|n}$ is similar.

The above argument also shows that the triangular relation (6.4.2) satisfies all $g_{B,A} \in \mathcal{Z} = \mathbb{Z}[v, v^{-1}]$. Hence, the basis assertion follows. In other words, both BLM and monomial bases are (integral) bases for $\dot{U}_{\mathcal{Z}}$.

Finally, repeatedly applying Proposition 3.2 yields

$$E_h^s.'X^{[\lambda]} = [s]^!{}'X^{[sE_{h,h+1} - s\mathbf{e}_{h+1} + \lambda]} \text{ and } F_h^s.'X^{[\lambda]} = [s]^!{}'X^{[sE_{h+1,h} - s\mathbf{e}_h + \lambda]}.$$

Thus, $E_h^{(s)} 1_\lambda = [sE_{h,h+1} - s\mathbf{e}_{h+1} + \lambda]$ and $F_h^{(s)} 1_\lambda = [sE_{h+1,h} - s\mathbf{e}_h + \lambda]$. Hence, every basis vector $\mathfrak{m}^{A^\pm,\mathbf{0}}1_{\mathrm{co}(A)}$ is a product of $[sE_{h,h+1}+\lambda], [sE_{h+1,h}+\mu]$, for some $s \in \mathbb{N}$, $\lambda, \mu \in \mathbb{Z}^{m+n}$ (i.e., is a monomial of the generators for $\dot{U}_{\mathcal{Z}}$). This proves the last assertion. □

REMARK 7.2. Observe from (7.1.1) that, if the hth component a_h of $\mathbf{a}$ is negative and $a_h + s \geq 0$, then $\left[{a_h+s \atop s}\right] = 0$ and so, $E_h^{(s)}.X^{(\mathbf{a})} = 0$. A similar statement hold for $F_h^{(s)}.X^{(\mathbf{a})}$. This implies the following fact: if $A \in \widetilde{M}(m|n)\backslash M(m|n)$, then both $E_h^{(s)}.'X^{[A]}$ and $F_h^{(s)}.'X^{[A]}$ are linear combinations of $'X^{[B]}$ with $B \in \widetilde{M}(m|n)\backslash M(m|n)$.

An immediate consequence of this result is the integral version of Corollary 6.5. For any $r \in \mathbb{Z}$, let

$$\dot{U}_{\mathcal{Z}}(r) := \bigoplus_{A \in \widetilde{M}(m|n)_r} \mathcal{Z}[A]. \tag{7.2.1}$$

Corollary 6.5 and its proof imply immediately the following.

COROLLARY 7.3. *Each $\dot{U}_{\mathcal{Z}}(r)$ is a $\mathcal{Z}$-algebra generated by $1_\lambda, [sE_{h,h+1} + \mu]$, $[sE_{h+1,h} + \mu]$, for all $s \in \mathbb{N}$, $\lambda \in (\mathbb{Z}^{m+n})_r$, and $\mu \in (\mathbb{Z}^{m+n})_{r-s}$ (with $s = 1$ for $h = m$). Moreover, $\dot{U}_{\mathcal{Z}}(r)$ is an ideal of $\dot{U}_{\mathcal{Z}}$, and the $\mathcal{Z}$-algebra $\dot{U}_{\mathcal{Z}}$ has the following decomposition:*

$$\dot{U}_{\mathcal{Z}} = \bigoplus_{r \in \mathbb{Z}} \dot{U}_{\mathcal{Z}}(r).$$

We are now ready to introduce the canonical basis of the (integral) modified quantum supergroup $\dot{U}_{\mathcal{Z}}$. The key ingredient is the triangular relation in (6.4.2) between the monomial and BLM bases. The construction is elementary (see [**DDPW**, §0.5]) and parallel to the one in [**DG2**, §5] for the positive part of $\mathbf{U}$.

Recall from the bar involution (2.0.5) on $\mathbf{U} = \mathbf{U}(\mathfrak{gl}_{m|n})$ (see also [**DG2**, §5]). Since $\overline{K^{\mathbf{j}} - v^{\lambda\bullet_s\mathbf{j}}} = K^{-\mathbf{j}} - v^{-\lambda\bullet_s\mathbf{j}} = -K^{-\mathbf{j}}v^{-\lambda\bullet_s\mathbf{j}}(K^{\mathbf{j}} - v^{\lambda\bullet_s\mathbf{j}})$, the bar map stabilises ${}_\lambda\mathbf{K}_\mu$. Thus, it induces a group automorphism on each ${}_\lambda\mathbf{U}_\mu$, for all $\lambda, \mu \in \mathbb{Z}^{m+n}$, and hence, a group automorphism on $\dot{\mathbf{U}}$.

LEMMA 7.4. *The (abelian) group automorphism $^- : \dot{\mathbf{U}} \to \dot{\mathbf{U}}$ is a ring automorphism of order 2, and $\overline{\dot{U}_{\mathcal{Z}}} = \dot{U}_{\mathcal{Z}}$.*

PROOF. By the definition, we have $\overline{\pi_{\lambda,\mu}(t)} = \pi_{\lambda,\mu}(\bar{t})$ on $\dot{\mathbf{U}}$. Thus, for $\lambda, \mu, \lambda', \mu' \in \mathbb{Z}^{m+n}, t, s \in \mathbf{U}$, we have

$$\begin{aligned}\overline{\pi_{\lambda,\mu}(t)\pi_{\lambda',\mu'}(s)} &= \overline{\delta_{\mu,\lambda'}\pi_{\lambda,\mu'}(ts)} = \delta_{\mu,\lambda'}\pi_{\lambda,\mu'}(\overline{ts}) = \delta_{\mu,\lambda'}\pi_{\lambda,\mu'}(\bar{t}\bar{s})\\ &= \pi_{\lambda,\mu}(\bar{t})\pi_{\lambda',\mu'}(\bar{s}) = \overline{\pi_{\lambda,\mu}(t)}\ \overline{\pi_{\lambda',\mu'}(s)},\end{aligned}$$

proving the first assertion. Since the bar involutions on $\mathbf{U}$ and on $\dot{\mathbf{U}}$ fix $\mathfrak{m}^{A,\mathbf{0}}$ and $1_{\mathrm{co}(A)}$, respectively, it follows that $\overline{\mathfrak{m}^{A^\pm,\mathbf{0}}1_{\mathrm{co}(A)}} = \mathfrak{m}^{A^\pm,\mathbf{0}}1_{\mathrm{co}(A)}$, proving the last assertion. □

Recall from (4.0.1) and (4.0.2) the preorder relation $\preceq$ defined on $\widetilde{M}(m|n)$.

THEOREM 7.5. *The basis $\{[A]\}_{A \in \widetilde{M}(m|n)}$ for $\dot{U}_{\mathcal{Z}}$, the bar involution on $\dot{U}_{\mathcal{Z}}$, and the preorder $\preceq$ on $\widetilde{M}(m|n)$ define uniquely a canonical basis $\mathscr{C} = \{\mathsf{C}_A \mid A \in \widetilde{M}(m|n)\}$ for the (integral) modified quantum supergroup $\dot{U}_{\mathcal{Z}}$. In other words, the elements C_A are uniquely defined by the conditions $\bar{\mathsf{C}}_A = \mathsf{C}_A$ and*

$$\mathsf{C}_A - [A] \in \sum_{B \in \widetilde{M}(m|n), B \prec A} v^{-1}\mathbb{Z}[v^{-1}][B].$$

PROOF. By the triangular relation (6.4.2) between the monomial and BLM bases, we may write the basis $\{[A]\}_{A\in\widetilde{M}(m|n)}$ in terms of the monomial basis:

$$[A] = \mathfrak{m}^{A^{\pm},\mathbf{0}}1_{\mathrm{co}(A)} + \sum_{B\in\widetilde{M}(m|n),B\prec A} h_{B,A}\mathfrak{m}^{B^{\pm},\mathbf{0}}1_{\mathrm{co}(A)},$$

where $h_{B,A}\in\mathcal{Z}$. Since $\overline{\mathfrak{m}^{A^{\pm},\mathbf{0}}1_{\mathrm{co}(A)}} = \mathfrak{m}^{A^{\pm},\mathbf{0}}1_{\mathrm{co}(A)}$, applying the bar involution to the expression above yields

$$\begin{aligned}\overline{[A]} &= \mathfrak{m}^{A^{\pm},\mathbf{0}}1_{\mathrm{co}(A)} + \sum_{B\in\widetilde{M}(m|n),B\prec A} \overline{h_{B,A}}\,\mathfrak{m}^{B^{\pm},\mathbf{0}}1_{\mathrm{co}(A)}\\ &= [A] + \sum_{C\in\widetilde{M}(m|n),C\prec A} f_{C,A}\,[C] \quad (f_{C,A}\in\mathcal{Z})\ (\text{by } (6.4.2)).\end{aligned}$$

Thus, a standard construction (see, e.g., [**DDPW**, §0.5]) shows that there exist unique polynomials $p_{B,A}$ with $p_{A,A}=1$ and $p_{B,A}\in v^{-1}\mathbb{Z}[v^{-1}]$ if $B\prec A$ such that the elements

$$\mathsf{C}_A = [A] + \sum_{B\in\widetilde{M}(m|n),B\prec A} p_{B,A}[B] \quad (A\in\widetilde{M}(m|n)) \tag{7.5.1}$$

are bar invariants. □

REMARK 7.6. It should be noted that Benkart, Kang, and Kashiwara [**BKK**] constructed the crystal bases for polynomial representations of $\mathbf{U}(\mathfrak{gl}_{m|n})$, and subsequently Kwon [**Kw**] constructed crystal bases for Kac modules of $\mathbf{U}(\mathfrak{gl}_{m|n})$; none of these authors constructed the global crystal or canonical bases. The latter was constructed for $\mathbf{U}^{\pm}(\mathfrak{gl}_{m|n})$ in [**CHW**, §7.2] or [**DG2**, Th. 5.2] and for $\dot{\mathbf{U}}(\mathfrak{gl}_{m|n})$ in this section.

8. Canonical bases for v-Schur superalgebras

In this section, we bring v-Schur superalgebras into the game. We first construct them via $\dot{U}_{\mathcal{Z}}$ and then make a comparison between the canonical basis for $\dot{U}_{\mathcal{Z}}$ and canonical basis for v-Schur superalgebras.

Recall from the last section the Lusztig form $U_{\mathcal{Z}}$, the $U_{\mathcal{Z}}$-module $\mathcal{K}_{\mathcal{Z}}^{m|n}$, and the ideal $\dot{U}_{\mathcal{Z}}(r)$ in (7.2.1). With (6.4.4), let

$$\dot{U}_{\mathcal{Z}}(r)^{<0} := \bigoplus_{A\in\widetilde{M}(m|n)_r\backslash M(m|n)_r} \mathcal{Z}[A]. \tag{8.0.1}$$

LEMMA 8.1. *The $\mathcal{Z}$-submodule $\dot{U}_{\mathcal{Z}}(r)^{<0}$ is an ideal of $\dot{U}_{\mathcal{Z}}(r)$.*

PROOF. By Theorem 6.2, γ restricts to a linear isomorphism $\gamma_r^{<0}:\dot{U}_{\mathcal{Z}}(r)^{<0}\to\mathcal{K}_{\mathcal{Z}}^{m|n}(r)^{<0}$, where

$$\mathcal{K}_{\mathcal{Z}}^{m|n}(r)^{<0} := \bigoplus_{A\in\widetilde{M}(m|n)_r\backslash M(m|n)_r} \mathcal{Z}'X^{[A]}$$

For $A\in\widetilde{M}(m|n)_r\backslash M(m|n)_r$, by Remark 7.2, we have $E_h^{(s)}.'X^{[A]}, F_h^{(s)}.'X^{[A]}\in\mathcal{K}_{\mathcal{Z}}^{m|n}(r)^{<0}$. Thus, $\mathcal{K}_{\mathcal{Z}}^{m|n}(r)^{<0}$ is a $U_{\mathcal{Z}}$-subsupermodule. Hence, it is a $\dot{U}_{\mathcal{Z}}$-subsupermodule. In other words, $\dot{U}_{\mathcal{Z}}(r)^{<0}$ is a left ideal of $\dot{U}_{\mathcal{Z}}(r)$ and, consequently, a two-sided ideal, since the anti-involution $\dot{\tau}$ in (5.5.2) stabilises $\dot{U}_{\mathcal{Z}}(r)^{<0}$. □

Recall the Lusztig form $U_{\mathbb{Z}}$ defined in the proof of Proposition 7.1. For any $\lambda \in \mathbb{Z}^{m|n}$ and $\mu \in \mathbb{N}^{m|n}$, let $\left[\begin{smallmatrix}\lambda\\ \mu\end{smallmatrix}\right] = \prod_{a=1}^{m+n} \left[\begin{smallmatrix}\lambda_a\\ \mu_a\end{smallmatrix}\right]$ and $\left[\begin{smallmatrix}K\\ \mu\end{smallmatrix}\right] = \prod_{a=1}^{m+n} \left[\begin{smallmatrix}K_a;0\\ \mu_a\end{smallmatrix}\right]$. Let J_r be the ideal of $U_{\mathbb{Z}}$ generated by

$$(8.1.1)\quad 1-\sum_{\lambda\in(\mathbb{N}^{m+n})_r}\begin{bmatrix}K\\ \lambda\end{bmatrix},\quad K_a^{\epsilon}\begin{bmatrix}K\\ \lambda\end{bmatrix}-v_a^{\epsilon\lambda_a}\begin{bmatrix}K\\ \lambda\end{bmatrix},\quad \begin{bmatrix}K_a;c\\ t\end{bmatrix}\begin{bmatrix}K\\ \lambda\end{bmatrix}-\begin{bmatrix}\lambda_a+c\\ t\end{bmatrix}\begin{bmatrix}K\\ \lambda\end{bmatrix},$$

where $\epsilon \in \{-1,1\}$, $1 \le a \le m+n, t \in \mathbb{N}, c \in \mathbb{Z}, \lambda \in \Lambda(m|n,r)$. By [**DLZ1**, Th. 4.3], we have a $\mathbb{Z}$-algebra isomorphism: for any $r \ge 0$

$$(8.1.2)\qquad U_{\mathbb{Z}}/J_r \cong \mathcal{S}_{\mathbb{Z}}(m|n,r),$$

where $\mathcal{S}_{\mathbb{Z}}(m|n,r)$ is the v-Schur superalgebra defined in [**DR**, 5.6, 9.8.] (cf. [**DZ**, Lemma 3.3 and Theorem 5.7]). See [**EK**] for a simpler presentation for the v-Schur superalgebra defined over $\mathbb{Q}(v)$.

PROPOSITION 8.2. *For any integer $r > 0$, the quotient algebra $\dot{U}_{\mathbb{Z}}(r)/\dot{U}_{\mathbb{Z}}(r)^{<0}$ is isomorphic to the (integral) v-Schur superalgebra $\mathcal{S}_{\mathbb{Z}}(m|n,r)$.*

PROOF. Since $\dot{U}_{\mathbb{Z}}(r)/\dot{U}_{\mathbb{Z}}(r)^{<0}$ has a finite basis labelled by $M(m|n)_r$, this algebra has the identity element $1_r = \sum_{\lambda\in(\mathbb{N}^{m+n})_r} 1_\lambda$. Thus, there is a well defined algebra homomorphism

$$\eta_{r,\mathbb{Z}} : U_{\mathbb{Z}} \longrightarrow \dot{U}_{\mathbb{Z}}(r)/\dot{U}_{\mathbb{Z}}(r)^{<0},\quad u \longmapsto u1_r.$$

We first prove that this induces an epimorphism $\bar{\eta}_{r,\mathbb{Z}}:\mathcal{S}_{\mathbb{Z}}(m|n,r)\to\dot{U}_{\mathbb{Z}}(r)/\dot{U}_{\mathbb{Z}}(r)^{<0}$. By (8.1.2), it suffices to prove that $\eta_{r,\mathbb{Z}}$ sends every element in (8.1.1) to 0. Equivalently, these elements act on $\mathcal{K}_{\mathbb{Z}}^{m|n}(r)/\mathcal{K}_{\mathbb{Z}}^{m|n}(r)^{<0}$ trivially, where

$$\mathcal{K}_{\mathbb{Z}}^{m|n}(r) := \bigoplus_{A\in\widetilde{M}(m|n)_r} \mathbb{Z}'X^{[A]}.$$

Since, for any $A \in M(m|n)_r$ and $\lambda \in (\mathbb{N}^{m+n})_r$, we have $K_a^{\pm1}.'X^{[A]} = v_a^{\pm\,\mathrm{ro}(A)_a}{}'X^{[A]}$, $\left[\begin{smallmatrix}K\\ \lambda\end{smallmatrix}\right].'X^{[A]} = \left[\begin{smallmatrix}\mathrm{ro}(A)\\ \lambda\end{smallmatrix}\right]'X^{[A]} = \delta_{\mathrm{ro}(A),\lambda}{}'X^{[A]}$, and $\left[\begin{smallmatrix}\lambda\\ \lambda\end{smallmatrix}\right] = 1$, and noting $\left[\begin{smallmatrix}\mathrm{ro}(A)\\ \lambda\end{smallmatrix}\right] = 0$ whenever $\mathrm{ro}(A) \ne \lambda$, then we have

$$\Big(1-\sum_{\lambda\in(\mathbb{N}^{m+n})_r}\begin{bmatrix}K\\ \lambda\end{bmatrix}\Big).'X^{[A]} = (1-\sum_{\lambda\in(\mathbb{N}^{m+n})_r}\delta_{\mathrm{ro}(A),\lambda})'X^{[A]} = 0;$$

$$\Big(K_a^{\pm1}\begin{bmatrix}K\\ \lambda\end{bmatrix}-v_a^{\pm\lambda_a}\begin{bmatrix}K\\ \lambda\end{bmatrix}\Big).'X^{[A]} = (K_a^{\pm1}-v_a^{\pm\lambda_a})\delta_{\mathrm{ro}(A),\lambda}{}'X^{[A]} = 0;$$

$$\Big(\begin{bmatrix}K_a;c\\ t\end{bmatrix}\begin{bmatrix}K\\ \lambda\end{bmatrix}-\begin{bmatrix}\lambda_a+c\\ t\end{bmatrix}\begin{bmatrix}K\\ \lambda\end{bmatrix}\Big).'X^{[A]} = \Big(\begin{bmatrix}K_a;c\\ t\end{bmatrix}-\begin{bmatrix}\lambda_a+c\\ t\end{bmatrix}\Big)\delta_{\mathrm{ro}(A),\lambda}{}'X^{[A]} = 0.$$

Thus, we obtain the desired epimorphism $\bar{\eta}_{r,\mathbb{Z}} : \mathcal{S}_{\mathbb{Z}}(m|n,r) \to \dot{U}_{\mathbb{Z}}(r)/\dot{U}_{\mathbb{Z}}(r)^{<0}$.

Finally, $\bar{\eta}_{r,\mathbb{Z}}$ sends a monomial basis described in the proof of [**DG1**, Cor. 8.5] to the monomial basis, the image of $\{\mathfrak{m}^{A^{\pm},\mathbf{0}}1_{\mathrm{co}(A)} \mid A \in M(m|n)_r\}$. Hence, $\bar{\eta}_{r,\mathbb{Z}}$ is an isomorphism. □

We obtained an isomorphism $\bar{\eta}_{r,\mathbb{Z}} : \mathcal{S}_{\mathbb{Z}}(m|n,r) \to \dot{U}_{\mathbb{Z}}(r)/\dot{U}_{\mathbb{Z}}(r)^{<0}$ using the presentation of the integral v-Schur superalgebra. However, a v-Schur superalgebra is also an endomorphism superalgebra; see (B.0.1). It would be interesting to know what is the inverse image $\bar{\eta}_{r,\mathbb{Z}}^{-1}([A] + \dot{U}_{\mathbb{Z}}(r)^{<0})$ as a module endomorphism? The following result answers the question by setting a direct isomorphism between the two.

Recall the endomorphism $[A]^\sharp$ mentioned in Remark 6.6(1) (see Appendix B for more details). For any $A \in M(m|n)_r$, let $[A, r] := [A] + \dot{U}_{\mathcal{Z}}(r)^{<0}$.

PROPOSITION 8.3. *The $\mathcal{Z}$-linear map*

$$\zeta_{r,\mathcal{Z}} : \dot{U}_{\mathcal{Z}}(r)/\dot{U}_{\mathcal{Z}}(r)^{<0} \longrightarrow \mathcal{S}_{\mathcal{Z}}(m|n,r), \quad [A,r] \longmapsto [A]^\sharp$$

is a superalgebra isomorphism.

PROOF. First, $\{[A,r] \mid A \in M(m|n)_r\}$ forms a basis for $\dot{U}_{\mathcal{Z}}(r)/\dot{U}_{\mathcal{Z}}(r)^{<0}$ and the map $\zeta_{r,\mathcal{Z}}$ is a $\mathcal{Z}$-module isomorphism. Base change to $\mathbb{Q}(v)$ induces a linear isomorphism

$$\zeta_r : \dot{\mathbf{U}}(r)/\dot{\mathbf{U}}(r)^{<0} \longrightarrow \mathbf{S}(m|n,r),\ [A] + \dot{\mathbf{U}}(r)^{<0} \longmapsto [A]^\sharp.$$

Now, the multiplication formulas in Theorem 6.4 in $\dot{\mathbf{U}}$ induces multiplication formulas in $\dot{\mathbf{U}}(r)/\dot{\mathbf{U}}(r)^{<0}$ which are mapped to the corresponding formulas in Proposition B.2. For example, in Theorem 6.4(2), if the entry $a_{h+1,h+1} = 0$, then the $i = h+1$ term is in $\dot{\mathbf{U}}(r)^{<0}$ and, hence, is 0 in $\dot{\mathbf{U}}(r)/\dot{\mathbf{U}}(r)^{<0}$. Thus, in $\dot{\mathbf{U}}(r)/\dot{\mathbf{U}}(r)^{<0}$, this formula has the form

$$\begin{aligned}&[E_{h,h+1} + \mathrm{ro}(A) - \mathbf{e}_{h+1}] \cdot [A] \\ &\quad = \sum_{\substack{1\le i\le m+n \\ a_{h+1,i}\ge 1}} (-1)^{\tilde{\sigma}_{h,m}(i,A)} v_h^{f(i,A)} [a_{h,i}+1][A + E_{h,i} - E_{h+1,i}] \pmod{\dot{\mathbf{U}}^{<0}},\end{aligned}$$

whose image under ζ_r is exactly the one in Proposition B.2(2). Hence, ζ_r is a $\mathbb{Q}(v)$-superalgebra isomorphism. Clearly, restriction gives a $\mathcal{Z}$-algebra isomorphism

$$\zeta_r : (\dot{U}_{\mathcal{Z}}(r) + \dot{\mathbf{U}}(r)^{<0})/\dot{\mathbf{U}}(r)^{<0} \longrightarrow \mathcal{S}_{\mathcal{Z}}(m|n,r).$$

Since $(\dot{U}_{\mathcal{Z}}(r) + \dot{\mathbf{U}}(r)^{<0})/\dot{\mathbf{U}}(r)^{<0} \cong \dot{U}_{\mathcal{Z}}(r)/(\dot{U}_{\mathcal{Z}}(r) \cap \dot{\mathbf{U}}(r)^{<0}) = \dot{U}_{\mathcal{Z}}(r)/\dot{U}_{\mathcal{Z}}(r)^{<0}$, the restriction map induces the desired $\mathcal{Z}$-algebra isomorphism $\zeta_{r,\mathcal{Z}}$. □

Let

$$\dot{U}_{\mathcal{Z}}^{<0} = \mathrm{span}_{\mathcal{Z}}\{[A] \mid A \in \widetilde{M}(m|n) \backslash M(m|n)\}.$$

Then, by (8.0.1). $\dot{U}_{\mathcal{Z}}^{<0} = \bigoplus_{r\in\mathbb{Z}} \dot{U}_{\mathcal{Z}}(r)^{<0}$. Thus, by Lemma 8.1, $\dot{U}_{\mathcal{Z}}^{<0}$ is an ideal of $\dot{U}_{\mathcal{Z}}$.

COROLLARY 8.4. *There is a $\mathcal{Z}$-superalgebra isomorphism*

$$\zeta_{\mathcal{Z}} = \oplus_r \zeta_{r,\mathcal{Z}} : \dot{U}_{\mathcal{Z}}/\dot{U}_{\mathcal{Z}}^{<0} \longrightarrow \bigoplus_{r\ge 0} \mathcal{S}_{\mathcal{Z}}(m|n,r).$$

Moreover, the quotient morphism $\pi_{\mathcal{Z}}^{\ge 0} : \dot{U}_{\mathcal{Z}} \to \dot{U}_{\mathcal{Z}}/\dot{U}_{\mathcal{Z}}^{<0}$ induces a $\mathcal{Z}$-superalgebra epimorphism

$$\dot{\eta}_{\mathcal{Z}}^{\ge 0} := \zeta_{\mathcal{Z}} \circ \pi_{\mathcal{Z}}^{\ge 0} = \oplus_{r\ge 0} \dot{\eta}_{r,\mathcal{Z}} : \dot{U}_{\mathcal{Z}} \longrightarrow \bigoplus_{r\ge 0} \mathcal{S}_{\mathcal{Z}}(m|n,r),$$

where $\dot{\eta}_{r,\mathcal{Z}} : \dot{U}_{\mathcal{Z}}(r) \to \mathcal{S}_{\mathcal{Z}}(m|n,r)$, for all $r \ge 0$, with

$$\dot{\eta}_{r,\mathcal{Z}}([A]) = \begin{cases} [A]^\sharp, & \text{if } A \in M(m|n)_r; \\ 0, & \text{otherwise.} \end{cases} \tag{8.4.1}$$

PROOF. All statements are clear except (8.4.1). Since

$$\dot{U}_{\mathbb{Z}}(r) + \dot{U}_{\mathbb{Z}}^{<0}/\dot{U}_{\mathbb{Z}}(r)^{<0} \cong \dot{U}_{\mathbb{Z}}(r)/\dot{U}_{\mathbb{Z}}(r) \cap \dot{U}_{\mathbb{Z}}^{<0} = \dot{U}_{\mathbb{Z}}(r)/\dot{U}_{\mathbb{Z}}(r)^{<0},$$

we may identify $\pi_{\mathbb{Z}}([A])$ with $[A,r]$, for all $A \in M(m|n)$. By Proposition 8.3, $\zeta_{r,\mathbb{Z}}([A,r]) = [A]^\sharp$, proving (8.4.1). □

Since the bar involution on $\dot{U}_{\mathbb{Z}}$ fixes all idempotents 1_λ, and $\dot{U}_{\mathbb{Z}}^{<0}$ has a monomial basis with basis vectors bar-fixed, it follows that $\overline{\dot{U}_{\mathbb{Z}}^{<0}} = \dot{U}_{\mathbb{Z}}^{<0}$. Thus, it induces a bar involution on the quotient $\dot{U}_{\mathbb{Z}}/\dot{U}_{\mathbb{Z}}^{<0}$ and $\dot{U}_{\mathbb{Z}}(r)/\dot{U}_{\mathbb{Z}}(r)^{<0}$, for all $r \geq 0$. In other words, for any $A \in M(m|n)_r$, we have

$$\overline{[A,r]} = \overline{[A]} + \dot{U}_{\mathbb{Z}}(r)^{<0}.$$

Recall from Corollary 8.4 that the epimorphism $\dot{\eta}_{\mathbb{Z}}$ is compatible with the bar involution by the argument above.

COROLLARY 8.5. *For any $r > 0$, let $\mathsf{C}_A^{(r)} = \dot{\eta}_{r,\mathbb{Z}}(\mathsf{C}_A)$ for all $A \in M(m|n)_r$. Then $\{\mathsf{C}_A^{(r)}\}_{A\in M(m|n)_r}$ forms the canonical basis for $\mathcal{S}_{\mathbb{Z}}(m|n,r)$ relative to the basis $\{[A]^\sharp \mid A \in M(m|n)_r\}$. Moreover, we have*

$$\dot{\eta}_{r,\mathbb{Z}}(\mathsf{C}_A) = \begin{cases} \mathsf{C}_A^{(r)}, & \textit{if } A \in M(m|n)_r; \\ 0, & \textit{if } A \in \widetilde{M}(m|n)\backslash M(m|n)_r. \end{cases} \tag{8.5.1}$$

PROOF. We first observe that the bar involution on $\dot{U}_{\mathbb{Z}}$ induces a bar involution on $\dot{U}_{\mathbb{Z}}(r)/\dot{U}_{\mathbb{Z}}(r)^{<0}$. The construction of the required canonical basis for $\dot{U}_{\mathbb{Z}}(r)/\dot{U}_{\mathbb{Z}}(r)^{<0}$ can then be done similarly as given in the proof of Theorem 7.5: one simply drop out the "tail", which is a linear combination of $[B]$ with $B \in \widetilde{M}(m|n)_r\backslash M(m|n)_r$, in each step of the construction. Hence, $\mathsf{C}_A^{(r)}$ is obtained from C_A in (7.5.1) just by dropping out the tail. We finally interpret the basis for $\dot{U}_{\mathbb{Z}}(r)/\dot{U}_{\mathbb{Z}}(r)^{<0}$ as the canonical basis for $\mathcal{S}_{\mathbb{Z}}(m|n,r)$ via the isomorphism $\zeta_{r,\mathbb{Z}}$. □

REMARK 8.6. There is another approach to construct the canonical basis for $\mathcal{S}_{\mathbb{Z}}(m|n,r)$ in [**DR**], using the Bruhat ordering. It should be pointed out that the bar involution used here should be the same bar involution coming from the bar involution on Hecke algebras. To see this, we need to recover Hecke algebras from certain monomial basis elements; see the construction in [**DP**, §9]. Thus, up to an adjustment by the sign $\overline{A}$, the canonical basis $\{\mathsf{C}_A^{(r)}\}_{A\in M(m|n)_r}$ agree with the canonical basis constructed in [**DG2**, Cor. 2.4], building on [**DR**] (see also [**DG2**, Rem. 4.3]).

Appendix A. A construction of $\mathbf{U}(\mathfrak{gl}_{m|n})$ from $\dot{\mathbf{U}}$

In this Appendix, we show how the BLM basis $\{[A] \mid A \in \widetilde{M}(m|n)\}$ for $\dot{\mathbf{U}}$ in (6.4.1) is in turn used to obtain a new realization of $\mathbf{U} = \mathbf{U}(\mathfrak{gl}_{m|n})$. This limiting procedure was first developed in [**BLM**, §5]. Unlike the modified approach developed in [**DF1**], where the quantum supergroup is embedded in the direct product of v-Schur algebras, we now embed the quantum supergroup $\mathbf{U}$ into a kind of completion algebra $\widehat{\dot{\mathbf{U}}}$ of $\dot{\mathbf{U}}$.

Let $\widehat{\dot{\mathbf{U}}}$ be the superspace consisting of all formal (possibly infinite) sums

$$\sum_{A\in\widetilde{M}(m|n)} f_A[A],$$

satisfying the following properties: for any $\lambda,\mu\in\mathbb{Z}^{m+n}$, both sums $\sum_{A\in\widetilde{M}(m|n)} f_A 1_\lambda[A]$ and $\sum_{A\in\widetilde{M}(m|n)} f_A[A]1_\mu$ are finite. Clearly, $\sum_{\lambda\in\mathbb{Z}^{m+n}} 1_\lambda\in\widehat{\dot{\mathbf{U}}}$. By Theorem 6.4(1), $\sum_{\lambda\in\mathbb{Z}^{m+n}} 1_\lambda$ is the identity of $\widehat{\dot{\mathbf{U}}}$. Thus, following [**BLM**, 5.1], we may define a multiplication on $\widehat{\dot{\mathbf{U}}}$ by setting

$$\Big(\sum_A f_A[A]\Big)\Big(\sum_B g_B[B]\Big) := \sum_{A,B} f_A g_B[A][B].$$

Then, $\widehat{\dot{\mathbf{U}}}$ becomes an associative algebra (with 1).

For $\mathbf{j}\in\mathbb{Z}^{m+n}$ and $A\in M(m|n)^\pm$, define

$$A(\mathbf{j}) = \sum_{\lambda\in\mathbb{Z}^{m+n}} v^{\lambda\centerdot_s\mathbf{j}}[A+\lambda]\in\widehat{\dot{\mathbf{U}}}. \tag{A.0.1}$$

It should be pointed out that the same notation $A(\mathbf{j})$ is also used in [**DG1**, (5.3.1)] and [**DZ**, (4.0.2)] for formal (possibly infinite) sums in different constructions. Let

$$\mathfrak{U}(m|n) = \mathrm{span}_{\mathbb{Q}(v)}\{A(\mathbf{j})\mid A\in M(m|n)^\pm,\mathbf{j}\in\mathbb{Z}^{m+n}\}. \tag{A.0.2}$$

Recall from Corollary 8.4 the $\mathcal{Z}$-superalgebra epimorphism

$$\dot{\eta}^{\geq 0}_{\mathcal{Z}} : \dot{U}_{\mathcal{Z}} \longrightarrow \bigoplus_{r\geq 0} \mathcal{S}_{\mathcal{Z}}(m|n,r).$$

This map induces a map (after base change to $\mathbb{Q}(v)$)

$$\widehat{\eta} : \widehat{\dot{\mathbf{U}}} \longrightarrow \widehat{\mathbf{S}},\quad \widehat{u}\longmapsto \widehat{u}^{\geq 0},$$

where $\widehat{\mathbf{S}} = \widehat{\mathbf{S}}(m|n) := \prod_{r\geq 0}\mathbf{S}(m|n,r)$, and, if $\widehat{u} = \sum_{A\in\widetilde{M}(m|n)} f_A[A]$ then, with the notation in (8.4.1),

$$\widehat{u}^{\geq 0} = \sum_{A\in\widetilde{M}(m|n)} f_A\dot{\eta}_{\mathcal{Z}}([A]) = \sum_{r\geq 0}\sum_{A\in M(m|n)_r} f_A\zeta_{\mathcal{Z}}([A,r]) = \sum_{r\geq 0}\sum_{A\in M(m|n)_r} f_A[A]^\sharp.$$

LEMMA A.1. *The map $\widehat{\eta}$ is a superalgebra homomorphism with*

$$\ker(\widehat{\eta}) = \Big\{\widehat{u}\in\widehat{\dot{\mathbf{U}}}\mid \widehat{u} = \sum_{A\in\widetilde{M}(m|n)\setminus M(m|n)} f_A[A]\Big\}.$$

PROOF. For each element $\widehat{u} = \sum_{A\in\widetilde{M}(m|n)} f_A[A], \widehat{v}\in\widehat{\dot{\mathbf{U}}}$, and $s\in\mathbb{Z}, r\in\mathbb{N}$, let $\widehat{u}_s = \sum_{A\in\widetilde{M}(m|n)_s} f_A[A]$ and $\widehat{u}^{\geq 0}_r = \sum_{A\in M(m|n)_r} f_A[A]$. Then $\widehat{u} = \sum_{s\in\mathbb{Z}}\widehat{u}_s$, $\widehat{u}^{\geq 0} = \sum_{r\in\mathbb{Z}}\widehat{u}^{\geq 0}_r$ and $\widehat{u}\cdot\widehat{v} = \sum_{s\in\mathbb{Z}}(\widehat{u}_s\cdot\widehat{v}_s)$, $\widehat{u}^{\geq 0}\cdot\widehat{v}^{\geq 0} = \sum_{r\in\mathbb{N}}(\widehat{u}^{\geq 0}_r\cdot\widehat{v}^{\geq 0}_r)$. Thus, by Lemma 8.1,

$$\widehat{\eta}(\widehat{u}\cdot\widehat{v}) = \sum_{r\in\mathbb{N}}\widehat{\eta}(\widehat{u}_r\cdot\widehat{v}_r) = \sum_{r\in\mathbb{N}}(\widehat{u}^{\geq 0}_r\cdot\widehat{v}^{\geq 0}_r) = \sum_{r\in\mathbb{N}}\widehat{u}^{\geq 0}_r\cdot\sum_{r\in\mathbb{N}}\widehat{v}^{\geq 0}_r = \widehat{\eta}(\widehat{u})\cdot\widehat{\eta}(\widehat{v}). \qquad\square$$

Recall from [**DG1**, (5.3.1)] that $\widehat{\mathbf{S}}$ contains a subsuperalgebra $\mathfrak{A}(m|n)$ spanned by

$$A(\mathbf{j})^\sharp = \sum_{r\geq 0}\sum_{\lambda\in(\mathbb{N}^{m+n})_{r-|A|}} v^{\lambda\centerdot_s\mathbf{j}}[A+\lambda]^\sharp \quad (A\in M(m|n)^\pm,\ \mathbf{j}\in\mathbb{Z}^{m+n}).$$

The subsuperalgebra $\mathfrak{A}(m|n)$ has a presentation as described in [**DG1**, Th. 8.1&8.4] and is isomorphic to $\mathbf{U}(\mathfrak{gl}_{m|n})$. We are now ready to obtain similar results for $\mathfrak{U}(m|n)$.

Recall $\alpha_h = \mathbf{e}_h - \mathbf{e}_{h+1}$, and set $-\beta_h = \mathbf{e}_h + \mathbf{e}_{h+1}$. Recall also from (3.0.4) the row/column sum vectors $\mathrm{ro}(A), \mathrm{co}(A)$ and denote their ith component by $\mathrm{ro}(A)_i$, $\mathrm{co}(A)_i$.

THEOREM A.2. *The superspace $\mathfrak{U}(m|n)$ is a subsuperalgebra of $\widehat{\mathbf{U}}$ with basis*

$$\{A(\mathbf{j}) \mid A\in M(m|n)^\pm, \mathbf{j}\in\mathbb{Z}^{m+n}\},$$

and the restriction of $\widehat{\eta}$ to $\mathfrak{U}(m|n)$ induces a superalgebra isomorphism

$$\widehat{\eta}: \mathfrak{U}(m|n) \longrightarrow \mathfrak{A}(m|n), \quad A(\mathbf{j}) \longmapsto A(\mathbf{j})^{\geq 0},$$

for all $A\in M(m|n)^\pm$ and $\mathbf{j}\in\mathbb{Z}^{m+n}$. Moreover, $\mathfrak{U}(m|n)$ is generated by

$$O(\pm\mathbf{e}_i), \quad (E_{h,h+1})(\mathbf{0}), \quad (E_{h+1,h})(\mathbf{0}),$$

whose matrix representations on $\mathfrak{U}(m|n)$ are given as follows:

(1) $O(\pm\mathbf{e}_i)\cdot A(\mathbf{j}) = v_i^{\pm\,\mathrm{ro}(A)_i}A(\mathbf{j}\pm\mathbf{e}_i), \quad A(\mathbf{j})\cdot O(\pm\mathbf{e}_i) = v_i^{\pm\,\mathrm{co}(A)_i}A(\mathbf{j}\pm\mathbf{e}_i)$;

(2) $(E_{h,h+1})(\mathbf{0}).A(\mathbf{j})$

$$\begin{aligned}
&= \sum_{\substack{i<h;\\ a_{h+1,i}\geq 1}} (-1)^{\tilde\sigma_{h,m}(i,A)} v_h^{f(i,A)}[a_{h,i}+1](A+E_{h,i}-E_{h+1,i})(\mathbf{j}+\alpha_h)\\
&+ \sum_{\substack{i>h+1;\\ a_{h+1,i}\geq 1}} (-1)^{\tilde\sigma_{h,m}(i,A)} v_h^{f(i,A)}[a_{h,i}+1](A+E_{h,i}-E_{h+1,i})(\mathbf{j})\\
&+ (-1)^{\tilde\sigma_{h,m}(i,A)} v_h^{f(h,A)-j_h}\frac{(A-E_{h+1,h})(\mathbf{j}+\alpha_h)-(A-E_{h+1,h})(\mathbf{j}+\beta_h)}{v_h-v_h^{-1}}\\
&+ (-1)^{\tilde\sigma_{h+1,m}(i,A)} v_h^{f(h+1,A)+(-1)^{\delta_{h,m}}j_{h+1}}[a_{h,h+1}+1](A+E_{h,h+1})(\mathbf{j});
\end{aligned}$$

(3) $(E_{h+1,h})(\mathbf{0}).A(\mathbf{j})$

$$\begin{aligned}
&= \sum_{i<h;a_{h+1,i}\geq 1} (-1)^{\tilde\sigma_{h,m}(i,A)} v_h^{g(i,A)}[a_{h+1,i}+1](A-E_{h,i}+E_{h+1,i})(\mathbf{j})\\
&+ \sum_{\substack{i>h+1;\\ a_{h+1,i}\geq 1}} (-1)^{\tilde\sigma_{h,m}(i,A)} v_h^{g(i,A)}[a_{h+1,i}+1](A-E_{h,i}+E_{h+1,i})(\mathbf{j}-\alpha_h)\\
&+ (-1)^{\tilde\sigma_{h,m}(h,A)} v_{h+1}^{g(h,A)+(-1)^{\delta_{h,m}}j_h}[a_{h+1,h}+1](A+E_{h+1,h})(\mathbf{j})\\
&+ (-1)^{\tilde\sigma_{h,m}(h+1,A)} v_{h+1}^{g(h+1,A)-j_{h+1}}\frac{(A-E_{h,h+1})(\mathbf{j}-\alpha_h)-(A-E_{h,h+1})(\mathbf{j}+\beta_h)}{v_{h+1}-v_{h+1}^{-1}}.
\end{aligned}$$

PROOF. Since

$$A(\mathbf{j})^{\geq 0} = \sum_{r\geq 0}\sum_{\lambda\in(\mathbb{N}^{m+n})_{r-|A|}} v^{\lambda\centerdot_s\mathbf{j}}[A]^\sharp = A(\mathbf{j})^\sharp,$$

for all $A \in M(m|n)^{\pm}$ and $\mathbf{j} \in \mathbb{Z}^{m+n}$, the map $\widehat{\eta}$ is a well-defined linear isomorphism preserving the grading for the super structure. Since $\mathfrak{A}(m|n)$ is a superalgebra by [**DG1**, Th. 8.1], its inverse image $\mathfrak{U}(m|n)$ is a superalgebra by Lemma A.1.

The multiplication formulas can be derived by directly applying these "short elements" multiplication formulas in Theorem 6.4. They may also be obtained by taking the inverse image of the corresponding formulas in [**DG1**, Th. 8.4] via [**DG1**, Props. 5.5&5.6]. For example, to compare [**DG1**, Prop. 5.6(2)] and (2) above, observe, by [**DG1**, (5.5.2)], $(-1)^{\varepsilon\sigma(k)}$ there equals $(-1)^{\tilde{\sigma}_{h,m}(k,A)}$ here, and $v_h^{f(k)}$ there equals $v_h^{f(k,A)+a_{h,k}}$ here. Also, $v_h^{a_{h,k}}\overline{[\![a_{h,k}+1]\!]}_{v_h^2} = [a_{h,k}+1]$, for all $k \neq h$. Hence, (2) follows from [**DG1**, Prop. 5.6(2)]. □

By this theorem and [**DG1**, Th. 8.4], we may identify $\mathbf{U}(\mathfrak{gl}_{m|n})$ with $\mathfrak{U}(m|n)$ under the map

$$\eta : \mathbf{U}(\mathfrak{gl}_{m|n}) \longrightarrow \mathfrak{U}(m|n), \quad E_h \mapsto (E_{h,h+1})(\mathbf{0}),\ F_h \mapsto (E_{h+1,h})(\mathbf{0}),\ K_i^{\pm 1} \mapsto O(\pm\mathbf{e}_i).$$

COROLLARY A.3. *We may present the quantum supergroups* $\mathbf{U}(\mathfrak{gl}_{m|n})$ *by the generators* $K_i^{\pm 1} = O(\pm\mathbf{e}_i)$, $E_h = (E_{h,h+1})(\mathbf{0})$, *and* $F_h = (E_{h+1,h})(\mathbf{0})$ *and the basis*

$$\{A(\mathbf{j}) \mid A \in M(m|n)^{\pm}, \mathbf{j} \in \mathbb{Z}^{m+n}\}$$

together with the relations in Theorem A.2*(1)–(3).*

This completes the construction of $\mathbf{U}(\mathfrak{gl}_{m|n})$ by using the elements $A(\mathbf{j})$ similar to those in [**BLM**]. We summarize this approach by the top trapezium of the following diagram: starting with the $\mathbf{U}$-module $\mathcal{K}^{m|n}$, which is the regular $\dot{\mathbf{U}}$-module, one gets the BLM basis for $\dot{\mathbf{U}}$ and, hence, the completion algebra $\widehat{\dot{\mathbf{U}}}$ from which the quantum supergroup $\mathbf{U}$ is extracted. Here the dashed arrow represents a module action

The bottom trapezium of the diagram shows the approach used in [**DG1**]. In this approach, all arrows are algebra homomorphisms. Theorem A.2 shows that this completion algebra ($\widehat{\dot{\mathbf{U}}}$) approach and the v-Schur algebra ($\widehat{\mathbf{S}}(m|n)$) approach are equivalent.

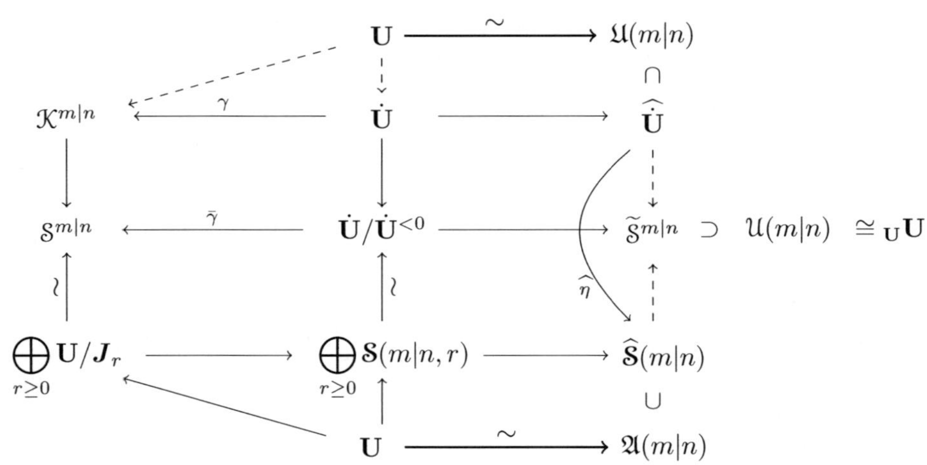

The central horizontal diagram shows the approach via v-differential operators developed in [**DZ**]. We remark that the multiplication formulas given in

[**DZ**, Th. 5.7] are also the "same" as those given in Theorem A.2. Thus, the three approaches to the realization problem for $\mathbf{U}(\mathfrak{gl}_{m|n})$ result in the same presentation.

Arrows in the left column represent some module homomorphisms, while those in the central column are algebra homomorphisms. Finally, two dashed arrows in the right column should also be viewed as module actions—the top action is the inflation of the bottom via $\widehat{\eta}:\widehat{\dot{\mathbf{U}}}\to\widehat{\mathbf{S}}$.

Appendix B. v-Schur superalgebras and some multiplication formulas

Following [**DR**, Def. 5.6], the v-Schur superalgebra is the following endomorphism

$$\mathcal{S}_{\mathcal{Z}}(m|n,r)=\mathrm{End}_{\mathcal{H}_{\mathcal{Z}}}\Big(\bigoplus_{\lambda\in(\mathbb{N}^{m+n})_r}x_{\lambda^{(0)}}y_{\lambda^{(1)}}\mathcal{H}_{\mathcal{Z}}\Big), \tag{B.0.1}$$

where $\mathcal{H}_{\mathcal{Z}}$ is the Hecke algebra over $\mathcal{Z}$ associated with the symmetric group $\mathfrak{S}_r$ and $\lambda=(\lambda^{(0)},\lambda^{(1)})$ with $\lambda^{(0)}=(\lambda_1,\dots,\lambda_m)$ and $\lambda^{(1)}=(\lambda_{m+1},\dots,\lambda_{m+n})$.

Let $\mathbf{S}(m|n,r)=\mathcal{S}_{\mathcal{Z}}(m|n,r)\otimes_{\mathcal{Z}}\mathbb{Q}(v)$. There is also a tensor-space definition together with a basis $\{\xi_A\mid A\in M(m|n)_r\}$; see [**DR**, Cor. 8.4] and [**DG1**, (4.2.1)]. The relationship between this basis and the double-coset basis defined in [**DR**, (6.2.1)] or [**DG2**, p.635] can be found in [**DG2**, Lem. 2.3].

For any $A\in M(m|n)_r$, let $[A]^\sharp=(-1)^{\bar{A}}\xi_A$, where

$$\bar{A}=\sum_{i,j=m+1}^{m+n}a_{i,j}\Big(\sum_{k=1}^{m}\sum_{l=j+1}^{m+n}a_{k,l}\Big).$$

Obviously, $\{[A]^\sharp\mid A\in M(m|n)_r\}$ is a basis of $\mathcal{S}_{\mathcal{Z}}(m|n,r)$. The following formulas is a modified version of [**DG1**, Prop. 4.4] with $[A]^\sharp$ replacing ξ_A. This is possible since for any matrix appearing on the right hand side of the formulas, we have $\bar{A}=\bar{B}$ and $\bar{U}_{h,p}=\bar{L}_{h,p}=0$.

PROPOSITION B.1 (The $h\neq m$ case).
Let $A=(a_{i,j})\in M(m|n)_r$. Assume $1\le h<m+n$ and $h\neq m$. Define

$$U_{h,p}:=pE_{h,h+1}-pE_{h+1,h+1}+\mathrm{ro}(A),\ L_{h,p}:=pE_{h+1,h}-pE_{h,h}+\mathrm{ro}(A)$$

with $p\ge 1$. The following multiplication formulas hold in $\mathcal{S}_{\mathcal{Z}}(m|n,r)$:

(1)

$$[U_{h,p}]^\sharp\cdot[A]^\sharp=\sum_{\substack{\nu\in(\mathbb{N}^{m+n})_p\\ \nu\le row_{h+1}(A)}}v_h^{f_h(\nu,A)}\prod_{k=1}^{m+n}\overline{\left[\!\!\left[{a_{h,k}+\nu_k\atop \nu_k}\right]\!\!\right]}_{v_h^2}[A+\sum_l\nu_l(E_{h,l}-E_{h+1,l})]^\sharp,$$

(2)

$$[L_{h,p}]^\sharp\cdot[A]^\sharp=\sum_{\substack{\nu\in(\mathbb{N}^{m+n})_p\\ \nu\le row_{h}(A)}}v_{h+1}^{f'_h(\nu,A)}\prod_{k=1}^{m+n}\overline{\left[\!\!\left[{a_{h+1,k}+\nu_k\atop \nu_k}\right]\!\!\right]}_{v_{h+1}^2}[A-\sum_l\nu_l(E_{h,l}-E_{h+1,l})]^\sharp,$$

where $f_h(\nu,A)$ and $f'_h(\nu,A)$ are defined in [**DG1**, (4.4.1-2)].

With this new basis, the υ-Schur superalgebra $\mathbf{S}(m|n,r)$ can be presented by generators

$$[\lambda]^\sharp, \quad [E_{h,h+1}+\mu]^\sharp, \quad [E_{h+1,h}+\mu]^\sharp, \text{ for all } \lambda \in (\mathbb{N}^{m+n})_r, \mu \in (\mathbb{N}^{m+n})_{r-1}$$

and their matrix representation (relative to the new basis) described below.

PROPOSITION B.2. *Let $A = (a_{i,j}) \in M(m|n)_r$. Assume $1 \le h < m+n$. The following multiplication formulas hold in the υ-Schur superalgebra $\mathcal{S}_{\mathbb{Z}}(m|n,r)$:*

(1) $[\lambda]^\sharp \cdot [A]^\sharp = \delta_{\lambda,\mathrm{ro}(A)}[A]^\sharp, \quad [A]^\sharp \cdot [\lambda]^\sharp = \delta_{\lambda,\mathrm{co}(A)}[A]^\sharp$;

(2) $[E_{h,h+1}+\mu]^\sharp \cdot [A]^\sharp = \sum_{a_{h+1,i}\ge 1}(-1)^{\tilde\sigma_{h,m}(i,A)}\upsilon_h^{f(i,A)}[a_{h,i}+1][A+E_{h,i}-E_{h+1,i}]^\sharp$ *with* $\mu = -\mathbf{e}_{h+1}+\mathrm{ro}(A)$;

(3) $[E_{h+1,h}+\nu]^\sharp \cdot [A]^\sharp = \sum_{a_{h,i}\ge 1}(-1)^{\tilde\sigma_{h,m}(i,A)}\upsilon_{h+1}^{g(i,A)}[a_{h+1,i}+1][A-E_{h,i}+E_{h+1,i}]^\sharp$ *with* $\nu = -\mathbf{e}_h+\mathrm{ro}(A)$.

PROOF. The $h \ne m$ case is the $p = 1$ case in Proposition B.1. We now prove the $h = m$ case. By definition, $\overline{A+E_{m,i}-E_{m+1,i}} = \overline{A}$ if $i \le m$, and $\overline{A+E_{m,i}-E_{m+1,i}} = \overline{A} + \sum_{s>m,m<t<i} a_{s,t} - \sum_{s\le m,t>i} a_{s,t}$ if $i > m$. So for $i \le m$,

$$\sum_{s>m,t<i} a_{s,t} + \overline{A+E_{m,i}-E_{m+1,i}} - \overline{A} = \sum_{s>m,t<i} a_{s,t} = \tilde\sigma_{m,m}(i,A),$$

and for $i > m$,

$$\begin{aligned}
&\sum_{s>m,t<i} a_{s,t} + \overline{A+E_{m,i}-E_{m+1,i}} - \overline{A}\\
&= \sum_{s>m,t<i} a_{s,t} + \sum_{s>m,m<t<i} a_{s,t} - \sum_{s\le m,t>i} a_{s,t}\\
&= \sum_{s>m,t\le m} a_{s,t} + 2\sum_{s>m,m<t<i} a_{s,t} - \sum_{s\le m,t>i} a_{s,t}\\
&=\tilde\sigma_{m,m}(i,A) + 2\sum_{s>m,m<t<i} a_{s,t} - 2\sum_{s\le m,t>i} a_{s,t}.
\end{aligned}$$

Substituting in [**DG1**, Prop.4.5] yields formula(2). A similar argument proves formula (3). □

As mentioned in Remark 6.6(1), the two sets of multiplication formulas can be used to derive the stablisation property, similar to the discussion in [**BLM**, §4]. In other words, the original construction for the algebra $\mathbf{K}$ in [**BLM**] can now be easily generalized to the quantum linear supergroup via a stabilisation property over the multiplication formulas given here; see [**BLM**, §5].

References

[BKLW] Huanchen Bao, Jonathan Kujawa, Yiqiang Li, and Weiqiang Wang, *Geometric Schur duality of classical type*, Transform. Groups **23** (2018), no. 2, 329–389, DOI 10.1007/s00031-017-9447-4. MR3805209

[BLM] A. A. Beilinson, G. Lusztig, and R. MacPherson, *A geometric setting for the quantum deformation of* GL_n, Duke Math. J. **61** (1990), no. 2, 655–677, DOI 10.1215/S0012-7094-90-06124-1. MR1074310

[DDPW] Bangming Deng, Jie Du, Brian Parshall, and Jianpan Wang, *Finite dimensional algebras and quantum groups*, Mathematical Surveys and Monographs, vol. 150, American Mathematical Society, Providence, RI, 2008, DOI 10.1090/surv/150. MR2457938

[BKK] Georgia Benkart, Seok-Jin Kang, and Masaki Kashiwara, *Crystal bases for the quantum superalgebra $U_q(\mathfrak{gl}(m,n))$*, J. Amer. Math. Soc. **13** (2000), no. 2, 295–331, DOI 10.1090/S0894-0347-00-00321-0. MR1694051

[CHW] Sean Clark, David Hill, and Weiqiang Wang, *Quantum shuffles and quantum supergroups of basic type*, Quantum Topol. **7** (2016), no. 3, 553–638, DOI 10.4171/QT/81. MR3551173

[DG1] Jie Du and Haixia Gu, *A realisation of the quantum supergroup $\mathbf{U}(\mathfrak{gl}_{m|n})$*, J. Algebra **404** (2014), 60–99, DOI 10.1016/j.jalgebra.2013.12.034. MR3177887

[DG2] Jie Du and Haixia Gu, *Canonical bases for the quantum supergroups $\mathrm{U}(\mathfrak{gl}_{m|n})$*, Math. Z. **281** (2015), no. 3-4, 631–660, DOI 10.1007/s00209-015-1499-3. MR3421634

[DF1] Jie Du and Qiang Fu, *A modified BLM approach to quantum affine $\mathfrak{gl}_n$*. Math. Z. **266** (2010), no. 4, 747–781, DOI 10.1007/s00209-009-0596-6. MR2729289

[DF2] Jie Du and Qiang Fu, *The integral quantum loop algebra of $\mathfrak{gl}_n$*, Int. Math. Res. Not. IMRN **20** (2019), 6179–6215, DOI 10.1093/imrn/rnx300. MR4031235

[DLZ1] Jie Du, Yanan Lin, and Zhongguo Zhou, *Polynomial super representations of $U_q^{\mathrm{res}}(\mathfrak{gl}_{m|n})$ at roots of unity*, J. Algebraic Combin. **51** (2020), no. 2, 195–235, DOI 10.1007/s10801-019-00872-z. MR4069340

[DLZ2] Jie Du, Yanan Lin, and Zhongguo Zhou, *Quantum queer supergroups via υ-differential operators*, J. Algebra **599** (2022), 48–103, DOI 10.1016/j.jalgebra.2022.01.017. MR4390456

[DP] Jie Du and Brian Parshall, *Monomial bases for q-Schur algebras*, Trans. Amer. Math. Soc. **355** (2003), no. 4, 1593–1620, DOI 10.1090/S0002-9947-02-03188-4. MR1946407

[DR] Jie Du and Hebing Rui, *Quantum Schur superalgebras and Kazhdan-Lusztig combinatorics*, J. Pure Appl. Algebra **215** (2011), no. 11, 2715–2737, DOI 10.1016/j.jpaa.2011.03.015. MR2802161

[DZ] Jie Du and Zhongguo Zhou, *The regular representation of $U_\upsilon(\mathfrak{gl}_{m|n})$*, Proc. Amer. Math. Soc. **148** (2020), no. 1, 111–124, DOI 10.1090/proc/14688. MR4042835

[EK] Houssein El Turkey and Jonathan R. Kujawa, *Presenting Schur superalgebras*, Pacific J. Math. **262** (2013), no. 2, 285–316, DOI 10.2140/pjm.2013.262.285. MR3069063

[GV] Victor Ginzburg and Éric Vasserot, *Langlands reciprocity for affine quantum groups of type A_n*, Internat. Math. Res. Notices **3** (1993), 67–85, DOI 10.1155/S1073792893000078. MR1208827

[Jan] Jens Carsten Jantzen, *Lectures on quantum groups*, Graduate Studies in Mathematics, vol. 6, American Mathematical Society, Providence, RI, 1996, DOI 10.1090/gsm/006. MR1359532

[Kw] Jae-Hoon Kwon, *Crystal bases of q-deformed Kac modules over the quantum superalgebras $U_q(\mathfrak{gl}(m|n))$*, Int. Math. Res. Not. IMRN **2** (2014), 512–550, DOI 10.1093/imrn/rns221. MR3159079

[Lu] G. Lusztig, *Aperiodicity in quantum affine $\mathfrak{gl}_n$*, Asian J. Math. **3** (1999), no. 1, 147–177, DOI 10.4310/AJM.1999.v3.n1.a7. Sir Michael Atiyah: a great mathematician of the twentieth century. MR1701926

[Lubk] George Lusztig, *Introduction to quantum groups*, Progress in Mathematics, vol. 110, Birkhäuser Boston, Inc., Boston, MA, 1993. MR1227098

[Zh] R. B. Zhang, *Finite-dimensional irreducible representations of the quantum supergroup $U_q(\mathrm{gl}(m/n))$*, J. Math. Phys. **34** (1993), no. 3, 1236–1254, DOI 10.1063/1.530198. MR1207983

School of Mathematics and Statistics, University of New South Wales, Sydney New South Wales 2052, Australia
Email address: j.du@unsw.edu.au

School of Science, Huzhou University, Huzhou, People's Republic of China
Email address: ghx@zjhu.edu.cn

College of Science, Hohai University, Nanjing, People's Republic of China
Email address: zhgzhou@hhu.edu.cn

Proceedings of Symposia in Pure Mathematics
Volume **108**, 2024
https://doi.org/10.1090/pspum/108/01971

An exact category approach to Hecke endomorphism algebras

Jie Du, Brian Parshall, and Leonard Scott,
with an appendix by Veronica Shalotenko

Abstract. Let G be a finite group of Lie type. In studying the nondefining characteristic (or cross-characteristic) representation theory of G, the (specialized) Hecke algebra $H = \mathrm{End}_G(\mathrm{Ind}_B^G 1_B)$ has played an important role. In particular, when $G = GL_n(\mathbb{F}_q)$ is a finite general linear group, this approach led to the Dipper-James theory of q-Schur algebras A. These algebras can be constructed over $\mathcal{Z} := \mathbb{Z}[t, t^{-1}]$ as the q-analog (with $q = t^2$) of an endomorphism algebra larger than H, involving parabolic subgroups. The algebra A is quasi-hereditary over $\mathcal{Z}$. An analogous algebra, still denoted A, can always be constructed in other types. However, these algebras have so far been less useful than in the GL_n case, in part because they are not generally quasi-hereditary.

Several years ago, reformulating a 1998 conjecture, the authors proposed (for all types) the existence of a $\mathcal{Z}$-algebra A^+ having a stratified derived module category, with strata constructed via Kazhdan-Lusztig cell theory. The algebra A is recovered as $A = eA^+e$ for an idempotent $e \in A^+$. A main goal of this monograph is to prove this conjecture completely. The proof involves several new homological techniques using exact categories. Following the proof, we show that A^+ does become quasi-hereditary after the inversion of the bad primes. Some first applications of the result—e.g., to decomposition matrices—are presented, together with several open problems.

Contents

0. Introduction
A special note
1. Preliminaries
2. Some exact categories
3. The conjecture and its proof
4. The quasi-heredity of $A^{+\natural}$
5. Some related applications
Appendix A. Projective coverings for Steinberg modules by Veronica Shalotenko
Appendix B. Height functions
Appendix C. Decomposition numbers and Krull dimension ≤ 2

2020 *Mathematics Subject Classification.* Primary .

This work was supported by grants from the ARC (DP120101436, Jie Du), and the Simons Foundation (#359360, Brian Parshall; #359363, Leonard Scott). The last author would like to thank UNSW for support from the Distinguished Research Visitor Scheme.

0. Introduction

Representations of the symmetric group $\mathfrak{S}_r$ are the first examples of group representations systematically investigated by the pioneers in representation theory such as F.G. Frobenius, A. Young, and W. Burnside. When I. Schur used the symmetric group representations to determine polynomial representations of the complex general linear group $\mathrm{GL}_n(\mathbb{C})$, certain finite-dimensional algebras, known as Schur algebras, played a bridging role between the two. The well-known Schur–Weyl duality summarizes the relation between the representations of $\mathrm{GL}_n(\mathbb{C})$ and $\mathfrak{S}_r$.

A Schur algebra is the endomorphism algebra of a direct sum of certain permutation modules associated with Young (or parabolic) subgroups of $\mathfrak{S}_r$. These permutation modules are filtered by Specht modules (with multiplicities given by Young's rule), and such filtrations result in a standard stratification for the Schur algebra so that its representation category is a so-called highest weight category, a notion naturally arising from the study of representations of Lie groups/algebras and algebraic groups. The Schur algebra itself is an example of a quasi-hereditary algebra, a notion introduced by E. Cline together with the last two authors of this monograph (CPS) in the late 1980s.

The endomorphism algebra of a permutation representation on a Borel subgroup $B(q)$, consisting of all upper triangular matrices, of a finite general linear group $\mathrm{GL}_r(q)$ is called an (Iwahori–)Hecke algebra which is a q-deformation of the group algebra of $\mathfrak{S}_r$. Such Hecke algebras and their generic versions are all defined for finite groups G of Lie type with a parameter q [**Iwa64, C85**]. Often q is a prime power and the associated Hecke algebras are q-deformations of the group algebra of the Weyl group W of G. Now, using q-permutation modules associated with parabolic subgroups of W, the notion of Schur algebras is generalized not only to q-Schur algebras in the type A case, but also to Hecke endomorphism algebras (Hecke endo-algebras) in general.

More precisely, let $\mathcal{H}$ be a generic Hecke algebra over $\mathcal{Z} := \mathbb{Z}[t, t^{-1}]$ associated to a finite group G of Lie type with Weyl group W. For each subset J of the set S of fundamental reflections for W, form the associated q-permutation module $x_J\mathcal{H}$. Put $T := \bigoplus_J x_J\mathcal{H}$ and call the endomorphism algebra $A := \mathrm{End}_{\mathcal{H}}(T)$ a generic Hecke endo-algebra. If G is a finite general linear group and $\mathscr{K}$ is a field, $A_{\mathscr{K}} = \mathscr{K} \otimes A$ is (Morita equivalent to) a q-Schur algebra, as introduced by Dipper and James [**DJ89**], and used effectively in the study of cross-characteristic representation theory of G (so that q is nonzero in $\mathscr{K}$). The definition makes sense for G of any type, but A itself, without modification, appears less useful (compared to q-Schur algebras), partly for structural and homological reasons. For simplicity, we will assume that G is a fixed-point group of a Frobenius map on a *simple* algebraic group $\mathbf{G}$ throughout the monograph.

The structure of q-Schur algebras is a natural quantization from the structure of Schur algebras, thanks to the existence of filtrations of q-permutation modules by q-Specht modules. What is a replacement of Specht or q-Specht modules in the general case?

In 1996, CPS developed a theory of stratifying endomorphism algebras. They considered abstract Specht modules and a set of stratification hypotheses with which the associated endomorphism algebra is a standardly stratified algebra, a notion that is slightly weaker than that of a quasi-hereditary algebra. For the aforementioned generic Hecke algebras, the authors used the modules associated with Kazhdan–Lusztig left cells as "Specht modules" to filter these q-permutation modules. This results in a standard stratification for q-Schur algebras, but a coarser stratification for a generic Hecke endo-algebra A of other Lie types; see [**DPS98b**, **DPS98a**]. However, in the rank-two case, if we enlarge T to $T^+ = T \oplus X$ where X is a certain module filtered by certain cell modules, we obtain a finer stratification in terms of two-sided cells. In other words, if an enlargement T^+ of T is used, then $A^+ = \mathrm{End}_{\mathcal{H}}(T^+)$ could be standardly stratified. This motivated the conjecture [**DPS98a**, Conj. 2.5.2], later modified (slightly) in [**DPS15**, Conj. 1.2].

The conjecture is about the construction of an extended (generic) Hecke endo-algebra A^+ with good homological properties. It proposes the existence of a $\mathcal{Z}$-algebra A^+ with a nicely stratified derived module category, with all strata constructed via Kazhdan-Lusztig cell theory. It is required by the conjecture to extend the $\mathcal{H}$-module T to $T^+ = T \oplus X$, for some right $\mathcal{H}$–module X with a suitable dual left-cell (module) filtration. Note, this implies that $A = eA^+e$, where e is the idempotent projection of T^+ onto T along X. The conjecture also specifies "standard modules," parametrized by left cells, that filter a projective generator for the category A^+–mod of finitely generated A^+-modules.

More precisely, the q-permutation module $x_J\mathcal{H}$, filtered by dual left-cell modules S_ω, has a bottom section $S_{\omega^{0,J}}$, where $\omega^{0,J}$ is the left cell that contains the longest element $w_{0,J}$ of the parabolic subgroup W_J. For a left cell ω that satisfies $S_\omega \ncong S_{\omega^{0,J}}$ for all $J \subseteq S$, we want to construct a right $\mathcal{H}$-module X_ω filtered by dual left-cell modules and with bottom section S_ω. We require X_ω to satisfy a certain homological property, the Ext^1 vanishing property, which forms part of the Stratification Hypothesis [**DPS98a**, (1.2.9)]. Note that, for the symmetric group, every two-sided cell contains some $w_{0,J}$ and every $S_\omega \cong S_{\omega^{0,J}}$ for some J. So, we take $X = 0$ in this case. In general, we may take X to be a direct sum of such X_ω. Thus, the proof of the conjecture relies on the existence of such X_ω.

In [**DPS15**], we studied a local case of the conjecture, under the equal parameter assumption for $\mathcal{H}$ in order to use a result from [**GGOR03**]. In the DPS work, the module X_ω was constructed inductively, using a new order on left cells, to satisfy the required Ext^1 vanishing condition. This order is strictly compatible with the two-sided cell order $\leq_{LR}$, and is defined by a height function on left cells, which is called a sorting function in [**GGOR03**]. Note that the height function is motivated by Lusztig's a-function on W that takes constant values on two-sided cells. In this way, we proved a special case ([**DPS15**, Th. 5.6]) of the conjecture [**DPS15**, Conj. 1.2].

To attack the conjecture in general, the authors developed in [**DPS17**] an exact category approach to control the Ext^1 vanishing property when constructing certain modules (such as X_ω) with some desired filtrations. Applying this general construction to $\mathcal{H}$-modules yields the exact category $(\mathscr{A}(\mathscr{L}^*), \mathscr{E}(\mathscr{L}^*))$, where $\mathscr{A}(\mathscr{L}^*)$ is a full (additive) subcategory of mod-$\mathcal{H}$, and $\mathscr{E}(\mathscr{L}^*)$ is a subcategory of the category of short exact sequences in the category mod-$\mathcal{H}$ of finite right $\mathcal{H}$-modules. However, this general construction does not decide if the q-permutation modules, though belonging to $\mathscr{A}(\mathscr{L}^*)$, satisfy the required Ext^1 vanishing property. To overcome the hurdle, two more exact subcategories are constructed to obtain an exact category $(\mathscr{A}_\flat(\mathscr{L}^*), \mathscr{E}_\flat(\mathscr{L}^*))$ in which X_ω may be constructed with a required filtration and both X_ω and $x_J\mathcal{H}$ satisfy the Ext^1 vanishing property. Though the process is quite sophisticated, we eventually prove the conjecture [**DPS15**, Conj. 1.2]; it is a consequence of Theorem 3.3.1 whose proof is outlined in Section 0.2 below.

0.1. Further results and applications. After preliminaries in Chapters 1, we begin in Chapter 2 with a construction of a sequence of exact categories, consisting of objects with filtrations of left-cell modules, inside the category $\mathcal{H}$-mod of left $\mathcal{H}$-modules. By applying a standard duality theory to the construction in Section 2.2, we obtain in Section 2.3 the exact categories $(\mathscr{A}_\flat(\mathscr{L}^*), \mathscr{E}_\flat(\mathscr{L}^*))$, required in the proof of the conjecture, of right $\mathcal{H}$-modules with filtrations of dual left-cell modules and the homological property satisfied by $x_J\mathcal{H}$. Section 2.4 reviews and reformulates an abstract setting developed in [**DPS17**] for the construction of these aforementioned extra modules X_ω.

The conjecture is stated and proved in Chapter 3; see Theorem 3.3.1. The proof is technically quite difficult, so we provide at the end of this introduction a conceptual framework for the way A^+ and T^+ evolve in these sections.

Theorem 3.3.1 involves *no mention of primes whatsoever*, good, bad, or otherwise. Nevertheless, the theorem proves that the category A^+-mod has a natural (standard) stratifying system. Section 3.4 is devoted to a "uniqueness theory." (See similar uniqueness results for quasi-hereditray covers, discussed by Rouquier in [**Ro08**, §4].) Its original motivation was to sharpen Theorem 3.3.1: A main step in the proof of that theorem involved the construction of certain $\mathcal{H}$-modules X_ω associated to left cells ω; see Construction 2.4.12. By Theorem 2.4.10, the X_ω have remarkable homological properties, which are required in realizing A^+ as an endomorphism algebra. Although no claim is made that the modules X_ω are unique, in Theorem 3.4.5, the following uniqueness result is proved: *the Morita class of A^+ is independent of the choice of the modules X_ω.* The uniqueness theory is formulated in such a way (see Theorem 3.4.2) that it might be useful for other suitably filtered endomorphism algebras. In Section 3.5, we turn the categorical property — i.e., the stratifying system for A^+-mod, established in Theorem 3.3.1 — into the structural property for A^+ by explicitly displaying a defining sequence, called a standard stratification, for A^+ with stratifying ideals as sections; see Theoren 3.4.5.

REMARK 0.1.1 (Good and bad primes). The "bad primes" are the only integer primes that divide the denominators of generic character degree formulas; see [**DPS98a**, (4.1.12)] or the discussion of display (4.1.14) in the appendix to Section 4.1, where only the ${}^2F_4$ case is considered. They form a subset of $\{2, 3, 5\}$. We declare all such prime divisors to be "bad" (for W, G, or the Lie series to which G belongs) and all others to be "good." Recall the classical definition that a prime is good for $\mathbf{G}$ if it does not divide any coefficient of the maximal root in the root system of $\mathbf{G}$. By [**DPS98a**, 4.1], all primes that are good for $\mathbf{G}$ are good for G.

Suppose that $A^{+\natural} := \mathcal{Z}^\natural \otimes_{\mathcal{Z}} A^+$ is obtained from A^+ by inverting all the bad primes for W. Theorem 4.3.4 then proves that $A^{+\natural}$ is quasi-hereditary of separable type in all cases (and split quasi-hereditary except in type ${}^2F_4$, becoming split upon the adjunction of $\sqrt{2}$)). Sections 4.1 and 4.2 prepare for the above results in Section 4.3. They also provide an interesting study in themselves of the base-changed version $\mathcal{J} := \mathcal{Z}\otimes\mathsf{J}$ of the Lusztig asymptotic $\mathbb{Z}$–algebra J. There is a natural inclusion of the Hecke algebra $\mathcal{H}$ into (a copy of) $\mathcal{J}$, and the latter may be regarded as an order in $\mathcal{H}_{\mathbb{Q}(t)}$ containing $\mathcal{H}$. One approach to understanding the representation theory of the generic Hecke algebra $\mathcal{H}$ is to relate it to the simpler algebra $\mathcal{J}$. Any $\mathcal{J}$-module M can be restricted to $\mathcal{H}$, and our Theorem 4.2.1 says every left-cell module for $\mathcal{H}$ arises this way. Once again there is no mention of primes. However, if bad primes are inverted in $\mathbb{Z}^\natural \subseteq \mathcal{Z}^\natural$, we show the resulting order $\mathcal{J}^\natural$ is maximal (in the sense of Reiner [**Reiner03**]). We then prove that every stratifying ideal arising from the defining sequence for A^+ in Theorem 3.4.5 is, after base change to $\mathcal{Z}^\natural$, a hereditary ideal of separable type (and of split type, if ${}^2F_4$ is excluded), completing the proof of Theorem 4.3.4. In the rest of Section 4.3, we focus on a refinement of the defining sequence for $A^{+\natural}$ such that its sections are indecomposable split hereditary ideals, a notion used in [**Ro08**]. In particular, standard modules and Rouquier's isomorphism are given in Corollary 4.3.5. Corollary 4.3.6 brings in the idempotents that define these indecomposable split hereditary ideals. (Note that the approach developed in [**Ro08**] is an idempotent-free approach.) Finally, in Corollary 4.3.7, we prove that $A^{+\natural}$-mod is a highest weight category in the sense of [**DS94**] and [**Ro08**].

Chapter 5 is devoted to applications to several related topics. In Section 5.1, we consider the algebras A and A^+ from the point of view of standardly based algebras, as introduced by Du and Rui [**DuR98**]. For example, Theorem 5.1.8 proves that the algebras $A^\natural$, $A^{+\natural}$ and $\mathcal{H}^\natural$ are all standardly based. The latter case was previously known (see [**G07**]), though our proof is new.

Section 5.2 focuses on a new study of the decomposition numbers of specialized Hecke endo-algebras relevant to those for finite groups of Lie type in cross-characteristic representation theory. It is not a new theme, but our results are quite general and often deeper than those found in the literature. Theorems 5.2.2 and 5.2.8 identify a large collection of decomposition numbers which remain constant for all groups in a given series, independent of their q-parameter, the latter parameter, defined by [**C85**], is a general version of the q in $SL_n(q)$ for the SL_n-series. The uni-triangularity of the associated decomposition matrix is an easy consequence of the split quasi-hereditary structure.

Section 5.3 links the decomposition matrix for a Hecke endo-algebra A with that of the corresponding finite groups of Lie type. We prove in Theorem 5.3.1 that, if A is the Hecke endo-algebra associated with the standard finite Coxeter system (W, S, L) and G is a finite group in the Lie type series associated with (W, S, L), then the decomposition matrix for the specialized $A_\mathcal{O}$ is part of the decomposition matrix of $\mathcal{O}G$. In particular, if r is good, then this part of the decomposition matrix is unitriangular. This provides evidence for possible improvement of Geck's conjecture [**G12**] proved recently in [**BDT20**]. See comments around footnote 45 in Chapter 5. Part of our proof in the adjoint case is based on a note of Veronica Shalotenko; see Appendix A. We thank her for allowing us to include this work in our monograph. We also point out in a Remark 5.3.2(1) that her proof continues to work if the hypothesis on the connected center of $\mathbf{G}$ is removed, due to a result of Dipper and Fleischmann [**DF92**, Lem. (3.8)].

Section 5.4 is motivated by Schur's original work, the Schur–Weyl duality, as mentioned at the beginning of this Introduction, and its quantum analog. In their study of quantum symmetric pairs of type AIII, H. Bao and W. Wang [**BW18**] established a new type of the Schur–Weyl duality between $\imath$-quantum groups $\mathbf{U}^\jmath$ and $\mathbf{U}^\imath$ and Hecke algebras of types B and C, respectively. Wu and the first author have partially lifted Bao–Wang's duality to the integral level (see [**DW22a, DW22b**]).

In these works, certain Hecke endomorphism algebras, which are centralizer subalgebras of $A^\ddagger$, played the bridging role. In particular, the standard basis theory in Section 5.1 and the discovery about the decomposition matrix for A^+ or $A^\ddagger$ in Section 5.2 can be applied to representations of $\imath$-quantum groups and their hyperalgebras. See Theorems 5.4.4 and 5.4.16.

In the final section of Chapter 5, we pose some problems, suggested by this monograph, on decomposition numbers for finite groups of Lie type. Each of the problems can be viewed as a step toward understanding the decomposition numbers in reductions mod p of ordinary irreducible representations in the unipotent series. The problems are formulated at a "starter" level, assuming the underlying field characteristic is a good prime. We hope this assumption might be eventually removed in some future reformulation. The framework of Theorem 3.3.1, which treats all primes alike, makes such a reworking of "good prime" formulations possible, and provides a general starting point. Note that Theorem 3.3.1 itself is a kind of reformulation of the "good prime" result Theorem 4.3.4.

Finally, the monograph closes with several appendices which cover a number of preparatory topics used in its main body.

0.2. Outline and guide for the proof of Theorem 3.3.1. Let Ω be the set of Kazhdan-Lusztig left cells in a finite standard Coxeter system (W, S, L) as defined in [**Lus03**], using parameters from an ambient algebraic group and a fixed-point group of a Frobenius map (see [**C85**, §1.17]).[1] In [**DPS17**], the authors constructed an early version $A^\dagger$ of A^+ based on the existence of suitable modules X_ω, for all $\omega \in \Omega$. Each X_ω has a filtration with sections isomorphic to dual left-cell modules S_ν, $\nu \in \Omega$, and a bottom section S_ω (defined precisely in display (1.2.10)), which, in addition, satisfy

$$\text{(0.2.1)} \qquad \operatorname{Ext}^1_{\mathscr{E}(\mathscr{S})}(S_\tau, X_\omega) = 0, \quad \forall \tau, \omega \in \Omega.$$

Here S_τ and X_ω are objects in a suitable exact category $(\mathscr{A}(\mathscr{S}), \mathscr{E}(\mathscr{S}))$. The category $\mathscr{A}(\mathscr{S})$ is a full subcategory of the category of $\mathbb{Z}[t, t^{-1}]$-free, finite right $\mathcal{H}$-modules. In addition, it is required that each object in $\mathscr{A}(\mathscr{S})$ have a certain filtration with sections direct sums of dual left-cell modules. Also, $A^\dagger := \operatorname{End}_{\mathscr{A}(\mathscr{S})}(T^\dagger)$, where $T^\dagger := \bigoplus_{\omega\in\Omega} X_\omega^{\oplus m_\omega}$ with each m_ω a positive integer. This definition of $T^\dagger$, together with the $\operatorname{Ext}^1_{\mathscr{E}(\mathscr{S})}$-vanishing (0.2.1), implies that $\operatorname{Ext}^1_{\mathscr{E}(\mathscr{S})}(S_\tau, T^\dagger) = 0$, for all $\tau \in \Omega$. This is a key property for managing filtrations of $A^\dagger$ that arise from filtrations of $T^\dagger$.

Generally speaking, the pair $(A^\dagger, T^\dagger)$ has many desirable properties, except we want $T^\dagger$ to contain, as an $\mathscr{A}(\mathscr{S})$-summand, each of the q-permutation modules $x_J\mathcal{H}$, as well as their direct sum (denoted T). This problem presents a major difficulty in passing from [**DPS17**] to the set-up of this monograph. It is solved by introducing a new exact category $(\mathscr{A}_\flat, \mathscr{E}_\flat) = (\mathscr{A}_\flat(\mathscr{S}), \mathscr{E}_\flat(\mathscr{S}))$ inside $(\mathscr{A}(\mathscr{S}), \mathscr{E}(\mathscr{S}))$ with $x_J\mathcal{H} \in \mathscr{A}_\flat$ and $\operatorname{Ext}^1_{\mathscr{E}_\flat}(S_\tau, x_J\mathcal{H}) = 0$, for all τ.

The construction of $(\mathscr{A}_\flat, \mathscr{E}_\flat)$ is highly nontrivial. It begins in Sections 1.3 and 2.1 which study a dual category $(\mathscr{A}^\flat, \mathscr{E}^\flat)$ in the category of $\mathcal{Z}$-free left $\mathcal{H}$-modules. Of course, it is possible to pass between right and left $\mathcal{Z}$-free, finite $\mathcal{H}$-modules by simply applying the dual functor $(-)^* = \operatorname{Hom}_{\mathcal{Z}}(-, \mathcal{Z})$. However, this process interchanges left exact functors with right exact functors, while we need to use

[1] See also footnote 2 in Section 1.1 below.

the left-module versions to insure the various functors sending an exact sequence $0 \to M \to N \to P \to 0$ to the fixed-point sequences $0 \to M^{\mathcal{H}_J} \to N^{\mathcal{H}_J} \to P^{\mathcal{H}_J}$, $J \subseteq S$, are left exact. This allows us to build an exact subcategory by focusing on the short exact sequences $0 \to M \to N \to P \to 0$ whose associated fixed-point sequences are also all short exact sequences [**DPS17**, Lem. 3.1]. A version of this approach is used in Proposition 2.1.3 to create $\mathscr{E}^\flat = \mathscr{E}^\flat(\mathscr{L})$ inside an exact category $(\mathscr{A}(\mathscr{L}), \mathscr{E}(\mathscr{L})$ dual to $(\mathscr{A}(\mathscr{S}), \mathscr{E}(\mathscr{S}))$. The category $\mathscr{A}^\flat$ is then defined to consist of objects with certain "internal" candidate exact sequences belonging to $\mathscr{E}^\flat$. Remarkably, the objects $\mathcal{H}x_J$ belong to $\mathscr{A}^\flat$; see Corollary 2.1.8. This implies that each dual object $x_J\mathcal{H}$ belongs to $\mathscr{A}_\flat$. We want more than that, however. We want $x_J\mathcal{H}$ to be a direct summand of a new version of $T^\dagger$ with the key vanishing (0.2.1) preserved. This forces us to prove that $\mathrm{Ext}^1_{\mathscr{E}_\flat}(S_\omega, x_J\mathcal{H}) = 0$, for each $\omega \in \Omega$ and $J \subseteq S$. This vanishing is proved in its dual incarnation, Theorem 2.1.6. The short exact sequence (2.1.8) displayed in the proof of Theorem 2.1.6 is the key ingredient. It is available from the exactness of $(-)^{\mathcal{H}_J}$ there.

Section 2.3 can be viewed as dualizing the constructions of Section 2.1. Of course, Section 2.1 was formulated with these duals in mind.

A main step in completing the proof of the conjecture, at this point, is carried out in Section 2.4, namely, in Theorem 2.4.10, which imitates the construction of X_ω in [**DPS17**, Th. 4.5].

Section 3.1 is also preparatory, providing more detail on dual left-cell module filtrations of q-permutation modules. It refines the analysis of q-permutation module filtrations begun above Corollary 2.1.5. The discussion draws from earlier work in [**DPS98a**].

The main goal of Section 3.3 is to prove Theorem 3.3.1. This implies Conjecture 3.2.1, equivalent to [**DPS15**, Conj. 1.2], a modification of [**DPS98a**, Conj. 2.5.2]. A further consequence of the proof is the Ext^1-vanishing result Corollary 3.3.2. It is used later in Section 3.4 in constructing a uniqueness Morita theory complementary to Theorem 3.3.1, and discussed further below.

For further discussions in outline of this monograph, see the previous two sections.

Sydney and Charlottesville
September 2022

Jie Du
Brian Parshall
Leonard Scott

A special note

Our lifelong collaborator and friend, Brian Parshall, passed away on January 17, 2022. This was a few days after a Zoom meeting we had held, generally weekly, since mid 2020. This last meeting was very successful and lasted 2 hours. The last question, one Jie had for Brian, was about using "Hecke endo-algebra" as a short form for "Hecke endomorphism algebra". Brian answered "it sounds good. I will think about it". Unfortunately, these words from Brian became the last words we heard from him. He passed away within a week of this last meeting. It was a shock and surprise for everyone, certainly for Brian's collaborators. Most of the work on this monograph, however, had been done with Brian's participation. Its remaining co-authors, Jie Du and Leonard Scott, wish to take this opportunity to acknowledge Brian Parshall as being a co-author of this monograph. In a variation on the words above that Brian used in our last meeting, "we didn't have to think about it."

Jie Du and Leonard Scott

1. Preliminaries

In this chapter, we briefly review some notions that will be used throughout. Topics include finite groups of Lie type, generic Hecke algebras, exact categories and stratified algebras.

If B is a ring, then B-mod (respectively, mod-B) will denote the category of all finite (i.e., finitely generated) left (respectively, right) B-modules. Usually, B will be a finitely generated algebra over a commutative Noetherian ring $\mathscr{K}$. In this case, and unless otherwise noted, objects in B-mod will be assumed to be finite $\mathscr{K}$-modules (with a similar convention for mod-B).

1.1. Finite groups of Lie type. We review some basic notions concerning finite groups of Lie type. More material on this topic will be introduced in Chapter 5. We closely follow the treatment given in Carter's book [**C85**]. In particular, we will assume the reader is familiar with the notion [**C85**, pp.16–17] of a connected, reductive algebraic group over an algebraically closed field of positive characteristic. Such a group $\mathbf{G}$ may be characterized as the internal product $\mathbf{Z}^\circ\mathbf{G}'$, where $\mathbf{Z}^\circ$ is a torus, the connected component of the identity in the center $\mathbf{Z} = Z(\mathbf{G})$ of $\mathbf{G}$, and $\mathbf{G}'$ is a semisimple algebraic group, defined by Carter to be nontrivial and an "almost direct product" (a "central product" in traditional terminology) of simple algebraic groups. The latter are defined as having a maximal normal closed subgroup which is finite. Note that $\mathbf{G}'$ is necessarily the commutator subgroup of $\mathbf{G}$.

Now let $\mathbf{G}$ be a (connected) reductive algebraic group over an algebraically closed field of positive characteristic p. Recall that a Frobenius map $F : \mathbf{G} \to \mathbf{G}$ in Carter's sense [**C85**, §1.17] is a morphism, some power of which equals a standard Frobenius map. Fix a Frobenius map F, and let $G := \mathbf{G}^F$ be the group of fixed points of F. Necessarily, G is finite. It is called a *finite group of Lie type*, and all finite groups of Lie type arise this way.[2]

The finite groups of Lie type most studied here arise in the context of the previous paragraph when $\mathbf{G}$ is taken to be simple.[3] In addition to the defining (algebraically closed) field, the (simple) algebraic group $\mathbf{G}$ is determined by its root datum and its isogeny class (see [**C85**, §§1.9, 1.11]), and $\mathbf{G}^F$ is discussed in [**C85**, §1.19].

It is known that $\mathbf{G}$ has an F-stable Borel subgroup $\mathbf{B}$. In addition, $\mathbf{B}$ contains a maximal torus $\mathbf{T}$ of $\mathbf{G}$ which is F-stable. The Weyl group $\mathbf{W} := N_{\mathbf{G}}(\mathbf{T})/\mathbf{T}$ is then part of a finite Coxeter system $(\mathbf{W}, \mathbf{S})$ for a (finite) subset $\mathbf{S}$ of elements of

[2]Our notation differs slightly from that in [**C85**], where $\mathbf{G}$ is sometimes written G. Also, we sometimes call a Frobenius map a Frobenius endomorphism or a Frobenius morphism. See [**C85**, p. 31] for a detailed definition. A Frobenius endomorphism is sometimes called a Steinberg endomorphism in other sources (see, e.g., [**MT11**]). We remark from [**MT11**, p.184, top] that Carter's assertion near the bottom of [**C85**, p. 31] (just above the definition) regarding surjective endomorphisms is not correct. However, this does not affect the validity of his definition.

[3]This assumption is convenient for the description of the Lie type series and weight function L discussed below. Note also that sometimes it is useful to study the more general case where $\mathbf{G}'$ is simple, either to place $\mathbf{G}$ in a familiar context (e.g., $\mathbf{G}$ is a general linear group), or, after beginning a study using a simple group $\mathbf{G}$, viewing it as a commutator group of a larger reductive group $\widetilde{\mathbf{G}}$ with favorable properties, such as a connected center. See Lemma 5.2.4 in Chapter 5.

order 2 and $\mathbf{W}$ is irreducible (i.e., the Coxeter diagram is connected) under the assumption that $\mathbf{G}$ is simple. In addition, the pair $(\mathbf{B}, \mathbf{N})$, where $\mathbf{N} = \mathbf{N_G}(\mathbf{T})$, give $\mathbf{G}$ the structure of a BN-pair. The group $G = \mathbf{G}^F$ also has a natural BN-pair structure B, N, where $B = \mathbf{B}^F$ and $N = \mathbf{N}^F$. Its Weyl group is $W = \mathbf{W}^F$. In fact, W is generated by the set S of all the long words in the various $\mathbf{W}_J$, as stated in [**C85**, p. 34] and proved in [**MT11**, Lem. C.1]. Moreover, the proof in [**MT11**] shows that every element $w \in W$ can be written as a product of elements of S with lengths additive. One can use these facts to complete Carter's sketch [**C85**, p.34] that W is the Weyl group in a (split) BN-pair structure on G, so, in particular, it is a Coxeter group (with S as its set of fundamental generators).

Let Φ be the root system of $\mathbf{G}$, and let Δ be the set of simple roots associated to $\mathbf{B}$. The Frobenius map F on $\mathbf{G}$ is closely related to a certain permutation ρ of Δ. In fact, ρ is a graph automorphism of the undirected Dynkin diagram associated to Δ. (See [**C85**, §1.19].) Two finite groups G_1 and G_2 are said to be in the same *Lie type series* if there exist a connected reductive group $\mathbf{G}$ and Frobenius morphisms F_1 and F_2, inducing the same graph automorphism ρ, and such that $G_i = \mathbf{G}^{F_i}$, for $i = 1, 2$. By inspection, this amounts to saying that G_1 and G_2 are in one of the groupings on [**C85**, pp. 39–41]; see also [**MT11**, p.193]. Further, if ρ is the trivial (or identity) graph automorphism, then G is called a *split* (or untwisted) group of Lie type; if ρ is non-trivial, then G is called a *quasi-split* (or twisted) group of Lie type. Here, following Carter's classification in [**C85**, p.41], we exclude the Ree and Suzuki groups in the quasi-split case.

There is a natural action of F on the character group $\mathbf{X}$ (respectively, cocharacter group $\mathbf{Y}$) of $\mathbf{T}$. The maximum of the absolute values of the eigenvalues of F as an operator on $\mathbf{X}$ is denoted q. It is a (possibly fractional) power $q = p^d$, where p is the characteristic of the defining field of $\mathbf{G}$. For example, if G is a classical finite (split) Chevalley group over a finite field k of order r (an integral power of the defining characteristic p), then $q = r$. In this case, $\rho = 1$, the identity map.

Returning to the general case, a useful construction of W as a subgroup of $\mathbf{W}$ using only the Dynkin diagram permutation ρ associated to F (rather than F itself) can be based on [**C85**, p. 34]: First, $\mathbf{W}$ may be viewed as constructed from the Dynkin diagram Γ, with the fundamental reflections associated to the nodes of Γ. Thus, we can associate to each orbit J of ρ on Γ a parabolic subgroup $\mathbf{W}_J$ of $\mathbf{W}$. Now, the fundamental reflections s in W may be defined as the longest words of the various subgroups $\mathbf{W}_J$. Next, define the "weight function" L on W by taking it to be the restriction of the length function $\boldsymbol{\ell}$ on $\mathbf{W}$ to W. That is, $L(w) := \boldsymbol{\ell}(w)$, for $w \in W \subseteq \mathbf{W}$. Then, (W, S, L) is a finite (irreducible) Coxeter system with weight function $L : W \to \mathbb{Z}$ in the sense of [**Lus03**, §3.1]. Thus, if $w = xy$ with $\ell(w) = \ell(x) + \ell(y)$, then $L(xy) = L(x) + L(y)$, for $x, y, w \in W$.

Within its Lie type series, the group $G = \mathbf{G}^F$ is determined by (and determines) a parameter q [**C85**, p. 35] which is described as follows. If δ is the smallest positive integer such that F^δ is a standard Frobenius endomorphism (as per [**C85**, p. 31]), then q^δ (rather than q itself) is defined to be the power of p associated to the latter endomorphism. The number δ is also the order of the Dynkin diagram graph automorphism ρ. If G is a (split) Chevalley group, $\delta = 1$ and q can be any positive power of the defining characteristic p. In the quasi-split cases which are not split, δ is the order of a non-trivial automorphism of the Dynkin diagram, and q is still any positive power of p. However, in the Ree and Suzuki cases, $\delta = 2$ and $q^2 = p^{2n+1}$

with $p = 2$ for the ${}^2B_2$ and ${}^2F_4$ cases and $p = 3$ for the ${}^2G_2$ case. In particular, q is a positive power of $\sqrt{p}$. See [**C85**, p. 41]. Remarkably, in all cases, q together with the weight function L gives

$$q^{L(s)} = [B : B \cap B^s], \tag{1.1.1}$$

where B is a Borel subgroup of G and s is a fundamental reflection in its Weyl group W. Here $B = \mathbf{B}^F$, where $\mathbf{B}$ is an F-stable Borel subgroup of $\mathbf{G}$; see [**C85**, pp. 34-35, and p. 74] for further details. This formula holds in the more general case where $\mathbf{G}$ is connected and reductive. (See the partial discussion in [**C85**, p. 74] and also [**St68**, 11.8].)

Throughout this work, W will be, unless otherwise noted, the Weyl group of a finite group G of Lie type (i.e., the Weyl group of the BN-pair associated to G). We will use the natural positive weight function L, constructed above and illustrated in [**Lus03**, Lem.16.2]. The resulting weighted Coxeter system (W, S, L) is irreducible and will be called *standard finite*.[4] We also ignore the rank one case, where W has order 2, unless otherwise noted.[5] Note that standard finite Coxeter systems (W, S, L) can also be divided into the *split case*, where the weight function L equals the length function ℓ on W, and the *quasi-split case*, where $L \neq \ell$. (The only non-trivial case for the Ree and Suzuki groups is the ${}^2F_4$ case, where W is the dihedral group of order 16.) See [**Lus03**, Chs. 15&16].

EXAMPLE 1.1.1. Let $\mathbb{F}_q$ be the finite field of q elements, and let n be a positive integer. The finite general linear group $\mathrm{GL}_n(\mathbb{F}_q)$ is the fixed-point group of the following (standard) Frobenius map

$$F_q : \mathrm{GL}_n(k) \longrightarrow \mathrm{GL}_n(k),\ (a_{i,j}) \longmapsto (a_{i,j}^q), \tag{1.1.2}$$

where k is the algebraic closure of $\mathbb{F}_q$. Note that $\mathrm{GL}_n(k)$ is a connected reductive group and the subgroup $\mathrm{SL}_n(k)$ of determinant 1 matrices is simple with F_q-fixed-point group $\mathrm{SL}_n(\mathbb{F}_q)$.

For any field k, $\mathrm{GL}_n(k)$ has the following split BN-pair:

$$B_n(k) := \text{subgroup of all upper triangular matrices in } \mathrm{GL}_n(k),$$
$$N_n(k) := \text{subgroup of all monomial matrices in } \mathrm{GL}_n(k).$$

Let $T_n(k) = B_n(k) \cap N_n(k)$ be the subgroup of diagonal matrices. If n_i denotes the permutation matrix obtained by interchanging the i-th and $(i+1)$-th rows of I_n, then the Weyl group

$$\widetilde{W} := N_n(\mathbb{F})/T_n(\mathbb{F}) \cong \mathfrak{S}_n \text{ under the correspondence } n_i \mapsto \tilde{s}_i = (i, i+1).$$

See, e.g., [**G03**, 1.6.9] for more details.

The other finite classical groups can be constructed by fixed-point groups of certain group automorphisms on $\mathrm{GL}_n(\mathbb{F}_q)$. In Example 1.1.2, we consider the symplectic and orthogonal groups. They are relevant to our studies of $\imath$-quantum groups in Section 5.4. Unitary groups are taken up in Example 1.1.3. These both help to

[4]The Hecke algebras associated to these triples are especially important for the modular representation theory of unipotent blocks for finite groups of Lie type. According to a theory put forward by Bonafé and Rouquier [**BR03**], [**BR17**], the case of general blocks (for finite groups of Lie type) could, eventually, reduce to the unipotent case. In any event, all irreducible or indecomposable modules with nontrivial cohomology belong to the unipotent principal block.

[5]This connectedness assumption is a matter of convenience and is unnecessary in most cases, e.g., if no component is dihedral of order 16, the Weyl group of ${}^2F_4$.

complete the classical group picture and to provide examples of "unequal parameter" weight functions.

EXAMPLE 1.1.2. For any field k and $J \in \mathrm{GL}_n(k)$, consider the group isomorphism

$$\vartheta = \vartheta_J : \mathrm{GL}_n(k) \longrightarrow \mathrm{GL}_n(k),\ x \longmapsto J^{-1}(x^t)^{-1}J, \tag{1.1.3}$$

and define orthogonal and symplectic groups as the fixed-point groups of some ϑ:

$$\mathrm{O}_n(k) := \{x \in \mathrm{GL}_n(k) \mid J_n = x^t J_n x\},$$
$$\mathrm{Sp}_{2m}(k) := \{x \in \mathrm{GL}_{2m}(k) \mid J_{2m}^- = x^t J_{2m}^- x\},$$

where

$$J_n := \begin{pmatrix} 0 & 0 & \cdots & 0 & 1 \\ 0 & 0 & \cdots & 1 & 0 \\ & \cdots & \cdots & \cdots & \\ 0 & 1 & \cdots & 0 & 0 \\ 1 & 0 & \cdots & 0 & 0 \end{pmatrix}, \quad J_{2m}^- := \begin{pmatrix} 0 & J_m \\ -J_m & 0 \end{pmatrix}. \tag{1.1.4}$$

Let $\mathrm{SO}_n(k) = \mathrm{O}_n(k) \cap \mathrm{SL}_n(k)$ and let $\mathbf{G} \subseteq \mathrm{GL}_n(k)$ be one of the classical groups:

$$\mathbf{G} = \begin{cases} \mathrm{SO}_{2m+1}(k), & n = 2m+1, \mathrm{char}(k) \neq 2; \\ \mathrm{Sp}_{2m}(k), & n = 2m, \text{any characteristic}; \\ \mathrm{SO}_{2m}(k), & n = 2m, \mathrm{char}(k) \neq 2. \end{cases}$$

These are known as split twisted groups in [**St67**, Ch. 11].[6]

If k is the algebraic closure of $\mathbb{F}_q$, then $\mathbf{G}$ is a connected simple group and has a split BN-pair $\mathbf{B} = \mathbf{G} \cap B_n(k)$, $\mathbf{N} = \mathbf{G} \cap N_n(k)$.

For the Frobenius map F_q as defined in (1.1.2) and $J = J_n$ or J_{2m}^-, since $J = x^t J x$ implies $J = (F_q(x))^t J F_q(x)$, it follows that F_q induces the (standard) Frobenius maps

$$F : \mathbf{G} \longrightarrow \mathbf{G},$$

and $\mathbf{G}^F = \mathrm{SO}_n(\mathbb{F}_q)$ $(n = 2m+1,\, 2m)$ or $\mathrm{Sp}_{2m}(\mathbb{F}_q)$. Note also that $\mathbf{G}^F = \mathrm{GL}_n(\mathbb{F}_q)^\vartheta$ has BN-pair $\mathbf{B}^F = B_n(\mathbb{F}_q)^\vartheta$ and $\mathbf{N}^F = N_n(\mathbb{F}_q)^\vartheta$ since ϑ stabilizes the split BN-pair $(B_n(k), N_n(k))$. In particular, ϑ induced a group automorphism $\sigma : \widetilde{W} \longrightarrow \widetilde{W}$, and there is a group isomorphism

$$\mathbf{W} := \mathbf{N}/\mathbf{B} \cap \mathbf{N} \cong \widetilde{W}^\sigma = N_n(\mathbb{F}_q)^\vartheta / T_n(\mathbb{F}_q)^\vartheta.$$

Note that the map induced by F on $\mathbf{W}$ is the identity map. Note also that, for $\mathbf{G} = \mathrm{SO}_{2m+1}(k)$ or $\mathrm{Sp}_{2m}(k)$, $W := \mathbf{W}^F = \mathbf{W}$ is the Weyl group of type B/C with generators $S = \{s_1, \ldots, s_{m-1}, t\}$, where

$$s_i = \tilde{s}_{m-i}\tilde{s}_{m+1+i}, \text{ and } t = \tilde{s}_m\tilde{s}_{m+1}\tilde{s}_m, \text{ for } n = 2m+1;$$
$$s_i = \tilde{s}_{m-i}\tilde{s}_{m+i}, \quad \text{and } t = \tilde{s}_m, \text{ for } n = 2m.$$

Finally, t is the homomorphic image of the elements in $N_n(k)^\vartheta$:

$$\begin{pmatrix} I_{m-1} & 0 & 0 \\ 0 & J & 0 \\ 0 & 0 & I_{m-1} \end{pmatrix}, \text{ where } J = \begin{pmatrix} 0 & 0 & 1 \\ 0 & -1 & 0 \\ 1 & 0 & 0 \end{pmatrix} \text{ or } \begin{pmatrix} 0 & 1 \\ -1 & 0 \end{pmatrix}.$$

[6]See [**MT11**, p.8] for the definition for $\mathrm{char}(k) = 2$.

For more details, see, e.g., [**G03**, 1.7.3].

EXAMPLE 1.1.3. Now, following [**DVV19**, §4.1], we combine F_q in (1.1.2) with ϑ in (1.1.3), using the choice $J = J_n$, to obtain a new Frobenius morphism

$$F := F_q\vartheta = \vartheta F_q : \mathrm{GL}_n(k) \longrightarrow \mathrm{GL}_n(k).$$

Since ϑ^2 is the identity map, it follows that F^2 is the standard Frobenius map F_{q^2} and $\mathrm{GL}_n(k)^F \subseteq \mathrm{GL}_n(\mathbb{F}_{q^2})$. We have

$$\mathrm{GU}_n(\mathbb{F}_q) := \mathrm{GL}_n(k)^F = \{x \in \mathrm{GL}_n(\mathbb{F}_{q^2}) \mid (F_q(x))^t J_n x = J_n\}.$$

This is called the *general unitary group* associated with the hermitian form defined by J_n. Replacing "G" by "S" in the definition above defines the special unitary group. (See also [**MT11**, Ex. 21.14(2)] or [**G03**, 4.1.10(c)].)

For $\mathrm{SU}_4(\mathbb{F}_{q^2})$, the F-orbits of the simple roots $\alpha_{1,2}, \alpha_{2,3}, \alpha_{3,4}$ for $\mathrm{SL}_4(k)$, with associated root subgroups $U_{i,j} = \{I_4 + aE_{i,j} \mid a \in \mathbb{F}_{q^2}\}$, are $\{\alpha_{1,2}, \alpha_{3,4}\}$, $\alpha_{2,3}$. The root subgroups $\mathrm{SU}_4(\mathbb{F}_{q^2}) = \mathrm{SL}_4(k)^F$ are

$$(U_{1,2}U_{3,4})^F = \{I_4 + aE_{1,2} - a^qE_{3,4} \mid a \in \mathbb{F}_{q^2}\}, \quad U_{2,3}^F = \{I_4 + aE_{2,3} \mid a + a^q = 0\}.$$

The values $q^{L(s)}$ $(s \in S)$ associated to the weight function L on $\mathbf{W}^F = W(B_2)$ in this case are q^2, q. In general, the values $q^{L(s)}$ for $\mathrm{SU}_{2m}(\mathbb{F}_{q^2})$ are listed in the Coxeter diagram:

For $\mathrm{SU}_3(\mathbb{F}_{q^2})$, $\{\alpha_{1,2}, \alpha_{2,3}\}$ is the F-orbit of the simple roots $\alpha_{1,2}, \alpha_{2,3}$ for $\mathrm{SL}_3(k)$, and we have the root subgroup (cf. [**G03**, 4.5.12(b)])

$$(U_{1,2}U_{2,3}U_{1,3})^F = \left\{ \begin{pmatrix} 1 & a & b \\ 0 & 1 & -a^q \\ 0 & 0 & 1 \end{pmatrix} \mid a, b \in \mathbb{F}_q, b^q + b + a^{q+1} = 0 \right\}.$$

So, $q^{L(s)} = q^3$. In general, the values $q^{L(s)}$ $(s \in S)$ for $\mathrm{SU}_{2m+1}(\mathbb{F}_{q^2})$ take the form:

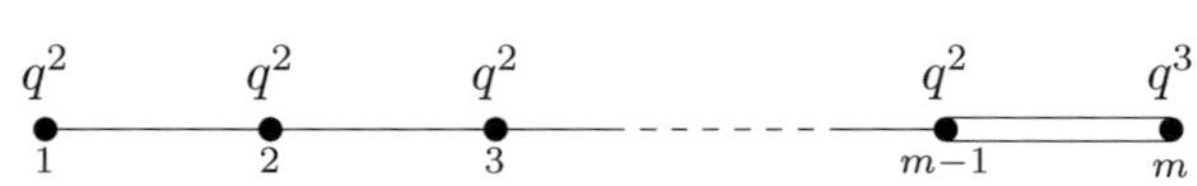

1.2. (Iwahori-)Hecke algebras. We review the theory of Hecke algebras in [**Lus03**]. Each weighted Coxeter system (W, S, L) defines an algebra $\mathcal{H}$, called an *Iwahori-Hecke algebra* in [**Lus03**] and called a *generic Hecke algebra* in this monograph. See (1.2.1) and (1.2.2) for more details.

Many results in [**Lus03**] are contingent on a series P1—P15 of conjectures (involving (W, S, L) and $\mathcal{H}$), all of which have been proved in the standard finite case. See [**Lus03**, Chs. 15, 16], revisited in [**GJ11**, Ch. 2] (especially [**GJ11**, 2.4.1(b)] for the case ${}^2F_4$). Thus, we can assume Lusztig's conjectures P1–P15 hold for the triples (W, S, L) considered in this monograph. *Henceforth, we assume that* (W, S, L) *is standard finite unless otherwise noted. In particular,* L *is positive and Lusztig's conjectures P1-P15 hold, as does condition* (1.2.11) *below.* In fact, we mention that in Chapters 1–3 which include the main theorem, we need only assume conjectures P4 and P9, which, together with the rank 2 discussion in [**DPS98a**, p. 204],[7] imply that property (1.2.11) holds.

Let $\mathcal{Z} = \mathbb{Z}[t, t^{-1}]$ be the ring of integer Laurent polynomials in a variable t. Let $\mathcal{H}$ be the generic Hecke algebra over $\mathcal{Z}$ associated with (W, S, L). Recall that $\mathcal{H}$ has basis $\{T_w\}_{w\in W}$ and defining relations ($s \in S$, $w \in W$):

$$(1.2.1) \qquad T_sT_w = \begin{cases} T_{sw}, & sw > w; \\ t_s^2T_{sw} + (t_s^2 - 1)T_w, & sw < w, \end{cases}$$

where $t_s := t^{L(s)}$ (and, more generally, $t_w = t^{L(w)}$). Also, for $x, y \in W$, $x < y$ denotes the Chevalley-Bruhat partial order. The algebra $\mathcal{H}$ is defined just using t^2, but, in practice, it is convenient to have its square root t available.[8]

We often use a different basis $\{\widetilde{T}_w\}$ also indexed by W defined by putting

$$(1.2.2) \qquad \widetilde{T}_w := t^{-L(w)}T_w, \quad w \in W.$$

(Our notation here differs slightly from that in [**Lus03**], where our $\widetilde{T}_w$ is denoted T_w.)

The *bar operator* $^- : \mathcal{H} \to \mathcal{H}$ sending t to t^{-1} and T_w to $(T_{w^{-1}})^{-1}$ is a ring involution on $\mathcal{H}$ (sending t to t^{-1}), and $\mathcal{H}$ admits a Kazhdan-Lusztig (or "canonical" in the sense of [**Lus90**]; see also [**DDPW08**, §0.5]) basis $\{\mathrm{c}_w\}_{w\in W}$. This basis is defined in [**Lus03**, Chap.5] using the formula

$$\mathrm{c}_w = \sum_{y\leq w} p_{y,w}\widetilde{T}_y.$$

In addition to the condition $\overline{\mathrm{c}}_w = \mathrm{c}_w$, for all w, the Laurent polynomials $p_{y,w} \in \mathcal{Z}$ satisfy (1) $p_{y,w} = 0$ unless $y \leq w$, (2) $p_{w,w} = 1$, and (3) $p_{y,w} \in \mathcal{Z}_{<0} := t^{-1}\mathbb{Z}[t^{-1}]$, if $y < w$. The polynomials $p_{y,w}$ are determined by these properties. For example,

$$(1.2.3) \qquad \mathrm{c}_s = t_s^{-1} + \widetilde{T}_s = t_s^{-1}(1 + T_s).$$

(Note that the element c_x is denoted c_x in [**Lus03**].)

If $J \subseteq S$ is non-empty, let $W_J := \langle s\rangle_{s\in J}$, a parabolic subgroup of W; if $J = \emptyset$, set $W_J := \{1\}$. Then, W_J has a unique longest (reduced) word, denoted $w_{0,J}$, and

$$(1.2.4) \qquad \mathrm{c}_{w_{0,J}} = t^{-L(w_{0,J})} \sum_{y\in W_J} T_y =: t^{-L(w_{0,J})}x_J$$

(which defines x_J). See [**Lus03**, Cor. 12.2] (or [**DDPW08**, (7.2.5)] in the split case). In the literature, the endomorphism algebra

$$(1.2.5) \qquad A := \mathrm{End}_{\mathcal{H}}\left(\bigoplus_{J\subseteq S} x_J\mathcal{H}\right)$$

(or any algebra Morita equivalent to it) is called "the" generalized generic q-Schur algebra or a *generic Hecke endomorphism algebra* (or *generic Hecke endo-algebra* for short) by analogy with our naming of $\mathcal{H}$. Various specializations of A and its extensions will also be considered in later chapters.

---

[7] See footnote 27 in Chapter 3 which corrects a typo.

[8] It is well known that the algebra $\mathbb{Q}(t) \otimes_{\mathcal{Z}} \mathcal{H}$ is a semisimple $\mathbb{Q}(t)$-algebra. See [**CR81**, Cor. 68.12] as well as Remark 3.3.4(b) below.

Let $h_{x,y,z} \in \mathcal{Z}$ denote the structure constants for $\mathcal{H}$, defined by the equations

$$\mathrm{c}_x \mathrm{c}_y = \sum_{z \in W} h_{x,y,z} \mathrm{c}_z. \tag{1.2.6}$$

In particular, we have [**Lus03**, Th. 6.6]

$$\mathrm{c}_s \mathrm{c}_w = \begin{cases} \mathrm{c}_{sw} + \sum_{x:sx<x<w} \mu^s_{x,w} \mathrm{c}_x, & \text{if } w < sw; \\ (t_s + t_s^{-1}) \mathrm{c}_w, & \text{if } w > sw. \end{cases} \tag{1.2.7}$$

Here $\mu^s_{x,w} \in \mathcal{Z}$ is fixed under the bar operator. In the case of split groups, $\mu^s_{x,w}$ is constant (in $\mathbb{Z}$). Also, (1.2.7) implies that the c_s, $s \in S$, generate the $\mathcal{Z}$-algebra $\mathcal{H}$.

If c_y appears in $\mathrm{c}_s\mathrm{c}_w$ with a nonzero coefficient, we set $y \leftarrow_L w$, following [**Lus03**, 8.1]. Define a *preorder* (that is, a relation which is reflexive and transitive) on W by setting, for $w, w' \in W$, $w \leq_L w'$ if there exists $w = w_0, w_1, \ldots, w_n = w'$ in W such that $w_{i-1} \leftarrow_R w_i$, for all $i = 1, 2, \ldots, n$, and define $w \leq_R w'$ if $w^{-1} \leq_L (w')^{-1}$.

We now consider the preorders $\leq_L$ and $\leq_R$ defined on W, and the preorder $\leq_{LR}$ generated by $\leq_L$ and $\leq_R$ and their associated *quasi-poset* $(W, \leq_L)$ etc. See, e.g., [**Lus03**, Ch. 8]. Let $\sim_L, \sim_R$ and $\sim_{LR}$ be the equivalence relations associated with $\leq_L, \leq_R$, and $\leq_{LR}$, respectively. The equivalence classes for $\sim_L, \sim_R$, and $\sim_{LR}$ are called the *left, right*, and *two-sided cells*, respectively.

For a quasi-poset, there is an associated poset (see Section B2 in Appendix B). The poset associated with the quasi-poset $(W, \leq_L)$ is the set Ω of all left cells, while the poset associated with the quasi-poset $(\Omega, \leq_{LR})$ is the set $\overline{\Omega}$ of all two-sided cells. Note that if a left cell $\omega \in \Omega$ contains an element $w_{0,J}$, then J is a uniquely determined subset of S. See [**Lus03**, Lem. 8.6] and the argument above Lemma 3.1.2 in this monograph.

We record that

$$h_{x,y,z} \neq 0 \implies z \leq_L y, z \leq_R x. \tag{1.2.8}$$

See [**Lus03**, 13.1(d)]. In particular, for any $w \in W$,

$$\mathcal{H}_{\leq_L w} := \sum_{\substack{y \in W \\ y \leq_L w}} \mathcal{Z}\mathrm{c}_y, \quad \mathcal{H}_{\leq_R w} := \sum_{\substack{y \in W \\ y \leq_R w}} \mathcal{Z}\mathrm{c}_y, \quad \mathcal{H}_{\leq_{LR} w} := \sum_{\substack{y \in W \\ y \leq_{LR} w}} \mathcal{Z}\mathrm{c}_y \tag{1.2.9}$$

is a left, right, or two-sided ideal of $\mathcal{H}$, respectively (see [**Lus03**, Lem. 8.2].). We may also define similar ideals by using $<_L, <_R, <_{LR}$.

Each left cell $\omega \in \Omega$ defines a *left-cell module* $S(\omega) := \mathcal{H}_{\leq_L \omega}/\mathcal{H}_{<_L \omega}$—it is explicitly defined in [**Lus03**, §8.3] (but with a different notation). Here $\mathcal{H}_{\leq_L \omega} = \mathcal{H}_{\leq_L y}$ and $\mathcal{H}_{<_L \omega} = \mathcal{H}_{<_L y}$, for any $y \in \omega$. Similarly, we may define right- or two-sided-cell modules. Also, there are base-change versions associated with the Hecke alegbra $\mathscr{K} \otimes_{\mathcal{Z}} \mathcal{H}$ and the basis $\{1 \otimes \mathrm{c}_w\}_{w \in W}$ over any commutative ring $\mathscr{K}$ which is a $\mathcal{Z}$-module.

In addition, there is a corresponding

$$\textit{dual left-cell module } S_\omega := S(\omega)^* = \mathrm{Hom}_{\mathcal{Z}}(S(\omega), \mathcal{Z}). \tag{1.2.10}$$

It is a *right* $\mathcal{H}$-module. Both $S(\omega)$ and S_ω are free $\mathcal{Z}$-modules; see [**Lus03**, §8.3].

REMARK 1.2.1. If $\mathcal{H}$ is the generic Hecke algebra for a symmetric group $W = \mathfrak{S}_r$, then it can be proved that the dual left-cell module S_ω is a "classical" Specht module for $\mathcal{H}$, as defined more generally by Dipper-James; see [**DJ86**, p 34], taking

R in the latter reference to be the ring $\mathbb{Z}$ of integers. On the other hand, the right cell $\gamma := \omega^{-1}$, defines a right-cell module K_γ, which is related to S_ω by the isomorphism $\mathfrak{D}_{\mathcal{H}}S_\omega \cong K_\gamma$, $\gamma = \omega^{-1}$. (Here, $\mathfrak{D}_{\mathcal{H}}$ is the contravariant duality functor on mod-$\mathcal{H}$ defined in [**DPS98c**, (1.2)].) We will not make use of any of the theory of Specht modules in this monograph.

The preorders $\leq_L, \leq_{LR}$ on W induce partial orders $\leq_L, \leq_R$ on Ω. Thus, for $\omega, \omega' \in \Omega$ and $X \in \{L, LR\}$, $\omega \leq_X \omega' \iff x \leq_X x'$, for some $x \in \omega, x' \in \omega'$. Also, $\omega \in \Omega$ defines a unique two-sided cell $\bar{\omega}$ with the property that $\omega \subseteq \bar{\omega}$. It is extremely important in our context here (see the second paragraph of Section 1.2 or Proposition B2.3 in Appendix B) that, given $\omega_1, \omega_2 \in \Omega$,

(1.2.11) $$\omega_1 <_L \omega_2 \implies \overline{\omega}_1 <_{LR} \overline{\omega}_2.$$

Thus, $\leq_L$ strictly dominates $\leq_{LR}$ in the sense of Definition B2.1 in Appendix B. As discussed at the beginning of the section, this property holds when (W, S, L) is standard finite, our standing assumption.

For each two-sided cell $\mathbf{c}$ of W, there is an $\mathcal{H}$-bimodule $M(\mathbf{c})$. The two-sided cell decomposition of W induces a decomposition of $\mathcal{H}_{\mathbb{Q}(t)}$ into two-sided ideals $M(\mathbf{c})_{\mathbb{Q}(t)}$ and, hence, a decomposition of $\Lambda = \operatorname{Irr}(\mathcal{H}_{\mathbb{Q}(t)})$. Further, the partial order $\leq_{LR}$ on the set $\bar{\Omega}$ of two-sided cells induces a preorder on Λ by setting, for $E, E' \in \Lambda$,

(1.2.12) $$E \leq_{LR} E' \text{ if } \mathbf{c}_E \leq_{LR} \mathbf{c}_{E'}.$$

Here, $\mathbf{c}_E$ denotes the two-sided cell $\mathbf{c}$ such that $E | M(\mathbf{c})_{\mathbb{Q}(t)}$ in the sense that E is a direct summand of the left $\mathcal{H}_{\mathbb{Q}(t)}$-module defined by $M(\mathbf{c})_{\mathbb{Q}(t)}$.

We end this section with a brief discussion about the origin of the generic Hecke algebra $\mathcal{H}$ associated with a standard finite Coxeter system (W, S, L). Here $W = \mathbf{W}^F$ is the Weyl group of a finite group $G = \mathbf{G}^F$ of Lie type that has a Borel subgroup $B = \mathbf{B}^F$; see Section 1.1. The following result, which will not be used until Section 5.2, was first discovered by Iwahori [**Iwa64**] for (split) Chevalley groups.

LEMMA 1.2.2. *Let $\mathcal{H}^o$ be the subalgebra over $\mathbb{Z}[t^2]$ generated by all $T_s \in \mathcal{H}$, $s \in S$.*[9] *Suppose $\mathcal{O}$ is a commutative ring and $q \in \mathcal{O}$, and view $\mathcal{O}$ as a $\mathbb{Z}[t^2]$-algebra by the homomorphism $\mathbb{Z}[t^2] \to \mathcal{O}$ sending $t^2 \mapsto q$. Then,*

$$\mathcal{H}^o_{\mathcal{O}} \cong \operatorname{End}_{\mathcal{O}G}(\mathcal{O}G/B),$$

where $\mathcal{O}G/B$ is the permutation module for the action of G on the set of left cosets of B.

PROOF. A proof of display (1.1.1) is outlined in [**C85**, p.74], beginning with a general root subgroup cardinality formula

$$|X^F_\Psi| = q^{|\Psi|},$$

where Ψ is a certain subset of the positive roots Φ^+ of $\mathbf{G}$. For each fundamental reflection s in the Weyl group W of G, one of the root subgroups X^F_Ψ can be seen, using [**C85**, Prop. 2.5.11], to satisfy $|B| = |X^F_\Psi(B \cap B^s)| = |X^F_\Psi| \cdot |B \cap B^s|$. The set Ψ here is the set of positive roots in Φ made negative by the action of s. Thus,

[9]Thus, $\mathcal{H}^o$ is a form for $\mathcal{H}$ over $\mathbb{Z}[t^2]$.

$|\Psi| = L(s)$ for this Ψ, where $L(s)$ is the length of s in the Weyl group $\mathbf{W}$ of $\mathbf{G}$; see Section 1.1. Thus, we obtain the expression (1.1.1) above from the expression

$$q^{L(s)} = q^{|\Psi|} = |B/B \cap B^s| = [B : B \cap B^s].$$

(Carter actually gives in [**C85**, p. 74] an analogous formula for $q^{L(w)}$, for any w in the Weyl group of $\mathbf{G}$.)

As in [**CPS73**, p. 188], the elements

$$\frac{1}{|B|}\underline{BwB}, \quad w \in W \tag{1.2.13}$$

span a $\mathbb{Z}$-algebra H, well known to be the endomorphism algebra of the permutation module $\mathbb{Z}G/B$ for the action of G on coset space G/B. This algebra base changes well, so that we can replace $\mathbb{Z}$ by $\mathcal{O}$, obtaining a basis for $\mathrm{End}_{\mathcal{O}G}(\mathcal{O}G/B)$. See, e.g., [**DDPW08**, Exercise 9.3]. In [**CPS73**, fn. 31], a multiplication relation is showed to be analogous to the right-hand version of the relation $T_sT_w = T_{sw}$ in (1.2.1), when $\ell(sw) = 1 + \ell(w)$. Here, we think of $\frac{1}{|B|}\underline{BwB}$ as analogous to T_w in (1.2.1). There is also the "quadratic relation"

$$\left(\frac{1}{|B|}\underline{BsB}\right)^2 = a\frac{1}{|B|}\underline{BsB} + b1$$

with $b = [B : B \cap B^s]$ and $a = b - 1$. Such a relation holds in any finite doubly transitive group, such as $\langle B, s\rangle$ acting on the coset space $\langle B, s\rangle/B$.

Hence $b = q^{L(s)}$ (and $a = b - 1$). The two types of relations are analogous to (1.2.1) with $w = s$, replacing the t_s^2 there with $q^{L(s)}$. The two types of relations we have considered clearly form special cases of (1.2.1), sufficient to provide generators and relations for $\mathcal{H}^o_{\mathcal{O}}$. Similarly, we have obtained from $\mathrm{End}_G(\mathcal{O}G/B)$ analogs with enough information to do all multiplications of the basis elements $\frac{1}{|B|}\underline{BwB}$. □

1.3. Exact categories and stratified algebras. Beginning in this section, considerable use is made of exact categories $(\mathscr{A}, \mathscr{E})$ as formulated in Keller [**K90**, Appendix A] and revisited in Bühler's exposition [**Bu10**]. The axioms given in these papers are equivalent to (but shorter than) the original axioms of Quillen [**Q73**]. A version[10] is stated in [**DPS17**, Appendix] which follows [**K90**, Appendix A] quite closely (as quoted in [**DRSS99**, Appendix A, Set 2] written by Keller). Briefly, $\mathscr{A}$ is an additive category, while $\mathscr{E}$ consists of certain pairs (i, d) of composable morphisms $i : X \to Y$ and $d : Y \to Z$ in $\mathscr{A}$, such that i is the kernel of d and d is the cokernel of i. The pairs (i, d) which belong to $\mathscr{E}$ are required to satisfy certain axioms; see explicitly [**K90**, Appendix A], [**DRSS99**, Appendix A, Set 2], [**Bu10**, Def. 2.1], or [**DPS17**, Appendix A]. As in [**K90**], we sometimes refer to i as an *inflation*, d as a *deflation*, and the pair (i, d) as a *conflation*. Conflations (i, d) are usually denoted by $(X \to Y \to Z)$.[11] As in [**Bu10**], we sometimes call conflations "short exact sequences" in $\mathscr{E}$. When it is convenient and clear from context, $(X \to Y \to Z)$ is denoted $0 \to X \to Y \to Z \to 0$ and called a short exact sequence (in $\mathscr{E}$).

Notice that $\mathscr{E}$ is itself an additive category, using the usual mappings between the categorical diagrams formed by composable pairs of morphisms in $\mathscr{A}$; see

[10]There are several different versions of exact categories in the literature. In particular, following Quillen, we do not assume any kind of idempotent splitting.

[11]Such a sequence is denoted by $X \rightarrowtail Y \twoheadrightarrow Z$ in [**Bu10**]. There are also other notations in the literature used with the same meaning.

[**Bu10**, Cor. 2.10]. As such, $\mathscr{E}$ is required to be closed under isomorphisms as part of the exactness axioms.

One can consider a pair $(\mathscr{A}, \mathscr{E})$ consisting of an additive category $\mathscr{A}$ together with a collection $\mathscr{E}$ of composable morphisms (i, d) in $\mathscr{A}$ and ask if it defines an exact category. Of course, one needs any (i, d) in $\mathscr{E}$ to be a kernel-cokernel pair (in the terminology of [**Bu10**]). Expanding on [**DPS17**], Chapters 2 and 3 present some important examples of exact categories $(\mathscr{A}, \mathscr{E})$ inspired by Hecke algebra representation theory, especially involving the construction and control of endomorphism rings, as in Theorem 3.3.1 below. Often $\mathscr{A}$ is a full subcategory of an additive category $\mathscr{B}$, associated to an already introduced exact category $(\mathscr{B}, \mathscr{F})$. The theme of constructing new exact categories from old ones is continued in Sections 2.2 and 2.3 using duality.

Finally, for any exact category $(\mathscr{A}, \mathscr{E})$, we can consider the abelian group $\mathrm{Ext}^1_{\mathscr{E}}(X, Y)$ for any pair of objects $X, Y \in \mathscr{A}$. In case $\mathscr{A}$ is an additive $\mathscr{K}$-category for a commutative ring $\mathscr{K}$, $\mathrm{Ext}^1_{\mathscr{E}}(X, Y)$ carries a natural structure as a $\mathscr{K}$-module; see [**DPS17**, Prop. A.2]. The action of $a \in \mathscr{K}$ on $\mathrm{Ext}^1_{\mathscr{E}}(X, Y)$ is given by either pull-back of multiplication by a on X or (equally, as it turns out) by push-out of multiplication of a on Y.

1.3A. Exact subcategories: some terminology and a new strategy. A note on exact category terminology: Let $(\mathscr{A}, \mathscr{E})$ and $(\mathscr{B}, \mathscr{F})$ be exact categories. Suppose $\mathscr{A}$ is a subcategory of $\mathscr{B}$, and $\mathscr{E} \subseteq \mathscr{F}$, viewing $\mathscr{E}$ and $\mathscr{F}$ as classes of composable pairs of morphisms in $\mathscr{A}$ and $\mathscr{B}$, respectively. We say that $(\mathscr{A}, \mathscr{E})$ is an *exact subcategory*[12] of $(\mathscr{B}, \mathscr{F})$. It is a *full exact subcategory* if $\mathscr{A}$ is also a full subcategory of $\mathscr{B}$—that is, $\mathrm{Hom}_{\mathscr{A}}(A_1, A_2) = \mathrm{Hom}_{\mathscr{B}}(A_1, A_2)$ whenever A_1, A_2 are objects in $\mathscr{A}$.

With this terminology in hand, we can now describe our new approach and some of its motivation.

STRATEGY 1.3.1. *If $(\mathscr{A}, \mathscr{E})$ is a full exact subcategory of $(\mathscr{B}, \mathscr{F})$, then the fullness implies that the inclusion $\mathscr{E} \subseteq \mathscr{F}$ induces an inclusion, i.e., an injective map,* $\mathrm{Ext}^1_{\mathscr{E}}(A, A') \hookrightarrow \mathrm{Ext}^1_{\mathscr{F}}(A, A')$ *for each pair (A, A') of objects in $\mathscr{A}$.*[13] *The inclusion is even $\mathscr{K}$-linear if $\mathscr{K}$ is a commutative Noetherian ring and the inclusion $\mathscr{A} \subseteq \mathscr{B}$ is also an inclusion of $\mathscr{K}$-categories (with $\mathscr{A}$ full in $\mathscr{B}$ as before.) See* [**DPS17**, Prop. A.2].

As in traditional homological algebra, the vanishing of a group $\mathrm{Ext}^1_{\mathscr{F}}(A_1, A_2)$ *leads to the exactness of* $\mathrm{Hom}_{\mathscr{F}}(A_1, -)$ *when applied to any given short exact sequence γ in $\mathscr{F}$ that begins with A_2. Assume this exactness is desirable for some set Γ of such γ's, but the proposed* Ext^1 *vanishing is too much to ask for. In that case,*

---

[12]Some authors note that the set $\mathscr{E}$ has a natural category structure, using "morphisms" constructed from one short exact sequence to another via commutative diagrams. This does not, however, affect our definition of an exact subcategory.

[13]The members of exact category Ext^1 groups are defined as equivalence classes of short exact sequences under a well-known equivalence relation. See [**DPS17**, p.249]. In the presence of fullness, each equivalence class in $\mathscr{F}$ intersects at most one of the $\mathscr{E}$ equivalence classes, resulting in the inclusions of this sentence.

We take this opportunity to note that our notion of "full exact subcategory" is quite different from the notion "fully exact subcategory" in [**Bu10**, Def. 10.21], apparently aimed in a different direction.

we can try to build a full exact subcategory $(\mathscr{A}, \mathscr{E})$ *as above, but with* $\Gamma \subseteq \mathscr{E}$ *and* $\mathrm{Ext}^1_{\mathscr{E}}(A_1, A_2) = 0$. *This gives the desired exactness.*

Though the discussion above is only an outline, we are able to use it when constructing stratifying systems, as we now show.

1.3B. Stratifying Systems. Let $\mathscr{K}$ be a Noetherian commutative ring. Let B be an algebra which is finite (= finitely generated) and projective as a $\mathscr{K}$-module. Let Λ be a quasi-poset, i.e., a finite set Λ with a preorder, denoted $\leq$ here. The associated poset is $\bar{\Lambda}$, and $\bar{\lambda}$ is the element of $\bar{\Lambda}$ associated to any given $\lambda \in \Lambda$.

For $\lambda \in \Lambda$, assume that there is given both a B-module $\Delta(\lambda)$, which is finite and projective over $\mathscr{K}$, and a finite and projective B-module $P(\lambda)$, together with an epimorphism $P(\lambda) \twoheadrightarrow \Delta(\lambda)$. The following conditions are assumed to hold:

(SS1) For $\lambda, \mu \in \Lambda$,

$$\mathrm{Hom}_B(P(\lambda), \Delta(\mu)) \neq 0 \implies \lambda \leq \mu.$$

(SS2) Every irreducible B-module is a homomorphic image of some $\Delta(\lambda)$.

(SS3) For $\lambda \in \Lambda$, the B-module $P(\lambda)$ has a finite filtration by B-submodules with top section $\Delta(\lambda)$ and other sections of the form $\Delta(\mu)$ with $\bar{\mu} > \bar{\lambda}$.

When these conditions hold, the data

$$\{\Delta(\lambda), P(\lambda)\}_{\lambda \in \Lambda} \tag{1.3.1}$$

form (by definition) a *stratifying system*[14] for the category B-mod of finite B-modules. Note that if (1.3.1) is a stratifying system for B-mod, then the system $\{\Delta(\lambda)_{\mathscr{K}'}, P(\lambda)_{\mathscr{K}'}\}_{\lambda\in\Lambda}$ is a stratifying system for $B_{\mathscr{K}'}$-mod, *for any base change* $\mathscr{K} \to \mathscr{K}'$, provided $\mathscr{K}'$ is a Noetherian commutative ring. (See [**DPS15**, p. 231]. For more details and further results, see [**DPS98a**, Lem. 1.2.5].)

REMARK 1.3.2. If B has a stratifying system as explained above, it has a "standard stratification" by idempotent ideals $0 \subseteq J_0 \subseteq J_1 \subseteq \cdots \subseteq J_n = B$ much like a defining sequence in the theory of quasi-hereditary algebras. Thus, B is a kind of generalization of a quasi-hereditary algebra, or, equivalently, if B-mod has a stratifying system, then B-mod is a generalization of a highest weight category. We call B a *standardly stratified algebra*; see Section 3.5 for more details. We do not pursue this further here, but refer to [**DPS98a**, §1.2] and [**DPS18**] for a more complete discussion.

1.3C. Exact categories and the stratification hypothesis. Stratifying systems can be constructed in an endomorphism-ring setting. Originally, the construction was based on assuming a "stratification hypothesis" in [**DPS98a**, Hyp. 1.2.9, Th. 1.2.10], which required a difficult Ext^1-vanishing condition (see [**DPS98a**, Th. 2.3.9, Th. 2.4.4]). This problem was partly circumvented in [**DPS17**, Hyp. 2.5 & Th. 2.6] by introducing "designer" exact categories which had, a priori, smaller Ext^1-groups, and thus were more likely to vanish. However, the new version was somewhat awkward to use in the Quillen axiom system. See [**DPS17**, Rem. 2.7]. The updated version below, given in the Stratification Hypothesis 1.3.3, removes this issue, by replacing all Ext^1-vanishings with existence assertions for needed short exact sequences. As noted in Remark 1.3.4, these assertions are at least implied by

[14] Following [**DPS17**], we drop the word "strict" in "strict stratifying system" as used in [**DPS98a**]. I.e., all stratifying systems in this monograph are strict in the sense of [**DPS98a**], thus, we insist on the strict comparison $\bar{\mu} > \bar{\lambda}$ in (SS3).

a natural exact category Ext^1-vanishing condition. This is the way they are verified in this monograph.

Let R be a finite and projective $\mathscr{K}$-algebra. Our aim will be to provide ingredients in mod-R, including a right R-module T with suitable filtrations, which enable the construction of a stratifying system for B-mod, where $B := \mathrm{End}_R(T)$. Clearly, T is naturally a left B-module.

Construction. We will construct T as an object in a full subcategory $\mathscr{A}$ of mod-R, with the following assumptions. First, the subcategory $\mathscr{A}$ is part of an exact category $(\mathscr{A}, \mathscr{E})$. Second, the exact sequences $(X \to Y \to Z) \in \mathscr{E}$ are among the short exact sequences $0 \to X \to Y \to Z \to 0$ in mod-R. Thus, $(\mathscr{A}, \mathscr{E})$ is a full exact subcategory of mod-R. (In our terminology, "full exact" is just an abbreviation for "full and exact.") Third, we have a collection of objects $S_\lambda, T_\lambda \in \mathscr{A}$ indexed by the elements λ of a finite quasi-poset Λ, where T_λ is a finite and projective $\mathscr{K}$-module.[15] For each $\lambda \in \Lambda$, S_λ is a subobject of T_λ, with the inclusion $S_\lambda \hookrightarrow T_\lambda$ an inflation in $\mathscr{E}$. For each $\lambda \in \Lambda$, fix a positive integer m_λ. Define

$$T := \bigoplus_{\lambda \in \Lambda} T_\lambda^{\oplus m_\lambda}. \tag{1.3.2}$$

With this notation, we can now state the following:

STRATIFICATION HYPOTHESIS 1.3.3. *Assume the set-up introduced in the previous paragraph. In particular, $(\mathscr{A}, \mathscr{E})$ is an exact category, a full exact subcategory of mod-R, and $T \in \mathscr{A}$ with $B = \mathrm{End}_R(T) = \mathrm{End}_{\mathscr{A}}(T)$. There are also indexed objects S_λ, T_λ ($\lambda \in \Lambda$), and the following statements are required as part of the hypothesis:*

(1) *For $\lambda \in \Lambda$, there is an increasing filtration of T_λ:*

$$0 = T_\lambda^{-1} \subseteq T_\lambda^0 \subseteq \cdots \subseteq T_\lambda^{\ell(\lambda)} = T_\lambda$$

in which each inclusion $T_\lambda^{i-1} \subseteq T_\lambda^i$ defines a conflation

$$(T_\lambda^{i-1} \longrightarrow T_\lambda^i \longrightarrow T_\lambda^i/T_\lambda^{i-1}) \in \mathscr{E}. \tag{1.3.3}$$

In addition, $T_\lambda^0 \cong S_\lambda$, and, for $i > 0$, the "section"[16] $T_\lambda^i/T_\lambda^{i-1}$ is a direct sum of various S_μ, $\mu \in \Lambda$ and $\bar{\mu} > \bar{\lambda}$ (repetitions allowed).

(2) *For $\lambda, \mu \in \Lambda$, $\mathrm{Hom}_{\mathscr{A}}(S_\mu, T_\lambda) \neq 0 \implies \lambda \leq \mu$.*

(3) *For all $\lambda \in \Lambda$ and integers $i \geq 0$, applying the contravariant "diamond" functor[17] $(-)^\diamond := \mathrm{Hom}_R(-, T)$ to (1.3.3) gives the exact sequence*

$$0 \to (T_\lambda^i/T_\lambda^{i-1})^\diamond \longrightarrow (T_\lambda^i)^\diamond \longrightarrow (T_\lambda^{i-1})^\diamond \to 0$$

of B-modules.

REMARK 1.3.4. Notice that condition (3) *is implied by the vanishing of* the abelian group $\mathrm{Ext}^1_{\mathscr{E}}(S_\lambda, T)$, for all $\lambda \in \Lambda$; see [**DPS17**, Prop. A.2]. This is the way that we will use the (updated) Stratification Hypothesis 1.3.3 in this monograph.

---

[15]This assumption is not often used directly, but can be a step toward the required projectivity over $\mathscr{K}$ of the modules $\Delta(\lambda)$; see Remark 1.3.6(b). The assumption for T often comes for free, since the S_ν are typically projective (or even free) $\mathscr{K}$-modules and each T_λ is constructed to have a filtration with each section a direct sum of various S_ν.

[16]Observe that $T_\lambda^i \to T_\lambda^i/T_\lambda^{i-1}$ is the unique (up to isomorphism) cokernel in $\mathscr{A}$ of the inflation $T_\lambda^{i-1} \to T_\lambda^i$ and, thus, is a deflation. We generally leave such exact category exercises to the reader.

[17]This functor will also be used later in the monograph.

We mention, without proof, that it is implied by the less flexible earlier [**DPS17**, Hyp. 2.5], which was modeled closely on [**DPS98a**, Hyp. 1.2.9] and which is harder to check, as written, in the Quillen exact category setting.

The following theorem updates [**DPS98a**, Th. 1.3.3] and [**DPS17**, Th. 2.6].

THEOREM 1.3.5. *Assume that the Stratification Hypothesis* 1.3.3 *holds. For* $\lambda \in \Lambda$, *put*

$$\begin{cases} \Delta(\lambda) := \mathrm{Hom}_{\mathscr{A}}(S_\lambda, T) = \mathrm{Hom}_R(S_\lambda, T) \in B\text{-mod}, \\ P(\lambda) := \mathrm{Hom}_{\mathscr{A}}(T_\lambda, T) = \mathrm{Hom}_R(T_\lambda, T) \in B\text{-mod}. \end{cases} \tag{1.3.4}$$

If each $\Delta(\lambda)$ *is projective over* $\mathscr{K}$, *then* $\{\Delta(\lambda), P(\lambda)\}_{\lambda\in\Lambda}$ *is a stratifying system for* B-mod.

PROOF. First, we check (SS1). Observe that the contravariant functor $(-)^\diamond$ provides maps

$$\mathrm{Hom}_R(S_\mu, T_\lambda) \longrightarrow \mathrm{Hom}_B(P(\lambda), \Delta(\mu)).$$

Passing to $T = \bigoplus_{\lambda\in\Lambda} T_\lambda^{\oplus m_\lambda}$ on the left and ${}_BB = \bigoplus_{\lambda\in\Lambda} P(\lambda)^{\oplus m_\lambda}$ on the right, we obtain an isomorphism

$$\mathrm{Hom}_R(S_\mu, T) \xrightarrow{\sim} \mathrm{Hom}_B({}_BB, \Delta(\mu)) \cong \Delta(\mu).$$

Consequently, the original maps must also be isomorphisms. Thus, (SS1) follows from Condition (2) (of Hypothesis 1.3.3).

Next, we will prove (SS2) assuming (SS3) has been proved. By the construction of $B = \mathrm{End}_R(T)$, we have that ${}_BB \cong \bigoplus_{\lambda\in\Lambda} P(\lambda)^{\oplus m_\lambda}$. Using (SS3), ${}_BB$ has a filtration with sections various $\Delta(\nu)$, $\nu \in \Lambda$. Any irreducible B-module is, of course, a homomorphic image of ${}_BB$, and so (using irreducibility) a homomorphic image of some $\Delta(\nu)$. This proves (SS2) in the presence of (SS3).

Now consider (SS3) itself for a given λ. Let n be an integer such that $T_\lambda^n = T_\lambda$. Let $j \geq i \geq -1$ be integers. Thus, T_λ^i is a submodule of T_λ^j.

CLAIM. The natural map $(T_\lambda^j)^\diamond \to (T_\lambda^i)^\diamond$ is a surjection.

To see this, observe that the inclusion $T_\lambda^i \hookrightarrow T_\lambda^j$ factors into a composition of inclusions $T_\lambda^i \hookrightarrow T_\lambda^{i+1} \hookrightarrow T_\lambda^{i+2} \hookrightarrow \cdots \hookrightarrow T_\lambda^j$. By Condition (3), for $i \leq a < j$, the inclusion $T_\lambda^a \hookrightarrow T_\lambda^{a+1}$ becomes a surjection $(T_\lambda^{a+1})^\diamond \twoheadrightarrow (T_\lambda^a)^\diamond$ upon applying $(-)^\diamond$. The Claim follows by taking compositions.

We return to the proof of (SS3) for a given λ. Let $n \geq j \geq 0$. The maps $(T_\lambda^n)^\diamond \to (T_\lambda^j)^\diamond$ and $(T_\lambda^n)^\diamond \to (T_\lambda^{j-1})^\diamond$ are both surjective by the Claim. This gives the two exact sequences forming the rows of the natural commutative diagram

$$\begin{array}{ccccccccc} 0 & \longrightarrow & (T_\lambda^n/T_\lambda^j)^\diamond & \longrightarrow & (T_\lambda^n)^\diamond & \longrightarrow & (T_\lambda^j)^\diamond & \longrightarrow & 0 \\ & & \alpha\big\downarrow & & \big\| & & \beta\big\downarrow & & \\ 0 & \longrightarrow & (T_\lambda^n/T_\lambda^{j-1})^\diamond & \longrightarrow & (T_\lambda^n)^\diamond & \longrightarrow & (T_\lambda^{j-1})^\diamond & \longrightarrow & 0 \end{array} \tag{1.3.5}$$

Take a sufficiently large integer n, so that $T_\lambda^n = T_\lambda$. Then, $P(\lambda) = T_\lambda^\diamond$ has a filtration by submodules of the form $(T_\lambda/T_\lambda^j)^\diamond$, and the map α above is an inclusion between adjacent filtration terms. The Snake Lemma shows that the section $\mathrm{Coker}(\alpha)$ of these two filtration terms is isomorphic to $\mathrm{Ker}(\beta) = (T_\lambda^j/T_\lambda^{j-1})^\diamond$. Thus, $P(\lambda)$ has a filtration with sections isomorphic to various $(T_\lambda^j/T_\lambda^{j-1})^\diamond$. The module

$(T^j_\lambda/T^{j-1}_\lambda)^\diamond$ is of the form $S^\diamond_\nu = \mathrm{Hom}_R(S_\nu, T) = \Delta(\nu)$, where $\nu > \lambda$ if $j > 0$, and $\nu = \lambda$ if $j = 0$. In the $j = 0$ case, $\mathrm{Coker}(\alpha)$ is the "top" section of the given filtration on $P(\lambda)$, and it is isomorphic to $\mathrm{Ker}(\beta) = (T^0_\lambda)^\diamond = S^\diamond_\lambda = \Delta(\lambda)$. This completes the proof of (SS3).

Finally, because it is assumed that each $\Delta(\nu)$, $\nu \in \Lambda$, is a projective $\mathscr{K}$-module, it follows that each $P(\lambda)$, $\lambda \in \Lambda$, is also a projective $\mathscr{K}$-module. Since B is a direct sum of the $P(\lambda)$ (repetitions allowed), B is a projective $\mathscr{K}$-module. This completes the proof. □

REMARKS 1.3.6.

(a) Correspondingly, $\mathscr{A}$ can be one of the categories $\mathscr{A}(\mathscr{L}^*)$ in Section 2.3 below (or a full subcategory, such as $\mathscr{A}_\flat(\mathscr{L}^*)$ there.) A similar remark holds for the corresponding exact category in Section 2.2; however, the notation above is chosen to be compatible with the applications in Section 2.4 as developed in Section 2.3.

(b) If $\mathscr{K}$ is a regular ring of Krull dimension at most 2, then, in the statement of Theorem 1.3.5, the sentence "Assume that each $\Delta(\lambda)$ is projective over $\mathscr{K}$" can be omitted, since $\Delta(\lambda)$ is automatically projective over $\mathcal{Z}$. This is shown in the proof of [**DPS98a**, Cor. 1.2.12)], using earlier work of Auslander-Goldman together with the projectivity of T as a $\mathscr{K}$-module (assumed in the preamble to the Stratification Hypothesis 1.3.3). For a detailed account, see [**DDPW08**, §C.3]. A similar result holds if the stratifying system is "defined over Z" where Z is a regular ring of Krull dimension ≤ 2. This means that the stratifying system over $\mathscr{K}$ is obtained by base change from the regular ring Z.

(c) As an example in (b), we will be mainly interested in the special case in which $R = \mathcal{H}$ is the Hecke algebra defined over $\mathcal{Z} = \mathbb{Z}[t, t^{-1}]$ (and its localizations $\mathcal{Z}^\natural$ at multiplicative subsets, not containing 0, of $\mathbb{Z}$). Since $\mathcal{Z}$ is a regular ring of dimension 2, (b) above applies to show that each $\Delta(\lambda)$ is projective over $\mathcal{Z}$. In fact, by a result of Swan [**Sw78**, p.111, first paragraph], $\Delta(\lambda)$ is then $\mathcal{Z}$-free. (Swan's paragraph also implies that any finite, projective $\mathcal{Z}^\natural$-module is free. In addition, if $\mathcal{Z}^{\natural\oplus m} \cong \mathcal{Z}^{\natural\oplus n}$ for integers m, n, then $m = n$.)

2. Some exact categories

Section 1.3 in Chapter 1 provides a general theory of stratifying endomorphism algebras using exact categories, reformulating that developed in [**DPS17**]. To apply the theory in the context of the conjecture [**DPS98a**, Conj. 2.5.2] by the authors, we need to construct a nested sequence of exact categories involving Hecke algebra modules with cell module filtrations. We eventually prove the Ext^1 vanishing condition (see Remark 1.3.4) in one of the exact categories and, then, prove the (slightly modified) conjecture in the next chapter.

2.1. Three exact categories. *In this section, we work exclusively with left modules for an algebra.* Let $\mathcal{H}$ be the generic Hecke algebra over $\mathcal{Z}$, associated to a standard finite Coxeter system (W, S, L) (as defined in Section 1.1).

We begin with the following variation on [**DPS15**, Lem. 4.3]. If V is a left $\mathcal{H}$-module, the "fixed-point module" for the action of $\mathcal{H}$ on V is defined to be

$$V^{\mathcal{H}} := \{v \in V \mid \mathrm{c}_s v = (t_s + t_s^{-1})v, \quad \forall s \in S\} \in \mathcal{Z}\text{-mod},$$

where t_s, c_s are defined in (1.2.1) and (1.2.3), respectively. Equivalently,

$$v \in V^{\mathcal{H}} \iff T_s v = t_s^2 v, \text{ for all } s \in S.$$

The assignment $V \mapsto V^{\mathcal{H}}$ is part of an evident left-exact endo-functor. Similarly, $V^{\mathcal{H}_J}$ is defined for any parabolic subalgebra $\mathcal{H}_J$ of $\mathcal{H}$ ($J \subseteq S$). The next result follows from [**DPS15**, Lem.4.3] and the above definition of fixed points, using (1.2.1) and (1.2.3).

LEMMA 2.1.1. *Let $\mathfrak{I} \subseteq \mathfrak{K}$ be left ideals in the Hecke algebra $\mathcal{H}$ over $\mathcal{Z}$. Assume that both $\mathfrak{I}$ and $\mathfrak{K}$ are spanned by the Kazhdan-Lusztig basis elements c_x, $x \in W$, that they contain. Then, for any $J \subseteq S$, the natural map*

$$\mathfrak{K}^{\mathcal{H}_J} \longrightarrow (\mathfrak{K}/\mathfrak{I})^{\mathcal{H}_J} \tag{2.1.1}$$

is surjective.

NOTATION 2.1.2. At this point, the reader should review the notion of height functions $\mathfrak{h}$ discussed in Appendix B below. In this section, we fix

$$\mathfrak{h} = \mathfrak{h}_l : \Omega \to \mathbb{Z},$$

a function defined on the set Ω of left cells in W, and compatible with the preorder $\leq_{LR}$; see display (B2.1) in Appendix B for the meaning of "compatible." For another example, see the order defined in display (3.1.1). The subscript l is there to remind the reader of our left-module context. Such a height function $\mathfrak{h}$ is also compatible with $\leq_L$, as formalized in Corollary B2.6 (which uses property (1.2.11) via Proposition B2.3). In particular, $\mathfrak{h}$ is constant on left cells which belong to the same two-sided cell. It is also observed in Remark B2.7 that $\mathfrak{h}$ may be viewed as assigning heights to left-cell modules $S(\omega)$. Thus, if $\omega \in \Omega$, $\mathfrak{h}$ assigns the left-cell module $S(\omega)$ height $\mathfrak{h}(\omega)$. Also, the value $\mathfrak{h}(\omega) = \mathfrak{h}(\bar{\omega})$, where $\bar{\omega}$ is the two-sided cell containing ω, may be assigned without ambiguity as a "height" to the irreducible constituents of the base-changed module $S(\omega)_{\mathbb{Q}(t)}$. In particular, this means that $\mathfrak{h}$ uniquely determines a height function on the set Λ of irreducible modules for the semisimple algebra $\mathcal{H}_{\mathbb{Q}(t)}$. Since this height function also determines the original $\mathfrak{h} : \Lambda \to \mathbb{Z}$ (see Remark B2.7), we continue to denote it by $\mathfrak{h}$. As explained in Example B1.1, $\mathfrak{h}$ induces a canonical decreasing filtration

$$M = M_0 \supseteq M_1 \supseteq M_2 \supseteq \cdots \supseteq M_{n-1} \supseteq M_n = 0 \tag{2.1.2}$$

on each $\mathcal{H}$-module M. We call this filtration the *$\mathfrak{h}$-filtration* or the *height filtration,* when $\mathfrak{h}$ is understood on M. Similar terminology can be used for right modules in the next three sections, although the height filtrations are increasing and the subscripts here become superscripts there.

Finally, as a convenience, we assume in this section (unless noted otherwise) that $\mathfrak{h} = \mathfrak{h}_l : \Omega \to \mathbb{Z}$ only takes negative values. See Appendix B. Later, we make use of $\mathfrak{h}_r := -\mathfrak{h}_l : \Omega \to \mathbb{Z}$, which is, in particular, used to assign positive height values to dual left-cell modules $S_\omega = S(\omega)^* = \mathrm{Hom}_{\mathcal{Z}}(S(\omega), \mathcal{Z})$, $\omega \in \Omega$ (see (1.2.10)) which are naturally right $\mathcal{H}$-modules. See Remark B.8.

Relative to such a height function $\mathfrak{h} : \Omega \to \mathbb{Z}$ that induces $\mathfrak{h} : \Lambda \to \mathbb{Z}$ as above, there is an exact category $(\mathscr{A}, \mathscr{E})$, as constructed in Example B1.1. Thus, $(\mathscr{A}, \mathscr{E})$ consists of the full subcategory $\mathscr{A} = \mathscr{A}_l$ of $\mathcal{H}$-mod whose objects are left $\mathcal{H}$-modules, finite and torsion-free over $\mathcal{Z}$, and the (full) "subcategory" $\mathscr{E} = \mathscr{E}_l$ (in the sense of footnote 12 in Chapter 1) of short exact sequences $0 \to M' \to M \to M'' \to 0$ in $\mathscr{A}$ satisfying the condition that $0 \to M'_h \to M_h \to M''_h \to 0$ is exact, for all heights h. (Compare the right-module version given in [**DPS17**, Constr. 3.5] or Theorem 2.2.1 below.)

Below, building on $(\mathscr{A},\mathscr{E})=(\mathscr{A}_l,\mathscr{E}_l)$, we present three more examples of exact categories:

$$\mathcal{H}\text{-mod}\geq(\mathscr{A},\mathscr{E})\geq(\mathscr{A}(\mathscr{L}),\mathscr{E}(\mathscr{L}))\geq(\mathscr{A}(\mathscr{L}),\mathscr{E}^\flat(\mathscr{L}))\geq(\mathscr{A}^\flat(\mathscr{L}),\mathscr{E}^\flat(\mathscr{L})),$$

which will be important in the sequel. Here "$\geq$" indicates "a full subcategory". In particular, $\mathrm{Hom}_{\mathscr{C}}(M,N)=\mathrm{Hom}_{\mathcal{H}}(M,N)$ for objects M,N in $\mathscr{C}$ and every $\mathscr{C}=\mathscr{A},\mathscr{A}(\mathscr{L})$, etc.

2.1A. The exact category $(\mathscr{A}(\mathscr{L}),\mathscr{E}(\mathscr{L}))$. Its construction involves the category $\mathcal{H}$-mod of (finite) left $\mathcal{H}$-modules, but it closely follows the more abstract discussion in [**DPS17**, Assump. 3.2 and Constr. 3.8] which worked with the category mod-H of right modules for a suitable algebra H more general than the generic Hecke algebra $\mathcal{H}$.

For any integer h, let $\mathscr{L}_h$ be the full subcategory of $\mathscr{A}$ consisting of modules N which are a finite direct sum of modules isomorphic to left-cell modules $S(\omega)$ of height h. Let $\mathscr{L}$ be the set-theoretic union of the $\mathscr{L}_h$. Next, let $\mathscr{A}(\mathscr{L})$ be the full additive subcategory of $\mathcal{H}$-mod with objects M (necessarily in $\mathscr{A}_l$) whose $\mathfrak{h}$-filtration (2.1.2) has the property that, for any integer h, M_{h-1}/M_h is an object in $\mathscr{L}_{-h}$.[18] We will also informally call such an $\mathfrak{h}$-filtration *a left-cell filtration.*[19]

By analogy with [**DPS17**, §3], let $\mathscr{E}(\mathscr{L})$ consist of those sequences $(A\to B\to C)$ of objects in $\mathscr{A}(\mathscr{L})$ such that, for every h,

$$0\longrightarrow A_{h-1}/A_h\longrightarrow B_{h-1}/B_h\longrightarrow C_{h-1}/C_h\longrightarrow 0$$

is a *split* short exact sequence in the additive category $\mathscr{L}_{-h}$ (or, equivalently, in the category $\mathcal{H}$–mod). Then,

(2.1.3) $\qquad(\mathscr{A}(\mathscr{L}),\mathscr{E}(\mathscr{L}))$ is an exact category.

This was proved in the more abstract setting of right H-modules in [**DPS17**, Th. 3.9] for an analogous category $(\mathscr{A}(\mathscr{S}),\mathscr{E}(\mathscr{S}))$ associated to H. The proof in the present situation is entirely similar. The result can also be deduced from the special case $H=\mathcal{H}^{\rm op}$ of the cited theorem.

Note that left-cell modules are, by definition, finite and free over $\mathcal{Z}$. This implies that any object in $\mathscr{A}(\mathscr{L})$ is also finite and free over $\mathcal{Z}$. Looking ahead to Section 2.2, this implies that $\mathscr{A}(\mathscr{L})\subseteq\mathscr{A}_l'$, the left-hand version of the category $\mathscr{A}'$ introduced in Theorem 2.2.2. Thus, $\mathscr{A}_l'$ denotes the full additive subcategory of objects $M\in\mathscr{A}_l$ for which M_h/M_{h+1} is projective as a $\mathscr{K}$-module, for each $h\in\mathbb{Z}$, and $\mathscr{E}_l'=\mathscr{E}_l|_{\mathscr{A}_l'}$. Also, $\mathscr{E}(\mathscr{L})\subseteq\mathscr{E}_l'$, so that $(\mathscr{A}(\mathscr{L}),\mathscr{E}(\mathscr{L}))$ is a (full) exact subcategory of $(\mathscr{A}_l',\mathscr{E}_l')$; see the definition at the beginning of Subsection 1.3A.

Note also that, for each object M in $\mathscr{A}(\mathscr{L})$ with the $\mathfrak{h}$-filtration given in display (2.1.2), each inclusion $M_h\subseteq M_{h-1}$ gives a conflation $(M_h\to M_{h-1}\to M_{h-1}/M_h)\in\mathscr{E}(\mathscr{L})$, for all $1\leq h\leq n$, so that its $\mathcal{Z}$-dual M^* satisfies the condition (1) in the Stratification Hypothesis 1.3.3.

[18] The minus sign is chosen to be compatible with the duality theory developed later in Sections 2.2 and 2.3. In particular, we will consider dual left-cell modules S_ω, for each $\omega\in\Omega$, and assign to each S_ω a positive height, negative to that of $S(\omega)$. See the beginning paragraphs of Section 2.3 and also (2.2.2). Note that, with Notation 2.1.2, this height filtration coincides with the one defined in Example B1.1.

[19] Clearly, every $\mathfrak{h}$-filtration here may be refined to a filtration with sections isomorphic to left-cell modules. This refined filtration is a left-cell filtration in the sense of [**DPS98a**, §2.2].

2.1B. The exact category $(\mathscr{A}(\mathscr{L}), \mathscr{E}^\flat(\mathscr{L}))$. With the notation above, let $\mathscr{E}^\flat(\mathscr{L})$ denote the full subcategory of $\mathscr{E}(\mathscr{L})$ with objects $(M \to N \to P) \in \mathscr{E}(\mathscr{L})$ such that, for any integer h and any subset J of S, the sequence

$$0 \longrightarrow (M_h)^{\mathcal{H}_J} \longrightarrow (N_h)^{\mathcal{H}_J} \longrightarrow (P_h)^{\mathcal{H}_J} \longrightarrow 0 \tag{2.1.4}$$

of fixed points is exact. Note that, for any h, each $0 \to M_h/M_{h-1} \to N_h/N_{h-1} \to P_h/P_{h-1} \to 0$ is split in $\mathcal{H}$-mod, since $(M \to N \to P) \in \mathscr{E}(\mathscr{L})$, but this does not mean that the short exact sequence $0 \to M_h \to N_h \to P_h \to 0$ is split. In particular, the exactness of (2.1.4) is a nontrivial property.

We have:

PROPOSITION 2.1.3. *$(\mathscr{A}(\mathscr{L}), \mathscr{E}^\flat(\mathscr{L}))$ is an exact category.*

PROOF. Let $\mathscr{C}$ be the abelian category of all finite $\mathcal{Z}$-modules. Put

$$F(-) := \bigoplus_{J,h}((-)_h)^{\mathcal{H}_J} : \mathscr{A}(\mathscr{L}) \to \mathscr{C}.$$

Then, F is an $\mathscr{E}(\mathscr{L})$-left exact functor. Now we apply [**DPS17**, Lem. 3.1], with $\mathscr{E}$ there taken to be $\mathscr{E}^\flat(\mathscr{L})$ as defined above. □

For $J \subseteq S$, if W_J is the parabolic subgroup of W defined by J, then $\mathcal{H}x_J$ is the usual (left) q-permutation module determined by any $J \subseteq S$, where

$$x_J := \sum_{w \in W_J} T_w \tag{2.1.5}$$

was defined in display (1.2.4). For future use in Section 5.2, we also define

$$y_J := \sum_{w \in W_J} \operatorname{sgn}(w)(t^2)^{-L(w)} T_w, \tag{2.1.6}$$

where $\operatorname{sgn}(w) := (-1)^{\ell(w)}$ is the sign of w. See [**DPS98c**, pp. 324–326], taking $q = t^2$ in our current notation.

PROPOSITION 2.1.4. *The $\mathcal{H}$-module $\mathcal{H}x_J$ belongs to $\mathscr{A}(\mathscr{L})$.*

PROOF. Let M denote $\mathcal{H}x_J$. Using display (1.2.8), we find M has a basis of elements c_v, with $v \in W$, in a certain quasi-poset ideal V for $\leq_L$, necessarily closed under $\sim_L$. In fact, we have $V = \{w \in W \mid ws < w, \forall s \in J\} = \{w \in W \mid w \leq_L w_{0,J}\}$; see [**DPS98a**, (2.3.3)] or Lemma 3.1.2 in this monograph. The quasi-poset ideal V is, thus, a union of left cells. For any integer h, let V_{h-1} be the set of all $v \in V$ belonging to a left cell ω with $\mathfrak{h}(\omega) \leq -h$. Then, each V_{h-1} is, by construction, a union of left cells, and is a quasi-poset ideal, since $\mathfrak{h}$ is compatible with $\leq_L$ on Ω. (This was observed in Notation 2.1.2 as a consequence of (1.2.11).) The $\mathcal{Z}$-span M_{h-1} of V_{h-1} is, thus, a left $\mathcal{H}$-module, as follows from the definition of $\leq_L$ [**Lus03**, p. 43]. Similar considerations identify M_{h-1}/M_h as a direct sum of left-cell modules of height $-h$, one summand for each left cell in V of height $-h$. Thus, $M \in \mathscr{A}(\mathscr{L})$, proving the assertion. □

We remark that improvements of Proposition 2.1.4 occur in several places in this monograph. For example, Corollary 2.1.8 gives the sharper result that, for $J \subseteq S$, $\mathcal{H}x_J \in \mathscr{A}^\flat(\mathscr{L})$. See Theorem 3.1.3 for a right-module version involving the exact category $(\mathscr{A}_\flat(\mathscr{L}^*), \mathscr{E}_\flat(\mathscr{L}^*))$.

COROLLARY 2.1.5. *Let* $M = \mathcal{H}x_J$, *and consider the decreasing filtration*

$$M = M_0 \supseteq M_1 \supseteq \cdots \supseteq M_{l_J} = 0$$

defined by the height function $\mathfrak{h}$. *For any* h, *the short exact sequence* $0 \to M_h \to M_{h-1} \to M_{h-1}/M_h \to 0$ *belongs of* $\mathscr{E}^\flat(\mathscr{L})$.

PROOF. By the discussion at the end of the proof of Proposition 2.1.4, each inclusion $M_h \subseteq M_{h-1}$ is an inflation. Thus, each $(M_h \to M_{h-1} \to M_{h-1}/M_h) \in \mathscr{E}(\mathscr{L})$. By Lemma 2.1.1, the sequence

$$0 \to (M_h)^{\mathcal{H}_J} \to (M_{h-1})^{\mathcal{H}_J} \to (M_{h-1}/M_h)^{\mathcal{H}_J} \to 0$$

is exact, for all $J \subseteq S$. Hence, by definition (2.1.4), the short exact sequence $0 \to M_h \to M_{h-1} \to M_{h-1}/M_h \to 0$ belongs of $\mathscr{E}^\flat(\mathscr{L})$. □

In the following theorem, the $\mathrm{Ext}^1_{\mathscr{E}^\flat(\mathscr{L})}$-group is calculated in the exact category $(\mathscr{A}(\mathscr{L}), \mathscr{E}^\flat(\mathscr{L}))$. Subsection 2.1C below introduces a second exact category $(\mathscr{A}^\flat(\mathscr{L}), \mathscr{E}^\flat(\mathscr{L}))$, possibly setting up a conflict of notation. However, as pointed out in Proposition 2.1.7 (or the proof of its corollary), the $\mathrm{Ext}^1_{\mathscr{E}^\flat(\mathscr{L})}(M, N)$-groups can be calculated in either exact category, as long as M, N both belong to $\mathscr{A}^\flat(\mathscr{L})$. We also remark that the condition (2.1.1) plays an essential role in the proof of the theorem.

THEOREM 2.1.6. *For any* $J \subseteq S$, $\mathrm{Ext}^1_{\mathscr{E}^\flat(\mathscr{L})}(\mathcal{H}x_J, S(\omega)) = 0$, *for all* $\omega \in \Omega$.

PROOF. Consider an exact sequence

$$0 \longrightarrow S(\omega) \longrightarrow E \xrightarrow{g} \mathcal{H}x_J \longrightarrow 0 \tag{2.1.7}$$

in $\mathscr{E}^\flat(\mathscr{L})$. We want to prove that the sequence is split. By definition, we obtain an exact sequence

$$0 \longrightarrow S(\omega)^{\mathcal{H}_J} \longrightarrow E^{\mathcal{H}_J} \xrightarrow{g_J} (\mathcal{H}x_J)^{\mathcal{H}_J} \longrightarrow 0. \tag{2.1.8}$$

Consider the $\mathcal{H}_J$-submodule $\mathcal{Z}x_J$ of $(\mathcal{H}x_J)^{\mathcal{H}_J}$. The exactness of (2.1.8) implies that there exists $e \in E^{\mathcal{H}_J}$ such that $g_J(e) = x_J$. Thus, $\mathcal{Z}x_J \cong \mathcal{Z}e$ as $\mathcal{H}_J$-modules. Since $\mathcal{Z}e \subseteq E$, this induces an $\mathcal{H}_J$-module homomorphism $f_J : \mathcal{Z}x_J \to E$. By Frobenius reciprocity, there is an $\mathcal{H}$-module homomorphism $f : \mathcal{H}x_J \to E$ such that $f(x_J) = e$. Hence, $gf(x_J) = x_J$, and, consequently, gf is the identity map on $\mathcal{H}x_J$. In other words, the sequence (2.1.7) is split. □

Again, we note that $\mathscr{E}^\flat(\mathscr{L}) \subseteq \mathscr{E}(\mathscr{L}) \subseteq \mathscr{E}'_l$, the latter defined in Section 2.2 (see Theorem 2.2.2). Thus, $(\mathscr{A}(\mathscr{L}), \mathscr{E}^\flat(\mathscr{L}))$ is a full exact subcategory of $(\mathscr{A}'_l, \mathscr{E}'_l)$.

Note that, in the exact category $(\mathscr{A}(\mathscr{L}), \mathscr{E}^\flat(\mathscr{L}))$, not every object satisfies the property in Corollary 2.1.5. This motivates the definition of the following full subcategory.

2.1C. The exact category $(\mathscr{A}^\flat(\mathscr{L}), \mathscr{E}^\flat(\mathscr{L}))$. Let $\mathscr{A}^\flat(\mathscr{L})$ be the full subcategory of $\mathscr{A}(\mathscr{L})$ with objects M having the property that, for any integer h and any $J \subseteq S$, the map

$$(M_{h-1})^{\mathcal{H}_J} \longrightarrow (M_{h-1}/M_h)^{\mathcal{H}_J} \tag{2.1.9}$$

is surjective. In other words, each conflation $(M_h \to M_{h-1} \to M_{h-1}/M_h)$ in $\mathscr{E}(\mathscr{L})$ gives rise, for each $J \subseteq S$, to a short exact seqeunce $0 \to (M_h)^{\mathcal{H}_J} \to (M_{h-1})^{\mathcal{H}_J} \to (M_{h-1}/M_h)^{\mathcal{H}_J} \to 0$. Hence, $(M_h \to M_{h-1} \to M_{h-1}/M_h) \in \mathscr{E}^\flat(\mathscr{L})$.

PROPOSITION 2.1.7. *Let $M, P \in \mathscr{A}^\flat(\mathscr{L})$, and assume a given $(M \to N \to P) \in \mathscr{E}(\mathscr{L})$. Then, $N \in \mathscr{A}^\flat(\mathscr{L})$ if and only if $(M \to N \to P) \in \mathscr{E}^\flat(\mathscr{L})$. In particular, the full subcategory $\mathscr{A}^\flat(\mathscr{L})$ of $\mathscr{A}(\mathscr{L})$ is closed under extensions in $(\mathscr{A}(\mathscr{L}), \mathscr{E}^\flat(\mathscr{L}))$.*

PROOF. First, assume that $(M \to N \to P) \in \mathscr{E}^\flat(\mathscr{L})$. For an integer h and a subset $J \subseteq S$, form the commutative diagram

$$\begin{array}{ccccc} (M_{h-1}/M_h)^{\mathcal{H}_J} & \longrightarrow & (N_{h-1}/N_h)^{\mathcal{H}_J} & \longrightarrow & (P_{h-1}/P_h)^{\mathcal{H}_J} \\ \uparrow & & \uparrow & & \uparrow \\ (M_{h-1})^{\mathcal{H}_J} & \longrightarrow & (N_{h-1})^{\mathcal{H}_J} & \longrightarrow & (P_{h-1})^{\mathcal{H}_J} \\ \uparrow & & \uparrow & & \uparrow \\ (M_h)^{\mathcal{H}_J} & \longrightarrow & (N_h)^{\mathcal{H}_J} & \longrightarrow & (P_h)^{\mathcal{H}_J}. \end{array} \tag{2.1.10}$$

Since $M, P \in \mathscr{A}^\flat(\mathscr{L})$, columns 1 and 3 are short exact sequences. Since $(M \to N \to P) \in \mathscr{E}^\flat(\mathscr{L})$, the three rows are all short exact sequences as well. Row 1 is already, before taking fixed points, a split short exact sequence of $\mathcal{H}$-modules, because $(M \to N \to P) \in \mathscr{E}(\mathscr{L})$. Thus, it remains exact upon taking $\mathcal{H}_J$-fixed points. Hence, column 2 is also exact, using [**Mac94**, Ex. 2, p. 51], so that $N \in \mathscr{A}^\flat(\mathscr{L})$, as required.

Conversely, if $N \in \mathscr{A}^\flat(\mathscr{L})$, then all columns in (2.1.10) are short exact sequences. Row 1 is exact as in the previous paragraph. By induction on h, the bottom row is also a short exact sequence. By [**Mac94**, Ex. 2, p. 51] again, row 2 is exact, proving that $(M \to N \to P) \in \mathscr{E}^\flat(\mathscr{L})$, as required. □

If $\mathscr{F}$ is a category of composible morphism pairs from an additive category $\mathscr{B}$, and $\mathscr{C}$ is an additive subcategory of $\mathscr{B}$, let $\mathscr{F}|_{\mathscr{C}}$ denote the category of pairs from $\mathscr{F}$ whose objects belong to $\mathscr{C}$. In the corollary below and elsewhere, we abuse notation by writing $(\mathscr{A}^\flat(\mathscr{L}), \mathscr{E}^\flat(\mathscr{L}))$ for $(\mathscr{A}^\flat(\mathscr{L}), \mathscr{E}^\flat(\mathscr{L})|_{\mathscr{A}^\flat(\mathscr{L})})$.

COROLLARY 2.1.8. *With the above notation, $(\mathscr{A}^\flat(\mathscr{L}), \mathscr{E}^\flat(\mathscr{L}))$ is an exact category, and is a full exact subcategory of $(\mathscr{A}(\mathscr{L}), \mathscr{E}^\flat(\mathscr{L}))$. Moreover, for every $J \subseteq S$, $\mathcal{H}x_J \in \mathscr{A}^\flat(\mathscr{L})$. (In particular, Theorem 2.1.6 holds in the full exact subcategory $(\mathscr{A}^\flat(\mathscr{L}), \mathscr{E}^\flat(\mathscr{L}))$.)*

PROOF. Observe that the category $\mathscr{A}^\flat(\mathscr{L})$ is closed under extensions in the exact category $(\mathscr{A}(\mathscr{L}), \mathscr{E}^\flat(\mathscr{L}))$ by Proposition 2.1.7. So $(\mathscr{A}^\flat(\mathscr{L}), \mathscr{E}^\flat(\mathscr{L}))$ is itself exact by [**Bu10**, Lem. 10.20]. (The proof in [**Bu10**] is largely left as an exercise, in part using the diagram of [**Bu10**, Lem. 3.5].) The assertion of the corollary regarding $\mathcal{H}x_J$ follows from Lemma 2.1.1. Alternately, it follows from a downward induction on h by applying Proposition 2.1.7 to Corollary 2.1.5. □

LEMMA 2.1.9. *Let $M \in \mathscr{A}^\flat(\mathscr{L})$, $h < a$ be integers, and $J \subseteq S$. Then, M_h and M_h/M_a belong to $\mathscr{A}^\flat(\mathscr{L})$, and the natural map*

$$(M_h)^{\mathcal{H}_J} \longrightarrow (M_h/M_a)^{\mathcal{H}_J} \tag{2.1.11}$$

is surjective. The same result is true if M_h is replaced by M.

PROOF. Obviously, $M_h \in \mathscr{A}^\flat(\mathscr{L})$ by definition. Once the surjectivity of (2.1.11) is proven, it follows that $M_h/M_a \in \mathscr{A}^\flat(\mathscr{L})$, by writing the map $M_h \longrightarrow M_h/M_{h+1}$ as the composite $M_h \to M_h/M_a \to M_h/M_{h+1}$ and taking $\mathcal{H}_J$-fixed points. It remains to prove the surjectivity of (2.1.11). Of course, it holds for $a = h + 1$ by the definition of $\mathscr{A}^{\mathfrak{h}}(\mathscr{L})$. By induction on $a - h$, it holds if h is replaced by $h + 1$. The required surjectivity follows from, say, a diagram chase in $\mathcal{Z}$-mod, using the commutative diagram

$$\begin{array}{ccccc} M_{h+1}^{\mathcal{H}_J} & \longrightarrow & M_h^{\mathcal{H}_J} & \longrightarrow & (M_h/M_{h+1})^{\mathcal{H}_J} \\ {\scriptstyle f}\big\downarrow & & {\scriptstyle g}\big\downarrow & & \big\| \\ (M_{h+1}/M_a)^{\mathcal{H}_J} & \longrightarrow & (M_h/M_a)^{\mathcal{H}_J} & \longrightarrow & (M_h/M_{h+1})^{\mathcal{H}_J}, \end{array}$$

whose rows are short exact sequences. □

COROLLARY 2.1.10.

(a) *If $M \in \mathscr{A}^{\mathfrak{h}}(\mathscr{L})$ has the $\mathfrak{h}$-filtration* (2.1.2), *and $a \in \mathbb{Z}$ with $0 \leq a \leq n$, then*

$$(M_a \longrightarrow M \longrightarrow M/M_a) \in \mathscr{E}^{\mathfrak{h}}(\mathscr{L}).$$

In particular, $(M_h \to M_{h-1} \to M_{h-1}/M_h) \in \mathscr{E}^{\mathfrak{h}}(\mathscr{L})$, for all $1 \leq h \leq n$.

(b) *Suppose $(M \to N \to P) \in \mathscr{E}^{\mathfrak{h}}(\mathscr{L})$. If $h \in \mathbb{Z}$, then both $(M_h \to N_h \to P_h)$ and $(M/M_h \to N/N_h \to P/P_h)$ belong to $\mathscr{E}^{\mathfrak{h}}(\mathscr{L})$.*

PROOF.

(a) follows from Lemma 2.1.9, using M for M_h.

(b) Clearly, $(M_h \to N_h \to P_h) \in \mathscr{E}^{\mathfrak{h}}(\mathscr{L})$. The corresponding statement for the quotients follows from the 3×3-lemma for exact categories [**Bu10**, Cor. 3.6], using (a) to obtain the columns. □

Note that $\mathscr{A}^{\mathfrak{h}}(\mathscr{L})$ is, by definition, a full subcategory of $\mathscr{A}(\mathscr{L})$ and, thus, of $\mathscr{A}_l'$. The latter is defined, along with $\mathscr{E}_l'$, in Section 2.2. Also, $\mathscr{E}^{\mathfrak{h}}(\mathscr{L}) \subseteq \mathscr{E}_l'$ so that $(\mathscr{A}^{\mathfrak{h}}(\mathscr{L}), \mathscr{E}^{\mathfrak{h}}(\mathscr{L}))$ is a full exact subcategory of $(\mathscr{A}_l', \mathscr{E}_l')$.[20] By the discussion in Section 2.2 below Proposition 2.2.3 (including Remark 2.2.4), the contravariant dual $(\mathscr{A}^{\mathfrak{h}}(\mathscr{L}), \mathscr{E}^{\mathfrak{h}}(\mathscr{L}))^{\mathrm{op}}$ is realized as $(\mathscr{A}^{\mathfrak{h}}(\mathscr{L})^*, \mathscr{E}^{\mathfrak{h}}(\mathscr{L})^*)$ inside $(\mathscr{A}_l'^*, \mathscr{E}_l'^*) = (\mathscr{A}_r', \mathscr{E}_r') = (\mathscr{A}', \mathscr{E}')$.

Remarkably, it is this dual of the exact category in the corollary that interests us most. We pursue this in Section 2.3.

2.2. Some generalities on duality. This section uses the (more abstract) right-module setup of [**DPS17**, Assump. 3.2]; see Example B1.1 in Appendix B for the left-module setup. Thus, $\mathscr{K}$ is a Noetherian integral domain, K is its fraction field, and H is a finite and torsion-free algebra over $\mathscr{K}$ such that H_K is semisimple. The (isomorphism classes of) irreducible *right* H_K-modules are indexed by a given set Λ, and $\mathfrak{h} : \Lambda \to \mathbb{Z}$ is a height function.

It was also assumed "for convenience" in [**DPS17**, Assump. 3.2] that $\mathfrak{h}$ takes non-negative values. We do not make that assumption here. The "convenience" was in being able to use 0 as a lower bound for indexing elements of Λ via $\mathfrak{h}$. In no case is the validity of a stated result in [**DPS17**] affected by dropping the requirement that $\mathfrak{h}$ takes non-negative values. One such general result is the following.

THEOREM 2.2.1 ([**DPS17**, Th. 3.6]). *Let $\mathscr{A}$ denote the category of right H-modules which are finite and torsion-free over $\mathscr{K}$. For each M in $\mathscr{A}$, and $h \in \mathbb{Z}$, let M^h denote $M \cap (M_K)^{\leq h}$, where $(M_K)^{\leq h}$ is the sum in M_K of all irreducible*

[20]See the discussion above (1.3.2) for the terminology.

right H_K-submodules indexed by $\lambda \in \Lambda$ with $\mathfrak{h}(\lambda) \leq h$. Let $\mathscr{E}$ denote the set of short exact sequences $0 \to X \xrightarrow{i} Y \xrightarrow{d} Z \to 0$ in mod-H of objects $X, Y, Z \in \mathscr{A}$, such that, for each $h \in \mathbb{Z}$, the induced sequence

$$0 \longrightarrow X^h/X^{h-1} \longrightarrow Y^h/Y^{h-1} \longrightarrow Z^h/Z^{h-1} \longrightarrow 0 \tag{2.2.1}$$

is also exact in mod-H. Then, $(\mathscr{A}, \mathscr{E})$ is an exact category.

We also have

THEOREM 2.2.2. *Keep the notation above, but let $\mathscr{A}'$ denote the full additive subcategory of objects $M \in \mathscr{A}$ for which M^h/M^{h-1} is projective as a $\mathscr{K}$-module, for each $h \in \mathbb{Z}$. If $\mathscr{E}'$ denotes $\mathscr{E}|_{\mathscr{A}'}$, the short exact sequences in $\mathscr{E}$ whose objects all belong to $\mathscr{A}'$, then $(\mathscr{A}', \mathscr{E}')$ is an exact category.*

PROOF. The category $\mathscr{A}'$ is precisely the category $\mathscr{A}(\mathscr{S})$ in [**DPS17**, Constr. 3.8] with $\mathscr{S} = \bigcup_{i \in \mathbb{Z}}, \mathscr{S}_i$ in the special case that each $\mathscr{S}_i$ is the full additive category of $\mathscr{A}$ consisting of objects S projective over $\mathscr{K}$ with S_K a sum of irreducible right H_K-modules of height i, for all $i \in \mathbb{Z}$.

Note that $\mathscr{A}'$ is closed under extensions in $\mathscr{A}$, since the middle term of the exact sequence (2.2.1) is projective over $\mathscr{K}$ when the other terms are projective over $\mathscr{K}$. Consequently, [**Bu10**, Lem. 10.20] implies that $(\mathscr{A}', \mathscr{E}')$ is an exact category. □

Observe that the category $\mathscr{E}(\mathscr{S})$, as defined in [**DPS17**, Constr. 3.8], is not used in the proof of the above result. Nevertheless, $(\mathscr{A}(\mathscr{S}), \mathscr{E}(\mathscr{S}))$ is a (full) exact subcategory of $(\mathscr{A}', \mathscr{E}')$, while $(\mathscr{A}', \mathscr{E}')$ provides us with a general place to park duals, arising from the construction below. Those duals of most interest to us will actually belong to $(\mathscr{A}(\mathscr{S}), \mathscr{E}(\mathscr{S}))$ (with $\mathscr{S}$ restricted to suitable direct sums of dual left-cell modules).

2.2A. Duality. In Section 2.1, we emphasized the differences between left and right modules, because of the historical use of left-cell modules and their duals. Here, we want to do the opposite and try to view left and right modules on a more equal footing.

The discussion and results above in this section are formulated in terms of right H-modules. They all have analogs for left H-modules as may be seen from the right H-module case by substituting the opposite algebra $H^{\rm op}$ for H. In a context where both left and right H-modules are used, it is useful to use a subscript l or r to distinguish the two cases. Thus, $\mathscr{A}_l$, $(\mathscr{A}_l, \mathscr{E}_l)$, $(\mathscr{A}_l', \mathscr{E}_l')$ are $\mathscr{A}$, $(\mathscr{A}, \mathscr{E})$, $(\mathscr{A}', \mathscr{E}')$ analogs in the left H-module case, and a subscript r could be used similarly to give a balanced treatment of the (original) right H-module case.

Next, let $(-)^* = \operatorname{Hom}_{\mathscr{K}}(-, \mathscr{K})$ be the usual duality functor—actually, there are two functors, $\mathscr{A}_l \to \mathscr{A}_r$ and $\mathscr{A}_r \to \mathscr{A}_l$. Both are contravariant and are denoted by the same symbol. The functors apply also to $\mathscr{A}_l'$ and $\mathscr{A}_r'$ and induce equivalences of additive categories

$$\mathscr{A}_l' \xrightarrow{\sim} \mathscr{A}_r'^{\rm op} \quad \text{and} \quad \mathscr{A}_r' \xrightarrow{\sim} \mathscr{A}_l'^{\rm op}.$$

Similarly, we can apply $(-)^*$ to $\mathscr{E}_l'$ and $\mathscr{E}_r'$. Finally, $(-)^*$ can be applied to the pairs $(\mathscr{A}_l', \mathscr{E}_l')$ and $(\mathscr{A}_r', \mathscr{E}_r')$ as well as the opposite categories.

Our notation hides the height function $\mathfrak{h}$, which affects the definition of categories $\mathscr{A}'$, $\mathscr{E}'$, ... If we wish, we can always add a constant integer to any given $\mathfrak{h}$ function without affecting $\mathscr{A}'$, $\mathscr{E}'$, ... We assume in the proposition below, and unless otherwise noted, that the dual of any irreducible right H-module indexed by $\lambda \in \Lambda$ is also indexed by λ and that

$$\mathfrak{h}_l(\lambda) = -\mathfrak{h}(\lambda) = -\mathfrak{h}_r(\lambda), \tag{2.2.2}$$

for all $\lambda \in \Lambda$.

PROPOSITION 2.2.3. *The duality functor* $(-)^*$ *induces a contravariant equivalence between* $(\mathscr{A}'_l, \mathscr{E}'_l)$ *and* $(\mathscr{A}'_r, \mathscr{E}'_r)$.

We leave the proof to the reader, except for noting that the projectivity (over $\mathscr{K}$) property of objects in $\mathscr{A}'_l$ is preserved by $(-)^*$ and also guarantees that $(-)^*$ sends $\mathscr{E}'_l$ to $\mathscr{E}'_r$. Similar properties hold with r and l interchanged. Also, $(-)^{**} = ((-)^*)^*$ is naturally isomorphic to the identity functor.

REMARK 2.2.4. The functor $(-)^*$ above is more than a contravariant equivalence. It defines an isomorphism of one category with a full representative subcategory of another. (Recall that a representative subcategory contains a member of every isomorphism class of objects in the larger category. If it is also full, its inclusion is an equivalence.)

In view of the proposition, we may think of $(\mathscr{A}'_r, \mathscr{E}'_r)$ as a dual of $(\mathscr{A}'_l, \mathscr{E}'_l)$, via the contravariant functor $(-)^*$. More interesting to note is the possibility of using $(-)^*$ to construct new exact categories inside $(\mathscr{A}'_r, \mathscr{E}'_r)$ as duals (i.e., strict images under $(-)^*$) of exact subcategories of $(\mathscr{A}'_l, \mathscr{E}'_l)$. More precisely, if $(\mathscr{B}, \mathscr{F})$ is a full exact subcategory of $(\mathscr{A}'_l, \mathscr{E}'_l)$, let $\mathscr{B}^*$ denote the (strict) image of $\mathscr{B}$ under $(-)^*$ in $\mathscr{A}'_r$, and define $\mathscr{F}^* \subseteq \mathscr{E}'_r$ similarly. Then, $(-)^*$ induces a contravariant equivalence $(\mathscr{B}, \mathscr{F}) \xrightarrow{\sim} (\mathscr{B}^*, \mathscr{F}^*)$. That is, the subcategory $(\mathscr{B}^*, \mathscr{F}^*) \subseteq (\mathscr{A}', \mathscr{E}')$ is equivalent to the contravariant dual $(\mathscr{B}, \mathscr{F})^{\mathrm{op}}$ of $(\mathscr{B}, \mathscr{F})$.

In general, the notation M^h applies to both the left- and right-module case. If it is really needed to distinguish the left case from the right, subscripts may be used, with the convention $M_h = M^{-h}$. This notation is implicit in Section 2.1, and we use it below.

2.2B. Balanced subcategories. A property we are looking for in exact subcategories of $(\mathscr{A}', \mathscr{E}')$ $(= (\mathscr{A}'_r, \mathscr{E}'_r))$ is *balance*. More generally, if $(\mathscr{B}, \mathscr{F})$ is a full[21] exact subcategory of $(\mathscr{A}', \mathscr{E}')$, we say $(\mathscr{B}, \mathscr{F})$ is *balanced* provided, for each $M \in \mathscr{B}$ and $h \in \mathbb{Z}$, the objects M^h and M/M^h are both isomorphic to objects in $\mathscr{B}$, and the sequence $(M^h \to M \to M/M^h)$ (which is obviously in $\mathscr{E}'$) is isomorphic to a member of $\mathscr{F}$. Here, we are using the defining properties of $\mathscr{A}'$ and $\mathscr{E}'$ introduced in Theorems 2.2.1 and 2.2.2 above. Of course, $(\mathscr{A}', \mathscr{E}')$ itself is balanced.

We have not assumed in the definition above that $\mathscr{B}$ is closed under isomorphism in $\mathscr{A}'$. However, $(\mathscr{B}, \mathscr{F})$ is balanced if and only if its strict closure in $(\mathscr{A}', \mathscr{E}')$ is balanced. See footnote 21 above for further discussion.

A similar discussion applies for $(\mathscr{A}'_l, \mathscr{E}'_l)$.

---

[21] We could use a weaker condition than "full" in this definition. All we really need is that, for any $X, Y \in \mathscr{B}$, if $f \in \mathrm{Hom}_{\mathscr{A}'}(X, Y)$ is an isomorphism in $\mathscr{A}'$, then $f \in \mathrm{Hom}_{\mathscr{B}}(X, Y)$. One could call the embedding $\mathscr{B} \to \mathscr{A}'$ *isomorphism-full* under such a condition. Note that the strict closure $\overline{\mathscr{B}}$ of $\mathscr{B}$ in $\mathscr{A}'$ (throwing in all objects in $\mathscr{A}'$ isomorphic to an object in $\mathscr{B}$) makes sense as a category in the presence of this condition, using for morphisms $f : X' \to Y'$ between its objects those morphisms f in $\mathscr{A}'$ for which there are in $\mathscr{A}'$ isomorphisms $e : X \to X'$ and $g : Y' \to Y$ with objects $X, Y \in \mathscr{B}$ and composite efg a morphism in $\mathscr{B}$. As such, the inclusion of $\mathscr{B}$ into $\overline{\mathscr{B}}$ is full and an equivalence of categories, and $\mathscr{B}$ is a subcategory of $\mathscr{A}'$ closed under object isomorphisms.

PROPOSITION 2.2.5. *Suppose* $(\mathscr{B}, \mathscr{F})$ *is a balanced exact subcategory of* $(\mathscr{A}_l', \mathscr{E}_l')$. *If* $(\mathscr{B}^*, \mathscr{F}^*)$ *denotes its strict image in* $(\mathscr{A}_r', \mathscr{E}_r')$ *under* $(-)^*$, *then* $(\mathscr{B}^*, \mathscr{F}^*)$ *is balanced.*

We leave the proof of this result to the reader.

We have already observed in Corollary 2.1.10 and Proposition 2.3.3 below that the exact categories $(\mathscr{A}^\flat(\mathscr{L}), \mathscr{E}^\flat(\mathscr{L}))$ and $(\mathscr{A}_\flat(\mathscr{L}^*), \mathscr{E}_\flat(\mathscr{L}^*))$ are balanced. As a consequence, the following proposition applies to both. Note that $\mathscr{A}^\flat(\mathscr{L})$ and $\mathscr{A}_\flat(\mathscr{L}^*)$ are, respectively, closed under isomorphism of objects in $\mathscr{A}_l'$ and $\mathscr{A}_r'$.

PROPOSITION 2.2.6. *Let* $(\mathscr{A}', \mathscr{E}')$ *denote either* $(\mathscr{A}_r', \mathscr{E}_r')$ *or its left-hand counterpart. Suppose* $(\mathscr{B}, \mathscr{F})$ *is a balanced exact subcategory of* $(\mathscr{A}', \mathscr{E}')$. *If* $M \in \mathscr{B}$, $h \in \mathbb{Z}$, *then the sequence*

$$M^{h-1} \longrightarrow M^h \longrightarrow M^h/M^{h-1} \tag{2.2.3}$$

belongs to $\mathscr{E}'$ *and is isomorphic to a sequence in* $\mathscr{F}$. *In fact, the sequence belongs to* $\mathscr{F}$, *if* $\mathscr{B}$ *is closed under isomorphisms in* $\mathscr{A}'$.

PROOF. We treat the right-hand case $(\mathscr{A}', \mathscr{E}') = (\mathscr{A}_r', \mathscr{E}_r')$, which also implies the left-hand version of the proposition.

Fix M in $\mathscr{B}$ and let h be an integer. Applying the definition of balanced twice, we obtain that $M^h, (M^h)^{h-1}$, and $M^h/(M^h)^{h-1}$ all belong to $\mathscr{B}$. Also, $((M^h)^{h-1} \to M^h \to M^h/M^{h-1}) \in \mathscr{F}$. Using the definition of $\mathscr{A}$ in Theorem 2.2.1 we obtain the following commutative diagram in $\mathscr{A}$ with all vertical arrows isomorphisms.

$$\begin{array}{ccccc} (M^h)^{h-1} & \longrightarrow & M^h & \longrightarrow & M^h/(M^h)^{h-1} \\ \wr\downarrow & & \| & & \wr\downarrow \\ M^{h-1} & \longrightarrow & M^h & \longrightarrow & M^h/M^{h-1}. \end{array}$$

Since $\mathscr{A}'$ is closed under isomorphism (and full) in $\mathscr{A}$, this diagram and its isomorphic images are all in $\mathscr{A}'$. The top row is in $\mathscr{B}$ and, in fact, belongs to $\mathscr{F}$. The bottom rows belongs to $\mathscr{E}'$ and are isomorphic to the top row (as a diagram). This proves the first assertion of the proposition.

To prove the second, assume $\mathscr{B}$ is closed under isomorphism. Then, the objects and morphisms in the bottom row all belong to $\mathscr{B}$, so the bottom row belongs to $\mathscr{F}$. (Recall that the conflations in any exact category are required to be closed under isomorphism as part of their definition.) This completes the proof of the proposition. □

The last part of the proof above implicitly used the fact that $\mathscr{B}$ is full, assumed in the definition of balanced. It would have been enough to assume the weaker property in the footnote 21 of this chapter.

2.3. The dual category of $(\mathscr{A}^\flat(\mathscr{L}), \mathscr{E}^\flat(\mathscr{L}))$**.** We now turn back to the generic Hecke algebra $\mathcal{H}$ associated with a standard finite Coxeter system (W, S, L). We first consider the right $\mathcal{H}$-module analog $(\mathscr{A}_r(\mathscr{L}^*), \mathscr{E}_r(\mathscr{L}^*))$ of $(\mathscr{A}_l(\mathscr{L}), \mathscr{E}_l(\mathscr{L}))$ defined in display (2.1.3) (see (2.3.2) below).

Thus, $\mathscr{A}_r$ is the full subcategory of right $\mathcal{H}$-modules which are finite and torsion-free over $\mathscr{K} = \mathcal{Z}$. We often denote the category $\mathscr{A}_r$ simply as $\mathscr{A}$ when no confusion can arise.

NOTATION 2.3.1. It will be convenient in this section to use the height function $\mathfrak{h} = \mathfrak{h}_r$, which is the negative of the height function $\mathfrak{h}_l$ introduced in Notation 2.1.2. Thus, the symbol $\mathfrak{h}$ in this section, without further comment, denotes $\mathfrak{h}_r = -\mathfrak{h}_l$; it takes positive values. It is also compatible with the preorder $\leq^{\rm op}_{LR}$ and partial order $\leq^{\rm op}_{L}$ on Ω.

For an integer h, let $\mathscr{L}^*_h$ be the full subcategory of $\mathscr{A}$ consisting of right $\mathcal{H}$-modules N such that N is a finite direct sum of modules which are isomorphic to dual left-cell modules $S_\omega := S(\omega)^*$ of height h. Then let $\mathscr{L}^* := \bigcup_h \mathscr{L}^*_h$. We remind the reader that, by a slight abuse of notation, a left-cell module $S(\omega)$ has a negative height $h = \mathfrak{h}_l(\omega) < 0$, and the dual left-cell module S_ω has height $-h = \mathfrak{h}_r(\omega) > 0$.

Let $\mathscr{A}(\mathscr{L}^*)$ be the full additive subcategory of mod-$\mathcal{H}$ with objects M whose $\mathfrak{h}_r$-filtration

$$0 = M^0 \subseteq M^1 \subseteq M^2 \subseteq \cdots \subseteq M^{n-1} \subseteq M^n = M \tag{2.3.1}$$

has the property that, for any integer $h > 0$, M^h/M^{h-1} is an object in $\mathscr{L}^*_h$. See [**DPS17**, §3], which also allowed $h = 0$ (but did not use this case). Like the refinement in footnote 19, such a filtration will be informally called *a dual left-cell filtration.*

By analogy with [**DPS17**, §3], let $\mathscr{E}(\mathscr{L}^*)$ consist of those sequences $(M \to N \to P)$ in $\mathscr{A}$ such that, for every h,

$$0 \to M^h/M^{h-1} \to N^h/N^{h-1} \to P^h/P^{h-1} \to 0$$

is a split short exact sequence in the additive category $\mathscr{L}^*_h$. Then, as a special case of [**DPS17**, Th. 3.9], we have

(2.3.2) $(\mathscr{A}(\mathscr{L}^*), \mathscr{E}(\mathscr{L}^*))$ is an exact category.

Recall from Section 2.2 the standard $\mathcal{Z}$-duality functor $(-)^*$ discussed in the context of the exact categories used here. Thus, if $N \in \mathscr{A}_l(\mathscr{L})$, then $N^* \in \mathscr{A}_r(\mathscr{L}^*)$ and $(N_{h-1}/N_h)^* \cong (N^*)^h/(N^*)^{h-1}$, for each $h \in \mathbb{Z}$. More precisely, if N has a (left-cell) $\mathfrak{h}_l$-filtration as in (2.1.2), then N^* has a (dual left-cell) $\mathfrak{h}_r$-filtration

$$0 = (N^*)^0 \subseteq (N^*)^1 \subseteq (N^*)^2 \subseteq \cdots \subseteq (N^{n-1})^* \subseteq (N^*)^n = N$$

as given in (2.3.1), where $(N^*)^i = (N/N_i)^*$ so that $(N^*)^h/(N^*)^{h-1} \cong (N_{h-1}/N_h)^* \in (\mathscr{L}_{-h})^* = (\mathscr{L}^*)_h$, for all $1 \leq h \leq n$. Hence, N^* is an object of $\mathscr{A}_r(\mathscr{L}^*)$. Similarly, for a conflation $(M \to N \to P) \in \mathscr{E}_l(\mathscr{L})$, we have $(P^* \to N^* \to M^*) \in \mathscr{E}_r(\mathscr{L}^*)$, since $\big((P^*)^h/(P^*)^{h-1} \to (N^*)^h/(N^*)^{h-1} \to (M^*)^h/(M^*)^{h-1}\big)$, which is isomorphic to $\big((P_{h-1}/P_h)^* \to (N_{h-1}/N_h)^* \to (M_{h-1}/M_h)^*\big)$, is split exact, for all $1 \leq h \leq n$.

Hence, the exact category $(\mathscr{A}(\mathscr{L}^*), \mathscr{E}(\mathscr{L}^*))$ is naturally isomorphic to the dual category $(\mathscr{A}_l(\mathscr{L})^*, \mathscr{E}_l(\mathscr{L})^*)$ of $(\mathscr{A}_l(\mathscr{L}), \mathscr{E}_l(\mathscr{L}))$. As in Section 2.2, $\mathscr{A}_l(\mathscr{L})^*$ and $\mathscr{E}_l(\mathscr{L})^*$ both denote strict images (closed under isomorphism).

Let $\mathscr{E}_\flat(\mathscr{L}^*) = \mathscr{E}^\flat(\mathscr{L})^*$ and let $\mathscr{A}_\flat(\mathscr{L}^*) = \mathscr{A}^\flat(\mathscr{L})^*$. We easily prove the following key proposition, motivated by the weaker [**DPS15**, Cor. 4.5] and [**DPS98a**, Th. 2.3.9]. For $J \subseteq S$,

$$x_J\mathcal{H} \cong (\mathcal{H}x_J)^* \tag{2.3.3}$$

by [**DPS98a**, Lem. 2.3.9]. In particular,

$$x_J\mathcal{H} \quad \text{belongs to } \mathscr{A}_\flat(\mathscr{L}^*) \tag{2.3.4}$$

by Corollary 2.1.8.

PROPOSITION 2.3.2. *With the notation above, $(\mathscr{A}(\mathscr{L}^*), \mathscr{E}_\flat(\mathscr{L}^*))$ is an exact category. Also, $(\mathscr{A}_\flat(\mathscr{L}^*), \mathscr{E}_\flat(\mathscr{L}^*))$ is a full exact subcategory of $(\mathscr{A}(\mathscr{L}^*), \mathscr{E}_\flat(\mathscr{L}^*))$ which is closed under extensions. Moreover, for any left cell $\omega \in \Omega$ and any $J \subseteq S$, there is a vanishing*

$$\text{Ext}^1_{\mathscr{E}_\flat(\mathscr{L}^*)}(S_\omega, x_J\mathcal{H}) = 0 \tag{2.3.5}$$

in both of these categories.

PROOF. The category $(\mathscr{A}(\mathscr{L}^*), \mathscr{E}_\flat(\mathscr{L}^*))$ is the (contravariant) dual of the exact category $(\mathscr{A}(\mathscr{L}), \mathscr{E}^\flat(\mathscr{L}))$. Hence, it is exact, since $(\mathscr{A}(\mathscr{L}), \mathscr{E}^\flat(\mathscr{L}))$ is exact by Proposition 2.1.3. Similarly, $(\mathscr{A}_\flat(\mathscr{L}^*), \mathscr{E}_\flat(\mathscr{L}^*))$ is exact, by Corollary 2.1.8. Clearly, $(\mathscr{A}_\flat(\mathscr{L}^*), \mathscr{E}_\flat(\mathscr{L}^*))$ is a full exact subcategory of $(\mathscr{A}(\mathscr{L}^*), \mathscr{E}_\flat(\mathscr{L}^*))$ and is closed under extensions by Proposition 2.1.7. The displayed vanishing, in the case of $(\mathscr{A}(\mathscr{L}^*), \mathscr{E}_\flat(\mathscr{L}^*))$, follows from Theorem 2.1.6 and duality. However, we have already noted that $(\mathscr{A}_\flat(\mathscr{L}^*), \mathscr{E}_\flat(\mathscr{L}^*))$ is an exact subcategory of $(\mathscr{A}(\mathscr{L}^*), \mathscr{E}_\flat(\mathscr{L}^*))$, and $\mathscr{A}_\flat(\mathscr{L}^*)$ is full in $\mathscr{A}(\mathscr{L}^*)$, so the displayed vanishing also holds for the subcategory $(\mathscr{A}_\flat(\mathscr{L}^*), \mathscr{E}_\flat(\mathscr{L}^*))$. This completes the proof of the proposition. □

We conclude this section with the following proposition, a version of Corollary 2.1.10 in our present dual setting. The proposition is essentially equivalent to that corollary, and can be deduced from it using the duality functor $(-)^*$. We omit the details. An alternative argument can be given using the duality functors for part (a); then using the 3×3-Lemma. See the proof of Proposition 2.1.10(b).

PROPOSITION 2.3.3.

(a) *If $M \in \mathscr{A}_\flat(\mathscr{L}^*)$ has the $\mathfrak{h}$-filtration (2.3.1), and $a \in \mathbb{Z}$ with $0 \leq a \leq n$. Then,*

$$(M^a \to M \to M/M^a) \in \mathscr{E}_\flat(\mathscr{L}^*).$$

In particular, each inclusion $M^{h-1} \subseteq M^h$ gives a conflation $(M^{h-1} \to M^h \to M^h/M^{h-1}) \in \mathscr{E}^\flat(\mathscr{L})$.

(b) *Suppose $(M \to N \to P) \in \mathscr{E}_\flat(\mathscr{L}^*)$. If $h \in \mathbb{Z}$, then both $(M^h \to N^h \to P^h)$ and $(M/M^h \to N/N^h \to P/P^h)$ belong to $\mathscr{E}_\flat(\mathscr{L}^*)$.*

2.4. Another abstract setting. Our main goal in this section is to prove Theorem 2.4.10, which builds a kind of injective envelope for certain exact categories, improving upon [**DPS17**, Th. 4.7]. This goal is achieved by refining the abstract framework used in the earlier paper. At least some abstraction is desirable. The category $(\mathscr{A}_\flat(\mathscr{L}^*), \mathscr{E}_\flat(\mathscr{L}^*))$ used in the main Theorem 3.3.1 is not easy to investigate directly, but it inherits many properties from its "dual" $(\mathscr{A}^\flat(\mathscr{L}), \mathscr{E}^\flat(\mathscr{L}))$. Those dual properties are often most easily checked in the abstract setting of an exact category satisfying Conditions 2.4.2 below.

We follow the notation of [**DPS17**, Assump. 3.2 and Constr. 3.8].[22] Thus, $\mathscr{K}$ is a Noetherian integral domain with fraction field K. Also, H is a finite and torsion-free $\mathscr{K}$-algebra such that H_K is semisimple. Thus, H_K has a finite set Λ of isomorphism classes of irreducible modules. In this section, "modules" for H and H_K are taken as right modules, We are given a "height function" $\mathfrak{h} : \Lambda \to \mathbb{Z}$, as

[22]In [**DPS17**, Const.3.8] at the end of the first paragraph, it should be assumed that M^j/M^{j-1} is isomorphic to, not necessarily equal to, an object in $\mathscr{S}$ for all j. This is not a problem usually, since one often requires that $\mathscr{S}$ be closed under isomorphism.

defined in Appendix B. We will assume $\mathfrak{h}$ takes positive values only on Λ, unless noted otherwise.

Often $\mathscr{K}$ is the algebra $\mathcal{Z} = \mathbb{Z}[t, t^{-1}]$ of Laurent polynomials, and $H = \mathcal{H}$ is a generic Hecke algebra (as defined in Section 1.2 above and used in Sections 2.1 and 2.3). To pass from the height function domain Ω used in Sections 2.1 and 2.3 to the set Λ above, see Notation 2.1.2 and Remark B2.7. We will later apply the abstract results of this section to Section 2.3, where the height function $\mathfrak{h} = \mathfrak{h}_r$ also takes positive values.

Let $(\mathscr{A}, \mathscr{E})$ be the exact category given in Theorem 2.2.1. In particular, $\mathscr{A}$ is the full subcategory of mod-H consisting of right H-modules which are finite and torsion-free over $\mathscr{K}$. For each integer i, let $\mathscr{S}_i$ be a set of objects in a full additive subcategory of $\mathscr{A}$. For each $S \in \mathscr{S}_i$, it is required that S_K be a direct sum of finitely many irreducible modules of the same height i. If i is not in the image of $\mathfrak{h}$, set $\mathscr{S}_i = [0]$, a full subcategory of zero objects in $\mathscr{A}$. Let $\mathscr{S} := \bigcup_i \mathscr{S}_i$. Finally, there is an exact category $(\mathscr{A}(\mathscr{S}), \mathscr{E}(\mathscr{S}))$ which is constructed in [**DPS17**, Constr. 3.8]. (This construction with some details has already been referenced in a left-module setting in (2.1.8). See also (2.3.2) for a right-module setting.) The construction shows that $\mathscr{A}(\mathscr{S})$ is a full subcategory of $\mathscr{A}$ (and thus of mod-H), and it shows that $\mathscr{E}(\mathscr{S})$ is contained in $\mathscr{E}$.

REMARK 2.4.1. For convenience, we will assume that each category $\mathscr{S}_i$, and thus $\mathscr{A}(\mathscr{S})$, is closed under isomorphism in mod-H.

In particular, $(\mathscr{A}(\mathscr{L}^*), \mathscr{E}(\mathscr{L}^*))$, in the Section 2.3 setting, is an example of an exact category $(\mathscr{A}(\mathscr{S}), \mathscr{E}(\mathscr{S}))$.

We will be interested in exact categories $(\mathscr{A}_\flat, \mathscr{E}_\flat)$, which satisfy the following

CONDITIONS 2.4.2. $(\mathscr{A}_\flat, \mathscr{E}_\flat)$ is a full exact subcategory of $(\mathscr{A}(\mathscr{S}), \mathscr{E}(\mathscr{S}))$, with $\mathscr{A}_\flat \hookrightarrow \mathscr{A}(\mathscr{S})$ an inclusion of $\mathscr{K}$-categories, such that the following hold:

(1) $\mathscr{A}_\flat$ is object-closed under isomorphisms in $\mathscr{A}(\mathscr{S})$ (and thus, by Remark 2.4.1 above, in mod-H).
(2) All objects of $\mathscr{S}$ are contained in $\mathscr{A}_\flat$.
(3) If $M \in \mathscr{A}_\flat$ and h is an integer, then M^h, M^{h-1} and M^h/M^{h-1} belong to $\mathscr{A}_\flat$, and $(M^{h-1} \to M^h \to M^h/M^{h-1})$ belongs to $\mathscr{E}_\flat$.
(4) For $M, N, P \in \mathscr{A}_\flat$, if $(M \to N \to P) \in \mathscr{E}_\flat$, then $(M^h \to N^h \to P^h) \in \mathscr{E}_\flat$ for all integers h.

EXAMPLE 2.4.3.

(a) The exact category $(\mathscr{A}(\mathscr{S}), \mathscr{E}(\mathscr{S}))$, a full exact subcategory of itself, satisfies the conditions of Conditions 2.4.2. In particular, in the setting of Section 2.3, $(\mathscr{A}(\mathscr{L}^*), \mathscr{E}(\mathscr{L}^*))$ provides an example.

(b) Another example is $(\mathscr{A}_\flat(\mathscr{L}^*), \mathscr{E}_\flat(\mathscr{L}^*))$, constructed in Section 2.3 and viewed as a full exact subcategory of $(\mathscr{A}(\mathscr{L}^*), \mathscr{E}(\mathscr{L}^*))$. Conditions (1) and (2) are clear, while conditions (3) and (4) (for $H = \mathcal{H}$) are explicit in Proposition 2.3.3.

(c) The exact category $(\mathscr{A}(\mathscr{L}^*), \mathscr{E}_\flat(\mathscr{L}^*))$ discussed in Proposition 2.3.2 does **not** satisfy the Conditions 2.4.2. In fact, condition (3) fails to hold in general.

We now state several lemmas on $\mathrm{Ext}^1_{\mathscr{E}_\flat}$-vanishing which hold for an exact category $(\mathscr{A}_\flat, \mathscr{E}_\flat)$ satisfying the above conditions. Lemma 2.4.5 is difficult, while the remaining ones are more routine generalizations of known results for the exact categories $(\mathscr{A}(\mathscr{S}), \mathscr{E}(\mathscr{S}))$ studied and proved in [**DPS17**, §4]. Thus, we give a detailed proof of Lemma 2.4.5, and just brief comments on the remaining ones. Everything is pulled together in Theorem 2.4.10.

Throughout the rest of this section, $(\mathscr{A}_\flat, \mathscr{E}_\flat)$ denotes a fixed exact category which satisfies Conditions 2.4.2.

LEMMA 2.4.4. *If $M, N \in \mathscr{A}_\flat$, then the natural maps*

$$\operatorname{Ext}^1_{\mathscr{E}_\flat}(M,N) \to \operatorname{Ext}^1_{\mathscr{E}(\mathscr{S})}(M,N) \quad \textit{and} \quad \operatorname{Ext}^1_{\mathscr{E}_\flat}(M,N) \to \operatorname{Ext}^1_H(M,N),$$

as well as $\operatorname{Ext}^1_{\mathscr{E}(\mathscr{S})}(M,N) \to \operatorname{Ext}^1_H(M,N)$, are inclusions of $\mathscr{K}$-modules.

The above lemma follows from the discussion of Strategy 1.3.1.

LEMMA 2.4.5. *Let h be an integer. Let $S \in \mathscr{S}_h$ and $M \in \mathscr{A}_\flat$, and suppose the map*

$$\operatorname{Hom}_{\mathscr{A}_\flat}(S, M^h/M^{h-1}) \longrightarrow \operatorname{Ext}^1_{\mathscr{E}_\flat}(S, M^{h-1})$$

defined by pull-back is surjective. Then, $\operatorname{Ext}^1_{\mathscr{E}_\flat}(S, M) = 0$.

PROOF. First, we claim it is sufficient to treat the case $M = M^h$: Given $(M \to E \to S) \in \mathscr{E}_\flat$, $(M^h \to E^h \to S) \in \mathscr{E}_\flat$ by Condition 2.4.2(4). The commutative diagram

$$\begin{array}{ccccc} M & \longrightarrow & E & \longrightarrow & S \\ \cup\uparrow & & \uparrow & & \| \\ M^h & \longrightarrow & E^h & \longrightarrow & S \end{array} \tag{2.4.1}$$

identifies the original sequence $(M \to E \to S)$ as a push-out of $(M^h \to E^h \to S)$ in mod-H. Since we have full subcategory inclusions $\mathscr{A}_\flat \subseteq \mathscr{A}(\mathscr{S}) \subseteq \mathscr{A} \subseteq$ mod-H, each map in the diagram (2.4.1) above belongs to $\mathscr{A}_\flat$. Also, each row is a conflation in $(\mathscr{A}_\flat, \mathscr{E}_\flat)$. Thus, the diagram (2.4.1) defines the original sequence $(M \to E \to S)$ as a push-out in $\mathscr{A}_\flat$. That is, the top row $(M \to E \to S)$ is isomorphic to the push-out of the bottom row, regarded as a conflation in $(\mathscr{A}_\flat, \mathscr{E}_\flat)$. Obviously, this push-out is 0 in $\operatorname{Ext}^1_{\mathscr{E}^\flat}(S, M)$ if it is a push-out of 0 in $\operatorname{Ext}^1_{\mathscr{E}_\flat}(S, M^h)$. Thus, from now on, we can assume that $M = M^h$. (A somewhat shorter argument can be made using [**Bu10**, Prop. 2.12], which relates push-outs to rectangles like (2.4.1) in general, in an exact category.)

Let $\chi \in \operatorname{Ext}^1_{\mathscr{E}_\flat}(S, M)$ and let $M \overset{\iota}{\to} E \overset{\delta}{\to} S$ be a representative conflation sequence for χ. Without loss of generality, we can assume M is a (right) H-submodule of E, with i its inclusion. Also, M^{h-1} identifies with E^{h-1}. Form the push-out diagram of the deflation $M = M^h \to M^h/M^{h-1}$ and the sequence $M \overset{\iota}{\to} E \overset{\delta}{\to} S$, to get a commutative diagram in the abelian category mod-H:

$$\begin{array}{ccccc} M^h/M^{h-1} & \longrightarrow & E/E^{h-1} & \longrightarrow & S \\ \uparrow & & \uparrow & & \| \\ M & \xrightarrow{\subseteq} & E & \longrightarrow & S \\ \uparrow & & \uparrow & & \\ M^{h-1} & \xrightarrow{=} & E^{h-1} & & \end{array} \tag{2.4.2}$$

with exact rows and columns (after adding 0 maps at the beginning and end of each row). Also, all objects in the diagram are in $\mathscr{A}_\flat$, and each-three term row or column is in $\mathscr{E}_\flat$. (The only non-obvious case, given our hypothesis on $(\mathscr{A}_\flat, \mathscr{E}_\flat)$, is the top row. However, this row is also in $\mathscr{E}_\flat$, since the push-out of an inflation is an inflation; see [**DPS17**, A.1(a)] which is based on [**K90**].)

In particular, the top row is defined by an element of

$$\mathrm{Ext}^1_{\mathscr{E}_\flat}(S, M^h/M^{h-1}) \subseteq \mathrm{Ext}^1_{\mathscr{E}(\mathscr{S})}(S, M^h/M^{h-1}) = 0.$$

It is, thus, split by an H-module map $\alpha : S \to E/E^{h-1}$. Let j denote the map $M^h/M^{h-1} \to E/E^{h-1}$ in the top row, and let $\beta : S \to M^h/M^{h-1}$ be any (right) H-module homomorphism. Fullness of $\mathscr{A}_\flat$ in $\mathscr{A}(\mathscr{S})$, and of the latter in mod-H, guarantees that β is a map in the additive category defined by $\mathscr{A}_\flat$. In particular, the pull-back through β of the element of $\mathrm{Ext}^1_{\mathscr{E}_\flat}(M^h/M^{h-1}, M^{h-1})$ defining the left-hand column is an element of $\mathrm{Ext}^1_{\mathscr{E}_\flat}(S, M^{h-1})$. Indeed, this process defines the natural homomorphism

$$\mathrm{Hom}_{\mathscr{A}_\flat}(S, M^h/M^{h-1}) \longrightarrow \mathrm{Ext}^1_{\mathscr{E}_\flat}(S, M^{h-1}),$$

which our hypothesis in the lemma guarantees to be surjective. Notice that the image of β may also be described as the image of $j \circ \beta$ under a similarly defined map

$$\mathrm{Hom}_{\mathscr{A}_\flat}(S, E/E^{h-1}) \longrightarrow \mathrm{Ext}^1_{\mathscr{E}_\flat}(S, E^{h-1}) = \mathrm{Ext}^1_{\mathscr{E}_\flat}(S, M^{h-1}).$$

(The top left square in diagram (2.4.2) defines M as a pull-back with top row j. Extend it to the left by a pull-back square with top row β, and consider the composite $j \circ \beta$.)

Observe that the map $\alpha + j \circ \beta : S \to E/E^{h-1}$ is still a splitting map of the top row in (2.4.2). Now choose β so that its image in $\mathrm{Ext}^1_{\mathscr{E}_\flat}(S, M^{h-1})$ is the negative of the image of α under the display above, using its identification of E^{h-1} with M^{h-1}. With this choice, the image of $\alpha + j \circ \beta$ is 0. This implies that the map $\alpha + j \circ \beta : S \to E/E^{h-1}$ can be lifted to a map $\alpha' : S \to E$ of right H-modules. That is, the composition of α' with $E \to E/E^{h-1}$ is equal to $\alpha + j \circ \beta$. Composing further with $E/E^{h-1} \to S$ gives the identity map of S to itself.

We have thus constructed a splitting map of the map $E \to S$ in mod-H. That is, the short exact sequence defined by χ is 0 in $\mathrm{Ext}^1_H(S, M)$. It is, therefore, also 0 in $\mathrm{Ext}^1_{\mathscr{E}_\flat}(S, M)$.

This completes the proof of the lemma. □

The following result is a generalization of [**DPS17**, Lem. 4.1]. See [**DPS17**, Prop. A.2] for the definition of $\mathrm{Ext}^1_{\mathscr{E}}$-groups.

LEMMA 2.4.6. *Let $M, N \in \mathscr{A}_\flat$ and $h \in \mathbb{Z}$.*

(a) *There is a natural isomorphism* $\mathrm{Ext}^1_{\mathscr{E}_\flat}(N^h, M) \cong \mathrm{Ext}^1_{\mathscr{E}_\flat}(N^h, M^h)$.

(b) *In particular, if $S \in \mathscr{S}_h$, we have*

$$\mathrm{Ext}^1_{\mathscr{E}_\flat}(S, M) \cong \mathrm{Ext}^1_{\mathscr{E}_\flat}(S, M^h).$$

(c) *Assume that $S \in \mathscr{S}_h$. If $M = M^h$ and $M^{h-1} = 0$, then* $\mathrm{Ext}^1_{\mathscr{E}_\flat}(S, M) = 0$.

PROOF. Note that part (b) follows from part (a). Part (a) follows from [**DPS17**, Prop. 4.1] and its description of the isomorphism

$$\mathrm{Ext}^1_{\mathscr{E}(\mathscr{S})}(N^h, M) \cong \mathrm{Ext}^1_{\mathscr{E}(\mathscr{S}}(N^h, M^h),$$

together with Condition 2.4.2(4) and Lemma 2.4.4. Since the proof will be quoted later, we provide some of the details.

First, we give a map

$$\mathrm{Ext}^1_{\mathscr{E}(\mathscr{S})}(N^h, M) \xrightarrow{\phi} \mathrm{Ext}^1_{\mathscr{E}(\mathscr{S})}(N^h, M^h)$$

and show that it is an isomorphism by explicitly giving an inverse. The map ϕ is obtained by applying the functor $(-)^h$ to a conflation. The inverse of ϕ is defined by push-out through $M^h \to M$. The restriction of ϕ to $\mathscr{E}_\flat$ makes sense using Condition 2.4.2(4) and Lemma 2.4.4.

This push-out also make sense in the exact category $(\mathscr{A}_\flat, \mathscr{E}_\flat)$ and agrees with the push-out in $(\mathscr{A}(\mathscr{S}), \mathscr{E}(\mathscr{S}))$. (This is an elementary argument. The agreement of push-outs also follows from [**Bu10**, Prop. 2.12].) Thus, the isomorphism ϕ induces a corresponding isomorphism when $\mathscr{E}(\mathscr{S})$ is replaced by $\mathscr{E}_\flat$.

Finally, part (c) (which is independent of (a) and (b)) follows from [**DPS17**, Lem. 4.1(c)] and Lemma 2.4.4 above. □

LEMMA 2.4.7. *Let $M \in \mathscr{A}_\flat$. If $S \in \mathscr{S}_h$, then*

$$\mathrm{Ext}^1_{\mathscr{E}_\flat}(S, M^{h-1}) \hookrightarrow \mathrm{Ext}^1_{\mathscr{E}(\mathscr{S})}(S, M^{h-1}) \cong \mathrm{Ext}^1_H(S, M^{h-1}).$$

PROOF. The containment $\mathrm{Ext}^1_{\mathscr{E}_\flat} \hookrightarrow \mathrm{Ext}^1_{\mathscr{E}(\mathscr{S})}$ in the display is obtained from Lemma 2.4.4. The isomorphism in the display is from [**DPS17**, Lem. 4.2]. □

LEMMA 2.4.8. *Let $S \in \mathscr{S}_h$, and let $M \in \mathscr{A}_\flat$. Suppose that $\mathrm{Ext}^1_{\mathscr{E}_\flat}(S, M^{h-1})$ is generated as a $\mathscr{K}$-module by $\epsilon_1, \ldots, \epsilon_n$ (regarded as elements $\mathrm{Ext}^1_H(S, M^{h-1})$ using the inclusion above). Let $M^{h-1} \to N \to S^{\oplus n}$ represent the element of $\mathrm{Ext}^1_H(S^{\oplus n}, M^{h-1})$ corresponding to $\chi := \epsilon_1 \oplus \cdots \oplus \epsilon_n$ above. Suppose there is a commutative diagram in* mod-H

$$\begin{array}{ccccc} M^{h-1} & \longrightarrow & N & \longrightarrow & S^{\oplus n} \\ \| & & \downarrow & & \downarrow f \\ M^{h-1} & \longrightarrow & M^h & \longrightarrow & M^h/M^{h-1}. \end{array}$$

Then, $\mathrm{Ext}^1_{\mathscr{E}_\flat}(S, M) = 0$.

PROOF. By the construction of addition in $\mathrm{Ext}^1_H(S, M^{h-1})$, for $1 \le j \le n$, there is an inclusion $\iota_j : S \to S^{\oplus n}$, inducing by pull-back a map $\mathrm{Ext}^1_H(S, M) \longrightarrow \mathrm{Ext}^1_H(S^{\oplus n}, M^{h-1})$ taking χ to ϵ_j. This gives the top rectangle in the commutative diagram below, with top row representing ϵ_j.

$$\begin{array}{ccccc} M^{h-1} & \longrightarrow & N_j & \longrightarrow & S \\ \| & & \downarrow & & \downarrow \\ M^{h-1} & \longrightarrow & N & \longrightarrow & S^{\oplus n} \\ \| & & \downarrow & & \downarrow f \\ M^{h-1} & \longrightarrow & M^h & \longrightarrow & M^h/M^{h-1}. \end{array}$$

Thus, the conflation corresponding to ϵ_j may be obtained by a pull-back of $M^{h-1} \to M^h \to M^h/M^{h-1}$ using the evident composite

$$g_j : S \to S^{\oplus n} \xrightarrow{f} M^h/M^{h-1}.$$

Consequently, the image of $g_j \in \mathrm{Hom}_{\mathscr{A}_\flat}(S, M^h/M^{h-1}) \subseteq \mathrm{Hom}_H(S, M^h/M^{h+1})$ under the connecting homomorphism to $\mathrm{Ext}^1_{\mathscr{E}_\flat}(S, M^{h-1}) \subseteq \mathrm{Ext}^1_H(S, M^{h-1})$ is ϵ_j. Since j was arbitrary, the connecting homomorphism is surjective, so $\mathrm{Ext}^1_{\mathscr{E}_\flat}(S, M) = 0$ by Lemma 2.4.5. □

The next result will assume the following hypothesis for $\mathscr{S}$:

FINITENESS HYPOTHESIS 2.4.9. *For any integer h, $\mathscr{S}_h$ is strictly generated as an additive category by a finite set $F(h)$, i.e., every object in $\mathscr{S}_h$ is isomorphic to a finite direct sum of objects chosen from $F(h)$.*

THEOREM 2.4.10. *Suppose $(\mathscr{A}_\flat, \mathscr{E}_\flat)$ is a full exact subcategory of $(\mathscr{A}(\mathscr{S}), \mathscr{E}(\mathscr{S}))$ satisfying conditions (1)–(4) in Conditions 2.4.2. Assume $\mathscr{S}$ satisfies Hypothesis 2.4.9. Then, given $M \in \mathscr{A}_\flat$, there exists an object $X = X_M$ in $\mathscr{A}_\flat$ and an inflation $M \xrightarrow{i} X$ in $(\mathscr{A}_\flat, \mathscr{E}_\flat)$ such that $\mathrm{Ext}^1_{\mathscr{E}_\flat}(S, X) = 0$, for all $S \in \mathscr{S}$. In addition, if h is chosen minimal such that $M^{h-1} \neq 0$, it may be assumed that the inflation $M \xrightarrow{i} X$ induces an isomorphism $M^{h-1} \cong X^{h-1}$.*

PROOF. Without loss of generality, we can assume that $M \neq 0$ and that $\mathrm{Ext}^1_{\mathscr{E}_\flat}(S, M) \neq 0$, for some $S \in \mathscr{S}$. Choose an integer h minimal with such a non-vanishing occurring for some $S \in \mathscr{S}_h$. Note that $M^{h-1} \neq 0$ by Lemma 2.4.6(c). We will enlarge M to an object X, closer to the X required in the theorem.

Let $S_1, \ldots, S_m$ be generators for $\mathscr{S}_h$. For each index i, let $\epsilon_{i,1}, \ldots, \epsilon_{i,n_i}$ be a finite set of generators for $\mathrm{Ext}^1_{\mathscr{E}_\flat}(S_i, M^{h-1})$. Form conflations $0 \to M^{h-1} \to Y_i \to S_i^{\oplus n_i} \to 0$ corresponding to $\chi_i := \epsilon_{i,1} \oplus \cdots \oplus \epsilon_{i,n_i} \in \mathrm{Ext}^1_{\mathscr{E}_\flat}(S_i^{\oplus n_i}, M^{h-1})$. Put $\chi := \chi_1 \oplus \cdots \oplus \chi_m$, and let $\chi' \in \mathrm{Ext}^1_{\mathscr{E}_\flat}(M^h/M^{h-1}, M^{h-1})$ correspond to the conflation $M^{h-1} \to M^h \to M^h/M^{h-1}$. Put $Z := \oplus_i S_i^{\oplus n_i} \oplus M^h/M^{h-1}$, and let $M^{h-1} \to X^h \to Z$ correspond to $\chi \oplus \chi'$. Observe that there is a commutative diagram

$$\begin{array}{ccccc} M^{h-1} & \longrightarrow & Y_i & \longrightarrow & S_i^{\oplus n_i} \\ \| & & \downarrow & & \downarrow \\ M^{h-1} & \longrightarrow & X^h & \longrightarrow & Z, \end{array}$$

in which the top row corresponds to χ_i and the bottom row to $\chi \oplus \chi'$. Note that the bottom row belongs to $\mathscr{E}_\flat$. Comparison with Lemma 2.4.8, allowing for the differences in notation, shows $\mathrm{Ext}^1_{\mathscr{E}_\flat}(S_i, X^h) = 0$, for all i. Thus, $\mathrm{Ext}^1_{\mathscr{E}_\flat}(S, X^h) = 0$, for all S in $\mathscr{S}_h$. Note that we have the same vanishing for $S \in \mathscr{S}_j$ with $j < h$, by our choice of h. In all cases, we can replace X^h with any X' containing it with $(X')^h = X^h$ by Lemma 2.4.6.

So far, we have not constructed an object X, only X^h. However, the latter may be viewed as the middle term of an exact sequence of right H-modules $0 \to M^h \to X^h \to S' \to 0$, where $S' := \bigoplus S_i^{\oplus n_i} \in \mathscr{S}_h$. This sequence clearly corresponds to a conflation in $\mathscr{E}_\flat$. (Notice that $X^h \to S'$ is the composition of the deflation $X^h \to Z$ and the split deflation $Z \to S'$, both for $\mathscr{E}_\flat$. It is therefore a deflation for $\mathscr{E}_\flat$. The morphism $M^h \to X^h$ belongs to $\mathscr{A}_\flat$ and is a kernel of $X^h \to S'$ in mod-H, hence is a kernel in $\mathscr{A}_\flat$.) Applying a push-out construction using $M^h \to M$ (see Lemma 2.4.6(a,b) and their proof), we obtain an object X in $\mathscr{A}_\flat$ which contains a copy of M under an inflation and has our constructed X^h as its image under the functor $(-)^h$. In addition, $X^j = M^j$ for $j \leq h-1$.

As noted above, $\mathrm{Ext}^1_{\mathscr{E}_\flat}(-,X)$ vanishes on all objects in $\mathscr{S}_j$ with $j \leq h$. Now repeat the argument with X in the role of M. This requires a bigger h, unless $\mathrm{Ext}^1_{\mathscr{E}_\flat}(S,X)$ already vanishes, for all $S \in \mathscr{S}$. Eventually the process stops, at which point X satisfies all requirements of the theorem. □

REMARKS 2.4.11.

(1) Note that [**DPS17**, Th. 4.7] is the special case of the theorem with $\mathscr{A}_\flat = \mathscr{A}(\mathscr{S})$ and $\mathscr{E}_\flat = \mathscr{E}(\mathscr{S})$. The result was applied in the main theorem [**DPS17**, Th. 4.9] there to build, for each dual left-cell module S_ω, an object X_ω in $\mathscr{A}(\mathscr{S})$ such that $S_\omega \subseteq X_\omega$ (part of the filtration structure required for $\mathscr{A}(\mathscr{S})$) and such that $\mathrm{Ext}^1_{\mathscr{E}(\mathscr{S})}(S, X_\omega) = 0$, for all $S \in \mathscr{S}$. We can apply the same construction here.

(2) The abstract construction of X_M is known as "universal extension" which has been used in [**HN04**, p.635] for a construction of Young modules. We thank the referee for pointing out this reference.

In preparation for the next chapter, we now assume that $H = \mathcal{H}$ is the generic Hecke algebra over $\mathcal{Z} = \mathbb{Z}[t,t^{-1}]$ as discussed in Section 1.2. Thus, the Noetherian domain $\mathscr{K}$ will be specialized to $\mathcal{Z}$.

CONSTRUCTION 2.4.12. *Let $(\mathscr{A}_\flat, \mathscr{E}_\flat) = (\mathscr{A}_\flat(\mathscr{L}^*), \mathscr{E}_\flat(\mathscr{L}^*))$, using a height function $\mathfrak{h} = \mathfrak{h}_r$ as in Notation 2.3.1. (Thus, $\mathfrak{h}$ is compatible with the preorder $\leq^{\mathrm{op}}_{LR}$.) Let Ω be the set of (Kazhdan-Lusztig) left cells in W.*

(a) *Let Ω' be the set of left cells that do not contain the longest element $w_{0,J}$ in a parabolic subgroup W_J, $J \subseteq S$. For $\omega \in \Omega'$, define (as given in Theorem 2.4.10) $X_\omega = X_M$, for $M = S_\omega$, and let $X := \bigoplus_{\omega\in\Omega'} X_\omega$.*

(b) *Define right $\mathcal{H}$-modules*

$$\begin{cases} T := \bigoplus_{J\subseteq S} x_J\mathcal{H} = \bigoplus_{\omega\in\Omega\backslash\Omega'} T_\omega, \\ T^+ := T \bigoplus X, \end{cases} \tag{2.4.3}$$

and form the generic Hecke endo-algebra and its extended version

$$\begin{cases} A := \mathrm{End}_{\mathcal{H}}(T), \\ A^+ := \mathrm{End}_{\mathcal{H}}(T^+). \end{cases} \tag{2.4.4}$$

(c) *Finally, for $\omega \in \Omega\backslash\Omega'$, let $T^+_\omega := x_J\mathcal{H}$, where J is the unique subset of S with $w_{0,J} \in \omega$, and $T^+_\omega = X_\omega$, for $\omega \in \Omega'$. Define the following left A^+-modules*

$$\begin{cases} \Delta(\omega) := \mathrm{Hom}_{\mathcal{H}}(S_\omega, T^+), \\ P(\omega) := \mathrm{Hom}_{\mathcal{H}}(T^+_\omega, T^+), \ \textit{for any } \omega \in \Omega. \end{cases} \tag{2.4.5}$$

As considered in (1.3.2) and also for later application in Section 5.4, we define, for any poset ideal Θ of $(\Omega, \leq_L)$,

$$T^\ddagger_\Theta := \bigoplus_{\omega\in\Theta} (T^+_\omega)^{m_\omega} \quad \text{and} \quad A^\ddagger_\Theta = \mathrm{End}_{\mathcal{H}}(T^\ddagger_\Theta), \tag{2.4.6}$$

and

$$A^\ddagger = \mathrm{End}_{\mathcal{H}}(T^\ddagger_\Omega), \tag{2.4.7}$$

where $\{m_\omega\}_{\omega\in\Theta}$ is any fixed set of positive integers.[23]

[23] We view $T^\ddagger$ as a modification of T^+, rather than T.

REMARK 2.4.13. It should be mentioned that the "enlargement" idea has occurred in various context. For example, Auslander [**A71**] showed that, for any finite-dimensional algebra $R \cong \mathrm{End}_R(R_R)$ over a field (or more generally any semiprimary ring R), the enlarged endomorphism ring $A = \mathrm{End}_R(\oplus_{i=1}^n R/J^i)$, where J is the radical of R and n is the nilpotency of J, has a finite global dimension and $R \cong eAe$ for some idempotent $e \in A$. Dlab and Ringel [**DR89**] further proved that A is quasi-hereditary.[24]

3. The conjecture and its proof

The conjecture in the title of this chapter is Conjecture 3.2.1, which is [**DPS15**, Conj.1.2] and also captures essentially the earlier conjecture [**DPS98a**, Conj. 2.5.2], as discussed in Remarks 3.3.4. A special local case [**DPS15**, Th. 3.6] of the [**DPS15**] conjecture was proved in a Cherednik algebra context, using an Ext^1 vanishing result [**DPS15**, Cor. 5.3] obtained in part from [**GGOR03**]. The latter results seemed to be hard to generalize. Nevertheless, in Chapter 2 here, we have used exact categories to formulate by analogy a tractable global version of the Ext^1 vanishing condition, not mentioning localization or Cherednik algebras. Using it, we are now ready to prove this new version of the conjecture.

For this chapter, we continue the notation in the previous chapter, but, as in Construction 2.4.12, *return to the special case where H is the generic Hecke algebra $\mathcal{H}$ (so that $\mathscr{K} = \mathcal{Z}$) associated to a standard finite Coxeter system (W, S, L).*

3.1. q-permutation modules and their filtrations. Let $\mathcal{H}$ be a generic Hecke algebra over $\mathcal{Z} = \mathbb{Z}[t, t^{-1}]$ associated to a standard finite Coxeter system (W, S, L). Let Ω be the set of left cells of W. The preorders $\leq_L$, $\leq_{LR}$ on W induce partial order $\leq_L$ and preorder $\leq_{LR}$ on Ω; see (1.2.11). We take this opportunity to repeat and highlight the following result, mentioned after Proposition B2.3.

PROPOSITION 3.1.1. *Each two-sided cell module is a direct sum of left-cell modules.*

PROOF. Let $\mathfrak{c}$ be a two-sided cell of W, and set $M(\mathfrak{c}) := \mathcal{H}_{\leq_{LR}\mathfrak{c}}/\mathcal{H}_{<_{LR}\mathfrak{c}}$, the associated two-sided cell module, where $\mathcal{H}_{\leq_{LR}\mathfrak{c}} = \mathcal{H}_{\leq_{LR}w}$ for $w \in \mathfrak{c}$ is defined in (1.2.9). Write $\mathfrak{c} = \cup_{i=1}^m \omega_i$ (a disjoint union) with $\omega_i \in \Omega$. Then, the free $\mathcal{Z}$-module $M(\mathfrak{c})$ is the direct sum $\bigoplus_{j=1}^m S'(\omega_j)$, where each submodule $S'(\omega_j) := \sum_{y\in\omega_j}(\mathcal{Z}c_y + \mathcal{H}_{<_{LR}\mathfrak{c}})/\mathcal{H}_{<_{LR}\mathfrak{c}}$ is a free $\mathcal{Z}$-module. It remains to prove that each $S'(\omega_j)$ is isomorphic to the left-cell module $S(\omega_j) = \mathcal{H}_{\leq_L\omega_j}/\mathcal{H}_{<_L\omega_j}$. This is clear, since the inclusion $\mathcal{H}_{\leq_L\omega_j} \subseteq \mathcal{H}_{\leq_{LR}\mathfrak{c}}$ induces by (1.2.11) an $\mathcal{H}$-module homomorphism $\mathcal{H}_{\leq_L\omega_j} \to M(\mathfrak{c})$ with image $S'(\omega_j)$ and with kernel $\mathcal{H}_{<_L\omega_j} = \mathcal{H}_{\leq_L\omega_j} \cap \mathcal{H}_{<_{LR}\mathfrak{c}}$. (Observe that, for some $z \in \omega_j$, by (1.2.11), if $y <_L z$, then $y <_{LR} z(\in \mathfrak{c})$, and if $y \leq_L z$ and $y <_{LR} w$ for some $w \in \mathfrak{c}$, then $y <_L z$.) $\square$

Observe that the argument for this result uses display (1.2.11). A similar fact holds for right-cell modules or dual left-cell modules. Also, the q-permutation modules $\mathcal{H}x_J$ in Proposition 2.1.4 have natural sections arising from $\mathfrak{h}_l$ which are direct sums of left-cell modules. Once more, this goes back to (1.2.11), as discussed at the end of the proof of Proposition 2.1.4. The (right) q-permutation modules $x_J\mathcal{H}$ have, similarly, filtrations by the direct sum of dual left-cell modules $S_\omega := S(\omega)^*$ defined as the $\mathcal{Z}$-linear dual of the left-cell module $S(\omega)$; see Notation 2.1.2.

[24] We thank Bangming Deng for the references.

We now define a new preorder $\preceq = \preceq_\Omega$ on Ω by setting, for, $\omega, \omega' \in \Omega$,

$$\omega \preceq \omega' \iff \mathfrak{h}(\omega) < \mathfrak{h}(\omega'), \text{ or } \mathfrak{h}(\omega) = \mathfrak{h}(\omega') \text{ and } \omega \sim_{LR} \omega'. \tag{3.1.1}$$

Here, the height function $\mathfrak{h} = \mathfrak{h}_r = -\mathfrak{h}_l$ is chosen so that it is compatible with $\leq_{LR}^{\rm op}$ on Ω. See Notation 2.3.1 and display (B2.1) in Appendix B. In particular, $\mathfrak{h}$ is constant on two-sided cells. Thus, $\preceq$ above is well-defined on Ω and even induces a partial order on the set $\overline{\Omega}$ of two-sided cells.

It also follows from the definition in display (3.1.1) that the equivalence classes of the preorder $\preceq$ are precisely the two-sided cells, satisfying a requirement in [**DPS15**, Conj. 1.2]. More is true along these lines: the preorders $\leq_{LR}^{\rm op}$ and $\preceq$ share a compatible height function $\mathfrak{h}$. In addition, $\leq_{LR}^{\rm op}$ strictly dominates $\preceq$ (see Definition B2.1 in Appendix B)[25] so that any height function compatible with $\preceq$ is also compatible with $\leq_{LR}^{\rm op}$ (as well as with $\leq_{L}^{\rm op}$). See Proposition B2.4 (and Proposition B2.3.) Again, the discussion traces back to (1.2.11).

The *right set* of an element $w \in W$ is defined as the set $\mathscr{R}(x) = \{s \in S \mid xs < x\}$. Then, if $x \leq_L y$, it is known (and elementary) that $\mathscr{R}(x) \supseteq \mathscr{R}(y)$ [**Lus03**, Lem. 8.6]. For example, if $J \subseteq S$, let $w_{0,J}$ be the longest word in W_J. Clearly, $\mathscr{R}(w_{0,J}) = J$. Thus, if $\omega \in \Omega$ contains $w_{0,I}$ and $w_{0,J}$, for subsets I, J of S, then, if $w_{0,I} \leq_L w_{0,J} \leq_L w_{0,I}$, so $I = J$.

We have the following elementary result.

LEMMA 3.1.2. *Let $J \subseteq S$, and let $w_{0,J}$ be the longest word in the parabolic subgroup W_J. The following sets A_J, B_J, C_J are equal:*

(a) $A_J = \{w \in W \mid \mathscr{R}(w) \supseteq J\}$.
(b) $B_J = \{w = uw_{0,J} \mid \ell(w) = \ell(u) + \ell(w_{0,J})\}$.
(c) $C_J = \{w \in W \mid w \leq_L w_{0,J}\}$.

PROOF. Let $w \in A_J$, and write $w = uv$, where $v \in W_J$ and u is a distinguished left W_J-coset representative. For $s \in J$, it follows that $ws < w$, so $vs < v$. Thus, $v = w_{0,J}$ and $w \in B_J$. Conversely, if $w = uw_{0,J} \in B_J$, then it follows easily that u is a distinguished left W_J-coset representative. So $J \subseteq \mathscr{R}(uw_{0,J})$, and $B_J \subseteq A_J$. Thus, $A_J = B_J$. Next, $w \leq_L w_{0,J}$ implies $\mathscr{R}(w) \supseteq \mathscr{R}(w_{0,J})$, so $C_J \subseteq A_J$. Given $w = uw_{0,J} \in B_J$, let $s_{i_1} \cdots s_{i_r}$ be a reduced expression for u. Because $\ell(w) = \ell(u) + \ell(w_{0,J})$, it follows that, if $1 < m \leq r$ and $u_m := s_{i_m} \cdots s_{i_r}$, then $s_{-_{m-1}} u_m w_{0,J} > u_m w_{0,J}$, so that $s_{i_{m-1}} u_m w_{0,J} \leq_L u_m w_{0,J}$. Thus, $w \leq_L w_{0,J}$ and $w \in C_J$. □

THEOREM 3.1.3. *Let $\omega \in \Omega$ and assume that, for some $J \subseteq S$, $w_{0,J} \in \omega$, where $w_{0,J}$ is the longest word in W_J. Then $x_J\mathcal{H}$ has an increasing filtration in the exact category $(\mathscr{A}_\flat(\mathscr{L}^*), \mathscr{E}_\flat(\mathscr{L}^*))$ which satisfies Condition 2.4.2. The sections of the filtration are direct sums of dual left-cell modules $S_{\omega'}$, starting with $S_\omega \subseteq x_J\mathcal{H}$. The other sections have summands $S_{\omega'}$ satisfying $\omega <_L \omega'$ (equivalently, $\omega' >_L^{\rm op} \omega$).*

PROOF. We consider the left module $\mathcal{H}x_J$ and then apply the contravariant $\mathcal{Z}$-duality $(-)^*$ in Section 2.2. We use the construction of the height filtration for $\mathcal{H}x_J$ in the proof of Proposition 2.1.4. The quasi-poset ideal V used in the construction from [**DPS98a**, Lem. 2.3.5] is, by inspection of the latter lemma, the

[25] To check this, suppose that $\omega, \omega' \in \Omega$ and $\omega <_{LR}^{\rm op} \omega'$. Since $\mathfrak{h}$ is compatible with $\leq_{LR}^{\rm op}$, we have $\mathfrak{h}(\omega) < \mathfrak{h}(\omega')$. Thus, $\omega \preceq \omega'$. However, it is not possible for $\omega \sim_{LR} \omega'$, since $\mathfrak{h}$ is constant on two-sided cells. So $\omega \prec \omega'$. Since we already know for $\omega, \omega' \in \Omega$, that $\omega \sim_{LR} \omega'$ implies $\omega \sim_\prec \omega'$, we have proved that $\leq_{LR}^{\rm op}$ strictly dominates $\preceq$.

set A_J in Lemma 3.1.2, which is also the set C_J. In particular, ω generates V as a quasi-poset ideal with preorder $\leq_L$. The construction uses a height function which we will take to be $\mathfrak{h}_l = -\mathfrak{h}$ for a given height function $\mathfrak{h} = \mathfrak{h}_r$ compatible with $\leq_{LR}^{\rm op}$. Thus, $\mathfrak{h}_l = -\mathfrak{h}$ is compatible with $\leq_{LR}$ and, therefore, with $\leq_L$. Hence, ω has a unique maximal height among all left cells in the construction. The left-cell module $S(\omega)$ is the top section in this height filtration of $\mathcal{H}x_J$, and the left-cell modules $S(\omega')$, which must satisfy $\omega' <_L \omega$, must appear in lower sections. The theorem now follows by duality. □

3.2. The conjecture. A main theorem in this monograph is Theorem 3.3.1, which, we will see, implies Conjecture 3.2.1 below. As noted in the introduction to this chapter, this latter conjecture was stated in [**DPS15**, Conj. 1.2], which revised slightly[26] an earlier formulation [**DPS98a**, Conj. 2.5.2]. In assertion (2), the "strict" stratifying system in [**DPS15**] is simply called a stratifying system here (with the same meaning).

Recall that the set Ω of left cells (unequal parameter in the sense of Lusztig [**Lus03**]) in the Weyl group W has a natural preorder $\leq_{LR}^{\rm op}$ which partitions Ω into its "two-sided Kazhdan-Lusztig cells." That is, ω and ω' belong to the same two-sided cell provided $\omega \leq_{LR}^{\rm op} \omega'$ and $\omega' \leq_{LR}^{\rm op} \omega$. Similarly, any preorder $\leq$ on Ω also partitions Ω into "cells." Let $\widetilde{\Omega}$ be the corresponding set of cells. We say that $\leq$ is *strictly compatible* with $\leq_{LR}^{\rm op}$ provided $\overline{\Omega} = \widetilde{\Omega}$. The preorder $\preceq$ in display (3.1.1) is clearly an example of such a preorder.

Recall also that given a left cell $\omega \in \Omega$, S_ω denotes the corresponding dual left-cell module $S(\omega)^* = \mathrm{Hom}_{\mathcal{Z}}(S(\omega), \mathcal{Z})$; see Notation 2.1.2. For the definition of the element x_J, see (2.1.5).

CONJECTURE 3.2.1. *There exist a preorder $\leq$ on the set Ω of left cells in W, strictly compatible with its partition $\overline{\Omega}$ into two-sided cells, and a right $\mathcal{H}$-module X such that the following statements hold, where $T = \bigoplus_{J\subseteq S} x_J\mathcal{H}$:*

(1) *X has a finite filtration with sections of the form S_ω, $\omega \in \Omega$.*
(2) *Let $T^+ := T \oplus X$, $A^+ := \mathrm{End}_{\mathcal{H}}(T^+)$ and, for any $\omega \in \Omega$, put*

$$\begin{cases} \Delta(\omega) := \mathrm{Hom}_{\mathcal{H}}(S_\omega, T^+), \\ P(\omega) := \mathrm{Hom}_{\mathcal{H}}(T^+_\omega, T^+), \ \text{as in (2.4.5)}. \end{cases}$$

Then, for any commutative, Noetherian $\mathcal{Z}$-algebra R, the set $\{\Delta(\omega)_R, P(\omega)_R\}_{\omega\in\Omega}$ is a stratifying system for A^+_R-mod relative to the quasi-poset $(\Omega, \leq)$.

In [**DPS15**, p. 231], in the notion of a stratifying system for A^+-mod, it is assumed, for each $\Delta(\omega)$, that there is a certain projective A^+-module $P(\omega)$ satisfying certain properties. In the notation of the present monograph, reviewed in Section 1.3, the projective modules $P(\omega)$ are explicitly mentioned as part of the data for a stratifying system. The $P(\omega)$'s are required to satisfy the same properties. Finally, as already noted above Remark 1.3.2, the notion of a stratifying system behaves well with respect to base change, so it suffices to prove the conjecture for $R = \mathcal{Z}$ and, thus, for $A^+_R = A^+$.

We remark that the conjecture was checked to be true for all rank 2 cases in [**DPS98a**, §3]. See Remarks 3.3.4 below for more details.

---

[26] See Remarks 3.3.4.

3.3. Proof of the conjecture. Before proceeding, we state and prove our main result, Theorem 3.3.1. The preorder $\preceq$ in the statement of the theorem is defined explicitly in display (3.1.1). It depends upon a suitable height function $\mathfrak{h}$. Also, the exact category $(\mathscr{A}_\flat(\mathscr{L}^*), \mathscr{E}_\flat(\mathscr{L}^*))$ depends on the same height function, as discussed, as does Construction 2.4.12.

THEOREM 3.3.1. *Let* $(\mathscr{A}_\flat, \mathscr{E}_\flat) = (\mathscr{A}_\flat(\mathscr{L}^*), \mathscr{E}_\flat(\mathscr{L}^*))$, T, X, T^+, *and* $\mathfrak{h} = \mathfrak{h}_r$ *be as in Construction* 2.4.12. *For* $\omega \in \Omega$, *let*

$$T^+_\omega = \begin{cases} X_\omega, & \text{if } \omega \in \Omega', \\ x_J\mathcal{H}, & \text{if } w_{0,J} \in \omega. \end{cases}$$

(Recall that X_ω *is constructed from* S_ω*, possibly non-uniquely, as in Theorem* 2.4.10. *Here* Ω' *is defined in Construction* 2.4.12*(a).)*

Form the endomorphism algebra

$$A^+ := \mathrm{End}_{\mathcal{H}}(T^+). \tag{3.3.1}$$

Similarly, for $\omega \in \Omega$, $\Delta(\omega) := \mathrm{Hom}_{\mathcal{H}}(S_\omega, T^+)$ *and* $P(\omega) := \mathrm{Hom}_{\mathcal{H}}(T^+_\omega, T^+)$.

Then $\{\Delta(\omega), P(\omega)\}_{\omega\in\Omega}$ *is a stratifying system for the category* A^+-mod *with respect to the quasi-poset* $(\Omega, \preceq)$.

PROOF. By Remarks 1.3.6(b)&(c), the left A^+-modules $\Delta(\omega)$, $\omega \in \Omega$, are automatically $\mathcal{Z}$-projective, even $\mathcal{Z}$-free. If $\omega \in \Omega'$, X_ω is, by construction, a free $\mathcal{Z}$-module. Also, for $\omega \in \Omega\backslash\Omega'$, $T^+_\omega \cong x_J\mathcal{H}$, for some $J \subseteq S$. By Theorem 3.1.3, T^+_ω has a filtration by dual left-cell modules S_τ, $\tau \in \Omega$. (This follows already from (2.3.4).) Each S_τ is, by construction, $\mathcal{Z}$-free, so T^+_ω is also $\mathcal{Z}$-free.

Now we must verify conditions (1), (2), and (3) in the Stratification Hypothesis 1.3.3. Given $\omega \in \Omega'$, Condition (1) follows from Theorem 2.4.10. If $\omega \in \Omega\backslash\Omega'$, then Condition (1) follows from Theorem 3.1.3.

Suppose that $\omega, \tau \in \Omega$, and $\mathrm{Hom}_{\mathscr{A}_\flat(\mathscr{L}^*)}(S_\omega, T^+_\tau) \neq 0$. Then, setting $K = \mathbb{Q}(t)$, $\mathrm{Hom}_{\mathcal{H}_K}(S_{\omega,K}, S_{\nu,K}) \neq 0$, so for some $\tau \in \Omega$, $\tau \preceq \nu$. Since $S_{\omega,K}$ and $S_{\nu,K}$ share a composition factor, they belong to the same (dual) two-sided cell module. Thus, ω and ν belong to the same two-sided cell for the preorder $\leq_{LR}$, or, equivalently for $\leq^{\mathrm{op}}_{LR}$. Since $\preceq$ is strictly compatible with $\leq^{\mathrm{op}}_{LR}$, ω and ν belong to the same two-sided cell for $\preceq$. It follows that $\tau \preceq \omega$. This proves (2).

Next, we prove condition (3). In case $\omega \in \Omega'$, we can apply the 4-term exact sequence in [**DPS17**, Prop. A.2] and the vanishing of $\mathrm{Ext}^1_{\mathscr{E}_\flat(\mathscr{L}^*)}(S, X)$ in Theorem 2.4.10 to conclude that condition (3) holds in this case. If $\omega \in \Omega\backslash\Omega'$, we can argue in the same way, but using Theorem 3.1.3 and Proposition 2.3.2. □

The following is a corollary of the proof of the theorem, and Construction 2.4.12, using Theorem 2.4.10, Theorem 3.1.3, Proposition 2.3.2, displays (2.3.4), and (3.1.1). The last assertion of the corollary follows from Theorem 2.4.10 ($\omega \in \Omega'$) and the last assertion of Theorem 3.1.3 (when $\omega \notin \Omega'$).

COROLLARY 3.3.2. *Let* $(\mathscr{A}_\flat, \mathscr{E}_\flat) = (\mathscr{A}_\flat(\mathscr{L}^*), \mathscr{E}_\flat(\mathscr{L}^*))$ *and keep the notation of the theorem above. Then,* S_ω *and* T^+_τ *belong to* $\mathscr{A}_\flat$*, for all* $\omega, \tau \in \Omega$. *Also,*

$$\mathrm{Ext}^1_{\mathscr{E}_\flat}(S_\omega, T^+_\tau) = 0.$$

In addition, S_ω *is a submodule (via an inflation in* $\mathscr{E}_\flat$*) of* T^+_ω *and is the lowest nonzero term in its height filtration.*

If Θ is a poset ideal of $(\Omega, \leq_L)$, then $(\Theta, \preceq)$ is a quasi-poset co-ideal of $(\Omega, \preceq)$. Thus, $T^\ddagger_\Theta := \bigoplus_{\omega\in\Theta}(T^+_\omega)^{m_\omega} \in \mathscr{A}_\flat$ with all $m_\omega > 0$, as defined in (2.4.6), satisfies Stratification Hypothesis 1.3.3 relative to the exact category $(\mathscr{A}_\flat, \mathscr{E}_\flat)$. Theorems 1.3.5 and 3.3.1 immediately imply the following.

COROLLARY 3.3.3. *Maintain the notation introduced above and form the endomorphism algebra* $A^\ddagger_\Theta = \mathrm{End}_{\mathcal{H}}(T^\ddagger_\Theta)$ *and* $A^\ddagger_\Theta$*-modules* $\Delta^+(\omega) = \mathrm{Hom}_{\mathcal{H}}(S_\omega, T^\ddagger_\Theta)$ *and* $P^+(\omega) = \mathrm{Hom}_{\mathcal{H}}(T^+_\omega, T^\ddagger_\Theta)$. *Then,* $\{\Delta^+(\omega), P^+(\omega)\}_{\omega\in\Theta}$ *is a stratifying system for the category* $A^\ddagger_\Theta$-mod *with respect to the quasi-poset* $(\Theta, \preceq)$.

We now complete the proof of Conjecture 3.2.1. The above theorem proves part (1) of Conjecture 3.2.1. Part (1) of the conjecture deals with the filtration of X by dual left-cell modules and follows easily from the construction given above. Our construction here is, in fact, much more informative than that envisioned by part (1) of the conjecture. For example, X_ω is built as an object in $\mathscr{A}_\flat(\mathscr{L}^*)$. As such, it has a height filtration with sections that are direct sums of dual left-cell modules. Also, left cells associated to distinct sections always lie in distinct two-sided cells.

Part (2) of the conjecture is phrased in the language of base ring extensions of $\mathcal{Z}$, but it is sufficient to check it over $\mathcal{Z}$, by [**DPS98a**, Lem. 1.2.5(4)]. (Stratifying systems behave well under base change.) Now taking $R = \mathcal{Z}$, part (2) of the conjecture follows from Theorem 3.3.1, provided we know the preorder $\preceq$ has the same equivalence classes as $\leq^{\rm op}_{LR}$. The actual preorder $\preceq$ in Theorem 3.3.1 has this property (and is constructed from $\leq^{\rm op}_{LR}$, sharing the height function $\mathfrak{h}$ with it; see display (3.1.1)).

This completes the proof. □

The following remarks discuss further the difference between the Conjecture proved in Theorem 3.3.1 and its original version given in [**DPS98a**]. It can be skipped by readers interested only in the more recent version as stated in Theorem 3.3.1 or in [**DPS15**, Conj. 1.2].

REMARKS 3.3.4.

(a) When [**DPS98a**, Conj. 2.5.2] was written and checked for all rank 2 cases,[27] the best choice of a preorder $\leq$ seemed to be $\leq^{\rm op}_{LR}$. In the revised conjecture [**DPS15**, Conj.1.2] (identical to Conjecture 3.2.1 above), we realized that this choice of a preorder was too strong, and so it was weakened to require only a preorder with the same equivalence classes as those of $\leq^{\rm op}_{LR}$. As discussed in (a), the preorder $\preceq$ here in Theorem 3.3.1 satisfies this condition and more, sharing also a height function $\mathfrak{h}$ with $\leq^{\rm op}_{LR}$. Moreover, $\preceq$ is a preorder completely determined by its equivalence classes and the height function $\mathfrak{h}$, a specificity giving it some advantage over $\leq^{\rm op}_{LR}$. For example, suppose $\mathfrak{h}$ is given by Lusztig's a-function

[27] We correct some typos around the third display on [**DPS98a**, p. 203]: ... Each module $\widetilde{S}^{\mathscr{R}}_{\{u\}} = S_{0'} \oplus S_{0''}$ and $\widetilde{S}^{\mathscr{R}}_{\{v\}} = S_0 \oplus S_{0''}$ has a length two filtration with two dual left-cell modules $\widetilde{S}_{\omega k}$ as sections, corresponding to left cells $\omega = B, C_1, C_2$, and D (in the notation of (3.5) below). The filtrations are of the form:

$$\widetilde{S}^{\mathscr{R}}_{\{u\}} = \boxed{\begin{matrix} C_1 \\ D \end{matrix}} \quad \text{and} \quad \widetilde{S}^{\mathscr{R}}_{\{v\}} = \boxed{\begin{matrix} B \\ C_2 \end{matrix}},$$

and ...

(see (4.1.2) below), a choice of height function we can certainly make, provided conjectures P4 and P11 hold; see [**Lus03**, 13.6, 14.1]. If ω, ω' belong to distinct 2-sided cells, to decide if $\omega \preceq \omega'$, (equivalent here to $\omega \prec \omega'$), we need only check that $a(\omega) < a(\omega')$—which is a much more computable condition than $\omega \leq^{\rm op}_{LR} \omega'$ or $\omega <^{\rm op}_{LR} \omega'$.

(b) The other main difference between the conjecture in [**DPS98a**] and its version in [**DPS15**] is that we generally worked over $\mathbb{Z}[t^2, t^{-2}]$ in [**DPS98a**] rather than $\mathcal{Z}$. A case for using $\mathcal{Z}$ rather than $\mathbb{Z}[t^2, t^{-2}]$ is that Lusztig's explicit isomorphism $\mathcal{H}_{\mathbb{Q}(t)} \cong \mathbb{Q}(t)W$ holds without exception (for the standard finite Coxeter system, even when G has type ${}^2F_4$). These Hecke algebras are all semisimple and, in fact, split semisimple except for the case ${}^2F_4$. We mention that the explicit isomorphisms can be obtained by applying [**Lus03**, 18.12] to the cases $R = \mathbb{Q}$ and $R = \mathbb{Q}(t)$. This works for any finite W satisfying P1–P15, and semisimplicity follows. The splitting, except for the ${}^2F_4$-case, follows from the isomorphisms $\mathcal{H}_{\mathbb{Q}(t)} \cong \mathbb{Q}(t)W$ above, and the fact that $\mathbb{Q}(t)W$ (in fact, $\mathbb{Q}W$) is split semisimple; see [**B71**], [**BC71**]. The latter reference discusses the Weyl groups arising in the "split" (Chevalley group) cases, also including those W for the "Steinberg twisted" cases. The remaining groups, associated to the Ree and Suzuki groups, are finite dihedral groups, and the only new group W arising, that has not already been considered, is the dihedral group $I_2(8)$ of order 16 associated to ${}^2F_4$. It has $\mathbb{Q}[\sqrt{2}]$ as its unique minimal splitting field.

(c) [**DPS98a**, Conj. 2.5.2] left an ambiguity, removed in [**DPS15**, Conj. 1.2]. In detail, the second sentence of part (2) of [**DPS98a**, Conj. 2.5.2] contains an expression "$\widetilde{A}_{\mathcal{Z}'} = \mathrm{End}_{\widetilde{H}}(T^+_{\mathcal{Z}'})$" which is a misprint, and should have been written "$\widetilde{A}_{\mathcal{Z}'} = \mathrm{End}_{\widetilde{H}}(T^+)_{\mathcal{Z}'}$." However, it is shown in [**DPS98a**, Th. 2.4.4(b)] that "$\mathrm{End}_{\mathcal{H}}(T)_{\mathcal{Z}'} \cong \mathrm{End}_{\mathcal{H}}(T_{\mathcal{Z}'})$," not using the "+" as a superscript on T. Possibly the same isomorphism holds using +, but we only know this to be true if $\mathcal{Z}'$ is flat over $\mathcal{Z}$.

(d) Finally, we observe that while the methods of [**DPS98a**] are useful in studying low rank examples of the Conjecture, they do not provide much help in the ultimate proof of the Conjecture!

3.4. A uniqueness theory. The main aim of this section is to provide a companion uniqueness theorem for (the Morita classes of) the algebras A^+ resulting from Construction 2.4.12. If A^+ is quasi-hereditary, such uniqueness result is available in Rouquier's theory of quasi-hereditary covers (see [**Ro08**, §4]) as well as in Dlab–Ringel's standardization construction of quasi-hereditary algebras (see [**DR92**, Th. 2 in §3]).

However, A^+ may not be quasi-hereditary. The method we use instead applies in a quite general, but natural, context of endomorphism algebras of suitably filtered injective–like objects in an exact category (e.g., the object T^+ in Theorem 3.3.1). We begin with a description of this general context (with a notation that is independent of previous sections, though chosen to be suggestive).

Let $\mathscr{K}$ be a commutative Noetherian ring, and let H be a finite and projective $\mathscr{K}$-algebra. We view H and its modules as described by $\mathscr{K}$-module structure. Let $(\mathscr{A}, \mathscr{E})$ be a strict and full exact subcategory of mod–H. Recall this means that $\mathscr{A}$ is strict (closed under isomorphisms) and full in mod–H and that all conflations in $\mathscr{E}$ are also short exact sequences in mod$-H$. Given $M \in \mathscr{A}$, a filtration $0 = M^{-1} \subseteq M^0 \subseteq \cdots \subseteq M^\ell = M$ in mod–H is called an $\mathscr{E}$-filtration provided each short exact sequence $0 \to M^{i-1} \to M^i \to M^i/M^{i-1} \to 0$ $(0 \le i \le \ell)$ in mod-H is a conflation in $\mathscr{E}$. (In particular, each M^i belongs to $\mathscr{A}$.)

LEMMA 3.4.1. *Suppose $M \in \mathscr{A}$ has an $\mathscr{E}$-filtration, indexed as above, and that T is an object in $\mathscr{A}$ such that $\mathrm{Ext}^1_{\mathscr{E}}(N,T) = 0$ for each section $N = M^i/M^{i-1}$, $0 \le i \le \ell$. Then,*

$$0 \to \mathrm{Hom}_H(M^j/M^i, T) \to \mathrm{Hom}_H(M^j, T) \to \mathrm{Hom}_H(M^i, T) \to 0$$

is a short exact sequence, for all $0 \le i < j \le \ell$. In particular, the map

$$\mathrm{Hom}_H(M, T) \to \mathrm{Hom}_H(M^0, T)$$

is surjective.

For the proof, see the Claim in the proof of Theorem 1.3.5.

Now let $\mathscr{S}$ be a (strict) full additive subcategory of $\mathscr{A}$ generated by a finite set $\{S_\omega\}_{\omega\in\Omega}$ of nonzero objects indexed by a finite set Ω. That is, every object in $\mathscr{S}$ is a finite direct sum of objects S_ω, $\omega \in \Omega$. Define $\mathscr{A}_{\mathscr{E}}(\mathscr{S})$ to be the subcategory of objects M in $\mathscr{A}$ which have an $\mathscr{E}$-filtration with every section M^i/M^{i-1} in $\mathscr{S}$.

Note that, by Corollary 3.3.2, the theorem below applies in particular to the algebra A^+ in the conclusion of the main Theorem 3.3.1, using A^+ for $A^\dagger$.

THEOREM 3.4.2. *Assume that, for each $\omega \in \Omega$, there exists an object $X_\omega \in \mathscr{A}_{\mathscr{E}}(\mathscr{S})$ such that*

(1) *$X_\omega \in \mathscr{A}_{\mathscr{E}}(\mathscr{S})$ has an $\mathscr{E}$-filtration with lowest section S_ω (and all other sections in $\mathscr{S}$), and*
(2) *$\mathrm{Ext}^1_{\mathscr{E}}(S_{\omega'}, X_\omega) = 0$, for all $\omega, \omega' \in \Omega$.*

If $A^\dagger = \mathrm{End}_H(T^\dagger)$ where $T^\dagger = \bigoplus_{\omega\in\Omega} X_\omega$, then the Morita class of $A^\dagger$ is independent of the choices of the X_ω satisfying (1) and (2).

As a corollary, we have the following result, whose statement will also be used in the proof of the theorem.

COROLLARY 3.4.3. *Suppose that, for each $\omega \in \Omega$, there are given objects $X_\omega, Y_\omega \in \mathscr{A}_{\mathscr{E}}(\mathscr{S})$ such that conditions (1) and (2) in Theorem 3.4.2 hold for each X_ω and maintain the notations $T^\dagger$ and $A^\dagger$. Assume also that each Y_ω satisfies condition (2) in the sense that $\mathrm{Ext}^1_{\mathscr{E}}(S_{\omega'}, Y_\omega) = 0$, for all $\omega' \in \Omega$.*

Put $T^{\dagger\prime} = \bigoplus_{\omega\in\Omega}(X_\omega \oplus Y_\omega)$ and $A^{\dagger\prime} := \mathrm{End}_H(T^{\dagger\prime})$. Then, $A^\dagger$ and $A^{\dagger\prime}$ are Morita equivalent.

Note that, as a special case of this corollary, $A^\ddagger := A^\ddagger_\Omega$ in (2.4.7) is Morita equivalent to A^+.

PROOF. It is clear that the theorem implies the corollary by using $X_\omega \oplus Y_\omega$ for X_ω satisfying (1) and (2). The filtration of the direct sum starts with that of X_ω, then uses that of Y_ω. However, it is also true that the corollary implies the theorem. In fact, if Y_ω also satisfies condition (1), then the roles of X_ω and Y_ω can be interchanged in the corollary, thus giving Morita equivalences (designated here by $\sim$)

$$(3.4.1)\qquad \begin{cases} \mathrm{End}_H(\bigoplus_\omega Y_\omega) \sim \mathrm{End}_H(\bigoplus_\omega (Y_\omega \oplus X_\omega)) \\ \cong \mathrm{End}_H(\bigoplus_\omega (X_\omega \oplus Y_\omega)) \sim \mathrm{End}_H(\bigoplus_\omega X_\omega). \end{cases}$$

Thus, the corollary and the theorem are equivalent, so it remains to prove the corollary directly. We follow the strategy in [**DPS18**, proof of Th. 4.1].

Let $e \in A^{\dagger\prime} = \mathrm{End}_H(T^{\dagger\prime})$ be the idempotent projection of $T^{\dagger\prime}$ onto $T^\dagger \subseteq T^{\dagger\prime}$ along $\bigoplus_\omega Y_\omega$. Since $A^\dagger \cong eA^{\dagger\prime}e$, it suffices to show that the projective module $A^{\dagger\prime}e$ is a progenerator, or, equivalently, that every irreducible $A^{+\prime}$-module L is a homomorphic image of $A^{\dagger\prime}e$. (A better-known criterion is that $A^{\dagger\prime}eA^{\dagger\prime} = A^{\dagger\prime}$. However, if this fails, then the two-sided ideal generated by e is contained in a maximal left ideal $\mathfrak{m}$ of $A^{\dagger\prime}$. Take $L = A^{\dagger\prime}/\mathfrak{m}$, and note that $eL = 0$, so that L cannot be a homomorphic image of $A^{\dagger\prime}e$.)

However, $A^{\dagger\prime}$ is filtered by modules $\mathrm{Hom}_H(S_\omega, T^{\dagger\prime})$ by Lemma 3.4.1, and each of these is a homomorphic image of $\mathrm{Hom}_H(T^\dagger, T^{\dagger\prime})$ again by Lemma 3.4.1. The corollary now follows from the identification $A^{\dagger\prime}e \cong \mathrm{Hom}_H(T^\dagger, T^{\dagger\prime})$ of $A^{\dagger\prime}$-modules. This completes the proof of the corollary and, as noted, the theorem. □

REMARK 3.4.4. The proof of Corollary 3.4.3 can be used to give an explicit description of a Morita ($\mathscr{K}$-category) equivalence $A^\dagger\text{-mod} \xrightarrow{\sim} A^{\dagger\prime}\text{-mod}$, where $A^\dagger$ and $A^{\dagger\prime}$ are as stated in the corollary. First, identify $A^\dagger$-mod with $eA^{\dagger\prime}e$-mod as above. Then, use the well-known equivalence

$$A^{\dagger\prime} \otimes_{eA^{\dagger\prime}e} (-) : eA^{\dagger\prime}e\text{-mod} \longrightarrow A^{\dagger\prime}\text{-mod}.$$

THEOREM 3.4.5. *Let A^+ be a $\mathcal{Z}$-algebra as constructed in the hypothesis of Theorem* 3.3.1 *and let $A^{+\prime}$ be a $\mathcal{Z}$-algebra also constructed to satisfy this hypothesis. (Here we use the same height function for A^+ and $A^{+\prime}$, but allow the X_ω to vary.) Then A^+ is Morita equivalent to $A^{+\prime}$.*

The proof is immediate from Corollary 3.3.2 using Theorem 3.4.2. Further details on the Morita equivalence can be obtained from the proof of the latter theorem and its corollary.

REMARKS 3.4.6. We now mention several useful applications of Theorem 3.4.2 and its corollary:

(a) The identification $A^{\dagger\prime}e \cong \mathrm{Hom}_H(T^\dagger, T^{\dagger\prime})$ in the proof of the corollary, together with the corresponding identification for $eA^{\dagger\prime}$ and the equality $A^{\dagger\prime}eA^{\dagger\prime} = A^{\dagger\prime} = \mathrm{End}_H(T^{\dagger\prime})$, shows:

(i) the trace of $T^\dagger$ is all of $T^{\dagger\prime}$. In particular,

(ii) any H-projective direct summand of $T^{\dagger\prime}$ is isomorphic to a direct summand of $(T^\dagger)^{\oplus m}$, for a sufficiently large $m \in \mathbb{N}$.

(b) Part (ii) of (a) may be used to prove the claim made in [**DPS17**, Remark 4.10(a)] that $H = \mathcal{H}$ is a direct summand of $T^\dagger$ there in the conclusion of [**DPS17**, Th. 4.9]. (In the context of [**DPS17**, Remark 4.10(a)], we are allowed to replace $T^\dagger$ here by $(T^\dagger)^{\oplus m}$.) In fact, it can also be proved that $T^\dagger$ on [**DPS17**, p. 245], taking the multiplicities m_ω all equal to 1, always has the regular representation $H = \mathcal{H}$ as a direct summand.

(c) Other Morita equivalences in the spirit of Corollary 3.4.3 are developed in [**DPS18**, §4]. Conversely, Theorem 3.4.2 can be used to prove [**DPS18**, Th. 4.3]. The theorem gives a Morita equivalence between localizations of stratified algebras in [**DPS17**] with algebras in [**DPS15**] which base change to algebras arising in a Cherednik category $\mathcal{O}$ context (with equal parameters).

3.5. The standard stratification of A^+. We now use the stratifying system $\{\Delta(\omega), P(\omega)\}_{\omega\in\Omega}$ in the category A^+-mod of finite A^+-modules given in Theorem 3.3.1 to construct a sequence of stratifying ideals in A^+ and to show that A^+ is a standardly stratified algebra.

Let $\mathscr{K}$ be a commutative Noetherian ring and let B be a $\mathscr{K}$-algebra which is finite and projective as a $\mathscr{K}$-module.

DEFINITION 3.5.1 ([**DPS18**, Defs. 3.1&3.2]). An ideal J in B is called a *standard stratifying ideal* if the following conditions hold:

(HI1) B/J is projective over $\mathscr{K}$;
(HI2) ${}_BJ$ is B-projective as a left B-module;
(HI3) $J^2 = J$.

The algebra B over $\mathscr{K}$ is called a *standardly stratified algebra* (SSA) provided there exists a finite "defining sequence" of ideals in B:

$$0 = J_0 \subseteq J_1 \subseteq J_2 \subseteq \cdots \subseteq J_m = B \tag{3.5.1}$$

such that, for $0 < i \leq m$, J_i/J_{i-1} is a standard stratifying ideal in B/J_{i-1}. We also call (3.5.1) a *standard stratification.*

We will see in Section 4.3 that stratifying ideals and standardly stratified algebras are weakened (more general) versions of heredity ideals and quasi-hereditary algebras.

We first observe the following nice properties.

PROPOSITION 3.5.2.

(1) *Every ideal in the standard stratification* (3.5.1) *is an idempotent ideal.*

(2) *Suppose that J is a stratifying ideal in a $\mathscr{K}$-algebra B. If M, N are B/J-modules which are also regarded as B-modules by inflation, then, for any integer $n \geq 0$, there is a $\mathscr{K}$-module isomorphism*

$$\mathrm{Ext}^n_{B/J}(M,N) \longrightarrow \mathrm{Ext}^n_B(M,N).$$

PROOF. A proof for (2) can be found in [**DPS17**, Appendix B]. We now prove (1). Since $J_1^2 = J_1$ and $(J_2/J_1)^2 = J_2/J_1$, it follows that

$$J_2^2 = J_2^2 + J_1^2 = J_2^2 + J_1 = J_2 + J_1 = J_2.$$

Hence, J_2 is an idempotent ideal. An induction argument proves the first assertion. □

We now prove that the Hecke endomorphism $\mathcal{Z}$-algebra A^+ is a standardly stratified algebra. There is a general proof of such a fact for any $\mathscr{K}$-algebra B which is finite projective over $\mathscr{K}$, provided that the category B-mod has a stratifying system; see [**DPS18**, §3]. The proof for A^+ below is independent and more direct, laying down a foundation for a further proof of the quasi-heredity of a base-changed $A^{+\natural}$ in the next chapter.

THEOREM 3.5.3. *If A^+ is the (extended) generic Hecke endo-algebra associated with a standard finite Coxeter system (W, S, L) as defined in* (2.4.4)*, then the $\mathcal{Z}$-algebra A^+ is a standardly stratified algebra.*

PROOF. We maintain the notation of Theorem 3.3.1, so, in particular, $A^+ = \mathrm{End}_{\mathcal{H}}(T^+)$ is a certain endomorphism algebra over $\mathcal{Z}$, where $T^+ = \bigoplus_{\omega\in\Omega} T^+_\omega$, and the category A^+-mod has a stratifying system $\{\Delta(\omega), P(\omega)\}_{\omega\in\Omega}$. Recall also from Theorem 3.3.1 that, for each $\omega \in \Omega$, the left A^+-modules $\Delta(\omega)$ and $P(\omega)$ are defined as $\mathrm{Hom}_{\mathcal{H}}(S_\omega, T^+)$ and $\mathrm{Hom}_{\mathcal{H}}(T^+_\omega, T^+)$, respectively, with the action of $A^+ := \mathrm{Hom}_{\mathcal{H}}(T^+, T^+)$ through compositions of morphisms (between right $\mathcal{H}$-modules).

Now, put $P := \bigoplus_{\omega\in\Omega} P(\omega)$ and observe that $P \cong A^+$ as left A^+-modules. (Use the equality $\bigoplus_{\omega\in\Omega} T^+_\omega = T^+$ of right $\mathcal{H}$-modules.)

Let $\mathfrak{h} : \Omega \to \mathbb{Z}$ and $\preceq$ be as discussed above Proposition 3.1.1, and used in Theorem 3.3.1. Here Ω is the set of (Kazhdan–Lusztig) left cells of W, and $(\Omega, \preceq)$ is the quasi-poset used in Theorem 3.3.1. For more details on our set-up here, see the paragraph above the statement of Theorem 3.3.1 and the internal references it provides.

Our immediate strategy is to construct a tractable "defining sequence" of ideals in A^+ (see Def. 3.5.1)

$$0 = J_0 \subseteq J_1 \subseteq \cdots \subseteq J_m = A^+ \tag{3.5.2}$$

such that all the conditions (HI1)–(HI3) in Definition 3.5.1, for every $J = J_i/J_{i-1}$, hold as an ideal of $B = A^+/J_{i-1}$.

We construct the defining sequence for A^+ as follows. First, for any $\omega \in \Omega$ and $j \in \mathbb{Z}$, we define a submodule $P(\omega)_j \subseteq P(\omega)$ as the sum of all images in $P(\omega)$ of all maps in $\mathrm{Hom}_{A^+}(P(\mu), P(\omega))$ with $\mathfrak{h}(\mu) \geq j$. Using the ring-theoretic language of traces, we may write this as

$$P(\omega)_j = \sum_{\mathfrak{h}(\mu)\geq j} \mathrm{trace}_{P(\omega)}(P(\mu)) = \mathrm{trace}_{P(\omega)}\Big(\bigoplus_{\mathfrak{h}(\mu)\geq j} P(\mu)\Big) := \sum_{f\in \mathrm{Hom}_{A^+}(X, P(\omega))} \mathrm{Im}(f),$$

where $X = \bigoplus_{\mathfrak{h}(\mu)\geq j} P(\mu)$. Observe that $P(\omega)_{j+1} \subseteq P(\omega)_j$. It is proved[28] in [**DPS17**, Prop. 2.2] and [**DPS18**, Prop. 3.9] that $P(\omega)_j/P(\omega)_{j+1}$ is a direct sum of modules $\Delta(\nu)$ with $\nu \in \Omega$ and $\mathfrak{h}(\nu) = j$. Multiplicities are allowed. That is, isomorphic copies of the same $\Delta(\nu)$ may appear in the direct sum.

Next, we may repeat the above discussion of $P(\omega)_j$ with P replacing $P(\omega)$. That is, we define

$$P_j = \mathrm{trace}_P\Big(\bigoplus_{\mathfrak{h}(\mu)\geq j} P(\mu)\Big). \tag{3.5.3}$$

Clearly, $P_j = \bigoplus_{\omega\in\Omega} P(\omega)_j$, since $P = \bigoplus_{\omega\in\Omega} P(\omega)$ as a left A^+-module. From this we obtain $P_{j+1} \subseteq P_j$ and deduce also that P_j/P_{j+1} is a direct sum of A^+-modules isomorphic to $\Delta(\nu)$, for some $\nu \in \Omega$, with $\mathfrak{h}(\nu) = j$. If the equality holds for a given ν, then some $\Delta(\nu)$ does occur.

We are now ready to return to A^+ and its ideals. We have observed near the beginning of this proof that $P \cong A^+$ as left A^+-modules. For each $j \in \mathbb{Z}$, let A^+_j denote the left ideal in A^+ identifying with $P_j \subseteq P$.

Claim. Each A^+_j is a two-sided ideal of A^+. It is idempotent (allowing zero), and A^+/A^+_j is free as a $\mathcal{Z}$-module.

To prove the claim, we first show that each A^+_j is a two-sided ideal in A^+. By definition, it is the left ideal of A^+ corresponding to the left A^+-submodule P_j of

[28]The method of proof reorders the sections $\Delta(\nu)$ in any given filtration of $P(\omega)$ that has sections given by various $\Delta(\nu)$'s with $\nu \in \Omega$. The multiplicities of any given $\Delta(\nu)$ does change in this process.

P under the natural identification $P \cong A^+$ of left A^+-modules discussed above. Similarly, A^+/A^+_j is isomorphic to P/P_j as left A^+-modules. As such, it is filtered by copies of various modules $\Delta(\nu)$ with $\mathfrak{h}(\nu) < j$, and A^+_j is similarly filtered by modules $\Delta(\tau)$ with $\mathfrak{h}(\tau) \geq j$. (In both cases, "filtered by" refers informally to a filtration by left A^+-modules, with the given Δ's as the only nonzero sections.)

If A^+_j were not a two-sided ideal in A^+, it would not be a right ideal, and there would be an element $a \in A^+$ with $A^+_j a \not\subseteq A^+_j$. This implies that the composition

$$A^+_j \longrightarrow A^+_j a \longrightarrow (A^+_j a + A^+_j)/A^+_j \subseteq A^+/A^+_j$$

of left A^+-module homomorphisms would be non-zero. Combining this with the above paragraph, we find from this that there is a nonzero map $\Delta(\tau) \to \Delta(\nu)$ with $\mathfrak{h}(\tau) > \mathfrak{h}(\nu)$. However, this is impossible. It leads to a nonzero composite $P(\tau) \twoheadrightarrow \Delta(\tau) \to \Delta(\nu)$, implying $\tau \preceq \nu$, by the axiom (SS2) for stratifying systems. But $\tau \preceq \nu$ and $\mathfrak{h}(\tau) > \mathfrak{h}(\nu)$ cannot both hold, by the compatibility of $\mathfrak{h}$ with $\preceq$; see (3.1.1). This contradiction shows that each A^+_j is a two-sided ideal.

Next, we show each ideal A^+_j is idempotent. In fact, we will show it is generated by an idempotent element $e_j \in A^+_j$ (if $A^+_j \neq 0$).

First, we need some refinements of our notation. Define

$$P_{j,0} = \bigoplus_{\mathfrak{h}(\omega)\geq j} P(\omega)_j \quad \text{and} \quad P_{j,1} = \bigoplus_{\mathfrak{h}(\omega)<j} P(\omega)_j.$$

Note that the trace definition $P(\omega)_j = \mathrm{trace}_{P(\omega)}\left(\bigoplus_{\mathfrak{h}(\mu)\geq j} P(\mu)\right)$ implies $P(\omega)_j \subseteq P(\omega)$, for any $\omega \in \Omega$, and $P(\omega)_j = P(\omega)$ if $\mathfrak{h}(\omega) \geq j$. (In particular, $P_{j,0} = \bigoplus_{\mathfrak{h}(\omega)\geq j} P(\omega)$.) Recalling the identity $P_j = \bigoplus_{\omega\in\Omega} P(\omega)_j$, we have $P_j = P_{j,0} \oplus P_{j,1}$. It is also useful to set $\widehat{P}_{j,1} = \bigoplus_{\mathfrak{h}(\omega)<j} P(\omega)$; so that $P_{j,1} \subseteq \widehat{P}_{j,1}$ and $P = P_{j,0} \oplus \widehat{P}_{j,1}$.

We are almost ready to define e_j. First, let A^+, A^+_j, $A^+_{j,0}$, $A^+_{j,1}$, $\widehat{A}^+_{j,1}$, ... correspond to P, P_j, $P_{j,0}$, $P_{j,1}$, $\widehat{P}_{j,1}$, ... under our standard isomorphism $A^+ \cong P$ of left A^+-modules. Next, let $f_j : P \twoheadrightarrow P_{j,0} \subseteq P_j \subseteq P$ be the obvious idempotent endomorphism of P passing through P_j. Then, there is a unique idempotent $e_j \in A^+$ such that right multiplication by e_j gives the map $A^+ \twoheadrightarrow A^+_{j,0} \subseteq A^+_j \subseteq A^+$ corresponding to f_j. Thus, e_j is idempotent (since $e_j^2 = 1 \cdot e_j^2 = f_j^2(1) = f_j(1) = 1 \cdot e_j = e_j$), and $e_j \in A^+_j$ (since $e_j = 1 \cdot e_j \in A^+_{j,0} \subseteq A^+_j$).

To complete the proof that A^+_j is an idempotent ideal in A^+, it is enough now (and even necessary and sufficient) to show $A^+ e_j A^+ = A^+_j$. This is actually fairly easy. The submodule $A^+ e_j$ of A^+_j corresponds, by construction, to the submodule $P_{j,0}$ of P. Also, we have the equality $P_j = \mathrm{trace}_P(P_{j,0})$ by definition. It follows, passing to A^+, that

(3.5.4)
$$A^+_j = \mathrm{trace}_{A^+}(A^+_{j,0}) = \mathrm{trace}_{A^+}(A^+ e_j) = \sum_{f\in \mathrm{Hom}_{A^+}(A^+e_j, A^+)} \mathrm{Im}(f) = A^+ e_j A^+.$$

Here the last equality follows from a well-known identification of $\mathrm{Hom}_{A^+}(A^+e_j, A^+)$ with $e_j A^+$, which leads to an identification of $\mathrm{trace}_{A^+}(A^+e_j)$ with $A^+e_jA^+$.

To prove the last remaining assertion of the claim that A^+/A^+_j is free as a $\mathcal{Z}$-module, it is enough to show A^+_i/A^+_{i+1} is free for every $i < j$. However, we have already observed that the isomorphic module P_i/P_{i+1} is a direct sum of A^+-modules isomorphic to $\Delta(\nu)$ for some $\nu \in \Omega$. Thus, it is projective (of finite rank) over $\mathcal{Z}$, hence free, by Swan's theorem. See Remark 1.3.6(c). The claim is now proved.

We are now able to construct a candidate for the defining sequence (3.5.2) previewed earlier in this proof.

Observe that $A_j^+ = 0$, for all sufficiently large integers j. In fact, it suffices to take $j > \mathfrak{h}(\omega)$, for all $\omega \in \Omega$, using the trace formula for P_j. Put $J_0 = 0$, and define $J_1 = A^+_{j(1)}$, where $j(1)$ denotes the largest value of $\mathfrak{h}$ on Ω. Continuing, define $J_2 = A^+_{j(2)}$, where $j(2)$ denotes the largest value of $\mathfrak{h}$ on $\Omega\backslash\Omega_{j(1)}$ with $\Omega_{j(1)} = \{\omega \in \Omega \mid \mathfrak{h}(\omega) = j(1)\}$. This process continues until $j(1), j(2), \ldots, j(m)$ have been defined as all the elements of $\mathfrak{h}(\Omega)$, listed in decreasing order. For each integer i with $1 \le i \le m$, we set $J_i = A^+_{j(i)}$, generalizing our definitions for J_1 and J_2. We note that $A^+_{j(m)} = A^+$, since $P_{j(m)} = P$ by the trace formula. Thus, $J_m = A^+_{j(m)} = A^+$, the terminal member of our ascending sequence of J_i's:

$$0 = J_0 \subseteq J_1 \subseteq \cdots \subseteq J_m = A^+ \quad \text{with} \quad j(1) > j(2) > \cdots > j(m). \tag{3.5.5}$$

(The general inclusion $P_j \subseteq P_{j-1}$, which holds for any $j \in \mathbb{Z}$, also implies that $P_j \subseteq P_{j-a}$, for any positive $a \in \mathbb{Z}$. Consequently, $A_j^+ \subseteq A^+_{j-a}$. So $A^+_{j(i)} \subseteq A^+_{j(i+1)}$, using $j = j(i)$ and $a = j(i) - j(i+1)$. Here, i is an integer with $1 \le i \le m-1$. Since we have defined $J_i = A^+_{j(i)}$ for each i, as well as $i = m$, we have the desired sequence $0 = J_0 \subseteq J_1 \subseteq \cdots \subseteq J_m = A^+$.)

Continuing the discussion above, we now wish to analyze more closely the left A^+-modules $J_1 = J_1/J_0$ and J_i/J_{i-1} for $2 \le i \le m$. We start with $J_1 = A^+_{j(1)}$. Recall that $j(1)$ is the largest value of the height function so that $A_j^+ = 0$, for $j > j(1)$. (Use the trace formula for P_j.) Thus, $A^+_{j(1)} = A^+_{j(1)}/A^+_{j(1)+1}$ is a direct sum of various left A^+-modules $\Delta(\omega)$ with $\mathfrak{h}(\omega) = j(1)$. These modules are projective for A^+, by the maximality of $\mathfrak{h}(\omega)$. (Just observe that $P(\omega) = \Delta(\omega)$ by axiom (SS3).)

We now consider J_i/J_{i-1} for $2 \le i \le m$ (the case $i = 1$ having been just discussed). We note

$$J_i/J_{i-1} = A^+_{j(i)}/A^+_{j(i-1)} = A^+_{j(i)}/A^+_{j(i)+1}. \tag{3.5.6}$$

The right-hand equality holds because $j(i) < j(i-1)$ are adjacent values of the height function. Consequently, for any integer j with $j(i)+1 \le j < j(i-1)$, the A^+-module A_j^+/A^+_{j+1} must be 0. (Use our description, earlier in this proof, of P_j/P_{j+1} as a direct sum of modules $\Delta(\omega)$ with $j = \mathfrak{h}(\omega)$.) But this gives $A_j^+ = A^+_{j+1}$, with repeated application giving $A^+_{j(i)+1} = A^+_{j(i-1)}$, as desired.

We are now in a position to prove condition (HI2), the projectivity of $J := J_i/J_{i-1}$ as a left B-module, for $B := A^+/J_{i-1}$, $1 \le i \le m$. (The case $i = 1$ has already been treated.) We have $J = A^+_{j(i)}/A^+_{j(i)+1}$ by (3.5.6). Since $j(i)$ is a height function value, we know that J in this latter quotient form is the direct sum of various $A^+/A^+_{j(i)+1}$-modules isomorphic to $\Delta(\omega)$ with $\mathfrak{h}(\omega) = j(i)$. (Such a module may be regarded as a left A^+- or $A^+/A^+_{j(i)+1}$-module.) Recall that we have direct sum formulas $P = \bigoplus_{\omega\in\Omega} P(\omega)$ and $P_j = \bigoplus_{\omega\in\Omega} P(\omega)_j$ for each integer j, and variations. In particular, note that $P/P_{j+1} = \bigoplus_{\omega\in\Omega} P(\omega)/P(\omega)_{j+1}$. When interpreted with A^+ replacing P and $j = j(i)$, we find that the left-hand side is $B = A^+/A^+_{j(i)+1}$, and the right-hand side, viewed a left A^+-module, has $\Delta(\omega)$ as a direct summand whenever $\mathfrak{h}(\omega) = j(i)$, as is the case for $\Delta(\omega)$ above. Thus, $\Delta(\omega)$, as a left B-module, is projective. Since the $\Delta(\omega)$ also represents the general direct summand of J, we have proved J is projective as a left B-module, as desired. (Note also the A^+ analog of the formula $P_j/P_{j+1} = \bigoplus_{\omega\in\Omega} P(\omega)_j/P(\omega)_{j+1}$. For $j = j(i)$, the left-hand side becomes $A^+_{j(i)}/A^+_{j(i)+1} = A^+_{j(i)}/A^+_{j(i-1)} = J_i/J_{i-1}$, which is J in the current context. On the right, we obtain a direct sum of modules $\Delta(\omega)$ with $\mathfrak{h}(\omega) = j(i)$, all now known to be projective as left B-modules.)

This completes the proof of (HI2) for B and J and all choices of i with $1 \leq i \leq m$. Properties (HI1) and (HI3) hold as immediate consequence of our claim studied several paragraphs above. This completes the proof of the theorem. □

In Chapter 4, we will consider the localization $\mathcal{Z}^\natural$ of $\mathcal{Z}$ at the multiplicative set generated by all bad primes for W; see Remark 0.1.1. By base change from $\mathcal{Z}$ to $\mathcal{Z}^\natural$, (3.5.2) gives rise to a sequence of ideals of $A^{+\natural} := \mathcal{Z}^\natural \otimes A^+$:

$$0 = J_0^\natural \subseteq J_1^\natural \subseteq \cdots \subseteq J_m^\natural = A^{+\natural}. \tag{3.5.7}$$

We will prove in Section 4.3 that every ideal $J^\natural := J_i^\natural/J_{i-1}^\natural$ $(1 \leq i \leq m)$ is a "heredity ideal of split type" in $B^\natural := A^{+\natural}/J_{i-1}^\natural$. More precisely, every $\mathrm{End}_{B^\natural}(J^\natural)$ is a direct product of matrix algebras $M_n(\mathcal{Z}^\natural)$. Consequently, if W is not of type ${}^2F_4$, $A^{+\natural}$ is a (split) quasi-hereditary algebra.

REMARKS 3.5.4.

(1) The equivalence classes associated with the preorder $\preceq$ defined in (3.1.1) are two-sided cells of W. Thus, the defining sequence (3.5.5) has sections labelled by two-sided cells. This labelling is the same as the labelling for Lusztig's families of irreducible unipotent characters of G (see [**C85**, §§12.1,12.3]). The latter was conjectured (see [**C85**, p. 387]) and proved by Lusztig in [**Lus86**]. We also note that these families play a key role for the proof of Geck's conjecture in [**BDT20**].

(2) Replacing Ω by a poset ideal Θ of $(\Omega, \leq_L)$, T^+ by the $\mathcal{H}$-module $T^\ddagger_\Theta$, A^+ by $A^\ddagger_\Theta$ as defined in (2.4.6), and using the stratifying system in Corollary 3.3.3 (thus, replacing $\Delta(\omega), P(\omega), P$, etc. by $\Delta^+(\omega), P^+(\omega), P^+$, etc.) throughout the proof of Theorem 3.5.3, we may construct a standard stratification

$$0 = J_0^+ \subseteq J_1^+ \subseteq \cdots \subseteq J_m^+ = A^\ddagger_\Theta. \tag{3.5.8}$$

for the standardly stratified algebra $A^\ddagger_\Theta$.

4. The quasi-heredity of $A^{+\natural}$

We proved in the last chapter that the enlarged Hecke endomorphism algebra A^+ is standardly stratified, a property reminiscent of quasi-hereditary algebras [**CPS88**]. In this chapter, we will prove that its base change $A^{+\natural} := \mathcal{Z}^\natural \otimes_{\mathcal{Z}} A^+$ over $\mathcal{Z}^\natural$ (in which all bad primes are invertible) is *integral* quasi-hereditary; see, e.g., [**CPS90**] (also [**DPS18**]) for integral quasi-hereditary algebras.

4.1. Lusztig's asymptotic forms J and maximal orders $\mathcal{J}^\natural := \mathcal{Z}^\natural \otimes_{\mathbb{Z}} \mathsf{J}$. Let (W, S, L) be a standard finite Coxeter system associated to a finite group G of Lie type, and let $\mathcal{H}$ be its generic Hecke algebra over $\mathcal{Z} := \mathbb{Z}[t, t^{-1}]$. As noted in Section 1.2, Lusztig's conjectures P1–P15 all hold for this situation. Also, W is, of course, finite, so the standing hypotheses of Lusztig used in [**Lus03**, §18] hold. *In this section, we temporarily exclude the case that G is of type* $^2F_4$*, i.e., we assume* W *is not the dihedral group* $I_2(8)$ *of order 16.* We will treat this case in a short appendix at the end of this section. (However, everything up through (4.1.6) holds for $^2F_4$. Recall that the smallest splitting field of $I_2(8)$ is $\mathbb{Q}(\sqrt{2})$, the field generated by its irreducible character values. The other Weyl groups have splitting field $\mathbb{Q}$.)

Let $\mathcal{Z}^\natural := \mathcal{S}^{-1}\mathcal{Z}$, the localization of $\mathcal{Z}$ at the multiplicative set $\mathcal{S}$ generated by the bad primes of W.[29] Set $\mathcal{H}^\natural := \mathcal{Z}^\natural \otimes_{\mathcal{Z}} \mathcal{H}$. Of course, $\mathcal{Z}^\natural$ is a unique factorization domain of Krull dimension 2 (which is regular). In particular, its height one prime ideals $\mathfrak{p}$ are all principal and are given (in this case) as follows: (i) $\mathfrak{p} = p\mathcal{Z}^\natural$, where $p \in \mathbb{Z}$ is a good prime for W; and (ii) $\mathfrak{p} = f(t)\mathcal{Z}^\natural$, where $f(t)$ is an irreducible polynomial in $\mathbb{Z}[t]$, of positive degree and not equal to $\pm t$. This description follows easily from [**Mum88**, Example H, p. 74], noting that the prime ideals of $\mathcal{Z}$ correspond naturally to those of $\mathbb{Z}[t]$ which do not contain t.

Before going further we introduce some general terminology on lattices and orders over a Noetherian integral domain $\mathscr{K}$. We have $\mathscr{K} = \mathcal{Z}$ or $\mathcal{Z}^\natural$ in mind. Often it will be convenient to let K be the fraction field of $\mathscr{K}$, e.g., $K = \mathbb{Q}(t)$, if $\mathscr{K} = \mathcal{Z}$ or $\mathcal{Z}^\natural$.

DEFINITIONS 4.1.1.

(a) For any finite-dimensional K-space V, a *(full)* $\mathscr{K}$*-lattice* in V is a finite $\mathscr{K}$-submodule M in V such that $KM = V$, where

$$KM := \Big\{\sum r_i m_i (\text{finite sum}) \mid r_i \in K, m_i \in M\Big\}.$$

Following [**Reiner03**, p. 108], a $\mathscr{K}$*-order* is a $\mathscr{K}$-subalgebra $\mathcal{A}$ of a finite-dimensional K-algebra $\mathfrak{A}$, having the same identity as $\mathfrak{A}$, such that $\mathcal{A}$ is a full $\mathscr{K}$-lattice in $\mathfrak{A}$. Unless otherwise noted, from now on, the term "lattice" refers to a "full lattice."

(b) Also, by some abuse of notation, we shall refer to a module M for a $\mathscr{K}$-order $\mathcal{A}$ as an $\mathcal{A}$-lattice if it inherits a $\mathscr{K}$-lattice structure from the action of $\mathcal{A}$. A $\mathscr{K}$-order $\mathcal{A}$ in $\mathfrak{A}$ is a *maximal* $\mathscr{K}$*-order* if it is not properly contained in a larger $\mathscr{K}$-order in $\mathfrak{A}$; see [**Reiner03**, p. 110].

To give a common example of a maximal order in a finite-dimensional K-algebra $\mathfrak{A}$, assume that $\mathscr{K}$ is integrally closed and $\mathfrak{A}$ is a direct product $\prod_i M_{n_i}(K)$ of a finite number of matrix algebras $M_{n_i}(K)$. Then $\mathcal{A} := \prod_i M_{n_i}(\mathscr{K})$ is a maximal $\mathscr{K}$-order in $\mathfrak{A}$. (See [**Reiner03**, Th. 8.7].)

We need some further notation, mostly taken from [**Lus03**]. For $w \in W$, let $\operatorname{sgn}(w) = (-1)^{\ell(w)}$. Recall the notation $\widetilde{T}_w$ from display (1.2.2). The map

$$\widetilde{T}_w \longmapsto \widetilde{T}_w^\dagger := \operatorname{sgn}(w)\widetilde{T}_{w^{-1}}^{-1} \tag{4.1.1}$$

defines a $\mathcal{Z}$-algebra involution $h \mapsto h^\dagger$ of $\mathcal{H}$. See [**Lus03**, 3.5]. Next, for $z \in W$, let $a(z)$ be the smallest non-negative integer such that

$$t^{a(z)} h_{x,y,z} \in \mathbb{Z}[t], \quad \text{for all } x, y \in W. \tag{4.1.2}$$

Here $h_{x,y,z}$ is the structure constant defined in (1.2.6). Then let $\gamma_{x,y,z} \in \mathbb{Z}$ be the coefficient of $t^{a(z)}$ in $h_{x,y,z}$.

[29] As noted with more detail in Remark 0.1.1, one could use "bad prime" in the traditional sense here, or, for our purposes, simply define it to be a prime divisor of the generic degree denominator m for a principal series character.

Obviously, $\mathcal{H}$ is a $\mathscr{K}$-order in $\mathcal{H}_{\mathbb{Q}(t)}$, using $\mathscr{K} = \mathcal{Z}$. Next, we define two further orders in (isomorphic copies of) $\mathcal{H}_{\mathbb{Q}(t)}$ in the sense of Definition 4.1.1. The first one $\mathcal{J} := \mathcal{Z} \otimes_{\mathbb{Z}} \mathsf{J}$ is the base change of a $\mathbb{Z}$-algebra J; the latter, defined below, is called *Lusztig's asymptotic form*. As we shall see, the order $\mathcal{J}$ contains a copy of $\mathcal{H}$, as discussed below. The second order $\mathcal{J}^\natural := \mathcal{Z}^\natural \otimes_{\mathbb{Z}} \mathsf{J} \cong \mathcal{Z}^\natural \otimes_{\mathcal{Z}} \mathcal{J}$, a $\mathcal{Z}^\natural$-order, is obtained from the first by inverting all bad primes for W.

The algebra J is a free $\mathbb{Z}$-algebra with $\mathbb{Z}$-basis $\{\mathsf{j}_x \mid x \in W\}$ and multiplication given by

$$\mathsf{j}_x\mathsf{j}_y = \sum_{z\in W} \gamma_{x,y,z^{-1}}\mathsf{j}_z, \quad x, y \in W. \tag{4.1.3}$$

Then J is associative and unital [**Lus03**, 18.3]. (Note that $\operatorname{rank}(\mathsf{J}) = |W|$.)

Recall that W has a subset $\mathcal{D}$ of distinguished involutions [**Lus03**, P6, 14.1]. Each left cell ω contains a unique element $d \in \mathcal{D}$. Also, for $d \in \mathcal{D}$, an integer $n_d = \pm 1$ is defined in [**Lus03**, 14.1, P5 & P13]. Given $z \in W$, let $d \in \mathcal{D}$ be the unique distinguished involution satisfying $z^{-1} \sim_L d$, and set $\hat{n}_z := n_d$. By [**Lus03**, Th. 18.9], there is a $\mathcal{Z}$-algebra homomorphism $\phi : \mathcal{H} \longrightarrow \mathcal{J}$ defined by

$$\phi(\mathrm{c}_w^\dagger) = \sum h_{w,d,z}\hat{n}_z\mathsf{j}_z. \tag{4.1.4}$$

In this expression, $w \in W$ and the sum is over all pairs $(d, z) \in \mathcal{D} \times W$ such that $a(z) = a(d)$. Then, for any commutative $\mathcal{Z}$-algebra R, ϕ induces an R-algebra homomorphism $\phi_R : \mathcal{H}_R \to \mathcal{J}_R$. By [**Lus03**, Prop. 18.12a], the kernel of ϕ_R is a nilpotent ideal. Thus, when R is a field and $\mathcal{H}_R$ is a semisimple algebra, ϕ_R is injective and, thus, an isomorphism by dimension considerations. In particular, $\phi_{\mathbb{Q}(t)}$ is an isomorphism. Of course, this means that ϕ is an injection, so that $\mathcal{J}$ and $\mathcal{J}^\natural$ both contain subalgebras isomorphic to $\mathcal{H}$. The case $R = \mathbb{Q}$ leads to a similar isomorphism $\mathcal{H}_{\mathbb{Q}} \xrightarrow{\sim} \mathcal{J}_{\mathbb{Q}}$, regarding $\mathbb{Q}$ as a $\mathcal{Z}$-algebra by means of the specialization $t \mapsto 1$. Note $\mathcal{J}_{\mathbb{Q}} \cong \mathsf{J}_{\mathbb{Q}}$ by construction. For later use in the proof of the theorem below, note also that $\mathcal{H}_{\mathbb{Q}} \cong \mathbb{Q}W$, and so, $\mathsf{J}_{\mathbb{Q}} \cong \mathcal{J}_{\mathbb{Q}} \cong \mathcal{H}_{\mathbb{Q}} \cong \mathbb{Q}W$.

REMARK 4.1.2. All the assertions of the above paragraph (below display (4.1.4)) hold as written if the homomorphism ϕ is replaced by the composition

$$\varpi := \phi \circ \dagger : \mathcal{H} \longrightarrow \mathcal{J}, \tag{4.1.5}$$

where $\dagger$ is defined in display (4.1.1). This is not needed right away, but will be useful in showing in the next section that all left-cell modules for $\mathcal{H}$ extend to $\mathcal{J}$-modules. For the convenience of the reader, we write down explicitly the expression for the composite $\varpi = \phi \circ \dagger$:

$$\varpi(\mathrm{c}_w) = \sum_{z\in W} \sum_{\substack{d\in\mathcal{D} \\ a(d)=a(z)}} h_{w,d,z}\hat{n}_z\mathsf{j}_z \quad (w \in W). \tag{4.1.6}$$

We will also use ϖ in the context of $\mathcal{H}^\natural$ and $\mathcal{J}^\natural$, in which case we define $\varpi^\natural$ to be the base-changed map

$$\varpi^\natural : \mathcal{H}^\natural = \mathcal{H}_{\mathcal{Z}^\natural} \xrightarrow{\varpi_{\mathcal{Z}^\natural}} \mathcal{J}_{\mathcal{Z}^\natural} = \mathcal{J}^\natural. \tag{4.1.7}$$

Moreover, the isomorphisms $\mathsf{J}_{\mathbb{Q}} \cong \mathcal{H}_{\mathbb{Q}} \cong \mathbb{Q}W$ and $\mathsf{J}_{\mathbb{Q}(t)} = \mathcal{J}_{\mathbb{Q}(t)} \cong \mathcal{H}_{\mathbb{Q}(t)}$ induce an isomorphism $\mathbb{Q}(t)W \cong \mathcal{H}_{\mathbb{Q}(t)}$ and, hence, a bijection

$$\operatorname{Irr}(\mathbb{Q}W) \longrightarrow \operatorname{Irr}(\mathcal{H}_{\mathbb{Q}(t)}),\ E \longmapsto E_{\mathbb{Q}(t)}. \tag{4.1.8}$$

If G is of type ${}^2F_4$, $\mathbb{Q}$ should be replaced by $\mathbb{Q}(\sqrt{2})$.

Next, we pursue some of the remarkable properties of the $\mathcal{Z}^\natural$-order $\mathcal{J}^\natural$. Here, $\mathscr{K} = \mathcal{Z}^\natural$ and $K = \mathbb{Q}(t)$.

Note that $\mathcal{J}^\natural_{\mathbb{Q}(t)}$ identifies with $\mathcal{J}_{\mathbb{Q}(t)}$ which is isomorphic (via the inverse of $\phi_{\mathbb{Q}(t)}$) to $\mathcal{H}_{\mathbb{Q}(t)}$. The latter is semisimple by Remark 3.3.4(b) and even split, since $\mathbb{Q}W$ is split semisimple. (Recall that we have postponed the case ${}^2F_4$ to an appendix at the end of this section.) Hence, there is a "natural" isomorphism

$$\mathcal{J}^\natural_{\mathbb{Q}(t)} \xrightarrow{\sim} \prod_i \mathrm{End}_{\mathbb{Q}(t)}(E_i), \tag{4.1.9}$$

where E_i ranges over the distinct irreducible $\mathcal{J}_{\mathbb{Q}(t)}$-modules. The map is obtained from the action of $\mathcal{J}^\natural_{\mathbb{Q}(t)} = \mathcal{J}_{\mathbb{Q}(t)}$ on the various E_i. We can assume that $E_i = KL_i$ where L_i is a $\mathcal{J}^\natural$-stable lattice in E_i. In fact, using Auslander-Goldman theory (see [**DDPW08**, Cor. C.18]), it can also be assumed (after replacing L_i by its double dual L_i^{**}) that all the L_i are projective over $\mathcal{Z}^\natural$. Then, by the result of Swan mentioned in Remark 1.3.6(c)), the L_i are free over $\mathcal{Z}^\natural$.

Thus, (4.1.9) may be obtained by base change from $\mathcal{Z}^\natural$ to $\mathbb{Q}(t)$ of an injection

$$\psi : \mathcal{J}^\natural \to \prod_i \mathrm{End}_{\mathcal{Z}^\natural}(L_i), \tag{4.1.10}$$

where each L_i is $\mathcal{Z}^\natural$-free. Note that each $\mathcal{J}^\natural$-module L_i is an $\mathcal{H}^\natural$-module by restriction through the map $\varpi^\natural$ in (4.1.7).

We can now prove:

THEOREM 4.1.3. *Maintain the assumptions on W, $\mathcal{H}$, etc. above. The map ψ in* (4.1.10) *is an isomorphism of $\mathcal{Z}^\natural$-algebras.*

PROOF. We have already noted that the map ψ is an injection (since its base change to $\mathbb{Q}(t)$ is obviously an injection). Clearly, $\mathcal{J}^\natural$ is a $\mathcal{Z}^\natural$-order (and, in particular, a full lattice) in $\mathcal{J}^\natural_{\mathbb{Q}(t)}$. To prove the theorem, it suffices to prove that $\mathcal{J}^\natural$ is a maximal $\mathcal{Z}^\natural$-order in $\mathcal{J}^\natural_{\mathbb{Q}(t)}$. This will show that ψ is an isomorphism, since it is an inclusion of orders. To do this, it is necessary and sufficient, by the Auslander-Goldman theorem (see [**Reiner03**, Th. 11.4]), to check two conditions:

(a) the $\mathcal{Z}^\natural$-order $\Xi := \mathcal{J}^\natural$ is reflexive in the sense that $\Xi^{**} \cong \Xi$ as $\mathcal{Z}^\natural$-modules;

(b) for any height one prime ideal $\mathfrak{p}$ in $\mathcal{Z}^\natural$, $\mathcal{J}^\natural_{\mathfrak{p}}$ is a maximal $\mathcal{Z}^\natural_{\mathfrak{p}}$-order in $\mathcal{J}^\natural_{\mathbb{Q}(t)}$.

But $\mathcal{J}^\natural$ is $\mathcal{Z}^\natural$-free, so it is certainly reflexive. Thus, (a) holds.

For part (b), it suffices to show, for any height one prime ideal $\mathfrak{p}$ in $\mathcal{Z}^\natural$, that there is an abstract algebra isomorphism

$$\mathcal{J}^\natural_{\mathfrak{p}} \cong \prod_i M_{n_i}(\mathcal{Z}^\natural_{\mathfrak{p}}) \tag{4.1.11}$$

for some finite list of positive integers n_i. See [**Reiner03**, Ths. 8.7, 10.5].

Consider first the case in which $\mathfrak{p} = (p)$, where p is a good prime for W. Then [**DPS98a**, Ths. 4.2.1, 4.2.2] and their proofs show that

$$\mathcal{H}_{\mathbb{F}_p(t)} \cong \prod_i \mathrm{End}_{\mathbb{F}_p(t)}(L_{i\mathbb{F}_p(t)}) \tag{4.1.12}$$

with L_i as above, and $L_{i\mathbb{F}_p(t)} = (L_i)_{\mathbb{F}_p(t)}$ denoting the base-changed module $\mathbb{F}_p(t)\otimes_{\mathcal{Z}^\natural} L_i$. Note that $\mathbb{F}_p(t)$ is the residue field of $\mathcal{Z}^\natural_{\mathfrak{p}}$. (Essentially, the argument in [**DPS98a**] uses Schur matrix relations and the fact that good primes avoid generic degree denominators [**DPS98a**, (4.1.2)] to show that all matrix units on the right have preimages on the left.) Since the action of $\mathcal{H}_{\mathcal{Z}^\natural}$ on L_i is by restriction through the algebra $\mathcal{J}^\natural$, we have a surjective composite map

$$(\mathcal{H}_{\mathcal{Z}^\natural})_{\mathfrak{p}} \hookrightarrow \mathcal{J}^\natural_{\mathfrak{p}} \longrightarrow M_{n_i}(\mathcal{Z}^\natural_{\mathfrak{p}}) \longrightarrow M_{n_i}(\mathbb{F}_p(t)),$$

where the two matrix algebras are the full endomorphism rings of $(L_i)_{\mathcal{Z}^\natural_{\mathfrak{p}}}$ and $(L_i)_{\mathbb{F}_p(t)}$, respectively. Consequently, the map

$$\mathcal{J}^\natural_{\mathfrak{p}} \longrightarrow \prod_i M_{n_i}(\mathbb{F}_p(t))$$

is surjective, as is the map

$$\mathcal{J}^\natural_{\mathfrak{p}} \longrightarrow \prod_i M_{n_i}(\mathcal{Z}^\natural_{\mathfrak{p}}) \qquad \text{(by Nakayama's Lemma).}$$

Since the latter map is an injection, it is also an isomorphism.

Finally, it is enough to show that an isomorphism (4.1.11) holds whenever $\mathfrak{p} = (f(t))$ with $f(t) \in \mathbb{Z}[t]$ an irreducible polynomial of positive degree and $\neq \pm t$. (These are the remaining height one primes in $\mathcal{Z}^\natural$.) In this case, no nonzero integer $z \neq 0$ is divisible by $f(t)$ in the unique factorization domain $\mathcal{Z}^\natural$. Thus, $\mathbb{Q} \subseteq \mathcal{Z}^\natural_{\mathfrak{p}}$. Since

$$\mathsf{J}_{\mathbb{Q}} \cong \mathbb{Q}W \cong \prod_j M_{n_i}(\mathbb{Q})$$

for some positive integers n_i,[30] we have

$$\begin{aligned}\mathcal{J}^\natural_{\mathfrak{p}} := \mathcal{Z}^\natural_{\mathfrak{p}} \otimes_{\mathcal{Z}^\natural} \mathcal{J}^\natural &\cong \mathcal{Z}^\natural_{\mathfrak{p}} \otimes_{\mathcal{Z}^\natural} \mathcal{Z}^\natural \otimes_{\mathcal{Z}} \mathcal{J} \cong \mathcal{Z}^\natural_{\mathfrak{p}} \otimes_{\mathcal{Z}^\natural} \mathcal{Z}^\natural \otimes_{\mathcal{Z}} \mathcal{Z} \otimes_{\mathbb{Z}} \mathsf{J} \\ &\cong \mathcal{Z}^\natural_{\mathfrak{p}} \otimes_{\mathbb{Z}} \mathsf{J} \cong \mathcal{Z}^\natural_{\mathfrak{p}} \otimes_{\mathbb{Q}} \mathbb{Q} \otimes_{\mathbb{Z}} \mathsf{J} \cong \mathcal{Z}^\natural_{\mathfrak{p}} \otimes_{\mathbb{Q}} \mathsf{J}_{\mathbb{Q}} \cong \prod_i M_{n_i}(\mathcal{Z}^\natural_{\mathfrak{p}}),\end{aligned}$$

completing part (b) of the proof. □

This has several consequences. The first corollary, which we call a theorem, follows either from the proof of Theorem 4.1.3 or from its statement. (Remember that $K = \mathbb{Q}(t)$ and G is not of type ${}^2F_4$.)

THEOREM 4.1.4.

(a) *The $\mathcal{Z}^\natural$-algebra $\mathcal{J}^\natural = \mathcal{J}_{\mathcal{Z}^\natural}$ is isomorphic to a maximal order in $\mathcal{H}_K$.*

(b) *Furthermore,*

$$\mathcal{J}^\natural = \mathcal{J}_{\mathcal{Z}^\natural} \cong \prod_i M_{n_i}(\mathcal{Z}^\natural), \tag{4.1.13}$$

for a finite sequence n_1, n_2, ... of positive integers.

REMARKS 4.1.5.

(1) The above theorem has an analogous version in which the $\mathcal{Z}^\natural$-algebra $\mathcal{J}^\natural$ is replaced by the $\mathbb{Z}^\natural$-algebra $\mathsf{J}^\natural$. In this context, another consequence of Theorem 4.1.4 is that $\mathsf{J}^\natural$ is a maximal order in $\mathcal{H}_{\mathbb{Q}} = \mathbb{Q}W$. Explicitly, $\mathsf{J}^\natural$ is isomorphic to a direct product of full matrix algebras over $\mathbb{Z}^\natural$. This follows by a base-change argument using the specialization $\mathcal{Z}^\natural \to \mathbb{Z}^\natural$ given by $t \mapsto 1$. Finally, all corollaries

[30] The first isomorphism is discussed above Remark 4.1.2. It may be regarded as one of the many well-known results of Lusztig generalized in [**Lus03**] to an unequal parameter setting.

below have an analogous version for $\mathsf{J}^\natural$. They are easily derived through a similar proof, or through a base-change argument together with the statements for $\mathcal{Z}$.

(2) By the theorem, we have $\mathcal{J}_K = \mathcal{J}^\natural_K \cong \prod_{i=1}^s M_{n_i}(K)$, for some s, when $K = \mathbb{Q}(t)$. Thus, analyzing projections, there are a total of s non-isomorphic irreducible $\mathcal{J}_K$-modules $L_1, \ldots, L_s$ such that, for each i with $1 \leq i \leq s$, $L_i \cong K^{n_i}$ as $\mathcal{J}_K$-modules, where K^{n_i} is the irreducible left $M_{n_i}(K)$-module via matrix multiplication. The discussion of the map ψ, in and above Theorem 4.1.3, implies that L_i has a $\mathcal{J}^\natural$-lattice $L_{i,\mathcal{Z}^\natural}$ isomorphic to $(\mathcal{Z}^\natural)^{n_i}$. In particular, $L_{i,\mathcal{Z}^\natural}$ is $\mathcal{Z}^\natural$-free.

(3) In the equal parameter case, Lusztig first proved in [**Lus87d**, (3.1j),(1.2a)] that, for a large enough field F, J_F is split semisimple. Further, in [**G98**, Remark 2.5], Geck observed that this result applies to the case where the characteristic of F is a good prime. See also [**GJ11**, Props. 1.5.7&2.3.16] for a similar result in general over a sufficiently large field $\mathbb{K}$ of characteristic 0.

We note that a maximal order copy of $\mathcal{J}^\natural$ in $\mathcal{H}_K$ is $\phi_K^{-1}(\mathcal{J}^\natural)$, in the notation of (4.1.4). Another useful copy is obtained by applying to this copy the involution on $\mathcal{H}_K$ induced by $\dagger$. The resulting copy may also be written $\varpi_K^{-1}(\mathcal{J}^\natural)$, with ϖ as in Remark 4.1.2.

Recall that if R is a commutative ring, an algebra B over R is separable if B is a projective $B \otimes_R B^{\rm op}$-module; see Auslander-Goldman [**AG60b**, p. 1].

COROLLARY 4.1.6.

(a) *The algebra $\mathcal{J}^\natural$ is separable over $\mathcal{Z}^\natural$.*

(b) *Every finite module for $\mathcal{J}^\natural$, which is projective over $\mathcal{Z}^\natural$, is projective over $\mathcal{J}^\natural$.*

PROOF. The reader may directly check that the direct product of a finite set of separable algebras is a separable algebra. Then, in view of Theorem 4.1.3, part (a) follows from [**AG60b**, Prop. 5.1]. Finally, part (b) follows from [**Hat63**, Th. 2.11]. (See also [**Reiner03**, Ex. 9, p. 106].) □

COROLLARY 4.1.7. *Let M and M' be finite $\mathcal{J}^\natural$-modules, projective over $\mathcal{Z}^\natural$. If $M_K \cong M'_K$ as $\mathcal{J}^\natural_K$-modules, then $M \cong M'$ as $\mathcal{J}^\natural$-modules.*

PROOF. Note that M and M' are projective over $\mathcal{J}^\natural$ by Corollary 4.1.6(b). By Theorem 4.1.4, we can reduce to the case where M and M' are finite projective modules for a matrix algebra $M_n(\mathcal{Z}^\natural)$ which become isomorphic over the fraction field K. By Morita theory, we can assume $n = 1$. (Let $F : \text{mod-}M_m(\mathcal{Z}^\natural) \to \text{mod-}\mathcal{Z}^\natural$ be an equivalence of $\mathcal{Z}^\natural$-categories, e.g., a suitable idempotent projection. Note that F induces by base change an equivalence $F_K : \text{mod-}M_m(K) \to \text{mod-}K$. For $X, Y \in \text{mod-}M_m(\mathcal{Z}^\natural)$, we clearly have $X \cong Y$ if and only if $F(X) \cong F(Y)$, and $X_K \cong Y_K$ in $\text{mod-}M_m(K)$ if and only if $F(X)_K \cong F(Y)_K$ in mod-K.) By Swan's theorem (see Remark 1.3.6(c)), we can assume that M and M' are free $\mathcal{Z}^\natural$-modules which are isomorphic over K. Obviously, this means $M \cong M'$. □

Finally, we have:

COROLLARY 4.1.8. *If M is a $\mathcal{J}^\natural$-module which is finite and projective over $\mathcal{Z}^\natural$, then*

(a) *M is a direct sum of $\mathcal{J}^\natural$-modules which are projective over $\mathcal{J}^\natural$ and which have irreducible base change to $\mathcal{J}^\natural_K$-modules.*

(b) *Moreover, M is uniquely determined (up to isomorphism) by the multiplicity of each isomorphism type of the irreducible summands of the $\mathcal{J}^{\natural}_K$-module M_K.*

PROOF. We first prove part (a). By Theorem 4.1.4, we may replace $\mathcal{J}^{\natural}$ by one of its direct factors $M_m(\mathcal{Z}^{\natural})$. Using Morita theory, we can even replace $M_m(\mathcal{Z}^{\natural})$ by $\mathcal{Z}^{\natural}$. Once again, applying Swan's theorem, the $\mathcal{Z}^{\natural}$-module M must be free, and so clearly is a direct sum of $\mathcal{Z}^{\natural}$-modules with irreducible base change to K. This proves (a).

For the proof of (b), the reduction is simply reversed. Further details are left to the reader. □

Appendix to Section 4.1*: the case of the Ree groups of type* $^2F_4$. The Weyl group W is the dihedral group $I_2(8)$ of order 16, generated by reflections s_1 and s_2 with s_1s_2 having order 8. The weights are determined from the longest words in each individual orbit of a graph "symmetry"; see [**C85**, pp. 36-38] or by analogy with the construction [**Lus03**, 16.1–16.2] formulated in the quasi-split case. We can assume $L(s_1) = 2$ and $L(s_2) = 4$ for the values of the weight function L on fundamental reflections s_1, s_2. See [**C85**, pp. 489-490] for the character degrees of the unipotent characters in the principal series. (Note that the degree associated to ρ_2 should be corrected to $(1/2)q^4\Phi_8^2\Phi_{24}$.)[31] There are 7 degrees in all, labeled $1, \epsilon', \epsilon'', \epsilon$ (for the degree 1 representations of $\mathbb{C}W$) and $\rho_2', \rho_2'', \rho_2$ (for the degree 2 representations of $\mathbb{C}W$). The degrees of all the unipotent characters are in terms of polynomial functions, elements of $\mathbb{Q}(\sqrt{2})[q]$, evaluated with q an odd power of $\sqrt{2}$. The generic degrees are obtained by replacing q by the variable t. From the corrected list, we see that the generic degrees have the form

$$\frac{1}{m}t^a\Phi'_{b_1}(t)\cdots\Phi'_{b_r}(t), \tag{4.1.14}$$

where a is a non-negative integer. Also, $\Phi'_{b_i}(t)$ is a factor in $\mathbb{Z}[\sqrt{2}][t]$ of the corresponding cyclotomic polynomial $\Phi_{b_i}(t)$ in $\mathbb{Z}[t]$. The integer m is equal to 1, 2, or 4.

The expression (4.1.14) is sufficiently similar to [**DPS98a**, (4.1.2)] that analogs of [**DPS98a**, Th. 4.2.1] and [**DPS98a**, Th. 4.2.2] may be proved with parallel arguments. It is only necessary to replace $\mathbb{Z}$ by $\mathbb{Z}' = \mathbb{Z}[\sqrt{2}]$, $\mathbb{Q}$ by $\mathbb{Q}' = \mathbb{Q}[\sqrt{2}]$, $\mathcal{H}$ by $\mathcal{H}' = \mathcal{H}_{\mathbb{Z}'[t,t^{-1}]}$, K by $K' = \mathbb{Q}'(t) = K[\sqrt{2}]$, etc.

It will also be useful to define

$$\mathcal{Z}' = \mathbb{Z}'[t,t^{-1}] = \mathbb{Z}[\sqrt{2},t,t^{-1}], \quad \mathcal{Z}'^{\natural} = \mathcal{Z}'[1/\sqrt{2},t,t^{-1}], \quad \mathbb{Z}^{\natural} = \mathbb{Z}[1/2],$$

and $\mathbb{Z}^{\natural\prime} = \mathbb{Z}[1/2][1/\sqrt{2}] = \mathbb{Z}[1/\sqrt{2}]$. Finally, defining $\mathcal{Z}^{\natural\prime} = \mathbb{Z}^{\natural\prime}[t,t^{-1}]$, we have $\mathcal{Z}^{\natural\prime} = \mathcal{Z}'^{\natural}$.

The only "bad prime" is $p = 2$ in $\mathbb{Z}$ (or $p = \sqrt{2}$ in $\mathbb{Z}'$) in the sense of dividing m in (4.1.14). Fortunately, $\mathbb{Z}' = \mathbb{Z}[\sqrt{2}]$ is a principal ideal domain. Using these facts, the argument in the proof of Theorem 4.1.3 establishing the intermediate isomorphism (4.1.12) determines an analogous isomorphism. In the analogy, $\mathcal{H}'$ replaces $\mathcal{H}$, and $\mathcal{Z}'^{\natural}$-free $\mathcal{J}'^{\natural}$-modules L_i' replaces L_i, for each i. Also, $\mathbb{F}_p(t)$ is

[31] The character degree given in [**C85**] does not appear in Lusztig's list of unipotent character degrees. The value we have given does appear in [**Lus84**, p. 374] but just as a unipotent character degree. It can be seen to be a generic/principal series degree by direct calculation using [**GP00**, pp. 270-272].

replaced by $\mathbb{F}'_p(t)$, the residue field of $\mathcal{Z}'_{\mathfrak{p}}$ (which equals the residue field of $\mathcal{Z}'^{\natural}_{\mathfrak{p}}$.) Here, $\mathfrak{p}$ is the ideal generated by $p \in \mathcal{Z}'$, where p is a prime not dividing $\sqrt{2}$. With this analog of the isomorphism (4.1.12) in hand, an evident version of the isomorphism (4.1.11) follows for each $\mathfrak{p}$, using the argument in the proof of Theorem 4.1.3. Consequently, each $\mathfrak{J}'^{\natural}_{\mathfrak{p}}$ is a maximal order. Thus, $\mathfrak{J}'^{\natural}$, which is clearly reflexive, is also a maximal order, and the map

$$\psi : \mathfrak{J}'^{\natural} \to \prod_i \operatorname{End}_{\mathcal{Z}'^{\natural}}(L'_i), \tag{4.1.15}$$

analogous to (4.1.10), is an isomorphism of $\mathcal{Z}'^{\natural}$-algebras. This proves the following theorem, an analog of Theorem 4.1.3 above, in the Ree group setting of this appendix.

THEOREM 4.1.9. *The map ψ in* (4.1.15) *is an isomorphism of $\mathcal{Z}'^{\natural}$-algebras.*

Each of the endomorphism ("matrix") components of (4.1.15) is a maximal order in some $M_n(K')$ by [**Reiner03**, Th. 8.7], since $\mathcal{Z}'^{\natural}$ is integrally closed in $K' = \mathbb{Q}'(t)$. The result below is a corollary of the above theorem, but given its importance, we upgrade it to a theorem.

THEOREM 4.1.10. *The $\mathcal{Z}'^{\natural}$-algebra $\mathfrak{J}'^{\natural} = \mathfrak{J}_{\mathcal{Z}'^{\natural}}$ is a maximal order in $\mathcal{H}_{\mathbb{Q}'(t)}$. Moreover,*

$$\mathfrak{J}'^{\natural} = \mathfrak{J}_{\mathcal{Z}'^{\natural}} \cong \prod_i M_{m_1}(\mathcal{Z}'^{\natural}) \tag{4.1.16}$$

for a finite sequence m_1, m_2, ... of positive integers.

REMARKS 4.1.11.

(a) Similarly, Corollaries 4.1.6–4.1.8 above have analogs for ${}^2F_4$ in the $\mathcal{Z}^{\natural\prime} = \mathcal{Z}'^{\natural}$-setting. We leave the details and proofs, which follow the original arguments, to the reader. Theorem 4.1.10 above is similarly analogous to Theorem 4.1.4.

(b) Also, Corollaries 4.1.6 to 4.1.8 and Theorem 4.1.4(a) remain true in the ${}^2F_4$-case, as stated, that is, using $\mathcal{Z}^{\natural} := \mathbb{Z}[\frac{1}{2}, t, t^{-1}]$, and $\mathfrak{J}^{\natural} := \mathfrak{J}_{\mathcal{Z}^{\natural}}$. (However, Theorem 4.1.4(b) is not true without further modification; see display (4.1.17) below.) The proofs below, in this new setting for the ${}^2F_4$ case, are somewhat more involved.

To prove Theorem 4.1.4(a) in this new setting, note that if $\mathfrak{J}^{\natural}$ were properly contained in a $\mathcal{Z}^{\natural}$-order—call it $\mathfrak{J}^{\clubsuit}$—then the $\mathcal{Z}^{\natural\prime}$-order $\mathfrak{J}^{\clubsuit}_{\mathcal{Z}^{\natural\prime}}$ would properly contain $\mathfrak{J}^{\natural\prime}$, contradicting Theorem 4.1.10 above.

Next, to prove Corollary 4.1.6 in the new setting, observe first that part (b) follows from part (a), using the argument already given in the original context. Part (a) follows from the following faithfully flat descent theorem:

THEOREM 4.1.12 ([**F17**, Th. 8.1.20(1)]). *Let R be a commutative ring, B a finite R-algebra. Suppose S is a commutative faithfully flat R-algebra such that $B \otimes_R S$ is separable over S. Then, B is separable over R.*

For a more elementary proof of Corollary 4.1.6(a), first show, using the maximality of $\mathfrak{J}^{\natural}$ (Theorem 4.1.4(a)), that $\mathfrak{J}^{\natural}$ is a direct product of 1×1 and 2×2 full matrix algebras over $\mathcal{Z}^{\natural}$, and a 2×2 full matrix algebra over $\mathcal{Z}^{\natural\prime}$:

$$\mathfrak{J}^{\natural} \cong \mathcal{Z}^{\natural} \oplus \mathcal{Z}^{\natural} \oplus \mathcal{Z}^{\natural} \oplus \mathcal{Z}^{\natural} \oplus M_2(\mathcal{Z}^{\natural}) \oplus M_2(\mathcal{Z}^{\natural\prime}). \tag{4.1.17}$$

Here, the factor $M_2(\mathcal{Z}^{\natural\prime})$ arises, using the field automorphism $\sqrt{2} \mapsto -\sqrt{2}$, as a "diagonal" $\mathcal{Z}$-subalgebra of $M_2(\mathcal{Z}^{\natural\prime}) \oplus M_2(\mathcal{Z}^{\natural\prime})$. It suffices to prove the separability of this factor over $\mathcal{Z}^{\natural}$, which reduces to the separability of $\mathcal{Z}^{\natural\prime}$ over $\mathcal{Z}^{\natural}$, and then to the separability of $\mathbb{Z}^{\natural\prime}$ over $\mathbb{Z}^{\natural}$. A splitting of the (bimodule) multiplication surjection $\mathbb{Z}^{\natural\prime} \otimes_{\mathbb{Z}^{\natural}} \mathbb{Z}^{\natural'} \longrightarrow \mathbb{Z}^{\natural\prime}$ is given by sending $1 \in \mathbb{Z}^{\natural\prime}$ to $(1/4)(\sqrt{2} \otimes \sqrt{2}) + (1/2)(1 \otimes 1)$. This proves the needed separability and completes the elementary proof.

To prove Corollary 4.1.7 in the new setting, note that proving a similar statement for each component algebra of $\mathcal{J}^{\natural}$ in (4.1.17) is equivalent. The original proof obviously works for all components except for the $\mathcal{Z}^{\natural}$-algebra $M_n(\mathcal{Z}^{\natural\prime})$. Here, it is necessary and sufficient to show (call it a claim) that projective modules M, M' for this component, which are isomorphic after base change to K, are themselves isomorphic. If we view $M_n(\mathcal{Z}^{\natural\prime})$ as a $\mathcal{Z}^{\natural\prime}$-algebra in this claim, and replace K by K', its validity follows from the proof of the original Corollary 4.1.7, in the spirit of (a) above. However, the module categories for this matrix component, whether viewed as a $\mathcal{Z}^{\natural}$- or $\mathcal{Z}^{\natural\prime}$-algebra, are the same. Moreover, the respective base-change functors, $K \otimes_{\mathcal{Z}^{\natural}} (-)$ and $K' \otimes_{\mathcal{Z}^{\natural\prime}} (-)$ are naturally isomorphic, even in the larger category of $\mathcal{Z}^{\natural\prime}$-modules. In view of these identifications, the cited argument proves the claim, and the version of Corollary 4.1.6 in new setting follows.

Corollary 4.1.8 follows in the new setting from its current proof together with the decomposition of $\mathcal{J}^{\natural}$ given in (4.1.17) above. The above identification of $K-$ and $K'-$base changes for $M_n(\mathcal{Z}^{\natural\prime})$-modules is also useful here.

Note, too, that specializing $t \mapsto 1$ in the expression (4.1.17) gives an analogous and very precise version of Remark 4.1.5 in the ${}^2F_4$ case.

Finally, we alert the reader that Theorem 4.1.12 will be needed in Section 4.3.

4.2. Left cells and the order $\mathcal{J} := \mathcal{Z} \otimes_{\mathbb{Z}} \mathsf{J}$. Recall from (4.1.5) that $\varpi : \mathcal{H} \to \mathcal{J}$ is the $\mathcal{Z}$-algebra homomorphism $\varpi = \phi \circ \dagger$, where $\phi : \mathcal{H} \to \mathcal{J}$ is defined in (4.1.4) and $\dagger : \mathcal{H} \to \mathcal{H}$ is defined in (4.1.1). Given a $\mathcal{J}$-module M, there is a unique $\mathcal{H}$-module $M^{\varpi} = \{m^{\varpi} \mid m \in M\}$ obtained by letting $\mathcal{H}$-act on M through ϖ, i.e., $hm^{\varpi} = (\varpi(h)m)^{\varpi}$, for each $h \in \mathcal{H}$, $m \in M$. In other words, M^{ϖ} is the restriction of the $\mathcal{J}$-module M to $\mathcal{H}$ through ϖ. (Throughout this section, modules are taken to be left modules, unless otherwise indicated.)

One approach to understanding the representation theory of a generic Hecke algebra $\mathcal{H}$ is to relate it to the representation theory of the "simpler" asymptotic algebra $\mathcal{J}$, by means of the inclusion $\varpi : \mathcal{H} \to \mathcal{J}$. For example, as just explained, any $\mathcal{J}$-module M can be restricted to $\mathcal{H}$, giving rise to a functor $M \mapsto M^{\varpi}$. Conversely, Theorem 4.2.1 below shows that (under the standing hypotheses on $\mathcal{H}$ of Section 1.2) for certain $\mathcal{H}$-modules S, there exists a unique (up to isomorphism) $\mathcal{J}$-module V such that $V^{\varpi} \cong S$. In fact, we prove that this is true whenever S is a finite direct sum of left-cell modules for $\mathcal{H}$.

If bad primes are inverted as in Section 4.1, then $\mathcal{Z}$ becomes $\mathcal{Z}^{\natural}$, and $\mathcal{J}$ becomes a maximal (and separable) order $\mathcal{J}^{\natural}$ over $\mathcal{Z}^{\natural}$. See Theorem 4.1.4, and Remark 4.1.11(b) in the case ${}^2F_4$. This implies that every finite and $\mathcal{Z}^{\natural}$-projective $\mathcal{J}^{\natural}$-module is a direct sum of finite projective $\mathcal{J}^{\natural}$-modules, each of which becomes irreducible upon base change to $\mathcal{J}_{\mathbb{Q}(t)}$. Such a decomposition is unique up to isomorphism. Also, because $\mathcal{Z}^{\natural}$ is regular of Krull dimension ≤ 2 (it is dimension 2, in fact), every irreducible $\mathcal{J}_{\mathbb{Q}(t)}$-module is the base change of a $\mathcal{J}^{\natural}$-module which is projective over

$\mathcal{Z}^\natural$. In fact, applying Swan's theorem, any such module is free over $\mathcal{Z}^\natural$. In effect, this provides a new refinement of the set of left-cell modules.

THEOREM 4.2.1. *Let S be a left-cell module (or even a finite direct sum of left-cell modules) for $\mathcal{H}$. Then, there exists a $\mathfrak{J}$-module M, unique up to isomorphism, such that $S \cong M^{\varpi}$.*

PROOF. We use several results from [**Lus03**, Ch. 18]. Let $*$ be as defined in [**Lus03**, 18.10]; it is a left-module action of $\mathfrak{J}$ on $\mathcal{H}$. Now define a new action $\boxdot$ of $\mathfrak{J}$ on $\mathcal{H}$ by putting $j \boxdot h := (j * h^\dagger)^\dagger$, for $j \in \mathfrak{J}$ and $h \in \mathcal{H}$. For each integer a, let

$$\mathcal{H}_a = \oplus_{w;a(w)=a}\mathcal{Z}\mathrm{c}_w^\dagger \quad \text{and} \quad \mathcal{H}_{\geq a} = \oplus_{w;a(w)\geq a}\mathcal{Z}\mathrm{c}_w^\dagger \tag{4.2.1}$$

as in [**Lus03**, 18.10], where $a(-)$ is the Lusztig a-function defined in (4.1.2); see [**Lus03**, 13.6]. Lusztig observes that $\mathrm{j}_x * c_w^\dagger \in \mathcal{H}_{a(w)}$, for all $x, w \in W$. In particular, $\mathcal{H}_{\geq a}$ is stable under the $*$ action of $\mathfrak{J}$. He proves [**Lus03**, 18.10(a)] that

$$h\mathrm{c}_w^\dagger \equiv \phi(h) * \mathrm{c}_w^\dagger \bmod \mathcal{H}_{\geq a(w)+1}, \quad \forall w \in W.$$

Replacing h by $h^\dagger$, and then applying $\dagger$ to both sides, we get

$$h\mathrm{c}_w = (\phi(h^\dagger) * \mathrm{c}_w^\dagger)^\dagger = \varpi(h) \boxdot \mathrm{c}_w \mod (\mathcal{H}_{\geq a(w)+1})^\dagger. \tag{4.2.2}$$

It follows from Lusztig's observation above that $(\mathcal{H}_{\geq a})^\dagger$ is stable under the $\boxdot$-action of $\mathfrak{J}$, as is $(\mathcal{H}_{\geq q})^\dagger/(\mathcal{H}_{\geq a+1})^\dagger$. Observe that $-a(-)$ is a height function in the sense of Appendix B compatible with $\leq_{LR}$ by [**Lus03**, Ch. 14, P4]. By Corollary 2.1.8 above, the left $\mathcal{H}$-module $\mathcal{H} = {}_{\mathcal{H}}\mathcal{H}$ has a height filtration with respect to the (compatible and negatively-valued) height function $-a(-)-1$. Note that $(\mathcal{H}_{\geq a})^\dagger/(\mathcal{H}_{\geq a+1})^\dagger$ is a section of such a filtration. The equation (4.2.2) shows that this natural $\mathcal{H}$-module is also the restriction through ϖ of a module for $\mathfrak{J}$, the latter acting via $\boxdot$ on the same $\mathcal{Z}$-module. Since the $\mathcal{H}$-module ${}_{\mathcal{H}}\mathcal{H}$ belongs to $\mathscr{A}(\mathscr{L})$ using Corollary 2.1.8, its filtration sections are direct sums of left-cell modules. Also, it is clear that every left-cell module for $\mathcal{H}$ appears as a direct summand of some such section. The theorem now follows from the elementary Lemma 4.2.2 below, which proves uniqueness (see part (c)) as well as completing the proof of existence (see part (d)—the notation differs slightly). □

Temporarily, let $\mathscr{K}$ be any Noetherian integral domain, and let K be its fraction field. The following lemma is inspired by the arguments in [**Reiner03**, below (8.2), p. 109]. Note the definition of $\mathfrak{B}\text{-mod}^f$ is given in part (a). A similar definition applies for $\mathfrak{A}\text{-mod}^f$. Note also that part (b) will be used in the proof of Theorem 4.3.4.

LEMMA 4.2.2. *Let $\mathfrak{A}$ be a finite and torsion-free $\mathscr{K}$-algebra and let $\mathfrak{B}$ be an order in the K-algebra $\mathfrak{A}_K$ containing $\mathfrak{A}$. Then,*

(a) *The restriction functor $\mathfrak{B}\text{-mod}^f \to \mathfrak{A}$-mod defines a fully faithful $\mathscr{K}$-linear functor from the $\mathscr{K}$-category $\mathfrak{B}\text{-mod}^f$ of $\mathscr{K}$-finite and $\mathscr{K}$-torsion-free $\mathfrak{B}$-modules to the $\mathscr{K}$-category of $\mathfrak{A}$-modules.*
(b) *In particular, for any $M \in \mathfrak{B}\text{-mod}^f$, we have*

$$\mathrm{End}_{\mathfrak{B}}(M) \cong \mathrm{End}_{\mathfrak{A}}(M|_{\mathfrak{A}})$$

as $\mathscr{K}$-algebras. Also, $M|_{\mathfrak{A}} \in \mathfrak{A}\text{-mod}^f$.

(c) *Suppose that* $M, M' \in \mathfrak{B}\text{-mod}^f$, *and let* N, N' *denote the* $\mathfrak{A}$*-modules which are their respective restrictions to* $\mathfrak{A}$. *Then,* $M \cong M'$ *as* $\mathfrak{B}$*-modules if and only if* $N \cong N'$ *as* $\mathfrak{A}$*-modules.*

(d) *Suppose* $M \in \mathfrak{B}\text{-mod}^f$. *Assume that* $N := M|_{\mathfrak{A}}$ *can be written as a direct sum* $N = N_1 \oplus N_2$ *in* $\mathfrak{A}$*-mod. Then,* $M = M_1 \oplus M_2$ *in* $\mathfrak{B}\text{-mod}^f$, *where* $M_i|_{\mathfrak{A}} \cong N_i$, $i = 1, 2$. *(Also,* N_1, N_2 *are in* $\mathfrak{A}\text{–mod}^f$*.)*

PROOF. We first prove part (a). Obviously, restriction is a $\mathscr{K}$-linear functor. We have to prove it induces an isomorphism on Hom-spaces between objects. For $M, M' \in \mathfrak{B}\text{-mod}^f$, note that $\mathrm{Hom}_{\mathfrak{B}}(M, M')$ identifies with the $\mathscr{K}$-submodule of $\mathrm{Hom}_{\mathfrak{B}_K}(M_K, M'_K)$ consisting of the elements of the latter that send $M \subseteq M_K$ to $M' \subseteq M'_K$. A similar identification holds for $\mathrm{Hom}_{\mathfrak{A}}(M|_{\mathfrak{A}}, M'|_{\mathfrak{A}})$ in $\mathrm{Hom}_{\mathfrak{A}_K}(M_K, M'_K)$. Since $\mathfrak{B}_K = K\mathfrak{B} = K\mathfrak{A} = \mathfrak{A}_K$, the isomorphism required in part (a) holds.

Part (b) follows from the argument above with $M = M'$ (using the fact that restriction is a functor, hence it induces an algebra homomorphism). The fact that $M|_{\mathfrak{A}}$ belongs to $\mathfrak{A}\text{–mod}^f$ is obvious. Part (c) also follows from a functorial argument using (a). Finally, (d) follows from the functorial properties of (a) and (b). It is also necessary to observe for part (d) that idempotents split in $\mathfrak{B}\text{–mod}^f$. More precisely, in terms of the argument for (a), if $e \in \mathrm{End}_{\mathfrak{B}}(M)$ is an idempotent, with $M \in \mathfrak{B}\text{-mod}^f$, then eM also belongs to $\mathfrak{B}\text{–mod}^f$. □

Now return to the setting of Theorem 4.2.1, but consider modules over $\mathcal{Z}^\natural$, rather than $\mathcal{Z}$. For each $E \in \mathrm{Irr}(\mathbb{Q}W)$, let $E_{\mathbb{Q}(t)}$ be the corresponding irreducible $\mathbb{Q}(t)W$-module, regarded as a $\mathcal{J}_{\mathbb{Q}(t)}$-module through the isomorphism $\mathbb{Q}(t)W \cong \mathcal{J}_{\mathbb{Q}(t)}$. (In the ${}^2F_4$ case, E and $E_{\mathbb{Q}(t)}$ may not be absolutely irreducible.) Thus, by Remark 4.1.5(2), $E_{\mathbb{Q}(t)} \cong L_i$, for some i. Choose a $\mathcal{J}^\natural$-lattice $E_\natural$ in $E_{\mathbb{Q}(t)}$ that is a projective as a $\mathcal{Z}^\natural$-module (thus free by Swan's theorem [**Sw78**, first paragraph, p. 111]). For example, for a type other than ${}^2F_4$, we may choose $E_\natural \cong L_{i\,\mathcal{Z}^\natural}$, for some i, which is $\mathcal{Z}^\natural$-free in the notation of Remark 4.1.5(2). Since $\mathcal{J}^\natural$ is projective over $\mathcal{Z}^\natural$, the $\mathcal{J}^\natural$-module $E_\natural$ is projective over $\mathcal{J}^\natural$; see Corollary 4.1.6(b) and its analog in Remark 4.1.11(b).

We remark that $E_\natural$ is uniquely determined (up to isomorphism) by Corollary 4.1.7 and its twisted ${}^2F_4$ analog in Remark 4.1.11(b). Define

$$S^\natural(E) := (E_\natural)^{\varpi^\natural}, \quad S_E^\natural := \mathrm{Hom}_{\mathcal{Z}^\natural}(S^\natural(E), \mathcal{Z}^\natural), \tag{4.2.3}$$

which belong to $\mathcal{H}^\natural$-mod and mod-$\mathcal{H}^\natural$, respectively. Note that $S^\natural(E)$ and $S_E^\natural$ are natural $\mathcal{H}^\natural$-forms for the $\mathcal{H}_{\mathbb{Q}[t,t^{-1}]}$-modules $S(E)$ and S_E defined in [**DPS15**, §3]. Note also that $S^\natural(E)_{\mathbb{Q}(t)} \cong E_{\mathbb{Q}(t)} \cong L_i$.

REMARK 4.2.3. Since $\{S^\natural(E)_K \mid E \in \mathrm{Irr}(\mathbb{Q}W)\}$, where $K = \mathbb{Q}(t)$, forms a complete set of irreducible left $\mathcal{H}_K$-modules (see Remark 4.1.5(2)), it follows that $\{(S_E^\natural)_K \mid E \in \mathrm{Irr}(\mathbb{Q}W)\}$ is a complete set of irreducible right $\mathcal{H}_K$-modules. Thus, since every $(S_E^\natural)_K$ is an irreducible constituent of T_K^+, the set

$$\{\mathrm{Hom}_{\mathcal{H}_K^\natural}((S_E^\natural)_K, T_K^{+\natural}) \mid E \in \mathrm{Irr}(\mathbb{Q}W)\}$$

forms a complete set of irreducible left A_K^+-modules. (Note that $A_K^+ = \mathrm{End}_{\mathcal{H}^\natural}(T^{+\natural})_K$. So, A_K^+ is split semisimple if ${}^2F_4$ is excluded.) Now, if $\Delta(E) := \mathrm{Hom}_{\mathcal{H}^\natural}(S_E^\natural, T^{+\natural})$, then $\Delta(E) \subseteq \mathrm{Hom}_{\mathcal{H}_K^\natural}((S_E^\natural)_K, T_K^{+\natural})$ is an $A^{+\natural}$-submodule, which is a torsion-free

$\mathcal{Z}^\natural$-module. Hence, $\Delta(E)_K = K\Delta(E)$ is a (nonzero) submodule of the A_K^+-module $\mathrm{Hom}_{\mathcal{H}_K^\natural}((S_E^\natural)_K, T_K^{+\natural})$. Irreducibility (treated above) forces

$$\Delta(E)_K = \mathrm{Hom}_{\mathcal{H}_K^\natural}((S_E^\natural)_K, T_K^{+\natural}).$$

Therefore, $\{\Delta(E)_K \mid E \in \mathrm{Irr}(\mathbb{Q}W)\}$ forms a complete set of irreducible left A_K^+-modules. To include the ${}^2F_4$ case, we may simply replace $\mathcal{Z}^\natural$ by $\mathcal{Z}^{\natural\prime} = \mathcal{Z}^\natural[\sqrt{2}]$, K by K', and $\Delta(E)$ by $\Delta'(E) := \Delta(E)_{\mathcal{Z}^{\natural\prime}}$ to get a complete set $\{\Delta'(E)_{K'} \mid E \in \mathrm{Irr}(\mathbb{Q}'W)\}$ of irreducible left $A_{K'}^+$-modules. Here, the fields K', $\mathbb{Q}'$ are obtained by adjoining $\sqrt{2}$ to K, $\mathbb{Q}$, respectively.

THEOREM 4.2.4.

(a) *Let N be a finite direct sum of left-cell modules for $\mathcal{H}$, and let $N^\natural := N_{\mathcal{Z}^\natural}$ be the corresponding (base-changed) module for $\mathcal{H}^\natural$. If M is a direct summand of the $\mathcal{H}^\natural$-module $N^\natural$, then M admits a direct sum decomposition $M = \oplus_{i=1}^r M_i$ in $\mathcal{H}^\natural$-mod such that each $M_i \cong S^\natural(E_i)$, for some $E_i \in \mathrm{Irr}(\mathbb{Q}W)$. The modules M_i and their multiplicities in M are determined up to isomorphism by M_K, where $K = \mathbb{Q}(t)$.*

(b) *Assume that (W, S, L) is of type ${}^2F_4$ (so the base-change notations $(-)^{\natural\prime}$ and $(-)'$ are defined). Then, (a) also holds after substituting the top row for the bottom row in the display (see the appendix to Section 4.1):*

$$\text{(4.2.4)}\qquad \begin{array}{cccccc} \mathcal{H}' & \mathcal{Z}^{\natural\prime} & \mathcal{H}^{\natural\prime} & \mathbb{Q}' & K' & N^{\natural\prime} \\ \mathcal{H} & \mathcal{Z}^\natural & \mathcal{H}^\natural & \mathbb{Q} & K & N^\natural \end{array}$$

PROOF. (a) First, we establish some projectivity properties. Note that all left-cell modules for $\mathcal{H}$ are projective (even free) over $\mathcal{Z}$. Thus, N is projective over $\mathcal{Z}$, and both $N^\natural$ and its direct summand M are projective over $\mathcal{Z}^\natural$. Consequently, if $M \cong X^{\varpi^\natural}$ (the restriction of X to $\mathcal{H}^\natural$ through $\varpi_{\mathcal{Z}^\natural}$) for a given $\mathcal{J}^\natural$-module X, then X is projective over $\mathcal{Z}^\natural$. Consequently, X is also projective over $\mathcal{J}^\natural$, by Corollary 4.1.6(b).

By Theorem 4.2.1, N is (isomorphic to) the restriction (through ϖ) to $\mathcal{H}$ of a module for $\mathcal{J}$. Thus, $N^\natural$ is the restriction to $\mathcal{H}^\natural$ of a module Y for the order $\mathcal{J}^\natural$, and the analogous statement holds for M by Lemma 4.2.2(d). (Use $\mathfrak{A} = \mathcal{H}^\natural$ and let $\mathfrak{B} = \varpi_K^{-1}(\mathcal{J}^\natural)$.) Accordingly, let X be a $\mathcal{J}^\natural$-module such that M is the restriction $X^{\varpi^\natural}$, and note that X is projective by the paragraph above. (Alternately, Lemma 4.2.2(d) gives an X which is a direct summand of Y, and Y can also be shown to be projective over $\mathcal{J}^\natural$.) Write $X_K \cong L_1 \oplus \cdots \oplus L_r$, where the L_i are irreducible $\mathcal{J}_K^\natural$-modules. Then, $L_i = (L_i')_K$, for some $\mathcal{J}^\natural$-lattice L_i' in L_i. In fact, we may take $L_i \cong (E_i)_{\mathbb{Q}(t)}$, $E_i \in \mathrm{Irr}(\mathbb{Q}W)$, and $L_i' \cong E_{i\natural}$ as discussed above the theorem. Note also from that discussion that $E_{i\natural}$ is projective over $\mathcal{J}^\natural$ and restricts though $\varpi^\natural$ to $S^\natural(E_i)$. The direct sum of all of the $E_{i\natural}$ remains projective and has the same base change to $\mathcal{J}_K^\natural$ as X_K. Using the projectivity of X and Corollary 4.1.7, we deduce that X is the direct sum of all the $E_{i\natural}$. Applying the functor $(-)^{\varpi^\natural}$ to X and its summands gives the desired description of M in part (a)—all but the last sentence, which is again a consequence of Corollary 4.1.7. The L_i' and their multiplicities are uniquely determined by Corollary 4.1.7 as well.

This completes the proof of (a), while the proof of (b) is similar, *mutatis mutandis*. Remark 4.1.11(a) can be used as a guide. □

COROLLARY 4.2.5. *The following statements hold for $\mathcal{H}$-mod:*

(a) *Let N_1, N_2 be left-cell modules for $\mathcal{H}$, and suppose M_1, M_2 are direct summands of $N_1^\natural, N_2^\natural$, respectively. Then, $M_1 \cong M_2$ as $\mathcal{H}^\natural$-modules if and only if $(M_1)_{\mathbb{Q}(t)} \cong (M_2)_{\mathbb{Q}(t)}$ as $\mathcal{H}_{\mathbb{Q}(t)}$-modules.*

(b) *In the type ${}^2F_4$ case, (a) also holds if changes in notation analogous to those in Theorem 4.2.4(b) are made.*

Part (a) is clear from the above theorem and its preamble, using the split semsimplicity of $\mathbb{Q}(t)W$ and algebras isomorphic to it. Part (b) is similar, replacing $\mathbb{Q}(t)W$ by $\mathbb{Q}'(t)W$.

4.3. The quasi-heredity of $A^{+\natural}$. Quasi-hereditary algebras (over a field) were first introduced by CPS (E. Cline and the second and third authors) [**CPS88**] in order to deal with highest weight categories arising naturally in the representation theory of semisimple Lie algebras and algebraic groups. Examples of quasi-hereditary algebras include Schur (or, more generally, q-Schur) algebras and Auslander algebras which arise rather naturally. With the maximal order theory developed in the previous sections, we are now ready to prove that the base change from $\mathcal{Z}$ to $\mathcal{Z}^\natural$ of the standardly stratified algebra A^+ is, in fact, (integrally) quasi-hereditary.

We begin with some terminology about (integral) quasi-hereditary algebras. A good source for much of this material is [**DPS18**, §2].

DEFINITION 4.3.1.

(1) Let B be an algebra over a commutative Noetherian ring $\mathscr{K}$. Assume that B is projective and finite over $\mathscr{K}$. An ideal J of B is called a *heredity ideal* in B if the following four conditions hold:

(HI1) B/J is projective over $\mathscr{K}$;
(HI2) ${}_BJ$ is B-projective;
(HI3) $J^2 = J$; and
(HI4) The endomorphism algebra $E := \mathrm{End}_B({}_BJ)$ is $\mathscr{K}$-semisimple.[32]

(2) The algebra B, still assumed to be projective over $\mathscr{K}$, is called a *quasi-hereditary algebra* (QHA) provided there exists a finite "defining sequence"

$$0 = J_0 \subseteq J_1 \subseteq \cdots \subseteq J_m = B$$

of ideals in B such that, for $0 < i \leq m$, J_i/J_{i-1} is a heredity ideal in B/J_{i-1}.

(3) If, in addition, every J_i/J_{i-1} is a heredity ideal of *split* type, in the sense that the $\mathscr{K}$-semisimple algebra $\mathrm{End}_{B/J_{i-1}}(J_i/J_{i-1})$ is isomorphic to a direct product of $\mathrm{End}_{\mathscr{K}}(P)$, where P is a finite (faithful) $\mathscr{K}$-projective module, we call B a *split* quasi-hereditary algebra (or a quasi-hereditary algebra of *split* type).

Notice that, since P is necessarily a projective generator for $\mathscr{K}$-mod, $\mathrm{End}_{\mathscr{K}}(P)$ is Morita equivalent to $\mathscr{K}$.

Clearly, a heredity ideal is a stratifying ideal and every quasi-hereditary algebra is a standardly stratified algebra. The "split" quasi-hereditary algebras defined above are quite popular and useful. Some mild variations have been collected in Remark 4.3.2 below.

REMARK 4.3.2.

[32]This means that, for all $\mathfrak{p} \in \mathrm{Spec}(\mathscr{K})$, $E(\mathfrak{p})$ is a semisimple algebra over the residue field $\mathscr{K}(\mathfrak{p}) := \mathscr{K}_\mathfrak{p}/\mathfrak{p}\mathscr{K}_\mathfrak{p}$. Equivalently, any short exact sequence of left E-modules which is split over $\mathscr{K}$, is split over E. (See [**CPS90**, §2].)

(1) Recall that an algebra E is *separable* over $\mathscr{K}$, provided the multiplication map $E \otimes_{\mathscr{K}} E^{\rm op} \to E$ splits as an $(E, E^{\rm op})$-bimodule map. By definition, E is *semisplit* if it is a direct product of algebras, each of which is separable with center $\mathscr{K}$ (i.e., an Azumaya algebra). If the $\mathscr{K}$-semisimple algebra $E := \mathrm{End}_B({}_BJ)$ in (HI4) is separable (respectively, semisplit) over $\mathscr{K}$, we say that J is of *separable* (respectively, *semisplit) type* over $\mathscr{K}$.

Then, if each J_i/J_{i-1} is of separable (respectively, semisplit) type over $\mathscr{K}$, we say that the quasi-hereditary algebra B is of *separable* (respectively, *semisplit*) *type* over $\mathscr{K}$. See [**CPS90**] for more details.

(2) Following [**Ro08**, Def. 4.1.4], a heredity ideal J of split type is said to be *indecomposable*, or J is an *indecomposable split heredity ideal*, if (HI4) is replaced by (HI4)$'$:

(HI4)$'$ The algebra $E := \mathrm{End}_B({}_BJ)$ is Morita equivalent to $\mathscr{K}$.

Rouquier established a very nice relation between indecomposable split heredity ideals J and "standard (projective) B-modules" L that are $\mathscr{K}$-faithful and satisfy one of the equivalent conditions in [**Ro08**, Lem. 4.5]:

Given L, the coresponding J satisfies the following B-B-bimodule isomorphism

$$J \cong L \otimes_{\mathrm{End}_B(L)} \mathrm{Hom}_B(L, A), \tag{4.3.1}$$

while, given J, the corresponding L satisfies the left B-module isomorphism

$$L \cong J \otimes_{\mathrm{End}_B(J)} P, \tag{4.3.2}$$

where P is a progenerator for $\mathrm{End}_B(J)$ with $\mathrm{End}_{\mathrm{End}_B(J)}(P) \cong \mathscr{K}$.

The construction of L from J in (4.3.2) is also given in [**CPS90**, p.157]. This, in particular, implies that $\mathrm{End}_B(L) \cong \mathscr{K}$. We will call the isomorphism (4.3.1) *Rouquier's isomorphism*. This isomorphism is useful in the construction of a standard basis for the split quasi-hereditary algebra; see Section 5.1.

(3) Note also that the definition of split quasi-hereditary algebras in [**Ro08**, 4.1.4], using indecomposable split heredity ideals ([**Ro08**, Def. 4.1]) and a poset, is stronger (at least in formulation) than Definition 4.3.1(3).

The following result is needed in the proof of the theorem below and is obtained by adapting an argument of Hochster [**Ho07**] (compare also [**Bo72**, Prop. 11, p.23]).

LEMMA 4.3.3. *Let* $\mathtt{H}$ *be a* $\mathcal{Z}$*-algebra which is finite and projective as a* $\mathcal{Z}$*-module, and let* M, N *be finite left* $\mathtt{H}$*-modules. If* M *is a finitely presented* $\mathtt{H}$*-module, then there exists a* $\mathcal{Z}^\natural$*-module isomorphism*

$$\phi^{\mathcal{Z}^\natural,\mathtt{H}}_{M,N} : \mathcal{Z}^\natural \otimes_{\mathcal{Z}} \mathrm{Hom}_{\mathtt{H}}(M, N) \xrightarrow{\sim} \mathrm{Hom}_{\mathtt{H}^\natural}(M^\natural, N^\natural), \tag{4.3.3}$$

defined, for $u \in \mathcal{Z}^\natural$ *and* $f \in \mathrm{Hom}_{\mathtt{H}}(M, N)$, *by*

$$\phi^{\mathcal{Z}^\natural,\mathtt{H}}_{M,N}(u \otimes f)(v \otimes m) = uvf(m) \quad (v \in \mathcal{Z}^\natural, m \in M).$$

Moreover, when $M = N$, $\phi^{\mathcal{Z}^\natural,\mathtt{H}}_{N,N} : \mathcal{Z}^\natural \otimes \mathrm{End}_{\mathtt{H}}(N) \to \mathrm{End}_{\mathtt{H}^\natural}(N^\natural)$ *is a natural isomorphism of* $\mathcal{Z}^\natural$*-algebras.*

PROOF. Let $\phi = \phi^{\mathcal{Z}^\natural,\mathtt{H}}_{M,N}$. It is helpful to think of ϕ as a composition $\nu \circ \psi$ of two simpler maps ν and ψ. The map

$$\psi : \mathcal{Z}^\natural \otimes_{\mathcal{Z}} \mathrm{Hom}_{\mathtt{H}}(M, N) \longrightarrow \mathrm{Hom}_{\mathtt{H}}(M, N^\natural)$$

is given by $\psi(u\otimes f)(m) = uf(m)$ $(u\in\mathcal{Z}^\natural,\ f\in \mathrm{Hom}_{\mathtt{H}}(M,N),\ m\in M)$. We will discuss ν in a moment, but, for the next two paragraphs, we continue to focus on ψ.

We will, however, let M vary. First, take $M={}_{\mathtt{H}}\mathtt{H}$, the left $\mathtt{H}$-module which is $\mathtt{H}$ itself. We abbreviate ${}_{\mathtt{H}}\mathtt{H}$ to $\mathtt{H}$ when usage is clear from context. There are well-known isomorphisms $\mathrm{Hom}_{\mathtt{H}}(\mathtt{H},N)\cong N$ (here N could be any left $\mathtt{H}$-module) in which $f\in\mathrm{Hom}_{\mathtt{H}}(\mathtt{H},N)$ corresponds to $f(1)\in N$. Equivalently, $n\in N$ corresponds to $f_n\in\mathrm{Hom}_{\mathtt{H}}(\mathtt{H},N)$, where $f_n(h)=hn$ $(h\in\mathtt{H})$. Using these isomorphisms and correspondences, we obtain a commutative diagram

$$\begin{array}{ccc}
\mathcal{Z}^\natural\otimes N & \xrightarrow{\ =\ } & N^\natural\\
\big\downarrow & & \big\uparrow\\
\mathcal{Z}^\natural\otimes\mathrm{Hom}_{\mathtt{H}}(\mathtt{H},N) & \xrightarrow{\ \psi\ } & \mathrm{Hom}_{\mathtt{H}}(\mathtt{H},N^\natural).
\end{array}$$

with the top arrow equality and each vertical map an isomorphism. As a consequence, the map ψ is an isomorphism, under the $M={}_{\mathtt{H}}\mathtt{H}$ hypothesis. This carries over to the weaker assumption that M is a direct sum of finitely many copies of the $\mathtt{H}$-modules ${}_{\mathtt{H}}\mathtt{H}$. Finally, ψ is an isomorphism whenever M is a finitely presented over $\mathtt{H}$.

We sketch the argument for this last assertion, which follows the spirit of [**Ho07**]: Let $E\to G\to M\to 0$ be a presentation of M in terms of finite direct sums E and G of the $\mathtt{H}$-modules ${}_{\mathtt{H}}\mathtt{H}$. Use it to form the columns of the diagram below

$$\begin{array}{ccc}
\mathcal{Z}^\natural\otimes\mathrm{Hom}_{\mathtt{H}}(E,N) & \longrightarrow & \mathrm{Hom}_{\mathtt{H}}(E,N^\natural)\\
\big\uparrow & & \big\uparrow\\
\mathcal{Z}^\natural\otimes\mathrm{Hom}_{\mathtt{H}}(G,N) & \longrightarrow & \mathrm{Hom}_{\mathtt{H}}(G,N^\natural)\\
\big\uparrow & & \big\uparrow\\
\mathcal{Z}^\natural\otimes\mathrm{Hom}_{\mathtt{H}}(M,N) & \longrightarrow & \mathrm{Hom}_{\mathtt{H}}(M,N^\natural)\\
\big\uparrow & & \big\uparrow\\
0 & & 0
\end{array}$$

The columns are exact, by virtue of partial exactness properties of $\mathrm{Hom}_{\mathtt{H}}$, together with flatness of $\mathcal{Z}^\natural$ over $\mathcal{Z}$. The bottom row is ψ and the other two rows are the versions of ψ obtained by using G and E, respectively, in the role of M. The squares are both commutative, as can be checked with the formulas for ψ which apply in each row. Finally, since the top two rows are isomorphisms, we conclude that the bottom row is an isomorphism (by the 5-Lemma), as asserted.

We now return to the map $\nu:\mathrm{Hom}_{\mathtt{H}}(M,N^\natural)\to\mathrm{Hom}_{\mathtt{H}^\natural}(M^\natural,N^\natural)$ and define it as follows. First, define

$$\tilde\nu:\mathrm{Hom}_{\mathcal{Z}}(M,N^\natural)\longrightarrow\mathrm{Hom}_{\mathcal{Z}^\natural}(M^\natural,N^\natural)$$

on $f\in\mathrm{Hom}_{\mathcal{Z}}(M,N^\natural)$ by $\tilde\nu(f)(v\otimes f)=vf(m)$ $(m\in M,\ v\in\mathcal{Z}^\natural)$. From this formula, it is easily seen that $\tilde\nu(f)\in\mathrm{Hom}_{\mathcal{Z}^\natural}(M^\natural,N^\natural)$. Moreover, if f belongs to the Hom space $\mathrm{Hom}_{\mathtt{H}}(M,N^\natural)$, then $\tilde\nu(f)$ belongs to $\mathrm{Hom}_{\mathtt{H}^\natural}(M^\natural,N^\natural)$. Starting from $f\in\mathrm{Hom}_{\mathtt{H}}(M,N^\natural)$, we have $hf(m)=f(hm)$, for $h\in\mathtt{H}, m\in M$. If $v\in\mathcal{Z}^\natural$, we have

$$h\tilde\nu(f)(v\otimes m)=hvf(m)=vhf(m)=vf(hm)=\tilde\nu(f)(v\otimes hm)=\tilde\nu(f)h(v\otimes m).$$

Thus, $\tilde{\nu}(f) \in \mathrm{Hom}_{\mathtt{H}}(M^\natural, N^\natural)$. We already know that $\tilde{\nu}(f) \in \mathrm{Hom}_{\mathcal{Z}^\natural}(M^\natural, N^\natural)$, so now we have $\tilde{\nu}(f) \in \mathrm{Hom}_{\mathcal{Z}^\natural\mathtt{H}}(M^\natural, N^\natural) = \mathrm{Hom}_{\mathtt{H}^\natural}(M^\natural, N^\natural)$. At this point, we can (and do) define ν to be the restriction of $\tilde{\nu}$ to $\mathrm{Hom}_{\mathtt{H}}(M, N^\natural)$:

$$\nu : \mathrm{Hom}_{\mathtt{H}}(M, N^\natural) \longrightarrow \mathrm{Hom}_{\mathtt{H}^\natural}(M^\natural, N^\natural).$$

Our next aim is to show that ν is an isomorphism, assuming M is finitely presented (as we assumed when dealing with ψ). As before, the case $M = {}_{\mathtt{H}}\mathtt{H}$ (which we abbreviate to $\mathtt{H}$) leads to a standard identification $\mathrm{Hom}_{\mathtt{H}}(\mathtt{H}, N^\natural) \cong N^\natural$, and there is a similar identification $\mathrm{Hom}_{\mathtt{H}^\natural}(\mathtt{H}^\natural, N^\natural) \cong N^\natural$. A calculation, much as we carried out for ψ, leads to a commutative diagram

$$\begin{array}{ccc} N^\natural & \xrightarrow{=} & N^\natural \\ \downarrow & & \downarrow \\ \mathrm{Hom}_{\mathtt{H}}(\mathtt{H}, N^\natural) & \xrightarrow{\nu} & \mathrm{Hom}_{\mathtt{H}^\natural}(\mathtt{H}^\natural, N^\natural). \end{array}$$

We provide a few more details about the diagram and its commutativity. Here, the two vertical maps are "standard", sending $n^\natural \in N^\natural$ to the unique function in the Hom space drawn below it, and taking the value $n^\natural$ at 1. We let $f_{n^\natural} \in \mathrm{Hom}_{\mathtt{H}}(\mathtt{H}, N^\natural)$ be the unique function on the left, and $f^\natural_{n^\natural} \in \mathrm{Hom}_{\mathtt{H}^\natural}(\mathtt{H}^\natural, N^\natural)$ on the right (keeping $n^\natural$ the same). Then, $\nu(f_{n^\natural})(1 \otimes 1) = f_{n^\natural}(1) = n^\natural$. Consequently, $\nu(f_{n^\natural}) = f^\natural_{n^\natural}$, and the diagram (which has equality as its top row, and the map ν as its bottom row) is now shown to be commutative. Since the columns are isomorphisms, as is the top row, it follows that the bottom row ν is also an isomorphism.

We can also use a similar argument to show that ν is an isomorphism when M is a direct sum $\mathtt{H}^{\oplus r}$ of finitely many copies of $\mathtt{H}$ (say r of them). In the diagram used for $M = \mathtt{H}$, replace the variable $\mathtt{H}$ inside the left "Hom" term with $\mathtt{H}^{\oplus r}$ and replace $\mathtt{H}^\natural$ inside the right "Hom" term with $(\mathtt{H}^\natural)^{\oplus r}$. Both terms $N^\natural$ in the top row are replaced by $(N^\natural)^{\oplus r}$. Remaining details, showing ν is an isomorphism in this $M = \mathtt{H}^{\oplus r}$ case, are left to the reader.

We can now show that ν is an isomorphism whenever the associated $\mathtt{H}$-module M has a presentation $E \to G \to M \to 0$ by finite direct sums G and E of copies of $\mathtt{H}$. The relevant diagram is displayed below.

$$\begin{array}{ccc} \mathrm{Hom}_{\mathtt{H}}(E, N^\natural) & \longrightarrow & \mathrm{Hom}_{\mathtt{H}^\natural}(E^\natural, N^\natural) \\ \uparrow & & \uparrow \\ \mathrm{Hom}_{\mathtt{H}}(G, N^\natural) & \longrightarrow & \mathrm{Hom}_{\mathtt{H}^\natural}(G^\natural, N^\natural) \\ \uparrow & & \uparrow \\ \mathrm{Hom}_{\mathtt{H}}(M, N^\natural) & \longrightarrow & \mathrm{Hom}_{\mathtt{H}^\natural}(M^\natural, N^\natural) \\ \uparrow & & \uparrow \\ 0 & & 0 \end{array}$$

The right column is obtained by applying $\mathcal{Z}^\natural \otimes (-)$ and, then, $\mathrm{Hom}_{\mathtt{H}^\natural}(-, N^\natural)$ to $E \to G \to M \to 0$. The left column is obtained similarly, omitting the $\mathcal{Z}^\natural \otimes (-)$ step. Both columns are exact. The rows are given by the maps ν associated with M, G, E on their respective rows. By our previous discussion, this implies that the top two rows are isomorphisms. Both squares are commutative, as can be checked with the formulas for ν that apply. By the 5-Lemma, the bottom row ν is also an isomorphism.

It now follows that the composite

$$\phi = \nu \circ \psi : \mathcal{Z}^\natural \otimes_{\mathcal{Z}} \mathrm{Hom}_{\mathbf{H}}(M, N) \xrightarrow{\sim} \mathrm{Hom}_{\mathbf{H}^\natural}(M^\natural, N^\natural)$$

is an isomorphism, proving the first assertion of the lemma.

We now assume $M = N$ and prove the last assertion for $\phi = \phi_{N,N}^{\mathcal{Z}^\natural,\mathbf{H}}$. We begin by showing that ϕ, as a $\mathcal{Z}^\natural$-linear map, induces a natural homomorphism of $\mathcal{Z}^\natural$-modules between its domain $\mathcal{Z}^\natural \otimes \mathrm{Hom}_{\mathbf{H}}(N, N)$ and its image $\mathrm{Hom}_{\mathbf{H}^\natural}(N^\natural, N^\natural)$. More precisely, for any $u \in \mathcal{Z}^\natural$, we show that $\phi(u \otimes 1_N)$ is an element of $\mathrm{Hom}_{\mathbf{H}^\natural}(M^\natural, N^\natural) = \mathrm{Hom}_{\mathbf{H}^\natural}(\mathcal{Z}^\natural \otimes N, \mathcal{Z}^\natural \otimes N)$ that acts on $\mathcal{Z}^\natural \otimes N$ via multiplication by $u \otimes 1_N$: If $v \in \mathcal{Z}^\natural$ and $n \in N$, then

$$\phi(u \otimes 1_N)(v \otimes n) = uvn = (u \otimes 1_N)(v \otimes n),$$

proving the naturality assertion. Since v and n were arbitrary, this equation proves the assertion regarding $\phi(u \otimes 1_N)$.

To establish the isomorphism of $\mathcal{Z}^\natural$-algebras, we need only show that ϕ preserves multiplication. Let $u_1, u_2 \in \mathcal{Z}^\natural$ and $f_1, f_2 \in \mathrm{Hom}_{\mathbf{H}}(N, N)$. Then, for $v \in \mathcal{Z}^\natural$ and $n \in N$, $\phi((u_1 \otimes f_1)(u_2 \otimes f_2))(v \otimes n) = \phi(u_1u_2 \otimes f_1f_2)(v \otimes n) = u_1u_2vf_1f_2(n)$, and $\phi(u_1 \otimes f_1) \cdot \phi(u_2 \otimes f_2)(v \otimes n) = \phi(u_1 \otimes f_1)(u_2v \otimes f_2(n)) = u_1u_2vf_1f_2(n)$, as desired. This completes our arguments for $\phi = \phi_{N,N}^{\mathcal{Z}^\natural,\mathbf{H}}$ and, hence, the proof of the lemma. □

We now prove a result which strengthens Theorem 3.5.3 upon localizing $\mathcal{Z}$ away from bad primes. In the ${}^2F_4$ case, we take 2 as the only bad prime (thus, $\mathcal{Z}^\natural := \mathcal{Z}[1/2]$ in this case). We maintain the notation of Theorem 3.3.1, so, in particular, $A^+ = \mathrm{End}_H(T^+)$ is a certain endomorphism algebra over $\mathcal{Z}$ which has a stratifying system $\{\Delta(\omega), P(\omega)\}_{\omega\in\Omega}$. Now, base change to $\mathcal{Z}^\natural$ or $\mathcal{Z}^{\natural\prime} := \mathcal{Z}^\natural[\sqrt{2}]$, and set $A^{+\natural} := \mathcal{Z}^\natural \otimes A^+$, $A^{+\natural\prime} = \mathcal{Z}^{\natural\prime} \otimes A^+$.

THEOREM 4.3.4.

(a) *The algebra $A^{+\natural}$ over $\mathcal{Z}^\natural$ is quasi-hereditary of separable type in all cases.*

(b) *Except in the case of ${}^2F_4$, the algebra $A^{+\natural}$ over $\mathcal{Z}^\natural$ is quasi-hereditary of split type.*

(c) *In the ${}^2F_4$ case, the algebra $A^{+\natural\prime}$ over $\mathcal{Z}^{\natural\prime}$ is quasi-hereditary of split type.*

PROOF. First, we prove (b). Thus, we are not in the case of ${}^2F_4$. It has already been shown in Theorem 3.3.1 that A^+ has a stratifying system $\{\Delta(\omega), P(\omega)\}_{\omega\in\Omega}$, where $\Delta(\omega) = \mathrm{Hom}_{\mathcal{H}}(S_\omega, T^+)$ and $P(\omega) = \mathrm{Hom}_{\mathcal{H}}(T_\omega^+, T^+)$. Likewise, Theorem 3.5.3 gives us that A^+ is standardly stratified with the standard stratification

$$0 = J_0 \subseteq J_1 \subseteq \cdots \subseteq J_m = A^+,$$

where $J_i = A_{j(i)}^+$ is an idempotent ideal, which is the image of the submodule $P_{j(i)} := \mathrm{trace}_P\left(\bigoplus_{\mathfrak{h}(\mu)\geq j(i)} P(\mu)\right)$ of $P = \oplus_{\omega\in\Omega} P(\omega)$ (see (3.5.3)) under the left A^+-module isomorphism $A^+ \cong P$. Thus, each J_i/J_{i-1} is an ideal of A^+/J_{i-1} satisfying (HI1)–(HI3).

It remains to study (HI4) under the base change to $\mathcal{Z}^\natural$ from $\mathcal{Z}$, but we will not do that yet. Instead, we study some further interesting properties that do not require base change.

The first is the isomorphism $\mathrm{Hom}_{\mathcal{H}}(S_\mu, T^+_\lambda) \xrightarrow{\sim} \mathrm{Hom}_{A^+}(P(\lambda), \Delta(\mu))$, obtained (in greater generality) at the beginning of the proof of Theorem 1.3.5. Restricting λ and μ to having the same height, we quickly find that $\mathrm{Hom}_{A^+}(P(\lambda), \Delta(\mu)) \cong \mathrm{Hom}_{A^+}(\Delta(\lambda), \Delta(\mu))$. (Use (SS3) and (SS1).) Next, we observe an isomorphism $\mathrm{Hom}_{\mathcal{H}}(S_\mu, S_\lambda) \cong \mathrm{Hom}_{\mathcal{H}}(S_\mu, T^+_\lambda)$ from the exact sequence $0 \to S_\lambda \to T^+_\lambda \to T^+_\lambda/S_\lambda \to 0$ in mod-$\mathcal{H}$ and the vanishing of $\mathcal{Z}$-torsion-free module $\mathrm{Hom}_{\mathcal{H}}(S_\mu, T^+_\lambda/S_\lambda)$. (Base change from $\mathcal{Z}$ to $\mathbb{Q}(t)$ to get the vanishing. Then, observe that the kernel of the restricting map $\mathrm{Hom}_{\mathcal{H}}(S_\mu, T^+_\lambda) \to \mathrm{Hom}_{\mathcal{H}}(S_\mu, T^+_\lambda/S_\lambda) = 0$ is the image of the injective map $\mathrm{Hom}_{\mathcal{H}}(S_\mu, S_\lambda) \to \mathrm{Hom}_{\mathcal{H}}(S_\mu, T^+_\lambda)$, which is, therefore, both injective and surjective.) Putting together isomorphisms and inverses of isomorphisms (which are also isomorphisms), we obtain the ($\mathcal{Z}$-linear) composite isomorphism, for λ, μ, of the same height:

$$\mathrm{Hom}_{\mathcal{H}}(S_\mu, S_\lambda) \xrightarrow{\sim} \mathrm{Hom}_{\mathcal{H}}(S_\mu, T^+_\lambda) \xrightarrow{\sim} \mathrm{Hom}_{A^+}(P(\lambda), \Delta(\mu)) \cong \mathrm{Hom}_{A^+}(\Delta(\lambda), \Delta(\mu)).$$

Reading from left to right, these isomorphisms fit in nicely with the commutative diagram

$$\begin{array}{ccc} \mathrm{Hom}_{\mathcal{H}}(S_\mu, T^+_\lambda) & \longrightarrow & \mathrm{Hom}_{A^+}(P(\lambda), \Delta(\mu)) \\ \uparrow & & \uparrow \\ \mathrm{Hom}_{\mathcal{H}}(S_\mu, S_\lambda) & \longrightarrow & \mathrm{Hom}_{A^+}(\Delta(\lambda), \Delta(\mu)). \end{array}$$

Here, the top row and vertical columns are isomorphisms we have seen before, in which the map in the bottom row arises from the ($\mathcal{Z}$-linear) contravariant functor $(-)^\diamond = \mathrm{Hom}_{\mathcal{H}}(-, T^+)$ from mod-$\mathcal{H}$ to A^+-mod. We leave it to the reader to check that the diagram is commutative.

The remarkable consequence is that the bottom row map is an isomorphism. Preceding it with the isomorphism $\mathrm{Hom}_{\mathcal{H}}(S(\lambda), S(\mu)) = \mathrm{Hom}_{\mathcal{H}}((S_\lambda)^*, (S_\mu)^*) \cong \mathrm{Hom}_{\mathcal{H}}(S_\mu, S_\lambda)$, it remains an isomorphism. In other words, the composite is a $\mathcal{Z}$-linear isomorphism

$$\mathrm{Hom}_{\mathcal{H}}(S(\lambda), S(\mu)) \xrightarrow{\sim} \mathrm{Hom}_{A^+}(\Delta(\lambda), \Delta(\mu)),$$

now the specialization of a covariant $\mathcal{Z}$-linear functor from $\mathcal{H}$-mod to A^+-mod. (The functor in question is the composite $(-)^\diamond \circ (-)^*$, which is being "specialized" in its domain category, to the full subcategory of $\mathcal{H}$-mod formed by the objects $S(\tau)$ with τ of the same height as λ, ν.)

Now, return to the A^+-module J_i/J_{i-1}, which we have shown in (3.5.6) to be isomorphic to $A^+_{j(i)}/A^+_{j(i)+1}$. In turn, the latter A^+-module is a direct sum, with various multiplicities, of modules $\Delta(\omega)$, with $\omega \in \Omega$ having height $\mathfrak{h}(\omega) = j(i)$. Ignoring uniqueness issues, fix a decomposition

$$J_i/J_{i-1} \cong \bigoplus_{\omega\in\Omega,\, \mathfrak{h}(\omega)=j(i)} \Delta(\omega)^{\oplus n_\omega},$$

where $j(1) > j(2) > \cdots > j(m)$ by (3.5.5). Using the same multiplicities n_ω of $\Delta(\omega)$ as a direct summand, as in the formula above, define a module N for $\mathcal{H}$ by the formula

$$N := \bigoplus_{\omega\in\Omega,\, \mathfrak{h}(\omega)=j(i)} S(\omega)^{\oplus n_\omega}. \tag{4.3.4}$$

The reader will immediately notice that we have an isomorphism $\mathrm{Hom}_{\mathcal{H}}(N, N) \xrightarrow{\sim} \mathrm{Hom}_{A^+}(J_i/J_{i-1}, J_i/J_{i-1})$, via the functor $(-)^\diamond \circ (-)^*$, as discussed above. More is true, however. Because the isomorphism is induced by a $\mathcal{Z}$-linear functor, it induces an isomorphism of $\mathcal{Z}$-algebras

$$\text{(4.3.5)} \qquad \mathrm{End}_{\mathcal{H}}(N) \xrightarrow{\sim} \mathrm{End}_{A^+}(J_i/J_{i-1}) \cong \mathrm{End}_B(J),$$

where, for notational simplicity in the sequel, we set $J = J_i/J_{i-1}$ and $B = A^+/J_{i-1}$ when i is fixed.

We now make a base change from $\mathcal{Z}$ to $\mathcal{Z}^\natural$ to find a verifiable version of (HI4). Note that the base-changed version of (HI1), (HI2), and (HI3) all hold, since base change takes projectives over $\mathcal{Z}$ (or B) to projectives over $\mathcal{Z}^\natural$ (or $B^\natural$). Also, base change takes idempotent ideals to idempotent ideals.

It remains to consider the base-changed version of (HI4). Here, we have to use the flatness of $\mathcal{Z}^\natural$ over $\mathcal{Z}$ (which we have) and the existence of finite presentations of N (and $J = J_i/J_{i-1}$) over $\mathcal{H}$ (and $B = A^+/J_{i-1}$). We may apply Lemma 4.3.3 to the $\mathcal{Z}$-algebras $\mathtt{H} = \mathcal{H}$ and B, and the $\mathcal{H}$-module N and B-module J, respectively. We thus obtain algebra isomorphisms:

$$\text{(4.3.6)} \qquad \begin{aligned} \phi_{N,N}^{\mathcal{Z}^\natural,\mathcal{H}} &: \mathcal{Z}^\natural \otimes \mathrm{End}_{\mathcal{H}}(N) \xrightarrow{\sim} \mathrm{End}_{\mathcal{H}^\natural}(N^\natural) \\ \phi_{J,J}^{\mathcal{Z}^\natural,B} &: \mathcal{Z}^\natural \otimes \mathrm{End}_B(J) \xrightarrow{\sim} \mathrm{End}_{B^\natural}(J^\natural). \end{aligned}$$

Combining the two isomorphisms with the one in (4.3.5) gives a string of $\mathcal{Z}^\natural$-algebra isomorphisms:

$$\text{(4.3.7)} \qquad \mathrm{End}_{B^\natural}(J^\natural) \xrightarrow{\sim} \mathcal{Z}^\natural \otimes \mathrm{End}_B(J) \xrightarrow{\sim} \mathcal{Z}^\natural \otimes \mathrm{End}_{\mathcal{H}}(N) \xrightarrow{\sim} \mathrm{End}_{\mathcal{H}^\natural}(N^\natural).$$

Reading from left to right, the first isomorphism is the inverse of $\phi_{J,J}^{\mathcal{Z}^\natural,B}$ as given in (4.3.6), the second is the inverse of the base change from $\mathcal{Z}$ to $\mathcal{Z}^\natural$ of (4.3.5), the latter map arising from our construction of N using left-cell modules (see (4.3.5)). The third and last isomorphism in the string is $\phi_{N,N}^{\mathcal{Z}^\natural,\mathcal{H}}$, the map in (4.3.3), for $M = N$. To complete the proof in progress of Theorem 4.3.4(b), we must show that the $\mathcal{Z}^\natural$-algebra $\mathrm{End}_{B^\natural}(J^\natural)$ on the left is a direct sum (categorical direct product) of (a finite number of) full matrix algebras over $\mathcal{Z}^\natural$. By the string of isomorphisms above, it is enough to prove that property for the $\mathcal{Z}^\natural$-algebra $\mathrm{End}_{\mathcal{H}^\natural}(N^\natural)$ on the right.

Applying Theorem 4.2.1, we find that $N^\natural$ is the restriction through $\varpi^\natural$ (see (4.1.7)) of a module $N^\natural_{\text{lift}}$ for $\mathcal{J}^\natural$. Of course, the endomorphism algebra $\mathrm{End}_{\mathcal{H}^\natural}(N^\natural)$ is isomorphic to $\mathrm{End}_{\varpi^\natural\mathcal{H}^\natural}(N^\natural_{\text{lift}})$, where the latter endomorphism algebra is isomorphic to $\mathrm{End}_{\mathcal{J}^\natural}(N^\natural_{\text{lift}})$ by the Reiner observation Lemma 4.2.2(b).

Next, note from Theorem 4.1.4(b) that $\mathcal{J}^\natural$ is isomorphic to a direct product of full matrix algebras $M_{m_i}(\mathcal{Z}^\natural)$, for a finite sequence $m_1, m_2, \ldots, m_s$ of positive integers; see (4.1.13). There is a corresponding decomposition of the identity $1 \in \mathcal{J}^\natural$ into a sum $1 = e_1 + e_2 + \cdots + e_s$ of pairwise orthogonal central idempotents e_i of $\mathcal{J}^\natural$ with $e_i\mathcal{J}^\natural = \mathcal{J}^\natural e_i \cong M_{m_i}(\mathcal{Z}^\natural)$, $i = 1, 2, \ldots, s$. There is also a decomposition of the $\mathcal{J}^\natural$-module $N^\natural_{\text{lift}}$ into $\oplus_i e_i N^\natural_{\text{lift}}$, each $e_i N^\natural_{\text{lift}}$ is a left $\mathcal{J}^\natural$-module, the latter acting through $e_i\mathcal{J}^\natural$ (and all $e_j \neq e_i$ acting as 0). Clearly, $\mathrm{Hom}_{\mathcal{J}^\natural}(e_iM, e_jM) = 0$, for all left $\mathcal{J}^\natural$-modules M, whenever $i \neq j$. Thus, $\mathrm{End}_{\mathcal{J}^\natural}(M) \cong \bigoplus_i \mathrm{End}_{e_i\mathcal{J}^\natural}(e_iM)$. (Here and below, we use $\bigoplus$ instead of Π, with some abuse of notation.) Taking $M = N^\natural_{\text{lift}}$ and using the previous paragraph, we have the reduction (all through $\mathcal{Z}^\natural$-algebra isomorphisms)

$$\text{(4.3.8)} \qquad \mathrm{End}_{\mathcal{H}^\natural}(N^\natural) \cong \mathrm{End}_{\mathfrak{J}^\natural}(N^\natural_{\mathrm{lift}}) \cong \bigoplus_{i=1}^{s} \mathrm{End}_{e_i\mathfrak{J}^\natural}(e_iN^\natural_{\mathrm{lift}}).$$

It is useful to note at this point that $N^\natural_{\mathrm{lift}}$ is a projective $\mathfrak{J}^\natural$-module, since it is projective (even free) as $\mathcal{Z}^\natural$-module. See Corollary 4.1.6(b). It follows easily that each $\mathfrak{J}^\natural$-module $e_jN^\natural_{\mathrm{lift}}$ is projective. Recall that the $\mathcal{Z}^\natural$-algebra $e_i\mathfrak{J}^\natural$ is isomorphic to the full matrix algebra $M_{m_i}(\mathcal{Z}^\natural)$, well-known to be Morita equivalent to $\mathcal{Z}^\natural$. A $\mathcal{Z}^\natural$-linear Morita equivalence which achieves this is a multiplication by an idempotent $f_i \in e_i\mathfrak{J}^\natural$, corresponding to an idempotent matrix unit $e_{1,1}$ in $M_{m_i}(\mathcal{Z}^\natural)$, on $e_i\mathfrak{J}^\natural$-modules. Such an equivalence takes projectives to projectives and preserves endomorphism algebras up to $\mathcal{Z}^\natural$-algebra isomorphism.

Since (finitely generated) projective $\mathcal{Z}^\natural$-modules are free (see Remark 1.3.6(c)), the $\mathcal{Z}^\natural$-module $f_ie_iN^\natural_{\mathrm{lift}}$ is free and isomorphic to ${\mathcal{Z}^\natural}^{\oplus n_i}$, for some integer $n_i \geq 0$, depending on $e_iN^\natural_{\mathrm{lift}}$. We mention here that the ring $\mathcal{Z}^\natural$ used as the base ring here is $e_{1,1}M_{m_i}(\mathcal{Z}^\natural)e_{1,1} \cong f_ie_i\mathfrak{J}^\natural f_i$, the target of the Morita equivalence given by left multiplication by f_i. We find now that the Morita equivalence induces an isomorphism

$$\text{(4.3.9)} \qquad \mathrm{End}_{e_i\mathfrak{J}^\natural}(e_iN^\natural_{\mathrm{lift}}) \cong \mathrm{End}_{\mathcal{Z}^\natural}({\mathcal{Z}^\natural}^{\oplus n_i}) \cong M_{n_i}(\mathcal{Z}^\natural).$$

Finally, combining these isomorphisms with many others demonstrated above, tracing our way from the left to the right of (4.3.7) and (4.3.8) and to the direct sum over i of (4.3.9), we obtain (again with some abuse of notation, using $\bigoplus$ instead of Π)

$$\text{(4.3.10)} \qquad \mathrm{End}_{B^\natural}(J^\natural) \cong \bigoplus_i M_{n_i}(\mathcal{Z}^\natural).$$

This establishes (HI4) and completes the proof of part (b).

Note that (a) follows from (b) except in the ${}^2F_4$ case. Assume we are in the latter case. After making the same changes in the appendix to Section 4.1 ($\mathcal{Z} \to \mathcal{Z}'$, $\mathbb{Q} \to \mathbb{Q}'$, etc.), the above argument shows, *mutadis mutandis,* that $A^{+\natural'}$ is quasi-hereditary of split type. In particular, it is of separable type. Therefore, as a faithfully flat descent, Theorem 4.1.12 implies that $A^{+\natural}$ itself is of separable type. (Note that each $J^{\natural'}_i/J^{\natural'}_{i-1}$ is the base change of the projective module $J^\natural_i/J^\natural_{i-1}$, hence its endomorphism algebra is a corresponding base change of the latter module. Also, note that $\mathcal{Z}^{\natural'}$ is faithfully flat over $\mathcal{Z}^\natural$.) This not only completes the proof of (a) but also establishes (c). □

We remark that the heredity ideals $J^\natural_i/J^\natural_{i-1}$ involved in the defining sequence above are not indecomposable in Rouquier's sense (see Remark 4.3.2(2)). The reader may ask if the defining sequence in the theorem can be refined so that only indecomposable split heredity ideals will occur. This is possible due to the fact that each $\Delta^\natural(\omega)$ is a direct sum of various $\Delta(E)$'s (Theorem 4.2.4), with the E's all of the same height as ω. Recall $\Lambda := \mathrm{Irr}(\mathbb{Q}W)$ and the preorder $\preceq$ on Ω given in (3.1.1), which induces a partial order $\preceq$ (using the same name) defined via (4.1.8) by

$$\text{(4.3.11)} \qquad E \preceq E' \iff E_{\mathbb{Q}(t)} = E'_{\mathbb{Q}(t)} \quad \text{or} \quad \mathfrak{h}(E_{\mathbb{Q}(t)}) < \mathfrak{h}(E'_{\mathbb{Q}(t)}), \quad E, E' \in \Lambda.$$

For each $E \in \Lambda$, $\Delta(E) := (S^\natural_E)^\diamond = \mathrm{Hom}_{\mathcal{H}^\natural}(S^\natural_E, T^{+\natural})$, where the $\mathcal{H}^\natural$-module $S^\natural_E := \mathrm{Hom}_{\mathcal{Z}^\natural}(S^\natural(E), \mathcal{Z}^\natural)$ is defined in the display (4.2.3).

The following corollary continues the argument from the proof of Theorem 4.3.4(b). We leave the counterpart for the $^2F_4$ case in Theorem 4.3.4(c) to the reader.

COROLLARY 4.3.5. *The defining sequence $0 = J^\natural_0 \subseteq J^\natural_1 \subseteq \cdots \subseteq J^\natural_m = A^{+\natural}$ in the proof may be refined to a defining sequence of length $s = |\Lambda|$:*

$$0 = I_0 \subseteq I_1 \subseteq \cdots \subseteq I_{s-1} \subseteq I_s = A^{+\natural}, \tag{4.3.12}$$

which is called an indecomposable *defining sequence (partly our terminology). Moreover, it may be assumed that, for every $1 \leq j \leq s$,*

(1) *there exists an idempotent e in $B' := A^{+\natural}/I_{j-1}$ such that $J' := I_j/I_{j-1} = B'eB'$.*
(2) *$J' \cong \Delta(E)^{\oplus l_j} \cong \Delta(E) \otimes_{\mathcal{Z}^\natural} \mathrm{Hom}_{B'}(\Delta(E), B')$, for some $E \in \Lambda$ and $l_j > 0$. (We will see that the l_j is equal to some n_i in the display (4.3.9)).*

Hence, $A^{+\natural}$ is also a split quasi-hereditary algebra in the sense of [**Ro08**, 4.1.4].

PROOF. Continuing the argument from the proof of Theorem 4.3.4(b), we refine each section $J^\natural_{h-1} \subseteq J^\natural_h$ $(1 \leq h \leq m)$. For a fixed h, let $J^\natural = J^\natural_h/J^\natural_{h-1}$ and $B^\natural = A^{+\natural}/J^\natural_{h-1}$. Recall from Notation 2.1.2 that the height function $\mathfrak{h} : \Omega \to \mathbb{Z}$ induces a height function $\mathfrak{h} : \Lambda \to \mathbb{Z}$. Recall also the idempotent decompsition $e_1 + \cdots + e_s = 1 \in \mathcal{J}^\natural$, where $s = |\Lambda|$. By displays (4.3.7), (4.3.8), and (4.3.9),

$$\mathcal{E} := \mathrm{End}_{B^\natural}(J^\natural) \cong \mathrm{End}_{\mathcal{H}^\natural}(N^\natural) \cong \mathrm{End}_{\mathcal{J}^\natural}(N^\natural_{\mathrm{lift}}) \cong \bigoplus_{i=1}^{s_h} \mathrm{End}_{e_i\mathcal{J}^\natural}(e_i N^\natural_{\mathrm{lift}}) \cong \bigoplus_{i=1}^{s_h} M_{n_i}(\mathcal{Z}^\natural).$$

Here, after relabelling, we assume $n_i > 0$, for all $1 \leq i \leq s_h$, and $n_i = 0$, for $s_h < i \leq s$. Note that the numbers s_h are related to a partition (see (4.3.4))

$$\Lambda = \bigcup_{i=1}^{m} \Lambda_{j(i)}, \text{ where } \Lambda_{h'} := \{E \in \Lambda \mid \mathfrak{h}(E) = h'\} \text{ and } |\Lambda_{j(h)}| = s_h. \tag{4.3.13}$$

If $L := N^\natural_{\mathrm{lift}}$ and $L_i := e_i N^\natural_{\mathrm{lift}}$ $(1 \leq i \leq s_h)$, then $N^\natural = L^{\varpi^\natural} = \oplus_{i=1}^{s_h} L_i^{\varpi^\natural}$. Thus, denoting the canonical projection map $N^\natural \to L_i^{\varpi^\natural} \subseteq N^\natural$ by e_i again, we have $e_i(N^\natural) = L_i^{\varpi^\natural}$ and the identity map $1_{N^\natural} = e_1 + \cdots + e_{s_h}$ is decomposed into a sum of orthogonal central idempotents. Since the isomorphism $\phi : \mathrm{End}_{\mathcal{H}^\natural}(N^\natural) \xrightarrow{\sim} \mathcal{E}$ is induced by the functor $(\ \ )^{*\diamond} = \mathrm{Hom}_{\mathcal{H}^\natural}(-, T^{+\natural}) \circ \mathrm{Hom}_{\mathcal{Z}^\natural}(-, \mathcal{Z}^\natural)$, the canonical projection e_i induces the canonical projection $\tilde{e}_i = (e_i)^{*\diamond} : J^\natural \to J^\natural$. Hence, the element $1_{J^\natural}$ of $\mathcal{E}$ (i.e., the identity map on $J^\natural$) is decomposed into a sum of orthogonal central idempotents $\tilde{e}_i \in \mathcal{E}$: $1 = \tilde{e}_1 + \tilde{e}_2 + \cdots + \tilde{e}_s$, where $\tilde{e}_i = (e_i)^{*\diamond}$. Putting $J^{\natural(i)} = \tilde{e}_i J^\natural$, we have $J^\natural = \oplus_{i=1}^{s_h} J^{\natural(i)}$.

Note that $J^\natural$ is a $B^\natural$-$B^\natural$-bimodule. Also, each $J^{\natural(i)}$ is a left $B^\natural$-module, since the action of $\tilde{e}_i \in \mathcal{E}$ commutes with that of $B^\natural$. We claim that each $J^{\natural(i)}$ $(1 \leq i \leq s_h)$ is also a right $B^\natural$-module. Indeed, suppose there exists $b \in B^\natural$ such that $J^{\natural(i)}b \not\subseteq J^{\natural(i)}$. Then, the composition of the following homomorphisms

$$J^{\natural(i)} \longrightarrow J^{\natural(i)}b \longrightarrow (J^{\natural(i)}b + J^{\natural(i)})/J^{\natural(i)} \subseteq J^\natural/J^{\natural(i)}$$

is non-zero. This is impossible, since $\mathrm{Hom}_{B^\natural}(\tilde{e}_i J^\natural, \tilde{e}_j J^\natural) = \tilde{e}_j \mathcal{E} \tilde{e}_i = 0$, for all $i \neq j$. Hence, each $J^{\natural(i)}$ is a $B^\natural$-$B^\natural$-bimodule.

Let $J^\natural_{h-1,j} = \pi^{-1}(\oplus_{i=1}^j J^{\natural(i)})$, where $\pi : J^\natural_h \to J^\natural_h/J^\natural_{h-1}$ is the obvious surjective homomorphism. Each $J^\natural_{h-1,j}$ is an ideal of $A^{+\natural}$ and the section $J^\natural_{h-1} \subseteq J^\natural_h$ is refined as

$$J^\natural_{h-1} = J^\natural_{h-1,0} \subseteq J^\natural_{h-1,1} \subseteq J^\natural_{h-1,2} \subseteq \cdots \subseteq J^\natural_{h-1,s_h} = J^\natural_h.$$

Note that such a refinement can be done for any order on $\Lambda_{h'}$. Combining these refinements for all h results in a candidate defining sequence which has length $\sum_{h=1}^m s_h = s = |\Lambda|$.

We now prove that it does give a defining sequence by checking (HI1)–(HI4) for each $J' := J^\natural_{h-1,i}/J^\natural_{h-1,i-1} \cong J^{\natural(i)}$ and $B' := A^{+\natural}/J^\natural_{h-1,i-1} \cong B^\natural/\oplus_{j=1}^{i-1} J^{\natural(j)}$.

The proof for (HI1) and (HI2) is clear, since the $\mathcal{Z}^\natural$-module

$$B'/J' \cong A^{+\natural}/J^\natural_{h-1,i} \cong A^{+\natural}/J^\natural_h \oplus J^\natural_h/J^\natural_{h-1,i}$$

is a direct sum of two projective $\mathcal{Z}^\natural$-modules, and ${}_{B'}J' = {}_{B^\natural}J'$ is a direct summand of the projective $B^\natural$-module ${}_{B^\natural}J^\natural$. (HI4) may be verified as follows:

$$\text{(4.3.14)} \qquad \mathrm{End}_{B'}(J') \cong \mathrm{End}_{B'}(\tilde{e}_i J^\natural) \cong \mathrm{End}_{B^\natural}(\tilde{e}_i J^\natural) \cong \tilde{e}_i \mathcal{E} \cong M_{n_i}(\mathcal{Z}^\natural).$$

For (HI3), we first note from the display (3.5.4) that the ideal $J^\natural_h = A^{+\natural}_{j(h)}$ is equal to $A^{+\natural} f A^{+\natural}$, for some idempotent f. Then the ideal $J^\natural = B^\natural \bar{f} B^\natural$, where $\bar{f} = f + J^\natural_{h-1}$ defines an idmpotent in $J^\natural$. If $\tilde{f}_i = \tilde{e}_i(\bar{f})$, for each i $(1 \le i \le s_h)$, then $\tilde{f}_i^2 = \tilde{e}_i(\bar{f})\tilde{e}_i(\bar{f}) = \tilde{e}_i(\tilde{e}_i(\bar{f})\bar{f}) = \tilde{e}_i(\tilde{e}_i(\bar{f}^2)) = \tilde{f}_i$ and $J^{\natural(i)} = \tilde{e}_i(B^\natural \bar{f} B^\natural) = B^\natural \tilde{f}_i B^\natural$. Hence, $J' = B'\tilde{f}_i B'$ is idempotent, proving (HI3) and (1) as well.

It remains to prove (2). The assertion that J' is isomorphic to a direct sum of $\Delta(E)$, for some $E \in \Lambda$, can be seen as follows. Since, by the proof of the theorem and recalling the notation $L_i = e_i N^\natural_{\mathrm{lift}}$, $f_i L_i \cong \mathcal{Z}^{\natural \oplus n_i}$ (see the discussion above the diaplay (4.3.9)), it follows from the Morita equivalence (between $M_{n_i}(\mathcal{Z}^\natural)$ and $\mathcal{Z}^\natural$) that $L_i \cong E_\natural^{\oplus n_i}$, where $E = E_{h',i} \in \Lambda_{h'} := \{E_{h',i} \mid 1 \le i \le s_h\}$ for some height $h' = j(h)$ (see (4.3.4)), and the $\mathcal{H}^\natural$-module $L_i^{\varpi^\natural} \cong S^\natural(E)^{\oplus n_i}$; see (4.2.3). (Here we remind the reader that $E_\natural$ $(E \in \Lambda)$ is a $\mathcal{J}^\natural$-lattice of the irreducible $\mathcal{J}_{\mathbb{Q}(t)}$-module $E_{\mathbb{Q}(t)}$.) Hence, $(e_i(N^\natural))^{*\diamond} = (L_i^{\varpi^\natural})^{*\diamond} \cong \Delta(E_{h',i})^{\oplus n_i}$. Finally, we have

$$J^{\natural(i)} = \tilde{e}_i(J^\natural) = (e_i(N^\natural))^{*\diamond} \cong \Delta(E_{h',i})^{\oplus n_i}.$$

Since $\mathrm{Hom}_{B'}(J', B'/J') = 0$ (see [**Ro08**, Lem. 4.4]), it follows that

$$\mathrm{Hom}_{B'}(\Delta(E_{h'i}), B') = \mathrm{Hom}_{B'}(\Delta(E_{h',i}), J') \cong (\mathcal{Z}^\natural)^{\oplus n_i}.$$

This implies the last isomorphism in (2).

It remains to verify Rouquier's definition [**Ro08**, 4.1.4] for split quasi-hereditary algebras. By the definition in display (4.3.11), each coideal Γ of Λ has the form $\Gamma = \big(\cup_{h''>h'} \Lambda_{h''}\big) \cup \Phi$, where $\Phi \subseteq \Lambda_{h'}$ for some $h' = j(h)$. Thus, each indecomposable defining sequence of the form (4.3.12) defines a sequence of coideals $\{\Gamma_i \mid 1 \le i \le s\}$ of Λ such that the set of ideals $\{I_{\Gamma_i} := I_i \mid 1 \le i \le s\}$ satisfies the two conditions in [**Ro08**, 4.1.4]. This proves that $A^{+\natural}$ is a split quasi-hereditary algebra in the sense of Rouquier, completing the proof of the corollary. □

COROLLARY 4.3.6. *Maintain the notations of Corollary 4.3.5. There exists an idempotent $e_1 \in B'$ such that $B'e_1 \cong \Delta(E)$ as left B'-modules.*

PROOF. By Corollary 4.3.5(1) and [**DPS18**, Prop. 2.2], we have

$$J' = B'eB' \cong B'e \otimes_{eB'e} eB',$$

$\mathrm{End}_{B'}(J')$ is Morita equivalent to $eB'e$, and eB' is a projective $eB'e$-module. By (4.3.14), we see that $eB'e$ is Morita equivalent to $\mathcal{Z}^\natural$. Since projective finite $\mathcal{Z}^\natural$-modules are free, it follows from the Morita equivalence that $eB'e \cong M_m(\mathcal{Z}^\natural)$, for some $m \geq 1$.

If $m = 1$, then $eB'e \cong \mathcal{Z}^\natural$ and $J' = B'eB' \cong B'e \otimes_{\mathcal{Z}^\natural} eB' \cong B'e \otimes_{\mathcal{Z}^\natural} \mathrm{Hom}_{B'}(B'e, B')$. Thus, the proof of [**Ro08**, Prop. 4.7] and Corollary 4.3.5(2) imply $B'e \cong \Delta(E)$. So, we take $e_1 = e$ in this case.

Suppose now $m > 1$. Then, there exist idempotents $e_1, \dots, e_m$ (corresponding to the matrix idempotents $e_{1,1}, \dots, e_{m,m}$ in $M_m(\mathcal{Z}^\natural)$) such that $e = e_1 + \cdots + e_m$ and $e_ie_j = \delta_{i,j}e_i$. Thus, $B'e = B'e_1 \oplus \cdots \oplus B'e_m$ and $\mathrm{Hom}_{B'}(B'e_i, B'e_j) \cong \mathcal{Z}^\natural$. (The right action of $eB'e \cong M_m(\mathcal{Z}^\natural)$ on $B'e$ may be interpreted as matrix multiplication.) Hence, $B'e \cong (B'e_1)^{\oplus m}$. Since the left $eB'e$-module eB' is projective, it follows (from the Morita equivalence between $eB'e$ and $e_1B'e_1 \cong \mathcal{Z}^\natural$) that the $e_1B'e_1$-module $e_1B' = e_1(eB')$ is projective (over $\mathcal{Z}^\natural$) and, hence, is free: $e_1B' \cong (\mathcal{Z}^\natural)^{\oplus m'}$ for some $m' > 0$. The Morita equivalence gives $eB' \cong (\mathcal{Z}^{\natural m})^{\oplus m'} \cong C_1^{\oplus m'}$, where $C_1 = M_m(\mathcal{Z}^\natural)e_{1,1}$. Consequently,

$$J' \cong B'e \otimes_{eB'e} eB' \cong (B'e_1)^{\oplus m} \otimes_{M_m(\mathcal{Z}^\natural)} C_1^{\oplus m'} \cong (B'e_1)^{\oplus m'}.$$

In other words, we may write $J' = P_1 \oplus \cdots \oplus P_{m'}$, where $B'e_1 \cong P_i$. Taking isomorphism $\phi_i : B'e_1 \to P_i$, we have $P_i = \phi_i(B'e_1) = B'e_1\phi_i(e_1) \subseteq B'e_1B'$. Hence, $J' \subseteq B'e_1B'$. Clearly, we have $B'e_1B' = B'ee_1B' \subseteq B'eB' = J'$. This shows $J' = B'e_1B'$. By the argument for the $m = 1$ case, we obtain $B'e_1 \cong \Delta(E)$. The corollary is proven. □

Corollary 4.3.5 together with [**Ro08**, Th. 4.16] gives the following result.

COROLLARY 4.3.7. *If $\mathcal{C} := A^{+\natural}$-mod is the category of finite $A^{+\natural}$-modules, then $(\mathcal{C}, \{\Delta(E)\}_{E\in\Lambda})$ is a highest weight category in the sense of* [**Ro08**, Def. 4.11], *where Λ is the poset with respect to the order defined in display* (4.3.11).

Note that the projective modules $P(E)$ used to form the stratifying system $\{\Delta(E), P(E)\}_{E\in\Lambda}$ may be recursively constructed; see the proof of [**Ro08**, Lem. 4.12].

Similarly to $A^{+\natural}$ and $A^{+\natural'}$ as defined above the statement of Theorem 4.3.4, we define, for a poset ideal Θ of $(\Omega, \leq_L)$, $A_\Theta^{\ddagger\natural} := \mathcal{Z}^\natural \otimes_{\mathcal{Z}} A_\Theta^\ddagger$ and $A_\Theta^{\ddagger\natural'} := \mathcal{Z}^{\natural'} \otimes_{\mathcal{Z}} A_\Theta^\ddagger$, where $A_\Theta^\ddagger$ is defined in (2.4.6). For later use in Sections 5.1 and 5.4, we need the quasi-hereditary structures of these algebras.

THEOREM 4.3.8. *All the statements* (a), (b), *and* (c) *in Theorem* 4.3.4 *are true if $A^{+\natural}$ and $A^{+\natural'}$ are replaced by $A_\Theta^{\ddagger\natural}$ and $A_\Theta^{\ddagger\natural'}$, respectively. Moreover, the base-changed sequence of the defining sequence* (3.5.8) *to $\mathcal{Z}^\natural$ (or $\mathcal{Z}^{\natural'}$) may be refined to a defining sequence*

$$0 = I_0^+ \subseteq I_1^+ \subseteq \cdots \subseteq I_s^+ = A_\Theta^{\ddagger\natural}$$

of length $s = |\Lambda_\Theta|$, where $\Lambda_\Theta := \{E \in \Lambda \mid E_{\mathbb{Q}(t)}|S(\omega)_{\mathbb{Q}(t)}$, for some $\omega \in \Theta\}$, such that $\bar{I}_j := I_j/I_{j-1}$ is an indecomposable heredity ideal of $B_j := A_\Theta^{\ddagger\natural}/I_{j-1}$. In particular, the highest weight category $A_\Theta^{\ddagger\natural}$-mod has standard objects

$$\Delta^+(E) := \mathrm{Hom}_{\mathcal{H}^\natural}(S_E^\natural, T^{\ddagger\natural}), \quad \forall E \in \Lambda$$

such that, if $\Delta^+(E) \otimes_{\mathcal{Z}^\natural} \mathrm{Hom}_{B_j}(\Delta^+(E), B_j) \cong \bar{I}_j$*, for some* $1 \le j \le s$*, then* $\Delta^+(E) \cong B_j e$ *for some idempotent* $e \in B_j$ *with* $eB_je \cong \mathcal{Z}^\natural$.

PROOF. Note that, by Remark 3.5.4(2), Theorem 1.3.5 is applicable in this case. Thus, the proof of Theorem 4.3.4 can be easily modified by simply adding a $^+$ to the superscripts of the various objects throughout its proof to get a proof for the first assertion. (The first assertion also follows easily from the Morita equivalence discussed in Sec 3.4.)

The remaining assertions follow from arguments similar to the proofs of Corollaries 4.3.5 and 4.3.6. □

Finally, in a different direction, the following remark will be useful in the next chapter.

REMARK 4.3.9. Let B be a split quasi-hereditary algebra over $\mathcal{Z}$. (For example, $B = A^{+\natural}$ as in Theorem 4.3.4(b) above.) *If* R *is a commutative Noetherian* $\mathcal{Z}$*-algebra,* $B_R := R \otimes_{\mathcal{Z}} B$ *is a split quasi-hereditary algebra over* R*.* This is verified by directly checking the axioms at the start of this section. For example, if J is a heredity ideal in B, then ${}_BJ$ is a projective B-module. It follows easily, using this projectivity, that the R-algebra $\mathrm{End}_{B_R}({}_{B_R}J)$ is isomorphic to the R-algebra $R\otimes_{\mathcal{Z}}\mathrm{End}_B({}_BJ)$ and so, is split over R. Thus, the operation $R\otimes_{\mathcal{Z}}-$ carries a defining sequence for B (exhibiting a splitting) into a similar defining sequence for B_R. The other steps in checking that B_R is split quasi-hereditary are straightforward.

The other parts (a) and (c) of Theorems 4.3.4 and 4.3.8 similarly behave well under base change.

5. Some related applications

We now present some applications of our theory to several related topics. First, the split quasi-heredity of $A^{+\natural}$ gives rise to a full standard basis for $A^{+\natural}$, making its centralizer subalgebras such as A and $\mathcal{H}$ into standardly based algebras. Second, the quasi-heredity of $A^{+\natural}$ gives a unitriangular property of several decomposition matrices, including the unipotent principal blocks of the corresponding finite groups of Lie type in cross characteristic. Third, using a recent Schur duality of Bao–Wang for certain $\imath$-quantum groups and Hecke algebras of type B/C, we relate representations of finite orthogonal/symplectic groups to those of $\imath$-quantum groups. Finally, we end this chapter with some open problems.

5.0. Introduction. The usual decomposition number framework for finite groups is the classical one of Brauer, involving integral representations over DVRs (discrete valuation rings, usually complete); see, e.g., [**Se77**]. However, the theory is not adequate for generic Hecke algebras $\mathcal{H}$ (or A or A^+), because the generic versions of these algebras are defined over $\mathcal{Z} = \mathbb{Z}[t, t^{-1}]$, and some authors introduce even more variables. Geck and Rouquier provided in [**GR97**] a general theory which applies to $\mathcal{Z}$ and many higher dimensional commutative rings. The authors of this monograph introduced in [**DPS98a**] a decomposition number theory applicable to $\mathcal{Z}$-algebras, which (though not applicable to commutative base rings of Krull dimension > 2) is closer to the original Brauer framework.

We will use the [**DPS98a**] theory in this section. An exposition, with several new results, is provided in Appendix C. We suggest that the reader look at this section first, then read some of Appendix C. After that, both the present section and Appendix C may be read in tandem.

Regular local triples. Consider a "local triple" (R, Q, F), with a local Noetherian integral domain R (with maximal ideal $\mathfrak{m}$), fraction field Q, and residue field $F = R/\mathfrak{m}$ of R. Several evident variations on this notational theme are useful. For example, if R is just a domain, call (R, Q, F) a ring triple. As before, Q is the fraction field of R, and there is given an identification of F with $R/\mathfrak{m}$ for an implicitly given maximal ideal $\mathfrak{m}$ of R. Also, if R is a DVR (automatically "local"), then (R, Q, F) is a "DVR triple." If R is a regular ring (in the sense of [**M86**]), then the ring triple (R, Q, F) is a "regular triple." Regular triples will play an important role in this discussion—and R will usually be a regular local ring of Krull dimension ≤ 2.

In case R is a regular local ring of Krull dimension d, we call (R, Q, F) a d-regular local triple.

Let B be an R-free and finite algebra. Also, assume that R is a regular local ring of Krull dimension ≤ 2. Let $B_Q := B \otimes_R Q$ be the corresponding finite-dimensional Q-algebra, and define B_F similarly. A basic result given in Theorem C1.2 (and based on work of Auslander-Goldman [**AG60a, AG60b**]) states that if $\mathfrak{X}$ is a finite B_Q-module, then $\mathfrak{X} \cong X_Q$, for some finite R-free B-submodule $X \subseteq \mathfrak{X}$. Furthermore, assume that X_Q is irreducible. If Y is a second R-free B-submodule such that $\mathfrak{X} \cong Y_Q$, then X_F and Y_F have the same composition factors (with the same multiplicities) as B_F-modules. This means the decomposition numbers of B defined below depend only on (R, Q, F).

Thus, under the hypotheses and notation of Theorem C1.2, there is a natural way to relate the representation theory of B_F (often a *modular theory*, usually meaning that F has positive characteristic,[33]) with that of B_Q (often a *characteristic zero theory*). In this regard, we are particularly interested in the *decomposition map* in display (C1.1). Note that the hypothesis that the Krull dimension of R is at most 2 is often satisfied in the present monograph, since R is usually closely related to the ring $\mathcal{Z}$ of integral Laurent polynomials in a variable t.

Decomposition matrices. Let (R, Q, F) be a regular local triple, where R has Krull dimension ≤ 2. (See above.) If $B = B_R$ is a finite and R-free algebra, we can use the triple to associate decomposition maps, decomposition numbers, and decomposition matrices to B, imitating the classical constructions. More explicitly, suppose that X is a B-module which is finite and projective (thus free) over R. Assume that X_Q is an irreducible B_Q-module. If L is an irreducible B_F-module, then the decomposition number d_{L,X_Q} is defined to be the multiplicity $[X_F : L]$ of L as a composition factor of X_F. This number depends only on X_Q not on the choice of X [**DPS98a**, §1.1]. Collectively, the decomposition numbers form the *decomposition matrix*, with rows indexed by "all non-isomorphic" irreducible B_F-modules, and columns indexed by "all non-isomorphic" irreducible B_K-modules.[34]

[33]For algebras over $\mathcal{Z}$, if the order of t in F is finite (that is, the image of t is a root of unity), that order is often called the quantum characteristic. "Modular theory" can occur in this case even when F has characteristic zero!

[34]Many authors take the decomposition matrix to be the transpose of our decomposition matrix. Thus, the rows of the decomposition matrix index the irreducible B_K-modules. This causes no difficulty, provided the reader is aware of the different conventions.

Notice that, as seen above, these decomposition numbers for B depend only on B and the underlying triple (R, Q, F). It is, however, common also to assume (although we will not do it) that the irreducible modules indexing the rows and columns are absolutely irreducible. Equivalently, both B_F and B_Q are split as algebras. In many cases, this can be achieved by enlarging R and replacing B by B_R. See Propositions C4.2 and C4.3, as well as Proposition C4.1 regarding uniqueness.

Although we often speak of "the" decomposition matrix, it is, in fact, only determined up to a permutation of its rows and a permutation of its columns, unless some specific (or restricted) ordering of irreducible modules is required. We will usually be somewhat informal about this and hope that it does not cause too much confusion.

Decomposition submatrices and unitriangularity. Given two (rectangular) matrices $D = (d_{ij})$ and $D' = (d'_{rs})$ (decomposition matrices or not), we say that D is a submatrix of D' if there are injective assignments $i \mapsto i'$ and $j \mapsto j'$ of the row (respectively, column) indices of D such that

$$d_{i,j} = d'_{i',j'},$$

for each row index i and column index j of D.

A rectangular matrix $D = (d_{ij})$ over any ring is said to be (upper) *unitriangular* if

(1) D has at least as many columns as rows;
(2) $d_{ij} = 0$ if $i > j$; and
(3) $d_{ii} = 1$ for each row index i.

5.1. Standardly based algebras. In this section, we consider the algebras A and A^+ from the point of view of the theory of standardly based algebras as developed in Du-Rui [**DuR98**] (see [**KM20**] for a new development). Such algebras are quite close to the Graham-Lehrer theory of cellular algebras [**GL96**]. For example, they have a similar family of bilinear forms (see (5.1.2) below), and all cellular algebras are standardly based. But, there is no requirement that a standardly based algebra have an involution as in [**GL96**].

DEFINITION 5.1.1. Let $\mathcal{A}$ be an algebra over a commutative ring $\mathscr{K}$. We say that $\mathcal{A}$ is a *standardly based algebra* (SBA) with respect to a poset $(\Lambda, \leq)$ if the following conditions hold:

(a) For any $\lambda \in \Lambda$, there are index sets $I(\lambda)$ and $J(\lambda)$ and subsets of $\mathcal{A}$,

$$\mathsf{B}^\lambda = \{a^\lambda_{i,j} \mid (i,j) \in I(\lambda) \times J(\lambda)\}$$

such that $\mathsf{B} := \bigcup_{\lambda \in \Lambda} \mathsf{B}^\lambda$ is a disjoint union and forms a $\mathscr{K}$-basis for $\mathcal{A}$.

(b) For any $a \in \mathcal{A}$, $a^\lambda_{i,j} \in \mathsf{B}$, we have

$$a \cdot a^\lambda_{i,j} \equiv \sum_{i' \in I(\lambda)} f_{i',\lambda}(a,i) a^\lambda_{i',j} \pmod{\mathcal{A}^{>\lambda}}$$

$$a^\lambda_{i,j} \cdot a \equiv \sum_{j' \in J(\lambda)} f^{\circ}_{\lambda,j'}(a,j) a^\lambda_{i,j'} \pmod{\mathcal{A}^{>\lambda}},$$

where $\mathcal{A}^{>\lambda}$ is the $\mathscr{K}$-submodule of $\mathcal{A}$ generated by B^{μ} with $\mu > \lambda$, and the coefficients $f_{i',\lambda}(a,i)$ and $f^{\circ}_{\lambda,j'}(a,j) \in \mathscr{K}$, defined by the display, are independent of j and i, respectively.

We also call B a *standard basis* for $\mathcal{A}$.

For any coideal Γ of Λ, the submodule $A^{\Gamma} := \mathrm{span}(\cup_{\lambda\in\Gamma}\mathsf{B}^{\lambda})$ is an ideal of $\mathcal{A}$. For a finite Λ, if we order $\Lambda = \{\lambda_1, \lambda_2, \ldots, \lambda_m\}$ such that $\Gamma_i = \{\lambda_1, \lambda_2, \ldots, \lambda_i\}$ is a coideal for every $1 \le i \le m$, then we obtain a sequence of ideals

$$0 = J_0 \subseteq J_1 \subseteq J_2 \subseteq \cdots \subseteq J_m = \mathcal{A}, \tag{5.1.1}$$

where $J_i = \mathcal{A}^{\Gamma_i}$. Thus, every section $A^{\lambda_i} := J_i/J_{i-1}$ is an $\mathcal{A}$-$\mathcal{A}$-bimodule.

Given a SBA as above, we may introduce, for each $\lambda \in \Lambda$, a function $f_\lambda : J(\lambda) \times I(\lambda) \to \mathscr{K}$ defined as follows: for any $a^{\lambda}_{ij}, a^{\lambda}_{i'j'} \in \mathsf{B}^{\lambda}$, there is a unique element $f_\lambda(j,i') \in \mathscr{K}$ (in other words, $f_\lambda(j,i') := f_{i,\lambda}(a^{\lambda}_{i,j}, i') = f^{\circ}_{\lambda,j'}(a^{\lambda}_{i',j'}, j)$) such that

$$a^{\lambda}_{ij}a^{\lambda}_{i'j'} \equiv f_\lambda(j,i')a^{\lambda}_{ij'} \pmod{\mathcal{A}^{>\lambda}}.$$

DEFINITION 5.1.2. For $\lambda \in \Lambda$, we define $\Delta(\lambda)$ (resp. $\Delta^{\circ}(\lambda)$) to be the left (resp. right) $\mathcal{A}$-module with $\mathscr{K}$-basis $\{a^{\lambda}_i \mid i \in I(\lambda)\}$ (resp. $\{b^{\lambda}_j \mid j \in J(\lambda)\}$) and the module action defined by

$$aa^{\lambda}_i = \sum_{i'\in I(\lambda)} f_{i',\lambda}(a,i)a^{\lambda}_{i'} \quad (\text{resp.}, b^{\lambda}_j a = \sum_{j'\in J(\lambda)} f^{\circ}_{\lambda,j'}(a,j)b^{\lambda}_{j'}.)$$

We shall call the modules $\Delta(\lambda)$ (respectively, $\Delta^{\circ}(\lambda)$) the *standard* modules in the category $\mathcal{A}$-mod (respectively, mod-$\mathcal{A}$).

One sees easily from the definition that there is an $\mathcal{A}$-$\mathcal{A}$-bimodule isomorphism $A^{\lambda} \cong \Delta(\lambda) \otimes \Delta^{\circ}(\lambda)$.

We define (compare [**GL96**, 2.3]) a bilinear pairing $\beta_\lambda : \Delta(\lambda) \times \Delta^{\circ}(\lambda) \to \mathscr{K}$ by putting

$$\beta_\lambda(a^{\lambda}_i, b^{\lambda}_j) = f_\lambda(j,i). \tag{5.1.2}$$

Clearly, β_λ is "associative" in a natural sense, see [**DuR98**, Lem. 2.3.1(a)]. A SBA $\mathcal{A}$ is called a *standardly full-based algebra* if $\mathrm{im}(\beta_\lambda) = \mathscr{K}$ for every λ.

When $\mathscr{K}$ is a field, these forms can be used to determine a classification of irreducible left $\mathcal{A}$-modules. In this case, we first observe that $\Delta(\lambda)$ is a cyclic $\mathcal{A}$-module if $\beta_\lambda \neq 0$. Let $\Lambda_1 = \Lambda_{1,\mathscr{K}} = \{\lambda \in \Lambda \mid \beta_\lambda \neq 0\}$ and, for $\lambda \in \Lambda_1$, let

$$L(\lambda) := \Delta(\lambda)/\mathrm{rad}\Delta(\lambda), \text{ where } \mathrm{rad}\Delta(\lambda) = \{v \in \Delta(\lambda) \mid \beta_\lambda(v,y) = 0\ \forall y \in \Delta^{\circ}(\lambda)\}$$

LEMMA 5.1.3 ([**DuR98**, Th. 2.4.1]). *Maintain the notation introduced above and assume that $\mathcal{A}$ is a finite-dimensional standardly based algebra over a field $\mathscr{K}$ with respect to a given poset Λ. Then:*

(a) *For each $\lambda \in \Lambda_1$, $L(\lambda)$ is absolutely irreducible, and $\{L(\lambda)\}_{\lambda\in\Lambda_1}$ forms a complete set of all non-isomorphic irreducible $\mathcal{A}$-modules.*
(b) *In particular, $|\Lambda_1|$ is the number of non-isomorphic irreducible $\mathcal{A}$-modules.*
(c) *If $L(\lambda)$ is a composition factor of $\Delta(\mu)$, then $\lambda \le \mu$.*
(d) *If $\lambda \in \Lambda_1$, then $[\Delta(\lambda) : L(\lambda)] = 1$.*

COROLLARY 5.1.4. *Let $(K, \mathcal{O}, k)$ be a local triple in which $\mathcal{O}$ is regular of Krull dimension ≤ 2 and suppose that $\mathcal{A}$ is a SBA over $\mathcal{O}$ associated with the poset Λ such that $\mathcal{A}_K$ is finite dimensional and semisimple. Then the decomposition matrix*

D of $\mathcal{A}$ is unitriangular. If, in addition, $\mathcal{A}$ is standardly full-based, then D is also square.

PROOF. Clearly, $\mathcal{A}_K$ is a SBA with standard modules $\Delta(\lambda)_K$, $\lambda \in \Lambda$, and has a filtration of ideals obtained by base changing (5.1.1) to K whose sections $I := \mathcal{A}_K^{\lambda_i}$ are ideals of $\mathcal{A}_K$ (since $\mathcal{A}_K$ is semisimple). Thus, $I^2 = I$ and, hence, there exists an idempotent $e \in I$ such that $I = \mathcal{A}_K e \mathcal{A}_K$. This implies $\Lambda_{1,K} = \Lambda$. Now the first assertion for the $|\Lambda_{1,k}| \times |\Lambda|$-matirx D follows from the lemma above. If $\mathcal{A}$ is standardly full-based, then $\Lambda_{1,k} = \Lambda$. Hence, D is a square matrix. □

The next proposition follows from a general result proved in [**DuR98**, (4.2.7)]. The assumption there that the base ring k be a commutative local ring is not necessary here when $k = \mathcal{Z}^\natural$ (or $\mathcal{Z}^{\natural\prime}$ if we are in the ${}^2F_4$ case). This is because Swan's result [**Sw78**, p.111] implies that any finite projective $\mathcal{Z}$-module is automatically free.[35] For completeness, we provide a constructive argument for the proposition.

PROPOSITION 5.1.5.

(a) *Assume that we are not in the ${}^2F_4$ case. Then, the split quasi-hereditary $\mathcal{Z}^\natural$-algebra $A^{+\natural}$ is a standardly full-based algebra.*

(b) *In the case of ${}^2F_4$, the split quasi-hereditary $\mathcal{Z}^{\natural\prime}$-algebra $A^{+\natural\prime}$ is a standardly full-based algebra.*

PROOF. We only give the proof in case (a). Let $\mathcal{A} := A^{+\natural}$ and let $\Lambda = \mathrm{Irr}(\mathbb{Q}W)$. Then, by Theorem 4.3.4(b), $\mathcal{A}$ is a split quasi-hereditary algebra, and Λ becomes a poset with the partial order given in (4.3.11) which is induced from the preorder (3.1.1). Consider the defining sequence of length $n = |\Lambda|$ given in Corollary 4.3.5:

$$0 = I_0 \subset I_1 \subset I_2 \subset \cdots \subset I_n = \mathcal{A}.$$

Assume that the chain is associated with the linear order $\lambda^{(1)}, \lambda^{(2)}, \ldots, \lambda^{(n)}$ on Λ, compatible with the partial order in Λ. For each $i = 1, \ldots, n$, $\mathcal{A}_i := \mathcal{A}/I_{i-1}$ is quasi-hereditary and $\bar{I}_i = I_i/I_{i-1}$ is a heredity ideal of $\mathcal{A}_i$. The split condition then implies, by Corollary 4.3.5(2), that there is an $\mathcal{A}_i$-$\mathcal{A}_i$-bimodule isomorphism

$$\bar{I}_i \overset{\phi_i}{\cong} \Delta(E) \otimes_{\mathcal{Z}^\natural} \mathrm{Hom}_{\mathcal{A}_i}(\Delta(E), \mathcal{A}_i), \tag{5.1.3}$$

where $\lambda^{(i)} = E$. Since both $\Delta(\lambda^{(i)}) := \Delta(E)$ and $\Delta^\circ(\lambda^{(i)}) := \mathrm{Hom}_{\mathcal{A}_i}(\Delta(E), \mathcal{A}_i)$ are $\mathcal{Z}^\natural$-projective, they are free $\mathcal{Z}^\natural$-modules.

Now, we may construct a standard basis by induction. The case for $i = n$ is clear, since bases for $\Delta(\lambda^{(n)})$ and $\Delta^\circ(\lambda^{(n)})$ give rise to a standard basis for $\mathcal{A}_n$. Assume now, for $1 < k \le n$, that a standard basis $\mathsf{B}_k = \bigcup_{l=k}^n \{b_{i,j}^{\lambda^{(l)}} \mid i \in I(\lambda^{(l)}), j \in J(\lambda^{(l)})\}$, for $\mathcal{A}_k = \mathcal{A}/I_{k-1}$ has been constructed. Use (5.1.3) and the bases $\{a_i\}$ and $\{b_j\}$ for $\Delta(\lambda^{(k-1)})$ and $\Delta^\circ(\lambda^{(k-1)})$, respectively, to get a basis $\{a_{i,j}^{\lambda^{(k-1)}} = \phi_{k-1}^{-1}(a_i \otimes b_j)\}_{i,j}$ for $\bar{I}_{k-1}$. This together with a pre-image $\{a_{i,j}^{\lambda^{(l)}}\}_{l,i,j}$ of the basis B_k under $\pi_k : \mathcal{A}_{k-1} \to \mathcal{A}_k$, the natural homomorphism, gives a basis

$$\mathsf{B}_{k-1} = \bigcup_{l=i-1}^{n} \{a_{i,j}^{\lambda^{(l)}} \mid i \in I(\lambda^{(l)}), j \in J(\lambda^{(l)})\},$$

[35]Note that $\mathcal{Z}^\natural = \mathbb{Z}^\natural[t, t^{-1}]$, where $\mathbb{Z}^\natural$ is a PID. Similarly, $\mathcal{Z}^{\natural\prime} = \mathcal{Z}'^\natural = \mathbb{Z}'^\natural[t, t^{-1}]$, where $\mathbb{Z}'^\natural$ is a PID. Thus, Swan's result [**Sw78**, p.111] applies.

for $\mathcal{A}_{k-1}$. Here, we require B_{k-1} to satisfy the condition $(a_{i,j}^{\lambda^{(l)}} + I_{k-1})/I_{k-1} = b_{i,j}^{\lambda^{(l)}}$, for all $l = k, \ldots, n$. Now, by induction, $\mathsf{B} = \mathsf{B}_0$ (for $i = 1$) is a desired standard basis.

Finally, the basis B is a full-standard basis, since every heredity ideal $\bar{I}_i$ has the form $\bar{I}_i = \mathcal{A}_i e_i \mathcal{A}_i$ with $\Delta(\lambda^{(i)}) \cong \mathcal{A}_i e_i$, $\Delta^{\circ}(\lambda^{(i)}) \cong e_i \mathcal{A}_i$ (see Corollary 4.3.6), and $e_i \mathcal{A}_i e_i \cong \mathcal{Z}^{\natural}$ for some idempotents $e_i \in \mathcal{A}_i$, where $i = 1, 2, \ldots, n$. □

We now use Theorem 4.3.8 to get a more general version of the proposition.

COROLLARY 5.1.6. *For a poset ideal Θ of $(\Omega, \leq_L)$, the statements* (a) *and* (b) *in Proposition* 5.1.5 *are true if $A^{+\natural}$ and $A^{+\natural'}$ are replaced by the quasi-hereditary algebras $A_{\Theta}^{\ddagger\natural}$ and $A_{\Theta}^{\ddagger\natural'}$, respectively.*

The following lemma follows from the proof of the above proposition. The base ring $\mathcal{Z}^{\natural}$ can be replaced by any commutative Noetherian ring with the property that any finite projective module is free.

LEMMA 5.1.7. *If $\mathcal{A}$ is a standardly based algebra over $\mathcal{Z}^{\natural}$ (or $\mathcal{Z}^{\natural'}$), and if $e \in \mathcal{A}$ is an idempotent, then $e\mathcal{A}e$ is standardly based.*

PROOF. Suppose $\mathsf{B} = \bigcup_{\lambda \in \Lambda} \mathsf{B}^{\lambda}$ is a standard basis for $\mathcal{A}$ and assume $\Lambda = \{\lambda^{(1)}, \lambda^{(2)}, \ldots, \lambda^{(m)}\}$ such that $i < j$ whenever $\lambda^{(i)} > \lambda^{(j)}$. Then, the two-sided ideals $J_i = \text{span}(\mathsf{B}^{\lambda^{(1)}} \cup \cdots \cup \mathsf{B}^{\lambda^{(i)}})$ of $\mathcal{A}$ gives a sequence in $\mathcal{A}$

$$0 = J_0 \subseteq J_1 \subseteq \cdots \subseteq J_m = \mathcal{A}.$$

This, in turn, gives a sequence of two-sided ideals in $e\mathcal{A}e$

$$0 = eJ_0e \subseteq eJ_1e \subseteq \cdots \subseteq eJ_me = e\mathcal{A}e.$$

Since all standard modules $\Delta(\lambda)$ and $\Delta^{\circ}(\lambda)$ of $\mathcal{A}$ are $\mathcal{Z}^{\natural}$-free, it follows that $e\Delta(\lambda)$ and $\Delta^{\circ}(\lambda)e$ are projective $\mathcal{Z}^{\natural}$-modules. Hence, they are $\mathcal{Z}^{\natural}$-free by Swan's result (if non-zero). Since $J_i/J_{i-1} \cong \Delta(\lambda^{(i)}) \otimes \Delta^{\circ}(\lambda^{(i)})$ as $\mathcal{A}_i$-$\mathcal{A}_i$-bimodules, where $\mathcal{A}_i = \mathcal{A}/J_{i-1}$, it follows that $eJ_ie/eJ_{i-1}e \cong e\Delta(\lambda^{(i)}) \otimes \Delta^{\circ}(\lambda^{(i)})e$ for all $i = 1, \ldots, m$. By induction, we may now construct a standard basis for $e\mathcal{A}e$ using the bases for $e\Delta(\lambda)$ and $\Delta^{\circ}(\lambda)e$. □

As a corollary, we obtain the following theorem.

THEOREM 5.1.8.

(a) *Assume that we are not in the case of ${}^2F_4$. The $\mathcal{Z}^{\natural}$-algebras $A^{\natural}$, $A^{+\natural}$ and $\mathcal{H}^{\natural}$ are all standardly based algebras.*

(b) *The analogous result holds in type ${}^2F_4$ for the corresponding $\mathcal{Z}^{\natural'}$–algebras. (See the appendix to Section* 4.1.*)*

This result for the $\mathcal{Z}^{\natural}$-algebra $A^{\natural}$ (and for $A^{+\natural}$, of course) is new. Geck [**G07**] obtained a stronger result in the case of $\mathcal{H}^{\natural}$, namely, that it has a cellular basis in the sense of [**GL96**]. As mentioned at the beginning of this section, cellular bases require the existence of an anti-involution which interacts with the algebra's basis and its bilinear forms.

By analyzing the structure of endomorphism algebras of tilting objects, G. Bellamy and U. Thiel proved that Hecke algebras associated to finite complex reflection groups are standardly based algebras over a field K. For real reflection groups, these bases are even cellular; see [**BT22**, Th. 3.8, Ex. 3.25].

5.2. Decomposition matrices, I. This and the next section present some applications to decomposition numbers for representations of finite groups of Lie type (generally in cross-characteristic) and associated algebras, such as the generic Hecke algebra $\mathcal{H}$ from Section 1.2 and the endomorphism algebras A and A^+ of display (2.4.4).

5.2A. Decomposition matrices associated with A, A^+, and $\mathcal{H}$. In this subsection, we begin our study of decomposition numbers of the endomorphism algebras A and A^+, defined in display (2.4.4), relative to a local triple $(\mathcal{O}, K, k)$. These algebras, as well as an underlying (generic) Hecke algebra $\mathcal{H}$, are built from a standard finite Coxeter system (W, S, L) arising. by the definition in Section 1.1, from a finite group G of Lie type. The "generic" aspect of this subsection comes from our naming of $\mathcal{H}$ in Section 1.2 as a generic Hecke algebra.

We do not directly use G in this subsection, although its underlying Lie theory context is in the background, especially the defining characteristic p of the ambient algebraic group.

In particular, finite groups of Lie type do appear informally throughout this section, so we digress slightly to introduce some further notation regarding them. Our assumption (in Section 1.2) that the Coxeter diagram of W is connected implies we may take $\mathbf{G}$ simple in the sense of algebraic groups [**C85**, p, 17, top]. (Alternatively, we can take $\mathbf{G}$ to be connected reductive, provided its semisimple part is simple. See [**C85**, §1.9]. Unless otherwise noted, we take $\mathbf{G}$ to be simple.) The group G may then be viewed as part of a Lie-type series associated to $\mathbf{G}$ and a permutation ρ (a graph automorphism) of its Dynkin diagram. There is also associated to G a power q of $\sqrt{p}$ described explicitly on [**C85**, pp. 39–42].[36]

As previously remarked, the group G plays a relatively minor role in the present subsection, but we will return to it in the next section. We do take this opportunity to remark that, if two groups G are in the same series, then they have the same standard finite Coxeter system (W, S, L), as shown by the construction [**C85**, pp. 34–36]. We also note that the prime p will remain fixed throughout this section (as the defining characteristic of $\mathbf{G}$), and the defining parameter q of G is a positive integer power of $\sqrt{p}$. In fact, if the Ree and Suzuki groups are excluded, q is a positive integer power of p.

We now return to the main task of studying decomposition numbers of A, A^+, and $\mathcal{H}$. The reader may recall the definition of a 2-regular local triple $(\mathcal{O}', K', k)$ from Section 5.0. We call such a triple *well-adapted* provided the following three conditions hold:

(1) $\mathcal{Z}$ is a subring of $\mathcal{O}'$;
(2) The field k has positive characteristic r distinct from p; and
(3) K' is a splitting field for $\mathcal{H}_{K'}$.

Note that (1) implies that $\mathbb{Q}(t)$ is contained in K', and so $\mathcal{H}_{K'}$ is always semisimple, even split, by (3). The split condition holds automatically except in the ${}^2F_4$ case. See footnote 8 in Chapter 1 for the semisimplicity of $\mathcal{H}_{\mathbb{Q}(t)}$.

Also, *the split semisimplicity of $\mathcal{H}_{K'}$ implies that of $A_{K'}$ and $A^+_{K'}$, as endomorphism algebras of modules for a split semisimple algebra.*

[36]In the split Chevalley group case, the group G itself may be written $\mathbf{G}(q)$. Within each series, the allowable parameters q determine a unique group G.

An example of a well-adapted 2-regular local triple is the Laurent triple associated with a DVR triple $(\mathcal{O}, K, k)$, where $\mathrm{char}(K) = 0$ and $\mathrm{char}(k) = r$ different from the defining characteristic of G. Also assume $\sqrt{2} \in K$ if G is of type ${}^2F_4$. See Section C3 in Appendix C.

Finally, we mention that, in the context of (1) above, condition (3) is always true when G is quasi-split or has rank 1. This covers all G except those of type ${}^2F_4$, in which case the field K' containing $\mathbb{Q}(t)$ is a splitting field if and only if $\sqrt{2} \in K'$. It is also the case that $\sqrt{2} \in K'$ if and only if $\sqrt{2} \in \mathcal{O}'$. This can be proved using the fact that $\mathcal{O}'$ is a regular local ring, hence a UFD (Auslander-Buchsbaum), hence integrally closed [**Jac89**, Ex. 1, p. 414]. In turn, $\sqrt{2} \in \mathcal{O}'$ if and only if

(1′) $\mathcal{Z}' \subseteq \mathcal{O}'$, where $\mathcal{Z}' := \mathbb{Z}'[t, t^{-1}]$ with $\mathbb{Z}' = \mathbb{Z}[\sqrt{2}]$.

(This condition is analogous to condition (1), and we call it condition (1′).) Consequently, a 2-regular local triple $(\mathcal{O}', K', k)$ is well-adapted when G is not of type ${}^2F_4$, provided conditions (1) and (2) hold and provided (1′) and (2) hold when G is of type ${}^2F_4$.

The reader may notice that $p = 2$ in the ${}^2F_4$ case, so that requiring $\sqrt{2} \in \mathcal{O}'$ is the same as requiring $\sqrt{p} \in \mathcal{O}'$. Later (such as in Lemma 5.2.7 below), we will need to consider similar requirements for other types, and even $\sqrt[4]{p} \in \mathcal{O}'$ in some cases, but this will not be necessary in Theorem 5.2.2 below.

In the subsection "Laurent triples" of Appendix C, we show how to construct many 2-regular local triples $(\mathcal{O}', K', k)$, starting with a DVR triple $(\mathcal{O}, K, k)$ and using for $\mathcal{O}'$ a localization of $\mathcal{O}[t, t^{-1}]$. If we take k to have positive characteristic $r \neq p$, as in (2) above, and K to have characteristic 0, then conditions (1) and (2) will be satisfied. Also, (1′) will be satisfied if $\sqrt{2} \in K$.

REMARK 5.2.1. Assume that the prime r is good for the group G. Then $\mathcal{Z}^\natural$ is contained in $\mathcal{O}'$, while in type ${}^2F_4$, the ring $\mathcal{Z}'^\natural$ is a subring of $\mathcal{O}'$. Thus, the algebra $A^+_{\mathcal{O}'}$ is a split quasi-hereditary algebra, as follows from the main result of the previous chapter, Theorem 4.3.4, together with Remark 4.3.9.

THEOREM 5.2.2. *Let $(\mathcal{O}', K', k)$ be a well-adapted 2-regular local triple with r (= the characteristic of k) good for G.*

Let $D_{\mathcal{O}'}$ be the decomposition matrix of $A_{\mathcal{O}'}$, and let $D^+_{\mathcal{O}'}$ be the decomposition matrix of $A^+_{\mathcal{O}'}$. Then, the following hold.

(a) *With respect to an ordering of the (non-isomorphic) irreducible modules for $A^+_{K'}$ and A^+_k, the decomposition matrix $D^+_{\mathcal{O}'}$ of $A^+_{\mathcal{O}'}$ is unitriangular and square.*
(b) *Let $D^+_{\mathcal{O}'}$ be arranged as in (a). Then, a version of $D_{\mathcal{O}'}$ can be obtained by removing some (possibly none) of the rows of $D^+_{\mathcal{O}'}$. Also, this puts $D_{\mathcal{O}'}$ in row echelon form with no row of zeros.*
(c) *In particular, the indices of the columns of $D_{\mathcal{O}'}$ in (b) above can be permuted so that $D_{\mathcal{O}'}$ has the form $[D_1, D_2]$, where D_1 is a square unitriangular matrix and D_2 is a rectangular matrix with the same number of rows as D_1. In particular, this version of $D_{\mathcal{O}'}$ is itself unitriangular.*
(d) *$D^+_{\mathcal{O}'}$ and $D_{\mathcal{O}'}$ have the same number of columns. Also, $A^+_{K'}$ and $A_{K'}$ have the same number of non-isomorphic irreducible representations as $\mathcal{H}_{K'}$.*
(e) *Parts (b) and (c) hold if $D_{\mathcal{O}'}$ is replaced by the decomposition matrix $D^\circ_{\mathcal{O}'}$ of $\mathcal{H}_{\mathcal{O}'}$. Hence, there exist submatrices D' and $D^{\circ\prime}$ of $D^+_{\mathcal{O}'}$, $D_{\mathcal{O}'}$,*

respectively, such that

$$D^+_{\mathcal{O}'} = \begin{pmatrix} D_{\mathcal{O}'} \\ D' \end{pmatrix} \text{ and } D_{\mathcal{O}'} = \begin{pmatrix} D^\circ_{\mathcal{O}'} \\ D^{\circ\prime} \end{pmatrix}.$$

PROOF. We will give arguments for the split, quasi-split, and some ${}^2F_4$ cases, leaving remaining ${}^2F_4$ arguments, and the rank 1 Ree and Suzuki groups to the reader.

We first prove (a),[37] noting that the $\mathcal{Z}^\natural$-algebra $A^{+\natural}$ in Corollary 4.3.7 gives rise to highest weight category $(\mathscr{C}, \{\Delta(E)\}_{E\in\Lambda})$ in the sense of Rouquier [**Ro08**, Def. 4.11]. Here, $\mathscr{C} = A^{+\natural}$-mod, the poset Λ is discussed in the paragraph surrounding (4.3.11), and the $\Delta(E)$'s (and $\Delta'(E)$'s) are discussed in Remark 4.2.3. (The $\Delta'(E)$'s are variations on the $\Delta(E)$'s. They are used in the ${}^2F_4$ case.)

The $\Delta(E)$'s above, in Remark 4.2.3, agree with those in Corollary 4.3.7, as may be shown by tracing (backwards) through the discussion above Corollary 4.3.5. (Here the discussion of the ${}^2F_4$ case is left to the reader.) A big advantage is that all $\Delta(E)$'s in Remark 4.2.3 are shown to have $\Delta(E)_{\mathbb{Q}(t)}$ irreducible. Applying this, and then base changing, we may conclude that $\Delta(E)_{K'}$ is irreducible, in our $(\mathscr{Q}', K', k)$ context, over $A^+_{K'} = (A^+_{\mathcal{O}'})_{K'}$.

We begin with the highest weight category $\mathscr{C}_{\mathcal{O}'} := A^{+\natural}_{\mathcal{O}'}$-mod by base changing $\mathscr{C}$ above from $\mathcal{Z}^\natural$ to $\mathcal{O}'$.

Using the notation in Corollary 4.3.7, together with base-change properties such as [**DPS98a**, Lem. 1.2.5(4)] and [**Ro08**, Prop. 4.14],[38] it is useful to define $V(\lambda) = \Delta(\lambda)_{\mathcal{O}'}$ in $\mathscr{C}_{\mathcal{O}'}$, modifying E to $\lambda \in \Lambda$ by abuse of notation. Here $\Lambda = \mathrm{Irr}(\mathbb{Q}W)$ (or $\Lambda = \mathrm{Irr}(\mathbb{Q}(\sqrt{2})W)$ if G is of type ${}^2F_4$), and $\Delta(\lambda) = (S^\natural_\lambda)^\diamond$ with $S^\natural_\lambda$ as defined in display (4.2.3). Thus, passing from $\mathcal{O}'$ to k, A^+_k–mod is a highest weight category in the original sense of [**CPS88**] for the poset $(\Lambda, \preceq)$, where $\preceq$ is defined in (4.3.11). In particular, Λ is in bijection with the distinct irreducible modules $\{L(\lambda)\}_{\lambda\in\Lambda}$ of A^+_k–mod, where $L(\lambda)$ is the head of the corresponding standard modules $V(\lambda)_k$.

Next, we note that each $V(\lambda)_{K'}$ is irreducible and that all irreducible objects in $A^+_{K'}$–mod have the form $V(\lambda)_{K'}$. (This follows from Remark 4.2.3.) Hence, all the decomposition numbers needed in (a) have the form $[V(\lambda)_k : L(\mu)]$, for any $\lambda, \mu \in \Lambda$. This gives a $|\Lambda| \times |\Lambda|$ square(!) matrix as well as the unitriangular property, since, by [**DS94**, Def. 2.1(i)], we have

$$[V(\lambda)_k : L(\mu)] = \begin{cases} 1, & \text{if } \lambda = \mu; \\ 0, & \text{unless } \mu \prec \lambda. \end{cases}$$

This completes the proof of (a). As noted, Remark 5.2.3(c) gives an alternate proof.

To prove Part (b) (and Part (d)),[39] we first observe from the notation in Construction 2.4.12 that $A^+ \cong \begin{pmatrix} \mathrm{Hom}_{\mathcal{H}}(T,T) & \mathrm{Hom}_{\mathcal{H}}(X,T) \\ \mathrm{Hom}_{\mathcal{H}}(T,X) & \mathrm{Hom}_{\mathcal{H}}(X,X) \end{pmatrix}$. This isomorphism induces isomorphisms $eA^+e \cong \begin{pmatrix} A & 0 \\ 0 & 0 \end{pmatrix} \cong A$, where $e \in A^+$ is the idempotent corresponding to $\begin{pmatrix} 1_T & 0 \\ 0 & 0 \end{pmatrix}$. Clearly, by noting that $T^+ = \begin{pmatrix} T \\ X \end{pmatrix}$, e defines the natural projection $T^+ \to T \subseteq T^+$. Thus, by combining the isomorphisms above, we may now identify A with eA^+e and $A_{\mathscr{K}}$ with $eA^+_{\mathscr{K}}e$ for any commutative ring $\mathscr{K}$. In the rest of the

[37] For a proof using standard basis theory, see Remark 5.2.3(c) below.

[38] Compare also Remark 4.3.9.

[39] This proof is inspired in part by arguments for [**DS00a**, Th. 6.3.2], but we include further details here.

proof, we will need only the cases $\mathscr{K} = k$ and $\mathscr{K} = K'$. These are fields, of course, allowing us to quote from [**Gr80**, Section 6.2] for algebras S over them such as $S = A_k^+$ and idempotents $e \in S$, taking, for instance, $eSe = eA_k^+e = A_k$,

If L is an irreducible A_k^+-module, then eL is either 0 or an irreducible A_k-module (observing that $A_k = eA_k^+e$). Further, the irreducible A_k-modules all arise this way. Also, if L, L' are irreducible A_k^+-modules such that $eL \neq 0 \neq eL'$, then $eL \cong eL'$ if and only if $L \cong L'$. For these assertions, see [**Gr80**, Th. (6.2g)]; their analogs also hold with K' replacing k. In fact, the latter case can be used to prove Part (d).[40] (Since the module $\mathcal{H}_{K'} = x_\emptyset \mathcal{H}_{K'}$, with $x_\emptyset$ defined in display (1.2.4), is a direct summand of both $T_{K'}^+$ and $T_{K'}$, we have, in this semisimple case, $eV(\lambda)_{K'} = e\Delta(\lambda)_{K'} \neq 0$, for all $\lambda \in \Lambda$ since $\mathrm{Hom}_{\mathcal{H}_K}((S_\lambda^\natural)_K, \mathcal{H}_K) \neq 0$. Thus, $\{eV(\lambda)_{K'} \mid \lambda \in \Lambda\}$ forms a complete set of irreducible $A_{K'}$-modules. Note that Λ also indexes irreducible modules of $\mathcal{H}_{K'}$; see (4.1.8). Hence, Part (d) follows.)

Continuing now the proof of Part (b), note that if $L = V_i/V_{i-1}$ is a composition factor of $V(\lambda)_k$, where $V_{i-1} \subseteq V_i$ are submodules of $V(\lambda)_k$, and $eL \neq 0$, then eL is irreducible and [**Gr80**, (6.2a)] now implies that $eL \cong eV_i/eV_{i-1}$ is a composition factor of $eV(\lambda)_k$. Also, every composition factor of $eV(\lambda)_k$ has such a form. This shows that a composition series $0 = V_0 \subseteq V_1 \subseteq \cdots \subseteq V_m = V(\lambda)_k$ of the left A_k^+-module $V(\lambda)_k$ gives rise to a composition series $0 = eV_0 \subseteq eV_1 \subseteq \cdots \subseteq eV_m = eV(\lambda)_k$ of the A_k-module $eV(\lambda)_k$ and, for every L with $eL \neq 0, [V(\lambda)_k : L] = [eV(\lambda)_k : eL]$. Hence, Part (b) follows except for its last assertion and that follows from Part (c).

We leave the proofs of (c) and (e) to the reader. This completes the proof of the theorem. □

REMARKS 5.2.3.

(a) These results appear to be new, certainly for A and A^+, but, because of our "generic" focus, also new for $\mathcal{H}$. Notice that $\mathcal{H}_{\mathcal{O}'}$ is a local version of the *generic* Hecke algebra $\mathcal{H}$, not one of its standard specializations (see Subsection 5.2B below), which typically require a replacement of the variable t or t^2 with a power of $\sqrt{p}$. For a variation of the theorem for such a standard specialization of $\mathcal{H}$, see [**GJ11**, Th. 4.4.1].

(b) Another variation of the theorem is obtained in [**DS00a**, (6.3.3)] in Lie type B, where a different version of A^+ is used, not involving Kazhdan-Lusztig cell modules.

(c) **Second proof of Theorem 5.2.2(a).** The fact that $\mathcal{O}'$-algebra $A_{\mathcal{O}'}^+$ is quasi-hereditary of split type (see Remark 5.2.1) implies that $A_{\mathcal{O}'}^+$ is a standardly full-based algebra associated with a poset Λ (see Proposition 5.1.5). Thus, every bilinear form associated with a cell module for $A_{K'}^+$ or A_k^+ is nonzero. Hence, we have $\Lambda_{1,K'} = \Lambda = \Lambda_{1,k}$ for both $A_{K'}^+$ and A_k^+ (see Lemma 5.1.3(b) and Corollary 5.1.4). This proves that the decomposition matrix $D_{\mathcal{O}'}^+$ is square. The unitriangular property follows from Lemma 5.1.3(c)&(d).

5.2B. From the generic case to standard specializations. With the exception of the more general Lemma 5.2.4 below, we continue with the notations $\mathbf{G}$ and G of the previous subsection. In particular, $\mathbf{G}$ is a simple algebraic group. We will fix $\mathbf{G}$ and the Lie-type series to which G belongs. In the setup of [**C85**, p. 31],

[40] See Remark 4.2.3 for an alternate proof of Part (d).

this means there is a "Frobenius endomorphism"[41] F of $\mathbf{G}$ with $G = \mathbf{G}^F$ and with F inducing a fixed permutation ρ of the Dynkin diagram of $\mathbf{G}$ [**C85**, pp.31–41]. Many of these—and other—details may be found in Section 1.1 of this monograph—and others.

We retain the notations $F, G, \mathbf{B}, B$ introduced in Section 1.1, which may also be used in a more general context such as a connected reductive group equipped with a Frobenius endomorphism.

It is useful to consider connected reductive groups $\widetilde{\mathbf{G}}$ which are central products[42] $\widetilde{\mathbf{Z}}\mathbf{G}$ of a torus $\widetilde{\mathbf{Z}}$ with $\mathbf{G}$. We assume, in the structures we seek, that F extends to a Frobenius endomorphism $\widetilde{F}$ of $\widetilde{\mathbf{G}}$ and write $\widetilde{G} = \widetilde{\mathbf{G}}^{\widetilde{F}}$. We also set $\widetilde{\mathbf{B}} = \widetilde{\mathbf{Z}}\mathbf{B}$, $\widetilde{B} = \widetilde{\mathbf{B}}^{\widetilde{F}}$, and $\widetilde{Z} = \widetilde{\mathbf{Z}}^{\widetilde{F}}$. Note that $\widetilde{\mathbf{Z}}$ is the connected component of the identity in the center of $\widetilde{\mathbf{G}}$, and, thus, is stable under $\widetilde{F}$. The lemma immediately below constructs functorially such a pair $\widetilde{\mathbf{G}}, \widetilde{F}$ in which $\widetilde{\mathbf{Z}}$ is the full center of $\widetilde{\mathbf{G}}$. (Compare [**Lus88**, pp.161–163] and [**DF94**, §§1,2].)

LEMMA 5.2.4. *If $\mathbf{G}$ is semisimple, then there exists a pair $\widetilde{\mathbf{G}}, \widetilde{F}$ in the setup of the above paragraph, in which $\widetilde{\mathbf{G}} = \widetilde{\mathbf{Z}}\mathbf{G}$ is connected reductive and $\widetilde{\mathbf{Z}}$ is the full center of $\widetilde{\mathbf{G}}$. (Equivalently, the center of $\widetilde{\mathbf{G}}$ is connected.) Also, $\widetilde{\mathbf{Z}}$ acts trivially on the coset space $\widetilde{\mathbf{G}}/\widetilde{\mathbf{B}}$. The G-sets $(\widetilde{\mathbf{G}}/\widetilde{\mathbf{B}})^{\widetilde{F}}$, $\widetilde{G}/\widetilde{B}$, $(\mathbf{G}/\mathbf{B})^F$, and G/B are all naturally isomorphic.*

PROOF. Observe that $Z(\mathbf{G})$ is finite and contained in all maximal tori $\mathbf{T}$ of $\mathbf{G}$. Take $\mathbf{T}$ to be F-stable. (See [**C85**, p. 33].) Let $\mathbf{T}_1$ be a copy of $\mathbf{T}$ independent of $\mathbf{G}$. Let $Z(\mathbf{G})_1 \subseteq \mathbf{T}_1$ correspond to $Z(\mathbf{G})$. Thus, the natural action of F on $\mathbf{T}$ gives rise to a corresponding action on $\mathbf{T}_1$ and on $Z(\mathbf{G})_1 \subseteq \mathbf{T}_1$. The natural maps $Z(\mathbf{G}) \cong Z(\mathbf{G})_1 \subseteq \mathbf{T}_1$ and $Z(\mathbf{G}) \subseteq \mathbf{G}$ give rise to a connected, reductive pushout group: $\widetilde{\mathbf{G}} = \mathbf{T}_1\mathbf{G}$ with (connected) center $\mathbf{T}_1$, and we now write $\widetilde{\mathbf{Z}} = \mathbf{T}_1$. Also, F extends to a Frobenius endomorphism on $\widetilde{\mathbf{G}}$, duplicating on $\mathbf{T}_1$ its action on $\mathbf{T}$.

Since the coset space $\widetilde{\mathbf{G}}/\widetilde{\mathbf{B}}$ identifies here, as a $\widetilde{\mathbf{G}}$-set, with the set of $\widetilde{\mathbf{G}}$-conjugates of $\widetilde{\mathbf{B}}$, the action of $\mathbf{T}_1 = \widetilde{\mathbf{Z}} := Z(\widetilde{\mathbf{G}})$ is trivial. The G-set identifications at the end of the lemma follow from elementary arguments and the Lang-Steinberg theorem, as illustrated in [**C85**, p. 33]. All the identifications are the "obvious" candidates. This completes the proof. □

REMARK 5.2.5. The lemma is also true if "semisimple" is replaced by "connected reductive." We sketch a proof in this case. First, change notation so that the group in question is $\mathbf{G}\mathbf{T}_2$, where $\mathbf{G}$ is semisimple and $\mathbf{T}_2$ is a (central) torus. Apply the proof of the lemma for the group $\mathbf{G}$ to show $\mathbf{T}_1\mathbf{G}$ has connected center $\mathbf{T}_1$. Then construct $\mathbf{T}_1\mathbf{G}\mathbf{T}_2$ and show it has center $\mathbf{T}_1\mathbf{T}_2$. The remaining details of the proof are left to the reader. (Later, we only use the lemma as stated, and only when $\mathbf{G}$ is simple.)

In the rest of subsection 5.2B, let $(\mathcal{O}, K, k)$ be a DVR triple (in some cases, a more general local triple could be used) in which

- the field k has positive characteristic r, distinct from the defining characteristic $p > 0$ of $\mathbf{G}$;

[41] Also called a "Frobenoius map" in Section 1.1. Thus, some power of F is a standard Frobenius endomorphism.

[42] See [**Go80**, p.29] for the definition of a central product.

- K has characteristic 0; and
- $\sqrt{2} \in \mathcal{O}$ if G is of type ${}^2F_4$.

These choices are natural when concerned with the representation theory of kG (i.e., the cross-characteristic theory)—see specifically [**DPS98a**, (1), p. 170].

We continue the assumption of this subsection that G is the group $\mathbf{G}^F$ in a fixed Lie-type series with $\mathbf{G}$ simple (as discussed above). Associated with G is a parameter q (which determines G in its series) and a parameter δ which equals 1, 2, or 3, depending on the series, with q^δ a power of p (an odd power, if the series is Ree or Suzuki).

The following lemma, included for completeness, provides largely well-known "standard specialization" properties of the generic Hecke algebra $\mathcal{H}$. Recall the $\mathbb{Z}[t^2]$-form $\mathcal{H}^o$ of $\mathcal{H}$ considered in Lemma 1.2.2.

LEMMA 5.2.6. *Let $(\mathcal{O}, K, k)$ be a DVR triple as detailed above. Suppose $\sqrt{q} \in \mathcal{O}$ is invertible in $\mathcal{O}$; this gives $\mathcal{O}$ a natural structure as a $\mathcal{Z}$-algebra (via $t \mapsto \sqrt{q}$). Then, the base-changed algebras $\mathcal{H}_\mathcal{O}$ and $\mathcal{H}^o_\mathcal{O}$ are naturally isomorphic, and the base-changed algebras $\mathcal{H}_K$ and A_K are split semisimple.*

PROOF. The isomorphism $\mathcal{H}_\mathcal{O} \cong \mathcal{H}^o_\mathcal{O}$ follows from the hypothesis. It remains to verify the final assertion of the lemma. Theorem 4.1.3 shows that $\mathcal{J}^\natural$ is a direct product of matrix algebras over $\mathcal{Z}^\natural$. Base change from $\mathcal{Z}^\natural$ to K gives that $\mathcal{J}_K$ is a similar direct product of full matrix algebras. Taking $R = K$ in the discussion below display (4.1.4), we have $\mathcal{H}_K \cong \mathcal{J}_K$. (Note that $\mathcal{H}_K \cong \mathcal{H}^o_K$ is semisimple, as the endomorphism ring of a KG-module.) The isomorphism with a direct product of matrix algebras shows that $\mathcal{H}_K$ is split semisimple.

Finally, A_K is the endomorphism algebra of an $\mathcal{H}_K$-module by [**DPS98a**, Th. 2.4.4(b)]. Since $\mathcal{H}_K$ is split semisimple, it follows that A_K is also split semisimple. □

Recall the definition of a Laurent triple associated with a DVR triple in Appendix C. (See the paragraph above Example C3.1 in Appendix C.)

LEMMA 5.2.7. *Let $(\mathcal{O}, K, k)$ be a DVR triple as detailed above. Also, assume $\sqrt{q} \in \mathcal{O}$.*[43] *Let $(\mathcal{O}', K(t), k)$ be the Laurent triple associated to the DVR triple $(\mathcal{O}, K, k)$ and the element $a = \sqrt{q}$.*

(a) *The decomposition matrix $D_{\mathcal{O}'}$ of $A_{\mathcal{O}'}$ (with respect to the triple $(\mathcal{O}', K(t), k)$) agrees with the decomposition matrix $D_\mathcal{O}$ of $A_\mathcal{O}$ with respect to $(\mathcal{O}, K, k)$.*

(b) *Similarly, the decomposition matrix of $\mathcal{H}_{\mathcal{O}'}$ agrees with that of $\mathcal{H}_\mathcal{O}$.*

PROOF. The reader may check that the diagram in display (C1.2) in Appendix C is interpreted as follows for $B = A$ or $\mathcal{H}$ and with the indicated prime ideal $\mathfrak{p}$

$$(5.2.1)\qquad \begin{array}{ccccc} B & \longrightarrow & B_\mathfrak{p} & \longrightarrow & B_Q \\ \downarrow & & \downarrow & & \\ B/\mathfrak{p}B & \longrightarrow & B(\mathfrak{p}) & & \\ \downarrow & & & & \\ B_F & & & & \end{array}$$

[43]We note that the latter assumption is implied by the condition that $\sqrt[4]{p} \in \mathcal{O}$. If the series of G is not Ree or Suzuki, this may be weakened to the condition that $\sqrt{p} \in \mathcal{O}$.

where in case (a), $B = A_{\mathcal{O}'}$ and $\mathfrak{p} = (t-a) \in \mathcal{O}'$, while $B(\mathfrak{p})$ identifies with $B_K = B_{(\mathcal{O}'/\mathfrak{p})_\mathfrak{p}}$. Now apply Theorem C1.4 and Lemma 5.2.6.

A similar argument applies in the Hecke algebra case (b), where $B = \mathcal{H}_{\mathcal{O}'}$. This proves the lemma. □

The notation $G = \mathbf{G}^F$ below continues as above. Recall the generic Hecke endo-algebra A defined in (1.2.5).

THEOREM 5.2.8. *Suppose G_1, G_2 are finite groups of Lie type in the same Lie-type series as G and let q_1^δ, q_2^δ be the powers of p associated to G_1, G_2, respectively (as in* [**C85**, pp. 38–41]*).*

Let $(\mathcal{O}, K, k)$ be a DVR triple as in (the first sentence of) Lemma 5.2.7, and assume that $\sqrt{p} \in \mathcal{O}$. If G is of Ree or Suzuki type, assume $\sqrt[4]{p} \in \mathcal{O}$.

Fix $B = A$ (the generic Hecke endo-algebra) or $B = \mathcal{H}$ (the generic Hecke algebra).

Suppose that D' and D'' are the decomposition matrices of $B_{\mathcal{O}'}$ and $B_{\mathcal{O}''}$, with respect to the Laurent triples $(\mathcal{O}', K', k)$ and $(\mathcal{O}'', K'', k)$ associated to $(\mathcal{O}, K, k)$ with $a_1 = \sqrt{q_1}$ and $a_2 = \sqrt{q_2}$, respectively.[44] *Also, viewing $\mathcal{O}$ as the $\mathcal{Z}$-algebra $\mathcal{O}_1$ or $\mathcal{O}_2$ via the specializations $t \mapsto a_1$ and $t \mapsto a_2$, respectively, let D_1, D_2 be the decomposition matrices for $B_{\mathcal{O}_1}, B_{\mathcal{O}_2}$, respectively, relative to the DVR triples $(\mathcal{O}_1, K_1, k)$, $(\mathcal{O}_2, K_2, k)$.*

Then, the decomposition matrices D', D'', D_1, and D_2 can all be identified with the same matrix D, where D depends only on the choice $B = A$ or $B = \mathcal{H}$.

PROOF. First, assume $B = A$ and G is not of Ree or Suzuki type.

Lemma 5.2.7 gives an identification of D_1 with D' and an identification of D_2 with D'', while Proposition C3.2 identifies D' and D''. It is useful to note here that there is, in the notation of Example C3.1, a Laurent triple $({}^a\mathcal{O}', K', k)$ with $a = \sqrt{p}$, even if this value of a is not a_1, a_2, or associated to any G in the Lie-type series. So, the relations $(\mathcal{O}', K', k) \leq ({}^a\mathcal{O}', K(t), k)$ and $(\mathcal{O}'', K'', k) \leq ({}^a\mathcal{O}', K(t), k)$ can be used with Proposition C3.2 to identify D'' with D' as follows: First, note that there is, in the notation of Example C3.1, a Laurent triple $({}^a\mathcal{O}', K(t), k)$ with $a = \sqrt{p}$, even if this value of a is not a_1, a_2 or associated to G in the Lie-type series. So, the relations $(\mathcal{O}', K', k) \leq ({}^a\mathcal{O}', K(t), k)$ and $(\mathcal{O}'', K'', k) \leq ({}^a\mathcal{O}', K(t), k)$ can be used with Proposition C3.2 to identify D'' with D'.

We leave the remaining cases, with either A replaced by $\mathcal{H}$ or G replaced by a group of Ree or Suzuki type, to the reader. □

5.3. Decomposition matrices, II. We have seen from Theorem 5.2.8 that the decomposition matrix D is an invariant relative to the standard specializations arising from finite groups G of Lie type in the same Lie-type series. We now prove that, for such a G and DVR triple $(\mathcal{O}, K, k)$, this matrix D is part of the decomposition matrix of $\mathcal{O}G$. This shows that D is, in fact, an invariant associated with the Lie type series.

We continue with the notation in the previous section. The unipotent representations in this section are the irreducible constituents of the permutation module $\mathbb{C}G/B$, where $B = \mathbf{B}^F$ for an F-stable Borel subgroup $\mathbf{B}$ of $\mathbf{G}$.

[44]Thus, $(\mathcal{O}', K', k)$ and $(\mathcal{O}'', K'', k)$ are defined in the paragraphs above Example C3.1 in Appendix C with a_1 and a_2 as the parameter a there. In particular, $K' = K'' = K(t)$.

In the following theorem, as in Theorem 5.2.8, $(\mathcal{O}, K, k)$ is a DVR triple with $\mathcal{O}$ complete, k is algebraically closed of characteristic $r \neq p$, and K has characteristic 0 (see Appendix A, Main Theorem). Also, by Hensel's lemma, $\mathcal{O}$ contains a primitive p^{th}-root of unity. Here p is the defining characteristic of $\mathbf{G}$. We also require, as in Theorem 5.2.8, that $\sqrt{p} \in \mathcal{O}$ and $\sqrt[4]{p} \in \mathcal{O}$ for G of Ree or Suzuki type; thus, we always have $\sqrt{q} \in \mathcal{O}$.

The proof of part (a) involves methods invented by Dipper many years ago (although we could not find an explicit statement in [**DG99**]). Part (b) seems to be a remarkable fact, not considered before in the literature. It is based on our study (in Appendix C) of Krull dimension 2 decomposition numbers. Part (c) depends on our main theorem in Subsection 5.2B, and provides evidence for the conjecture [**GJ11**, Conj. 4.5.2], which references work of Geck [**G90**, **G06**] and Geck–Hiss [**GH97**, Conj. 3.4]. Geck's conjecture [**G12**, Conj. 2.1], originating from his 1990 Ph.D. thesis [**G90**], on unitriangular shape of decomposition matrices of unipotent blocks has been proved recently in [**BDT20**], under the assumption that the defining characteristic for $\mathbf{G}$ is good and the non-defining characteristic r is *very good*.[45] The following result suggests that, for certain unipotent blocks, these conditions could be weakened by simply assuming that r is good. (Compare bottom line of [**BDT20**, p.587].)

THEOREM 5.3.1. *Let $G = \mathbf{G}^F$ be as above. (Thus, $\mathbf{G}$ is simple,[46] and we think of G as a finite group in a given Lie-type series.)*

(a) *Let the matrix D be as described above in Theorem 5.2.8 for the generic Hecke endo-algebra A. Then, D is a submatrix of the decomposition matrix of $\mathcal{O}G$. The columns of D are indexed by the irreducible constituents of the permutation module $\mathbb{C}G/B$.*

(b) *The same statements hold, with the same matrix D, if G is replaced by any group in the same finite Lie-type series.*

(c) *If r is a good prime for $\mathbf{G}$, then D above is unitriangular.*

PROOF. First, assuming that (a) holds, (b) follows from Theorem 5.2.8 and (c) follows from Theorem 5.2.2(c) and Theorem 5.2.8.

It remains to prove (a). This assertion contains two parts, the second of which, counting columns, follows from the argument for Theorem 5.2.2(d). It remains to prove the first part of (a), which asserts an inclusion of decomposition matrices. We divide the proof into two parts.

We give the proof when $\mathbf{G}$ is simple of adjoint type in Part 1. In this case, we may directly apply Theorem A4.1 in Appendix A. We also supply the main ingredient in the argument for the general case. In Part 2, we prove a variation on Theorem A4.1 for $\mathbf{G}$ simple, reducing the general case to the validity of Theorem A4.1 for the group $\widetilde{\mathbf{G}}$ in Lemma 5.2.4 (which has connected center).

PART 1 (The adjoint-type case). Let $\mathbf{G}$ be a simple algebraic group of adjoint type. Then, by the proof of [**MT11**, Prop. 9.15], $\mathbf{G}$ has a trivial center $Z(\mathbf{G}) = 1$; see also the proof of [**C85**, Prop. 8.1.2] on page 254 for this fact. Thus, Theorem A4.1 applies in the current (adjoint) case.

[45] Call r a very good prime if r is good and r does not divide the order of $Z(\mathbf{G})_F$, the largest (finite) quotient of $Z(\mathbf{G})$ on which F acts trivially. See [**BDT20**, Th. A].

[46] We continue to make this assumption for simplicity; see footnote 3 in Chapter 1. In fact, much of the argument applies if $\mathbf{G}$ is just semisimple, or even connected reductive.

Let

$$\mathcal{M} = \mathcal{M}_{\mathcal{O}} := \bigoplus_{J\subseteq S} R^G_{L_J}(\mathscr{S}_\sigma^{L_J}),$$

using, on the right-hand side, the notation of the proof of Theorem A4.1. Then, Theorem A4.1 holds for $\mathcal{M}_K$. This provides a projective cover of $\mathcal{M}$ satisfying [**Dip91**, 2.3]. Equivalently, in the context of [**Dip90**, §4], the cover satisfies [**Dip90**, 2.6]; see [**Dip90**, Th. 4.7]. Thus, by [**Dip90**, Cor. 4.10], the decomposition matrix of $\mathrm{End}_{\mathcal{O}G}(\mathcal{M})$ is a submatrix of the decomposition matrix of $\mathcal{O}G$. (See the formulation of [**GJ11**, Th. 4.1.14] which uses a slightly different hypothesis on $(\mathcal{O}, K, k)$ but also references the same result [**Dip90**, Cor. 4.10].) Now, assertion (a) follows from the algebra isomorphism $\mathrm{End}_{\mathcal{O}G}(\mathcal{M}) \cong A_{\mathcal{O}}$, which is proved below.

For an F-stable maximal torus $\mathbf{T} \subseteq \mathbf{B}$, let $T = \mathbf{T}^F$. Let $M_\emptyset := R^G_T\mathcal{O} = \mathcal{O}G/B$, where $\mathcal{O}$ is the trivial T-module. By remarks before the statement of the theorem, $\mathcal{O}$ contains a primitive p^{th}-root of unity. Thus, the underlying regular character σ of Γ_σ is defined over $\mathcal{O}$. (The reader may wish to review properties of Gelfand-Graev characters and their underlying modules in [**C85**, §8.1] or the helpful summary in [**G17**, §4.3].) As an $\mathcal{O}G$-module, Γ_σ is induced to G from σ on U, and the latter is not trivial on any fundamental root subgroup of U, nor are any of the conjugates of σ by elements of T. We mention that [**C85**, §2.1–§2.6] gives an excellent treatment of root subgroups and root systems in the broad context of algebraic groups with split BN-pairs, suitable for our use here. So, if $w \in W$ is not the longest word, neither is its inverse, and there are no nontrivial B-homomorphisms from $\mathrm{Ind}_U^B\,\mathcal{O}_\sigma$ to $\mathrm{Ind}^B_{{}^wB\cap B}\,\mathcal{O}$. (The point is that ${}^wB \cap B$ must contain a fundamental root group X, on which σ and its conjugates through elements of T are all nontrivial.) Mackey decomposition arguments now show that $\mathrm{Hom}_{\mathcal{O}G}(\Gamma_\sigma, M_\emptyset) \cong \mathcal{O}$, generated by a map $\beta_\emptyset : \Gamma_\sigma \longrightarrow M_\emptyset$ with image $\mathcal{O}$-pure. By [**G17**, 4.1], $\dim_K \mathrm{Hom}_{KG}(K\Gamma_\sigma, \mathrm{St}_K) = 1$. As is well known, St_K is a submodule of $KM_\emptyset$. Thus, the image of $\beta_\emptyset$ satisfies $K\,\mathrm{Im}\,\beta_\emptyset = \mathrm{St}_K \subseteq KM_\emptyset$. So the kernel $\mathrm{Ker}\,\beta_\emptyset$ is just the intersection of Γ_σ with the sum of all irreducible KG-submodules of $K\Gamma_\sigma$ not isomorphic to St_K. (Recall that $K\Gamma_\sigma$ is multiplicity free by [**C85**, Th. 8.1.3]. This reference requires that $\mathbf{G}$ be connected reductive with connected center, a condition that holds for $\mathbf{G}$ with adjoint type and also for $\widetilde{\mathbf{G}}$ in the more general case discussed in Part 2. Corresponding results for Levi subgroups also hold in each case. The multiplicity freeness may then be deduced for $\widetilde{G}$ and its Levi subgroups by using [**DF94**, Lem. 2.3(a),(c)].) A similar kernel analysis applies to the surjective map $\Gamma_\sigma \longrightarrow \mathscr{S}_\sigma$ discussed above (A3.2). Thus, the latter map has the same kernel as $\beta_\emptyset$. Consequently,

$$\mathrm{Im}\,\beta_\emptyset \cong \mathscr{S}_\sigma.$$

The remaining discussion in Part 1 does not assume that $\mathbf{G}$ is of adjoint type.

The Steinberg KG-submodule of $KM_\emptyset$ can also be written as $y_SKM_\emptyset$, where

$$y_S = \sum_{w\in W} \mathrm{sgn}(w)t^{-2L(w)}T_w = \sum_{w\in W} \mathrm{sgn}(w)t^{-L(w)}\widetilde{T}_w \in \mathcal{H}_K.$$

This is just the module-theoretic realization of the correspondence between irreducible KG-constituents of $KM_\emptyset$ and irreducible characters of its endomorphism algebra. The Steinberg character corresponds to the sign character; see [**G17**, Th. 2.1, Lem. 2.2], and Lemma 5.2.6 above. Note that $T_sx = -x$, for $x = y_S$, and, consequently, for any $x \in y_SKM_\emptyset$. For the $x = y_S$ case, see [**DPS98c**, Lem. 1.1(b)].

There is thus a unique pure $\mathcal{O}G$-submodule of $M_\emptyset$ which base changes over K to St_K, namely, $M_\emptyset \cap y_S K M_\emptyset$. The latter module may also be written as $\sqrt{y_S M_\emptyset}$ in the (now common) terminology of [**Dip90**, p. 251]. We have proved $\mathscr{S}_\sigma \cong \sqrt{y_S M_\emptyset}$.

The argument for this isomorphism has not used the assumption that $\mathbf{G}$ is simple of adjoint type. There are similar characterizations of the modules $\mathscr{S}_\sigma^{L_J}$ introduced above Theorem A4.1. We may write

$$\mathscr{S}_\sigma^{L_J} = \sqrt{y_J M_\emptyset^{L_J}},$$

where $y_J = \sum_{w \in W_J} (-1)^{\ell(w)} (t_s^2)^{-\ell(w)} T_w$, and $\sqrt{y_J M_\emptyset} \cong R_T^G(\mathscr{S}_\sigma^{L_J})$, the submodule of the $\mathcal{O}G$-module $M_\emptyset$ generated by $\sqrt{y_J M_\emptyset^{L_J}} = \mathscr{S}_\sigma^{L_J}$.

Applying [**Dip98**, Ths. 1.30&1.36] with $R = \mathcal{O}$, and $M_\mathcal{O} = M_\emptyset$ in Theorem 1.30 there, we obtain[47]

$$\mathrm{End}_{\mathcal{O}G}\left(\bigoplus_J \sqrt{y_J M_\emptyset}\right) \cong \mathrm{End}_{\mathcal{H}_\mathcal{O}}\left(\bigoplus_J y_J \mathcal{H}_\mathcal{O}\right).$$

However, there is a well-known identification

$$\mathrm{Hom}_{\mathcal{H}_\mathcal{O}}(y_J\mathcal{H}_\mathcal{O}, y_I\mathcal{H}_\mathcal{O}) \cong \mathrm{Hom}_{\mathcal{H}_\mathcal{O}}(x_J\mathcal{H}_\mathcal{O}, x_I\mathcal{H}_\mathcal{O})$$

for each $I, J \subseteq S$. See [**DPS98c**, Lem. 1.1(c)]. The left-hand side of this display sums to the right-hand side of the display above it. The lower display's own righth-hand-side sums to the algebra $A_\mathcal{O}$, using the base-change result [**DPS98a**, Th. 2.4.4]. The left-hand side of the higher display has decomposition matrix a submatrix of that of $\mathcal{O}G$, as noted in our discussion of $\mathrm{End}_{\mathcal{O}G}(M)$ in the opening paragraph of this proof. However, by Theorem 5.2.8 (or Lemma 5.2.7), the decomposition matrix D of $A_{\mathcal{O}'}$ agrees with that of $A_\mathcal{O}$. The theorem is now completely proved for $\mathbf{G}$ simple of adjoint type.

PART 2 (The general case). Next, we consider the general case where $\mathbf{G}$ is simple, but possibly not adjoint. (Much of the argument applies if $\mathbf{G}$ is just semisimple, or even connected reductive.) Possibly, $\mathbf{G}$ does not have connected center. In particular, Theorem A4.1 is not directly available, but does apply to the group $\widetilde{\mathbf{G}}$ as constructed in Lemma 5.2.4. The theorem can at least be (re)stated for G (that is, for $G = \mathbf{G}^F$), using a fixed regular character $\sigma : U \to K^\times$ and drawing on the notation and discussion in Appendix A, which in turn draws on [**G17**, §4]. We next give the main details and then use the version of Theorem A4.1 in Appendix A to prove the modified version.

In particular, an $\mathcal{O}G$-lattice Γ_σ is defined as in [**G17**, 4.1] via an idempotent in $\mathcal{O}U \subseteq \mathcal{O}G$ constructed from σ. (The homomorphism σ^{-1} applied to the idempotent is 1.) Equivalently, Γ_σ is the $\mathcal{O}G$-module induced by the rank 1 $\mathcal{O}U$-module determined by σ^{-1} (a module we call $\mathcal{O}_\sigma$ by abuse of notation). Briefly, $\Gamma_\sigma := \mathrm{Ind}_U^G \mathcal{O}_\sigma$. This definition makes sense for $G = \mathbf{G}^F$ (or for any finite group of Lie type).

Next, keeping the same level of generality, let $P_J = U_J L_J$ be the standard parabolic subgroup of G corresponding to $J \subseteq S$. A specific analog $\Gamma_\sigma^{L_J}$ of Γ_σ is constructed "by restriction" in Appendix A. This makes sense here, since $U \cap L_J$ is the analog for L of U, and the restriction of σ to $U \cap L_J$ is regular. We write

$$\Gamma_\sigma^{L_J} := \mathrm{Ind}_{U\cap L_J}^{L_J} \mathrm{Res}_{U\cap L_J}^{U} \mathcal{O}_\sigma.$$

[47]See also [**Ca98**, Th. 1] and [**DG99**, Th. 1.10].

Continuing, note from [**G17**, 4.1] that the Steinberg module St_K appears in $K\Gamma_\sigma$ with multiplicity 1. It follows (see [**G17**, Prop. 4.2], attributed to Hiss, and its proof) that there is a unique pure $\mathcal{O}G$-sublattice Γ'_σ such that $K(\Gamma_\sigma/\Gamma'_\sigma) \cong \mathrm{St}_K$. Both [**G17**] and Appendix A take this property to define $\mathscr{S}_\sigma := \Gamma_\sigma/\Gamma'_\sigma$, and we make the same definition here. As in Appendix A, we also make the parallel definition

$$\mathscr{S}_\sigma^{L_J} := \Gamma_\sigma^{L_J}/(\Gamma_\sigma^{L_J})'.$$

This does not require the connected center hypothesis and makes sense in the general context $G = \mathbf{G}^F$ with $\mathbf{G}$ simple (or even just connected reductive).

To summarize, we have now defined for $\mathbf{G}$ all of the ingredients needed to define β_K for the formal statement of Theorem A4.1.

We claim that Theorem A4.1 *holds for* $\mathbf{G}$ *with this notation.* We intend to prove this claim by using the fact that it holds for $\widetilde{\mathbf{G}}$ with obvious changes in notation from $G, L_J, \beta, \Gamma_\sigma^{L_J}, \mathscr{S}_\sigma^{L_J}, \mathrm{St}_K^{L_J}$ to $\widetilde{G}, \widetilde{L}_J, \widetilde{\beta}, \Gamma_\sigma^{\widetilde{L}_J}, \mathscr{S}_\sigma^{\widetilde{L}_J}, \mathrm{St}_K^{\widetilde{L}_J}$, for any $J \subseteq S$.

Theorem A4.1 as currently stated requires that we prove that the kernel and image of

$$\beta_K : \bigoplus_{J\subseteq S} R_{L_J}^G(K\Gamma_\sigma^{L_J}) \longrightarrow \bigoplus_{J\subseteq S} R_{L_J}^G(St_K^{L_J}) \tag{5.3.1}$$

have no common nonzero constituents, a property that we know (and will prove) to be true in our context.

For each $J \subseteq S$, let $\gamma^J : K\Gamma_\sigma^{L_J} \to K\mathscr{S}_\sigma^{L_J} \cong \mathrm{St}_K^{L_J}$, discussed in the proof of Theorem A4.1. We take L_J to be the standard Levi subgroup of G associated to J, and all the terms $\Gamma_\sigma^{L_J}, \mathscr{S}_\sigma^{L_\sigma}, \mathrm{St}_K^{L_J}$ as discussed above. Similarly, we have a map $\widetilde{\gamma}^J$, replacing G with $\widetilde{G}$, and using $\Gamma_\sigma^{\widetilde{L}_J}$, $\mathscr{S}_\sigma^{\widetilde{L}_J}$, and $\mathrm{St}_K^{\widetilde{L}_J}$ in place of their counterparts for G.

We pause here to note that

$$\beta_K = \bigoplus_{J\subseteq S} R_{L_J}^G(\gamma^J) \quad \text{and} \quad \widetilde{\beta}_K = \bigoplus_{J\subseteq S} R_{\widetilde{L}_J}^{\widetilde{G}}(\widetilde{\gamma}^J).$$

Recall that $\widetilde{G}$ is defined in the notation and setting of Lemma 5.2.4 as $\widetilde{G} = \widetilde{\mathbf{G}}^{\widetilde{F}}$, where $\widetilde{\mathbf{G}} = \mathbf{G}\widetilde{\mathbf{Z}}$. Here, $\widetilde{\mathbf{Z}}$ is an isomorphic copy of the "maximally split" torus $\mathbf{T}$ of $\mathbf{G}$ in the terminology of [**C85**, §1.18] and $\widetilde{\mathbf{Z}} = Z(\widetilde{\mathbf{G}})$. The torus $\mathbf{T}\widetilde{\mathbf{Z}}$ is maximally split in $\widetilde{\mathbf{G}}$, and in $\widetilde{G} = G(\mathbf{T}\widetilde{\mathbf{Z}})^{\widetilde{F}}$, as may be shown using the Lang-Steinberg theorem. There is also an F-stable Borel subgroup $\mathbf{B}$ of $\mathbf{G}$ containing $\mathbf{T}$ in the setting of Lemma 5.2.4. A closed subgroup $\mathbf{N}$ normalizing $\mathbf{T}$ may then be chosen so that $\mathbf{B}$, $\mathbf{N}$ give a split BN-pair structure to $\mathbf{G}$ as per the axioms (specialized for algebraic groups) of [**C85**, p.23], with $\mathbf{B} \cap \mathbf{N} = \mathbf{T}$. (The choice in [**C85**, p.22] of $\mathbf{N}$ is $\mathbf{N} = N_{\mathbf{G}}(\mathbf{T})$, but this is not part of the axioms. We note that this choice of $\mathbf{N}$ is F-stable, a useful property.) We now have three additional candidates for split BN-pairs: the pair $\mathbf{B}\widetilde{\mathbf{Z}}, \mathbf{N}\widetilde{\mathbf{Z}}$ with intersection $\mathbf{T}\widetilde{\mathbf{Z}}$, the pair $B = \mathbf{B}^F, N = \mathbf{N}^F$ with intersection $T = \mathbf{T}^F$, and, finally, the pair $(\mathbf{B}\widetilde{\mathbf{Z}})^{\widetilde{F}}, (\mathbf{N}\widetilde{\mathbf{Z}})^{\widetilde{F}}$ with intersection $(\mathbf{T}\widetilde{\mathbf{Z}})^{\widetilde{F}}$. The first of these candidate pairs satisfies Carter's axioms by an easy inspection, while the last two are covered by [**C85**, pp. 33-34], which also discusses their respective Weyl groups.

We now have four "algebraic groups with split BN-pairs" in the sense of [**C85**], namely $\mathbf{G}$, $\widetilde{\mathbf{G}} = \mathbf{G}\widetilde{\mathbf{Z}}$, $G = \mathbf{G}^F$, and $\widetilde{G} = \widetilde{\mathbf{G}}^{\widetilde{F}} = G(\mathbf{T}\widetilde{\mathbf{Z}})^{\widetilde{F}}$. The first two groups are connected reductive, and the second two are the respective finite fixed-point groups under Frobenius endomorphisms. According to [**C85**, p. 61 bottom] these groups all "satisfy the commutator relations" of Chevalley in the form displayed there. This is needed for the full Levi decomposition theory developed in [**C85**, §2.6], quoted in [**G17**, §4]. The standard Levi subgroup L_J of G is introduced above [**C85**, Prop. 2.6.3]. Implicit in its definitions is the split BN-pair context provided by the pair $\mathbf{B}^F, \mathbf{N}^F$ discussed above. If that pair is changed to $(\mathbf{B}\widetilde{\mathbf{Z}})^{\widetilde{F}}, (\mathbf{N}\widetilde{\mathbf{Z}})^{\widetilde{F}}$, the analogous definition gives the Levi subgroup $\widetilde{L}_J$ of $\widetilde{G} = (\mathbf{G}\widetilde{\mathbf{Z}})^{\widetilde{F}}$.

In more detail, L_J is defined as the group generated by the BN-pair intersection $B \cap N = T$ and certain "root groups" X_α, $\alpha \in \Phi_J$, determined by the interaction of the unipotent radical $U = (\mathbf{U})^F$ of (the B-part of) the pair with its Weyl group, $W = N/T$. See [**C85**, p. 43, §2.2, Prop. 2.3.1, p. 57]. We take the opportunity to observe that W also identifies with the Weyl group $(\mathbf{N}\widetilde{\mathbf{Z}})^{\widetilde{F}}/(\mathbf{T}\widetilde{\mathbf{Z}})^{\widetilde{F}}$ of the split BN-pair $(\mathbf{B}\widetilde{\mathbf{Z}})^{\widetilde{F}}$, $(\mathbf{N}\widetilde{\mathbf{Z}})^{\widetilde{F}}$. Briefly, using the Lang-Steinberg theorem and the observation that $\mathbf{N} \cap \widetilde{\mathbf{Z}} \subseteq \mathbf{G} \cap \widetilde{\mathbf{Z}} \subseteq \mathbf{T}$, we have

$$\begin{aligned} N/T \cong \mathbf{N}^F/\mathbf{T}^F &\cong (\mathbf{N}/\mathbf{T})^F = (\mathbf{N}/\mathbf{T}(\mathbf{N} \cap \widetilde{\mathbf{Z}}))^F = (\mathbf{N}/\mathbf{N} \cap \mathbf{T}\widetilde{\mathbf{Z}})^F \\ &\cong (\mathbf{N}\mathbf{T}\widetilde{\mathbf{Z}}/\mathbf{T}\widetilde{\mathbf{Z}})^{\widetilde{F}} \cong (\mathbf{N}\widetilde{\mathbf{Z}})^{\widetilde{F}}/(\mathbf{T}\widetilde{\mathbf{Z}})^{\widetilde{F}}. \end{aligned}$$

Order considerations now imply that the natural map

$$N/T \longrightarrow (\mathbf{N}\widetilde{\mathbf{Z}})^{\widetilde{F}}/(\mathbf{T}\widetilde{\mathbf{Z}})^{\widetilde{F}}$$

is an isomorphism. (The map is injective, since

$$N \cap (\mathbf{T}\widetilde{\mathbf{Z}})^{\widetilde{F}} = (\mathbf{N} \cap \mathbf{T}\widetilde{\mathbf{Z}})^{\widetilde{F}} = (\mathbf{T}(\mathbf{N} \cap \widetilde{\mathbf{Z}}))^{\widetilde{F}} \subseteq (\mathbf{T}(\mathbf{G} \cap \widetilde{\mathbf{Z}}))^{\widetilde{F}} = \mathbf{T}^{\widetilde{F}} = \mathbf{T}^F = T.)$$

Surjectivity gives the useful identity $(\mathbf{N}\widetilde{\mathbf{Z}})^{\widetilde{F}} = N(\mathbf{T}\widetilde{\mathbf{Z}})^{\widetilde{F}}$.

Next, we note that the following statements all hold.

(1) T normalizes U.
(2) If V is any subgroup of G or $\widetilde{G}$ normalized by T, then ${}^nV := nVn^{-1}$ and $V^n := n^{-1}Vn$ are also normalized by T, for any $n \in N$.
(3) If V_1 and V_2 are subgroups of G or $\widetilde{G}$, both normalized by T, then $V_1 \cap V_2$ and $\langle V_1, V_2\rangle$, the subgroup generated by V_1, V_2, are also normalized by T.

The proofs are all elementary, and the same properties (1), (2), (3) hold if T is replaced by $(\mathbf{T}\widetilde{\mathbf{Z}})^{\widetilde{F}}$. (To see that (1) holds in this case, note that $U = \mathbf{U}^F$, and $\mathbf{T}\widetilde{\mathbf{Z}}$ normalizes $\mathbf{U}$.)

The above paragraph contains an implicit recipe for computing root groups, in the sense of [**C85**, p. 50], for G or $\widetilde{G}$, using [**C85**, §2.5]. The split BN-pair context we use for G is $\mathbf{B}^F, \mathbf{N}^F$ and $(\mathbf{B}\widetilde{\mathbf{Z}})^{\widetilde{F}}, (\mathbf{N}\widetilde{\mathbf{Z}})^{\widetilde{F}}$ for $\widetilde{G}$. First, we use the first two lines of the display [**C85**, p. 50] to construct a "fundamental" root group X_i from U and a fundamental reflection s_i in W, represented by $n_i \in N$. A representative n_0 for any longest word is also needed. We then write

$$X_i = U \cap U^{n_0 n_i},$$

which is the same group whether working with G or $\widetilde{G}$. All the other root groups for G have the form nX_in^{-1} for some fundamental reflection s_i and some element $n \in N = \mathbf{N}^F$ [**C85**, Prop. 2.5.15]. A similar construction applies for $\widetilde{G}$, using

$(\mathbf{N}\widetilde{\mathbf{Z}})^{\widetilde{F}}$ in place of $\mathbf{N}^F$. However, it leads to the same root groups, since each X_i is normalized by $(\mathbf{T}\widetilde{\mathbf{Z}})^{\widetilde{F}}$ and $(\mathbf{N}\widetilde{\mathbf{Z}})^{\widetilde{F}} = N(\mathbf{T}\widetilde{\mathbf{Z}})^{\widetilde{F}}$.

It will now be convenient to write nX_in^{-1} above as $wX_iw^{-1}(= {}^wX_i = X_i^{w^{-1}})$, where $w \in W$ is represented by $n \in N$. More generally, if X, Y are root groups and $Y = nXn^{-1}$, with n representing w ($n \in N$, $w \in W$), we write $Y = wXw^{-1}$ ($= {}^wX = X^{w^{-1}}$). This gives a permutation action of W on the set of root subgroups. This action agrees with the permutation action on the set of roots Φ in a "root system" constructed from W as a finite Coxeter group in [**C85**, §2.2]. See [**C85**, Prop. 2.5.15] and its proof, which gives the equivariant bijection. The notation Φ_J is introduced in [**C85**, Prop. 2.3.1], using as J a subset of the indices of the fundamental reflections S of W. Our J denotes the subset of these reflections themselves. With this modification, root groups X_α "with $\alpha \in \Phi_J$" are those of the form $w(X_i)$ with $s_i \in J$ and $w \in W_J := \langle J \rangle$. (The key to proving this is observing that the correspondence given in the proof of [**C85**, Prop. 2.5.15] is W-equivariant. It can then be applied to the definition of $\Phi_J = W_J(\Delta_J)$ in [**C85**, Prop. 2.3.1], noting that Δ_J corresponds to the fundamental reflections in J.)

We can now return to our discussion of L_J and $\widetilde{L}_J$ earlier in the proof. Let

$$X_J = \langle w(X_i) \mid s_i \in J \quad \& \quad w \in W_J \rangle.$$

We now know $L_J = \langle T, X_J \rangle$, from the definition in [**C85**, §2.6] in terms of Φ_J and the reformulation above. This is in the context of $G = \mathbf{G}^F$ and its split BN-pair $\mathbf{B}^F, \mathbf{N}^F$. If instead we work with $\widetilde{G} = (\mathbf{G}\widetilde{\mathbf{Z}})^{\widetilde{F}}$, we obtain $\widetilde{L}_J$ in place of L_J. Since the root groups for G and $\widetilde{G}$ are the same, with the same Weyl group actions on them, we have

$$\widetilde{L}_J = \langle (\mathbf{T}\widetilde{\mathbf{Z}})^{\widetilde{F}}, X_J \rangle = (\mathbf{T}\widetilde{\mathbf{Z}})^{\widetilde{F}} X_J = (\mathbf{T}\widetilde{\mathbf{Z}})^{\widetilde{F}} \mathbf{T}^F X_J = (\mathbf{T}\widetilde{\mathbf{Z}})^{\widetilde{F}} L_J,$$

taking into account the normalization of all the root groups by both T and $(\mathbf{T}\widetilde{\mathbf{Z}})^{\widetilde{F}}$. We may also write

$$\widetilde{L}_J = L_J(\mathbf{T}\widetilde{\mathbf{Z}})^{\widetilde{F}}.$$

Notice that the unipotent radical of $\widetilde{B}_J$ and the unipotent radical of B_J are the same. Hence, $\Gamma_\sigma^{\widetilde{L}_J} \cong \mathrm{Ind}_{L_J}^{\widetilde{L}_J}(\Gamma_\sigma^{L_J})$ by definition. This is a good place to mention that L_J (resp., $\widetilde{L}_J$) inherits a split BN-pair structure from G (resp., $\widetilde{G}$). See [**C85**, Prop. 2.6.3]. We denote the pair as B_J, N_J (resp., $\widetilde{B}_J, \widetilde{N}_J$). We remark that $B_J \subseteq B$ and $N_J \subseteq N$ (resp., $\widetilde{B}_J \subseteq \widetilde{B}$ and $\widetilde{N}_J \subseteq \widetilde{N} = \widetilde{\mathbf{N}}^{\widetilde{F}}$, where $\widetilde{\mathbf{N}} = \mathbf{N}\widetilde{\mathbf{Z}}$). Also, we find, again using [**C85**, Prop. 2.6.3] with our discussions above (of root subgroups and of W), that

$$B_J \cap N_J = T, \quad \widetilde{B}_J \cap \widetilde{N}_J = (\mathbf{T}\widetilde{\mathbf{Z}})^{\widetilde{F}}, \quad N_J/T = W_J \subseteq W = N/T.$$

The group W_J in this display is the group generated by the reflections in J, in our notation, and by the reflections indexed by J in the notation of [**C85**, Prop. 2.6(4)]. With some abuse of notation, we also have

$$\widetilde{N}_J/(\mathbf{T}\widetilde{\mathbf{Z}})^{\widetilde{F}} = W_J \subseteq W = (\mathbf{N}\widetilde{\mathbf{Z}})^{\widetilde{F}}/(\mathbf{T}\widetilde{\mathbf{Z}})^{\widetilde{F}}.$$

Technically, we should use groups $\widetilde{W_J}$, $\widetilde{W}$, naturally isomorphic to W_J, W, respectively, in the display above. Here, $\widetilde{W_J} = \overline{\eta}(W_J)$ and $\widetilde{W} = \overline{\eta}(W)$, where

$$\overline{\eta} : N/T \longrightarrow (\mathbf{N}\widetilde{\mathbf{Z}})^{\widetilde{F}}/(\mathbf{T}\widetilde{\mathbf{Z}})^{\widetilde{F}}$$

is the natural isomorphism discussed above in this proof. It is induced by the natural inclusion map $\eta : N \longrightarrow (\mathbf{N}\widetilde{\mathbf{Z}})^{\widetilde{F}}$ that sends any element of N to itself.

For use in the next paragraph, observe that if $w \in W = N/T$ is represented by $n_w \in N$, then $\overline{\eta}(w) \in \widetilde{W} = (\mathbf{N}\widetilde{\mathbf{Z}})^{\widetilde{F}}/(\mathbf{T}\widetilde{\mathbf{Z}})^{\widetilde{F}}$ is represented by the same element n_w, viewed as a member of $(\mathbf{N}\widetilde{\mathbf{Z}})^{\widetilde{F}}$. If $w \in W_J = N_J/T$, then $n_w \in N_J$, and $\overline{\eta}(w) \in \overline{\eta}(W_J) = \widetilde{W}_J$ is, as before, represented by n_w. Since the isomorphism $\overline{\eta}$ takes W_J onto $\widetilde{W}_J$, this process can be reversed, showing that any element $\widetilde{w} \in \widetilde{W}_J$ is representable by an element $n_w \in N_J$, where $\overline{\eta}(w) = \widetilde{w}$. We remark that the element n_w belongs to $\widetilde{N}_J$, since the latter group is the preimage of $\widetilde{W}_J$ under the natural map $\widetilde{N} = (\mathbf{N}\widetilde{\mathbf{Z}})^{\widetilde{F}} \longrightarrow \widetilde{W}$.

We next note, using the previous paragraph, that the Steinberg module $\mathrm{St}_K^{\widetilde{L}_J}$ has $\mathrm{St}_K^{L_J}$ as its restriction to L_J, up to isomorphism: To see this, consider the description of $\mathrm{St}_K^{L_J}$ (and of $\mathrm{St}_K^{\widetilde{L}_J}$ in [**G17**, Th. 2.1], attributed to Steinberg):

$$\mathrm{St}_K^{L_J} = KL_J\mathfrak{e} \subseteq KL_J\underline{\mathfrak{b}},$$

where $\mathfrak{e} := \sum_{w\in W_J}(-1)^{\ell(w)}n_w\underline{\mathfrak{b}}$ and $\underline{\mathfrak{b}}$ is the sum of the elements in the "B"-part of a given split BN-pair structure for L_J. We use the pair B_J, N_J above for L_J, and $\widetilde{B}_J, \widetilde{N}_J$ for $\widetilde{L}_J$ and an analogous description of its Steinberg module $\mathrm{St}_K^{\widetilde{L}_J}$ over K. The elements n_w are required to be in the "N" part N_J of the pair for L_J, with n_w a preimage of $w \in W_J$ under the natural map $N_J \twoheadrightarrow W_J$. Since N_J is the full preimage of W_J under the natural map $N \to W$, it is enough to know that $n_w \in N$ and its image w lies in W_J. Passing to the formula for $\mathrm{St}_K^{\widetilde{L}_J}$, we find that the expression $\sum_{w\in W_J}(-1)^{\ell(w)}n_w$ in the formula for $\mathrm{St}_K^{L_J}$ can be used in the same place for $\mathrm{St}_K^{\widetilde{L}_J}$ in the formula for the analog for $\widetilde{\mathfrak{e}}$ of $\mathfrak{e}$. Let $\mathfrak{c}$ be a set of representatives for all the right cosets of $\mathfrak{b} := B_J$ in $\widetilde{\mathfrak{b}} := \widetilde{B}_J$, and let $\underline{\mathfrak{c}}$ denote the sum of all the elements in $\mathfrak{c}$ viewed as members of $K\widetilde{L}_J$. Regard also the sums $\underline{\mathfrak{b}}$ and $\underline{\widetilde{\mathfrak{b}}}$ as taking place in $K\widetilde{L}_J$. Then

$$\underline{\mathfrak{b}} \cdot \underline{\mathfrak{c}} = \underline{\widetilde{\mathfrak{b}}}, \quad \mathfrak{e} \cdot \underline{\mathfrak{c}} = \widetilde{\mathfrak{e}}.$$

Right multiplication by $\underline{\mathfrak{c}}$ thus provides, inside $K\widetilde{L}_J$, a nonzero map of KL_J-modules $\mathrm{St}_K^{L_J} \longrightarrow \mathrm{St}_K^{\widetilde{L}_J}|_{L_J}$. The map is injective, since $\mathrm{St}_K^{L_J}$ is irreducible [**G17**, Th. 2.1(ii)]. It is an isomorphism, since both domain and target have the same dimension $|B_J|_p = |\widetilde{B}_J|_p$ by [**G17**, Th. 2.1(i)]. (The two given p-parts represent the equal cardinalities of Sylow p-subgroups of the underlying groups B and $\widetilde{B}$, and the quoted result implies these cardinalities give the dimensions of the relevant Steinberg modules.)

We now return to the task of proving that the kernel $\mathrm{Ker}\,\beta_K$ and the image $\mathrm{Im}\,\beta_K$ of β_K have no nonzero constituents in common. (Recall that this will prove that Theorem A4.1 holds for our $\mathbf{G}$ here, i.e., one that may not satisfy the connected center hypothesis). Equivalently, we must show that $\mathrm{Ker}\,R^G_{L_J}(\gamma^J)$ and $\mathrm{Im}\,R^G_{L_{J'}}(\gamma^{J'})$ have no common nonzero constituents, for each J, J' contained in S. Recall that γ^J is a surjective map $K\Gamma_\sigma^{L_J} \twoheadrightarrow \mathrm{St}_K^{L_J}$ with analogous notation for $\widetilde{\gamma}^J$ (and $\gamma^{J'}, \widetilde{\gamma}^{J'}$). As such, it is unique up to nonzero scalar multiples. A similar uniqueness holds for its analogous maps.

We next **claim** that $R^{\widetilde{G}}_{L_J}(\operatorname{Ker}\gamma^J)$ is isomorphic to a submodule of $\operatorname{Ker} R^{\widetilde{G}}_{\widetilde{L}_J}(\widetilde{\gamma}^J)$. To see this, first induce the KL_J exact sequences

$$0 \longrightarrow \operatorname{Ker}\gamma^J \longrightarrow K\Gamma^{L_J}_{\sigma} \xrightarrow{\gamma^J} \operatorname{St}^{L_J}_K \to 0$$

to $K\widetilde{L}_J$, obtaining

$$0 \to \operatorname{Ind}^{\widetilde{L}_J}_{L_J}(\operatorname{Ker}\gamma^J) \longrightarrow K\Gamma^{\widetilde{L}_J}_{\sigma} \longrightarrow \operatorname{St}^{\widetilde{L}_J}_K \otimes K(\widetilde{L}_J/L_J) \to 0.$$

Here, we view $\Gamma^{L_J}_{\sigma}$ as the induction of σ to L_J from the "unipotent radical" of B_J, and $\Gamma^{\widetilde{L}_J}_{\sigma}$ as its further induction to $\widetilde{L}_J$. (We first learned this viewpoint from [**DF94**, Lem. 2.3(a)].) The tensor product expression on the right is obtained from the well-known tensor identity, using the fact that $\operatorname{St}^{\widetilde{L}_J}_K|_{L_J} \cong \operatorname{St}^{L_J}_K$ (proved above). Since the $K\widetilde{L}_J$-module $K(\widetilde{L}_J/L_J)$ has the trivial module K as a homomorphic image, we have an epimorphism $\operatorname{St}^{\widetilde{L}_J}_K \otimes K(\widetilde{L}_J/L_J) \twoheadrightarrow \operatorname{St}^{\widetilde{L}_J}_K$. After possibly multipling by a nonzero scalar, we can compose this map with the map $K\Gamma^{\widetilde{L}_J}_{\sigma} \longrightarrow \operatorname{St}^{\widetilde{L}_J}_K \otimes K(\widetilde{L}_J/L_J)$ above to obtain $\widetilde{\gamma}^J$. Consequently, there is a natural inclusion

$$\operatorname{Ind}^{\widetilde{L}_J}_{L_J}(\operatorname{Ker}\gamma^J) \subseteq \operatorname{Ker}\widetilde{\gamma}^J.$$

Next, apply $R^{\widetilde{G}}_{\widetilde{L}_J}$ to both sides (an exact functor). The right-hand side becomes $R^{\widetilde{G}}_{\widetilde{L}_J}(\operatorname{Ker}\widetilde{\gamma}^J) \cong \operatorname{Ker} R^{\widetilde{G}}_{\widetilde{L}_J}(\widetilde{\gamma}^J)$. To understand the left-hand side, note first that the standard parabolic subgroups P_J and $\widetilde{P}_J$ associated to J (and containing L_J and $\widetilde{L}_J$, respectively) have the same unipotent radical, called U_J in [**C85**, §2.6]. As a consequence, inflation commutes with induction, in the sense that $\operatorname{Inf}^{\widetilde{P}_J}_{\widetilde{L}_J} \circ \operatorname{Ind}^{\widetilde{L}_J}_{L_J} \cong \operatorname{Ind}^{\widetilde{P}_J}_{P_J} \circ \operatorname{Inf}^{P_J}_{L_J}$ as functors on KL_J-modules. Apply this isomorphism of functors to the object $\operatorname{Ker}\gamma^J$, and then apply $\operatorname{Ind}^{\widetilde{G}}_{\widetilde{P}_J}$ to obtain

$$R^{\widetilde{G}}_{\widetilde{L}_J}(\operatorname{Ind}^{\widetilde{L}_J}_{L_J}(\operatorname{Ker}\gamma^J)) \cong R^{\widetilde{G}}_{L_J}(\operatorname{Ker}\gamma^J).$$

The **claim** which began this paragraph follows.

As a corollary to the claim, using the validity of Theorem A4.1 for $\widetilde{\mathbf{G}}$ (and its consequences for $\widetilde{\beta}_K$ and the various $R^{\widetilde{G}}_{\widetilde{L}_J}(\widetilde{\gamma}^J)$, we now know that each $\operatorname{Ker} R^{\widetilde{G}}_{L_J}(\gamma^J) = R^{\widetilde{G}}_{L_J}(\operatorname{Ker}\gamma^J)$ has no nonzero constituents in common with each

$$\operatorname{Im} R^{\widetilde{G}}_{\widetilde{L}_{J'}}(\widetilde{\gamma}^{J'}) \cong R^{\widetilde{G}}_{\widetilde{L}_{J'}}(\operatorname{St}^{\widetilde{L}_{J'}}_K).$$

Applying Frobenius reciprocity (for ordinary induction) and writing $R^{\widetilde{G}}_{\widetilde{L}'_J} \cong \operatorname{Ind}^{\widetilde{G}}_{P_J} \circ \operatorname{Inf}^{P_J}_{L_J}$, where Inf is inflation, we find that $R^{G}_{L_J}(\operatorname{Ker}\gamma^J)$ has no nonzero constituents in common with the restriction

$$\begin{aligned} R^{\widetilde{G}}_{\widetilde{L}_{J'}}(\operatorname{St}^{\widetilde{L}_{J'}}_K)|_G &\cong \left(\operatorname{Ind}^{\widetilde{G}}_{\widetilde{P}_{J'}}(\operatorname{Inf}^{\widetilde{P}_{J'}}_{\widetilde{L}_{J'}} \operatorname{St}^{\widetilde{L}_{J'}}_K)\right)\Big|_G \\ &\cong \operatorname{Ind}^{G}_{\widetilde{P}_{J'}\cap G}(\operatorname{Inf}^{\widetilde{P}_{J'}}_{\widetilde{L}_{J'}} \operatorname{St}^{\widetilde{L}_{J'}}_K)|_{\widetilde{P}_{J'}\cap G} \\ &\cong \operatorname{Ind}^{G}_{P_{J'}}(\operatorname{Inf}^{P_{J'}}_{L_{J'}}(\operatorname{St}^{\widetilde{L}_{J'}}_K|_{L_{J'}})) \\ &\cong \operatorname{Ind}^{G}_{P_{J'}}(\operatorname{Inf}^{P_{J'}}_{L_{J'}} \operatorname{St}^{L_{J'}}_K) \cong R^{G}_{L_{J'}}(\operatorname{St}^{L_{J'}}_K). \end{aligned}$$

In this display, the first line holds by definition. The second line follows from the Mackey decomposition theory, using $\widetilde{G} = \widetilde{P}_{J'}G$. The third line uses $P_{J'} \subseteq \widetilde{P}_{J'} \cap G \subseteq N_G(P_{J'}) \subseteq P_{J'}$; see [**C85**, Prop. 2.1.6]. In particular, $P_{\widetilde{J}'} = \widetilde{P}_{J'} \cap G$. The agreement of $P_{J'}$ actions inside the outer parentheses, in lines two and three, can be checked on $L_{J'}$ and $U_{J'}$. The resulting $L_{J'}$-module in both lines is $\mathrm{St}^{\widetilde{L}_K^{J'}}|_{L_{J'}}$, and the $U_{J'}$-actions are both trivial. Passage from line three to four is achieved directly from the Steinberg module restriction result earlier in this proof. This concludes the proof of the claimed variation on Theorem A4.1.

The remaining details in the proof of the general case of Theorem 5.3.1 (**G** is simple) parallel those given for **G** of adjoint type and are left to the reader. □

REMARKS 5.3.2.

(1) If we use a result of Dipper and Fleischmann, the proof for Theorem A4.1 in Appendix A can be generalized to the general case where the center of **G** is not necessarily connected. To show that the kernel and image of β_K in (5.3.1) have no irreducible constituents in common, it suffices to show that $\dim \mathrm{Hom}_{KG}(P, M) = \dim \mathrm{Hom}_{KG}(M, M)$, where $P = \bigoplus_{J\subseteq S} R^G_{L_J}(K\Gamma^{L_J}_\sigma)$ and $M = \bigoplus_{J\subseteq S} R^G_{L_J}(St^{L_J}_K)$. This can be easily reduced to prove that

(5.3.2)
$$\dim \mathrm{Hom}_{KG}\big(R^G_{L_I}(K\Gamma^{L_I}_\sigma), R^G_{L_J}(St^{L_J}_K)\big) = \dim \mathrm{Hom}_{KG}\big(R^G_{L_I}(St^{L_I}_K), R^G_{L_J}(St^{L_J}_K)\big),$$

for all $I, J \subseteq S$. An application of Frobenius reciprocity and the Mackey decomposition theorem for Harish-Chandra inductions and restrictions as in the proof of Theorem A4.1 requires the fact that ${}^*R^{{}^nL_I}_{{}^nL_I\cap L_J}({}^nK\Gamma^{L_I}_\sigma)$ is isomorphic to a Gelfand–Graev module $K\Gamma^{{}^nL_I\cap L_J}_{\sigma'}$ for some regular character σ', where $n = n_w \in N$ with $w \in \mathscr{D}_{J,I}$, the set of (W_J, W_I)-double coset representatives in W. This result is proved in [**DF92**, Lem. (3.6)] via [**DF92**, Lem. (1.8)]. (Note that if the center of **G** is connected, then we may replace $K\Gamma^{{}^nL_I\cap L_J}_{\sigma'}$ by $K\Gamma^{{}^nL_I\cap L_J}_{\sigma}$, the case considered in Appendix A.) Thus, by applying Frobenius reciprocity again together with the fact that ${}^*R^G_{L_J}(St^G_K) \cong St^{L_J}_K$ (see [**CR87**, (71.6)]) and $[\Gamma_\sigma : St_K] = 1$ (see [**G17**, 4.1]), we obtain LHS $= \#\mathscr{D}_{J,I}$ in (5.3.2). A similar argument shows that RHS $= \#\mathscr{D}_{J,I}$ as well. This proves (5.3.2) and, hence, the generalized version of Theorem A4.1 .

(2) We conclude this section by again calling attention to Theorem 5.3.1(b), which shows that the decomposition matrix D relative to a given cross-characteristic r can be evaluated by working with the smallest groups (as measured by a power q of p) in any Lie-type series as parametrized by Carter in [**C85**, §1.1.9].

5.4. $\imath$-Quantum groups. The q-Schur algebras [**DJ89**] are certain Hecke endo-algebras of type A which play a central role in linking representations of quantum $\mathfrak{gl}_n$ and Hecke algebras of the symmetric groups as well as finite general linear groups (see, e.g., [**DD97**]). In this section, we will look at how certain Hecke endo-algebras of type B/C can be introduced to play a similar role in connecting representations of finite orthogonal/symplectic groups with those of the $\imath$-quantum groups introduced in [**BW18**]. For this purpose, we will modify the definition of the algebra A to introduce q-Schur algebras of type B/C with bidegree (n, r), where n is the rank of the $\imath$-quantum group and r is the rank of the Hecke algebra.

5.4A. The q-Schur algebras of type B/C. Let $\mathbb{F}_q$ be the finite field of q elements and r a positive integer. (In this section, r is not required to be the cross-characteristic associated with G.) Consider the following two types of the orthogonal/symplectic groups introduced in Example 1.1.2:

$$(5.4.1)\qquad G(r_*,q)=\begin{cases}\mathrm{O}_{2r+1}(\mathbb{F}_q), & \text{if } *=\jmath\\ \mathrm{Sp}_{2r}(\mathbb{F}_q), & \text{if } *=\imath,\end{cases}\quad\text{where}\quad r_*=\begin{cases}2r+1, & \text{if } *=\jmath\\ 2r. & \text{if } *=\imath.\end{cases}$$

Then, $G(r_*,q)$ is the fixed-point subgroup of $\mathrm{GL}_{r_*}(\mathbb{F}_q)$ under the graph automorphism in (1.1.3):

$$\vartheta_{r_*}:\mathrm{GL}_{r_*}(\mathbb{F}_q)\longrightarrow \mathrm{GL}_{r_*}(\mathbb{F}_q),\quad x\longmapsto J_*^{-1}(x^t)^{-1}J_*,$$

where $J_\jmath=J_{2r+1}$ and $J_\imath=J_{2r}^-$ as in (1.1.4).

The graph automorphism ϑ_{r_*} induces an automorphism $\sigma=\sigma_{r_*}$ on the Weyl group $\widetilde{W}^*$ of $\mathrm{GL}_{r_*}(\mathbb{F}_q)$, which is identified as the symmetric group $\mathfrak{S}_{r_*}$. More precisely, we have an involution:

$$(5.4.2)\qquad \sigma:\mathfrak{S}_{r_*}\to\mathfrak{S}_{r_*},(i,j)\mapsto(r_*+1-i,r_*+1-j),\ \text{for all } i,j\in[1,r_*],$$

where $[1,r_*].=\{1,2,\dots,r_*\}$, such that the Weyl group W^* of $G(r_*,q)$ is isomorphic to the fixed-point group $(\mathfrak{S}_{r_*})^\sigma$.

If $S=\{s_1,s_2,\dots,s_r\}$ is the set of the Coxeter generators for W^*, then we have the following identifications:

For $*=\jmath$, $W^\jmath=(\mathfrak{S}_{2r+1})^\sigma$ and

$$s_i=(i,i+1)(2r+2-i,2r+1-i)\ (1\le i<r),\quad s_r=(r,r+1)(r+1,r+2)(r,r+1),$$

while, for $*=\imath$, $W^\imath=(\mathfrak{S}_{2r})^\sigma$ and

$$s_i=(i,i+1)(2r+1-i,2r-i)\ (1\le i<r),\quad s_r=(r,r+1).$$

See Example 1.1.2. Hence, $W^\jmath\cong W^\imath$ and we set $W_r=W^*$.

Note that W_r is the subgroup of $\mathfrak{S}_{r_*}$ consisting of permutations

$$(5.4.3)\qquad \begin{pmatrix}1 & 2 & \cdots & r & r+1 & \cdots & r_*\\ i_1 & i_2 & \cdots & i_r & i_{r+1} & \cdots & i_{r_*}\end{pmatrix}$$

satisfying $i_j+i_{r_*+1-j}=r_*+1$.

Let $\widetilde{B}(r_*,\mathbb{F}_q)$ be the subgroup of $\mathrm{GL}_{r_*}(\mathbb{F}_q)$ consisting of upper triangular matrices, where $r_*\in\{2r,2r+1\}$ and, for $*\in\{\jmath,\imath\}$, let

$$B(r_*,q)=G(r_*,q)\cap\widetilde{B}(r_*,\mathbb{F}_q).$$

Then, $B(r_*,q)$ is a Borel subgroup of $G(r_*,q)$ and, by Lemma 1.2.2,

$$\mathrm{End}_{\mathbb{Z}G(r_*,q)}\left(\mathrm{Ind}_{B(r_*,q)}^{G(r_*,q)}1_{\mathbb{Z}}\right)\cong\mathcal{H}^o(W_r)_{\mathbb{Z}},$$

where $1_{\mathbb{Z}}$ is the trivial representation of $B(r_*,q)$, $\mathcal{H}^o(W_r)_{\mathbb{Z}}$ is the specialization via $t^2\to q$ of the generic Hecke algebra of W_r with equal parameters.

Let $\Lambda(m,r)=(\mathbb{N}^m)_r$ be the set of sequences $(a_1,a_2\dots,a_m)\in\mathbb{N}^m$ with entries summing to r, and let

$$(5.4.4)\qquad \Lambda_\jmath(n,r)=\Lambda(n+1,r),\quad \Lambda_\imath(n,r)=\Lambda(n,r).$$

Note that we often regard $\Lambda_\imath(n,r)$ as a subset of $\Lambda_\jmath(n,r)$ via the embedding $\lambda\mapsto(\lambda,0)$. Define

(5.4.5)

$$\widetilde{\ }:\Lambda_\jmath(n,r)\longrightarrow\Lambda(2n+1,2r+1),\quad \lambda\longmapsto\widetilde{\lambda}=(\lambda_1,\dots,\lambda_n,2\lambda_{n+1}+1,\lambda_n,\dots,\lambda_1)$$

$$\widetilde{\ }:\Lambda_\imath(n,r)\longrightarrow\Lambda(2n,2r),\quad \lambda\longmapsto\widetilde{\lambda}=(\lambda_1,\dots,\lambda_n,\lambda_n,\dots,\lambda_1).$$

For $\lambda \in \Lambda_*(n,r)$ with $* \in \{\jmath, \imath\}$, let $\widetilde{P}_\lambda(r_*,q)$ be the standard parabolic subgroup of $\mathrm{GL}_{r_*}(\mathbb{F}_q)$ associated with $\widetilde{\lambda}$, consisting of upper triangular matrices with blocks of sizes $\widetilde{\lambda}_i$ on the diagonal. Let

$$P_\lambda(r_*,q) = \widetilde{P}_\lambda(r_*,q) \cap G(r_*,q),$$

and define, for any commutative ring R with $q = q1_R \in R$,

$$\mathcal{E}_R^*(n,r) = \mathrm{End}_{RG(r_*,q)}\Big(\bigoplus_{\lambda\in\Lambda_*(n,r)} \mathrm{Ind}_{P_\lambda(r_*,q)}^{G(r_*,q)} 1_R\Big),$$

where 1_R denotes the trivial representation of $P_\lambda(r_*,q)$.

The Weyl group of the Levi subgroup $L_\lambda(r_\jmath,q)$ of $P_\lambda(r_\jmath,q)$ has the form

$$W_\lambda = W_\lambda^\jmath = \langle S\backslash\{s_{\lambda_1+\cdots+\lambda_i} \mid 1\le i\le n\}\rangle, \quad (\text{Here } \lambda\in\Lambda(n+1,r).)$$

while the Weyl group of the Levi subgroup $L_\lambda(r_\imath,q)$ of $P_\lambda(r_\imath,q)$ has the form

$$W_\lambda = W_\lambda^\imath = \langle S\backslash\{s_{\lambda_1+\cdots+\lambda_i} \mid 1\le i\le n\}\rangle. \quad (\text{Here } \lambda\in\Lambda(n,r).)$$

They are both the fixed-point subgroups of the corresponding standard Young subgroups $\mathfrak{S}_{\widetilde{\lambda}}$ under the graph automorphism σ in (5.4.2): $W_\lambda = (\mathfrak{S}_{\widetilde{\lambda}})^\sigma$.

Let $\mathcal{H}(W_r)$ be the Hecke algebra over $\mathcal{Z}$ associated with W_r. It is generated by $T_i = T_{s_i}$, for $1\le i\le r$, with basis $\{T_w\}_{w\in W}$. The subalgebra generated by T_1, $\ldots$, T_{r-1} is the Hecke algebra $\mathcal{H}(\mathfrak{S}_r)$ associated with the symmetric group $\mathfrak{S}_r$.

For $* \in \{\jmath,\imath\}$ and $\lambda\in\Lambda_*(n,r)$, let $x_\lambda = \sum_{w\in W_\lambda^*} T_w$ (perhaps, x_λ^* is more accurate than x_λ), $T_{n,r}^* := \bigoplus_{\lambda\in\Lambda_*(n,r)} x_\lambda\mathcal{H}(W_r)$, and define a Hecke endomorphism $\mathcal{Z}$-algebra of bidegree (n,r):

$$\text{(5.4.6)} \qquad \mathcal{S}^*(n,r) = \mathcal{S}_{t,\mathcal{Z}}^*(n,r) = \mathrm{End}_{\mathcal{H}(W_r)}(T_{n,r}^*).$$

The algebra $\mathcal{S}^\jmath(n,r)$ (resp., $\mathcal{S}^\imath(n,r)$) is called the *q-Schur algebra of type B* (resp, *type C*). See Proposition 5.4.2 below for a justification of the terminology.

REMARK 5.4.1.

(1) Note that if we replace T in displays (2.4.3) and (2.4.4) by

$$T^\dagger = \bigoplus_{J\subseteq S} (x_J\mathcal{H})^{m_J}$$

with all $m_J > 0$, and define $A^\dagger := \mathrm{End}_{\mathcal{H}}(T^\dagger)$, then, for $n\ge r$ and a suitable choice of m_J, $\mathcal{S}^\jmath(n,r) = A^\dagger$. However, all $\mathcal{S}^\imath(n,r)$ or $\mathcal{S}^\jmath(n,r)$ for $n<r$ cannot be of the form $A^\dagger$, since $T_{n,r}^\imath$ does not contain $x_J\mathcal{H}$ (as a direct summand) with $s_r\in J$ and $T_{n,r}^\jmath$ does not contain $\mathcal{H}$ if $n<r$. But each of them has the form $eA^\dagger e$ for some $A^\dagger$ and some idempotent $e\in A^\dagger$, i.e., is a centralizer subalgebra of $A^\dagger$.

(2) The algebras $\mathcal{S}^\jmath(n,r)$ $(n\ge r)$ and $\mathcal{S}^\imath(n,r)$ $(n\ge r)$ are Morita equivalent to the q-Schur^2 algebra of parabolic and Young type, respectively (see [**DS00a**, p.4337]).

PROPOSITION 5.4.2. *Let R be a commutative ring and assume $\sqrt{q}, q^{-1}\in R$. By specializing t to $\sqrt{q}$, there is an algebra isomorphism*

$$\mathcal{S}^*(n,r)_R \cong \mathcal{E}_R^*(n,r) \quad (*\in\{\jmath,\imath\}).$$

PROOF. The isomorphism is an extension of the Iwahori isomorphism (see Lemma 1.2.2):

$$\mathcal{H}(W_r)|_{t^2=q} \xrightarrow{\sim} \mathrm{End}_{\mathbb{Z}'G^*}(\mathrm{Ind}_{B^*}^{G^*} 1_{\mathbb{Z}'}), \quad T_w \mapsto (\underline{B^*} \mapsto \underline{B^*wB^*}),$$

where $\mathbb{Z}' = \mathbb{Z}[\sqrt{q}, \sqrt{q}^{-1}]$, $G^* = G(r_*, q)$, $B^* = B(r_*, q)$, and $\underline{X} = \sum_{x\in X} x$. Compare [**DDPW08**, Th. 4.37]. Now, an argument similar to the proof of [**DDPW08**, Th. 13.15] proves the isomorphism. $\square$

Recall the ring $\mathcal{Z}^\natural$ with all bad primes invertible. For $* = \jmath, \imath$, let $\mathcal{S}^*(n,r)^\natural := \mathcal{Z}^\natural \otimes_{\mathcal{Z}} \mathcal{S}^*(n,r)$.

PROPOSITION 5.4.3. *There exist split quasi-hereditary $\mathcal{Z}^\natural$-algebras $\mathcal{S}^{\jmath+}(n,r)^\natural$ and $\mathcal{S}^{\imath+}(n,r)^\natural$, which are endomorphism algebras of $\mathcal{H}(W_r)$-modules with dual left-cell filtrations, such that $\mathcal{S}^\jmath(n,r)^\natural$ (resp., $\mathcal{S}^\imath(n,r)^\natural$) is a centralizer subalgebra of $\mathcal{S}^{\jmath+}(n,r)^\natural$ (resp., $\mathcal{S}^{\imath+}(n,r)^\natural$).*

PROOF. If $n \geq r$, then, for any $J \subseteq S$, there exists $\lambda \in \Lambda(n+1,r)$ such that $W_\lambda = W_J$ (e.g., $W_{(1^r,0)} = W_\emptyset$ for $n=r$). Thus, $\mathcal{S}^\jmath(n,r)^\natural = A^{\dagger\natural}$ with a suitable choice of $m_J, J \subseteq S$, as seen in Remark 5.4.1. Thus, in this case, $\mathcal{S}^\jmath(n,r)^\natural$ is a centraliser subalgebra of some $A^{\ddagger\natural}$ defined in display (2.4.7)) with base change to $\mathcal{Z}^\natural$. Here the associated finite standard system (W,S,L) is $W = W_r$ and $L = \ell$, the usual length function. In this case, we choose $\mathcal{S}^{\jmath+}(n,r)^\natural = A^{\ddagger\natural}$. Now, the split quasi-heredity of $A^{\ddagger\natural}$ follows from Theorem 4.3.8 in our Chapter 4.

If $n < r$, let

$$\Theta = \Theta_{n,r} = \{\omega \in \Omega \mid \omega \leq_L \omega_\lambda, \text{ for some } \lambda \in \Lambda(n+1,r)\},$$

where ω_λ is the left cell that contains the longest word of W_λ. Then, Θ is a poset ideal of $(\Omega, \leq_L)$ and $A^{\ddagger\natural}_\Theta$ is a split quasi-hereditary algebra by Theorem 4.3.8. For a suitable choice $\{m_\omega \in \mathbb{Z}_+ \mid \omega \in \Theta\}$, $\mathcal{S}^\jmath(n,r)^\natural$ is a centralizer subalgebra of $A^{\ddagger\natural}_\Theta$. We then choose $\mathcal{S}^{\jmath+}(n,r)^\natural := A^{\ddagger\natural}_\Theta$ as the desired algebra.

By Remark 5.4.1, the $\imath$ case may be proved similarly as in the $n<r$ case above. $\square$

By Corollary 5.1.6 and Lemma 5.1.7 and noting the remark right above Proposition 5.4.2, we immediately have:

THEOREM 5.4.4. *The algebras $\mathcal{S}^\jmath(n,r)^\natural$ and $\mathcal{S}^\imath(n,r)^\natural$ are standardly based algebras. In particular, their decomposition matrices are unitriangular.*

5.4B. i-Quantum groups and the associated Schur–Weyl duality. We now introduce the i-quantum groups arising from the study of quantum symmetric pairs; see [**Le03**], [**BW18**] and the references therein.

Symmetric pairs refer to pairs of the form $(\mathfrak{g}, \mathfrak{g}^\theta)$ or $(\mathcal{U}(\mathfrak{g}), \mathcal{U}(\mathfrak{g}^\theta))$, where $\mathfrak{g}$ is a Lie algebra with its enveloping algebra $\mathcal{U}(\mathfrak{g})$, $\theta : \mathfrak{g} \to \mathfrak{g}$ is an involution, and $\mathfrak{g}^\theta$ is the fixed-point subalgebra. Note that $\mathcal{U}(\mathfrak{g}^\theta)$ is a Hopf subalgebra of $\mathcal{U}(\mathfrak{g})$.

For example, if $\mathfrak{g} = \mathfrak{gl}_N$ with basis consisting of matrix units $\{e_{i,j}\}_{1\leq, i,j\leq N}$, and

$$\theta : \mathfrak{gl}_N \longrightarrow \mathfrak{gl}_N, \qquad e_{i,j} \longmapsto e_{\tau(i),\tau(j)},$$

where $\tau(i) = N+1-i$, then (see, e.g., [**LZ**, 2.4.2, 3.5.1])

$$\mathfrak{gl}_N^\theta \cong \begin{cases} \mathfrak{gl}_{n+1} \oplus \mathfrak{gl}_n, & \text{if } N = 2n+1 \\ \mathfrak{gl}_n \oplus \mathfrak{gl}_n, & \text{if } N = 2n. \end{cases}$$

Thus,

$$\mathcal{U}(\mathfrak{gl}_N^\theta) \cong \begin{cases} \mathcal{U}(\mathfrak{gl}_{n+1}) \otimes \mathcal{U}(\mathfrak{gl}_n), & \text{if } N = 2n+1 \\ \mathcal{U}(\mathfrak{gl}_n) \otimes \mathcal{U}(\mathfrak{gl}_n), & \text{if } N = 2n. \end{cases} \tag{5.4.7}$$

Note that this involution θ is different from the differentiation γ of ϑ in (1.1.3). Here, for $* \in \{\jmath, \imath\}$, $\gamma : \mathfrak{gl}_N \to \mathfrak{gl}_N$, $x \mapsto -J_* x^t J_*$, with $J_\jmath = J_{2n+1}$, $J_\imath = J_{2n}^-$ being given in (1.1.4), is used to define $\mathfrak{o}_{2n+1}$ and $\mathfrak{sp}_{2n}$.

We now define the quantum analogs of $(\mathcal{U}(\mathfrak{gl}_N), \mathcal{U}(\mathfrak{gl}_N^\theta))$.

DEFINITION 5.4.5. The quantum enveloping algebra $\mathbf{U}(\mathfrak{gl}_N)$ over $\mathbb{Q}(t)$ is generated by

$$E_a, F_a, K_i^{\pm 1},\ 1 \le a, i \le N, a \ne N.$$

These elements are subject to the following relations:

(QG1) $K_i K_j = K_j K_i$, $K_i K_i^{-1} = K_i^{-1} K_i = 1$;
(QG2) $K_i E_a = t^{\delta_{i,a} - \delta_{i,a+1}} E_a K_i$, $K_i F_a = t^{-\delta_{i,a} + \delta_{i,a+1}} F_a K_i$;
(QG3) $[E_a, F_b] = \delta_{a,b} \frac{\widetilde{K}_a - \widetilde{K}_a^{-1}}{t - t^{-1}}$, where $\widetilde{K}_a = K_a K_{a+1}^{-1}$;
(QG4) $E_a E_b = E_b E_a$, $F_a F_b = F_b F_a$, if $|a - b| > 1$; and
(QG5) $E_a^2 E_b - (t + t^{-1}) E_a E_b E_a + E_b E_a^2 = 0$, $F_a^2 F_b - (t + t^{-1}) F_a F_b F_a + F_b F_a^2 = 0$, if $|a - b| = 1$.

A Hopf algebra structure on $\mathbf{U}(\mathfrak{gl}_N)$ is defined (see [**Jan96**, p. 199]) by:

$$\begin{aligned} &\Delta(K_i) = K_i \otimes K_i, \\ &\Delta(E_a) = E_a \otimes \widetilde{K}_a^{-1} + 1 \otimes E_a, \quad \Delta(F_a) = F_a \otimes 1 + \widetilde{K}_a \otimes F_a, \\ &\varepsilon(K_i) = 1, \quad \varepsilon(E_a) = \varepsilon(F_a) = 0, \\ &S(K_i) = K_i^{-1}, \quad S(E_a) = -E_a \widetilde{K}_a, \quad S(F_a) = -\widetilde{K}_a^{-1} F_a. \end{aligned} \tag{5.4.8}$$

The quantum $\mathfrak{sl}_N$ is the Hopf subalgebra of $\mathbf{U}(\mathfrak{gl}_N)$

$$\mathbf{U}_N = \mathbf{U}(\mathfrak{sl}_N) = \langle E_i, F_i, \widetilde{K}_i \mid 1 \le i \le N-1 \rangle.$$

We now follow [**BW18**] to introduce two subalgebras in the cases $N = 2n+1$ and $N = 2n$ via the corresponding graph automorphism of the Dynkin diagram of $\mathfrak{sl}_N$. For this purpose, we label the vertices of the Dynkin diagrams in a more symmetric fashion, i.e., simple roots are labeled symmetrically about 0. See [**BW18**, (1.1)].

For integers $m < n$, let

$$\begin{aligned} &[m, n] = \{i \in \mathbb{Z} \mid m \le i \le n\}, \\ &[m, n]_{\frac{1}{2}} = \{i \in \mathbb{Z} + \frac{1}{2} \mid m \le i \le n\}. \end{aligned}$$

Note that $[m, n]$ (resp., $[m, n]_{\frac{1}{2}}$) consists of all endpoints (resp., midpoints) of intervals $[i, i+1]$ for $m \le i < n$.

Let V_N be the natural representation of $\mathbf{U}_N = \mathbf{U}(\mathfrak{sl}_N)$. Following the notation used in [**BW18**, Chs. 2&6], we now relabel the generators of $\mathbf{U}_N$ by simple roots associated with $\mathfrak{sl}_N$.

For $N = 2n+1$, we label the basis for V_{2n+1} by

$$v_{-n}, \dots, v_{-2}, v_{-1}, v_0, v_1, v_2, \dots, v_n \tag{5.4.9}$$

and the corresponding simple roots $\alpha_i = \varepsilon_{i-\frac{1}{2}} - \varepsilon_{i+\frac{1}{2}}$ by all $i \in [-n, n]_{\frac{1}{2}}$. Here, $\varepsilon_j, j \in [-n, n]$ are the fundamental weights for $\mathfrak{sl}_{2n+1}$.

For $N = 2n$, we label the basis for V_{2n} by

(5.4.10) $$v_{-n+\frac{1}{2}}, v_{-n+\frac{1}{2}}, \ldots, v_{-\frac{1}{2}}, v_{\frac{1}{2}}, v_{1+\frac{1}{2}}, \ldots, v_{n-\frac{1}{2}}$$

and the corresponding simple roots $\alpha_i = \varepsilon_{i-\frac{1}{2}} - \varepsilon_{i+\frac{1}{2}}$ by all $i \in [-n+1, n-1]$.

DEFINITION 5.4.6.

(1) The $\mathbb{Q}(t)$-subalgebra $\mathbf{U}_n^\jmath$ of $\mathbf{U}_{2n+1}$ is generated by

$$\mathbf{k}_i = \widetilde{K}_i\widetilde{K}_{-i}^{-1}, \quad \mathbf{e}_i = E_i + \widetilde{K}_i^{-1}F_{-i}, \quad \mathbf{f}_i = F_i\widetilde{K}_{-i}^{-1} + E_{-i}, \quad \forall i \in [0, n]_{\frac{1}{2}}.$$

(2) The $\mathbb{Q}(t)$-subalgebra $\mathbf{U}_n^\imath$ of $\mathbf{U}_{2n}$ is generated by, for $1 \le i \le n-1$,

$$\mathbf{k}_i = \widetilde{K}_i\widetilde{K}_{-i}^{-1}, \quad \mathbf{e}_i = E_i + \widetilde{K}_i^{-1}F_{-i}, \quad \mathbf{f}_i = F_i\widetilde{K}_{-i}^{-1} + E_{-i}, \quad \mathbf{t} = E_0 + tF_0\widetilde{K}_0^{-1} + \widetilde{K}_0^{-1}.$$

REMARK 5.4.7. Note that $\mathbf{U}_n^\jmath, \mathbf{U}_n^\imath$ are not Hopf subalgebras, but coideal subalgebras, and $(\mathbf{U}_{2n+1}, \mathbf{U}_n^\jmath)$, $(\mathbf{U}_{2n}, \mathbf{U}_n^\imath)$ form some quantum symmetric pairs,[48] which are the quantum analogs of the symmetric pairs $(\mathcal{U}(\mathfrak{gl}_N), \mathcal{U}(\mathfrak{gl}_N^\theta))$ with $N = 2n+1, 2n$. Thus, they are the quantum analog of $\mathcal{U}(\mathfrak{gl}_N^\theta)$, $N \in \{2n+1, 2n\}$, and will be called *i-quantum group*.

Note that, by (5.4.7), there is another quantization of $\mathcal{U}(\mathfrak{gl}_N^\theta)$ as a tensor product of two quantum $\mathfrak{gl}_m$ which cannot be embedded into $\mathcal{U}(\mathfrak{gl}_N)$.

We summarize the two cases in the following table:

$\mathbf{U}_N = \mathbf{U}(\mathfrak{sl}_N)$	Basis for V_N	Simple roots	i-Quanutm groups
$N = 2n+1$	$v_i, i \in [-n, n]$	$\alpha_i, i \in [-n, n]_{\frac{1}{2}}$	$\mathbf{U}_n^\jmath \subset \mathbf{U}_{2n+1}$
$N = 2n$	$v_i, i \in [-n, n]_{\frac{1}{2}}$	$\alpha_i, i \in [-n+1, n-1]$	$\mathbf{U}_n^\imath \subset \mathbf{U}_{2n}$

For $* \in \{\jmath, \imath\}$, let

$$\epsilon_* = \begin{cases} 1, & \text{if } * = \jmath \\ 0, & \text{if } * = \imath \end{cases}$$

and let

(5.4.11)
$$\begin{aligned} \mathbb{I}(2n + \epsilon_\jmath, r) &= \{(i_1, i_2, \ldots, i_r) \mid i_j \in [-n, n]\} \\ \mathbb{I}(2n + \epsilon_\imath, r) &= \{(i_1, i_2, \ldots, i_r) \mid i_j \in [-n, n]_{\frac{1}{2}}\}. \end{aligned}$$

Both sets index bases for the *tensor spaces* $V_{2n+1}^{\otimes r}$ and $V_{2n}^{\otimes r}$, respectively.

The new symmetric notation above gives rise to an adjustment of the associated Coxeter system (W_0, S_0) with $S_0 = \{s_0, s_1, \ldots, s_{r-1}\}$. Here, we relabel the generators s_i $(1 \le i \le r)$ considered in Subsection 5.4A by $s_{r-i}, 1 \le i \le r$. Thus, the Dynkin diagram of type B_r takes the form

(5.4.12)

Note that this labeling reverses the labeling in Examples 1.1.2 and 1.1.3.

Let

(5.4.13) $$\mathcal{H}' := \mathcal{H}_{t^{-1}}(W_0) \quad \text{and} \quad \boldsymbol{\mathcal{H}}' := \mathcal{H}'_{\mathbb{Q}(t)}$$

be the Hecke algebras over $\mathcal{Z}$ and $\mathbb{Q}(t)$, respectively (relative to parameter t^{-1}) of the standard finite Coxeter system (W_0, S_0, L) with constant weight function

[48] $\mathbf{U}_n^\jmath$, $\mathbf{U}_n^\imath$ are denoted $\mathbf{U}^\jmath, \mathbf{U}^\imath$, respectively, in [**BW18**].

$L(s_i) = 1$. Then, $\mathcal{H}'$ is generated by $T_{s_0}, T_{s_1}, \ldots, T_{s_{r-1}}$. The subalgebra $\mathcal{H}'(\mathfrak{S}_r)$ generated by $T_{s_1}, \ldots, T_{s_{r-1}}$ is the Hecke algebra of the symmetric group $\mathfrak{S}_r$.

The $\mathbf{U}_{2n+\epsilon_*}$-module $V_{2n+\epsilon_*}^{\otimes r}$ commutes with the following action by $\mathcal{H}'(\mathfrak{S}_r)$ on the basis $v_{\underline{i}} := v_{i_1} \otimes \cdots \otimes v_{i_r}$, for $\underline{i} \in \mathbb{I}(2n+\epsilon_*, r)$:

$$v_{\underline{i}} T_{s_j} = \begin{cases} v_{\underline{i}s_j}, & \text{if } i_j < i_{j+1}; \\ t^{-2} v_{\underline{i}}, & \text{if } i_j = i_{j+1}; \\ (t^{-2}-1)v_{\underline{i}} + t^{-2} v_{\underline{i}s_j}, & \text{if } i_j > i_{j+1}. \end{cases} \tag{5.4.14}$$

The $\mathbf{U}_{2n+\epsilon_*}$-$\boldsymbol{\mathcal{H}}'(\mathfrak{S}_r)$-bimodule $V_{2n+\epsilon_*}^{\otimes r}$ satisfies the double centralizer property—the quantum Schur–Weyl duality.

We extend the action to the Hecke algebra $\mathcal{H}'$ of W_0. We first do it integrally. Consider the free $\mathcal{Z}$-submodule of $V_{2n+\epsilon_*}^{\otimes r}$ spanned by all $v_{\underline{i}} := v_{i_1} \otimes \cdots \otimes v_{i_r}$:

$$T(2n+\epsilon_*, r) = \mathrm{span}_{\mathcal{Z}}\{v_{\underline{i}} \mid \underline{i} \in \mathbb{I}(2n+\epsilon_*, r)\}. \tag{5.4.15}$$

This is clearly a right $\mathcal{H}'(\mathfrak{S}_r)$-module with the above action in (5.4.14). Extend this action to $\mathcal{H}'$ by setting (cf. [**BW18**, (6.8), (5.2)])

$$v_{\underline{i}} T_{s_0} = \begin{cases} v_{\underline{i}s_0}, & \text{if } i_1 > 0; \\ t^{-2} v_{\underline{i}}, & \text{if } i_1 = 0; \\ (t^{-2}-1)v_{\underline{i}} + t^{-2} v_{\underline{i}s_0}, & \text{if } i_1 < 0. \end{cases} \quad \text{for } N = 2n+1 \tag{5.4.16}$$

$$v_{\underline{i}} T_{s_0} = \begin{cases} v_{\underline{i}s_0}, & \text{if } i_1 > 0; \\ (t^{-2}-1)v_{\underline{i}} + t^{-2} v_{\underline{i}s_0}, & \text{if } i_1 < 0. \end{cases} \quad \text{for } N = 2n. \tag{5.4.17}$$

For $\lambda \in \Lambda_*(n,r)$, we will also relabel the components as $\lambda = (\lambda_0, \lambda_1, \ldots, \lambda_n)$, if $\lambda \in \Lambda_\jmath(n,r) = \Lambda(n+1, r)$, and the generators for the parabolic subgroup $W_{0,\lambda}$ of W_0. See Lemma 5.4.10 below. Let $\mathscr{D}_\lambda$ be the shortest representatives of the right cosets of $W_{0,\lambda}$ in W_0. It is easily checked that, for $\underline{i} \in \mathbb{I}(2n+\epsilon_*, r)$, if $\underline{i} = \underline{i}_\lambda d$ for some $d \in \mathscr{D}_\lambda$, then $v_{\underline{i}} = v_{\underline{i}_\lambda} T_d$.

Note that, putting $e_{\underline{i}} = v_{\underline{i}_\lambda} \widetilde{T}_d$, where $\widetilde{T}_d = t^{\ell(d)} T_d$, then the induced action from (5.4.14) and (5.4.16) (resp., (5.4.17)) coincides with the action in [**BW18**, (6.8)] (resp., [**BW18**, (5.2)]).

For $* \in \{\jmath, \imath\}$, define *$\imath$-quantum Schur algebras* over $\mathcal{Z}$:

$$U_{n,r}^* := \mathrm{End}_{\mathcal{H}'}\big(T(2n+\epsilon_*, r)\big), \qquad \mathbf{U}_{n,r}^* := \mathrm{End}_{\boldsymbol{\mathcal{H}}'}(V_{2n+\epsilon_*}^{\otimes r}).$$

The following theorem gives a new Schur–Weyl duality between $\mathbf{U}_n^*$ and $\boldsymbol{\mathcal{H}}' := \mathbb{Q}(t) \otimes_{\mathcal{Z}} \mathcal{H}'$ for $* \in \{\jmath, \imath\}$.

THEOREM 5.4.8 ([**BW18**, Ths. 5.4&6.27]). *Maintaining the notation above and letting $* \in \{\jmath, \imath\}$, we have:*

(1) *The restricted $\mathbf{U}_n^*$-action on $V_{2n+\epsilon_*}^{\otimes r}$ commutes with the $\boldsymbol{\mathcal{H}}'$-action.*
(2) *The $\mathbf{U}_n^*$-$\boldsymbol{\mathcal{H}}'$-bimodule $V_{2n+\epsilon_*}^{\otimes r}$ satisfies the double centralizer property. Hence, there is a Schur–Weyl duality over $\mathbb{Q}(t)$ in each of the choices of $*$.*
(3) *In particular, there is an algebra epimorphism over $\mathbb{Q}(t)$:*

$$\phi_* : \mathbf{U}_n^* \longrightarrow \mathbf{U}_{n,r}^* := \mathrm{End}_{\boldsymbol{\mathcal{H}}'}(V_{2n+\epsilon_*}^{\otimes r}).$$

REMARKS 5.4.9.

(1) We will see below that $U^{\jmath}_{n,r}$ and $U^{\imath}_{n,r}$ are isomorphic to $\mathcal{S}^{\jmath}_{t^{-1},\mathcal{Z}}(n,r)$ and $\mathcal{S}^{\imath}_{t^{-1},\mathcal{Z}}(n,r)$, respectively. Hence, they are Hecke endo-algebras associated with a finite orthogonal group $O_{2r+1}(\mathbb{F}_q)$ or a finite symplectic group $\mathrm{Sp}_{2r}(\mathbb{F}_q)$.

(2) By (5.4.7), there is another quantum analog of $\mathcal{U}(\mathfrak{gl}_N^{\theta})$, which is a tensor product of some quantum $\mathfrak{gl}_m$. The quantum Schur–Weyl duality relative to these quantum analogs and Hecke algebra of type B is discussed in [**SS99**] and [**Hu01**].

5.4C. Place permutations and the tensor space decomposition. We now establish isomorphism between the Hecke endo-algebras $\mathcal{S}^{\jmath}(n,r)$ and $\mathcal{S}^{\imath}(n,r)$ introduced in Subsection 5.4A and $U^{\jmath}_{n,r}$ and $U^{\imath}_{n,r}$ introduced in Subsection 5.4B. The isomorphisms are largely known; see, e.g., [**LL21, LNX20, LW22**]. We include a proof using place permutations for completion.

We may regard an element $\underline{i} \in \mathbb{I}(2n+\epsilon_{\jmath}, r)$ as a function

$$\underline{i} : [-r,r] \longrightarrow [-n,n], \quad k \longmapsto i_k \text{ with } i_{-k} = \begin{cases} -i_k, & \text{if } i_k \neq 0 \\ 0, & \text{if } i_k = 0, \end{cases}$$

and, similarly, regard an element $\underline{i} \in \mathbb{I}(2n+\epsilon_{\imath}, r)$ as a function

$$\underline{i} : [-r,-1] \cup [1,r] \longrightarrow [-n,n]_{\frac{1}{2}}, \quad k \longmapsto i_k \text{ with } i_{-k} = -i_k.$$

We now introduce W_0-actions on the sets $\mathbb{I}(2n+\epsilon_*, r)$, for $* \in \{\jmath, \imath\}$. Recall from (5.4.12) that W_0 is the Weyl group of type B_r with Coxeter generators $S_0 = \{s_0, s_1, \ldots, s_{r-1}\}$, where $S_0 \backslash \{s_0\}$ generates a parabolic subgroup isomorphic to the symmetric group $\mathfrak{S}_r$. Note that we have relabeled the Coxeter generators $s_i, 1 \leq i \leq r$ in Subsection 5.4A by $s_{r-i}, 1 \leq i \leq r$. By regarding $\mathfrak{S}_{2r+1}$ as the permutation group $\mathfrak{S}_{[-r,r]}$ of the set $[-r,r]$, W_0 is the subgroup of $\mathfrak{S}_{[-r,r]}$ (resp., $\mathfrak{S}_{[-r,r]\backslash\{0\}}$, regarded as a subgroup of $\mathfrak{S}_{[-r,r]}$) consisting of permutations

$$\begin{pmatrix} -r & \cdots & -2 & -1 & 0 & 1 & 2 & \cdots & r \\ i_{-r} & \cdots & i_{-2} & i_{-1} & 0 & i_1 & i_2 & \cdots & i_r \end{pmatrix} \tag{5.4.18}$$

satisfying $i_{-j} = -i_j$, for all $j \in [1,r]$ (cf. (5.4.3)). Hence, W_0 is generated by

$$s_0 = (-1,0)(0,1)(-1,0) = (-1,1), \quad s_i = (i,i+1)(-i,-i-1), \quad 1 \leq i \leq r-1.$$

Since every element w in W_0 satisfies $w(-j) = -w(j)$, for $j \in [-r,r]\backslash\{0\}$ and $w(0) = 0$, the following action, defined by composition of functions, is well-defined

$$\underline{i}.w = (i_{w(1)}, i_{w(2)}, \ldots, i_{w(r)}),$$

for all $\underline{i} \in \mathbb{I}(2n+\epsilon_*, r)$. This action generalizes the well-known *place permutation* from $\mathfrak{S}_r$ to the Weyl group W_0. So, W_0 acts on $\mathbb{I}(2n+\epsilon_*, r)$.

For $\underline{i} = (i_1, i_2, \ldots, i_r) \in \mathbb{I}(2n+\epsilon_{\jmath}, r)$ (here $* = \jmath$), let

$$\lambda_k = \#\{i_j \mid 1 \leq j \leq r, k = |i_j|\}, \text{ for all } k \in [0,n]$$

and define $\mathrm{wt}(\underline{i}) = \lambda := (\lambda_0, \lambda_1, \ldots, \lambda_n)$.

For $\underline{i} = (i_1, i_2, \ldots, i_r) \in \mathbb{I}(2n+\epsilon_{\imath}, r)$ (here $* = \imath$), by (5.4.11), $|i_k| + \frac{1}{2} \in [1,n]$, for all $k \in [1,r]$. Let

$$\lambda_k = \#\{i_j \mid 1 \leq j \leq r, k = |i_j| + \frac{1}{2}\}, \text{ for all } k \in [1,n]$$

and define $\mathrm{wt}(\underline{i}) = \lambda := (\lambda_1, \ldots, \lambda_n)$.

Thus, for $* \in \{\jmath, \imath\}$, the sets $\Lambda_*(n,r)$ in (5.4.4) have the following interpretation

$$\Lambda_*(n,r) = \{\mathrm{wt}(\underline{i}) \mid \underline{i} \in \mathbb{I}(2n+\epsilon_*, r)\}.$$

For $\lambda \in \Lambda_\jmath(n,r)$, let

$$\underline{i_\lambda} = (\underbrace{0,\dots,0}_{\lambda_0}, \underbrace{1,\dots,1}_{\lambda_1}, \dots, \underbrace{n,\dots,n}_{\lambda_n})$$

and, for $\lambda \in \Lambda_\imath(n,r)$, let

$$\underline{i_\lambda} = (\underbrace{1_{\frac12},\dots,1_{\frac12}}_{\lambda_1}, \underbrace{2_{\frac12},\dots,2_{\frac12}}_{\lambda_2}, \dots, \underbrace{n_{\frac12},\dots,n_{\frac12}}_{\lambda_n}),$$

where $a_{\frac12} = a - \frac12$, for all $a \in [1,n]$.

LEMMA 5.4.10. *For $* \in \{\jmath, \imath\}$ and $\lambda \in \Lambda_*(n,r)$, the orbit*

$$\underline{i_\lambda}.W_0 = \{\underline{i} \in \mathbb{I}(2n+\epsilon_*, r), \mathrm{wt}(\underline{i}) = \lambda\},$$

and $\mathrm{Stab}_{W_0}(\underline{i_\lambda}) = W_{0,\lambda}$ *is the subgroup generated by*

(1) $S\backslash\{s_{\lambda_0+\cdots+\lambda_i} \mid 0 \le i < n\}$, *if* $\lambda \in \Lambda_\jmath(n,r)$ *and*
(2) $S\backslash\{s_0, s_{\lambda_1+\cdots+\lambda_i} \mid 1 \le i < n\}$, *if* $\lambda \in \Lambda_\imath(n,r)$.

PROOF. We first assume $\lambda \in \Lambda_\jmath(n,r)$. Consider the place permutation of $\mathfrak{S}_{[-r,r]}$ on

$$I(2n+1, 2r+1) = \{(i_{-r},\dots,i_{-1},i_0,i_1,\dots,i_r) \mid i_j \in [-n,n],\ \forall j \in [-r,r]\}.$$

For $\underline{i} = (i_1, i_2, \dots, i_r) \in \mathbb{I}(2n+\epsilon_\jmath, r)$, let

$$-\underline{i}^\tau \circ \underline{i} := (i_{-r},\dots,i_{-2},i_{-1},0,i_1,i_2,\dots,i_r),$$

where $i_0 = 0$. (Recall $i_{-j} = -i_j$.) By regarding W_0 as a subgroup of $\mathfrak{S}_{[-r,r]}$ via (5.4.18), it is clear that $\mathrm{Stab}_{W_0}(\underline{i}) = W_0 \cap \mathrm{Stab}_{\mathfrak{S}_{[-r,r]}}(-\underline{i}^\tau \circ \underline{i})$. In particular, $\mathrm{Stab}_{W_0}(\underline{i_\lambda}) = W_0 \cap \mathrm{Stab}_{\mathfrak{S}_{[-r,r]}}(-\underline{i_\lambda}^\tau \circ \underline{i_\lambda})$. Since

$$\mathrm{Stab}_{\mathfrak{S}_{[-r,r]}}(-\underline{i_\lambda}^\tau \circ \underline{i_\lambda}) = \mathfrak{S}_{\widetilde{\lambda}} \cong \mathfrak{S}_{\lambda_n} \times \cdots \times \mathfrak{S}_{\lambda_1} \times \mathfrak{S}_{2\lambda_0+1} \times \mathfrak{S}_{\lambda_1} \times \dots \times \mathfrak{S}_{\lambda_n},$$

it follows that $\mathrm{Stab}_{W_0}(\underline{i_\lambda}) = W_0 \cap \mathfrak{S}_{\widetilde{\lambda}} = W_{0,\lambda}$.

The orbit equality can be seen as follows. First, for any $\underline{i} = (i_1, i_2, \dots, i_r) \in \mathbb{I}(2n+\epsilon_\jmath, r)$ with $\mathrm{wt}(\underline{i}) = \lambda$, there is $x \in \mathfrak{S}_r$ such that $\underline{i_\lambda} x = (|i_1|, |i_2|, \dots, |i_n|)$. Now, assume $i_{j_1}, \dots, i_{j_t}$ are negative, then $(|i_1|, |i_2|, \dots, |i_n|)t_{j_1}, \dots, t_{j_t} = (i_1, \dots, i_r)$, where $t_i = s_{i-1}\cdots s_1 s_0 s_1 \cdots s_{i-1}$ is the permutation swapping i and $-i$ and fixing others. Hence, $\underline{i_\lambda} x t_{j_1}, \dots, t_{j_t} = (i_1, \dots, i_r)$.

The case for $\lambda \in \Lambda_\imath(n,r)$ is similar. □

We now consider the $\mathcal{Z}$-submodules of $T(2n+\epsilon_*, r)$ defined in (5.4.15):

$$T(2n+\epsilon_\jmath, r)_\lambda = \mathrm{span}_{\mathcal{Z}}\{v_{\underline{i}} \mid \underline{i} \in \mathbb{I}(2n+\epsilon_\jmath, r), \mathrm{wt}(\underline{i}) = \lambda\} = \mathrm{span}_{\mathcal{Z}}\{v_{\underline{i}_\lambda d} \mid d \in \mathscr{D}_\lambda\},$$
$$T(2n+\epsilon_\imath, r)_\lambda = \mathrm{span}_{\mathcal{Z}}\{v_{\underline{i}} \mid \underline{i} \in \mathbb{I}(2n+\epsilon_\imath, r), \mathrm{wt}(\underline{i}) = \lambda\} = \mathrm{span}_{\mathcal{Z}}\{v_{\underline{i}_\lambda d} \mid d \in \mathscr{D}_\lambda\}.$$

Recall from (5.4.13) that $\mathcal{H}' = \mathcal{H}_{t^{-1}}(B_r)$.

LEMMA 5.4.11. *For $* \in \{\jmath, \imath\}$ and any $\lambda \in \Lambda_*(n,r)$, $T(2n+\epsilon_*,r)_\lambda$ is an $\mathcal{H}'$-submodule of $T(2n+\epsilon_*,r)$ and, as right $\mathcal{H}'$-modules,*

$$T(2n+\epsilon_*,r)_\lambda \cong x_\lambda \mathcal{H}', \text{ where } x_\lambda = \sum_{w\in W_\lambda} T_w.$$

Thus, we have an algebra isomorphism

$$U^*_{n,r} \cong \mathrm{End}_{\mathcal{H}'}\Big(\bigoplus_{\lambda\in\Lambda_*(n,r)} x_\lambda\mathcal{H}'\Big) = \mathcal{S}^*_{t^{-1},\mathcal{Z}}(n,r).$$

PROOF. By definition, $\mathcal{Z}v_{\underline{i_\lambda}}$ is an $\mathcal{H}'_\lambda$-module isomorphic to $\mathcal{Z}x_\lambda$, where $\mathcal{H}'_\lambda$ is the subalgebra of $\mathcal{H}'$ associated with the parabolic subgroup $W_{0,\lambda}$. By the Frobenius reciprocity, we have an isomorphism

$$\mathrm{Hom}_{\mathcal{H}'}(x_\lambda\mathcal{H}', T(2n+\epsilon_*,r)_\lambda) \cong \mathrm{Hom}_{\mathcal{H}'_\lambda}(\mathcal{Z}x_\lambda, T(2n+\epsilon_*,r)_\lambda).$$

Thus, the $\mathcal{H}'_\lambda$-module homomorphism $\mathcal{Z}x_\lambda \to T(2n+\epsilon_*,r), x_\lambda \mapsto v_{\underline{i_\lambda}}$ induces an $\mathcal{H}'$-module isomorphism from $x_\lambda\mathcal{H}'$ to $T(2n+\epsilon_*,r)_\lambda$. □

REMARKS 5.4.12. For the $*=\imath$ case, this isomorphism is also given in [**Grn97**, Prop. 4.1.4]. As noted in Remark 5.4.1, Morita equivalent versions of $\mathcal{S}^\jmath(n,r)$ and $\mathcal{S}^\imath(n,r)$ are considered in [**DS00a**].

5.4D. $\imath$-quantum groups and finite orthogonal/symplectic groups. In order to link $\imath$-quantum groups with the finite orthogonal/symplectic groups discussed in Subsection 5.4A, we now recover the standard labeling used in Definition 5.4.5. The $\imath$-quantum groups $\mathbf{U}^*_n$ introduced in Definition 5.4.6 are subalgebras of $\mathbf{U}(\mathfrak{sl}_N)$. We present them in generators and relations as a counterpart of $\mathbf{U}(\mathfrak{gl}_N)$. The following definition is taken from [**BKLW18**].

DEFINITION 5.4.13.

(1) The algebra $\mathbf{U}^\jmath(n)$ is defined to be the associative algebra over $\mathbb{Q}(t)$ generated by e_i, f_i, d_a, d_a^{-1}, $i\in[1,n]$, $a\in[1,n+1]$ subject to the relations (QG1)–(QG5) of Definition 5.4.5 with e_i, f_i, d_a, d_a^{-1} replacing E_i, F_i, K_a, K_a^{-1}, where, for (QG1) and (QG4), $N=n+1$ and, for (QG2), (QG3), and (QG5), $N=n$, together with the following additional relations:

(iQG1) $d_{n+1}e_jd_{n+1}^{-1} = v^{-2\delta_{n,j}}e_j$, $d_{n+1}f_jd_{n+1}^{-1} = v^{2\delta_{n,j}}f_j$, $j\in[1,n]$;

(iQG2) $f_n^2e_n + e_nf_n^2 = [2]\big(f_ne_nf_n - (vd_nd_{n+1}^{-1} + v^{-1}d_n^{-1}d_{n+1})f_n\big)$; and

(iQG3) $e_n^2f_n + f_ne_n^2 = [2]\big(e_nf_ne_n - e_n(vd_nd_{n+1}^{-1} + v^{-1}d_n^{-1}d_{n+1})\big)$.

(2) The algebra $\mathbf{U}^\imath(n)$ is defined to be the associative algebra over $\mathbb{Q}(t)$ generated by e_i, f_i, d_a, d_a^{-1}, $\mathfrak{t}$, $i\in[1,n)$, $a\in[1,n]$ subject to the relations (QG1)–(QG5) with e_i, f_i, d_a, d_a^{-1} replacing E_i, F_i, K_a, K_a^{-1} and $N=n$ together with the following relations involving $\mathfrak{t}$:

(iQG1′) $\mathfrak{t}d_a = d_a\mathfrak{t}$, $e_i\mathfrak{t} = \mathfrak{t}e_i$, $f_j\mathfrak{t} = \mathfrak{t}f_j$ for $i\neq n-1$;

(iQG2′) $\mathfrak{t}^2e_{n-1} + e_{n-1}\mathfrak{t}^2 = [2]\mathfrak{t}e_{n-1}\mathfrak{t} + e_{n-1}$, $e_{n-1}^2\mathfrak{t} + \mathfrak{t}e_{n-1}^2 = [2]e_{n-1}\mathfrak{t}e_{n-1}$; and

(iQG3′) $\mathfrak{t}^2f_{n-1} + f_{n-1}\mathfrak{t}^2 = [2]\mathfrak{t}f_{n-1}\mathfrak{t} + f_{n-1}$, $f_{n-1}^2\mathfrak{t} + \mathfrak{t}f_{n-1}^2 = [2]f_{n-1}\mathfrak{t}f_{n-1}$.

Like the embedding $\mathbf{U}(\mathfrak{sl}_N)$ into $\mathbf{U}(\mathfrak{gl}_N)$, we have the various algebra embeddings. Recall the notations in Definition 5.4.6.

PROPOSITION 5.4.14. *The following are $\mathbb{Q}(t)$-algebra isomorphisms:*

$$\eta^\jmath : \mathbf{U}^\jmath_n \longrightarrow \mathbf{U}^\jmath(n),\ \mathbf{e}_{n-i+\frac{1}{2}} \longmapsto e_i, \mathbf{f}_{n-i+\frac{1}{2}} \longmapsto f_i,$$
$$\mathbf{k}_{n-i+\frac{1}{2}} \longmapsto d_i d_{i+1}^{-1}, \mathbf{k}_{\frac{1}{2}} \longmapsto t^{-1} d_n d_{n+1}^{-1},$$

for all $i \in [1, n]$ (see [**BKLW18**, Remark 4.3]*), and*

$$\eta^\imath : \mathbf{U}^\imath_n \longrightarrow \mathbf{U}^\imath(n),\ \mathbf{e}_{n-h} \longmapsto e_h, \mathbf{f}_{n-h} \longmapsto f_h, \mathbf{k}_{n-h} \longmapsto d_h d_{h+1}^{-1}, \mathbf{t} \longmapsto \mathfrak{t},$$

for all $h \in [1, n)$.

Theorem 5.4.8 continues to hold for these enlarged algebras.[49] Let $\mathbf{S}^\jmath(n,r) = \mathcal{S}^\jmath(n,r)_{\mathbb{Q}(t)}$ and $\mathbf{S}^\imath(n,r) = \mathcal{S}^\imath(n,r)_{\mathbb{Q}(t)}$. Thus, we have the following corollary.

COROLLARY 5.4.15. *The category $\mathbf{S}^\jmath(n,r)$-mod of finite-dimensional $\mathbf{S}^\jmath(n,r)$-modules is a full subcategory of $\mathbf{U}^\jmath(n)$-mod; while the category $\mathbf{S}^\imath(n,r)$-mod of finite-dimensional $\mathbf{S}^\imath(n,r)$-modules is a full subcategory of $\mathbf{U}^\imath(n)$-mod.*

For certain classes of classical weight modules of $\mathbf{U}^*(n)$ over $\mathbb{Q}(t)$ and their crystal bases, see [**W1, W2**].

In order to link representations of finite orthogonal/symplectic groups in cross characteristics, we need to consider $\imath$-quantum hyperalgebras. This is done in [**DW22a, DW22b**].

Let

$$\begin{bmatrix} d_i \\ s \end{bmatrix} = \begin{bmatrix} d_i \\ s \end{bmatrix}_t = \prod_{i=1}^{s} \frac{d_i t^{-i+1} - d_i^{-1} t^{i-1)}}{t^i - t^{-i}} \ (s \geq 1), \quad \text{and} \quad \begin{bmatrix} d_i \\ 0 \end{bmatrix} = 1.$$

Let $U^\jmath_{\mathcal{Z}}(n)$ be the $\mathcal{Z}$-subalgebra of $\mathbf{U}^\jmath(n)$ generated by

$$d_i, \begin{bmatrix} d_i \\ s \end{bmatrix}, d_{n+1}, \begin{bmatrix} d_{n+1} \\ s \end{bmatrix}_{t^2}, \quad e_i^{(m)}, f_i^{(m)}\ (1 \leq i \leq n, s, m \in \mathbb{N}).$$

See [**DW22a**]. The definition of an integral form $U^\imath_{\mathcal{Z}}(n)$ for $\mathbf{U}^\imath(n)$ is a bit tricky since the $\mathcal{Z}$-subalgebra generated by

$$d_i, \begin{bmatrix} d_i \\ s \end{bmatrix}, \quad e_h^{(m)}, f_h^{(m)}, \mathfrak{t}^{(m)} \quad (1 \leq i \leq n, 1 \leq h \leq n-1, s, m \in \mathbb{N})$$

does not map onto $\mathcal{S}^\imath(n,r)$. In fact, replacing all $\mathfrak{t}^{(m)}$ by the image of ${}^{(m)}g := (mE^\theta_{n,n+1})(\mathbf{0})$ via a new realization of $\mathbf{U}^\imath(n)$ can fix the problem. See [**DW22b**] for more details.

For any field k and a specialisation $\mathcal{Z} \to k$, the base-changed algebras $U^\jmath_{\mathcal{Z}}(n)_k$ and $U^\imath_{\mathcal{Z}}(n)_k$ are called *$\imath$-quantum hyperalgebras.*

By Theorem 5.4.8, Lemma 5.4.11, and results from [**DW22a**], and [**DW22b**], we have the following.

THEOREM 5.4.16.

(1) *There is an algebra epimorphism $\eta^\jmath_r : U^\jmath_{\mathcal{Z}}(n) \to \mathcal{S}^\jmath(n,r)$, for all $r \geq 0$. Thus, for any base change from $\mathcal{Z}$ to any field k, it induces a surjective map. Hence, the category $\mathcal{S}^\jmath(n,r)_k$-mod of finite-dimensional $\mathcal{S}^\jmath(n,r)_k$-modules is a full subcategory of $U^\jmath_{\mathcal{Z}}(n)_k$-mod.*

---

[49] After an index shift similar to the one in Proposition 5.4.14 and an adjustment on the coalgebra structure on $\mathbf{U}(\mathfrak{gl}_N)$, the parameters for $\mathbf{U}^*(n)$ and $\mathcal{S}^*(n,r)_{\mathbb{Q}(t)}$ can be made the same t.

(2) *Similarly, with* $\imath$ *replacing* $\jmath$, *we obtain that the category* $\mathcal{S}^\imath(n,r)_k$-mod *of finite-dimensional* $\mathcal{S}^\imath(n,r)_k$*-modules is a full subcategory of* $U^\imath_{\mathbb{Z}}(n)_k$-mod.

Since the q-Schur algebras of type B or C are not necessarily quasi-hereditary, it is unlikely that the representation categories of these i-quantum hyperalgebras are highest weight categories. However, since the decomposition matrices of $\mathcal{S}^*(n,r)$, for $* \in \{\jmath,\imath\}$, are unitriangular by Theorem 5.4.4, it seems to indicate that the representation category of $U^\jmath_{\mathbb{Z}}(n)_k$ or $U^\imath_{\mathbb{Z}}(n)_k$ has a sort of connection to a highest weight category.

REMARKS 5.4.17.

(1) Lai, Nakano and Xiang considered the representation theory of $\mathcal{S}^*(n,r)_k$ over a *field* k. They used a coordinate algebra approach, giving a natural polynomial representation theory as discussed in [**Gr80**]. In particular, under a certain invertibility condition, it has been shown in [**LNX20**, Cors. 5.2.1&6.1.1)] that these algebras are cellular and quasi-hereditary.

(2) Theorem 5.4.16 provides a method for investigating the representation theory of $U^*_{\mathbb{Z}}(n)_k$ ($* \in \{\jmath,\imath\}$) arising from those of $\mathcal{S}^*(n,r)_k$. A representation of $U^*_{\mathbb{Z}}(n)_k$ obtained by inflating a representation of $\mathcal{S}^*(n,r)_k$ is also called a *polynomial representation* of $U^*_{\mathbb{Z}}(n)_k$. The theorem establishes a link between polynomial representations of $U^*_{\mathbb{Z}}(n)_k$ and representations of finite orthogonal/symplectic groups. The first step toward understanding the *modular* representation theory of $U^*_{\mathbb{Z}}(n)_k$ would be the development of the polynomial representation theory.

5.5. Open problems. We formulate several open problems inspired by this monograph. We first recall some notation.

Let G be a finite group of Lie type with standard finite Coxeter system (W,S,L). We maintain the notation of Section 5.2 (and so of Section 4.1, too). Moreover, let $(K,\mathcal{O},k)$ be a triple as in Section 5.0 with $\mathcal{O}$ sufficiently large to define the decomposition matrices $D = D_{\mathcal{O}}$ and $D^+ = D^+_{\mathcal{O}}$ of $A_{\mathcal{O}}$ and $A^+_{\mathcal{O}}$, respectively. Recall that r is the characteristic of k, and we consider representations of kG in the cases where r does not equal the defining characteristic p of $\mathbf{G}$. We also assume in this section that r is good. Using Theorem 5.2.2(c), we see that the rank of D is the number of modular irreducible A_k-modules.

PROBLEM 1. What is the rank of D in terms of r and W?

The rank of D always satisfies

$$\operatorname{rank}(D) \le \operatorname{rank}(D^+) = |\operatorname{Irr}(\mathbb{Q}W)|,$$

by Theorem 5.2.2(b),(d). When equality holds, it also follows that $D = D^+$. In particular, D^+ is a submatrix of the decomposition matrix of G in this case, at least when the center of $\mathbf{G}$ is connected, by Theorem 5.3.1. Notice that, in any given case, the rank can be determined as the number of nonzero bilinear forms in the standardly based algebra A_k. See Lemma 5.1.3.

PROBLEM 2. Is the decomposition matrix $D^+ = D^+_{\mathcal{O}}$ of $A^+ = A^+_{\mathcal{O}}$ a submatrix of the decomposition matrix of $\mathcal{O}G$?

This is related to the realization problem of the modules X_ω ($\omega \in \Omega'$) constructed in Theorem 2.4.10. A positive answer to Problem 2 would determine many decomposition numbers of G associated to complex irreducible characters in the permutation module $\mathrm{Ind}_{B(q)}^{G}\mathbb{C}$. These characters are all "unipotent," and much is known about the remaining unipotent characters from Lusztig's theory of "families" and how they are parametrized. See [**C85**, 12.3, 12.4, 13.2, 13.8, 13.9].

It is a consequence of Lusztig's theory that each irreducible complex character in the "unipotent series" lies in the same "series" as one of the irreducible complex characters in the paragraph above. A great deal is known about characters in a series and characters of the Weyl group, see [**Lus93**, p. 191] and [**C85**, §§12.3-12.4].

PROBLEM 3. Assume that the center of G is connected and recall that $r =$ characteristic of k is good. Is D a submatrix of a square unitriangular decomposition matrix $D^{++} = D^{++}_{\mathcal{O}}$ (for, say, an $\mathcal{O}$-algebra $A^{++}_{\mathcal{O}}$) with columns indexed by the set of irreducible unipotent characters?

We would hope that D^{++} would be the decomposition matrix arising from a stratified $\mathcal{Z}$-algebra A^{++} in the spirit of this monograph, and that it is a submatrix of the decomposition matrix of $\mathcal{O}G$.

Related to this problem is the Geck-Hiss conjecture [**GJ11**, Conj. 4.5.2] (see also [**GH97**, Conj. 3.4]) that the decomposition matrix associated to all the unipotent characters is square unitriangular (when r is good). So, an informal version of Problem 3 might be: Could the now known validity of our Conjecture 3.2.1 (Theorem 3.3.1) above lead to progress on the Geck-Hiss conjecture? Theorem 5.3.1 may be viewed as a step in this direction. Moreover, if the Geck-Hiss conjecture could be proved in this way, then it would be natural to investigate if Theorem 3.3.1 might be used, in the general case (where bad primes are allowed), to formulate some kind of approximate analog (of that conjecture).

Appendix A. Projective coverings for Steinberg modules by Veronica Shalotenko

A1. Preliminaries. Let p be a prime number and let $\mathbf{G}$ be a connected reductive algebraic group over the algebraic closure $\bar{\mathbb{F}}_p$ of the finite field $\mathbb{F}_p$. If F is an endomorphism of G such that some power of F is a (standard) Frobenius morphism, then the fixed-point subgroup $G = \mathbf{G}^F$ of $\mathbf{G}$ under the action of F is a finite group of Lie type. (Finite groups of Lie type are discussed in detail in [**1**].) We will assume that the center of $\mathbf{G}$ is connected. (We will use this assumption when describing the Gelfand-Graev module for G below.) The finite group of Lie type G is equipped with a BN-pair which is split in the sense of [**1**, §2.5], i.e. this BN-pair satisfies the additional axioms (a) $B = UT$, where $T = B \cap N$ and U is the largest normal p-subgroup of B, and (b) $\cap_{n\in N} nBn^{-1} = T$. Let (W, S) be the Coxeter system corresponding to the BN-pair structure on G, where $W = N/T$. There is a length function l on W, where for any $w \in W$, $l(w)$ is the minimal length of an expression of w in the generators $s \in S$.

Harish-Chandra induction and restriction. We will briefly describe several key aspects of modular Harish-Chandra theory; for a detailed treatment, see [**5**, §4.2]. In our description of Harish-Chandra induction and restriction, we will assume that k is an algebraically closed field of characteristic $r \geq 0$, $r \neq p$ (thus, k may be a field of characteristic 0).

Let P be a parabolic subgroup of G and let L be a Levi complement in P; in this case, $P = U_P \rtimes L$. Let kL-mod denote the category of (finite-dimensional) left kL-modules, and let kG-mod denote the category of (finite-dimensional) left kG-modules. There is a Harish-Chandra induction functor

$$R^G_{L\subseteq P} : kL\text{-mod} \to kG\text{-mod},$$

defined by $R^G_{L\subseteq P}(X) = \mathrm{Ind}^G_P(\tilde{X})$ for all $X \in kL$-mod, where $\tilde{X}$ denotes the inflation of X from L to P via the surjective homomorphism $P \to L$ with kernel U_P.

There is also a Harish-Chandra restriction functor

$$^*R^G_{L\subseteq P} : kG\text{-mod} \to kL\text{-mod}.$$

Given a kG-module Y, $^*R^G_{L\subseteq P}(Y) = Y^{U_P}$ (which has the structure of a kL-module since U_P is a normal subgroup of P).

Howlett and Lehrer [**6**] and Dipper and Du [**3**] proved that the Harish-Chandra induction and restriction functors $R^G_{L\subseteq P}$ and $^*R^G_{L\subseteq P}$ are independent of the choice of parabolic subgroup P containing L. Thus, we will omit the parabolic subgroup P and write R^G_L and $^*R^G_L$ for the Harish-Chandra induction and restriction functors.

We will use the following properties of Harish-Chandra induction and restriction in the proof of Theorem A4.1.

<u>Adjointness</u>. For any Levi subgroup L, R^G_L and $^*R^G_L$ are exact. The functors R^G_L and $^*R^G_L$ are each other's two-sided adjoints.

<u>Mackey decomposition</u>. Let L be a Levi subgroup of G. Given an element $n \in N$ and a (left) kL-module X, we can define a $k({}^nL)$-module structure on X by setting $nln^{-1}.x = l.x$, for any $l \in L$ and $x \in X$. The resulting $k({}^nL)$-module will be denoted by ${}^nX$.

Now, let L and M be two Levi subgroups of G, and let P and Q be parabolic subgroups containing L and M, respectively. Let X be a kL-module, and let $D(Q,P)$ denote a full set of (Q,P)-double coset representatives in G. The Mackey formula provides the following direct sum decomposition of the kM-module $^*R^G_M(R^G_L(X))$:

$$^*R^G_M(R^G_L(X)) \cong \bigoplus_{n\in D(Q,P)} R^M_{{}^nL\cap M}({}^*R^{{}^nL}_{{}^nL\cap M}({}^nX)).$$

A2. The Steinberg module and the Gelfand-Graev module. Let G be a finite group of Lie type defined in characteristic p, and let R be a commutative ring (with unity).

<u>The Steinberg module</u>. We define an element $\mathfrak{e} \in RG$ by $\mathfrak{e} = \sum_{w\in W} (-1)^{l(w)} n_w \mathfrak{b}$, where $\mathfrak{b} = \sum_{b\in B} b$ and $n_w \in N_G(T)$ is a representative of the coset $w \in W = N/T \leq N_G(T)/T$. The left RG-module $\mathrm{St}_R = RG\mathfrak{e}$ is the Steinberg module for G over R [**7**].

<u>An r-modular system</u>. For the remainder of this appendix, k will be an algebraically closed field of characteristic $r > 0$, $r \neq p$. We will work with an r-modular system $(\mathcal{O}, K, k)$, where $\mathcal{O}$ is a complete discrete valuation ring with residue field k and K is the fraction field of $\mathcal{O}$. We will assume that $\mathrm{char}(K) = 0$. We will call an $\mathcal{O}G$-module M a lattice if M is finitely generated and free over $\mathcal{O}$.

Let $\mathrm{St}_{\mathcal{O}}=\mathcal{O}G\mathfrak{e}$ be the Steinberg module for G over $\mathcal{O}$ ($\mathrm{St}_{\mathcal{O}}$ is finitely generated and free over $\mathcal{O}$, so it is a lattice). Then, $K\otimes_{\mathcal{O}}\mathrm{St}_{\mathcal{O}}\cong\mathrm{St}_K$ is an irreducible KG-module [**7**, Theorem 3].

<u>The Gelfand-Graev module</u>. We will construct the Gelfand-Graev module for G following [**4**, §4]. Let $\sigma: U\to K^\times$ be a group homomorphism, and define $\mathfrak{u}_\sigma := \sum_{u\in U}\sigma(u)u\in KG$. Since U is a p-subgroup of G, $r\nmid |U|$ and $\sigma(u)\in\mathcal{O}^\times$, for all $u\in U$, which means $\mathfrak{u}_\sigma\in\mathcal{O}G$. We have

$$\mathfrak{u}_\sigma^2=\sum_{u,u'\in U}\Big(\sigma(u)u\Big)\Big(\sigma(u')u'\Big)=\sum_{u\in U}\sum_{u'\in U}\sigma(uu')uu'=\sum_{u\in U}\mathfrak{u}_\sigma=|U|\mathfrak{u}_\sigma.$$

Since $|U|$ is a unit in $\mathcal{O}$, $\frac{1}{|U|}\mathfrak{u}_\sigma$ is an idempotent in $\mathcal{O}G$ and the $\mathcal{O}G$-lattice $\Gamma_\sigma := \mathcal{O}G\mathfrak{u}_\sigma$ is projective.

We will now assume that $\sigma: U\to K^\times$ is a fixed regular character (defined in [**4**, §4.3]). As above, we assume that $G=\mathbf{G}^F$ and that the algebraic group $\mathbf{G}$ has a connected center. In this case, the projective $\mathcal{O}G$-lattice Γ_σ is called a Gelfand-Graev module for G. Since the center of $\mathbf{G}$ is connected, the Gelfand-Graev module is unique up to isomorphism [**1**, Prop. 8.1.2]. We will denote $K\otimes_{\mathcal{O}}\Gamma_\sigma$ (the Gelfand-Graev module for G over K) by $K\Gamma_\sigma$; we will denote $k\otimes_{\mathcal{O}}\Gamma_\sigma$ (the Gelfand-Graev module for G over k) by $\overline{\Gamma}_\sigma$. Since Γ_σ is a projective $\mathcal{O}G$-module, $\overline{\Gamma}_\sigma$ is a projective kG-module.

A3. Properties of the Steinberg and Gelfand-Graev modules. In the proof of Theorem A4.1, we will use several facts about the Steinberg and Gelfand-Graev modules which can be found in [**4**]. We begin with a summary of the relevant results.

By [**4**, §4.1],

$$\dim \mathrm{Hom}_{KG}(K\Gamma_\sigma,\mathrm{St}_K)=1. \tag{A3.1}$$

Since St_K is irreducible, it follows that there is a surjective KG-module homomorphism from $K\Gamma_\sigma$ to St_K.

A similar result holds over $\mathcal{O}$. In [**4**, Prop. 4.2], Geck defines an $\mathcal{O}G$-lattice $\rho_\sigma(\mathscr{S}_\sigma)$ in St_K (which we will denote by $\mathscr{S}_\sigma$ to simplify the notation) such that $K\otimes_{\mathcal{O}}\mathscr{S}_\sigma\cong\mathrm{St}_K$. The lattice $\mathscr{S}_\sigma$ is a homomorphic image of the Gelfand-Graev lattice Γ_σ [**4**, Prop. 4.2]. Thus, there is a surjective homomorphism $\Gamma_\sigma\twoheadrightarrow\mathscr{S}_\sigma$, which, upon reduction to k, yields a surjective homomorphism

$$\overline{\Gamma}_\sigma\twoheadrightarrow\overline{\mathscr{S}}_\sigma \tag{A3.2}$$

from the projective kG-module $\overline{\Gamma}_\sigma$ to the reduction $\overline{\mathscr{S}}_\sigma$ of the lattice $\mathscr{S}_\sigma$. (By [**4**, Prop. 4.7], there are many groups G and fields k for which $\overline{\mathscr{S}}_\sigma\cong\mathrm{St}_k$.)

Statements analogous to (A3.1) and (A3.2) hold for any Levi subgroup L of G. Since a Levi subgroup L is itself a finite group of Lie type, there is a Steinberg module St_K^L for L over K. By [**4**, §4.3], there is also a Gelfand-Graev lattice Γ_σ^L for L (obtained using a restriction of the regular character σ). Thus, we have a Gelfand-Graev module $K\Gamma_\sigma^L$ for L over K. By (A3.1), $\dim_{KL}(K\Gamma_\sigma^L,\mathrm{St}_K^L)=1$; by (A3.2), there is an $\mathcal{O}L$-lattice $\mathscr{S}_\sigma^L$ in St_K^L with $K\otimes_{\mathcal{O}}\mathscr{S}_\sigma^L\cong\mathrm{St}_K^L$ and a surjective homomorphism $\overline{\Gamma}_\sigma^L\twoheadrightarrow\overline{\mathscr{S}}_\sigma^L$.

If L and M are Levi subgroups of G with $L \subseteq M$, we have the following isomorphisms of KL-modules (these isomorphisms are recorded in [**4**, Lemma 4.4]):

(A3.3) $$^*R_L^M(K\Gamma_\sigma^M) \cong K\Gamma_\sigma^L,$$

(A3.4) $$^*R_L^M(\mathrm{St}_K^M) \cong \mathrm{St}_K^L, \text{ and}$$

(A3.5) if $L = {}^nL'$, for some $n \in N$, then ${}^nK\Gamma_\sigma^{L'} \cong K\Gamma_\sigma^L$ and ${}^n\mathrm{St}_K^{L'} \cong \mathrm{St}_K^L$.

A4. Main result. As above, let G be a finite group of Lie type defined in characteristic p (arising from an algebraic group with a connected center). Let k be an algebraically closed field of characteristic $r > 0$, $r \neq p$, and let $(\mathcal{O}, K, k)$ be an r-modular system, where $\mathcal{O}$ is a complete discrete valuation ring and K (the fraction field of $\mathcal{O}$) has characteristic 0. Let (W, S) be the Coxeter system associated to the BN-pair of G. Given a subset $J \subseteq S$, let L_J denote the standard Levi subgroup of G corresponding to J. By (A3.2), for all $J \subseteq S$, there is a surjective homomorphism

$$\overline{\Gamma}_\sigma^{L_J} \twoheadrightarrow \overline{\mathscr{S}}_\sigma^{L_J}$$

from the projective kL_J-module $\overline{\Gamma}_\sigma^{L_J}$ to the kL_J-module $\overline{\mathscr{S}}_\sigma^{L_J}$, where $\mathscr{S}_\sigma^{L_J}$ is an $\mathcal{O}L_J$-lattice in $\mathrm{St}_K^{L_J}$ with the property that $K \otimes_{\mathcal{O}} \mathscr{S}_\sigma^{L_J} \cong \mathrm{St}_K^{L_J}$. Applying the exact Harish-Chandra induction functor $R_{L_J}^G$, we obtain a surjective homomorphism

$$R_{L_J}^G(\overline{\Gamma}_\sigma^{L_J}) \twoheadrightarrow R_{L_J}^G(\overline{\mathscr{S}}_\sigma^{L_J}).$$

Thus, taking direct sums over all $J \subseteq S$, we obtain a surjective homomorphism

$$\beta_k : \bigoplus_{J\subseteq S} R_{L_J}^G(\overline{\Gamma}_\sigma^{L_J}) \twoheadrightarrow \bigoplus_{J\subseteq S} R_{L_J}^G(\overline{\mathscr{S}}_\sigma^{L_J})$$

from the projective kG-module $\bigoplus_{J\subseteq S} R_{L_J}^G(\overline{\Gamma}_\sigma^{L_J})$ to the kG-module $\bigoplus_{J\subseteq S} R_{L_J}^G(\overline{\mathscr{S}}_\sigma^{L_J})$.

We can construct an analogous homomorphism β_K over the characteristic 0 field K. For any $J \subseteq S$, the surjective homomorphism $\overline{\Gamma}_\sigma^{L_J} \twoheadrightarrow \overline{\mathscr{S}}_\sigma^{L_J}$ of (A3.2) is induced by a surjective homomorphism $\Gamma_\sigma^{L_J} \twoheadrightarrow \mathscr{S}_\sigma^{L_J}$ from the Gelfand-Graev lattice $\Gamma_\sigma^{L_J}$ to the lattice $\mathscr{S}_\sigma^{L_J}$ of $\mathrm{St}_K^{L_J}$. It follows that the surjective homomorphism β_k defined above is induced by a surjective homomorphism

$$\beta : \bigoplus_{J\subseteq S} R_{L_J}^G(\Gamma_\sigma^{L_J}) \to R_{L_J}^G(\mathscr{S}_\sigma^{L_J})$$

of $\mathcal{O}G$-lattices. Taking tensor products with K, we obtain a surjective homomorphism

$$\beta_K : \bigoplus_{J\subseteq S} K \otimes_{\mathcal{O}} R_{L_J}^G(\Gamma_\sigma^{L_J}) \twoheadrightarrow \bigoplus_{J\subseteq S} K \otimes_{\mathcal{O}} R_{L_J}^G(\mathscr{S}_\sigma^{L_J})$$

of KG-modules. Since Harish-Chandra induction is "compatible" with the r-modular system (See [**4**, §4]), we have $K \otimes_{\mathcal{O}} R_{L_J}^G(\Gamma_\sigma^{L_J}) \cong R_{L_J}^G(K\Gamma_\sigma^{L_J})$ and $K \otimes_{\mathcal{O}} R_{L_J}^G(\mathscr{S}_\sigma^{L_J}) \cong R_{L_J}^G(K \otimes_{\mathcal{O}} \mathscr{S}_\sigma^{L_J}) \cong R_{L_J}^G(\mathrm{St}_K^{L_J})$, for any $J \subseteq S$. Therefore, there is a surjective KG-module homomorphism

$$\beta_K : \bigoplus_{J\subseteq S} R_{L_J}^G(K\Gamma_\sigma^{L_J}) \twoheadrightarrow \bigoplus_{J\subseteq S} R_{L_J}^G(\mathrm{St}_K^{L_J}).$$

THEOREM A4.1. *If* $\beta_K : \underset{J\subseteq S}{\oplus} R^G_{L_J}(K\Gamma^{L_J}_\sigma) \twoheadrightarrow \underset{J\subseteq S}{\oplus} R^G_{L_J}(\mathrm{St}^{L_J}_K)$ *is the surjective homomorphism described above, then the kernel of* β_K *has no composition factors in common with* $\underset{J\subseteq S}{\oplus} R^G_{L_J}(\mathrm{St}^{L_J}_K)$.[50]

PROOF. We claim that it suffices to show that[51]

$$\begin{aligned}\dim \mathrm{Hom}_{KG}(\underset{J\subseteq S}{\oplus} R^G_{L_J}(K\Gamma^J_\sigma), \underset{J\subseteq S}{\oplus} R^G_{L_J}(\mathrm{St}^{L_J}_K))\\ = \dim \mathrm{Hom}_{KG}(\underset{J\subseteq S}{\oplus} R^G_{L_J}(\mathrm{St}^{L_J}_K), \underset{J\subseteq S}{\oplus} R^G_{L_J}(\mathrm{St}^{L_J}_K)).\end{aligned} \tag{A4.1}$$

For suppose that the equality in (A4.1) holds. Since K is a field of characteristic 0, every short exact sequence of KG-modules splits and we have a direct sum decomposition

$$\underset{J\subseteq S}{\oplus} R^G_{L_J}(K\Gamma^{L_J}_\sigma) \cong \mathrm{Ker}(\beta_K) \oplus \mathrm{Im}(\beta_K) \cong \mathrm{Ker}(\beta_K) \oplus \Big(\underset{J\subseteq S}{\oplus} R^G_{L_J}(\mathrm{St}^{L_J}_K)\Big)$$

(the homomorphism β_K is surjective, so $\mathrm{Im}(\beta_K) \cong \underset{J\subseteq S}{\oplus} R^G_{L_J}(\mathrm{St}^{L_J}_K)$). It follows that $\mathrm{Hom}_{KG}(\underset{J\subseteq S}{\oplus} R^G_{L_J}(K\Gamma^J_\sigma), \underset{J\subseteq S}{\oplus} R^G_{L_J}(\mathrm{St}^{L_J}_K))$ decomposes as a direct sum of

$$\mathrm{Hom}_{KG}(\mathrm{Ker}(\beta_K), \underset{J\subseteq S}{\oplus} R^G_{L_J}(\mathrm{St}^{L_J}_K)) \text{ and } \mathrm{Hom}_{KG}(\underset{J\subseteq S}{\oplus} R^G_{L_J}(\mathrm{St}^{L_J}_K), \underset{J\subseteq S}{\oplus} R^G_{L_J}(\mathrm{St}^{L_J}_K)).$$

Therefore, for equality to hold in (A4.1), we must have

$$\dim \mathrm{Hom}_{KG}(\mathrm{Ker}(\beta_K), \underset{J\subseteq S}{\oplus} R^G_{L_J}(\mathrm{St}^{L_J}_K)) = 0.$$

Since $\mathrm{char}(K) = 0$ and all KG-modules are completely reducible, this is possible if and only if the KG-modules $\mathrm{Ker}(\beta_K)$ and $\underset{J\subseteq S}{\oplus} R^G_{L_J}(\mathrm{St}^{L_J}_K)$ have no composition factors in common.

We will now establish the equality

$$\begin{aligned}\dim \mathrm{Hom}_{KG}(\underset{J\subseteq S}{\oplus} R^G_{L_J}(K\Gamma^J_\sigma), \underset{J\subseteq S}{\oplus} R^G_{L_J}(\mathrm{St}^{L_J}_K))\\ = \dim \mathrm{Hom}_{KG}(\underset{J\subseteq S}{\oplus} R^G_{L_J}(\mathrm{St}^{L_J}_K), \underset{J\subseteq S}{\oplus} R^G_{L_J}(\mathrm{St}^{L_J}_K)).\end{aligned}$$

In fact, since

$$\mathrm{Hom}_{KG}(\underset{J\subseteq S}{\oplus} R^G_{L_J}(K\Gamma^{L_J}_\sigma), \underset{J\subseteq S}{\oplus} R^G_{L_J}(\mathrm{St}^{L_J}_K)) \cong \bigoplus_{I,J\subseteq S} \mathrm{Hom}_{KG}(R^G_{L_I}(K\Gamma^{L_I}_\sigma), R^G_{L_J}(\mathrm{St}^{L_J}_K))$$

and

$$\mathrm{Hom}_{KG}(\underset{J\subseteq S}{\oplus} R^G_{L_J}(\mathrm{St}^{L_J}_K), \underset{J\subseteq S}{\oplus} R^G_{L_J}(\mathrm{St}^{L_J}_K)) \cong \bigoplus_{I,J\subseteq S} \mathrm{Hom}_{KG}(R^G_{L_I}(\mathrm{St}^{L_I}_K), R^G_{L_J}(\mathrm{St}^{L_J}_K)),$$

it is enough to show that

$$\dim \mathrm{Hom}_{KG}(R^G_{L_I}(K\Gamma^{L_I}_\sigma), R^G_{L_J}(\mathrm{St}^{L_J}_K)) = \dim \mathrm{Hom}_{KG}(R^G_{L_I}(\mathrm{St}^{L_I}_K), R^G_{L_J}(\mathrm{St}^{L_J}_K)), \tag{A4.2}$$

for all $I, J \subseteq S$.

---

[50]A related open question is whether or not this result holds in positive characteristic.

[51]Our use of this condition on the dimension of Hom spaces was inspired by a reading of [**5**, §4.1]. (A more general version of this condition appears in [**5**, Lemma 4.1.7].)

We will now compute $\dim \operatorname{Hom}_{KG}(R^G_{L_I}(K\Gamma^{L_I}_\sigma), R^G_{L_J}(\mathrm{St}^{L_J}_K))$. Since the Harish-Chandra restriction functor ${}^*R^G_{L_J}$ is adjoint to the Harish-Chandra induction functor $R^G_{L_J}$, we have

$$\operatorname{Hom}_{KG}(R^G_{L_I}(K\Gamma^{L_I}_\sigma), R^G_{L_J}(\mathrm{St}^{L_J}_K)) \cong \operatorname{Hom}_{KL_J}({}^*R^G_{L_J} R^G_{L_I}(K\Gamma^{L_I}_\sigma), \mathrm{St}^{L_J}_K).$$

Let P_J and P_I denote the standard parabolic subgroups of G containing L_J and L_I, respectively, and let $D(P_J, P_I) \subseteq N$ be a full set of (P_J, P_I)-double coset representatives in G. By Mackey decomposition, we have

$$\begin{aligned}
&\operatorname{Hom}_{KL_J}({}^*R^G_{L_J} R^G_{L_I}(K\Gamma^{L_I}_\sigma), \mathrm{St}^{L_J}_K) \\
&\qquad \cong \operatorname{Hom}_{KL_J}\Big(\bigoplus_{n\in D(P_J,P_I)} R^{L_J}_{{}^nL_I\cap L_J}\, {}^*R^{{}^nL_I}_{{}^nL_I\cap L_J}({}^nK\Gamma^{L_I}_\sigma), \mathrm{St}^{L_J}_K \Big) \\
&\qquad \cong \bigoplus_{n\in D(P_J,P_I)} \operatorname{Hom}_{KL_J}(R^{L_J}_{{}^nL_I\cap L_J}\, {}^*R^{{}^nL_I}_{{}^nL_I\cap L_J}({}^nK\Gamma^{L_I}_\sigma), \mathrm{St}^{L_J}_K).
\end{aligned}$$

Now, by (A3.3) and (A3.5), the $K({}^nL_I \cap L_J)$-module ${}^*R^{{}^nL_I}_{{}^nL_I\cap L_J}({}^nK\Gamma^{L_I}_\sigma)$ is the Gelfand-Graev module $K\Gamma^{{}^nL_I\cap L_J}_\sigma$, for ${}^nL_I \cap L_J$ over K. Thus, continuing the computation above (and, using the adjointness of $R^{L_J}_{{}^nL_I\cap L_J}$ and ${}^*R^{L_J}_{{}^nL_I\cap L_J}$), we have

$$\begin{aligned}
&\bigoplus_{n\in D(P_J,P_I)} \operatorname{Hom}_{KL_J}(R^{L_J}_{{}^nL_I\cap L_J}\, {}^*R^{{}^nL_I}_{{}^nL_I\cap L_J}({}^nK\Gamma^{L_I}_\sigma), \mathrm{St}^{L_J}_K) \\
&\cong \bigoplus_{n\in D(P_J,P_I)} \operatorname{Hom}_{KL_J}(R^{L_J}_{{}^nL_I\cap L_J}(K\Gamma^{{}^nL_I\cap L_J}_\sigma), \mathrm{St}^{L_J}_K) \\
&\cong \bigoplus_{n\in D(P_J,P_I)} \operatorname{Hom}_{K({}^nL_I\cap L_J)}(K\Gamma^{{}^nL_I\cap L_J}_\sigma, {}^*R^{L_J}_{{}^nL_I\cap L_J}(\mathrm{St}^{L_J}_K)).
\end{aligned}$$

By (A3.4), ${}^*R^{L_J}_{{}^nL_I\cap L_J}(\mathrm{St}^{L_J}_K)$ is the Steinberg module $\mathrm{St}^{{}^nL_I\cap L_J}_K$, for ${}^nL_I \cap L_J$ over K. Therefore,

$$\begin{aligned}
&\operatorname{Hom}_{K({}^nL_I\cap L_J)}(K\Gamma^{{}^nL_I\cap L_J}_\sigma, {}^*R^{L_J}_{{}^nL_I\cap L_J}(\mathrm{St}^{L_J}_K)) \\
&\qquad\qquad \cong \operatorname{Hom}_{K({}^nL_I\cap L_J)}(K\Gamma^{{}^nL_I\cap L_J}_\sigma, \mathrm{St}^{{}^nL_I\cap L_J}_K),
\end{aligned}$$

which has dimension 1 by (A3.1). Thus,

$$\dim \operatorname{Hom}_{KG}(R^G_{L_I}(K\Gamma^{L_I}_\sigma), R^G_{L_J}(\mathrm{St}^{L_J}_K)) = \sum_{n\in D(P_J,P_I)} 1 = |D(P_J, P_I)|.$$

We can similarly compute $\dim \mathrm{Hom}_{KG}(R^G_{L_I}(\mathrm{St}^{L_I}_K), R^G_{L_J}(\mathrm{St}^{L_J}_K))$. We have

$$\begin{aligned}
\mathrm{Hom}_{KG}(R^G_{L_I}(\mathrm{St}^{L_I}_K), R^G_{L_J}(\mathrm{St}^{L_J}_K)) &\cong \mathrm{Hom}_{KL_J}({}^*R^G_{L_J} R^G_{L_I}(\mathrm{St}^{L_I}_K), \mathrm{St}^{L_J}_K) \\
&\cong \mathrm{Hom}_{KL_J}\Big(\bigoplus_{n\in D(P_J,P_I)} R^{L_J}_{{}^nL_I\cap L_J}\, {}^*R^{{}^nL_I}_{{}^nL_I\cap L_J}({}^n\mathrm{St}_K{}^{L_I}), \mathrm{St}^{L_J}_K\Big) \\
&\cong \bigoplus_{n\in D(P_J,P_I)} \mathrm{Hom}_{KL_J}(R^{L_J}_{{}^nL_I\cap L_J}\, {}^*R^{{}^nL_I}_{{}^nL_I\cap L_J}({}^n\mathrm{St}_K{}^{L_I}), \mathrm{St}^{L_J}_K) \\
&\cong \bigoplus_{n\in D(P_J,P_I)} \mathrm{Hom}_{KL_J}(R^{L_J}_{{}^nL_I\cap L_J}(\mathrm{St}^{{}^nL_I\cap L_J}_K), \mathrm{St}^{L_J}_K) \\
&\cong \bigoplus_{n\in D(P_J,P_I)} \mathrm{Hom}_{K({}^nL_I\cap L_J)}(\mathrm{St}^{{}^nL_I\cap L_J}_K, {}^*R^{L_J}_{{}^nL_I\cap L_J}(\mathrm{St}^{L_J}_K)) \\
&\cong \bigoplus_{n\in D(P_J,P_I)} \mathrm{Hom}_{K({}^nL_I\cap L_J)}(\mathrm{St}^{{}^nL_I\cap L_J}_K, \mathrm{St}^{{}^nL_I\cap L_J}_K).
\end{aligned}$$

Since each Steinberg module $\mathrm{St}^{{}^nL_I\cap L_J}_K$ is absolutely irreducible (see [**4**, Theorem 2.1]), we have

$$\dim \mathrm{Hom}_{K({}^nL_I\cap L_J)}(\mathrm{St}^{{}^nL_I\cap L_J}_K, \mathrm{St}^{{}^nL_I\cap L_J}_K) = 1,$$

for all $n \in D(P_J, P_I)$. Therefore,

$$\dim \mathrm{Hom}_{KG}(R^G_{L_I}(\mathrm{St}^{L_I}_K), R^G_{L_J}(\mathrm{St}^{L_J}_K)) = |D(P_J, P_I)|,$$

proving that the equality in (A4.2) holds. □

References

[1] Roger W. Carter, *Finite groups of Lie type: Conjugacy classes and complex characters*, Wiley Classics Library, John Wiley & Sons, Ltd., Chichester, 1993. Reprint of the 1985 original; A Wiley-Interscience Publication. MR1266626

[2] Richard Dipper, *On quotients of Hom-functors and representations of finite general linear groups. I*, J. Algebra **130** (1990), no. 1, 235–259, DOI 10.1016/0021-8693(90)90111-Z. MR1045747

[3] Richard Dipper and Jie Du, *Harish-Chandra vertices*, J. Reine Angew. Math. **437** (1993), 101–130, DOI 10.1515/crll.1993.437.101. MR1212254

[4] Meinolf Geck, *On the modular composition factors of the Steinberg representation*, J. Algebra **475** (2017), 370–391, DOI 10.1016/j.jalgebra.2015.11.005. MR3612476

[5] Meinolf Geck and Nicolas Jacon, *Representations of Hecke algebras at roots of unity*, Algebra and Applications, vol. 15, Springer-Verlag London, Ltd., London, 2011, DOI 10.1007/978-0-85729-716-7. MR2799052

[6] R. B. Howlett and G. I. Lehrer, *On Harish-Chandra induction and restriction for modules of Levi subgroups*, J. Algebra **165** (1994), no. 1, 172–183, DOI 10.1006/jabr.1994.1104. MR1272585

[7] Robert Steinberg, *Prime power representations of finite linear groups. II*, Canadian J. Math. **9** (1957), 347–351, DOI 10.4153/CJM-1957-041-1. MR87659

Appendix B. Height functions

We collect together some terminology on height functions and related matters which are used repeatedly. Given a non-empty finite set Λ, a *height function* on Λ is a function $\mathfrak{h} : \Lambda \to \mathbb{Z}$. Often, but not always, the set Λ indexes the irreducible modules for an algebra. This is the case in the following.

B1. Height filtrations. We will consider here decreasing filtrations associated with a negative-valued height function as set in the following example.

EXAMPLE B1.1. Let $\mathscr{K}$ be a Noetherian domain with fraction field K, and let H be a finite and torsion-free $\mathscr{K}$-algebra. Let $\mathscr{A}$ be the additive category of H-finite, $\mathscr{K}$-torsion-free left H-modules. Assume that H_K is semisimple, and let Λ be a set indexing the distinct irreducible left H_K-modules: $\lambda \in \Lambda$ corresponds to an irreducible H_K-module $L_K(\lambda)$. Fix a height function $\mathfrak{h}$ on Λ which takes negative values. (The assumption that $\mathfrak{h}$ only takes negative values is a convenience; see Remark B1.2 immediately below.) Given a module $M \in \mathscr{A}$, $\mathfrak{h}$ defines a natural (decreasing) filtration $M = M_0 \supseteq M_1 \supseteq \cdots \supseteq M_n = 0$ of M by $\mathscr{A}$-submodules: let $(M_K)_h$ be the H_K-submodule of M_K which is the sum of all irreducible submodules isomorphic to $L_K(\lambda)$ with $\mathfrak{h}(\lambda) < -h$. The decreasing filtration of M is obtained by putting $M_h := M \cap (M_K)_h$, for all h. Notice that $(M_{h-1}/M_h)_K$ is a direct sum of irreducible modules $L_K(\lambda)$ with $\mathfrak{h}(\lambda) = -h$. Also, a short exact sequence $0 \to M \to N \to P \to 0$ in $\mathscr{A}$ gives, for any integer h, an exact sequence $0 \to M_h \to N_h \to P_h$; see [**DPS17**, Prop. 3.3(a)]. If $\mathscr{E}$ denotes the category (in the sense of footnote 12 of Chapter 1) of those exact sequences $0 \to M \to N \to P \to 0$ such that, for any h, $0 \to M_h \to N_h \to P_h \to 0$ *is* exact, then $(\mathscr{A}, \mathscr{E})$ is an example of an exact category. This fact may be proved in a way similar to the proof of [**DPS17**, Th. 3.6], though the latter is given in a right-module context, using superscript notation.

For more important examples of exact categories in this spirit, see [**DPS17**, §3] as well as earlier sections of this monograph. While $(\mathscr{A}, \mathscr{E})$ depends on the fixed height function $\mathfrak{h}$, we prefer not to include $\mathfrak{h}$ as part of its name. Also, replacing $\mathfrak{h}$ by $\mathfrak{h} + c$, $c \in \mathbb{Z}$, does not change the filtrations on $\mathscr{A}$ or the exact category $(\mathscr{A}, \mathscr{E})$.

REMARK B1.2. Suppose we drop the assumption that $\mathfrak{h}$ takes only negative values. Then, one should replace the filtration $M = M_0 \supseteq M_1 \supseteq \cdots \supseteq M_n = 0$ with a filtration $M = M_{-m} \supseteq \cdots \supseteq M_n = 0$ for sufficiently large (positive) n and m. This can be done uniformly for all $M \in \mathscr{A}$, provided all $\mathfrak{h}(\lambda)$ lie in the interval $(-n, m]$. If we modify such a height function by adding m' with $m' < -m$ to all its values and reindex M_i, we return to the setting assumed in Example B1.1, and it is evident that the two height functions lead to canonically isomorphic categories $(\mathscr{A}, \mathscr{E})$. The extra flexibility, however, is useful:

Suppose we take a possibly positive-valued height function $\mathfrak{h} : \Lambda \to \mathbb{Z}$, and construct M^h according to the recipe $M^h = M \cap M_K^{\leq h}$ in Theorem 2.2.1, by giving (in the category of right H-modules) a recipe for constructing an exact category through *increasing filtrations*

$$\cdots \subseteq M^{h-1} \subseteq M^h \subseteq \cdots ,$$

where the irreducible (right) H_K-modules appearing in $(M^h/M^{h-1})_K$ are indexed by $\lambda \in \Lambda$ with $\mathfrak{h}(\lambda) = h$. Obviously, such an increasing filtration canonically identifies (for any M) with a decreasing filtration

$$\cdots \subseteq M_{-(h-1)} \subseteq M_{-h} \subseteq \cdots , \tag{B1.1}$$

where $M_{-h} := M^h$ for every h. Moreover, a second height function can be given (assigning $\lambda \in \Lambda$ to $\mathfrak{h}(\lambda) - 1$) so that the filtration of M constructed from Example B1.1 is precisely that given in (B1.1), which is, in turn, "the same" as that in Theorem 2.2.1 (using $\mathfrak{h}$). This is true at least up to the left-right-module discrepancy, but that disappears when we use H^{op}, instead of H, in one of the cases.

The passage here from increasing to decreasing filtrations via the convention $M_{-h} = M^h$ is quite natural but does not, in general, lead to a negative-valued height function without further adjustment of the subscripts.

The notational scheme in Example B1.1 has, however, been chosen to behave well under a particularly important passage from left to right modules, namely, the passage from left to right arising from the contravariant functor $(-)^* = \mathrm{Hom}_{\mathcal{Z}}(-, \mathcal{Z})$ — the duality functor; see Section 2.2 for more details.

B2. Height functions compatible with a preorder. In practice, though some useful set Λ may be in the background (see Remark B2.7), we also consider a set Ω with a preorder $\leq$.

A relation $\leq$ on a set Ω that is reflexive and transitive is called a *preorder*. We will call $\Omega = (\Omega, \leq)$ a *quasi-poset* in the monograph.

For such a preorder $\leq$, define an equivalence relation $\sim$ on Ω by declaring that $\omega \sim \nu \iff \omega \leq \nu$ and $\nu \leq \omega$, and a strict version $<$ of $\leq$ by setting $\omega < \nu$ provided $\omega \leq \nu$ and $\omega \not\sim \nu$ (or $\omega \not\geq \nu$).

If $\overline{\Omega} = \Omega/\sim$ denotes the set of equivalence classes, then $\leq$ induces a partial order, denoted again by $\leq$, on $\overline{\Omega}$. We call $(\overline{\Omega}, \leq)$ the *poset associated with* $(\Omega, \leq)$. Thus, for any $\lambda, \mu \in \Omega$, if $\overline{\lambda}$ denotes the equivalence class containing λ, then we have $\overline{\omega} < \overline{\nu} \iff \omega < \nu$.

By definition, a height function $\mathfrak{h}$ on Ω is *compatible* with a preorder $\leq$ on Ω provided that

$$\text{(B2.1)} \qquad \begin{cases} (1) \quad \omega < \nu \implies \mathfrak{h}(\omega) < \mathfrak{h}(\nu), & \text{and} \\ (2) \quad \omega \sim \nu \implies \mathfrak{h}(\omega) = \mathfrak{h}(\nu). \end{cases}$$

Note that if $\mathfrak{h}$ is compatible with $\leq$, then the function $-\mathfrak{h} : \Omega \to \mathbb{Z}$ is compatible with $\leq^{\mathrm{op}}$.

Given a preorder $\leq$ on Ω, there are many choices for a compatible height function $\mathfrak{h}$. For example, the *standard height function* $\mathfrak{h}_{st}$ is defined by setting $\mathfrak{h}_{st}(\omega)$ to be the maximal length of a proper chain $\omega_0 < \omega_1 < \cdots < \omega_n = \omega$. It is compatible with the preorder on Ω.

DEFINITION B2.1. Let $\leq$ and $\leq'$ be two preorders on the same finite set Ω. We say that $\leq$ *dominates* $\leq'$ provided $\omega \leq \nu \implies \omega \leq' \nu$, for all $\omega, \nu \in \Omega$. If, in addition, $\omega < \nu \implies \omega <' \nu$, for all $\omega, \nu \in \Omega$, then we say that $\leq$ *strictly dominates* $\leq'$.

Recall that $\omega < \nu$ means that $\omega \leq \nu$, but $\omega \not\sim \nu$. Observe that if $\leq$ also dominates (respectively, strictly dominates) $\leq'$, then $\leq^{\mathrm{op}}$ dominates (respectively, strictly dominates) $\leq'^{\mathrm{op}}$.

EXAMPLE B2.2. This example explains the notation above, in which the quasi-poset Λ has been renamed Ω. Let (W, S, L) be a standard finite Coxeter system. We will make use of the (generalized) Kazhdan-Lusztig cell theory mentioned in Section 1.2. See [**Lus03**, Ch. 8]. Let Ω be the set of left Kazhdan-Lusztig cells in W. There is a preorder $\leq_L$ on W whose equivalence classes are the left cells $\omega \in \Omega$. Similarly, there is a preorder $\leq_{LR}$ on W whose equivalence classes are the two-sided Kazhdan-Lusztig cells. Clearly, $\leq_L$ also defines a preorder (still denoted $\leq_L$) on Ω, which is a partial order. Also, $\leq_{LR}$ defines a preorder on Ω (still denoted $\leq_{LR}$).

PROPOSITION B2.3. *As preorders on Ω, $\leq_L$ strictly dominates $\leq_{LR}$.*

PROOF. This (nontrivial) result is immediate from the statement (1.2.11) in Section 1.2 which follows from (P4) and (P9) of [**Lus03**, Ch.14]. Note that all conjectures (P1)–(P15) hold for the standard finite Coxeter system (W, S, L). □

An important consequence is that a two-sided cell module is a direct sum of left-cell modules. (See Proposition 3.1.1.)

An important property of strict dominance in general is the following consequence for height functions:

PROPOSITION B2.4. *Suppose $\leq, \leq'$ are preorders on a set Ω and $\leq$ strictly dominates $\leq'$.*[52] *If $\mathfrak{h}$ is any height function compatible with $\leq'$, then $\mathfrak{h}$ is also compatible with $\leq$.*

PROOF. Let $\omega, \nu \in \Omega$. Suppose first that $\omega \sim \nu$. Then, $\omega \sim' \nu$ just from the dominance property. So $\mathfrak{h}(\omega) = \mathfrak{h}(\nu)$. Next, suppose $\omega < \nu$. Then, $\omega <' \nu$ by strict dominance, and so $\mathfrak{h}(\omega) < \mathfrak{h}(\nu)$. The proposition follows. □

REMARK B2.5. For an alternate proof of the proposition, introduce a third preorder $\leq_{\mathfrak{h}}$ on Ω by declaring $\omega \leq_{\mathfrak{h}} \mu \iff \mathfrak{h}(\omega) \leq \mathfrak{h}(\mu)$. The compatability of $\mathfrak{h}$ with $\leq'$ is equivalent to the strict dominance of $\leq_{\mathfrak{h}}$ by $\leq'$. Obviously, strict dominance is a transitive relation between preorders, and the proposition follows. (Use the fact that there is a similar equivalence with $\leq'$ replaced by $\leq$.)

In the context of Example B2.2, we consider a height function $\mathfrak{h} : \Omega \to \mathbb{Z}$. Using Proposition B2.3, the following result follows immediately from Proposition B2.4.

COROLLARY B2.6. *Any height function $\mathfrak{h}$ compatible (see display* (B2.1)*) with the preorder $\leq_{LR}$ on the set Ω of left cells in W is compatible with $\leq_L$.*

This result—now seen to be easy—plays an important role in the application of Theorem 2.4.10. See, e.g., Theorem 3.3.1. We must use $\leq_{LR}$ or a related preorder in the statement of the latter theorem in order for it to prove the conjecture under discussion there. However, the proof of Theorem 3.3.1 requires properties of q-permutation modules $\mathcal{H}x_J$ developed in Section 1.3, which depend on a height function compatible with $\leq_L$.

REMARK B2.7. We conclude with a few words about height functions associated to the Hecke algebras of standard finite Coxeter systems. In Example B1.1, the height functions arise as functions $\mathfrak{h} : \Lambda \to \mathbb{Z}$, where Λ indexes the irreducible (left) modules (of $H_K = \mathcal{H}_K$). In favorable circumstances, $\mathfrak{h}$ is constant on indices that index irreducible modules associated to the same two-sided cell. Thus, $\mathfrak{h}$ defines, in particular, a function $\mathfrak{h} : \Omega \to \mathbb{Z}$. This procedure can be reversed. If we have a height function $\mathfrak{h} : \bar{\Omega} \to \mathbb{Z}$, where $\bar{\Omega}$ is the set of two-sided cells, then $\mathfrak{h}$ arises from functions (to which we give the same name) $\mathfrak{h} : \Omega \to \mathbb{Z}$ and $\mathfrak{h} : \Lambda \to \mathbb{Z}$. If $\mathfrak{h}$ is compatible with $(\Omega, \leq_{LR})$, then it is compatible with $(\Omega, \leq_L)$; see Corollary B2.6. This method gives a rich supply of height functions $\mathfrak{h} : \Omega \to \mathbb{Z}$ in the "cell theory" context of Example B2.2 above, all compatible with the $\Lambda \to \mathbb{Z}$ context of Example B1.1.

---

[52]In practice, Ω will often be the set of left cells as described above, but, in general, it might just be a quasi-poset. It will be clear from context which is which.

Further abusing notation, sometimes we think of *modules* indexed by Ω as having a height. For example, a left-cell module $S(\omega)$ would have height $\mathfrak{h}_l(\omega)$, while the dual left-cell module S_ω has height $\mathfrak{h}_r(\omega) = -\mathfrak{h}_l(\omega)$. (We often denote $\mathfrak{h}_l$ or $\mathfrak{h}_r$ by simply $\mathfrak{h}$ if no confusion results.)

Appendix C. Decomposition numbers and Krull dimension ≤ 2

Following [**DPS98a**], but modifying the notation slightly, a local triple (R, Q, F) consists of a commutative local domain R, having fraction field Q and residue field $F = R/\mathfrak{m}$, where $\mathfrak{m}$ denotes the maximal ideal of R. In the modular representation theory of a finite group G, the local domain R will usually be a discrete (sometimes complete) valuation ring (DVR), while F and Q will typically be large enough to be splitting fields for some designated subgroups of G (for example, G itself, or all of its subgroups). There are also many variations involving specific subgroups.

Given an irreducible QG-module L, it is elementary that there exists a (full) R-lattice L_0 in L in the sense of Definition 4.1.1. Because R is a DVR, we can easily assume that L_0 is G-stable. When QG is also semisimple, a classical result states that there is a map

$$\mathscr{G}(QG) \longrightarrow \mathscr{G}(FG) \tag{C0.1}$$

from the Grothendieck group of QG to that of FG. The map (C0.1) is called the *decomposition map* of the triple (R, Q, F). For an irreducible QG-module L, it sends $[L] \in \mathscr{G}(QG)$ to $\sum_i m_i[Z_i] \in \mathscr{G}(FG)$, where Z_1, Z_2, ... is a listing of the irreducible FG-modules, and m_i is the multiplicity of Z_i as a composition factor of the FG-module $F \otimes_R L_0$. The point is that the multiplicity m_i is well-defined. The various m_i are called *decomposition numbers* and are often collected together to give matrix versions (*decomposition matrices*) of the decomposition maps.

C1. Regular local rings of Krull dimension ≤ 2. In this monograph, we often work with the category mod-$\mathcal{H}$ of finite (right) modules over the generic Hecke algebra $\mathcal{H}$; see (1.2.1). In this case, fix a triple (R, Q, F), where R is not a discrete valuation ring but is a regular local ring of Krull dimension ≤ 2. For example, R is the localization of $\mathcal{Z}$ or $\mathcal{Z}^\natural$ at some prime ideal; see Subsection C3 below for more details (also see the proof of [**M86**, Th. 19.5]). The group algebra RG is replaced by the generic Hecke algebra $\mathcal{H}$ (or sometimes $\mathcal{H}^\natural$). In this case, the construction of the lattice L_0 in the previous paragraph is problematic. We briefly indicate how this works in the discussion below, following [**DPS98a**, §1]. We start with the following result of Auslander-Goldman:

LEMMA C1.1 ([**AG60a**, p. 17], [**DDPW08**, Th. C.17]). *Let R be a regular local commutative ring of Krull dimension ≤ 2, and let M be a finite R-module. Then, the dual R-module $M^* := \mathrm{Hom}_R(M, R)$ is R-projective (and hence free, since R is local).*

The following consequence is contained in [**DPS98a**, Prop. 1.1.1 & Cor. 1.1.2b], taking into account slight differences in notation. In particular, we have avoided the lattice terminology of [**AG60a**], since we use the terminology of [**Reiner03**]. Of course, if R is a DVR and $A = RG$ for a finite group G, this result is classical.

THEOREM C1.2. *Let (R, Q, F) be a local triple with R a regular local domain of Krull dimension ≤ 2. Let B be an algebra which is finite and free over R.*

Suppose that $\mathfrak{X}$ is an irreducible B_Q-module. Then $\mathfrak{X} = X_Q$, for some R-projective B-submodule $X \subseteq \mathfrak{X}$. If Y is another R-projective B-submodule of $\mathfrak{X}$ such that $\mathfrak{X} = Y_Q$, then X_F and Y_F have the same composition factors over B_F with the same multiplicities.[53]

This theorem gives an evident way to define a decomposition map

$$\mathscr{G}(B_Q) \longrightarrow \mathscr{G}(B_F) \tag{C1.1}$$

entirely analogous to the map (C0.1). This decomposition map often has natural factorizations in terms of decomposition maps associated with DVR triples. We discuss such a factorization, leading to Theorem C1.3 below.

In general, define a prime ideal $\mathfrak{p}$ in a domain R to be *coregular* if $R/\mathfrak{p}$ is a regular ring. We think of the condition of coregularity as one that can be checked by construction. We give several examples later in this section.

Let (R, Q, F) be a local triple, but assume that R is a regular local domain of Krull dimension 2, and again let B be as in Theorem C1.2. Let $\mathfrak{p}$ be a height one coregular prime ideal in R. This implies that both the rings $R/\mathfrak{p}$ and $R_\mathfrak{p}$ are DVRs. The fraction field of $R/\mathfrak{p}$ is the residue field $R(\mathfrak{p})$ of $R_\mathfrak{p}$. The residue field of $R/\mathfrak{p}$ is the same as the residue field F of R. Consider the following commutative diagram of rings. Here $B(\mathfrak{p}) := B_{R(\mathfrak{p})}$.

$$\begin{array}{ccccc} B & \longrightarrow & B_\mathfrak{p} & \longrightarrow & B_Q \\ \downarrow & & \downarrow & & \\ B/\mathfrak{p}B & \longrightarrow & B(\mathfrak{p}) & & \\ \downarrow & & & & \\ B_F & & & & \end{array} \tag{C1.2}$$

Observe that the natural ring maps associated with B and the triple (R, Q, F) can be found as compositions along the perimeter. They may be used to construct the decomposition map $\mathscr{G}(B_Q) \longrightarrow \mathscr{G}(B_F)$. (One starts with a B_Q-module, finds in it a full B-module lattice, then base changes the latter to B_F.) Analogous ring maps associated with $B_\mathfrak{p}$ and the DVR triple $(R_\mathfrak{p}, Q, R(\mathfrak{p}))$ can be found as individual maps in the right-hand portion of the diagram. These can be used to construct a decomposition map $\mathscr{G}(B_Q) \longrightarrow \mathscr{G}(B(\mathfrak{p}))$. The ring maps associated with $B/\mathfrak{p}B$ and the DVR triple $(R/\mathfrak{p}, R(\mathfrak{p}), F)$ are similarly found at the lower left. They can be used to construct a decomposition map $\mathscr{G}(B(\mathfrak{p})) \longrightarrow \mathscr{G}(B_F)$. Easy arguments, starting with a B-module and various base changes of it, now give the following result.

THEOREM C1.3. *Let (R, Q, F) be a regular local triple. Assume that R has Krull dimension ≤ 2, and let $\mathfrak{p} \in \operatorname{Spec} R$ be coregular of height 1. Finally, let B be an R-algebra as in Theorem* C1.2. *Then, the decomposition map $\mathscr{G}(B_Q) \longrightarrow \mathscr{G}(B_F)$ is the composition of the decomposition maps*

$$\mathscr{G}(B_Q) \longrightarrow \mathscr{G}(B(\mathfrak{p})) \longrightarrow \mathscr{G}(B_F). \tag{C1.3}$$

[53]We remark that the same theorem holds without the requirement that R be local, provided the condition that X is "R-projective" is replaced by the condition that X is "R-free," with a similar replacement regarding Y. The proof comes down to showing that Theorem C1.1 holds without assuming that R is local. This fact follows from the local version, which implies, in the non-local case, that $\operatorname{Hom}_R(M, R)_\mathfrak{p}$ is projective over $R(\mathfrak{p})$, for every $\mathfrak{p} \in \operatorname{Spec} R$.

The theorem is especially interesting when one of the decomposition map factors in display (C1.3) induces a bijective correspondence on isomorphism classes of irreducible modules. When this occurs, the theorem gives an identification of a decomposition matrix for the remaining factor with one for the decomposition map $\mathscr{G}(B_Q) \longrightarrow \mathscr{G}(B_F)$.

The following theorem involves such ingredients and is used in Section 5.2. The proof shows that the left-hand decomposition map factor in display (C1.3) induces the desired bijective correspondence. Thus, the decomposition matrix for $\mathscr{G}(B_Q) \longrightarrow \mathscr{G}(B_F)$ may be identified with that for the right-hand factor $\mathscr{G}(B(\mathfrak{p})) \longrightarrow \mathscr{G}(B_F)$ in this case.

THEOREM C1.4. *Assume the hypotheses of Theorem* C1.3. *Also, assume that the following two conditions hold:*

(1) B_Q *is split, and*
(2) $B(\mathfrak{p})$ *is split semisimple.*

Then, any decomposition matrix for B, with respect to the regular local triple (R, Q, F), agrees with one for $B/\mathfrak{p}B$ with respect to the DVR triple $(R/\mathfrak{p}, R(\mathfrak{p}), F)$.

The proof of Theorem C1.4 is given at the end of the subsection **C2**. We also recall that the term "split" applied to a finite (dimensional) algebra over a given base field means that every irreducible module for the algebra is absolutely irreducible. Equivalently, the algebra mod its Jacobson radical is a direct product of matrix algebras over the base field.

We first prove some independent results which will be useful in its proof. The notation, especially as regards (R, Q, F), is independent of that used in the statement of the theorem.

C2. On semiperfect rings. The following discussion makes use of the theory of semiperfect rings. Recall that a ring E is semiperfect provided $E/J(E)$ is semisimple (automatic in many cases) and idempotents in $E/J(E)$ "lift" to idempotents in E. (See [**AF74**, p. 303]. Here $J(E)$ denotes the Jacobson radical of E.) The lifting condition implies that any complete finite set of orthogonal idempotents in $E/J(E)$ lifts to a complete finite set of orthogonal idempotents in E. See [**AF74**, Prop. 27.4] and [**AF74**, pp. 303–304]. Another useful criterion, given in [**AF74**, Th. 27.6], states that E is semiperfect provided every irreducible E-module has a projective cover. Finally, [**AF74**, Cor. 27.8] states that E is semiperfect if some Morita equivalent ring is semiperfect.

PROPOSITION C2.1. *Let (R, Q, F) be a DVR triple and let E be a finite and R-free algebra. Then, condition (a) below implies condition (b):*

(a) E_Q *is split semisimple over* Q,
(b) E *is semiperfect as a ring.*

PROOF. See [**CR81**, §6, Ex. 16]. □

The following proposition is less trivial and could be used as a kind of converse of the proposition above. It may be viewed as a version of the Tits deformation theorem in the spirit of [**GP00**, Th. 7.4.6], which uses a different set of hypotheses.[54]

[54]The authors of [**GP00**] only require that R be integrally closed, but assume that E_Q is split; in particular they do not include the semiperfect case (which holds when R is complete). They discuss further variations of their hypotheses.

PROPOSITION C2.2. *Maintain the notation in the hypothesis of Proposition C2.1. Assume*

(1) *E is semiperfect, and*
(2) *E_F is split semisimple.*

Then, E is a direct product of full matrix algebras over R, and E_Q is split semisimple.

PROOF. Let π generate the maximal ideal of R. Since $E/\pi E \cong E_F$ is a semisimple algebra by hypothesis, it follows that $J(E) \subseteq \pi E$. Conversely, let L be an irreducible (left) E-module. Thus, L is an irreducible $E/\pi E = E_F$-module. Notice that every irreducible E-module L is finitely generated over R. Hence, πL is a proper submodule of L over R or E, so it must be 0. Thus, $\pi E \subseteq \ker(\phi)$, for every E-module homomorphism $\phi : {}_EE \to L$. Hence, $\pi E \subseteq J(E)$, giving $\pi E = J(E)$.

Since E is semiperfect, for the irreducible E-modules L and L', there are projective E-modules P and P' covering L and L', respectively.

CLAIM 1. If $L \ncong L'$, then $\mathrm{Hom}_E(P, P') = 0$.

PROOF OF CLAIM 1. We argue by contradiction. Let $0 \neq f$ be in $\mathrm{Hom}_E(P, P')$. Since R is a DVR, $\bigcap \pi^n R = 0$, and it follows that $\bigcap \pi^n P' = 0$. Hence, $f(P) \nsubseteq \pi^n P'$, for some $n > 0$. Let n be the smallest non-negative integer with this property. Then, f induces a nonzero map

$$P/\pi P \to \pi^{n-1}P'/\pi^n P'.$$

However, $P/\pi P \cong L$ and $\pi^{n-1}P'/\pi^n P' \cong P'/\pi P' \cong L'$. But, such a nonzero map does not exist, proving Claim 1.

CLAIM 2. $\mathrm{Hom}_E(P, P) \cong R$.

PROOF OF CLAIM 2. First, note that any non-zero $f \in \mathrm{Hom}_R(P, P)$ induces a nonzero map $P/\pi P \longrightarrow \pi^{n-1}P/\pi^n P$, if $n-1$ is chosen maximal with $f(P) \subseteq \pi^{n-1}P$. Thus, we can write f as $\pi^{n-1}g$, where $g \in \mathrm{Hom}_R(P, P)$ and $g(P) \nsubseteq \pi P$. (Take $g = \pi^{-n+1}f$.) Thus, g induces a nonzero $E/\pi E$-module homomorphism to $P/\pi P \to P/\pi P$. Such a map must be a scalar multiple of the identity $1_{P/\pi P}$ since $P/\pi P \cong L$ is absolutely irreducible. Thus, g belongs to $R \cdot 1_P + \pi \,\mathrm{Hom}_E(P, P)$ as does $f = \pi^{n-1}g$. However, f was an arbitrary non-zero element in $\mathrm{Hom}_E(P, P)$. Thus, $R \cdot 1_P = \mathrm{Hom}_E(P, P)$ by Nakayama's Lemma. This proves Claim 2. □

Next, we show that E is a direct product of full matrix algebras over R. Note that it follows immediately that E_Q is split semisimple, so the proposition will also follow.

Since E is semiperfect, the decomposition of E_F into a direct sum of irreducible (left) modules

$$E_F = \bigoplus_i L_i^{\oplus n_i},$$

where $n_i = \dim_F L_i$, gives a corresponding decomposition of ${}_EE$. We have

$${}_EE = \bigoplus P_i^{n_i},$$

where P_i is the projective cover of L_i as above. Using the (remarkable!) fact of Claim 1,

$$E \cong \mathrm{End}_E(E)^{\mathrm{op}} \cong \bigoplus_i \mathrm{End}_E(P_i^{\oplus n_i})^{\mathrm{op}}.$$

By Claim 2, the latter direct sum can be written as

$$\bigoplus_i M_{n_i}(R). \tag{C2.1}$$

Here the direct sum gives a direct product in the category of rings or R-algebras. This proves the first part of the assertion, and the rest follows. □

The next corollary is an obvious consequence of Proposition C2.2. Notice that each of the conclusions (a), (b), (c) of the corollary (we only need (a) later) easily reduces to the case where E is a full matrix algebra over R.

COROLLARY C2.3. *Let E and the DVR triple (R, Q, F) satisfy the hypotheses of Proposition* C2.2.

(a) *The natural decomposition map*

$$\mathscr{G}(E_Q) \longrightarrow \mathscr{G}(E_F) \tag{C2.2}$$

on Grothedieck groups induces a natural one-to-one correspondence between the (isomorphism types of) irreducible modules for E_Q and those of E_F.

(b) *Let M be any finite R-free E-module. Then, M_Q is irreducible over E_Q if and only if M_F is irreducible over E_F. In this case, $[M_Q]$ corresponds to $[M_F]$ in (a).*

(c) *Finally, given any ordering of the distinct irreducible E_Q-modules (resp., E_F-modules), an ordering of the irreducible E_F-modules (resp., E_Q-modules) can be chosen so that the corresponding decomposition matrix is the identity matrix.*

This completes our discussion of semiperfect rings. They will be used only in the proof below of Theorem C1.4.

We now return to the assumption, as part of the hypotheses shared by Theorems C1.3 and C1.4, that (R, Q, F) is a *regular local triple* with R of Krull dimension ≤ 2, and B is a finite and R-free algebra. Thus, associated to a coregular $\mathfrak{p} \in \operatorname{Spec} R$ having height 1, there are two local triples $(R_\mathfrak{p}, Q, R(\mathfrak{p}))$ and $(R/\mathfrak{p}, R(\mathfrak{p}), F)$; see display (C1.2). Finally, reading more from the hypothesis of Theorem C1.4, we note that B_Q is split, and $B(\mathfrak{p})$ is split semisimple.

PROOF OF THEOREM C1.4. We first apply Proposition C2.1 with the DVR triple $(R_\mathfrak{p}, Q, R(\mathfrak{p}))$ playing the role of (R, Q, F) there, and $B_\mathfrak{p}$ playing the role of E. We use this framework to prove that $B_Q = (B_\mathfrak{p})_Q$ is semisimple:

Since $R_\mathfrak{p}$ is a DVR, we can fix a generator π for its maximal ideal. By the semisimplicity of $B(\mathfrak{p}) = B_{R(\mathfrak{p})}$, we have $\pi B_\mathfrak{p} = J(B_\mathfrak{p})$ (see the first paragraph of the proof of Proposition C2.2). Thus, $B_\mathfrak{p}/\pi B_\mathfrak{p} \cong B_{R(\mathfrak{p})}$ implies $\operatorname{rank}(B_\mathfrak{p}) = \dim(B_{R(\mathfrak{p})})$. We now show that $B_\mathfrak{p} \cap J(B_Q) = 0$; equivalently, $J(B_Q) = 0$, implying B_Q is semisimple. Note that $B_\mathfrak{p} \cap J(B_Q)$ is a nilpotent ideal in $B_\mathfrak{p}$, hence contained in $J(B_\mathfrak{p})$. So, there is an epimorphism $B_\mathfrak{p}/B_\mathfrak{p} \cap J(B_Q) \twoheadrightarrow B_{R(\mathfrak{p})}$. But, $B_\mathfrak{p}/B_\mathfrak{p} \cap J(B_Q)$, embedded in $B_Q/J(B_Q)$, is $R_\mathfrak{p}$-torsion-free, hence free for the DVR $R_\mathfrak{p}$. Also $B_\mathfrak{p} \cap J(B_Q)$ is torsion-free and finitely generated over $R_\mathfrak{p}$, hence free. Rank considerations show $B_\mathfrak{p} \cap J(B_Q) = 0$.

By the hypothesis of the theorem, we conclude that $B_Q = (B_\mathfrak{p})_Q$ is split semisimple. Thus, $B_\mathfrak{p}$ is semiperfect by Proposition C2.1. We may now apply Proposition C2.2 (and Corollary C2.3) to the DVR triples $(R_\mathfrak{p}, Q, R(\mathfrak{p}))$ for $E = B_\mathfrak{p}$ to conclude that the decomposition matrix associated with the left decomposition map in display (C1.3) is an identity matrix (up to labeling of irreducible modules). Theorem C1.4 now follows from Theorem C1.3. □

Before proceeding, we introduce some further terminology that will be useful in the rest of this appendix. Let (R_1, Q_1, F_1) and (R, Q, F) be two local triples. Write

$$(R, Q, F) \leq (R_1, Q_1, F_1) \tag{C2.3}$$

if there is given an inclusion $R \subseteq R_1$ which also induces both inclusions $Q \subseteq Q_1$ and $F \subseteq F_1$ of rings. Unlike $Q \subseteq Q_1$, the inclusion $F \subseteq F_1$ is not automatic (from $R \subseteq R_1$) but exists precisely when $\mathfrak{m}_1 \cap R = \mathfrak{m}$, where $\mathfrak{m}_1$ and $\mathfrak{m}$ are the maximal ideals of R_1 and R, respectively. (Equivalently, $\mathfrak{m} \subseteq \mathfrak{m}_1$, since $\mathfrak{m}$ is maximal in R.)

It will also be convenient to use the following notion:

DEFINITION C2.4. Consider two local triples related as in display (C2.3).

(a) We say the triples are *residually equal* if the inclusion $F \subseteq F_1$ is an equality.

(b) We say that the local ring R_1 is *unramified* over the local ring R if $\mathfrak{m}_1 = \mathfrak{m}R_1$.

C3. Laurent triples. Suppose that $(\mathcal{O}, K, k)$ is a DVR triple, so that $\mathcal{O}$ is a discrete valuation ring with fraction field K and residue field k. We always assume in this subsection (and in this appendix) that K is a field of characteristic 0, so that all of its extensions are separable and $\mathbb{Z}$ is contained in K. Assume also that k has characteristic equal to a prime integer $r > 0$.[55] Form the localization

$$\mathcal{O}' := \mathcal{O}[t, t^{-1}]_{\mathfrak{m}}, \tag{C3.1}$$

where $\mathfrak{m}$ is a maximal ideal in $\mathcal{O}[t, t^{-1}] := \mathcal{O} \otimes_{\mathbb{Z}} \mathcal{Z}$ containing the maximal ideal of $\mathcal{O}$. Such an ideal $\mathfrak{m}$ may be constructed as the ideal generated by π and $t - a$, where $\pi \in \mathcal{O}$ generates the maximal ideal of $\mathcal{O}$ and $a \in \mathcal{O}$ does not belong to $\pi\mathcal{O}$. (We will only consider maximal ideals $\mathfrak{m}$ of this form.) Let $K' = K(t)$ be the fraction field of $\mathcal{O}'$, and note that the residue field of the local ring $\mathcal{O}'$ identifies naturally with $\mathcal{O}[t, t^{-1}]/\mathfrak{m} \cong \mathcal{O}/\pi\mathcal{O} = k$. (This property will result in the "residual" equality noted in Proposition C3.2 below.)

We call $(\mathcal{O}', K', k)$ the *Laurent triple* associated to the DVR triple $(\mathcal{O}, K, k)$ and the element $a \in \mathcal{O}\backslash\pi\mathcal{O}$. It follows from [**M86**, Th. 19.5] that the local triple $(\mathcal{O}', K', k)$ is a regular local triple in which $\mathcal{O}'$ has Krull dimension at most 2 (in fact, exactly 2).

EXAMPLE C3.1. When it is useful to identify a, we will write ${}^a\mathcal{O}'$ for the ring $\mathcal{O}'$. We give an important example of this setup.

Start with a DVR triple $(\mathcal{O}, K, k)$ in which K is the fraction field of $\mathcal{O}$ and k is the residue field of $\mathcal{O}$. Now, let $a \in \mathcal{O}\backslash\pi\mathcal{O}$ be arbitrary, and form two discrete valuation rings $\mathcal{O}' = \mathcal{O}[t, t^{-1}]_{\mathfrak{m}'}$ and $\mathcal{O}'' = \mathcal{O}[t, t^{-1}]_{\mathfrak{m}''}$, where $\mathfrak{m}' := (t - a, \pi)$ and $\mathfrak{m}'' := (t - a^n, \pi)$ are maximal ideals. Here, n can be any positive integer.

We can regard $\mathcal{O}'' \subseteq \mathcal{O}'$ by identifying $t \in \mathcal{O}'' = \mathcal{O}[t, t^{-1}]_{\mathfrak{m}''}$ with $t^n \in \mathcal{O}' = \mathcal{O}[t, t^{-1}]_{\mathfrak{m}'}$. This also results in an inclusion $K(t) \subseteq K(t)$, where t on the left is identified with t^n on the right.

[55] In typical applications, $\mathcal{O}$ is complete and k is algebraically closed, but we do not make these assumptions here. Also, the assumption that the characteristic of k is positive is a matter of notational convenience.

The following proposition is used in Section 5.2 in the proof of Theorem 5.2.8.

PROPOSITION C3.2. *With the above (slight abuse of) notation,*

$$(\mathcal{O}'', K(t), k) \leq (\mathcal{O}', K(t), k),$$

and these triples are residually equal (as defined in Definition C2.4*(a)). Finally, suppose that* B *is a finite and free* $\mathcal{O}''$*-algebra. Assume that* $B_{K(t)} := K(t) \otimes_{\mathcal{O}''} B$ *is split. Then, the decomposition matrix of* $B_{\mathcal{O}''}$ *with respect to* $(\mathcal{O}'', K(t), k)$ *agrees with that for* $B_{\mathcal{O}'}$ *with respect to* $(\mathcal{O}', K(t), k)$.

PROOF. The first assertion is proved using the discussion above the proposition: Observe that, in the notation for $\mathcal{O}'$, $\mathfrak{m}''$ is generated as an ideal in $\mathcal{O}''$ by π and the element $t^n - a^n \in \mathcal{O}'$. The factorization

$$t^n - a^n = (t-a)[(t^n - a^n)/(t-a)]$$

now shows that $\mathfrak{m}'' \subseteq \mathfrak{m}'$, and the first assertion follows.

The second assertion is a consequence of Proposition C4.1 below. □

C4. Some uniqueness/comparison results. Let (R, Q, F) be a local triple in which R is a regular local domain of Krull dimension ≤ 2. Let B be an R-algebra which is finite and free over R. We have seen that decomposition maps and matrices can be associated to B using such a triple. The "decomposition number" entries of the decomposition matrix are defined for a pair $(\mathfrak{X}, L)$ from mod-B_Q and mod-B_F, respectively (or their left-hand analogs) as the multiplicity $[X_F : L]$ of L in the base change to F of an R-free B-lattice X with $X_Q \cong \mathfrak{X}$. These multiplicities do not depend on the choice of X by Theorem C1.2.

This provides a kind of internal uniqueness for the decomposition maps and matrices defined using the local triple (R, Q, F). But, what if we change the triple?

PROPOSITION C4.1. *Let* $(R, Q, F) \leq (R_1, Q_1, F_1)$ *be regular local triples with* R, R_1 *regular of Krull dimension* ≤ 2. *Let* B *be a finite and* R*-free algebra. (These two lines are part of the hypotheses for each of the assertions (a), (b) below, and, in an evident modified form, for part (c).)*

(a) *Assume that* B_Q *and* B_F *are split. Then,* B_{Q_1} *and* B_{F_1} *are also split and the decomposition matrices associated to* B_{R_1} *and* (R_1, Q_1, F_1) *agree via base-change indexing with those associated to* B *and* (R, Q, F).

(b) *Similarly, if* B_Q *is split and the triples above are residually equal, then* B_{Q_1} *is split, and the decomposition numbers above agree via base-change indexing.*

(c) *More generally, all claims above hold if the inequality hypothesis* $(R, Q, F) \leq (R_1, Q_1, F_1)$ *is replaced by an injective ring homomorphism* $i : R \longrightarrow R_1$ *that induces an inequality* $(i(R), i(Q), i(F)) \leq (R_1, Q_1, F_1)$. *(Note that the isomorphism* $R \cong i(R)$ *enables natural definitions of* $i(Q)$ *and* $i(F)$ *as the fraction and residue fields of* $i(R)$*, respectively.)*

PROOF. We briefly sketch the proof, partly to make precise the agreement of decomposition numbers. Recall that the term "split" applied to a finite (dimensional) algebra over a given base field means that every irreducible module for the algebra is absolutely irreducible. Thus, if B_Q is split over Q, so is $B_{Q_1} = B_Q \otimes_Q Q_1$ over Q_1. In addition, if L, L', ... is a complete (finite) list of distinct irreducible B_Q-modules, then $L_{Q_1}, L'_{Q_1}, \ldots$ is a corresponding complete list for B_{Q_1}. A similar discussion applies for F and F_1.

Let X be any R-finite B-lattice, projective over R, with X_Q irreducible. Let L be any irreducible B_F-module. Then, the decomposition number associated with the pair (X_Q, L) is the multiplicity of L as a B_F-composition factor of X_F. This is also the multiplicity of L_{F_1} as a B_{F_1}-composition factor of

$$\begin{aligned}(X_F)_{F_1} &= (X \otimes_R F)_{F_1} \cong (X \otimes_R F) \otimes_F F_1 \cong X \otimes_R (F \otimes_F F_1) \\ &\cong X \otimes_R (R_1 \otimes_{R_1} F_1) \cong (X \otimes_R R_1)_{F_1}.\end{aligned}$$

The last module on the right is the B_{F_1}-module obtained by base change of the B_R-lattice $X \otimes_R R_1$. This lattice has irreducible base change to B_{Q_1}, namely,

$$(X \otimes_R R_1) \otimes_{R_1} Q_1 \cong (X \otimes_R Q) \otimes_Q Q_1 \cong (X_Q)_{Q_1}.$$

So, the given decomposition number associated to the pair (X_Q, L) is also the decomposition number associated to the pair $((X_Q)_{Q_1}, L_{F_1})$. This proves all the claims of part (a). The assertions of part (b) are proved similarly. We leave the assertions of part (c) to the reader. □

If we have a directed system of such triples, such that "eventually" B_Q and B_F are split, Proposition C4.1 implies that the associated decomposition matrix is uniquely determined by B. This assumes, of course, that we are able to define the system of triples naturally. We have two useful types of systems in mind, one for Krull dimension ≤ 1 (the DVR triples) and one for Krull dimension 2 (the Laurent triples). Comparisons of Laurent triples can often be reduced to considerations for underlying DVR triples, using Theorem C1.4. One important comparison, where a different approach is used, is given in Proposition C3.2.

We now give two useful results for constructing comparable DVR triples:

PROPOSITION C4.2 (Boosting Q). *Let (R, Q, F) be a DVR triple and let Q_1 be a finite extension of the field Q. Then, there exists a DVR triple (R_1, Q_1, F_1) such that $(R, Q, F) \leq (R_1, Q_1, F_1)$.*

PROOF. Let Z_1 be the integral closure of R in Q_1. Then, Z_1 is a Dedekind domain since R is a Dedekind domain; see [**Jac89**, Th. 10.7] or [**Reiner03**, Th. 4.4]. Let $\mathfrak{p}_1$ be a prime ideal in Z_1 such that $\mathfrak{p}_1 \cap R = \mathfrak{m}$, where $\mathfrak{m}$ is the maximal ideal of R. (Such a $\mathfrak{p}_1$ exists; see [**M86**, Lem. 2, p. 66] or [**AM69**, Th. 5.10, p.62]. Note that $\mathfrak{p}_1$ is maximal by [**Jac89**, Th. 10.3]) Set $R_1 = (Z_1)_{\mathfrak{p}_1}$ and $F_1 = Z_1/\mathfrak{p}_1$. Note that R_1 is a DVR by [**AM69**, Th. 9.3] or [**Jac89**, Th. 10.6]. The proposition follows. □

PROPOSITION C4.3 (Boosting F). *Let (R, Q, F) be a DVR triple with the field Q of characteristic 0 and the field F finite. Let F_1 be a finite extension field of F. Then, there is a DVR triple (R_1, Q_1, F_1) such that $(R, Q, F) \leq (R_1, Q_1, F_1)$, and the local ring R_1 is unramified over R (as defined in Definition* C2.4*(b)).*

PROOF. Put $m := |F_1| - 1$. Let ϵ be a primitive mth root of 1, arbitrary for the moment. Now, apply Proposition C4.2 to the field inclusion $Q \subseteq Q[\epsilon]$, which results in a DVR triple (R_1, Q_1, F_1), where $Q_1 = Q[\epsilon]$.

We can interpret $Q[\epsilon]$ as $Q[x]/f(x)Q(x)$, where $f(x)$ is an irreducible R-integral monic polynomial depending on ϵ and dividing the cyclotomic polynomial $\Phi_m(x)$ (Gauss). Let $g(x)$ be the minimal polynomial over F of a generator θ of F_1 with order $|F_1| - 1 = m$. Thus, $\bar{\Phi}_m(\theta) = 0$, where $\bar{\Phi}_m$ is the reduction of Φ_m mod r, where r is the characteristic of F (and $r \nmid m$). We may also regard $\bar{\Phi}_m$ as a (polynomial) factor of $x^m - 1 \in F[x]$ with distinct roots in F. As such, it is a product of the reduction $\bar{f}(x)$ mod $\mathfrak{m}$ (where $F = R/\mathfrak{m}$) of the various candidates for $f(x)$ dividing $\Phi_m(x)$ in $R[x]$. The reduction of one of these $f(x)$ must have $g(x)$ as a factor in $F[x]$. Choose such an $f(x)$, and identify $R[\epsilon]$ with $R[x]/f(x)R[x]$.

CLAIM. $R[\epsilon]$ is integrally closed in its fraction field $Q[\epsilon] = Q[x]/f(x)Q[x]$.

Assuming this claim, it follows that $R[\epsilon]$ is a Dedekind domain (see [**Jac89**, Th. 10.7] or [**Reiner03**, Th. 4.4]). Consequently, if $\mathfrak{m}_1$ is any one of its maximal ideals, the localization $R_1 = R[\epsilon]_{\mathfrak{m}_1}$ is a DVR. Now, let $\mathfrak{m}_1$ be the kernel of the composite (surjective) map

$$R[\epsilon] = R[x]/f(x)R[x] \longrightarrow F[x]/\bar{f}(x)F[x] \longrightarrow F[x]/g(x)F[x] \cong F_1.$$

It remains to prove the Claim. Let R' be the integral closure of R in $Q(\epsilon)$. Clearly, $R[\epsilon] \subseteq R'$, and both rings have the same base change as R-modules to Q. It follows that both have the same rank as (free) R-modules. To prove that they are equal, it is sufficient, by Nakayama's lemma, to prove $\pi R' + R[\epsilon] = R'$, where $\pi \in R$ generates its maximal ideal. Since both sides contain $\pi R'$, it is sufficient to show that this equality holds if $\pi R'$ is factored out of each side. The resulting left-hand side $(\pi R' + R[\epsilon])/\pi R'$ is the image of the obvious map

$$i : R[\epsilon]/\pi R[\epsilon] \longrightarrow R'/\pi R',$$

which has kernel $I/\pi R[\epsilon]$, where $I := R[\epsilon] \cap \pi R'$ is an ideal in $R[\epsilon]$. Some power of I is contained in $\pi R[\epsilon]$, since the R-module $\pi R'/\pi R[\epsilon]$ contains no nonzero π-torsion-free R-submodule (and so is killed by some nth power of π or of $\pi R'$). Thus, $I/\pi R[\epsilon]$ is a nilpotent ideal in the direct product of fields $R[\epsilon]/\pi R[\epsilon] \cong F[x]/\bar{f}(x)F[x]$ and must be zero. Now, we know that the map i above must be injective. However, both target and domain have the same rank as $R/\pi R$-modules. Since $R/\pi R = F$ is a field, i must also be surjective. This proves the claim, and the proof of Proposition C4.3 is now complete. □

REMARKS C4.4. We conclude this appendix with two further examples of triples that are useful.

(a) **Completions.** Let (R, Q, F) be a regular local triple. Then, R is a local ring, so let $\widehat{R}$ be its completion at its maximal ideal; necessarily, $\widehat{R}$ is a regular local ring having the same Krull dimension as R. ([**AM69**, Ch. 10] has a general treatment of completions. See also [**M86**, Th. 17.5, and the proof of Th. 17.8] and the preceding discussion.) If $\widehat{Q}$ is the fraction field of $\widehat{R}$, then $(\widehat{R}, \widehat{Q}, F)$ is a "complete regular local triple."

(b) **Direct unions.** Suppose that (R_n, Q_n, F_n) is a sequence of DVR triples with

$$(R_n, Q_n, F_n) \leq (R_{n+1}, Q_{n+1}, F_{n+1}), \quad n \geq 0.$$

Assume also, for sufficiently large n, that R_{n+1} is unramified over R_n. Then,

$$(\varinjlim R_n, \varinjlim Q_n, \varinjlim F_n).$$

is a DVR triple. We leave the proof of this as an exercise, but do record the following corollary.

COROLLARY C4.5. *Let F be the algebraic closure of a prime field $\mathbb{F}_r$, let R be the ring of Witt vectors over F, and let Q be the fraction field R. Then, (R, Q, F) is a DVR triple.*

This fact follows from Remark C4.4(b) and the construction of Witt vectors (see [**Jac89**, §8.10]).

Acknowledgments

We would like to thank our colleague Weiqiang Wang for several discussions and special assistances during the writing of this monograph. Section 5.4 was heavily influenced by his work with H. Bao.

Part of the work was written while Jie Du was visiting the University of Virginia in 2016 and 2022, using Simons grants there. Leonard Scott visited the University of New South Wales in 2015 and 2019 with support there through an ARC grant and the Distinguished Research Visitor Scheme. Both authors want to thank the two universities for their support and hospitality.

Finally, we wish to thank Karen Parshall (University of Virginia) for reading the entire manuscript and making many helpful comments and thank the referee for the reference [**DR92**] and the comments about the works of Auslander, Dlad–Ringel, and Hemmer–Nakano in [**A71, DR89, HN04**].

References

[AF74] Frank W. Anderson and Kent R. Fuller, *Rings and categories of modules*, Graduate Texts in Mathematics, Vol. 13, Springer-Verlag, New York-Heidelberg, 1974. MR417223

[AM69] M. F. Atiyah and I. G. Macdonald, *Introduction to commutative algebra*, Addison-Wesley Publishing Co., Reading, Mass.-London-Don Mills, Ont., 1969. MR242802

[A71] M. Auslander, *Representation dimension of Artin algebras*, Queen Mary College Mathematical Notes (London), 1971.

[AG60a] Maurice Auslander and Oscar Goldman, *Maximal orders*, Trans. Amer. Math. Soc. **97** (1960), 1–24, DOI 10.2307/1993361. MR117252

[AG60b] Maurice Auslander and Oscar Goldman, *The Brauer group of a commutative ring*, Trans. Amer. Math. Soc. **97** (1960), 367–409, DOI 10.2307/1993378. MR121392

[BW18] Huanchen Bao and Weiqiang Wang, *A new approach to Kazhdan-Lusztig theory of type B via quantum symmetric pairs* (English, with English and French summaries), Astérisque **402** (2018), vii+134. MR3864017

[BKLW18] Huanchen Bao, Jonathan Kujawa, Yiqiang Li, and Weiqiang Wang, *Geometric Schur duality of classical type*, Transform. Groups **23** (2018), no. 2, 329–389, DOI 10.1007/s00031-017-9447-4. MR3805209

[BT22] Gwyn Bellamy and Ulrich Thiel, *Cellularity of endomorphism algebras of tilting objects*. part A, Adv. Math. **404** (2022), no. part A, Paper No. 108387, 38, DOI 10.1016/j.aim.2022.108387. MR4413216

[B71] Mark Benard, *On the Schur indices of characters of the exceptional Weyl groups*, Ann. of Math. (2) **94** (1971), 89–107, DOI 10.2307/1970736. MR297887

[BC71] C. T. Benson and C. W. Curtis, *On the degrees and rationality of certain characters of finite Chevalley groups*, Trans. Amer. Math. Soc. **165** (1972), 251–273, DOI 10.2307/1995885. MR304473

[BM75] Stephen Berman and Robert V. Moody, *Extensions of Chevalley groups*, Israel J. Math. **22** (1975), no. 1, 42–51, DOI 10.1007/BF02757272. MR390077

[BR03] Cédric Bonnafé and Raphaël Rouquier, *Catégories dérivées et variétés de Deligne-Lusztig* (French), Publ. Math. Inst. Hautes Études Sci. **97** (2003), 1–59, DOI 10.1007/s10240-003-0013-3. MR2010739

[BR17] Cédric Bonnafé, Jean-François Dat, and Raphaël Rouquier, *Derived categories and Deligne-Lusztig varieties II*, Ann. of Math. (2) **185** (2017), no. 2, 609–670, DOI 10.4007/annals.2017.185.2.5. MR3612005

[Bo72] N. Bourbaki, *Commutative algebras*, Hermann, Paris, 1972.

[BDT20] Olivier Brunat, Olivier Dudas, and Jay Taylor, *Unitriangular shape of decomposition matrices of unipotent blocks*, Ann. of Math. (2) **192** (2020), no. 2, 583–663, DOI 10.4007/annals.2020.192.2.7. MR4151085

[Bu10] Theo Bühler, *Exact categories*, Expo. Math. **28** (2010), no. 1, 1–69, DOI 10.1016/j.exmath.2009.04.004. MR2606234

[Ca98] Marc Cabanes, *Algèbres de Hecke comme algèbres symétriques et théorème de Dipper* (French, with English and French summaries), C. R. Acad. Sci. Paris Sér. I Math. **327** (1998), no. 6, 531–536, DOI 10.1016/S0764-4442(98)89158-1. MR1647300

[C85] Roger W. Carter, *Finite groups of Lie type: Conjugacy classes and complex characters*, Pure and Applied Mathematics (New York), John Wiley & Sons, Inc., New York, 1985. A Wiley-Interscience Publication. MR794307

[CPS73] Edward Cline, Brian Parshall, and Leonard Scott, *Cohomology of finite groups of Lie type. I*, Inst. Hautes Études Sci. Publ. Math. **45** (1975), 169–191. MR399283

[CPS88] E. Cline, B. Parshall, and L. Scott, *Finite-dimensional algebras and highest weight categories*, J. Reine Angew. Math. **391** (1988), 85–99. MR961165

[CPS90] E. Cline, B. Parshall, and L. Scott, *Integral and graded quasi-hereditary algebras. I*, J. Algebra **131** (1990), no. 1, 126–160, DOI 10.1016/0021-8693(90)90169-O. MR1055002

[CPS96] Edward Cline, Brian Parshall, and Leonard Scott, *Stratifying endomorphism algebras*, Mem. Amer. Math. Soc. **124** (1996), no. 591, viii+119, DOI 10.1090/memo/0591. MR1350891

[CR81] Charles W. Curtis and Irving Reiner, *Methods of representation theory. Vol. I*, Pure and Applied Mathematics, John Wiley & Sons, Inc., New York, 1981. With applications to finite groups and orders; A Wiley-Interscience Publication. MR632548

[CR87] Charles W. Curtis and Irving Reiner, *Methods of representation theory. Vol. II*, Pure and Applied Mathematics (New York), John Wiley & Sons, Inc., New York, 1987. With applications to finite groups and orders; A Wiley-Interscience Publication. MR892316

[DDPW08] Bangming Deng, Jie Du, Brian Parshall, and Jianpan Wang, *Finite dimensional algebras and quantum groups*, Mathematical Surveys and Monographs, vol. 150, American Mathematical Society, Providence, RI, 2008, DOI 10.1090/surv/150. MR2457938

[Dip90] Richard Dipper, *On quotients of Hom-functors and representations of finite general linear groups. I*, J. Algebra **130** (1990), no. 1, 235–259, DOI 10.1016/0021-8693(90)90111-Z. MR1045747

[Dip91] Richard Dipper, *Polynomial representations of finite general linear groups in nondescribing characteristic*, Representation theory of finite groups and finite-dimensional algebras (Bielefeld, 1991), Progr. Math., vol. 95, Birkhäuser, Basel, 1991, pp. 343–370. MR1112168

[Dip98] Richard Dipper, *On quotients of Hom-functors and representations of finite general linear groups. II*, J. Algebra **209** (1998), no. 1, 199–269, DOI 10.1006/jabr.1998.7541. MR1652126

[DD97] Richard Dipper and Jie Du, *Harish-Chandra vertices and Steinberg's tensor product theorems for finite general linear groups*, Proc. London Math. Soc. (3) **75** (1997), no. 3, 559–599, DOI 10.1112/S0024611597000439. MR1466661

[DF92] Richard Dipper and Peter Fleischmann, *Modular Harish-Chandra theory. I*, Math. Z. **211** (1992), no. 1, 49–71, DOI 10.1007/BF02571417. MR1179779

[DF94] Richard Dipper and Peter Fleischmann, *Modular Harish-Chandra theory. II*, Arch. Math. (Basel) **62** (1994), no. 1, 26–32, DOI 10.1007/BF01200434. MR1249581

[DG99] Richard Dipper and Jochen Gruber, *Generalized q-Schur algebras and modular representation theory of finite groups with split (BN)-pairs*, J. Reine Angew. Math. **511** (1999), 145–191, DOI 10.1515/crll.1999.511.145. MR1695794

[DJ86] Richard Dipper and Gordon James, *Representations of Hecke algebras of general linear groups*, Proc. London Math. Soc. (3) **52** (1986), no. 1, 20–52, DOI 10.1112/plms/s3-52.1.20. MR812444

[DJ89] Richard Dipper and Gordon James, *The q-Schur algebra*, Proc. London Math. Soc. (3) **59** (1989), no. 1, 23–50, DOI 10.1112/plms/s3-59.1.23. MR997250

[DPS98a] Jie Du, Brian Parshall, and Leonard Scott, *Stratifying endomorphism algebras associated to Hecke algebras*, J. Algebra **203** (1998), no. 1, 159–210, DOI 10.1006/jabr.1997.7325. MR1620729

[DPS98b] J. Du, B. Parshall, and L. Scott, *Cells and q-Schur algebras*, Transform. Groups **3** (1998), no. 1, 33–49, DOI 10.1007/BF01237838. MR1603810

[DPS98c] Jie Du, Brian Parshall, and Leonard Scott, *Quantum Weyl reciprocity and tilting modules*, Comm. Math. Phys. **195** (1998), no. 2, 321–352, DOI 10.1007/s002200050392. MR1637785

[DPS15] Jie Du, Brian J. Parshall, and Leonard L. Scott, *Extending Hecke endormorphism algebras*, Pacific J. Math. **279** (2015), no. 1-2, 229–254, DOI 10.2140/pjm.2015.279.229. MR3437777

[DPS17] Jie Du, Brian J. Parshall, and Leonard L. Scott, *Stratifying endomorphism algebras using exact categories*, J. Algebra **475** (2017), 229–250, DOI 10.1016/j.jalgebra.2016.11.002. MR3612470

[DPS18] Jie Du, Brian J. Parshall, and Leonard L. Scott, *Local and global methods in representations of Hecke algebras*, Sci. China Math. **61** (2018), no. 2, 207–226, DOI 10.1007/s11425-017-9171-2. MR3755776

[DR89] Vlastimil Dlab and Claus Michael Ringel, *Every semiprimary ring is the endomorphism ring of a projective module over a quasihereditary ring*, Proc. Amer. Math. Soc. **107** (1989), no. 1, 1–5, DOI 10.2307/2048026. MR943793

[DR92] Vlastimil Dlab and Claus Michael Ringel, *The module theoretical approach to quasi-hereditary algebras*, Representations of algebras and related topics (Kyoto, 1990), London Math. Soc. Lecture Note Ser., vol. 168, Cambridge Univ. Press, Cambridge, 1992, pp. 200–224. MR1211481

[DRSS99] Peter Dräxler, Idun Reiten, Sverre O. Smalø, and Øyvind Solberg, *Exact categories and vector space categories*, Trans. Amer. Math. Soc. **351** (1999), no. 2, 647–682, DOI 10.1090/S0002-9947-99-02322-3. With an appendix by B. Keller. MR1608305

[DuR98] Jie Du and Hebing Rui, *Based algebras and standard bases for quasi-hereditary algebras*, Trans. Amer. Math. Soc. **350** (1998), no. 8, 3207–3235, DOI 10.1090/S0002-9947-98-02305-8. MR1603902

[DS94] Jie Du and Leonard Scott, *Lusztig conjectures, old and new. I*, J. Reine Angew. Math. **455** (1994), 141–182, DOI 10.1515/crll.1994.455.141. MR1293877

[DS00a] Jie Du and Leonard Scott, *The q-*Schur^2 *algebra*, Trans. Amer. Math. Soc. **352** (2000), no. 9, 4325–4353, DOI 10.1090/S0002-9947-00-02262-5. MR1655991

[DS00b] Jie Du and Leonard Scott, *Stratifying q-Schur algebras of type D*, Representations and quantizations (Shanghai, 1998), China High. Educ. Press, Beijing, 2000, pp. 167–197. MR1802172

[DW22a] Jie Du and Yadi Wu, *A new realisation of the i-quantum group* $\mathbf{U}^{\jmath}(n)$, J. Pure Appl. Algebra **226** (2022), no. 1, Paper No. 106793, 27, DOI 10.1016/j.jpaa.2021.106793. MR4262075

[DW22b] Jie Du and Yadi Wu, *The i-quantum group* $U^{\imath}(n)$, Pacific J. Math. **320** (2022), no. 1, 61–101, DOI 10.2140/pjm.2022.320.61. MR4496094

[DM18] Olivier Dudas and Gunter Malle, *Modular irreducibility of cuspidal unipotent characters*, Invent. Math. **211** (2018), no. 2, 579–589, DOI 10.1007/s00222-017-0753-1. MR3748314

[DVV19] O. Dudas, M. Varagnolo, and E. Vasserot, *Categorical actions on unipotent representations of finite unitary groups*, Publ. Math. Inst. Hautes Études Sci. **129** (2019), 129–197, DOI 10.1007/s10240-019-00104-x. MR3949029

[F17] Timothy J. Ford, *Separable algebras*, Graduate Studies in Mathematics, vol. 183, American Mathematical Society, Providence, RI, 2017, DOI 10.1090/gsm/183. MR3618889

[G90] M. Geck, *Verallgemeinerte Gelfand–Graev Charaktere und Zerlegungszahlen endlicher Gruppen vom Lie-Typ*, Ph.D. thesis, RWTH Aachen (1990).

[G91] Meinolf Geck, *On the decomposition numbers of the finite unitary groups in non-defining characteristic*, Math. Z. **207** (1991), no. 1, 83–89, DOI 10.1007/BF02571376. MR1106814

[G92] Meinolf Geck, *Brauer trees of Hecke algebras*, Comm. Algebra **20** (1992), no. 10, 2937–2973, DOI 10.1080/00927879208824499. MR1179271

[G98] Meinolf Geck, *Kazhdan-Lusztig cells and decomposition numbers*, Represent. Theory **2** (1998), 264–277, DOI 10.1090/S1088-4165-98-00042-9. MR1628035

[G03] Meinolf Geck, *An introduction to algebraic geometry and algebraic groups*, Oxford Graduate Texts in Mathematics, vol. 10, Oxford University Press, Oxford, 2003. MR2032320

[G06] Meinolf Geck, *Modular principal series representations*, Int. Math. Res. Not., posted on 2006, Art. ID 41957, 20, DOI 10.1155/IMRN/2006/41957. MR2219230

[G07] Meinolf Geck, *Hecke algebras of finite type are cellular*, Invent. Math. **169** (2007), no. 3, 501–517, DOI 10.1007/s00222-007-0053-2. MR2336039

[G12] Meinolf Geck, *Remarks on modular representations of finite groups of Lie type in non-defining characteristic*, Algebraic groups and quantum groups, Contemp. Math., vol. 565, Amer. Math. Soc., Providence, RI, 2012, pp. 71–80, DOI 10.1090/conm/565/11160. MR2932422

[G17] Meinolf Geck, *On the modular composition factors of the Steinberg representation*, J. Algebra **475** (2017), 370–391, DOI 10.1016/j.jalgebra.2015.11.005. MR3612476

[G18] Meinolf Geck, *A first guide to the character theory of finite groups of Lie type*, Local representation theory and simple groups, EMS Ser. Lect. Math., Eur. Math. Soc., Zürich, 2018, pp. 63–106. MR3821138

[GH97] Meinolf Geck and Gerhard Hiss, *Modular representations of finite groups of Lie type in non-defining characteristic*, Finite reductive groups (Luminy, 1994), Progr. Math., vol. 141, Birkhäuser Boston, Boston, MA, 1997, pp. 195–249. MR1429874

[GJ11] Meinolf Geck and Nicolas Jacon, *Representations of Hecke algebras at roots of unity*, Algebra and Applications, vol. 15, Springer-Verlag London, Ltd., London, 2011, DOI 10.1007/978-0-85729-716-7. MR2799052

[GP00] Meinolf Geck and Götz Pfeiffer, *Characters of finite Coxeter groups and Iwahori-Hecke algebras*, London Mathematical Society Monographs. New Series, vol. 21, The Clarendon Press, Oxford University Press, New York, 2000. MR1778802

[GR97] Meinolf Geck and Raphaël Rouquier, *Centers and simple modules for Iwahori-Hecke algebras*, Finite reductive groups (Luminy, 1994), Progr. Math., vol. 141, Birkhäuser Boston, Boston, MA, 1997, pp. 251–272. MR1429875

[GGOR03] Victor Ginzburg, Nicolas Guay, Eric Opdam, and Raphaël Rouquier, *On the category $\mathcal{O}$ for rational Cherednik algebras*, Invent. Math. **154** (2003), no. 3, 617–651, DOI 10.1007/s00222-003-0313-8. MR2018786

[GL96] J. J. Graham and G. I. Lehrer, *Cellular algebras*, Invent. Math. **123** (1996), no. 1, 1–34, DOI 10.1007/BF01232365. MR1376244

[Go80] Daniel Gorenstein, *Finite groups*, 2nd ed., Chelsea Publishing Co., New York, 1980. MR569209

[Gr80] James A. Green, *Polynomial representations of* GL_n, Lecture Notes in Mathematics, vol. 830, Springer-Verlag, Berlin-New York, 1980. MR606556

[Grn97] R. M. Green, *Hyperoctahedral Schur algebras*, J. Algebra **192** (1997), no. 1, 418–438, DOI 10.1006/jabr.1996.6935. MR1449968

[Hat63] Akira Hattori, *Semisimple algebras over a commutative ring*, J. Math. Soc. Japan **15** (1963), 404–419, DOI 10.2969/jmsj/01540404. MR158903

[HN04] David J. Hemmer and Daniel K. Nakano, *Specht filtrations for Hecke algebras of type A*, J. London Math. Soc. (2) **69** (2004), no. 3, 623–638, DOI 10.1112/S0024610704005186. MR2050037

[Ho07] M. Hochster, *Flat base change and Hom*, University of Michigan Math711 Lecture, September 24, 2007.

[Hu01] Jun Hu, *Schur-Weyl reciprocity between quantum groups and Hecke algebras of type* $G(r,1,n)$, Math. Z. **238** (2001), no. 3, 505–521, DOI 10.1007/s002090100264. MR1869696

[Iwa64] Nagayoshi Iwahori, *On the structure of a Hecke ring of a Chevalley group over a finite field*, J. Fac. Sci. Univ. Tokyo Sect. I **10** (1964), 215–236 (1964). MR165016

[Jac89] Nathan Jacobson, *Basic algebra. II*, 2nd ed., W. H. Freeman and Company, New York, 1989. MR1009787

[Jan96] Jens Carsten Jantzen, *Lectures on quantum groups*, Graduate Studies in Mathematics, vol. 6, American Mathematical Society, Providence, RI, 1996, DOI 10.1090/gsm/006. MR1359532

[K90] Bernhard Keller, *Chain complexes and stable categories*, Manuscripta Math. **67** (1990), no. 4, 379–417, DOI 10.1007/BF02568439. MR1052551

[K96] Bernhard Keller, *Derived categories and their uses*, Handbook of algebra, Vol. 1, Handb. Algebr., vol. 1, Elsevier/North-Holland, Amsterdam, 1996, pp. 671–701, DOI 10.1016/S1570-7954(96)80023-4. MR1421815

[KM20] Alexander Kleshchev and Robert Muth, *Based quasi-hereditary algebras*, J. Algebra **558** (2020), 504–522, DOI 10.1016/j.jalgebra.2019.04.034. MR4102115

[LL21] Chun-Ju Lai and Li Luo, *Schur algebras and quantum symmetric pairs with unequal parameters*, Int. Math. Res. Not. IMRN **13** (2021), 10207–10259, DOI 10.1093/imrn/rnz110. MR4283577

[LNX20] Chun-Ju Lai, Daniel K. Nakano, and Ziqing Xiang, *q-Schur algebras corresponding to Hecke algebras of type* B, Transform. Groups **27** (2022), no. 3, 983–1024, DOI 10.1007/s00031-020-09628-7. MR4475484

[Le03] Gail Letzter, *Quantum symmetric pairs and their zonal spherical functions*, Transform. Groups **8** (2003), no. 3, 261–292, DOI 10.1007/s00031-003-0719-9. MR1996417

[LZ] Yiqiang Li and Jieru Zhu, *Quasi-split symmetric pairs of $U(\mathfrak{gl}_N)$ and their Schur algebras*, Nagoya Math. J. **245** (2022), 1–27, DOI 10.1017/nmj.2020.16. MR4413360

[LN97] Rudolf Lidl and Harald Niederreiter, *Finite fields*, 2nd ed., Encyclopedia of Mathematics and its Applications, vol. 20, Cambridge University Press, Cambridge, 1997. With a foreword by P. M. Cohn. MR1429394

[LW22] Li Luo and Weiqiang Wang, *The q-Schur algebras and q-Schur dualities of finite type*, J. Inst. Math. Jussieu **21** (2022), no. 1, 129–160, DOI 10.1017/S1474748020000031. MR4366335

[Lus81] George Lusztig, *On a theorem of Benson and Curtis*, J. Algebra **71** (1981), no. 2, 490–498, DOI 10.1016/0021-8693(81)90188-5. MR630610

[Lus83] G. Lusztig, *Left cells in Weyl groups*, Lie group representations, I (College Park, Md., 1982/1983), Lecture Notes in Math., vol. 1024, Springer, Berlin, 1983, pp. 99–111, DOI 10.1007/BFb0071433. MR727851

[Lus84] George Lusztig, *Characters of reductive groups over a finite field*, Annals of Mathematics Studies, vol. 107, Princeton University Press, Princeton, NJ, 1984, DOI 10.1515/9781400881772. MR742472

[Lus86] George Lusztig, *Sur les cellules gauches des groupes de Weyl* (French, with English summary), C. R. Acad. Sci. Paris Sér. I Math. **302** (1986), no. 1, 5–8. MR827096

[Lus87a] George Lusztig, *Cells in affine Weyl groups*, Algebraic groups and related topics (Kyoto/Nagoya, 1983), Adv. Stud. Pure Math., vol. 6, North-Holland, Amsterdam, 1985, pp. 255–287, DOI 10.2969/aspm/00610255. MR803338

[Lus87b] George Lusztig, *Cells in affine Weyl groups. II*, J. Algebra **109** (1987), no. 2, 536–548, DOI 10.1016/0021-8693(87)90154-2. MR902967

[Lus87c] George Lusztig, *Cells in affine Weyl groups. III*, J. Fac. Sci. Univ. Tokyo Sect. IA Math. **34** (1987), no. 2, 223–243. MR914020

[Lus87d] G. Lusztig, *Leading coefficients of character values of Hecke algebras*, The Arcata Conference on Representations of Finite Groups (Arcata, Calif., 1986), Proc. Sympos. Pure Math., vol. 47, Amer. Math. Soc., Providence, RI, 1987, pp. 235–262, DOI 10.1090/pspum/047.2/933415. MR933415

[Lus88] G. Lusztig, *On the representations of reductive groups with disconnected centre*, Astérisque **168** (1988), 10, 157–166. Orbites unipotentes et représentations, I. MR1021495

[Lus90] G. Lusztig, *Canonical bases arising from quantized enveloping algebras*, J. Amer. Math. Soc. **3** (1990), no. 2, 447–498, DOI 10.2307/1990961. MR1035415

[Lus93] George Lusztig, *Appendix: Coxeter groups and unipotent representations*, Astérisque **212** (1993), 191–203. Représentations unipotentes génériques et blocs des groupes réductifs finis. MR1235835

[Lus03] G. Lusztig, *Hecke algebras with unequal parameters*, CRM Monograph Series, vol. 18, American Mathematical Society, Providence, RI, 2003, DOI 10.1090/crmm/018. MR1974442

[MT11] Gunter Malle and Donna Testerman, *Linear algebraic groups and finite groups of Lie type*, Cambridge Studies in Advanced Mathematics, vol. 133, Cambridge University Press, Cambridge, 2011, DOI 10.1017/CBO9780511994777. MR2850737

[Mac94] Saunders MacLane, *Homology*, 1st ed., Die Grundlehren der mathematischen Wissenschaften, Band 114, Springer-Verlag, Berlin-New York, 1967. (Fourth printing 1994). MR349792

[M86] Hideyuki Matsumura, *Commutative ring theory*, Cambridge Studies in Advanced Mathematics, vol. 8, Cambridge University Press, Cambridge, 1986. Translated from the Japanese by M. Reid. MR879273

[Mum88] David Mumford, *The red book of varieties and schemes*, Lecture Notes in Mathematics, vol. 1358, Springer-Verlag, Berlin, 1988, DOI 10.1007/978-3-662-21581-4. MR971985

[Reiner03] I. Reiner, *Maximal orders*, London Mathematical Society Monographs. New Series, vol. 28, The Clarendon Press, Oxford University Press, Oxford, 2003. Corrected reprint of the 1975 original; With a foreword by M. J. Taylor. MR1972204

[Q73] Daniel Quillen, *Higher algebraic K-theory. I*, Algebraic K-theory, I: Higher K-theories (Proc. Conf., Battelle Memorial Inst., Seattle, Wash., 1972), Lecture Notes in Math., Vol. 341, Springer, Berlin-New York, 1973, pp. 85–147. MR338129

[Ro08] Raphaël Rouquier, *q-Schur algebras and complex reflection groups* (English, with English and Russian summaries), Mosc. Math. J. **8** (2008), no. 1, 119–158, 184, DOI 10.17323/1609-4514-2008-8-1-119-158. MR2422270

[SS99] Masahiro Sakamoto and Toshiaki Shoji, *Schur-Weyl reciprocity for Ariki-Koike algebras*, J. Algebra **221** (1999), no. 1, 293–314, DOI 10.1006/jabr.1999.7973. MR1722914

[St67] Robert Steinberg, *Lectures on Chevalley groups*, Yale University, New Haven, CT, 1968. Notes prepared by John Faulkner and Robert Wilson. MR466335

[St68] Robert Steinberg, *Endomorphisms of linear algebraic groups*, Memoirs of the American Mathematical Society, No. 80, American Mathematical Society, Providence, RI, 1968. MR230728

[Se77] Jean-Pierre Serre, *Linear representations of finite groups*, Graduate Texts in Mathematics, Vol. 42, Springer-Verlag, New York-Heidelberg, 1977. Translated from the second French edition by Leonard L. Scott. MR450380

[Sw78] Richard G. Swan, *Projective modules over Laurent polynomial rings*, Trans. Amer. Math. Soc. **237** (1978), 111–120, DOI 10.2307/1997613. MR469906

[W1] Hideya Watanabe, *Classical weight modules over $\imath$quantum groups*, J. Algebra **578** (2021), 241–302, DOI 10.1016/j.jalgebra.2021.02.023. MR4234802

[W2] Hideya Watanabe, *Global crystal bases for integrable modules over a quantum symmetric pair of type AIII*, Represent. Theory **25** (2021), 27–66, DOI 10.1090/ert/556. MR4198491

Index

A, generic Hecke endo-algebra, 171
W, a Weyl group, 167
  (W, S, L), a standard finite $\sim$, 168
Ω, set of left cells, 172
$\mathcal{Z} = \mathbb{Z}[t, t^{-1}]$, 171
$\mathfrak{h}$-filtration, 180
$\mathscr{K}$-order, 210
  maximal $\sim$, 210
J, asymptotic form, 211
  $\mathcal{J} := \mathcal{Z} \otimes_{\mathbb{Z}} \mathsf{J}$, 211
  $\mathcal{J}^{\natural} := \mathcal{Z}^{\natural} \otimes_{\mathbb{Z}} \mathsf{J}$, 211
$\overline{\Omega}$, set of two-sided cells, 172
ϖ, the embedding $\mathcal{H} \to \mathcal{J}$, 211, 217
a-function, 202, 210, 218
$\imath$-quantum group
  $\sim \mathbf{U}^{\jmath}(n)$, $\mathbf{U}^{\imath}(n)$, 262
  $\sim \mathbf{U}^{\jmath}_{n}$, $\mathbf{U}^{\imath}_{n}$, 258
$\imath$-quantum groups, 258
q-Schur algebra, 160
  $\sim$ of type B, 255
  $\sim$ of type C, 255
  generic $\sim$, 171

cell, 172
  left $\sim$, 172
  right $\sim$, 172
  two-sided $\sim$, 172
cell filtration, 181
  dual left $\sim$, 189
  left $\sim$, 181
cell module
  dual left $\sim$, S_{ω}, 172
  left $\sim$, $S(\omega)$, 172, 180, 275
Coxeter system, 167
  standard finite $\sim$, 168
  weighted $\sim$, 168

decomposition map, 233, 276
decomposition matrix, 233
  unitriangular $\sim$, 234

exact category, 174
  $\sim (\mathscr{A}(\mathscr{L}), \mathscr{E}(\mathscr{L}))$, 181
  $\sim (\mathscr{A}(\mathscr{L}), \mathscr{E}^{\flat}(\mathscr{L}))$, 182
  $\sim (\mathscr{A}^{\flat}(\mathscr{L}), \mathscr{E}^{\flat}(\mathscr{L}))$, 183
  dual $\sim (\mathscr{A}_{\flat}(\mathscr{L}^{*}), \mathscr{E}_{\flat}(\mathscr{L}^{*}))$, 189

finite group of Lie type, 166
  Lie type series of $\sim$, 167
  quasi-split $\sim$, 167
  split $\sim$, 167

Gelfand-Graev module, 267
generic Hecke endo-algebra, 171
  $\sim A$, 196
  extended $\sim A^{+}$, 196
  multiply extended $\sim A^{\ddagger}$, 196

Harish-Chandra induction, 266
Harish-Chandra restriction, 266
Hecke algebra, 170
  generic $\sim$, 171
height filtration, 180, 272
height function, 180, 271
heredity ideal, 221
  $\sim$ of semisplit type, 222
  $\sim$ of separable type, 222
  $\sim$ of split type, 221
  indecomposable split $\sim$, 222, 228

Laurent triple, 280
local triple, 233
  DVR triple, 233
  regular $\sim$, 233

Mackey decomposition, 266

preorder, 273
  dominates, 273
  strictly dominates, 198, 273
preorder $\preceq$, 198

quasi-hereditary algebra, 221
  split $\sim$, 221
quasi-poset, 176, 273
  poset associated with $\sim$, 176, 273

Rouquier's isomorphism, 222

standard basis, 235
standard module, 235
standardly based algebra, 234
standardly full-based algebra, 235
standardly stratified algebra, 176
Steinberg module, 266
stratification hypothesis, 177
stratifying system, 176

trace of a module, 206

unipotent representation, 244

School of Mathematics and Statistics, University of New South Wales, Sydney, New South Wales 2052, Australia
Email address: j.du@unsw.edu.au

Department of Mathematics, University of Virginia, Charlottesville, Virginia 22903

Department of Mathematics, University of Virginia, Charlottesville, Virginia 22903
Email address: lls2l@virginia.edu

Proceedings of Symposia in Pure Mathematics
Volume **108**, 2024
https://doi.org/10.1090/pspum/108/01963

On two constructions of Gelfand-Tsetlin modules with arbitrary characters

Vyacheslav Futorny, Dimitar Grantcharov, Luis Enrique Ramirez, and Pablo Zadunaisky

ABSTRACT. Recently two different constructions of Gelfand-Tsetlin modules with arbitrary character and full support were introduced. In this paper we prove that the two constructions are equivalent.

1. Introduction

In the last few years, two different constructions of Gelfand-Tsetlin modules with arbitrary characters and full support have been provided - one by Ramirez and Zadunaisky in [**RZ18**] and one by Early, Mazorchuk and Vishnyakova in [**EMV20**]. For the first construction, one first defines a certain "big tableaux module" which after specializing at a fixed character leads to the desired module. This construction relies on the celebrated tableaux realization of simple finite-dimensional representations of $\mathfrak{gl}(n, \mathbb{C})$ introduced by Gelfand and Tsetlin in [**GC50**], and more precisely on the generalization of this realization in terms of derivative tableaux introduced in [**FGR16**]. The second construction is based on the realization of $\mathfrak{gl}(n, \mathbb{C})$ as an algebra of operators over the linear dual of the OGZ algebra Γ (see §2.3 for details). It then uses Demazure operators to describe the module containing the desired character χ, which itself is an element of Γ^*.

The main result of this paper is Theorem 3.4, which provides an explicit isomorphism between the two constructions. The comparison is obtained as follows: it is shown that every OGZ algebra is contained in an affine nilHecke algebra, while the big tableaux module is shown to be a sub OGZ-module of this larger algebra. Once the isomorphism is established, an evaluation at a point leads to the second construction, while specialization at that point gives the first one. It is easy to see that these operations give isomorphic modules. Using this isomorphism, and the computation of the matrix coefficients for the EMV constructionin [**FGRZ20**], one obtains matrix coefficients for the tableaux modules.

Throughout the document we work over $\mathbb{C}$. Given $n \in \mathbb{N}$ we set $[\![n]\!] = \{1, 2, \ldots, n\}$.

2020 *Mathematics Subject Classification.* Primary 16G99, 17B10.

Key words and phrases. Gelfand-Tsetlin modules, Gelfand-Tsetlin bases, nilHecke algebras.

The third author was supported in part by CNPq (316163/2021-0). The second author was supported in part by Simons Collaboration Grant 855678. The fourth author is a CONICET researcher and was supported in part by BID PICT 0099/2019.

2. OGZ algebras

2.1. Root systems. Let $n \in \mathbb{Z}_{\geq 2}$, and let $\mathbf{v} = (v_1, \ldots, v_n) \in (\mathbb{Z}_{>0})^n$. Set $\Omega := \Omega(n, \mathbf{v}) = \{(k,i) \mid k \in [\![n]\!], i \in [\![v_k]\!]\}$ and $V = \mathbb{C}^{\Omega}$. We use the unusual notation $[k,i]$ for the unique vector of V whose (k,i)-coordinate is 1 and all other coordinates are zero. Thus $\{[k,i] \mid (k,i) \in \Omega\}$ is a basis of V, and we denote the corresponding dual basis of V^* by $\{x_{k,i} \mid (k,i) \in \Omega\}$. Set $\mathsf{Z} \subset V$ to be the abelian group spanned by $\{[k,i] \mid k \in [\![n-1]\!], i \in [\![v_k]\!]\}$.

Inside V^* we have the root system

$$\Phi = \{x_{k,i} - x_{k,j} \mid k \in [\![n-1]\!],\ i,j \in [\![v_k]\!](i \neq j)\}$$

which is a decomposable root system of type $A_{v_1} \times \cdots \times A_{v_{n-1}}$. The set of positive roots Φ^+ is given by the roots $x_{k,i} - x_{k,j}$ with $i < j$, and the corresponding standard basis is denoted by Σ. We denote by W the corresponding Coxeter group, which we identify with $S_{v_1} \times \cdots \times S_{v_{n-1}}$ in the natural way.

For each $v \in V$ we set

$$\Phi(v) = \{\alpha \in \Phi \mid \alpha(v) \in \mathbb{Z}\};$$
$$\Phi_0(v) = \{\alpha \in \Phi \mid \alpha(v) = 0\}.$$

We define a set of positive roots by $\Phi(v)^+ = \Phi(v) \cap \Phi^+$. We denote by $W(v)$ the Weyl group of the root system $\Phi(v)$. The Weyl group of $\Phi_0(v)$ is then $W_v = W(v)_v$. Given an element $z \in \mathsf{Z}$ we denote by $W(v)^z$ the set of minimal length representatives of cosets in $W(v)/W(v)_z$.

A point $v \in V$ is called *generic* if $\Phi(v) = \emptyset$, i.e. if $v_{k,i} - v_{k,j} \notin \mathbb{Z}$ for all $i,j \in [\![v_k]\!]$ with $i \neq j$, and all $k \in [\![n-1]\!]$, otherwise it is called *singular*. It is *critical* if $\Phi_0(v)$ is not empty, i.e. if $v_{k,i} = v_{k,j}$ for some k and some $i,j \in [\![v_k]\!]$ with $i \neq j$, and *fully critical* if $\Phi(v) = \Phi_0(v)$. It is *positive* if $\alpha \in \Phi(v)^+$ if and only if $\alpha(v) \in \mathbb{Z}_{\geq 0}$. Finally, we say v is in *normal form* if $\Phi(v)$ is a parabolic subsystem of Φ i.e. $\Sigma \cap \Phi(v)$ is a base of $\Phi(v)$. For every $v \in V$ there exists $\sigma \in W$ such that $\sigma(v)$ is in normal form and positive, and $z \in \mathsf{Z}$ such that $\sigma(v) + z$ is fully critical and in normal form.

The group $W \times S_{v_n}$ acts on Ω by permutations in the obvious way, and hence on V. The restriction of this action to W clearly coincides with the one described above.

2.2. Affine nilHecke algebras. Set $\Lambda_{\mathbf{v}} = \mathbb{C}[x_{k,i} \mid (k,i) \in \Omega]$, and denote by L the fraction field of $\Lambda_{\mathbf{v}}$. The group $W \times S_{v_n}$ acts on $\Lambda_{\mathbf{v}}$ and on L by algebra automorphisms, and we set $\Gamma_{\mathbf{v}} = \Lambda_{\mathbf{v}}^{W \times S_{v_n}}$.

Let $\alpha \in \Sigma$ and let s be the corresponding simple transposition. The Demazure operator ∂_s on $\Lambda_{\mathbf{v}}$ is defined as $\partial_s(f) = \frac{1}{\alpha}(f - s(f))$ [**KK86**, section 4]. If $\sigma \in W$ is an arbitrary element, and $\sigma = st \cdots u$ is a reduced decomposition then the operator $\partial_\sigma = \partial_s \circ \partial_t \circ \cdots \circ \partial_u$ is independent of the chosen decomposition.

The affine nilHecke algebra associated to Φ (see [**KK86**, section 4]) is the algebra $\mathcal{H}_W \subset \operatorname{End}_{\mathbb{C}}(\Lambda_{\mathbf{v}})$ generated by all Demazure operators along with $\Lambda_{\mathbf{v}}$, which we see as acting on itself by multiplication. It is free both as a left and right $\Lambda_{\mathbf{v}}$-module, with the Demazure operators forming a basis. The center of $\mathcal{H}_W$ is $\Gamma_{\mathbf{v}}$, and so $\mathcal{H}_W$ is free over its center with rank $|W|^2$. Each Demazure operator is $\Gamma_{\mathbf{v}}$-linear, and $\mathcal{H}_W$ coincides with $\operatorname{End}_{\Gamma_{\mathbf{v}}}(\Lambda_{\mathbf{v}})$. In particular $W \subset \mathcal{H}_W$, and in fact the smash product $\Lambda_{\mathbf{v}} \# \mathbb{C}[W]$ is a subalgebra of $\mathcal{H}_W$.

The polynomial $\Delta = \prod_{\alpha \in \Phi^+} \alpha$ is normal in both $\Lambda_{\mathbf{v}} \# \mathbb{C}[W]$ and $\mathcal{H}_W$, and there is an isomorphism $\Lambda_{\mathbf{v}}[\Delta^{-1}] \# \mathbb{C}[W] \cong \mathcal{H}_W[\Delta^{-1}]$. Let M be a $\Lambda_{\mathbf{v}}[\Delta^{-1}]$-module with an equivariant W-action. Then M is an $\mathcal{H}_W$-module, in particular we can apply Demazure operators to its elements.

2.3. Orthogonal Gelfand-Zetlin subalgebras. For each $z \in \mathsf{Z}$ we denote by t_z the unique algebra automorphism of $\Lambda_{\mathbf{v}}$ such that $t_z(x_{k,i}) = x_{k,i} + z_{k,i}$. This defines an action of Z on $\Lambda_{\mathbf{v}}$ by automorphisms. Each map t_z extends to an automorphism of L, and hence we can form the smash product $L \# \mathsf{Z}$. The obvious map $L \# V \longrightarrow \operatorname{End}_{\mathbb{C}}(L)$ is an embedding [**FGRZ20**, Lemma 4.1].

Let A be the localization of $\Lambda_{\mathbf{v}}$ at the multiplicatively closed set generated by $\{\alpha - d \mid \alpha \in \Phi, d \in \mathbb{Z}\}$. Then A is stable by the automorphisms t_z, and we can form the smash product $A \# \mathsf{Z}$, which is a subalgebra of $L \# \mathsf{Z}$. It is a free A-module with basis $\{t_z \mid z \in \mathsf{Z}\}$, and embedds into $\operatorname{End}_{\mathbb{C}}(A)$. Notice that both $A \# \mathsf{Z}$ and $\operatorname{End}_{\mathbb{C}}(A)$ are $\mathcal{H}_W$-modules, and the embedding is a map of $\mathcal{H}_W$-modules.

The Orthogonal Gelfand-Zeitlin algebra (OGZ algebra for short) associated to $\mathbf{v}$, denoted $U_{\mathbf{v}}$, is the subalgebra of $A \# \mathsf{Z}$ generated by $\Gamma_{\mathbf{v}}$ and the operators

$$E_k = \sum_{i=1}^{v_k} e_{k,i} t_{+[k,i]}; \qquad F_k = \sum_{i=1}^{v_k} f_{k,i} t_{-[k,i]}; \qquad k \in [\![n-1]\!],$$

where

$$e_{k,i} = -\frac{\prod_{j=1}^{v_{k+1}} (x_{k,i} - x_{k+1,j})}{\prod_{j \neq i} (x_{k,i} - x_{k,j})}; \qquad f_{k,i} = \frac{\prod_{j=1}^{v_{k-1}} (x_{k,i} - x_{k-1,j})}{\prod_{j \neq i} (x_{k,i} - x_{k,j})}.$$

The definition is taken from [**EMV20**, subsection 2.2] (also, see Remark 3.3). The subalgebra $\Gamma_{\mathbf{v}} \subset U_{\mathbf{v}}$ is called the *Gelfand-Tsetlin* subalgebra of $U_{\mathbf{v}}$.

The algebra $A \# \mathsf{Z}$ has an antiautomorphism, which we denote by $\dagger$, given by $f^\dagger = f$ and $t_z^\dagger = t_{-z}$ for every $f \in A$ and $z \in \mathsf{Z}$. Clearly $\dagger$ is its own inverse. We identify $U_{\mathbf{v}}^\dagger$ with $U_{\mathbf{v}}^{\mathsf{op}}$. By definition $\Gamma_{\mathbf{v}}$ is fixed pointwise by $\dagger$ so $U_{\mathbf{v}}^\dagger$ also contains $\Gamma_{\mathbf{v}}$.

2.4. OGZ algebras and Demazure operators. We now introduce some combinatorial notions that, along with the $\mathcal{H}_W$-module structure of $A \# \mathsf{Z}$, will allow us to give new presentations for $U_{\mathbf{v}}$. An *interval* in Ω is a set of the form $I = \{(k,a), (k,a+1), \ldots, (k,b) \mid 1 \leq a < b \leq v_k\}$. We set $a(I) = a$ and $b(I) = b$, and denote by $\alpha(I), \beta(I) \in S_{v_k}$ the permutations $\alpha(I) = (b\ b-1\ \ldots\ a)$ and $\beta(I) = (a\ a+1\ \ldots\ b)$. We regard these as elements of W in the obvious way.

Given an interval $I \subset \Omega$ we also set

$$e_{k,I} = -\frac{\prod_{j=1}^{v_{k+1}} \left(x_{k,a(I)} - x_{k+1,j}\right)}{\prod_{(k,j) \notin I} \left(x_{k,a(I)} - x_{k,j}\right)}; \qquad f_{k,I} = \frac{\prod_{j=1}^{v_{k-1}} \left(x_{k,b(I)} - x_{k-1,j}\right)}{\prod_{(k,j) \notin I} \left(x_{k,b(I)} - x_{k,j}\right)}.$$

Any $v \in V$ defines an equivalence relation in Ω, where $(k,i) \sim_v (k,j)$ if and only if $v_{k,i} - v_{k,j} \in \mathbb{Z}$, or in other words $x_{k,i} - x_{k,j} \in \Phi(v)$. If v is in normal form then each equivalence class is an interval. Given v in normal form we set $\overline{\Omega}(v) = \Omega / \sim_v$, and denote by $\overline{\Omega}(v)[k]$ the equivalence classes contained in $\{(k,i) \mid i \in [\![v_k]\!]\}$. It follows from [**EMV20**, Proposition 4.1] that we can write

$$E_k = \sum_{I \in \overline{\Omega}(v)[k]} \partial_{\alpha(I)}(e_{k,I} \circ t_{[k,a(I)]}); \qquad F_k = \sum_{I \in \overline{\Omega}(v)[k]} \partial_{\beta(I)}(f_{k,I} \circ t_{-[k,b(I)]}).$$

3. Representations of OGZ algebras

3.1. Gelfand-Tsetlin modules. Evaluation at a point $v \in V$ defines a character $\mathsf{ev}_v : \Lambda_{\mathbf{v}} \longrightarrow \mathbb{C}$, which in turn restricts to a character of $\Gamma_{\mathbf{v}}$. Of course if $w = \sigma(v)$ for some $\sigma \in W \times S_{v_n}$ the characters ev_v and ev_w restrict to the same character of $\Gamma_{\mathbf{v}}$. Given a $U_{\mathbf{v}}$-module M and $v \in V$ we denote by $M[v]$ the subspace formed by the generalized eigenvectors for the action of $\Gamma_{\mathbf{v}}$ with eigenvalue ev_v.

A Gelfand-Tsetlin module over an OGZ algebra $U_{\mathbf{v}}$ is a (left or right) $U_{\mathbf{v}}$-module M such that the action of $\Gamma_{\mathbf{v}}$ is locally finite on M, that is $\dim \Gamma_{\mathbf{v}} m < \infty$ for any $m \in M$. Equivalently, a module M is a Gelfand-Tsetlin module if and only if it is the direct sum of the spaces $M[v]$ with v running over a set of representatives of $V/W \times S_{v_n}$. It follows from [**FO14**] that for every $v \in V$ there exists a Gelfand-Tsetlin module M with $M[v] \neq 0$. We will now review the explicit constructions of such modules given in [**RZ18, EMV20**], and show that they coincide. We point out that the setup of [**RZ18**] assumes $\mathbf{v} = (1, 2, \dots, n)$ where $U_{\mathbf{v}}^{\dagger} \cong \mathcal{U}(\mathfrak{gl}(n, \mathbb{C}))$; as pointed out below, the construction easily generalizes.

We introduce some more notation. Given $v \in V$ which is fully critical and in normal form we set

$$\mathsf{Z}(v)_{\geq 0} = \{z \in \mathsf{Z} \mid \alpha(z) \geq 0 \text{ for all } \alpha \in \Phi(v)^+\}.$$

In other words, $\mathsf{Z}(v)$ is the set of points in the integral lattice Z lying in the positive Weyl chamber of the root system $\Phi(v)$. We also set $\Pi(v) = \Phi \setminus \Phi(v)$, i.e. the set of all roots such that $\alpha(v) \neq 0$, and write $A(v) = \Pi(v)^{-1}\Lambda_{\mathbf{v}} \subset A$. This is a subalgebra of A consisting of rational functions that can be evaluated at v. Thus $A(v)$ has a 1-dimensional module $\mathbb{C}_v$ given by evaluation at v. We denote by 1_v the canonical generator of this module.

3.2. The EMV construction. Since $U_{\mathbf{v}} \subset A\#\mathsf{Z}$, the algebra A is naturally a left $U_{\mathbf{v}}$-module. The subalgebra $\Gamma_{\mathbf{v}} \subset U_{\mathbf{v}}$ acts on A by multiplication. It is a remarkable fact that $\Gamma_{\mathbf{v}}$ is a $U_{\mathbf{v}}$-submodule of A, and hence $\Gamma_{\mathbf{v}}^*$ is a right $U_{\mathbf{v}}$-module. This was used by Early, Mazorchuk and Vishnyakova in [**EMV20**] to produce a Gelfand-Tsetlin module M containing an arbitrary character χ in its support. Since $\chi \in \Gamma_{\mathbf{v}}^*$, it is natural to consider the submodule generated by this character, but the structure of this module is not well-understood in general. There is, however, a larger submodule which is easier to describe as shown in the next theorem (for the relevant definitions see §2.1).

THEOREM 3.1 ([**EMV20**, Theorem 4.4]). *Let v be fully critical and in normal form. The set*

$$\mathbf{B}(v) = \{\mathsf{ev}_v \circ \partial_\sigma \circ t_z \mid z \in \mathsf{Z}(v)_{\geq 0} \text{ and } \sigma \in W(v)^z\}$$

is free over $\mathbb{C}$, and its span M_v is a right $U_{\mathbf{v}}$-submodule of $\Gamma_{\mathbf{v}}^*$.

Since $M_v[v+z] \neq 0$ for any z, this solves the problem of finding modules with arbitrary characters in their support. We will refer to M_v as the *EMV construction* at v. The structure of the module M_v was further analyzed in [**FGRZ20**], and from the results found in section 8 of this reference we obtain the matrix coefficients for the action of $U_{\mathbf{v}}$ on the basis $\mathbf{B}(v)$.

THEOREM 3.2. *The action of the canonical generators of $U_{\mathbf{v}}$ on M_v is given by the formulas*

$$(\mathsf{ev}_v \circ \partial_\sigma \circ t_z)E_k = - \sum_{I\in\overline{\Omega}(v)[k]} \sum_{\tau\leq\sigma\alpha(I)} \mathfrak{D}^{v+z}_{\tau,\sigma\alpha(I)}(e_{k,I})(\mathsf{ev}_v \circ \partial_\tau \circ t_{z+[k,a(I)]}),$$

$$(\mathsf{ev}_v \circ \partial_\sigma \circ t_z)F_k = \sum_{I\in\overline{\Omega}(v)[k]} \sum_{\tau\leq\sigma\beta(I)} \mathfrak{D}^{v+z}_{\tau,\sigma\beta(I)}(f_{k,I})(\mathsf{ev}_v \circ \partial_\tau \circ t_{z-[k,b(I)]}),$$

$$(\mathsf{ev}_v \circ \partial_\sigma \circ t_z)\gamma = \sum_{\tau<\sigma} \mathfrak{D}^{v+z}_{\tau,\sigma}(\gamma)(\mathsf{ev}_v \circ \partial_\tau \circ t_z).$$

for each $\gamma \in \Gamma_{\mathbf{v}}$. Here $\mathfrak{D}_{\tau,\sigma}$ is the Postnikov-Stanley operator introduced in [**FGRZ20**, Definition 3.1].

3.3. The universal tableaux module. Set $\mathcal{B} = \{\mathcal{T}(x+z) \mid z \in \mathsf{Z}\}$, on which W acts by permutation of z. The symbol $\mathcal{T}(x+z)$ is meant to represent a Gelfand-Tsetlin tableaux with generic entries x shifted by z, but for the moment is purely formal. The group W also acts on A, and hence on the free A-module spanned by $\mathcal{B}$ which we denote $\mathbb{V}_A$. This module has a right $U_{\mathbf{v}}$-module structure, analogous to the one introduced in [**RZ18**, 3.4 Proposition]; the simplest way to describe this structure is to use the isomorphism $\mathbb{V}_A \cong A\#\mathsf{Z}$, and use that $A\#\mathsf{Z}$ is naturally a right $U_{\mathbf{v}}$-module.

REMARK. The module introduced in [**RZ18**, 3.4 Proposition] is actually a *left* $U^\dagger_{\mathbf{v}}$-module in the special case where $\mathbf{v} = (1,2,\ldots,n)$; as we have already mentioned, in this case $U^{\mathsf{op}}_{\mathbf{v}}$ is isomorphic to the enveloping algebra of $\mathfrak{gl}(n,\mathbb{C})$. Since $U^\dagger_{\mathbf{v}} \cong U^{\mathsf{op}}_{\mathbf{v}}$, this defines a right $U_{\mathbf{v}}$-module structure on $\mathbb{V}_A$.

Given $\sigma \in W(v)$ we denote by $\mathcal{D}_\sigma$ the element $\mathrm{sym}_{W(v)}\,\partial_\sigma \in \mathcal{H}_W$, where $\mathrm{sym}_{W(v)} = \frac{1}{|W(v)|}\sum_{\tau\in W(v)}\tau \in \mathbb{C}[W] \subset \mathcal{H}_W$. Since $\mathbb{V}_A \cong A\#\mathsf{Z}$ is an $\mathcal{H}_W$-module we can apply the operator $\mathcal{D}_\sigma = \mathrm{sym}_{W(v)}\,\partial_\sigma$ to the element $\mathcal{T}(x{+}z)$. The arguments from [**RZ18**, sections 4 and 5] can be adapted to show the following.

THEOREM 3.3. *Let v be fully critical and in normal form. The set*

$$\mathcal{B}(v) = \{\mathcal{D}_\sigma(\mathcal{T}(x+z)) \mid z \in \mathsf{Z}(v)_{\geq 0} \text{ and } \sigma \in W(v)^z\}$$

is free over A, and its $A(v)$-span $\mathcal{L}(v)$ is a right $U_{\mathbf{v}}$-submodule of $\mathbb{V}_A$.

Thus for each fully critical $v \in V$ in normal form we obtain a right $U_{\mathbf{v}}$-module $\mathbb{V}(T(v)) = \mathbb{C}_v \otimes_{A(v)} \mathcal{L}(v)$, which we call the *universal tableaux module at v*. The family $\{D_\sigma(v+z) = 1_v \otimes_{A(v)} \mathcal{D}_\sigma(\mathcal{T}(x+z)) \mid z \in \mathsf{Z}(v)_{\geq 0}, \sigma \in W(v)^z\}$ is a basis of $\mathbb{V}(T(v))$, and its elements are called *derivative tableaux*. Each derivative tableau is a generalized $\Gamma_{\mathbf{v}}$ eigenvector with eigenvalue ev_{v+z}, so this also solves the problem of finding Gelfand-Tsetlin modules with an arbitrary character in its support.

REMARK. The OGZ algebras were originally introduced by Mazorchuk in [**Maz99**] as algebras of operators acting on the left on spaces with tableaux-like bases. These modules can be identified with the big tableaux modules at generic points. Thus OGZ algebras as originally defined naturally identify with $U^{\mathsf{op}}_{\mathbf{v}} \cong U^\dagger_{\mathbf{v}}$.

3.4. Isomorphism of M_v and $\mathbb{V}(T(v))$. We now show that the EMV construction and the universal tableaux module at v are isomorphic. The link is given by the reinterpretation of the module $\mathbb{V}_A$ as the smash product $A\#\mathsf{Z}$.

THEOREM 3.4. *Let $v \in V$ be fully critical and in normal form. The modules M_v and $\mathbb{V}(T(v))$ are isomorphic as right $U_{\mathbf{v}}$-modules.*

PROOF. As mentioned earlier, the algebra $A\#\mathsf{Z}$ embeds in $\mathrm{End}_{\mathbb{C}}(A)$ and the embedding is a map of left $A\#W$-modules. Since $\mathcal{H}_W \cong A\#W$, it follows that we can act by divided differences on $\mathbb{C}$-linear endomorphisms of A. With this in mind $\mathcal{L}(v)$ can be realized as the $A(v)$-submodule of $\mathrm{End}_{\mathbb{C}}(A)$ spanned by the operators $(\mathrm{sym}_{W(v)}\, \partial_\sigma) \cdot t_z$ with $z \in \mathsf{Z}(v)$ and $\sigma \in W(v)^z$.

Now, an induction on the length of σ shows that $(\mathrm{sym}_{W(v)}\, \partial_\sigma \cdot t_z) \in \mathrm{End}_{\mathbb{C}}(A)$ maps a symmetric polynomial $\gamma \in \Gamma_{\mathbf{v}}$ to $\mathrm{sym}_{W(v)}(\partial_\sigma(\gamma(x+z))) \in \Lambda_{\mathbf{v}}$, and hence the endomorphisms in $\mathcal{L}(v)$ map $\Gamma_{\mathbf{v}}$ into $A(v)$. We thus have a commutative diagram of right $U_{\mathbf{v}}$-modules

$$\begin{array}{ccccc}
A\#Z & \hookrightarrow & \mathrm{End}_{\mathbb{C}}(A) & \xrightarrow{\mathsf{res}} & \mathrm{hom}_{\mathbb{C}}(\Gamma_{\mathbf{v}}, A) \\
\uparrow & & & & \uparrow \\
\mathcal{L}(v) & & \longrightarrow & & \mathrm{hom}_{\mathbb{C}}(\Gamma_{\mathbf{v}}, A(v)) \\
\downarrow & & & & \downarrow{\scriptstyle \mathsf{ev}_v \circ -} \\
\mathbb{V}(T(v)) & & \longrightarrow & & \Gamma_{\mathbf{v}}^*
\end{array}$$

Using standard diagram chasing we show that the bottom map sends $D_\sigma(v+z) \in \mathbb{V}(T(v))$ to the linear functional $\mathsf{ev}_v \circ \mathrm{sym}_{W(v)} \circ \partial_\sigma \circ t_z \in \Gamma_{\mathbf{v}}^*$. Since $\mathsf{ev}_v \circ \mathrm{sym}_{W(v)} = \mathsf{ev}_v$, the basis of derivative tableaux of $\mathbb{V}(T(v))$ maps to the basis $\mathbf{B}(v)$ of the module M_v, so the bottom map restricts to an isomorphism of $U_{\mathbf{v}}$-modules $\mathbb{V}(T(v)) \cong M_v$. $\square$

3.5. The enveloping algebra of $\mathfrak{gl}(n, \mathbb{C})$ as an OGZ algebra. Take $\mathbf{v} = (1, 2, \ldots, n)$. The A-module $\mathbb{V}_A$ is then isomorphic to the module V_A introduced in [**Zad17**, Proposition 1], with t_z sent to the symbol $T(z)$ for all $z \in \mathbb{Z}$. As we have seen $\mathbb{V}_A$ is a left $U_{\mathbf{v}}^\dagger$-module, while V_A is a left $\mathfrak{gl}(n, \mathbb{C})$-module. The isomorphism intertwines the action of the elements $E_k^\dagger, F_k^\dagger$ and $H_k^\dagger$ with those of $E_{k,k+1}, E_{k+1,k}$ and $E_{k,k}$, respectively. Both representations are faithful, and hence it follows that there is an algebra isomorphism $\mathcal{U}(\mathfrak{gl}(n, \mathbb{C})) \longrightarrow U_{\mathbf{v}}^\dagger$ given by $E_{k,k+1} \mapsto E_k^\dagger, E_{k+1,k} \mapsto F_k^\dagger, E_{k,k} \mapsto H_k^\dagger$.

Several variants of this embedding appear in the literature. Ours coincides with those of [**DFO94, Maz99, RZ18**], but seems to be different from the ones in [**FO10, FO14**]. These variants can be reconciled through various automorphisms or antiautomorphisms of U, and variations of the Gelfand-Tsetlin formulas; for these last point see [**Gra04**]. Notice in particular that any antiautomorphism of $\mathfrak{gl}(n, \mathbb{C})$ gives us an isomorphism $U \cong U_{\mathbf{v}}$.

References

[DFO94] Yu. A. Drozd, V. M. Futorny, and S. A. Ovsienko, *Harish-Chandra subalgebras and Gel′fand-Zetlin modules*, Finite-dimensional algebras and related topics (Ottawa, ON,

1992), NATO Adv. Sci. Inst. Ser. C: Math. Phys. Sci., vol. 424, Kluwer Acad. Publ., Dordrecht, 1994, pp. 79–93, DOI 10.1007/978-94-017-1556-0_5. MR1308982

[EMV20] N. Early, V. Mazorchuk, and E. Vishnyakova, *Canonical Gelfand-Zeitlin modules over orthogonal Gelfand-Zeitlin algebras*, Int. Math. Res. Not. IMRN **20** (2020), 6947–6966, DOI 10.1093/imrn/rnz002. MR4172674

[FGR16] V. Futorny, D. Grantcharov, and L. E. Ramirez, *Singular Gelfand-Tsetlin modules of* $\mathfrak{gl}(n)$, Adv. Math. **290** (2016), 453–482, DOI 10.1016/j.aim.2015.12.001. MR3451929

[FGRZ20] V. Futorny, D. Grantcharov, L. E. Ramirez, and P. Zadunaisky, *Gelfand-Tsetlin theory for rational Galois algebras*, Israel J. Math. **239** (2020), no. 1, 99–128, DOI 10.1007/s11856-020-2048-2. MR4160883

[FO10] V. Futorny and S. Ovsienko, *Galois orders in skew monoid rings*, J. Algebra **324** (2010), no. 4, 598–630, DOI 10.1016/j.jalgebra.2010.05.006. MR2651560

[FO14] V. Futorny and S. Ovsienko, *Fibers of characters in Gelfand-Tsetlin categories*, Trans. Amer. Math. Soc. **366** (2014), no. 8, 4173–4208, DOI 10.1090/S0002-9947-2014-05938-2. MR3206456

[GC50] I. M. Gel′fand and M. L. Cetlin, *Finite-dimensional representations of the group of unimodular matrices* (Russian), Doklady Akad. Nauk SSSR (N.S.) **71** (1950), 825–828. MR35774

[Gra04] M. I. Graev, *Infinite-dimensional representations of the Lie algebra* $\mathfrak{gl}(n,\mathbb{C})$ *related to complex analogs of the Gelfand-Tsetlin patterns and general hypergeometric functions on the Lie group* $\mathrm{GL}(n,\mathbb{C})$, Acta Appl. Math. **81** (2004), no. 1-3, 93–120, DOI 10.1023/B:ACAP.0000024196.16709.f6. MR2069333

[KK86] B. Kostant and S. Kumar, *The nil Hecke ring and cohomology of G/P for a Kac-Moody group G*, Advances in Mathematics **62** (1986), no. 3, 187-237, `https://www.sciencedirect.com/science/article/pii/0001870886901015`, DOI https://doi.org/10.1016/0001-8708(86)90101-5.

[Maz99] V. Mazorchuk, *Orthogonal Gelfand-Zetlin algebras. I*, Beiträge Algebra Geom. **40** (1999), no. 2, 399–415. MR1720114

[RZ18] L. E. Ramírez and P. Zadunaisky, *Gelfand-Tsetlin modules over* $\mathfrak{gl}(n)$ *with arbitrary characters*, J. Algebra **502** (2018), 328–346, DOI 10.1016/j.jalgebra.2018.01.034. MR3774896

[Zad17] P. Zadunaisky, *A new way to construct 1-singular Gelfand-Tsetlin modules*, Algebra Discrete Math. **23** (2017), no. 1, 180–193. MR3634504

Shenzhen International Center for Mathematics, Southern University of Science and Technology, Shenzhen, People's Republic of China; and Instituto de Matemática e Estatística, Universidade de São Paulo, São Paulo, Brasil"
Email address: `futorny@ime.usp.br`

Department of Mathematics, University of Texas at Arlington, Arlington, Texas 76019
Email address: `grandim@uta.edu`

Universidade Federal do ABC, Santo André-SP, Brasil
Email address: `luis.enrique@ufabc.edu.br`

Instituto de Investigaciones Matemáticas Luis A. Santaló (IMAS)
Email address: `pzadub@dm.uba.ar`

Proceedings of Symposia in Pure Mathematics
Volume **108**, 2024
https://doi.org/10.1090/pspum/108/01964

Lie algebra cohomology of the positive part of twisted affine Lie algebras

Jiuzu Hong and Shrawan Kumar

ABSTRACT. The explicit Verlinde formula for the dimension of conformal blocks, attached to a marked projective curve Σ, a simple Lie algebra $\mathfrak{g}$ over $\mathbb{C}$ and integrable highest weight modules of a fixed central charge of the corresponding affine Lie algebra $\hat{L}(\mathfrak{g})$ attached to the marked points, requires (among several other important ingredients) a Lie algebra cohomology vanishing result due to C. Teleman for the positive part $\hat{L}^+(\mathfrak{g})$ with coefficients in the tensor product of an integrable highest weight module with copies of finite dimensional evaluation modules. The aim of this paper is to extend this result of Teleman to a twisted setting where $\mathfrak{g}$ is endowed with a special automorphism σ and the curve Σ is endowed with the action of σ. In this general setting, the affine Lie algebra gets replaced by twisted affine Lie algebras. The crucial ingredient (as in Teleman) is to prove a certain Nakano Identity.

1. Introduction

We take the base field in this paper to be the field of complex numbers $\mathbb{C}$. The dimension of the space of conformal blocks can be computed by the celebrated Verlinde formula (cf. [**Ku2**, Theorem 4.2.19]). This formula was originally conjectured by Verlinde in Conformal Field Theory. It was mathematically derived by combining the efforts of several mathematicians (cf. [**Ku2**, 4.C]). Some of the main ingredients going into its proof consisted of reducing the problem from a genus g curve Σ with any number of marked points to a genus 0 curve (i.e. $\mathbb{P}^1$) with three marked points (by using the Factorization Theorem and the existence of a flat projective connection on the bundle of conformal blocks over the moduli space of curves [**TUY**]). The problem for three marked points on $\mathbb{P}^1$ is encoded in so called the *Fusion Algebra*. Faltings obtained a certain result on the classification of characters of the fusion algebra for the classical simple Lie algebras $\mathfrak{g}$ as well as $\mathfrak{g}$ of type G_2. This allows one to derive the explicit dimension formula for these Lie algebras. Then, Teleman proved the cohomology vanishing result mentioned in the Abstract (see below for a precise statement) for any simple $\mathfrak{g}$, thereby one obtains the explicit dimension formula for the space of conformal blocks for any simple $\mathfrak{g}$.

Let Γ be a finite group acting on a simple Lie algebra $\mathfrak{g}$ and a projective algebraic curve Σ. One may define the space of twisted conformal blocks in this equivariant setting. A generalization of the work of Tsuchiya-Ueno-Yamada [**TUY**]

The first author was partially supported by the NSF grant DMS-2001365 and the second author was partially supported by the NSF grant DMS-1802328.

in this setting was developed in [**Da**, **HK1**]. Subsequently, in [**HK2**] the authors proved that the calculation of the dimension of twisted conformal blocks can be reduced to the situation that Γ acts on $\mathfrak{g}$ by diagram automorphisms. Moreover, an explicit formula for the dimension of the twisted conformal blocks (akin to the classical Verlinde dimension formula) can be derived assuming a twisted analogue of Teleman's vanishing theorem on the Lie algebra (co)homology (cf. [**HK2**, Theorem 6.9]). It should be mentioned that, using the machinery of modular tensor categories, under some restrictions, the dimension formula for the twisted conformal blocks was obtained in [**DM**] earlier. However, our approach to the same problem is quite different. In fact, combining the works of [**HK1**, **HK2**, **DM**], one can derive the dimension formula for the twisted conformal blocks for a general Γ-curve Σ and any Γ action on $\mathfrak{g}$.

Let $\Gamma = \langle\sigma\rangle$ be the cyclic group acting on $\mathbb{P}^1$ with ramified points only at $\{0, \infty\}$, and σ acting on $\mathfrak{g}$ as a special automorphism (cf. Definition 3.5). Let $\hat{L}(\mathfrak{g}, \sigma)$ be the associated twisted affine Lie algebra, the central extension of the twisted loop algebra $(\mathfrak{g} \otimes \mathbb{C}((t)))^\sigma$ and let $L^+(\mathfrak{g}, \sigma) := (t\mathfrak{g}[t])^\sigma$ be its Lie subalgebra (called the positive part of $\hat{L}(\mathfrak{g}, \sigma)$). Let $\mathscr{H}_c(\lambda)$ be the integrable highest weight module of $\hat{L}(\mathfrak{g}, \sigma)$ of level c and highest weight λ. We choose $s + 1$ points $\vec{z}$ in $\mathbb{C} \subset \mathbb{P}^1$ such that their Γ-orbits are distinct, where $\vec{z} = (z_0 = 0, z_1, \ldots, z_s)$. At each point z_i, we attach the finite dimensional irreducible representation $V(\mu_i)$ of the fixed point Lie subalgebra $\mathfrak{g}^{\Gamma_i}$ with highest weight μ_i, where Γ_i is the stabilizer group of Γ at z_i. Thus, with our choice, $V(\mu_0)$ is an irreducible representation of $\mathfrak{g}^\sigma$, whereas $V(\mu_i)$ (for $1 \leq i \leq s$) is an irreducible representation of $\mathfrak{g}$. The choice of z_i gives rise to an evaluation representation $V(\mu_i)_{z_i}$ of $L^+(\mathfrak{g}, \sigma)$ at z_i. Thus, $L^+(\mathfrak{g}, \sigma)$ naturally acts on the tensor product

$$\hat{\mathscr{H}}_c(\lambda) \otimes V(\vec{\mu})_{\vec{z}} := \hat{\mathscr{H}}_c(\lambda) \otimes V(\mu_0)_{z_0} \otimes \cdots \otimes V(\mu_s)_{z_s},$$

where $\hat{\mathscr{H}}_c(\lambda)$ is the direct *product* of the energy eigenspaces of $\mathscr{H}_c(\lambda)$.

Let $H^q(L^+(\mathfrak{g}, \sigma), \hat{\mathscr{H}}_c(\lambda) \otimes V(\vec{\mu})_{\vec{z}})$ be the q-th Lie algebra cohomology of $L^+(\mathfrak{g}, \sigma)$ with coefficients in $\hat{\mathscr{H}}_c(\lambda) \otimes V(\vec{\mu})_{\vec{z}}$. Our main result of this paper is Theorem 3.17, which asserts the following.

THEOREM. *Let σ be a special automorphism of $\mathfrak{g}$ of order m and let $(\mathfrak{g}, m)$ be different from $(D_4, 3)$. For any $q \geq 1$, the Lie algebra cohomology*

$$H^q(L^+(\mathfrak{g}, \sigma), \hat{\mathscr{H}}_c(\lambda) \otimes V(\vec{\mu})_{\vec{z}})^{\mathfrak{g}^\sigma} = 0, \qquad \text{if } d_{\vec{\mu}} < \frac{c + 2\check{h}}{m},$$

and the points $\vec{z} = \{z_0 = 0, z_1, \ldots, z_s\}$ in the open unit disc are far apart in the hyperbolic metric. Here $\check{h}$ is the dual Coxeter number of $\mathfrak{g}$ and $d_{\vec{\mu}}$ is defined by the identity (29).

The above theorem, in the special case σ is the identity automorphism, recovers Teleman's higher cohomology vanishing result (cf. [**Te**, Theorem 0 (c)]).

To prove this vanishing theorem, we generally follow the strategy of Teleman's proof in the untwisted case. However, there is one major difference in that the Möbius transformations do not preserve the twisted affine Lie algebras (unlike the untwisted affine case). In the untwisted case, this allowed Teleman to reduce the problem of any marked point z in $\mathbb{C}$ to the point 0, which made some of the computations much easier to deal with. We establish a twisted analogue of Nakano identity in Theorems 3.11 and 3.12. The proof of this identity requires a cumbersome computation of Laplacian operators, which is carried out in Section 4. Some results on Casimir operators on graded Lie algebras in Section 3.2, especially Proposition 3.8 is used crucially in Section 4. We would like to stress that, in our above vanishing theorem, the restriction on $\vec{\mu}$ is not optimal.

As a corollary of the above theorem, we get the following (cf. Corollary 3.20).

COROLLARY. *Under the assumptions of the above theorem, for any $q \geq 1$ we have*

$$H_q(L^-(\mathfrak{g},\sigma^{-1}), \mathscr{H}_c(\lambda^*) \otimes V(\vec{\mu}^*)_{\vec{z}^{-1}})^{\mathfrak{g}^\sigma} = 0, \qquad \text{if } d_{\vec{\mu}} < \frac{c+2\check{h}}{m},$$

where $\vec{z}^{-1} = (\infty, z_1^{-1}, \cdots, z_s^{-1})$, $\mathscr{H}_c(\lambda^*)$ is the highest weight module of the twisted affine Lie algebra $\hat{L}(\mathfrak{g},\sigma^{-1})$, and $V(\vec{\mu}^*)_{\vec{z}^{-1}} := V(\mu_0^*)_\infty \otimes \cdots \otimes V(\mu_s^*)_{z_s^{-1}}$ represents the evaluation representation of $L^-(\mathfrak{g},\sigma^{-1})$ in the usual sense.

The above corollary implies the analogue of the Verlinde dimension formula for twisted conformal blocks on $\mathbb{P}^1$ with three marked points under the restriction on $\vec{\mu}$ and σ as in the above corollary (cf. [**HK2**, Theorem 6.9]).

2. Preliminaries

Let $\mathfrak{g}$ be a simple Lie algebra over $\mathbb{C}$. Let σ be an automorphism of order m of $\mathfrak{g}$. Let $\mathcal{K}$ be the field of Laurent series in the parameter t, such that $\sigma(t) = \epsilon^{-1}t$ where $\epsilon = e^{\frac{2\pi i}{m}}$ and σ acts on $\mathbb{C}$ trivially. Let $\mathcal{O}$ be the field of formal power series in t. We now define a central extension $\hat{L}(\mathfrak{g},\sigma) := \mathfrak{g}(\mathcal{K})^\sigma \oplus \mathbb{C}C$ of $\mathfrak{g}(\mathcal{K})^\sigma$ under the bracket

$$(1) \quad [x[P] + zC, x'[P'] + z'C] = [x,x'][PP'] + m^{-1}\,\mathrm{Res}_{t=0}\big((dP)P'\big)\langle x, x'\rangle C,$$

for $x[P], x'[P'] \in \mathfrak{g}(\mathcal{K})^\sigma$, $z, z' \in \mathbb{C}$; where $\mathrm{Res}_{t=0}$ denotes the coefficient of $t^{-1}dt$ and $\langle\, ,\, \rangle$ denotes the normalized invariant form on $\mathfrak{g}$ so that the induced form on $\mathfrak{g}^*$ satisfies $\langle \theta, \theta \rangle = 2$ for the highest root θ of $\mathfrak{g}$.

Throughout the paper, we fix a positive integer (called the level) $c > 0$. We also fix an integer $s > 0$ denoting the number of marked points.

We use $D_{c,\sigma}$ to denote the set of highest weights of $\mathfrak{g}^\sigma$ which parametrizes the integrable highest weight modules of $\hat{L}(\mathfrak{g},\sigma)$ of level c, where the level denotes the action of C, see [**HK1**, Section 2]. When σ is trivial, we also use D_c to denote this set for brevity. For each $\lambda \in D_{c,\sigma}$, we will denote by $(\mathscr{H}_c(\lambda), \rho_\lambda)$ (or for simplicity $\mathscr{H}_c(\lambda)$) the associated integrable highest weight module of $\hat{L}(\mathfrak{g},\sigma)$ of level c.

There exists a 'compatible' Cartan subalgebra $\mathfrak{h}$ and a 'compatible' Borel subalgebra $\mathfrak{b} \supset \mathfrak{h}$ of $\mathfrak{g}$ both stable under the action of σ such that

$$(2) \qquad \sigma = \tau\epsilon^{\mathrm{ad}h},$$

where τ is a (possibly trivial) diagram automorphism of $\mathfrak{g}$ of order r preserving $\mathfrak{h}$ and $\mathfrak{b}$, $\alpha(h) \in \mathbb{Z}$ for any root α of $\mathfrak{g}$ and $\epsilon^{\mathrm{ad}h}$ is the inner automorphism of $\mathfrak{g}$ such that for any root α of $\mathfrak{g}$, $\epsilon^{\mathrm{ad}h}$ acts on the root space $\mathfrak{g}_\alpha$ by the multiplication $\epsilon^{\alpha(h)}$, and $\epsilon^{\mathrm{ad}h}$ acts on $\mathfrak{h}$ by the identity. Here h is an element in $\mathfrak{h}^\tau$. In particular, τ and $\epsilon^{\mathrm{ad}h}$ commute. Moreover, r divides m, $\alpha(h) \in \mathbb{Z}^{\geq 0}$ for any positive root α of $\mathfrak{g}^\tau$

and $\theta_0(h) \leq \frac{m}{r}$ where $\theta_0 \in (\mathfrak{h}^\tau)^*$ denotes the following weight of $\mathfrak{g}^\tau$:

$$\theta_0 = \begin{cases} \text{highest root of } \mathfrak{g}, \text{ if } r = 1 \\ \text{highest short root of } \mathfrak{g}^\tau, \text{ if } r > 1 \text{ and } (\mathfrak{g}, r) \neq (A_{2n}, 2) \\ 2 \cdot \text{highest short root of } \mathfrak{g}^\tau, \text{ if } (\mathfrak{g}, r) = (A_{2n}, 2). \end{cases}$$

Let $\hat{L}(\mathfrak{g}, \tau)$ denote the Lie algebra with the construction similar to $\hat{L}(\mathfrak{g}, \sigma)$ where σ is replaced by τ, m is replaced by r and ϵ is replaced by $\epsilon^{\frac{m}{r}}$. There exists an isomorphism of Lie algebras (cf. [**Ka**, Theorem 8.5]):

$$\Psi_\sigma : \hat{L}(\mathfrak{g}, \tau) \simeq \hat{L}(\mathfrak{g}, \sigma) \tag{3}$$

given by $C \mapsto C$ and $x[t^j] \mapsto x[t^{\frac{m}{r}j+k}]$ modulo $\mathbb{C}C$, for any x an $\epsilon^{\frac{m}{r}j}$-eigenvector of τ, and x also a k-eigenvector of $\operatorname{ad} h$. Then, the isomorphism Ψ_σ induces a bijection

$$D_{c,\sigma} \simeq D_{c,\tau}, \quad \lambda \mapsto \bar{\lambda}. \tag{4}$$

REMARK 2.1. The explicit description of $D_{c,\sigma}$ is given in [**HK1**, Lemma 2.1] in terms of $\{n_{\lambda,i} \mid i \in \hat{I}(\mathfrak{g}, \sigma)\}$ defined there. Also, $\bar{\lambda}$ can be expressed in terms of numbers $a_i, a_i^\vee$ which can be read from [**Ka**, p.54-55] via [**Ka**, Theorem 8.7]. For the convenience of readers, we would like to point out that there is a typo in the formula for $\alpha_i^\vee$ in [**Ka**, Theorem 8.7]. The correct expression is: $\alpha_i^\vee = 1 \otimes H_i + \frac{a_i s_i r}{a_i^\vee m} K$.

3. Vanishing theorem for Lie algebra cohomology

In this section, we consider the twisted affine Lie algebra $\hat{L}(\mathfrak{g}, \sigma)$ for a special automorphism σ. The main technical result is the Nakano identity (Theorem 3.11, equivalently Theorem 3.12) for $\hat{L}(\mathfrak{g}, \sigma)$. To formulate Nakano identity, we introduce Laplacians in Section 3.3 and Section 3.4. The proof of Nakano identity will be given in Section 4, where Proposition 3.8 in this section will be crucially used. Using Nakano identity and the estimate in Proposition 3.16, we obtain the main result of this paper in Theorem 3.17, which asserts that certain Lie algebra cohomology of $L^+(\mathfrak{g}, \sigma)$ vanishes. An equivalent form in terms of Lie algebra homology vanishing is given in Corollary 3.20.

3.1. Contravariant Hermitian form on twisted affine algebras. Let σ be an automorphism of $\mathfrak{g}$ of order m. As in (2), we write $\sigma = \tau \epsilon^{adh}$ with respect to a compatible Borel subalgebra $\mathfrak{b}$ and a Cartan sublagebra $\mathfrak{h}$ contained in $\mathfrak{b}$, where $\epsilon = e^{\frac{2\pi i}{m}}$. With respect to $\mathfrak{b}$ and $\mathfrak{h}$, we choose a set of Chevalley generators $\{e_i, f_i, h_i \mid i \in I\}$ such that they are permuted by τ.

Let $\mathfrak{g}_\mathbb{R}$ be the real split form of $\mathfrak{g}$ generated by $\{e_i, f_i, h_i \mid i \in I\}$. Let ω be the Cartan involution of $\mathfrak{g}$ defined by

$$\omega(e_i) = -f_i, \omega(f_i) = -e_i, \omega(h_i) = -h_i, \quad \text{for any } i \in I.$$

Define a *compact (conjugate-linear) involution* κ of $\mathfrak{g}$:

$$\kappa = \omega \circ J, \tag{5}$$

where J is the complex conjugation of $\mathfrak{g}$ with respect to $\mathfrak{g}_\mathbb{R}$.

LEMMA 3.1. *The equality $\sigma \circ \kappa = \kappa \circ \sigma$ holds on $\mathfrak{g}$.*

PROOF. For each $i \in I$, we have

$$\sigma(e_i) = \epsilon^{\alpha_i(h)} e_{\tau(i)}, \sigma(f_i) = \epsilon^{-\alpha_i(h)} f_{\tau(i)}, \sigma(h_i) = h_{\tau(i)},$$

where e_i belongs to the root space $\mathfrak{g}_{\alpha_i}$. It is now easy to see that the equality $\sigma \circ \kappa = \kappa \circ \sigma$ holds on $\{e_i, f_i, h_i\}$. Thus, the equality holds on $\mathfrak{g}$. □

Consider the decomposition $\mathfrak{g} = \oplus_{\underline{n} \in \mathbb{Z}/m\mathbb{Z}} \mathfrak{g}_{\underline{n}}$, where $\mathfrak{g}_{\underline{n}}$ is the ϵ^n-eigenspace of σ of $\mathfrak{g}$. Then, $\kappa(\mathfrak{g}_{\underline{n}}) = \mathfrak{g}_{-\underline{n}}$ and $\mathfrak{g}_{\underline{0}} = \mathfrak{g}^\sigma$.

As before, let $\langle , \rangle$ be the normalized invariant form on $\mathfrak{g}$ so that $\langle \theta, \theta \rangle = 2$ for the highest root θ of $\mathfrak{g}$. Then, $\langle , \rangle : \mathfrak{g}_{\underline{n}} \times \mathfrak{g}_{-\underline{n}} \to \mathbb{C}$ is non-degenerate for each $\underline{n}$, and the eigenspaces $\mathfrak{g}_{\underline{n}}$ and $\mathfrak{g}_{-\underline{n}'}$ for $\underline{n} \neq \underline{n}'$ are orthogonal with respect to $\langle , \rangle$.

Define a Hermitian form $\{ , \}$ on $\mathfrak{g}$ as follows:

$$\{x, y\} = -\langle x, \kappa(y) \rangle, \quad \text{for any } x, y \in \mathfrak{g}. \tag{6}$$

Then, $\{ , \}$ is a positive-definite contravariant Hermitian form on $\mathfrak{g}$, where contravariance means

$$\{[z, x], y\} = -\{x, [\kappa(z), y]\}, \text{ for all } x, y, z \in \mathfrak{g}.$$

Moreover, $\mathfrak{g} = \oplus_{\underline{n} \in \mathbb{Z}/m\mathbb{Z}} \mathfrak{g}_{\underline{n}}$ is an orthogonal decomposition with respect to $\{ , \}$. Thus, we can choose an orthonormal basis $\mathcal{B} = \mathcal{B}_\sigma$ with respect to $\{ , \}$ such that $\mathcal{B}_{\underline{n}} := \mathcal{B} \cap \mathfrak{g}_{\underline{n}}$ is a basis of $\mathfrak{g}_{\underline{n}}$. We define a function $[\cdot] : \mathcal{B} \to \mathbb{Z}/m\mathbb{Z}$ given by $[b] = \underline{n}$, if $b \in \mathcal{B}_{\underline{n}}$. For each $b \in \mathcal{B}$, set $\check{b} = -\kappa(b)$. Then, $\{\check{b}\}_{b \in \mathcal{B}}$ is the dual basis of $\mathcal{B}$ with respect to $\langle , \rangle$. Moreover, $[\check{b}] = -\underline{n}$, if $[b] = \underline{n}$.

For $x \in \mathfrak{g}$ and $n \in \mathbb{Z}$, let $x(n) := x[t^n]$. The action of σ on $\mathfrak{g}(\mathcal{K})$ is given in Section 2 and $\hat{L}(\mathfrak{g}, \sigma)$ defined there. Moreover, σ acts trivially on C.

Let $\hat{L}^f(\mathfrak{g})$ denote the Lie subalgebra of $\hat{L}(\mathfrak{g}) := \hat{L}(\mathfrak{g}, I)$ obtained by replacing the function field $\mathcal{K}$ by the ring of Laurent polynomials $\mathbb{C}[t, t^{-1}]$. Similarly,

$$\hat{L}^f(\mathfrak{g}, \sigma) := \left(\mathfrak{g} \otimes_{\mathbb{C}} \mathbb{C}[t, t^{-1}]\right)^\sigma \oplus \mathbb{C}C.$$

Extend the compact involution κ of $\mathfrak{g}$ to $\hat{L}^f(\mathfrak{g})$ as the conjugate-linear involution defined by

$$\kappa(x(n)) = \kappa(x)(-n), \text{ for any } x(n) \in \hat{L}^f(\mathfrak{g}), \text{ and } \kappa(C) = -C. \tag{7}$$

We take

$$(\mathfrak{g}_{\mathbb{R}} \otimes \mathbb{R}[t^\pm]) \oplus \mathbb{R}C$$

as the split real form of $\hat{L}^f(\mathfrak{g})$.

Similarly, we extend the contravariant Hermitian form $\{ , \}$ on $\mathfrak{g}$ to $\hat{L}^f(\mathfrak{g})$ by

$$\{x(n), y(n')\} = \{x, y\}\delta_{n,n'}, \quad \{x(n), C\} = 0 \text{ and } \{C, C\} = 0. \tag{8}$$

Similar to Lemma 3.1, σ and κ commute on $\hat{L}^f(\mathfrak{g})$ since, for any $x(n) \in \hat{L}^f(\mathfrak{g})$,

$$\sigma\kappa(x(n)) = \sigma(\kappa(x)(-n)) = \epsilon^n (\sigma\kappa)(x)(-n), \text{ and } \sigma\kappa(C) = -C,$$

and

$$\kappa\sigma(x(n)) = \kappa(\epsilon^{-n}\sigma(x)(n)) = \epsilon^n (\kappa\sigma)(x)(-n), \text{ and } \kappa\sigma(C) = -C.$$

Thus, κ gives rise to a (conjugate-linear) *compact involution* of $\hat{L}^f(\mathfrak{g}, \sigma)$, and the induced Hermitian form $\{ , \}$ on $\hat{L}^f(\mathfrak{g}, \sigma)$ is contravariant, and also positive-definite on $\mathfrak{g}[t, t^{-1}]^\sigma$ (cf. [**Ka**, §§7.6, 8.4]). Moreover, $\kappa : L^\pm(\mathfrak{g}, \sigma) \to L^\mp(\mathfrak{g}, \sigma)$ is an anti-linear isomorphism, where $L^\pm(\mathfrak{g}, \sigma) := (t^\pm \mathfrak{g}[t^\pm])^\sigma$, and $L^\pm(\mathfrak{g}, \sigma)$ are Lie subalgebras of $\hat{L}^f(\mathfrak{g}, \sigma)$. Moreover, the restriction of $\{ , \}$ to $L^\pm(\mathfrak{g}, \sigma)$ is positive-definite. We now introduce two different Hermitian forms $\{,\}_0$ and $\{,\}_1$ on $L^-(\mathfrak{g},\sigma)$ modified from $\{,\}$ by defining

$$(9)\quad \{x(-n),y(-n')\}_0=\frac{1}{n}\{x(-n),y(-n')\},\qquad \text{for any } x(-n),y(-n')\in L^-(\mathfrak{g},\sigma),$$

and

$$(10)\quad \{x(-n),y(-n')\}_1=n\{x(-n),y(-n')\},\qquad \text{for any } x(-n),y(-n')\in L^-(\mathfrak{g},\sigma).$$

We introduce the following set

$$(11)\qquad \hat{\mathcal{B}}_\sigma=\{b(k)\mid k\in\mathbb{Z}_{<0}, b\in\mathcal{B} \text{ with } [b]=\underline{k}\}.$$

Then, $\hat{\mathcal{B}}_\sigma$ is an orthonormal basis of $L^-(\mathfrak{g},\sigma)$ with respect to the Hermitian form $\{,\}$. Moreover, $\{\sqrt{-k}b(k)\}_{b(k)\in\hat{\mathcal{B}}_\sigma}$ is an orthonormal basis of $L^-(\mathfrak{g},\sigma)$ with respect to the Hermitian form $\{,\}_0$, and $\{\frac{1}{\sqrt{-k}}b(k)\}_{b(k)\in\hat{\mathcal{B}}_\sigma}$ is an orthonormal basis of $L^-(\mathfrak{g},\sigma)$ with respect to $\{,\}_1$.

NOTATION 3.2. Fix a level $c>0$. Let $\lambda\in D_{c,\sigma}$, $\mu_0\in D_{c,\sigma^{-1}}$ and $\mu_1,\mu_2,\dots,\mu_s\in D_c$. For $1\le i\le s$, let $V(\mu_i)$ be the irreducible representation of $\mathfrak{g}$ with highest weight μ_i and let $V(\mu_0)$ be the irreducible representation of $\mathfrak{g}^\sigma$ with highest weight μ_0. Let $(\mathbb{P}^1,\vec{z})$ be a (stable) $s+2$-pointed Γ-curve (for $s\ge 1$), where $\Gamma=\langle\sigma\rangle$, σ acts on $\mathbb{P}^1$ via $\sigma z=e^{\frac{2\pi i}{m}}z$ for $z\in\mathbb{P}^1$, $\vec{z}=(\infty,z_0,z_1,\cdots,z_s)$ with $z_0=0$ and $0<|z_i|<1$ for other i. For $0\le i\le s$, let $V(\mu_i)_{z_i}$ be the evaluation representation of $\mathfrak{g}[t^{-1}]^\sigma$ with the underlying space $V(\mu_i)$, defined by

$$x(-n)\cdot v=\bar{z}_i^n x\cdot v,\quad \text{for any } x(-n)\in\mathfrak{g}[t^{-1}]^\sigma,$$

where $\bar{z}_i$ is the complex conjugate of z_i. For each $0\le i\le s$, we define a Lie algebra action of $L^+(\mathfrak{g},\sigma)$ on $V(\mu_i)_{z_i}$ via

$$x(n)\cdot v=z_i^n x\cdot v,\quad \text{for any } x(n)\in L^+(\mathfrak{g},\sigma).$$

We further define $C\cdot v=0$. Observe that these actions *do not combine to give a Lie algebra action of* $\hat{L}^f(\mathfrak{g},\sigma)$.

Set

$$V(\vec{\mu})_{\vec{z}}=V(\mu_0)_{z_0}\otimes V(\mu_1)_{z_1}\otimes\cdots\otimes V(\mu_s)_{z_s}.$$

We take the positive-definite contravariant Hermitian form on the $\hat{L}(\mathfrak{g},\sigma)$-module $\mathscr{H}_c(\lambda)$ (cf. [**Ku1**, Theorem 2.3.13]), and the $\mathfrak{g}^{\Gamma_{z_i}}$-module $V(\mu_i)$ (still denoted by $\{,\}$), where Γ_{z_i} is the isotropy of Γ at z_i.

Now, the following lemma is a restatement of $\{\,,\,\}$ being contravariant.

LEMMA 3.3. *For any $v,w\in\mathscr{H}_c(\lambda)$ (resp. $V(\mu_i)_{z_i}$), we have*

$$\{x(n)v,w\}=-\{v,\kappa(x(n))\cdot w\},\qquad \textit{for any } x(n)\in L^f(\mathfrak{g},\sigma).$$

From the Hermitian form $\{,\}$ on $\hat{L}^f(\mathfrak{g},\sigma)$, one introduces a Hermitian form on $\wedge^q\hat{L}^f(\mathfrak{g},\sigma)$ in the standard way as follows. For any $\vec{y}=y_1\wedge\cdots\wedge y_q$, $\vec{u}=u_1\wedge\cdots\wedge u_q$ in $\wedge^q\hat{L}^f(\mathfrak{g},\sigma)$, define

$$\{\vec{y},\vec{u}\}:=\det(\{y_i,u_j\}).$$

Then, we have the follow lemma, which is easy to prove.

LEMMA 3.4. *The Hermitian form* $\{,\}$ *on* $\wedge^q \hat{L}^f(\mathfrak{g},\sigma)$ *with respect to the adjoint action of* $\hat{L}^f(\mathfrak{g},\sigma)$ *is contravariant, i.e.,*

$$\{ad_x(\vec{y}),\vec{u}\} = -\{\vec{y}, ad_{\kappa(x)}(\vec{u})\},$$

for any $x \in \hat{L}^f(\mathfrak{g},\sigma)$, *and* $\vec{y},\vec{u} \in \wedge^q \hat{L}^f(\mathfrak{g},\sigma)$.

Similarly,

$$\langle ad_x(\vec{y}),\vec{u}\rangle = -\langle \vec{y}, ad_x(\vec{u})\rangle.$$

Let $\bar{L}^-(\mathfrak{g})$ (resp. $\bar{L}^-(\mathfrak{g},\sigma)$ be the L^2-completion of $L^-(\mathfrak{g})$ (resp. $L^-(\mathfrak{g},\sigma)$) with respect to the positive definite Hermitian form $\{,\}_0$ on $L^-(\mathfrak{g})$ (resp. $L^-(\mathfrak{g},\sigma)$).

3.2. Casimir operators on graded Lie algebras. Recall that $\langle,\rangle$ is the normalized invariant form on $\mathfrak{g}$. It induces an invariant form on $\mathfrak{g}^\sigma$ via restriction, which is still denoted by $\langle,\rangle$.

DEFINITION 3.5. An automorphism σ of $\mathfrak{g}$ is called **special** if σ is a diagram automorphism (including the identity automorphism), or the order 4 automorphism of $\mathfrak{g}$ or its inverse when $\mathfrak{g}$ is of type A_{2n}, which is defined as follows. Let e_i, f_i, h_i, $i = 1, \dots, 2n$, be the set of Chevalley generators. The automorphism σ of $\mathfrak{g}$ is defined such that

$$\begin{cases} \sigma(e_i) = e_{\tau(i)}, & \text{if } i \neq n, n+1; \\ \sigma(e_i) = \sqrt{-1}e_{\tau(i)}, & \text{if } i \in \{n, n+1\}; \\ \sigma(f_\theta) = f_\theta, \end{cases} \quad , \tag{12}$$

where θ is the highest root of $\mathfrak{g}$, $\sqrt{-1} := i$, and τ is the nontrivial diagram automorphism. In fact, we can write

$$\sigma = \tau\sqrt{-1}^{\,\mathrm{ad}h}, \tag{13}$$

where $h \in \mathfrak{h}$ is such that $\alpha_i(h) = 0$ if $i \neq n, n+1$ and $\alpha_i(h) = 1$ if $i = n, n+1$. As mentioned above, σ^{-1} is also a special automorphism.

The following table describes the fixed point Lie algebra for all the nontrivial special automorphisms (cf. [**BH**, Section 2.1]):

(14)

$(\mathfrak{g}, m)$	$(A_{2n-1}, 2)$	$(A_{2n}, 4)$	$(A_{2n}, 2)$	$(D_{n+1}, 2)$	$(D_4, 3)$	$(E_6, 2)$
$\mathfrak{g}^\sigma$	C_n	C_n	B_n	B_n	G_2	F_4

where by convention C_1 and B_1 are A_1 and $n \geq 3$ for D_{n+1}.

LEMMA 3.6. *Let* σ *be a special automorphism of* $\mathfrak{g}$ *of order* m *and let* θ_σ *be the element defined below:*

$$\theta_\sigma = \begin{cases} \text{highest short root of } \mathfrak{g}^\sigma, & (\mathfrak{g}, m) \neq (A_{2n}, 4), (A_{2n}, 2) \\ \frac{1}{2} \text{ highest root of } \mathfrak{g}^\sigma, & (\mathfrak{g}, m) = (A_{2n}, 4) \\ 2 \text{ highest short root of } \mathfrak{g}^\sigma, & (\mathfrak{g}, m) = (A_{2n}, 2) \end{cases} \tag{15}$$

Then, $\langle \theta_\sigma, \theta_\sigma + 2\rho_\sigma\rangle = \frac{2\check{h}}{m}$, *where* $\check{h}$ *as earlier is the dual Coxeter number of* $\hat{L}(\mathfrak{g},\sigma)$ *(or* $\mathfrak{g}$*), and* ρ_σ *is half the sum of positive roots of* $\mathfrak{g}^\sigma$.

PROOF. Let $(,)$ be the standard pairing between $\mathfrak{h}^\sigma$ and $(\mathfrak{h}^\sigma)^*$. Then,

$$\langle \theta_\sigma, \theta_\sigma + 2\rho_\sigma\rangle = \frac{(2\check{\theta}_\sigma, \theta_\sigma + 2\rho_\sigma)}{\langle \check{\theta}_\sigma, \check{\theta}_\sigma\rangle},$$

where $\check{\theta}_\sigma$ is the element defined below:

$$(16)\qquad \check{\theta}_\sigma = \begin{cases} \text{highest coroot of } \mathfrak{g}^\sigma, & (\mathfrak{g}, m) \neq (A_{2n}, 4), (A_{2n}, 2) \\ 2 \cdot \text{ highest short coroot of } \mathfrak{g}^\sigma, & (\mathfrak{g}, m) = (A_{2n}, 4). \\ \frac{1}{2} \cdot \text{highest coroot of } \mathfrak{g}^\sigma, & (\mathfrak{g}, m) = (A_{2n}, 2). \end{cases}$$

When $(\mathfrak{g}, m)$ is different from $(A_{2n}, 4)$ and $(A_{2n}, 2)$, as $\check{\theta}_\sigma$ is the highest coroot, $\langle \check{\theta}_\sigma, \check{\theta}_\sigma \rangle = 2m$. When $(\mathfrak{g}, m)$ is $(A_{2n}, 4)$, from an explicit realization of $\mathfrak{g}^\sigma$, one can see that for any short coroot $\check{\beta}$, $\langle \check{\beta}, \check{\beta} \rangle = 2$. Since $\check{\theta}_\sigma$ is twice the highest short coroot, $\langle \check{\theta}_\sigma, \check{\theta}_\sigma \rangle = 8 = 2m$. When $(\mathfrak{g}, m) = (A_{2n}, 2)$,

$$\langle 2\theta_s, 2\theta_s + 2\rho_\sigma \rangle = \frac{4(\check{\theta}_s, 2\theta_s + 2\rho_\sigma)}{\langle \check{\theta}_s, \check{\theta}_s \rangle} = \frac{(\check{\theta}_s, 2\theta_s + 2\rho_\sigma)}{2} = 2 + (\check{\theta}_s, \rho_\sigma) = 2n + 1 = \frac{2\check{h}}{m},$$

where $\langle \check{\theta}_s, \check{\theta}_s \rangle = 8$, since $\check{\theta}_s$ is conjugate to $2(\check{\alpha}_n + \check{\alpha}_{n+1})$ (both are long coroots of $\mathfrak{g}^\sigma$) (cf. [**Ka**, Case 5, p.129]). □

Let σ be a special automorphism of $\mathfrak{g}$ of order m. Let θ_l denote the highest root of $\mathfrak{g}^\sigma$ and let θ_s denote the highest short root of $\mathfrak{g}^\sigma$. Let $\theta_{\underline{i}}$ be the unique maximal weight (with respect to the Bruhat-Chevalley order on the set of weights) of $\mathfrak{g}^\sigma$ that appears in $\mathfrak{g}_{\underline{i}}$.

LEMMA 3.7. *We have*

$$\theta_{\underline{i}} = \begin{cases} \theta_l & \textit{if } \underline{i} = \underline{0} \\ \theta_s & \textit{if } \mathfrak{g} \neq A_{2n} \textit{ and } (m, \underline{i}) = (2, \underline{1}) \textit{ or } (3, \underline{1}) \textit{ or } (3, \underline{2}) \\ 0 & \textit{if } \mathfrak{g} = A_2 \textit{ and } (m, \underline{i}) = (4, \underline{2}) \\ \theta_s & \textit{if } \mathfrak{g} = A_{2n} (n > 1) \textit{ and } (m, \underline{i}) = (4, \underline{2}) \\ \frac{\theta_l}{2} & \textit{if } \mathfrak{g} = A_{2n} (n \geq 1) \textit{ and } (m, \underline{i}) = (4, \underline{1}), (m, \underline{i}) = (4, \underline{3}) \\ 2\theta_s & \textit{if } (\mathfrak{g}, m) = (A_{2n}, 2) \textit{ and } \underline{i} = \underline{1}. \end{cases}$$

If $(\mathfrak{g}, m) \neq (A_{2n}, 4)$ or $\underline{i} \neq \underline{2}$, $\mathfrak{g}_{\underline{i}}$ is an irreducible representation of $\mathfrak{g}^\sigma$ of highest weight $\theta_{\underline{i}}$. If $(\mathfrak{g}, m) = (A_{2n}, 4) (n > 1)$, $\mathfrak{g}_{\underline{2}} \simeq V(\theta_s) \oplus \mathbb{C}$. If $(\mathfrak{g}, m) = (A_2, 4)$, $\mathfrak{g}_{\underline{2}} \simeq \mathbb{C}$.

PROOF. This lemma follows from [**Ka**, Proposition 8.3] for σ a diagram automorphism. For the special automorphism σ of order 4 of A_{2n}, it follows from the calculation given in the proof of Proposition 3.8. □

PROPOSITION 3.8. *Let σ be a special automorphism of order m of $\mathfrak{g}$. Then, for any $x(-p) \in L^-(\mathfrak{g}, \sigma)$,*

$$(17)\qquad \sum_{k \geq 0, b \in \mathcal{B} \textit{ with } [b] = -\underline{k}} ad_{b(-k)} \overline{ad}_{\check{b}(k)} \cdot x(-p) = \frac{2p\check{h}}{m} x(-p).$$

PROOF. Observe that

$$S := \sum_{k \geq 0, b \in \mathcal{B} \text{ with } [b] = -\underline{k}} ad_{b(-k)} \overline{ad}_{\check{b}(k)} \cdot x(-p) = \sum_{k=0}^{p-1} C_{-\underline{k}} \cdot x(-p),$$

where we set

$$(18)\qquad C_{-\underline{k}} := \sum_{b \in \mathcal{B} \text{ with } [b] = -\underline{k}} ad_b ad_{\check{b}}.$$

Then, $C = \sum_{k=0}^{m-1} C_{-\underline{k}}$ is the Casimir operator of $\mathfrak{g}$ with respect to the normalized invariant form. Its action on any irreducible $\mathfrak{g}$-module of highest weight λ is given by the scalar $\langle \lambda, \lambda + 2\rho \rangle$ (cf. [**Ku1**, Lemma 2.1.16]).

CASE 1. $(\mathfrak{g}, m) = (A_{2n-1}, 2)(n \geq 2); (D_{n+1}, 2)(n \geq 3); (E_6, 2)$. When $p = 2i$ for $i \geq 1$,

$$S = iC \cdot x(-p) = 2i\check{h}x(-p) = \frac{2p\check{h}}{m}x(-p), \text{ since } C \text{ acts on } \mathfrak{g} \text{ via multiplication by } 2\check{h}.$$

When $p = 2i + 1 (i \geq 0)$,

$$S = \sum_{k=0}^{2i-1} C_{-\underline{k}} \cdot x(-p) + C_{\underline{0}} \cdot x(-p) = 2i\check{h}x(-p) + C_{\underline{0}} \cdot x(-p).$$

Since $x \in \mathfrak{g}_{\underline{1}}$ and $\mathfrak{g}_{\underline{1}}$ is an irreducible representation of $\mathfrak{g}^\sigma$ with highest weight θ_s, by Lemma 3.6, the formula (17) holds.

CASE 2. $(\mathfrak{g}, m) = (D_4, 3)$. When $p = 3i$ or $p = 3i + 1$, the proof is similar to the Case 1, where we use Lemma 3.6 for $p = 3i + 1$. When $p = 3i + 2$,

$$S = \sum_{k=0}^{3i+2} C_{-\underline{k}} \cdot x(-p) - C_{\underline{1}} \cdot x(-p) = \frac{2(p+1)}{3}\check{h}x(-p) - C_{\underline{1}} \cdot x(-p),$$

where $x \in \mathfrak{g}_{\underline{1}}$. The operator $C_{\underline{1}}$ on $\mathfrak{g}_{\underline{1}}$ commutes with the action of $\mathfrak{g}^\sigma$, as $\langle , \rangle : \mathfrak{g}_{\underline{1}} \times \mathfrak{g}_{\underline{2}} \to \mathbb{C}$ is $\mathfrak{g}^\sigma$-invariant and $\{b\}_{b \in \mathcal{B}_{\underline{1}}}$ and $\{\check{b}\}_{b \in \mathcal{B}_{\underline{1}}}$ are dual bases with respect to $\langle , \rangle$. Since $\mathfrak{g}_{\underline{1}}$ is an irreducible representation of $\mathfrak{g}^\sigma$ with highest weight θ_s, by Schur lemma, $C_{\underline{1}}$ acts by a scalar $c_{\underline{1}}$ on $\mathfrak{g}_{\underline{1}}$. We compute this number by choosing a basis $\mathcal{B}_{\underline{1}}$ of $\mathfrak{g}_{\underline{1}}$ and a specific vector v in $\mathfrak{g}_{\underline{1}}$. Following the Bourbaki labelling [**Bo**, Planche IV], let $\{\alpha_1, \alpha_2, \alpha_3, \alpha_4\}$ be the set of simple roots of $\mathfrak{g}$, and σ be the cycle (134) on the set of vertices $\{1, 2, 3, 4\}$. Choose a Chevalley basis $\{e_\alpha, h_i\}$ of $\mathfrak{g}$. For each positive root α we denote the associated sl_2-triple by $\{e_\alpha, e_{-\alpha}, h_\alpha\}$ satisfying $\sigma(e_{\pm\alpha}) = e_{\pm\sigma(\alpha)}$, and $\langle e_\alpha, e_{-\alpha} \rangle = 1$ (cf. [**Ka**, §7.9]). We choose $\mathcal{B}_{\underline{1}}$ to be the set of the following elements:

(19)

$$h_{\alpha_1} + \epsilon^2 h_{\alpha_3} + \epsilon h_{\alpha_4}, \quad e_{\pm\alpha_1} + \epsilon^2 e_{\pm\alpha_3} + \epsilon e_{\pm\alpha_4}, \quad e_{\pm(\alpha_1+\alpha_2)} + \epsilon^2 e_{\pm(\alpha_3+\alpha_2)} + \epsilon e_{\pm(\alpha_4+\alpha_2)},$$
$$e_{\pm(\alpha_1+\alpha_2+\alpha_3)} + \epsilon^2 e_{\pm(\alpha_3+\alpha_2+\alpha_4)} + \epsilon e_{\pm(\alpha_4+\alpha_2+\alpha_1)},$$

where $\epsilon = e^{\frac{2\pi i}{3}}$. Then, the dual basis in $\mathfrak{g}_{\underline{2}}$ is given by

(20)
$$\frac{1}{6}(h_{\alpha_1} + \epsilon h_{\alpha_3} + \epsilon^2 h_{\alpha_4}), \quad \frac{1}{3}(e_{\mp\alpha_1} + \epsilon e_{\mp\alpha_3} + \epsilon^2 e_{\mp\alpha_4}),$$
$$\frac{1}{3}(e_{\mp(\alpha_1+\alpha_2)} + \epsilon e_{\mp(\alpha_3+\alpha_2)} + \epsilon^2 e_{\mp(\alpha_4+\alpha_2)}),$$
$$\frac{1}{3}(e_{\mp(\alpha_1+\alpha_2+\alpha_3)} + \epsilon e_{\mp(\alpha_3+\alpha_2+\alpha_4)} + \epsilon^2 e_{\mp(\alpha_4+\alpha_2+\alpha_1)}).$$

We choose the vector $v = h_{\alpha_1} + \epsilon^2 h_{\alpha_3} + \epsilon h_{\alpha_4}$ in $\mathfrak{g}_{\underline{1}}$. Then, $C_{\underline{1}}v = \sum_{[b]=\underline{1}} ad_b ad_{\check{b}} v = 4v$. In fact, for the first basis element b in (19), $ad_b ad_{\check{b}} v = 0$, and for any other b there, $ad_b ad_{\check{b}} v = \frac{2}{3}v$. Thus, $c_{\underline{1}} = 4 = \frac{2\check{h}}{3}$, as the dual Coxeter number of D_4 is 6. Thus, (17) holds.

CASE 3. $(\mathfrak{g}, m) = (A_{2n}, 4)$. Let $\{e_{i,j}\}_{1\leq i\neq j\leq 2n+1} \sqcup \{e_{i,i} - e_{i+1,i+1}\}_{1\leq i\leq 2n}$ be the standard basis of $\mathfrak{g} = sl_{2n+1}$, where $e_{i,j}$ is the $(2n+1)\times(2n+1)$-matrix with (i,j)-entry equal to 1 and zero elsewhere. Then, $\mathfrak{g}^\sigma \simeq sp_{2n}$, and as representations of sp_{2n}, $\mathfrak{g}_{\underline{1}}$ and $\mathfrak{g}_{\underline{3}}$ are both isomorphic to the standard representation, which is of highest weight θ_σ (half of the highest root of $\mathfrak{g}^\sigma$) (cf. [**Bo**, Planche III]). When $p = 4i$ or $p = 4i+1$, the proof is similar to Case 1, where Lemma 3.6 is used for $p = 4i+1$.

When $p = 4i+3$, the operator $C_{\underline{1}}$ acts on $\mathfrak{g}_{\underline{1}}$ by a scalar $c_{\underline{1}}$ as $\mathfrak{g}_{\underline{1}}$ is an irreducible $\mathfrak{g}^\sigma$-representation and $C_{\underline{1}}$ commutes with the action of $\mathfrak{g}^\sigma$. We choose a basis $\mathcal{B}_{\underline{1}}$ of $\mathfrak{g}_{\underline{1}}$ as follows:

$$\{e_{i,n+1} + (-1)^{n-i} e_{n+1,2n+2-i}, e_{n+1,i} - (-1)^{n-i} e_{2n+2-i,n+1}\}_{1\leq i\leq n}. \tag{21}$$

The corresponding dual basis of $\mathfrak{g}_{\underline{3}}$ is given by:

$$\{\frac{1}{2}(e_{n+1,i} + (-1)^{n-i} e_{2n+2-i,n+1}),\quad \frac{1}{2}(e_{i,n+1} - (-1)^{n-i} e_{n+1,2n+2-i})\}_{1\leq i\leq n}. \tag{22}$$

We choose the vector $v = e_{1,n+1} + (-1)^{n-1} e_{n+1,2n+1}$ of $\mathfrak{g}_{\underline{1}}$. Then, when $b = v$, $ad_b ad_{\check{b}} v = v$, and for any other $b \in \mathcal{B}_{\underline{1}}$, $ad_b ad_{\check{b}} v = \frac{1}{2} v$. Thus, the scalar $c_{\underline{1}} = \frac{2n+1}{2} = \frac{2\check{h}}{m}$.

When $p = 4i+2$, it suffices to show that the operator $C_{\underline{0}} + C_{\underline{-1}}$ acts on $\mathfrak{g}_{\underline{2}}$ by the scalar $2n+1$.

When $n > 1$, $\mathfrak{g}_{\underline{2}}$ is not irreducible as a representation of $\mathfrak{g}^\sigma$; in fact, we have $\mathfrak{g}_{\underline{2}} = W \oplus \mathbb{C}v_0$, where W is an irreducible representation of $\mathfrak{g}^\sigma$ with highest weight the highest short root θ_s of $\mathfrak{g}^\sigma$, and

$$v_0 := \sum_{i=1}^{n} (e_{i,i} + e_{2n+2-i,2n+2-i}) - 2n e_{n+1,n+1}$$

is an $\mathfrak{g}^\sigma$-invariant vector.

We show that $C_{\underline{0}} + C_{\underline{-1}}$ acts on W and v_0 by the same scalar $2n+1$. First notice that $C_{\underline{0}}$ and $C_{\underline{-1}}$ act on W by scalars. Of course, $C_{\underline{0}}$ acts on W by the scalar

$$\langle \theta_s, \theta_s + 2\rho_\sigma\rangle = \frac{2(\check{\theta}_s, \theta_s + 2\rho_\sigma)}{\langle \check{\theta}_s, \check{\theta}_s\rangle} = 1 + (\check{\theta}_s, \rho_\sigma) = 2n.$$

Further, $C_{\underline{-1}}$ acts on W by a scalar since the action of $C_{\underline{-1}}$ on W commutes with the $\mathfrak{g}^\sigma$-action. Moreover, one computes that $C_{\underline{-1}}$ acts on W by 1 by using the basis given in (22) and its dual basis in (21) and fixing the vector $e_{1,2} - e_{2n,2n+1} \in W$. We now calculate the action of $C_{\underline{-1}}$ on v_0. It turns out that it acts by the scalar $2n+1$. Furthermore, $C_{\underline{0}}$ clearly acts on v_0 by 0. Thus, $C_{\underline{0}} + C_{\underline{-1}}$ acts on $\mathfrak{g}_{\underline{2}}$ by $2n+1$.

But, when $n = 1$, $\mathfrak{g}_{\underline{2}}$ is the trivial one dimensional representation of $\mathfrak{g}^\sigma$ spanned by v_0. Hence, $C_{\underline{0}}$ acts by zero on $\mathfrak{g}_{\underline{2}}$. Moreover, $C_{\underline{-1}}$ acts on v_0 by $3 = 2n+1$. Thus, in the case $n = 1$ as well, $C_{\underline{0}} + C_{\underline{-1}}$ acts on $\mathfrak{g}_{\underline{2}}$ by the scalar $2n+1$.

CASE 4. $(\mathfrak{g}, m) = (A_{2n}, 2)$. In this case, σ is a diagram automorphism and $\mathfrak{g}^\sigma$ is of type B_n. Similar to Case 1, we need to compute the action of $C_{\underline{0}}$ on $\mathfrak{g}_{\underline{1}}$. Note that $\mathfrak{g}_{\underline{1}}$ is an irreducible representation of $\mathfrak{g}^\sigma$ with highest weight $2\theta_s$, where θ_s is the highest short root of $\mathfrak{g}^\sigma$. Thus, $C_{\underline{0}}$ acts on $\mathfrak{g}_{\underline{1}}$ by the scalar $\langle 2\theta_s, 2\theta_s + 2\rho_\sigma\rangle = \frac{2\check{h}}{m}$ (by Lemma 3.6). Thus, the formula (17) holds. $\square$

LEMMA 3.9. *Let $V(\mu)$ be an irreducible $\mathfrak{g}^\sigma$-module with highest weight μ, where σ is a special automorphism of $\mathfrak{g}$ of order m. Consider the operator*

$$C_\mu^{\underline{i}} : \mathfrak{g}_{\underline{i}} \otimes V(\mu) \to \mathfrak{g}_{\underline{i}} \otimes V(\mu)$$

defined by $C_\mu^{\underline{i}}(x \otimes v) = \sum_{b \in \mathcal{B}_{\underline{0}}} [b, x] \otimes \check{b} \cdot v$, where $\mathcal{B}_{\underline{0}} := \{b \in \mathcal{B} : [b] = \underline{0}\}$. Then, except for $(\mathfrak{g}, m) = (D_4, 3)$ in the case $\underline{i} = \underline{0}$, the eigenvalues $e_\mu^{\underline{i}}$ of $C_\mu^{\underline{i}}$ are bounded by

$$-\langle \mu + 2\rho_\sigma, \theta_{\underline{i}} \rangle \leq e_\mu^{\underline{i}} \leq \langle \mu, \theta_{\underline{i}} \rangle. \tag{23}$$

PROOF. Let $C_{\mathfrak{g}^\sigma}$ be the Casimir operator of $\mathfrak{g}^\sigma$ with respect to the invariant form induced from the normalized invariant form $\langle , \rangle$ on $\mathfrak{g}$. Then, for $x \in \mathfrak{g}_{\underline{i}}(\theta_{\underline{i}})$ (where $\mathfrak{g}_{\underline{i}}(\theta_{\underline{i}})$ denotes the $\theta_{\underline{i}}$-isotypic $\mathfrak{g}^\sigma$-module component of $\mathfrak{g}_{\underline{i}}$) and $v \in V(\mu)$,

$$\begin{aligned} -2C_\mu^{\underline{i}}(x \otimes v) &= -C_{\mathfrak{g}^\sigma}(x \otimes v) + (C_{\mathfrak{g}^\sigma} x) \otimes v + x \otimes C_{\mathfrak{g}^\sigma} v \\ &= -C_{\mathfrak{g}^\sigma}(x \otimes v) + \langle \theta_{\underline{i}}, \theta_{\underline{i}} + 2\rho_\sigma \rangle x \otimes v + \langle \mu, \mu + 2\rho_\sigma \rangle x \otimes v. \end{aligned}$$

Now, any irreducible $\mathfrak{g}^\sigma$-submodule of $\mathfrak{g}_{\underline{i}}(\theta_{\underline{i}}) \otimes V(\mu)$ is of the form $V(\mu + \gamma)$, for some weight γ of $\mathfrak{g}_{\underline{i}}(\theta_{\underline{i}})$ such that $\mu + \gamma$ is $\mathfrak{g}^\sigma$-dominant. The eigenvalue of $C_{\mathfrak{g}^\sigma}$ on $V(\mu + \gamma)$ is given by $\langle \mu + \gamma, \mu + \gamma + 2\rho_\sigma \rangle$. Now, for any such $V(\mu + \gamma)$,

$$\langle \mu - \theta_{\underline{i}}, \mu - \theta_{\underline{i}} + 2\rho_\sigma \rangle \leq \langle \mu + \gamma, \mu + \gamma + 2\rho_\sigma \rangle \leq \langle \mu + \theta_{\underline{i}}, \mu + \theta_{\underline{i}} + 2\rho_\sigma \rangle, \tag{24}$$

where the second inequality can be easily seen since $\mu + \gamma$ is $\mathfrak{g}^\sigma$-dominant, and $\theta_{\underline{i}} - \gamma$ is a non-negative linear combination of the positive roots of $\mathfrak{g}^\sigma$. To prove the first inequality, by Lemma 3.7, we observe that $\mathfrak{g}_{\underline{i}}(\theta_{\underline{i}})$ is self-dual, and thus γ can be written as $-\theta_{\underline{i}} + \beta$, where β is a non-negative linear comination of positive roots of $\mathfrak{g}^\sigma$. Moreover, $2\rho_\sigma - \theta_{\underline{i}}$ is a dominant weight, since $(\theta_{\underline{i}}, \check{\beta}) \leq 2$ for any simple coroot $\check{\beta}$ of $\mathfrak{g}^\sigma$ except when $(\mathfrak{g}, m) = (A_2, 2)$ and $\underline{i} = \underline{1}$. So assume that $(\mathfrak{g}, m) \neq (A_2, 2)$ when $\underline{i} = \underline{1}$. Thus, any eigenvalue $e_\mu^{\underline{i}}$ of $C_\mu^{\underline{i}}$ acting on $\mathfrak{g}_{\underline{i}} \otimes V(\mu)$ has lower bound

$$\frac{1}{2}\left(\langle \mu - \theta_{\underline{i}}, \mu - \theta_{\underline{i}} + 2\rho_\sigma \rangle - \langle \theta_{\underline{i}}, \theta_{\underline{i}} + 2\rho_\sigma \rangle - \langle \mu, \mu + 2\rho_\sigma \rangle\right)$$

and upper bound

$$\frac{1}{2}\left(\langle \mu + \theta_{\underline{i}}, \mu + \theta_{\underline{i}} + 2\rho_\sigma \rangle - \langle \theta_{\underline{i}}, \theta_{\underline{i}} + 2\rho_\sigma \rangle - \langle \mu, \mu + 2\rho_\sigma \rangle\right),$$

i.e., the inequality (23) holds.

If we take $x \in \mathfrak{g}_{\underline{i}}(0)$ (which exists only for $(\mathfrak{g}, m) = (A_{2n}, 4)$ and $\underline{i} = \underline{2}$), then $C_\mu^{\underline{i}}(x \otimes v) = 0$. Thus, the inequality (23) is clearly satisfied.

For $(\mathfrak{g}, m) = (A_2, 2)$ and $\underline{i} = \underline{1}$, the equation (23) is still satisfied. To prove this, if $\mu = n\omega, n \geq 2$ (ω being the fundamental weight of $\mathfrak{g}^\sigma = A_1$), the equation (23) is satisfied by the same proof. If $\mu = 0$, then $e_\mu^{\underline{1}} = 0$ and if $\mu = \omega$, then $e_\mu^{\underline{1}} = \frac{1}{2}, -\frac{3}{4}$. Thus, the equation (23) is satisfied in this case as well. □

3.3. Chain complex. Set $M_q := \wedge^q(L^-(\mathfrak{g}, \sigma)) \otimes \mathscr{H}_c(\lambda) \otimes V(\vec{\mu})_{\vec{z}}$ as a representation of $\mathfrak{g}[t^{-1}]^\sigma$. We first consider the standard chain complex (C_*, ∂_*) of $L^-(\mathfrak{g}, \sigma)$ with coefficients in M_q (cf. [**Ku1**, §3.1.1]). We calculate the formal adjoint operator ∂_1^* of ∂_1 with respect to the Hermitian form $\{,\}_1$ on $L^-(\mathfrak{g}, \sigma)$, the Hermitian form $\{,\}_0$ on $\wedge^q L^-(\mathfrak{g}, \sigma)$ (tensor component of M_q), and the contravariant Hermitian forms $\{,\}$ on $\mathscr{H}_c(\lambda)$ and $V(\vec{\mu})_{\vec{z}}$.

Consider the standard differential $\partial_p : \wedge^p(L^-(\mathfrak{g}, \sigma)) \otimes M_q \to \wedge^{p-1}(L^-(\mathfrak{g}, \sigma)) \otimes M_q$. In particular, when $p = 1$, $\partial_1(x(-n) \otimes m) = -x(-n) \cdot m$, for any $x(-n) \in$

$L^-(\mathfrak{g},\sigma)$ and $m \in M_q$. Take $\vec{x}(-\vec{p}) := x_1(-p_1)\wedge\cdots\wedge x_q(-p_q) \in \wedge^q L^-(\mathfrak{g},\sigma)$, and $v \in \mathscr{H}_c(\lambda)\otimes V(\vec{\mu})_{\vec{z}}$. Also, take an orthonormal basis $\{v_\phi\}_\phi$ of $\mathscr{H}_c(\lambda)\otimes V(\vec{\mu})_{\vec{z}}$. Then, we get (where $\hat{\mathcal{B}}^q_\sigma$ is an orthonormal basis of $\wedge^q L^-(\mathfrak{g},\sigma)$ with respect to the Hermitian form $\{\,,\,\}$ obtained as wedge product of an orthonormal basis of $L^-(\mathfrak{g},\sigma)$ and for $\vec{n} = (n_1,\dots,n_q)$, $|\vec{n}| := n_1\cdots n_q$)

$$
\begin{aligned}
&\partial_1^*(\vec{x}(-\vec{p})\otimes v)\\
&= \sum_{\substack{\vec{a}(-\vec{n})\in\hat{\mathcal{B}}^q_\sigma,\\ b(-k)\in\hat{\mathcal{B}}_\sigma, v_\phi}} \frac{|\vec{n}|}{k}\big\{\partial_1^*(\vec{x}(-\vec{p})\otimes v), b(-k)\boxtimes\vec{a}(-\vec{n})\otimes v_\phi\big\} b(-k)\boxtimes\vec{a}(-\vec{n})\otimes v_\phi\\
&= -\sum_{\substack{\vec{a}(-\vec{n})\in\hat{\mathcal{B}}^q_\sigma,\\ b(-k)\in\hat{\mathcal{B}}_\sigma, v_\phi}} \Big(\frac{|\vec{n}|}{k}\big\{\vec{x}(-\vec{p})\otimes v, ad_{b(-k)}(\vec{a}(-\vec{n}))\otimes v_\phi\big\} b(-k)\boxtimes\vec{a}(-\vec{n})\otimes v_\phi\\
&\quad + \frac{|\vec{n}|}{k}\big\{\vec{x}(-\vec{p})\otimes v, \vec{a}(-\vec{n})\otimes b(-k)\cdot v_\phi\big\} b(-k)\boxtimes\vec{a}(-\vec{n})\otimes v_\phi\Big)\\
&= -\sum_{\substack{\vec{a}(-\vec{n})\in\hat{\mathcal{B}}^q_\sigma,\\ b(-k)\in\hat{\mathcal{B}}_\sigma, v_\phi}} \Big(\frac{|\vec{n}|}{k}\big\{\vec{x}(-\vec{p}), ad_{b(-k)}(\vec{a}(-\vec{n}))\big\}_0\big\{v, v_\phi\big\} b(-k)\boxtimes\vec{a}(-\vec{n})\otimes v_\phi\\
&\quad + \frac{|\vec{n}|}{k}\big\{\vec{x}(-\vec{p}), \vec{a}(-\vec{n})\big\}_0\big\{v, b(-k)\cdot v_\phi\big\} b(-k)\boxtimes\vec{a}(-\vec{n})\otimes v_\phi\Big)\\
&= -\sum_{\substack{\vec{a}(-\vec{n})\in\hat{\mathcal{B}}^q_\sigma,\\ b(-k)\in\hat{\mathcal{B}}_\sigma, v_\phi}} \Big(\frac{|\vec{n}|}{k|\vec{p}|}\big\{ad_{\check{b}(k)}(\vec{x}(-\vec{p})), \vec{a}(-\vec{n})\big\}\big\{v, v_\phi\big\} b(-k)\boxtimes\vec{a}(-\vec{n})\otimes v_\phi\\
&\quad + \frac{|\vec{n}|}{k}\big\{\vec{x}(-\vec{p}), \vec{a}(-\vec{n})\big\}_0\big\{\check{b}(k)\cdot v, v_\phi\big\} b(-k)\boxtimes\vec{a}(-\vec{n})\otimes v_\phi\Big), \text{ by Lemmas 3.3 and 3.4}
\end{aligned}
$$

(25)

$$
= -\sum_{b(-k)\in\hat{\mathcal{B}}_\sigma}\Big(\sum_{i=1}^{q}\Big(\frac{p_i-k}{kp_i} b(-k)\boxtimes\overline{ad}^{\,i}_{\check{b}(k)}(\vec{x}(-\vec{p}))\otimes v\Big) + \frac{b(-k)}{k}\boxtimes\vec{x}(-\vec{p})\otimes\check{b}(k)\cdot v\Big),
$$

where $ad^i_{\check{b}(k)}(\vec{x}(-\vec{p}))$ denotes the action only on the i-th factor, and $\overline{ad}$ denotes the projection onto $L^-(\mathfrak{g},\sigma)$. Observe that the second term of the above sum is an infinite sum taking values in $\bar{L}^-(\mathfrak{g},\sigma)\otimes M_q$.

From the above equation (25) and the definition of ∂_1, the following proposition follows easily.

PROPOSITION 3.10. *Define* $\Box = \partial_1\circ\partial_1^*$. *Then, as an operator from* M_q *to* $\bar{M}_q := \overline{\wedge^q(L^-(\mathfrak{g},\sigma))\otimes\mathscr{H}_c(\lambda)\otimes V(\vec{\mu})_{\vec{z}}}$,

$$
\begin{aligned}
&\Box(\vec{x}(-\vec{p})\otimes v)\\
&= \sum_{b(-k)\in\hat{\mathcal{B}}_\sigma}\sum_{i=1}^{q}\Big(\frac{p_i-k}{kp_i}\Big(ad_{b(-k)}\big(\overline{ad}^{\,i}_{\check{b}(k)}(\vec{x}(-\vec{p}))\big)\otimes v + \overline{ad}^{\,i}_{\check{b}(k)}(\vec{x}(-\vec{p}))\otimes b(-k)\cdot v\Big)\Big)\\
&\quad + \sum_{b(-k)\in\hat{\mathcal{B}}_\sigma}\frac{1}{k}\Big(ad_{b(-k)}(\vec{x}(-\vec{p}))\otimes\check{b}(k)\cdot v + \vec{x}(-\vec{p})\otimes b(-k)\check{b}(k)\cdot v\Big),
\end{aligned}
$$

where $\bar{M}_q$ is the L^2-completion of M_q with respect to the metric $\{\, , \}_0$ on $L^-(\mathfrak{g},\sigma)$ and the standard metrics on $\mathscr{H}_c(\lambda)$ and $V(\vec{\mu})_{\vec{z}}$ as in Subsection 3.3.

Observe that the above sum makes sense as an element of $\bar{M}_q$ since each $|z_i| < 1$ by assumption.

3.4. Cochain complex. We now consider the standard cochain complex $C^* := C^*(L^+(\mathfrak{g},\sigma), \hat{\mathscr{H}}_c(\lambda) \otimes V(\vec{\mu})_{\vec{z}})$ of the Lie algebra $L^+(\mathfrak{g},\sigma)$ with coefficients in $\hat{\mathscr{H}}_c(\lambda) \otimes V(\vec{\mu})_{\vec{z}}$, where $\hat{\mathscr{H}}_c(\lambda)$ is the formal completion of $\mathscr{H}_c(\lambda)$ with respect to the energy grading (i.e., it is the direct *product* of the energy eigenspaces of $\mathscr{H}_c(\lambda)$),

$$C^q = \operatorname{Hom}_{\mathbb{C}}(\wedge^q L^+(\mathfrak{g},\sigma), \hat{\mathscr{H}}_c(\lambda) \otimes V(\vec{\mu})_{\vec{z}}),$$

and we denote $d^q : C^q \to C^{q+1}$ the standard differential in C^* (cf. [**Ku1**, §3.1.2]). Observe that

$$\hat{\mathscr{H}}_c(\lambda) \simeq \left(\mathscr{H}^{\text{low}}_{-c}(\lambda^{*_\sigma})\right)^* \text{ the full vector space dual,} \tag{26}$$

where λ^{*_σ} is the highest weight of the dual $\mathfrak{g}^\sigma$-module $V(\lambda)^*$ and $\mathscr{H}^{\text{low}}_{-c}(\lambda^{*_\sigma})$ is the integrable *lowest weight* $\hat{L}^f(\mathfrak{g},\sigma)$-module with lowest energy $\mathfrak{g}^\sigma$-module the dual of the $\mathfrak{g}^\sigma$-module $V(\lambda)$.

We take the Hermitian form $\{,\}_0$ on $L^-(\mathfrak{g},\sigma)$ as defined before, and the $\hat{L}(\mathfrak{g},\sigma)$ (resp. $\mathfrak{g}^{\Gamma_{z_i}}$)-contravariant positive-definite Hermitian form $\{,\}$ on $\mathscr{H}_c(\lambda)$ (resp. $V(\mu_i)_{z_i}$).

Define the invariant form $\langle\, , \rangle$ on $\hat{L}(\mathfrak{g},\sigma)$ by

$$\langle x[P], y[Q]\rangle = \langle x, y\rangle \operatorname{Res}_{t=0}(t^{-1}PQ), \quad \langle x[P], C\rangle = \langle C, C\rangle = 0, \text{ for } x[P], y[Q] \in \mathfrak{g}(\mathcal{K})^\sigma.$$

The invariant form $\langle,\rangle$ on $\hat{L}(\mathfrak{g},\sigma)$ gives rise to an embedding $\xi : L^-(\mathfrak{g},\sigma) \to L^+(\mathfrak{g},\sigma)^*$ given by $\xi(X)(Y) = \langle X, Y\rangle$, for $X \in L^-(\mathfrak{g},\sigma)$ and $Y \in L^+(\mathfrak{g},\sigma)$. This induces an embedding

$$\bar{\xi} : \wedge^q(L^-(\mathfrak{g},\sigma)) \otimes \mathscr{H}_c(\lambda) \otimes V(\vec{\mu})_{\vec{z}} \longrightarrow \operatorname{Hom}_{\mathbb{C}}(\wedge^q L^+(\mathfrak{g},\sigma), \hat{\mathscr{H}}_c(\lambda) \otimes V(\vec{\mu})_{\vec{z}}).$$

From now on we will often identify $\wedge^q(L^-(\mathfrak{g},\sigma)) \otimes \mathscr{H}_c(\lambda) \otimes V(\vec{\mu})_{\vec{z}}$ as a subspace of $\operatorname{Hom}_{\mathbb{C}}(\wedge^q L^+(\mathfrak{g},\sigma), \hat{\mathscr{H}}_c(\lambda) \otimes V(\vec{\mu})_{\vec{z}})$ via the embedding $\bar{\xi}$.

Let $(d^{q-1})^* : \wedge^q(L^-(\mathfrak{g},\sigma)) \otimes \mathscr{H}_c(\lambda) \otimes V(\vec{\mu})_{\vec{z}} \to C^{q-1}$ be the restriction of the formal adjoint of d^{q-1} to $\wedge^q(L^-(\mathfrak{g},\sigma)) \otimes \mathscr{H}_c(\lambda) \otimes V(\vec{\mu})_{\vec{z}}$ (via the embedding $\bar{\xi}$). Then, $(d^{q-1})^*$ extends continuously (under the L^2-metric on $\bar{L}^-(\mathfrak{g},\sigma)$) to a linear map (still denoted by)

$$(d^{q-1})^* : \wedge^q(\bar{L}^-(\mathfrak{g},\sigma)) \otimes \mathscr{H}_c(\lambda) \otimes V(\vec{\mu})_{\vec{z}} \to \wedge^{q-1}(\bar{L}^-(\mathfrak{g},\sigma)) \otimes \mathscr{H}_c(\lambda) \otimes V(\vec{\mu})_{\vec{z}}.$$

By Lemma 4.1, $d^q\left(\wedge^q(L^-(\mathfrak{g},\sigma)) \otimes \mathscr{H}_c(\lambda) \otimes V(\vec{\mu})_{\vec{z}}\right)$ lies in $\wedge^{q+1}(\bar{L}^-(\mathfrak{g},\sigma)) \otimes \mathscr{H}_c(\lambda) \otimes V(\vec{\mu})_{\vec{z}}$ and by Lemma 4.4,

$$(d^{q-1})^*\left(\wedge^q(L^-(\mathfrak{g},\sigma)) \otimes \mathscr{H}_c(\lambda) \otimes V(\vec{\mu})_{\vec{z}}\right)$$

lies in $\wedge^{q-1}(L^-(\mathfrak{g},\sigma)) \otimes \mathscr{H}_c(\lambda) \otimes V(\vec{\mu})_{\vec{z}}$. Let $\overline{\Box} := dd^* + d^*d$ be the Laplacian acting on $\wedge^q(L^-(\mathfrak{g},\sigma)) \otimes \mathscr{H}_c(\lambda) \otimes V(\vec{\mu})_{\vec{z}}$ with image in C^q. We prove the following theorem giving an expression for the difference $\overline{\Box} - \Box$ restricted to $[\wedge^q(L^-(\mathfrak{g},\sigma)) \otimes \mathscr{H}_c(\lambda) \otimes V(\vec{\mu})_{\vec{z}}]^{\mathfrak{g}_0}$.

THEOREM 3.11 (Nakano identity). *Let σ be a special automorphism of $\mathfrak{g}$ of order m. For $\vec{x}(-\vec{p}) \otimes v_1 \otimes v_2 \in [\wedge^q(L^-(\mathfrak{g},\sigma)) \otimes \mathscr{H}_c(\lambda) \otimes V(\vec{\mu})_{\vec{z}}]^{\mathfrak{g}^\sigma}$, where $\vec{x}(-\vec{p}) = x_1(-p_1) \wedge \cdots \wedge x_q(-p_q)$,*

$$
\begin{aligned}
&\left(\overline{\square}-\square\right)\left(\vec{x}(-\vec{p})\otimes v_1\otimes v_2\right)\\
&=\frac{(c+2\check{h})q}{m}\vec{x}(-\vec{p})\otimes v_1\otimes v_2\\
&\quad+\sum_{1\le\ell\le q;b(-k)\in\hat{\mathcal{B}}_\sigma}\frac{1}{p_\ell}ad^\ell_{b(-k)}\vec{x}(-\vec{p})\otimes v_1\otimes\left((1-(z\bar{z})^{p_\ell})z^k\check{b}\right)\cdot v_2\\
&\quad+\sum_{\ell;b(-k)\in\hat{\mathcal{B}}_\sigma}\frac{1}{p_\ell}\overline{ad}^\ell_{\check{b}(k)}\vec{x}(-\vec{p})\otimes v_1\otimes\left((1-(z\bar{z})^{p_\ell-k})(\bar{z})^k b\right)\cdot v_2\\
&\quad+\sum_{\ell;b_0\in\mathcal{B}_{\underline{0}}}\frac{1}{p_\ell}ad^\ell_{b_0}\vec{x}(-\vec{p})\otimes v_1\otimes\left((1-(z\bar{z})^{p_\ell})\check{b}_0\right)\cdot v_2,
\end{aligned}
$$

where

$$
(z^{k_1}\bar{z}^{k_2}\check{b})\cdot v_2:=\sum_{i=0}^{s}v_2^0\otimes\cdots\otimes z_i^{k_1}\bar{z}_i^{k_2}\check{b}\cdot v_2^i\otimes\cdots\otimes v_2^s,
$$

$$
\text{for } v_2=v_2^0\otimes\cdots\otimes v_2^s \text{ with } v_2^i\in V(\mu_i).
$$

Here c is the scalar by which C of $\hat{L}(\mathfrak{g},\sigma)$ acts on $\mathscr{H}_c(\lambda)$ and $\check{h}$ is the dual Coxeter number of $\mathfrak{g}$. Observe that, from the above expression, we see that

$$
(\overline{\square}-\square)\left(\left[\wedge^q\left(L^-(\mathfrak{g},\sigma)\right)\otimes\mathscr{H}_c(\lambda)\otimes V(\vec{\mu})_{\vec{z}}\right]^{\mathfrak{g}^\sigma}\right)\subset\left[\wedge^q\left(\bar{L}^-(\mathfrak{g},\sigma)\right)\otimes\mathscr{H}_c(\lambda)\otimes V_{\vec{z}}(\vec{\mu})\right]^{\mathfrak{g}^\sigma}.
$$

Its proof is given in Section 4.

3.5. Difference of Laplacians. Recall that $\bar{L}^-(\mathfrak{g})$ (resp. $\bar{L}^-(\mathfrak{g},\sigma)$) is the L^2-completion of $L^-(\mathfrak{g})$ (resp. $L^-(\mathfrak{g},\sigma)$) with respect to the positive definite Hermitian form $\{,\}_0$ on $L^-(\mathfrak{g})$ (resp. $L^-(\mathfrak{g},\sigma)$).

Define the operator $T_z: L^-(\mathfrak{g},\sigma)\otimes V(\mu)_z\to\bar{L}^-(\mathfrak{g},\sigma)\otimes V(\mu)_z$ for any $z\in\mathbb{D}:=\{z\in\mathbb{C},|z|<1\}$, and $V(\mu)$ an irreducible $\mathfrak{g}$-module with highest weight μ if $z\neq0$ and $V(\mu)$ an irreducible $\mathfrak{g}^\sigma$-module if $z=0$, by

(27)

$$
T_z(x(-p)\otimes v)=\begin{cases}\sum_{b(-k)\in\hat{\mathcal{B}}_\sigma}\frac{1}{p}\Big(ad_{b(-k)}x(-p)\otimes(1-(z\bar{z})^p)z^k\check{b}\cdot v\\ \qquad+\overline{ad}_{\check{b}(k)}x(-p)\otimes(1-(z\bar{z})^{p-k})(\bar{z})^k b\cdot v\Big)\\ \qquad+\sum_{b_0\in\mathcal{B}_{\underline{0}}}\frac{1}{p}ad_{b_0}x(-p)\otimes(1-(z\bar{z})^p)\,\check{b}_0\cdot v, & \text{if } z\neq0,\\ \sum_{b_0\in\mathcal{B}_{\underline{0}}}\frac{1}{p}[b_0,x](-p)\otimes\check{b}_0\cdot v, & \text{if } z=0.\end{cases}
$$

Then, the operator T_z extends to a multilinear operator

$$
T_{\vec{z}}:\wedge^q(L^-(\mathfrak{g},\sigma))\otimes\mathscr{H}_c(\lambda)\otimes V(\vec{\mu})_{\vec{z}}\to\wedge^q(\bar{L}^-(\mathfrak{g},\sigma))\otimes\mathscr{H}_c(\lambda)\otimes V(\vec{\mu})_{\vec{z}} \tag{28}
$$

defined by

$$
\begin{aligned}
&T_{\vec{z}}\left((x_1(-p_1)\wedge\cdots\wedge x_q(-p_q))\otimes h\otimes v_0\otimes\cdots\otimes v_s\right)\\
&=\sum_{k=0}^{s}\sum_{\ell=1}^{q}x_1(-p_1)\wedge\cdots\wedge T_{z_k}(x_\ell(-p_\ell)\otimes v_k)\wedge\cdots\wedge x_q(-p_q)\otimes h\otimes v_0\otimes\cdots\otimes\hat{v_k}\otimes\cdots\otimes v_s,
\end{aligned}
$$

for any $h\otimes v_0\otimes\cdots\otimes v_s\in\mathscr{H}_c(\lambda)\otimes V(\vec{\mu})_{\vec{z}}$. In view of the operator $T_{\vec{z}}$, the Nakano's identity (Theorem 3.11) clearly takes the following form:

THEOREM 3.12. *Follow the notation and assumptions from Theorem 3.11. In particular, σ is a special automorphism of $\mathfrak{g}$ of order m. Then, for $\vec{x}(-\vec{p}) \otimes v \in [\wedge^q(L^-(\mathfrak{g},\sigma)) \otimes \mathscr{H}_c(\lambda) \otimes V(\vec{\mu})_{\vec{z}}]^{\mathfrak{g}^\sigma}$, where $\vec{x}(-\vec{p}) := x_1(-p_1) \wedge \cdots \wedge x_q(-p_q) \in \wedge^q L^-(\mathfrak{g},\sigma)$ and $v \in \mathscr{H}_c(\lambda) \otimes V(\vec{\mu})_{\vec{z}}$, we have*

$$(\bar{\square} - \square)(\vec{x}(-\vec{p}) \otimes v) = q\frac{c+2\check{h}}{m}\vec{x}(-\vec{p}) \otimes v + T_{\vec{z}}(\vec{x}(-\vec{p}) \otimes v).$$

Let $V(\mu)$ be an irreducible representation of $\mathfrak{g}^\sigma$. Now, consider the self-adjoint operator $S_0 : \mathfrak{g} \otimes V(\mu) \to \mathfrak{g} \otimes V(\mu)$ given by

$$S_0(x \otimes v) = \sum_{b \in \mathcal{B}_{\underline{0}}} [b, x] \otimes \check{b} \cdot v.$$

Then, by Lemma 3.9, for $(\mathfrak{g}, m)$ different from $(D_4, 3)$, eigenvalues of S_0 are bounded above in absolute value by

$$d_\mu := \max_{i=0,\cdots,m-1} \{\langle \mu + 2\rho_\sigma, \theta_{\underline{i}} \rangle\}. \tag{29}$$

Let σ be a special automorphism of $\mathfrak{g}$ of order m and let σ act on $\mathbb{P}^1$ by $\sigma(z) = e^{\frac{2\pi i}{m}} z$ for $z \in \mathbb{P}^1$. Let $\vec{z} = (z_0 = 0, z_1, \cdots, z_s)$ be distinct points in the unit disc $\mathbb{D}$ with distinct Γ-orbits, and $\vec{\mu} = (\mu_i)_{0 \le i \le s}$ be the weights associated to $\vec{z}$ respectively, where Γ is the cyclic group of order m generated by σ, μ_0 is a dominant weight of $\mathfrak{g}^\sigma$ and μ_i $(i \geq 1)$ is a dominant weight of $\mathfrak{g}$.

Let

$$d_{\vec{\mu}} := \max_{k=1,\cdots,s} \{d_{\mu_0}, \langle \mu_k + 2\rho, \theta \rangle\}, \tag{30}$$

where d_{μ_0} is defined in (29), and θ is the highest root of $\mathfrak{g}$.

Consider the $\mathbb{C}$-linear map (for any $z \in \mathbb{D}$),

$$\eta_z : t^{-1}\mathbb{C}[t^{-1}] \to t^{-1}\mathbb{C}[[t^{-1}]]$$

defined by (for any $p > 0$)

$$\eta_z(t^{-p}) := \frac{1}{p} \sum_{0<n<p} (\bar{z}^{p-n}(1 - |z|^{2n})t^{-n}) + \frac{1}{p}(1 - |z|^{2p}) \sum_{n \geq 0} z^n t^{-n-p}. \tag{31}$$

As before, $L^-(\mathfrak{g}) := t^{-1}\mathfrak{g}[t^{-1}]$ with the Hermitian form $\{\,\}_0$ and $\bar{L}^-(\mathfrak{g})$ is its L^2-completion. Then, it is easy to see that for $z \neq 0$,

(32)

$$T_z(x(-p)\otimes v) = \pi\left(\sum_{b\in\mathcal{B}} ([b,x]\otimes\eta_z(t^{-p}))\right)\otimes\check{b}\cdot v, \text{ for } x(-p)\in L^-(\mathfrak{g},\sigma) \text{ and } v\in V(\mu)_z,$$

where $\pi : \bar{L}^-(\mathfrak{g}) \to \bar{L}^-(\mathfrak{g}, \sigma)$ is the orthogonal projection onto the σ-invariants.

DEFINITION 3.13. For any $z \in \mathbb{D}$, choose a Möbius transformation $\phi = \phi_z$ of the unit disc $\mathbb{D}$:

$$\phi(t) = \frac{t+z}{1+\bar{z}t}.$$

Then, $\phi^{-1}(t) = \frac{t-z}{1-\bar{z}t}$. Let H_0^{-1} be the operator acting on $t\mathbb{C}[[t]]$ such that $H_0^{-1}(t^k) = \frac{t^k}{k}$ for any $k \geq 1$. Then, the inverse of H_0^{-1} on $t\mathbb{C}[[t]]$ is $H_0 = t\frac{d}{dt}$. We extend H_0^{-1} to $\hat{H}_0^{-1}$ acting on $\mathbb{C}[[t]]$ defined by

$$\hat{H}_0^{-1}(f(t)) := H_0^{-1}(f(t) - f(0)), \text{ for } f(t) \in \mathbb{C}[[t]]. \tag{33}$$

Following Teleman (cf. [**Te**, Proposition 2.5.4]), let $H_z^{-1} : L^-(\mathfrak{g}) \to L^+(\mathfrak{g})^*$ be the operator defined by

$$\begin{aligned}(34)\quad &\left(H_z^{-1}(x(-p))\right)(y(n))\\ &= \langle y \otimes \phi^{-1} \circ \hat{H}_0^{-1} \circ \phi(t^n), x(-p)\rangle, \text{ for } x(-p) \in L^-(\mathfrak{g}) \text{ and } y(n) \in L^+(\mathfrak{g}).\end{aligned}$$

Then, the image of H_z^{-1} is contained in the completion $\bar{L}^-(\mathfrak{g})$ under the identification of $\bar{L}^-(\mathfrak{g})$ as a subspace of $L^+(\mathfrak{g})^*$ under ξ, where $\xi : L^-(\mathfrak{g}) \to L^+(\mathfrak{g})^*$ is induced from the invariant form $\langle\, ,\, \rangle$ on $\hat{L}(\mathfrak{g})$ (cf. Subsection 3.4). Thus, we will think of the operator H_z^{-1} as a linear map $L^-(\mathfrak{g}) \to \bar{L}^-(\mathfrak{g})$.

PROPOSITION 3.14. *For any $z \in \mathbb{D}$, the operator H_z^{-1} coincides with the operator $x(-p) \in L^-(\mathfrak{g}) \mapsto x \otimes \eta_z(t^{-p}) \in \bar{L}^-(\mathfrak{g})$.*

PROOF. We first show that, for any $n \geq 1$, we have the following equality:

$$(35)\qquad (\phi^{-1} \circ \hat{H}_0^{-1} \circ \phi)(t^n) = \sum_{k \geq 0} a_k^n t^k,$$

where

$$(36)\qquad a_k^n = \begin{cases} \frac{1}{k}(1 - |z|^{2k})z^{n-k}, & \text{when } 1 \leq k \leq n \\ \frac{1}{k}\bar{z}^{k-n}(1 - |z|^{2n}), & \text{when } k \geq n\ . \end{cases}$$

Observe that when $k = n$, both of the above cases give the same value.

By (33), it suffices to show that

$$\begin{aligned}\phi(t)^n - z^n &= \sum_{k \geq 1} a_k^n t \frac{d}{dt}(\phi(t)^k)\\ &= \sum_{k \geq 1} k a_k^n \phi(t)^{k-1} t \phi(t)'\\ &= \sum_{k \geq 1} k a_k^n \phi(t)^{k-1}(1 - |z|^2)\frac{t}{(1 + \bar{z}t)^2}.\end{aligned}$$

Equivalently,

(37)

$$t^n - z^n = \sum_{k \geq 1} k a_k^n t^{k-1}(1 - |z|^2)\frac{\phi^{-1}(t)}{(1 + \bar{z}\phi^{-1}(t))^2} = \sum_{k \geq 1} \frac{k a_k^n}{1 - |z|^2} t^{k-1}(t - z)(1 - \bar{z}t).$$

We compare this identity on each degree of t. When the degree of t is 0, we get $a_1^n = z^{n-1}(1 - |z|^2)$. For any $1 \leq k \leq n$, we now prove the identity $a_k^n = \frac{1}{k}(1 - |z|^{2k})z^{n-k}$. When $k = 1$, we have just confirmed it. When $k \geq 2$, by comparing the coefficients of t in (37), we get

$$a_2^n = \frac{1}{2}z^{n-2}(1 - |z|^4) - \delta_{n,1}\frac{1 - |z|^2}{2z}.$$

This proves (35) for $k = 2$ and any $n \geq 1$.

Thus, we can assume $n \geq 3$. We induct on k. When $3 \leq k \leq n$, comparing the coefficients of t^{k-1}, we have

$$\begin{aligned}k a_k^n &= \frac{1}{z}(-\bar{z}(k-2)a_{k-2}^n + (1 + |z|^2)(k-1)a_{k-1}^n)\\ &= \frac{1}{z}(-\bar{z}(1 - |z|^{2k-4})z^{n-k+2} + (1 + |z|^2)(1 - |z|^{2k-2})z^{n-k+1})\end{aligned}$$

$$= (1-|z|^{2k})z^{n-k}.$$

Now, we consider the case $k > n$. When $k = n+1$, comparing the coefficients of t^n in (37), we get

$$\begin{aligned}(n+1)a^n_{n+1} &= \frac{1}{z}(-(1-|z|^2) + (n-1)a^n_{n-1}(-\bar{z}) + na^n_n(1+|z|^2)) \\ &= \frac{1}{z}(-(1-|z|^2) + (1-|z|^{2n-2})(-|z|^2) + (1-|z|^{2n})(1+|z|^2)) \\ &= \bar{z}(1-|z|^{2n}).\end{aligned}$$

When $k = n+2$, we have

$$(n+2)a^n_{n+2} = \frac{1}{z}(na^n_n(-\bar{z}) + (n+1)a^n_{n+1}(1+|z|^2)) = \bar{z}^2(1-|z|^{2n}).$$

When $k \geq n+3$, inducting on k, we have

$$\begin{aligned}ka^n_k &= \frac{1}{z}((k-2)a^n_{k-2}(-\bar{z}) + (k-1)a^n_{k-1}(1+|z|^2)) \\ &= \frac{1}{z}(-\bar{z}^{k-1-n}(1-|z|^{2n}) + \bar{z}^{k-1-n}(1-|z|^{2n})(1+|z|^2)) \\ &= \bar{z}^{k-n}(1-|z|^{2n}).\end{aligned}$$

This finishes the proof of the identity (35).

Write

$$H_z^{-1}(x(-p)) = \sum_{n>0; b\in\mathcal{B}} c_{b(-n)} b(-n), \text{ for some } c_{b(-n)} \in \mathbb{C}.$$

Pairing the right side of the above identity with $\check{b}(n)$ under $\langle\, ,\, \rangle$, we get

$$c_{b(-n)} = \langle \check{b}, x\rangle \operatorname{Res}_{t=0}\left(\phi^{-1}\circ \hat{H}_0^{-1}\circ\phi(t^n)t^{-p-1}\right).$$

Thus,

$$\begin{aligned}H_z^{-1}(x(-p)) &= \sum_{n>0; b\in\mathcal{B}} \langle \check{b}, x\rangle \operatorname{Res}_{t=0}\left(\phi^{-1}\circ \hat{H}_0^{-1}\circ\phi(t^n)t^{-p-1}\right) b(-n) \\ &= \sum_{n>0} \operatorname{Res}_{t=0}\left(\phi^{-1}\circ \hat{H}_0^{-1}\circ\phi(t^n)t^{-p-1}\right) x(-n) \\ &= \sum_{n>0} a^n_p(x(-n)), \text{ where } a^n_p \text{ is given by the equation (36)} \\ &= \frac{1}{p}\sum_{0<n<p} \bar{z}^{p-n}(1-|z|^{2n})x(-n) + \frac{1}{p}(1-|z|^{2p})\sum_{n\geq p} z^{n-p}x(-n). \qquad (38)\end{aligned}$$

Thus, from the above identity, the operator H_z^{-1} coincides with the operator

$$x(-p) \mapsto x\otimes \eta_z(t^{-p}), \text{ for } x(-p)\in L^-(\mathfrak{g}).$$

This proves the proposition. □

In view of Proposition 3.14, the following proposition is a restatement of Teleman ([**Te**, Proposition 2.5.5]).

PROPOSITION 3.15. *For any $A = \sum_{p\geq 1} x_p(-p) \in L^-(\mathfrak{g})$ and any $\epsilon > 0$,*

$$\| \sum_{k=0}^{s} \sum_{p\geq 1} x_p \otimes \eta_{z_k}(t^{-p}) \|_0 \leq (1+\epsilon)\|A\|_0,$$

provided the points $\{z_0, \cdots, z_s\}$ are sufficiently far apart (only depending upon ϵ) in the hyperbolic metric on the unit disc $\mathbb{D}$, where $\| - \|_0$ is the norm with respect to the Hermitian form $\{-\}_0$ (defined by the equation (9) by taking $\sigma = \mathrm{Id}$). Such points $\{z_0, \cdots, z_s\}$ always exist.

PROPOSITION 3.16. *Let σ be a special automorphism of $\mathfrak{g}$ of order m and let $(\mathfrak{g}, m)$ be different from $(D_4, 3)$. For any given $\epsilon > 0$, there exist points $\{z_0 = 0, z_1, \ldots, z_s\}$ in $\mathbb{D}$ far apart in the hyperbolic metric such that the operator $T_{\vec{z}}$ defined in (28) satisfies $| \{T_{\vec{z}}(A), A\}_0 | \leq q d_{\vec{\mu}}(1+\epsilon)\|A\|_0^2$, for any A in $\wedge^q(L^-(\mathfrak{g},\sigma)) \otimes \mathscr{H}_c(\lambda) \otimes V(\vec{\mu})_{\vec{z}}$.*

PROOF. By the definition of $T_{\vec{z}}$ (cf. (28)),

$$\{T_{\vec{z}}(A), A\}_0 = \sum_{\ell=1}^{q} \{\sum_{k=0}^{s} T^{\ell}_{z_k}(A), A\}_0, \tag{39}$$

where $T^{\ell}_{z_k}(A)$ denotes the action only on the ℓ-th factor of $\wedge^q L^-(\mathfrak{g},\sigma)$ and the factor $V(\mu_k)_{z_k}$.

Let ρ^k denote the representation $V(\mu_k)$ of $\mathfrak{g}^{\Gamma_{z_k}}$. Define the operator $S_k := \sum_{b\in\mathcal{B}} ad_b \boxtimes \rho^k_{\check{b}}$ on $L^-(\mathfrak{g}) \otimes V(\mu_k)$ for $k \geq 1$ (which clearly extends to an operator on $\bar{L}^-(\mathfrak{g}) \otimes V(\mu_k)$); and when $k = 0$, denote $S_0 = \sum_{b\in\mathcal{B}_{\underline{0}}} ad_b \boxtimes \rho^0_{\check{b}}$ on $L^-(\mathfrak{g}) \otimes V(\mu_0)$. Consider the operator

$$\bar{\eta}_{z_k} : L^-(\mathfrak{g}) \otimes V(\mu_k) \to \bar{L}^-(\mathfrak{g}) \otimes V(\mu_k)$$

defined by

$$\bar{\eta}_{z_k}(x(-p) \otimes v) = (x \otimes \eta_{z_k}(t^{-p})) \otimes v, \qquad \text{for any } x \in \mathfrak{g}, p \geq 1 \text{ and } v \in V(\mu_k),$$

where η_{z_k} is defined in (31), and extend it multi-linearly on $\wedge^q(L^-(\mathfrak{g})) \otimes V(\mu_k)$. By Proposition 3.15, $\bar{\eta}_{z_k}$ extends uniquely to a continuous operator on $\bar{L}^-(\mathfrak{g}) \otimes V(\mu_k)$ with respect to the Hermitian metric $\{,\}_0$ on $\bar{L}^-(\mathfrak{g})$. Then, the operator S_k is self-adjoint and it commutes with $\bar{\eta}_{z_k}$ for any $0 \leq k \leq s$. We extend both the operators S^{ℓ}_k and $\bar{\eta}^{\ell}_{z_k}$ to $\wedge^q(\bar{L}^-(\mathfrak{g})) \otimes \mathscr{H}_c(\lambda) \otimes V(\vec{\mu})_{\vec{z}}$ by defining them to be the identity operator on the $\mathscr{H}_c(\lambda)$ factor, where S^{ℓ}_k is the operator S_k acting only on the ℓ-th factor of $\wedge^q \bar{L}^-(\mathfrak{g})$ and the k-th factor $V(\mu_k)$ of $V(\vec{\mu})_{\vec{z}}$ and similarly for $\bar{\eta}^{\ell}_{z_k}$. Moreover, for $A, B \in \wedge^q(L^-(\mathfrak{g},\sigma)) \otimes \mathscr{H}_c(\lambda) \otimes V(\vec{\mu})_{\vec{z}}$,

$$\{T^{\ell}_{z_k} A, B\}_0 = \{S^{\ell}_k \circ \bar{\eta}^{\ell}_{z_k}(A), B\}_0, \tag{40}$$

since for any $x(-p), x'(-p') \in L^-(\mathfrak{g},\sigma)$, any $1 \leq k \leq s$ and $v \in V(\mu_k)$,

$$\{\sum_{b\in\mathcal{B}} \pi([b,x]\otimes\eta_{z_k}(t^{-p}))\otimes\check{b}\cdot v, x'(-p')\otimes v\}_0 = \{\sum_{b\in\mathcal{B}} ([b,x]\otimes\eta_{z_k}(t^{-p}))\otimes\check{b}\cdot v, x'(-p')\otimes v\}_0,$$

and similarly for $k = 0$, where π is the orthogonal projection onto the σ-invariants (as defined below equation (32)).

We next show that for any $1 \leq \ell \leq q$ and $A \in \wedge^q(L^-(\mathfrak{g},\sigma)) \otimes \mathscr{H}_c(\lambda) \otimes V(\vec{\mu})_{\vec{z}}$,

$$| \{\sum_{k=0}^{s} S^{\ell}_k \circ \bar{\eta}^{\ell}_{z_k}(A), A\}_0 | \leq d_{\vec{\mu}} \sum_{k=0}^{s} \{\bar{\eta}^{\ell}_{z_k}(A), A\}_0. \tag{41}$$

To prove this, observe first that any eigenvalue of the operator S_k, acting on $\mathfrak{g} \otimes V(\mu_k)$ for $k \geq 1$, is bounded above in absolute value by $\langle \mu_k + 2\rho, \theta \rangle$ (by Lemma 3.9 for $\sigma = \text{Id}$). Further, for $z_0 = 0$, i.e., $k = 0$, by Lemma 3.9, the operator S_0 acting on $\mathfrak{g} \otimes V(\mu_0)$ has all its eigenvalues in absolute value bounded above by d_{μ_0} defined in (29). Thus, all the eigenvalues in absolute value of any of the operators $\{S_k\}_{0 \leq k \leq s}$ are bounded above by $d_{\vec{\mu}}$.

We now prove (41). Decompose $\wedge^q(L^-(\mathfrak{g})) \otimes \mathscr{H}_c(\lambda) \otimes V(\vec{\mu})_{\vec{z}}$ into the direct sum of eigenspaces under S_k^ℓ as follows:

$$\wedge^q(L^-(\mathfrak{g})) \otimes \mathscr{H}_c(\lambda) \otimes V(\vec{\mu})_{\vec{z}} = \perp_{|n| \leq d_{\vec{\mu}}} E_n^k,$$

where E_n^k is the eigenspace of the self-adjoint operator S_k^ℓ with eigenvalue n. Since the operator $\bar{\eta}_{z_k}^\ell$ commutes with S_k^ℓ for any $0 \leq k \leq s$ and any $1 \leq \ell \leq q$, the eigenspaces E_n^k are stable under the action of $\bar{\eta}_{z_k}^\ell$. Write

$$A = \sum_n A_n^k, \text{ for } A_n^k \in E_n^k.$$

By orthogonality, $A_n^k \perp A_m^k$ for $n \neq m$. Thus, we get (for any fixed $0 \leq k \leq s$ and $1 \leq \ell \leq q$)

$$\begin{aligned} | \{S_k^\ell \circ \bar{\eta}_{z_k}^\ell(A), A\}_0 | &= | \{\sum_n n\bar{\eta}_{z_k}^\ell(A_n^k), A\}_0 | = | \sum_{|n| \leq d_{\vec{\mu}}} n\{\bar{\eta}_{z_k}^\ell(A_n^k), A_n^k\}_0 | \\ &\leq d_{\vec{\mu}} \sum_n \{\bar{\eta}_{z_k}^\ell(A_n^k), A_n^k\}_0 = d_{\vec{\mu}}\{\bar{\eta}_{z_k}^\ell(A), A\}_0, \end{aligned} \tag{42}$$

since $\bar{\eta}_{z_k}^\ell$ is a positive operator as it is diagonalizable with all its eigenvalues $\{\frac{1}{p}\}_{p \in \mathbb{Z}_{>0}}$ (cf. [**R**, Theorem 12.32]). To prove this, observe that $\bar{\eta}_{z_k}^\ell$ is conjugate of the operator $\bar{\eta}_0^\ell$ under the action of ϕ_{z_k} on the holomorphic functions on the unit disc which keeps the inner product $\{,\}_0$ invariant (use Proposition 3.14). Thus, by identity (42), for any $1 \leq \ell \leq q$, the inequality (41) holds.

Thus, by Proposition 3.15, identities (39) and (40) (for $B = A$) and the inequality (41), we have

$$| \{T_{\vec{z}}A, A\}_0 | = | \sum_{\ell=1}^q \{\sum_{k=0}^s T_{z_k}^\ell(A), A\}_0 | \leq q d_{\vec{\mu}}(1 + \epsilon)\|A\|_0^2,$$

provided the points $\{z_k\}$ are far apart under the hyperbolic metric in the unit disc. This proves our proposition. □

We now come to the main theorem of this paper.

THEOREM 3.17. *Let σ be a special automorphism of $\mathfrak{g}$ of order m and let $(\mathfrak{g}, m)$ be different from $(D_4, 3)$. For any $q \geq 1$, the Lie algebra cohomology*

$$H^q(L^+(\mathfrak{g}, \sigma), \hat{\mathscr{H}}_c(\lambda) \otimes V(\vec{\mu})_{\vec{z}})^{\mathfrak{g}^\sigma} = 0, \quad \text{if } d_{\vec{\mu}} < \frac{c + 2\check{h}}{m},$$

and the points $\vec{z} = \{z_0 = 0, z_1, \ldots, z_s\}$ in $\mathbb{D}$ are far apart in the hyperbolic metric.

PROOF. Let A be a nonzero element in $[\wedge^q(L^-(\mathfrak{g}, \sigma)) \otimes \mathscr{H}_c(\lambda) \otimes V(\vec{\mu})_{\vec{z}}]^{\mathfrak{g}^\sigma}$ such that $\bar{\Box} A = 0$. Then, by Theorem 3.12,

$$0 = \bar{\Box} A = \Box A + \frac{q(c + 2\check{h})}{m} A + T_{\vec{z}}A.$$

This gives

$$-\{T_{\vec{z}}A, A\}_0 = \{\Box A, A\}_0 + \frac{q(c+2\check{h})}{m}\|A\|_0^2,$$

which gives, since $\{\Box A, A\}_0 \geq 0$,

$$-\{T_{\vec{z}}A, A\}_0 \geq \frac{q(c+2\check{h})}{m}\|A\|_0^2.$$

But, by Proposition 3.16,

$$| \{T_{\vec{z}}A, A\}_0 | \leq q d_{\vec{\mu}}(1+\epsilon)\|A\|_0^2.$$

Thus, we get $\frac{c+2\check{h}}{m} \leq d_{\vec{\mu}}(1+\epsilon)$. By assumption, $d_{\vec{\mu}} < \frac{c+2\check{h}}{m}$, which violates the last inequality for some small enough ϵ. This proves that $A = 0$. Thus, $\bar{\Box}$ has no nonzero kernel in $[\wedge^q(L^-(\mathfrak{g},\sigma)) \otimes \mathscr{H}_c(\lambda) \otimes V(\vec{\mu})_{\vec{z}}]^{\mathfrak{g}^\sigma}$. Rest of the proof of the theorem follows from [**Te**, Proposition 3.3.1], since by loc. cit., for any element in the cohomology H^q, there exists a representative A such that $\bar{\Box}A = 0$. □

REMARK 3.18.

(a) In fact, the above Theorem 3.17 remains true under no restrictions on $\vec{z} \in \mathbb{C}^{s+1}$ (except that they are distinct) by following the same argument as that of the proof of [**Te**, Proposition 3.4.2].

(b) Theorem 3.17 extends [**Te**, Theorem 0 (c)] to the *twisted* situation.

Let σ be an automorphism of $\mathfrak{g}$ of order m. Further, as in Notation 3.2, let σ act on $\mathbb{P}^1$ via $\sigma(z) = e^{\frac{2\pi i}{m}} z$ for $z \in \mathbb{P}^1$. Take $s+2$ points $(\infty, \vec{z})$ in $\mathbb{P}^1$, where $\vec{z} = (z_0 = 0, z_1, \ldots, z_s)$, such that their σ orbits are distinct. Let $\lambda \in D_{c,\sigma^{-1}}, \mu_0 \in D_{c,\sigma}$ and $\mu_1, \ldots, \mu_s \in D_c$ be attached successively to the points $(\infty, \vec{z})$. Let Γ be the cyclic group generated by σ, and let Γ_z be the stabilizer group of Γ at $z \in \mathbb{P}^1$. Let λ^* (resp. μ_i^*) denote the highest weight of the dual representation $V(\lambda)^*$ (resp. $V(\mu_i)^*$) of $\mathfrak{g}^\sigma$ (resp. $\mathfrak{g}^{\Gamma_{z_i}}$).

LEMMA 3.19. *Follow the notation as above.*

(a) *The space of vacua*

$$\mathscr{V}_{\mathbb{P}^1,\Gamma}\left((\infty,\vec{z}),(\lambda,\mu_0,\mu_1,\ldots,\mu_s)\right)^\dagger \simeq H^0\left(L^+(\mathfrak{g},\sigma), \hat{\mathscr{H}}_c(\lambda^*) \otimes V(\vec{\mu}^*)_{\vec{z}}\right)^{\mathfrak{g}^\sigma},$$

where $\vec{\mu}^* = (\mu_0^*, \mu_1^*, \ldots, \mu_s^*)$.

(b) *For any* $\lambda \in D_{c,\sigma^{-1}}, \mu_0$ *a dominant weight of* $\mathfrak{g}^\sigma$ *and* $\mu_1, \ldots, \mu_s$ *any dominant weights of* $\mathfrak{g}$ *(we do not require* $\mu_0 \in D_{c,\sigma}$ *nor do we require* $\mu_1, \ldots, \mu_s \in D_c$*) and any* $q \geq 0$,

$$H^q\left(L^+(\mathfrak{g},\sigma), \hat{\mathscr{H}}_c(\lambda^*) \otimes V(\vec{\mu}^*)_{\vec{z}}\right) \simeq \left(H_q(L^+(\mathfrak{g},\sigma), \mathscr{H}_c(\lambda)_\infty \otimes V(\vec{\mu})_{\vec{z}})\right)^*$$

as $\mathfrak{g}^\sigma$*-modules, where* $\mathscr{H}_c(\lambda)_\infty$ *is the highest weight integrable module of* $\hat{L}(\mathfrak{g},\sigma^{-1})$ *attached to* $\infty \in \mathbb{P}^1$.

PROOF.

(a) To prove the (a)-part, by the Propagation of Vacua (cf. [**HK1**, Theorem 4.3]),

$$
\begin{aligned}
&\mathscr{V}_{\mathbb{P}^1,\Gamma}\left((\infty,\vec{z}),(\lambda,\mu_0,\mu_1,\dots,\mu_s)\right)^{\dagger}\\
&\quad=\left[\mathscr{H}_c(\lambda)^*_{\infty}\otimes V(\mu_0)^*_{z_0}\otimes\cdots\otimes V(\mu_s)^*_{z_s}\right]^{(\mathfrak{g}\otimes\mathbb{C}[\mathbb{P}^1\backslash\infty])^{\Gamma}}\\
&\quad\simeq\left[\hat{\mathscr{H}}_c(\lambda^*)\otimes V(\mu_0^*)_{z_0}\otimes V(\mu_1^*)_{z_1}\otimes\cdots\otimes V(\mu_s^*)_{z_s}\right]^{\mathfrak{g}[t]^{\sigma}},\ \text{by using (26)}\\
&\quad=H^0\left(L^+(\mathfrak{g},\sigma),\hat{\mathscr{H}}_c(\lambda^*)\otimes V(\vec{\mu}^*)_{\vec{z}}\right)^{\mathfrak{g}^{\sigma}}.
\end{aligned}
$$

Observe that $\mathscr{H}^{\mathrm{low}}_{-c}(\lambda)$ is exactly $\mathscr{H}_c(\lambda)_\infty$ with respect to the action of $\hat{L}(\mathfrak{g},\sigma^{-1})$, where $L^+(\mathfrak{g},\sigma)=(t\mathfrak{g}[t])^{\sigma}$ is naturally the negative part of $\hat{L}(\mathfrak{g},\sigma^{-1})$ with $t_\infty=t^{-1}$ as the coordinate at ∞. This proves (a)

(b) By [**Ku1**, Lemmas 3.1.9 and 3.1.13]

$$
\begin{aligned}
(43)\quad &H^q\left(L^+(\mathfrak{g},\sigma),\hat{\mathscr{H}}_c(\lambda^*)\otimes V(\vec{\mu}^*)_{\vec{z}}\right)\\
&\qquad\simeq\left(H_q\left(L^+(\mathfrak{g},\sigma),\mathscr{H}_c(\lambda)_\infty\otimes V(\vec{\mu})_{\vec{z}}\right)\right)^*,\ \text{as } \mathfrak{g}^{\sigma}\text{-modules.}
\end{aligned}
$$

This conclude the proof of part (b) of the lemma. □

We now assume σ is a special automorphism of $\mathfrak{g}$.

COROLLARY 3.20. *Under the assumptions of Theorem 3.17, for any $q\geq 1$ we have*

$$H_q(L^-(\mathfrak{g},\sigma^{-1}),\mathscr{H}_c(\lambda^*)\otimes V(\vec{\mu}^*)_{\vec{z}^{-1}})^{\mathfrak{g}^{\sigma}}=0,\qquad \textit{if } d_{\vec{\mu}}<\frac{c+2\check{h}}{m},$$

where $\vec{z}^{-1}=(\infty,z_1^{-1},\cdots,z_s^{-1})$, $\mathscr{H}_c(\lambda^*)$ is the highest weight module of $\hat{L}(\mathfrak{g},\sigma^{-1})$, and $V(\vec{\mu}^*)_{\vec{z}^{-1}}:=V(\mu_0^*)_\infty\otimes\cdots\otimes V(\mu_s^*)_{z_s^{-1}}$ represents the evaluation representation of $L^-(\mathfrak{g},\sigma^{-1})$ in the usual sense.

PROOF. Observe that there is an isomorphism of Lie algebras $\gamma: L^-(\mathfrak{g},\sigma)\to L^+(\mathfrak{g},\sigma^{-1})$ given by $x\otimes f(t)\mapsto x\otimes f(t^{-1})$. Note that in this isomorphism, by convention on LHS $\sigma(t)=e^{\frac{-2\pi i}{m}}t$ and on RHS $\sigma(t)=e^{\frac{2\pi i}{m}}t$. With respect to the isomorphism γ, in part (b) of Lemma 3.19, the action of $L^+(\mathfrak{g},\sigma)$ on $\mathscr{H}_c(\lambda)_\infty$ corresponds to the action of $L^-(\mathfrak{g},\sigma^{-1})$ on $\mathscr{H}_c(\lambda)$ as a highest weight module of $\hat{L}(\mathfrak{g},\sigma^{-1})$; the evaluation representation $V(\mu_i)_{z_i}$ of $L^+(\mathfrak{g},\sigma)$ corresponds to the evaluation representation $V(\mu_i)_{z_i^{-1}}$ of $L^-(\mathfrak{g},\sigma^{-1})$. Then, from part (b) of Lemma 3.19, we get

$$H^q\left(L^+(\mathfrak{g},\sigma),\hat{\mathscr{H}}_c(\lambda^*)\otimes V(\vec{\mu}^*)_{\vec{z}}\right)\simeq\left(H_q(L^-(\mathfrak{g},\sigma^{-1}),\mathscr{H}_c(\lambda)\otimes V(\vec{\mu},_{\vec{z}^{-1}})\right)^*.$$

Finally, our corollary follows from Theorem 3.17. (Also, see Remark 3.18 (a).) □

REMARK 3.21. Corollary 3.20 verifies the conjecture in [**HK2**, Conjecture 5.6] under some strong constraints.

4. Laplacian computation

Throughout this section σ is any automorphism of $\mathfrak{g}$ of order m except in the last part of subsection 4.1, where we restrict σ to be a special automorphism. We abbreviate $\hat{\mathcal{B}}_\sigma$ (resp. $\hat{\mathcal{B}}^q_\sigma$) by $\hat{\mathcal{B}}$ (resp. $\hat{\mathcal{B}}^q$).

We follow the notation from §3.4 and first calculate the differential $d^q : C^q \to C^{q+1}$ restricted to $\wedge^q (L^-(\mathfrak{g},\sigma)) \otimes \mathscr{H}_c(\lambda) \otimes V(\vec{\mu})_{\vec{z}}$ with image inside

$$\wedge^{q+1} \left(\bar{L}^-(\mathfrak{g},\sigma)\right) \otimes \mathscr{H}_c(\lambda) \otimes V(\vec{\mu})_{\vec{z}}.$$

LEMMA 4.1. *For $\vec{x}(-\vec{p}) \in \wedge^q (L^-(\mathfrak{g},\sigma))$ and $v \in \mathscr{H}_c(\lambda) \otimes V(\vec{\mu})_{\vec{z}}$,*

$$d^q\left(\vec{x}(-\vec{p}) \otimes v\right) = \frac{1}{2}\sum_{b(-k)\in\hat{\mathcal{B}}} b(-k) \wedge \left(\overline{ad}_{\check{b}(k)}\vec{x}(-\vec{p})\right) \otimes v + \sum_{b(-k)\in\hat{\mathcal{B}}} b(-k) \wedge \vec{x}(-\vec{p}) \otimes \check{b}(k)\cdot v,$$

where, as before, $\overline{ad}$ is the projection of ad to $L^-(\mathfrak{g},\sigma)$.

PROOF. Let d^q_1 be the same differential d^q but with the trivial action of $L^+(\mathfrak{g},\sigma)$ on $\mathscr{H}_c(\lambda) \otimes V(\vec{\mu})_{\vec{z}}$. Then,

$$\begin{aligned} &d_1^q\left(\vec{x}(-\vec{p}) \otimes v\right)\left(\check{b}_1(k_1) \wedge \cdots \wedge \check{b}_{q+1}(k_{q+1})\right) \\ &\quad = \left(\vec{x}(-\vec{p}) \otimes v\right)\left(\frac{1}{2}\sum_{i=1}^{q+1}(-1)^i \,\mathrm{ad}_{\check{b}_i(k_i)}\cdot\left(\check{b}_1(k_1)\wedge\cdots\wedge\widehat{\check{b}_i(k_i)}\wedge\cdots\wedge\check{b}_{q+1}(k_{q+1})\right)\right). \end{aligned}$$

Thus,

$$\begin{aligned} &d_1^q\left(\vec{x}(-\vec{p}) \otimes v\right) \\ &\quad = \frac{1}{2}\sum_{i;\vec{b}(-\vec{k})\in\hat{\mathcal{B}}^{q+1}} (-1)^i \left\langle \vec{x}(-\vec{p}), ad_{\check{b}_i(k_i)}\left(\check{b}_1(k_1)\wedge\ldots\wedge\widehat{\check{b}_i(k_i)}\wedge\ldots\wedge\check{b}_{q+1}(k_{q+1})\right)\right\rangle \\ &\quad\quad \cdot b_1(-k_1)\wedge\cdots\wedge b_{q+1}(-k_{q+1})\otimes v \\ &\quad = \frac{1}{(q+1!)2}\sum_{i;b_j(-k_j)\in\hat{\mathcal{B}}} \left\langle\left(ad_{\check{b}_i(k_i)}\vec{x}(-\vec{p})\right), \check{b}_1(k_1)\wedge\cdots\wedge\widehat{\check{b}_i(k_i)}\wedge\cdots\wedge\check{b}_{q+1}(k_{q+1})\right\rangle \\ &\quad\quad \cdot b_i(-k_i)\wedge b_1(-k_1)\wedge\cdots\wedge\widehat{b_i(-k_i)}\wedge\cdots\wedge b_{q+1}(-k_{q+1})\otimes v, \text{ by Lemma 3.4} \end{aligned}$$

(44)

$$= \frac{1}{2}\sum_{b(-k)\in\hat{\mathcal{B}}} b(-k)\wedge\left(\overline{ad}_{\check{b}(k)}\vec{x}(-\vec{p})\right)\otimes v$$

Further,

$$\begin{aligned} &(d^q - d^q_1)\left(\vec{x}(-\vec{p}) \otimes v\right)\left(\check{b}_1(k_1) \wedge \cdots \wedge \check{b}_{q+1}(k_{q+1})\right) \\ &\quad = \sum_{i=1}^{q+1}(-1)^{i+1}\left\langle \vec{x}(-\vec{p}), \check{b}_1(k_1)\wedge\cdots\wedge\widehat{\check{b}_i(k_i)}\wedge\cdots\wedge\check{b}_{q+1}(k_{q+1})\right\rangle \check{b}_i(k_i)\cdot v \end{aligned}$$

Thus,

$$\begin{aligned} &(d^q - d^q_1)\left(\vec{x}(-\vec{p}) \otimes v\right) = \\ &\quad \sum_{i;\vec{b}(-\vec{k})\in\hat{\mathcal{B}}^{q+1}} (-1)^{i+1}\left\langle \vec{x}(-\vec{p}), \check{b}_1(k_1)\wedge\cdots\wedge\widehat{\check{b}_i(k_i)}\wedge\cdots\wedge\check{b}_{q+1}(k_{q+1})\right\rangle \\ &\quad\quad \cdot b_1(-k_1)\wedge\cdots\wedge b_i(-k_i)\wedge\cdots\wedge b_{q+1}(-k_{q+1})\otimes\left(\check{b}_i(k_i)\cdot v\right) \end{aligned}$$

$$= \sum_{i;b_i(-k_i)\in\hat{\mathcal{B}}} \frac{q!}{(q+1)!} b_i(-k_i) \wedge \vec{x}(-\vec{p}) \otimes \check{b}_i(k_i) \cdot v$$

$$= \sum_{b(-k)\in\hat{\mathcal{B}}} b(-k) \wedge \vec{x}(-\vec{p}) \otimes \left(\check{b}(k) \cdot v\right). \tag{45}$$

Combining the equations (44) and (45), we get the lemma. □

LEMMA 4.2. *For any* $Y \in \wedge^{q+1}(L^-(\mathfrak{g},\sigma))$, $Z \in \wedge^q(L^-(\mathfrak{g},\sigma))$ *and* $x \in L^-(\mathfrak{g},\sigma)$,

$$\{i_x Y, Z\} = \{Y, x \wedge Z\},$$

where $i_x Y$ *is the graded derivation such that*

$$i_x(x') = \{x', x\}, \text{ for } x' \in L^-(\mathfrak{g},\sigma).$$

PROOF. We can assume that $Y = y_1 \wedge \cdots \wedge y_{q+1}, y_i \in L^-(\mathfrak{g},\sigma)$ and $Z = z_1 \wedge \cdots \wedge z_q$. Then,

$$\{i_x Y, Z\} = \sum_{p=1}^{q+1} (-1)^{p-1} \{y_p, x\} \det\left(\{\overline{y_i}, z_j\}_{1\le i,j\le q}\right), \tag{46}$$

where

$$\overline{y_i} = y_i \text{ for } i < p, \text{ and}$$
$$= y_{i+1} \text{ for } i \ge p.$$

Similarly,

$$\{Y, x\wedge Z\} = \det\begin{pmatrix} \{y_1, x\}, & \{y_1, z_1\}, & \dots, & \{y_1, z_q\} \\ \{y_2, x\}, & \{y_2, z_1\}, & \dots, & \{y_2, z_q\} \\ \vdots & & & \vdots \\ \{y_{q+1}, x\}, & \{y_{q+1}, z_1\}, & \dots, & \{y_{q+1}, z_q\} \end{pmatrix}. \tag{47}$$

Combining the equations (46) and (47). we get the lemma. □

LEMMA 4.3. *For* $x \in L^-(\mathfrak{g},\sigma)$ *and* $y \in \hat{L}^f(\mathfrak{g},\sigma)$,

$$i_x(\overline{ad}_y^{\ell}) = (\overline{ad}_y^{\ell}) i_x + i^{\ell}_{[x,\kappa(y)]_-}, \quad \text{for } 1 \le \ell \le q, \text{ acting on } \wedge^q(L^-(\mathfrak{g},\sigma)),$$

where i^ℓ *is the contraction action only on the* ℓ*-th factor with sign* $(-1)^{\ell-1}$ *and* $[\,]_-$ *denotes the projection onto* $L^-(\mathfrak{g},\sigma)$.

PROOF. Take $Z = z_1 \wedge \cdots \wedge z_q, z_i \in L^-(\mathfrak{g},\sigma)$. Then,

$$i_x(\overline{ad}_y^{\ell})(Z) = \sum_{j\ne\ell} (-1)^{j-1} z_1 \wedge \cdots \wedge \{z_j, x\} \wedge \cdots \wedge [y, z_\ell]_- \wedge \cdots \wedge z_q$$
$$+ (-1)^{\ell-1} z_1 \wedge \cdots \wedge \{[y, z_\ell], x\} \wedge \cdots \wedge z_q.$$

Further,

$$\left(\overline{ad}_y^{\ell}\right) i_x(Z) = \sum_{j\ne\ell} (-1)^{j-1} z_1 \wedge \cdots \wedge \{z_j, x\} \wedge \cdots \wedge [y, z_\ell]_- \wedge \cdots \wedge z_q. \tag{48}$$

By Lemma 3.4, the lemma follows. □

We consider the Hermitian form $\{\,,\,\}_0$ on $\wedge^*(L^-(\mathfrak{g},\sigma))$ and the standard Hermitian forms on $\mathscr{H}_c(\lambda)$ and $V(\vec{\mu})_{\vec{z}}$ as in §3.4.

LEMMA 4.4. *For* $\vec{x}(-\vec{p}) = x_1(-p_1) \wedge \cdots \wedge x_{q+1}(-p_{q+1}) \in \wedge^{q+1}(L^-(\mathfrak{g},\sigma))$ *and* $v \in \mathscr{H}_c(\lambda) \otimes V(\vec{\mu})_{\vec{z}}$,

$$d^*\left(\vec{x}(-\vec{p}) \otimes v\right) = \frac{1}{2} \sum_{j;b(-k)\in\hat{\mathcal{B}}} \left(\frac{1}{p_j} + \frac{1}{k}\right) i_{b(-k)} \left(ad^j_{b(-k)}\vec{x}(-\vec{p})\right) \otimes v$$
$$+ \sum_{b(-k)\in\hat{\mathcal{B}}} \frac{1}{k}\left(i_{b(-k)}\vec{x}(-\vec{p})\right) \otimes b(-k)\cdot v.$$

PROOF. As in §3.3, $\{v_\phi\}$ is an orthonormal basis of $\mathscr{H}_c(\lambda) \otimes V(\vec{\mu})_{\vec{z}}$.

$$\begin{aligned}
&d^*\left(\vec{x}(-\vec{p}) \otimes v\right)\\
&= \sum_{v_\phi;\vec{a}(-\vec{n})\in\hat{\mathcal{B}}^q} |\vec{n}| \left\{d^*\left(\vec{x}(-\vec{p})\otimes v\right), \vec{a}(-\vec{n})\otimes v_\phi\right\} \vec{a}(-\vec{n})\otimes v_\phi\\
&= \sum \frac{|\vec{n}|}{|\vec{p}|} \left\{\vec{x}(-\vec{p})\otimes v, d\left(\vec{a}(-\vec{n})\otimes v_\phi\right)\right\} \vec{a}(-\vec{n})\otimes v_\phi\\
&= \frac{1}{2} \sum_{v_\phi;\vec{a}(-\vec{n})\in\hat{\mathcal{B}}^q;b(-k)\in\hat{\mathcal{B}}} \frac{|\vec{n}|}{|\vec{p}|} \left\{\vec{x}(-\vec{p})\otimes v, b(-k)\wedge\left(\overline{ad}_{\check{b}(k)}\vec{a}(-\vec{n})\right)\otimes v_\phi\right\} \vec{a}(-\vec{n})\otimes v_\phi\\
&\quad+ \sum_{v_\phi;\vec{a}(-\vec{n})\in\hat{\mathcal{B}}^q;b(-k)\in\hat{\mathcal{B}}} \frac{|\vec{n}|}{|\vec{p}|} \left\{\vec{x}(-\vec{p})\otimes v, b(-k)\wedge\vec{a}(-\vec{n})\otimes\check{b}(k)\cdot v_\phi\right\} \vec{a}(-\vec{n})\otimes v_\phi,\\
&\quad\text{by Lemma 4.1}\\
&= \frac{1}{2}\sum \frac{|\vec{n}|}{|\vec{p}|} \left\{\left(i_{b(-k)}\vec{x}(-\vec{p})\right)\otimes v, \overline{ad}_{\check{b}(k)}\vec{a}(-\vec{n})\otimes v_\phi\right\} \vec{a}(-\vec{n})\otimes v_\phi\\
&\quad+ \sum \frac{|\vec{n}|}{|\vec{p}|} \left(\left\{i_{b(-k)}\vec{x}(-\vec{p})\right)\otimes v, \vec{a}(-\vec{n})\otimes\check{b}(k)\cdot v_\phi\right\} \vec{a}(-\vec{n})\otimes v_\phi, \text{ by Lemma 4.2}\\
&= \frac{1}{2}\sum \frac{|\vec{n}|}{|\vec{p}|} \left\{ad_{b(-k)}i_{b(-k)}\vec{x}(-\vec{p})\otimes v, \vec{a}(-\vec{n})\otimes v_\phi\right\} \vec{a}(-\vec{n})\otimes v_\phi\\
&\quad+ \sum \frac{|\vec{n}|}{|\vec{p}|} \left\{i_{b(-k)}\vec{x}(-\vec{p}), \vec{a}(-\vec{n})\right\} \vec{a}(-\vec{n})\otimes\left\{v, \check{b}(k)\cdot v_\phi\right\} v_\phi, \text{ by Lemma 3.4}\\
&= \frac{1}{2} \sum_{j;b(-k)\in\hat{\mathcal{B}}} \left(\frac{1}{p_j}+\frac{1}{k}\right) i_{b(-k)} \left(ad^j_{b(-k)}\vec{x}(-\vec{p})\right)\otimes v\\
&\quad+ \sum_{b(-k)\in\hat{\mathcal{B}}} \frac{1}{k}\left(i_{b(-k)}\vec{x}(-\vec{p})\right)\otimes b(-k)\cdot v, \text{ by Lemmas 4.3 and 3.3.}
\end{aligned}$$

This proves the lemma. □

Combining the Lemmas 4.1 and 4.4, we get the following.

PROPOSITION 4.5.
$\overline{\Box} := dd^* + d^*d : \wedge^q\left(L^-(\mathfrak{g},\sigma)\right)\otimes\mathscr{H}_c(\lambda)\otimes V(\vec{\mu})_{\vec{z}} \to \wedge^q\left(\bar{L}^-(\mathfrak{g},\sigma)\right)\otimes\mathscr{H}_c(\lambda)\otimes V(\vec{\mu})_{\vec{z}}$ *is given as follows.*

For $\vec{x}(-\vec{p}) \in \wedge^q\left(L^-(\mathfrak{g},\sigma)\right)$ *and* $v = v_1\otimes v_2 \in \mathscr{H}_c(\lambda)\otimes V(\vec{\mu})_{\vec{z}}$,

$$\overline{\Box}\left(\vec{x}(-\vec{p})\otimes v\right)$$
$$= \frac{1}{4} \sum_{\substack{1\le\ell\le q;\\ 1\le j\le q+1;\\ b(-k),a(-n)\in\hat{\mathcal{B}}}} \frac{p_j(\ell)+n}{np_j(\ell)} i_{a(-n)} ad^j_{a(-n)} \left(b(-k)\wedge\left(\overline{ad}^{\ell}_{\check{b}(k)}\vec{x}(-\vec{p})\right)\right)\otimes v \tag{49}$$

$$+\frac{1}{2}\sum_{b(-k),a(-n)\in\hat{\mathcal{B}}}\frac{1}{n}i_{a(-n)}\left(b(-k)\wedge\left(\overline{ad}_{\check{b}(k)}\vec{x}(-\vec{p})\right)\right)\otimes a(-n)\cdot v \tag{50}$$

$$-\frac{1}{2}\sum_{b(-k),a(-n)\in\hat{\mathcal{B}}}\frac{k+n}{kn}\left[a(-n),b(-k)\right]\wedge\left(i_{a(-n)}\cdot\vec{x}(-\vec{p})\right)\otimes\check{b}(k)\cdot v \tag{51}$$

$$+\frac{1}{2}\sum_{j>1;b(-k),a(-n)\in\hat{\mathcal{B}}}\frac{p_{j-1}+n}{np_{j-1}}i_{a(-n)}ad^{j}_{a(-n)}\left(b(-k)\wedge\vec{x}(-\vec{p})\right)\otimes\check{b}(k)\cdot v \tag{52}$$

$$+\sum_{b(-k),a(-n)\in\hat{\mathcal{B}}}\frac{1}{n}i_{a(-n)}\left(b(-k)\wedge\vec{x}(-\vec{p})\right)\otimes a(-n)\check{b}(k)\cdot v \tag{53}$$

$$+\frac{1}{4}\sum_{j;b(-k),a(-n)\in\hat{\mathcal{B}}}\frac{p_j+n}{np_j}b(-k)\wedge\overline{ad}_{\check{b}(k)}\left(i_{a(-n)}ad^{j}_{a(-n)}\vec{x}(-\vec{p})\right)\otimes v \tag{54}$$

$$+\frac{1}{2}\sum_{j;b(-k),a(-n)\in\hat{\mathcal{B}}}\frac{p_j+n}{np_j}b(-k)\wedge i_{a(-n)}\left(ad^{j}_{a(-n)}\vec{x}(-\vec{p})\right)\otimes\check{b}(k)\cdot v \tag{55}$$

$$+\frac{1}{2}\sum_{b(-k),a(-n)\in\hat{\mathcal{B}}}\frac{1}{n}b(-k)\wedge\left(\overline{ad}_{\check{b}(k)}i_{a(-n)}\vec{x}(-\vec{p})\right)\otimes a(-n)\cdot v \tag{56}$$

$$+\sum_{b(-k),a(-n)\in\hat{\mathcal{B}}}\frac{1}{n}b(-k)\wedge\left(i_{a(-n)}\vec{x}(-\vec{p})\right)\otimes\check{b}(k)a(-n)\cdot v, \tag{57}$$

where $p_1(\ell)=k$; $p_j(\ell)=p_{j-1}$ *for* $j>1$ *and* $j\neq\ell+1$; $p_{\ell+1}(\ell)=p_\ell-k$ *if* $p_\ell>k$ *otherwise* 0.

PROOF.

$$\begin{aligned}\overline{\square}&\left(\vec{x}(-\vec{p})\otimes v\right)\\&=\frac{1}{2}d^*\sum_{b(-k)\in\hat{\mathcal{B}}}b(-k)\wedge\left(\overline{ad}_{\check{b}(k)}\vec{x}(-\vec{p})\right)\otimes v+d^*\sum_{b(-k)\in\hat{\mathcal{B}}}b(-k)\wedge\vec{x}(-\vec{p})\otimes\check{b}(k)\cdot v\\&\quad+\frac{1}{2}d\sum_{j\neq\ell;a(-n)\in\hat{\mathcal{B}}}\left(\frac{1}{p_j}+\frac{1}{n}\right)i^{\ell}_{a(-n)}(ad^{j}_{a(-n)}\vec{x}(-\vec{p}))\otimes v\\&\quad+d\sum_{a(-n)\in\hat{\mathcal{B}}}\frac{1}{n}i_{a(-n)}\vec{x}(-\vec{p})\otimes a(-n)\cdot v,\ \text{by Lemmas 4.1 and 4.4.}\end{aligned}$$

Again using the Lemmas 4.1 and 4.4, we get the proposition. □

We now simplify the terms (49)–(57) of Proposition 4.5.

LEMMA 4.6. *With the notation as in Proposition* 4.5, *we have the following, where* m *is the order of the automorphism* σ *and* $\mathcal{B}_{\underline{0}}$ *is an orthonormal basis of* $\mathfrak{g}_{\underline{0}}$ *as in Section* 3.1.

$$(50)+(56)=\sum_{b(-k)\in\hat{\mathcal{B}}}\frac{1}{k}\overline{ad}_{\check{b}(k)}\vec{x}(-\vec{p})\otimes b(-k)\cdot v.$$

$$\begin{aligned}&(53)+(57)\\&=-\sum_{\ell;a(-n)\in\hat{\mathcal{B}}}\frac{1}{p_\ell}\left(ad^{\ell}_{a(-n)}\vec{x}(-\vec{p})\right)\otimes(\check{a}(n)\cdot v_1)\otimes v_2\end{aligned}$$

$$
\begin{aligned}
&- \sum_{\ell; a(-n)\in\hat{\mathcal{B}}} \frac{1}{p_\ell}\left(\overline{ad}^{\ell}_{\check{a}(n)}\vec{x}(-\vec{p})\right)\otimes (a(-n)\cdot v_1)\otimes v_2 \\
&+ \vec{x}(-\vec{p})\otimes \frac{q}{m}c\cdot v_1\otimes v_2 - \sum_{\ell; b_0\in\mathcal{B}_{\underline{0}}} \frac{1}{p_\ell}\left(ad^{\ell}_{\check{b}_o}\vec{x}(-\vec{p})\right)\otimes(\check{b}_0\cdot v_1)\otimes v_2 \\
&- \sum_{\ell; a(-n)\in\hat{\mathcal{B}}} \frac{1}{p_\ell}\left(ad^{\ell}_{a(-n)}\vec{x}(-\vec{p})\right)\otimes v_1\otimes (z\bar{z})^{p_\ell} z^n \check{a}\cdot v_2 \\
&- \sum_{\ell; a(-n)\in\hat{\mathcal{B}}} \frac{1}{p_\ell}\left(\overline{ad}^{\ell}_{\check{a}(n)}\vec{x}(-\vec{p})\right)\otimes v_1\otimes (z\bar{z})^{p_\ell - n}\bar{z}^n a\cdot v_2 \\
&- \sum_{\ell; b_0\in\mathcal{B}_{\underline{0}}} \frac{1}{p_\ell}\left(ad^{\ell}_{\check{b}_o}\vec{x}(-\vec{p})\right)\otimes v_1\otimes (z\bar{z})^{p_\ell}\check{b}_0\cdot v_2 \\
&+ \sum_{a(-n)\in\hat{\mathcal{B}}} \frac{1}{n}\vec{x}(-\vec{p})\otimes a(-n)\check{a}(n)\cdot v.
\end{aligned}
$$

$$
(51)+(52)+(55) = \sum_{\ell; b(-k)\in\hat{\mathcal{B}}} \frac{p_\ell + k}{p_\ell\cdot k} ad^{\ell}_{b(-k)}\vec{x}(-\vec{p})\otimes \check{b}(k)\cdot v.
$$

$$
\begin{aligned}
&(49)+(54) \\
&= \sum_{j\neq\ell; b(-k)\in\hat{\mathcal{B}}} \left(\frac{1}{k}-\frac{1}{p_\ell}\right) ad^{j}_{b(-k)}\overline{ad}^{\ell}_{\check{b}(k)}\vec{x}(-\vec{p})\otimes v \\
&\quad + \frac{1}{2}\sum_{\ell; b(-k)\in\hat{\mathcal{B}}} \frac{p_\ell}{(p_\ell - k)\cdot k} ad^{\ell}_{b(-k)}\overline{ad}^{\ell}_{\check{b}(k)}\vec{x}(-\vec{p})\otimes v \\
&\quad - \frac{1}{2}\sum_{j\neq\ell; b_0\in\mathcal{B}_{\underline{0}}} \left(\frac{1}{p_j}+\frac{1}{p_\ell}\right) ad^{j}_{b_0} ad^{\ell}_{\check{b}_0}\vec{x}(-\vec{p})\otimes v.
\end{aligned}
$$

PROOF.

$$
\begin{aligned}
&(50)+(56) \\
&= \frac{1}{2}\sum_{b(-k)\in\hat{\mathcal{B}}} \frac{1}{k}\overline{ad}_{\check{b}(k)}\vec{x}(-\vec{p})\otimes b(-k)\cdot v + \frac{1}{2}\sum_{a(-n)\in\hat{\mathcal{B}}} \frac{1}{n}\overline{ad}_{\check{a}(n)}\vec{x}(-\vec{p})\otimes a(-n)\cdot v \\
&= \sum_{b(-k)\in\hat{\mathcal{B}}} \frac{1}{k}\overline{ad}_{\check{b}(k)}\vec{x}(-\vec{p})\otimes b(-k)\cdot v.
\end{aligned}
$$

$$
\begin{aligned}
&(53)+(57) \\
&= \sum_{a(-n), b(-k)\in\hat{\mathcal{B}}} \frac{1}{n} i_{a(-n)}\left(b(-k)\wedge\vec{x}(-\vec{p})\right)\otimes a(-n)\check{b}(k)\cdot v \\
&\quad + \sum_{a(-n), b(-k)\in\hat{\mathcal{B}}} \frac{1}{n} b(-k)\wedge\left(i_{a(-n)}\vec{x}(-\vec{p})\right)\otimes \check{b}(k)a(-n)\cdot v \\
&= \sum_{a(-n)\in\hat{\mathcal{B}}} \frac{1}{n}\vec{x}(-\vec{p})\otimes a(-n)\check{a}(n)\cdot v
\end{aligned}
$$

$$- \sum_{a(-n),b(-k)\in\hat{\mathcal{B}}} \frac{1}{n} b(-k) \wedge \left(i_{a(-n)}\vec{x}(-\vec{p})\right) \otimes a(-n)\check{b}(k) \cdot v$$

$$+ \sum_{a(-n),b(-k)\in\hat{\mathcal{B}}} \frac{1}{n} b(-k) \wedge \left(i_{a(-n)}\vec{x}(-\vec{p})\right) \otimes \check{b}(k)a(-n) \cdot v$$

$$= \sum_{a(-n)\in\hat{\mathcal{B}}} \frac{1}{n}\vec{x}(-\vec{p}) \otimes a(-n)\check{a}(n) \cdot v$$

$$- \sum_{a(-n),b(-k)\in\hat{\mathcal{B}}} \frac{1}{n} b(-k) \wedge \left(i_{a(-n)}\vec{x}(-\vec{p})\right) \otimes \left(a(-n)\check{b}(k) \cdot v - \check{b}(k)a(-n) \cdot v\right)$$

$$= \sum_{a(-n)\in\hat{\mathcal{B}}} \frac{1}{n}\vec{x}(-\vec{p}) \otimes a(-n)\check{a}(n) \cdot v$$

$$- \sum_{\ell;b(-k),a(-n)\in\hat{\mathcal{B}}} \frac{1}{n} x_1(-p_1) \wedge \quad \wedge b(-k)^{[\ell]} \wedge \cdots \wedge x_q(-p_q)$$

$$\otimes \{x_\ell(-p_\ell), a(-n)\} \left(a(-n)\check{b}(k) \cdot v - \check{b}(k)a(-n) \cdot v\right)$$

where $(\,)^{[\ell]}$ denotes the enclosed item placed in ℓ-th place

$$= \sum_{a(-n)\in\hat{\mathcal{B}}} \frac{1}{n}\vec{x}(-\vec{p}) \otimes a(-n)\check{a}(n) \cdot v$$

$$- \sum_{\ell;b(-k)\in\hat{\mathcal{B}}} \left(\frac{1}{p_\ell} x_1(-p_1) \wedge \cdots \wedge b(-k)^{[\ell]} \wedge \cdots \wedge x_q(-p_q)\right.$$

$$\left.\otimes \left(\left[x_\ell(-p_\ell), \check{b}(k)\right] \cdot v_1\right) \otimes v_2 + v_1 \otimes \left[\theta(x_\ell(-p_\ell)), \theta(\check{b}(k))\right] \cdot v_2\right),$$

where θ denotes the action on $V_{\vec{z}}(\vec{\mu})$

$$= \sum_{a(-n)\in\hat{\mathcal{B}}} \frac{1}{n}\vec{x}(-\vec{p}) \otimes a(-n)\check{a}(n) \cdot v$$

$$- \sum_{\ell;b(-k)\in\hat{\mathcal{B}}} \frac{1}{p_\ell} x_1(-p_1) \wedge \cdot \wedge b(-k)^{[\ell]} \wedge \cdot \wedge x_q(-p_q)$$

$$\otimes \Bigg(\sum_{a(-n)\in\hat{\mathcal{B}}} \left\{[x_\ell(-p_\ell), \check{b}(k)], \check{a}(n)\right\} (\check{a}(n) \cdot v_1) \otimes v_2$$

$$+ \sum_{a(-n)\in\hat{\mathcal{B}}} \left\{[x_\ell(-p_\ell), \check{b}(k)], a(-n)\right\} (a(-n) \cdot v_1) \otimes v_2$$

$$+ v_1 \otimes [\theta(x_\ell(-p_\ell)), \theta(\check{b}(k)] \cdot v_2 + (\delta_{k,p_\ell}([x_\ell, \check{b}] - \frac{p_\ell}{m}\langle x_\ell, \check{b}\rangle C) \cdot v_1) \otimes v_2\Bigg)$$

where in the last sum only those $\check{b}$ appear which belong to $\mathfrak{g}_{p_\ell}$ and $\hat{\mathcal{B}}_{p_\ell}$ is an orthonormal basis of $\mathfrak{g}_{\underline{p}_\ell}$

$$= - \sum_{\ell;b(-k)\in\hat{\mathcal{B}}} \frac{1}{p_\ell} x_1(-p_1) \wedge \cdot \wedge b(-k)^{[\ell]} \wedge \cdots \wedge x_q(-p_q)$$

$$\otimes \Bigg(\sum_{a(-n)\in\hat{\mathcal{B}}} \{[a(-n), x_\ell(-p_\ell)], b(-k)\} (\check{a}(n)\cdot v_1)\otimes v_2$$

$$+ \sum_{a(-n)\in\hat{\mathcal{B}}} \{[\check{a}(n), x_\ell(-p_\ell)], b(-k)\} (a(-n)\cdot v_1)\otimes v_2$$

$$+ \Big(\delta_{k,p_\ell}\Big([x_\ell, \check{b}] - \frac{p_\ell}{m}\langle x_\ell, \check{b}\rangle C\Big)\cdot v_1\Big)\otimes v_2$$

$$+ v_1 \otimes \big[\theta(x_\ell(-p_\ell)), \theta(\check{b}(k))\big]\cdot v_2\Bigg),$$

$$\text{by Lemma 3.4 } + \sum_{a(-n)\in\hat{\mathcal{B}}} \frac{1}{n}\vec{x}(-\vec{p})\otimes a(-n)\check{a}(n)\cdot v$$

$$= - \sum_{\ell;a(-n)\in\hat{\mathcal{B}}} \frac{1}{p_\ell}\Big(ad^\ell_{a(-n)}\vec{x}(-\vec{p})\Big)\otimes(\check{a}(n)\cdot v_1)\otimes v_2$$

$$- \sum_{\ell;a(-n)\in\hat{\mathcal{B}}} \frac{1}{p_\ell}\Big(\overline{ad}^\ell_{\check{a}(n)}\vec{x}(-\vec{p})\Big)\otimes(a(-n)\cdot v_1)\otimes v_2$$

$$+ \frac{qc}{m}\vec{x}(-\vec{p})\otimes v_1\otimes v_2 - \sum_{\ell;b_0\in\mathcal{B}_{\underline{0}}} \frac{1}{p_\ell} ad^\ell_{b_0}\vec{x}(-\vec{p})\otimes(\check{b}_0\cdot v_1)\otimes v_2$$

$$- \sum_{\ell;a(-n)\in\hat{\mathcal{B}}} \frac{1}{p_\ell}\Big(ad^\ell_{a(-n)}\vec{x}(-\vec{p})\Big)\otimes v_1\otimes((z\bar{z})^{p_\ell}z^n\check{a})\cdot v_2$$

$$- \sum_{\ell;a(-n)\in\hat{\mathcal{B}}} \frac{1}{p_\ell}\Big(\overline{ad}^\ell_{\check{a}(n)}\vec{x}(-\vec{p})\Big)\otimes v_1\otimes\big((z\bar{z})^{p_\ell-n}\bar{z}^n a\big)\cdot v_2$$

$$- \sum_{\ell;b_0\in\mathcal{B}_{\underline{0}}} \frac{1}{p_\ell}\big(ad^\ell_{b_o}\vec{x}(-\vec{p})\big)\otimes v_1\otimes\big((z\bar{z})^{p_\ell}\check{b}_0\big)\cdot v_2$$

$$+ \sum_{a(-n)\in\hat{\mathcal{B}}} \frac{1}{n}\vec{x}(-\vec{p})\otimes a(-n)\check{a}(n)\cdot v.$$

(51) + (52) + (55)

$$= -\frac{1}{2}\sum_{\ell;b(-k),a(-n)\in\hat{\mathcal{B}}} \frac{p_\ell+k}{p_\ell\cdot k}[a(-n), b(-k)]\wedge\Big(i^\ell_{a(-n)}\vec{x}(-\vec{p})\Big)\otimes\check{b}(k)\cdot v$$

$$+ \frac{1}{2}\sum_{\ell;b(-k)\in\hat{\mathcal{B}}} \frac{p_\ell+k}{p_\ell\cdot k} ad^\ell_{b(-k)}\vec{x}(-\vec{p})\otimes\check{b}(k)\cdot v$$

$$= \sum_{\ell;b(-k)\in\hat{\mathcal{B}}} \frac{p_\ell+k}{p_\ell\cdot k} ad^\ell_{b(-k)}\vec{x}(-\vec{p})\otimes\check{b}(k)\cdot v.$$

(49) + (54)

$$= \frac{1}{4}\sum_{\substack{1\le\ell\le q;\\ 1\le j\le q+1;\\ b(-k),a(-n)\in\hat{\mathcal{B}}}} \frac{p_j(\ell)+n}{p_j(\ell)\cdot n} i_{a(-n)} ad^j_{a(-n)}\Big(b(-k)\wedge\overline{ad}^\ell_{\check{b}(k)}\vec{x}(-\vec{p})\Big)\otimes v$$

$$+\frac{1}{4}\sum_{j;b(-k),a(-n)\in\hat{\mathcal{B}}}\frac{n+p_j}{n\cdot p_j}b(-k)\wedge\overline{ad}_{\check{b}(k)}\left(i_{a(-n)}ad^j_{a(-n)}\vec{x}(-\vec{p})\right)\otimes v$$

$$=\frac{1}{4}\sum_{\ell;b(-k),a(-n)\in\hat{\mathcal{B}}}\frac{k+n}{kn}i_{a(-n)}\left([a(-n),b(-k)]\wedge\overline{ad}^\ell_{\check{b}(k)}\vec{x}(-\vec{p})\right)\otimes v$$

$$+\frac{1}{4}\sum_{\ell\neq j;b(-k),a(-n)\in\hat{\mathcal{B}}}\frac{p_j+n}{p_jn}i_{a(-n)}\left(b(-k)\wedge\mathrm{ad}^j_{a(-n)}\overline{ad}^\ell_{\check{b}(k)}\vec{x}(-\vec{p})\right)\otimes v$$

$$+\frac{1}{4}\sum_{\ell;b(-k),a(-n)\in\hat{\mathcal{B}}}\frac{p_\ell-k+n}{(p_\ell-k)n}i_{a(-n)}\left(b(-k)\wedge\mathrm{ad}^\ell_{a(-n)}\overline{ad}^\ell_{\check{b}(k)}\vec{x}(-\vec{p})\right)\otimes v$$

$$+\frac{1}{4}\sum_{j;b(-k),a(-n)\in\hat{\mathcal{B}}}\frac{n+p_j}{n\cdot p_j}b(-k)\wedge\overline{ad}_{\check{b}(k)}\left(i_{a(-n)}ad^j_{a(-n)}\vec{x}(-\vec{p})\right)\otimes v$$

$$=-\frac{1}{4}\sum_{\ell;b(-k),a(-n)\in\hat{\mathcal{B}}}\frac{k+n}{kn}\left([a(-n),b(-k)]\wedge i_{a(-n)}\overline{ad}^\ell_{\check{b}(k)}\vec{x}(-\vec{p})\right)\otimes v$$

$$+\frac{1}{4}\sum_{\ell\neq j;b(-k)\in\hat{\mathcal{B}}}\frac{p_j+k}{p_jk}\left(\mathrm{ad}^j_{b(-k)}\overline{ad}^\ell_{\check{b}(k)}\vec{x}(-\vec{p})\right)\otimes v$$

$$-\frac{1}{4}\sum_{\ell\neq j;b(-k),a(-n)\in\hat{\mathcal{B}}}\frac{p_j+n}{p_jn}b(-k)\wedge\left(i_{a(-n)}\mathrm{ad}^j_{a(-n)}\overline{ad}^\ell_{\check{b}(k)}\vec{x}(-\vec{p})\right)\otimes v$$

$$+\frac{1}{4}\sum_{\ell;b(-k)\in\hat{\mathcal{B}}}\frac{p_\ell}{(p_\ell-k)k}\left(\mathrm{ad}^\ell_{b(-k)}\overline{ad}^\ell_{\check{b}(k)}\vec{x}(-\vec{p})\right)\otimes v$$

$$-\frac{1}{4}\sum_{\ell;b(-k),a(-n)\in\hat{\mathcal{B}}}\frac{p_\ell-k+n}{(p_\ell-k)n}b(-k)\wedge\left(i_{a(-n)}\mathrm{ad}^\ell_{a(-n)}\overline{ad}^\ell_{\check{b}(k)}\vec{x}(-\vec{p})\right)\otimes v$$

$$+\frac{1}{4}\sum_{j;b(-k),a(-n)\in\hat{\mathcal{B}}}\frac{n+p_j}{n\cdot p_j}b(-k)\wedge\overline{ad}_{\check{b}(k)}\left(i_{a(-n)}ad^j_{a(-n)}\vec{x}(-\vec{p})\right)\otimes v$$

$$=\frac{1}{4}\sum_{j\neq\ell;b(-k)\in\hat{\mathcal{B}}}\frac{k+p_j}{k\cdot p_j}\mathrm{ad}^j_{b(-k)}\overline{ad}^\ell_{\check{b}(k)}\vec{x}(-\vec{p})\otimes v$$

$$+\frac{1}{4}\sum_{\ell;b(-k)\in\hat{\mathcal{B}}}\frac{p_\ell}{(p_\ell-k)k}\mathrm{ad}^\ell_{b(-k)}\overline{ad}^\ell_{\check{b}(k)}\vec{x}(-\vec{p})\otimes v$$

$$+\frac{1}{4}\sum_{j\neq\ell;b(-k)\in\hat{\mathcal{B}}}\frac{k+p_j}{kp_j}ad^j_{b(-k)}\overline{ad}^\ell_{\check{b}(k)}\vec{x}(-\vec{p})\otimes v$$

$$+\frac{1}{4}\sum_{\ell\neq j;b(-k)\in\hat{\mathcal{B}}}\frac{p_j+p_\ell-k}{p_j(p_\ell-k)}x_1(-p_1)\wedge\cdots\wedge b(-k)^{[\ell]}\wedge\ldots$$

$$\wedge\left(ad_{x_j(-p_j)}\overline{ad}_{\check{b}(k)}(x_\ell(-p_\ell))\right)^{[j]}\wedge\cdots\wedge x_q(-p_q)\otimes v$$

$$-\frac{1}{4}\sum_{\ell\neq j;i\notin\{j,\ell\};b(-k)\in\hat{\mathcal{B}}}\frac{p_j+p_i}{p_j\cdot p_i}x_1(-p_1)\wedge\cdots\wedge\left(\overline{ad}_{\check{b}(k)}x_\ell(-p_\ell)\right)^{[\ell]}\wedge\ldots$$

$$\wedge[x_i(-p_i),x_j(-p_j)]^{[j]}\wedge\cdots\wedge b(-k)^{[i]}\wedge\cdots\wedge x_q(-p_q)\otimes v$$

$$
\begin{aligned}
&+\frac{1}{4}\sum_{\ell;b(-k)\in\hat{\mathcal{B}}}\frac{p_\ell}{k(p_\ell-k)}ad^\ell_{b(-k)}\overline{ad}^\ell_{\check{b}(k)}\vec{x}(-\vec{p})\otimes v\\
&-\frac{1}{4}\sum_{\ell\neq j;b(-k)\in\hat{\mathcal{B}}}\frac{p_j+p_\ell-k}{p_j(p_\ell-k)}x_1(-p_1)\wedge\cdots\wedge\Big(ad_{x_j(-p_j)}\overline{ad}_{\check{b}(k)}(x_\ell(-p_\ell))\Big)^{[\ell]}\\
&\wedge\cdots\wedge b(-k)^{[j]}\wedge\cdots\wedge x_q(-p_q)\otimes v\\
&+\frac{1}{4}\sum_{j\neq\ell;b(-k)\in\hat{\mathcal{B}}}\frac{p_j+p_\ell}{p_j\cdot p_\ell}\overline{ad}_{\check{b}(k)}\Big(x_1(-p_1)\wedge\cdots\wedge\big(ad_{x_\ell(-p_\ell)}x_j(-p_j)\big)^{[j]}\\
&\wedge\cdots\wedge b(-k)^{[\ell]}\wedge\cdots\wedge x_q(-p_q)\Big)\otimes v\\
=&\frac{1}{2}\sum_{j\neq\ell;b(-k)\in\hat{\mathcal{B}}}\frac{k+p_j}{k\cdot p_j}\,\mathrm{ad}^j_{b(-k)}\,\overline{ad}^\ell_{\check{b}(k)}\vec{x}(-\vec{p})\otimes v\\
&+\frac{1}{2}\sum_{\ell;b(-k)\in\hat{\mathcal{B}}}\frac{p_\ell}{k(p_\ell-k)}\,\mathrm{ad}^\ell_{b(-k)}\,\overline{ad}^\ell_{\check{b}(k)}\vec{x}(-\vec{p})\otimes v\\
&+\frac{1}{4}\sum_{\ell\neq j;b(-k)\in\hat{\mathcal{B}}}\frac{p_j+p_\ell-k}{p_j(p_\ell-k)}x_1(-p_1)\wedge\cdots\wedge b(-k)^{[\ell]}\wedge\dots\\
&\wedge\Big(ad_{x_j(-p_j)}\overline{ad}_{\check{b}(k)}(x_\ell(-p_\ell))\Big)^{[j]}\wedge\cdots\wedge x_q(-p_q)\otimes v\\
&-\frac{1}{4}\sum_{\ell\neq j;b(-k)\in\hat{\mathcal{B}}}\frac{p_j+p_\ell-k}{p_\ell(p_j-k)}x_1(-p_1)\wedge\cdots\wedge b(-k)^{[\ell]}\wedge\dots\\
&\wedge\Big(ad_{x_\ell(-p_\ell)}\overline{ad}_{\check{b}(k)}(x_j(-p_j))\Big)^{[j]}\wedge\cdots\wedge x_q(-p_q)\otimes v\\
&+\frac{1}{4}\sum_{\ell\neq j;b(-k)\in\hat{\mathcal{B}}}\frac{p_j+p_\ell}{p_j\cdot p_\ell}x_1(-p_1)\wedge\cdots\wedge\Big(\overline{ad}_{\check{b}(k)}\,\mathrm{ad}_{x_j(-p_j)}(x_\ell(-p_\ell))\Big)^{[\ell]}\\
&\wedge\cdots\wedge b(-k)^{[j]}\wedge x_q(-p_q)\otimes v\\
=&\frac{1}{2}\sum_{j\neq\ell;b(-k)\in\hat{\mathcal{B}}}\frac{k+p_j}{k\cdot p_j}\,\mathrm{ad}^j_{b(-k)}\,\overline{ad}^\ell_{\check{b}(k)}\vec{x}(-\vec{p})\otimes v\\
&+\frac{1}{2}\sum_{\ell;b(-k)\in\hat{\mathcal{B}}}\frac{p_\ell}{k(p_\ell-k)}\,\mathrm{ad}^\ell_{b(-k)}\,\overline{ad}^\ell_{\check{b}(k)}\vec{x}(-\vec{p})\otimes v\\
&+\frac{1}{4}\sum_{\ell\neq j;a(-n)\in\hat{\mathcal{B}}}\frac{p_j+n}{n\cdot p_j}\overline{ad}^\ell_{\check{a}(n)}\,\mathrm{ad}^j_{a(-n)}\,\vec{x}(-\vec{p})\otimes v\\
&-\frac{1}{4}\sum_{\ell\neq j;a(-n)\in\hat{\mathcal{B}}}\frac{p_\ell+n}{n\cdot p_\ell}x_1(-p_1)\wedge\dots\\
&\wedge\big(\overline{ad}_{\check{a}(n)}(x_j(-p_j))\big)^{[\ell]}\wedge\cdots\wedge\big(ad_{a(-n)}x_\ell(-p_\ell)\big)^{[j]}\wedge\cdots\wedge x_q(-p_q)\otimes v\\
&+\frac{1}{4}\sum_{\ell\neq j;b(-k)\in\hat{\mathcal{B}}}\frac{p_j+p_\ell}{p_j\cdot p_\ell}x_1(-p_1)\wedge\cdots\wedge\Big(\overline{ad}_{\check{b}(k)}\,\mathrm{ad}_{x_j(-p_j)}(x_\ell(-p_\ell))\Big)^{[\ell]}\wedge\dots\\
&\wedge b(-k)^{[j]}\wedge\cdots\wedge x_q(-p_q)\otimes v,\ \text{by Lemma 4.7}
\end{aligned}
$$

$$= \sum_{j\neq \ell; b(-k)\in \hat{\mathcal{B}}} \frac{p_\ell - k}{k\cdot p_\ell} \,\mathrm{ad}^j_{b(-k)}\, \overline{ad}^{\ell}_{\check{b}(k)} \vec{x}(-\vec{p}) \otimes v$$
$$+ \frac{1}{2} \sum_{\ell; b(-k)\in \hat{\mathcal{B}}} \frac{p_\ell}{k(p_\ell - k)} \,\mathrm{ad}^{\ell}_{b(-k)}\, \overline{ad}^{\ell}_{\check{b}(k)} \vec{x}(-\vec{p}) \otimes v$$
$$- \frac{1}{2} \sum_{j\neq \ell; b_0\in \mathcal{B}_{\underline{0}}} \frac{p_\ell + p_j}{p_\ell p_j} ad^j_{b_0} ad^{\ell}_{\check{b}_0} \vec{x}(-\vec{p}) \otimes v, \text{ by following Lemma 4.8.}$$

This completes the proof of the lemma. □

Lemma 4.7. *For $j \neq \ell$ and $\vec{x}(-\vec{p}) = x_1(-p_1) \wedge \cdots \wedge x_q(-p_q) \in \wedge^q (L^-(\mathfrak{g},\sigma))$, we have*

$$\sum_{b(-k)\in \hat{\mathcal{B}}} \frac{p_j + p_\ell - k}{p_j(p_\ell - k)} x_1(-p_1)\wedge \ldots \wedge b(-k)^{[\ell]} \wedge \ldots \wedge \Big(ad_{x_j(-p_j)} \overline{ad}_{\check{b}(k)}(x_\ell(-p_\ell))\Big)^{[j]} \wedge \ldots \wedge x_q(-p_q)$$
$$= \sum_{a(-n)\in \hat{\mathcal{B}}} \frac{p_j + n}{p_j \cdot n} \overline{ad}^{\ell}_{\check{a}(n)}\, \mathrm{ad}^j_{a(-n)}\, \vec{x}(-\vec{p}).$$

Proof.

$$\sum_{b(-k)\in \hat{\mathcal{B}}} \frac{p_j + p_\ell - k}{p_j(p_\ell - k)} x_1(-p_1)\wedge \ldots \wedge b(-k)^{[\ell]} \wedge \ldots \wedge \Big(ad_{x_j(-p_j)} \overline{ad}_{\check{b}(k)}(x_\ell(-p_\ell))\Big)^{[j]} \wedge \ldots \wedge x_q(-p_q)$$
$$= \sum_{b(-k), a(-n)\in \hat{\mathcal{B}}} \frac{p_j + p_\ell - k}{p_j(p_\ell - k)} x_1(-p_1) \wedge \cdots \wedge \{[\check{b}(k), x_\ell(-p_\ell)], a(-n)\} b(-k)^{[\ell]} \wedge \ldots$$
$$\wedge \left(ad_{x_j(-p_j)}(a(-n))\right)^{[j]} \wedge \cdots \wedge x_q(-p_q)$$
$$= \sum_{b(-k), a(-n)\in \hat{\mathcal{B}}} \frac{p_j + p_\ell - k}{p_j(p_\ell - k)} x_1(-p_1) \wedge \cdots \wedge \{[\check{a}(n), x_\ell(-p_\ell], b(-k)\} b(-k)^{[\ell]} \wedge \ldots$$
$$\wedge \left(ad_{a(-n)} x_j(-p_j)\right)^{[j]} \wedge \cdots \wedge x_q(-p_q), \text{by Lemma 3.4}$$
$$= \sum_{a(-n)\in \hat{\mathcal{B}}} \frac{p_j + n}{p_j \cdot n} x_1(-p_1) \wedge \cdots \wedge \left(\overline{ad}_{\check{a}(n)} x_\ell(-p_\ell)\right)^{[\ell]} \wedge \ldots$$
$$\wedge \left(ad_{a(-n)} x_j(-p_j)\right)^{[j]} \wedge \cdots \wedge x_q(-p_q).$$

This proves the lemma. □

Lemma 4.8. *For $\vec{x}(-\vec{p}) = x_1(-p_1) \wedge \cdots \wedge x_q(-p_q) \in \wedge^q (L^-(\mathfrak{g},\sigma))$, we have*

$$\sum_{\substack{j\neq \ell;\\ b(-k)\in \hat{\mathcal{B}}}} \left(\frac{1}{p_\ell} + \frac{1}{p_j}\right) x_1(-p_1) \wedge \cdots \wedge b(-k)^{[j]} \wedge \cdots \wedge \left[\left[\check{b}(k), x_j(-p_j)\right], x_\ell(-p_\ell)\right]_-^{[\ell]} \wedge \cdots \wedge x_q(-p_q)$$
$$= -2 \sum_{j\neq \ell; a(-n)\in \hat{\mathcal{B}}} \left(\frac{1}{p_\ell} + \frac{1}{p_j}\right) ad^j_{a(-n)}\, \overline{ad}^{\ell}_{\check{a}(n)}\, \vec{x}(-\vec{p})$$
$$- \sum_{j\neq \ell; b_0\in \mathcal{B}_{\underline{0}}} \left(\frac{1}{p_\ell} + \frac{1}{p_j}\right) ad^j_{b_0}\, ad^{\ell}_{\check{b}_0}\, \vec{x}(-\vec{p}),$$

where $[\,]_-$ denotes the projection to $L^-(\mathfrak{g},\sigma)$. Similarly,

$$\sum_{j\neq \ell; b(-k)\in \hat{\mathcal{B}}} \left(\frac{1}{p_\ell} + \frac{1}{p_j}\right) x_1(-p_1) \wedge \cdots \wedge b(-k)^{[j]}$$

$$\wedge \cdots \wedge \left[x_j(-p_j), [\check{b}(k), x_\ell(-p_\ell)]\right]_-^{[\ell]} \wedge \cdots \wedge x_q(-p_q)$$

$$= -2 \sum_{j\neq \ell; a(-n)\in \hat{\mathcal{B}}} \left(\frac{1}{p_j} + \frac{1}{p_\ell}\right) ad^j_{a(-n)}\, \overline{ad}^\ell_{\check{a}(n)}\, \vec{x}(-\vec{p})$$

$$- \sum_{j\neq \ell; b_0\in \mathcal{B}_{\underline{0}}} \left(\frac{1}{p_j} + \frac{1}{p_\ell}\right) ad^j_{b_0}\, ad^\ell_{\check{b}_0}\, \vec{x}(-\vec{p}).$$

PROOF.

$$\sum_{\substack{j\neq \ell;\\ b(-k)\in \hat{\mathcal{B}}}} \left(\frac{1}{p_\ell} + \frac{1}{p_j}\right) x_1(-p_1) \wedge \cdots \wedge b(-k)^{[j]} \wedge \cdots$$

$$\wedge \left[\left[\check{b}(k), x_j(-p_j)\right], x_\ell(-p_\ell)\right]_-^{[\ell]} \wedge \cdots \wedge x_q(-p_q)$$

$$= \sum_{\substack{j\neq \ell;\\ a(-n), b(-k)\in \hat{\mathcal{B}}}} \left(\frac{1}{p_\ell} + \frac{1}{p_j}\right) x_1(-p_1) \wedge \cdots \wedge b(-k)^{[j]} \wedge \cdots$$

$$\wedge \left\{\left[\check{b}(k), x_j(-p_j)\right], a(-n)\right\} [a(-n), x_\ell(-p_\ell)]^{[\ell]} \wedge \cdots \wedge x_q(-p_q)$$

$$+ \sum_{j\neq \ell; a(-n), b(-k)\in \hat{\mathcal{B}}} \left(\frac{1}{p_\ell} + \frac{1}{p_j}\right) x_1(-p_1) \wedge \cdots \wedge b(-k)^{[j]} \wedge \cdots$$

$$\wedge \left\{\left[\check{b}(k), x_j(-p_j)\right], \check{a}(n)\right\} [\check{a}(n), x_\ell(-p_\ell)]_-^{[\ell]} \wedge \cdots \wedge x_q(-p_q)$$

$$+ \sum_{j\neq \ell; b_0\in \mathcal{B}_{\underline{0}}; b(-k)\in \hat{\mathcal{B}}} \left(\frac{1}{p_\ell} + \frac{1}{p_j}\right) x_1(-p_1) \wedge \cdots \wedge b(-k)^{[j]} \wedge \cdots$$

$$\wedge \left\{\left[\check{b}(k), x_j(-p_j)\right], b_0\right\} [b_0, x_\ell(-p_\ell)]^{[\ell]} \wedge \cdots \wedge x_q(-p_q)$$

$$= \sum_{j\neq \ell; a(-n)\in \hat{\mathcal{B}}} \left(\frac{1}{p_\ell} + \frac{1}{p_j}\right) x_1(-p_1) \wedge \cdots \wedge [x_j(-p_j), \check{a}(n)]_-^{[j]}$$

$$\wedge \cdots \wedge [a(-n), x_\ell(-p_\ell)]^{[\ell]} \wedge \cdots \wedge x_q(-p_q)$$

$$+ \sum_{j\neq \ell; a(-n)\in \hat{\mathcal{B}}} \left(\frac{1}{p_\ell} + \frac{1}{p_j}\right) x_1(-p_1) \wedge \cdots \wedge [x_j(-p_j), a(-n)]^{[j]}$$

$$\wedge \cdots \wedge [\check{a}(n), x_\ell(-p_\ell)]_-^{[\ell]} \wedge \cdots \wedge x_q(-p_q)$$

$$+ \sum_{j\neq \ell; b_0\in \mathcal{B}_{\underline{0}}} \left(\frac{1}{p_\ell} + \frac{1}{p_j}\right) x_1(-p_1) \wedge \cdots \wedge \left[x_j(-p_j), \check{b}_0\right]^{[j]}$$

$$\wedge \cdots \wedge [b_0, x_\ell(-p_\ell)]^{[\ell]} \wedge \cdots \wedge x_q(-p_q), \text{ by Lemma 3.4}$$

$$= -2 \sum_{\substack{j\neq \ell;\\ a(-n)\in \hat{\mathcal{B}}}} \left(\frac{1}{p_\ell} + \frac{1}{p_j}\right) ad^j_{a(-n)}\, \overline{ad}^\ell_{\check{a}(n)}\, \vec{x}(-\vec{p}) - \sum_{\substack{j\neq \ell;\\ b_o\in \mathcal{B}_{\underline{0}}}} \left(\frac{1}{p_\ell} + \frac{1}{p_j}\right) ad^j_{b_0}\, ad^\ell_{\check{b}_0}\, \vec{x}(-\vec{p}).$$

□

LEMMA 4.9. *If $\vec{x}(-\vec{p})\otimes v \in \wedge^q(L^-(\mathfrak{g},\sigma))\otimes\mathscr{H}_c(\lambda)\otimes V(\vec{\mu})_{\vec{z}}$ is $\mathfrak{g}^\sigma$-invariant, then*

$$\frac{1}{2}\sum_{1\le j,\ell\le q;b_0\in\mathcal{B}_{\underline{0}}}\left(\frac{1}{p_j}+\frac{1}{p_\ell}\right)ad^j_{b_0}ad^\ell_{\check{b}_0}\vec{x}(-\vec{p})\otimes v+\sum_{\ell;b_0\in\mathcal{B}_{\underline{0}}}\frac{1}{p_\ell}ad^\ell_{b_0}\vec{x}(-\vec{p})\otimes\check{b}_0\cdot v=0,$$

where $\mathfrak{g}_{\underline{0}}=\mathfrak{g}^\sigma$ is the $+1$ eigenspace with respect to the operator σ.

PROOF. For any $x\in\mathfrak{g}_{\underline{0}}$

$$x\cdot\left(\sum_\ell\frac{1}{p_\ell}ad^\ell_x\vec{x}(-\vec{p})\otimes v\right)=0. \tag{58}$$

To prove this, observe that

$$x\cdot\left(\sum_\ell\frac{1}{p_\ell}ad^\ell_x\vec{x}(-\vec{p})\otimes v\right)=\sum_{j,\ell}\frac{1}{p_\ell}ad^\ell_x ad^j_x\vec{x}(-\vec{p})\otimes v+\sum_\ell\frac{1}{p_\ell}ad^\ell_x\vec{x}(-\vec{p})\otimes x\cdot v,$$

since ad^ℓ_x and ad^j_x commute

$$\begin{aligned}&=\left(\sum_\ell\frac{1}{p_\ell}ad^\ell_x\right)\cdot\left(ad_x\vec{x}(-\vec{p})\otimes v+\vec{x}(-\vec{p})\otimes x\cdot v\right)\\&=0,\qquad\text{since } \vec{x}(-\vec{p})\otimes v \text{ is } \mathfrak{g}_{\underline{0}}\text{-invariant.}\end{aligned}$$

This proves (58).

Thus, for any fixed b_0,

$$\sum_\ell\frac{1}{p_\ell}ad^\ell_{\check{b}_0}\vec{x}(-\vec{p})\otimes b_0\cdot v=-\sum_\ell\frac{1}{p_\ell}\left(ad_{b_0}ad^\ell_{\check{b}_0}\vec{x}(-\vec{p})\right)\otimes v.$$

Thus, summing over b_0, we get the following equation:

$$\sum_{\ell;b_0\in\mathcal{B}_{\underline{0}}}\frac{1}{p_\ell}ad^\ell_{\check{b}_0}\vec{x}(-\vec{p})\otimes b_0\cdot v=-\sum_{j,\ell;b_0\in\mathcal{B}_{\underline{0}}}\frac{1}{p_\ell}ad^j_{b_0}ad^\ell_{\check{b}_0}\vec{x}(-\vec{p})\otimes v. \tag{59}$$

Moreover,

$$\begin{aligned}&\frac{1}{2}\sum_{j,\ell;b_0\in\mathcal{B}_{\underline{0}}}\left(\frac{1}{p_j}+\frac{1}{p_\ell}\right)\left(ad^j_{b_0}ad^\ell_{\check{b}_0}\vec{x}(-\vec{p})\right)\otimes v\\&=\frac{1}{2}\sum_{j,\ell;b_0\in\mathcal{B}_{\underline{0}}}\frac{1}{p_j}\left(ad^j_{b_0}ad^\ell_{\check{b}_0}\vec{x}(-\vec{p})\right)\otimes v+\frac{1}{2}\sum_{j,\ell;b_0\in\mathcal{B}_{\underline{0}}}\frac{1}{p_\ell}\left(ad^j_{b_0}ad^\ell_{\check{b}_0}\vec{x}(-\vec{p})\right)\otimes v\\&=\frac{1}{2}\sum_{j,\ell;b_0\in\mathcal{B}_{\underline{0}}}\frac{1}{p_\ell}\left(ad^\ell_{b_0}ad^j_{\check{b}_0}\vec{x}(-\vec{p})\right)\otimes v+\frac{1}{2}\sum_{j,\ell;b_0\in\mathcal{B}_{\underline{0}}}\frac{1}{p_\ell}\left(ad^j_{b_0}ad^\ell_{\check{b}_0}\vec{x}(-\vec{p})\right)\otimes v,\end{aligned}$$

by interchanging j and ℓ in the first factor,

$$=\frac{1}{2}\sum_{j,\ell;b_0\in\mathcal{B}_{\underline{0}}}\frac{1}{p_\ell}\left(ad^j_{\check{b}_0}ad^\ell_{b_0}\vec{x}(-\vec{p})\right)\otimes v+\frac{1}{2}\sum_{j,\ell;b_0\in\mathcal{B}_{\underline{0}}}\frac{1}{p_\ell}\left(ad^j_{b_0}ad^\ell_{\check{b}_0}\vec{x}(-\vec{p})\right)\otimes v, \tag{60}$$

since $\sum_{b_0\in\mathcal{B}_{\underline{0}}}[b_o,[\check{b}_0,y]]=\sum_{b_0\in\mathcal{B}_{\underline{0}}}[\check{b}_o,[b_0,y]]$ for any $y\in\mathfrak{g}$.

Combining (59) and (60), we get the lemma. □

4.1. Proof of the Nakano's identity Theorem 3.11. We first calculate $\overline{\square}$. By Proposition 4.5 and Lemma 4.6,

$$\begin{aligned}
&\overline{\square}\,(\vec{x}(-\vec{p})\otimes v) = \\
&\quad + \sum_{b(-k)\in\hat{\mathcal{B}}} \frac{1}{k}\overline{ad}_{\check{b}(k)}\cdot \vec{x}(-\vec{p})\otimes b(-k)\cdot v + \sum_{b(-k)\in\hat{\mathcal{B}}} \frac{1}{k}\vec{x}(-\vec{p})\otimes b(-k)\check{b}(k)\cdot v \\
&\quad - \sum_{\ell;b(-k)\in\hat{\mathcal{B}}} \frac{1}{p_\ell} ad^\ell_{b(-k)}\vec{x}(-\vec{p})\otimes \big(\check{b}(k)\cdot v_1\big)\otimes v_2 - \sum_{\ell;b_0\in\mathcal{B}_{\underline{0}}} \frac{1}{p_\ell} ad^\ell_{b_0}\vec{x}(-\vec{p})\otimes\big(\check{b}_0\cdot v_1\big)\otimes v_2 \\
&\quad - \sum_{\ell;b(-k)\in\hat{\mathcal{B}}} \frac{1}{p_\ell}\overline{ad}^\ell_{\check{b}(k)}\vec{x}(-\vec{p})\otimes (b(-k)\cdot v_1)\otimes v_2 + \frac{qc}{m}\vec{x}(-\vec{p})\otimes v_1\otimes v_2 \\
&\quad - \sum_{\ell;b_0\in\mathcal{B}_{\underline{0}}} \frac{1}{p_\ell} ad^\ell_{b_0}\vec{x}(-\vec{p})\otimes v_1\otimes\big((z\bar{z})^{p_\ell}\check{b}_0\big)\cdot v_2 \\
&\quad - \sum_{\ell;a(-n)\in\hat{\mathcal{B}}} \frac{1}{p_\ell}\Big(ad^\ell_{a(-n)}\vec{x}(-\vec{p})\Big)\otimes v_1\otimes((z\bar{z})^{p_\ell}z^n\check{a})\cdot v_2 \\
&\quad - \sum_{\ell;a(-n)\in\hat{\mathcal{B}}} \frac{1}{p_\ell}\Big(\overline{ad}^\ell_{\check{a}(n)}\vec{x}(-\vec{p})\Big)\otimes v_1\otimes\big((z\bar{z})^{p_\ell-n}\bar{z}^n a\big)\cdot v_2 \\
&\quad + \sum_{\ell;b(-k)\in\hat{\mathcal{B}}}\left(\frac{1}{p_\ell}+\frac{1}{k}\right) ad^\ell_{b(-k)}\vec{x}(-\vec{p})\otimes\check{b}(k)\cdot v \\
&\quad + \sum_{j\neq\ell;b(-k)\in\hat{\mathcal{B}}}\left(\frac{1}{k}-\frac{1}{p_\ell}\right) ad^j_{b(-k)}\overline{ad}^\ell_{\check{b}(k)}\vec{x}(-\vec{p})\otimes v \\
&\quad + \frac{1}{2}\sum_{\ell;b(-k)\in\hat{\mathcal{B}}}\left(\frac{1}{p_\ell-k}+\frac{1}{k}\right) ad^\ell_{b(-k)}\overline{ad}^\ell_{\check{b}(k)}\vec{x}(-\vec{p})\otimes v \\
&\quad - \frac{1}{2}\sum_{j\neq\ell;b_0\in\mathcal{B}_{\underline{0}}}\left(\frac{1}{p_j}+\frac{1}{p_\ell}\right) ad^j_{b_0} ad^\ell_{\check{b}_0}\vec{x}(-\vec{p})\otimes v.
\end{aligned}$$

Observe that for any $1\le\ell\le q$,

$$\sum_{b(-k)\in\hat{\mathcal{B}}}\left(\frac{1}{p_\ell-k}+\frac{1}{k}\right) ad^\ell_{b(-k)}\overline{ad}^\ell_{\check{b}(k)}\vec{x}(-\vec{p})\otimes v = \sum_{b(-k)\in\hat{\mathcal{B}}}\left(\frac{1}{2k}\right) ad^\ell_{b(-k)}\overline{ad}^\ell_{\check{b}(k)}\vec{x}(-\vec{p})\otimes v.$$

Thus, using the expression for $\square$ as in Proposition 3.10, we get the following:

For $\vec{x}(-\vec{p})\otimes v_1\otimes v_2\in\wedge^q\left(L^-(\mathfrak{g},\sigma)\right)\otimes\mathscr{H}_c(\lambda)\otimes V(\vec{\mu})_{\vec{z}}$, we have

$$\begin{aligned}
&\left(\overline{\square}-\square\right)(\vec{x}(-\vec{p})\otimes v_1\otimes v_2) \\
&\quad = \frac{qc}{m}(\vec{x}(-\vec{p})\otimes v_1\otimes v_2) + \sum_{\ell;b(-k)\in\hat{\mathcal{B}}}\frac{1}{p_\ell}\Big(ad^\ell_{b(-k)}\vec{x}(-\vec{p})\Big)\otimes v_1\otimes\check{b}(k)\cdot v_2 \\
&\qquad - \sum_{\ell;b_0\in\mathcal{B}_{\underline{0}}}\frac{1}{p_\ell} ad^\ell_{b_0}\vec{x}(-\vec{p})\otimes v_1\otimes\big((z\bar{z})^{p_\ell}\check{b}_0\big)\cdot v_2
\end{aligned}$$

$$
\begin{aligned}
&- \sum_{\ell; a(-n)\in\hat{\mathcal{B}}} \frac{1}{p_\ell}\left(ad^\ell_{a(-n)}\vec{x}(-\vec{p})\right)\otimes v_1\otimes\left((z\bar{z})^{p_\ell}z^n\check{a}\right)\cdot v_2\\
&- \sum_{\ell; a(-n)\in\hat{\mathcal{B}}} \frac{1}{p_\ell}\left(\overline{ad}^\ell_{\check{a}(n)}\vec{x}(-\vec{p})\right)\otimes v_1\otimes\left((z\bar{z})^{p_\ell-n}\bar{z}^n a\right)\cdot v_2\\
&+ \sum_{\substack{\ell;\\ b(-k)\in\hat{\mathcal{B}}}} \frac{1}{p_\ell} ad^\ell_{b(-k)}\overline{ad}^\ell_{\check{b}(k)}\vec{x}(-\vec{p})\otimes v - \frac{1}{2}\sum_{\substack{j\neq\ell;\\ b_0\in\mathcal{B}_{\underline{0}}}}\left(\frac{1}{p_j}+\frac{1}{p_\ell}\right) ad^j_{b_0} ad^\ell_{\check{b}_0}\vec{x}(-\vec{p})\otimes v\\
&+ \sum_{\substack{\ell;\\ b(-k)\in\hat{\mathcal{B}}}} \frac{1}{p_\ell}\overline{ad}^\ell_{\check{b}(k)}\vec{x}(-\vec{p})\otimes v_1\otimes b(-k)\cdot v_2 - \sum_{\substack{\ell;\\ b_0\in\mathcal{B}_{\underline{0}}}}\frac{1}{p_\ell} ad^\ell_{b_0}\vec{x}(-\vec{p})\otimes\check{b}_0\cdot v_1\otimes v_2,
\end{aligned}
$$

since, for any $1\le\ell\le q$,

$$
\sum_{b(-k)\in\hat{\mathcal{B}}}\frac{1}{p_\ell-k}\left(ad^\ell_{b(-k)}\overline{ad}^\ell_{\check{b}(k)}\vec{x}(-\vec{p})\right) = \sum_{a(-n)\in\hat{\mathcal{B}}}\frac{1}{n}\left(ad^\ell_{a(-n)}\overline{ad}^\ell_{\check{a}(n)}\vec{x}(-\vec{p})\right),
$$

as can be easily seen.

Now, from the above expression of $\overline{\Box}-\Box$, using Lemma 4.9 and Proposition 3.8, we get the following, which proves the Nakano's identity Theorem 3.11.

For $\vec{x}(-\vec{p})\otimes v_1\otimes v_2\in\left[\wedge^q\left(L^-(\mathfrak{g},\sigma)\right)\otimes\mathscr{H}_c(\lambda)\otimes V(\vec{\mu})_{\vec{z}}\right]^{\mathfrak{g}^\sigma}$,

$$
\begin{aligned}
\left(\overline{\Box}-\Box\right)\left(\vec{x}(-\vec{p})\otimes v_1\otimes v_2\right) &= \frac{(c+2\check{h})q}{m}\vec{x}(-\vec{p})\otimes v\\
&+ \sum_{\ell; b(-k)\in\hat{\mathcal{B}}}\frac{1}{p_\ell} ad^\ell_{b(-k)}\vec{x}(-\vec{p})\otimes v_1\otimes\left((1-(z\bar{z})^{p_\ell})z^k\check{b}\right)\cdot v_2\\
&+ \sum_{\ell; b(-k)\in\hat{\mathcal{B}}}\frac{1}{p_\ell}\overline{ad}^\ell_{\check{b}(k)}\vec{x}(-\vec{p})\otimes v_1\otimes\left((1-(z\bar{z})^{p_\ell-k})(\bar{z})^k b\right)\cdot v_2\\
&+ \sum_{\ell; b_0\in\mathcal{B}_{\underline{0}}}\frac{1}{p_\ell} ad^\ell_{b_0}\vec{x}(-\vec{p})\otimes v_1\otimes\left((1-(z\bar{z})^{p_\ell})\check{b}_0\right)\cdot v_2.
\end{aligned}
$$

Observe that we have obtained Theorem 3.11 for *special* automorphisms σ of $\mathfrak{g}$ since its proof given above uses Proposition 3.8, which is proved only for special automorphisms σ.

Acknowledgments

We thank Constantin Teleman for many helpful correspondences and conversations. We thank the referee for helpful comments to improve the exposition.

References

[BH] M. Besson and J. Hong, *Smooth locus of twisted affine Schubert varieties and twisted affine Demazure modules*, `arXiv:2010.11357`, 2021.

[Bo] N. Bourbaki, *Groupes et Algèbres de Lie, Ch. 4-6*, Masson, Paris, 1981.

[Da] C. Damiolini, *Conformal blocks attached to twisted groups*, Math. Z. **295** (2020), no. 3-4, 1643–1681, DOI 10.1007/s00209-019-02414-6. MR4125705

[DM] T. Deshpande and S. Mukhopadhyay, *Crossed modular categories and the Verlinde formula for twisted conformal blocks*, Camb. J. Math. **11** (2023), no. 1, 159–297. MR4600543

[Fa] G. Faltings, *A proof for the Verlinde formula*, J. Algebraic Geom. **3** (1994), no. 2, 347–374. MR1257326

[HK1] J. Hong and S. Kumar, *Conformal blocks for Galois covers of algebraic curves*, Compos. Math. **159** (2023), no. 10, 2191–2259, DOI 10.1112/s0010437x23007418. MR4634665

[HK2] J. Hong and S. Kumar, *Twisted conformal blocks and their dimension*, Math. Z. **306** (2024), no. 4, Paper No. 76, 29, DOI 10.1007/s00209-024-03461-4. MR4719989

[HS] G. Hochschild and J.-P. Serre, *Cohomology of Lie algebras*, Ann. of Math. (2) **57** (1953), 591–603, DOI 10.2307/1969740. MR54581

[Hum] J. E. Humphreys, *Linear algebraic groups*, Graduate Texts in Mathematics, No. 21, Springer-Verlag, New York-Heidelberg, 1975. MR396773

[Ka] V. G. Kac, *Infinite-dimensional Lie algebras*, 3rd ed., Cambridge University Press, Cambridge, 1990, DOI 10.1017/CBO9780511626234. MR1104219

[Ko] B. Kostant, *Powers of the Euler product and commutative subalgebras of a complex simple Lie algebra*, Invent. Math. **158** (2004), no. 1, 181–226, DOI 10.1007/s00222-004-0370-7. MR2090363

[Ku1] S. Kumar, *Kac-Moody groups, their flag varieties and representation theory*, Progress in Mathematics, vol. 204, Birkhäuser Boston, Inc., Boston, MA, 2002, DOI 10.1007/978-1-4612-0105-2. MR1923198

[Ku2] S. Kumar, *Conformal blocks, generalized theta functions and the Verlinde formula*, New Mathematical Monographs, vol. 42, Cambridge University Press, Cambridge, 2022. MR4327093

[R] W. Rudin, *Functional analysis*, McGraw-Hill Series in Higher Mathematics, McGraw-Hill Book Co., New York-Düsseldorf-Johannesburg, 1973. MR365062

[Te] C. Teleman, *Lie algebra cohomology and the fusion rules*, Comm. Math. Phys. **173** (1995), no. 2, 265–311. MR1355626

[TUY] A. Tsuchiya, K. Ueno, and Y. Yamada, *Conformal field theory on universal family of stable curves with gauge symmetries*, Integrable systems in quantum field theory and statistical mechanics, Adv. Stud. Pure Math., vol. 19, Academic Press, Boston, MA, 1989, pp. 459–566, DOI 10.2969/aspm/01910459. MR1048605

Department of Mathematics, University of North Carolina at Chapel Hill, Chapel Hill, North Carolina 27599-3250

Email address: jiuzu@email.unc.edu

Department of Mathematics, University of North Carolina at Chapel Hill, Chapel Hill, North Carolina 27599-3250

Email address: shrawan@email.unc.edu

Proceedings of Symposia in Pure Mathematics
Volume **108**, 2024
https://doi.org/10.1090/pspum/108/01966

Realizing rings of regular functions via the cohomology of quantum groups

Zongzhu Lin and Daniel K. Nakano

In memory of Georgia M. Benkart and Brian J. Parshall

ABSTRACT. Let G be a complex reductive group and P be a parabolic subgroup of G. In this paper, the authors address questions involving the realization of the G-module of the global sections of the (twisted) cotangent bundle over the flag variety G/P via the cohomology of the small quantum group. Our main results generalize the important computation of the cohomology ring for the small quantum group by Ginzburg and Kumar [Duke Math. J. 69 (1993) 179–198], and provide a generalization of well-known calculations by Kumar, Lauritzen, and Thomsen [Invent. Math. 136 (1999) 603–621] to the quantum case and the parabolic setting. As an application we answer the question (first posed by Friedlander and Parshall for Frobenius kernels [Invent. Math. 86 (1986), 553–562, (3.2)] about the realization of coordinate rings of Richardson orbit closures for complex semisimple groups via quantum group cohomology. Formulas will be provided which relate the multiplicities of simple G-modules in the global sections with the dimensions of extension groups over the large quantum group.

1. Introduction

1.1. Let G be a complex simple algebraic group, $\mathfrak{g}$ be its Lie algebra, and Φ be the associated root system. Fix ζ an ℓth root of unity in $\mathbb{C}$. Using divided powers, Lusztig defined a $\mathbb{Z}[q,q^{-1}]$-form, from which one can specialize q to an ℓ-th primitive root $\zeta \in \mathbb{C}$ of 1 to get $U_\zeta(\mathfrak{g})$. We will call $U_\zeta(\mathfrak{g})$ the large quantum group. The small quantum group $u_\zeta(\mathfrak{g})$ is a finite-dimensional normal Hopf subalgebra of $U_\zeta(\mathfrak{g})$ whose quotient $U_\zeta(\mathfrak{g})/\!\!/u_\zeta(\mathfrak{g})$ identifies with the ordinary universal enveloping algebra $\mathcal{U}(\mathfrak{g})$. This basic fact implies that for any $U_\zeta(\mathfrak{g})$-module, M, the cohomology group $\mathrm{H}^\bullet(\mathfrak{u}_\zeta(\mathfrak{g}), M)$ is a rational G-module. When $M \cong \mathbb{C}$ (i.e., the trivial module), Ginzburg and Kumar [**GK**] demonstrated that for $\ell > h$ (h is the Coxeter number for Φ) the odd degree cohomology vanishes and $\mathrm{H}^{2\bullet}(\mathfrak{u}_\zeta(\mathfrak{g}), \mathbb{C}) \cong \mathbb{C}[\mathcal{N}]$ where $\mathcal{N}$ is the variety of nilpotent elements of $\mathfrak{g}$. Arkhipov, Bezrukavnikov and Ginzburg [**ABG**] have used this computation as a starting point to show that there are beautiful connections between representations for quantum groups and the complex algebraic geometry of $\mathcal{N}$. As an application of their work, they provided a proof of Lusztig character formula for simple modules in the quantum group setting.

2020 *Mathematics Subject Classification.* Primary 20G42, 20G10; Secondary 17B56.
Research of the second author was supported in part by NSF grant DMS-2101941.

In [**BNPP**], Bendel, Nakano, Parshall, and Pillen have also exploited the powerful tools available in complex algebraic geometry to compute the cohomology ring $\mathrm{H}^\bullet(\mathfrak{u}_\zeta(\mathfrak{g}),\mathbb{C})$ when $\ell \leq h$. Their computation verified that the cohomology ring is finitely generated and allowed them to develop a theory of support varieties. Furthermore, they computed the support varieties of quantum Weyl modules in the case when $(p,\ell)=1$ for every bad prime p for Φ which proves a quantum version of a conjecture of Jantzen (cf. [**NPV**, (6.2.1) Theorem]).

1.2. Let G_k be a reductive algebraic group over an algebraically closed field k of positive characteristic $p>0$ with a rational structure over the prime field $\mathbb{F}_p$. In 1986, Friedlander and Parshall [**FP1**, (3.2)] posed several questions about (i) the realization of orbit closures via support varieties for finite-dimensional rational G_k-modules and (ii) the realization of the coordinate algebras of the closures of nilpotent G_k-orbits in the Lie algebra $\mathfrak{g}_k=\mathrm{Lie}(G_k)$ via the cohomology of the first Frobenius kernel of the reductive group with coefficients in a suitable algebra. In the quantum group setting, the analog of (i) was verified by Bezrukavnikov ([**Be**]) by computing the support varieties of tilting modules. The quantum analog of (ii) can be stated as follows:

QUESTION 1.2.1. Let $\ell>h$ and $J\subseteq\Delta$ with Levi decomposition of $\mathfrak{g}=\mathfrak{u}_J\oplus\mathfrak{l}_J\oplus\mathfrak{u}_J^+$. Does there exist a $U_\zeta(\mathfrak{g})$-algebra A such that $\mathrm{H}^{\mathrm{odd}}(u_\zeta(\mathfrak{g}),A)=0$ and $\mathrm{H}^{2\bullet}(u_\zeta(\mathfrak{g}),A)=\mathbb{C}[G\cdot\mathfrak{u}_J]$?

We will also be interested in another related question:

QUESTION 1.2.2. Let $\ell>h$ and $J\subseteq\Delta$ with Levi decomposition of $\mathfrak{g}=\mathfrak{u}_J\oplus\mathfrak{l}_J\oplus\mathfrak{u}_J^+$. Does there exist a $U_\zeta(\mathfrak{g})$-module M such that $\mathrm{H}^\bullet(u_\zeta(\mathfrak{g}),M)$ identifies as a G-module with $\mathrm{ind}_{P_J}^G(S^\bullet(\mathfrak{u}_J^*)\otimes\mathsf{k}_\gamma)$ where $\gamma\in X_{P_J}\cap X^+$?

The paper is devoted to answering Questions 1.2.1 and 1.2.2 for quantum groups. We show that Question 1.2.2 has a positive answer in full generality. When Φ is of type A_n, Question 1.2.1 has an affirmative answer for all $J\subseteq\Delta$. For other root systems one needs to impose conditions involving the moment map and the normality of the orbit closure to guarantee the realization. In the process of answering Questions 1.2.1 and 1.2.2 we construct modules over the small quantum group whose cohomology identifies with the global sections of the twisted cotangent bundle on G/P_J. The resulting theorem can be viewed as a generalization of the aforementioned result of Ginzburg and Kumar [**GK**] and of Kumar, Lauritzen, and Thomsen [**KLT**] to the quantum and parabolic settings. The computation of cohomologies in [**GK**] is a special case of what we computed here. However, [**GK**] deals with the Borel subalgebra, for which the Levi subalgebra $\mathfrak{l}_J$ is the Cartan subalgebra $\mathfrak{h}$. When the Levi subalgebra is not torus, the towers of algebras by regular sequences in the center of $\mathcal{U}_\zeta(\mathfrak{u}_J)$ are not $U_\zeta(\mathfrak{l}_J)$-modules in contrast to case of $\mathfrak{l}_J=\mathfrak{h}$. We had to search for a completely different approach to deal with $U_\zeta(\mathfrak{p}_J)$-module structures on the cohomology by proving a theorem that, for two $U_\zeta(\mathfrak{p}_J)$-modules V_1 and V_2, a linear isomorphism $\phi: V_1\to V_2$ is a $U_\zeta(\mathfrak{p}_J)$-module isomorphism if and only if it is a $U_\zeta(\mathfrak{b})$-module isomorphism. This property is more general than what we need in this paper and uses deep properties of the generalized tensor identity and the completeness of the flag varieties.

We should also mention that our calculations involve computing the ring structure of $\mathrm{Ext}^\bullet_{u_\zeta(\mathfrak{g})}(D,D)$ for a specific $U_\zeta(\mathfrak{g})$-modules D. There is very little known about these Ext-algebras in general and we hope that our techniques and explicit computations will lead to a better understanding of these structures. It should be noted that the presence of strong cohomological vanishing results in the complex reductive group case allows us to make computations for the Ext-algebra in the quantum setting.

It should be observed that Questions 1.2.1 and 1.2.2 are still not completely resolved for reductive groups in positive characteristic. For Richardson orbits, Question 1.2.1 holds by [**CLNP**], and for arbitrary orbits in classical groups in [**NT**] when the field has good characteristics. Partial results for exceptional groups are provided in [**NT**]. We will indicate the necessary vanishing results needed to establish positive answers to Questions 1.2.1 and 1.2.2 over fields of positive characteristics.

1.3. The outline of the paper is as follows. In Section 2, we review the construction of quantum groups for $\mathfrak{g}$ and its parabolic subalgebras. For our purposes these are important considerations to construct suitable $U_\zeta(\mathfrak{g})$-modules. In Section 3, we prove a general theorem which gives sufficient conditions to realize the global sections of the twisted cotangent bundle as the cohomology of the small quantum group with coefficients in a $U_\zeta(\mathfrak{g})$-module N. In Section 4 we produce a module N which gives a non-shifted realization of these global sections and as an application we provide an affirmative answer to Question 1.2.1. Later in Section 5, we apply the theory of tilting modules to yield a shifted realization of the global sections. This shifted realization result is a generalization of both the quantum and the parabolic versions of the work of Kumar, Lauritzen and Thomsen [**KLT**].

Section 6 is devoted to investigating the connections between G-composition factors in the coordinate rings with quantum group cohomology. We demonstrate that computing certain Ext-groups for modules over the large quantum groups is equivalent to determining the G-composition factors in the modules $\text{ind}_{P_J}^G S^\bullet(\mathfrak{u}_J^*) \otimes \mathsf{k}_\gamma$ for a character γ of P. It is well-known that the later problem involves Kostant's Partition Function which can be related to the coefficients of certain Kazhdan-Lusztig Polynomials (cf. [**Br2**]). In Section 7, we indicate how one can extend our results to Frobenius kernels under suitable cohomological vanishing assumptions in positive characteristic cases. Finally, in Section 8, we include an appendix which gives details on Hopf algebra actions on cohomology rings. The results are of independent interest and form the foundational basis for the work in the paper. For instance, in Section 8.4, we show that two modules for the algebraic group (quantum group, or more general Hopf algebras) are isomorphic if and only if they are isomorphic when restricted to a particular subalgebra provided the (rational) induced functor sends the trivial module for the subalgebra to the trivial module for the larger algebra. Thus many equivariant questions can be reduced to equivariant questions for a Borel subgroups (subalgebras).

2. Quantum groups and Hopf algebra actions

2.1. We will follow the conventions as described in [**BNPP**, Section 2]. Let $G := G_{\mathbb{Q}}$ be a simply connected simple algebraic group which is defined and split over $\mathbb{Q}$ with Lie algebra $\mathfrak{g}$ over $\mathbb{Q}$ with a fixed Chevalley basis. Let Φ be the irreducible root system associated to $\mathfrak{g}$ with respect to a fixed maximal split torus T of G. Let Δ be the corresponding set of simple roots. The set Φ spans a real vector space $\mathbb{E}$ with positive definite inner product $\langle u, v\rangle$, $u, v \in \mathbb{E}$, adjusted so that $\langle\alpha,\alpha\rangle = 2$ if $\alpha \in \Phi$ is a short root. If $\alpha \in \Phi$ then let $\alpha^\vee = \frac{2}{\langle\alpha,\alpha\rangle}\alpha$ be the coroot. For $J \subseteq \Delta$, let $\Phi_J = \Phi \cap \mathbb{Z}J$ be the root system of Φ generated by J. Let W be the Weyl group corresponding to T. For $J \subset \Delta$, let W_J be the Weyl group of Φ_J, viewed as a subgroup of W, and ${}^JW$ be the set of minimal length coset representatives for $W_J\backslash W$. Let $w_0 \in W$ be the unique longest element. We will use $w_{0,J} \in W_J$ to denote the unique longest element of W_J.

Define the fundamental dominant weights $\omega_{\alpha_1}, \ldots, \omega_{\alpha_n}$ by $\langle\omega_{\alpha_i}, \alpha_j^\vee\rangle = \delta_{i,j}$ for all $\alpha_j \in \Delta$, so that the weight lattice $X = X(T) = \mathbb{Z}\omega_{\alpha_1} \oplus \cdots \oplus \mathbb{Z}\omega_{\alpha_n}$ and the set of dominant weights $X^+ = \mathbb{N}\omega_{\alpha_1} \oplus \cdots \oplus \mathbb{N}\omega_{\alpha_n}$.

Let $\mathfrak{t}$ be a Lie algebra of T which is a Cartan subalgebra of $\mathfrak{g}$. Given $\alpha \in \Phi$, let $\mathfrak{g}_\alpha$ be the α-root space. Put $\mathfrak{b}^+ = \mathfrak{t} \oplus \bigoplus_{\alpha\in\Phi^+} \mathfrak{g}_\alpha$ (the positive Borel subalgebra), and $\mathfrak{b} = \mathfrak{t} \oplus \bigoplus_{\alpha\in\Phi^-} \mathfrak{g}_\alpha$ (the Borel subalgebra opposite to $\mathfrak{b}^+$). We will denote by B^+ and B the corresponding Borel subgroups of G containing the maximal torus T. More generally, given a subset $J \subseteq \Delta$, one can consider the parabolic subgroup P_J of G containing B with a fixed Levi subgroup L_J (containing T) such that their Lie algebras are $\mathrm{Lie}(L_J) = \mathfrak{l}_J$ and $\mathrm{Lie}(P_J) = \mathfrak{p}_J = \mathfrak{l}_J \oplus \mathfrak{u}_J$ of $\mathfrak{g}$. Note that Φ_J is the root system of L_J. In this case, set $X_{P_J} = \{\lambda \in X : \langle\lambda,\alpha^\vee\rangle = 0 \text{ for all } \alpha \in J\}$. Observe that X_{P_J} identifies with the one-dimensional representations of L_J or the the character group $\mathrm{Hom}(L_J, G_m)$.

2.2. Throughout this paper let $\ell > 1$ be a fixed odd positive integer. If Φ has type G_2, then we assume that 3 does not divide ℓ. Let $\mathcal{A} = \mathbb{Q}[q, q^{-1}]$ with fraction field $\mathbb{Q}(q)$. Let $\zeta \in \mathbb{C}$ be a primitive ℓth root of unity and $\mathsf{k} = \mathbb{Q}(\zeta)$. One can regard k as an $\mathcal{A}$-algebra by means of the homomorphism $\mathbb{Q}[q,q^{-1}] \to \mathsf{k}$ where $q \mapsto \zeta$.

The quantized enveloping algebra $\mathbb{U}_q(\mathfrak{g})$ of $\mathfrak{g}$ is the $\mathbb{Q}(q)$-algebra with generators E_α, $K_\alpha^{\pm 1}$, F_α, $\alpha \in \Delta$ and relations (R1)—(R6) listed in [**Jan2**, (4.3)]. The algebra $\mathbb{U}_q(\mathfrak{g})$ has two $\mathcal{A}$-forms. One is $U_q^{\mathcal{A}}(\mathfrak{g})$ defined by Lusztig [**Lu1**] as the $\mathcal{A}$-subalgebra of $\mathbb{U}_q(\mathfrak{g})$ generated by $E_\alpha^{(n)} = E_\alpha^n/[n]!$, $K_\alpha^{\pm 1}$, $F_\alpha^{(n)} = F_\alpha^n/[n]!$. The other is $\mathcal{U}_q^{\mathcal{A}}(\mathfrak{g})$ defined De Concini and Kac [**DK**]) as the $\mathcal{A}$-subalgebra of $\mathbb{U}_q(\mathfrak{g})$ generated by $\{E_\alpha, K_\alpha^{\pm 1}, F_\alpha \mid \alpha \in \Delta\}$. Both are free as $\mathcal{A}$-modules and Hopf $\mathcal{A}$-subalgebras of $\mathbb{U}_q(\mathfrak{g})$. Since $\mathcal{U}_q^{\mathcal{A}}(\mathfrak{g}) \subseteq U_q^{\mathcal{A}}(\mathfrak{g})$, applying the functor $\mathsf{k} \otimes_{\mathcal{A}} -$, one obtains a Hopf k-algebra homomorphism $\mathcal{U}_{\mathsf{k}}(\mathfrak{g}) = \mathsf{k} \otimes_{\mathcal{A}} \mathcal{U}_q^{\mathcal{A}}(\mathfrak{g}) \to \mathsf{k} \otimes_{\mathcal{A}} U_q^{\mathcal{A}}(\mathfrak{g}) = U_{\mathsf{k}}(\mathfrak{g})$. These two Hopf k-algebras $\mathcal{U}_{\mathsf{k}}(\mathfrak{g})$ and $U_{\mathsf{k}}(\mathfrak{g})$ play the roles analogous to the the universal enveloping algebra of its Lie algebra, respectively and hyperalgebra (algebra of distributions) of a reductive group over fields of positive characteristics [**Jan1**].

Let $\mathbb{U}_q^0$ be the subalgebra of $\mathbb{U}_q(\mathfrak{g})$ generated by the $\{K_\alpha^{\pm 1} : \alpha \in \Delta\}$. Then $\mathbb{U}_q^0$ is exactly the Laurent polynomial ring $\mathbb{Q}(q)[K_\alpha^{\pm 1} \mid \alpha \in \Delta]$. Let $\mathcal{U}^{0,\mathcal{A}} = \mathcal{U}^{\mathcal{A}}(\mathfrak{g}) \cap \mathbb{U}_q^0$ and $U^{0,\mathcal{A}} = U^{\mathcal{A}}(\mathfrak{g}) \cap \mathbb{U}_q^0$. We have $\mathcal{U}^{0,\mathcal{A}} \subseteq U^{0,\mathcal{A}}$. Similarly, one has a Hopf algebra homomorphism $\mathcal{U}_{\mathsf{k}}^0 \to U_{\mathsf{k}}^0$.

The elements E_α, F_α, $\alpha \in \Delta$, in $U_{\mathsf{k}}(\mathfrak{g})$ generate a finite dimensional Hopf subalgebra, denoted by $u_\zeta(\mathfrak{g})$, of $U_{\mathsf{k}}(\mathfrak{g})$. The Hopf algebra $u_\zeta(\mathfrak{g})$ will be referred to as the *small quantum group*. We remark that $u_\zeta(\mathfrak{g})$ is a normal Hopf subalgebra of $U_{\mathsf{k}}(\mathfrak{g})$ such that $U_{\mathsf{k}}(\mathfrak{g}) /\!\!/ u_\zeta(\mathfrak{g}) = U_{\mathsf{k}}(\mathfrak{g})/\langle \mathrm{Ker}(\epsilon : u_\zeta(\mathfrak{g}) \to k)\rangle$ is a Hopf algebra [**Lin2**] and the $(U_{\mathsf{k}}(\mathfrak{g}) /\!\!/ u_\zeta(\mathfrak{g}))/\langle K_\alpha^\ell - 1 \mid \alpha \in \Delta\rangle \cong \mathcal{U}(\mathfrak{g}_{\mathsf{k}})$, where $\mathcal{U}(\mathfrak{g}_k)$ is the ordinary universal enveloping algebra $\mathfrak{g}_k = \mathfrak{g} \otimes_{\mathbb{Q}} k$ over k, see [**Lu1, Lu2**] for more details.

We will also modify some of the algebras a little and define

$$\mathcal{U}_\zeta(\mathfrak{g}) = \mathcal{U}_{\mathsf{k}}(\mathfrak{g})/\langle K_\alpha^\ell - 1, \alpha \in \Delta\rangle$$

and denote $U_\zeta(\mathfrak{g}) = U_\mathsf{k}(\mathfrak{g})$. In this paper we will use the same letters E_α, F_α, etc, for elements in any one of the algebra $\mathcal{U}_\mathsf{k}(\mathfrak{g})$, $\mathcal{U}_\zeta(\mathfrak{g})$, $U_\zeta(\mathfrak{g})$, or in $u_\zeta(\mathfrak{g})$ once the context is understood.

The (Hopf) algebra $\mathcal{U}_\zeta(\mathfrak{g})$ has a central subalgebra $\mathcal{Z}$ which is generated by $\{E_\alpha^\ell, F_\alpha^\ell \mid \alpha \in \Delta\}$ and is a normal Hopf k-subalgebra such that $u_\zeta(\mathfrak{g}) \cong \mathcal{U}_\zeta(\mathfrak{g}) /\!\!/ \mathcal{Z}$ (cf. [**DK**, Cor. 3.1] for more details).

We will assume throughout the paper that the $U_\zeta(\mathfrak{g})$-modules M are *integrable of type 1* (cf. [**BNPP**, Section 2.2]). Let $\mathrm{Fr} : U_\zeta(\mathfrak{g}) \to \mathcal{U}(\mathfrak{g}_\mathsf{k})$ be the (quantum) Frobenius homomorphism. If N is a (locally finite) $\mathcal{U}(\mathfrak{g}_\mathsf{k})$-module then the $U_\zeta(\mathfrak{g})$-module $N^{[1]}$ (integrable, type 1) is the inflation of N by Fr. Conversely, for any $U_\zeta(\mathfrak{g})$-module M which is $u_\zeta(\mathfrak{g})$-trivial, then $M \cong N^{[1]}$ for some $\mathcal{U}(\mathfrak{g})$-module N.

2.3. Levi and parabolic subalgebra. For each $\alpha \in \Delta$, Lusztig has defined an automorphism T_α of $\mathbb{U}_q(\mathfrak{g})$ (cf. [**Jan3**, Ch. 8], [**Lu2**]). If s_α is a simple reflection in W, let $T_{s_\alpha} := T_\alpha$. More generally, given any $w \in W$, let $w = s_{\beta_1}s_{\beta_2}\cdots s_{\beta_n}$ be a reduced expression, and define $T_w := T_{\beta_1}\cdots T_{\beta_n} \in \mathrm{Aut}(\mathbb{U}_q(\mathfrak{g}))$. The automorphism T_w is independent of the reduced expression of w [**Lu1**, Thm 3.2]. Now for each $\gamma \in \Phi^+$, one can define $E_\gamma = T_w(E_\beta)$, $F_\gamma = T_w(F_\beta)$ for $w \in W$ with $w(\beta) \in \Phi^+$ and $\beta \in \Delta$ [**Lu2**, Prop. 1.8] and [**BNPP**, Section 2.4]. Note that E_γ has weight γ, and F_γ has weight $-\gamma$.

Now let $J \subseteq \Delta$ and fix a reduced expression $w_0 = s_{\beta_1}\cdots s_{\beta_N}$ that starts with a reduced expression for the long element $w_{0,J}$ for W_J. If $w_{0,J} = s_{\beta_1}\cdots s_{\beta_M}$, then $s_{\beta_{M+1}}\cdots s_{\beta_N}$ is a reduced expression for $w_J = w_{0,J}w_0$. Let $\mathfrak{l}_J$ be the Levi subalgebra corresponding to J and $\mathfrak{p}_J$ the parabolic subalgebra containing $\mathfrak{b}$ and $\mathfrak{l}_J$. The universal enveloping algebras (over the field k) will be denoted by $\mathcal{U}(\mathfrak{l}_J)$ and $\mathcal{U}(\mathfrak{p}_J)$. The Lie algebras considered here will be over the field k without adding the subscript k (despite the fact it was assumed that $\mathfrak{g}$ is a Lie algebra over $\mathbb{Q}$).

One can naturally define corresponding quantized enveloping algebras $\mathbb{U}_q(\mathfrak{l}_J)$ and $\mathbb{U}_q(\mathfrak{p}_J)$ over $\mathbb{Q}(q)$ as subalgebras of $\mathbb{U}_q(\mathfrak{g})$. The subalgebra $\mathbb{U}_q(\mathfrak{l}_J)$ is generated by $\{E_\alpha, F_\alpha : \alpha \in J\} \cup \{K_\alpha^{\pm 1} : \alpha \in \Pi\}$, and $\mathbb{U}_q(\mathfrak{p}_J)$ is the subalgebra generated by $\{E_\alpha : \alpha \in J\} \cup \{F_\alpha, K_\alpha^{\pm 1} : \alpha \in \Pi\}$. Using the PBW type of basis constructed by Lusztig in [**Lu2**], each of these subalgebras has two different $\mathcal{A}$-forms as pure subalgebras of $U_q^{\mathcal{A}}(\mathfrak{g})$ and $\mathcal{U}_q^{\mathcal{A}}(\mathfrak{g})$ respectively. Upon specialization of q to ζ as in Section 2.1, one obtains the subalgebras $U_\zeta(\mathfrak{l}_J)$, $U_\zeta(\mathfrak{p}_J)$, $u_\zeta(\mathfrak{l}_J)$, $u_\zeta(\mathfrak{p}_J)$ of $U_\zeta(\mathfrak{g})$, and $\mathcal{U}_\zeta(\mathfrak{l}_J)$ and $\mathcal{U}_\zeta(\mathfrak{p}_J)$ of $\mathcal{U}_\zeta(\mathfrak{g})$. One can also make analogous constructions with the opposite parabolic $\mathfrak{p}_J^+$. The subalgebras $U_\zeta(\mathfrak{l}_J)$ and $U_\zeta(\mathfrak{p}_J)$ were first constructed in [**APW**].

We remark that the algebras $U_\zeta(\mathfrak{l}_J)$, $U_\zeta(\mathfrak{p}_J)$, $\mathcal{U}_\zeta(\mathfrak{l}_J)$, and $\mathcal{U}_\zeta(\mathfrak{p}_J)$ are Hopf algebras. The inclusion $\mathcal{U}_q^{\mathcal{A}}(\mathfrak{g}) \subseteq U_q^{\mathcal{A}}(\mathfrak{g})$ of $\mathcal{A}$-algebras induces a homomorphism of Hopf algebras: $\mathcal{U}_\zeta(\mathfrak{g}) \to U_\zeta(\mathfrak{g})$ with image being $u_\zeta(\mathfrak{g})$. This Hopf algebra homomorphism induces Hopf algebra homomorphisms $\mathcal{U}_\zeta(\star) \to U_\zeta(\star)$ with $\star = \mathfrak{l}_J, \mathfrak{p}_J$

The above reduced expression for w_0 (beginning with one for $w_{0,J}$) defines an ordering on positive roots $\Phi^+ = \{\gamma_1, \cdots \gamma_N\}$, with $\{\gamma_1, \cdots, \gamma_{N_J}\} = \Phi_J^+$ being the positive roots of the reductive Lie algebra $\mathfrak{l}_J$ and $\{\gamma_{N_J+1}, \cdots, \gamma_N\}$ being the roots in the nilpotent radical $\mathfrak{u}_J^+$ of $\mathfrak{p}_J^+$. With the PBW basis $\{F_{\gamma_1}^{(a_1)}\cdots F_{\gamma_N}^{(a_N)} \mid (a_1, \cdots, a_N) \in \mathbb{N}^N\}$ of $\mathbb{U}_q(\mathfrak{n})$ as described in [**Lu1**, Appendix] using this reduced expression of w_0, one can define a subalgebra $\mathbb{U}_q(\mathfrak{u}_J)$ of $\mathbb{U}_q(\mathfrak{n}) \subseteq \mathbb{U}_q(\mathfrak{p}_J)$ generated by all F_{γ_i} with $i > N_J$. $\mathbb{U}_q(\mathfrak{u}_J)$ is an augmented normal subalgebra of $\mathbb{U}_q(\mathfrak{p}_J)$

(although not a Hopf subalgebra of $\mathbb{U}_q(\mathfrak{g})$). The algebra $\mathbb{U}_q(\mathfrak{u}_J)$ is analogous to $\mathcal{U}(\mathfrak{u}_J) \subseteq \mathcal{U}(\mathfrak{p}_J)$.

The $\mathbb{Q}(q)$-subalgebra $\mathbb{U}_q(\mathfrak{u}_J)$ in $\mathbb{U}_q(\mathfrak{g})$ is spanned by the $F_{\gamma_{N_J+1}}^{(a_{N_J+1})} \cdots F_{\gamma_N}^{(a_N)}$, $a_i \in \mathbb{N}$. According to [**BNPP**, Lemma 2.4.1] $\mathbb{U}_q(\mathfrak{u}_J)$ is a subalgebra of $\mathbb{U}_q(\mathfrak{p}_J)$ and independent of the choice of reduced expression for w_0. Again by specializing q to ζ, one obtains algebras $U_\zeta(\mathfrak{u}_J)$ and $u_\zeta(\mathfrak{u}_J)$ as subalgebras of $U_\zeta(\mathfrak{g})$ and $\mathcal{U}_\zeta(\mathfrak{u}_J)$ as a subalgebra of $\mathcal{U}_\zeta(\mathfrak{p}_J)$. Again, the Hopf algebra homomorphism $\mathcal{U}_\zeta(\mathfrak{g}) \to U_\zeta(\mathfrak{g})$ restricts to an algebra homomorphism $\mathcal{U}_\zeta(\mathfrak{u}_J) \to U_\zeta(\mathfrak{u}_J)$ with image being $u_\zeta(\mathfrak{u}_J)$.

2.4. Adjoint action. Since $\mathcal{U}_\zeta(\mathfrak{g})$ is a Hopf algebra, and $E_\beta^\ell, F_\beta^\ell$, and K_i^ℓ are central elements, then the adjoint action of $\mathcal{U}_\zeta(\mathfrak{g})$ on the central elements is trivial in the sense that $u \cdot z = \epsilon(u)z$ for all $u \in \mathcal{U}_\zeta(\mathfrak{g})$ and z central in $\mathcal{U}_\zeta(\mathfrak{g})$. Thus, the adjoint action of elements E_β^ℓ and F_β^ℓ on $\mathcal{U}_\zeta(\mathfrak{g})$ is zero and K_i^ℓ acts on $\mathcal{U}_\zeta(\mathfrak{g})$ as the identity. Therefore, the adjoint action of $\mathcal{U}_\zeta(\mathfrak{g})$ on $\mathcal{U}_\zeta(\mathfrak{g})$ factors through $u_\zeta(\mathfrak{g})$. In particular, an element $u \in u_\zeta(\mathfrak{p}_J)$ acts on f^ℓ via the counit $\epsilon(u)$ for $f = E_\beta, F_\beta, K_i$. Furthermore, for any subalgebra A of $\mathcal{U}_\zeta(\mathfrak{g})$ we can also construct the adjoint action of A on certain subspaces of $\mathcal{U}_\zeta(\mathfrak{g})$. This action by A always factors through its image in $u_\zeta(\mathfrak{g})$.

To extend the adjoint action of $u_\zeta(\mathfrak{g})$ on $\mathcal{U}_\zeta(\mathfrak{g})$ we have to return to the $\mathcal{A}$-forms. Both $\mathcal{A}$-forms $U_q^{\mathcal{A}}(\mathfrak{g})$ and $\mathcal{U}_q^{\mathcal{A}}(\mathfrak{g})$ are Hopf $\mathcal{A}$-subalgebras of $\mathbb{U}_q(\mathfrak{g})$. The following proposition is from [**ABG**, Prop. 2.9.2]. For the definition and properties of the adjoint action of Hopf algebras we refer the reader to Section 8.

PROPOSITION 2.4.1. *Let $\mathcal{A}_\zeta$ be the localization of $\mathcal{A}$ at the maximal ideal generated by $q - \zeta$. The algebra $\mathcal{U}_q^{\mathcal{A}_\zeta}(\mathfrak{g})$ is invariant under the left adjoint action of $U_q^{\mathcal{A}_\zeta}(\mathfrak{g})$. In particular, the algebra $\mathcal{U}_\zeta(\mathfrak{g})$ is a left $U_\zeta(\mathfrak{g})$-module algebra.*

The proof in [**Lin1**, Prop. 5.3] shows that $u_\zeta(\mathfrak{u}_J)$ is invariant under the left adjoint action of $U_\zeta(\mathfrak{p}_J)$. The above proposition defines a $U_\zeta(\mathfrak{p}_J)$-module algebra structure on $\mathcal{U}_\zeta(\mathfrak{u}_J)$ such that the homomorphism $\mathcal{U}_\zeta(\mathfrak{u}_J) \to u_\zeta(\mathfrak{u}_J)$ is a homomorphism of $U_\zeta(\mathfrak{p}_J)$-module algebras. Since the algebra $\mathcal{U}_\zeta(\mathfrak{u}_J)$ is generated by F_{γ_i} for $i = N_J + 1, \ldots, N$, one only needs to define the action on these elements. The algebra $U_\zeta(\mathfrak{p}_J)$ is generated as an algebra by $E_\alpha, E_\alpha^{(\ell)}$ and F_β, $F_\beta^{(\ell)}$ with $\alpha \in J$ and $\beta \in \Delta$ together with U_ζ^0. The action of $E_\alpha, F_\beta, K_\beta$ on $\mathcal{U}_\zeta(\mathfrak{u}_J)$ can be expressed by writing down the comultiplication. We only consider the action of $F_\beta^{(\ell)}$ and $E_\alpha^{(\ell)}$. In the formal definition of the action of $F_\beta^{(\ell)}$, one follows the argument in [**Lin1**, Prop. 5.3] together with the argument of [**Lu1**, Lemma 8.5]. The action of $E_\alpha^{(\ell)}$ follows a similar argument in [**Lin2**, 5.1].

2.5. The Hopf subalgebras U_ζ^0 and $U_\zeta^0(\mathfrak{p}_J)$ act on the algebras $\mathcal{U}_\zeta(\mathfrak{u}_J)$ and $U_\zeta(\mathfrak{u}_J)$. Moreover, the algebra homomorphism $\mathcal{U}_\zeta(\mathfrak{u}_J) \to U_\zeta(\mathfrak{u}_J)$ is a homomorphism of $U_\zeta(\mathfrak{p}_J)$-module algebras. In this paper, we will need to discuss U_ζ^0, $U_\zeta(\mathfrak{l}_J)$, and $U_\zeta(\mathfrak{p}_J)$-equivariant $\mathcal{U}_\zeta(\mathfrak{u}_J)$-modules and $U_\zeta(\mathfrak{u}_J)$-modules. This is discussed in Section 8. By U_ζ^0 and $U_\zeta(\mathfrak{l}_J)$-modules, we mean the integrable of type 1 modules, which are locally finite and with a weight space decomposition as described in [**APW**] and [**Lin1**].

3. Cohomological calculations

3.1. In this section we will present a general method for making cohomological computations for quantum groups given two key assumptions. The first assumption appears in the next proposition and involves computing cohomology for $\mathcal{U}_\zeta(\mathfrak{u}_J)$. Note that although $\mathcal{U}_\zeta(\mathfrak{u}_J)$ is not a Hopf algebra, it is an augmented algebra and the cohomology we consider here is the cohomology of the augmented (supplemented) algebra in the sense of Cartan-Eilengberg [**CE**, Ch X]. Once this assumption is satisfied one can compute the corresponding cohomology for $u_\zeta(\mathfrak{u}_J)$. The proof uses the constructions presented in Section 2 and employs the techniques outlined in the proofs of [**BNPP**, Theorem 5.3.1, Lemma 5.4.1]. The proof will proceed by induction on successive quotients of $U_\zeta(\mathfrak{u}_J)$ by regular sequences (cf. [**GK**, 2.4]) and can be described as follows.

Let $N^J = |\Phi^+\backslash\Phi^+_J|$ (thus $N_J+N^J = N$) and choose a fixed ordering of root vectors $f_1, f_2, \ldots, f_{N^J}$ in $\mathcal{U}_\zeta(\mathfrak{u}_J)$ corresponding to the positive roots $\{\gamma_1, \gamma_2, \ldots, \gamma_{N^J}\}$ in $\Phi^+\backslash\Phi^+_J$ such that for each i, the subalgebra generated by $\langle f_1, \cdots, f_i\rangle$ is $U_\zeta(\mathfrak{b})$-stable through the action of $U_\zeta(\mathfrak{g})$ on $\mathcal{U}_\zeta(\mathfrak{g})$. Each f_i^ℓ is central in $\mathcal{U}_\zeta(\mathfrak{g})$ and is contained in the augmented ideal of $\mathcal{U}_\zeta(\mathfrak{u}_J)$. Let Z_i be the subalgebra of $\mathcal{U}_\zeta(\mathfrak{g})$ generated by $\langle f_1^\ell, f_2^\ell, \ldots, f_i^\ell\rangle$ (with $i = 1, \ldots, N^J$). Then Z_i is a central subalgebra of $\mathcal{U}_\zeta(\mathfrak{u}_J)$ and $u_\zeta(\mathfrak{u}_J) = \mathcal{U}_\zeta(\mathfrak{u}_J) /\!/ Z_{N^J}$. We note that for each i, Z_i are $U_\zeta(\mathfrak{b})$-stable subalgebras of $\mathcal{U}_\zeta(\mathfrak{u}_J)$, but in general, Z_i are not $U_\zeta(\mathfrak{p}_J)$-stable. However, Z_{N^J} is $U_\zeta(\mathfrak{p}_J)$-stable.

For $0 \leq i \leq N^J$, let $A_i = \mathcal{U}_\zeta(\mathfrak{u}_J)/\langle f_1^\ell, f_2^\ell, \ldots, f_i^\ell\rangle$with $A_{N^J} = u_\zeta(\mathfrak{u}_J)$. Note that each A_i is a $U_\zeta(\mathfrak{b})$-module algebra. Furthermore, for $1 \leq i \leq N^J$, let $B_i = \langle f_i^\ell\rangle \subseteq A_{i-1}$ be the augmented subalgebra generated by f_i^ℓ. Each B_i is a polynomial algebra in one variable, central in A_{i-1} and stable under $U_\zeta(\mathfrak{b})$. Thus B_i is a normal augmented subalgebra of A_{i-1} in the sense of [**CE**, XVI. §6], and $A_{i-1} /\!/ B_i \cong A_i$ as U^0_ζ-modules. By the discussion in Section 2.4, the ideals $\langle f_1^\ell, f_2^\ell, \ldots, f_i^\ell\rangle$ are $u_\zeta(\mathfrak{p}_J)$-stable which induces an action of $U_\zeta(\mathfrak{b})u_\zeta(\mathfrak{p}_J)$ on A_i.

For $0 \leq i \leq N^J$, let V_i be a vector space with basis $\{x_1, x_2, \ldots, x_i\}$ considered as a $U_\zeta(\mathfrak{b})$-module by letting x_i have weight $-\gamma_i$ contained as a $U_\zeta(\mathfrak{b})$-submodule in $V_N \cong \mathfrak{u}_J^*$ with $V_0 = \{0\}$. For each integrable $U_\zeta(\mathfrak{p}_J)$-module M of type 1 (in the sense of [**APW**]), the space $\operatorname{Hom}_{u_\zeta(\mathfrak{u}_J)}(\mathsf{k}, M)$ is a rational P_J-module for the algebraic group P_J (the parabolic subgroup of G defined and split over k) with differential being the $\mathcal{U}(\mathfrak{p}_J)$-module arising from the Hopf algebra isomorphism $\mathcal{U}(\mathfrak{p}_J) \cong U_\zeta(\mathfrak{p}_J) /\!/ u_\zeta(\mathfrak{p}_J)$. Any $U_\zeta(\mathfrak{p}_J)$-module M restricts to a $u_\zeta(\mathfrak{p}_J)$-module. The module can then be inflated to a $\mathcal{U}_\zeta(\mathfrak{p}_J)$-module via the Hopf algebra homomorphism $\mathcal{U}_\zeta(\mathfrak{p}_J) \to U_\zeta(\mathfrak{p}_J)$. Then the module can be regarded as a $\mathcal{U}_\zeta(\mathfrak{u}_J)$-module where $Z_J^+M = 0$. Let $Z_J = Z_{N^J}$ be the polynomial algebra with generators f_i^ℓ and $Z_J^+ = \langle f_1^\ell, \cdots, f_{N^J}^\ell\rangle$ be the maximal ideal of Z_J.

3.2. We recall the following general setting. For a given Hopf algebra H and a H-module algebra A, we can consider the category of HA-modules as the A-modules in the tensor category of H-modules. The objects are H-modules M with an A-module structure $A \otimes_{\mathsf{k}} M \to M$, which is an H-module homomorphism. The morphisms are the k-linear maps that are homomorphisms for both A-modules and H-modules. If M, N are two HA-modules, then $\operatorname{Hom}_A(M, N)$ is an H-module. If A is an augmented algebra with the augmentation $A \to \mathsf{k}$ being an H-module homomorphism, then H acts on the cohomology groups $\mathrm{H}^k(A, M) = \mathrm{Ext}^k_A(\mathsf{k}, M)$ for any HA-module M.

For $H = U_\zeta(\mathfrak{p}_J)$, $u_\zeta(\mathfrak{u}_J)$ is an augmented $U_\zeta(\mathfrak{p}_J)$-module algebra. According to Lemma 2.4.1, $\mathcal{U}_\zeta(\mathfrak{u}_J)$ is also a $U_\zeta(\mathfrak{p}_J)$-module algebra and the augmented algebra homomorphism $\mathcal{U}_\zeta(\mathfrak{u}_J) \to u_\zeta(\mathfrak{u}_J)$ is also a homomorphism of $U_\zeta(\mathfrak{p}_J)$-module algebras. We have two types of cohomology groups $\mathrm{H}^*(\mathcal{U}_\zeta(\mathfrak{u}_J), M)$ and $\mathrm{H}^*(u_\zeta(\mathfrak{u}_J), M)$ for each $U_\zeta(\mathfrak{p}_J)$-module M. Thus both $\mathrm{H}^*(\mathcal{U}_\zeta(\mathfrak{u}_J), M)$ and $\mathrm{H}^*(u_\zeta(\mathfrak{u}_J), M)$ are $U_\zeta(\mathfrak{p}_J)$-modules, on which $u_\zeta(\mathfrak{u}_J)$ acts trivially. But the subalgebras A_i constructed above are only $U_\zeta(\mathfrak{b})$-module algebras. By noting that $U_\zeta(\mathfrak{b})//u_\zeta(\mathfrak{b}) \cong \mathcal{U}(\mathfrak{b})$, we know that $\mathrm{Hom}_{u_\zeta(\mathfrak{l}_J)}(\mathsf{k}, \mathrm{H}^n(A_i, M) \otimes Q))$ is a rational B-module for any (integrable of type 1) $U_\zeta(\mathfrak{p}_J)$-modules M and Q as stated in the following proposition.

PROPOSITION 3.2.1. *Let $J \subseteq \Delta$. Let M and Q be $U_\zeta(\mathfrak{p}_J)$-modules such that Q is an injective $u_\zeta(\mathfrak{l}_J)$-module and is trivial as $U_\zeta(\mathfrak{u}_J)$-module. If there is an integer $t \geq 0$ such that there are rational P_J-module isomorphisms*

$$\mathrm{Hom}_{u_\zeta(\mathfrak{l}_J)}(\mathsf{k}, \mathrm{H}^n(\mathcal{U}_\zeta(\mathfrak{u}_J), M) \otimes Q) \cong \begin{cases} \mathsf{k} & n = t, \\ 0 & \textit{otherwise}, \end{cases}$$

then we have the following rational P_J-module isomorphisms:

(a)

$$\mathrm{Hom}_{u_\zeta(\mathfrak{l}_J)}(\mathsf{k}, \mathrm{H}^n(u_\zeta(\mathfrak{u}_J), M) \otimes Q)) \cong \begin{cases} S^{\frac{n-t}{2}}(\mathfrak{u}_J^*) & n \equiv t \pmod 2, \\ 0 & \textit{otherwise}; \end{cases}$$

(b)

$$\mathrm{H}^n(u_\zeta(\mathfrak{p}_J), M \otimes Q) \cong \begin{cases} S^{\frac{n-t}{2}}(\mathfrak{u}_J^*) & n - t \equiv 0 \pmod 2, \\ 0, & \textit{otherwise}. \end{cases}$$

PROOF.

(a) Note that each $U_\zeta(\mathfrak{p}_J)$-module M becomes a $\mathcal{U}_\zeta(\mathfrak{u}_J)$-module satisfying $Z_J M = 0$. By Proposition 2.4.1, $\mathcal{U}_\zeta(\mathfrak{u}_J)$ is a right $U_\zeta(\mathfrak{p}_J)$-module algebra and the left $\mathcal{U}_\zeta(\mathfrak{u}_J)$-module structure on M is compatible with the left $U_\zeta(\mathfrak{p}_J)$-module structure on M by Section 3.1. Hence, the space $\mathrm{Hom}_{\mathcal{U}_\zeta(\mathfrak{u}_J)}(\mathsf{k}, M)$ is a $U_\zeta(\mathfrak{l}_J)$-module (and a $\mathcal{U}(\mathfrak{u}_J)$-module). In particular, $\mathrm{H}^i(\mathcal{U}_\zeta(\mathfrak{u}_J), M)$ is a $U_\zeta(\mathfrak{p}_J)$-module such that $u_\zeta(\mathfrak{u}_J)$ acts trivially by Proposition 8.3.1.

Set $\mathcal{G}(-) = \mathrm{Hom}_{u_\zeta(\mathfrak{l}_J)}(\mathsf{k}, - \otimes Q)$. For each $U_\zeta(\mathfrak{p}_J)$-module E (integrable of type 1) on which $u_\zeta(\mathfrak{u}_J)$-acts trivially, $\mathcal{G}(E)$ is a rational P_J-module. Thus $\mathcal{G}$ is a functor from the full subcategory of $U_\zeta(\mathfrak{p}_J)$-modules on which $u_\zeta(\mathfrak{u}_J)$ acts trivially to the category of rational P_J-modules. Since the module Q is injective as a $u_\zeta(\mathfrak{l}_J)$-module and $E \otimes Q$ is injective for any $u_\zeta(\mathfrak{l}_J)$-module E, the functor $\mathcal{G}$ is an exact functor. However, the Hopf algebra $U_\zeta(\mathfrak{p}_J)$ does not necessarily act on A_i, thus we cannot assume that $\mathrm{H}^s(A_i, M)$ are $U_\zeta(\mathfrak{p}_J)$-modules. Thus we let $\mathcal{G}'(-)$ be the restriction of $\mathcal{G}(-)$ to the category of (integrable of type 1) $U_\zeta(\mathfrak{b})$-modules to the category of rational B-modules.

We first note that for each $0 \leq i \leq N^J$, each $\mathrm{H}^s(A_i, M)$ is a $U_\zeta(\mathfrak{b})$-module (cf. 2.4) which is $u_\zeta(\mathfrak{u}_J)$-trivial. Thus we can apply the functor $\mathcal{G}'(-)$ to get a B-module $\mathcal{G}'(\mathrm{H}^s(A_i, M))$.

We will first prove by induction on i that for $0 \le i \le N^J$ as $\mathcal{U}(\mathfrak{t})$-modules:

$$\mathcal{G}'(\mathrm{H}^s(A_i, M)) \cong \begin{cases} S^r(V_i) & \text{if } s = 2r + t, \\ 0 & \text{otherwise.} \end{cases} \tag{3.2.1}$$

For $i = 0$, this follows from the hypothesis. The statement of this proposition (as B-modules) is the case when $i = N^J$.

Assume that (3.2.1) holds for $i - 1$. We will prove that it is also valid for i. By using the PBW basis Lusztig constructed for $\mathbb{U}_q(\mathfrak{u})$, one can construct a PBW basis for $\mathcal{U}_\zeta(\mathfrak{u}_J)$ by removing the denominators in the divided powers. Then by using induction on i, one can show that A_{i-1} is a projective B_i-module (see Section 3.1 for the definition of B_i). Now we can consider the Lyndon-Hochschild-Serre spectral sequence [**CE**, Thm. 6.1]:

$$E_2^{a,b} = \mathrm{H}^a(A_i, \mathrm{H}^b(B_i, M)) \Rightarrow \mathrm{H}^{a+b}(A_{i-1}, M).$$

Since $Z_J^+.M = 0$, then B_i acts on M trivially and $\mathrm{H}^*(B_i, M) = M \otimes \mathrm{H}^*(B_i, \mathsf{k})$ by Proposition 8.3.2. The algebra A_i acts trivially on B_i via the adjoint action induced from the right adjoint action of the Hopf algebra $\mathcal{U}_\zeta(\mathfrak{p}_J)$. In particular, $u_\zeta(\mathfrak{u}_J)$ acts on B_i trivially. Thus $A_i = \mathcal{U}_\zeta(\mathfrak{u}_J)/\!\!/ Z_i$ acts trivially on $\mathrm{H}^*(B_i, \mathsf{k})$ using the argument following Proposition 8.3.2. The A_i-action on $\mathrm{H}^*(B_i, M) = M \otimes \mathrm{H}^*(B_i, \mathsf{k})$ is via the action on M. Thus, the spectral sequence can be rewritten as

$$E_2^{a,b} = \mathrm{H}^a(A_i, M) \otimes \mathrm{H}^b(B_i, \mathsf{k}) \Rightarrow \mathrm{H}^{a+b}(A_{i-1}, M). \tag{3.2.2}$$

The cohomology of B_i is an exterior algebra with one generator in degree 1 and, as $U_\zeta(\mathfrak{b})$-modules, we have

$$\mathrm{H}^b(B_i, \mathsf{k}) = \begin{cases} \mathsf{k} & \text{if } b = 0, \\ \mathsf{k}_{\gamma_i} & \text{if } b = 1, \\ 0 & \text{otherwise.} \end{cases}$$

Since the functor $\mathcal{G}'(-)$ is exact, we can apply this functor to the above spectral sequence and obtain a new spectral sequence using the fact that $u_\zeta(\mathfrak{l}_J)$-action on the cohomology of B_i is trivial to get the following spectral sequence:

$$E_2^{a,b} = \mathcal{G}'(\mathrm{H}^a(A_i, M)) \otimes \mathrm{H}^b(B_i, \mathsf{k}) \Rightarrow \mathcal{G}'(\mathrm{H}^{a+b}(A_{i-1}, M)).$$

The spectral sequence consists of at most two non-zero rows (i.e., when $b = 0, 1$). Therefore, the spectral sequence collapses after the E_3-page and the abutment can be obtained by calculating the differential d_2. Thus, we have a short exact sequence

$$0 \to E_3^{u-1,1} \to \mathcal{G}'(\mathrm{H}^u(A_{i-1}, M)) \to E_3^{u,0} \to 0 \tag{3.2.3}$$

for all u. On the other hand, we have

$$E_3^{u,1} = \mathrm{Ker}(d_2 : E_2^{u,1} \to E_2^{u+2,0}), \qquad E_3^{u,0} = \mathrm{coker}(d_2 : E_2^{u-2,1} \to E_2^{u,0}). \tag{3.2.4}$$

Observe that from the induction hypothesis the abutment is

$$\mathcal{G}'(\mathrm{H}^{a+b}(A_{i-1}, M)) \cong \begin{cases} S^r(V_{i-1}), & \text{if } a + b = 2r + t, \\ 0, & \text{otherwise.} \end{cases} \tag{3.2.5}$$

The first row $E_2^{a,0} = \mathcal{G}'(\mathrm{H}^a(A_i, M))$ is what we are trying to determine. Note also that $E_2^{a,1} \cong E_2^{a,0} \otimes \mathsf{k}_{\gamma_i}$.

We observe that $E_2^{u,0} \cong E_2^{u,1} = 0$ for $u < t$. Note that $\mathcal{G}'(\mathrm{H}^u(A_{i-1}, M)) = 0$ for $u < t$. Therefore, $E_3^{u,0} = E_\infty^{u,0} = 0$ and $E_3^{u,1} = E_\infty^{u,1} = 0$ for $u + 1 < t$. Thus, for $u < t$, $d_n^{u-2,1}$ is an isomorphism by (3.2.4) and we have

$$E_2^{u,0} \overset{d_2}{\cong} E_2^{u-2,1} \cong E_2^{u-2,0} \otimes \mathsf{k}_{\gamma_i}.$$

One can conclude inductively that $E_2^{u,0} = 0$ for $u < t$.

When $a - t$ is odd, then by (3.2.5) and (3.2.3), we have $E_3^{a,0} = 0 = E_3^{a-1,1}$ and that $d_2^{a-2,1}$ is surjective and $d_2^{a-1,1}$ is injective. If $a = t + 1$, the surjectivity of $d_2^{t-1,1}$ and the fact that $E_2^{t-1,1} = 0$ implies that $E_2^{t+1,0} = 0$, which further implies $E_2^{t+1,1} = 0$.

We now assume that we have proved that $E_2^{a,0} = 0$ for an a with $a - t$ odd. Then $E_2^{a,1} = 0$. Since $\mathcal{G}'(\mathrm{H}^{a+2}(A_{i-1}, M)) = 0$, we have $E_3^{a+2,0} = 0$. Similarly, the surjectivity of $d_2^{a,1} : E_2^{a,1} \to E_2^{a+2,0}$ implies that $E_2^{a+2,0} = 0$. Therefore, by induction on a, we conclude that $E_2^{a,0} = 0 = E_2^{a,1}$ whenever $a - t$ is odd. In particular we have $E_3^{a,1} = 0$ for a with $a - t$ odd. Thus we have $E_3^{a,1} = 0$ for all a by (3.2.3) and (3.2.5). Consequently, we have $E_3^{a,0} = \mathcal{G}'(\mathrm{H}^a(A_{i-1}, M))$ for all a.

Next we analyze the case when $a - t$ is even. First observe that $d_2 : E_2^{a,1} \to E_2^{a+2,0}$ is a monomorphism, and for $a - t$ even,

$$E_2^{a,0}/E_2^{a-2,1} = E_3^{a,0} = \mathcal{G}'(\mathrm{H}^a(A_{i-1}, M)).$$

This yields a short exact sequence of $U_\zeta(\mathfrak{b})$-modules:

$$0 \to \mathcal{G}'(\mathrm{H}^{a-2}(A_i, M)) \otimes \mathsf{k}_{\gamma_i} \to \mathcal{G}'(\mathrm{H}^a(A_i, M)) \to S^{(a-t)/2}(V_{i-1}) \to 0. \tag{3.2.6}$$

We can now use induction on $a - t$ (even) to determine $\mathcal{G}'(\mathrm{H}^{a-t}(A_i, M))$. For even $a - t$, assume that

$$\mathcal{G}'(\mathrm{H}^{a-2}(A_i, M)) = S^{(a-t-2)/2}(V_i).$$

Then the short exact sequence (3.2.6) can be written as

$$0 \to S^{(a-t-2)/2}(V_i) \otimes \mathsf{k}_{\gamma_i} \to \mathcal{G}'(\mathrm{H}^a(A_i, M)) \to S^{(a-t)/2}(V_{i-1}) \to 0.$$

Now set $a = 2r + t$, thus we have isomorphism of $U_\zeta(\mathfrak{t})$-modules

$$\mathcal{G}'(\mathrm{H}^a(A_i, M)) \cong (S^{r-1}(V_i) \otimes \mathsf{k}_{\gamma_i}) \oplus S^r(V_{i-1}) \cong S^r(V_i). \tag{3.2.7}$$

Note that the above isomorphism is as T-modules.

Next one can apply the argument in [**BNPP**, Lemma 5.4.1] to verify (3.2.1), and therefore statement (a) with the isomorphism

$$\mathcal{G}'(\mathrm{H}^a(A_i, M)) \cong S^r(V_i) \tag{3.2.8}$$

as B-modules.

Next we need to verify that the identifications given in the statement (a) holds as $\mathcal{U}(\mathfrak{p}_J)$-modules. Recall Z_J is a central subalgebra of $\mathcal{U}_\zeta(\mathfrak{u}_J)$ which is $U_\zeta(\mathfrak{p}_J)$-stable under the right adjoint action on $\mathcal{U}_\zeta(\mathfrak{u}_J)$ (cf. [**BNPP**, Cor. 2.7.4]). We have a spectral sequence of $U_\zeta(\mathfrak{p}_J)$-modules:

$$E_2^{a,b} = \mathrm{H}^a(u_\zeta(\mathfrak{u}_J), \mathrm{H}^b(Z_J, M)) \Rightarrow \mathrm{H}^{a+b}(\mathcal{U}_\zeta(\mathfrak{u}_J), M)$$

The algebra $u_\zeta(\mathfrak{p}_J)$-action (the restriction from the $U_\zeta(\mathfrak{p}_J)$-action which is the same as that factored through the adjoint action of $\mathcal{U}_\zeta(\mathfrak{p}_J)$) on Z_J is trivial on since it Z_J is central in $\mathcal{U}_\zeta(\mathfrak{g})$. Thus $u_\zeta(\mathfrak{p}_J)$ also acts trivially on $\mathrm{H}^b(Z_J, \mathsf{k})$. Also Z_J (as augmented algebra) acts trivially on the $U_\zeta(\mathfrak{p}_J)$-module M. Therefore, the above spectral sequence becomes

$$E_2^{a,b} = \mathrm{H}^a(u_\zeta(\mathfrak{u}_J), M) \otimes \mathrm{H}^b(Z_J, \mathsf{k}) \Rightarrow \mathrm{H}^{a+b}(\mathcal{U}_\zeta(\mathfrak{u}_J), M).$$

The restriction of the $U_\zeta(\mathfrak{p}_J)$-action to the $u_\zeta(\mathfrak{u}_J)$-action is trivial on each term of the spectral sequence by Proposition 8.3.2. Now we apply the functor $\mathcal{G}(-)$ to obtain a new spectral sequence of $U_\zeta(\mathfrak{p}_J)$-modules:

$$E_2^{a,b} = \mathcal{G}(\mathrm{H}^a(u_\zeta(\mathfrak{u}_J), M)) \otimes \mathrm{H}^b(Z_J, \mathbb{C}) \Rightarrow \mathcal{G}(\mathrm{H}^{a+b}(\mathcal{U}_\zeta(\mathfrak{u}_J), M)). \tag{3.2.9}$$

Observe that $U_\zeta(\mathfrak{p}_J)$ acts on the terms in the spectral sequence with both $u_\zeta(\mathfrak{l}_J)$ and $u_\zeta(\mathfrak{u}_J)$ acting trivially on each term. Thus this is a spectral sequence of $\mathcal{U}(\mathfrak{p}_J)$-modules. Since we are working in the category of integrable $U_\zeta(\mathfrak{p}_J)$-modules of type 1, the terms of the spectral sequence are rational P_J-modules as well. By the assumption, the abutment $\mathcal{G}(\mathrm{H}^{a+b}(\mathcal{U}_\zeta(\mathfrak{u}_J), M))$ of this spectral sequence is nonzero only when $a + b = t$ in which case it is the trivial module k. Moreover, by (3.2.8), as $\mathcal{U}(\mathfrak{b})$-modules,

$$\mathcal{G}(\mathrm{H}^a(u_\zeta(\mathfrak{u}_J), M)) = \mathcal{G}'(\mathrm{H}^a(u_\zeta(\mathfrak{u}_J), M)) \cong S^{\frac{a-t}{2}}(\mathfrak{u}_J^*). \tag{3.2.10}$$

We note that both are $\mathcal{U}(\mathfrak{p}_J)$-modules. They are actually rational P_J-modules. The isomorphism is as B-modules. Proposition 8.4.2 implies that

$$\mathcal{G}(\mathrm{H}^a(u_\zeta(\mathfrak{u}_J), M)) \cong S^{\frac{a-t}{2}}(\mathfrak{u}_J^*)$$

as P_J-modules.

(b) Apply the Lyndon-Hochschild-Serre (LHS) spectral sequence to $u_\zeta(\mathfrak{u}_J)$ as a normal Hopf subalgebra in $u_\zeta(\mathfrak{p}_J)$:

$$E_2^{i,j} = \mathrm{H}^i(u_\zeta(\mathfrak{l}_J), \mathrm{H}^j(u_\zeta(\mathfrak{u}_J), M) \otimes Q) \Rightarrow \mathrm{H}^{i+j}(u_\zeta(\mathfrak{p}_J), M \otimes Q).$$

Since Q is injective over $u_\zeta(\mathfrak{l}_J)$ (thus tensor product with any $u_\zeta(\mathfrak{l}_J)$-module remains injective), this spectral sequence collapses and yields:

$$\mathrm{H}^n(u_\zeta(\mathfrak{p}_J), M \otimes Q) = \mathrm{Hom}_{u_\zeta(\mathfrak{l}_J)}(\mathsf{k}, \mathrm{H}^n(u_\zeta(\mathfrak{u}_J), M) \otimes Q).$$

The result now follows by part (a). □

3.3. In the next theorem we identify important axiomatic conditions which allow one to realize a G-module isomorphism between $\mathrm{ind}_{P_J}^G(S^\bullet(\mathfrak{u}_J^*) \otimes \mathsf{k}_\gamma)$ and cohomology for modules over the small quantum group in order to answer Question 1.2.2.

THEOREM 3.3.1. *Let $J \subseteq \Delta$, M be a $U_\zeta(\mathfrak{g})$-module with $Z_J^+.M = 0$ and Q be a $U_\zeta(\mathfrak{p}_J)$-module such that Q is injective as $u_\zeta(\mathfrak{l}_J)$-module and trivial as $U_\zeta(\mathfrak{u}_J)$-module. Assume that the following two conditions hold:*

(i) *As a rational B-module, and for some $t \geq 0$,*

$$\mathrm{Hom}_{u_\zeta(\mathfrak{l}_J)}(\mathsf{k}, \mathrm{H}^n(\mathcal{U}_\zeta(\mathfrak{u}_J), M) \otimes Q) \cong \begin{cases} \mathsf{k} & n = t, \\ 0 & \textit{otherwise}. \end{cases} \tag{3.3.1}$$

(ii) *For $\gamma \in X_{P_J} \cap X^+$ we have $R^i \mathrm{ind}_{U_\zeta(\mathfrak{p}_J)}^{U_\zeta(\mathfrak{g})}(Q \otimes \mathsf{k}_{\ell\gamma}) = 0$ for $i > 0$.*

Then there exists an isomorphism of rational G-modules:
(3.3.2)
$$\mathrm{H}^n(u_\zeta(\mathfrak{g}), M \otimes \operatorname{ind}_{U_\zeta(\mathfrak{p}_J)}^{U_\zeta(\mathfrak{g})}(Q \otimes \mathsf{k}_{\ell\gamma})) \cong \begin{cases} \operatorname{ind}_{P_J}^G(S^{\frac{n-t}{2}}(\mathfrak{u}_J^*) \otimes \mathsf{k}_\gamma) & n-t \equiv 0 \pmod 2, \\ 0 & \textit{otherwise.} \end{cases}$$

PROOF. We first consider the functors (cf. [**Jan1**, I 6.12])
$$\mathcal{F}_1(-) = \operatorname{Hom}_{u_\zeta(\mathfrak{g})}(\mathsf{k}, \operatorname{ind}_{U_\zeta(\mathfrak{p}_J)}^{U_\zeta(\mathfrak{g})}(-)) \text{ and } \mathcal{F}_2(-) = \operatorname{ind}_{P_J}^G \operatorname{Hom}_{u_\zeta(\mathfrak{p}_J)}(\mathsf{k}, -)$$
with $\mathcal{F}_1, \mathcal{F}_2 : U_\zeta(\mathfrak{p}_J)$-mod $\to U_\zeta(\mathfrak{g})$-mod (or G-mod). Observe that we are using the Frobenius map and the following identification of functors:
$$\operatorname{ind}_{P_J}^G(-) \cong \operatorname{ind}_{U_\zeta(\mathfrak{p}_J)/\!\!/u_\zeta(\mathfrak{p}_J)}^{U_\zeta(\mathfrak{g})/\!\!/u_\zeta(\mathfrak{g})}(-).$$

Set $D = M \otimes Q \otimes \mathsf{k}_{\ell\gamma}$. The functors $\mathcal{F}_1$ and $\mathcal{F}_2$ are naturally isomorphic and there exist two spectral sequences:
$$'E_2^{i,j} = \mathrm{H}^i(u_\zeta(\mathfrak{g}), R^j \operatorname{ind}_{U_\zeta(\mathfrak{p}_J)}^{U_\zeta(\mathfrak{g})} D) \Rightarrow (R^{i+j}\mathcal{F}_1)(D) \quad \text{and}$$
$$E_2^{i,j} = R^i \operatorname{ind}_{P_J}^G \mathrm{H}^j(u_\zeta(\mathfrak{p}_J), D) \Rightarrow (R^{i+j}\mathcal{F}_2)(D) \tag{3.3.3}$$
which converge to the same abutment. However, by the tensor identity and condition (ii),
$$R^i \operatorname{ind}_{U_\zeta(\mathfrak{p}_J)}^{U_\zeta(\mathfrak{g})} D \cong M \otimes R^i \operatorname{ind}_{U_\zeta(\mathfrak{p}_J)}^{U_\zeta(\mathfrak{g})}[Q \otimes \mathsf{k}_{\ell\gamma}] \cong 0$$
for $i > 0$. Consequently, the first spectral sequence collapses and we can combine this with the second spectral sequence to obtain a first quadrant spectral sequence:
(3.3.4)
$$E_2^{i,j} = R^i \operatorname{ind}_{P_J}^G(\mathrm{H}^j(u_\zeta(\mathfrak{p}_J), M \otimes Q) \otimes \mathsf{k}_\gamma) \Rightarrow \mathrm{H}^{i+j}(u_\zeta(\mathfrak{g}), M \otimes \operatorname{ind}_{U_\zeta(\mathfrak{p}_J)}^{U_\zeta(\mathfrak{g})}(Q \otimes \mathsf{k}_{\ell\gamma})).$$
One can rewrite the spectral sequence (3.3.4) using Proposition 3.2.1(b) as
$$E_2^{i,j} = R^i \operatorname{ind}_{P_J}^G(S^{\frac{j-t}{2}}(\mathfrak{u}_J^*) \otimes \mathsf{k}_\gamma) \Rightarrow \mathrm{H}^{i+j}(u_\zeta(\mathfrak{g}), M \otimes \operatorname{ind}_{U_\zeta(\mathfrak{p}_J)}^{U_\zeta(\mathfrak{g})}(Q \otimes \mathsf{k}_{\ell\gamma})). \tag{3.3.5}$$
Since $\gamma \in X_{P_J} \cap X^+$, the Grauert-Riemenschneider vanishing theorem implies that
$$R^i \operatorname{ind}_{P_J}^G(S^\bullet(\mathfrak{u}_J^*) \otimes \mathsf{k}_\gamma) = 0$$
for $i > 0$. The spectral sequence (3.3.5) collapses and yields the isomorphism stated in the theorem. □

4. Non-shifted realization of the coordinate ring of the twisted cotangent bundle

4.1. Let $J \subseteq \Delta$ and X_J^+ be the J-dominant weights (i.e., $\langle \mu, \alpha^\vee \rangle \geq 0$ for all $\alpha \in J$). For $\mu \in X_J^+$, let
$$H_J^0(\mu) = \operatorname{ind}_{U_\zeta(\mathfrak{b}_J)}^{U_\zeta(\mathfrak{l}_J)} \mathsf{k}_\mu.$$
In particular when $J = \varnothing$, we obtain the ordinary induced module $H^0(\mu) = \operatorname{ind}_{U_\zeta(\mathfrak{b})}^{U_\zeta(\mathfrak{g})} \mathsf{k}_\mu$ where $\mu \in X^+$. For $\mu \in X_J^+$, let $L_J(\mu)$ be the finite dimensional simple $U_\zeta(\mathfrak{l}_J)$-module which appears in the socle of $H_J^0(\mu)$.

Next we observe that, for any $w \in {}^JW$, the weights $w \cdot 0$ and $-w_{0,J}(w \cdot 0)$ are in X_J^+, and in fact in the bottom alcove of X_J^+ when $\ell > h$ [**UGA1**, Prop. 3.6.1]. Therefore,
$$L_J(w \cdot 0) \cong H_J^0(w \cdot 0)$$

and

$$L_J(w\cdot 0)^* = L_J(-w_{0,J}(w\cdot 0)) = H^0_J(-w_{0,J}(w\cdot 0)).$$

We will also be using the fact that $w\cdot 0 = -\sum_{\alpha\in\Phi^+(w)}\alpha$. Here $\Phi^+(w) = \{\alpha\in\Phi^+ \mid w^{-1}(\alpha)\in\Phi^-\}$. If $w\in {}^JW$, then $\Phi^+(w)\subseteq \Phi^+\setminus\Phi^+_J$. If $w\in W_J$ then $\Phi^+(w)\subseteq\Phi_J$ and $w(\Phi_J)\subseteq\Phi_J$.

Let $\epsilon_J := (\ell-1)\sum_{\alpha\in J}\omega_\alpha$. In this case $H^0_J(\epsilon_J) = L_J(\epsilon_J)$ is the Steinberg representation for $U_\zeta(\mathfrak{l}_J)$. We also note that $H^0_J(\epsilon_J)$ is the unique composition factor of $H^0(\epsilon_J)$ as $U_\zeta(\mathfrak{l}_J)$-module with highest weight ϵ_J. In fact

$$H^0(\epsilon_J)|_{U_\zeta(\mathfrak{l}_J)} \cong H^0_J(\epsilon_J)\oplus E$$

with E being a $U_\zeta(\mathfrak{l}_J)$-module without weight ϵ_J. In particular,

$$\operatorname{Hom}_{U_\zeta(\mathfrak{l}_J)}(H^0(\epsilon_J), H^0_J(\epsilon_1)) \cong \mathsf{k}.$$

LEMMA 4.1.1. *Assume $\ell\geq h-1$. Then for any $U_\zeta(\mathfrak{p}_J)$-module M and any $U_\zeta(\mathfrak{l}_J)$-module Q such that Q is injective as $u_\zeta(\mathfrak{l}_J)$-module, $\mathrm{H}^n(\mathcal{U}_\zeta(\mathfrak{u}_J), M)\otimes Q$ is isomorphic, as a $u_\zeta(\mathfrak{l}_J)$-module, to a direct summand of*

$$\bigoplus_{y\in {}^JW,\ l(y)=n} M\otimes L_J(y\cdot 0)^*\otimes Q.$$

In particular, for any $u_\zeta(\mathfrak{l}_J)$-module E, $\operatorname{Hom}_{u_\zeta(\mathfrak{l}_J)}(E, \mathrm{H}^n(\mathcal{U}_\zeta(\mathfrak{u}_J), M)\otimes Q) = 0$ if

$$\operatorname{Hom}_{u_\zeta(\mathfrak{l}_J)}(E, M\otimes L_J(y\cdot 0)\otimes Q) = 0$$

for all $y\in {}^JW$ with $l(y)=n$. Similar statements can be formulated for the functor $\operatorname{Hom}_{u_\zeta(\mathfrak{l}_J)}(-,E)$.

PROOF. We can first apply [**UGA2**, Theorem 3.5.2] to deduce that $\mathrm{H}^n(\mathcal{U}_\zeta(\mathfrak{u}_J), M)$ is a $u_\zeta(\mathfrak{l}_J)$-subquotient of

$$M\otimes \mathrm{H}^n(\mathcal{U}_\zeta(\mathfrak{u}_J),\mathsf{k}).$$

According to [**UGA2**, Theorem 6.4.1] since $\ell\geq h-1$, we have

$$\mathrm{H}^n(\mathcal{U}_\zeta(\mathfrak{u}_J),\mathsf{k}) \cong \bigoplus_{y\in {}^JW,\ l(y)=n} L_J(-w_{0,J}(y\cdot 0)).$$

Since Q is $u_\zeta(\mathfrak{l}_J)$-injective, any short exact sequence of $u_\zeta(\mathfrak{l}_J)$-modules tensored by Q becomes split as $u_\zeta(\mathfrak{l}_J)$-modules. Then $\mathrm{H}^n(\mathcal{U}_\zeta(\mathfrak{u}_J, M)\otimes Q$ is a direct $u_\zeta(\mathfrak{l}_J)$-direct summand of $M\otimes \mathrm{H}^n(\mathcal{U}_\zeta(\mathfrak{u}_J,\mathsf{k})\otimes Q$. Now the second part of the lemma follows by applying the functor additive functors $\operatorname{Hom}_{u_\zeta(\mathfrak{l}_J)}(E,-)$ and $\operatorname{Hom}_{u_\zeta(\mathfrak{l}_J)}(-,E)$. □

4.2. We can now apply the theory developed in Section 3 in order to give a precise realization of the coordinate algebra of the twisted cotangent bundle.

Let $M = H^0(\epsilon_J)^*$ *(a $U_\zeta(\mathfrak{g})$-module) and* $Q_w = H^0_J(\epsilon_J)\otimes H^0_J(-w_{0,J}(w\cdot 0))$ *for a fixed $w\in {}^JW$.*

THEOREM 4.2.1. *Let $\ell > 2h-1$, $J\subseteq\Delta$ and $\epsilon_J = \sum_{\alpha\in J}(\ell-1)\omega_\alpha$ with $w\in {}^JW$ and $\gamma\in X_{P_J}\cap X^+$. Set*

$$N_w := H^0(\epsilon_J)^*\otimes \operatorname{ind}_{U_\zeta(\mathfrak{p}_J)}^{U_\zeta(\mathfrak{g})}[H^0_J(\epsilon_J)\otimes H^0_J(-w_{0,J}(w\cdot 0))\otimes \mathsf{k}_{\ell\gamma}],$$

and assume that

(a) $\operatorname{Hom}_{u_\zeta(\mathfrak{l}_J)}(\mathsf{k}, \mathrm{H}^0(\mathcal{U}_\zeta(\mathfrak{u}_J), M)\otimes Q_w)\cong\mathsf{k}$,
(b) $\epsilon_J - w_{0,J}(w\cdot 0)+\ell\gamma\in X^+$.

Then there exists an isomorphism of rational G-modules:

$$\mathrm{H}^n(u_\zeta(\mathfrak{g}), N_w) \cong \begin{cases} \mathrm{ind}_{P_J}^G(S^{\frac{n}{2}}(\mathfrak{u}_J^*) \otimes \mathsf{k}_\gamma) & n \equiv 0 \pmod 2, \\ 0 & \textit{otherwise.} \end{cases} \tag{4.2.1}$$

Furthermore, in the case when $w = \mathrm{id}$ for N_w, (4.2.1) holds for $\ell > h$ with the condition (b). The condtion (a) is satisfied in the case when $w = \mathrm{id}$.

PROOF. The module Q_w is a $U_\zeta(\mathfrak{l}_J)$-module inflated to a $U_\zeta(\mathfrak{p}_J)$. Since $L_J(\epsilon_J) = H_J^0(\epsilon_J)$ is an injective $u_\zeta(\mathfrak{l}_J)$-module, thus Q_w is an injective $u_\zeta(\mathfrak{l}_J)$-module and also injective $U_\zeta(\mathfrak{l}_J)$. Therefore, M and Q_w satisfy the conditions of Theorem 3.3.1. Since ϵ_J is an ℓ-restricted weight, M is a quotient of the baby Verma module $u_\zeta(\mathfrak{g}) \otimes_{u_\zeta(\mathfrak{b})} \mathsf{k}_{-w_0\epsilon_J}$, thus $Z_J^+.M = 0$ when M considered a $\mathcal{U}_\zeta(\mathfrak{g})$-module. We need to verify conditions (i) and (ii) of Theorem 3.3.1 in order to conclude our result.

For condition (i), we need to show that as a rational B-module,

$$\mathrm{Hom}_{u_\zeta(\mathfrak{l}_J)}(\mathsf{k}, \mathrm{H}^n(\mathcal{U}_\zeta(\mathfrak{u}_J), M) \otimes Q_w) \cong \begin{cases} \mathsf{k} & n = 0, \\ 0 & \text{otherwise.} \end{cases} \tag{4.2.2}$$

Assume the left hand side of (4.2.2)

$$\mathrm{Hom}_{u_\zeta(\mathfrak{l}_J)}(\mathsf{k}, \mathrm{H}^n(\mathcal{U}_\zeta(\mathfrak{u}_J), H^0(\epsilon_J)^*) \otimes L_J(\epsilon_J) \otimes L_J(-w_{0,J}(w \cdot 0))) \neq 0. \tag{4.2.3}$$

Then, by Lemma 4.1.1, there is $y \in {}^JW$ with $l(y) = n$ such that

$$\mathrm{Hom}_{u_\zeta(\mathfrak{l}_J)}(\mathsf{k}, H^0(\epsilon_J)^* \otimes L_J(-w_{0,J}(y \cdot 0)) \otimes L_J(\epsilon_J) \otimes L_J(-w_{0,J}(w \cdot 0))) \neq 0$$

which is equivalent to

$$\mathrm{Hom}_{u_\zeta(\mathfrak{l}_J)}(L_J(-w_{0,J}\epsilon_J), H^0(\epsilon_J)^* \otimes L_J(-w_{0,J}(y \cdot 0)) \otimes L_J(-w_{0,J}(w \cdot 0))) \neq 0.$$

All the modules under consideration are $U_\zeta(\mathfrak{l}_J)$-modules, so we can conclude that

$$\mathrm{Hom}_{U_\zeta(\mathfrak{l}_J)}(L_J(-w_{0,J}\epsilon_J) \otimes L(\nu)^{[1]}, H^0(\epsilon_J)^* \otimes L_J(-w_{0,J}(y \cdot 0)) \otimes L_J(-w_{0,J}(w \cdot 0))) \neq 0 \tag{4.2.4}$$

for some $\nu \in X_J^+$. Since

$$L_J(-w_{0,J}(y \cdot 0)) \cong H_J^0(-w_{0,J}(y \cdot 0)) \cong \mathrm{ind}_{U_\zeta(\mathfrak{b}_J)}^{U_\zeta(\mathfrak{l}_J)} \mathsf{k}_{-w_{0,J}(y \cdot 0)},$$

it follows by Frobenius reciprocity that

$$\mathrm{Hom}_{U_\zeta(\mathfrak{b}_J)}(L_J(-w_{0,J}\epsilon_J) \otimes L_J(\nu)^{[1]}, H^0(\epsilon_J)^* \otimes \mathsf{k}_{-w_{0,J}(y \cdot 0)} \otimes L_J(-w_{0,J}(w \cdot 0))) \neq 0. \tag{4.2.5}$$

The head of $L_J(-w_{0,J}\epsilon_J) \otimes L(\nu)^{[1]}$ as a $U_\zeta(\mathfrak{b}_J)$-module is one-dimensional and isomorphic to $-w_{0,J}\epsilon_J + \ell\nu$. We can now deduce from (4.2.5) that

$$-w_{0,J}\epsilon_J + \ell\nu = \mu - w_{0,J}(y \cdot 0) + \sigma \tag{4.2.6}$$

for some $\nu \in X_J^+$, some weight μ of $H^0(\epsilon_J)^*$, and a weight σ of $L_J(-w_{0,J}(w \cdot 0)) = L_J(w \cdot 0)^*$. We note that for any $U_\zeta(\mathfrak{l}_J)$-module V, μ is a weight of V if any only if $-w(\mu)$ is a weight of V^* for any $w \in W_J$. Applying $-w_{0,J}$ to (4.2.6) justifies revising the weight condition as follows:

$$\epsilon_J + \ell\nu = \mu + y \cdot 0 + \sigma \tag{4.2.7}$$

where $\nu \in X_J^+$, μ an weight of $H^0(\epsilon_J)$ and σ a weight of $L_J(w \cdot 0)$.

The Weyl group for $\Delta - J$, $W_{\Delta - J}$ is generated by the simple reflections s_β such that $\beta \in \Delta - J$. Note that $s_\beta(\omega_\alpha) = \omega_\alpha$ when $\beta \in \Delta - J$, $\alpha \in J$, thus

$\tilde{w}(\omega_\alpha) = \omega_\alpha$ for all $\tilde{w} \in W_{\Delta-J}$, $\alpha \in J$. Therefore, we can find some $\widetilde{y} \in W_{\Delta-J}$ such that $\nu' = \widetilde{y}\nu \in X^+$. Applying $\widetilde{y}$ to (4.2.7) yields:

$$\epsilon_J + \ell\nu' = \mu' + \widetilde{y}(y \cdot 0) + \widetilde{y}\sigma \tag{4.2.8}$$

where μ' is a weight of $H^0(\epsilon_J)$. Rewriting this equation, one has

$$\epsilon_J - \mu' + \ell\nu' = \widetilde{y}(y \cdot 0) + \widetilde{y}\sigma = \widetilde{y}(y \cdot 0 + \sigma). \tag{4.2.9}$$

Taking the inner product with $\alpha_0^\vee$ (with α_0 being the highest short root) and using the facts that $\langle \beta, \alpha_0^\vee \rangle \geq 0$ and $\langle \delta, \alpha_0^\vee \rangle \leq 2(h-1)$ for all $\beta \in \Phi^+$ and $\delta \leq z \cdot 0$ where $z \in W$, one has

$$0 \leq \langle \epsilon_J - \mu', \alpha_0^\vee \rangle + \ell\langle \nu', \alpha_0^\vee \rangle \leq 4(h-1) < 2\ell. \tag{4.2.10}$$

The equation (4.2.10) implies that $\langle \nu', \alpha_0^\vee \rangle = 0, 1$.

From (4.2.9) it follows that $\ell\nu'$ is in the root lattice by noting that $\mu - \epsilon_J$ as well as all weights of $L_J(y \cdot 0)$ and $L_J(w \cdot 0)$ are in the root lattice. Since $\ell > h$, $\ell \nmid |X/\mathbb{Z}\Phi|$, and ν' must be in the root lattice (i.e., cannot be minuscule). Consequently, $\nu = 0$. Since $y \cdot 0, \sigma \in -\mathbb{N}\Phi^+$, (4.2.7) implies that $y = \text{id}$, $\sigma = 0$, $n = 0$, and $\mu = \epsilon_J$.

Note when $w = \text{id}$ for N_w with $\ell > h$, one has $\sigma = 0$, and one can replace (4.2.10) with

$$0 \leq \langle \epsilon_J - \mu', \alpha_0^\vee \rangle + \ell\langle \nu', \alpha_0^\vee \rangle \leq 2(h-1) < 2\ell. \tag{4.2.11}$$

The arguments from the preceding paragraph can then be repeated to draw the same conclusions (i.e., $\nu = 0$ and $y = \text{id}$). Our analysis proves that for $n > 0$

$$\text{Hom}_{u_\zeta(\mathfrak{l}_J)}(\mathsf{k}, \text{H}^n(\mathcal{U}_\zeta(\mathfrak{u}_J), M) \otimes Q_w) \cong 0.$$

Now we can use the condition (a) in the statement of the theorem to conclude that (i) holds.

Next we show that condition (a) holds in the case when $w = \text{id}$. Since $Q_{\text{id}} = L_J(\epsilon_J)$ is trivial as $u_\zeta(\mathfrak{u}_J)$-module and $\text{H}^0(\mathcal{U}_\zeta(\mathfrak{u}_J), M) = \text{Hom}_{\mathcal{U}_\zeta(\mathfrak{u}_J)}(H^0(\epsilon_J), \mathsf{k})$, we have

$$\begin{aligned}
\text{Hom}_{u_\zeta(\mathfrak{l}_J)}(\mathsf{k}, \text{H}^n(\mathcal{U}_\zeta(\mathfrak{u}_J), M) \otimes Q_{\text{id}}) &\cong \text{Hom}_{u_\zeta(\mathfrak{l}_J)}(\mathsf{k}, L_J(\epsilon_J) \otimes \text{Hom}_{\mathcal{U}_\zeta(\mathfrak{u}_J)}(H^0(\epsilon_J), \mathsf{k})) \\
&\cong \text{Hom}_{u_\zeta(\mathfrak{l}_J)}(\mathsf{k}, \text{Hom}_{u_\zeta(\mathfrak{u}_J)}(H^0(\epsilon_J), L_J(\epsilon_J)) \\
&\cong \text{Hom}_{u_\zeta(\mathfrak{p}_J)}(H^0(\epsilon_J), L_J(\epsilon_J)).
\end{aligned}$$

Set $S := \text{Hom}_{u_\zeta(\mathfrak{p}_J)}(H^0(\epsilon_J), L_J(\epsilon_J)) \cong \text{Hom}_{u_\zeta(\mathfrak{p}_J)}(H^0(\epsilon_J), H^0_J(\epsilon_J))$. By using a slight variation of the spectral sequences in the proof of Theorem 3.3.1 (cf. [**Jan1**, II 12.7(2)]), we have

$$S \cong \text{ind}_B^{P_J} \text{Hom}_{u_\zeta(\mathfrak{b}_J)}(H^0(\epsilon_J), \mathsf{k}_{\epsilon_J}). \tag{4.2.12}$$

Now set $V = \text{Hom}_{u_\zeta(\mathfrak{b}_J)}(H^0(\epsilon_J), \mathsf{k}_{\epsilon_J})$. The verification of (4.2.2) for $w = \text{id}$ will be done once we show that $V \cong \mathsf{k}$ as a B-module. This will imply that $S \cong \mathsf{k}$ as a P_J-module using (4.2.12).

We will next analyze the structure of V as a B-module. First, we consider the B-socle of V. Suppose that $0 \neq \text{Hom}_B(\mathsf{k}_\nu, V) \cong \text{Hom}_{U_\zeta(\mathfrak{b})}(H^0(\epsilon_J), \mathsf{k}_{\epsilon_J+\ell\nu})$. By using Frobenius reciprocity, it follows that $\epsilon_J + \ell\nu$ is in X^+. Since

$$\text{Hom}_{U_\zeta(\mathfrak{b})}(H^0(\epsilon_J), \mathsf{k}_{\epsilon_J+\ell\nu}) \neq 0,$$

it follows that $\epsilon_J + \ell\nu \leq \epsilon_J$, thus $\ell\nu \leq 0$. Moreover, ν has to be dominant because $\epsilon_J + \ell\nu$ is dominant (use the definition of ϵ_J). Therefore, $\ell\nu \in \mathbb{N}\Phi^- \cap X^+ = \{0\}$, and $\nu = 0$. Furthermore,

$$\mathrm{Hom}_B(\mathsf{k}, V) \cong \mathrm{Hom}_{U_\zeta(\mathfrak{b})}(H^0(\epsilon_J), \mathsf{k}_{\epsilon_J}) \cong \mathrm{Hom}_{U_\zeta(\mathfrak{g})}(H^0(\epsilon_J), H^0(\epsilon_J)) \cong \mathsf{k}.$$

This shows that $\mathrm{soc}_B V \cong \mathsf{k}$.

The next step is to show that $V \cong \mathsf{k}$. First, one has $V \subseteq \mathrm{Hom}_{u_\zeta(\mathfrak{t})}(H^0(\epsilon_J), \mathsf{k}_{\epsilon_J})$. By using a weight space decomposition for $H^0(\epsilon_J)$, one sees that any B-composition factor of V must be of the form k_ν where $\mathsf{k}_{\epsilon_J - \ell\nu}$ is a $U_\zeta(\mathfrak{b})$-composition factor of $H^0(\epsilon_J)$. Since ϵ_J is the highest weight of $H^0(\epsilon_J)$, it follows that $\ell\nu \in \mathbb{N}\Phi^+$.

If $V/\mathrm{soc}_B V \neq 0$ then there exists a B-composition factor k_ν of V such that $\mathrm{Ext}^1_B(\mathsf{k}_\nu, \mathsf{k}) \neq 0$. As a consequence of the Borel-Bott-Weil Theorem (cf. [**Jan1**, II 5.5]), $\nu = -\alpha$ where $\alpha \in \Delta$. This contradicts the prior paragraph that $\ell\nu \in \mathbb{N}\Phi^+$. Hence, $V \cong \mathsf{k}$ as B-module.

We will now verify condition (ii) of Theorem 3.3.1 by showing that

$$R^i \,\mathrm{ind}_{U_\zeta(\mathfrak{p}_J)}^{U_\zeta(\mathfrak{g})}[H^0_J(\epsilon_J) \otimes H^0_J(-w_{0,J}(w \cdot 0)) \otimes \mathsf{k}_{\ell\gamma}] = 0 \tag{4.2.13}$$

for $i > 0$. Set $\widehat{S} := H^0_J(-w_{0,J}(w \cdot 0))$. There exists a spectral sequence

$$E_2^{i,j} = R^i \,\mathrm{ind}_{U_\zeta(\mathfrak{p}_J)}^{U_\zeta(\mathfrak{g})} R^j \,\mathrm{ind}_{U_\zeta(\mathfrak{b})}^{U_\zeta(\mathfrak{p}_J)}[\mathsf{k}_{\epsilon_J} \otimes \widehat{S} \otimes \mathsf{k}_{\ell\gamma}] \Rightarrow R^{i+j} \,\mathrm{ind}_{U_\zeta(\mathfrak{b})}^{U(\mathfrak{g})}[\mathsf{k}_{\epsilon_J} \otimes \widehat{S} \otimes \mathsf{k}_{\ell\gamma}].$$

Observe that as a $U_\zeta(\mathfrak{l}_J)$-module one has by the generalized tensor identity

$$\begin{aligned} R^j \,\mathrm{ind}_{U_\zeta(\mathfrak{b})}^{U_\zeta(\mathfrak{p}_J)}[\mathsf{k}_{\epsilon_J} \otimes \widehat{S} \otimes \mathsf{k}_{\ell\gamma}] &\cong R^j \,\mathrm{ind}_{U_\zeta(\mathfrak{b}_J)}^{U_\zeta(\mathfrak{l}_J)}[H^0_J(\epsilon_J) \otimes \widehat{S} \otimes \mathsf{k}_{\ell\gamma}] \\ &\cong (R^j \,\mathrm{ind}_{U_\zeta(\mathfrak{b}_J)}^{U_\zeta(\mathfrak{l}_J)} \mathsf{k}) \otimes [H^0_J(\epsilon_J) \otimes \widehat{S} \otimes \mathsf{k}_{\ell\gamma}] \\ &= 0 \end{aligned}$$

for $j > 0$. Therefore, this spectral sequence collapses and gives the following isomorphism for $i \geq 0$:

$$R^i \,\mathrm{ind}_{U_\zeta(\mathfrak{p}_J)}^{U_\zeta(\mathfrak{g})}[H^0_J(\epsilon_J) \otimes \widehat{S} \otimes \mathsf{k}_{\ell\gamma}] \cong R^i \,\mathrm{ind}_{U_\zeta(\mathfrak{b})}^{U(\mathfrak{g})}[\mathsf{k}_{\epsilon_J} \otimes \widehat{S} \otimes \mathsf{k}_{\ell\gamma}].$$

Let μ be a weight of $\widehat{S} = H^0_J(-w_{0,J}(w \cdot 0))$. Then (4.2.13) will follow by Kempf's Vanishing Theorem, if we demonstrate that $\epsilon_J + \mu + \ell\gamma \in X^+$. Observe that $\langle \alpha, \beta^\vee \rangle \leq 0$ for all $\alpha \in J$ and $\beta \in \Delta - J$. This implies that for $\beta \in \Delta - J$,

$$\langle \mu, \beta^\vee \rangle \geq \langle -w_{0,J}(w \cdot 0), \beta^\vee \rangle$$

because $-w_{0,J}(w \cdot 0) - \mu \in \mathbb{N}\Phi_J^+$. Consequently, for $\beta \in \Delta - J$,

$$\langle \epsilon_J + \mu + \ell\gamma, \beta^\vee \rangle \geq \langle \epsilon_J + -w_{0,J}(w \cdot 0) + \ell\gamma, \beta^\vee \rangle \geq 0 \tag{4.2.14}$$

since $\epsilon_J - w_{0,J}(w \cdot 0) + \ell\gamma \in X^+$. On the other hand, let $J = J_1 \cup J_2 \cup \cdots \cup J_t$ be the decomposition of J corresponding to decomposing Φ_J into a union of irreducible root systems. Let J_s be one of the components, and δ be either the highest long root or the highest short root of Φ_{J_s}. The lowest weight of $L_J(-w_{0,J}(w \cdot 0))$ is $-w \cdot 0$, and since $w \in {}^J W$,

$$\langle -w \cdot 0, \delta^\vee \rangle = -\langle \rho, w^{-1}\delta^\vee \rangle + \langle \rho, \delta^\vee \rangle \geq -(h-1).$$

Now $\mu - (-w \cdot 0) \in \mathbb{N}\Phi_J^+$ so $\langle \mu - (-w \cdot 0), \delta^\vee \rangle \geq 0$, thus $\langle \mu, \delta^\vee \rangle > -(h-1)$. If β is a simple root in J_s there exists $\tilde{w} \in W_{J_s}$ such that $\tilde{w}\beta = \delta$ (where δ depends on whether β is a short or long root). Using the fact that $\tilde{w}\mu$ is a weight of $L_J(w\cdot 0)$, we have

(4.2.15) $\langle \epsilon_J+\mu+\ell\gamma, \beta^\vee\rangle = \langle \epsilon_J, \beta^\vee\rangle + \langle \tilde{w}\mu, \tilde{w}\beta^\vee\rangle + \langle \ell\gamma, \beta^\vee\rangle \geq (\ell-1)-(h-1)+0 \geq 0.$

Consequently, $\epsilon_J+\mu+\ell\gamma \in X^+$ follows by (4.2.14) and (4.2.15). □

4.3. We can now give a positive answer to the 1986 question posed by Friedlander and Parshall (cf. [**FP1**, (3.2)]) in the case of quantum groups.

THEOREM 4.3.1. *Let $\ell > h$ and $\mathcal{O} := \mathcal{O}_J$ be a Richardson orbit in $\mathcal{N}$ with moment map $\Gamma : G\times_{P_J}\mathfrak{u}_J \to G\cdot \mathfrak{u}_J = \overline{\mathcal{O}}$. Then there exists a finite-dimensional $U_\zeta(\mathfrak{g})$-algebra $A_{\mathcal{O}}$ (which is not necessarily commutative) such that as rational G-algebras:*

(a) $\mathrm{H}^{2\bullet+1}(u_\zeta(\mathfrak{g}), A_{\mathcal{O}}) = 0$;
(b) $\mathrm{H}^{2\bullet}(u_\zeta(\mathfrak{g}), A_{\mathcal{O}}) \cong \mathsf{k}[\mathcal{O}] \cong \mathsf{k}[G\times_{P_J}\mathfrak{u}_J] \cong \mathrm{ind}_{P_J}^G S^\bullet(\mathfrak{u}_J^*)$.

Furthermore, if Γ is a resolution of singularities and $\overline{\mathcal{O}}$ is normal then

$$\mathrm{H}^{2\bullet}(u_\zeta(\mathfrak{g}), A_{\mathcal{O}}) \cong \mathsf{k}[\overline{\mathcal{O}}].$$

PROOF. We apply Theorem 4.2.1 in the case when $w = \mathrm{id}$ and $\gamma = 0$. Then if we set $A_{\mathcal{O}} := Q_w = H^0(\epsilon_J)^*\otimes H^0(\epsilon_J) \cong \mathrm{End}_{\mathsf{k}}(H^0(\epsilon_J))$ which is a $U_\zeta(\mathfrak{g})$-algebra. The statements (a) and (b) now follow by Theorem 4.2.1 by noting that in this specific setting the isomorphisms are now as rational G-algebras.

For the last statement of the theorem, the conditions that Γ is a resolution of singularities and $\overline{\mathcal{O}}$ is normal imply that $\mathsf{k}[\overline{\mathcal{O}}] \cong \mathsf{k}[\mathcal{O}]$ (cf. [**BNPP**, §3.5]). □

We remark that we have assumed that k is of characteristic 0 and not necessarily algebraically closed. Since the group G is assumed to be defined and split over k, the Richardson orbit is defined over any fields and thus the statement of the theorem makes sense. Friedlander and Parshall's question is for algebraic groups in positive characteristic. However, the question of which closures of nilpotent orbits are normal is still open in types E_7 and E_8. For type A_n all nilpotent orbit closures are normal and for types B_n, C_n, D_n, E_6, F_2, and G_2 this question has been resolved (cf. [**KP**], [**So1, So2**], [**Br1**]).

Let $\Phi_\lambda = \{\alpha\in\Phi : \langle \lambda+\rho, \alpha^\vee\rangle \in \ell\mathbb{Z}\}$. Under the condition that $(\ell, p) = 1$ for any bad prime p of Φ, one can find $w\in W$ such that $w(\Phi_\lambda) = \Phi_J$. Moreover, one can apply [**BNPP**, Theorem 1.3.5] to see that the support variety of $A_{\mathcal{O}}$ is the closure of the Richardson orbit associated to J (i.e., $\mathcal{V}_{\mathfrak{g}}(A_{\mathcal{O}}) = G\cdot\mathfrak{u}_J$). In general the rate of growth of $\dim \mathrm{H}^p(u_\zeta(\mathfrak{g}), A_{\mathcal{O}})$ will equal the $\dim G\cdot \mathfrak{u}_J$. A priori, $\mathrm{H}^\bullet(u_\zeta(\mathfrak{g}), A_{\mathcal{O}})$ could be a non-commutative algebra, but Theorem 4.3.1 demonstrates that the Yoneda product in this case is commutative.

4.4. When $\Phi = A_{n-1}$, the $G = GL_n$-orbits are labelled by partitions of n. If λ is a partition of n, let x_λ be the nilpotent matrix in $\mathfrak{g} = \mathfrak{gl}_n$ with Jordan blocks with sizes matching up with the parts of λ. Set $\mathcal{O}_\lambda = G\cdot x_\lambda$. Given $J\subseteq\Delta$, one can associate a partition $\sigma(J)$ to the subroot system generated by J (cf. [**NPV**, Section 4.5]). According to [**Kr**],

$$\overline{\mathcal{O}_{\sigma(J)^t}} = G\cdot \mathfrak{u}_J.$$

Here $(-)^t$ denotes the transposed partition.

The centralizers of nilpotent elements under G are connected so Γ is a desingularization. Moreover, from [**KP**] all G-orbits have normal orbit closures. Consequently, Theorem 4.3.1 can be stated in terms of partitions indexing the various nilpotent orbits.

COROLLARY 4.4.1. *Let $\mathfrak{g} = \mathfrak{gl}_n$, $\ell > n$ and λ be a partition of n. Let J be a subset of simple roots corresponding to the partition λ^t. Then there exists a finite dimensional $U_\zeta(\mathfrak{g})$-algebra $A_{\mathcal{O}_\lambda}$ such that*

(a) $\mathrm{H}^{2\bullet+1}(u_\zeta(\mathfrak{g}), A_{\mathcal{O}_\lambda}) = 0$;
(b) $\mathrm{H}^{2\bullet}(u_\zeta(\mathfrak{g}), A_{\mathcal{O}_\lambda}) \cong \mathsf{k}[G \cdot \mathfrak{u}_J] \cong \mathsf{k}[\overline{\mathcal{O}}]$.

5. Shifted realization of the coordinate ring of the twisted cotangent bundle

5.1. In this section we will provide a computation involving the cohomology of certain tilting modules for $U_\zeta(\mathfrak{l}_J)$ and demonstrate that this yields another realization of the coordinate ring of the twisted cotangent bundle. This result is quite striking in the sense that the computation of the characters of the tilting modules is only known conjecturally or via p-Kazhdan-Lusztig polynomials (cf. [**Jan1**, E.10 (3) Conjecture] and [**AMRW**, Theorem 7.6]), yet we will show that without assuming knowledge about the character formula one can still compute non-trivial cohomology.

For $\mu \in X_J^+$, let $T_J(\mu)$ be the unique indecomposable tilting $U_\zeta(\mathfrak{l}_J)$-module with highest weight μ. Recall $\epsilon_J = \sum_{\alpha \in J}(\ell - 1)\omega_\alpha$. Let $\mu \in X_{J,\mathrm{res}}$ be the set of ℓ-restricted weights of X_J^+. Using the work of Pillen [**P**, Section 2, Corollary A], one can show that for $\mu \in X_{J,\mathrm{res}}$ the tilting module $T_J(\epsilon_J - w_{0,J}\epsilon_J + w_{0,J}\mu)$ can be realized as a multiplicity one $U_\zeta(\mathfrak{l}_J)$-summand in $L_J(\epsilon_J) \otimes L_J(-w_{0,J}\epsilon_J + w_{0,J}\mu)$. Set $\mu_J = -w_{0,J}\epsilon_J + w_{0,J}\mu$. We also review the important fact (cf. [**Jan1**, II 10.15 Lemma]) that

$$\mathrm{Hom}_{u_\zeta(\mathfrak{l}_J)}(L(\mu), L_J(\epsilon_J) \otimes L_J(\mu_J)) \cong \mathrm{Hom}_{U_\zeta(\mathfrak{l}_J)}(L(\mu), L_J(\epsilon_J) \otimes L_J(\mu_J)). \tag{5.1.1}$$

THEOREM 5.1.1. *Let $\ell > h$ and $J \subseteq \Delta$. Set $\sigma_{w,J} = \epsilon_J - w_{0,J}\epsilon_J + w_{0,J}(w \cdot 0)$ for $w \in {}^JW$. Moreover, let $\gamma \in X^+$ such that all weights of $\widehat{T} := T_J(\sigma_{w,J}) \otimes \mathsf{k}_{\ell\gamma}$ are in X^+. Then there exists an isomorphism of rational G-modules:*
(5.1.2)
$$\mathrm{H}^n(u_\zeta(\mathfrak{g}), \mathrm{ind}_{U_\zeta(\mathfrak{p}_J)}^{U(\mathfrak{g})} \widehat{T}) \cong \begin{cases} \mathrm{ind}_{P_J}^G S^{\frac{n-l(w)}{2}}(\mathfrak{u}_J^*) \otimes \mathsf{k}_\gamma & n - l(w) \equiv 0 \pmod 2, \\ 0 & \textit{otherwise.} \end{cases}$$

PROOF. Let $M = \mathsf{k}$ and $Q_w = T_J(\epsilon_J - w_{0,J}\epsilon_J + w_{0,J}(w \cdot 0))$. Recall that if $\ell > h$ then $w \cdot 0 \in X_{J,\mathrm{res}}$ by [**UGA1**, Prop. 3.6.1]. Therefore, Q_w is a $U_\zeta(\mathfrak{l}_J)$-summand in $L_J(\epsilon_J) \otimes L_J(-w_{0,J}\epsilon_J + w_{0,J}(w \cdot 0))$, and thus Q_w is injective over $u_\zeta(\mathfrak{l}_J)$. The statement of the theorem will follow if we can verify conditions (i) and (ii) of Theorem 3.3.1.

Condition (ii) holds because we have assumed that all weights of $T_J(\epsilon_J - w_{0,J}\epsilon_J + w_{0,J}(w \cdot 0)) \otimes \mathsf{k}_{\ell\gamma}$ are in X^+. Thus, we only need to verify condition (i) which is equivalent to showing that

$$\mathrm{Hom}_{u_\zeta(\mathfrak{l}_J)}(L_J(\nu)^{[1]}, \mathrm{H}^n(\mathcal{U}_\zeta(\mathfrak{u}_J), \mathsf{k}) \otimes Q_w) \cong \begin{cases} \mathsf{k} & n = l(w) \text{ and } \nu = 0, \\ 0 & \text{otherwise.} \end{cases}$$

If this Hom-space is non-zero, by Lemma 4.1.1 there exists $y \in {}^JW$ with $l(y) = n$ such that $\operatorname{Hom}_{u_\zeta(\mathfrak{l}_J)}(L_J(y \cdot 0) \otimes L_J(\nu)^{[1]}, Q_w) \neq 0$. The simple $U_\zeta(\mathfrak{l}_J)$-module $L_J(w \cdot 0)$ appears in the socle (and head) of Q_w. This implies that $y \cdot 0 + \ell\nu$ is linked to $w \cdot 0$ under the action of the affine Weyl group associated to Φ_J. Therefore,

$$y \cdot 0 + \ell\nu = x \cdot (w \cdot 0) + \ell\sigma,$$

where $\sigma \in \mathbb{Z}\Phi_J$ and $x \in W_J$. Rewriting this equation we have

$$y \cdot 0 = (xw) \cdot 0 + \ell(\sigma - \nu).$$

We can use the arguments given in the proof of Theorem 4.2.1 to conclude that $\sigma = \nu$ and $y = xw$ (or see [**DNP**, Lemmas 2.1.1 and 2.1.2]). But y and w are in ${}^JW$ so $y = w$, and $x = \text{id}$.

Now by (5.1.1) and the fact that $L_J(w \cdot 0)$ appears in the $U_\zeta(\mathfrak{l}_J)$-socle and the $u_\zeta(\mathfrak{l}_J)$-socle of $L_J(\epsilon_J) \otimes L_J(-w_{0,J}\epsilon_J + w_{0,J}(w \cdot 0))$ exactly once we can conclude that $L_J(w \cdot 0 + \ell\nu)$ is a composition factor in Q_w only if $\nu = 0$. This verifies condition (i). □

5.2. We will first indicate how Theorem 5.1.1 can be viewed as a quantum analog of the computation of the cohomology of the first Frobenius kernel with coefficients in an induced module. This result was first proved by Andersen and Jantzen in most cases and by Kumar, Lauritzen, and Thomsen in general (cf. [**AJ**, Corollary 3.7(b)] [**KLT**, Theorem 8]).

COROLLARY 5.2.1. *If $\ell > h$ and $w \in W$ such that $w \cdot 0 + \ell\gamma \in X^+$. Then there exists an isomorphism of rational G-modules*
(5.2.1)

$$\mathrm{H}^n(u_\zeta(\mathfrak{g}), \operatorname{ind}_{U_\zeta(\mathfrak{b})}^{U_\zeta(\mathfrak{g})}[\mathsf{k}_{w\cdot 0+\ell\gamma}]) \cong \begin{cases} \operatorname{ind}_B^G S^{\frac{n-l(w)}{2}}(\mathfrak{u}^*) \otimes \mathsf{k}_\gamma & \text{if } n - l(w) \equiv 0 \pmod 2, \\ 0 & \text{otherwise.} \end{cases}$$

PROOF. The statement of the corollary is a special case of Theorem 5.1.1 in the case when $J = \varnothing$. Then $\mathfrak{p}_J = \mathfrak{b}$, $\epsilon_J = 0$, $w_{0,J} = \text{id}$, and $w \in W$. Then $T_J(\epsilon_J - w_{0,J}\epsilon_J + w_{0,J}(w \cdot 0)) = w \cdot 0$. □

6. Connections with cohomology for quantum groups

6.1. In this section we will indicate the strong interplay between the structure of geometric objects in the setting of complex Lie theory and homological information for modules over the quantum group. The first result of this section connects the multiplicities of simple G-modules in the global sections of the twisted cotangent bundle with the cohomology for modules over the large quantum group.

THEOREM 6.1.1. *Let $J \subseteq \Delta$, M be a $U_\zeta(\mathfrak{g})$-module with $Z_J.M = 0$ and Q be a $U_\zeta(\mathfrak{p}_J)$-module such that Q is an injective $u_\zeta(\mathfrak{l}_J)$-module which is trivial as $U_\zeta(\mathfrak{u}_J)$-module. Assume that the following two conditions hold:*

(i) *As a rational B-module,*

$$\operatorname{Hom}_{u_\zeta(\mathfrak{l}_J)}(\mathsf{k}, \mathrm{H}^n(\mathcal{U}_\zeta(\mathfrak{u}_J), M) \otimes Q) \cong \begin{cases} k & \text{if } n = t, \\ 0 & \text{otherwise.} \end{cases} \tag{6.1.1}$$

(ii) *For $\gamma \in X_{P_J} \cap X^+$, $R^i \operatorname{ind}_{U_\zeta(\mathfrak{p}_J)}^{U_\zeta(\mathfrak{g})}(Q \otimes \mathsf{k}_{\ell\gamma}) = 0$ for $i > 0$. Then for $n \geq 0$ and $\sigma \in X^+$, the composition factor multiplicities as G-modules are given by*

$$[\operatorname{ind}_{P_J}^G(S^{\frac{n-t}{2}}(\mathfrak{u}_J^*)\otimes \mathsf{k}_\gamma):L(\sigma)]=\begin{cases}\dim \operatorname{Ext}^{n-t}_{U_\zeta(\mathfrak{g})}(L(\sigma)^{[1]},\widehat{M}) & \text{if } n-t\equiv 0 \pmod 2,\\ 0 & \text{otherwise}\end{cases}$$

where $\widehat{M} = M \otimes \operatorname{ind}_{U_\zeta(\mathfrak{p}_J)}^{U_\zeta(\mathfrak{g})}(Q\otimes \mathsf{k}_{\ell\gamma})$.

PROOF. One can apply the Lyndon-Hochschild-Serre spectral sequence for $u_\zeta(\mathfrak{g})$, a normal Hopf subalgebra in $U_\zeta(\mathfrak{g})$, to obtain:

$$E_2^{i,j} = \operatorname{Ext}^i_G(L(\sigma),\operatorname{Ext}^j_{u_\zeta(\mathfrak{g})}(\mathsf{k},\widehat{M}))\Rightarrow \operatorname{Ext}^{i+j}_{U_\zeta(\mathfrak{g})}(L(\sigma)^{[1]}, M\otimes \operatorname{ind}_{U_\zeta(\mathfrak{p}_J)}^{U_\zeta(\mathfrak{g})}(Q\otimes \mathsf{k}_{\ell\gamma})).$$

The finite-dimensional G-modules are completely reducible so there are no higher extensions between G-modules. This implies that this spectral sequence collapses and yields:

$$\begin{aligned}[\mathrm{H}^\bullet(u_\zeta(\mathfrak{g}),\widehat{M}):L(\sigma)] &= \dim \operatorname{Hom}_G(L(\sigma),\mathrm{H}^\bullet(u_\zeta(\mathfrak{g}),M\otimes \operatorname{ind}_{U_\zeta(\mathfrak{p}_J)}^{U_\zeta(\mathfrak{g})}(Q\otimes \mathsf{k}_{\ell\gamma})))\\ &= \dim \operatorname{Ext}^\bullet_{U_\zeta(\mathfrak{g})}(L(\sigma)^{[1]}, M\otimes \operatorname{ind}_{U_\zeta(\mathfrak{p}_J)}^{U_\zeta(\mathfrak{g})}(Q\otimes \mathsf{k}_{\ell\gamma})).\end{aligned}$$

The result now follows from Theorem 3.3.1. □

We note that the multiplicities $[\operatorname{ind}_{P_J}^G S^\bullet(\mathfrak{u}_J^*)\otimes \mathsf{k}_\gamma : L(\sigma)]$ are given in terms of classical formulas involving values of Kostant's partition function (cf. [**Jan3**, 8.18]). These values can also be interpreted as certain coefficients of Kazhdan-Lusztig polynomials. The theorem above can now be used to give an interpretation of the dimensions of the extension groups

$$\operatorname{Ext}^\bullet_{U_\zeta(\mathfrak{g})}(L(\sigma)^{[1]}, M\otimes \operatorname{ind}_{U_\zeta(\mathfrak{p}_J)}^{U_\zeta(\mathfrak{g})}(Q\otimes \mathsf{k}_{\ell\gamma}))$$

via values on the Kostant's partition function.

6.2. Theorem 6.1.1 can be combined with Theorem 4.2.1 to yield the following result.

COROLLARY 6.2.1. *Let* $\ell > h$, $J\subseteq \Delta$, $\epsilon_J = \sum_{\alpha\in J}(\ell-1)\omega_\alpha$, *and* $M = H^0(\epsilon_J)^*$. *For* $w\in {}^JW$ *and* $\gamma\in X_{P_J}\cap X^+$, *set*

$$N_w(\gamma) := H^0(\epsilon_J)^*\otimes \operatorname{ind}_{U_\zeta(\mathfrak{p}_J)}^{U_\zeta(\mathfrak{g})}(H^0_J(\epsilon_J)\otimes H^0_J(-w_{0,J}w\cdot 0)\otimes \mathsf{k}_{\ell\gamma}).$$

Let Q_w *be as in Section* 4.2, *and assume that*

(a) $\operatorname{Hom}_{u_\zeta(\mathfrak{l}_J)}(\mathsf{k},\mathrm{H}^0(\mathcal{U}_\zeta(\mathfrak{u}_J),M)\otimes Q_w)\cong \mathsf{k}$,
(b) $\epsilon_J - w_{0,J}(w\cdot 0)+\ell\gamma\in X^+$.

Then for $n\geq 0$

$$[\operatorname{ind}_{P_J}^G(S^{\frac{n}{2}}(\mathfrak{u}_J^*)^{[1]}\otimes \mathsf{k}_\gamma):L(\sigma)]=\begin{cases}\dim \operatorname{Ext}^n_{U_\zeta(\mathfrak{g})}(L(\sigma)^{[1]},N_w(\gamma)) & \text{if } w=\mathrm{id} \text{ and}\\ & n\equiv 0 \pmod 2,\\ 0 & \text{otherwise.}\end{cases}$$

As a special case of the result above one sees that the multiplicities of simple G-modules in the graded coordinate algebra of (conical) orbit closures are directly related to the cohomology of quantum groups.

COROLLARY 6.2.2. *Let* $\ell > h$ *and* $\mathcal{O} := \mathcal{O}_J$ *be a Richardson orbit in* $\mathcal{N}$ *with moment map* $\Gamma : G\times_{P_J}\mathfrak{u}_J\to G\cdot \mathfrak{u}_J$. *Set* $\epsilon_J = (\ell-1)\sum_{\alpha\in J}\omega_\alpha$. *Then*

(a) $[\mathsf{k}[G\times_{P_J}\mathfrak{u}_J]_\bullet : L(\sigma)] = \dim \operatorname{Ext}^{2\bullet}_{U_\zeta(\mathfrak{g})}(H^0(\epsilon_J)\otimes L(\sigma)^{[1]}, H^0(\epsilon_J))$.

(b) *Furthermore, if* Γ *is a resolution of singularities and* $\overline{\mathcal{O}}$ *is normal then*

$$[\mathsf{k}[\overline{\mathcal{O}}]_{\bullet} : L(\sigma)] = \dim \operatorname{Ext}^{2\bullet}_{U_\zeta(\mathfrak{g})}(H^0(\epsilon_J) \otimes L(\sigma)^{[1]}, H^0(\epsilon_J))$$

where $k[\overline{\mathcal{O}}]_{\bullet}$ *is the (graded) coordinate algebra of the orbit closure* $\overline{\mathcal{O}}$ *which is a conical variety.*

(c) *In particular* $[\mathsf{k}[\mathcal{N}]_{\bullet} : L(\sigma)] = \dim \operatorname{Ext}^{2\bullet}_{U_\zeta(\mathfrak{g})}(L(\sigma)^{[1]}, \mathsf{k})$.

We can also use Theorem 6.1.1 in conjunction with Theorem 5.1.1 to realize composition factor multiplicities in a shifted version of quantum group cohomology.

COROLLARY 6.2.3. *Let* $\ell > h$, $J \subseteq \Delta$, $w \in {}^J W$ *and* $\epsilon_J = (\ell - 1)\sum_{\alpha \in J} \omega_\alpha$. *Moreover, let* $\gamma \in X^+$ *such that all weights of* $T_J(\epsilon_J - w_{0,J}\epsilon_J + w_{0,J}(w \cdot 0)) \otimes \mathsf{k}_{\ell\gamma}$ *are in* X^+. *Set*

$$N_w = \operatorname{ind}_{U_\zeta(\mathfrak{p}_J)}^{U_\zeta(\mathfrak{g})}(T_J(\epsilon_J - w_{0,J}\epsilon_J + w_{0,J}(w \cdot 0)) \otimes \mathsf{k}_{\ell\gamma}).$$

Then

$$[\widehat{I} : L(\sigma)] = \begin{cases} \dim \operatorname{Ext}^n_{U_\zeta(\mathfrak{g})}(L(\sigma)^{[1]}, N_w(\gamma)) & \text{if } n - l(w) \equiv 0 \pmod 2, \\ 0 & \text{otherwise} \end{cases}$$

where $\widehat{I} = \operatorname{ind}_{P_J}^G(S^{\frac{n-l(w)}{2}}(\mathfrak{u}_J^*) \otimes \mathsf{k}_\gamma)$.

7. Frobenius kernels

7.1. We will demonstrate which results we have proved for quantum groups will hold for Frobenius kernels (under suitable cohomological vanishing conditions). Let G be a reductive algebraic group scheme defined and split over $\mathbb{F}_p$ and k be an algebraically closed field of characteristic $p > 0$. Moreover, in this section, for any algebraic group H defined over $\mathbb{F}_p$, let H_1 be the scheme theoretic kernel of the Frobenius map $\mathrm{Fr} : H \to H$. If $J \subseteq \Delta$, we will use $P_J = L_J \ltimes U_J$ to denote the parabolic subgroup of G corresponding to J with a Levi subgroup L_J and the unipotent radical U_J. All of these subgroups are defined over $\mathbb{F}_p$ and are Fr-stable. As before we use $\mathfrak{p}_J$, $\mathfrak{u}_J$, and $\mathfrak{l}_J$ to denote the corresponding Lie algebras over k. Following [**BNPP**, §7.2] we state the following assumptions on $J \subseteq \Delta$. The first assumption entails the Grauert-Riemenschneider vanishing result:

(A1) $R^i \operatorname{ind}_{P_J}^G(S^\bullet(\mathfrak{u}_J^*) \otimes k_\gamma) = 0$ for $i > 0$ and $\gamma \in X_{P_J} \cap X^+$, and $\operatorname{ind}_{P_J}^G(S^\bullet(\mathfrak{u}_J^*) \otimes k_\gamma)$ has a good filtration.

The second assumption on J is a condition on the normality of $G \cdot \mathfrak{u}_J$.

(A1) The variety $G \cdot \mathfrak{u}_J$ is normal.

We note that (A1) have been shown when p is a good prime and γ satisfies the additional condition that $\langle \gamma, \alpha^\vee \rangle > 0$ for all $\alpha \in \Delta - J$ [**KLT**, Theorem 5]. The condition (*A2*), for example, has been verified when $\Phi = A_n$ (see [**Don**]).

7.2. Next we indicate how one can prove the analogue of Theorem 3.3.1. For Frobenius kernels the proof is simpler because one has natural actions of the Levi subgroup on the exterior algebra and the ordinary Lie algebra cohomology.

We recall that $\mathrm{H}^*(\mathfrak{h}, -)$ is the Lie algebra cohomology while $\mathrm{H}^*(H_1, -) = \operatorname{Ext}^*_{H_1}(k, -)$ for any algebraic group H defined over $\mathbb{F}_p$.

THEOREM 7.2.1. *Let* $p \geq 3$, $J \subseteq \Delta$, M *be a* G*-module, and* Q *be a* P_J*-module such that* Q *is an injective* $(L_J)_1$*-module and trivial as a* U_J*-module. Assume that (A1) and the following two conditions hold:*

(i) *As a rational $P_J/(P_J)_1$-module, and for some $t \geq 0$,*

$$\operatorname{Hom}_{(L_J)_1}(k, \mathrm{H}^n(\mathfrak{u}_J, M) \otimes Q) \cong \begin{cases} k & \text{if } n = t, \\ 0 & \text{otherwise.} \end{cases} \tag{7.2.1}$$

(ii) *For $\gamma \in X_{P_J} \cap X^+$, $R^i \operatorname{ind}_{P_J}^G (Q \otimes k_{p\gamma}) = 0$ for $i > 0$.*

Then there exists an isomorphism of rational G-modules:

(7.2.2)

$$\mathrm{H}^n(G_1, M \otimes \operatorname{ind}_{P_J}^G (Q \otimes p\gamma)) \cong \begin{cases} \operatorname{ind}_{P_J}^G (S^{\frac{n-t}{2}}(\mathfrak{u}_J^*) \otimes k_\gamma)^{[1]} & n - t \equiv 0 \pmod 2, \\ 0 & \text{otherwise.} \end{cases}$$

PROOF. In order to prove the theorem, we apply the LHS spectral sequence to $(U_J)_1$ embedded in $(P_J)_1$ as a normal subgroup scheme with quotient $(L_J)_1$ using the fact that Q is U_J-trivial:

$$E_2^{i,j} = \mathrm{H}^i((L_J)_1, \mathrm{H}^j((U_J)_1, M) \otimes Q) \Rightarrow \mathrm{H}^{i+j}((P_J)_1, M \otimes Q).$$

Since Q is projective over $(L_J)_1$, this spectral sequence collapses and yields:

$$\mathrm{H}^n((P_J)_1, M \otimes Q) \cong \operatorname{Hom}_{(L_J)_1}(k, \mathrm{H}^n((U_J)_1, M) \otimes Q).$$

For $p \geq 3$, there exists a first quadrant spectral sequence [**FP2**, (1.3) Proposition]:

$$E_2^{2i,j} = S^i(\mathfrak{u}_J^*)^{[1]} \otimes \mathrm{H}^j(\mathfrak{u}_J, M \otimes Q) \Rightarrow \mathrm{H}^{2i+j}((U_J)_1, M \otimes Q).$$

Since the functor $\operatorname{Hom}_{(L_J)_1}(k, - \otimes Q)$ is exact, we can compose it with the spectral sequence above and use the fact that $S^*(\mathfrak{u}_J^*)^{[1]}$ is a trivial $(P_J)_1$-module and Q is a trivial module when restricted to U_J to get another spectral sequence:

$$E_2^{2i,j} = S^i(\mathfrak{u}_J^*)^{[1]} \otimes \operatorname{Hom}_{(L_J)_1}(k, \mathrm{H}^j(\mathfrak{u}_J, M) \otimes Q) \Rightarrow \mathrm{H}^{2i+j}((P_J)_1, M \otimes Q). \tag{7.2.3}$$

Condition (i) implies that the E_2 only lives one (horizontal) line, thus the spectral sequence collapses and yields

$$\operatorname{Hom}_{(L_J)_1}(k, \mathrm{H}^n((U_J)_1, M) \otimes Q)) \cong \begin{cases} S^{\frac{n-t}{2}}(\mathfrak{u}_J^*)^{[1]} & \text{if } n \equiv t \pmod 2, \\ 0 & \text{otherwise,} \end{cases}$$

and

$$\mathrm{H}^n((P_J)_1, M \otimes Q) \cong \begin{cases} S^{\frac{n-t}{2}}(\mathfrak{u}_J^*)^{[1]} & \text{if } n - t \equiv 0 \pmod 2, \\ 0 & \text{otherwise.} \end{cases}$$

Now one can apply the same techniques given in the proof of Theorem 3.3.1 and (A1) to yield the desired results. □

7.3. By using Theorem 7.2.1 one can prove Frobenius kernel analogs (using the same weight arguments as given in the quantum case) of Theorems 4.2.1, 4.3.1, and 5.1.1.

THEOREM 7.3.1. *Let $p > 2h - 1$, $J \subseteq \Delta$, $\epsilon_J = \sum_{\alpha \in J}(p-1)\omega_\alpha$, $M = H^0(\epsilon_J)^*$. For $w \in {}^J W$ and $\gamma \in X_{P_J} \cap X^+$ set*

$$N_w(\gamma) := H^0(\epsilon_J)^* \otimes \operatorname{ind}_{P_J}^G (H_J^0(\epsilon_J) \otimes H_J^0(-w_{0,J}w \cdot 0) \otimes k_{p\gamma}).$$

Assume (A1) holds and

(a) $\operatorname{Hom}_{(L_J)_1}(k, \mathrm{H}^0((U_J)_1, M) \otimes Q)) \cong k$,

(b) $\epsilon_J - w_{0,J}(w \cdot 0) + p\gamma \in X^+$.

Then there exists an isomorphism of rational G-modules:
(7.3.1)

$$\mathrm{H}^n(G_1, N_w(\gamma)) \cong \begin{cases} \mathrm{ind}_{P_J}^G(S^{\frac{n}{2}}(\mathfrak{u}_J^*)\otimes k_\gamma)^{[1]} & \text{if } w = \mathrm{id} \text{ and } n \equiv 0 \pmod 2, \\ 0 & \text{otherwise.}\end{cases}$$

In the case when $w = \mathrm{id}$ *for* $N_w(\gamma)$, *the condition on* p *can be replaced by* $p > h$.

This theorem specializes to the case when $\gamma = 0$ and $w = \mathrm{id}$ to answer the question posed by Friedlander and Parshall for G_1 assuming that $(A1)$ and $(A2)$ hold.

THEOREM 7.3.2. *Let* $p > h$, $J \subseteq \Delta$ *and assume that (A1) holds. Let* $\mathcal{O} := \mathcal{O}_J$ *be a Richardson orbit in* $\mathcal{N}$ *with moment map* $\Gamma : G\times_P \mathfrak{u}_J \to G\cdot \mathfrak{u}_J$. *Then there exists a finite-dimensional G-algebra* $A_{\mathcal{O}}$ *such that*

(a) $\mathrm{H}^{2\bullet+1}(G_1, A_{\mathcal{O}}) = 0$;

(b) $\mathrm{H}^{2\bullet}(G_1, A_{\mathcal{O}}) \cong k[\mathcal{O}]^{[1]} \cong k[G\times_{P_J}\mathfrak{u}_J]^{[1]} \cong \mathrm{ind}_{P_J}^G S^\bullet(\mathfrak{u}_J^*)^{[1]}$.

Furthermore, if Γ *is a resolution of singularities and* $(A2)$ *holds then*

$$\mathrm{H}^{2\bullet}(u_\zeta(\mathfrak{g}), A_{\mathcal{O}}) \cong k[\overline{\mathcal{O}}]^{[1]}.$$

In the following theorem, we will use the notation $T_J(\lambda)$ to denote the partial tilting L_J-module of highest weight $\lambda \in X_J^+$.

THEOREM 7.3.3. *Let* $p > h$, $J \subseteq \Delta$, $w \in {}^JW$ *and* $\epsilon_J = (p-1)\sum_{\alpha\in J}\omega_\alpha$. *Assume that (A1) holds. Moreover, let* $\gamma \in X_{P_J}\cap X^+$ *such that all weights of* $T_J(\epsilon_J - w_{0,J}\epsilon_J + w_{0,J}(w\cdot 0))\otimes k_{p\gamma}$ *are in* X^+ *Then there exists an isomorphism of rational G-modules:*
(7.3.2)

$$\mathrm{H}^n(G_1, \mathrm{ind}_{P_J}^G(\widehat{T}\otimes k_{p\gamma})) \cong \begin{cases} \mathrm{ind}_{P_J}^G(S^{\frac{n-l(w)}{2}}(\mathfrak{u}_J^*)\otimes k_\gamma)^{[1]} & n - l(w) \equiv 0 \pmod 2, \\ 0 & \text{otherwise}\end{cases}$$

where $\widehat{T} = T_J(\epsilon_J - w_{0,J}\epsilon_J + w_{0,J}(w\cdot 0))$.

7.4. Corollaries 6.2.1, 6.2.2, and 6.2.3 have analogs in the case of Frobenius kernels via the use of good filtrations. We recall that a G-module E is said to have a good filtration if E has a filtration with sections isomorphism to modules of the form $H^0(\sigma)$ with $\sigma \in X^+$. The number of times $H^0(\sigma)$ appears as the sections is independent of the choice of the good filtration and denoted by $[E : H^0(\sigma)]$. This notation (not to be confused with composition factor multiplicities) will only be used in this section. Let $V(\sigma)$ be the Weyl module of highest weight $\sigma \in X^+$.

COROLLARY 7.4.1. *Let* $p > h$, $J \subseteq \Delta$, $\epsilon_J = \sum_{\alpha\in J}(p-1)\omega_\alpha$, *and* $M = H^0(\epsilon_J)^*$. *For* $w \in {}^JW$ *and* $\gamma \in X_{P_J}\cap X^+$, *set*

$$N_w(\gamma) := H^0(\epsilon_J)^*\otimes \mathrm{ind}_{P_J}^G(H_J^0(\epsilon_J)\otimes H_J^0(-w_{0,J}w\cdot 0)\otimes k_{p\gamma}).$$

Assume that (A1) holds and

(a) $\mathrm{Hom}_{(L_J)_1}(k, \mathrm{H}^0((U_J)_1, M)\otimes Q)) \cong k$,

(b) $\epsilon_J - w_{0,J}(w\cdot 0) + p\gamma \in X^+$.

Then

$$[\widehat{I} : H^0(\sigma)] = \begin{cases} \dim \mathrm{Ext}_G^n(V(\sigma)^{[1]}, N_w(\gamma)) & \text{if } w = \mathrm{id} \text{ and } n \equiv 0 \pmod 2, \\ 0 & \text{otherwise}\end{cases}$$

where $\widehat{I} = \operatorname{ind}_{P_J}^G(S^{\frac{n}{2}}(\mathfrak{u}_J^*)^{[1]} \otimes k_\gamma)$.

COROLLARY 7.4.2. *Assume that (A1) holds. Let* $p > h$ *and* $\mathcal{O} := \mathcal{O}_J$ *be a Richardson orbit in* $\mathcal{N}$ *with moment map* $\Gamma : G \times_{P_J} \mathfrak{u}_J \to G \cdot \mathfrak{u}_J$. *Set* $\epsilon_J = (p-1)\sum_{\alpha \in J} \omega_\alpha$. *Then*

(a) $[k[G \times_{P_J} \mathfrak{u}_J] : H^0(\sigma)] = \dim \operatorname{Ext}_G^{2\bullet}(V(\sigma)^{[1]}, H^0(\epsilon_J)^* \otimes H^0(\epsilon_J))$.

(b) *Furthermore, if* Γ *is a resolution of singularities and (A2) holds then*

$$[k[\overline{\mathcal{O}}]_\bullet : H^0(\sigma)] = \dim \operatorname{Ext}_G^{2\bullet}(V(\sigma)^{[1]}, H^0(\epsilon_J)^* \otimes H^0(\epsilon_J)).$$

In particular when $J = \varnothing$, *(A1) and (A2) are satisfied, and the moment map is a resolution of singularities, then*

$$[k[\mathcal{N}]_\bullet : H^0(\sigma)] = \dim \operatorname{Ext}_G^{2\bullet}(V(\sigma)^{[1]}, k).$$

COROLLARY 7.4.3. *Let* $p > h$, $J \subseteq \Delta$, $w \in {}^JW$ *and* $\epsilon_J = (p-1)\sum_{\alpha \in J} \omega_\alpha$. *Assume that (A1) holds. Moreover, let* $\gamma \in X^+$ *such that all weights of* $T_J(\epsilon_J - w_{0,J}\epsilon_J + w_{0,J}(w \cdot 0)) \otimes k_{p\gamma}$ *are in* X^+. *Set*

$$N_w = \operatorname{ind}_{U_\zeta(\mathfrak{p}_J)}^{U(\mathfrak{g})}(T_J(\epsilon_J - w_{0,J}\epsilon_J + w_{0,J}(w \cdot 0)) \otimes k_{p\gamma}).$$

Then

$$[\widehat{I} : H^0(\sigma)] = \begin{cases} \dim \operatorname{Ext}_G^n(V(\sigma)^{[1]}, N_w) & \text{if } n - l(w) \equiv 0 \pmod 2, \\ 0 & \text{otherwise} \end{cases}$$

where $\widehat{I} = \operatorname{ind}_{P_J}^G(S^{\frac{n-l(w)}{2}}(\mathfrak{u}_J^*) \otimes k_\gamma)$.

8. Appendix: Hopf algebra actions on cohomology

8.1. Let H be an arbitrary Hopf algebra over a commutative ring k. As in [**Lin1**, 4.1] and [**LN**], the left adjoint action of H on itself is defined by

$$h \cdot x = \sum_{(h)} h_{(1)} x S(h_{(2)})$$

for all $h \in H$ and $x \in R$ with $\Delta(h) = \sum_{(h)} h_{(1)} \otimes h_{(2)}$ (using the Sweedler notation). Here S is the antipode. It is easily checked that

$$h \cdot (x_1 x_2) = \sum_{(h)} (h_{(1)} \cdot x_1)(h_{(2)} \cdot x_2)$$

making H into an H-module algebra. Any (unital) k-subalgebra $D \subseteq H$ stable under this action is a normal subalgebra of H regarded as an augmented (supplemented) algebra in sense of [**CE**] with the augmentation $\epsilon : D \to R$ being the restriction of the counit of H. One can similarly define the right adjoint action making H as a right H-module.

8.2. In [**LN**], a Hopf algebra action on cohomology of a Hopf algebra is defined. More general Hopf algebra actions on cohomology of augmented algebras were given in [**BNPP**]. We outline the setup here for later use in this section (and in this paper). For simplicity, all algebras in the rest of this section are over an arbitrary field k. Let H be a Hopf algebra and A be an augmented algebra with a *right* H-module structure, i.e., A is a right H-module such that both the multiplication $A \otimes A \to A$, the augmentation $\epsilon : A \to \mathsf{k}$, and the unit $u : \mathsf{k} \to A$ are H-module homomorphisms. In such case, we call A a right H-module augmented algebra.

This is equivalent that A is an augmented algebra object in the tensor category of right H-modules. One forms the smash product algebra $H\#A$ which is the vector space $H \otimes A$ with multiplication given by

$$(h'\#a)(h\#b) = \sum_{(h)} h'h_{(1)}\#(a \cdot h_{(2)})b.$$

For a left H-module augmented algebra A, one can similarly define $A\#H$. If A is a Hopf algebra with the comultiplication $A \to A \otimes A$ being an H-module homomorphism and H is cocommutative, then $H\#A$ is a Hopf algebra containing both A and H as Hopf subalgebras with A being normal. For a general augmented algebra A, $H\#A$ is an augmented algebra and A is a normal subalgebra [**ABG**, 2.1] such that the quotient $(H\#A) /\!\!/ A := (H\#A)/\langle A^+\rangle \cong H$ as augmented algebras. Note that $(H\#A) /\!\!/ A \cong (H\#A) \otimes_A \mathsf{k}$ as augmented algebras.

A left A-module X is said to have a compatible left H-module action if it extends to a left $H\#A$-module. However, the condition is slightly different from the right module structure

$$a(h(x)) = \sum_{(h)} h_{(1)}((a \cdot h_{(2)})x).$$

This is equivalent to M being a left $H\#A$-module. Similarly, one can define a right A-module X with a compatible right H-module action. This is equivalent to X being a right $H\#A$-module. One notes that right $H\#A$-modules are just the right module objects for the algebra object A in the tensor category of right H-modules. One can also define a left A module X with a compatible right H-module structure. This is equivalent to X being a left module object of the algebra object A in the tensor category of right H-modules.

If A and A' are two right H-module algebras, we say that an A'-A-bimodule X has a compatible right H-module structure if

$$(a'xa) \cdot h = \sum_{(h)} (a' \cdot h_{(1)})(x \cdot h_{(2)})(a \cdot h_{(3)}).$$

This is equivalent to X being a bimodule object of the two algebra objects A' and A in the tensor category of H-modules. In case A has a trivial right H-module structure, $H\#A = H \otimes A$ is simply the tensor product algebra. On any left (or right) $H\#A$-module X, the actions of H and A commute.

Let X be a left A-module with a compatible right H-module structure, i.e., the structure map $A \otimes X \to X$ is a homomorphism of H-modules. This means that X is a left A-module in the tensor category of right H-modules. If Y is an A'-A-bimodule with a compatible right H-module structure (i.e., $A' \otimes Y \otimes A \to Y$ is a homomorphism of right H-modules, then there is a right H-module structure on $Y \otimes_A X$ given via the tensor product of right H-modules. This setting is equivalent to defining the tensor product over an algebra object A of a left A-module object X and a right A-module Y in the tensor category of right H-modules. In particular, $Y \otimes_A X$ also has a left A'-module structure compatible with the right H-module structure.

We now consider the case $B = H\#A$. Then $Y = B$ is a B-A-bimodule with left and right multiplications in B with A being a subalgebra of B. The Hopf algebra

H also acts on Y by right multiplication. Thus,

$$\begin{aligned}((h'\#a')a)h &= (h'\#(a'a))h \\ &= \sum_{(h)} h'h_{(1)}\#((a'a)\cdot h_{(2)}) \\ &= \sum_{(h)} h'h_{(1)}\#((a'\cdot h_{(2)})(a\cdot h_{(3)}) \\ &= \sum_{(h)} (h'\#a')h_{(1)})(1\#(a\cdot h_{(2)})) \\ &= \sum_{(h)} (h'\#a')h_{(1)})(a\cdot h_{(2)}).\end{aligned}$$

We now consider the standard Hochschild complex $S_n(A) = A\otimes A^{\otimes n}\otimes A$ of A-A-bimodules [**CE**, IX.6]. For any left A-module N, the complex of left A-modules $C_\bullet(N) = S_\bullet(A)\otimes_A N$ is an A-projective resolution of the left A-module N (using [**CE**, IX.6 (2)]).

The Hopf algebra H acts on $S_n(A)$ from the right as the tensor product of right H-modules making $A\otimes A^{\otimes n}\otimes A$ compatible with the A-A-bimodule structure. By writing down the differentials $d_n : S_n(A)\to S_{n-1}(A)$ in the standard Hochschild complex, one notices that each $h\in H$ commutes with the differential d_n. By applying the functor $-\otimes_A \mathsf{k}$, we get an A-projective resolution $C_\bullet(\mathsf{k}) = A\otimes A^{\otimes\bullet}\to \mathsf{k}$ of the trivial A-module k (assuming A is an augmented algebra). The H-module structure on $C_n(\mathsf{k})$ is still given by the tensor product structure.

Applying the exact functor $B\otimes_A -$ to the complex $C_\bullet(\mathsf{k})$, one gets a complex of B-modules $D_\bullet = B\otimes_A C_\bullet(\mathsf{k}) = B\otimes A^{\otimes n}$ which is a B-projective resolution of $B\otimes_A \mathsf{k}$.

The right H-module structure on $C_n(\mathsf{k})$ does not commute with the left A-action. With H acting on B via right multiplication (identifying H with the subalgebra $H\#1$ of B), the tensor product H-module structure on $B\otimes C_n(\mathsf{k})$ is A-balanced, i.e., the quotient map $\pi : B\otimes C_n(\mathsf{k})\to B\otimes_A C_n(\mathsf{k})$ satisfy the following

$$\begin{aligned}\pi((b\otimes ac)\cdot h) &= \pi(\sum_{(h)}(bh_{(1)})\otimes((ac)\cdot h)) \\ &= \pi(\sum_{(h)}(bh_{(1)})\otimes((a\cdot h_{(2)})(c\cdot h_{(3)}))) \\ &= \pi(\sum_{(h)}(bh_{(1)}(a\cdot h_{(2)}))\otimes(c\cdot h)) \\ &= \pi(\sum_{(h)}((ba)\cdot h_{(1)}))\otimes(c\cdot h_{(2)})) \\ &= \pi(((ba)\otimes c)\cdot h).\end{aligned}$$

Hence, the H-right module structure on the tensor product $B\otimes_A C_n(\mathsf{k})$ is well defined and commutes with the left B-module structure via left multiplication. Then H acts on each term of D_n from right commuting with the left B-module structure and compatible with the right A-module structure. Moreover, the action of $h\in H$ commutes with the differentials, and one has a B-projective resolution

$$D_\bullet\to B\otimes_A \mathsf{k}.$$

With k being a trivial H-module, H acts on each term of the complex and commutes with the differentials and with the left B-module structure. Note that the right action of H on $B \otimes_A \mathsf{k}$ is given by

$$(b \otimes_A 1) \cdot h = \sum_{(h)} (b \cdot h_{(1)}) \otimes_A (1 \cdot h_{(2)}) = \sum_{(h)} (bh_{(1)} \otimes_A (\epsilon(h_{(2)})1) = bh \otimes_A 1$$

such that each $h \in H$ acts on the B-projective resolution $D_\bullet \to B \otimes_A \mathsf{k}$ as chain map of left B-modules. In particular, for any left B-module M, each $h \in H$ acts on the complex $\mathrm{Hom}_B(D_n, M)$ from left given by

$$(h\phi)(x) = \phi(x \cdot h), \quad \forall h \in H,\ x \in D_n.$$

This defines a left action of H on $\mathrm{Ext}^n_B(B \otimes_A \mathsf{k}, M)$ making it a left H-module. We claim that this action of H is exactly the left action of $B /\!\!/ A$-action on $\mathrm{Ext}^n_B(B /\!\!/ A, M)$ as described in [**CE**, XVI Thm 6.1]. Each $h \in H$ defines a left B-module homomorphism $B /\!\!/ A \to B /\!\!/ A$ by $(b \otimes_A 1) \mapsto (b \otimes_A 1) \cdot h = bh \otimes_A 1$. For any B-projective resolution $P_\bullet \to B \otimes_A \mathsf{k}$, there is a lift $\tilde{h}_\bullet : P_\bullet \to P_\bullet$ as a chain map of B-module, which defines the map $h : \mathrm{Ext}^\bullet_B(B /\!\!/ A, M) \to \mathrm{Ext}^\bullet_B(B /\!\!/ A, M)$, which is independent of the choice of the projective resolution and the lift $\tilde{h}_\bullet$ by using the Comparison Theorem for Resolutions [**W**]. The h action on the complex $B \otimes_A C_\bullet(\mathsf{k})$ from the right H-module is such a lift and thus defines the same action on $\mathrm{Ext}^\bullet_B(B /\!\!/ A, M)$.

On the other hand, we have the natural isomorphism of H-module homomorphism

$$\mathrm{Hom}_B(B \otimes_A C_n(\mathsf{k}), M) \overset{\rho}{\cong} \mathrm{Hom}_A(C_n(\mathsf{k}), M) = \mathrm{Hom}_k(A^{\otimes n}, M)$$

such that

$$\begin{aligned}\rho(h\phi)(x) &= \phi((1 \otimes_A x)h) \\ &= \sum_{(h)} \phi(h_{(1)} \otimes_A (x \cdot h_{(2)})) \\ &= \sum_{(h)} h_{(1)}\phi(1 \otimes_A (x \cdot h_{(2)})) \\ &= \sum_{(h)} h_{(1)}(\rho(\phi)(x \cdot h_{(2)})).\end{aligned}$$

The right hand side is exactly the H-action described in [**BNPP**, 2.8] on the reduced bar resolution. This verifies the claim in [**BNPP**] that the H-action using the reduced bar resolution on $\mathrm{Ext}^\bullet_B(B /\!\!/ A, M)$ is the same as $H = B /\!\!/ A$ action described in [**CE**, XVI Thm 6.1].

The advantage of using H-action on the bar resolution (or standard) resolution of A is that the Hopf algebra action of H respects the cup product in cohomology. We summarize our observations as follows.

THEOREM 8.2.1. *Let H be a Hopf algebra and A be an augmented algebra which is a right H-module algebra over a field k. Then for any left A-module M with a compatible left H-module structure there is a natural (functorial in M) left H action on $\mathrm{H}^*(A, M)$. Furthermore, under the conditions that A is a Hopf algebra which is also a right H-module algebra such that the co-multiplication of A is also a homomorphism of H-modules and M is a left A-module algebra extending to a* *left $H\#A$-module. Then under the cup product* [**CE**, XI.4], $\mathrm{H}^*(A, M)$ *a graded left H-module algebra.*

8.3. Actions by subalgebras of Hopf algebras on cohomology. We now apply the above Hopf algebra actions on cohomology groups to cases which are not Hopf algebras. Let P be a subalgebra of a Hopf algebra H. Then P is an augmented algebra and the restriction to P of the H-action on A, makes A into a P-module (although one cannot technically call it a P-module algebra because P is not necessarily a Hopf algebra). Then P also acts on the cohomology $\mathrm{Ext}^n_A(k, M)$ for any left A-module M with a compatible left H-module structure.

Note that H is a right H-module algebra under the right adjoint action defined by

$$h' \cdot h = \sum_{(h)} S(h_{(1)}) h' h_{(2)}.$$

In this case, we have [**Lin2**, Prop. 4.2]

$$h'h = \sum_{(h)} h_{(1)}(h' \cdot h_{(2)}). \tag{8.3.1}$$

If A is a subalgebra of H and is H-stable under the right adjoint action, then A is a normal subalgebra of H as an augmented algebra. If M is a left H-module, then the restriction to A makes M a left A-module which is compatible with H-action since

$$a(h(m)) = \sum_{(h)} h_{(1)}((a \cdot h_{(2)})m)$$

by using (8.3.1). In this case, we can say more about the H-action on $\mathrm{H}^n(A, M)$. In the construction of the H-action on $\mathrm{H}^n(A, M)$, instead of forming the algebra $B = H\#A$ which includes A as a subalgebra and applying the functor $B \otimes_A -$ to the complex $C_\bullet(\mathsf{k})$, we simply replace B by H and apply the functor $H \otimes_A -$ again with the right H-module structure on H by right multiplication. Assuming that H is flat as a right A-module, every computation in Section 8.2 still applies. In this case, $H \otimes \mathsf{k} = H /\!/ A$. Thus we get an H-action on $\mathrm{Ext}^n_A(\mathsf{k}, M) = \mathrm{Ext}^n_H(H \otimes_A \mathsf{k}, M)$ which is the same as the H-action described in Section 8.2 via the chain map $\phi \otimes_A 1 : (H\#A) \otimes_A C_n(\mathsf{k}) \to H \otimes_A C_n(\mathsf{k})$ with $\phi : H\#A \to H$ defined by $\phi(h\#a) = ha \in H$. Note that ϕ is a homomorphism of k-algebras, in particular, it is a right A-module homomorphism. On the other hand, the H-action on $\mathrm{Ext}^n_H(H /\!/ A, M)$ defined in [**CE**] is obtained by lifting the right multiplication of H on $H /\!/ A$ to an H-projective resolution of $H /\!/ A$. In this case, we consider the standard complex $S_\bullet(H /\!/ A)$ (but with a change of the differentials to $(-1)^n d_n$ and the homotopy map $s'_n : a_0 \otimes \cdots \otimes a_{n+1} \mapsto a_0 \otimes \cdots \otimes a_{n+1} \otimes 1$. Then consider the complex $H \otimes_H S_\bullet(H /\!/ A) \to H \otimes_H H /\!/ A$, which is an H-projective resolution of the H-module $H /\!/ A$. The right $H /\!/ A$-module structure (also an H-module structure) on the resolution defines an H-action on $\mathrm{Ext}^\bullet_H(H /\!/ A, M)$ which identifies with the H-action on $\mathrm{Ext}^\bullet_H(H /\!/ A, M) \cong \mathrm{Ext}^\bullet_A(\mathsf{k}, M)$. Hence, the restriction of the H-module structure on $\mathrm{Ext}^\bullet_A(\mathsf{k}, M)$ to the subalgebra $A \subseteq H$ is trivial.

PROPOSITION 8.3.1. *If A is a subalgebra of H which is stable under the right adjoint H-action and H is a right flat A-module, then the subalgebra A action on $\mathrm{H}^\bullet(A, M)$ is trivial when considered as restriction from the action of H.*

If further $A \subseteq B$ are subalgebras of H and both are invariant under the right adjoint action of H, then A is a normal subalgebra of B. In this case, $B \mathbin{/\!/} A$ is also a right H-module algebra. Then for any left H-module M, there are H-actions on each term of the spectral sequence described in [**BNPP**, Lemma 2.8.1]

$$E_2^{i,j} = \mathrm{H}^i(B \mathbin{/\!/} A, H^j(A, M)) \Rightarrow \mathrm{H}^{i+j}(B, M),$$

where H-modules and the differentials are H-module homomorphisms.

If A is a Noetherian central subalgebra of H then the right H-adjoint action on A is trivial. Hence, the action of H on any projective A-module P is trivial and on $\mathrm{Hom}_{\mathsf{k}}(P, M)$ is given by $(h \cdot f)(x) = h \cdot (f(x))$ for all $f \in \mathrm{Hom}_{\mathsf{k}}(P, M)$ and $x \in P$. In particular, if M is finitely generated and projective over k then

$$\mathrm{Hom}_{\mathsf{k}}(P, M) \cong M \otimes \mathrm{Hom}_{\mathsf{k}}(P, \mathsf{k}) \tag{8.3.2}$$

as H-modules. If M is trivial as A-module and P is finitely generated and projective over A, then one has an isomorphism of H-modules

$$\mathrm{Hom}_A(P, M) \cong M \otimes \mathrm{Hom}_A(P, \mathsf{k})$$

for all H-modules which is a direct limit of submodules which are finitely generated and projective over k.

PROPOSITION 8.3.2. *Let k be field. If A is a finitely generated central subalgebra of a Hopf algebra H and M is a locally finite left H-module such that its restriction to A is trivial, then one has an H-module isomorphism*

$$\mathrm{Ext}^n_A(\mathsf{k}, M) \cong M \otimes \mathrm{Ext}^n_A(\mathsf{k}, \mathsf{k}). \tag{8.3.3}$$

Let $N \subseteq A \subseteq B \subseteq H$ be subalgebras which are stable under the right adjoint action of H on H. Then H acts on $A' = A \mathbin{/\!/} N \subseteq B' = B \mathbin{/\!/} N$. Then H acts on each of the term of the spectral sequence

$$E_2^{i,j} = \mathrm{H}^i(B' \mathbin{/\!/} A', \mathrm{H}^j(A', M)) \Rightarrow H^{i+j}(B', M)$$

for each H-module M whose restriction to N is trivial. If A is a central subalgebra of H, then H acts trivially on both N and A, and $H \mathbin{/\!/} N$ also acts on $A' = A \mathbin{/\!/} N$ trivially. Furthermore, if M is A-trivial then

$$\mathrm{Ext}^n_{A'}(\mathsf{k}, M) \cong M \otimes \mathrm{Ext}^n_{A'}(\mathsf{k}, \mathsf{k}) \tag{8.3.4}$$

is an isomorphism of $H \mathbin{/\!/} N$-module. In particular it is an isomorphism of B'-module. Therefore, the action of $H \mathbin{/\!/} N$ on the tensor product module arises from the H-action on the tensor product.

8.4. Module isomorphisms for parabolic subgroups. Let G be a connected algebraic group over a field k of any characteristic. Moreover, let G-mod be the category of finite dimensional rational G-modules. Let $B \subseteq G$ be a Borel subgroup. Then G/B is a complete algebraic k-variety. Thus the global sections of the trivial line bundle is k. In particular the induced representation $\mathrm{ind}_B^G(\mathsf{k}) \cong \mathsf{k}$. Let M be any rational G-module. Then we have $\mathrm{ind}_B^G(M) \cong M \otimes \mathrm{ind}_B^G(\mathsf{k}) \cong M$. If $\mathcal{C}$ is any category, we use $\mathrm{Gpd}(\mathcal{C})$ to denote the groupoid of $\mathcal{C}$, which has the same objects but taking isomorphisms only.

PROPOSITION 8.4.1. *For any connected algebraic k-group G and B be a Borel subgroup. If M and M' are two rational G-modules, then $M \cong M'$ in the category* *of G-modules if and only if* $\operatorname{res}_B^G(M) \cong \operatorname{res}_B^G(M')$ *in the category of rational B-modules. In particular, the restriction functor* $\operatorname{res}_B^G : G\text{-mod} \to B\text{-mod}$, *which is not full in general, defines a fully faithful functor of groupoids* $\operatorname{Gpd}(G\text{-mod}) \to \operatorname{Gpd}(B\text{-mod})$.

PROOF. If $\phi : M \to M'$ is a B-module isomorphism, then $\operatorname{Ind}_B^G(\phi) : \operatorname{Ind}_B^G(M) \to \operatorname{Ind}_B^G(M')$ is an isomorphism of G-modules. Since both M and M' are G-modules, the tensor identity implies that $\operatorname{Ind}_B^G(M) \cong \operatorname{Ind}_B^G(\mathsf{k}) \otimes M \cong M$ and $\operatorname{Ind}_B^G(M') \cong M'$. Here we used the tensor identity [**Jan1**, I. 3.6] and the fact that $\operatorname{Ind}_B^G(\mathsf{k}) = \mathsf{k}$. □

This can be extended to representations of Hopf algebras. Assume that H is a Hopf algebra over k and let $\mathcal{C}(H)$ be a full tensor subcategory of H-modules which is closed under tensor product (defined by a topology as described in [**Lin1**]). If D is a Hopf subalgebra of H, consider the full subcategory $\mathcal{C}(D)$ of D-modules such that the restriction functor $\operatorname{res}_D^H : \mathcal{C}(H) \to \mathcal{C}(D)$ has a left adjoint functor $\operatorname{ind}_D^H : \mathcal{C}(D) \to \mathcal{C}(H)$. The argument in [**Lin1**] implies that tensor identity holds if H has a bijective antipode. In fact this follows from the adjointness of functors by using the following:

(1) $\operatorname{Hom}_H(M, N) = \operatorname{Hom}_{\mathsf{k}}(M, N)^H$;

(2) $\operatorname{Hom}_H(M \otimes P^*, N) \cong \operatorname{Hom}_H(M, N \otimes P)$ and $\operatorname{Hom}_H({}^*P \otimes M, N) \cong \operatorname{Hom}_H(M, P \otimes N)$ for all H-modules P which are finitely generated and projective over k. Here the H-module structures on ${}^*P = \operatorname{Hom}_{\mathsf{k}}(P, \mathsf{k})$ (resp. $P^* = \operatorname{Hom}_{\mathsf{k}}(P, \mathsf{k})$) are defined by $(hf)(p) = f(S(h)p)$ (resp. $(hf)(p) = f(S^{-1}(h)p)$), where M, N can be any H-modules.

(3) ind_D^H commutes with direct limits and every object in the category $\mathcal{C}(H)$ is a direct limit of objects that are finitely generated over k. The tensor identity can be stated in the following two ways: for any D-module N and H-module V in $\mathcal{C}(H)$,

$$\operatorname{ind}_D^H(N \otimes V) \cong \operatorname{ind}_D^H(N) \otimes V \quad \text{and} \quad \operatorname{ind}_D^H(V \otimes N) \cong V \otimes \operatorname{ind}_D^H(N).$$

This is a generalization of [**Jan1**, I. 6.13] from groups to Hopf algebras. The proof of [**Lin1**, Prop. 3.7] reduces to continuous dual Hopf algebra and realizes ind_D^H as cotensor product of comodules and uses [**PW**].

PROPOSITION 8.4.2. *Let* k *be a field. Assume that* H *is a Hopf algebra with bijective antipode and the category* $\mathcal{C}(H)$ *has the property that every object is locally finite. If* D *is a Hopf subalgebra of* H *such that* $\operatorname{ind}_D^H(\mathsf{k}) \cong \mathsf{k}$, *then any two modules* $V_1, V_2, \in \mathcal{C}(H)$ *are isomorphic if and only if* $\operatorname{res}_D^H(V_1) \cong \operatorname{res}_D^H(V_2)$ *as D-modules.*

PROOF. Clearly, if V_1 is isomorphic to V_2 over H then these modules must be isomorphic over D. Suppose that V_1 and V_2 are isomorphic when restricted to D. Then by the tensor identity and the fact that $\operatorname{ind}_D^H(\mathsf{k}) \cong \mathsf{k}$, one has that these modules must be isomorphic over H. □

The proposition holds for various Hopf algebras such as Lusztig's divided power quantum groups and their Borel subalgebras. The tensor identity holds and the induced representation of the trivial module from a Borel subalgebra remains trivial (cf. [**APW**]).

Acknowledgments

The authors dedicate this paper to Georgia Benkart and Brian Parshall for their many contributions in the area of Lie and representation theory. The authors (along with Jens C. Jantzen) were honored to serve with Georgia and Brian on the organizing committee for the 2004 AMS Summer Research Conference in Snowbird, Utah, celebrating James E. Humphreys' 65th birthday. Unfortunately, he also passed away during the pandemic in 2020. The authors thank the referee for a careful reading of the paper.

References

[AJ] Henning Haahr Andersen and Jens Carsten Jantzen, *Cohomology of induced representations for algebraic groups*, Math. Ann. **269** (1984), no. 4, 487–525, DOI 10.1007/BF01450762. MR766011

[APW] Henning Haahr Andersen, Patrick Polo, and Ke Xin Wen, *Representations of quantum algebras*, Invent. Math. **104** (1991), no. 1, 1–59, DOI 10.1007/BF01245066. MR1094046

[ABG] Sergey Arkhipov, Roman Bezrukavnikov, and Victor Ginzburg, *Quantum groups, the loop Grassmannian, and the Springer resolution*, J. Amer. Math. Soc. **17** (2004), no. 3, 595–678, DOI 10.1090/S0894-0347-04-00454-0. MR2053952

[AMRW] Pramod N. Achar, Shotaro Makisumi, Simon Riche, and Geordie Williamson, *Koszul duality for Kac-Moody groups and characters of tilting modules*, J. Amer. Math. Soc. **32** (2019), no. 1, 261–310, DOI 10.1090/jams/905. MR3868004

[BNPP] Christopher P. Bendel, Daniel K. Nakano, Brian J. Parshall, and Cornelius Pillen, *Cohomology for quantum groups via the geometry of the nullcone*, Mem. Amer. Math. Soc. **229** (2014), no. 1077, x+93. MR3204911

[Be] Roman Bezrukavnikov, *Cohomology of tilting modules over quantum groups and t-structures on derived categories of coherent sheaves*, Invent. Math. **166** (2006), no. 2, 327–357, DOI 10.1007/s00222-006-0514-z. MR2249802

[Br1] Abraham Broer, *Normal nilpotent varieties in F_4*, J. Algebra **207** (1998), no. 2, 427–448, DOI 10.1006/jabr.1998.7458. MR1644246

[Br2] Bram Broer, *Normality of some nilpotent varieties and cohomology of line bundles on the cotangent bundle of the flag variety*, Lie theory and geometry, Progr. Math., vol. 123, Birkhäuser Boston, Boston, MA, 1994, pp. 1–19, DOI 10.1007/978-1-4612-0261-5_1. MR1327529

[CLNP] Jon F. Carlson, Zongzhu Lin, Daniel K. Nakano, and Brian J. Parshall, *The restricted nullcone*, Combinatorial and geometric representation theory (Seoul, 2001), Contemp. Math., vol. 325, Amer. Math. Soc., Providence, RI, 2003, pp. 51–75, DOI 10.1090/conm/325/05664. MR1988985

[CE] Henri Cartan and Samuel Eilenberg, *Homological algebra*, Princeton University Press, Princeton, NJ, 1956. MR77480

[Ca] Roger W. Carter, *Finite groups of Lie type*, Pure and Applied Mathematics (New York), John Wiley & Sons, Inc., New York, 1985. Conjugacy classes and complex characters; A Wiley-Interscience Publication. MR794307

[C] A. L. Christophersen, *A Classification of the Normal Nilpotent Varieties for Groups of Type E_6*, Ph.D. Thesis, University of Aarhus, 2006.

[CM] David H. Collingwood and William M. McGovern, *Nilpotent orbits in semisimple Lie algebras*, Van Nostrand Reinhold Mathematics Series, Van Nostrand Reinhold Co., New York, 1993. MR1251060

[DK] Corrado De Concini and Victor G. Kac, *Representations of quantum groups at roots of 1*, Operator algebras, unitary representations, enveloping algebras, and invariant theory (Paris, 1989), Progr. Math., vol. 92, Birkhäuser Boston, Boston, MA, 1990, pp. 471–506. MR1103601

[Don] Stephen Donkin, *The normality of closures of conjugacy classes of matrices*, Invent. Math. **101** (1990), no. 3, 717–736, DOI 10.1007/BF01231523. MR1062803

[DNP] Christopher M. Drupieski, Daniel K. Nakano, and Nham V. Ngo, *Cohomology for infinitesimal unipotent algebraic and quantum groups*, Transform. Groups **17** (2012), no. 2, 393–416, DOI 10.1007/s00031-012-9181-x. MR2921071

[FP1] Eric M. Friedlander and Brian J. Parshall, *Support varieties for restricted Lie algebras*, Invent. Math. **86** (1986), no. 3, 553–562, DOI 10.1007/BF01389268. MR860682

[FP2] Eric M. Friedlander and Brian J. Parshall, *Geometry of p-unipotent Lie algebras*, J. Algebra **109** (1987), no. 1, 25–45, DOI 10.1016/0021-8693(87)90161-X. MR898334

[GK] Victor Ginzburg and Shrawan Kumar, *Cohomology of quantum groups at roots of unity*, Duke Math. J. **69** (1993), no. 1, 179–198, DOI 10.1215/S0012-7094-93-06909-8. MR1201697

[Jan1] Jens Carsten Jantzen, *Representations of algebraic groups*, 2nd ed., Mathematical Surveys and Monographs, vol. 107, American Mathematical Society, Providence, RI, 2003. MR2015057

[Jan2] Jens Carsten Jantzen, *Lectures on quantum groups*, Graduate Studies in Mathematics, vol. 6, American Mathematical Society, Providence, RI, 1996, DOI 10.1090/gsm/006. MR1359532

[Jan3] Jens Carsten Jantzen, *Nilpotent orbits in representation theory*, Lie theory, Progr. Math., vol. 228, Birkhäuser Boston, Boston, MA, 2004, pp. 1–211. MR2042689

[Kr] Hanspeter Kraft, *Parametrisierung von Konjugationsklassen in* $\mathfrak{sl}_n$ (German), Math. Ann. **234** (1978), no. 3, 209–220, DOI 10.1007/BF01420644. MR491855

[KP] Hanspeter Kraft and Claudio Procesi, *On the geometry of conjugacy classes in classical groups*, Comment. Math. Helv. **57** (1982), no. 4, 539–602, DOI 10.1007/BF02565876. MR694606

[KLT] Shrawan Kumar, Niels Lauritzen, and Jesper Funch Thomsen, *Frobenius splitting of cotangent bundles of flag varieties*, Invent. Math. **136** (1999), no. 3, 603–621, DOI 10.1007/s002220050320. MR1695207

[Lin1] Zong Zhu Lin, *Induced representations of Hopf algebras: applications to quantum groups at roots of* 1, J. Algebra **154** (1993), no. 1, 152–187, DOI 10.1006/jabr.1993.1010. MR1201918

[Lin2] Zong Zhu Lin, *A Mackey decomposition theorem and cohomology for quantum groups at roots of* 1, J. Algebra **166** (1994), no. 1, 100–129, DOI 10.1006/jabr.1994.1143. MR1276819

[LN] Zongzhu Lin and Daniel K. Nakano, *Algebraic group actions in the cohomology theory of Lie algebras of Cartan type*, J. Algebra **179** (1996), no. 3, 852–888, DOI 10.1006/jabr.1996.0040. MR1371747

[Lu1] George Lusztig, *Quantum groups at roots of* 1, Geom. Dedicata **35** (1990), no. 1-3, 89–113, DOI 10.1007/BF00147341. MR1066560

[Lu2] George Lusztig, *Finite-dimensional Hopf algebras arising from quantized universal enveloping algebra*, J. Amer. Math. Soc. **3** (1990), no. 1, 257–296, DOI 10.2307/1990988. MR1013053

[NPV] Daniel K. Nakano, Brian J. Parshall, and David C. Vella, *Support varieties for algebraic groups*, J. Reine Angew. Math. **547** (2002), 15–49, DOI 10.1515/crll.2002.049. MR1900135

[NT] Daniel K. Nakano and Toshiyuki Tanisaki, *On the realization of orbit closures as support varieties*, J. Pure Appl. Algebra **206** (2006), no. 1-2, 66–82, DOI 10.1016/j.jpaa.2005.04.014. MR2220082

[PW] Brian Parshall and Jian Pan Wang, *Quantum linear groups*, Mem. Amer. Math. Soc. **89** (1991), no. 439, vi+157, DOI 10.1090/memo/0439. MR1048073

[P] Cornelius Pillen, *Tensor products of modules with restricted highest weight*, Comm. Algebra **21** (1993), no. 10, 3647–3661, DOI 10.1080/00927879308824754. MR1231622

[So1] Eric Sommers, *Normality of nilpotent varieties in* E_6, J. Algebra **270** (2003), no. 1, 288–306, DOI 10.1016/S0021-8693(03)00416-2. MR2016663

[So2] Eric Sommers, *Normality of very even nilpotent varieties in* D_{2l}, Bull. London Math. Soc. **37** (2005), no. 3, 351–360, DOI 10.1112/S0024609304003935. MR2131388

[Th] Jesper Funch Thomsen, *Normality of certain nilpotent varieties in positive characteristic*, J. Algebra **227** (2000), no. 2, 595–613, DOI 10.1006/jabr.1999.8240. MR1759837

[UGA1] University of Georgia VIGRE Algebra Group, *On Kostant's theorem for Lie algebra cohomology*, Representation theory, Contemp. Math., vol. 478, Amer. Math. Soc., Providence, RI, 2009, pp. 39–60, DOI 10.1090/conm/478/09318. University of Georgia VIGRE Algebra Group: Irfan Bagci, Brian D. Boe, Leonard Chastkofsky, Benjamin Connell, Bobbe J. Cooper, Mee Seong Im, Tyler Kelly, Jonathan R. Kujawa, Wenjing Li, Daniel K. Nakano, Kenyon J. Platt, Emilie Wiesner, Caroline B. Wright and Benjamin Wyser. MR2513265

[UGA2] University of Georgia VIGRE Algebra Group, *An analog of Kostant's theorem for the cohomology of quantum groups*, Proc. Amer. Math. Soc. **138** (2010), no. 1, 85–99, DOI 10.1090/S0002-9939-09-10039-4. University of Georgia VIGRE Algebra Group: Irfan Bagci, Brian D. Boe, Leonard Chastkofsky, Benjamin Connell, Benjamin Jones, Wenjing Li, Daniel K. Nakano, Kenyon J. Platt, Jae-Ho Shin, and Caroline B. Wright. MR2550172

[W] Charles A. Weibel, *An introduction to homological algebra*, Cambridge Studies in Advanced Mathematics, vol. 38, Cambridge University Press, Cambridge, 1994, DOI 10.1017/CBO9781139644136. MR1269324

Department of Mathematics, Kansas State University, Manhattan, Kansas 66506
Email address: zlin@ksu.edu

Department of Mathematics, University of Georgia, Athens, Georgia 30602
Email address: nakano@math.uga.edu

Proceedings of Symposia in Pure Mathematics
Volume **108**, 2024
https://doi.org/10.1090/pspum/108/01967

$\imath$Hall algebras and $\imath$quantum groups

Ming Lu and Weiqiang Wang

In memory of Brian Parshall

ABSTRACT. We survey some recent development on the theory of $\imath$Hall algebras. Starting from $\imath$quivers (aka quivers with involutions), we construct a class of 1-Gorenstein algebras called $\imath$quiver algebras, whose semi-derived Hall algebras give us $\imath$Hall algebras. We then use these $\imath$Hall algebras to realize quasi-split $\imath$quantum groups arising from quantum symmetric pairs. Relative braid group symmetries on $\imath$quantum groups are realized via reflection functors. In case of Jordan $\imath$quiver, the $\imath$Hall algebra is commutative and connections to $\imath$Hall-Littlewood symmetric functions are developed. In case of $\imath$quivers of diagonal type, our construction amounts to a reformulation of Bridgeland-Hall algebra realization of the Drinfeld double quantum groups (which in turn generalizes Ringel-Hall algebra realization of halves of quantum groups). Many rank 1 and rank 2 computations are supplied to illustrate the general constructions. We also briefly review $\imath$Hall algebras of weighted projective lines, and use them to realize Drinfeld type presentations of $\imath$quantum loop algebras.

CONTENTS

1. Introduction
2. Quantum groups and $\imath$quantum groups
3. Hall algebras
4. Semi-derived Ringel-Hall algebras
5. $\imath$Quiver algebras
6. $\imath$Hall algebra for the Jordan quiver
7. $\imath$Hall algebras for Dynkin $\imath$quivers
8. $\imath$Hall algebras for acyclic $\imath$quivers
9. Reflection functors in $\imath$Hall algebras and relative braid group symmetries
10. $\imath$Hall algebras of weighted projective lines and $\imath$quantum loop algebras
11. Discussions and open problems
Acknowledgment
References

2020 *Mathematics Subject Classification.* Primary 17B37, 16E45, 18G80.

Key words and phrases. Hall algebras, Quantum groups, Quantum symmetric pairs.

The first author was partially supported by the Science and Technology Commission of Shanghai Municipality (grant No. 18dz2271000) and the National Natural Science Foundation of China (grant No. 12171333). The second author was partially supported by the NSF grant DMS-2001351.

1. Introduction

1.1. Hall algebras. The theory of Hall algebras has a rich history and diverse applications. We give a quick overview of 3 (interconnected) classes of Hall algebras.

This first Hall algebra (which was popularized by Macdonald's book [**Mac95**] but goes back to Steinitz decades earlier) is an associative algebra with a (Hall) basis parameterized by the isoclasses of finite modules over a discrete valuation ring (such as the ring of p-adic integers or the power series ring $\mathbf{k}[[x]]$ over a finite field $\mathbf{k}$ of q elements), or equivalently, parameterized by partitions. The structure constants arise from counting the extensions between these modules. In this case, the Hall algebra is commutative and isomorphic to the ring Λ of symmetric functions. The Hall basis corresponds to a distinguished basis of Λ known as Hall-Littlewood functions. From a modern point of view, the Steinitz-Hall algebra is associated to the Jordan quiver (consisting of a vertex and a loop).

Ringel-Hall algebra associated to a Dynkin quiver was introduced by [**Rin90**], whose structure constants count the extensions of quiver representations over $\mathbf{k}$. Ringel showed that this version of Hall algebra is isomorphic to half a quantum group associated to the underlying Dynkin diagram of the quiver. Lusztig [**Lus90, Lus93**] developed a geometric version of Ringel-Hall algebra, which allows him to construct a canonical basis of half the quantum group with remarkable positivity property. Ringel's construction was extended by J.A. Green [**Gr95**] to acyclic quivers in connection to quantum groups of Kac-Moody type, who also provided a categorical interpretation of the comultiplication. Reflection functors on Hall algebras provide a realization of braid group symmetries on quantum groups, and in addition, they help to construct PBW bases through Hall algebra approach; see [**Lus93, Rin96, CX99, Z00, XY01**].

In spite of various attempts, the search for a Hall algebra realization of the whole quantum group $\mathbf{U}$ (or the Drinfeld double $\widetilde{\mathbf{U}}$) was largely unsuccessful until the work of Bridgeland [**Br13**] 20 years later. We mention some of these attempts. One first (non-categorical) approach was to use the Drinfeld double of the twisted extended Hall algebras from hereditary categories to realize the whole quantum groups (see [**X97, SV99**]). Also relevant is the (not entirely categorical either) construction of the Hall type Lie algebras from root categories (see [**PX00, Hub06**]) to realize the Kac-Moody Lie algebras. Yet another construction is the derived Hall algebras from bounded derived categories or more general triangulated categories with certain finiteness conditions (see [**T06, XX08**], for a different version see [**Ka98**]). Bridgeland's construction starts with the Hall algebra of the exact category of $\mathbb{Z}_2$-graded complexes of projective modules of quivers and then applies localizations to the acyclic complexes.

Before describing a third class of Hall algebras, let us digress briefly to make some preparation. Affine Lie algebras can be viewed as a special class of Kac-Moody algebras and also as central extensions of loop algebras. Amazingly, the affine quantum groups admit a current presentation (known as Drinfeld presentation) besides the Serre presentation given by Drinfeld and Jimbo; see [**Dr87, Be94, Da15**].

Ringel's construction can be formally extended to hereditary abelian categories, which contain as distinguished examples the categories of quiver representations and the categories of coherent sheaves of (weighted) projective lines over $\mathbf{k}$. The Hall algebra of the projective line was shown in a pioneering work of Kaparanov [**Ka97**] to be isomorphic to half quantum affine $\widehat{\mathfrak{sl}}_2$ in its Drinfeld (current) presentation.

Kaparanov's work has been extended in [**Sch04**] (also [**BS13**]) who developed the connection between Hall algebras of weighted projective lines and halves of quantum loop algebras.

The versions of Hall algebras following Steinitz, Ringel, and Bridgeland are reviewed in Section 3; we refer to [**DDPW08**] and [**Sch06**] for in-depth expositions and extensive references on Hall algebras. Other important classes of Hall algebras, such as derived Hall algebras and cohomological Hall algebras, will not be considered in this survey paper.

1.2. Quantum groups. Drinfeld-Jimbo quantum groups $\mathbf{U}$ or Drinfeld double $\widetilde{\mathbf{U}}$ are by now well-known subjects [**Dr87, Lus90**]. Some fundamental constructions on quantum groups which are more relevant to this survey include PBW bases, canonical bases, Hall algebra realizations, braid group actions, and Drinfeld presentation for affine quantum groups.

Quantum groups admit a natural generalization known as ıquantum groups.

1.3. ıQuantum groups. Symmetric pairs can be classified in terms of Satake diagrams denoted by $(\mathbb{I} = \mathbb{I}_\bullet \cup \mathbb{I}_\circ, \tau)$, where $\mathbb{I} = \mathbb{I}_\bullet \cup \mathbb{I}_\circ$ is a suitable bicolored partition of the Dynkin diagram $\mathbb{I}$ together with a diagram automorphism τ such that $\tau^2 = \mathrm{Id}$; cf. [**OV90**]. Starting with Satake diagrams, Letzter [**Let99, Let02**] introduced quantum symmetric pairs $(\mathbf{U}, \mathbf{U}^\imath)$ of finite type, where $\mathbf{U}^\imath = \mathbf{U}^\imath_{\varsigma}$ (called an ıquantum group) is a coideal subalgebra of a Drinfeld-Jimbo quantum group $\mathbf{U}$ depending on parameters ς; Kolb [**Ko14**] improved and extended the constructions of quantum symmetric pairs to Kac-Moody type. In this survey, we shall restrict ourselves to the quasi-split ıquantum groups $\mathbf{U}^\imath$, associated to $\mathbb{I} = \mathbb{I}_\circ$ and $\mathbb{I}_\bullet = \emptyset$. The authors introduced in [**LW22a**] the universal (quasi-split) ıquantum group $\widetilde{\mathbf{U}}^\imath$, which is a coideal subalgebra of the Drinfeld double $\widetilde{\mathbf{U}}$. The ıquantum groups $\mathbf{U}^\imath_{\varsigma}$ for various parameters are obtained from $\widetilde{\mathbf{U}}^\imath$ by central reductions.

Drinfeld-Jimbo quantum groups can be viewed as ıquantum groups associated to diagonal Satake diagrams (recall a diagonal Satake diagram is a union of 2 identical copies of a Dynkin diagram with an involution swapping corresponding vertices). This viewpoint has been fruitful in the ıprogram (cf. [**BW18a**]) which seeks to generalize all fundamental (algebraic, geometric, and categorical) constructions in quantum groups to ıquantum groups; see [**W22**] for a survey.

In contrast to quantum groups (for which we have $\mathbf{U} = \mathbf{U}^-\mathbf{U}^0\mathbf{U}^+$), the ıquantum groups in general do not admit a natural triangular decomposition. For quantum groups, there is a unique rank 1 case (i.e., the quantum $\mathfrak{sl}_2$) and there is a uniform Serre relation associated to all rank 2 Dynkin diagrams. For quasi-split ıquantum groups, there are 3 (real) rank 1 cases, and accordingly there are several Serre type relations which have to be formulated separately.

ıDivided powers arise as ıcanonical basis for the split rank 1 ıquantum group (with a distinguished parameter) [**BW18a, BeW18**], and they have been generalized to $\mathbf{U}^\imath$ for arbitrary parameters [**CLW21a**] and also to the universal ıquantum groups $\widetilde{\mathbf{U}}^\imath$ [**LW20**]. A Serre presentation for quasi-split ıquantum groups $\widetilde{\mathbf{U}}^\imath$ of arbitrary Kac-Moody type was obtained in [**CLW21a**] (generalizing the earlier presentation of Letzter [**Let02**] in finite type and some relation in [**BK15**]), where a new key ingredient is the so-called ıSerre relation in terms of ıdivided powers.

The basics on quantum groups and ıquantum groups are reviewed in Section 2.

1.4. $\imath$Hall algebras. This survey is mainly concerned about the constructions of $\imath$Hall algebras and their use in realizing the quasi-split $\imath$quantum groups. The reflection functors on $\imath$Hall algebras are then used to realize the relative braid group actions on $\imath$quantum groups. We also develop the connection between the $\imath$Hall algebra of the Jordan quiver and ring of symmetric functions.

1.4.1. *Semi-derived Hall algebras.* With the help of the Riedtmann-Peng formula, the definition of Hall algebras has been extended to exact categories by Hubery [**Hub06**] (also cf. [**PX00**]). Inspired by Bridgeland's construction [**Br13**], M. Gorsky [**Gor18**] constructed semi-derived Hall algebras for Frobenius categories; see also [**Gor13**] for the semi-derived Hall algebras constructed for $\mathbb{Z}_2$-graded complexes of an exact category admitting enough projective objects.

More recently, motivated by the works of Bridgeland and Gorsky, the first author and Peng [**LP21**] formulated the *semi-derived Ringel-Hall algebras* for hereditary abelian categories, which may not have enough projectives. In [**Lu22**], the construction in [**LP21**] is further extended to weakly 1-Gorenstein exact categories. Note that the module categories of $\imath$quiver algebras below are 1-Gorenstein but not hereditary.

We give an overview of semi-derived Hall algebras in Section 4.

1.4.2. *$\imath$Quiver algebras.* Starting with $\imath$quivers (Q, τ) (aka quivers Q with involutions τ preserving the arrows of Q, where $\tau = \mathrm{Id}$ is allowed), we constructed in [**LW22a**] a class of 1-Gorenstein algebras $\Lambda^\imath$; in case $\tau = \mathrm{Id}$, the $\imath$quiver algebras can be identified with $\mathbf{k}Q \otimes \mathbf{k}[\varepsilon]/(\varepsilon^2)$, and these algebras also appeared in [**RZ17**] and [**GLS17**] from very different perspectives. The $\imath$quiver algebras can be infinite-dimensional if the quiver Q is not acyclic; but in all cases, they are 1-Gorenstein and admit favorable homological properties, cf. [**LW22a, LW20**]. The $\imath$Hall algebra $\widetilde{\mathcal{H}}(\mathbf{k}Q, \tau)$ is by definition the twisted semi-derived Hall algebra of (the module category of) the $\imath$quiver algebra $\Lambda^\imath$.

The $\imath$quiver algebra associated to the diagonal $\imath$quiver $(Q \cup Q, \mathrm{swap})$ is isomorphic to $\mathbf{k}Q \otimes R_2$, where R_2 is the exterior $\mathbf{k}$-algebra in 2 variables $\varepsilon, \varepsilon'$; see §5.2. We show that the $\imath$Hall algebra in this case is isomorphic to Drinfeld double $\widetilde{\mathbf{U}}$, and this provides a reformulation of the main theorem of Bridgeland [**Br13**].

The $\imath$quiver algebras and their basic homological properties were presented in Section 5.

1.4.3. *$\imath$Hall algebras for Jordan quiver.* A general theory of quiver representations is developed often for quivers without loops, and so representations of the Jordan quiver require some separate discussions. The framework of $\imath$quivers does allow to formulate Jordan $\imath$quiver naturally; see Example 5.4(5) and Figure 5.10.

It is shown in [**LRW21**] that the $\imath$Hall algebra of Jordan quiver is commutative and it is isomorphic to the ring of symmetric functions (now in 2 variables t and θ), extending the classic Steinitz-Hall isomorphism. The $\imath$Hall basis now corresponds to $\imath$Hall-Littlewood ($\imath$HL) functions $Q^\imath_\lambda$ (or its modified version $H^\imath_\lambda$), which admit a formulation via raising and lowering operators. One may view the (modified) HL functions as the leading term in the variable θ of the (modified) $\imath$HL functions, as $Q^\imath_\lambda$ and $H^\imath_\lambda$ at $\theta = 0$ reduce to Q_λ and H_λ. When setting $\theta = 1$, the $\imath$HL functions are identified with type C deformed universal characters introduced in [**SZ06**]. Several $\imath$Pieri rules for $\imath$HL functions were obtained through mixed combinatorial and $\imath$Hall approaches; the combinatorial approach follows and generalizes [**Tam11**].

We present the ıHall algebra of Jordan quiver and its connection to ring of symmetric functions in Section 6.

1.4.4. *ıHall algebras for ıquivers.* The ıHall algebra $\widetilde{\mathcal{H}}(\mathbf{k}Q,\tau)$ for a Dynkin ıquiver is shown by the authors [**LW22a**] to be isomorphic to the universal ıquantum group $\widetilde{\mathbf{U}}^\imath$. The relations (including the Serre relations) of $\widetilde{\mathbf{U}}^\imath$ can be verified directly in the framework of $\widetilde{\mathcal{H}}(\mathbf{k}Q,\tau)$, as there are only a few local rank 2 cases to consider. The connection between ıHall algebras for Dynkin ıquivers and ıquantum groups of finite type is developed in Section 7.

It requires some major work to extend the above construction for Dynkin ıquivers to a class of virtually acyclic (including all acyclic) ıquivers; in this generality, we obtain an injective homomorphism from the ıquantum group $\widetilde{\mathbf{U}}^\imath$ to the corresponding ıHall algebra [**LW20**]. The verification of ıSerre relations (among others) in an ıHall algebra is highly nontrivial. To that end, we first need to understand the ıdivided powers in terms of an ıHall basis. Then we express the ıSerre formula in terms of an ıHall basis, through nontrivial homological computations. Finally, we reduce the proof of the vanishing of various coefficients in the ıHall basis for the ıSerre formula into some new q-binomial identities.

The connection between ıHall algebras for virtually acyclic ıquivers and ıquantum groups of Kac-Moody type is developed in Section 8.

1.4.5. *Reflection functors.* Quiver representations have been generalized to representations of modulated graphs by Dlab-Ringel, and the latter can be adapted to ıquivers. BGP type reflection functors can also be formulated in the setting of representations of modulated graphs, and they can be realized in terms of tilting modules; see [**Li12, GLS17, LW21a, LW22b**].

For a sink ℓ of Q_0, let $\boldsymbol{s}_\ell Q$ be the quiver obtained from Q by reversing the arrows ending at ℓ and $\tau\ell$. Denote by $\Lambda^\imath$ (respectively, $\boldsymbol{s}_\ell\Lambda^\imath$) the ıquiver algebra for the ıquiver (Q,τ) (respectively, $(\boldsymbol{s}_\ell Q,\tau)$). A reflection functor from $\mathrm{mod}(\Lambda^\imath)$ to $\mathrm{mod}(\boldsymbol{s}_\ell\Lambda^\imath)$ can then be defined [**LW21a, LW22b**]. This induces an isomorphism of ıHall algebras, $\Gamma_\ell : \widetilde{\mathcal{H}}(\mathbf{k}Q,\tau) \xrightarrow{\sim} \widetilde{\mathcal{H}}(\mathbf{k}\boldsymbol{s}_\ell Q,\tau)$, with explicit formulas on generators; and when composed with a Fourier transformation, this further induces an automorphism of $\widetilde{\mathcal{H}}(\mathbf{k}Q,\tau)$.

Under the ıHall algebra realization of the ıquantum group $\widetilde{\mathbf{U}}^\imath$, we translate the (reflection functor) automorphisms of $\widetilde{\mathcal{H}}(\mathbf{k}Q,\tau)$ to (relative braid group) symmetries on $\widetilde{\mathbf{U}}^\imath$, which are given by explicit formulas on generators; this confirms substantial cases of [**CLW21a**, Conjecture 6.5] and [**CLW21b**, Conjecture 3.7]. These conjectures have been most recently completely settled by Weinan Zhang [**Z23**] who developed further the approach initiated in [**WZ23**]. Existence of relative braid group symmetries on ıquantum groups was originally conjectured in [**KP11**].

Reflection functors on ıHall algebras and relative braid group actions on $\widetilde{\mathbf{U}}^\imath$ are presented in Section 9.

1.4.6. *ıHall algebras for (weighted) projective lines.* Drinfeld (current) presentation of affine quantum groups were obtained in [**Dr87, Be94, Da15**]. Recently, the Drinfeld type presentations of split affine ıquantum groups were obtained in [**LW21c**] for ADE type and then in [**Z21**] for BCFG type (also cf. [**LWZ23**]). The affine ıquantum group of split type $A_1^{(1)}$ is also known as q-Onsager algebra.

For a hereditary abelian category $\mathcal{A}$, let $\mathcal{C}_{\mathbb{Z}_1}(\mathcal{A})$ be the category of 1-periodic complexes of $\mathcal{A}$. Denote its $\imath$Hall algebra by ${}^\imath\widetilde{\mathcal{H}}(\mathcal{A})$, which is the twisted semi-derived Ringel-Hall algebra of $\mathcal{C}_{\mathbb{Z}_1}(\mathcal{A})$.

When $\mathcal{A}$ is the category of coherent sheaves on the projective line $\mathbb{P}^1_{\mathbf{k}}$, we show in [**LRW20**] that the $\imath$Hall algebra ${}^\imath\widetilde{\mathcal{H}}(\mathcal{A})$ is isomorphic to the (universal) q-Onsager algebra in its Drinfeld type presentation. The isomorphism between two presentations of q-Onsager algebra is conceptually explained by the derived equivalence between category of modules over split Kronecker $\imath$quiver algebra and category $\mathcal{C}_{\mathbb{Z}_1}(\mathcal{A})$.

The above picture admits a vast generalization [**LR21**] when $\mathcal{A}$ is the category of coherent sheaves on a weighted projective line. In this case, one needs to replace q-Onsager algebra by $\imath$quantum loop algebras, which are by definition a generalization of Drinfeld presentation of split affine $\imath$quantum groups.

We review the Drinfeld type presentation of affine $\imath$quantum groups and $\imath$Hall algebras of weighted projective lines in Section 10.

1.4.7. *Open problems.* In the final Section 11, we formulate several open problems in $\imath$Hall algebras and put them in contexts.

1.5. Notations. We list the notations which are often used throughout the paper.

- ▷ $\mathbb{N}, \mathbb{Z}, \mathbb{Q}$, $\mathbb{C}$ – sets of nonnegative integers, integers, rational and complex numbers,
- ▷ $\mathbf{k}$ – a finite field of q elements,
- ▷ $|\mathcal{S}|$ – cardinality of a finite set $\mathcal{S}$.

For a quiver algebra $A = \mathbf{k}Q/I$ (not necessarily finite-dimensional), we always identify left A-modules with representations of Q satisfying relations in I. A representation $V = (V_i, V(\alpha))_{i \in Q_0, \alpha \in Q_1}$ of A is called *nilpotent* if for each oriented cycle $\alpha_m \cdots \alpha_1$ at a vertex i, the $\mathbf{k}$-linear map $V(\alpha_m) \cdots V(\alpha_1) : V_i \to V_i$ is nilpotent. We denote

- ▷ $\operatorname{proj}(A)$ – category of finitely generated projective A-modules,
- ▷ $\operatorname{mod}(A)$ – category of finite-dimensional nilpotent A-modules,
- ▷ $\mathcal{D}^b(\operatorname{mod}(A))$ – bounded derived category for $\operatorname{mod}(A)$,
- ▷ $\operatorname{proj.dim}_A M$ – projective dimension of an A-module M,
- ▷ $\operatorname{inj.dim}_A M$ – injective dimension of M,
- ▷ $\mathcal{P}^{\leq m}(A)$ – subcategory of $\operatorname{mod}(A)$ with projective dimension $\leq m$, for $m \geq 0$,
- ▷ $\mathcal{P}^{<\infty}(A)$ – subcategory of $\operatorname{mod}(A)$ with finite projective dimension,
- ▷ $\operatorname{Gproj}(A)$ – category of Gorenstein-projective A-modules.

Let $Q = (Q_0, Q_1)$ be a quiver, and $A = \mathbf{k}Q/I$ be a quiver algebra. For $i \in Q_0$, we denote

- ▷ e_i – the primitive idempotent supported at i of A,
- ▷ S_i – the 1-dimensional simple nilpotent A-module supported at i,
- ▷ P_i – the projective cover of S_i,
- ▷ I_i – the injective hull of S_i.

For an additive category $\mathcal{A}$ and $M \in \mathcal{A}$, we denote

- ▷ $\operatorname{add} M$ – subcategory of $\mathcal{A}$ whose objects are the direct summands of finite direct sums of copies of M,
- ▷ $\operatorname{Ind}(\mathcal{A})$ – set of the isoclasses of indecomposable objects in $\mathcal{A}$,

▷ $\mathrm{Iso}(\mathcal{A})$ – set of the isoclasses of objects in $\mathcal{A}$,
▷ $\mathrm{Aut}(M)$ – automorphism group of M.

For an exact category $\mathcal{A}$, we denote

▷ $K_0(\mathcal{A})$ – Grothendieck group of $\mathcal{A}$,
▷ $\widehat{A}$ – the class in $K_0(\mathcal{A})$ of $A \in \mathcal{A}$,
▷ $\mathcal{C}_{\mathbb{Z}_m}(\mathcal{A})$ – the category of $\mathbb{Z}_m$-graded complexes,
▷ $\mathcal{D}_{sg}(\mathcal{A})$ – the singularity category of $\mathcal{A}$.

For various Hall algebras, we denote

▷ $\mathcal{H}(\mathcal{A})$ – Hall algebra of the exact category $\mathcal{A}$,
▷ $\mathcal{H}(\mathbf{k}Q)$ – Hall algebra of the path algebra $\mathbf{k}Q$,
▷ $\widetilde{\mathcal{H}}(\mathbf{k}Q)$ – twisted Hall algebra of the path algebra $\mathbf{k}Q$,
▷ $D\widetilde{\mathcal{H}}(\mathbf{k}Q)$ – Bridgeland's Hall algebra of the path algebra $\mathbf{k}Q$ (with Q acyclic).

Associated to the ıquiver (Q,τ) (aka quiver with involution), we denote

▷ $\Lambda^\imath$ – ıquiver algebra,
▷ $\mathcal{SDH}(\Lambda^\imath)$ – semi-derived Hall algebra for $\Lambda^\imath$ (i.e., for $\mathrm{mod}(\Lambda^\imath)$),
▷ $\mathcal{T}(\Lambda^\imath)$ – quantum torus,
▷ $\mathcal{SD}\widetilde{\mathcal{H}}(\Lambda^\imath) = {}^\imath\widetilde{\mathcal{H}}(\mathbf{k}Q,\tau)$ – ıHall algebra (=twisted semi-derived Hall algebra for $\Lambda^\imath$),
▷ $\widetilde{\mathcal{T}}(\Lambda^\imath)$ – twisted quantum torus.
▷ Q_{J}–Jordan quiver,
▷ $\mathbb{P}^1_{\mathbf{k}}$ – the projective line over $\mathbf{k}$,
▷ $\mathrm{coh}(\mathbb{P}^1_{\mathbf{k}})$ – category of coherent sheaves of $\mathbb{P}^1_{\mathbf{k}}$,
▷ $\mathbb{X}$ – weighted projective line over $\mathbf{k}$,
▷ $\mathrm{coh}(\mathbb{X})$ – category of coherent sheaves of $\mathbb{X}$.

For quantum algebras, we denote

▷ $\mathbf{U}$ – quantum group,
▷ $\widetilde{\mathbf{U}}$ – Drinfeld double (a variant of $\mathbf{U}$ with doubled Cartan subalgebra),
▷ $\mathbf{U}^\imath = \mathbf{U}^\imath_{\boldsymbol{\varsigma}}$ – a right coideal subalgebra of $\mathbf{U}$, depending on parameter $\boldsymbol{\varsigma} \in (\mathbb{Q}(v)^\times)^{\mathbb{I}}$,
▷ $(\mathbf{U},\mathbf{U}^\imath)$ – quantum symmetric pair,
▷ $(\widetilde{\mathbf{U}},\widetilde{\mathbf{U}}^\imath)$ – a universal quantum symmetric pair, such that $\mathbf{U}^\imath$ is obtained from $\widetilde{\mathbf{U}}^\imath$ by a central reduction,
▷ ${}^{\mathrm{Dr}}\widetilde{\mathbf{U}}^\imath$ – ıquantum loop algebra.

2. Quantum groups and ıquantum groups

In this section, we recall Drinfeld double quantum groups, and review the (quasi-split) universal ıquantum groups from [**LW22a**] whose central reductions are the ıquantum groups à la Letzter and Kolb [**Ko14**]. We then give a Serre presentation of the (universal) ıquantum groups via the ıdivided powers, following [**CLW21a**].

2.1. Quantum groups and Drinfeld doubles. Let $Q = (Q_0, Q_1)$ be a quiver (without loops) with vertex set $Q_0 = \mathbb{I}$. The underlying graph of Q with orientations of edges forgotten is called a Dynkin diagram, denoted by D_Q. Let n_{ij} be the number of edges connecting vertex i and j. Let $C = (c_{ij})_{i,j\in\mathbb{I}}$ be the symmetric generalized Cartan matrix of D_Q, defined by $c_{ij} = 2\delta_{ij} - n_{ij}$. Let $\mathfrak{g}$ be the Kac-Moody Lie algebra associated to C, with simple roots denoted by α_i $(i \in \mathbb{I})$; for this survey of Hall algebras, we can simply take $\mathfrak{g}$ to be the derived subalgebra of the Kac-Moody algebra. The root lattice is denoted by $\mathbb{Z}^{\mathbb{I}} := \bigoplus_{i\in\mathbb{I}} \mathbb{Z}\alpha_i$. The *simple reflection* $s_i : \mathbb{Z}^{\mathbb{I}} \to \mathbb{Z}^{\mathbb{I}}$ is defined to be $s_i(\alpha_j) = \alpha_j - c_{ij}\alpha_i$, for $i, j \in \mathbb{I}$. Denote the Weyl group by $W = \langle s_i \mid i \in \mathbb{I}\rangle$.

Let v be an indeterminant. Write $[A, B] = AB - BA$. Denote, for $r \in \mathbb{N}$ and $m \in \mathbb{Z}$,

$$[r] = \frac{v^r - v^{-r}}{v - v^{-1}}, \quad [r]! = \prod_{i=1}^{r}[i], \quad \begin{bmatrix} m \\ r \end{bmatrix} = \frac{[m][m-1]\dots[m-r+1]}{[r]!}.$$

Then $\widetilde{\mathbf{U}} = \widetilde{\mathbf{U}}_v(\mathfrak{g})$ is the $\mathbb{Q}(v)$-algebra generated by $E_i, F_i, \widetilde{K}_i, \widetilde{K}_i'$, $i \in \mathbb{I}$, where $\widetilde{K}_i, \widetilde{K}_i'$ are invertible, subject to the following relations:

$$[E_i, F_j] = \delta_{ij}\frac{\widetilde{K}_i - \widetilde{K}_i'}{v - v^{-1}}, \qquad [\widetilde{K}_i, \widetilde{K}_j] = [\widetilde{K}_i, \widetilde{K}_j'] = [\widetilde{K}_i', \widetilde{K}_j'] = 0, \tag{2.1}$$

$$\widetilde{K}_i E_j = v^{c_{ij}} E_j \widetilde{K}_i, \qquad \widetilde{K}_i F_j = v^{-c_{ij}} F_j \widetilde{K}_i, \tag{2.2}$$

$$\widetilde{K}_i' E_j = v^{-c_{ij}} E_j \widetilde{K}_i', \qquad \widetilde{K}_i' F_j = v^{c_{ij}} F_j \widetilde{K}_i', \tag{2.3}$$

and the quantum Serre relations, for $i \neq j \in \mathbb{I}$,

$$\sum_{r=0}^{1-c_{ij}} (-1)^r E_i^{(r)} E_j E_i^{(1-c_{ij}-r)} = 0, \tag{2.4}$$

$$\sum_{r=0}^{1-c_{ij}} (-1)^r F_i^{(r)} F_j F_i^{(1-c_{ij}-r)} = 0. \tag{2.5}$$

Here we have used the divided powers

$$F_i^{(n)} = F_i^n/[n]!, \quad E_i^{(n)} = E_i^n/[n]!, \qquad \text{for } n \geq 1, i \in \mathbb{I}.$$

Note that $\widetilde{K}_i\widetilde{K}_i'$ are central in $\widetilde{\mathbf{U}}$ for all i. We endow $\widetilde{\mathbf{U}}$ a Hopf algebra structure with the comultiplication $\Delta : \widetilde{\mathbf{U}} \longrightarrow \widetilde{\mathbf{U}} \otimes \widetilde{\mathbf{U}}$ given by

$$\begin{aligned} \Delta(E_i) &= E_i \otimes 1 + \widetilde{K}_i \otimes E_i, \quad \Delta(F_i) = 1 \otimes F_i + F_i \otimes \widetilde{K}_i', \\ \Delta(\widetilde{K}_i) &= \widetilde{K}_i \otimes \widetilde{K}_i, \quad \Delta(\widetilde{K}_i') = \widetilde{K}_i' \otimes \widetilde{K}_i'. \end{aligned} \tag{2.6}$$

The Chevalley involution ω on $\widetilde{\mathbf{U}}$ is given by

$$\omega(E_i) = F_i, \quad \omega(F_i) = E_i, \quad \omega(\widetilde{K}_i) = \widetilde{K}_i', \quad \omega(\widetilde{K}_i') = \widetilde{K}_i, \quad \forall i \in \mathbb{I}. \tag{2.7}$$

Analogously as for $\widetilde{\mathbf{U}}$, the quantum group $\mathbf{U}$ is defined to be the $\mathbb{Q}(v)$-algebra generated by E_i, F_i, K_i, K_i^{-1}, $i \in \mathbb{I}$, subject to the relations modified from (2.1)–(2.5) with $\widetilde{K}_i$ and $\widetilde{K}_i'$ replaced by K_i and K_i^{-1}, respectively. The comultiplication Δ and Chevalley involution ω on $\mathbf{U}$ are obtained by modifying (2.6)–(2.7) with $\widetilde{K}_i$ and $\widetilde{K}_i'$ replaced by K_i and K_i^{-1}, respectively (cf. [**Lus93**]; beware that our K_i has a different meaning from $K_i \in \mathbf{U}$ therein).

The algebra $\mathbf{U}$ can be obtained from $\widetilde{\mathbf{U}}$ by a central reduction; that is, $\mathbf{U}$ is isomorphic to a quotient algebra of $\widetilde{\mathbf{U}}$ by the ideal $(\widetilde{K}_i\widetilde{K}_i' - 1 \mid i \in \mathbb{I})$.

Let $\widetilde{\mathbf{U}}^+$ be the subalgebra of $\widetilde{\mathbf{U}}$ generated by E_i $(i \in \mathbb{I})$, $\widetilde{\mathbf{U}}^0$ be the subalgebra of $\widetilde{\mathbf{U}}$ generated by $\widetilde{K}_i, \widetilde{K}_i'$ $(i \in \mathbb{I})$, and $\widetilde{\mathbf{U}}^-$ be the subalgebra of $\widetilde{\mathbf{U}}$ generated by F_i $(i \in \mathbb{I})$, respectively. The subalgebras $\mathbf{U}^+$, $\mathbf{U}^0$ and $\mathbf{U}^-$ of $\mathbf{U}$ are defined similarly.

Then both $\widetilde{\mathbf{U}}$ and $\mathbf{U}$ have triangular decompositions: $\widetilde{\mathbf{U}} = \widetilde{\mathbf{U}}^+ \otimes \widetilde{\mathbf{U}}^0 \otimes \widetilde{\mathbf{U}}^-$, $\mathbf{U} = \mathbf{U}^+ \otimes \mathbf{U}^0 \otimes \mathbf{U}^-$. Clearly, $\mathbf{U}^+ \cong \widetilde{\mathbf{U}}^+$, $\mathbf{U}^- \cong \widetilde{\mathbf{U}}^-$, and $\mathbf{U}^0 \cong \widetilde{\mathbf{U}}^0/(\widetilde{K}_i\widetilde{K}_i' - 1 \mid i \in \mathbb{I})$.

REMARK 2.1.

(1) $\widetilde{\mathbf{U}}$ (respectively, $\mathbf{U}$) is a $\mathbb{Z}^{\mathbb{I}}$-graded algebra with $\mathrm{wt}\, E_i = \alpha_i$, $\mathrm{wt}\, F_i = -\alpha_i$, and $\mathrm{wt}\, \widetilde{K}_i = 0 = \mathrm{wt}\, \widetilde{K}_i'$ (respectively, $\mathrm{wt}\, K_i = 0 = \mathrm{wt}\, K_i^{-1}$).

(2) There is a second $\mathbb{Z}^{\mathbb{I}}$-grading on $\widetilde{\mathbf{U}}$ (which do not descend to $\mathbf{U}$) by setting $\deg E_i = \alpha_i = \det F_i$, and $\deg \widetilde{K}_i = \alpha_i + \alpha_{\tau i} = \deg \widetilde{K}_i'$.

EXAMPLE 2.2 (Rank 1). There is a unique quantum group of rank 1 corresponding to $\mathbb{I}$ which consists of a single vertex. In this case, $\mathbf{U} = \mathbf{U}_v(\mathfrak{sl}_2)$ is generated by $E, F, K^{\pm 1}$ such that $KK^{-1} = K^{-1}K = 1$, $KE = v^2EK$, $KF = v^{-2}FK$, and $[E,F] = \frac{K-K^{-1}}{v-v^{-1}}$.

EXAMPLE 2.3 (Rank 2). The quantum groups $\mathbf{U}$ of rank 2 are associated to the Dynkin diagram $\mathbb{I}$ consisting of two vertices 1 and 2 with a edges between them; in this case $c_{12} = -a$, and we draw the diagram as

$$1 \overset{a}{\text{——}} 2$$

In particular, if $a = 1$, then $\mathbf{U} = \mathbf{U}_v(\mathfrak{sl}_3)$ is of type A_2; and if $a = 2$, then $\mathbf{U} = \mathbf{U}_v(\widehat{\mathfrak{sl}}_2)$ is of affine type $A_1^{(1)}$.

2.2. (Universal) ıquantum groups $\mathbf{U}^\imath$ and $\widetilde{\mathbf{U}}^\imath$. For a generalized Cartan matrix $C = (c_{ij})$, let $\mathrm{Aut}(C)$ be the group of all permutations τ of the set $\mathbb{I}$ such that $c_{ij} = c_{\tau i, \tau j}$. An element in $\mathrm{Aut}(C)$ is the same as a diagram automorphism of the Dynkin diagram associated to C. An element $\tau \in \mathrm{Aut}(C)$ is called an *involution* if $\tau^2 = \mathrm{Id}$; we allow $\tau = \mathrm{Id}$. Such a τ induces an automorphism of $\mathfrak{g}$, denoted again by τ.

Let τ be an involution in $\mathrm{Aut}(C)$. We define $\widetilde{\mathbf{U}}^\imath$ to be the $\mathbb{Q}(v)$-subalgebra of $\widetilde{\mathbf{U}}$ generated by

$$B_i = F_i + E_{\tau i}\widetilde{K}_i', \qquad \widetilde{k}_i = \widetilde{K}_i\widetilde{K}_{\tau i}', \quad \forall i \in \mathbb{I}.$$

Indeed, $\widetilde{\mathbf{U}}^\imath$ is a right coideal subalgebra of $\widetilde{\mathbf{U}}$, i.e., $\Delta(\widetilde{\mathbf{U}}^\imath) \subset \widetilde{\mathbf{U}}^\imath \otimes \widetilde{\mathbf{U}}$; this can be checked on generators of $\widetilde{\mathbf{U}}^\imath$. Let $\widetilde{\mathbf{U}}^{\imath 0}$ be the commutative $\mathbb{Q}(v)$-subalgebra (called the Cartan subalgebra) of $\widetilde{\mathbf{U}}^\imath$ generated by $\widetilde{k}_i$, for $i \in \mathbb{I}$. The elements

$$\widetilde{k}_i \ (i = \tau i) \qquad \widetilde{k}_i\widetilde{k}_{\tau i} \ (i \neq \tau i) \tag{2.8}$$

are central in $\widetilde{\mathbf{U}}^\imath$.

Let

$$\varsigma = (\varsigma_i \mid i \in \mathbb{I}) \in (\mathbb{Q}(v)^\times)^{\mathbb{I}}$$

be such that $\varsigma_i = \varsigma_{\tau i}$ whenever $i \in \mathbb{I}$ satisfies $c_{i,\tau i} = 0$. Set

$$\mathbb{I}_\tau = \{\text{the chosen representatives of } \tau\text{-orbits in } \mathbb{I}\}. \tag{2.9}$$

Let $\mathbf{U}^\imath_\varsigma$ be the $\mathbb{Q}(v)$-subalgebra of $\mathbf{U}$ generated by

$$B_i = F_i + \varsigma_i E_{\tau i} K_i^{-1}, \quad k_j = K_j K_{\tau j}^{-1}, \qquad \forall i \in \mathbb{I}, j \in \mathbb{I} \setminus \mathbb{I}_\tau.$$

We often write $\mathbf{U}^\imath$ for $\mathbf{U}^\imath_\varsigma$ by dropping the parameter ς when there is no confusion. It is known [**Let99, Ko14**] that $\mathbf{U}^\imath$ is a right coideal subalgebra of $\mathbf{U}$, i.e., $\Delta(\mathbf{U}^\imath) \subset \mathbf{U}^\imath \otimes \mathbf{U}$; and $(\mathbf{U}, \mathbf{U}^\imath)$ is called a (quasi-split) *quantum symmetric pair*, and it specializes at $v = 1$ to (the enveloping algebra version of) the symmetric pair $(\mathfrak{g}, \mathfrak{g}^\theta)$, where $\theta = \omega \circ \tau$. (The Chevalley involution ω and the involution τ on $\mathfrak{g}$ commute.)

We call $\mathbf{U}^\imath$ a (quasi-split) $\imath$quantum group and $\widetilde{\mathbf{U}}^\imath$ a (quasi-split) universal $\imath$quantum group. A quasi-split $\imath$quantum group is called *split* if in addition we have $\tau = \mathrm{Id}$.

REMARK 2.4. More general $\imath$quantum groups can be defined associated to Satake diagrams $(\mathbb{I} = \mathbb{I}_\bullet \cup \mathbb{I}_\circ, \tau)$ [**Let99**, **Ko14**], and so are the universal $\imath$quantum groups (cf. [**WZ23**]). For the purpose of $\imath$Hall algebra realization in this paper, we only consider quasi-split universal $\imath$quantum groups as defined above, which are associated to Satake diagrams with $\mathbb{I}_\bullet = \emptyset$.

The algebras $\mathbf{U}^\imath_{\varsigma}$, for various parameters ς, are obtained from $\widetilde{\mathbf{U}}^\imath$ by central reductions.

PROPOSITION 2.5 ([**LW22a**, Propositon 6.2]). *The $\mathbb{Q}(v)$-algebra $\mathbf{U}^\imath$ is isomorphic to the quotient of $\widetilde{\mathbf{U}}^\imath$ by the ideal $I^\imath$ generated by $\widetilde{k}_i - \varsigma_i$ (for $i = \tau i$) and $\widetilde{k}_i\widetilde{k}_{\tau i} - \varsigma_i\varsigma_{\tau i}$ (for $i \neq \tau i$). The isomorphism $\mathbf{U}^\imath \to \widetilde{\mathbf{U}}^\imath/I^\imath$ is given by sending*

$$B_i \mapsto B_i, \quad k_j \mapsto \varsigma_{\tau j}^{-1}\widetilde{k}_j, \quad k_j^{-1} \mapsto \varsigma_j^{-1}\widetilde{k}_{\tau j}, \quad \forall i \in \mathbb{I},\, j \in \mathbb{I} \setminus \mathbb{I}_\tau.$$

The *rank* of $\widetilde{\mathbf{U}}^\imath$ is by definition the number of τ-orbits in $\mathbb{I}$.

EXAMPLE 2.6 (Split type of rank $1, 2$).

(1) The split (universal) $\imath$quantum groups $\widetilde{\mathbf{U}}^\imath$ and $\mathbf{U}^\imath$ of rank 1 are associated to $\mathbb{I}$ which consists of a single vertex 1. In this case, $\widetilde{\mathbf{U}}^\imath$ is a commutative algebra generated by $B, \widetilde{k}^{\pm 1}$, while $\mathbf{U}^\imath$ is isomorphic to the polynomial algebra $\mathbb{Q}(v)[B]$.

(2) The split $\imath$quantum groups $\widetilde{\mathbf{U}}^\imath$ and $\mathbf{U}^\imath$ of rank 2 are associated to the following (Satake) diagram:

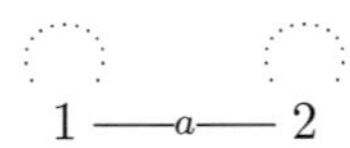

For $a = 2$, the $\imath$quantum groups $\mathbf{U}^\imath$ and $\widetilde{\mathbf{U}}^\imath$ are known as q-Onsager algebra.

EXAMPLE 2.7 (Quasi-split type of rank $1, 2$).

(1) The quasi-split $\imath$quantum groups $\widetilde{\mathbf{U}}^\imath$ and $\mathbf{U}^\imath$ of rank 1 are associated to the following (Satake) diagram:

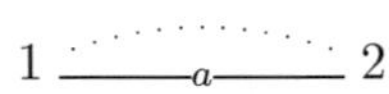

(2) The quasi-split $\imath$quantum groups $\widetilde{\mathbf{U}}^\imath$ and $\mathbf{U}^\imath$ of rank 2 are associated to the following (Satake) diagrams:

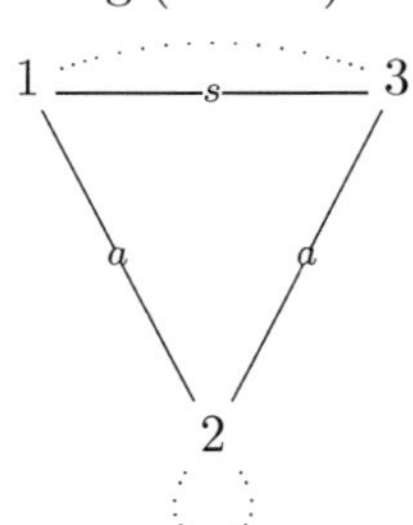

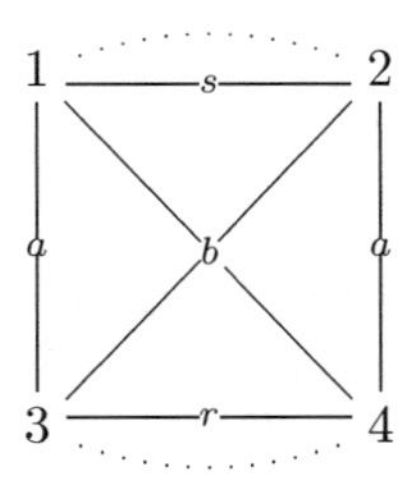

In the second diagram above, the number of edges between 1 and 4 (and respectively, between 2 and 3) is b.

2.3. A Serre presentation of $\widetilde{\mathbf{U}}^\imath$. For $\ell \in \mathbb{I}$ with $\tau\ell \neq \ell$, we define the ıdivided power of B_ℓ as

$$B_\ell^{(m)} := B_\ell^m/[m]!, \quad \forall m \geq 0, \qquad (\text{if } \ell \neq \tau\ell). \tag{2.10}$$

Let $i \in \mathbb{I}$ with $\tau i = i$. The *ıdivided power* of B_i in $\mathbf{U}^\imath_\varsigma$ for a distinguished parameter $\varsigma_i = v^{-1}$ was originally introduced in [**BW18a**, **BeW18**], and then generalized for a general parameter ς_i in [**CLW21a**]. Replacing ς_i by $\widetilde{k}_i$ (see Proposition 2.5) and following [**CLW21b**, (2.20)-(2.21)], we define the *ıdivided powers* of B_i in $\widetilde{\mathbf{U}}$ to be

$$B_{i,\bar{1}}^{(m)} = \frac{1}{[m]!}\begin{cases} B_i\prod_{s=1}^k (B_i^2 - [2s-1]^2 v\widetilde{k}_i) & \text{if } m = 2k+1,\\ \prod_{s=1}^k (B_i^2 - [2s-1]^2 v\widetilde{k}_i) & \text{if } m = 2k;\end{cases} \tag{2.11}$$

$$B_{i,\bar{0}}^{(m)} = \frac{1}{[m]!}\begin{cases} B_i\prod_{s=1}^k (B_i^2 - [2s]^2 v\widetilde{k}_i) & \text{if } m = 2k+1,\\ \prod_{s=1}^k (B_i^2 - [2s-2]^2 v\widetilde{k}_i) & \text{if } m = 2k.\end{cases} \tag{2.12}$$

Denote

$$(a;x)_0 = 1, \qquad (a;x)_n = (1-a)(1-ax)\cdots(1-ax^{n-1}), \quad n \geq 1.$$

The following theorem is an upgrade of [**CLW21a**, Theorem 3.1] for $\mathbf{U}^\imath$ to the setting of a universal ıquantum group $\widetilde{\mathbf{U}}^\imath$. In the setting of $\mathbf{U}^\imath$, the relations (2.16) and (2.17) below were obtained earlier in [**Let99**]-[**Ko14**] and [**BK15**], respectively, while the ıSerre relation (2.15) involving ıdivided powers (2.11)–(2.12) was established in [**CLW21a**].

THEOREM 2.8 (Serre presentation for $\widetilde{\mathbf{U}}^\imath$, [**LW20**, Theorem 4.2]). *Fix $\overline{p}_i \in \mathbb{Z}_2$ for each $i \in \mathbb{I}$. The $\mathbb{Q}(v)$-algebra $\widetilde{\mathbf{U}}^\imath$ has a presentation with generators B_i, $\widetilde{k}_i$ ($i \in \mathbb{I}$) and the relations* (2.13)–(2.17) *below: for $\ell \in \mathbb{I}$, and $i \neq j \in \mathbb{I}$,*

$$\widetilde{k}_i\widetilde{k}_\ell = \widetilde{k}_\ell\widetilde{k}_i, \quad \widetilde{k}_i B_\ell = v^{c_{\tau i,\ell} - c_{i\ell}} B_\ell \widetilde{k}_i, \tag{2.13}$$

$$B_iB_j - B_jB_i = 0, \qquad \text{if } c_{ij} = 0 \text{ and } \tau i \neq j, \tag{2.14}$$

$$\sum_{n=0}^{1-c_{ij}} (-1)^n B_{i,\overline{p_i}}^{(n)} B_j B_{i,\overline{c_{ij}}+\overline{p}_i}^{(1-c_{ij}-n)} = 0, \qquad \text{if } \tau i = i \neq j, \tag{2.15}$$

$$\sum_{n=0}^{1-c_{ij}} (-1)^n B_i^{(n)} B_j B_i^{(1-c_{ij}-n)} = 0, \qquad \text{if } i \neq j \neq \tau i \neq i, \tag{2.16}$$

(2.17)
$$\sum_{n=0}^{1-c_{i,\tau i}} (-1)^{n+c_{i,\tau i}} B_i^{(n)} B_{\tau i} B_i^{(1-c_{i,\tau i}-n)} = \frac{1}{v-v^{-1}} \times$$
$$\big(v^{c_{i,\tau i}}(v^{-2};v^{-2})_{-c_{i,\tau i}} B_i^{(-c_{i,\tau i})}\widetilde{k}_i - (v^2;v^2)_{-c_{i,\tau i}} B_i^{(-c_{i,\tau i})}\widetilde{k}_{\tau i}, \quad \text{if } \tau i \neq i.$$

(Note for $\widetilde{\mathbf{U}}^\imath$ of split type, only the relations (2.13)–(2.15) *are needed.)*

By the Serre presentation of $\widetilde{\mathbf{U}}^\imath$ in Theorem 2.8, letting $\deg B_i = \alpha_i$ and $\deg \widetilde{k}_i = 0$, for $i \in \mathbb{I}$, endows $\widetilde{\mathbf{U}}^\imath$ a $\mathbb{N}\mathbb{I}$-filtered algebra structure. Let $\widetilde{\mathbf{U}}^{\imath,\mathrm{gr}}$ be the associated graded algebra (with generators $\bar{B}_i$ ($i \in \mathbb{I}$) and so on).

REMARK 2.9. There exists a natural algebra monomorphism $\phi : \mathbf{U}^- \longrightarrow \widetilde{\mathbf{U}}^{\imath,\mathrm{gr}}$ which maps $F_i \mapsto \bar{B}_i$, for $i \in \mathbb{I}$. Moreover, the multiplication map $\operatorname{Im}\phi \times \widetilde{\mathbf{U}}^{\imath 0} \to \widetilde{\mathbf{U}}^{\imath,\mathrm{gr}}$ is an isomorphism of vector spaces. These are $\widetilde{\mathbf{U}}^\imath$-variants of some basic results on the structure of $\mathbf{U}^\imath$, which were established originally by Letzter [**Let02**] and Kolb [**Ko14**].

3. Hall algebras

In this section, we introduce (Steinitz-)Hall algebra of the Jordan quiver and its connection to symmetric functions. We formulate Ringel-Hall algebras of quivers to realize halves of quantum groups, and then formulate Bridgeland-Hall algebras to realize the Drinfeld double quantum groups.

3.1. Steinitz-Hall and Hall-Littlewood functions. Let $\mathcal{O}$ be a discrete valued ring with the maximal principal ideal $\mathfrak{p}$ and a finite residue field $\mathbf{k}$ of q elements.

Let $\mathcal{A}$ be the category of all finite $\mathcal{O}$-modules (i.e., modules with finitely many elements). Then $\mathcal{A}$ has only one simple module, $\mathbf{k}$ (which is identified with $\mathcal{O}/\mathfrak{p}$). The indecomposable objects of $\mathcal{A}$ are $\mathcal{O}/\mathfrak{p}^m$ ($m \geq 1$) up to isomorphisms. Given a partition $\lambda = (\lambda_1, \ldots, \lambda_n)$, we denote

$$M(\lambda) := \mathcal{O}/\mathfrak{p}^{\lambda_1} \oplus \cdots \oplus \mathcal{O}/\mathfrak{p}^{\lambda_n}.$$

Then every object $M \in \mathcal{A}$ is isomorphic to $M(\lambda)$ for some unique partition λ.

For any partitions λ, μ, ν, let $F^\lambda_{\mu,\nu}$ denote the number of submodules N of $M(\lambda)$ such that $N \cong M(\nu)$, and $M(\lambda)/N \cong M(\mu)$. Define $\mathcal{H} = \bigoplus_\lambda \mathbb{Q}\mathfrak{u}_\lambda$, a $\mathbb{Q}$-linear space with a basis indexed by partitions. We endow $\mathcal{H}$ with an algebra structure given by

$$\mathfrak{u}_\mu * \mathfrak{u}_\nu := \sum_\lambda F^\lambda_{\mu,\nu} \mathfrak{u}_\lambda, \quad \forall \mu, \nu.$$

We call $\mathcal{H}$ a (Steinitz-) Hall algebra, following [**Mac95**]; it will be understood as Hall algebra of the Jordan quiver, see §6.

Let Λ be the $\mathbb{Q}$-algebra of symmetric functions in infinitely many variables. It is well known that Λ is a polynomial ring $\mathbb{Q}[e_r \mid r \geq 1]$, where e_r denotes the rth elementary symmetric function. Let t be an indeterminant. Denote by $\Lambda[t] := \Lambda \otimes_{\mathbb{Q}} \mathbb{Q}[t]$.

We recall the Hall-Littlewood (HL) functions in variables $x = (x_1, x_2, \ldots)$, following [**Mac95**, Pages 208–211]. Denote by h_r the rth complete symmetric function, for $r \geq 1$. We also set $h_0 = 1$. For an indeterminate u, denote

$$H(u) = \sum_{r=0}^{\infty} h_r u^r = \prod_{i \geq 1} \frac{1}{1 - u x_i}, \tag{3.1}$$

$$Q(u) = \sum_{r=0}^{\infty} q_r u^r = H(u)/H(tu). \tag{3.2}$$

We set $q_\lambda = q_{\lambda_1} q_{\lambda_2} \cdots$, for any composition $\lambda = (\lambda_1, \lambda_2, \ldots)$. Let u_1, u_2, ... be independent indeterminates. Let

$$F(u) = \frac{1-u}{1-tu} = 1 + (1 - t^{-1}) \sum_{r \geq 1} t^r u^r. \tag{3.3}$$

Then the HL symmetric function Q_λ is the coefficient of $u^\lambda = u_1^{\lambda_1}u_2^{\lambda_2}\cdots$ in

$$Q(u_1,u_2,\dots) = \prod_{i\geq 1} Q(u_i)\prod_{i<j} F(u_iu_j^{-1}). \tag{3.4}$$

By definition, we have $Q_{(r)} = q_r$, which is also denoted by Q_r.

Let $1 \leq i < j$. The raising operator R_{ij} acts on integer sequences by

$$R_{ij}\alpha = (\dots,\alpha_i+1,\dots,\alpha_j-1,\dots).$$

The actions of raising operators on q_α are given by letting

$$R_{ij}q_\alpha = q_{R_{ij}\alpha}.$$

According to [**Mac95**, p.212], the generating function (3.4) for HL functions Q_λ can be restated that

$$Q_\lambda = \prod_{i<j}\frac{1-R_{ij}}{1-tR_{ij}}q_\lambda = \Big(1+(t-1)\sum_{r\geq 1} t^{r-1}R_{ij}^r\Big)q_\lambda. \tag{3.5}$$

Then $\{Q_\lambda \mid \lambda \text{ is a partition}\}$ forms a $\mathbb{Q}[t]$-basis of $\Lambda[t]$.

For a composition α, we denote

$$|\alpha| = \alpha_1+\alpha_2+\cdots, \qquad n(\lambda) = \sum_i (i-1)\lambda_i, \tag{3.6}$$

and denote by $m_i(\lambda)$ the number of times i occurring as a part of λ.

For any partition λ, define $P_\lambda \in \Lambda[t]$ such that $Q_\lambda = b_\lambda(t)P_\lambda$, where

$$\varphi_r(t) = (1-t)(1-t^2)\cdots(1-t^r); \tag{3.7}$$

$$b_\lambda(t) = \prod_{i\geq 1}\varphi_{m_i(\lambda)}(t). \tag{3.8}$$

In fact, $P_\lambda \in \Lambda[t]$.

We view all functions in $\Lambda[t]$ to be in Λ by setting $t = q^{-1}$.

THEOREM 3.1 (Steinitz, Hall, [**Mac95**, II, III]).

(1) *$\mathcal{H}$ is a commutative associative algebra with the identity $\mathfrak{u}_\emptyset$.*
(2) *There is an isomorphism of algebras $\phi_{(q)} : \mathcal{H} \to \Lambda$ which maps $\mathfrak{u}_{(1^r)} \mapsto q^{-\binom{r}{2}}e_r$, for $r \geq 1$.*
(3) *For any partition λ, we have $\phi_{(q)}(\mathfrak{u}_\lambda) = q^{-n(\lambda)}P_\lambda$.*

In fact, $|\operatorname{Aut}(M(\lambda))| = q^{|\lambda|+2n(\lambda)}b_\lambda(q^{-1})$; see e.g. [**Sch06**, Lemma 2.8]. Then we have

$$\phi_{(q)}(|\operatorname{Aut}(M(\lambda))|\cdot \mathfrak{u}_\lambda) = q^{|\lambda|+n(\lambda)}Q_\lambda, \tag{3.9}$$

for any partition λ.

3.2. Ringel-Hall and halves of quantum groups. Let $\mathcal{E}$ be an essentially small exact category in the sense of Quillen, linear over $\mathbf{k} = \mathbb{F}_q$. For the basics on exact categories, we refer to [**Buh10**] and references therein. In particular, abelian categories are exact categories.

Assume that $\mathcal{E}$ has finite morphism and extension spaces, i.e.,

$$|\operatorname{Hom}(M,N)| < \infty, \quad |\operatorname{Ext}^1(M,N)| < \infty, \quad \forall M,N\in\mathcal{E}.$$

Given objects $M,N,L \in \mathcal{E}$, define $\operatorname{Ext}^1(M,N)_L \subseteq \operatorname{Ext}^1(M,N)$ as the subset parameterizing extensions whose middle term is isomorphic to L. We define the

Ringel-Hall algebra $\mathcal{H}(\mathcal{E})$ (or *Hall algebra* for short) to be the $\mathbb{Q}$-vector space whose basis is formed by the isoclasses $[M]$ of objects M in $\mathcal{E}$, with the multiplication defined by (see [**Br13**])

$$[M] \diamond [N] = \sum_{[L] \in \mathrm{Iso}(\mathcal{E})} \frac{|\operatorname{Ext}^1(M,N)_L|}{|\operatorname{Hom}(M,N)|}[L]. \tag{3.10}$$

For any three objects L, M, N, let

$$F_{MN}^L := \big|\{X \subseteq L \mid X \cong N, L/X \cong M\}\big|. \tag{3.11}$$

The Riedtmann-Peng formula states that

$$F_{MN}^L = \frac{|\operatorname{Ext}^1(M,N)_L|}{|\operatorname{Hom}(M,N)|} \cdot \frac{|\operatorname{Aut}(L)|}{|\operatorname{Aut}(M)||\operatorname{Aut}(N)|},$$

where $\operatorname{Aut}(M)$ denotes the automorphism group of M. For any object M, let

$$[\![M]\!] := \frac{[M]}{|\operatorname{Aut}(M)|}. \tag{3.12}$$

Then the Hall multiplication (3.10) can be reformulated to be

$$[\![M]\!] \diamond [\![N]\!] = \sum_{[\![L]\!]} F_{M,N}^L [\![L]\!], \tag{3.13}$$

which is the version of Hall multiplication used in [**Rin90**].

Let $Q = (\mathbb{I}, Q_1)$ be a quiver (without loops), and $\mathbf{k} = \mathbb{F}_q$. Let $\mathcal{A} = \operatorname{mod}(\mathbf{k}Q)$ be the category of finite-dimensional nilpotent $\mathbf{k}Q$-modules. Then $\mathcal{A}$ is a hereditary abelian category. Let $\langle -, - \rangle$ be the Euler form of $\mathcal{A}$, i.e.,

$$\langle M, N \rangle = \dim_{\mathbf{k}} \operatorname{Hom}(M,N) - \dim_{\mathbf{k}} \operatorname{Ext}^1(M,N), \quad \forall M, N \in \mathcal{A}.$$

Throughout the paper we shall denote

$$\mathbf{v} = \sqrt{q}. \tag{3.14}$$

The twisted Hall algebra $\widetilde{\mathcal{H}}(\mathcal{A})$ is defined over $\mathcal{H}(\mathcal{A}) \otimes_{\mathbb{Q}} \mathbb{Q}(\mathbf{v})$ with multiplication twisted by Euler form:

$$[L] * [M] = \mathbf{v}^{\langle L, M \rangle} [L] \diamond [M].$$

Denote by S_i the 1-dimensional simple nilpotent module supported at $i \in \mathbb{I}$.

Recall the $\mathbb{Q}(v)$-algebra $\mathbf{U}^-$, and consider the base change $\mathbf{U}_{\mathbf{v}}^- := \mathbb{Q}(\mathbf{v}) \otimes_{\mathbb{Q}(v)} \mathbf{U}^-$.

THEOREM 3.2 (Ringel [**Rin90**], Green [**Gr95**]). *Let $Q = (\mathbb{I}, Q_1)$ be a quiver, and $\mathcal{A} = \operatorname{mod}(\mathbf{k}Q)$. Then there exists an injective homomorphism of $\mathbb{Q}(\mathbf{v})$-algebras $\phi : \mathbf{U}_{\mathbf{v}}^- \longrightarrow \widetilde{\mathcal{H}}(\mathcal{A})$, which maps F_i to $[\![S_i]\!]$, for all $i \in \mathbb{I}$. Moreover, ϕ is an isomorphism if and only if Q is of Dynkin type.*

PROOF. The verification that $\phi(F_i)$ satisfy the quantum Serre relations is given below in Example 3.5 (also cf. Examples 3.3–3.4). Hence we have a morphism $\phi : \mathbf{U}_{\mathbf{v}}^- \to \widetilde{\mathcal{H}}(\mathcal{A})$ which maps F_i to $[\![S_i]\!]$, for all $i \in \mathbb{I}$.

We now sketch the proof of injectivity of ϕ.

STEP 1. We identify $\mathbf{U}^-$ with $\mathbf{f}$ in [**Lus93**, §1.2]. Then $\mathbf{U}^- = \oplus_{\mu\in\mathbb{N}\mathbb{I}}\mathbf{U}^-_\mu$ is a graded bialgebra with $\deg F_i = \alpha_i$ and a comultiplication $r : \mathbf{U}^- \to \mathbf{U}^- \otimes \mathbf{U}^-$ such that $r(F_i) = F_i \otimes 1 + 1 \otimes F_i$, for each $i \in \mathbb{I}$. By [**Lus93**, §1.2], there exists a non-degenerate symmetric bilinear form $(\cdot,\cdot)$ on $\mathbf{U}^-_\mathbf{v}$ (which is slightly renormalized here) such that

(1) $(1,1) = 1$, and $(F_i, F_j) = \delta_{i,j}\frac{1}{\mathbf{v}^2-1}$, for $i, j \in \mathbb{I}$.
(2) $(x, yz) = (r(x), y\otimes z)$ for $x, y, z \in \mathbf{U}^-_\mathbf{v}$; it is understood that $(a\otimes a', b\otimes b') = (a, b)(a', b')$.

STEP 2. $\widetilde{\mathcal{H}}(\mathcal{A})$ is a bialgebra with comultiplication r' defined by

$$[\![A]\!] \mapsto \sum_{[B],[C]} \mathbf{v}^{\langle B,C\rangle}\frac{|\operatorname{Ext}^1_\mathcal{A}(B,C)_A|}{|\operatorname{Hom}_\mathcal{A}(B,C)|}[\![B]\!]\otimes[\![C]\!].$$

There is a non-degenerate bilinear pairing $(\cdot,\cdot)'$ on $\widetilde{\mathcal{H}}(\mathcal{A})$ defined by

$$([\![M]\!], [\![N]\!])' := \delta_{[M],[N]}\frac{1}{|\operatorname{Aut}(M)|}.$$

Note that $([\![S_i]\!], [\![S_j]\!])' = \delta_{ij}\frac{1}{q-1} = \delta_{ij}\frac{1}{\mathbf{v}^2-1}$.

STEP 3. We have $(\phi(x), \phi(y))' = (x, y)$ for any $x, y \in \mathbf{U}^-_\mathbf{v}$.

STEP 4. For any $x \in \ker\phi$, we have $(x, y) = 0$ for any $y \in \mathbf{U}^-_\mathbf{v}$, and then $x = 0$ since $(\cdot,\cdot)$ is non-degenerate. □

EXAMPLE 3.3. Let Q be the quiver consisting of a single vertex with no arrow. Then $\mathbf{k}Q = \mathbf{k}$. Denote by S the (1-dimensional) simple $\mathbf{k}Q$-module. For any $l \geq 1$, we shall write

$$[lS] = [\underbrace{S\oplus\cdots\oplus S}_{l}], \qquad [S]^{*l} = \underbrace{[S]*\cdots*[S]}_{l}, \qquad [\![S]\!]^{*l} = \underbrace{[\![S]\!]*\cdots*[\![S]\!]}_{l}.$$

One shows by induction on l that

$$[S]^{*l} = \mathbf{v}^{-\frac{l(l-1)}{2}}[lS]. \tag{3.15}$$

Note that $|\operatorname{Aut}(S^{\oplus l})| = (q-1)^l\mathbf{v}^{\frac{3l(l-1)}{2}}[l]^!_\mathbf{v}$ for any $l \geq 1$. So

$$[\![S]\!]^{*l} = \mathbf{v}^{-\frac{l(l-1)}{2}}\frac{|\operatorname{Aut}(S^{\oplus l})|}{(q-1)^l}[\![lS]\!] = \mathbf{v}^{l(l-1)}][l]^!_\mathbf{v}[\![lS]\!].$$

and then

$$[S]^{(l)} := \frac{[S]^{*l}}{[l]^!_\mathbf{v}} = \mathbf{v}^{-\frac{l(l-1)}{2}}\frac{[lS]}{[l]^!_\mathbf{v}}, \qquad [\![S]\!]^{(l)} := \frac{[\![S]\!]^{*l}}{[l]^!_\mathbf{v}} = \mathbf{v}^{l(l-1)}[\![lS]\!]. \tag{3.16}$$

Before tackling the general rank 2 quivers, let us first warm up with the simplest cases.

EXAMPLE 3.4.

(1) Assume Q is the $A_1 \times A_1$ quiver with vertices labelled by 1 and 2 and no arrows. Then we have

$$[S_1] * [S_2] = [S_1 \oplus S_2] = [S_2] * [S_1],$$

i.e., $[S_1]$ and $[S_2]$ commute with each other, since

$$\operatorname{Hom}_{\mathbf{k}Q}(S_1, S_2) = 0 = \operatorname{Hom}_{\mathbf{k}Q}(S_2, S_1), \qquad \operatorname{Ext}^1_{\mathbf{k}Q}(S_1, S_2) = 0 = \operatorname{Ext}^1_{\mathbf{k}Q}(S_2, S_1).$$

(2) Assume Q is the A_2 quiver with vertices labelled by 1 and 2 and one arrow from 1 to 2. Then $\dim_{\mathbf{k}} \mathrm{Ext}^1_{\mathbf{k}Q}(S_1, S_2) = 1$, and $\dim_{\mathbf{k}} \mathrm{Ext}^1_{\mathbf{k}Q}(S_2, S_1) = 0$. Let I_2 be the indecomposable injective module which fits into the following short exact sequence of $\mathbf{k}Q$-modules:

$$0 \longrightarrow S_2 \longrightarrow I_2 \longrightarrow S_1 \longrightarrow 0.$$

Note I_2 is also projective (i.e., it is the projective cover of S_1). Then we have

$$[S_1] * [S_2] = \mathbf{v}^{-1}[S_1 \oplus S_2] + \mathbf{v}^{-1}(q-1)[I_2],$$
$$[S_2] * [S_1] = [S_1 \oplus S_2],$$

and thus

$$\mathbf{v}[S_1] * [S_2] - [S_2] * [S_1] = (q-1)[I_2].$$

We have

$$[I_2] * [S_1] = \mathbf{v}^{-1}[S_1 \oplus I_2], \quad [S_1] * [I_2] = [S_1 \oplus I_2];$$
$$[I_2] * [S_2] = [S_2 \oplus I_2], \quad [S_2] * [I_2] = \mathbf{v}^{-1}[S_2 \oplus I_2],$$

by noting that

$$\dim_{\mathbf{k}} \mathrm{Hom}_{\mathbf{k}Q}(I_2, S_1) = 1 = \dim_{\mathbf{k}} \mathrm{Hom}_{\mathbf{k}Q}(S_2, I_2),$$
$$\dim_{\mathbf{k}} \mathrm{Hom}_{\mathbf{k}Q}(I_2, S_2) = 0 = \dim_{\mathbf{k}} \mathrm{Hom}_{\mathbf{k}Q}(S_1, I_2).$$

Therefore we have

$$\begin{aligned}&[\![S_1]\!] * [\![S_1]\!] * [\![S_2]\!] - (\mathbf{v}+\mathbf{v}^{-1})[\![S_1]\!] * [\![S_2]\!] * [\![S_1]\!] + [\![S_2]\!] * [\![S_1]\!] * [\![S_1]\!]\\ &= \frac{1}{(q-1)^3}\Big([S_1] * \big([S_1] * [S_2] - \mathbf{v}^{-1}[S_2] * [S_1]\big) - \big(\mathbf{v}[S_1] * [S_2] - [S_2] * [S_1]\big) * [S_1]\Big)\\ &= \frac{1}{(q-1)^2}\big(\mathbf{v}^{-1}[S_1] * [I_2] - [I_2] * [S_1]\big) = 0.\end{aligned}$$

An entirely similar computation shows that

$$[\![S_2]\!] * [\![S_2]\!] * [\![S_1]\!] - (\mathbf{v}+\mathbf{v}^{-1})[\![S_2]\!] * [\![S_1]\!] * [\![S_2]\!] + [\![S_1]\!] * [\![S_2]\!] * [\![S_2]\!] = 0.$$

That is, $[\![S_1]\!]$ and $[\![S_2]\!]$ satisfy the quantum Serre relation of type A_2.

Now we treat the general rank 2 quivers.

EXAMPLE 3.5 (General rank 2 quivers). Consider a rank 2 quiver Q with a arrows from vertex 1 to vertex 2, and b arrows from 2 to 1:

$$Q = (\, 1 \underset{\longleftarrow b \longrightarrow}{\overset{\longrightarrow a \longrightarrow}{\rightleftarrows}} 2 \,), \qquad \text{where } a+b=-c_{12}.$$

The arrows $1 \to 2$ are denoted by α_i, $1 \le i \le a$, and $2 \to 1$ are denoted by β_j, $1 \le j \le b$.

For any $\mathbf{k}Q$-module $M = (M_i, M(\alpha_j), M(\beta_k),)_{i=1,2;1\le j\le a,1\le k\le b}$ such that $\dim_{\mathbf{k}} M_2 = 1$, we define

$$U_M := \bigcap_{1\le i\le a} \mathrm{Ker}\, M(\alpha_i), \qquad W_M := \sum_{j=1}^{b} \mathrm{Im}\, M(\beta_j), \tag{3.17}$$

and let

$$u_M := \dim_{\mathbf{k}} U_M, \qquad w_M := \dim_{\mathbf{k}} W_M. \tag{3.18}$$

For any $r \geq 0$, define

$$\mathcal{I}_r := \{[M] \in \operatorname{Iso}(\operatorname{mod}(\mathbf{k}Q)) \mid \widehat{M} = r\widehat{S_1} + \widehat{S_2}, W_M \subseteq U_M\}, \tag{3.19}$$

$$\mathcal{J}_r := \{[M] \in \operatorname{Iso}(\operatorname{mod}(\mathbf{k}Q)) \mid \widehat{M} = r\widehat{S_1} + \widehat{S_2}, S_2 \in \operatorname{add}\operatorname{top}(M)\}. \tag{3.20}$$

We shall write $[sS_1] = [S_1^{\oplus s}]$. For any $s, t \geq 0$, we have

$$\begin{aligned}
&[sS_1] * [S_2] * [tS_1] \\
&= \mathbf{v}^{-tb} \sum_{[U] \in \mathcal{J}_t} |\operatorname{Ext}^1(S_2, S_1^{\oplus t})_U| [S_1^{\oplus s}] * [U] \\
&= \mathbf{v}^{-tb} \sum_{[U] \in \mathcal{J}_t} |\operatorname{Ext}^1(S_2, S_1^{\oplus t})_U| \sum_{[M] \in \mathcal{I}_{s+t}} \mathbf{v}^{sa-st} q^{s(t-a)} \frac{|\operatorname{Ext}^1(S_1^{\oplus s}, U)_M|}{|\operatorname{Hom}(S_1^{\oplus s}, U)|} \cdot [M].
\end{aligned}$$

It is clear that $F^U_{S_2, S_1^{\oplus t}} \neq 0$ if and only if $[U] \in \mathcal{J}_t$; in this case, $F^U_{S_2, S_1^{\oplus t}} = 1$. By the Riedtmann-Peng formula we have

$$|\operatorname{Ext}^1(S_2, S_1^{\oplus t})_U| = \frac{|\operatorname{Aut}(S_2)| \cdot |\operatorname{Aut}(S_1^{\oplus t})|}{|\operatorname{Aut}(U)|} F^U_{S_2, S_1^{\oplus t}} = \frac{|\operatorname{Aut}(S_2)| \cdot |\operatorname{Aut}(S_1^{\oplus t})|}{|\operatorname{Aut}(U)|},$$

$$|\operatorname{Ext}^1(S_1^{\oplus s}, U)_M| = F^M_{S_1^{\oplus s}, U} \cdot |\operatorname{Hom}(S_1^{\oplus s}, U)| \frac{|\operatorname{Aut}(S_1^{\oplus s})| \cdot |\operatorname{Aut}(U)|}{|\operatorname{Aut}(M)|}.$$

Therefore we have

$$\begin{aligned}
&[sS_1] * [S_2] * [tS_1] \\
&= \mathbf{v}^{st-sa-tb}(q-1) \sum_{[M] \in \mathcal{I}_{s+t}} |\operatorname{Aut}(S_1^{\oplus s})||\operatorname{Aut}(S_1^{\oplus t})| \Big(\sum_{[U] \in \mathcal{J}_t} F^M_{S_1^{\oplus s}, U} \Big) [\![M]\!].
\end{aligned}$$

A simple computation shows that

$$\begin{aligned}
\sum_{[U] \in \mathcal{J}_t} F^M_{S_1^{\oplus s}, U} &= |\{f : S_1^{\oplus s} \hookrightarrow M\}| \cdot \frac{1}{|\operatorname{Aut}(S_1^{\oplus s})|} \\
&= |\operatorname{Gr}(t - w_M, u_M - w_M)| \\
&= \mathbf{v}^{(u_M - t)(t - w_M)} \begin{bmatrix} u_M - w_M \\ t - w_M \end{bmatrix}_{\mathbf{v}}.
\end{aligned}$$

Hence, we can rewrite the previous formula as

$$[\![sS_1]\!] * [\![S_2]\!] * [\![tS_1]\!] = \mathbf{v}^{st-sa-tb} \sum_{[M] \in \mathcal{I}_{s+t}} \mathbf{v}^{(u_M - t)(t - w_M)} \begin{bmatrix} u_M - w_M \\ t - w_M \end{bmatrix}_{\mathbf{v}} [\![M]\!]. \tag{3.21}$$

Then by (3.16) and (3.21), we have

$$\begin{aligned}
&\sum_{t=0}^{a+b+1} (-1)^t [\![S_1]\!]^{(a+b+1-t)} * [\![S_2]\!] * [\![S_1]\!]^{(t)} \\
&= \sum_{[M] \in \mathcal{I}_{a+b+1}} \mathbf{v}^{(a+b+1)b - u_M w_M} \sum_{t=0}^{a+b+1} (-1)^t \mathbf{v}^{-(2b+1-u_M-w_M)t} \begin{bmatrix} u_M - w_M \\ t - w_M \end{bmatrix}_{\mathbf{v}} [\![M]\!].
\end{aligned}$$

Since $u_M \geq b+1 > w_M$ for any $[M] \in \mathcal{I}_{a+b+1}$, we have $1 - u_M - w_M \leq 2b + 1 - u_M - w_M \leq u_M + w_M - 1$, and by a standard v-binomial identity (cf.

[**Lus93**, 1.3.1(c)])

$$\sum_{t=0}^{a+b+1} (-1)^t \mathbf{v}^{-(2b+1-u_M-w_M)t} \begin{bmatrix} u_M - w_M \\ t - w_M \end{bmatrix}_{\mathbf{v}} = 0.$$

Therefore, we have concluded that

$$\sum_{t=0}^{a+b+1} (-1)^t [\![S_1]\!]^{(a+b+1-t)} * [\![S_2]\!] * [\![S_1]\!]^{(t)} = 0.$$

The analogous identity with $[\![S_1]\!], [\![S_2]\!]$ switched holds since the above identity is symmetric with respect to a, b.

3.3. Bridgeland-Hall algebras and quantum groups. Let $\mathcal{A}$ be a $\mathbf{k}$-linear exact category with finite-dimensional Hom-spaces and Ext^1-spaces. Let $\mathcal{C}_{\mathbb{Z}_2}(\mathcal{A})$ be the exact category of $\mathbb{Z}_2$-graded complexes over $\mathcal{A}$. Namely, an object M of this category is a diagram with objects and morphisms in $\mathcal{A}$:

$$M^0 \underset{d^1}{\overset{d^0}{\rightleftarrows}} M^1 , \quad d^1 d^0 = d^0 d^1 = 0.$$

All indices of components of $\mathbb{Z}_2$-graded objects will be understood modulo 2. A morphism $s = (s^0, s^1) : M \to N$ is a diagram

$$\begin{array}{ccc} M^0 & \underset{d^1}{\overset{d^0}{\rightleftarrows}} & M^1 \\ \downarrow s^0 & & \downarrow s^1 \\ N^0 & \underset{e^1}{\overset{e^0}{\rightleftarrows}} & N^1 \end{array}$$

with $s^{i+1} d^i = e^i s^i$ for $i = 0, 1$.

The shift functor on complexes is an involution $*$ on $\mathcal{C}_{\mathbb{Z}_2}(\mathcal{A})$ which shifts the grading and changes the sign of the differential as follows:

$$\left(M^0 \underset{d^1}{\overset{d^0}{\rightleftarrows}} M^1\right) \xrightarrow{\ *\ } \left(M^1 \underset{-d^0}{\overset{-d^1}{\rightleftarrows}} M^0\right).$$

Similar to ordinary complexes, we can define the ith homology group for M, denoted by $H^i(M)$, for $i \in \mathbb{Z}_2$. A complex is called acyclic if each of its homology groups is zero. The subcategory of all acyclic $\mathbb{Z}_2$-graded complexes over $\mathcal{A}$ is denoted by $\mathcal{C}_{\mathbb{Z}_2,ac}(\mathcal{A})$.

For any object $X \in \mathcal{A}$, we define

$$\begin{aligned} &K_X := \left(X \underset{0}{\overset{1}{\rightleftarrows}} X \right), && K_X^* := \left(X \underset{1}{\overset{0}{\rightleftarrows}} X \right), \\ &C_X := \left(0 \rightleftarrows X \right), && C_X^* := \left(X \rightleftarrows 0 \right) \end{aligned} \tag{3.22}$$

in $\mathcal{C}_{\mathbb{Z}_2}(\mathcal{A})$. Note that K_X, K_X^* are acyclic complexes. Let $\mathrm{res} : \mathcal{C}_{\mathbb{Z}_2}(\mathcal{A}) \to \mathcal{A} \times \mathcal{A}$ be the forgetful functor, which maps $M = \left(M^0 \underset{d^1}{\overset{d^0}{\rightleftarrows}} M^1 \right)$ to (M^0, M^1).

From now on, we shall impose a stronger assumption that $\mathcal{A}$ is a hereditary abelian category.

LEMMA 3.6 ([**LP21**, Proposition 2.3]). *Let $\mathcal{A}$ be a hereditary abelian category. For any $K \in \mathcal{C}_{\mathbb{Z}_2}(\mathcal{A})$ with K acyclic, we have* $\operatorname{proj.dim}_{\mathcal{C}_{\mathbb{Z}_2}(\mathcal{A})}(K) \leq 1$, *and* $\operatorname{inj.dim}_{\mathcal{C}_{\mathbb{Z}_2}(\mathcal{A})}(K) \leq 1$.

By Lemma 3.6, for any $K, M \in \mathcal{C}_{\mathbb{Z}_2}(\mathcal{A})$ with K acyclic, the following pairings are well defined:

$$\langle K, M\rangle = \dim \operatorname{Hom}_{\mathcal{C}_{\mathbb{Z}_2}(\mathcal{A})}(K, M) - \dim \operatorname{Ext}^1_{\mathcal{C}_{\mathbb{Z}_2}(\mathcal{A})}(K, M)$$

and

$$\langle M, K\rangle = \dim \operatorname{Hom}_{\mathcal{C}_{\mathbb{Z}_2}(\mathcal{A})}(M, K) - \dim \operatorname{Ext}^1_{\mathcal{C}_{\mathbb{Z}_2}(\mathcal{A})}(M, K).$$

They induce bilinear pairings (call *Euler forms*) between the Grothendieck groups $K_0(\mathcal{C}_{\mathbb{Z}_2,ac}(\mathcal{A}))$ and $K_0(\mathcal{C}_{\mathbb{Z}_2}(\mathcal{A}))$:

$$\langle \cdot, \cdot \rangle : K_0(\mathcal{C}_{\mathbb{Z}_2,ac}(\mathcal{A})) \times K_0(\mathcal{C}_{\mathbb{Z}_2}(\mathcal{A})) \longrightarrow \mathbb{Z},$$

and

$$\langle \cdot, \cdot \rangle : K_0(\mathcal{C}_{\mathbb{Z}_2}(\mathcal{A})) \times K_0(\mathcal{C}_{\mathbb{Z}_2,ac}(\mathcal{A})) \longrightarrow \mathbb{Z}.$$

We can use the same notation above, since the two forms coincide on $K_0(\mathcal{C}_{\mathbb{Z}_2,ac}(\mathcal{A})) \times K_0(\mathcal{C}_{\mathbb{Z}_2,ac}(\mathcal{A}))$.

Let $K_0(\mathcal{A})$ be the Grothendieck group of $\mathcal{A}$. For any $A \in \mathcal{A}$, we denote by $\widehat{A}$ the corresponding element in the Grothendieck group $K_0(\mathcal{A})$. We also use $\langle \cdot, \cdot \rangle$ to denote the Euler form of $\mathcal{A}$, i.e.,

$$\langle \widehat{A}, \widehat{B}\rangle = \dim \operatorname{Hom}_{\mathcal{A}}(A, B) - \dim \operatorname{Ext}^1_{\mathcal{A}}(A, B), \text{ for any } A, B \in \mathcal{A}.$$

Let $(\cdot, \cdot)$ be the symmetrized Euler form of $\mathcal{A}$, i.e., $(\widehat{A}, \widehat{B}) = \langle \widehat{A}, \widehat{B}\rangle + \langle \widehat{B}, \widehat{A}\rangle$.

PROPOSITION 3.7 ([**LP21**, Proposition 2.4, Corollary 2.5]). *For any $A, B \in \mathcal{A}$, we have*

$$\langle C_A, K_B\rangle = \langle C_A^*, K_B^*\rangle = \langle \widehat{A}, \widehat{B}\rangle, \quad \langle K_B, C_A^*\rangle = \langle K_B^*, C_A\rangle = \langle \widehat{B}, \widehat{A}\rangle; \tag{3.23}$$

$$\langle K_B, C_A\rangle = \langle C_A^*, K_B\rangle = \langle C_A, K_B^*\rangle = \langle K_B^*, C_A^*\rangle = 0; \tag{3.24}$$

$$\langle K_A, K_B\rangle = \langle K_A^*, K_B^*\rangle = \langle K_A, K_B^*\rangle = \langle K_A^*, K_B\rangle = \langle \widehat{A}, \widehat{B}\rangle. \tag{3.25}$$

In the remaining of this subsection, we shall assume that $\mathcal{A}$ is a hereditary abelian category with enough projective objects. Let $\mathcal{P}$ be the full subcategory of $\mathcal{A}$ formed by projective objects. Let $\mathcal{C}_{\mathbb{Z}_2}(\mathcal{P})$ be the category of $\mathbb{Z}_2$-graded complexes over $\mathcal{P}$. Then K_P, K_P^* are projective objects in $\mathcal{C}_{\mathbb{Z}_2}(\mathcal{A})$, and $\mathcal{C}_{\mathbb{Z}_2}(\mathcal{P})$ is a Frobenius category with K_P, K_P^* $(P \in \mathcal{P})$ as projective-injective objects; see §4.1 for the definition of Frobenius categories. For any $X \in \mathcal{A}$, let $0 \to P_X \xrightarrow{f} Q_X \to X \to 0$ be the minimal projective resolution of X. Define

$$G_X := \big(P_X \underset{0}{\overset{f}{\rightleftarrows}} Q_X \big), \qquad G_X^* := \big(Q_X \underset{f}{\overset{0}{\rightleftarrows}} P_X \big). \tag{3.26}$$

We consider $\widetilde{\mathcal{H}}(\mathcal{C}_{\mathbb{Z}_2}(\mathcal{P}))$, the twisted Ringel-Hall algebra of $\mathcal{C}_{\mathbb{Z}_2}(\mathcal{P})$, i.e.,

$$\widetilde{\mathcal{H}}(\mathcal{C}_{\mathbb{Z}_2}(\mathcal{P})) = \bigoplus_{[M]\in \operatorname{Iso}(\mathcal{C}_{\mathbb{Z}_2}(\mathcal{P}))} \mathbb{Q}(\mathbf{v})[M], \tag{3.27}$$

with multiplication given by

$$[L] * [M] = \mathbf{v}^{\langle \operatorname{res} L, \operatorname{res} M\rangle} \sum_{[N]} \frac{|\operatorname{Ext}^1(L,M)_N|}{|\operatorname{Hom}(L,M)|}[N].$$

Since $K_P, K_P^* \in \mathcal{C}_{\mathbb{Z}_2}(\mathcal{P})$ are projective-injective objects, we have

$$\begin{aligned}[K_P] * [M] &= \mathbf{v}^{\langle P, M^0 \oplus M^1\rangle} q^{-\langle K_P, M\rangle}[K_P \oplus M] \\ &= \mathbf{v}^{\langle P, M^0\rangle - \langle P, M^1\rangle}[K_P \oplus M],\end{aligned}$$

and similarly

$$[M] * [K_P] = \mathbf{v}^{\langle M^1, P\rangle - \langle M^0, P\rangle}[K_P \oplus M].$$

Therefore, we have

$$[K_P] * [M] = \mathbf{v}^{(P, M^0) - (P, M^1)}[M] * [K_P]. \tag{3.28}$$

So one can show that $\mathcal{S}_{\mathcal{P}} := \{a[K_P] \mid a \in \mathbb{Q}(\mathbf{v})^\times, P \in \mathcal{P}\}$ is a multiplicatively closed subset with the identity [0], which is a right Ore, right reversible subset.

By definition, the *Bridgeland-Hall algebra*, denoted by $\mathcal{D}\widetilde{\mathcal{H}}(\mathcal{A})$, is the right localization of $\widetilde{\mathcal{H}}(\mathcal{C}_{\mathbb{Z}_2}(\mathcal{P}))$ with respect to $\mathcal{S}_{\mathcal{P}}$.

Denote by

$$E_X = \mathbf{v}^{\langle P, X\rangle}[K_{P_X}]^{-1} * [G_X], \quad F_X = E_X^*,$$

and

$$[K_X] := [K_{Q_X}] * [K_{P_X}]^{-1}, \quad [K_X^*] := [K_{Q_X}^*] * [K_{P_X}^*]^{-1},$$

where $0 \to P_X \to Q_X \to X \to 0$ is a projective resolution of X.

Note that E_X, F_X, $[K_X]$, $[K_X^*]$ do not depend on the projective resolutions of X (not necessarily minimal). In fact, $[K_X]$ and $[K_X^*]$ only depend on the class $\widehat{X} \in K_0(\mathcal{A})$. So for any $\alpha \in K_0(\mathcal{A})$, one can define $[K_\alpha], [K_\alpha^*]$ in $\mathcal{D}\widetilde{\mathcal{H}}(\mathcal{A})$.

In case $\mathcal{A} = \operatorname{mod}(\mathbf{k}Q)$ for an acyclic quiver Q, we shall denote its Bridgeland-Hall algebra by $\mathcal{D}\widetilde{\mathcal{H}}(\mathbf{k}Q)$. We shall also consider the base change $\widetilde{\mathbf{U}}_{\mathbf{v}} = \mathbb{Q}(\mathbf{v}) \otimes_{\mathbb{Q}(v)} \widetilde{\mathbf{U}}$.

THEOREM 3.8 ([**Br13**]). *There is an injective homomorphism of* $\mathbb{Q}(\mathbf{v})$*-algebras* $\Psi : \widetilde{\mathbf{U}}_{\mathbf{v}} \longrightarrow \mathcal{D}\widetilde{\mathcal{H}}(\mathbf{k}Q)$ *such that*

$$\Psi(E_i) = \frac{\mathbf{v}}{q-1}E_{S_i}, \; \Psi(F_i) = \frac{-1}{q-1}F_{S_i}, \; \Psi(K_i) = [K_{S_i}], \; \Psi(K_i') = [K_{S_i}^*], \; \forall i \in \mathbb{I}. \tag{3.29}$$

PROOF. We give a sketched proof here. First, to show that Ψ is a homomorphism, we prove that Ψ preserves the defining relations for $\widetilde{\mathbf{U}}$.

For the Serre relations, we only prove (2.5) for F's, as the same proof applies for the Serre relations for E's. With the help of Theorem 3.2, we only need to check that

$$I^- : \widetilde{\mathcal{H}}(\mathbf{k}Q) \longrightarrow \mathcal{D}\widetilde{\mathcal{H}}(\mathbf{k}Q), \quad [M] \mapsto F_M$$

is an algebra homomorphism.

For any $X, Z \in \operatorname{mod}(\mathbf{k}Q)$, we have two (minimal) projective resolution:

$$0 \longrightarrow P_X \longrightarrow Q_X \longrightarrow X \longrightarrow 0, \qquad 0 \longrightarrow P_Z \longrightarrow Q_Z \longrightarrow Z \longrightarrow 0,$$

which give rise to two exact sequences:

$$(3.30)\quad 0 \longrightarrow K_{P_X} \to G_X \longrightarrow C_X \longrightarrow 0, \qquad 0 \longrightarrow K_{P_Z} \to G_Z \longrightarrow C_Z \longrightarrow 0.$$

We make the following

CLAIM. We have $\mathrm{Ext}^1_{\mathcal{A}}(X,Z) \cong \mathrm{Ext}^1_{\mathcal{C}_{\mathbb{Z}_2}(\mathcal{A})}(C_X, C_Z) \cong \mathrm{Ext}^1_{\mathcal{C}_{\mathbb{Z}_2}(\mathcal{A})}(G_X, G_Z)$.

Let us prove the Claim. Using (3.30), we have the following exact sequences:

$$0 \longrightarrow \mathrm{Hom}_{\mathcal{C}_{\mathbb{Z}_2}(\mathcal{A})}(K_{P_X}, C_Z) \longrightarrow \mathrm{Hom}_{\mathcal{C}_{\mathbb{Z}_2}(\mathcal{A})}(C_X, C_Z) \longrightarrow \mathrm{Hom}_{\mathcal{C}_{\mathbb{Z}_2}(\mathcal{A})}(G_X, C_Z) \longrightarrow 0,$$
$$0 \longrightarrow \mathrm{Ext}^1_{\mathcal{C}_{\mathbb{Z}_2}(\mathcal{A})}(G_X, G_Z) \longrightarrow \mathrm{Ext}^1_{\mathcal{C}_{\mathbb{Z}_2}(\mathcal{A})}(G_X, C_Z) \longrightarrow 0,$$

since K_{P_X} is projective in $\mathcal{C}_{\mathbb{Z}_2}(\mathcal{A})$, and K_{P_Z} is projective-injective in $\mathcal{C}_{\mathbb{Z}_2}(\mathcal{P})$. It follows that $\mathrm{Ext}^1_{\mathcal{C}_{\mathbb{Z}_2}(\mathcal{A})}(C_X, C_Z) \cong \mathrm{Ext}^1_{\mathcal{C}_{\mathbb{Z}_2}(\mathcal{A})}(G_X, G_Z)$ by $\mathrm{Hom}_{\mathcal{C}_{\mathbb{Z}_2}(\mathcal{A})}(K_{P_X}, C_Z) = 0$. The Claim follows.

It follows that $I^-([X]*[Z]) = I^-([X]) * I^-([Z])$ by definition and the Claim above. Hence the Serre relations are preserved by Ψ.

Similarly, one checks that

$$\mathrm{Ext}^1_{\mathcal{C}_{\mathbb{Z}_2}(\mathcal{A})}(G_X, G_Z^*) \cong \mathrm{Hom}_{\mathcal{A}}(X,Z) \cong \mathrm{Ext}^1_{\mathcal{C}_{\mathbb{Z}_2}(\mathcal{A})}(G_X^*, G_Z).$$

So we have $[E_{S_i}, F_{S_j}] = 0$, for any $i \neq j$.

On the other hand, we have $\mathrm{Ext}^1_{\mathcal{C}_{\mathbb{Z}_2}(\mathcal{A})}(G_{S_i}, G_{S_i}^*) \cong \mathbf{k}$, for any $i \in \mathbb{I}$. If an exact sequence $0 \to G_{S_i}^* \to M \to G_{S_i} \to 0$ is non-split, then M is acyclic and $M \cong K_{P_{S_i}}^* \oplus K_{Q_{S_i}}^*$. So we have $[E_{S_i}, F_{S_i}] = (q-1)([K_{S_i}^*] - [K_{S_i}])$, as desired.

It remains to prove the injectivity of Ψ. Any $M \in \mathcal{C}_{\mathbb{Z}_2}(\mathcal{P})$ is isomorphic to $C_A \oplus C_B^* \oplus K_P \oplus K_Q^*$; see [**Br13**, Lemma 4.2]. So $\mathcal{D}\widetilde{\mathcal{H}}(\mathbf{k}Q)$ has a basis consisting of elements

$$[C_A \oplus C_B^*] * [K_\alpha] * [K_\beta^*], \quad \forall [A],[B] \in \mathrm{Iso}(\mathrm{mod}(\mathbf{k}Q)), \alpha,\beta \in K_0(\mathrm{mod}(\mathbf{k}Q)).$$

It implies that $\mathcal{D}\widetilde{\mathcal{H}}(\mathbf{k}Q)$ has a basis consisting of elements

$$[C_A] * [C_B^*] * [K_\alpha] * [K_\beta^*], \quad \forall [A],[B] \in \mathrm{Iso}(\mathrm{mod}(\mathbf{k}Q)), \alpha,\beta \in K_0(\mathrm{mod}(\mathbf{k}Q));$$

see [**Br13**, Lemma 4.7] or [**LP21**, Theorem 3.25]. Then the injectivity of Ψ follows from this and Theorem 3.2. □

4. Semi-derived Ringel-Hall algebras

In this section, we formulate weakly 1-Gorenstein exact categories and provide several examples. We then introduce the semi-derived Ringel-Hall algebras for such categories.

4.1. Weakly 1-Gorenstein exact categories. Let $\mathcal{E}$ be an essentially small exact category in the sense of Quillen, linear over a finite field $\mathbf{k} = \mathbb{F}_q$. For an exact category $\mathcal{E}$, we introduce the following subcategories of $\mathcal{A}$:

$$\mathcal{P}^{\leq i}(\mathcal{E}) = \{X \in \mathcal{E} \mid \mathrm{proj.dim}\, X \leq i\},$$
$$\mathcal{I}^{\leq i}(\mathcal{E}) = \{X \in \mathcal{E} \mid \mathrm{inj.dim}\, X \leq i\}, \quad \forall i \in \mathbb{N},$$
$$\mathcal{P}^{<\infty}(\mathcal{E}) = \{X \in \mathcal{E} \mid \mathrm{proj.dim}\, X < \infty\},$$
$$\mathcal{I}^{<\infty}(\mathcal{E}) = \{X \in \mathcal{E} \mid \mathrm{inj.dim}\, X < \infty\}.$$

The category $\mathcal{E}$ is called *weakly Gorenstein* if $\mathcal{P}^{<\infty}(\mathcal{E}) = \mathcal{I}^{<\infty}(\mathcal{E})$, and $\mathcal{E}$ is a *weakly d-Gorenstein* exact category if $\mathcal{A}$ is weakly Gorenstein and $\mathcal{P}^{<\infty}(\mathcal{E}) = \mathcal{P}^{\leq d}(\mathcal{E})$, $\mathcal{I}^{<\infty}(\mathcal{E}) = \mathcal{I}^{\leq d}(\mathcal{E})$.

LEMMA 4.1 (Iwanaga's Theorem). *Let $\mathcal{E}$ be a weakly Gorenstein exact category with enough projectives and injectives. Then $\mathcal{P}^{<\infty}(\mathcal{E}) = \mathcal{P}^{\leq d}(\mathcal{E})$ if and only if $\mathcal{I}^{<\infty}(\mathcal{E}) = \mathcal{I}^{\leq d}(\mathcal{E})$.*

Throughout this section, we always assume that $\mathcal{E}$ is an exact category satisfying the following conditions:

(Ea) $\mathcal{E}$ is essentially small, with finite morphism spaces, and finite extension spaces,
(Eb) $\mathcal{E}$ is linear over $\mathbf{k} = \mathbb{F}_q$,
(Ec) $\mathcal{E}$ is weakly 1-Gorenstein,
(Ed) For any object $X \in \mathcal{A}$, there exists an object $P_X \in \mathcal{P}^{<\infty}(\mathcal{E})$ and a deflation $P_X \twoheadrightarrow X$.

In this case, we have $\mathcal{P}^{<\infty}(\mathcal{E}) = \mathcal{P}^{\leq 1}(\mathcal{E}) = \mathcal{I}^{<\infty}(\mathcal{E}) = \mathcal{I}^{\leq 1}(\mathcal{E})$.

EXAMPLE 4.2.

(1) Any $\mathbf{k}$-linear hereditary abelian category with finite-dimensional morphism spaces and extension spaces satisfies (Ea)-(Ed).
(2) For any finite-dimensional 1-Gorenstein algebra Λ over $\mathbf{k}$ (i.e., $\operatorname{inj.dim}{}_\Lambda\Lambda \leq 1$ and $\operatorname{inj.dim}\Lambda_\Lambda \leq 1$), $\mathcal{E} = \operatorname{mod}(\Lambda)$ satisfies (Ea)–(Ed).

An exact category $\mathcal{E}$ is called a *Frobenius category*, provided that the class of projective objects coincides with the class of injective objects and it has enough projective objects and enough injective objects. The importance of Frobenius categories lies in that they give rise naturally to triangulated categories; see [**Ha88**].

EXAMPLE 4.3. Any $\mathbf{k}$-linear Frobenius category with finite-dimensional morphism spaces and extension spaces satisfies (Ea)-(Ed).

Let A be a finite-dimensional algebra over $\mathbf{k}$. The notion of Gorenstein projective modules is introduced in [**EJ00**]. Let $\operatorname{Gproj}(A)$ be the category consisting of all finitely generated Gorenstein projective modules. Then $\operatorname{Gproj}(A)$ is a Frobenius category satisfies (Ea)-(Ed).

EXAMPLE 4.4. Let $\mathcal{A}$ be a hereditary abelian $\mathbf{k}$-category (not necessarily with enough projective objects). Let $\mathcal{C}_{\mathbb{Z}_n}(\mathcal{A})$ be the category of $\mathbb{Z}_n$-graded complexes (also called n-periodic complexes), for $n \geq 1$. Denote by $\mathcal{C}_{ac,\mathbb{Z}_n}(\mathcal{A})$ the subcategory of acyclic complexes in $\mathcal{C}_{\mathbb{Z}_n}(\mathcal{A})$. Then $\mathcal{C}_{\mathbb{Z}_n}(\mathcal{A})$ satisfies (Ea)-(Ed) with $\mathcal{P}^{<\infty}(\mathcal{C}_{\mathbb{Z}_n}(\mathcal{A})) = \mathcal{C}_{ac,\mathbb{Z}_n}(\mathcal{A}) = \mathcal{I}^{<\infty}(\mathcal{C}_{\mathbb{Z}_n}(\mathcal{A}))$; see [**LP21**, Proposition 2.3] and [**LRW20**, Corollary 2.4].

By definition, the *singularity category* $\mathcal{D}_{sg}(\mathcal{E})$ is the Verdier quotient of $\mathcal{D}^b(\mathcal{E})$ with respect to the thick subcategory generated by $\mathcal{P}^{<\infty}(\mathcal{E})$.

4.2. Definition of semi-derived Ringel-Hall algebra. Given an exact category $\mathcal{E}$ satisfying (Ea)-(Ed), its Ringel-Hall algebra $\mathcal{H}(\mathcal{E})$ can be too big to be interesting. We apply some suitable quotient and localization constructions below to $\mathcal{H}(\mathcal{E})$ to obtain a new algebra of suitable size, which has a natural (categorical) basis. As we shall see, for some distinguished classes of exact categories, the resulting algebras provide realizations of quantum groups and ıquantum groups.

Let I be the two-sided ideal of $\mathcal{H}(\mathcal{E})$ generated by all differences

$$[M]-[N], \text{ if } \widehat{M}=\widehat{N}\in K_0(\mathcal{E}), \text{ and } M\cong N \text{ in } \mathcal{D}_{sg}(\mathcal{E}). \tag{4.1}$$

Denote

$$\mathcal{S}_{\mathcal{E}} := \{a[K]\in \mathcal{H}(\mathcal{E})/I \mid a\in \mathbb{Q}^{\times}, K\in \mathcal{P}^{\leq 1}(\mathcal{E})\}. \tag{4.2}$$

PROPOSITION 4.5 ([**LP21**, **Lu22**]). *Let $\mathcal{E}$ be an exact category satisfying (Ea)-(Ed). Then the multiplicatively closed subset $\mathcal{S}_{\mathcal{E}}$ is a right Ore, right reversible subset of $\mathcal{H}(\mathcal{E})/I$. Equivalently, the right localization of $\mathcal{H}(\mathcal{E})/I$ with respect to $\mathcal{S}_{\mathcal{E}}$ exists, and will be denoted by $(\mathcal{H}(\mathcal{E})/I)[\mathcal{S}_{\mathcal{E}}^{-1}]$.*

DEFINITION 4.6 ([**LP21**, **Lu22**]). For any exact category $\mathcal{E}$ satisfying (Ea)-(Ed), $(\mathcal{H}(\mathcal{E})/I)[\mathcal{S}_{\mathcal{E}}^{-1}]$ is called the semi-derived Ringel-Hall algebra of $\mathcal{E}$, and denoted by $\mathcal{SDH}(\mathcal{E})$.

The quantum torus $\mathcal{T}(\mathcal{E})$ is defined to be the subalgebra of $\mathcal{SDH}(\mathcal{E})$ generated by $[M]$, for M in $\mathcal{P}^{\leq 1}(\mathcal{E})$.

EXAMPLE 4.7. Let $\mathcal{A}$ be a $\mathbf{k}$-linear hereditary abelian category with finite-dimensional morphism spaces and extension spaces. Then $\mathcal{SDH}(\mathcal{A})$ is the group algebra of $K_0(\mathcal{A})$ with its multiplication twisted by the Euler form, i.e.,

$$[M]\diamond[N]=q^{\langle M,N\rangle}[M\oplus N],$$

for any $M,N\in\mathcal{A}$.

EXAMPLE 4.8. Let $\mathcal{F}$ be a Frobenius category with $\mathcal{P}(\mathcal{F})$ the subcategory of projective-injective objects. The semi-derived Ringel-Hall algebra of $\mathcal{F}$ was introduced by M. Gorsky [**Gor18**].

EXAMPLE 4.9. Let Λ be a finite-dimensional 1-Gorenstein algebra. We denote by $\mathcal{SDH}(\Lambda)$ the semi-derived Ringel-Hall algebra of $\mathrm{mod}(\Lambda)$. Then the natural embedding $\mathrm{Gproj}(\Lambda)\subseteq \mathrm{mod}(\Lambda)$ induces an isomorphism

$$F:\mathcal{SDH}\big(\mathrm{Gproj}(\Lambda)\big)\xrightarrow{\cong}\mathcal{SDH}(\Lambda).$$

Moreover, for any Λ-module M, we have an exact sequence

$$0\longrightarrow P_M\longrightarrow G_M\longrightarrow M\longrightarrow 0,$$

where $G_M\in\mathrm{Gproj}(\Lambda)$ and P_M is projective. Then the inverse of F is given by $[M]\mapsto q^{-\langle M,P_M\rangle}[G_M]\diamond[P_M]^{-1}$; see [**Lu22**, Theorem A.15].

EXAMPLE 4.10. For a $\mathbf{k}$-linear hereditary abelian category $\mathcal{A}$, let $\mathcal{C}_{\mathbb{Z}_n}(\mathcal{A})$ be the category of $\mathbb{Z}_n$-graded complexes of modules over $\mathcal{A}$, for $n\geq 1$. Then $\mathcal{SDH}(\mathcal{C}_{\mathbb{Z}_n}(\mathcal{A}))$ was studied in [**LP21**, **LinP19**].

4.3. Bridgeland's theorem reformulated. Let $\mathcal{A}$ be a hereditary abelian $\mathbf{k}$-category. Recall that the category $\mathcal{C}_{\mathbb{Z}_2}(\mathcal{A})$ is defined in §3.3. Let $\mathcal{SD}\widetilde{\mathcal{H}}\big(\mathcal{C}_{\mathbb{Z}_2}(\mathcal{A})\big)$ be the twisted semi-derived Ringel-Hall algebra of $\mathcal{C}_{\mathbb{Z}_2}(\mathcal{A})$ over $\mathbb{Q}(\mathbf{v})$, i.e., its multiplication is given by

$$[M]*[N]:=\mathbf{v}^{\langle \mathrm{res}\, M,\mathrm{res}\, N\rangle}[M]\diamond[N],\quad \forall M,N\in\mathcal{C}_{\mathbb{Z}_2}(\mathcal{A}). \tag{4.3}$$

Let $\mathcal{P}$ be the subcategory of $\mathcal{A}$ consisting of projective objects. If $\mathcal{A}$ admits enough projectives, then the natural embedding $\mathcal{C}_{\mathbb{Z}_2}(\mathcal{P})\subseteq\mathcal{C}_{\mathbb{Z}_2}(\mathcal{A})$ induces an algebra isomorphism (cf. [**LP21**, Theorem 5.13])

$$\phi: D\widetilde{\mathcal{H}}(\mathcal{A})\xrightarrow{\cong}\mathcal{SD}\widetilde{\mathcal{H}}\big(\mathcal{C}_{\mathbb{Z}_2}(\mathcal{A})\big).$$

In fact, $D\widetilde{\mathcal{H}}(\mathcal{A})$ is just the twisted semi-derived Ringel-Hall algebra of $\mathcal{C}_{\mathbb{Z}_2}(\mathcal{P})$ by noting that $\mathcal{C}_{\mathbb{Z}_2}(\mathcal{P})$ is a Frobenius category.

We further specialize to the distinguished case when $\mathcal{A} = \mathrm{mod}(\mathbf{k}Q)$ for a finite quiver Q (not necessarily acyclic); recall $\mathrm{mod}(\mathbf{k}Q)$ may not have enough projectives. Denote by $\mathcal{SD}\widetilde{\mathcal{H}}_{\mathbb{Z}_2}(\mathbf{k}Q)$ the twisted semi-derived Ringel-Hall algebra of $\mathcal{C}_{\mathbb{Z}_2}(\mathrm{mod}(\mathbf{k}Q))$.

THEOREM 4.11 (Bridgeland's Theorem reformulated; see [**LW22a**]). *There exists an injective homomorphism of $\mathbb{Q}(\mathbf{v})$-algebras $\Psi : \widetilde{\mathbf{U}}_{\mathbf{v}} \longrightarrow \mathcal{SD}\widetilde{\mathcal{H}}_{\mathbb{Z}_2}(\mathbf{k}Q)$ such that*

(4.4)

$$\Psi(E_i) = \frac{\mathbf{v}}{q-1}[C_{S_i}], \Psi(F_i) = \frac{-1}{q-1}[C^*_{S_i}], \Psi(K_i) = [K_{S_i}], \Psi(K'_i) = [K^*_{S_i}], \quad \forall i \in \mathbb{I}.$$

PROOF. We remark that the verification that Ψ is an algebra homomorphism is much easier than in the setting of Theorem 3.8.

Let $\mathcal{A} = \mathrm{mod}(\mathbf{k}Q)$. The natural embeddings $\iota : \mathcal{A} \to \mathcal{C}_{\mathbb{Z}_2}(\mathcal{A})$ given by $X \mapsto C_X$ makes $\mathcal{A}$ a full subcategory of $\mathcal{C}_{\mathbb{Z}_2}(\mathcal{A})$ which is closed under taking extensions. So ι induces an algebra homomorphism $\widetilde{\mathcal{H}}(\mathcal{A}) \to \mathcal{SD}\widetilde{\mathcal{H}}_{\mathbb{Z}_2}(\mathbf{k}Q)$ sending $[X] \mapsto [C_X]$. So Ψ preserves the quantum Serre relation (2.4) by Theorem 3.2. Similarly, Ψ preserves the other quantum Serre relation (2.5).

Let $i \neq j$. Since $\dim_{\mathbf{k}} \mathrm{Ext}^1_{\mathcal{C}_{\mathbb{Z}_2}(\mathcal{A})}(C_{S_i}, C^*_{S_j}) = \dim_{\mathbf{k}} \mathrm{Hom}_{\mathcal{A}}(S_i, S_j) = 0$, we have $[C_{S_i}, C^*_{S_j}] = 0$.

For any non-split exact sequence $0 \to C^*_{S_i} \to M \to C_{S_i} \to 0$, we have $M \cong K^*_{S_i}$. So we have $[C_{S_i}, C^*_{S_i}] = (q-1)([K^*_{S_i}] - [K_{S_i}])$.

Finally, by Proposition 3.7, we see that the remaining relations (2.1)–(2.3) for $\widetilde{\mathbf{U}}$ are preserved by Ψ. □

5. $\imath$Quiver algebras

In this section, starting from quivers with involutions, we formulate a class of 1-Gorenstein algebras known as $\imath$quiver algebras. In case of acyclic quivers, these algebras are finite-dimensional. We present some basic homological properties of these algebras.

5.1. $\imath$Quivers. Let $\mathbf{k}$ be a field. Let $Q = (Q_0, Q_1)$ be a quiver. An *involution* of Q is defined to be an automorphism τ of the quiver Q such that $\tau^2 = \mathrm{Id}$. In particular, we allow the *trivial* involution $\mathrm{Id} : Q \to Q$. An involution τ of Q induces an involution of the path algebra $\mathbf{k}Q$, again denoted by τ. A quiver together with a specified involution τ, (Q, τ), will be called an *$\imath$quiver*. The *rank* of an $\imath$quiver (Q, τ) is by definition the number of τ-orbits in Q_0.

Let Z_m be the quiver with m vertices and m arrows which forms an oriented cycle. The vertex set of Z_m is $\{0, 1, \dots, m-1\}$. Let R_m be the radical square zero self-injective Nakayama algebra of Z_m, i.e., $R_m := \mathbf{k}Z_m/J$, where J denotes the ideal of $\mathbf{k}Z_m$ generated by all paths of length two. In particular,

- ▷ R_1 is isomorphic to the truncated polynomial algebra $\mathbf{k}[\varepsilon]/(\varepsilon^2)$;
- ▷ R_2 is the radical square zero of the path algebra of $1 \underset{\varepsilon'}{\overset{\varepsilon}{\rightleftarrows}} 1'$, i.e., $\varepsilon'\varepsilon = 0 = \varepsilon\varepsilon'$.

Recall that $\mathcal{C}_{\mathbb{Z}_m}(\text{mod}(\mathbf{k}Q))$ is the category of the $\mathbb{Z}_m$-graded complexes over $\text{mod}(\mathbf{k}Q)$ for any $m \geq 1$, see [**Br13,LP21**]. We have $\mathcal{C}_{\mathbb{Z}_m}(\text{mod}(\mathbf{k}Q)) \cong \text{mod}(\mathbf{k}Q \otimes_{\mathbf{k}} R_m)$ for any $m \geq 1$.

Define a $\mathbf{k}$-algebra

$$\Lambda = \mathbf{k}Q \otimes_{\mathbf{k}} R_2. \tag{5.1}$$

Let $Q^\sharp$ be the quiver such that

- the vertex set of $Q^\sharp$ consists of 2 copies of the vertex set Q_0, $\{i, i' \mid i \in Q_0\}$;
- the arrow set of $Q^\sharp$ is

$$\{\alpha : i \to j, \alpha' : i' \to j' \mid (\alpha : i \to j) \in Q_1\} \cup \{\varepsilon_i : i \to i', \varepsilon'_i : i' \to i \mid i \in Q_0\}.$$

We call $Q^\sharp$ the *double framed quiver* associated to the quiver Q.

Let $I^\sharp$ be the ideal of $\mathbf{k}Q^\sharp$ generated by

- (Nilpotent relations) $\varepsilon_i \varepsilon'_i$, $\varepsilon'_i \varepsilon_i$ for any $i \in Q_0$;
- (Commutative relations) $\varepsilon'_j \alpha' - \alpha \varepsilon'_i$, $\varepsilon_j \alpha - \alpha' \varepsilon_i$ for any $(\alpha : i \to j) \in Q_1$.

Then the algebra Λ can be realized as

$$\Lambda \cong \mathbf{k}Q^\sharp / I^\sharp. \tag{5.2}$$

Let Q (respectively, Q') be the full subquiver of $Q^\sharp$ formed by all vertices i (respectively, i') for $i \in Q_0$. Then $Q \sqcup Q'$ is a subquiver of $Q^\sharp$.

The category $\text{mod}(\Lambda)$ is isomorphic to $\mathcal{C}_{\mathbb{Z}_2}(\text{mod}(\mathbf{k}Q))$.

EXAMPLE 5.1.

(a) Let $Q = (1 \xrightarrow{\alpha} 2)$. Then its double framed quiver $Q^\sharp$ is

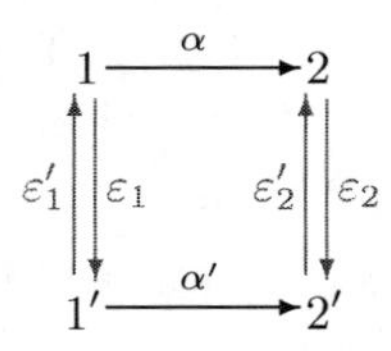

and $I^\sharp$ is generated by all possible quadratic relations

$$\varepsilon_1 \varepsilon'_1, \qquad \varepsilon'_1 \varepsilon_1, \qquad \varepsilon'_2 \varepsilon_2, \qquad \varepsilon_2 \varepsilon'_2, \alpha' \varepsilon_1 - \varepsilon_2 \alpha, \qquad \alpha \varepsilon'_1 - \varepsilon'_2 \alpha'.$$

(b) Let $Q = (1 \xrightarrow{\alpha} 2 \xleftarrow{\beta} 3)$. Then its double framed quiver $Q^\sharp$ is

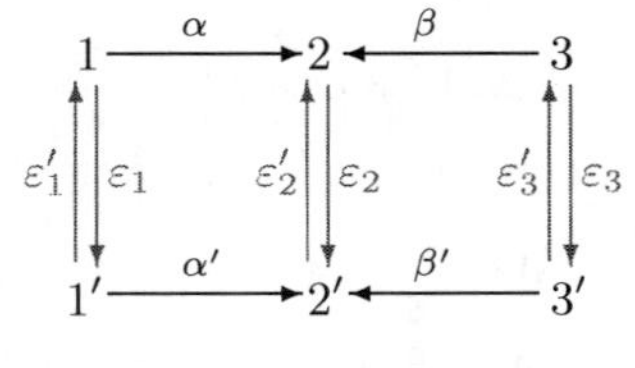

and $I^\sharp$ is generated by all possible quadratic relations

$$\varepsilon'_i \varepsilon_i, \qquad \varepsilon_i \varepsilon'_i, \quad \forall 1 \leq i \leq 3$$
$$\alpha' \varepsilon_1 - \varepsilon_2 \alpha, \qquad \alpha \varepsilon'_1 - \varepsilon'_2 \alpha', \qquad \beta \varepsilon'_3 - \varepsilon'_2 \beta', \qquad \beta' \varepsilon_3 - \varepsilon_2 \beta.$$

(c) Let Q be the quiver such that $Q_0 = \{1, 2\}$, and $Q_1 = \emptyset$. Then $\Lambda \cong R_2 \times R_2$.

(d) Let Q_{J} be the Jordan quiver. Then its double framed quiver $Q_{\mathrm{J}}^\sharp$ is

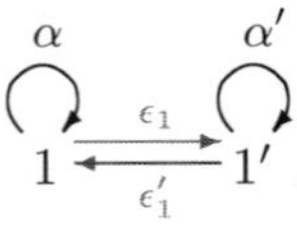

and $I^\sharp$ is generated by all possible quadratic relations $\alpha\epsilon_1' = \epsilon_1'\alpha'$, $\alpha'\epsilon_1 = \epsilon_1\alpha$.

(e) Let $Q_{\mathrm{Kr}} : 0 \overset{\alpha}{\underset{\beta}{\rightrightarrows}} 1$ be the Kronecker quiver. Then its double framed quiver $Q_{\mathrm{Kr}}^\sharp$ is

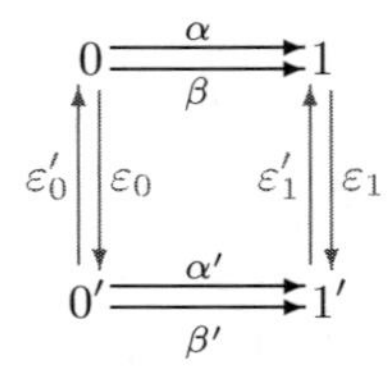

and $I^\sharp$ is generated by all possible quadratic relations

$$\varepsilon_0\varepsilon_0', \qquad \varepsilon_0'\varepsilon_0, \qquad \varepsilon_1'\varepsilon_1, \qquad \varepsilon_1\varepsilon_1',$$
$$\alpha'\varepsilon_0 - \varepsilon_1\alpha, \qquad \alpha\varepsilon_0' - \varepsilon_1'\alpha' \qquad \beta'\varepsilon_0 - \varepsilon_1\beta, \qquad \beta\varepsilon_0' - \varepsilon_1'\beta'.$$

EXAMPLE 5.2. Let $Q = (Q_0, Q_1)$ be a quiver, $Q^\sharp$ be its double framed quiver. Let $Q^{\mathrm{dbl}} = Q \sqcup Q^\diamond$, where $Q^\diamond$ is an identical copy of Q with a vertex set $\{i^\diamond \mid i \in Q_0\}$ and an arrow set $\{\alpha^\diamond \mid \alpha \in Q_1\}$. Let $\Lambda = \mathbf{k}Q \otimes_{\mathbf{k}} R_2$, $\Lambda^\diamond = kQ^\diamond \otimes_{\mathbf{k}} R_2$, and $\Lambda^{\mathrm{dbl}} := \mathbf{k}Q^{\mathrm{dbl}} \otimes R_2$. Then the double framed quiver $(Q^{\mathrm{dbl}})^\sharp$ of Q^{dbl} is $Q^\sharp \sqcup (Q^\sharp)^\diamond$, and $\Lambda^{\mathrm{dbl}} = \Lambda \times \Lambda^\diamond \cong \Lambda \times \Lambda$.

5.2. $\imath$Quiver algebras. The involution τ of a quiver Q induces an involution $\tau^\sharp$ of $Q^\sharp$ defined by

- $\tau^\sharp(i) = (\tau i)'$, $\tau^\sharp(i') = \tau i$ for any $i \in Q_0$;
- $\tau^\sharp(\varepsilon_i) = \varepsilon_{\tau i}'$, $\tau^\sharp(\varepsilon_i') = \varepsilon_{\tau i}$ for any $i \in Q_0$;
- $\tau^\sharp(\alpha) = (\tau\alpha)'$, $\tau^\sharp(\alpha') = \tau\alpha$ for any $\alpha \in Q_1$.

So starting from the $\imath$quiver (Q, τ) we have constructed a new $\imath$quiver $(Q^\sharp, \tau^\sharp)$.

The action of $\tau^\sharp$ preserves $I^\sharp$. Hence $\tau^\sharp$ induces an involution, again denoted by $\tau^\sharp$, of the algebra Λ.

DEFINITION 5.3. The fixed point subalgebra of Λ under $\tau^\sharp$,

$$\Lambda^\imath = \{x \in \Lambda \mid \tau^\sharp(x) = x\}, \tag{5.3}$$

is called the $\imath$quiver algebra of (Q, τ).

According to [**LW22a**, Proposition 2.7], we have $\Lambda^\imath \cong \mathbf{k}\overline{Q}/\overline{I}$, where

(i) $\overline{Q}$ is constructed from Q by adding a loop ε_i at the vertex $i \in Q_0$ if $\tau i = i$, and adding an arrow $\varepsilon_i : i \to \tau i$ for each $i \in Q_0$ if $\tau i \neq i$;
(ii) $\overline{I}$ is generated by
  (1) (Nilpotent relations) $\varepsilon_i\varepsilon_{\tau i}$ for any $i \in \mathbb{I}$;
  (2) (Commutative relations) $\varepsilon_i\alpha - \tau(\alpha)\varepsilon_j$ for any arrow $\alpha : j \to i$ in Q_1.

By [**LW22a**, Corollary 2.12], $\mathbf{k}Q$ is naturally a subalgebra and also a quotient algebra of $\Lambda^\imath$. Viewing $\mathbf{k}Q$ as a subalgebra of $\Lambda^\imath$, we have a restriction functor

$$\mathrm{res} : \mathrm{mod}(\Lambda^\imath) \longrightarrow \mathrm{mod}(\mathbf{k}Q).$$

Viewing $\mathbf{k}Q$ as a quotient algebra of $\Lambda^\imath$, we obtain a pullback functor

$$\iota : \mathrm{mod}(\mathbf{k}Q) \longrightarrow \mathrm{mod}(\Lambda^\imath). \tag{5.4}$$

Hence a simple module $S_i (i \in Q_0)$ of $\mathbf{k}Q$ is naturally a simple $\Lambda^\imath$-module.

For each $i \in Q_0$, define a $\mathbf{k}$-algebra (which can be viewed as a subalgebra of $\Lambda^\imath$)

$$\mathbb{H}_i := \begin{cases} \mathbf{k}[\varepsilon_i]/({\varepsilon_i}^2) & \text{if } i = \tau i, \\ \mathbf{k}(\, i \underset{\varepsilon_{\tau i}}{\overset{\varepsilon_i}{\rightleftarrows}} \tau i\,)/(\varepsilon_i\varepsilon_{\tau i}, \varepsilon_{\tau i}\varepsilon_i) & \text{if } \tau i \neq i. \end{cases} \tag{5.5}$$

Note that $\mathbb{H}_i = \mathbb{H}_{\tau i}$ for any $i \in Q_0$. Choose one representative for each τ-orbit on $\mathbb{I}$, and let

$$\mathbb{I}_\tau = \{\text{the chosen representatives of } \tau\text{-orbits in } \mathbb{I}\}. \tag{5.6}$$

Define the following subalgebra of $\Lambda^\imath$:

$$\mathbb{H} = \bigoplus_{i \in \mathbb{I}_\tau} \mathbb{H}_i. \tag{5.7}$$

Note that $\mathbb{H}$ is a radical square zero self-injective algebra. Denote by

$$\mathrm{res}_{\mathbb{H}} : \mathrm{mod}(\Lambda^\imath) \longrightarrow \mathrm{mod}(\mathbb{H}) \tag{5.8}$$

the natural restriction functor. On the other hand, as $\mathbb{H}$ is a quotient algebra of $\Lambda^\imath$, every $\mathbb{H}$-module can be viewed as a $\Lambda^\imath$-module.

Recall the algebra $\mathbb{H}_i$ for $i \in \mathbb{I}_\tau$ from (5.5). For $i \in Q_0 = \mathbb{I}$, define the indecomposable module over $\mathbb{H}_i$ (if $i \in \mathbb{I}_\tau$) or over $\mathbb{H}_{\tau i}$ (if $i \notin \mathbb{I}_\tau$)

$$\mathbb{K}_i = \begin{cases} \mathbf{k}[\varepsilon_i]/(\varepsilon_i^2), & \text{if } i = \tau i; \\ \mathbf{k} \underset{0}{\overset{1}{\rightleftarrows}} \mathbf{k} \text{ on the quiver } i \underset{\varepsilon_{\tau i}}{\overset{\varepsilon_i}{\rightleftarrows}} \tau i\,, & \text{if } i \neq \tau i. \end{cases} \tag{5.9}$$

Then $\mathbb{K}_i$, for $i \in Q_0$, can be viewed as a $\Lambda^\imath$-module and will be called a *generalized simple* $\Lambda^\imath$-module.

EXAMPLE 5.4.

(a) We continue Example 5.1(a), with $\tau = \mathrm{Id}$. Then $\Lambda^\imath$ is isomorphic to the algebra with its quiver $\overline{Q}$ and relations as follows:

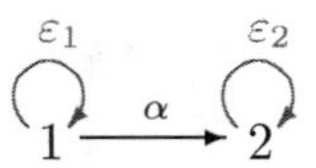

$$\varepsilon_1^2 = 0 = \varepsilon_2^2, \quad \varepsilon_2\alpha = \alpha\varepsilon_1.$$

Indeed, $\Lambda^\imath$ is isomorphic to $\mathbf{k}Q \otimes_{\mathbf{k}} R_1$. This class of algebras was independently studied by Ringel-Zhang and Geiss-Leclerc-Schröer [**GLS17**] from very different perspectives.

(b) We continue Example 5.1(b), with the involution τ of Q swapping vertices 1 and 3. Then $\Lambda^\imath$ is isomorphic to the algebra with its quiver $\overline{Q}$ and relations as follows:

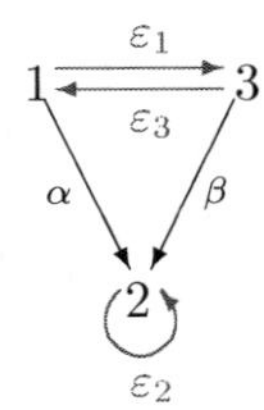

$$\varepsilon_1\varepsilon_3 = 0 = \varepsilon_3\varepsilon_1, \quad \varepsilon_2^2 = 0, \quad \varepsilon_2\beta = \alpha\varepsilon_3, \quad \varepsilon_2\alpha = \beta\varepsilon_1.$$

(c) We continue Example 5.1(c) with $\tau = \mathrm{Id}$. Then, $\Lambda^\imath \cong R_1 \times R_1$.

(d) We continue Example 5.1(d) of Jordan quiver with $\tau = \mathrm{Id}$. Then $\Lambda^\imath$ is isomorphic to the algebra with its quiver $\overline{Q}$ and relations as follows:

(5.10)

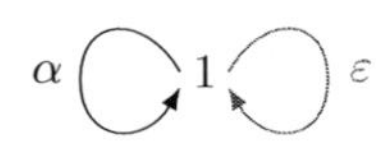

$$\alpha\epsilon = \epsilon\alpha, \quad \epsilon^2 = 0.$$

(e) We continue Example 5.1(e) of Kronecker quiver with $\tau = \mathrm{Id}$. Then $\Lambda^\imath_{\mathrm{Kr}}$ is isomorphic to the algebra with its quiver $\overline{Q_{\mathrm{Kr}}}$ and relations as follows:

(5.11)

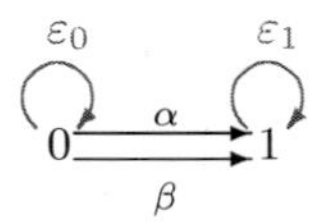

$$\varepsilon_0^2 = 0 = \varepsilon_1^2, \quad \varepsilon_1\alpha = \alpha\varepsilon_0, \varepsilon_1\beta = \beta\varepsilon_0.$$

EXAMPLE 5.5 ($\imath$quiver of diagonal type). Continuing Example 5.2, we let swap be the involution of Q^{dbl} uniquely determined by $\mathrm{swap}(i) = i^\diamond$ for any $i \in Q_0$. Then $(\Lambda^{\mathrm{dbl}})^\imath$ is isomorphic to Λ. Explicitly, let $(\overline{Q}^{\mathrm{dbl}}, \overline{I}^{\mathrm{dbl}})$ be the bound quiver of $(\Lambda^{\mathrm{dbl}})^\imath$. Then $(\overline{Q}^{\mathrm{dbl}}, \overline{I}^{\mathrm{dbl}})$ coincides with the double $\imath$quiver $(Q^\sharp, I^\sharp)$. So we just use $(Q^\sharp, I^\sharp)$ as the bound quiver of $(\Lambda^{\mathrm{dbl}})^\imath$ and identify $(\Lambda^{\mathrm{dbl}})^\imath$ with Λ.

5.3. Gorenstein properties of $\imath$quiver algebras. Let d be a positive integer. Recall [**Ha91**, **EJ00**] a noetherian algebra A is *d-Gorenstein* if $\mathrm{inj.dim}\, {}_A A \le d$ and $\mathrm{inj.dim}\, A_A \le d$.

PROPOSITION 5.6 ([**LW20**, Proposition 2.2]). *For a general $\imath$quiver (Q, τ), Λ and $\Lambda^\imath$ are 1-Gorenstein algebras.*

LEMMA 5.7 ([**LW20**, Lemma 2.4]). *For any $M \in \mathrm{mod}(\Lambda^\imath)$, there exist short exact sequences*

$$0 \longrightarrow M \longrightarrow H^M \longrightarrow X^M \longrightarrow 0 \tag{5.12}$$

$$0 \longrightarrow X_M \longrightarrow H_M \longrightarrow M \longrightarrow 0, \tag{5.13}$$

with $H^M, H_M \in \mathcal{P}^{\le 1}(\Lambda^\imath)$.

By definition of $(\overline{Q},\overline{I})$, $\Lambda^\imath$ is positively graded by $\deg \varepsilon_i = 1$, $\deg \alpha = 0$ for any $i \in Q_0$, $\alpha \in Q_1$. Note that $\Lambda^\imath_0 = \mathbf{k}Q$. Let $\mathrm{mod}^{\mathbb{Z}}(\Lambda^\imath)$ be the finite-dimensional nilpotent $\mathbb{Z}$-graded $\Lambda^\imath$-modules. The singularity category $\mathcal{D}_{sg}(\mathrm{mod}^{\mathbb{Z}}(\Lambda^\imath))$ is defined to be the Verdier quotient of $\mathcal{D}^b(\mathrm{mod}^{\mathbb{Z}}(\Lambda^\imath))$ modulo the thick subcategory generated by modules with finite projective dimensions.

Let G be the composition

$$\mathcal{D}^b(\mathrm{mod}(\mathbf{k}Q)) \xrightarrow{\mathcal{D}^b(\iota)} \mathcal{D}^b(\mathrm{mod}^{\mathbb{Z}}(\Lambda^\imath)) \xrightarrow{\pi} \mathcal{D}_{sg}(\mathrm{mod}^{\mathbb{Z}}(\Lambda^\imath)),$$

where $\mathcal{D}^b(\iota)$ is the derived functor of ι since ι is exact.

Let $\widehat{\tau}$ be the triangulated auto-equivalence of $D^b(\mathrm{mod}(\mathbf{k}Q))$ induced by τ. By [**LW22a**, Theorem 3.18], we have

$$(1) \circ G \simeq G \circ \Sigma \circ \widehat{\tau}, \tag{5.14}$$

where (1) is the degree shift functor, and Σ is the suspension functor of $\mathcal{D}^b(\mathrm{mod}(\mathbf{k}Q))$.

By [**LW20**, Lemma 2.10], the restriction of G provides us a triangulated equivalence

$$\mathcal{D}^b(\mathrm{mod}(\mathbf{k}Q)) \xrightarrow{\simeq} \mathcal{D}_{sg}(\mathrm{mod}^{\mathbb{Z}}(\Lambda^\imath)).$$

Combing with (5.14), we have the following theorem.

THEOREM 5.8 ([**LW20**, Theorem 2.11]). *Let (Q,τ) be an ıquiver. Then $D^b(\mathrm{mod}(\mathbf{k}Q))/\Sigma \circ \widehat{\tau}$ is a triangulated orbit category à la Keller* [**Ke05**], *and we have the following triangulated equivalence*

$$\mathcal{D}_{sg}(\mathrm{mod}(\Lambda^\imath)) \simeq \mathcal{D}^b(\mathrm{mod}(\mathbf{k}Q))/\Sigma \circ \widehat{\tau}.$$

As a corollary, for any $M \in D_{sg}(\mathrm{mod}(\Lambda^\imath))$, there exists a unique (up to isomorphisms) module $N \in \mathrm{mod}(\mathbf{k}Q) \subseteq \mathrm{mod}(\Lambda^\imath)$ such that $M \cong N$ in $D_{sg}(\mathrm{mod}(\Lambda^\imath))$; see [**LW22a**, Corollary 3.21].

COROLLARY 5.9 ([**LW20**, Corollary 2.13]). *For any $M \in \mathrm{mod}(\Lambda^\imath)$ the following are equivalent.*

(i) $\mathrm{proj.dim}\, M < \infty$;
(ii) $\mathrm{inj.dim}\, M < \infty$;
(iii) $\mathrm{proj.dim}\, M \leq 1$;
(iv) $\mathrm{inj.dim}\, M \leq 1$;
(v) $\mathrm{res}_{\mathbb{H}}(M)$ *is projective as an $\mathbb{H}$-module.*

Summarizing, we have established the following.

PROPOSITION 5.10. *The $\mathrm{mod}(\Lambda^\imath)$ is an abelian category satisfying (Ea)-(Ed).*

6. ıHall algebra for the Jordan quiver

In this short section, we formulate the connection between the ıHall algebra for the Jordan quiver and ring of symmetric functions.

Let Q_{J} be the Jordan quiver, i.e., a quiver with a single vertex 1 and a single loop $\alpha : 1 \to 1$. It is well known that $\mathrm{mod}(\mathbf{k}Q_{\mathrm{J}})$ is a uniserial category. Let S be the simple object in $\mathrm{mod}(\mathbf{k}Q_{\mathrm{J}})$. Then any indecomposable object of $\mathrm{mod}(\mathbf{k}Q_{\mathrm{J}})$ (up to isomorphisms) is of the form $S^{(n)}$ of length $n \geq 1$. Thus the set of isomorphism classes of $\mathrm{mod}(\mathbf{k}Q_{\mathrm{J}})$ is canonically isomorphic to the set $\mathcal{P}$ of all partitions, via the assignment

$$\lambda = (\lambda_1, \lambda_2, \cdots, \lambda_r) \mapsto S^{(\lambda)} = S^{(\lambda_1)} \oplus \cdots \oplus S^{(\lambda_r)}. \tag{6.1}$$

Let $\Lambda^\imath_{\mathrm{J}}$ be the $\imath$quiver algebra of the $\imath$quiver Q_{J} equipped with trivial involution [**LW22a, LW20**], see Example 5.4(d) and the $\imath$quiver in (5.10). In particular, $\Lambda^\imath_{\mathrm{J}}$ is a commutative $\mathbf{k}$-algebra.

The quotient algebra

$$(6.2)\qquad {}^\imath\widetilde{\mathcal{H}}(\mathbf{k}Q_{\mathrm{J}}) := \mathcal{H}(\Lambda^\imath_{\mathrm{J}})/I$$

is called the *ıHall algebra* of $\Lambda^\imath_{\mathrm{J}}$, where I is the two-sided ideal defined as in (4.1); comparing semi-derived Hall algebras defined in Section 4, we do not apply any localization here.

LEMMA 6.1 ([**LRW20**, Lemma 4.5], [**LRW21**, Proposition 4.4]). *The algebra* ${}^\imath\widetilde{\mathcal{H}}(\mathbf{k}Q_{\mathrm{J}})$ *is commutative, which is isomorphic to*

(1) *the polynomial $\mathbb{Q}$-algebra in the infinitely many generators* $[\mathbb{K}_1], [S^{(1^r)}]$, *for* $r \geq 1$;
(2) *the polynomial $\mathbb{Q}$-algebra in the infinitely many generators* $[\mathbb{K}_1], [S^{(r)}]$, *for* $r \geq 1$.

Similar to the Hall-Littlewood functions (see §3.1), we define the ıHall-Littlewood functions as follows.

Let $1 \leq i < j$. Similar to the raising operator R_{ij}, we define the lowering operator L_{ij} acting on integer sequences by

$$L_{ij}\alpha = (\ldots, \alpha_i - 1, \ldots, \alpha_j - 1, \ldots).$$

Recall that $q_r = Q_r$ for any $r \geq 1$; see §3.1. The actions of lowering operators on q_α are given by letting

$$L_{ij} q_\alpha = q_{L_{ij}\alpha}.$$

Note that all raising and lowering operators commute with each other.

Let t, θ be indeterminates. We consider the ring $\Lambda_{t,\theta} = \mathbb{Q}(t)[\theta][q_1, q_2, \ldots]$.

DEFINITION 6.2 (ıHL functions; [**LRW21**]). For any integer vector α, we define $Q^\imath_\alpha \in \Lambda_{t,\theta}$ by:

$$(6.3)\qquad Q^\imath_\alpha = \prod_{1\leq i<j} \frac{1-\theta L_{ij}}{1-\theta t L_{ij}} \frac{1-R_{ij}}{1-tR_{ij}} q_\alpha.$$

Note that $Q^\imath_{(r)} = Q_r = q_r$.

LEMMA 6.3 ([**LRW21**]). *For any composition λ, $Q^\imath_\lambda$ is the coefficient of $u^\lambda = u_1^{\lambda_1} u_2^{\lambda_2} \cdots$ in*

$$(6.4)\qquad \begin{aligned} Q^\imath(u_1, u_2, \ldots) &= \prod_{i\geq 1} Q(u_i) \prod_{i<j} F(u_i^{-1}u_j) F(\theta u_i u_j) \\ &= \prod_{i\geq 1} Q(u_i) \prod_{i<j} \frac{(1-u_i^{-1}u_j)(1-\theta u_i u_j)}{(1-tu_i^{-1}u_j)(1-t\theta u_i u_j)}. \end{aligned}$$

Denote $\Lambda_\theta := \Lambda_{t,\theta} \otimes_{\mathbb{Q}(t)} \mathbb{Q}$, where t acts on $\mathbb{Q}$ by q^{-1}.

THEOREM 6.4 ([**LRW21**, Theorem 4.9]). *There exists a $\mathbb{Q}$-algebra isomorphism* $\Phi_{(q)} : {}^\imath\widetilde{\mathcal{H}}(\mathbf{k}Q_{\mathrm{J}}) \to \Lambda_\theta$ *such that*

$$\Phi_{(q)}(\mathbb{K}_1) = q\theta, \qquad \Phi_{(q)}([S^{(r)}]) = q^r Q^\imath_{(r)} = q^r Q_r \quad (r \geq 1).$$

Moreover, for any partition λ, *we have*

$$\Phi_{(q)}([S^{(\lambda)}]) = q^{|\lambda|+n(\lambda)}Q^\imath_\lambda. \tag{6.5}$$

7. ıHall algebras for Dynkin ıquivers

In this section, we study the ıHall algebras for ıquivers without loops (i.e., the twisted semi-derived Hall algebras of ıquiver algebras), and establish their Hall bases. We show that ıHall algebra for a Dynkin ıquiver is isomorphic to the corresponding universal ıquantum group.

7.1. ıHall algebras and ıHall bases. Recall we have defined an ıquiver algebra $\Lambda^\imath$ (5.3) associated to an ıquiver (Q,τ); we shall assume that Q contains no loops in this section. According to §5.3, $\mathrm{mod}(\Lambda^\imath)$ is a 1-Gorenstein algebra satisfying (Ea)-(Ed). Thus, we can define the semi-derived Ringel-Hall algebra $\mathcal{SDH}(\Lambda^\imath)$ of $\Lambda^\imath$.

For $K, M \in \mathrm{mod}(\Lambda^\imath)$ with $K \in \mathcal{P}^{\leq 1}(\Lambda^\imath) = \mathcal{P}^{<\infty}(\Lambda^\imath)$, by Corollary 5.9, the following pairings are well defined:

(7.1)

$$\langle K, M\rangle = \sum_{i=0}^{\infty}(-1)^i \dim_{\mathbf{k}} \mathrm{Ext}^i(K,M) = \dim_{\mathbf{k}} \mathrm{Hom}(K,M) - \dim_{\mathbf{k}} \mathrm{Ext}^1(K,M),$$

(7.2)

$$\langle M, K\rangle = \sum_{i=0}^{\infty}(-1)^i \dim_{\mathbf{k}} \mathrm{Ext}^i(M,K) = \dim_{\mathbf{k}} \mathrm{Hom}(M,K) - \dim_{\mathbf{k}} \mathrm{Ext}^1(M,K).$$

These pairings induce bilinear pairings (called Euler forms) between the Grothendieck groups $K_0(\mathcal{P}^{\leq 1}(\Lambda^\imath))$ and $K_0(\mathrm{mod}(\Lambda^\imath))$:

$$\langle \cdot,\cdot\rangle : K_0(\mathcal{P}^{\leq 1}(\Lambda^\imath)) \times K_0(\mathrm{mod}(\Lambda^\imath)) \longrightarrow \mathbb{Z}, \tag{7.3}$$

$$\langle \cdot,\cdot\rangle : K_0(\mathrm{mod}(\Lambda^\imath)) \times K_0(\mathcal{P}^{\leq 1}(\Lambda^\imath)) \longrightarrow \mathbb{Z}. \tag{7.4}$$

Denote by $\langle\cdot,\cdot\rangle_Q$ the Euler form of $\mathbf{k}Q$, and $(\cdot,\cdot)_Q$ the symmetrized Euler form. Denote by S_i the simple $\mathbf{k}Q$-module (respectively, $\Lambda^\imath$-module) corresponding to the vertex $i \in Q_0$ (respectively, $i \in \overline{Q}_0$).

LEMMA 7.1. *For* $K, K' \in \mathcal{P}^{\leq 1}(\Lambda^\imath)$, $M \in \mathrm{mod}(\Lambda^\imath)$, $i, j \in \mathbb{I}$, *we have*

$$\langle K, M\rangle = \langle \mathrm{res}_{\mathbb{H}}(K), M\rangle, \qquad \langle M, K\rangle = \langle M, \mathrm{res}_{\mathbb{H}}(K)\rangle, \tag{7.5}$$

$$\langle \mathbb{K}_i, S_j\rangle = \langle S_i, S_j\rangle_Q, \qquad \langle S_j, \mathbb{K}_i\rangle = \langle S_j, S_{\tau i}\rangle_Q, \tag{7.6}$$

$$\langle K, K'\rangle = \frac{1}{2}\langle \mathrm{res}(K), \mathrm{res}(K')\rangle_Q. \tag{7.7}$$

Via the restriction functor $\mathrm{res} : \mathrm{mod}(\Lambda^\imath) \to \mathrm{mod}(\mathbf{k}Q)$, we define the twisted semi-derived Ringel-Hall algebra to be the $\mathbb{Q}(\mathbf{v})$-algebra on the same vector space as $\mathcal{SDH}(\Lambda^\imath)$ with twisted multiplication given by

$$[M] * [N] = \mathbf{v}^{\langle \mathrm{res}(M), \mathrm{res}(N)\rangle_Q}[M] \diamond [N]. \tag{7.8}$$

We shall denote this algebra $(\mathcal{SDH}(\Lambda^\imath), *)$ by $\mathcal{SD}\widetilde{\mathcal{H}}(\Lambda^\imath)$ or preferably, $\widetilde{\mathcal{H}}(\mathbf{k}Q,\tau)$, and call it the *Hall algebra associated to the ıquiver* (Q,τ), (or an *ıHall algebra*, for short). The *twisted quantum torus* $\widetilde{\mathcal{T}}(\Lambda^\imath)$ is defined to be the subalgebra of $\widetilde{\mathcal{H}}(\mathbf{k}Q,\tau)$ generated by $[K]$, $K\in\mathcal{P}^{\leq 1}(\Lambda^\imath)$. For any $\alpha=\sum_{i\in\mathbb{I}} a_i\widehat{S}_i\in K_0(\mathrm{mod}(\mathbf{k}Q))=\mathbb{Z}^{\mathbb{I}}$, define

$$\mathbb{K}_\alpha := [X]*[Y]^{-1}\in\widetilde{\mathcal{H}}(\mathbf{k}Q,\tau),$$

where $X=\bigoplus_{i\in\mathbb{I}:a_i\geq 0}\mathbb{K}_i^{\oplus a_i}$ and $Y=\bigoplus_{i\in\mathbb{I}:a_i<0}\mathbb{K}_i^{\oplus(-a_i)}$. Then $\widetilde{\mathcal{T}}(\Lambda^\imath)$ is a Laurent polynomial algebra generated by $[\mathbb{K}_i]$, for $i\in\mathbb{I}$; and $[\mathbb{K}_\alpha]*[\mathbb{K}_\beta]=[\mathbb{K}_{\alpha+\beta}]$ for any $\alpha,\beta\in\mathbb{Z}^{\mathbb{I}}$.

By definition and Lemma 7.1, we have

$$[M]*[\mathbb{K}_i]=\mathbf{v}^{\langle \mathrm{res}\, M,S_i\rangle_Q-\langle \mathrm{res}\, M,S_{\tau i}\rangle_Q}[M\oplus\mathbb{K}_i], \tag{7.9}$$

$$[\mathbb{K}_i]*[M]=\mathbf{v}^{\langle S_{\tau i},\mathrm{res}\, M\rangle_Q-\langle S_i,\mathrm{res}\, M\rangle_Q}[\mathbb{K}_i\oplus M], \tag{7.10}$$

and then

$$[\mathbb{K}_i]*[M]=\mathbf{v}^{(\mathrm{res}\, M,S_{\tau i})_Q-(\mathrm{res}\, M,S_i)_Q}[M]*[\mathbb{K}_i]. \tag{7.11}$$

for any $i\in\mathbf{I}$, $M\in\mathrm{mod}(\Lambda^\imath)$.

THEOREM 7.2 ($\imath$Hall basis; see [**LW20**, Theorem 3.6]). *Let (Q,τ) be an $\imath$quiver. Then*

$$\big\{[X]*\mathbb{K}_\alpha\mid [X]\in\mathrm{Iso}(\mathrm{mod}(\mathbf{k}Q))\subseteq\mathrm{Iso}(\mathrm{mod}(\Lambda^\imath)),\alpha\in\mathbb{Z}^{\mathbb{I}}\big\} \tag{7.12}$$

is a basis of $\widetilde{\mathcal{H}}(\mathbf{k}Q,\tau)$.

There is a filtered algebra structure on $\widetilde{\mathcal{H}}(\mathbf{k}Q,\tau)$ by setting $\deg[M]=\widehat{M}\in K_0(\mathrm{mod}(\mathbf{k}Q))=\mathbb{Z}^{\mathbb{I}}$, and $\deg\mathbb{K}_\alpha=0$ for any $M\in\mathrm{mod}(\mathbf{k}Q)$ and $\alpha\in\mathbb{Z}^{\mathbb{I}}$. We denote the associated graded algebra

$$\widetilde{\mathcal{H}}(\mathbf{k}Q,\tau)^{\mathrm{gr}}=\bigoplus_{\alpha\in K_0(\mathrm{mod}(\mathbf{k}Q))}\widetilde{\mathcal{H}}(\mathbf{k}Q,\tau)^{\mathrm{gr}}_\alpha.$$

It is natural to view the quantum torus $\widetilde{\mathcal{T}}(\Lambda^\imath)$ as a subalgebra of $\widetilde{\mathcal{H}}(\mathbf{k}Q,\tau)^{\mathrm{gr}}$. By [**LW22a**, Lemma 5.4(ii)], the linear map

$$\varphi:\widetilde{\mathcal{H}}(\mathbf{k}Q)\longrightarrow\widetilde{\mathcal{H}}(\mathbf{k}Q,\tau)^{\mathrm{gr}},\quad \varphi([M])=[M],\ \forall M\in\mathrm{mod}(\mathbf{k}Q), \tag{7.13}$$

is an algebra embedding. Then we have $\widetilde{\mathcal{H}}(\mathbf{k}Q,\tau)\cong\widetilde{\mathcal{H}}(\mathbf{k}Q)*\widetilde{\mathcal{T}}(\Lambda^\imath)$ as linear spaces.

Recall that $\langle\cdot,\cdot\rangle_Q$ is the Euler form of Q. Define

$$(x,y)=\langle x,y\rangle_Q+\langle y,x\rangle_Q.$$

In particular, $(S_i,S_j)=c_{ij}$ for any $i,j\in\mathbb{I}$, the entries for the Cartan matrix C.

7.2. $\imath$Hall algebras and $\imath$quantum groups of Dynkin type. Let (Q,τ) be a Dynkin $\imath$quiver. In this case, the presentation of the universal $\imath$quantum group $\widetilde{\mathbf{U}}^\imath$ in Theorem 2.8 can be simplified. That is, $\widetilde{\mathbf{U}}^\imath$ is a $\mathbb{Q}(v)$-algebra with generators $B_i,\widetilde{k}_i$ ($i\in\mathbb{I}$), where $\widetilde{k}_i$ are invertible, subject to the following relations

(7.14)–(7.18): for $\ell \in \mathbb{I}$, and $i \neq j \in \mathbb{I}$,

$$\widetilde{k}_i \widetilde{k}_\ell = \widetilde{k}_\ell \widetilde{k}_i, \quad \widetilde{k}_\ell B_i = v^{c_{\tau\ell,i} - c_{\ell i}} B_i \widetilde{k}_\ell, \tag{7.14}$$

$$B_i B_j - B_j B_i = 0, \quad \text{if } c_{ij} = 0 \text{ and } \tau i \neq j, \tag{7.15}$$

$$\sum_{s=0}^{1-c_{ij}} (-1)^s \begin{bmatrix} 1 - c_{ij} \\ s \end{bmatrix} B_i^s B_j B_i^{1-c_{ij}-s} = 0, \quad \text{if } j \neq \tau i \neq i, \tag{7.16}$$

$$B_{\tau i} B_i - B_i B_{\tau i} = \frac{\widetilde{k}_i - \widetilde{k}_{\tau i}}{v - v^{-1}}, \quad \text{if } \tau i \neq i \text{ (and hence } c_{i\,\tau i} = 0), \tag{7.17}$$

$$B_i^2 B_j - [2] B_i B_j B_i + B_j B_i^2 = v \widetilde{k}_i B_j, \quad \text{if } c_{ij} = -1 \text{ and } \tau i = i. \tag{7.18}$$

Denote $\widetilde{\mathbf{U}}^\imath_{\mathbf{v}} = \mathbb{Q}(\mathbf{v}) \otimes_{\mathbb{Q}(v)} \widetilde{\mathbf{U}}^\imath$. The following theorem provides an ıHall algebra realization of $\widetilde{\mathbf{U}}^\imath$ (which also shows why $\widetilde{\mathbf{U}}^\imath$ is more natural than $\mathbf{U}^\imath$ from a categorical viewpoint).

THEOREM 7.3. *Let (Q, τ) be a Dynkin ıquiver. Then there exists a $\mathbb{Q}(\mathbf{v})$-algebra isomorphism*

$$\widetilde{\psi} : \widetilde{\mathbf{U}}^\imath_{\mathbf{v}} \longrightarrow \widetilde{\mathcal{H}}(\mathbf{k}Q, \tau),$$

which sends

$$B_j \mapsto \frac{-1}{q-1}[S_j], \text{ if } j \in \mathbb{I}_\tau, \qquad \widetilde{k}_i \mapsto -q^{-1}[\mathbb{K}_i], \text{ if } \tau i = i \in \mathbb{I}; \tag{7.19}$$

$$B_j \mapsto \frac{\mathbf{v}}{q-1}[S_j], \text{ if } j \notin \mathbb{I}_\tau, \qquad \widetilde{k}_i \mapsto [\mathbb{K}_i], \quad \text{if } \tau i \neq i \in \mathbb{I}. \tag{7.20}$$

PROOF. It follows from (7.11) that $\widetilde{\psi}$ preserves the relation (7.14).

For (7.15), we have $\mathrm{Hom}_{\Lambda^\imath}(S_i, S_j) = 0 = \mathrm{Ext}^1_{\Lambda^\imath}(S_i, S_j)$ for any i, j such that $c_{ij} = 0$ and $\tau i \neq j$. So $[S_i] * [S_j] = [S_i \oplus S_j] = [S_j] * [S_i]$. Then $\widetilde{\psi}$ preserves the relation (7.15).

For (7.17), we have $\mathrm{Hom}_{\Lambda^\imath}(S_i, S_{\tau i}) = 0$ and $\mathrm{Ext}^1_{\Lambda^\imath}(S_i, S_{\tau i}) = \mathbf{k}$, and $K \cong \mathbb{K}_i$ for any non-split exact sequence $0 \to S_{\tau i} \to K \to S_i \to 0$. So $[S_{\tau i}] * [S_i] - [S_i] * [S_{\tau i}] = (q-1)([\mathbb{K}_{\tau i}] - [\mathbb{K}_i])$.

The verification that $\widetilde{\psi}$ preserves the relations (7.16) and (7.18) is reduced to the rank 2 Dynkin ıquivers, and the details will be given in §7.3–§7.4 below.

It remains to prove that $\widetilde{\psi} : \widetilde{\mathbf{U}}^\imath_{\mathbf{v}} \longrightarrow \widetilde{\mathcal{H}}(\mathbf{k}Q, \tau)$ is an isomorphism. The homomorphism $\widetilde{\psi} : \widetilde{\mathbf{U}}^\imath_{\mathbf{v}} \to \widetilde{\mathcal{H}}(\mathbf{k}Q, \tau)$ restricts to an algebra homomorphism

$$\widetilde{\psi} : \widetilde{\mathbf{U}}^{\imath 0}_{\mathbf{v}} \longrightarrow \widetilde{\mathcal{T}}(\Lambda^\imath),$$
$$\widetilde{k}_i \mapsto -q^{-1}[\mathbb{K}_i], \text{ if } \tau i = i, \qquad \widetilde{k}_i \mapsto [\mathbb{K}_i], \text{ if } \tau i \neq i.$$

Since both $\widetilde{\mathbf{U}}^{\imath 0}_{\mathbf{v}}$ and $\widetilde{\mathcal{T}}(\Lambda^\imath)$ are Laurent polynomial algebras in the same number of generators, $\widetilde{\psi} : \widetilde{\mathbf{U}}^{\imath 0}_{\mathbf{v}} \to \widetilde{\mathcal{T}}(\Lambda^\imath)$ is an isomorphism.

We observe that $\widetilde{\psi}$ is a morphism of filtered algebras. Let $\widetilde{\psi}^{\rm gr} : \widetilde{\mathbf{U}}^{\imath,{\rm gr}}_{\mathbf{v}} \longrightarrow \widetilde{\mathcal{H}}(\mathbf{k}Q,\tau)^{\rm gr}$ be its associated graded morphism, and we obtain the following commutative diagram

$$\begin{array}{ccc} \mathbf{U}^-_{\mathbf{v}} & \xrightarrow{\ \phi\ } & \widetilde{\mathbf{U}}^{\imath,{\rm gr}}_{\mathbf{v}} \\ \Big\downarrow R & & \Big\downarrow \widetilde{\psi}^{\rm gr} \\ \widetilde{\mathcal{H}}(\mathbf{k}Q) & \xrightarrow{\ \varphi\ } & \widetilde{\mathcal{H}}(\mathbf{k}Q,\tau)^{\rm gr} \end{array}$$

It follows that $\widetilde{\psi}^{\rm gr}\circ\phi$ is injective since φ and $R : \mathbf{U}^-_{\mathbf{v}} \to \widetilde{\mathcal{H}}(\mathbf{k}Q)$ are injective by Theorem 3.2 and (7.13).

We claim that $\widetilde{\psi}^{\rm gr}$ is injective. Indeed, any element in $\widetilde{\mathbf{U}}^{\imath,{\rm gr}}$ is of form $\sum_{\alpha\in\mathbb{Z}^{\mathbb{I}}}\phi(V_\alpha)\cdot\widetilde{k}_\alpha$, for $V_\alpha\in\mathbf{U}^-$. Here $\widetilde{k}_\alpha=\prod_{i\in\mathbb{I}}\widetilde{k}_i^{a_i}$ for $\alpha=\sum_{i\in\mathbb{I}}a_i\alpha_i$. Assume $\widetilde{\psi}^{\rm gr}(\sum_\alpha\phi(V_\alpha)\cdot\widetilde{k}_\alpha)=0$, i.e., $\sum_{\alpha\in\mathbb{Z}^{\mathbb{I}}}\widetilde{\psi}^{\rm gr}(\phi(V_\alpha))*\mathbb{K}_\alpha=0$. By Theorem 7.2, we obtain $\widetilde{\psi}^{\rm gr}(\phi(V_\alpha))*\mathbb{K}_\alpha=0$ for any α, and then $\widetilde{\psi}^{\rm gr}(\phi(V_\alpha))=0$ since $\mathbb{K}_\alpha$ is invertible. So $V_\alpha=0$ by using the above commutative diagram. It follows that $\widetilde{\psi}^{\rm gr}$ is injective.

Now by a standard filtered algebra argument, we obtain that $\widetilde{\psi} : \widetilde{\mathbf{U}}^\imath_{\mathbf{v}} \longrightarrow \widetilde{\mathcal{H}}(\mathbf{k}Q,\tau)$ is an algebra monomorphism.

Since Q is of Dynkin type, we have $\widetilde{\mathcal{H}}(\mathbf{k}Q)$ is generated by $[S_i]$ $(i\in\mathbb{I})$ by Theorem 3.2. Then $\widetilde{\mathcal{H}}(\mathbf{k}Q,\tau)$ is generated by $[S_i]$ $(i\in\mathbb{I})$ and $\mathbb{K}_\alpha$ $(\alpha\in\mathbb{Z}^{\mathbb{I}})$ since $\widetilde{\mathcal{H}}(\mathbf{k}Q,\tau)\cong\widetilde{\mathcal{H}}(\mathbf{k}Q)*\widetilde{\mathcal{T}}(\Lambda^\imath)$ as linear spaces. So $\widetilde{\psi}$ is surjective. The theorem is proved. □

7.3. Computations for rank 2 Dynkin $\imath$quivers, I.

PROPOSITION 7.4. *Let Q be the quiver $1\longrightarrow 2$, with $\tau=\mathrm{Id}$. Then in $\mathcal{SD}\widetilde{\mathcal{H}}(\Lambda^\imath)$ we have*

$$[S_2]*[S_1]*[S_1]-(\mathbf{v}+\mathbf{v}^{-1})[S_1]*[S_2]*[S_1]+[S_1]*[S_1]*[S_2] = -\frac{(q-1)^2}{\mathbf{v}}[S_2]*[\mathbb{K}_1], \tag{7.21}$$

$$[S_1]*[S_2]*[S_2]-(\mathbf{v}+\mathbf{v}^{-1})[S_2]*[S_1]*[S_2]+[S_2]*[S_2]*[S_1] = -\frac{(q-1)^2}{\mathbf{v}}[S_1]*[\mathbb{K}_2]. \tag{7.22}$$

PROOF. Recall from Example 5.4(a) the quiver and relations of $\Lambda^\imath$ for this $\imath$quiver. We shall only prove the first identity (7.21) while skipping a similar proof of the identity (7.22).

Denote by U_i the indecomposable projective $\Lambda^\imath$-module corresponding to i. Denote by X the unique indecomposable $\Lambda^\imath$-module with $\widehat{S_1}+\widehat{S_2}$ as its class in

$K_0(\mathrm{mod}(\Lambda^\imath))$. Then we have

$$\begin{aligned}
[S_2]*[S_1]*[S_1] &= \mathbf{v}^{\langle S_2,S_1\rangle_Q}[S_1\oplus S_2]*[S_1]\\
&= \mathbf{v}^{\langle S_2,S_1\rangle_Q}\mathbf{v}^{\langle S_1\oplus S_2,S_1\rangle_Q}\Big(\frac{1}{q}[S_2\oplus S_1\oplus S_1]+\frac{q-1}{q}[S_2\oplus\mathbb{K}_1]\Big)\\
&= \frac{1}{\mathbf{v}}[S_2\oplus S_1\oplus S_1]+\frac{q-1}{\mathbf{v}}[S_2\oplus\mathbb{K}_1];\\
[S_1]*[S_2]*[S_1] &= \mathbf{v}^{\langle S_1,S_2\rangle_Q}([S_1\oplus S_2]*[S_1]+(q-1)[X]*[S_1])\\
&= \mathbf{v}^{\langle S_1,S_2\rangle_Q}\mathbf{v}^{\langle S_1\oplus S_2,S_1\rangle_Q}\Big(\frac{1}{q}[S_1\oplus S_1\oplus S_2]+\frac{q-1}{q}[S_2\oplus\mathbb{K}_1]\Big)\\
&\quad+\mathbf{v}^{\langle S_1,S_2\rangle_Q}\mathbf{v}^{\langle S_1\oplus S_2,S_1\rangle_Q}\Big(\frac{q-1}{q}[X\oplus S_1]+\frac{(q-1)^2}{q}[U_1/S_2]\Big)\\
&= \frac{1}{q}[S_1\oplus S_1\oplus S_2]+\frac{q-1}{q}[S_2\oplus\mathbb{K}_1]\\
&\quad+\frac{q-1}{q}[X\oplus S_1]+\frac{(q-1)^2}{q}[U_1/S_2];\\
[S_1]*[S_1]*[S_2] &= [S_1]*\big(\mathbf{v}^{\langle S_1,S_2\rangle_Q}([S_1\oplus S_2]+(q-1)[X])\big)\\
&= \mathbf{v}^{\langle S_1,S_2\rangle_Q}\mathbf{v}^{\langle S_1,S_1\oplus S_2\rangle_Q}\\
&\quad\times\Big(\frac{1}{q}[S_1\oplus S_1\oplus S_2]+\frac{q-1}{q}[\mathbb{K}_1\oplus S_2]+\frac{q-1}{q}[S_1\oplus X]\Big)\\
&\quad+\mathbf{v}^{\langle S_1,S_2\rangle_Q}\mathbf{v}^{\langle S_1,S_1\oplus S_2\rangle_Q}\Big(\frac{(q-1)^2}{q}[U_1/S_2]+(q-1)[S_1\oplus X]\Big)\\
&= \frac{1}{q\mathbf{v}}[S_1\oplus S_1\oplus S_2]+\frac{q-1}{q\mathbf{v}}[\mathbb{K}_1\oplus S_2]\\
&\quad+\frac{q^2-1}{q\mathbf{v}}[S_1\oplus X]+\frac{(q-1)^2}{q\mathbf{v}}[U_1/S_2].
\end{aligned}$$

By the short exact sequence $0\to S_2\to U_1/S_2\to\mathbb{K}_1\to 0$ with $\mathbb{K}_1\in\mathcal{P}^{\leq 1}(\Lambda^\imath)$, we have $[U_1/S_2]=[\mathbb{K}_1\oplus S_2]$ in $\mathcal{SD}\widetilde{\mathcal{H}}(\Lambda^\imath)$, and then by Lemma 7.1, we have

$$[S_2]*[\mathbb{K}_1]=\mathbf{v}^{\langle S_2,S_1\oplus S_1\rangle_Q}q^{-\langle S_2,\mathbb{K}_1\rangle}[U_1/S_2]=[U_1/S_2].$$

Hence, in $\mathcal{SD}\widetilde{\mathcal{H}}(\Lambda^\imath)$ we obtain that

$$\begin{aligned}
[S_2]*[S_1]*[S_1]-&(\mathbf{v}+\mathbf{v}^{-1})[S_1]*[S_2]*[S_1]+[S_1]*[S_1]*[S_2]\\
&=-\frac{(q-1)^2}{\mathbf{v}}[U_1/S_2]=-\frac{(q-1)^2}{\mathbf{v}}[S_2]*[\mathbb{K}_1].
\end{aligned}$$

The proposition is proved. □

7.4. Computations for rank 2 Dynkin $\imath$quivers, II.

PROPOSITION 7.5. *Let Q be the quiver $1 \longrightarrow 2 \longleftarrow 3$ with τ being the nontrivial involution. Then in $\mathcal{SD}\widetilde{\mathcal{H}}(\Lambda^\imath)$ we have, for $i = 1, 3$,*

(7.23)
$$[S_i]*[S_i]*[S_2] - (\mathbf{v}+\mathbf{v}^{-1})[S_i]*[S_2]*[S_i]+[S_2]*[S_i]*[S_i] = 0,$$
(7.24)
$$[S_2]*[S_2]*[S_i] - (\mathbf{v}+\mathbf{v}^{-1})[S_2]*[S_i]*[S_2]+[S_i]*[S_2]*[S_2]=-\frac{(q-1)^2}{\mathbf{v}}[S_i]*[\mathbb{K}_2].$$

(Using the opposite quiver $1 \longleftarrow 2 \longrightarrow 3$ yields the same formulas above.)

PROOF. Recall from Example 5.4(b) the quiver and relations of $\Lambda^\imath$. Denote by U_i the indecomposable projective $\Lambda^\imath$-module corresponding to $i \in \{1, 2, 3\}$. Denote by X the unique indecomposable $\Lambda^\imath$-module with $\widehat{S_1} + \widehat{S_2}$ as its class in $K_0(\Lambda^\imath)$. We shall only prove the formulas for $i = 1$, as the remaining case with $i = 3$ follows by symmetry.

In $\mathcal{SD}\widetilde{\mathcal{H}}(\Lambda^\imath)$, we have

$$\begin{aligned}
[S_1] * [S_1] * [S_2] &= \mathbf{v}^{\langle S_1, S_2\rangle_Q}[S_1] * \big([S_1 \oplus S_2] + (q-1)[X]\big) \\
&= \mathbf{v}^{\langle S_1, S_2\rangle_Q}\mathbf{v}^{\langle S_1, S_1\oplus S_2\rangle_Q} \\
&\quad\times \Big(\frac{1}{q}[S_1 \oplus S_1 \oplus S_2] + \frac{q-1}{q}[S_1 \oplus X] + (q-1)[S_1 \oplus X]\Big) \\
&= \frac{1}{q\mathbf{v}}[S_1 \oplus S_1 \oplus S_2] + \frac{q^2-1}{q\mathbf{v}}[S_1 \oplus X]; \\
[S_1] * [S_2] * [S_1] &= \mathbf{v}^{\langle S_1, S_2\rangle_Q}\big([S_1 \oplus S_2] + (q-1)[X]\big) * [S_1] \\
&= \mathbf{v}^{\langle S_1, S_2\rangle_Q}\mathbf{v}^{\langle S_1\oplus S_2, S_1\rangle_Q}\Big(\frac{1}{q}[S_1 \oplus S_1 \oplus S_2] + \frac{q-1}{q}[S_1 \oplus X]\Big) \\
&= \frac{1}{q}[S_1 \oplus S_1 \oplus S_2] + \frac{q-1}{q}[S_1 \oplus X];
\end{aligned}$$

and

$$\begin{aligned}
[S_2] * [S_1] * [S_1] &= \mathbf{v}^{\langle S_2, S_1\rangle_Q}[S_1 \oplus S_2] * [S_1] \\
&= \mathbf{v}^{\langle S_2, S_1\rangle_Q}\mathbf{v}^{\langle S_1\oplus S_2, S_1\rangle_Q}\frac{1}{q}[S_1 \oplus S_1 \oplus S_2] \\
&= \frac{1}{\mathbf{v}}[S_1 \oplus S_1 \oplus S_2].
\end{aligned}$$

The first identity (7.23) follows from combining the above computations.

On the other hand, we have

$$\begin{aligned}
[S_2]*[S_2]*[S_1] &= \mathbf{v}^{\langle S_2,S_1\rangle_Q}[S_2]*[S_2\oplus S_1]\\
&= \mathbf{v}^{\langle S_2,S_1\rangle_Q}\mathbf{v}^{\langle S_2,S_1\oplus S_2\rangle_Q}\Big(\frac{1}{q}[S_2\oplus S_2\oplus S_1]+\frac{q-1}{q}[\mathbb{K}_2\oplus S_1]\Big)\\
&= \frac{1}{\mathbf{v}}[S_2\oplus S_2\oplus S_1]+\frac{q-1}{\mathbf{v}}[\mathbb{K}_2\oplus S_1];
\end{aligned}$$

(7.25)

$$\begin{aligned}
[S_2]*[S_1]*[S_2] &= \mathbf{v}^{\langle S_2,S_1\rangle_Q}[S_2\oplus S_1]*[S_2]\\
&= \mathbf{v}^{\langle S_2,S_1\rangle_Q}\mathbf{v}^{\langle S_2\oplus S_1,S_2\rangle_Q}\Big(\frac{1}{q}[S_2\oplus S_1\oplus S_2]+\frac{q-1}{q}[\mathbb{K}_2\oplus S_1]\\
&\quad+\frac{q-1}{q}[S_2\oplus X]+\frac{(q-1)^2}{q}[\mathrm{rad}(U_3)]\Big)\\
&= \frac{1}{q}[S_2\oplus S_1\oplus S_2]+\frac{q-1}{q}[\mathbb{K}_2\oplus S_1]+\frac{q-1}{q}[S_2\oplus X]\\
&\quad+\frac{(q-1)^2}{q}[\mathrm{rad}(U_3)]\\
&= \frac{1}{q}[S_2\oplus S_1\oplus S_2]+(q-1)[\mathbb{K}_2\oplus S_1]+\frac{q-1}{q}[S_2\oplus X],
\end{aligned}$$

where we have used $[\mathrm{rad}(U_3)]=[\mathbb{K}_2\oplus S_1]$ in $\mathcal{SD}\widetilde{\mathcal{H}}(\Lambda^\imath)$ thanks to a short exact sequence $0\to\mathbb{K}_2\to\mathrm{rad}(U_3)\to S_1\to 0$ with $\mathbb{K}_2\in P^{\leq 1}(\Lambda^\imath)$. In addition, (7.25) also implies that

$$[S_1\oplus S_2]*[S_2]=\frac{1}{q}[S_2\oplus S_1\oplus S_2]+(q-1)[\mathbb{K}_2\oplus S_1]+\frac{q-1}{q}[S_2\oplus X].$$

Then we have

$$\begin{aligned}
[S_1]*[S_2]*[S_2] &= \mathbf{v}^{\langle S_1,S_2\rangle_Q}([S_1\oplus S_2]*[S_2]+(q-1)[X]*[S_2])\\
&= \mathbf{v}^{\langle S_1,S_2\rangle_Q}\mathbf{v}^{\langle S_2\oplus S_1,S_2\rangle_Q}\Big(\frac{1}{q}[S_2\oplus S_1\oplus S_2]+(q-1)[\mathbb{K}_2\oplus S_1]\\
&\quad+\frac{q-1}{q}[S_2\oplus X]+(q-1)[X\oplus S_2]\Big)\\
&= \frac{1}{q\mathbf{v}}[S_2\oplus S_1\oplus S_2]+\frac{q-1}{\mathbf{v}}[\mathbb{K}_2\oplus S_1]+\frac{q^2-1}{q\mathbf{v}}[S_2\oplus X].
\end{aligned}$$

Summarizing, we have obtained

$$\begin{aligned}
&[S_2]*[S_2]*[S_1]-(\mathbf{v}+\mathbf{v}^{-1})[S_2]*[S_1]*[S_2]+[S_1]*[S_2]*[S_2]\\
&\qquad=-\frac{(q-1)^2}{\mathbf{v}}[\mathbb{K}_2\oplus S_1]=-\frac{(q-1)^2}{\mathbf{v}}[S_1]*[\mathbb{K}_2],
\end{aligned}$$

whence the identity (7.24). □

8. ıHall algebras for acyclic ıquivers

In this section, we establish the ıHall algebra realization of the universal ıquantum group of Kac-Moody type. The relations (2.15) and (2.17) in $\widetilde{\mathbf{U}}^\imath$ are highly nontrivial in contrast to the Dynkin case, and their verification in the ıHall algebra setting is highly nontrivial and requires different techniques. We outline the proof of the relation (2.15) for ıHall algebra, by first formulating the ıHall algebra counterpart of ıdivided powers.

8.1. Rank 1 and ıdivided powers. Consider the ıquiver consisting of a single vertex with a trivial involution. The associated ıquiver algebra is given by $\Lambda^\imath = \mathbf{k}[x]/(x^2)$, and the corresponding ıHall algebra $\widetilde{\mathcal{H}}(\mathbf{k}Q,\tau)$ is commutative and isomorphic to $\mathbb{Q}(\mathbf{v})[[S],\mathbb{K}^{\pm 1}]$. The split ıquantum group $\widetilde{\mathbf{U}}^\imath$ of rank one is the algebra $\mathbb{Q}(v)[B,\widetilde{k}^{\pm 1}]$. The $\mathbb{Q}(\mathbf{v})$-algebra isomorphism $\widetilde{\psi}:\widetilde{\mathbf{U}}^\imath_{\mathbf{v}}\longrightarrow\widetilde{\mathcal{H}}(\mathbf{k}Q,\tau)$ (of rank one) (see Theorem 7.3) sends

$$B\mapsto\frac{-1}{q-1}[S],\qquad \widetilde{k}\mapsto-\frac{\mathbb{K}}{q}. \tag{8.1}$$

Inspiring by (2.11)–(2.12) and (8.1), we define the ıdivided power of $[S]$ in $\widetilde{\mathcal{H}}(\mathbf{k}Q,\tau)$ to be

$$[S]_{\bar{1}}^{(m)}:=\frac{1}{[m]_{\mathbf{v}}!}\begin{cases}[S]\prod_{j=1}^{k}([S]^2+\mathbf{v}^{-1}(\mathbf{v}^2-1)^2[2j-1]_{\mathbf{v}}^2[\mathbb{K}]) & \text{if } m=2k+1,\\ \prod_{j=1}^{k}([S]^2+\mathbf{v}^{-1}(\mathbf{v}^2-1)^2[2j-1]_{\mathbf{v}}^2[\mathbb{K}]) & \text{if } m=2k;\end{cases}$$

$$[S]_{\bar{0}}^{(m)}:=\frac{1}{[m]_{\mathbf{v}}!}\begin{cases}[S]\prod_{j=1}^{k}([S]^2+\mathbf{v}^{-1}(\mathbf{v}^2-1)^2[2j]_{\mathbf{v}}^2[\mathbb{K}]) & \text{if } m=2k+1,\\ \prod_{j=1}^{k}([S]^2+\mathbf{v}^{-1}(\mathbf{v}^2-1)^2[2j-2]_{\mathbf{v}}^2[\mathbb{K}]) & \text{if } m=2k.\end{cases}$$

These ıdivided powers satisfy the following recursive relations:

$$[S]*[S]_{\bar{1}}^{(2m)}=[2m+1][S]_{\bar{1}}^{(2m+1)}, \tag{8.2}$$

$$[S]*[S]_{\bar{1}}^{(2m+1)}=[2m+2][S]_{\bar{1}}^{(2m+2)}-\mathbf{v}(\mathbf{v}-\mathbf{v}^{-1})^2[2m+1][S]_{\bar{1}}^{(2m)}*[\mathbb{K}], \tag{8.3}$$

$$[S]*[S]_{\bar{0}}^{(2m-1)}=[2m][S]_{\bar{0}}^{(2m)}, \tag{8.4}$$

$$[S]*[S]_{\bar{0}}^{(2m)}=[2m+1][S]_{\bar{0}}^{(2m+1)}-\mathbf{v}(\mathbf{v}-\mathbf{v}^{-1})^2[2m][S]_{\bar{0}}^{(2m-1)}*[\mathbb{K}]. \tag{8.5}$$

The following lemma follows by definition.

LEMMA 8.1. *The isomorphism $\widetilde{\psi}$ satisfies that, for $m\in\mathbb{N}$,*

$$\widetilde{\psi}(B_{\bar{1}}^{(m)})=\frac{[S]_{\bar{1}}^{(m)}}{(1-\mathbf{v}^2)^m},\qquad \widetilde{\psi}(B_{\bar{0}}^{(m)})=\frac{[S]_{\bar{0}}^{(m)}}{(1-\mathbf{v}^2)^m}. \tag{8.6}$$

We denote by $[0]_{\mathbf{v}}^{!!}=1$, and for any $k\in\mathbb{Z}_{\geq 1}$,

$$[2k]_{\mathbf{v}}^{!!}=[2k]_{\mathbf{v}}[2k-2]_{\mathbf{v}}\cdots[4]_{\mathbf{v}}[2]_{\mathbf{v}}.$$

We denote by $\lfloor x\rfloor$ the largest integer not exceeding x, for $x\in\mathbb{R}$.

PROPOSITION 8.2. *The following identity holds in $\widetilde{\mathcal{H}}(\mathbf{k}Q,\tau)$, for $n\in\mathbb{N}$:*

$$[S]_{\bar{0}}^{(n)}=\sum_{k=0}^{\lfloor\frac{n}{2}\rfloor}\frac{\mathbf{v}^{k(k-(-1)^n)-\binom{n-2k}{2}}\cdot(\mathbf{v}-\mathbf{v}^{-1})^k}{[n-2k]_{\mathbf{v}}^{!}[2k]_{\mathbf{v}}^{!!}}[(n-2k)S]*[\mathbb{K}]^k, \tag{8.7}$$

$$[S]_{\bar{1}}^{(n)}=\sum_{k=0}^{\lfloor\frac{n}{2}\rfloor}\frac{\mathbf{v}^{k(k+(-1)^n)-\binom{n-2k}{2}}\cdot(\mathbf{v}-\mathbf{v}^{-1})^k}{[n-2k]_{\mathbf{v}}^{!}[2k]_{\mathbf{v}}^{!!}}[(n-2k)S]*[\mathbb{K}]^k. \tag{8.8}$$

PROOF. The required Euler form is given by

$$\langle S,S^{\oplus m}\rangle_Q=m=\dim_{\mathbf{k}}\operatorname{Hom}_{\Lambda^\imath}(S,S^{\oplus m}).$$

For any non-split short exact sequence $0 \to S^{\oplus m} \to M \to S \to 0$ in $\operatorname{mod}(\Lambda^\imath)$, we have $M \cong S^{\oplus(m-1)} \oplus \mathbb{K}$. Note that $\operatorname{Ext}^1_{\Lambda^\imath}(S, S^{\oplus m}) = m$. Then we have

$$(8.9)\qquad \begin{aligned}[S] * [mS] &= \mathbf{v}^{-m}[(m+1)S] + \mathbf{v}^{-m}(q^m - 1)[(m-1)S \oplus \mathbb{K}] \\ &= \mathbf{v}^{-m}[(m+1)S] + (\mathbf{v}^m - \mathbf{v}^{-m})[(m-1)S] * [\mathbb{K}].\end{aligned}$$

The identities (8.7)–(8.8) can now be proved by induction on n using the recursive identities (8.2)–(8.5) and (8.9). □

8.2. Rank 2 computations. The following multiplication formula in the ıHall algebra is useful.

PROPOSITION 8.3 ([**LW20**, Proposition 3.10]). *Let (Q, τ) be an ıquiver with $\tau = \operatorname{Id}$. For any $A, B \in \operatorname{mod}(\mathbf{k}Q) \subset \operatorname{mod}(\Lambda^\imath)$, the following identity holds in $\widetilde{\mathcal{H}}(\mathbf{k}Q, \tau)$.*

(8.10)

$$[A] * [B] = \sum_{[L],[M],[N] \in \operatorname{Iso}(\operatorname{mod}(\mathbf{k}Q))} \mathbf{v}^{\langle A,B\rangle_Q} q^{\langle N,L\rangle_Q - \langle A,B\rangle_Q} \frac{|\operatorname{Ext}^1(N,L)_M|}{|\operatorname{Hom}(N,L)|}$$
$$\times |\{s \in \operatorname{Hom}(A,B) \mid \operatorname{Ker} s \cong N, \operatorname{Coker} s \cong L\}| \cdot [M] * [\mathbb{K}_{\widehat{A}-\widehat{N}}].$$

Consider the ıquiver

$$(8.11)\qquad Q = (\, 1 \underset{\longleftarrow b \longleftarrow}{\overset{\longrightarrow a \longrightarrow}{\rightleftarrows}} 2 \,), \quad \tau = \operatorname{Id}, \qquad \text{where } a + b = -c_{12}.$$

The arrows $1 \to 2$ are denoted by α_i, $1 \le i \le a$, and $2 \to 1$ are denoted by β_j, $1 \le j \le b$. Then the corresponding ıquiver algebra $\Lambda^\imath$ has its quiver $\overline{Q}$ as

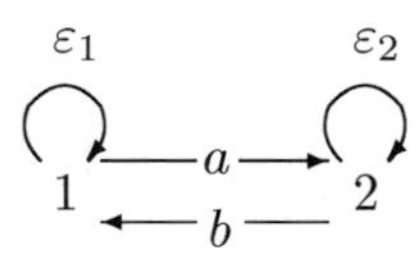

For any $\Lambda^\imath$-module $M = (M_i, M(\alpha_j), M(\beta_k), M(\varepsilon_i))_{i=1,2; 1\le j\le a, 1\le k\le b}$ such that $\dim_{\mathbf{k}} M_2 = 1$, we define

$$(8.12)\qquad U_M := \bigcap_{1\le i\le a} \operatorname{Ker} M(\alpha_i), \qquad W_M := \sum_{j=1}^{b} \operatorname{Im} M(\beta_j),$$

and let

$$(8.13)\qquad u_M := \dim U_M, \qquad w_M := \dim W_M.$$

For any $r \ge 0$, define

$$(8.14)\qquad \mathcal{I}_r := \{[M] \in \operatorname{Iso}(\operatorname{mod}(\mathbf{k}Q)) \mid \widehat{M} = r\widehat{S_1} + \widehat{S_2}, W_M \subseteq U_M\},$$

$$(8.15)\qquad \mathcal{J}_r := \{[M] \in \operatorname{Iso}(\operatorname{mod}(\mathbf{k}Q)) \mid \widehat{M} = r\widehat{S_1} + \widehat{S_2}, S_2 \in \operatorname{add}\operatorname{top}(M)\}.$$

The following proposition generalizes [**LW20**, Proposition 7.3], which treated the special cases when either a or b is 0.

PROPOSITION 8.4 (cf. [**LW20**, Proposition 7.3]). *The following identity holds in $\widetilde{\mathcal{H}}(\mathbf{k}Q,\tau)$, for $s,t\geq 0$:*

(8.16)
$$[sS_1]*[S_2]*[tS_1]=\mathbf{v}^{-tb}\sum_{r=0}^{\min\{s,t\}}\sum_{[M]\in\mathcal{I}_{s+t-2r}}\mathbf{v}^{\tilde{p}(a,b,r,s,t)}(\mathbf{v}-\mathbf{v}^{-1})^{s-r+t+1}\frac{[s]^!_{\mathbf{v}}[t]^!_{\mathbf{v}}}{[r]^!_{\mathbf{v}}}$$
$$\times\begin{bmatrix}u_M-w_M\\(t-r)-w_M\end{bmatrix}_{\mathbf{v}}\frac{1}{|\operatorname{Aut}(M)|}[M]*[\mathbb{K}_1]^r,$$

where

$$\tilde{p}(a,b,r,s,t)=2(s-r)(t-r-a)-s(t-a)+2br-2r(t-r)-r^2+rs+t^2+\binom{t}{2}+ (s-r)^2+\binom{s-r}{2}+\big(u_M-(t-r)\big)\big((t-r)-w_M\big)+1.$$

PROOF. The computation in cases either $a=0$ or $b=0$ can be found in the proof of [**LW20**, Proposition 7.3]. The computation required in the general case is similar, again by using crucially Proposition 8.3; we skip the details. □

We can formulate the ıSerre relations in the ıHall algebra, and we postpone its sketched proof in the next subsection.

THEOREM 8.5. *The following identity holds in $\widetilde{\mathcal{H}}(\mathbf{k}Q,\tau)$, for any $\overline{p}\in\mathbb{Z}_2$:*

$$\sum_{n=0}^{1+a+b}(-1)^n[S_1]^{(n)}_{\overline{p}}*[S_2]*[S_1]^{(1+a+b-n)}_{\overline{a}+\overline{b}+\overline{p}}=0. \tag{8.17}$$

8.3. Proof of Theorem 8.5. In this subsection, we sketch the proof of Theorem 8.5.

Note that (8.17) can be written as

$$\sum_{n=0}^{1+a+b}(-1)^n[S_1]^{(n)}_{\overline{0}}*[S_2]*[S_1]^{(1+a+b-n)}_{\overline{a}+\overline{b}}=0, \tag{8.18}$$

$$\sum_{n=0}^{1+a+b}(-1)^n[S_1]^{(n)}_{\overline{1}}*[S_2]*[S_1]^{(1+a+b-n)}_{\overline{a}+\overline{b}+\overline{1}}=0. \tag{8.19}$$

We only sketch the proof of (8.18) here, since the other one is similar.

Using (8.7)–(8.8) and Proposition 8.4, we obtain

(8.20)

$$\sum_{n=0}^{a+b+1} (-1)^n [S_1]_{\overline{0}}^{(n)} * [S_2] * [S_1]_{\overline{a}+\overline{b}}^{(1+a+b-n)}$$

$$= \sum_{n=0,2|n}^{a+b+1} \sum_{k=0}^{\frac{n}{2}} \sum_{m=0}^{\lfloor \frac{a+b+1-n}{2} \rfloor} \mathbf{v}^{-(1+a+b-n-2m)b} \sum_{r=0}^{\min\{n-2k,1+a+b-n-2m\}} \sum_{[M]\in\mathcal{I}_{1+a+b-2k-2m-2r}}$$

$$\frac{\mathbf{v}^{\tilde{z}}(\mathbf{v}-\mathbf{v}^{-1})^{2+a+b-k-m-r}}{[2k]^{!!}_{\mathbf{v}}[2m]^{!!}_{\mathbf{v}}[r]^{!}_{\mathbf{v}}} \begin{bmatrix} u_M - w_M \\ 1+a+b-n-2m-r-w_M \end{bmatrix}_{\mathbf{v}} \frac{[M]*[\mathbb{K}_1]^{k+m+r}}{|\operatorname{Aut}(M)|}$$

$$- \sum_{n=0,2\nmid n}^{a+b+1} \sum_{k=0}^{\frac{n-1}{2}} \sum_{m=0}^{\lfloor \frac{a+b+1-n}{2} \rfloor} \mathbf{v}^{-(1+a+b-n-2m)b} \sum_{r=0}^{\min\{n-2k,1+a+b-n-2m\}} \sum_{[M]\in\mathcal{I}_{1+a+b-2k-2m-2r}}$$

$$\frac{\mathbf{v}^{\tilde{z}+2k-2m}(\mathbf{v}-\mathbf{v}^{-1})^{2+a+b-k-m-r}}{[2k]^{!!}_{\mathbf{v}}[2m]^{!!}_{\mathbf{v}}[r]^{!}_{\mathbf{v}}} \begin{bmatrix} u_M - w_M \\ 1+a+b-n-2m-r-w_M \end{bmatrix}_{\mathbf{v}} \frac{[M]*[\mathbb{K}_1]^{k+m+r}}{|\operatorname{Aut}(M)|}.$$

Set

$$d = k + m + r.$$

Now we only need to prove that the coefficient of $\frac{[M]*[\mathbb{K}_1]^d}{|\operatorname{Aut}(M)|}$ in the RHS of (8.20) is zero, for any given $[M] \in \mathcal{I}_{1+a+b-2d}$ and any $d \in \mathbb{N}$. For simplicity, we denote by $u = u_M$, $w = w_M$ in the following.

Note the powers of $(\mathbf{v}-\mathbf{v}^{-1})$ in all terms are the same (equals to $2+a+b-d$). Denote

(8.21) $\widetilde{T}(a,b,d,u,w)$

$$= \sum_{n=0,2|n}^{a+b+1} \sum_{k=0}^{\frac{n}{2}} \sum_{m=0}^{\lfloor \frac{a+b+1-n}{2} \rfloor} \mathbf{v}^{-(1+a+b-n-2m)b} \delta\{0 \le r \le n-2k\}$$

$$\times \frac{\mathbf{v}^{\tilde{z}}}{[2k]^{!!}_{\mathbf{v}}[2m]^{!!}_{\mathbf{v}}[r]^{!!}_{\mathbf{v}}} \begin{bmatrix} u-w \\ 1+a+b-n-2m-r-w \end{bmatrix}_{\mathbf{v}}$$

$$- \sum_{n=0,2\nmid n}^{a+b+1} \sum_{k=0}^{\frac{n-1}{2}} \sum_{m=0}^{\lfloor \frac{a+b+1-n}{2}\, rfloor} \mathbf{v}^{-(1+a+b-n-2m)b} \delta\{0 \le r \le n-2k\}$$

$$\times \frac{\mathbf{v}^{\tilde{z}+2k-2m}}{[2k]^{!!}_{\mathbf{v}}[2m]^{!!}_{\mathbf{v}}[r]^{!!}_{\mathbf{v}}} \begin{bmatrix} u-w \\ 1+a+b-n-2m-r-w \end{bmatrix}_{\mathbf{v}},$$

where we set $\delta\{X\} = 1$ if the statement X holds and $\delta\{X\} = 0$ if X is false. Then the coefficient of $\frac{[M]*[\mathbb{K}_1]^d}{|\operatorname{Aut}(M)|}$ in the RHS of (8.20) is equal to

$$(\mathbf{v}-\mathbf{v}^{-1})^{2+a+b-d}\widetilde{T}(a,b,d,u,w).$$

So the identity (8.18) is equivalent to the identity $\widetilde{T}(a,b,d,u,w) = 0$, for any nonnegative integers a, b, d, u, w subject to the constraints

(8.22) $\quad 0 \le d \le (a+b+1)/2, \quad 0 \le w \le b, \quad b+1-2d \le u \le 1+a+b-2d,$

where u, d are not both zeros.

LEMMA 8.6 ([**LW20**, Proposition 8.1]). *For any nonnegative integers a, b, d, u, w satisfying the constraint* (8.22), *the following identity holds:*

$$\widetilde{T}(a, b, d, u, w) = 0. \tag{8.23}$$

The proof of Lemma 8.6 can be reduced to the following identities: for $p, d \geq 1$,

$$\sum_{k=0}^{p} v^{-k(p-k+1)} \begin{bmatrix} p \\ k \end{bmatrix} = \prod_{j=1}^{p} (1 + v^{-j}), \qquad \sum_{\substack{k,m,r \in \mathbb{N} \\ k+m+r=d}} (-1)^r \frac{v^{\binom{r+1}{2} - 2(k-1)m}}{[r]^! [2k]^{!!} [2m]^{!!}} = 0.$$

Thanks to Lemma 8.6, the proof of the identity (8.18) is complete. This proves Theorem 8.5.

8.4. $\imath$Hall algebra realization of $\widetilde{\mathbf{U}}^\imath$ of Kac-Moody type.

THEOREM 8.7 ([**LW20**, Theorem 9.6]). *Let (Q, τ) be an $\imath$quiver without loops. Then there exists a $\mathbb{Q}(\mathbf{v})$-algebra monomorphism*

$$\widetilde{\psi} : \widetilde{\mathbf{U}}^\imath_{\mathbf{v}} \longrightarrow \widetilde{\mathcal{H}}(\mathbf{k}Q, \tau),$$

which sends

$$B_j \mapsto \frac{-1}{q-1}[S_j], \ \text{if } j \in \mathbb{I}_\tau, \qquad \widetilde{k}_i \mapsto -q^{-1}[\mathbb{K}_i], \ \text{if } \tau i = i \in \mathbb{I}; \tag{8.24}$$

$$B_j \mapsto \frac{\mathbf{v}}{q-1}[S_j], \ \text{if } j \notin \mathbb{I}_\tau, \qquad \widetilde{k}_i \mapsto \mathbf{v}^{\frac{-c_{i,\tau i}}{2}}[\mathbb{K}_i], \quad \text{if } \tau i \neq i \in \mathbb{I}. \tag{8.25}$$

Due to the constraint arising from definition of $\imath$quivers, the Cartan matrix $(c_{ij})_{i,j \in \mathbb{I}}$ for Theorem 8.7 must satisfy the condition $c_{i,\tau i} \in 2\mathbb{Z}$, for all $i \in \mathbb{I}$.

PROOF. Let us outline the proof. First, we prove that $\widetilde{\psi}$ is an algebra homomorphism by verifying that it preserves the relations (2.13)–(2.17).

The verification of relations (2.13)–(2.14) is the same as for Theorem 7.3. The verification of relation (2.16) can be reduced to $\widetilde{\mathcal{H}}(\mathbf{k}Q)$, which holds by Theorem 3.2.

The verification of relation (2.17) requires long homological computations (compare to Theorem 3.2), and the detail can be found in [**LW20**, §5].

The verification of relation (2.15) can be reduced to the rank 2 $\imath$quivers, while the highly nontrivial rank 2 computation was sketched in §8.2–8.3.

The injectivity of $\widetilde{\psi}$ is proved in the same way as for Theorem 7.3. This completes the proof of the theorem. □

9. Reflection functors in $\imath$Hall algebras and relative braid group symmetries

In this section, we formulate and compute the reflection functors in the framework of $\imath$Hall algebras. The reflection functors give rise to (relative braid group) automorphisms of $\imath$quantum groups, and we provide explicit formulas for these symmetries on generators.

9.1. Modulated graphs for ıquivers. In this subsection, we study representations of modulated graphs associated to ıquivers, cf. [**LW20, LW22a**].

Let (Q,τ) be an ıquiver, and $\Lambda^\imath = \mathbf{k}\overline{Q}/\overline{I}$ with $(\overline{Q},\overline{I})$ as defined in §5.1. For each $i \in Q_0$, define a $\mathbf{k}$-algebra

(9.1)
$$\overline{\mathbb{H}}_i := \begin{cases} \mathbf{k}[\varepsilon_i]/(\varepsilon_i{}^2) & \text{if } i=\tau i, \\ \mathbf{k}\Big(i \underset{\varepsilon_{\tau i}}{\overset{\varepsilon_i}{\rightleftarrows}} \tau i \text{ with arrows } \xrightarrow{r_i},\ \xleftarrow{r_i} \Big)/(\varepsilon_i\varepsilon_{\tau i}, \varepsilon_{\tau i}\varepsilon_i, \alpha_j\varepsilon_{\tau i} - \varepsilon_i\beta_j, \beta_j\varepsilon_i - \varepsilon_{\tau i}\alpha_j \mid 1 \leq j \leq r) & \text{if } \tau i \neq i, n_{i,\tau i} = 2r_i. \end{cases}$$

Define the following subalgebra of $\Lambda^\imath$:

$$\overline{\mathbb{H}} = \bigoplus_{i \in \mathbb{I}_\tau} \overline{\mathbb{H}}_i. \tag{9.2}$$

Define

$$\Omega := \Omega(Q) = \{(i,j) \in Q_0 \times Q_0 \mid \exists (\alpha : i \to j) \in Q_1, \tau i \neq j\}. \tag{9.3}$$

Then Ω represents the orientation of Q. We also use $\Omega(i,-)$ to denote the subset $\{j \in Q_0 \mid \exists(\alpha : i \to j) \in Q_1\}$, and $\Omega(-,i)$ is defined similarly.

For any $(i,j) \in \Omega$, we define

$${}_j\overline{\mathbb{H}}_i := \overline{\mathbb{H}}_j \operatorname{Span}_{\mathbf{k}}\{\alpha, \tau\alpha \mid (\alpha : i \to j) \in Q_1 \text{ or } (\alpha : i \to \tau j) \in Q_1\}\overline{\mathbb{H}}_i. \tag{9.4}$$

Note that ${}_j\overline{\mathbb{H}}_i = {}_{\tau j}\overline{\mathbb{H}}_{\tau i} = {}_j\overline{\mathbb{H}}_{\tau i} = {}_{\tau j}\overline{\mathbb{H}}_i$ for any $(i,j) \in \Omega$.

Hence ${}_j\overline{\mathbb{H}}_i$ is an $\overline{\mathbb{H}}_j$-$\overline{\mathbb{H}}_i$-bimodule, which is free as a left $\overline{\mathbb{H}}_j$-module (and respectively, right $\overline{\mathbb{H}}_i$-module), with a basis ${}_j\mathtt{L}_i$ (and respectively, ${}_j\mathtt{R}_i$) defined in the following:

$${}_j\mathtt{L}_i = \begin{cases} \{\alpha \mid (\alpha : i \to j) \in Q_1\} & \text{if } i = \tau i, \tau j = j, \\ \{\alpha + \tau\alpha \mid (\alpha : i \to j) \in Q_1\} & \text{if } i = \tau i, \tau j \neq j, \\ \{\alpha, \tau\alpha \mid (\alpha : i \to j) \in Q_1\} & \text{if } i \neq \tau i, \tau j = j, \\ \{\alpha + \tau\alpha \mid (\alpha : i \to j) \text{ or } (\alpha : i \to \tau j) \in Q_1\} & \text{if } i \neq \tau i, \tau j \neq j; \end{cases} \tag{9.5}$$

$${}_j\mathtt{R}_i = \begin{cases} \{\alpha \mid (\alpha : i \to j) \in Q_1\} & \text{if } i = \tau i, \tau j = j, \\ \{\alpha, \tau\alpha \mid (\alpha : i \to j) \in Q_1\} & \text{if } i = \tau i, \tau j \neq j, \\ \{\alpha + \tau\alpha \mid (\alpha : i \to j) \in Q_1\} & \text{if } i \neq \tau i, \tau j = j, \\ \{\alpha + \tau\alpha \mid (\alpha : i \to j) \text{ or } (\alpha : i \to \tau j) \in Q_1\} & \text{if } i \neq \tau i, \tau j \neq j. \end{cases} \tag{9.6}$$

Denote

$$\overline{\Omega} := \{(i,j) \in \mathbb{I}_\tau \times \mathbb{I}_\tau \mid (i,j) \in \Omega \text{ or } (i,\tau j) \in \Omega\}. \tag{9.7}$$

Recall that $\mathbb{I}_\tau$ is a (fixed) subset of $Q_0 = \mathbb{I}$ consisting of the representatives of τ-orbits. The tuple $(\overline{\mathbb{H}}_i, {}_j\overline{\mathbb{H}}_i) := (\overline{\mathbb{H}}_i, {}_j\overline{\mathbb{H}}_i)_{i \in \mathbb{I}_\tau, (i,j) \in \overline{\Omega}}$ is called a *modulation* of (Q,τ) and is denoted by $\mathcal{M}(Q,\tau)$.

A representation $(N_i, N_{ji}) := (N_i, N_{ji})_{i \in \mathbb{I}_\tau, (i,j) \in \overline{\Omega}}$ of $\mathcal{M}(Q,\tau)$ is defined by assigning to each $i \in \mathbb{I}_\tau$ a finite-dimensional $\overline{\mathbb{H}}_i$-module N_i and to each $(i,j) \in \overline{\Omega}$ an $\overline{\mathbb{H}}_j$-morphism $N_{ji} : {}_j\overline{\mathbb{H}}_i \otimes_{\overline{\mathbb{H}}_i} N_i \to N_j$. A morphism $f : L \to N$ between representations $L = (L_i, L_{ji})$ and $N = (N_i, N_{ji})$ of $\mathcal{M}(Q,\tau)$ is a tuple $f = (f_i)_{i \in \mathbb{I}_\tau}$

of $\overline{\mathbb{H}}_i$-morphisms $f_i : L_i \to N_i$ such that the following diagram is commutative for each $(i,j) \in \overline{\Omega}$:

$$\begin{array}{ccc}
{}_j\overline{\mathbb{H}}_i \otimes_{\overline{\mathbb{H}}_i} L_i & \xrightarrow{1\otimes f_i} & {}_j\overline{\mathbb{H}}_i \otimes_{\overline{\mathbb{H}}_i} N_i \\
\downarrow{\scriptstyle L_{ij}} & & \downarrow{\scriptstyle N_{ij}} \\
L_j & \xrightarrow{\quad f_j \quad} & N_j
\end{array}$$

PROPOSITION 9.1 ([**LW22a**, Proposition 2.16]). *The categories* $\operatorname{rep}(\mathcal{M}(Q,\tau))$ *and* $\operatorname{rep}(\overline{Q},\overline{I})$ *are isomorphic.*

9.2. Reflection functors. In this subsection, we shall introduce the reflection functors in the setting of ıquivers; see [**LW21a, LW22b**].

Let Q^* be the quiver constructed from Q by reversing all the arrows $\alpha : i \to j$ such that $\tau i \neq j$. For any $i, j \in \mathbb{I}$ such that $\tau i \neq j$, we have $(i,j) \in \Omega$ if and only if $(j,i) \in \Omega^* := \Omega(Q^*)$. For any $\alpha : i \to j$ in Q such that $\tau i \neq j$, denote by $\widetilde{\alpha} : j \to i$ the corresponding arrow in Q^*. Then τ induces an involution τ^* of Q^*. Clearly, $\tau^* i = \tau i$ for any vertex $i \in Q_0$. Then similarly we can define $\Lambda^* = \mathbf{k}Q^* \otimes_{\mathbf{k}} R_2$, and an involution $\tau^{*\sharp}$ for Λ^*, and its $\tau^{*\sharp}$-fixed point subalgebra $(\Lambda^*)^\imath$. Note that $\overline{\mathbb{H}}$ is also a subalgebra of $(\Lambda^*)^\imath$.

It is worth noting that $\mathbb{I}_\tau$ is also a set of representatives of τ^*-orbits. In this way, one can define $\overline{\Omega}^*$ (cf. (9.7) for $\overline{\Omega}$).

For any $(j,i) \in \Omega^*$, we can define ${}_i\overline{\mathbb{H}}_j$ as follows:

$${}_i\overline{\mathbb{H}}_j := \overline{\mathbb{H}}_i \operatorname{Span}_{\mathbf{k}}\{\widetilde{\alpha}, \tau^*\widetilde{\alpha} \mid (\widetilde{\alpha} : j \to i) \in Q_1^* \text{ or } (\widetilde{\alpha} : j \to \tau^* i) \in Q_1^*\}\overline{\mathbb{H}}_j.$$

Recall from (9.5)–(9.6) the basis ${}_i\mathtt{L}_j$ (and respectively, ${}_i\mathtt{R}_j$) for ${}_i\overline{\mathbb{H}}_j$ as a left $\overline{\mathbb{H}}_i$-module (and respectively, right $\overline{\mathbb{H}}_j$-module). Let ${}_j\mathtt{L}_i^*$ and ${}_j\mathtt{R}_i^*$ be the dual bases of $\operatorname{Hom}_{\overline{\mathbb{H}}_j}({}_j\overline{\mathbb{H}}_i, \overline{\mathbb{H}}_j)$ and $\operatorname{Hom}_{\overline{\mathbb{H}}_i}({}_j\overline{\mathbb{H}}_i, \overline{\mathbb{H}}_i)$, respectively. Denote by b^* the corresponding dual basis vector for any $b \in {}_j\mathtt{L}_i$ or $b \in {}_j\mathtt{R}_i$.

Since ${}_i\overline{\mathbb{H}}_j$ and $\operatorname{Hom}_{\overline{\mathbb{H}}_j}({}_j\overline{\mathbb{H}}_i, \overline{\mathbb{H}}_j)$ are right free $\overline{\mathbb{H}}_j$-modules with bases given by ${}_i\mathtt{R}_j$ and ${}_j\mathtt{L}_i^*$ respectively, there is a right $\overline{\mathbb{H}}_j$-module isomorphism

$$\rho : {}_i\overline{\mathbb{H}}_j \longrightarrow \operatorname{Hom}_{\overline{\mathbb{H}}_j}({}_j\overline{\mathbb{H}}_i, \overline{\mathbb{H}}_j)$$

such that $\rho(\widetilde{b}) = b^*$ for any $b \in {}_j\mathtt{L}_i$. It is then routine to check that ρ is actually an $\overline{\mathbb{H}}_i$-$\overline{\mathbb{H}}_j$-bimodule isomorphism. Similarly, there is an $\overline{\mathbb{H}}_i$-$\overline{\mathbb{H}}_j$-bimodule isomorphism

$$\lambda : {}_i\overline{\mathbb{H}}_j \longrightarrow \operatorname{Hom}_{\overline{\mathbb{H}}_i}({}_j\overline{\mathbb{H}}_i, \overline{\mathbb{H}}_i).$$

These two isomorphisms satisfy that $\rho({}_i\mathtt{R}_j) = {}_j\mathtt{L}_i^*$ and $\lambda({}_i\mathtt{L}_j) = {}_j\mathtt{R}_i^*$. We sometimes identify the spaces $\operatorname{Hom}_{\overline{\mathbb{H}}_j}({}_j\overline{\mathbb{H}}_i, \overline{\mathbb{H}}_j)$, ${}_i\overline{\mathbb{H}}_j$ and $\operatorname{Hom}_{\overline{\mathbb{H}}_i}({}_j\overline{\mathbb{H}}_i, \overline{\mathbb{H}}_i)$ via ρ and λ.

If N_j is an $\overline{\mathbb{H}}_j$-module, then we have a natural isomorphism of $\overline{\mathbb{H}}_i$-modules

$$\operatorname{Hom}_{\overline{\mathbb{H}}_j}({}_j\overline{\mathbb{H}}_i, N_j) \longrightarrow {}_i\overline{\mathbb{H}}_j \otimes_{\overline{\mathbb{H}}_j} N_j$$

defined by

$$f \mapsto \sum_{b \in {}_j\mathtt{L}_i} b^* \otimes f(b).$$

Furthermore, for any $\overline{\mathbb{H}}_i$-module L_i, there is a natural isomorphism of $\mathbf{k}$-vector spaces:

$$\operatorname{Hom}_{\overline{\mathbb{H}}_j}({}_j\overline{\mathbb{H}}_i \otimes_{\overline{\mathbb{H}}_i} L_i, N_j) \longrightarrow \operatorname{Hom}_{\overline{\mathbb{H}}_i}(L_i, \operatorname{Hom}_{\overline{\mathbb{H}}_j}({}_j\overline{\mathbb{H}}_i, N_j)).$$

Composing the two maps above, we obtain the following.

LEMMA 9.2 (cf. [**LW21a**, Lemma 3.2]). *There exists a canonical* $\mathbf{k}$*-linear isomorphism*

$$\mathrm{ad}_{ji} = \mathrm{ad}_{ji}(L_i, N_j) : \mathrm{Hom}_{\overline{\mathbb{H}}_j}({}_j\overline{\mathbb{H}}_i \otimes_{\overline{\mathbb{H}}_i} L_i, N_j) \longrightarrow \mathrm{Hom}_{\overline{\mathbb{H}}_i}(L_i, {}_i\overline{\mathbb{H}}_j \otimes_{\overline{\mathbb{H}}_j} N_j)$$
$$\mathrm{ad}_{ji} : f \mapsto \big(f^\vee : l \mapsto \sum_{b\in {}_j\mathrm{L}_i} b^* \otimes f(b\otimes l)\big).$$

The inverse ad_{ji}^{-1} *is given by* $\mathrm{ad}_{ji}^{-1}(g) = \big(g^\vee : h\otimes l \mapsto \sum_{b\in_j \mathrm{L}_i} b^*(h)(g(l))_b\big)$, *where the elements* $(g(l))_b \in N_j$ *are uniquely determined by* $g(l) = \sum_{b\in_j\mathrm{L}_i} b^* \otimes (g(l))_b$.

Let (Q,τ) be an ıquiver. Without loss of generality, we assume Q to be connected and of rank ≥ 2. Recall $\Omega = \Omega(Q)$ is the orientation of Q. For any sink $\ell \in Q_0$, define the quiver $s_\ell(Q)$ by reversing all the arrows ending to ℓ. Note that in this case, we have $n_{\ell,\tau\ell} = 0$. By definition, ℓ is a sink of Q if and only if $\tau\ell$ is a sink of Q. Define the quiver

$$Q' := \boldsymbol{s}_\ell Q := \begin{cases} s_\ell(Q) & \text{if } \tau\ell = \ell, \\ s_\ell s_{\tau\ell}(Q) & \text{if } \tau\ell \neq \ell. \end{cases} \tag{9.8}$$

Note that $s_\ell s_{\tau\ell}(Q) = s_{\tau\ell}s_\ell(Q)$. Then τ also induces an involution τ on the quiver Q'. In this way, we can define $\Lambda' = \mathbf{k}Q' \otimes_{\mathbf{k}} R_2$ with an involution $\tau^\sharp$, and denote the $\tau^\sharp$-fixed point subalgebra by $\Lambda'^\imath = \boldsymbol{s}_\ell\Lambda^\imath$. Note that $\boldsymbol{s}_\ell\Lambda^\imath = \boldsymbol{s}_{\tau\ell}\Lambda^\imath$ for any sink $\ell \in Q_0$. The quiver $\overline{Q'}$ of $\boldsymbol{s}_\ell\Lambda^\imath$ can be constructed from $\overline{Q}$ by reversing all the arrows ending to ℓ and $\tau\ell$. Denote by $\Omega' := \Omega(Q')$ the orientation of Q'.

We shall define a reflection functor associated to a sink $\ell \in Q_0$

$$F_\ell^+ : \mathrm{mod}(\Lambda^\imath) := \mathrm{rep}(\overline{Q}, \overline{I}) \longrightarrow \mathrm{mod}(\boldsymbol{s}_\ell\Lambda^\imath) := \mathrm{rep}(\overline{Q'}, \overline{I'}), \tag{9.9}$$

in (9.11) below. Using Proposition 9.1, we shall identify the category $\mathrm{rep}(\overline{Q},\overline{I})$ with $\mathrm{rep}(\mathcal{M}(Q,\tau))$, and respectively, $\mathrm{rep}(\overline{Q'},\overline{I'})$ with $\mathrm{rep}(\mathcal{M}(Q',\tau))$.

Without loss of generality, we assume that the sink $\ell \in \overline{\mathbb{I}}_\tau$. Let $L = (L_i, L_{ji}) \in \mathrm{rep}(\mathcal{M}(Q,\tau))$. Then $L_{ji} : {}_j\overline{\mathbb{H}}_i \otimes_{\overline{\mathbb{H}}_i} L_i \to L_j$ is an $\overline{\mathbb{H}}_i$-morphism for any $(i,j) \in \overline{\Omega}$. Denote by

$$L_{\ell,\mathrm{in}} := (L_{\ell i})_i : \bigoplus_{i\in\overline{\Omega}(-,\ell)} {}_\ell\overline{\mathbb{H}}_i \otimes_{\overline{\mathbb{H}}_i} L_i \longrightarrow L_\ell.$$

Let $N_\ell := \ker(L_{\ell,\mathrm{in}})$. Recall $\mathbb{H}_\ell$ from (5.5), and note that $\overline{\mathbb{H}}_\ell = \mathbb{H}_\ell$ is finite-dimensional by our assumption that ℓ is a sink. We have $\dim N_\ell < \infty$. By definition, there exists an exact sequence

$$0 \longrightarrow N_\ell \longrightarrow \bigoplus_{i\in\overline{\Omega}(-,\ell)} {}_\ell\overline{\mathbb{H}}_i \otimes_{\overline{\mathbb{H}}_i} L_i \xrightarrow{L_{\ell,\mathrm{in}}} L_\ell. \tag{9.10}$$

Denote by $(N_{i\ell}^\vee)_i$ the inclusion map $N_\ell \to \bigoplus_{i\in\overline{\Omega}(-,\ell)} {}_\ell\overline{\mathbb{H}}_i \otimes_{\overline{\mathbb{H}}_i} L_i$.

For any $L \in \mathrm{rep}(\mathcal{M}(Q,\tau))$, define

$$F_\ell^+(L) = (N_r, N_{rs}) \in \mathrm{rep}(\mathcal{M}(Q',\tau)), \tag{9.11}$$

where

$$N_r := \begin{cases} L_r & \text{if } r \neq \ell, \\ N_\ell & \text{if } r = \ell, \end{cases} \qquad N_{rs} := \begin{cases} L_{rs} & \text{if } (s,r) \in \overline{\Omega} \text{ with } r \neq \ell, \\ (N_{r\ell}^\vee)^\vee & \text{if } (s,r) \in \overline{\Omega}^* \text{ and } s = \ell. \end{cases}$$

Here $(N_{r\ell}^{\vee})^{\vee} = \mathrm{ad}_{r\ell}^{-1}(N_{r\ell}^{\vee})$; see Lemma 9.2.

Dually, associated to any source $\ell \in Q_0$, we have a reflection functor

$$F_\ell^- : \mathrm{mod}(\Lambda^\imath) \longrightarrow \mathrm{mod}(\boldsymbol{s}_\ell \Lambda^\imath). \tag{9.12}$$

As proved in [**LW22b**, Proposition 2.4], the pair (F_ℓ^-, F_ℓ^+) is a pair of adjoint functors, i.e., there is a functorial isomorphism

$$\mathrm{Hom}_{\Lambda^\imath}(F_\ell^-(M), N) \cong \mathrm{Hom}_{\boldsymbol{s}_\ell \Lambda^\imath}(M, F_\ell^+(N)).$$

Let $\ell \in Q$ be a sink. For any $A \in \{\Lambda^\imath, \boldsymbol{s}_\ell \Lambda^\imath\}$ and $j \in Q_0$, let

$$\mathcal{T}_j^A := \{X \in \mathrm{mod}(A) \mid \mathrm{Hom}_A(X, S_j \oplus S_{\tau j}) = 0\},$$
$$\mathcal{S}_j^A := \{X \in \mathrm{mod}(A) \mid \mathrm{Hom}_A(S_j \oplus S_{\tau j}, X) = 0\}.$$

There is an equivalence of subcategories:

$$F_\ell^+ : \mathcal{T}_\ell^{\Lambda^\imath} \xrightarrow{\simeq} \mathcal{S}_\ell^{\boldsymbol{s}_\ell \Lambda^\imath}, \tag{9.13}$$

with its inverse given by F_ℓ^-; see [**LW22b**, Corollary 2.5].

Let $\mathcal{T} := \mathcal{T}_\ell^{\Lambda^\imath}$, and let $\mathcal{F}$ be the extension closed subcategory of $\mathrm{mod}(\Lambda^\imath)$ generated by S_ℓ and $S_{\tau\ell}$.

LEMMA 9.3 ([**LW22b**, Lemma 2.7]).
(a) *$(\mathcal{T}, \mathcal{F})$ is a torsion pair in* $\mathrm{mod}(\Lambda^\imath)$*;*
(b) *For any $M \in \mathrm{mod}(\Lambda^\imath)$, there exists a short exact sequence*

$$0 \longrightarrow M \longrightarrow T_M^0 \longrightarrow T_M^1 \longrightarrow 0$$

with $T_M^0, T_M^1 \in \mathcal{T}$ and $T_M^1 \in \mathcal{P}^{\leq 1}(\Lambda^\imath)$.

We have a $\mathbb{Z}$-linear isomorphism $\underline{\dim}\colon K_0(\mathrm{mod}(\mathbf{k}Q)) \to \mathbb{Z}^{\mathbb{I}}$, which sends an isoclass to its dimension vector. By identifying $i \in \mathbb{I}$ with a simple root α_i and thus $\mathbb{Z}^{\mathbb{I}}$ with the root lattice (of a Kac-Moody algebra $\mathfrak{g}$), we have a simple reflection s_i acting on $\mathbb{Z}^{\mathbb{I}}$. Then we denote

$$\boldsymbol{s}_i = \begin{cases} s_i, & \text{if } i = \tau i \\ s_i s_{\tau i}, & \text{if } i \neq \tau i. \end{cases}$$

LEMMA 9.4 ([**LW22b**, Lemma 2.9]). *Let (Q, τ) be an $\imath$quiver with a sink ℓ. Let $L \in \mathrm{mod}(\mathbf{k}Q) \subseteq \mathrm{mod}(\Lambda^\imath)$ be an indecomposable module. Then either $F_\ell^+(L) = 0$ (equivalently $L \cong S_\ell$ or $S_{\tau\ell}$) or $F_\ell^+(L)$ is indecomposable with $\underline{\dim} F_\ell^+(L) = \boldsymbol{s}_\ell(\underline{\dim} L)$.*

THEOREM 9.5 ([**LW21a**, Theorem 4.3]). *Let ℓ be a sink of Q. Then we have an isomorphism of algebras:*

$$\begin{aligned} \Gamma_\ell : \mathcal{S}\mathcal{D}\widetilde{\mathcal{H}}(\Lambda^\imath) &\xrightarrow{\cong} \mathcal{S}\mathcal{D}\widetilde{\mathcal{H}}(\boldsymbol{s}_\ell \Lambda^\imath), \\ [M] &\mapsto \mathbf{v}^{\langle \mathrm{res}(T_M), \mathrm{res}(M)\rangle_Q} q^{-\langle T_M, M\rangle} [F_\ell^+(T_M)]^{-1} * [F_\ell^+(X_M)]. \end{aligned} \tag{9.14}$$

where $M \in \mathrm{mod}(\Lambda^\imath)$ and $X_M, T_M \in \mathcal{T}$ ($T_M \in \mathcal{P}^{\leq 1}(\Lambda^\imath)$) fit into a short exact sequence $0 \to M \to X_M \to T_M \to 0$.

Dually, the functor F_ℓ^- for any source ℓ of Q induces an isomorphism of algebras:

$$\Gamma_\ell^- : \mathcal{S}\mathcal{D}\widetilde{\mathcal{H}}(\Lambda^\imath) \xrightarrow{\cong} \mathcal{S}\mathcal{D}\widetilde{\mathcal{H}}(\boldsymbol{s}_\ell \Lambda^\imath).$$

9.3. Actions of reflection functors on simple modules. We first have the following formulas for the isomorphism $\Gamma_\ell : \mathcal{SD}\widetilde{\mathcal{H}}(\Lambda^\imath) \to \mathcal{SD}\widetilde{\mathcal{H}}(\mathbf{s}_\ell\Lambda^\imath)$, which basically follow from the definitions:

$$\Gamma_\ell([M]) = [F_\ell^+(M)], \quad \forall M \in \mathcal{T}, \tag{9.15}$$

$$\Gamma_\ell([S_\ell]) = \begin{cases} \mathbf{v}[\mathbb{K}'_\ell]^{-1} * [S'_{\tau\ell}], & \text{if } \tau\ell \neq \ell, \\ [\mathbb{K}'_\ell]^{-1} * [S'_{\tau\ell}], & \text{if } \tau\ell = \ell, \end{cases} \tag{9.16}$$

$$\Gamma_\ell([S_{\tau\ell}]) = \mathbf{v}[\mathbb{K}'_{\tau\ell}]^{-1} * [S'_\ell], \qquad \text{if } \tau\ell \neq \ell, \tag{9.17}$$

$$\Gamma_\ell([\mathbb{K}_\alpha]) = [\mathbb{K}'_{\mathbf{s}_\ell\alpha}], \quad \forall \alpha \in K_0(\mathrm{mod}(\mathbf{k}Q)). \tag{9.18}$$

Next, we compute the formulas for Γ_ℓ on simple modules S_j, for $j \neq i, \tau i$.

THEOREM 9.6 ([**LW22b**, Theorem 3.2]). *Let (Q,τ) be an ıquiver. For any sink $i \in Q_0$ such that $i = \tau i \neq j$, we have*

(9.19)
$$\Gamma_i([S_j]) = \sum_{r+s=-c_{ij}} (-1)^r \mathbf{v}^r (1-\mathbf{v}^2)^{c_{ij}} [S'_i]_{\overline{p}}^{(r)} * [S'_j] * [S'_i]_{\overline{c_{ij}}+\overline{p}}^{(s)}$$
$$+ (-1)^p \sum_{t\geq 1} \sum_{\substack{r+s+2t=-c_{ij} \\ \overline{r}=\overline{p}}} \mathbf{v}^r (1-\mathbf{v}^2)^{c_{ij}+2t} [S'_i]_{\overline{p}}^{(r)} * [S'_j] * [S'_i]_{\overline{c_{ij}}+\overline{p}}^{(s)} * [\mathbb{K}'_i]^t.$$

The proof of Theorem 9.6 is reduced to the consideration of the following two types of rank 2 ıquivers (nevertheless, the computations are the same for both types):

(9.20)

(9.21)

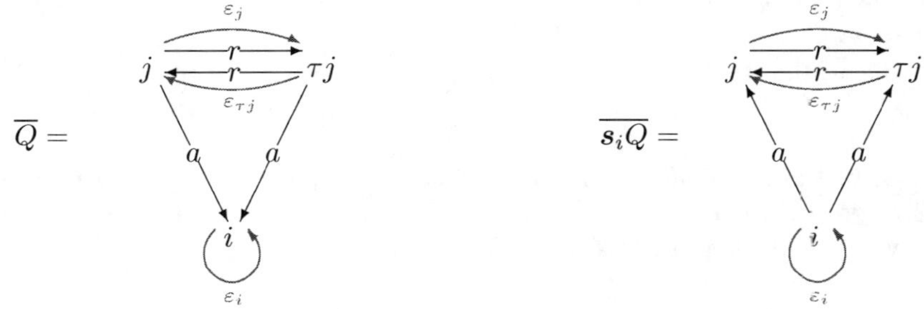

THEOREM 9.7 ([**LW22b**, Theorem 5.1]). *Let (Q,τ) be an ıquiver. For any sink $i \in Q_0$ such that $c_{i,\tau i} = 0$ and $j \neq i, \tau i$, we have*
(9.22)
$$\Gamma_i([S_j]) = \sum_{u=0}^{-\max(c_{ij},c_{\tau i,j})} \sum_{r=0}^{-c_{ij}-u} \sum_{s=0}^{-c_{\tau i,j}-u} (-1)^{r+s} \mathbf{v}^{-r-s+(r-s)u} (\mathbf{v}-\mathbf{v}^{-1})^{c_{ij}+c_{\tau i,j}+2u}$$
$$\times [S'_i]^{(-c_{ij}-r-u)} [S'_{\tau i}]^{(-c_{\tau i,j}-s-u)} * [S'_j] * [S'_{\tau i}]^{(s)} * [S'_i]^{(r)} * [\mathbb{K}'_{\tau i}]^u.$$

The proof of Theorem 9.7 is reduced to the consideration of the following two types of rank 2 $\imath$quivers (nevertheless, the computations are the same for both types):

(9.23)

$\overline{Q} =$ 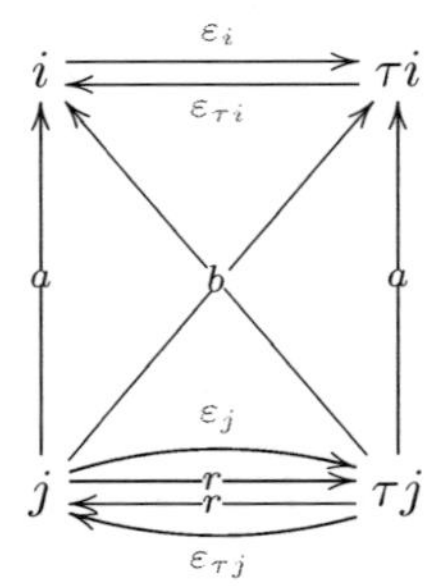

$\overline{\boldsymbol{s}_i Q} =$ 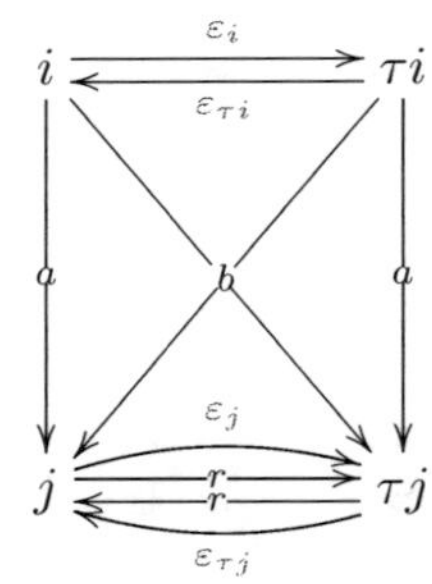

(9.24)

$\overline{Q} =$ 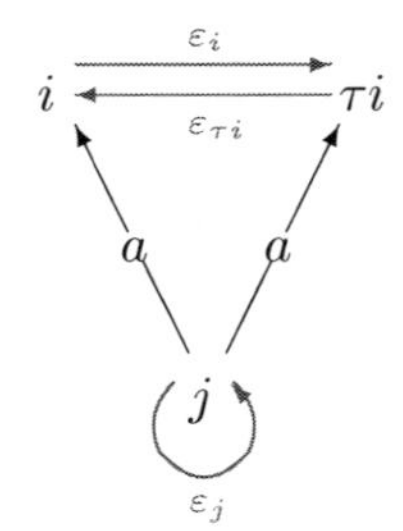

$\overline{\boldsymbol{s}_i Q} =$ 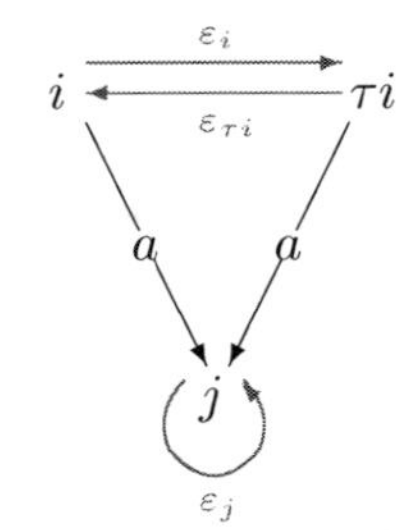

9.4. Relative braid group symmetries of $\imath$quantum groups. Let τ be an involution of Q, which induces an involution on $\mathfrak{g}$ again denoted by τ. We shall define the *relative (or restricted) Weyl group* associated to the quasi-split symmetric pair $(\mathfrak{g}, \mathfrak{g}^{\omega\tau})$ to be the following subgroup W^τ of W:

$$W^\tau = \{w \in W \mid \tau w = w\tau\} \tag{9.25}$$

where τ is regarded as an automorphism of $\mathrm{Aut}(C)$. In finite type, it is well known that the relative Weyl group defined in this way coincides with the one arising from real groups (cf., e.g., [**Lus03, KP11**]).

Recall the subset $\mathbb{I}_\tau$ of $\mathbb{I}$ from (5.6), and define

$$\overline{\mathbb{I}}_\tau := \{i \in \mathbb{I}_\tau \mid c_{i,\tau i} = 0 \text{ or } 2\}. \tag{9.26}$$

In our setting, $\overline{\mathbb{I}}_\tau$ consists of exactly those $i \in \mathbb{I}_\tau$ such that the τ-orbit of i is of finite type. Note that $\overline{\mathbb{I}}_\tau = \mathbb{I}_\tau$ if (Q,τ) is acyclic. We denote by $\boldsymbol{s}_i$, for $i \in \overline{\mathbb{I}}_\tau$, the following element of order 2 in the Weyl group W

$$\boldsymbol{s}_i = \begin{cases} s_i, & \text{if } i = \tau i \\ s_i s_{\tau i}, & \text{if } i \neq \tau i. \end{cases} \tag{9.27}$$

LEMMA 9.8 ([**Lus03**, Appendix]). *The relative Weyl group W^τ can be identified with a Coxeter group with $\boldsymbol{s}_i$ $(i \in \overline{\mathbb{I}}_\tau)$ as its generators.*

Let i be a sink of an $\imath$quiver (Q,τ). By Theorem 7.3, there are injective homomorphisms: $\widetilde{\Psi}_Q : \widetilde{\mathbf{U}}^\imath_{\mathbf{v}} \to \mathcal{SD}\widetilde{\mathcal{H}}(\Lambda^\imath)$, $\widetilde{\Psi}_{Q'} : \widetilde{\mathbf{U}}^\imath_{\mathbf{v}} \to \mathcal{SD}\widetilde{\mathcal{H}}(\boldsymbol{s}_i\Lambda^\imath)$.

Recall $\overline{\mathbb{I}}_\tau$ from (9.26). We define algebra automorphisms

$$\mathbf{T}''_{i,1} \in \mathrm{Aut}(\widetilde{\mathbf{U}}^\imath), \qquad \text{for } i \in \overline{\mathbb{I}}_\tau,$$

such that specializing at $v = \mathbf{v}$ we have the following commutative diagram:

$$\begin{array}{ccc} \widetilde{\mathbf{U}}^\imath_{\mathbf{v}} & \xrightarrow{\mathbf{T}''_{i,1}} & \widetilde{\mathbf{U}}^\imath_{\mathbf{v}} \\ \downarrow{\scriptstyle \widetilde{\Psi}_Q} & & \downarrow{\scriptstyle \widetilde{\Psi}_{Q'}} \\ \mathcal{SD}\widetilde{\mathcal{H}}(\Lambda^\imath) & \xrightarrow{\Gamma_i} & \mathcal{SD}\widetilde{\mathcal{H}}(\boldsymbol{s}_i\Lambda^\imath) \end{array} \tag{9.28}$$

Using Theorems 9.6–9.7, we have the following theorem.

THEOREM 9.9 ([**LW22b**, Theorem 6.8]). *We have a $\mathbb{Q}(v)$-algebra automorphism $\mathbf{T}''_{i,1}$ of $\widetilde{\mathbf{U}}^\imath$, for $i \in \mathbb{I}_\tau$, such that*

(1) $\underline{(i = \tau i)}$: $\mathbf{T}''_{i,1}(\widetilde{k}_j) = (-v^2\widetilde{k}_i)^{-c_{ij}}\widetilde{k}_j$, *and*

$$\mathbf{T}''_{i,1}(B_i) = (-v^2\widetilde{k}_i)^{-1}B_i,$$

$$\begin{aligned}\mathbf{T}''_{i,1}(B_j) = &\sum_{r+s=-c_{ij}} (-1)^r v^r B^{(r)}_{i,\overline{p}} B_j B^{(s)}_{i,\overline{c_{ij}}+\overline{p}} \\ &+ \sum_{u\geq 1} \sum_{\substack{r+s+2u=-c_{ij} \\ \overline{r}=\overline{p}}} (-1)^r v^r B^{(r)}_{i,\overline{p}} B_j B^{(s)}_{i,\overline{c_{ij}}+\overline{p}} (-v^2\widetilde{k}_i)^u, \qquad \text{for } j \neq i;\end{aligned}$$

(2) $\underline{(i \neq \tau i)}$: $\mathbf{T}''_{i,1}(\widetilde{k}_j) = \widetilde{k}_i^{-c_{ij}}\widetilde{k}_{\tau i}^{-c_{\tau i,j}}\widetilde{k}_j$,

$$\mathbf{T}''_{i,1}(B_j) = \begin{cases} -\widetilde{k}_i^{-1}B_{\tau i}, & \text{if } j = i \\ -B_i\widetilde{k}_{\tau i}^{-1}, & \text{if } j = \tau i, \end{cases}$$

and for $j \neq i, \tau i$,

$$\begin{aligned}\mathbf{T}''_{i,1}(B_j) = \sum_{u=0}^{-\max(c_{ij},c_{\tau i,j})} \sum_{r=0}^{-c_{i,j}-u} \sum_{s=0}^{-c_{\tau i,j}-u} &(-1)^{r+s} v^{r-s+(-c_{ij}-r-s-u)u} \\ &\times B_i^{(r)} B_{\tau i}^{(-c_{\tau i,j}-u-s)} B_j B_{\tau i}^{(s)} B_i^{(-c_{ij}-r-u)} \widetilde{k}_{\tau i}^u.\end{aligned}$$

The following lemma follows by inspection of the defining relations for $\widetilde{\mathbf{U}}^\imath$ in Theorem 2.8; cf. [**CLW23**].

LEMMA 9.10.

(a) *There exists a $\mathbb{Q}$-algebra involution $\psi_\imath : \widetilde{\mathbf{U}}^\imath \to \widetilde{\mathbf{U}}^\imath$ (called a bar involution) such that*

$$\psi_\imath(v) = v^{-1}, \quad \psi_\imath(\widetilde{k}_i) = v^{c_{i,\tau i}}\widetilde{k}_{\tau i}, \quad \psi_\imath(B_i) = B_i, \quad \forall i \in \mathbb{I}.$$

(b) *There exists a $\mathbb{Q}(v)$-algebra anti-involution $\sigma_\imath : \widetilde{\mathbf{U}}^\imath \to \widetilde{\mathbf{U}}^\imath$ such that*

$$\sigma_\imath(B_i) = B_i, \quad \sigma_\imath(\widetilde{k}_i) = \widetilde{k}_{\tau i}, \quad \forall i \in \mathbb{I}. \tag{9.29}$$

THEOREM 9.11 ([**LW22b**, Theorems 6.11–6.12]). *For $i \in \mathbb{I}_\tau$ and $e \in \{\pm 1\}$, there are automorphisms $\mathbf{T}'_{i,e}, \mathbf{T}''_{i,e}$ on $\widetilde{\mathbf{U}}^\imath$ such that*

$$\mathbf{T}'_{i,e} = \sigma_\imath \mathbf{T}''_{i,-e}\sigma_\imath, \qquad \psi_\imath \mathbf{T}'_{i,e}\psi_\imath = \mathbf{T}'_{i,-e}. \tag{9.30}$$

Moreover, we have $\mathbf{T}'_{i,e} = (\mathbf{T}''_{i,-e})^{-1}$, for any $i \in \mathbb{I}_\tau$ and $e = \{\pm 1\}$.

PROOF. We only need to prove the last statement.

It suffices to prove for $e = -1$ by (9.30). Given an $\imath$quiver (Q, τ) such that i is a sink, we have two isomorphisms, $\Gamma_i : \mathcal{SD}\widetilde{\mathcal{H}}(\Lambda^\imath) \xrightarrow{\cong} \mathcal{SD}\widetilde{\mathcal{H}}(\boldsymbol{s}_i\Lambda^\imath)$ and $\Gamma_i^- : \mathcal{SD}\widetilde{\mathcal{H}}(\boldsymbol{s}_i\Lambda^\imath) \xrightarrow{\cong} \mathcal{SD}\widetilde{\mathcal{H}}(\Lambda^\imath)$, which are inverses to each other. In particular, we have the following commutative diagram:

$$
\begin{array}{ccc}
\widetilde{\mathbf{U}}^\imath_{\mathbf{v}} & \xrightarrow{\mathbf{T}'_{i,-1}} & \widetilde{\mathbf{U}}^\imath_{\mathbf{v}} \\
\Big\downarrow \widetilde{\Psi}_{Q'} & & \Big\downarrow \widetilde{\Psi}_{Q} \\
\mathcal{SD}\widetilde{\mathcal{H}}(\boldsymbol{s}_i\Lambda^\imath) & \xrightarrow{\Gamma_i^-} & \mathcal{SD}\widetilde{\mathcal{H}}(\Lambda^\imath)
\end{array}
\tag{9.31}
$$

We conclude that $\mathbf{T}'_{i,-1}$ and $\mathbf{T}''_{i,1}$ are inverses to each other. □

REMARK 9.12. The results in Theorem 9.9 and Theorem 9.11 verify part of [**CLW21b**, Conjecture 6.5] in case $i = \tau i$ and [**CLW23**, Conjecture 3.7] in case $i \neq \tau i$, for quasi-split $\imath$quantum groups $\widetilde{\mathbf{U}}^\imath$ associated with symmetric generalized Cartan matrices such that all $c_{i,\tau i}$ even. (Note that in Theorem 9.9(1), our notation here $\mathbf{T}'_{i,e}, \mathbf{T}''_{i,e}$ follow [**LW22b**, Theorem **A**(1)], which was swapped in [**CLW21b**, Conjecture 6.5].)

A complete proof of [**CLW21b**, Conjecture 6.5] and [**CLW23**, Conjecture 3.7 & §3.5] is obtained by Zhang [**Z23**], who has further developed the approach initiated in [**WZ23**]. In addition, Zhang shows that $\mathbf{T}'_{i,e}$ (or $\mathbf{T}''_{i,e}$) satisfy the relative braid group relations.

10. $\imath$Hall algebras of weighted projective lines and $\imath$quantum loop algebras

In this section, we outline the connection between $\imath$Hall algebras of (weighted) projective lines and affine $\imath$quantum groups in Drinfeld type presentations.

10.1. Drinfeld type presentation of q-Onsager algebra. The (universal) $\imath$quantum group $\widetilde{\mathbf{U}}^\imath(\widehat{\mathfrak{sl}}_2)$ associated to the Satake diagram of split affine type $A_1^{(1)}$ is known as *(universal) q-Onsager algebra.* By definition, $\widetilde{\mathbf{U}}^\imath(\widehat{\mathfrak{sl}}_2)$ is generated by $B_i, \widetilde{k}_i$, for $i = 0, 1$, subject to relations (2.13)–(2.15); in $\imath$Hall algebra setting, the renormalized generators $\mathbb{K}_i := -v^2\widetilde{k}_i$ are more natural as they correspond to (generalized) simple modules.

DEFINITION 10.1 ([**LW21c**]). Let ${}^{\mathrm{Dr}}\widetilde{\mathbf{U}}^\imath(\widehat{\mathfrak{sl}}_2)$ be the $\mathbb{Q}(v)$-algebra generated by $\mathbb{K}_1^{\pm 1}$, $C^{\pm 1}$, H_m and $B_{1,r}$, where $m \geq 1$, $r \in \mathbb{Z}$, subject to the following relations, for $r, s \in \mathbb{Z}$ and $m, n \geq 1$:

$$\mathbb{K}_1\mathbb{K}_1^{-1} = 1, CC^{-1} = 1, \quad \mathbb{K}_1, C \text{ are central,} \tag{10.1}$$

$$[H_m, H_n] = 0, \tag{10.2}$$

$$[H_m, B_{1,r}] = \frac{[2m]}{m}B_{1,r+m} - \frac{[2m]}{m}B_{1,r-m}C^m, \tag{10.3}$$

(10.4)
$$[B_{1,r}, B_{1,s+1}]_{v^{-2}} - v^{-2}[B_{1,r+1}, B_{1,s}]_{v^2} = v^{-2}\Theta_{s-r+1}C^r\mathbb{K}_1 - v^{-4}\Theta_{s-r-1}C^{r+1}\mathbb{K}_1 + v^{-2}\Theta_{r-s+1}C^s\mathbb{K}_1 - v^{-4}\Theta_{r-s-1}C^{s+1}\mathbb{K}_1.$$

Here

$$1+\sum_{m\geq 1}(v-v^{-1})\Theta_m z^m = \exp\Big((v-v^{-1})\sum_{m\geq 1} H_m z^m\Big). \tag{10.5}$$

THEOREM 10.2 ([**LW21c**, Theorem 2.16]). *There is an isomorphism of* $\mathbb{Q}(v)$*-algebras*

$$\Phi : {}^{\mathrm{Dr}}\widetilde{\mathbf{U}}^\imath(\widehat{\mathfrak{sl}}_2) \longrightarrow \widetilde{\mathbf{U}}^\imath(\widehat{\mathfrak{sl}}_2).$$

We shall call ${}^{\mathrm{Dr}}\widetilde{\mathbf{U}}^\imath(\widehat{\mathfrak{sl}}_2)$ the Drinfeld type presentation of the q-Onsager algebra. To establish Theorem 10.2, one needs to construct the real root vectors $B_{1,r}$ and imaginary root vectors Θ_m in $\widetilde{\mathbf{U}}^\imath(\widehat{\mathfrak{sl}}_2)$ via braid group symmetries (cf. [**BK20, LW21c**]), and the map Φ matches the generators in the same notations and identifies $C = \mathbb{K}_0\mathbb{K}_1$. To show Φ is an algebra homomorphism amounts to showing that all the relations in ${}^{\mathrm{Dr}}\widetilde{\mathbf{U}}^\imath(\widehat{\mathfrak{sl}}_2)$ are satisfied by these root vectors in $\widetilde{\mathbf{U}}^\imath$. The surjectivity of Φ is easy, while the injectivity follows by some filtration argument.

10.2. ıHall algebra of the projective line. Recall $\mathbf{k}$ denotes a finite field of q elements. The coordinate ring of the projective line $\mathbb{P}^1_{\mathbf{k}}$ over $\mathbf{k}$ is the $\mathbb{Z}$-graded ring $\mathbf{S} = \mathbf{k}[X_0, X_1]$ with $\deg(X_0) = \deg(X_1) = 1$. A closed point x of $\mathbb{P}^1_{\mathbf{k}}$ is given by a prime homogeneous ideal of $\mathbf{S}$ generated by an irreducible polynomial in $\mathbf{S}$. The degree of x, denoted by $\deg(x)$ or d_x, is defined to be the degree of the defining irreducible polynomial associated to x. We denote

▷ $\mathrm{mod}^{\mathbb{Z}}(\mathbf{S})$ – category of finitely generated $\mathbb{Z}$-graded $\mathbf{S}$-modules

▷ $\mathrm{mod}^{\mathbb{Z}}_0(\mathbf{S})$– the full subcategory of $\mathrm{mod}^{\mathbb{Z}}(\mathbf{S})$ of finite-dimensional graded $\mathbf{S}$-modules

▷ $\mathrm{coh}(\mathbb{P}^1_{\mathbf{k}})$ – category of coherent sheaves on $\mathbb{P}^1_{\mathbf{k}}$

We can associate a coherent sheaf $\widetilde{M}$ on $\mathbb{P}^1_{\mathbf{k}}$ to any $M \in \mathrm{mod}^{\mathbb{Z}}(\mathbf{S})$. This gives rise to a category equivalence (which goes back to Serre): $\mathrm{mod}^{\mathbb{Z}}(\mathbf{S})/\mathrm{mod}^{\mathbb{Z}}_0(\mathbf{S}) \cong \mathrm{coh}(\mathbb{P}^1_{\mathbf{k}})$. The category $\mathrm{coh}(\mathbb{P}^1_{\mathbf{k}})$ is a finitary hereditary abelian Krull-Schmidt category.

Any indecomposable vector bundle on $\mathbb{P}^1_{\mathbf{k}}$ is a line bundle; more precisely [**Gro57**], it is of the form $\mathcal{O}(n) = \widetilde{\mathbf{S}[n]}$, for $n \in \mathbb{Z}$, where $\mathbf{S}[n]$ is the n-th shift of the trivial module $\mathbf{S}$, i.e., $\mathbf{S}[n]_i = \mathbf{S}_{n+i}$. The homomorphism between two line bundles are given by

$$\mathrm{Hom}(\mathcal{O}(m), \mathcal{O}(n)) \cong \mathbf{S}_{n-m}, \tag{10.6}$$

which then has dimension $n-m+1$ if $n \geq m$ and 0 otherwise.

For a hereditary abelian category $\mathcal{A}$, let $\mathcal{C}_{\mathbb{Z}_1}(\mathcal{A})$ be the category of 1-periodic complexes of $\mathcal{A}$. Recall that $\mathcal{SD}\widetilde{\mathcal{H}}(\mathcal{C}_{\mathbb{Z}_1}(\mathcal{A}))$ is the twisted semi-derived Ringel-Hall algebra of $\mathcal{C}_{\mathbb{Z}_1}(\mathcal{A})$.

For any $X \in \mathcal{A}$, denote the stalk complex in $\mathcal{C}_{\mathbb{Z}_1}(\mathcal{A})$ by

$$C_X = (X, 0)$$

(or just by X when there is no confusion), and denote by K_X the following acylic complex:

$$K_X := (X \oplus X, d), \qquad \text{where } d = \begin{pmatrix} 0 & \mathrm{Id} \\ 0 & 0 \end{pmatrix}.$$

For any $\alpha \in K_0(\mathcal{A})$, there exist $X, Y \in \mathcal{A}$ such that $\alpha = \widehat{X} - \widehat{Y}$. Define $[K_\alpha] := [K_X] * [K_Y]^{-1} \in \mathcal{SD}\widetilde{\mathcal{H}}(\mathcal{C}_{\mathbb{Z}_1}(\mathcal{A}))$.

Return to $\mathcal{A} = \mathrm{coh}(\mathbb{P}^1_{\mathbf{k}})$. We shall use a shorthand notation ${}^\imath\widetilde{\mathcal{H}}(\mathbb{P}^1_{\mathbf{k}})$ to denote the twisted semi-derived Ringel-Hall algebra of $\mathcal{C}_{\mathbb{Z}_1}(\mathrm{coh}(\mathbb{P}^1_{\mathbf{k}}))$.

For $m \geq 1$, we define

$$\widehat{\Theta}_m = \frac{1}{(q-1)^2 \mathbf{v}^{m-1}} \sum_{0 \neq f: \mathcal{O} \to \mathcal{O}(m)} [\mathrm{Coker}\, f]. \tag{10.7}$$

The classes $\widehat{\mathcal{O}}$ and $\widehat{\mathcal{O}(1)}$ form a basis of $K_0(\mathbb{P}^1_{\mathbf{k}}) := K_0(\mathrm{coh}(\mathbb{P}^1_{\mathbf{k}}))$. Denote by

$$\delta := \widehat{\mathcal{O}(1)} - \widehat{\mathcal{O}}. \tag{10.8}$$

Then $\{\widehat{\mathcal{O}}, \delta\}$ is also a basis.

THEOREM 10.3 ([**LRW20**, Theorem 4.2]). *There exists a $\mathbb{Q}(\mathbf{v})$-algebra homomorphism*

$$\Omega : {}^{\mathrm{Dr}}\widetilde{\mathbf{U}}^\imath_{\mathbf{v}}(\widehat{\mathfrak{sl}}_2) \longrightarrow {}^\imath\widetilde{\mathcal{H}}(\mathbb{P}^1_{\mathbf{k}}) \tag{10.9}$$

which sends, for all $r \in \mathbb{Z}$ and $m \geq 1$,

$$\mathbb{K}_1 \mapsto [K_{\mathcal{O}}], \quad C \mapsto [K_\delta], \quad B_{1,r} \mapsto -\frac{1}{q-1}[\mathcal{O}(r)], \quad \Theta_m \mapsto \widehat{\Theta}_m.$$

Moreover, $\widehat{H}_m := \Omega(H_m)$ is also described in [**LRW20**].

10.3. $\imath$Hall algebra interpretation of the isomorphism Φ. Consider the Kronecker quiver Q_{Kr} : $0 \overset{\alpha}{\underset{\beta}{\rightrightarrows}} 1$. Let $\Lambda^\imath_{\mathrm{Kr}}$ be the $\imath$quiver algebra of the split Kronecker $\imath$quiver $(Q_{\mathrm{Kr}}, \mathrm{Id})$ described as in Example 5.4(e). We can make sense of the $\imath$Hall algebra ${}^\imath\widetilde{\mathcal{H}}(\mathbf{k}Q_{\mathrm{Kr}}) = \widetilde{\mathcal{H}}(\mathbf{k}Q_{\mathrm{Kr}}, \mathrm{Id})$. Let $T = \mathcal{O} \oplus \mathcal{O}(1)$ and $B = \mathrm{End}_{\mathbb{P}^1_{\mathbf{k}}}(T)$. It is known that T is a tilting object, and $B^{op} \cong \mathbf{k}Q_{\mathrm{Kr}}$. It follows that

$$\mathrm{RHom}_{\mathbb{P}^1_{\mathbf{k}}}(T, -) : \mathcal{D}^b(\mathrm{coh}(\mathbb{P}^1_{\mathbf{k}})) \xrightarrow{\simeq} \mathcal{D}^b(\mathrm{rep}_{\mathbf{k}}(Q_{\mathrm{Kr}})) \tag{10.10}$$

is a derived equivalence. Let $\mathcal{V}$ be the subcategory of $\mathrm{coh}(\mathbb{P}^1_{\mathbf{k}})$ consisting of M such that $\mathrm{Hom}_{\mathbb{P}^1_{\mathbf{k}}}(T, M) = 0$. Denote $\mathcal{U} = \mathrm{Fac}\, T$, the full subcategory of $\mathrm{coh}(\mathbb{P}^1_{\mathbf{k}})$ consisting of homomorphic images of objects in $\mathrm{add}\, T$. Then $(\mathcal{U}, \mathcal{V})$ is a *torsion pair* of $\mathrm{coh}(\mathbb{P}^1_{\mathbf{k}})$. It is proved in [**LRW20**, Lemma 5.8] that $(\mathcal{C}_1(\mathcal{U}), \mathcal{C}_1(\mathcal{V}))$ is a torsion pair of $\mathcal{C}_1(\mathrm{coh}(\mathbb{P}^1_{\mathbf{k}}))$. In particular, any $M \in \mathcal{C}_1(\mathrm{coh}(\mathbb{P}^1_{\mathbf{k}}))$ admits a short exact sequence of the form

$$0 \longrightarrow M \longrightarrow X_M \longrightarrow T_M \longrightarrow 0 \tag{10.11}$$

where $X_M \in \mathcal{C}_1(\mathcal{U})$ and $T_M \in \mathrm{add}\, K_T$. Then applying [**Lu22**, Theorem A.22] to our setting gives us an algebra isomorphism:

$$\begin{aligned} F : {}^\imath\widetilde{\mathcal{H}}(\mathbb{P}^1_{\mathbf{k}}) &\xrightarrow{\sim} {}^\imath\widetilde{\mathcal{H}}(\mathbf{k}Q_{\mathrm{Kr}}) \\ [M] &\mapsto [F(T_M)]^{-1} * [F(X_M)], \end{aligned}$$

where $X_M \in \mathcal{C}_1(\mathcal{U})$ and $T_M \in \mathrm{add}\, K_T$, are defined in the short exact sequence (10.11).

THEOREM 10.4 ([**LRW20**, Theorem 5.11]). *We have the following commutative diagram of algebra homomorphisms*

$$
\begin{array}{ccc}
\widetilde{\mathbf{U}}^\imath_{\mathbf{v}}(\widehat{\mathfrak{sl}}_2) & \xrightarrow{\ \Phi^{-1}\ } & {}^{\mathrm{Dr}}\widetilde{\mathbf{U}}^\imath_{\mathbf{v}}(\widehat{\mathfrak{sl}}_2) \\
\big\downarrow{\scriptstyle \widetilde{\psi}} & & \big\downarrow{\scriptstyle \Omega} \\
{}^\imath\widetilde{\mathcal{H}}(\mathbf{k}Q_{\mathrm{Kr}}) & \xrightarrow{\ \digamma^{-1}\ } & {}^\imath\widetilde{\mathcal{H}}(\mathbb{P}^1_{\mathbf{k}})
\end{array}
$$

where $\Phi, \digamma$ are isomorphisms. In particular, the homomorphism Ω is injective.

10.4. ıQuantum loop algebras. Associated to a generalized Cartan matrix (GCM) $C = (c_{ij})_{i,j\in\mathbb{I}_0}$, the ıquantum loop algebra ${}^{\mathrm{Dr}}\widetilde{\mathbf{U}}^\imath$ of split type is the $\mathbb{Q}(v)$-algebra generated by $\mathbb{K}_i^{\pm 1}$, $C^{\pm 1}$, $H_{i,m}$, $\Theta_{i,l}$ and $B_{i,l}$, where $i \in \mathbb{I}_0$, $m \in \mathbb{Z}_+$, $l \in \mathbb{Z}$, subject to some relations.

To describe the relations in ${}^{\mathrm{Dr}}\widetilde{\mathbf{U}}^\imath$, we introduce some shorthand notations below. Let $k_1, k_2, l \in \mathbb{Z}$ and $i, j \in \mathbb{I}_0$. Set

$$
\begin{aligned}
S(k_1,k_2|l;i,j) &= B_{i,k_1}B_{i,k_2}B_{j,l} - [2]B_{i,k_1}B_{j,l}B_{i,k_2} + B_{j,l}B_{i,k_1}B_{i,k_2},\\
\mathbb{S}(k_1,k_2|l;i,j) &= S(k_1,k_2|l;i,j) + \{k_1 \leftrightarrow k_2\}.
\end{aligned}
\tag{10.12}
$$

Here and below, $\{k_1 \leftrightarrow k_2\}$ stands for repeating the previous summand with k_1, k_2 switched if $k_1 \neq k_2$, so the sums over k_1, k_2 are symmetric. We also denote

(10.13)

$$
\begin{aligned}
R(k_1,k_2|l;i,j) = \mathbb{K}_i C^{k_1}\Big(&-\sum_{p\geq 0} v^{2p}[2][\Theta_{i,k_2-k_1-2p-1}, B_{j,l-1}]_{v^{-2}}C^{p+1}\\
&-\sum_{p\geq 1} v^{2p-1}[2][B_{j,l}, \Theta_{i,k_2-k_1-2p}]_{v^{-2}}C^{p} - [B_{j,l}, \Theta_{i,k_2-k_1}]_{v^{-2}}\Big),\\
\mathbb{R}(k_1,k_2|l;i,j) = R(k_1,k_2|l;i,j) &+ \{k_1 \leftrightarrow k_2\}.
\end{aligned}
$$

Now we can give the definition of ıquantum loop algebras ${}^{\mathrm{Dr}}\widetilde{\mathbf{U}}^\imath$ for simply-laced generalized Cartan matrices C, following [**LW21c**, Remark 3.17]. If C is a Cartan matrix of simple Lie algebra $\mathfrak{g}$ of type ADE, then ${}^{\mathrm{Dr}}\widetilde{\mathbf{U}}^\imath$ provides a Drinfeld type presentation of the affine ıquantum group of split type (associated to the affinization of $\mathfrak{g}$). That is, there is a $\mathbb{Q}(v)$-algebra isomorphism $\Phi : {}^{\mathrm{Dr}}\widetilde{\mathbf{U}}^\imath \to \widetilde{\mathbf{U}}^\imath$; see [**LW21c**].

DEFINITION 10.5 (ıquantum loop algebras). Let $C = (c_{ij})_{i,j\in\mathbb{I}_0}$ be a simply-laced generalized Cartan matrix (GCM). The ıquantum loop algebra ${}^{\mathrm{Dr}}\widetilde{\mathbf{U}}^\imath$ of split type is the $\mathbb{Q}(v)$-algebra generated by $\mathbb{K}_i^{\pm 1}$, $C^{\pm 1}$, $H_{i,m}$ and $B_{i,l}$, where $i \in \mathbb{I}_0$,

$m \in \mathbb{Z}_+$, $l \in \mathbb{Z}$, subject to the following relations, for $m, n \in \mathbb{Z}_+$ and $k, l \in \mathbb{Z}$:

(10.14)
$$\mathbb{K}_i, C \text{ are central}, \quad \mathbb{K}_i\mathbb{K}_i^{-1} = 1, \quad CC^{-1} = 1,$$

(10.15)
$$[H_{i,m}, H_{j,n}] = 0,$$

(10.16)
$$[H_{i,m}, B_{j,l}] = \frac{[mc_{ij}]}{m} B_{j,l+m} - \frac{[mc_{ij}]}{m} B_{j,l-m}C^m,$$

(10.17)
$$[B_{i,k}, B_{j,l}] = 0, \text{ if } c_{ij} = 0,$$

(10.18)
$$[B_{i,k}, B_{j,l+1}]_{v^{-c_{ij}}} - v^{-c_{ij}}[B_{i,k+1}, B_{j,l}]_{v^{c_{ij}}} = 0, \text{ if } i \neq j;$$

(10.19)
$$\begin{aligned}[B_{i,k}, B_{i,l+1}]_{v^{-2}} - v^{-2}[B_{i,k+1}, B_{i,l}]_{v^2} &= v^{-2}\Theta_{i,l-k+1}C^k\mathbb{K}_i - v^{-4}\Theta_{i,l-k-1}C^{k+1}\mathbb{K}_i \\ &\quad + v^{-2}\Theta_{i,k-l+1}C^l\mathbb{K}_i - v^{-4}\Theta_{i,k-l-1}C^{l+1}\mathbb{K}_i,\end{aligned}$$

(10.20)
$$\mathbb{S}(k_1, k_2|l; i, j) = \mathbb{R}(k_1, k_2|l; i, j), \text{ if } c_{ij} = -1.$$

Here we set

$$\Theta_{i,0} = (v - v^{-1})^{-1}, \qquad \Theta_{i,m} = 0, \quad \text{for } m < 0; \tag{10.21}$$

and $\Theta_{i,m}$ ($m \geq 1$) are related to $H_{i,m}$ by the following equation:

$$1 + \sum_{m\geq 1}(v - v^{-1})\Theta_{i,m}u^m = \exp\Big((v - v^{-1})\sum_{m\geq 1} H_{i,m}u^m\Big). \tag{10.22}$$

10.5. Weighted projective lines and $\imath$quantum loop algebras. Fix a positive integer $\mathbf{t}$ such that $2 \leq \mathbf{t} \leq q$. Let $\mathbf{p} = (p_1, p_2, \dots, p_\mathbf{t}) \in \mathbb{Z}_+^\mathbf{t}$, and let $L(\mathbf{p})$ denote the rank one abelian group on generators $\vec{x}_1, \vec{x}_2, \dots, \vec{x}_\mathbf{t}$ with relations $p_1\vec{x}_1 = p_2\vec{x}_2 = \cdots = p_\mathbf{t}\vec{x}_\mathbf{t}$. We call $\vec{c} = p_i\vec{x}_i$ the *canonical element* of $L(\mathbf{p})$. Obviously, the polynomial ring $\mathbf{k}[X_1, \dots, X_\mathbf{t}]$ is a $L(\mathbf{p})$-graded algebra by setting $\deg X_i = \vec{x}_i$, which is denoted by $\mathbf{S}(\mathbf{p})$.

Let $\underline{\boldsymbol{\lambda}} = \{\boldsymbol{\lambda}_1, \dots, \boldsymbol{\lambda}_\mathbf{t}\}$ be a collection of distinguished closed points (of degree one) on the projective line $\mathbb{P}^1_\mathbf{k}$, normalized such that $\boldsymbol{\lambda}_1 = \infty$, $\boldsymbol{\lambda}_2 = 0$, $\boldsymbol{\lambda}_3 = 1$. Let $I(\mathbf{p}, \underline{\boldsymbol{\lambda}})$ be the $L(\mathbf{p})$-graded ideal of $\mathbf{S}(\mathbf{p})$ generated by

$$\{X_i^{p_i} - (X_2^{p_2} - \boldsymbol{\lambda}_i X_1^{p_1}) \mid 3 \leq i \leq \mathbf{t}\}.$$

Then $\mathbf{S}(\mathbf{p}, \underline{\boldsymbol{\lambda}}) := \mathbf{S}(\mathbf{p})/I(\mathbf{p}, \underline{\boldsymbol{\lambda}})$ is an $L(\mathbf{p})$-graded algebra. For $1 \leq i \leq \mathbf{t}$, denote by x_i the image of X_i in $\mathbf{S}(\mathbf{p}, \underline{\boldsymbol{\lambda}})$.

The *weighted projective line* $\mathbb{X} := \mathbb{X}_{\mathbf{p},\underline{\boldsymbol{\lambda}}}$ is the set of all non-maximal prime homogeneous ideals of $\mathbf{S} := \mathbf{S}(\mathbf{p}, \underline{\boldsymbol{\lambda}})$, which is also denoted by $\mathbb{X}_\mathbf{k}$ to emphasis the base field $\mathbf{k}$. The classification of the closed points in $\mathbb{X}$ is provided in [**GL87**, Proposition 1.3]. First, each $\boldsymbol{\lambda}_i$ corresponds to the prime ideal generated by x_i, called the *exceptional point*. Second, any other ideals are of the form $(f(x_1^{p_1}, x_2^{p_2}))$, where $f \in \mathbf{k}[y_1, y_2]$ is an irreducible homogeneous polynomial in y_1, y_2, which is different from y_1 and y_2, called the *ordinary point*.

Following [**GL87**], the category of coherent sheaves $\mathrm{coh}(\mathbb{X})$ over $\mathbb{X}$ is defined to be the Serre quotient $\mathrm{mod}^{L(\mathbf{p})}(\mathbf{S})/\mathrm{mod}_0^{L(\mathbf{p})}(\mathbf{S})$ of the category of finitely generated $L(\mathbf{p})$-graded $\mathbf{S}$-modules modulo the category of finite length graded $\mathbf{S}$-modules. Let $\mathcal{O}=\mathcal{O}_{\mathbb{X}}$ be the structure sheaf, i.e., the image of $\mathbf{S}$. A sheaf is called a *torsion sheaf* if it is a finite-length object in $\mathrm{coh}(\mathbb{X})$. Let $\mathscr{T}$ be the full subcategory consisting of all torsion sheaves.

In order to describe the category $\mathscr{T}$, we shall introduce the representation theory of cyclic quivers. We consider the oriented cyclic quiver Cy_n for $n\geq 2$ with its vertex set $\mathbb{Z}_n=\{0,1,2,\dots,n-2,n-1\}$:

(10.23)
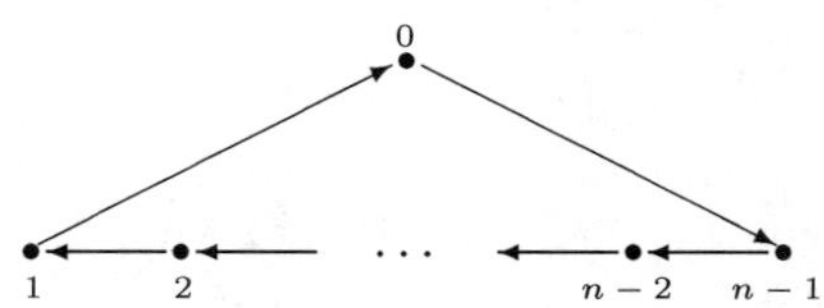

In particular, Cy_1 is just the Jordan quiver, i.e., the quiver with only one vertex and one loop arrow. Denote by $\mathrm{rep}_{\mathbf{k}}(\mathrm{Cy}_n)$ the category of finite-dimensional nilpotent representations of Cy_n over the field $\mathbf{k}$. Then the structure of $\mathscr{T}$ is described in the following.

LEMMA 10.6 ([**GL87**]).

(1) *The category $\mathscr{T}$ decomposes as a coproduct $\mathscr{T}=\coprod_{x\in\mathbb{X}}\mathscr{T}_x$, where $\mathscr{T}_x$ is the subcategory of torsion sheaves with support at x.*

(2) *For any ordinary point x of degree d, let $\mathbf{k}_x$ denote the residue field at x, i.e., $[\mathbf{k}_x:\mathbf{k}]=d$. Then $\mathscr{T}_x$ is equivalent to the category $\mathrm{rep}_{\mathbf{k}_x}(\mathrm{Cy}_1)$.*

(3) *For any exceptional point $\boldsymbol{\lambda}_i$ $(1\leq i\leq \mathbf{t})$, the category $\mathscr{T}_{\boldsymbol{\lambda}_i}$ is equivalent to $\mathrm{rep}_{\mathbf{k}}(\mathrm{Cy}_{p_i})$.*

For any ordinary point x of degree d, let π_x be the prime homogeneous polynomial corresponding to x. The multiplication by π_x gives the exact sequence

$$0\longrightarrow\mathcal{O}\xrightarrow{\pi_x}\mathcal{O}(d\vec{c})\longrightarrow S_x\to 0,$$

where S_x is the unique (up to isomorphism) simple sheaf in the category $\mathscr{T}_x$. For any exceptional point $\boldsymbol{\lambda}_i$, multiplication by x_i yields the short exact sequence

$$0\longrightarrow\mathcal{O}((j-1)\vec{x_i})\xrightarrow{x_i}\mathcal{O}(j\vec{x_i})\longrightarrow S_{ij}\to 0,\ \text{ for } 1\leq j\leq p_i;$$

where $\{S_{ij}\mid j\in\mathbb{Z}_{p_i}\}$ is a complete set of pairwise non-isomorphic simple sheaves in the category $\mathscr{T}_{\boldsymbol{\lambda}_i}$ for any $1\leq i\leq\mathbf{t}$.

Let $\mathcal{C}_{\mathbb{Z}_1}(\mathrm{coh}(\mathbb{X}_{\mathbf{k}}))$ be the category of 1-periodic complexes of $\mathrm{coh}(\mathbb{X}_{\mathbf{k}})$. We shall use a shorthand notation ${}^{\imath}\widetilde{\mathcal{H}}(\mathbb{X}_{\mathbf{k}})$ to denote the twisted semi-derived Ringel-Hall algebra of $\mathcal{C}_{\mathbb{Z}_1}(\mathrm{coh}(\mathbb{X}_{\mathbf{k}}))$.

For $\mathbf{p} = (p_1, \ldots, p_{\mathbf{t}}) \in \mathbb{Z}_+^{\mathbf{t}}$, let us consider the following star-shaped graph $\Gamma = \mathbb{T}_{p_1,\ldots,p_{\mathbf{t}}}$:

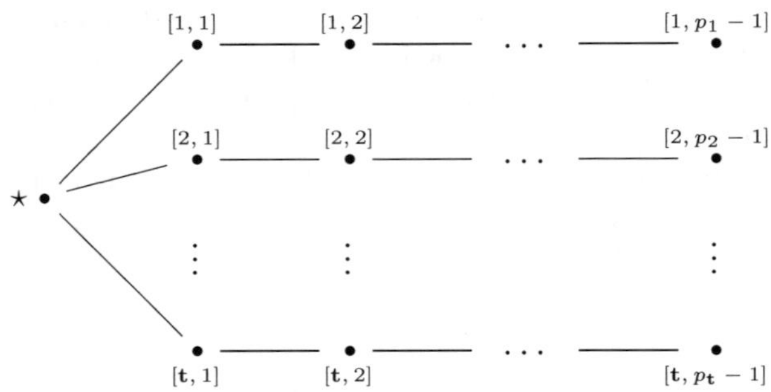

(10.24)

The set of vertices is denoted by $\mathbb{I}_0$. Let $C = (c_{ij})_{i,j\in\mathbb{I}_0}$ be the generalized Cartan matrix (GCM) of Γ. Then by Definition 10.5, we have an $\imath$quantum loop algebra ${}^{\mathrm{Dr}}\widetilde{\mathbf{U}}^\imath$ of split type associated to C.

Recall the cyclic quiver Cy_n from (10.23). We can make sense of the $\imath$Hall algebra ${}^\imath\widetilde{\mathcal{H}}(\mathbf{k}\mathrm{Cy}_n) = \widetilde{\mathcal{H}}(\mathbf{k}\mathrm{Cy}_n, \mathrm{Id})$. As a special case of [**LW20**, Theorem 9.6], we have an algebra monomorphism:

$$\widetilde{\psi}_{\mathrm{Cy}_n} : \widetilde{\mathbf{U}}^\imath_{\mathbf{v}}(\widehat{\mathfrak{sl}}_n) \longrightarrow {}^\imath\widetilde{\mathcal{H}}(\mathbf{k}\mathrm{Cy}_n) \tag{10.25}$$

$$B_j \mapsto \frac{-1}{q-1}[S_j], \quad \mathbb{K}_j \mapsto [K_{S_j}], \text{ for } 0 \le j \le n-1. \tag{10.26}$$

Define

$$\Omega_{\mathrm{Cy}_n} := \widetilde{\psi}_{\mathrm{Cy}_n} \circ \Phi : {}^{\mathrm{Dr}}\widetilde{\mathbf{U}}^\imath_{\mathbf{v}}(\widehat{\mathfrak{sl}}_n) \longrightarrow {}^\imath\widetilde{\mathcal{H}}(\mathbf{k}\mathrm{Cy}_n). \tag{10.27}$$

Then Ω_{Cy_n} sends

$$\mathbb{K}_j \mapsto [K_{S_j}], \qquad C \mapsto [K_\delta], \qquad B_{j,0} \mapsto \frac{-1}{q-1}[S_j], \qquad \forall \ 1 \le j \le n-1. \tag{10.28}$$

For any $1 \le j \le n-1$, $l \in \mathbb{Z}$ and $r \ge 1$, we define

(10.29)

$$\widehat{B}_{j,l} := (1-q)\Omega_{\mathrm{Cy}_n}(B_{j,l}), \qquad \widehat{\Theta}_{j,r} := \Omega_{\mathrm{Cy}_n}(\Theta_{j,r}), \qquad \widehat{H}_{j,r} := \Omega_{\mathrm{Cy}_n}(H_{j,r}).$$

In particular, $\widehat{B}_{j,0} = [S_j]$ for any $1 \le j \le n-1$.

Let $\mathbb{X}$ be a weighted projective line of weight type $(\mathbf{p}, \underline{\boldsymbol{\lambda}})$. Recall that $\boldsymbol{\lambda}_i$ is the exceptional closed point of $\mathbb{X}$ of weight p_i for any $1 \le i \le \mathbf{t}$. Recall that $\mathscr{T}_{\boldsymbol{\lambda}_i}$ is the Serre subcategory of $\mathrm{coh}(\mathbb{X})$ consisting of torsion sheaves supported at $\boldsymbol{\lambda}_i$, and we have an equivalence $\mathscr{T}_{\boldsymbol{\lambda}_i} \cong \mathrm{rep}_{\mathbf{k}}(\mathrm{Cy}_{p_i})$, which induces an embedding of $\imath$Hall algebras:

$$\iota_i : {}^\imath\widetilde{\mathcal{H}}(\mathbf{k}\mathrm{Cy}_{p_i}) \longrightarrow {}^\imath\widetilde{\mathcal{H}}(\mathbb{X}_{\mathbf{k}}). \tag{10.30}$$

Inspired by (10.29), we define

$$\widehat{B}_{[i,j],l} := \iota_i(\widehat{B}_{j,l}), \quad \widehat{\Theta}_{[i,j],r} := \iota_i(\widehat{\Theta}_{j,r}), \quad \widehat{H}_{[i,j],r} := \iota_i(\widehat{H}_{j,r}), \tag{10.31}$$

for any $1 \le j \le p_i - 1$, $l \in \mathbb{Z}$ and $r > 0$.

Let $\mathcal{C}$ be the Serre subcategory of $\mathrm{coh}(\mathbb{X})$ generated by those simple sheaves S satisfying $\mathrm{Hom}(\mathcal{O}, S) = 0$. Then the Serre quotient $\mathrm{coh}(\mathbb{X})/\mathcal{C}$ is equivalent to the category $\mathrm{coh}(\mathbb{P}^1)$ and the canonical functor $\mathrm{coh}(\mathbb{X}) \to \mathrm{coh}(\mathbb{X})/\mathcal{C}$ has an exact fully faithful right adjoint functor

$$\mathbb{F}_{\mathbb{X},\mathbb{P}^1} : \mathrm{coh}(\mathbb{P}^1) \to \mathrm{coh}(\mathbb{X}), \tag{10.32}$$

which sends

$$\mathcal{O}_{\mathbb{P}^1}(l) \mapsto \mathcal{O}(l\vec{c}), \quad S_{\boldsymbol{\lambda}_i}^{(r)} \mapsto S_{i,0}^{(rp_i)}, \quad S_x^{(r)} \mapsto S_x^{(r)}$$

for any $l \in \mathbb{Z}, r \geq 1$, $1 \leq i \leq \mathbf{t}$ and $x \in \mathbb{P}^1_{\mathbf{k}} \setminus \{\boldsymbol{\lambda}_1, \cdots, \boldsymbol{\lambda}_{\mathbf{t}}\}$. Then $\mathbb{F}_{\mathbb{X},\mathbb{P}^1}$ induces an exact fully faithful functor $\mathcal{C}_1(\mathrm{coh}(\mathbb{P}^1)) \to \mathcal{C}_1(\mathrm{coh}(\mathbb{X}))$, which is also denoted by $\mathbb{F}_{\mathbb{X},\mathbb{P}^1}$. This functor $\mathbb{F}_{\mathbb{X},\mathbb{P}^1}$ induces a canonical embedding of Hall algebras $\mathcal{H}(\mathcal{C}_1(\mathrm{coh}(\mathbb{P}^1))) \to \mathcal{H}(\mathcal{C}_1(\mathrm{coh}(\mathbb{X})))$, and then an embedding

$$F_{\mathbb{X},\mathbb{P}^1} : {}^{\imath}\widetilde{\mathcal{H}}(\mathbb{P}^1_{\mathbf{k}}) \longrightarrow {}^{\imath}\widetilde{\mathcal{H}}(\mathbb{X}_{\mathbf{k}}). \tag{10.33}$$

Inspired by (10.7), we define

$$\widehat{\Theta}_{\star,m} := F_{\mathbb{X},\mathbb{P}^1}(\widehat{\Theta}_m) = \frac{1}{(q-1)^2\mathbf{v}^{m-1}} \sum_{0\neq f:\mathcal{O}(\vec{c})\to\mathcal{O}(m\vec{c})} [\mathrm{Coker}\, f], \tag{10.34}$$

$$\widehat{H}_{\star,m} := F_{\mathbb{X},\mathbb{P}^1}(\widehat{H}_m). \tag{10.35}$$

Recall the star-shaped graph $\Gamma = T_{p_1,\dots,p_{\mathbf{t}}}$ in (10.24). We define

$$\delta := \widehat{\mathcal{O}(\vec{c})} - \widehat{\mathcal{O}} \in K_0(\mathrm{coh}(\mathbb{X})); \tag{10.36}$$

compare with (10.8).

THEOREM 10.7 ([**LR21**]). *For any star-shaped graph Γ, let C be the GCM and $\mathbb{X}$ be the weighted projective line associated to Γ. Then there exists a $\mathbb{Q}(\mathbf{v})$-algebra homomorphism*

$$\Omega : {}^{\mathrm{Dr}}\widetilde{\mathbf{U}}^{\imath}_{\mathbf{v}} \longrightarrow {}^{\imath}\widetilde{\mathcal{H}}(\mathbb{X}_{\mathbf{k}}), \tag{10.37}$$

which sends

$$\mathbb{K}_\star \mapsto [K_{\mathcal{O}}], \qquad \mathbb{K}_{[i,j]} \mapsto [K_{S_{ij}}], \qquad C \mapsto [K_\delta]; \tag{10.38}$$

$$B_{\star,l} \mapsto \frac{-1}{q-1}[\mathcal{O}(l\vec{c})], \qquad \Theta_{\star,r} \mapsto \widehat{\Theta}_{\star,r}, \qquad H_{\star,r} \mapsto \widehat{H}_{\star,r}; \tag{10.39}$$

$$B_{[i,j],l} \mapsto \frac{-1}{q-1}\widehat{B}_{[i,j],l}, \quad \Theta_{[i,j],r} \mapsto \widehat{\Theta}_{[i,j],r}, \quad H_{[i,j],r} \mapsto \widehat{H}_{[i,j],r} \tag{10.40}$$

for any $[i,j] \in \mathbb{I}_0 - \{\star\}$, $l \in \mathbb{Z}$, $r > 0$. Moreover, if Γ is of finite type or affine type, then Ω is an embedding.

REMARK 10.8. Assume that $\mathbb{X}$ is of weight type $\mathbf{p} = (p_1, p_2)$. Let $T = \bigoplus_{0\leq\vec{x}\leq\vec{c}} \mathcal{O}_{\mathbb{X}}(\vec{x})$ be the canonical tilting sheaf in $\mathrm{coh}(\mathbb{X})$; see [**GL87**]. Then $\mathrm{End}_{\mathbb{X}}(T) = \mathbf{k}Q$ is the canonical algebra with $Q = C_{p_1,p_2}$ as shown below.

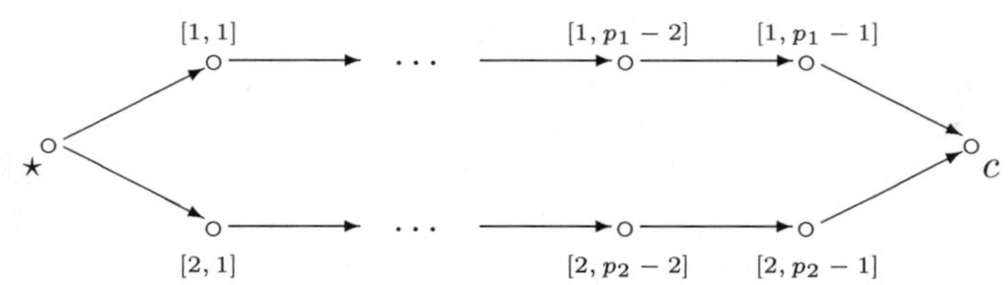

Then the corresponding result in Theorem 10.4 also holds for $\mathbb{X}$ in this case.

11. Discussions and open problems

11.1. Categorical symmetric pairs. Quantum group $\mathbf{U}$ and Drinfeld double $\widetilde{\mathbf{U}}$ are Hopf algebras. A categorical intepretation of the comultiplication was provided by Green [**Gr95**] (which is extended to Bridgeland's setup [**Br13**]).

We have a quantum symmetric pair $(\widetilde{\mathbf{U}}, \widetilde{\mathbf{U}}^\imath)$, where $\widetilde{\mathbf{U}}^\imath$ is a coideal subalgebra of $\widetilde{\mathbf{U}}$, i.e., $\Delta : \widetilde{\mathbf{U}}^\imath \to \widetilde{\mathbf{U}}^\imath \otimes \widetilde{\mathbf{U}}$. We have seen Hall algebra constructions of $\widetilde{\mathbf{U}}$ and $\widetilde{\mathbf{U}}^\imath$, separately.

PROBLEM 11.1. Find a categorical framework for the coideal subalgebra structure of $\widetilde{\mathbf{U}}^\imath$ in $\widetilde{\mathbf{U}}$ and for the quantum symmetric pair $(\widetilde{\mathbf{U}}, \widetilde{\mathbf{U}}^\imath)$.

See [**FL21**, **FL**^{3}**W20**] for a geometric interpretation of the quantum symmetric pair of (affine) type AIII.

11.2. $\imath$Hall basis. For Dynkin quivers, a suitable Hall basis (up to the normalization of some specific v-powers) for Ringel-Hall algebra is a PBW basis; cf. [**DDPW08**, Chapter 11].

PROBLEM 11.2. Let (Q, τ) be a Dynkin $\imath$quiver. Identify an $\imath$Hall basis in the $\imath$Hall algebra $\widetilde{\mathcal{H}}(\mathbf{k}Q, \tau)$ explicitly in terms of $\imath$quantum group $\widetilde{\mathbf{U}}^\imath$, under the isomorphism $\widetilde{\mathbf{U}}^\imath_{\mathbf{v}} \xrightarrow{\cong} \widetilde{\mathcal{H}}(\mathbf{k}Q, \tau)$ in Theorem 7.3.

11.3. Beyond quasi-split type. A Satake diagram consists of a pair $(\mathbb{I} = \mathbb{I}_\bullet \cup \mathbb{I}_\circ, \tau)$, where $\mathbb{I} = \mathbb{I}_\bullet \cup \mathbb{I}_\circ$ is a bi-colored partition of the Dynkin diagram $\mathbb{I}$, and τ a diagram involution (we allow $\tau = \mathrm{Id}$), subject to suitable conditions. In finite type, the Satake diagrams are classified by Araki; cf. [**BW18b**, Table 4]; they are in bijection with real forms of complex simple Lie algebras.

Associated to a general Satake diagram of Kac-Moody type, the $\imath$quantum groups $\mathbf{U}^\imath_\varsigma$ were defined in [**Ko14**] and the universal $\imath$quantum groups $\widetilde{\mathbf{U}}^\imath$ in [**LW22a**, **WZ23**], respectively.

PROBLEM 11.3. Provide an $\imath$Hall algebra realization of $\widetilde{\mathbf{U}}^\imath$ associated to a general Satake diagram.

11.4. Generalizing $\imath$quiver algebras. So far the $\imath$Hall algebras can only be used to realize quasi-split (i.e., $\mathbb{I}_\bullet = \emptyset$) $\imath$quantum groups under the additional assumption on the evenness of Cartan integers $c_{i,\tau i}$. (This assumption arises from the $\imath$quiver algebra construction in §5.)

This simplest example of quasi-split Satake diagrams which do not satisfy the evenness condition on $c_{i,\tau i}$ is the A_2 Dynkin diagram with nontrivial involution τ. Note that as τ does not preserve the orientation of the A_2 quiver, this is not an example of $\imath$quiver in the sense of §5.1.

PROBLEM 11.4. Extend the formulation of $\imath$quiver algebra and the construction of $\imath$Hall algebra to realize $\widetilde{\mathbf{U}}^\imath$ associated to A_2 with nontrivial involution τ.

11.5. Valued $\imath$quivers. In the realization of quantum groups via Hall algebras, Ringel and Green actually used the Hall algebras of valued quivers to realize quantum groups associated to symmetrizable (which may not be symmetric) GCM. The reflection functors of valued quivers were also introduced, and they were used to realize Lusztig's symmetries of quantum groups; we refer to [**DDPW08**] for a good survey on Hall algebras of valued quivers. We expect a good outcome for the following problem.

PROBLEM 11.5. Formulate valued ıquiver algebras and their homological properties. Use the ıHall algebras of valued ıquiver algebras to realize the ıquantum groups associated to symmetrizable GCM.

11.6. Geometric ıHall algebras. In [**LW21b**], we use (singular) Nakajima-Keller-Scherozke varieties [**KS16, S19**] to realize the ıquiver algebras, and then construct convolution algebras on the dual graded Grothendieck rings of perverse sheaves to realize the quasi-split ıquantum group of ADE type. As a result, the dual basis of perverse sheaves gives a basis of ıquantum groups, called dual ıcanonical basis, which has positive integral structure constants. This is an ıgeneralization of Qin's geometric realization [**Qin16**] of Drinfeld double quantum group $\widetilde{\mathbf{U}}$ and its dual canonical basis.

PROBLEM 11.6. Provide an algebraic construction of the dual canonical basis for $\widetilde{\mathbf{U}}$ and $\widetilde{\mathbf{U}}^\imath$, respectively. (The rank one cases including $\widetilde{\mathbf{U}}(\mathfrak{sl}_2)$ are already open.)

A precise relation (if any) between the ıcanonical basis on the modified ıquantum group $\dot{\mathbf{U}}^\imath$ [**BW18b,BW21**] and the dual canonical basis on the universal ıquantum group $\widetilde{\mathbf{U}}^\imath$ [**LW21b**] remains to be clarified.

11.7. Elliptic ıHall algebra. The ıconstructions are often related to various constructions of classical types, sometimes unexpectedly; see, e.g., [**BW18a**] for connections to Kazhdan-Lusztig theory, [**BKLW18, FL^{3}W20**] for connections to convolution algebras arising from (affine) flag varieties, and [**LRW21**] for connections to type C universal characters.

The definition of ıHall algebra of the projective line can be extended to higher genus curves. The ıHall algebra of an elliptic curve will be called the *elliptic ıHall algebra*. Stimulated by work of Schiffmann and Vasserot (see [**SV11**]), we post the following loosely phrased question.

PROBLEM 11.7. Is the elliptic ıHall algebra related to any of the following subjects? (i) spherical DAHA of type C; (ii) (inhomogeneous) ıMacdonald polynomials.

Acknowledgment

The first author thanks Liangang Peng for guidance and continuing encouragement, and thanks University of Virginia, Shanghai Key Laboratory of Pure Mathematics and Mathematical Practice, East China Normal University for hospitality and support.

References

[BK15] Martina Balagović and Stefan Kolb, *The bar involution for quantum symmetric pairs*, Represent. Theory **19** (2015), 186–210, DOI 10.1090/ert/469. MR3414769

[BW18a] Huanchen Bao and Weiqiang Wang, *A new approach to Kazhdan-Lusztig theory of type B via quantum symmetric pairs* (English, with English and French summaries), Astérisque **402** (2018), vii+134. MR3864017

[BW18b] Huanchen Bao and Weiqiang Wang, *Canonical bases arising from quantum symmetric pairs*, Invent. Math. **213** (2018), no. 3, 1099–1177, DOI 10.1007/s00222-018-0801-5. MR3842062

[BW21] Huanchen Bao and Weiqiang Wang, *Canonical bases arising from quantum symmetric pairs of Kac-Moody type*, Compos. Math. **157** (2021), no. 7, 1507–1537, DOI 10.1112/S0010437X2100734X. MR4277109

[BK20] Pascal Baseilhac and Stefan Kolb, *Braid group action and root vectors for the q-Onsager algebra*, Transform. Groups **25** (2020), no. 2, 363–389, DOI 10.1007/s00031-020-09555-7. MR4098883

[Be94] Jonathan Beck, *Braid group action and quantum affine algebras*, Comm. Math. Phys. **165** (1994), no. 3, 555–568. MR1301623

[BeW18] Collin Berman and Weiqiang Wang, *Formulae of ı-divided powers in $\mathbf{U}_q(\mathfrak{sl}_2)$*, J. Pure Appl. Algebra **222** (2018), no. 9, 2667–2702, DOI 10.1016/j.jpaa.2017.10.014. MR3783013

[Br13] Tom Bridgeland, *Quantum groups via Hall algebras of complexes*, Ann. of Math. (2) **177** (2013), no. 2, 739–759, DOI 10.4007/annals.2013.177.2.9. MR3010811

[Buh10] Theo Bühler, *Exact categories*, Expo. Math. **28** (2010), no. 1, 1–69, DOI 10.1016/j.exmath.2009.04.004. MR2606234

[BKLW18] Huanchen Bao, Jonathan Kujawa, Yiqiang Li, and Weiqiang Wang, *Geometric Schur duality of classical type*, Transform. Groups **23** (2018), no. 2, 329–389, DOI 10.1007/s00031-017-9447-4. MR3805209

[BS13] Igor Burban and Olivier Schiffmann, *The composition Hall algebra of a weighted projective line*, J. Reine Angew. Math. **679** (2013), 75–124, DOI 10.1515/crelle.2012.023. MR3065155

[CLW21a] Xinhong Chen, Ming Lu, and Weiqiang Wang, *A Serre presentation for the ıquantum groups*, Transform. Groups **26** (2021), no. 3, 827–857, DOI 10.1007/s00031-020-09581-5. MR4309551

[CLW21b] Xinhong Chen, Ming Lu, and Weiqiang Wang, *Serre-Lusztig relations for ıquantum groups*, Comm. Math. Phys. **382** (2021), no. 2, 1015–1059, DOI 10.1007/s00220-021-04035-9. MR4227167

[CLW23] Xinhong Chen, Ming Lu, and Weiqiang Wang, *Serre-Lusztig relations for ıquantum groups III*, J. Pure Appl. Algebra **227** (2023), no. 4, Paper No. 107253, 13, DOI 10.1016/j.jpaa.2022.107253. MR4510803

[CX99] Xueqing Chen and Jie Xiao, *Exceptional sequences in Hall algebras and quantum groups*, Compositio Math. **117** (1999), no. 2, 161–187, DOI 10.1023/A:1000947529874. MR1695853

[Da15] Ilaria Damiani, *From the Drinfeld realization to the Drinfeld-Jimbo presentation of affine quantum algebras: injectivity*, Publ. Res. Inst. Math. Sci. **51** (2015), no. 1, 131–171, DOI 10.4171/PRIMS/150. MR3367090

[DDPW08] Bangming Deng, Jie Du, Brian Parshall, and Jianpan Wang, *Finite dimensional algebras and quantum groups*, Mathematical Surveys and Monographs, vol. 150, American Mathematical Society, Providence, RI, 2008, DOI 10.1090/surv/150. MR2457938

[DJX12] Rujing Dou, Yong Jiang, and Jie Xiao, *The Hall algebra approach to Drinfeld's presentation of quantum loop algebras*, Adv. Math. **231** (2012), no. 5, 2593–2625, DOI 10.1016/j.aim.2012.07.026. MR2970461

[Dr87] V. G. Drinfel′d, *Quantum groups*, Proceedings of the International Congress of Mathematicians, Vol. 1, 2 (Berkeley, Calif., 1986), Amer. Math. Soc., Providence, RI, 1987, pp. 798–820. MR934283

[EJ00] Edgar E. Enochs and Overtoun M. G. Jenda, *Relative homological algebra*, De Gruyter Expositions in Mathematics, vol. 30, Walter de Gruyter & Co., Berlin, 2000, DOI 10.1515/9783110803662. MR1753146

[FL21] Zhaobing Fan and Yiqiang Li, *Positivity of canonical bases under comultiplication*, Int. Math. Res. Not. IMRN **9** (2021), 6871–6931, DOI 10.1093/imrn/rnz047. MR4251292

[FL^{3}W20] Zhaobing Fan, Chun-Ju Lai, Yiqiang Li, Li Luo, and Weiqiang Wang, *Affine flag varieties and quantum symmetric pairs*, Mem. Amer. Math. Soc. **265** (2020), no. 1285, v+123, DOI 10.1090/memo/1285. MR4080913

[GL87] Werner Geigle and Helmut Lenzing, *A class of weighted projective curves arising in representation theory of finite-dimensional algebras*, Singularities, representation of algebras, and vector bundles (Lambrecht, 1985), Lecture Notes in Math., vol. 1273, Springer, Berlin, 1987, pp. 265–297, DOI 10.1007/BFb0078849. MR915180

[GLS17] Christof Geiss, Bernard Leclerc, and Jan Schröer, *Quivers with relations for symmetrizable Cartan matrices I: Foundations*, Invent. Math. **209** (2017), no. 1, 61–158, DOI 10.1007/s00222-016-0705-1. MR3660306

[Gor13] M. Gorsky, *Semi-derived Hall algebras and tilting invariance of Bridgeland-Hall algebras*, `arXiv:1303.5879v2`, 2013

[Gor18] Mikhail Gorsky, *Semi-derived and derived Hall algebras for stable categories*, Int. Math. Res. Not. IMRN **1** (2018), 138–159, DOI 10.1093/imrn/rnv325. MR3800631

[Gr95] James A. Green, *Hall algebras, hereditary algebras and quantum groups*, Invent. Math. **120** (1995), no. 2, 361–377, DOI 10.1007/BF01241133. MR1329046

[Gro57] A. Grothendieck, *Sur la classification des fibrés holomorphes sur la sphère de Riemann* (French), Amer. J. Math. **79** (1957), 121–138, DOI 10.2307/2372388. MR87176

[Ha88] Dieter Happel, *Triangulated categories in the representation theory of finite-dimensional algebras*, London Mathematical Society Lecture Note Series, vol. 119, Cambridge University Press, Cambridge, 1988, DOI 10.1017/CBO9780511629228. MR935124

[Ha91] Dieter Happel, *On Gorenstein algebras*, Representation theory of finite groups and finite-dimensional algebras (Bielefeld, 1991), Progr. Math., vol. 95, Birkhäuser, Basel, 1991, pp. 389–404, DOI 10.1007/978-3-0348-8658-1_16. MR1112170

[Hub06] Andrew Hubery, *From triangulated categories to Lie algebras: a theorem of Peng and Xiao*, Trends in representation theory of algebras and related topics, Contemp. Math., vol. 406, Amer. Math. Soc., Providence, RI, 2006, pp. 51–66, DOI 10.1090/conm/406/07653. MR2258041

[Ka97] M. M. Kapranov, *Eisenstein series and quantum affine algebras*, J. Math. Sci. (New York) **84** (1997), no. 5, 1311–1360, DOI 10.1007/BF02399194. Algebraic geometry, 7. MR1465518

[Ka98] M. Kapranov, *Heisenberg doubles and derived categories*, J. Algebra **202** (1998), no. 2, 712–744, DOI 10.1006/jabr.1997.7323. MR1617651

[Ke90] Bernhard Keller, *Chain complexes and stable categories*, Manuscripta Math. **67** (1990), no. 4, 379–417, DOI 10.1007/BF02568439. MR1052551

[Ke94] Bernhard Keller, *Deriving DG categories*, Ann. Sci. École Norm. Sup. (4) **27** (1994), no. 1, 63–102. MR1258406

[Ke05] Bernhard Keller, *On triangulated orbit categories*, Doc. Math. **10** (2005), 551–581. MR2184464

[KS16] Bernhard Keller and Sarah Scherotzke, *Graded quiver varieties and derived categories*, J. Reine Angew. Math. **713** (2016), 85–127, DOI 10.1515/crelle-2013-0124. MR3483626

[Ko14] Stefan Kolb, *Quantum symmetric Kac-Moody pairs*, Adv. Math. **267** (2014), 395–469, DOI 10.1016/j.aim.2014.08.010. MR3269184

[KP11] Stefan Kolb and Jacopo Pellegrini, *Braid group actions on coideal subalgebras of quantized enveloping algebras*, J. Algebra **336** (2011), 395–416, DOI 10.1016/j.jalgebra.2011.04.001. MR2802552

[Let99] Gail Letzter, *Symmetric pairs for quantized enveloping algebras*, J. Algebra **220** (1999), no. 2, 729–767, DOI 10.1006/jabr.1999.8015. MR1717368

[Let02] Gail Letzter, *Coideal subalgebras and quantum symmetric pairs*, New directions in Hopf algebras, Math. Sci. Res. Inst. Publ., vol. 43, Cambridge Univ. Press, Cambridge, 2002, pp. 117–165. MR1913438

[Li12] Fang Li, *Modulation and natural valued quiver of an algebra*, Pacific J. Math. **256** (2012), no. 1, 105–128, DOI 10.2140/pjm.2012.256.105. MR2928543

[LinP19] Ji Lin and Liangang Peng, *Modified Ringel-Hall algebras, Green's formula and derived Hall algebras*, J. Algebra **526** (2019), 81–103, DOI 10.1016/j.jalgebra.2019.02.009. MR3914913

[Lu22] M. Lu, *Semi-derived Ringel-Hall algebras of 1-Gorenstein algebras, Appendix A to* [**LW22a**].

[LP21] Ming Lu and Liangang Peng, *Semi-derived Ringel-Hall algebras and Drinfeld double*, Adv. Math. **383** (2021), Paper No. 107668, 72, DOI 10.1016/j.aim.2021.107668. MR4232541

[LR21] Ming Lu and Shiquan Ruan, *ıHall algebras of weighted projective lines and quantum symmetric pairs*, Represent. Theory **28** (2024), 112–188, DOI 10.1090/ert/669. MR4712701

[LRW20] Ming Lu, Shiquan Ruan, and Weiqiang Wang, *ıHall algebra of the projective line and q-Onsager algebra*, Trans. Amer. Math. Soc. **376** (2023), no. 2, 1475–1505, DOI 10.1090/tran/8798. MR4531682

[LRW21] M. Lu, S. Ruan and W. Wang, *ıHall algebras of Jordan quiver and ıHall-Littlewood functions*, arXiv:2104.12336

[LW20] Ming Lu and Weiqiang Wang, *Hall algebras and quantum symmetric pairs of Kac-Moody type*, Adv. Math. **430** (2023), Paper No. 109215, 56, DOI 10.1016/j.aim.2023.109215. MR4619448

[LW21a] Ming Lu and Weiqiang Wang, *Hall algebras and quantum symmetric pairs II: Reflection functors*, Comm. Math. Phys. **381** (2021), no. 3, 799–855, DOI 10.1007/s00220-021-03965-8. MR4218672

[LW21b] Ming Lu and Weiqiang Wang, *Hall algebras and quantum symmetric pairs III: Quiver varieties*, Adv. Math. **393** (2021), Paper No. 108071, 70, DOI 10.1016/j.aim.2021.108071. MR4340223

[LW21c] Ming Lu and Weiqiang Wang, *A Drinfeld type presentation of affine ıquantum groups I: Split ADE type*, Adv. Math. **393** (2021), Paper No. 108111, 46, DOI 10.1016/j.aim.2021.108111. MR4340233

[LW22a] Ming Lu and Weiqiang Wang, *Hall algebras and quantum symmetric pairs I: Foundations*, Proc. Lond. Math. Soc. (3) **124** (2022), no. 1, 1–82, DOI 10.1112/plms.12423. With an appendix by Lu. MR4389313

[LW22b] Ming Lu and Weiqiang Wang, *Braid group symmetries on quasi-split ıquantum groups via ıHall algebras*, Selecta Math. (N.S.) **28** (2022), no. 5, Paper No. 84, 64, DOI 10.1007/s00029-022-00800-3. MR4486230

[LWZ23] Ming Lu, Weiqiang Wang, and Weinan Zhang, *Braid group action and quasi-split affine ıquantum groups I*, Represent. Theory **27** (2023), 1000–1040, DOI 10.1090/ert/657. MR4659754

[LZ21] Ming Lu and Bin Zhu, *Singularity categories of Gorenstein monomial algebras*, J. Pure Appl. Algebra **225** (2021), no. 8, Paper No. 106651, 39, DOI 10.1016/j.jpaa.2020.106651. MR4190607

[Lus90] G. Lusztig, *Canonical bases arising from quantized enveloping algebras*, J. Amer. Math. Soc. **3** (1990), no. 2, 447–498, DOI 10.2307/1990961. MR1035415

[Lus93] George Lusztig, *Introduction to quantum groups*, Progress in Mathematics, vol. 110, Birkhäuser Boston, Inc., Boston, MA, 1993. MR1227098

[Lus03] G. Lusztig, *Hecke algebras with unequal parameters*, CRM Monograph Series, vol. 18, American Mathematical Society, Providence, RI, 2003. enhanced version in arXiv:0208154v2, DOI 10.1090/crmm/018. MR1974442

[Mac95] I. G. Macdonald, *Symmetric functions and Hall polynomials*, 2nd ed., Oxford Mathematical Monographs, The Clarendon Press, Oxford University Press, New York, 1995. With contributions by A. Zelevinsky; Oxford Science Publications. MR1354144

[MY21] Hiroyuki Minamoto and Kota Yamaura, *The Happel functor and homologically well-graded Iwanaga-Gorenstein algebras*, J. Algebra **565** (2021), 441–488, DOI 10.1016/j.jalgebra.2020.08.021. MR4150804

[M06] Kent E. Morrison, *Integer sequences and matrices over finite fields*, J. Integer Seq. **9** (2006), no. 2, Article 06.2.1, 28. MR2217227

[OV90] A. L. Onishchik and È. B. Vinberg, *Lie groups and algebraic groups*, Springer Series in Soviet Mathematics, Springer-Verlag, Berlin, 1990. Translated from the Russian and with a preface by D. A. Leites, DOI 10.1007/978-3-642-74334-4. MR1064110

[PX00] Liangang Peng and Jie Xiao, *Triangulated categories and Kac-Moody algebras*, Invent. Math. **140** (2000), no. 3, 563–603, DOI 10.1007/s002220000062. MR1760751

[Qin16] Fan Qin, *Quantum groups via cyclic quiver varieties I*, Compos. Math. **152** (2016), no. 2, 299–326, DOI 10.1112/S0010437X15007551. MR3462555

[Rin90] Claus Michael Ringel, *Hall algebras and quantum groups*, Invent. Math. **101** (1990), no. 3, 583–591, DOI 10.1007/BF01231516. MR1062796

[Rin96] Claus Michael Ringel, *PBW-bases of quantum groups*, J. Reine Angew. Math. **470** (1996), 51–88, DOI 10.1515/crll.1996.470.51. MR1370206

[RZ17] Claus Michael Ringel and Pu Zhang, *Representations of quivers over the algebra of dual numbers*, J. Algebra **475** (2017), 327–360, DOI 10.1016/j.jalgebra.2016.12.001. MR3612474

[S19] Sarah Scherotzke, *Generalized quiver varieties and triangulated categories*, Math. Z. **292** (2019), no. 3-4, 1453–1478, DOI 10.1007/s00209-018-2140-z. MR3980299

[SS16] Sarah Scherotzke and Nicolo Sibilla, *Quiver varieties and Hall algebras*, Proc. Lond. Math. Soc. (3) **112** (2016), no. 6, 1002–1018, DOI 10.1112/plms/pdw016. MR3537330

[Sch04] Olivier Schiffmann, *Noncommutative projective curves and quantum loop algebras*, Duke Math. J. **121** (2004), no. 1, 113–168, DOI 10.1215/S0012-7094-04-12114-1. MR2031167

[Sch06] Olivier Schiffmann, *Lectures on Hall algebras* (English, with English and French summaries), Geometric methods in representation theory. II, Sémin. Congr., vol. 24, Soc. Math. France, Paris, 2012, pp. 1–141. MR3202707

[SV11] O. Schiffmann and E. Vasserot, *The elliptic Hall algebra, Cherednik Hecke algebras and Macdonald polynomials*, Compos. Math. **147** (2011), no. 1, 188–234, DOI 10.1112/S0010437X10004872. MR2771130

[SV99] Bert Sevenhant and Michel Van den Bergh, *On the double of the Hall algebra of a quiver*, J. Algebra **221** (1999), no. 1, 135–160, DOI 10.1006/jabr.1999.7958. MR1722908

[SZ06] Mark Shimozono and Mike Zabrocki, *Deformed universal characters for classical and affine algebras*, J. Algebra **299** (2006), no. 1, 33–61, DOI 10.1016/j.jalgebra.2006.03.008. MR2225764

[Tam11] Harry Tamvakis, *Giambelli, Pieri, and tableau formulas via raising operators*, J. Reine Angew. Math. **652** (2011), 207–244, DOI 10.1515/CRELLE.2011.023. MR2787357

[T06] Bertrand Toën, *Derived Hall algebras*, Duke Math. J. **135** (2006), no. 3, 587–615, DOI 10.1215/S0012-7094-06-13536-6. MR2272977

[W22] W. Wang, *Quantum symmetric pairs*, Proceedings of ICM 2022, `arXiv:2112.10911`

[WZ23] Weiqiang Wang and Weinan Zhang, *An intrinsic approach to relative braid group symmetries on ıquantum groups*, Proc. Lond. Math. Soc. (3) **127** (2023), no. 5, 1338–1423. MR4668526

[X97] Jie Xiao, *Drinfeld double and Ringel-Green theory of Hall algebras*, J. Algebra **190** (1997), no. 1, 100–144, DOI 10.1006/jabr.1996.6887. MR1442148

[XX08] Jie Xiao and Fan Xu, *Hall algebras associated to triangulated categories*, Duke Math. J. **143** (2008), no. 2, 357–373, DOI 10.1215/00127094-2008-021. MR2420510

[XY01] Jie Xiao and Shilin Yang, *BGP-reflection functors and Lusztig's symmetries: a Ringel-Hall algebra approach to quantum groups*, J. Algebra **241** (2001), no. 1, 204–246, DOI 10.1006/jabr.2001.8762. MR1838851

[Z00] Pu Zhang, *PBW-basis for the composition algebra of the Kronecker algebra*, J. Reine Angew. Math. **527** (2000), 97–116, DOI 10.1515/crll.2000.086. MR1794019

[Z21] Weinan Zhang, *A Drinfeld-type presentation of affine ıquantum groups II: split BCFG type*, Lett. Math. Phys. **112** (2022), no. 5, Paper No. 89, 33, DOI 10.1007/s11005-022-01583-6. MR4482099

[Z23] Weinan Zhang, *Relative braid group symmetries on ıquantum groups of Kac-Moody type*, Selecta Math. (N.S.) **29** (2023), no. 4, Paper No. 59, 54, DOI 10.1007/s00029-023-00861-y. MR4621136

Department of Mathematics, Sichuan University, Chengdu 610064, People's Republic of China

Email address: `luming@scu.edu.cn`

Department of Mathematics, University of Virginia, Charlottesville, Virginia 22904

Email address: `ww9c@virginia.edu`

Proceedings of Symposia in Pure Mathematics
Volume **108**, 2024
https://doi.org/10.1090/pspum/108/01972

Unipotent blocks and weighted affine Weyl groups

G. Lusztig

Contents

Introduction
1. Weighted Weyl groups
2. Sharp Weyl groups
3. Affine Weyl groups and the sets $\mathcal{C}_\omega(W)$
4. The function $c : \mathrm{Irr}(\bar{\mathcal{W}}) \to \mathbf{N}$
5. A weighted affine Weyl group
6. The set $\mathcal{G}_\omega(G)$ and the bijection $\mathcal{G}_\omega(G) \xrightarrow{\sim} \mathcal{C}_\omega(W)$
7. Cells in the weighted affine Weyl group $\mathcal{W}_J$
References

Introduction

0.1. Let G be a connected reductive algebraic group over $\mathbf{C}$. Let $un(G)$ be the set of unipotent conjugacy classes in G. Let $ls(G)$ be the set of all pairs $(\mathfrak{c}, \mathfrak{L})$ where $\mathfrak{c} \in un(G)$ and $\mathfrak{L}$ is an irreducible $\mathbf{C}$-local system on $\mathfrak{c}$, equivariant for the conjugation action of G. In [**Spr76**], Springer discovered a remarkable bijection between a certain subset $'ls(G)$ of $ls(G)$ and the set of irreducible representations of the Weyl group of G (up to isomorphism). In [**Lus84a**] I extended Springer result by defining the "generalized Springer correspondence" that is, a partition of $ls(G)$ into subsets (which could be called *unipotent blocks*) and a bijection, for each unipotent block β, between the set of objects in β and the set $\mathrm{Irr}(\mathfrak{W}_\beta)$ of irreducible representations (up to isomorphism) of a certain Weyl group $\mathfrak{W}_\beta$ associated to β. (The subset $'ls(G)$ is one of the unipotent blocks, namely the one containing $(\{1\}, \mathbf{C})$.) The arguments of [**Lus84a**] were based on the use of perverse sheaves on G.

0.2. In the remainder of this introduction we assume that G is almost simple, simply connected. Let W be the affine Weyl group of type dual to that of G. In this paper we try to show that various information on unipotent elements of G can be recovered from things which are more primitive than unipotent elements, namely from Weyl groups, their associated Hecke algebras and their representations. We want to recover such information from W and its subgroups, without use of the

This work was supported by NSF grant DMS-2153741.

geometry of G. Results of this type have been obtained in [**Lus23**] for the unipotent block $'ls(G)$ and this paper can be viewed as an attempt to extend [**Lus23**] to arbitrary unipotent blocks.

We now describe some earlier results in the same direction.

In [**Lus79**] it has been observed that there is a set defined purely in terms of W which is an indexing set for $un(G)$, via the ordinary Springer correspondence. More precisely, this set describes $un(G)$ in terms of truncated induction from special representations of the various finite standard parabolic subgroups of W, see 6.3. The same process allows one to recover the order of the group of components of the centralizer of a unipotent element in the adjoint group of G in terms of data attached to the various special representations in the previous sentence, see [**Lus09**]. In [**Lus80**] it was conjectured (and in [**Lus89**] proved) that the set of two-sided cells of W is an indexing set for $un(G)$. These results suggest that it may be possible to recover other properties of unipotent elements of G in terms of W.

0.3. Let Ω_W be the group of automorphisms of W as a Coxeter group which induce an inner automorphism of W modulo the subgroup of translations. In this paper, for any $\omega \in \Omega_W$, we define (without reference to G) a certain set $\mathcal{C}_\omega(W)$ of standard finite parabolic subgroups W_J of W, stable under ω, see §3. Moreover, we define a bijection between the set of unipotent blocks of G and the set $\cup_{\omega\in\Omega_W}\mathcal{C}_\omega(W)$, see 6.5. (This bijection is closely related to the arithmetic/geometric correspondence [L95] in the study of unipotent representations of a simple p-adic group.)

One of the properties that we impose on $W_J \in \mathcal{C}_\omega(W)$ implies that a finite reductive group over $\mathbf{F}_q$ with Weyl group W_J and with Frobenius acting on this Weyl group as ω should have a unipotent cuspidal representation (but we formulate this property without reference to finite reductive groups, see §2). This has the consequence that the subgroups W_J which appear are rather few. In §5 we attach to each $W_J \in \mathcal{C}_\omega(W)$ an affine Weyl group $\mathcal{W}_J$ with a weight function $\mathcal{L}_J : \mathcal{W}_J \to \mathbf{N}$. (This follows a procedure from [**Lus91**], [**Lus95**].) In this way to any unipotent block we have attached a weighted affine Weyl group. (We regard the group $\{1\}$ as a weighted affine Weyl group.) The weighted affine Weyl group $(\mathcal{W}_J, \mathcal{L}_J)$ is different in general from the one defined in [**Lus17**]. In the remainder of this introduction we fix a unipotent block β and we denote by $\omega, W_J, \mathcal{W}_J, \mathcal{L}_J$ the objects associated to β as above.

In Theorem 6.10 we show that the quotient $\bar{\mathcal{W}}_J$ of $\mathcal{W}_J$ by its group of translations (which is defined purely in terms of W) can be identified with the group $\mathfrak{W}_\beta$ of 0.1. In §4 we define a function $c : \mathrm{Irr}(\bar{\mathcal{W}}_J) \to \mathbf{N}$ purely in terms of W. The values of c are conjecturally related to dimensions of certain Springer fibres (see 6.11); this relation is unconditional for exceptional types. The function c is used in 6.12(a) to express conjecturally the generalized Green functions [**Lus86**, 24.8] purely in terms of W; again this is unconditional for exceptional types. In 7.3 we state a conjecture which provides an indexing set for the set of two-sided cells in $\mathcal{W}_J$ relative to $\mathcal{L}_J$; this is unconditional for exceptional types.

0.4. Notation. For a group L we denote by L_{der} the derived subgroup of L. For a finite group Γ let $\mathrm{Irr}(\Gamma)$ be the set of irreducible representations of Γ over $\mathbf{C}$ (up to isomorphism).

1. Weighted Weyl groups

1.1. Let $\mathfrak{W}$ be a Coxeter group and let $\mathfrak{S}$ be its set of simple reflections. Let $w \mapsto |w|$ be the length function on $\mathfrak{W}$. We say that $\mathfrak{W}$ is *weighted* if we are given a weight function $\mathcal{L} : \mathfrak{W} \to \mathbf{N}$ that is a function such that $\mathcal{L}(ww') = \mathcal{L}(w) + \mathcal{L}(w')$ for any w, w' in $\mathfrak{W}$ such that $|ww'| = |w| + |w'|$. Let $Cell(\mathfrak{W}, \mathcal{L})$ be the set of two sided cells of $(\mathfrak{W}, \mathcal{L})$ in the sense of [**Lus03**].

Let $v \in \mathbf{C} - \{0\}$ be a non-root of 1 and let $\mathfrak{H}_v$ be the Hecke algebra (over $\mathbf{C}$) associated to $\mathfrak{W}, \mathcal{L}, v$; thus $\mathfrak{H}_v$ is the $\mathbf{C}$-vector space with basis $\{T_w; w \in \mathfrak{W}\}$ with associative multiplication defined by the rules $T_w T_{w'} = T_{ww'}$ for any w, w' in $\mathfrak{W}$ such that $|ww'| = |w| + |w'|$ and $(T_\sigma + v^{-\mathcal{L}(\sigma)})(T_\sigma - v^{\mathcal{L}(\sigma)}) = 0$ for any $\sigma \in \mathfrak{S}$.

1.2. In the remainder of this section we assume that $\mathfrak{W}$ is a Weyl group with a given weight function $\mathcal{L} : \mathfrak{W} \to \mathbf{N}$. Let $\mathrm{Irr}(\mathfrak{H}_v)$ be the set of simple $\mathfrak{H}_v$-modules (up to isomorphism). It is known that $\mathrm{Irr}(\mathfrak{W}), \mathrm{Irr}(\mathfrak{H}_v)$ are in canonical bijection. (We can assume that $\mathfrak{W}$ is irreducible. If properties P1-P15 in [**Lus03**, §14] are assumed, then according to [**Lus03**], the algebras $\mathfrak{H}_v, \mathbf{C}[\mathfrak{W}]$ are canonically isomorphic and the result would follow. Now P1-P15 do hold when $\mathfrak{W}$ is of type E_7, E_8 or G_2. Thus we can assume that $\mathfrak{W}$ is of type other than E_7, E_8 or G_2. In this case, the result follows from the observation in [**BC72**] according to which the various $E \in \mathrm{Irr}(\mathfrak{W})$ are characterized by their multiplicities in representations of $\mathfrak{W}$ induced from the unit representation of various parabolic subgroups. Alternatively we can use [**Gec11**].) We denote this bijection by $E \leftrightarrow E_v$. For $E \in \mathrm{Irr}(\mathfrak{W})$ we set

$$f_{E,\mathcal{L},v} = (\dim E)^{-1} \sum_{w \in \mathfrak{W}} \mathrm{tr}(T_w, E_v)\, \mathrm{tr}(T_{w^{-1}}, E_v) \in \mathbf{C}^*$$

and

$$D_{E,\mathcal{L},v} = f_{E,\mathcal{L},v}^{-1} \sum_{w \in \mathfrak{W}} v^{2\mathcal{L}(w)} \in \mathbf{C}.$$

From the known explicit formulas for $D_{E,\mathcal{L},v}$ we see that $D_{E,\mathcal{L},v}$ is a nonzero rational function in v. We define $\mathbf{a}_\mathcal{L}(E) \in \mathbf{N}$ by the requirement that $D_{E,\mathcal{L},v} v^{-2\mathbf{a}_\mathcal{L}(E)}$ is a rational function in v whose value at $v = 0$ is $\neq 0$ and $\neq \infty$. In the case where $\mathcal{L} = ||$ we write $\mathbf{a}(E)$ instead of $\mathbf{a}_\mathcal{L}(E)$.

1.3. If $\mathfrak{W}$ is of type A_1 and the value of $\mathcal{L}_\mathfrak{S}$ is $a \in \mathbf{Z}_{>0}$ then setting $q = v^{2a}$ we have $D_{E,\mathcal{L},v} = 1$ if $E = 1$ and $D_{E,\mathcal{L}} = q$ if $E = \mathrm{sgn}$; the corresponding values of $\mathbf{a}_{E,\mathcal{L}}$ are $0; a$. We now assume that $\mathfrak{W}$ is of type A_2 and the values of $\mathcal{L}|_\mathfrak{S}$ are a, a in $\mathbf{Z}_{>0}$. Setting $q = v^{2a}$ we write below the values of $D_{E,\mathcal{L},v}$ for various E of dimension $1; 2; 1$:

$$1; q^2 + q, q^3.$$

The corresponding values of $\mathbf{a}_{E,\mathcal{L}}$ are $0; a; 3a$.

1.4. In this subsection we assume that $\mathfrak{W}$ is of type B_2 and the values of $\mathcal{L}|_\mathfrak{S}$ are a, b in $\mathbf{Z}_{>0}$. Setting $q = v^{2a}, y = v^{2b}$ we write below the values of $D_{E,\mathcal{L},v}$ for various E of dimension $1; 2; 1; 1; 1$:

$$1; qy(q+1)(y+1)/(q+y); q^2(qy+1)/(q+y); y^2(qy+1)/(q+y); q^2y^2.$$

The corresponding values of $\mathbf{a}_{E,\mathcal{L}}$ are:

$$0; a+b-m; 2a-m; 2b-m; 2a+2b$$

where $m = \min(a, b)$.

1.5. In this subsection we assume that $\mathfrak{W}$ is of type G_2 and the values of $\mathcal{L}|_{\mathfrak{S}}$ are a, b in $\mathbf{Z}_{>0}$. Setting $q = v^{2a}, y = v^{2b}, \sqrt{qy} = v^{a+b}$ we write below the values of $D_{E,\mathcal{L},v}$ for various E of dimension $1; 2; 2; 1; 1; 1$:

$$1; qy(q+1)(y+1)(qy+\sqrt{qy}+1)/(2(q+\sqrt{qy}+y));$$
$$qy(q+1)(y+1)(qy-\sqrt{qy}+1)/(2(q-\sqrt{qy}+y));$$
$$q^2(q^2y^2+qy+1)/(q^2+qy+y^2); y^2(q^2y^2+qy+1)/(q^2+qy+y^2);$$
$$q^3y^3.$$

The corresponding values of $\mathbf{a}_{E,\mathcal{L}}$ are:

$$0; a+b-m; a+b-m; 2a-2m; 2b-2m, 3a+3b$$

where $m = \min(a, b)$.

1.6. In this subsection we assume that $\mathfrak{W}$ is of type B_3 and the values of $\mathcal{L}|_{\mathfrak{S}}$ are a, a, b in $\mathbf{Z}_{>0}$. Setting $q = v^{2a}, y = v^{2b}$ we write below the values of $D_{E,\mathcal{L},v}$ for various E of dimension $1; 3; 2; 1; 3; 3; 2; 3; 1; 1$:

$$1; qy(q^2+q+1)(qy+1)/(q+y); q^2(q+1)(q^2y+1)/(q+y);$$
$$y^3(qy+1)(q^2y+1)/((q^2+y)(q+y));$$
$$qy^2(q^2+q+1)(q^2y+1)/(q^2+y); q^3y(q^2+q+1)(q^2y+1)/(q^2+y);$$
$$q^2y^3(q+1)(q^2y+1)/(q+y); q^3y^2(q^2+q+1)(qy+1)/(q+y);$$
$$q^6(qy+1)(q^2y+1)/((q^2+y)(q+y)); q^6y^3.$$

The corresponding values of $\mathbf{a}_{E,\mathcal{L}}$ are:

$$0; a+b-m; 2a-m; 3b-m-m'; a+2b-m'; 3a+b-m'; 2a+3b-m;$$
$$3a+2b-m; 6a-m-m'; 6a+3b$$

where $m = \min(a, b)$, $m' = \min(2a, b)$.

1.7. In [**Lus82**] a definition of a partition of $\mathrm{Irr}(\mathfrak{W})$ into subsets called families was given. Repeating that definition but using $\mathbf{a}_{\mathcal{L}}(E)$ instead of $\mathbf{a}(E)$ for $E \in \mathrm{Irr}(\mathfrak{W})$ we obtain a partition of $\mathrm{Irr}(\mathfrak{W})$ into subsets called $\mathcal{L}$-families. (This definition appears in [**Lus83**, no.7].) Thus the families of [**Lus82**] are the same as the $||$-families.

2. Sharp Weyl groups

2.1. Let $\mathfrak{W}, \mathfrak{S}, ||$ be as in 1.1. We assume that $\mathfrak{W}$ is a Weyl group. Let $A_{\mathfrak{W}}$ be the group of all automorphisms γ of $\mathfrak{W}$ preserving $\mathfrak{S}$ and such that whenever $\sigma \neq \sigma'$ are in the same γ-orbit in $\mathfrak{S}$, the product $\sigma\sigma'$ has order 2 or 3. For $\gamma \in A_{\mathfrak{W}}$ let $r(\gamma)$ be the number of γ-orbits on S; let $ord(\gamma)$ be the order of γ. We shall also write ${}^{\gamma}\mathfrak{W}$ instead of $(\mathfrak{W}, \gamma)$. (We sometimes write ${}^{d}\mathfrak{W}$ instead of ${}^{\gamma}W$ where $d = ord(\gamma)$; when $d = 1$ we write $\mathfrak{W}$ instead of ${}^{1}\mathfrak{W}$.)

Let $op \in A_{\mathfrak{W}}$ be given by conjugation by the longest element of $\mathfrak{W}$.

For any $E \in \mathrm{Irr}(\mathfrak{W})$ and $v \in \mathbf{C}^*$ a non-root of 1, $D_{E,||,v} \in \mathbf{C}$ is defined as in 1.2 with $\mathcal{L} = ||$. It is known that $D_{E,||,v}$ is a polynomial in v^2 with rational coefficients. Let $z(E)$ be the largest integer ≥ 0 such that $D_{E,||,v}/(v^2+1)^{z(E)}$ is a polynomial in v^2. From the known formulas for $D_{E,||,v}$ one can see that $z(E) \leq r(op)$.

Let $\gamma \in A_{\mathfrak{W}}$. Assuming that $\mathfrak{W}$ is $\{1\}$ or irreducible, we say that ${}^\gamma\mathfrak{W}$ is *sharp* if (a),(b),(c) below hold:

(a) there exists $E_0 \in \mathrm{Irr}(\mathfrak{W})$ such that $z(E_0) = r(op)$;
(b) $r(op)ord(op)$ is even;
(c) $ord(op\gamma)$ is odd.

Condition (b) is imposed to rule out a $\mathfrak{W}$ of type E_7. Condition (c) means that we have either $\gamma = op$ or $ord(\gamma) = 3$ (hence $\mathfrak{W}$ is of type D_4).

Note that the family of $\mathfrak{W}$ containing E_0 in (a) is necessarily unique (when such E_0 exists). But E_0 in (a) may not be unique: in type E_8 there are three such E_0; one of them is special.

For ${}^\gamma\mathfrak{W}$ sharp we set $\mathbf{a}[{}^\gamma\mathfrak{W}] = \mathbf{a}(E_0)$ where E_0 is as in (a).

2.2. Here is a complete list of the various sharp ${}^\gamma\mathfrak{W}$ and the corresponding $\mathbf{a}[{}^\gamma\mathfrak{W}]$:

(i) $\{1\}$, $\mathbf{a}[{}^\gamma\mathfrak{W}] = 0$:
(ii) ${}^2A_{(t^2-1)/8-1}$, $t \in \{5, 7, 9, \dots\}$, $\mathbf{a}[{}^\gamma\mathfrak{W}] = (t-3)(t-1)(t+1)/48$;
(iii) $B_{(t^2-1)/4}$, $t \in \{3, 5, 7, \dots\}$, $\mathbf{a}[{}^\gamma\mathfrak{W}] = (t-1)(t+1)(2t-3)/24$;
(iv) $D_{t^2/4}$, $t \in \{4, 8, 12, \dots\}$, $\mathbf{a}[{}^\gamma\mathfrak{W}] = (t-2)t(2t+1)/24$;
(v) ${}^2D_{t^2/4}$, $t \in \{6, 10, 14,$, $\mathbf{a}[{}^\gamma\mathfrak{W}] = (t-2)t(2t+1)/24$;
(vi) G_2, ${}^3D_4$, F_4, ${}^2E_6$, E_8, $\mathbf{a}[{}^\gamma\mathfrak{W}] = 1, 3, 4, 7, 16$ respectively.

For a general $\mathfrak{W}, \gamma$ we say that ${}^\gamma\mathfrak{W}$ is irreducible if $\mathfrak{W}$ is a product of $k \geq 1$ irreducible Weyl groups $\mathfrak{W}_1, \dots, \mathfrak{W}_k$ and γ permutes $\mathfrak{W}_1, \dots, \mathfrak{W}_k$ cyclically. In this case we say that ${}^\gamma\mathfrak{W}$ is sharp if ${}^{\gamma^k}\mathfrak{W}_1$ is sharp and we set $\mathbf{a}[{}^\gamma\mathfrak{W}] = k\mathbf{a}[{}^{\gamma^k}\mathfrak{W}_1]$.

For a general $\mathfrak{W}, \gamma$, we have that $\mathfrak{W}$ is a product $\mathfrak{W}'_1 \times \mathfrak{W}'_l$ of Weyl groups such that each $\mathfrak{W}'_j$ is γ-stable and ${}^\gamma\mathfrak{W}'_j$ is irreducible. We say that ${}^\gamma\mathfrak{W}$ is sharp if each ${}^\gamma\mathfrak{W}'_j$ is sharp; we set $\mathbf{a}[{}^\gamma\mathfrak{W}] = \sum_j \mathbf{a}[{}^\gamma\mathfrak{W}_j]$.

The objects in (i),(iii)-(v) can be viewed as vertices of a graph:

$$\{1\} --- B_{(3^2-1)/4} --- D_{4^2/4} --- B_{(5^2-1)/4} ---$$
$$--- {}^2D_{6^2/4} --- B_{(7^2-1)/4} --- D_{8^2/4} --- B_{(9^2-1)/4} ---$$
$$--- {}^2D_{10^2/4} --- B_{(11^2-1)/4} --- \dots. \tag{a}$$

We will attach to each vertex of this graph an index: the index of each of $B_{(t^2-1)/4}$, $D_{t^2/4}$, ${}^2D_{t^2/4}$ is t; the index of $\{1\}$ is 2.

From the objects in (i)-(v) we can form a second graph:

$$\{1\} \times \{1\} --- {}^2A_{(5^2-1)/8-1} --- {}^\gamma(B_{(6/2)^2-1)/4} \times B_{((6/2)^2-1)/4}) ---$$
$$--- {}^2A_{(7^2-1)/8-1} --- {}^\gamma(D_{(8/2)^2/4} \times D_{(8/2)^2/4}) ---$$
$$--- {}^2A_{(9^2-1)/8-1} --- {}^\gamma(B_{((10/2)^2-1)/4} \times B_{((10/2)^2-1)/4}) ---$$
$$--- {}^2A_{(11^2-1)/8-1} --- {}^\gamma(D_{(12/2)^2/4} \times D_{(12/2)^2/4}) ---$$
$$--- {}^2A_{(13^2-1)/8-1} --- \dots. \tag{b}$$

Here γ acts on $B_{((t/2)^2-1)/4} \times B_{((t/2)^2-1)/4}$ as an involution exchanging the two factors; it acts on $D_{(t/2)^2/4} \times D_{(t/2)^2/4}$ by permuting the two factors in such a way that $\gamma^2 = 1$ if $t \in \{8, 16, 24,$ and $\gamma^2 \neq 1$ if $t \in \{12, 20, 28,$. We will attach to each vertex of this graph an index: the index of each of

$${}^\gamma(D_{(t/2)^2/4} \times D_{(t/2)^2/4}), {}^2A_{(t^2-1)/8-1}, {}^\gamma(B_{(t/2)^2-1)/4} \times B_{((t/2)^2-1)/4})$$

is t; the index of $\{1\} \times \{1\}$ is 4.

3. Affine Weyl groups and the sets $\mathcal{C}_\omega(W)$

3.1. In this section W denotes an (irreducible) affine Weyl group. Let $\mathcal{T}$ be the set of all $w \in W$ such that the conjugacy class of w is finite. (Such w are said to be the translations of W.) Now $\mathcal{T}$ is a free abelian group of finite rank and of finite index in W. Let $w \mapsto |w|$ be the usual length function of W. Let S be the set of simple reflections of W. Let $S^!$ be the set of all $\sigma \in S$ such that the sum of labels of edges of the Coxeter graph of W which touch σ is ≥ 3. We have $\sharp(S^!) \leq 2$.

Let Ω_W be the (finite abelian) group of automorphisms of W preserving S whose restriction to $\mathcal{T}$ is given by conjugation by an element of W. If $S^! \neq \emptyset$ let Ω'_W be the set of all $\omega \in \Omega_W$ such that ω restricted to $S^!$ is the identity map (this is a subgroup of Ω_W); if $S^! = \emptyset$ we set $\Omega'_W = \Omega_W$. We set $\Omega''_W = \Omega_W - \Omega'_W$.

Let $\bar{W} = W/\mathcal{T}$ (a finite group). We show:

(a) *If $\omega \in \Omega_W$ then $\omega : W \to W$ induces an inner automorphism of $\bar{W}$.*

We can find $w \in W$ such that $\omega(\tau) = \mathrm{Ad}(w)(\tau)$ for all $\tau \in \mathcal{T}$. Let $\zeta = \mathrm{Ad}(w^{-1})\omega : W \to W$. We have $\zeta(\tau) = \tau$ for any $\tau \in \mathcal{T}$. Let $y \in W, \tau \in \mathcal{T}$. We have $y\tau y^{-1} \in \mathcal{T}$ hence $\zeta(y\tau y^{-1}) = y\tau y^{-1}$ that is $\zeta(y)\tau\zeta(y^{-1}) = y\tau y^{-1}$. Setting $y' = y^{-1}\zeta(y) \in W$ we have $y'\tau y'^{-1} = \tau$ for any $\tau \in W$. Now the action of $W/\mathcal{T}$ on $\mathcal{T}$ by conjugation is faithful hence $y' \in \mathcal{T}$. Thus $\zeta(y) \in y\mathcal{T}$ so that $w^{-1}\omega(y)w \in y\mathcal{T}$ and $\omega(y) \in wyw^{-1}\mathcal{T}$. This proves (a).

3.2. For any $J \subsetneqq S$ let W_J be the subgroup of W generated by J (a finite Weyl group).

Let S_* be the set of all $\sigma \in S$ such that $W_{S-\{\sigma\}} \to \bar{W}$ (restriction of the obvious map $W \to \bar{W}$) is an isomorphism. We have $S_* \neq \emptyset$ and the obvious action of Ω_W on S_* is simply transitive. If $J \subsetneqq S$ let $W_J \to \bar{W}$ be the restriction of the obvious homomorphism $W \to \bar{W}$; this is an imbedding, so that W_J can be viewed as a subgroup of $\bar{W}$. For any special representation $E \in \mathrm{Irr}(W_J)$ there is a unique $E' \in \mathrm{Irr}(\bar{W})$ such that E' appears in $\mathrm{Ind}_{W_J}^{\bar{W}}(E)$ and in the $\mathbf{a}(E)$-th symmetric power of the conjugation representation of $\bar{W}$ on $\mathbf{C} \otimes \mathcal{T}$ (with $\mathbf{a}(E)$ defined in terms of W_J); we set $E' = j_{W_J}^{\bar{W}}(E)$, see [**Lus09**, 1.3]. Let $\bar{S}(\bar{W})$ be the subset of $\mathrm{Irr}(\bar{W})$ consisting of representations of the form $j_{W_J}^{\bar{W}}(E)$ for some $J \subsetneqq S$ and some special $E \in \mathrm{Irr}(W_J)$.

3.3. Let $\omega \in \Omega_W$. We define a set $\mathcal{C}_\omega(W)$ of Weyl subgroups W_J of W with $J \subsetneqq S$. This set contains $\{1\}$. Now $\mathcal{C}_\omega(W) - \{1\}$ consists of the subgroups W_J with $J \subsetneqq S$, $J \neq \emptyset$ which satisfy the following requirements.

(i) W_J is ω'-stable for any $\omega' \in \Omega_W$.

(ii) ${}^\omega W_J$ is sharp.

(iii) ${}^\omega W_{S-J}$ is irreducible.

(iv) If $\sharp(S^!) = 2$ and $\Omega''_W = \emptyset$, then W_J is the product of two vertices of the graph 2.2(a) which are joined by an edge.

(v) if $\omega \in \Omega''_W$, then W_J is the product of two vertices of the graph 2.2(b) which are joined by an edge.

3.4. We now describe the set $\mathcal{C}_\omega(W)$ in each case. If W is of affine type A_{n-1}, $n \geq 2$, then Ω_W is cyclic of order n; for any $\omega \in \Omega_W$, $\mathcal{C}_\omega(W)$ consists of a single element: $\{1\}$ (with $\mathbf{a}[\{1\}] = 0$).

3.5. If W is of affine type E_6 then Ω_W is cyclic of order 3. Let $\omega \in \Omega_W$. If $\omega = 1$ then $\mathcal{C}_\omega(W)$ consists of a single element: $\{1\}$ (with $\mathbf{a}[\{1\}] = 0$). If $\omega \neq 1$ then $\mathcal{C}_\omega(W)$ consists of $\{1\}$ (with $\mathbf{a}[\{1\}] = 0$) and of the subgroup W_J of type D_4 (so that ${}^\omega W_J = {}^3D_4$ and $\mathbf{a}[{}^\omega W_J] = 3$).

3.6. If W is of affine type E_7 then Ω_W is cyclic of order 2. Let $\omega \in \Omega_W$. If $\omega = 1$ then $\mathcal{C}_\omega(W)$ consists of a single element: $\{1\}$ (with $\mathbf{a}[\{1\}] = 0$). If $\omega \neq 1$ then $\mathcal{C}_\omega(W)$ consists of $\{1\}$ (with $\mathbf{a}[\{1\}] = 0$) and of the subgroup W_J of type E_6 (so that ${}^\omega W_J = {}^2E_6$ and $\mathbf{a}[{}^\omega W_J] = 7$).

3.7. If W is of affine type E_8, F_4 or G_2 and $\omega \in \Omega_W$ then $\omega = 1$ and $\mathcal{C}_\omega(W)$ consists of $\{1\}$ (with $\mathbf{a}[\{1\}] = 0$) and of the subgroup W_J of non-affine type E_8, F_4 or G_2 (respectively), with $\mathbf{a}[W_J]$ equal to $16, 4, 1$ respectively.

3.8. In the remainder of this section we assume that W is of affine type $B_n (n \geq 3), C_n (n \geq 2)$ or $D_n (n \geq 4)$. Let $\omega \in \Omega_W$. Assume first that $\omega \in \Omega'_W$. Let ${}'\mathcal{C}_\omega(W)$ be the set of all pairs $(t, s) \in \mathbf{N}^2$ such that

$t - s = \pm 1$ (type B), $t = s$ (type C, D),

$t = 0 \mod 4$ (type B, D with $\omega = 1$), $t = 2 \mod 4$ (type B, D with $\omega \neq 1$) $t = 1 \mod 2$ (type C),

and for some $r \in \mathbf{N}$ we have

$t^2/4 + (s^2 - 1)/4 + r = n$, (type B),

$(t^2 - 1)/4 + (s^2 - 1)/4 + r = n$, (type C),

$t^2/4 + s^2/4 + r = n$, (type D),

that is, $ts + 2r = 2n$ (type B, D), $ts + 2r = 2n + 1$ (type C).

In type B we define a bijection $\mathcal{C}_\omega(W) \xrightarrow{\sim} {}'\mathcal{C}_\omega(W)$ by associating to $W_J \in \mathcal{C}_\omega(W)$ (assumed to be $\neq \{1\}$) the pair (t, s) formed by the indexes t, s of the two vertices attached to W_J in 3.3(iv) and by associating to $W_J = \{1\}$ the pair $(0, 1)$ (if $\omega = 1$) or $(2, 1)$ (if $\omega \neq 1$).

In type C, D, we define a bijection $\mathcal{C}_\omega(W) \xrightarrow{\sim} {}'\mathcal{C}_\omega(W)$ by associating to $W_J \in \mathcal{C}_\omega(W)$ (assumed to be $\neq \{1\}$) the pair (t, s) (with $t = s$) where W_J is the product of a vertex of index t in 2.2(a) with itself and by associating to $W_J = \{1\}$ the pair $(1, 1)$ (type C), $(0, 0)$ (type D with $\omega = 1$), $(2, 2)$ (type D with $\omega \neq 1$).

Assuming that W_J corresponds as above to (t, s) we can compute $\mathbf{a}[{}^\omega W_J]$ in each case.

If $s = t \in \{1, 3, 5,$ then

$$\mathbf{a}[{}^\omega W_J] = (t-1)(t+1)(2t-3)/24 + (t-1)(t+1)(2t-3)/24$$
$$= (t-1)(t+1)(2t-3)/12.$$

If $s = t \in \{0, 2, 4,$ then

$$\mathbf{a}[{}^\omega W_J] = (t-2)t(2t+1)/24 + (t-2)t(2t+1)/24 = (t-2)t(2t+1)/12.$$

If $t \in \{0, 2, 4, , s \in \{1, 3, 5, , t - s = \pm 1$ then

$$\mathbf{a}[{}^\omega W_J] = (t-2)t(2t+1)/24 + (s-1)s(2s+1)/24$$

and this equals $(t-1)t(t+1)/6$ if $s = t + 1$ and $(t-2)(t-1)t/6$ if $t = s + 1$.

3.9. Assume next that $\omega \in \Omega''_W$. Let ${}'\mathcal{C}_\omega(W)$ be the set of all pairs $(t,s) \in \mathbf{N}^2$ such that

$t - s = \pm 1$,
$t = 2 \mod 4$ (type C),
$t = 0 \mod 8$ (type D with $\omega^2 = 1$), $t = 4 \mod 8$ (type D with $\omega^2 \neq 1$),
and for some $r \in \mathbf{N}$ we have
$((t/2)^2 - 1)/2 + (s^2 - 1)/8 + 2r = n$ (type C),
$(t/2)^2/2 + (s^2 - 1)/8 + 2r = n$ (type D),
or equivalently $ts/2 + 4r = 2n + 1$ (type C), $ts/2 + 4r = 2n$ (type D).

Note that $(2,1) \in {}'\mathcal{C}_\omega(W)$ in type C with n even and $(2,3) \in {}'\mathcal{C}_\omega(W)$ in type C with n odd. We define a bijection $\mathcal{C}_\omega(W) \xrightarrow{\sim} {}'\mathcal{C}_\omega(W)$ by associating to $W_J \in \mathcal{C}_\omega(W)$ (assumed to be $\neq \{1\}$) the pair (t,s) formed by the indexes of the two vertices attached to W_J in 3.3(v) and by associating to $W_J = \{1\}$ the pair $(2,1)$ (type C with n even), the pair $(2,3)$ (type C with n odd), the pair $(0,1)$ (type D with $\omega^2 = 1$), the pair $(4,3)$ (type D with $\omega^2 \neq 1$).

Assuming that W_J corresponds as above to (t,s) we can compute $\mathbf{a}[{}^\omega W_J]$ in each case.

If $t \in \{2, 6, 10,$ then

$$\mathbf{a}[{}^\omega W_J] = (s-3)(s-1)(s+1)/48 + (t/2-1)(t/2+1)(t-3)/24+ \\ (t/2-1)(t/2+1)(t-3)/24$$

which equals $(t-2)(2t^2-5t-6)/48$ if $s = t-1$ and equals $(t-2)(t+2)(2t-3)/48$ if $s = t+1$.

If $t \in \{0, 4, 8, 10,$ then

$$\mathbf{a}[{}^\omega W_J] = (s-3)(s-1)(s+1)/48 + (t/2-2)(t/2)(t+1)/24 + (t/2-2)(t/2)(t+1)/24$$

which equals $(t-4)t(2t-1)/48$ if $s = t-1$ and equals $t(2t^2-3t-8)/48$ if $s = t+1$.

3.10. Let $\omega \in \Omega'_W$. Let ${}'\underline{\mathcal{C}}_\omega(W)$ be the set of all pairs $(\delta, r) \in \mathbf{N}^2$ such that

(type B) $\delta + 2r = 2n$, $\delta = 2 + 4 + 6 + \dots + (2\sigma)$, where $\sigma \in \mathbf{N}$ and $\sigma = 0 \mod 4$ or $\sigma = 3 \mod 4$ if $\omega = 1$; $\sigma = 1 \mod 4$ or $\sigma = 2 \mod 4$ if $\omega \neq 1$;

(type C) $\delta + 2r = 2n + 1$, $\delta = 1 + 3 + 5 + \dots + (2\sigma - 1)$ where $\sigma \in \mathbf{N}$ and $\sigma = 1 \mod 2$;

(type D) $\delta + 2r = 2n$, $\delta = 1 + 3 + 5 + \dots + (2\sigma - 1)$, where $\sigma \in \mathbf{N}$ and $\sigma = 0 \mod 4$ if $\omega = 1$, $\sigma = 2 \mod 4$ if $\omega \neq 1$.

We define a bijection

$${}'\mathcal{C}_\omega(W) \xrightarrow{\sim} {}'\underline{\mathcal{C}}_\omega(W)$$

by $(t,s) \mapsto (ts, r)$ where $r \in \mathbf{N}$ is as in 3.8.

3.11. Let $\Delta = \{t(t+1)/2; t \in \mathbf{N}\}$. Let $\omega \in \Omega''_W$. Let ${}'\underline{\mathcal{C}}_\omega(W)$ be the set of all pairs $(\delta, r) \in \Delta \times \mathbf{N}$ such that

$\delta + 4r = 2n + 1$ (type C),
$\delta + 4r = 2n$, $\delta = 0 \mod 4$ (type $D, \omega^2 = 1$),
$\delta + 4r = 2n$, $\delta = 2 \mod 4$ (type $D, \omega^2 \neq 1$).

We define a bijection

$${}'\mathcal{C}_\omega(W) \xrightarrow{\sim} {}'\underline{\mathcal{C}}_\omega(W)$$

by $(t,s) \mapsto (ts/2, r)$ where $r \in \mathbf{N}$ is as in 3.9.

4. The function $c : \mathrm{Irr}(\bar{\mathcal{W}}) \to \mathbf{N}$

4.1. In this section $\mathcal{W}$ denotes an irreducible affine Weyl group with a set $\mathcal{S}$ of simple reflections and with a given weight function $\mathcal{L} : \mathcal{W} \to \mathbf{N}$. Let $||$ be the length function of $\mathcal{W}$. Let $\mathcal{T}_{\mathcal{W}}$ be the group of translations of $\mathcal{W}$ (see 3.1) and let $\bar{\mathcal{W}} = \mathcal{W}/\mathcal{T}_{\mathcal{W}}$ (a finite group). For any $\mathcal{J} \subsetneqq \mathcal{S}$ we denote by $\mathcal{W}_{\mathcal{J}}$ the subgroup of $\mathcal{W}$ generated by $\mathcal{J}$ (a finite Weyl group). Let $\bar{\mathcal{W}}_{\mathcal{J}}$ be the image of $\mathcal{W}_{\mathcal{J}}$ under the obvious map $\mathcal{W} \to \bar{\mathcal{W}}$. Note that the obvious map $\mathcal{W}_{\mathcal{J}} \to \bar{\mathcal{W}}_{\mathcal{J}}$ is an isomorphism; we use this to identify $\mathcal{W}_{\mathcal{J}} = \bar{\mathcal{W}}_{\mathcal{J}}$.

For any $E \in \mathrm{Irr}(\bar{\mathcal{W}})$ we define $\mathfrak{T}(E)$ to be the set of all pairs $(\mathcal{J}, E')$ where $\mathcal{J} \subset \mathcal{S}$, $\sharp(\mathcal{J}) = \sharp(\mathcal{S}) - 1$ and $E' \in \mathrm{Irr}(\mathcal{W}_{\mathcal{J}})$ is such that E' appears in the restriction of E to $\mathcal{W}_j = \bar{\mathcal{W}}_{\mathcal{J}}$. We set

(a) $$c_E = \max_{(\mathcal{J},E') \in \mathfrak{T}(E)} \mathbf{a}_{\mathcal{L}}(E') \in \mathbf{N},$$

$$\mathfrak{T}^*(E) = \{(\mathcal{J}, E') \in \mathfrak{T}(E); \mathbf{a}_{\mathcal{L}}(E') = c_E\},$$

where $\mathbf{a}_{\mathcal{L}}(E')$ is defined as in 1.2 in terms of the Weyl group $\mathcal{W}_{\mathcal{J}}$ with the weight function obtained by restricting $\mathcal{L} : \mathcal{W} \to \mathbf{N}$ to $\mathcal{W}_{\mathcal{J}}$. We have $\mathfrak{T}^*(E) \neq \emptyset$.

4.2. The function $E \mapsto c_E$ has been computed explicitly in [**Lus23**] in the case where $\mathcal{W}$ is of exceptional type and $\mathcal{L} = ||$. We will now describe explicitly the map $\mathrm{Irr}(\bar{\mathcal{W}}) \to \mathbf{N}, E \mapsto c_E$ of 4.1 in several examples with $\mathcal{L} \neq ||$. Assume that either

(a) $\mathcal{W}$ is of affine type $C_r, r \geq 2$ and $\mathcal{L}|_{\mathcal{S}}$ takes the values $t, 1, 1, \ldots, 1, 1, s$ where $t > 0, s > 0$, or
(b) $\mathcal{W}$ is of affine type $C_r, r \geq 2$ and $\mathcal{L}|_{\mathcal{S}}$ takes the values $t, 2, 2, \ldots, 2, 2, s$ where $t > 0, s > 0$, or
(c) $\mathcal{W}$ is of affine type G_2 and $\mathcal{L}|_{\mathcal{S}}$ has values $3, 3, 1$.

In case (a),(b) we set $u = \max(t, s)$. In each of the examples below we give a table with two rows whose columns are indexed by the various $E \in \mathrm{Irr}(\bar{\mathcal{W}})$. The first row represents the numbers c_E. The second row represents the numbers $\mathbf{a}_{\mathcal{L}}(E')$ for various $E' \in \mathrm{Irr}(\mathcal{W}_{\mathcal{J}})$ where $\mathcal{J} = \mathcal{S} - \{s\}$ for some $s \in \mathcal{S}_*$ (see 3.2) which in case (a),(b) is chosen so that $\mathcal{L}(s) = u$. (We can identify $\mathcal{W}_{\mathcal{J}} = \bar{\mathcal{W}}$ hence E' can be identified with an $E \in \bar{\mathcal{W}}_J$.) Any entry e in the first row and column E is $\geq$ than the entry e' in the second row and column E. When $e > e'$ we indicate some other $\mathcal{J}' \subsetneqq \mathcal{S}$ such that some $E' \in \mathrm{Irr}(\mathcal{W}_{\mathcal{J}'})$ appears in the restriction of E to $\mathcal{W}_{\mathcal{J}'}$ and $\mathbf{a}_{\mathcal{L}}(E') = e$. (We will specify such $\mathcal{J}'$ by specifying the type of $\mathcal{W}_{\mathcal{J}'}$.)

Assume first (in case (a)) that $r = 2$ and $t = s = u \geq 1$. The table is

$$0; 1; u; 2u; 2u + 2$$
$$0; 1; t; 2u - 1; 2u + 2$$

with an additional $\mathcal{J}'$ with $\mathcal{W}_{\mathcal{J}'}$ of type $A_1 \times A_1$, contributing $2u$ to the fourth column.

Assume next (in case (a)) that $r = 2$ and $t = s \pm 1, u \geq 2$. The table is

$$0; 1; u; 2u - 1; 2u + 2$$
$$0; 1; u; 2u - 1; 2u + 2$$

In this case there is no need for an additional $\mathcal{J}'$.

Assume next (in case (b)) that $r = 2$ and $t = s \pm 1$, $u \geq 2$. The table is

$$0; u; 2; 2u - 1; 2u + 2$$
$$0; u; 2; 2u - 2; 2u + 2$$

with an additional $\mathcal{J}'$ with $\mathcal{W}_{\mathcal{J}'}$ of type $A_1 \times A_1$, contributing $2u - 1$ to the fourth column.

Assume next (in case (b)) that $r = 3$ and $t = s \pm 1$, $u \geq 2$. The table is

$$0; u; 2; 3u - 3; 2u - 1; u + 2; 3u + 3; 2u + 4; 6; 3u + 12$$
$$0; u; 2; 3u - 6; 2u - 2; u + 2, 3u + 2; 2u + 4; 6; 3u + 12$$

with additional $\mathcal{J}'$ with $\mathcal{W}_{\mathcal{J}'}$ of type $A_1 \times B_2$, contributing $3u - 3$ to the fourth column, $2u - 1$ to the fifth column and $3u + 3$ to the seventh column.

Finally assume that we are in case (c). The table is

$$0; 1; 3; 4; 9; 12$$
$$0; 1; 3; 3; 7; 12$$

with an additional $\mathcal{J}'$ with $\mathcal{W}_{\mathcal{J}'}$ of type A_2 contributing 9 to the fifth column and with an additional $\mathcal{J}'$ with $\mathcal{W}_{\mathcal{J}'}$ of type $A_1 \times A_1$ contributing 4 to the fourth column.

4.3. We set

$$\nu(\mathcal{W}, \mathcal{L}) = \max_{s \in \mathcal{S}_*} \mathcal{L}(w_{0,s})$$

where $w_{0,s}$ is the element of maximal length of $\mathcal{W}_{\mathcal{S}-\{s\}}$.

4.4. Let $\sim$ be the equivalence relation on $\mathrm{Irr}(\bar{\mathcal{W}})$ generated by the relation $E_1 \sim E_2$ when $c_{E_1} = c_{E_2}$ and there exist $(\mathcal{J}_1, E_1') \in \mathfrak{T}^*(E_1)$, $(\mathcal{J}_2, E_2') \in \mathfrak{T}^*(E_2)$ such that $\mathcal{J}_1 = \mathcal{J}_2$ and E_1', E_2' are in the same $\mathcal{L}$-family (see 1.7) of $\mathrm{Irr}(\mathcal{W}_{\mathcal{J}_1}) = \mathrm{Irr}(\mathcal{W}_{\mathcal{J}_2})$.

5. A weighted affine Weyl group

5.1. We preserve the setup of 3.1. For any $J \subsetneq S$ let w_0^J be the longest element of W_J. We now fix $\omega \in \Omega_W$, $J \subsetneq S$ such that $W_J \in \mathcal{C}_\omega(W)$. Following [**Lus91**] (see also [**Lus95**]), to W, ω, J we associate a weighted affine Weyl group $(\mathcal{W}_J, \mathcal{L}_J)$. (A similar procedure was used for finite Weyl groups in [**Lus78**].) The proofs of various statements in this section can be extracted from [**Lus95**]. Let

$$\mathcal{W}_J' = \{w \in W; wW_J = W_J w, w \text{ has minimal length in } wW_J = W_J w\};$$

this is a subgroup of W stable under ω; let $\mathcal{W}_J = \{w \in \mathcal{W}_J'; \omega(w) = w\}$ (another subgroup of W). Let $\mathcal{S}_J$ be the set of all ω-orbits on $S - J$.

When $\sharp(\mathcal{S}_J) \geq 2$ let $\mathcal{T}_J$ be the group of translations of the affine Weyl group $\mathcal{W}_J$ and let $\bar{\mathcal{W}}_J = \mathcal{W}_J/\mathcal{T}_J$, a finite group. When $\sharp(\mathcal{S}_J) = 1$ we have $\mathcal{W}_J = \{1\}$ and we set $\mathcal{T}_J = \{1\}, \bar{\mathcal{W}}_J = \{1\}$; in this case we regard $\mathcal{W}_J$ as a weighted affine Weyl group with weight function $\mathcal{L}_J = 0$.

We now assume that $\sharp(\mathcal{S}_J) \geq 2$. If $\theta \in \mathcal{S}_J$ we set $\tau_\theta = w_0^{J \cup \theta} w_0^J = w_0^J w_0^{J \cup \theta}$. (The last equality is a property of any $W_J \in \mathcal{C}_\omega(W)$.) We have $\tau_\theta \in \mathcal{W}_J, \tau_\theta^2 = 1$. Moreover $\mathcal{W}_J$ is a Coxeter group on the generators $\{\tau_\theta; \theta \in \mathcal{S}_J\}$ and with Coxeter relations $(t_\theta t_{\theta'})^{m_{\theta,\theta'}} = 1$ for any distinct $\theta \in \mathcal{S}_J, \theta' \in \mathcal{S}_J$ such that $J \cup \theta \cup \theta' \neq S$, where

$$m_{\theta,\theta'} = \frac{2(|w_0^{J \cup \theta \cup \theta'}| - |w_0^J|)}{|w_0^{J \cup \theta}| + |w_0^{J \cup \theta'}| - 2|w_0^J|}.$$

(When $J \cup \theta \cup \theta' = S$ then $\tau_\theta \tau_{\theta'}$ has infinite order.) Note that $\mathcal{W}_J$ is an (irreducible) affine Weyl group.

5.2. We define a weight function $\mathcal{L}_J : \mathcal{W}_J \to \mathbf{N}$. Let $\theta \in \mathcal{S}_J$. Let E_0 be the unique irreducible special representation of W_J such that $z(E_0) = r(op)$ (notation of 2.1 with $\mathfrak{W}$ replaced by W_J). Let

$$\mathcal{E} = \{\tilde{E} \in \mathrm{Irr}(W_{J\cup\theta}; \tilde{E} \text{ appears in } \mathrm{Ind}_{W_J}^{W_{J\cup\theta}}(E_0)\},$$

$$\mathcal{L}(\theta) = \max\{\mathbf{a}(\tilde{E}); \tilde{E} \in \mathcal{E}\} - \min\{\mathbf{a}(\tilde{E}); \tilde{E} \in \mathcal{E}\}.$$

(Here $\mathbf{a}(\tilde{E})$ is as in 1.2 with $\mathfrak{W}$ replaced by $W_{J\cup\theta}$.) This defines the function $\mathcal{L}_J : \mathcal{S}_J \to \mathbf{Z}_{>0}$. This extends uniquely to a weight function $\mathcal{W}_J \to \mathbf{N}$ denoted again by $\mathcal{L}_J$ so that $(\mathcal{W}_J, \mathcal{L}_J)$ is a weighted affine Weyl group.

5.3. Let $\omega \in \Omega_W$, $J \subsetneq S$ be such that $W_J \in \mathcal{C}_\omega(W)$. We now describe the pair $(\mathcal{W}, \mathcal{L}) = (\mathcal{W}_J, \mathcal{L}_J)$ in each case. We will write $\mathcal{S}$ instead of $\mathcal{S}_J$.

If W is of affine type $A_{n-1}, n \geq 2$, let $k = ord(\omega)$ (a divisor of n). We have $W_J = \{1\}$. If $k < n$, $\mathcal{W}$ is of affine type $A_{(n/k)-1}$ with $\mathcal{L}|_\mathcal{S}$ being constant equal to k. If $k = n$, we have $\mathcal{W} = \{1\}$.

Assume that W is of affine type E_6. If $\omega = 1$ then $W_J = \{1\}$ and $\mathcal{W} = W$ with $\mathcal{L}|_\mathcal{S}$ being constant equal to 1. If $\omega \neq 1$ and $W_J = \{1\}$ then $\mathcal{W}$ is of affine type G_2 with the values of $\mathcal{L}|_\mathcal{S}$ being $3, 3, 1$. If $\omega \neq 1$ and W_J is of type D_4 then $\mathcal{W} = \{1\}$.

Assume that W is of affine type E_7. If $\omega = 1$ then $W_J = \{1\}$ and $\mathcal{W} = W$ with $\mathcal{L}|_\mathcal{S}$ being constant equal to 1. If $\omega \neq 1$ and $W_J = \{1\}$ then $\mathcal{W}$ is of affine type F_4 with the values of $\mathcal{L}|_\mathcal{S}$ being $2, 2, 2, 1, 1$. If $\omega \neq 1$ and W_J is of type E_6 then $\mathcal{W} = \{1\}$.

Assume that W is of affine type E_8, F_4 or G_2. We have $\omega = 1$. If $W_J = \{1\}$ then $\mathcal{W} = W$ with $\mathcal{L}|_\mathcal{S}$ being constant equal to 1. If W_J is of the non-affine type E_8, F_4 or G_2 (respectively) then $\mathcal{W} = \{1\}$.

5.4. We now assume that W is of affine type $B_n (n \geq 3), C_n (n \geq 2)$ or $D_n (n \geq 4)$. Let $(\delta, r) \in {}'\underline{\mathcal{C}}_\omega(W)$ be the image of W_J under the composition $\mathcal{C}_\omega(W) \xrightarrow{\sim} {}'\mathcal{C}_\omega(W) \xrightarrow{\sim} {}'\underline{\mathcal{C}}_\omega(W)$ (the first map as in 3.8, 3.9, the second map as in 3.10, 3.11).

If $r = 0$ then $\mathcal{W} = \{1\}$. If $r > 0, \delta > 0$, then $\mathcal{W}$ is of affine type C_r, with Coxeter graph

(a) $\boxed{t} == \boxed{1} - - \boxed{1} - - \cdots - - \boxed{1} - - \boxed{1} == \boxed{s}$

(b) $\boxed{t} == \boxed{2} - - \boxed{2} - - \cdots - - \boxed{2} - - \boxed{2} == \boxed{s}$

where the boxed entries are the values of $\mathcal{L}|_\mathcal{S}$ and t, s in $\mathbf{N}$ are defined by

$ts = \delta$, $t - s \in \{0, 1, -1\}$ (in (a) with $\omega \in \Omega'_W$),

$ts/2 = \delta$, $t - s \in \{1, -1\}$ (in (b) with $\omega \in \Omega''_W$).

(Here affine of type C_1 is taken to be the same as affine of type A_1.)

If $\delta = 0$ (hence $r > 0$), then either:

W is of type B_n or D_n, $\omega = 1$ and $\mathcal{W} = W$ with $\mathcal{L}|_\mathcal{S}$ constant equal to 1, or

W is of type D_n, $n = 2r \geq 6$, $\omega \in \Omega''_W$, $\omega^2 = 1$ and $\mathcal{W}$ is of affine type B_r, with the values of $\mathcal{L}|_\mathcal{S}$ being $1, 2, 2, \ldots, 2$.

5.5. We now list the various $(\mathcal{W}, \mathcal{L})$ which are associated to various $W, \omega, W_J \in \mathcal{C}_\omega(W)$.

$\mathcal{W} = \{1\}$ or $\mathcal{W}$ is an irreducible affine Weyl groups of type E_6, E_7, E_8 or $D_m, m \geq 4$ with $\mathcal{L}|_\mathcal{S}$ constant equal to 1;

$\mathcal{W}$ of affine type $A_{n-1}, n \geq 3$ with $\mathcal{L}|_\mathcal{S}$ constant in $\mathbf{Z}_{>0}$;

$\mathcal{W}$ of affine type $B_m, m \geq 3$ with the values of $\mathcal{L}|_\mathcal{S}$ being $1, 1, \ldots, 1, 1$ or $1, 2, 2, \ldots 2$;

$\mathcal{W}$ of affine type $C_m, m \geq 1$ with the values of $\mathcal{L}|_\mathcal{S}$ being $t, 1, 1, \ldots, 1, 1, s$ with $(t, s) \in \mathbf{Z}_{>0} \times \mathbf{Z}_{>0}$ such that $t - s \in \{0, 1, -1\}$;

$\mathcal{W}$ of affine type $C_m, m \geq 2$ with the values of $\mathcal{L}|_\mathcal{S}$ being $t, 2, 2, \ldots, 2, 2, s$ with $(t, s) \in \mathbf{Z}_{>0} \times \mathbf{Z}_{>0}$ such that $t - s \in \{1, -1\}$;

$\mathcal{W}$ of affine type G_2 with the values of $\mathcal{L}|_\mathcal{S}$ being $1, 1, 1$ or $3, 3, 1$;

$\mathcal{W}$ of affine type F_4 with the values of $\mathcal{L}|_\mathcal{S}$ being $1, 1, 1, 1, 1$ or $2, 2, 2, 1, 1$.

(Here affine of type C_1 is taken to be the same as affine of type A_1.)

5.6. Let $\omega \in \Omega_W$, $J \subsetneq S$ be such that $W_J \in \mathcal{C}_\omega(W)$. Let $E \mapsto c_E$ be as in 4.1 with $\mathcal{W} = \mathcal{W}_J$. For $E, \tilde{E}$ in $\mathrm{Irr}(\bar{\mathcal{W}}_J)$ we write $E \leq \tilde{E}$ if $E = \tilde{E}$ or $c_E > c_{\tilde{E}}$. This is a partial order on $\mathrm{Irr}(\bar{\mathcal{W}}_J)$. For $E, \tilde{E}$ in $\mathrm{Irr}(\bar{\mathcal{W}}_J)$ we write $E \approx \tilde{E}$ if $c_E = c_{\tilde{E}}$. This is an equivalence relation on $\mathrm{Irr}(\bar{\mathcal{W}}_J)$.

Let $\mathbf{q}$ be an indeterminate. Let $\underline{\mathcal{T}}_J = \mathbf{Q}(\mathbf{q}) \otimes \mathcal{T}_J$. Now the obvious $\mathcal{W}_J$ action on $\mathcal{T}_J$ induces a linear action of $\mathcal{W}_J$ on $\underline{\mathcal{T}}_J$.

For $E, \tilde{E}$ in $\mathrm{Irr}(\bar{\mathcal{W}}_J)$ we define

$$\Omega'_{E,\tilde{E}} = \sharp(\bar{\mathcal{W}})^{-1} \sum_{w \in \bar{\mathcal{W}}_J} \mathrm{tr}(w^{-1}, E)\, \mathrm{tr}(w, \tilde{E}) \det(\mathbf{q} - w, \underline{\mathcal{T}}_J)^{-1} \mathbf{q}^{-c_E - c_{\tilde{E}}} \in \mathbf{Q}(\mathbf{q}).$$

The following result can be deduced from Lemma 2.1 in [**GM99**] (where it is attributed to the author).

PROPOSITION 5.7. *The system of equations*

$$\sum_{E', \tilde{E}' \text{ in } \bar{\mathcal{W}}_J; E' \leq E, \tilde{E}' \leq \tilde{E}, E' \sim \tilde{E}'} P_{E',E} \Lambda'_{E',\tilde{E}'} P_{\tilde{E}',\tilde{E}} = \Omega'_{E,\tilde{E}},$$

$$P_{E,E} = 1 \text{ for all } E,$$
$$P_{E',E} = 0 \text{ if } E' \approx E, E \neq E',$$
$$P_{E',E} = 0 \text{ if } E' \not\leq E,$$
$$\Lambda'_{E',E} = 0 \text{ unless } E' \approx E$$

with unknowns

$$P_{E',E} \in \mathbf{Q}(\mathbf{q}), \Lambda'_{E',E} \in \mathbf{Q}(\mathbf{q}), (E', E \text{ in } \mathrm{Irr}(\bar{\mathcal{W}}_J))$$

has a unique solution.

6. The set $\mathcal{G}_\omega(G)$ and the bijection $\mathcal{G}_\omega(G) \xrightarrow{\sim} \mathcal{C}_\omega(W)$

6.1. Let $le(G)$ be the set of subgroups of G (see 0.1) which are Levi subgroups of some parabolic subgroup of G. Let $ls^0(G)$ be the subset of $ls(G)$ (see 0.1) consisting of the pairs $(\mathfrak{c}, \mathfrak{L})$ each of which is a unipotent block by itself. Let $\underline{ls}(G)$ be the set of unipotent blocks of G. Note that $ls^0(G)$ can be identified with a subset $\underline{ls}^0(G)$ of $\underline{ls}(G)$. Let Z_G be the group of components of the centre of G and let $Z_G^* = \mathrm{Hom}(Z_G, \mathbf{C}^*)$. We have a partition $ls(G) = \sqcup_{\chi \in Z_G^*} ls_\chi(G)$ where $ls_\chi(G)$

consists of all $(\mathfrak{c}, \mathfrak{L}) \in ls(G)$ such that the natural action of Z_G on $\mathfrak{L}$ is through χ. For any $\chi \in Z_G^*$, $ls_\chi(C)$ is a union of unipotent blocks. Hence we have a partition

$$\underline{ls}(G) = \sqcup_{\chi \in Z_G^*} \underline{ls}_\chi(G)$$

where $\underline{ls}_\chi(G)$ is the set of unipotent blocks contained in $ls_\chi(G)$. In [**Lus84a**] it is shown that for any $\chi \in Z_G^*$

(a) the intersection $ls^0(G) \cap ls_\chi(G)$ consists of at most one element.
Let $\mathcal{G}'(G)$ be the set consisting of G-conjugacy classes of triples $(L, \mathfrak{c}_1, \mathfrak{L}_1)$ with $L \in le(G)$, $(\mathfrak{c}_1, \mathfrak{L}_1) \in ls^0(L)$. In [**Lus84a**] a bijection

(b) $\underline{ls}(G) \leftrightarrow \mathcal{G}'(G)$
is established. To $(L, \mathfrak{c}_1, \mathfrak{L}_1) \in \mathcal{G}'(G)$ we have associated in [**Lus84a**, 4.4] a perverse sheaf $\phi_! K$ (up to shift) on G whose cohomology sheaves restricted to unipotent classes are direct sums of local systems in a fixed unipotent block. (This defines the unipotent blocks and the bijection (b).) If $\chi \in Z_G^*$ and $\beta \in \underline{ls}_\chi(G)$, then the corresponding $(L, \mathfrak{c}_1, \mathfrak{L}_1) \in \mathcal{G}'(G)$ is such that $(\mathfrak{c}_1, \mathfrak{L}_1) \in ls_{\chi_1}(L)$ for some $\chi_1 \in Z_L^*$ which is uniquely determined by χ. (Note that χ is the image of χ_1 under the injective homomorphism $Z_L^* \to Z_G^*$ induced by the obvious (surjective) homomorphism $Z_G \to Z_L$.) Using (a) for L we see that $(\mathfrak{c}_1, \mathfrak{L}_1)$ is unique if it exists. Thus (b) gives rise to a bijection

(c) $\underline{ls}_\chi(G) \leftrightarrow \mathcal{G}_\chi(G)$
where $\mathcal{G}_\chi(G)$ is the set of all $L \in le(G)$ (up to G-conjugacy) such that there exists $(\mathfrak{c}_1, \mathfrak{L}_1) \in ls^0(L) \cap ls_{\chi_1}(L)$ where $\chi_1 \in Z_L^*$ maps to χ under $Z_L^* \to Z_G^*$. We set $\mathcal{G}(G) = \sqcup_{\chi \in Z_G^*} \mathcal{G}_\chi(G)$. Then (c) gives rise to a bijection

(d) $\underline{ls}(G) \leftrightarrow \mathcal{G}(G)$.
In [**Lus84a**] it is shown that the set of objects in a fixed unipotent block β is in bijection with the set of irreducible representations of the normalizer $\mathfrak{W}_\beta$ of L modulo L (with L corresponding to β) and that $\mathfrak{W}_\beta$ is naturally a Weyl group.

6.2. In the remainder of this section we assume that G in 0.1 is almost simple, simply connected and that W, S in 3.1 is the affine Weyl group associated with a simple algebraic group H over $\mathbf{C}$ of type dual to that of G. In particular, $\mathcal{T}$ (see 3.1) can be identified with the group of one parameter subgroups of a maximal torus T of H. For $h \in T$, the connected centralizer of h in H has Weyl group equal to a W-conjugate of W_J for some $J \subsetneqq S$. This gives a correspondence $h - - - J$ between T and $\{J; J \subsetneqq S\}$.

We can find an isomorphism $\iota : Z_G^* \xrightarrow{\sim} \Omega_W$ such that the following holds: if $G = Spin_N(\mathbf{C})$, with $N \geq 5$ odd or $N \geq 10$ even, the subset Ω''_W of Ω_W (see 3.1) corresponds to the set of characters of Z_G which do not factor through $SO_N(\mathbf{C})$. We shall identify $Z_G^* = \Omega_W$ via ι.

The following result appears in [**Lus79**], [**Lus09**].

THEOREM 6.3. *Define $un(G) \to \mathrm{Irr}(\bar{W})$ by $\mathfrak{c} \mapsto E$ where E is attached to $(\mathfrak{c}, \mathbf{C})$ under the Springer correspondence. This map defines a bijection $un(G) \xrightarrow{\sim} \bar{S}(\bar{W})$ (notation of 3.2).*

6.4. Let $\omega \in \Omega_W$. Let $L \in \mathcal{G}_\omega(G)$, let $(L, \mathfrak{c}_1, \mathfrak{L}_1)$ be the corresponding object of $\mathcal{G}'_G$ (see 6.1) and let $\phi_! K$ be the associated complex of sheaves on G (see 6.1); note that $\phi_! K[m]$ is a semisimple perverse sheaf for some m. Let K_1 be a simple perverse sheaf on G such that K_1 is a direct summand of $\phi_! K[m]$. Then K_1 is a character sheaf on G. Let C be the semisimple conjugacy class of H attached to K_1 by the known classification of character sheaves. Note that C is independent of the choice of K_1. There is a unique subset $J \subsetneq S$ such that for any $h \in T \cap C$ we have $h - - - J$ (see 6.2).

THEOREM 6.5. *For $L \in \mathcal{G}_\omega(G)$ we define $J \subsetneq S$ as in 6.4. We have $W_J \in \mathcal{C}_\omega(W)$ and $L \mapsto W_J$ is a bijection*

(a) $$\mathcal{G}_\omega(G) \xrightarrow{\sim} \mathcal{C}_\omega(W).$$

Note that the Theorem gives a parametrization of the set of unipotent blocks of G which is purely in terms of W and is thus independent of the geometry of G. The proof is given in the remainder of this section.

6.6. Let $\omega \in \Omega_W$. We now describe explicitly (and independently of 6.4) a bijection

(a) $$\mathcal{G}_\omega(G) \to \mathcal{C}_\omega(W).$$

The set $\mathcal{G}_\omega(G)$ is computed in [**Lus84a**]. Assume first that $G = SL_n(\mathbf{C})$, $n \geq 2$. Note that $Z_G^* = \Omega_W$ is a cyclic group of order n. Let $k = ord(\omega)$ (a divisor of n). Now $\mathcal{G}_\omega(G)$ consists of a single $L \in le(G)$ (up to conjugacy) such that $L_{der} \cong SL_k(\mathbf{C}) \times SL_k(\mathbf{C})$ (n/k copies). The bijection (a) is the obvious bijection between sets with one element.

Assume that G is of type E_6. Then Ω_W is cyclic of order 3. If $\omega = 1$ then $\mathcal{G}_\omega(G)$ consists of a single object, a maximal torus L. The bijection (a) is the obvious bijection between sets with one element. If $\omega \neq 1$ then $\mathcal{G}_\omega(G)$ consists of $L = G$ and of $L \in le(G)$ such that $L_{der} = SL_3(\mathbf{C}) \times SL_3(\mathbf{C})$ (up to conjugacy). Recall that $\mathcal{C}_\omega(W) = \{\{1\}, W_J\}$ where W_J is of type D_4. We define (a) by $G \mapsto W_J$, $L \mapsto \{1\}$ where $L \neq G$.

Assume that G is of type E_7. Then Ω_W is cyclic of order 2. If $\omega = 1$ then $\mathcal{G}_\omega(G)$ consists of a single object, a maximal torus L. The bijection (a) is the obvious bijection between sets with one element. If $\omega \neq 1$ then $\mathcal{G}_\omega(G)$ consists of $L = G$ and of $L \in le(G)$ (up to conjugacy) such that $L_{der} = SL_2(\mathbf{C}) \times SL_2(\mathbf{C}) \times SL_2(\mathbf{C})$ and L is contained in an $L' \in le(G)$ with $L'_{der} = SL_6(\mathbf{C})$ but L is not contained in an $L'' \in le(G)$ with $L''_{der} = SL_7(\mathbf{C})$. Recall that $\mathcal{C}_\omega(W) = \{\{1\}, W_J\}$ where W_J is of type E_6. We define (a) by $G \mapsto W_J$, $L \mapsto \{1\}$ where $L \neq G$.

Assume that G is of type E_8, F_4 or G_2. We have $\omega = 1$ and $\mathcal{G}_\omega(G)$ consists of two objects, an L which is a maximal torus and G. Recall that $\mathcal{C}_\omega(W) = \{\{1\}, W_J\}$ where W_J is of the non-affine type E_8, F_4 or G_2 (respectively). We define (a) by $G \mapsto W_J$, $L \mapsto \{1\}$ where $L \neq G$.

6.7. In this subsection we assume that G is $Sp_{2n}(\mathbf{C}), (n \geq 3)$, $Spin_{2n+1}(\mathbf{C})$, $(n \geq 2)$ or $Spin_{2n}(\mathbf{C}), (n \geq 4)$ so that W is of affine type $B_n (n \geq 3), C_n (n \geq 2)$ or $D_n (n \geq 4)$ (respectively).

From [**Lus84a**] we see that $\mathcal{G}_\omega(G)$ consists of all $L \in le(G)$ (up to conjugacy) such that

$L_{der} \cong Sp_\delta$, for various $(\delta, r) \in {}'\underline{\mathcal{C}}_\omega(W)$ (for W of type B),

$L_{der} \cong Spin_\delta$, for various $(\delta, r) \in {}'\underline{\mathcal{C}}_\omega(W)$ (for W of type C or D, $\omega \in \Omega'_W$),

$L_{der} \cong Spin_\delta \times SL_2(\mathbf{C})^r$, for various $(\delta, r) \in {}'\underline{\mathcal{C}}_\omega(W)$ (for W of type C or D, $\omega \in \Omega''_W$).

Then $L \mapsto (\delta, r)$ defines a bijection $\mathcal{G}_\omega(G) \xrightarrow{\sim} {}'\underline{\mathcal{C}}_\omega(W)$. Composing this with the inverses of the bijections ${}'\mathcal{C}_\omega(W) \xrightarrow{\sim} {}'\underline{\mathcal{C}}_\omega(W)$ (3.10, 3.11) and $\mathcal{C}_\omega(W) \xrightarrow{\sim} {}'\mathcal{C}_\omega(W)$ (3.8, 3.9) we obtain the bijection 6.6(a) in our case.

This completes the definition of the bijection 6.6(a) in all cases. It can be verified that this associates to $L \in \mathcal{G}_\omega(G)$ the same W_J as that defined in 6.5. Hence the map 6.5(a) is well defined and it is a bijection (the same as 6.6(a)).

6.8. Let $\omega \in \Omega_W, \beta \in \underline{ls}_\omega(G)$ with corresponding $L \in \mathcal{G}_\omega(G)$. Under 6.5(a), L corresponds to $W_J \in \mathcal{C}_\omega(W)$. Let $(\mathcal{W}, \mathcal{L}, \mathcal{S}, \bar{\mathcal{W}}) = (\mathcal{W}_J, \mathcal{L}_J, \mathcal{S}_J, \bar{\mathcal{W}}_J)$ be as in §5.

Let $(L, \mathfrak{c}_1, \mathfrak{L}_1) \in \mathcal{G}'(G)$ be a triple corresponding to L as in 6.1. Let $\mathfrak{c}_{max}$ be the unipotent class of G induced by $\mathfrak{c}_1$; let $\mathfrak{c}_{min}$ be the unipotent class of G that contains $\mathfrak{c}_1$.

For any $\mathfrak{c} \in un(G)$ we denote by $b_\mathfrak{c}$ the dimension of the variety of Borel subgroups of G that contain a fixed element of $\mathfrak{c}$. A case by case verification gives the following two results.

THEOREM 6.9. *We have*

(a) $b_{\mathfrak{c}_{max}} = \mathbf{a}[{}^\omega W_J]$ *(notation of 2.1).*
(b) $b_{\mathfrak{c}_{min}} - b_{\mathfrak{c}_{max}} = \nu(\mathcal{W}, \mathcal{L})$ *(notation of 4.3).*

THEOREM 6.10. *There exists a group isomorphism $\mathfrak{W}_\beta \xrightarrow{\sim} \bar{\mathcal{W}}$ well defined up to composition with an inner automorphism of $\bar{\mathcal{W}}$ given by the action of an element in $\Omega_\mathcal{W}$ (see 3.1). It carries the set of simple reflections of $\mathfrak{W}_\beta$ into the image of $\mathcal{S}$ under $\mathcal{W} \to \bar{\mathcal{W}}$.*

CONJECTURE 6.11. *Assume that $\mathcal{W} \neq \{1\}$. Let $(\mathfrak{c}, \mathfrak{L}), (\mathfrak{c}', \mathfrak{L}')$ be in β and let E, E' be in $\mathrm{Irr}(\mathfrak{W}_\beta) = \mathrm{Irr}(\bar{\mathcal{W}})$ (this equality follows from 6.10). Assume that the generalized Springer correspondence* [**Lus84a**] *associates E to $(\mathfrak{c}, \mathfrak{L})$ and E' to $(\mathfrak{c}', \mathfrak{L}')$. Then*

(a) $b_\mathfrak{c} - b_{\mathfrak{c}_{max}} = c_E$ *where c_E is defined as in 4.1 in terms of $(\mathcal{W}, \mathcal{L})$;*
(b) *we have $E \sim E'$ if and only if $\mathfrak{c} = \mathfrak{c}'$.*

This holds in the case where G is of exceptional type. (In the case where G is of type E_8, F_4 or G_2 this follows from [**Lus23**].) This can be also verified in the cases where $(\mathcal{W}, \mathcal{L})$ is as in the examples in 4.2.

6.12. For any $i = (\mathfrak{c}, \mathfrak{L}), i' = (\mathfrak{c}', \mathfrak{L}')$ in $\mathfrak{b}$ let $\Omega_{i,i'} \in \mathbf{Q}(\mathbf{q})$ be as in [**Lus86**, 24.7] and let $\Pi_{i',i} \in \mathbf{Q}(\mathbf{q})$ be as in [**Lus86**, 24.8]. Here $\mathbf{q}$ is an indeterminate. Let E, E' in $\mathrm{Irr}(\bar{\mathcal{W}})$ be corresponding to i, i' (respectively) under the generalized Springer correspondence. From the definitions we have $\Omega_{i,i'} = f(\mathbf{q})\Omega'_{E,E'}$ where $f(\mathbf{q}) \in \mathbf{Q}(\mathbf{q}) - \{0\}$ is independent of i, i'. Assuming that 6.11(a) holds we see from 5.7 and [**Lus86**, 24.8] that

(a)
$$\Pi_{i',i} = P_{E',E}$$

for any i, i' as above. In particular this holds in the case where G is of exceptional type. Note that $\Pi_{i',i}$ measures the stalks of the intersection cohomology sheaf on the closure of $\mathfrak{c}$ with coefficients in $\mathfrak{L}$ while $P_{E',E}$ is determined purely in terms of W.

7. Cells in the weighted affine Weyl group $\mathcal{W}_J$

7.1. Let

(a) $$\zeta : Cell(W, ||) \xrightarrow{\sim} un(G)$$

be the bijection defined in [**Lus89**].

7.2. We now fix $\omega \in \Omega_W$ and $\beta \in \underline{ls}_\omega(G)$ with corresponding $L \in \mathcal{G}_\omega(G)$. Under 6.5(a), L corresponds to $J \subsetneq S$ such that $W_J \in \mathcal{C}_\omega(W)$. Let $\mathcal{W}_J, \mathcal{S}_J, \mathcal{L}_J$ be as in §5. We assume that $\sharp(\mathcal{S}_J) \geq 2$. Let $un_\beta(G)$ be the set of all $\mathfrak{c} \in un(G)$ such that $(\mathfrak{c}, \mathfrak{L}) \in \beta$ for some $\mathfrak{L}$. We define a map

(a) $\mathfrak{A} : Cell(\mathcal{W}_J, \mathcal{L}_J) \to un(G)$

asssuming a conjecture in [**Lus03**, §25]. Let $\mathbf{c} \in Cell(\mathcal{W}_J, \mathcal{L}_J)$. Let E_0 be the unique irreducible special representation of W_J such that $z(E_0) = r(op)$ (notation of 2.1 with $\mathfrak{W}$ replaced by W_J) and let $\mathbf{c}_0 \in Cell(W_J, ||)$ be such that E_0 belongs to $\mathbf{c}_0$. According to [**Lus03**, Conj.25.3] there is a well defined $\tilde{\mathbf{c}} \in Cell(W, ||)$ which contains yx for any $y \in \mathbf{c}_0, x \in \mathbf{c}'$ (the product yx is taken in W). (We use the fact that, by results of [**Lus84b**], W_J satisfies the assumptions of [**Lus03**, 25.2].) We set $\mathfrak{A}(\mathbf{c}) = \zeta(\tilde{\mathbf{c}})$, with ζ as in 7.1(a).

CONJECTURE 7.3. *$\mathfrak{A}$ is injective with image equal to $un_\beta(G)$. Hence $\mathfrak{A}$ defines a bijection $Cell(\mathcal{W}_J, \mathcal{L}_J) \xrightarrow{\sim} un_\beta(G)$.*

This is a generalization of 7.1(a). Note that [**Lus03**, 25.2] holds if $J = \emptyset$ in which case it states that any $\mathbf{c} \in Cell(\mathcal{W}_J, \mathcal{L}_J)$ is contained in a two-sided cell of $(W, ||)$; this can be deduced from [**Lus03**, 10.14]. Using this one can define $\mathfrak{A}$ unconditionally when W is of exceptional type and verify the conjecture in that case.

7.4. Let $\mathbf{a} : \mathcal{W}_J \to \mathbf{N}$ be the a-function (see [**Lus03**]) of the weighted affine Weyl group $(\mathcal{W}_J, \mathcal{L}_J)$. Let $\mathbf{c} \in Cell(\mathcal{W}_J, \mathcal{L}_J)$ and let $\mathfrak{c} = \mathfrak{A}_J(\mathbf{c}) \in un_\beta(G)$. We expect that

(a) for any $w \in \mathbf{c}$ we have $\mathbf{a}(w) = b_\mathfrak{c} - b_{\mathfrak{c}_{max}}$.

References

[BC72] C. T. Benson and C. W. Curtis, *On the degrees and rationality of certain characters of finite Chevalley groups*, Trans. Amer. Math. Soc. **165** (1972), 251–273, DOI 10.2307/1995885. MR304473

[Gec11] Meinolf Geck, *On Iwahori-Hecke algebras with unequal parameters and Lusztig's isomorphism theorem*, Pure Appl. Math. Q. **7** (2011), no. 3, Special Issue: In honor of Jacques Tits, 587–620, DOI 10.4310/PAMQ.2011.v7.n3.a5. MR2848588

[GM99] Meinolf Geck and Gunter Malle, *On special pieces in the unipotent variety*, Experiment. Math. **8** (1999), no. 3, 281–290. MR1724160

[Lus78] George Lusztig, *Representations of finite Chevalley groups*, CBMS Regional Conference Series in Mathematics, vol. 39, American Mathematical Society, Providence, RI, 1978. Expository lectures from the CBMS Regional Conference held at Madison, Wis., August 8–12, 1977. MR518617

[Lus79] G. Lusztig, *A class of irreducible representations of a Weyl group*, Nederl. Akad. Wetensch. Indag. Math. **41** (1979), no. 3, 323–335. MR546372

[Lus80] George Lusztig, *Some problems in the representation theory of finite Chevalley groups*, The Santa Cruz Conference on Finite Groups (Univ. California, Santa Cruz, Calif., 1979), Proc. Sympos. Pure Math., vol. 37, Amer. Math. Soc., Providence, RI, 1980, pp. 313–317. MR604598

[Lus82] George Lusztig, *A class of irreducible representations of a Weyl group. II*, Nederl. Akad. Wetensch. Indag. Math. **44** (1982), no. 2, 219–226. MR662657

[Lus83] G. Lusztig, *Left cells in Weyl groups*, Lie group representations, I (College Park, Md., 1982/1983), Lecture Notes in Math., vol. 1024, Springer, Berlin, 1983, pp. 99–111, DOI 10.1007/BFb0071433. MR727851

[Lus84a] G. Lusztig, *Intersection cohomology complexes on a reductive group*, Invent. Math. **75** (1984), no. 2, 205–272, DOI 10.1007/BF01388564. MR732546

[Lus84b] George Lusztig, *Characters of reductive groups over a finite field*, Annals of Mathematics Studies, vol. 107, Princeton University Press, Princeton, NJ, 1984, DOI 10.1515/9781400881772. MR742472

[Lus86] George Lusztig, *Character sheaves. V*, Adv. in Math. **61** (1986), no. 2, 103–155, DOI 10.1016/0001-8708(86)90071-X. MR849848

[Lus89] George Lusztig, *Cells in affine Weyl groups. IV*, J. Fac. Sci. Univ. Tokyo Sect. IA Math. **36** (1989), no. 2, 297–328. MR1015001

[Lus91] George Lusztig, *Intersection cohomology methods in representation theory*, Proceedings of the International Congress of Mathematicians, Vol. I, II (Kyoto, 1990), Math. Soc. Japan, Tokyo, 1991, pp. 155–174. MR1159211

[Lus95] George Lusztig, *Classification of unipotent representations of simple p-adic groups*, Internat. Math. Res. Notices **11** (1995), 517–589, DOI 10.1155/S1073792895000353. MR1369407

[Lus03] G. Lusztig, *Hecke algebras with unequal parameters*, CRM Monograph Series, vol. 18, American Mathematical Society, Providence, RI, 2003. additional material in version 2 (2014), `arXiv:math/0208154`, DOI 10.1090/crmm/018. MR1974442

[Lus09] G. Lusztig, *Unipotent classes and special Weyl group representations*, J. Algebra **321** (2009), no. 11, 3418–3449, DOI 10.1016/j.jalgebra.2008.04.004. MR2510055

[Lus17] G. Lusztig, *Generalized Springer theory and weight functions*, Ann. Univ. Ferrara Sez. VII Sci. Mat. **63** (2017), no. 1, 159–167, DOI 10.1007/s11565-016-0249-8. MR3651646

[Lus23] George Lusztig, *From families in Weyl groups to Springer representations*, Bull. Inst. Math. Acad. Sin. (N.S.) **18** (2023), no. 1, 1–14. MR4582839

[Spr76] T. A. Springer, *Trigonometric sums, Green functions of finite groups and representations of Weyl groups*, Invent. Math. **36** (1976), 173–207, DOI 10.1007/BF01390009. MR442103

Department of Mathematics, M.I.T., Cambridge, Massachusetts 02139

Proceedings of Symposia in Pure Mathematics
Volume **108**, 2024
https://doi.org/10.1090/pspum/108/01968

From conjugacy classes in the Weyl group to representations

George Lusztig and Zhiwei Yun

1. Definition of the map Ψ

1.1. Let G be a connected reductive algebraic group defined and split over the finite field $\mathbf{F}_p$ with p elements where p is a prime number which we assume to be good for G. Let W be the Weyl group of G (a Coxeter group with length function $w \mapsto |w|$), let $\underline{W}$ be the set of conjugacy classes of W and let $\mathcal{R}_W$ be the Grothendieck group of (finite dimensional) representations of W over $\bar{\mathbf{Q}}_l$; here l is a fixed prime number, $l \neq p$. Let $\mathcal{U}$ be the variety of unipotent elements of G. Let $\underline{\mathcal{U}}$ be the set of G-conjugacy classes of unipotent elements. Let $\mathcal{B}$ be the variety of Borel subgroups of G. We fix $B^+ \in \mathcal{B}$ such that B^+ is defined over $\mathbf{F}_p$.

In [**12**] a (surjective) map $\Phi : \underline{W} \to \underline{\mathcal{U}}$ was defined based on the study of intersections of unipotent classes in G with (B^+, B^+) double cosets in G.

Let $\mathbf{q}$ be an indeterminate. In this paper we define a map

$$\Psi : \underline{W} \to \mathbf{Z}[\mathbf{q}] \otimes \mathcal{R}_W$$

whose image is denoted by Z. We also define a (surjective) map $\Theta : Z \to \underline{\mathcal{U}}$ such that $\Phi(C) = \Theta(\Psi(C))$ for any $C \in \underline{W}$. Thus, Ψ is a refinement of Φ. The definition of Ψ again involves the study of intersections of unipotent classes in G with (B^+, B^+) double cosets in G. We also describe Ψ explicitly for G of low rank (see §3).

A map closely related to Ψ appears in the work of Minh-Tam Trinh [**14**].

1.2. For $(B, B') \in \mathcal{B} \times \mathcal{B}$ we denote by $pos(B, B') \in W$ the relative position of B, B' (see [**4**]). For $w \in W$ let $B^+wB^+ = \{g \in G; pos(B^+, gB^+g^{-1}) = w\}$. Let $F_1 : G \to G$ be the Frobenius map relative to the $\mathbf{F}_p$-structure. Let $q = p^e$ $(e \geq 1)$ and let $F = F_1^e : G \to G$. This induces Frobenius maps $\mathcal{U} \to \mathcal{U}$, $\mathcal{B} \to \mathcal{B}$ denoted again by F. For $y \in W$, we set $X_y = \{B \in \mathcal{B}; pos(B, F(B)) = y\}$ (see [**4**]). If $i \in \mathbf{Z}$, the $\bar{\mathbf{Q}}_l$-cohomology space with compact support $H^i_c(X_y)$ has a natural action of the finite group G^F (see [**4**]); here $?^F$ denotes the fixed point set of $F : ? \to ?$. Let $R^1_y = \sum_i (-1)^i H^i_c(X_y)$, a virtual representation of G^F. For $g \in G^F$, $\mathrm{tr}(g, R^1_y)$ is an integer, see [**4**, §3.3]. For $w \in W, y \in W$ we set

$$a_{w,y} = \sum_{u \in \mathcal{U}^F \cap (B^+wB^+)} \mathrm{tr}(u, R^1_y) \in \mathbf{Z}. \tag{1.1}$$

The first author was supported by NSF grant DMS-1855773. The second author was supported by the Packard Fellowship and the Simons Investigatorship.

1.3. PROPOSITION. *Let* $w \in W$. *There is a unique element* $\zeta_{w,q} \in \mathcal{R}_W$ *such that for any* $y \in W$ *we have* $\mathrm{tr}(y, \zeta_{w,q}) = a_{w,y}$.

PROOF. For $u \in \mathcal{U}$ we set $\mathcal{B}_u = \{B \in \mathcal{B}; u \in B\}$. It is known (Springer) that W acts naturally on the l-adic cohomology space $H^i(\mathcal{B}_u)$. Let $\mathrm{Irr}(W)$ be a set of representatives for the isomorphism classes of irreducible representations of W over $\bar{\mathbf{Q}}_l$. We can identify $H^i(\mathcal{B}_u) = \oplus_{E \in \mathrm{Irr}(W)} E \otimes H^i(\mathcal{B}_u)_E$ where $H^i(\mathcal{B}_u)_E = \mathrm{Hom}_W(E, H^i(\mathcal{B}_u))$. From [**11**] (or [**7**] when $p \gg 0$), for $u \in \mathcal{U}^F$ and $y \in W$ we have

$$\mathrm{tr}(u, R_y^1) = \sum_i (-1)^i \mathrm{tr}(yF, H^i(\mathcal{B}_u)) = \sum_{E \in \mathrm{Irr}(W)} \mathrm{tr}(y, E)\,\mathrm{tr}(F, H^*(\mathcal{B}_u)_E). \tag{1.2}$$

Here $F : H^i(\mathcal{B}_u) \to H^i(\mathcal{B}_u)$ and $F : H^i(\mathcal{B}_u)_E \to H^i(\mathcal{B}_u)_E$ are induced by the restriction of $F : \mathcal{B} \to \mathcal{B}$ to $\mathcal{B}_u$ and $\mathrm{tr}(F, H^*(\mathcal{B}_u)_E)$ is defined to be

$$\sum_i (-1)^i \mathrm{tr}(F, H^i(\mathcal{B}_u)_E).$$

Thus we have

$$a_{w,y} = \sum_{u \in \mathcal{U}^F \cap (B^+ wB^+)} \sum_{E \in \mathrm{Irr}(W)} \mathrm{tr}(y, E)\,\mathrm{tr}(F, H^*(\mathcal{B}_u)_E).$$

For $E \in \mathrm{Irr}(W)$ we set

$$\zeta_{w,q;E} = \sum_{u \in \mathcal{U}^F \cap (B^+ wB^+)} \mathrm{tr}(F, H^*(\mathcal{B}_u)_E) \tag{1.3}$$

so that

$$a_{w,y} = \sum_{E \in \mathrm{Irr}(W)} \zeta_{w,q;E}\,\mathrm{tr}(y, E)$$

for any $y \in W$. Since $a_{w,y} \in \mathbf{Z}$ and the matrix $(\mathrm{tr}(y, E))$ (with y running through representatives for the conjugacy classes in W, and $E \in \mathrm{Irr}(W)$) has integer entries and nonzero determinant, we see that $\zeta_{w,q;E} \in \mathbf{Q}$ for any $E \in \mathrm{Irr}(W)$. On the other hand from the definition we see see that $\zeta_{w,q;E}$ is an algebraic integer. It follows that

$$\zeta_{w,q;E} \in \mathbf{Z}. \tag{1.4}$$

We set $\zeta_{w,q} = \sum_{E \in \mathrm{Irr}(W)} \zeta_{w,q;E} E \in \mathcal{R}_W$. This proves the existence statement in the proposition. The uniqueness is obvious. □

1.4. Let $\mathcal{F}$ be the vector space of functions $\mathcal{B}^F \to \bar{\mathbf{Q}}_l$. For $w \in W$ we define a linear map $T_w : \mathcal{F} \to \mathcal{F}$ by $T_w(f)(B) = \sum_{B' \in \mathcal{B}^F; pos(B,B')=w} f(B')$. Let $\mathcal{H}$ be the subspace of $End(\mathcal{F})$ with basis $\{T_w; w \in W\}$; this is a subalgebra of $End(\mathcal{F})$ (the Hecke algebra). Now G^F acts on $\mathcal{F}$ by $g : f \mapsto f'$ where $f'(B) = f(g^{-1}Bg)$. This action commutes with the $\mathcal{H}$-action so that we can identify $\mathcal{F} = \oplus_{\mathcal{E} \in \mathrm{Irr}(\mathcal{H})} \mathcal{E} \otimes \mathcal{F}_{\mathcal{E}}$ where $\mathcal{F}_{\mathcal{E}} = \mathrm{Hom}_{\mathcal{H}}(\mathcal{E}, \mathcal{F})$ and $\mathrm{Irr}(\mathcal{H})$ is a set of representatives for the simple $\mathcal{H}$-modules; here $\mathcal{F}_{\mathcal{E}}$ is an irreducible G^F-module. For $u \in \mathcal{U}^F$ we have

$$\sharp(B \in \mathcal{B}^F; pos(B, uBu^{-1}) = w) = \mathrm{tr}(uT_w, \mathcal{F}) = \sum_{\mathcal{E} \in \mathrm{Irr}(\mathcal{H})} \mathrm{tr}(T_w, \mathcal{E})\,\mathrm{tr}(u, \mathcal{F}_{\mathcal{E}}).$$

Hence for $y \in W$ we have

$$\sharp(\mathcal{B}^F)a_{w,y} = \sum_{u\in\mathcal{U}^F, B\in\mathcal{B}^F; pos(B,uBu^{-1})=w} \operatorname{tr}(u, R^1_y) \tag{1.5}$$
$$= \sum_{\mathcal{E}\in\operatorname{Irr}(\mathcal{H})} \sum_{u\in\mathcal{U}^F} \operatorname{tr}(u, \mathcal{F}_{\mathcal{E}}) \operatorname{tr}(u, R^1_y) \operatorname{tr}(T_w, \mathcal{E}).$$

From (1.5) and (1.2) for any $w \in W, E \in \operatorname{Irr}(W)$ we have:

$$\sharp(\mathcal{B}^F)\zeta_{w,q;E} = \sum_{u\in\mathcal{U}^F} \sum_{\mathcal{E}\in\operatorname{Irr}(\mathcal{H})} \operatorname{tr}(u, \mathcal{F}_{\mathcal{E}}) \operatorname{tr}(F, H^*(\mathcal{B}_u)_E) \operatorname{tr}(T_w, \mathcal{E}). \tag{1.6}$$

1.5. Let $w \in W, E' \in \operatorname{Irr}(W)$. We set

$$c_{w,E',q} = \sharp(W)^{-1} \sum_{\mathcal{E}\in\operatorname{Irr}(\mathcal{H})} \sum_{z\in W} (\mathcal{F}_{\mathcal{E}} : R^1_z)_{G^F} \operatorname{tr}(z, E') \operatorname{tr}(T_w, \mathcal{E}) \in \mathbf{Q}. \tag{1.7}$$

Here $(:)_{G^F}$ is the standard inner product of virtual representations of G^F. We shall now regard q as variable. We show:

(1.8) *$c_{w,E',q}$ is the value at q of a polynomial $\mathbf{c}_{w,E'}(\mathbf{q})$ in $\mathbf{q}$ with rational coefficients independent of q.*

We can assume that G is adjoint simple. The only part of the right hand side of (1.7) which depends on q is $\operatorname{tr}(T_w, \mathcal{E})$. This is known to be the value at q of a polynomial in $\mathbf{q}$ with integral coefficients independent of q except when G is of type E_7, $\dim E' = 512$ or G is of type E_8, $\dim E' = 4096$. Assume now that G is of type E_7, $\dim E' = 512$. Let $\mathcal{E}', \mathcal{E}''$ be the two objects of $\operatorname{Irr}(\mathcal{E})$ which have dimension 512. We have $c_{w,E',q} = (1/2)(\operatorname{tr}(T_w, \mathcal{E}') + \operatorname{tr}(T_w, \mathcal{E}''))$ (see the proof of [**9**, §7.12]). Although the quantities $\operatorname{tr}(T_w, \mathcal{E}'), \operatorname{tr}(T_w, \mathcal{E}'')$ are separately not necessarily polynomials in q their sum is. This proves (1.8) in the case where G is of type $E_7, \dim E' = 512$. The proof in the case where G is of type $E_8, \dim E' = 4096$, is entirely similar.

1.6. For $\mathcal{E} \in \operatorname{Irr}(\mathcal{H})$ there is a well defined class function $\xi_{\mathcal{E}}$ on G^F such that

$$\operatorname{tr}(g, \mathcal{F}_{\mathcal{E}}) = \sharp(W)^{-1} \sum_{z\in W} (\mathcal{F}_{\mathcal{E}} : R^1_z)_{G^F} \operatorname{tr}(g, R^1_z) + \xi_{\mathcal{E}}(g) \tag{1.9}$$

for all $g \in G^F$ and

(1.10) $\xi_{\mathcal{E}}$ is orthogonal to the character of any virtual representation R^θ_T of [**4**].

From (1.10) we deduce using results in [**10**] that for any $E \in \operatorname{Irr}(W)$ we have

$$\sum_{u\in\mathcal{U}^F} \xi_{\mathcal{E}}(u) \operatorname{tr}(F, H^*(\mathcal{B}_u)_E) = 0.$$

Using (1.9) and (1.6) we see that for any $w \in W, E \in \operatorname{Irr}(W)$ we have

$$\begin{aligned}&\sharp(\mathcal{B}^F)\zeta_{w,q;E}\\ &= \sharp(W)^{-1} \sum_{u\in\mathcal{U}^F} \sum_{\mathcal{E}\in\operatorname{Irr}(\mathcal{H})} \sum_{z\in W} (\mathcal{F}_{\mathcal{E}} : R^1_z)_{G^F} \operatorname{tr}(u, R^1_z) \operatorname{tr}(F, H^*(\mathcal{B}_u)_E) \operatorname{tr}(T_w, \mathcal{E}).\end{aligned}$$

Using (1.2) we deduce

$$\sharp(\mathcal{B}^F)\zeta_{w,q;E} = \sum_{E'\in\mathrm{Irr}(W)} c_{w,E',q} \sum_{u\in\mathcal{U}^F} \mathrm{tr}(F, H^*(\mathcal{B}_u)_{E'})\,\mathrm{tr}(F, H^*(\mathcal{B}_u)_E). \tag{1.11}$$

The right hand side of (1.11) is the value at $\mathbf{q} = q$ of a polynomial $\pi(\mathbf{q})$ (with the coefficients of π being rational numbers independent of q). This follows from (1.8) and the known properties of $\mathrm{tr}(F, H^i(\mathcal{B}_u)_E)$, see [**3**].

1.7. Let $w \in W, E \in \mathrm{Irr}(W)$. It is known that

(1.12) *$H^i(\mathcal{B}_u)_E$ can be interpreted as the stalk at $u \in \mathcal{U}$ of an intersection cohomology complex K_E on the closure of a unipotent class of G.*

(This was stated in [**8**, §3, Conj.2] and proved in [**1**].) We use this and the Grothendieck trace formula to rewrite the definition of $\zeta_{w,q;E}$ in the form

$$\zeta_{w,q;E} = \sum_i (-1)^i\,\mathrm{tr}(F, H^i(X, K_E|_X))$$

where $X = \mathcal{U} \cap (B^+wB^+)$ and the map $H^i(X, K_E|_X) \to H^i(X, K_E|_X)$ induced by $F : G \to G$ is denoted again by F. Replacing q by p^s with $s = 1, 2, \ldots$ we see that

$$\zeta_{w,p^s;E} = \sum_{\lambda\in V} n_\lambda \lambda^s$$

where V is a finite subset of $\bar{\mathbf{Q}}_l - \{0\}$ and n_λ are nonzero integers independent of s. It follows that

$$\left(\sum_{v\in W} p^{|v|s}\right)\zeta_{w,p^s;E} = \sum_{\lambda'\in V'} n'_{\lambda'}\lambda'^s$$

where V' is a finite subset of $\bar{\mathbf{Q}}_l - \{0\}$ and $n'_{\lambda'}$ are nonzero integers independent of s. By the results in §1.6 we have also

$$\left(\sum_{v\in W} p^{|v|s}\right)\zeta_{w,p^s;E} = \sum_{m\in[0,N]} n''_m p^{sm}$$

for some $N \geq 1$ where $n''_m \in \mathbf{Q}$ are independent of s. We deduce that

$$\sum_{\lambda'\in V'} n'_{\lambda'}\lambda'^s = \sum_{m\in[0,N]} n''_m p^{sm}$$

for $s = 1, 2, \ldots$. This forces V' to be a subset of $\{1, p, p^2, \ldots\}$. Thus we have the following result.

(1.13) *There exists a polynomial $\pi(\mathbf{q})$ in $\mathbf{q}$ with integer coefficients independent of s such that $(\sum_{v\in W} p^{|v|s})\zeta_{w,p^s;E} = \pi(p^s)$ for $s = 1, 2, \ldots$.*

1.8. For $C \in \underline{W}$ we denote by C_{min} the set of elements of minimal length in C. Let ρ be the reflection representation of W. For $w \in W$ let $n_w = \det(1 - w, \rho)$. We have $n_w \geq 0$. We say that w is elliptic if $n_w > 0$. Let $\underline{W}_{el}$ be the set of all $C \in \underline{W}$ such that for some/any $w \in C$, w is elliptic.

Let Z_G be the center of G. Let $\nu = \dim \mathcal{B}$. Let r be the rank of G/Z_G.

In this subsection we fix $C \in \underline{W}_{el}$ and $w \in C_{min}$. Let $n'_w \geq 1$ be the part prime to p of n_w. According to [**12**, §5.2], for any $g \in B^+wB^+$,

(1.14) *The group $\{b \in B^+; bgb^{-1} = g\}/Z_G$ is finite abelian of order dividing n'_w.*

(In the first line of [**12**, §5.1] one should add the sentence: *We fix $w \in C_{min}$.*) We show:

(1.15) *For any $E \in \mathrm{Irr}(W)$ we have $q^{-\nu}(q-1)^{-r}n'_w\zeta_{w,q;E} \in \mathbf{Z}$.*

We have $\zeta_{w,q;E} = \sum_S \sharp(S)\,\mathrm{tr}(F, H^*(\mathcal{B}_{u_S})_E)$ where S runs over the set of orbits of B^{+F} acting on $\mathcal{U}^F \cap B^+wB^+$ by conjugation and for each such S, u_S is an element of S. Here for any S in the sum, $\mathrm{tr}(F, H^*(\mathcal{B}_{u_S})_E)$ is an algebraic integer and, by (1.14), $\sharp(S) \in q^\nu(q-1)^r n'^{-1}_w\mathbf{Z}$. Hence $q^{-\nu}(q-1)^{-r}n'_w\zeta_{w,q;E}$ is an algebraic integer. By (1.4), this is also a rational number, hence an integer. This proves (1.15).

1.9. Proposition. *Let $C \in \underline{W}_{el}$ and $w \in C_{min}$. Let $E \in \mathrm{Irr}(W)$. Then there exists a polynomial $\pi_1(\mathbf{q}) \in \mathbf{Z}[\mathbf{q}]$ such that $\zeta_{w,p^s;E}/(p^{s\nu}(p^s-1)^r) = \pi_1(p^s)$ for $s = 1, 2, \ldots$. In particular, $\zeta_{w,q;E}/(q^\nu(q-1)^r) \in \mathbf{Z}$.*

Proof. Let $\pi_0(\mathbf{q}) = (\sum_{v \in W} \mathbf{q}^{|v|_s})\mathbf{q}^\nu(\mathbf{q}-1)^r$. This is a monic polynomial in $\mathbf{q}$ of degree $2\nu + r$ with integer coefficients. Let $\pi(\mathbf{q})$ be as in (1.13). We have $\pi(\mathbf{q}) = \pi_1(\mathbf{q})\pi_0(\mathbf{q}) + \pi_2(\mathbf{q})$ where $\pi_1(\mathbf{q}), \pi_2(\mathbf{q})$ are polynomials with integer coefficients and $\pi_2(\mathbf{q})$ is either 0 or has degree $< 2\nu + r$. By (1.13) and (1.15) for $s = 1, 2, \ldots$ we have $\pi(q^s)n'_w/\pi_0(q^s) \in \mathbf{Z}$ hence $\pi_1(q^s)n'_w + \pi_2(q^s)n'_w/\pi_0(q^s) \in \mathbf{Z}$ so that $\pi_2(q^s)n'_w/\pi_0(q^s) \in \mathbf{Z}$. If $\pi_2(\mathbf{q}) \neq 0$ this is impossible for large s since $\deg \pi_2 < \deg \pi_0$. We see that $\pi(\mathbf{q}) = \pi_1(\mathbf{q})\pi_0(\mathbf{q})$. Setting $\mathbf{q} = q$ we see that $\zeta_{w,q;E}/(q^\nu(q-1)^r) = \pi_1(q) \in \mathbf{Z}$. The proposition is proved. □

1.10. Let P be a parabolic subgroup of G containing B^+. Let L be an F-stable Levi subgroup of P and let U be the unipotent radical of P. Then $B_L^+ := B^+ \cap L$ is an F-stable Borel subgroup of L and we have $B^+ = B_L^+U$. Let $\mathcal{U}_L$ be the set of unipotent elements of L. Let W_L be the Weyl group of L. We can view W_L as a subgroup of W and as an indexing set for the (B_L^+, B_L^+) double cosets of L, so that for $w \in W_L$ the double coset $B_L^+wB_L^+$ satisfies $(B_L^+wB_L^+)U = B^+wB^+$. For $y' \in W_L$ we define $X_{y',L}$ in the same way as X_y, but replacing G, y by L, y'; then $R^1_{y',L} = \sum_i(-1)^iH^i_c(X_{y',L})$ is naturally a (virtual) L^F-module.

Let σ be an irreducible L^F-module. We can view σ as a P^F-module on which U^F acts trivially. Let $y \in W$. The following identity is a reformulation of a special case of a result in [**5**]:

$$(\mathrm{ind}_{P^F}^{G^F}(\sigma) : R^1_y)_{G^F} = \sharp(W_L)^{-1} \sum_{z \in W; zyz^{-1} \in W_L} (\sigma : R^1_{zyz^{-1},L})_{L^F}. \tag{1.16}$$

Here $(:)_{L^F}$ denote the standard inner product of virtual representations of L^F. The left hand side of (1.16) is equal to $(\sigma : \sum_i(-1)^iH^i_c(X_y)^{U^F})_{L^F}$ where $H^i_c(X_y)^{U^F}$ is the space of U^F-invariants on $H^i_c(X_y)$ viewed as an L^F-module. It follows that

$$\sharp(W_L)\sum_i(-1)^iH^i_c(X_y)^{U^F} = \sum_{z \in W; zyz^{-1} \in W_L} R^1_{zyz^{-1},L}$$

as virtual L^F-modules.

For $w \in W_L, y' \in W_L$ we define $a_{w,y';L}$ as in (1.1), in terms of L instead of G that is,

$$a_{w,y';L} = \sum_{u \in \mathcal{U}_L^F \cap (B_L^+ w B_L^+)} \operatorname{tr}(u, R^1_{y',L}) \in \mathbf{Z}.$$

Now let $w \in W_L, y \in W$. We have

$$\begin{aligned}
(1.17) \qquad a_{w,y} &= \sum_{u \in \mathcal{U}^F \cap (B_L^+ w B_L^+) U} \operatorname{tr}(u, R^1_y) \\
&= \sum_{(u',u'') \in \mathcal{U}_L^F \times U^F ; u' \in B_L^+ w B_L^+} \operatorname{tr}(u'u'', R^1_y) \\
&= \sharp(U^F) \sum_{u' \in \mathcal{U}_L^F (B_L^+ w B_L^+)} \sum_i (-1)^i \operatorname{tr}(u', H^i_c(X_y)^{U^F}) \\
&= \sharp(W_L)^{-1} \sharp(U^F) \sum_{u' \in \mathcal{U}_L^F (B_L^+ w B_L^+)} \sum_{z \in W; zyz^{-1} \in W_L} \operatorname{tr}(u', R^1_{zyz^{-1},L}) \\
&= \sharp(W_L)^{-1} \sharp(U^F) \sum_{z \in W; zyz^{-1} \in W_L} a_{w,zyz^{-1};L}.
\end{aligned}$$

Let $\mathcal{R}_{W_L}$ be the Grothendieck group of W_L-modules over $\bar{\mathbf{Q}}_l$. We define $\zeta_{w,L,q} \in \mathcal{R}_{W_L}$ as in Proposition 1.3 in terms of L instead of G. Using (1.17) we have

$$\operatorname{tr}(y, \zeta_{w,q}) = \sharp(W_L)^{-1} \sharp(U^F) \sum_{z \in W; zyz^{-1} \in W_L} \operatorname{tr}(zyz^{-1}, \zeta_{w,L,q}),$$

that is,

$$(1.18) \qquad \zeta_{w,q} = \sharp(U^F) \operatorname{ind}_{W_L}^W (\zeta_{w,L,q}).$$

1.11. Let $C \in \underline{W}$. For $w \in C_{min}, w' \in C_{min}$ we show:

$$(1.19) \qquad \zeta_{w,q} = \zeta_{w',q}.$$

It is enough to show that for any $y \in W$ we have $a_{w,y} = a_{w',y}$. We write (1.5) for w, y and for w', y. We see that it is enough to show that $\operatorname{tr}(T_w, \mathcal{E}) = \operatorname{tr}(T_{w'}, \mathcal{E})$ for any $\mathcal{E} \in \operatorname{Irr}(\mathcal{H})$; this follows from results in [**6**].

1.12. Let $C \in \underline{W}$. Let $m(C)$ be the multiplicity of the eigenvalue 1 of w in ρ (see §1.8) for some/any $w \in C$. We define an element $\underline{\zeta}_{C,q} \in \mathbf{Q} \otimes \mathcal{R}_W$ by

$$\underline{\zeta}_{C,q} = \zeta_{w,q} / (q^\nu (q-1)^{r-m(C)})$$

where $w \in C_{min}$. This is independent of the choice of w by 1.11. We show that

$$(1.20) \qquad \underline{\zeta}_{C,q} \in \mathcal{R}_W.$$

If $m(C) = 0$ (so that $C \in \underline{W}_{el}$) this follows from Proposition 1.9. Assume now that $m(C) > 0$. In this case C_{min} contains an element w which is contained and elliptic in W_L where P, U, L, W_L are as in §1.10 and $P \neq G$. By Proposition 1.9 for L instead of G we have $\zeta_{w,L,q}/(q^{\nu-\omega}(q-1)^{r-m(C)}) \in \mathcal{R}_{W_L}$ where $\sharp(U) = q^\omega$. It follows that $\operatorname{ind}_{W_L}^W(\zeta_{w,L,q}/(q^{\nu-\omega}(q-1)^{r-m(C)})) \in \mathcal{R}_W$. Using (1.18) we deduce $\zeta_{w,q}/(q^\omega q^{\nu-\omega}(q-1)^{r-m(C)}) \in \mathcal{R}_W$. This proves (1.20).

1.13. Let $C \in \underline{W}$. From Proposition 1.9 and the definitions we see that there is a well defined element $\Psi(C) \in \mathbf{Z}[\mathbf{q}] \otimes \mathcal{R}_W$ such that for any $s = 1, 2, \dots,$ the specialization $\Psi(C)|_{\mathbf{q}=p^s} \in \mathcal{R}_W$ is equal to $\underline{\zeta}_{C,p^s}$ as in §1.12. We can write $\Psi(C) = \sum_{i\ge 0} \mathbf{q}^i \Psi_i(C)$ where $\Psi_i(C) \in \mathcal{R}_W$ are zero for $i \gg 0$. Thus we have a map $\Psi : \underline{W} \to \mathbf{Z}[\mathbf{q}] \otimes \mathcal{R}_W$ and maps $\Psi_i : \underline{W} \to \mathcal{R}_W$.

2. Properties of the map Ψ

2.1. Let $w \in W, E \in \mathrm{Irr}(W)$. We show:

(2.1) $|\mathcal{B}^F|\zeta_{w,E,q}$ *is a polynomial in* q *of degree* $\le |w| + 2\nu$;

(2.2) $|\mathcal{B}^F|\zeta_{w,1,q}$ *is a monic polynomial in* q *of degree* $|w| + 2\nu$.

From (1.8) and the definitions, for any $E' \in \mathrm{Irr}(W)$,

(2.3) $c_{w,E',q}$ *is a polynomial in* q *of degree* $\le |w|$.

Moreover, we have

$$c_{w,1,q} = \sharp(W)^{-1} \sum_{\mathcal{E}\in\mathrm{Irr}(\mathcal{H})} \sum_{z\in W} (\mathcal{F}_{\mathcal{E}} : R_z^1)_{G^F} \operatorname{tr}(T_w, \mathcal{E}) \tag{2.4}$$
$$= \sum_{\mathcal{E}\in\mathrm{Irr}(\mathcal{H})} (\mathcal{F}_{\mathcal{E}} : 1)_{G^F} \operatorname{tr}(T_w, \mathcal{E}) = \operatorname{tr}(T_1, \mathcal{E}_1) = q^{|w|}$$

where $\mathcal{E}_1 \in \mathrm{Irr}(\mathcal{H})$ is such that T_y acts on it as $q^{|y|}$ for any $y \in W$.

For any $\gamma \in \underline{\mathcal{U}}$ and any $E' \in \mathrm{Irr}(W)$ we consider the sum

$$M_{E,E',\gamma} = \sum_{u\in\gamma^F} \operatorname{tr}(F, H^*(\mathcal{B}_u)_{E'}) \operatorname{tr}(F, H^*(\mathcal{B}_u)_E).$$

It is known that

(2.5) *This is a polynomial in* q *of degree* $\le 2\dim \mathcal{B}_u + \dim \gamma = 2\nu$ *(where* $u \in \gamma$*) with the inequality being strict unless* E, E' *appear in the top cohomology of* $\mathcal{B}_u, u \in \gamma$.

We have

$$\sharp(\mathcal{B}^F)\zeta_{w,q;E} = \sum_{E'\in\mathrm{Irr}(W)} c_{w,E',q} \sum_{\gamma\in\underline{\mathcal{U}}} M_{E,E',\gamma}.$$

(See (1.11).) Using this together with (2.5) and (2.3) we see that (2.1) holds. Now assume that $E = 1$. If γ is not the regular unipotent class then E does not appear in the top cohomology of $\mathcal{B}_u, u \in \gamma$ hence by (2.5), $M_{E,E',\gamma}$ is a polynomial in q of degree $< 2\nu$. If γ is the regular unipotent class and $E' \ne 1$ then $M_{E,E',\gamma} = 0$. If γ is the regular unipotent class and $E' = 1$ then $M_{E,E',\gamma} = |\gamma^F|$ is a monic polynomial in q of degree 2ν. Combining this with (2.4) we see that (2.2) holds.

Now let $C \in \underline{W}$ and let $w \in C_{min}$. From (2.1) and (2.2) we deduce

(2.6) *If* $i > |w| - (r - m(C))$ *then* $\Psi_i(C) = 0$.

(2.7) *If* $i = |w| - (r - m(C))$ *then* $\Psi_i(C) \ne 0$; *more precisely, the multiplicity of* 1 *in* $\Psi_i(C)$ *is* 1.

Note that $|w| - (r - m(C))$ is even.

We make the following conjecture.

2.2. CONJECTURE. *If $i, i' \in \mathbf{N}$ satisfy $i + i' = |w| - (r - m(C))$, then $\Psi_i(C) = \Psi_{i'}(C)$. In particular the multiplicity of 1 in $\Psi_0(C)$ is 1.*

This is supported by the examples in §3.

2.3. For $\gamma \in \underline{\mathcal{U}}$ we set $d(\gamma) = \dim \mathcal{B}_u$ where $u \in \gamma$. According to Springer, for any $E \in \mathrm{Irr}(W)$ there is a unique $\gamma \in \underline{\mathcal{U}}$ such that $H^{2d(\gamma)}(\mathcal{B}_u)_E \neq 0$ for some/any $u \in \gamma$; we set $\Xi(E) = \gamma$, $d'(E) = d(\gamma)$. The map $\Xi : \mathrm{Irr}(W) \to \underline{\mathcal{U}}$ is surjective. We shall need the following property of Ξ.

(2.8) *Let $\gamma \in \underline{\mathcal{U}}$, $u \in \gamma$ and let $E \in \mathrm{Irr}(W)$ be such that $H^i(\mathcal{B}_u)_E \neq 0$ for some i. Then $d'(E) = d(\Xi(E)) \leq d(\gamma)$. If in addition $i < 2d(\gamma)$ then $d'(E) < d(\gamma)$.*

This follows from (1.12).

2.4. Let $\Phi : \underline{W} \to \underline{\mathcal{U}}$ be the (surjective) map defined in [**12**]. By definition, if $C \in \underline{W}$ and $w \in C_{min}$, then $\Phi(C) \cap (B^+wB^+) \neq \emptyset$; moreover, if $\gamma' \in \underline{\mathcal{U}}$ satisfies $\gamma' \cap (B^+wB^+) \neq \emptyset$ then $\Phi(C)$ is contained in the closure of γ' in $\mathcal{U}$.

Let $C \in \underline{W}$ and let $\gamma = \Phi(C)$. Let $w \in C_{min}$. We show:

(2.9) *If $E \in \mathrm{Irr}(W)$ appears in $\underline{\zeta}_{C,q}$ then $d'(E) \leq d(\gamma)$.*

From (1.3) we see that $H^i(\mathcal{B}_u)_E \neq 0$ for some $u \in \mathcal{U} \cap (B^+wB^+)$. Now (2.9) follows from (2.8).

We show:

(2.10) *If $E \in \mathrm{Irr}(W)$ appears in $\underline{\zeta}_{C,q}$ and $\Xi(E) \neq \gamma$ then $d'(E) < d(\gamma)$.*

From (1.3) we see that $H^i(\mathcal{B}_u)_E \neq 0$ for some $\gamma' \in \underline{\mathcal{U}}$ and some $u \in \gamma' \cap (B^+wB^+)$. If $i < 2d(\gamma')$ then by (2.8) we have $d'(E) < d(\gamma')$. Using (2.9) we deduce $d(\gamma) < d(\gamma')$; but by the definition of Φ, γ is contained in the closure of γ' so that $d(\gamma) \geq d(\gamma')$, a contradiction. If $i = 2d(\gamma')$ then $\Xi(E) = \gamma'$ hence $d'(E) = d(\gamma')$. Since γ is contained in the closure of γ' and $\gamma' \neq \gamma$ we have $d(\gamma') < d(\gamma)$ hence $d'(E) < d(\gamma)$. This proves (2.10).

(2.11) *If $E_0 \in \mathrm{Irr}(W)$ is the part of $H^{2d(\gamma)}(\mathcal{B}_{u_0})$ which is fixed by the action of the centralizer of u_0 ($u_0 \in \gamma$) then E_0 appears in $\underline{\zeta}_{C,q}$.*

We must show that $\sum_{u \in \mathcal{U}^F \cap (B^+wB^+)} \mathrm{tr}(F, H^*(\mathcal{B}_u)_{E_0}) \neq 0$. Assume that

$$\sum_{u \in \mathcal{U}^F \cap (B^+wB^+)} \mathrm{tr}(F, H^*(\mathcal{B}_u)_{E_0}) = 0. \tag{2.12}$$

If $\gamma' \in \underline{\mathcal{U}}$ is such that γ is contained in the closure of γ' and $\gamma \neq \gamma'$ then $H^i(\mathcal{B}_u)_{E_0} = 0$ for all i hence from (2.12) we deduce $\sum_{u \in \gamma^F \cap (B^+wB^+)} \mathrm{tr}(F, H^*(\mathcal{B}_u)_{E_0}) = 0$. It follows that $\sharp(u \in \gamma^F \cap (B^+wB^+))q^{d(\gamma)} = 0$. This contradicts $\gamma^F \cap (B^+wB^+) \neq \emptyset$. This proves (2.11).

2.5. Let $Z = \Psi(\underline{W}) \subset \mathbf{Z}[\mathbf{q}] \otimes \mathcal{R}_W$. For $\xi \in Z$ there is a unique $\gamma \in \underline{\mathcal{U}}$ such that the following holds:

If $E \in \mathrm{Irr}(W)$ appears in ξ then $d'(E) \leq d(\gamma)$. If $E \in \mathrm{Irr}(W)$ appears in ξ and $\Xi(E) \neq \gamma$ then $d'(E) < d(\gamma)$. If $E_0 \in \mathrm{Irr}(W)$ is the part of $H^{2d(\gamma)}(\mathcal{B}_{u_0})$ which is fixed by the action of the centralizer of u_0 ($u_0 \in \gamma$), then E_0 appears in ξ.

The existence of γ follows from §2.4; the uniqueness is obvious. Thus $E \mapsto \gamma$ is a well defined map $\Theta : Z \to \underline{\mathcal{U}}$. From §2.4 we have that $\Phi(C) = \Theta(\Psi(C))$ for any $C \in \underline{W}$. Since Φ is surjective, we see that Θ is surjective.

2.6. Let $G_{ad} = G/Z_G$. From (1.2) for any $u \in \mathcal{U}^F$ and $E \in \mathrm{Irr}(W)$ we have

(2.13) $$\mathrm{tr}(F, H^*(\mathcal{B}_u)_E) = \sharp(W)^{-1} \sum_{y \in W} \mathrm{tr}(y, E)\,\mathrm{tr}(u, R^1_y).$$

For y, y' in W we have

$$\sum_{u \in \mathcal{U}^F} \mathrm{tr}(u, R^1_y)\,\mathrm{tr}(u, R^1_{y'}) = \sharp(z \in W; zyz^{-1} = y')\det(q - y, \rho)^{-1}\sharp(G^F_{ad}),$$

see [**4**, Theorem 6.9]. Using this and (2.13) we obtain

(2.14) $$\begin{aligned}&\sum_{u \in \mathcal{U}^F} \mathrm{tr}(F, H^*(\mathcal{B}_u)_{E'})\,\mathrm{tr}(F, H^*(\mathcal{B}_u)_E)\\ &= \sharp(W)^{-2} \sum_{y \in W, z \in W} \mathrm{tr}(y, E')\,\mathrm{tr}(zyz^{-1}, E)\det(q - y, \rho)^{-1}\sharp(G^F_{ad})\\ &= \sharp(W)^{-1} \sum_{y \in W} \mathrm{tr}(y, E \otimes E')\det(q - y, \rho)^{-1}\sharp(G^F_{ad}).\end{aligned}$$

Using this and (1.11) we see that if $C \in \underline{W}$, $w \in C_{min}$ then the coefficient of $E \in \mathrm{Irr}(W)$ in $\underline{\zeta}_{C,q}$ is

(2.15) $$\sum_{\mathcal{E} \in \mathrm{Irr}(\mathcal{H}), E' \in \mathrm{Irr}(W)} A_{C,\mathcal{E}} A'_{\mathcal{E},E'} A''_{E',E},$$

where

$$\begin{aligned}A_{C,\mathcal{E}} &= (q-1)^{m(C)}\,\mathrm{tr}(T_w, \mathcal{E}),\\ A'_{\mathcal{E},E'} &= \sharp(W)^{-1} \sum_{z \in W} (\mathcal{F}_{\mathcal{E}} : R^1_z)_{G^F}\,\mathrm{tr}(z, E'),\\ A''_{E',E} &= \sharp(W)^{-1} \sum_{y \in W} \mathrm{tr}(y, E \otimes E')\det(q - y, \rho)^{-1}.\end{aligned}$$

Thus this coefficient is an entry of a product of three square matrices of size $\sharp(\underline{W}) = \sharp(\mathrm{Irr}(W))$. The matrices $(A_{C,\mathcal{E}}), (A''_{E',E})$ are known to be invertible when regarded as matrices with entries in $\mathbf{Q}$. We see that

(2.16) *$\{\underline{\zeta}_{C,q}; C \in \underline{W}\}$ is a basis of $\mathbf{Q} \otimes \mathcal{R}_W$ if and only if the matrix $(A'_{\mathcal{E},E'})$ (with entries independent of q) is invertible.*

The entries of the last matrix are some of the entries of the nonabelian Fourier transform [**9**] for the various two-sided cells of W. This matrix is not necessarily invertible. For example if W is of type B_n or C_n this matrix is invertible if $n \le 11$ but is not invertible if $n = 12$. Also if W is of type E_6 this matrix is not invertible.

2.7. PROPOSITION. *For any $y \in W$ we have*

$$a_{w,y} = \sum_{\mathcal{E} \in \mathrm{Irr}(W)} (-1)^{|y|}(\mathcal{F}_{\mathcal{E}} : \dim(R^1_y)R^1_y)\,\mathrm{tr}(T_w, \mathcal{E})q^{\nu}\sharp(\mathcal{B}^F)^{-1}.$$

PROOF. Combining (1.7),(1.11) and (2.14), we obtain the identity

$$\sharp(\mathcal{B}^F)a_{w,y}=\sharp(W)^{-2}\sum_{E,E',\mathcal{E}\in\mathrm{Irr}(W)}\sum_{z,y'\in W}$$

$$(\mathcal{F}_{\mathcal{E}}:R^1_z)\,\mathrm{tr}(z,E')\,\mathrm{tr}(T_w,\mathcal{E})\,\mathrm{tr}(y',E)\,\mathrm{tr}(y',E')\det(q-y',\rho)^{-1}\sharp(G^F_{ad})\,\mathrm{tr}(y,E),$$

where $(\mathcal{F}_{\mathcal{E}}:?)$ is the multiplicity of $\mathcal{F}_{\mathcal{E}}$ in the virtual representation ? of G^F, ρ is the reflection representation of W and G_{ad} is the adjoint group of G.

We now replace $\sum_{E'\in\mathrm{Irr}(W)}\mathrm{tr}(z,E')\,\mathrm{tr}(y',E')$ by $\sharp(e\in W;eze^{-1}=y')$ and we obtain

$$\sharp(\mathcal{B}^F)a_{w,y}=\sharp(W)^{-1}\sum_{E,\mathcal{E}\in\mathrm{Irr}(W)}\sum_{z\in W}$$

$$(\mathcal{F}_{\mathcal{E}}:R^1_z)\,\mathrm{tr}(T_w,\mathcal{E})\,\mathrm{tr}(z,E)\det(q-z,\rho)^{-1}\sharp(G^F_{ad})\,\mathrm{tr}(y,E).$$

We now replace $\sum_{E\in\mathrm{Irr}(W)}\mathrm{tr}(z,E)\,\mathrm{tr}(y,E)$ by $\sharp(e'\in W;e'ze'^{-1}=y)$ and we obtain

$$a_{w,y}=\sum_{\mathcal{E}\in\mathrm{Irr}(W)}(\mathcal{F}_{\mathcal{E}}:R^1_y)\,\mathrm{tr}(T_w,\mathcal{E})\det(q-y,\rho)^{-1}\sharp(G^F_{ad})\sharp(\mathcal{B}^F)^{-1}.$$

We now replace $\det(q-y,\rho)^{-1}\sharp(G^F_{ad})$ by $\dim(R^1_y)q^{\nu}(-1)^{|y|}$ and we obtain the desired identity. $\square$

2.8. In this subsection we diverge from the setup of §1.1; we assume instead that W is a finite Coxeter group. Then all ingredients of (2.13) make sense. The (constant) matrix $(A'_{\mathcal{E},E'})$ makes sense as a matrix whose entries involve the appropriate generalization of the nonabelian Fourier transform. Hence the map Ψ can be defined in this generality (although the ring of coefficients $\mathbf{Z}$ may have to be increased).

3. Examples

3.1. We return to the setup in §1.1. We shall denote by sgn the sign representation of W, by 1 the unit representation of W. Let ρ be as in §1.8. In this section we shall sometime write w instead of C when w is an element of $C\in\underline{W}$.

Assume first that W is of type A_1. The elements of $\underline{W}$ are represented by $1,s$ where s is the simple reflection. The objects of $\mathrm{Irr}(W)$ are $1,\rho=\mathrm{sgn}$. We have

$\Psi(1)=1+\mathrm{sgn}$,
$\Psi(s)=1$.

3.2. We now assume that W is of type A_2. The elements of $\underline{W}$ are represented by $1,s_1,c$ where s_1,s_2 are the simple reflections and $c=s_1s_2$. The objects of $\mathrm{Irr}(W)$ are $1,\rho,\mathrm{sgn}$. We have

$\Psi(1)=1+2\rho+\mathrm{sgn}$,
$\Psi(s_1)=1+\rho$,
$\Psi(c)=1$.

3.3. We now assume that W is of type B_2. The elements of $\underline{W}$ are represented by $1,s_1,s_2,c,c^2$ where s_1,s_2 are the simple reflections and $c=s_1s_2$. The objects of $\mathrm{Irr}(W)$ are $1,\rho,\epsilon',\epsilon'',\mathrm{sgn}$ where ϵ',e'' are the one dimensional representations other than $1,\mathrm{sgn}$. The following result was obtained by making use of [**13**]. We can arrange notation so that:

$$\begin{aligned}
&\Psi(1) = 1 + 2\rho + \epsilon' + \epsilon'' + \mathrm{sgn},\\
&\Psi(s_1) = 1 + \rho + \epsilon',\\
&\Psi(s_2) = 1 + \rho + \epsilon'',\\
&\Psi(c) = 1,\\
&\Psi(c^2) = \mathbf{q}^2 1 + \mathbf{q}\rho + 1.
\end{aligned}$$

3.4. We now assume that W is of type G_2. The elements of $\underline{W}$ are represented by $1, s_1, s_2, c, c^2, c^3$ where s_1, s_2 are the simple reflections and $c = s_1 s_2$. The objects of $\mathrm{Irr}(W)$ are $1, \rho, \epsilon', \epsilon'', \rho' = \rho \otimes \epsilon' = \rho \otimes \epsilon'', \mathrm{sgn}$ where ϵ', e'' are the one dimensional representations other than $1, \mathrm{sgn}$. The following result was obtained by making use of [**2**]. We can arrange notation so that:

$$\begin{aligned}
&\Psi(1) = 1 + 2\rho + 2\rho' + \epsilon' + \epsilon'' + \mathrm{sgn},\\
&\Psi(s_1) = 1 + \rho + \rho' + \epsilon',\\
&\Psi(s_2) = 1 + \rho + \rho' + \epsilon'',\\
&\Psi(c) = 1,\\
&\Psi(c^2) = \mathbf{q}^2 1 + \mathbf{q}\rho + 1,\\
&\Psi(c^3) = \mathbf{q}^4 1 + \mathbf{q}^3 \rho + \mathbf{q}^2(\rho' + 1) + \mathbf{q}\rho + 1.
\end{aligned}$$

3.5. In the examples above, $\Psi_i(C)$ is always an actual representation of W. This is not so in higher rank. We have calculated $\Psi(C)$ for G of type F_4 on a computer using the formulas in §2.6, by the same method as in [**12**, §1.2]; we thank Gongqin Li for programming this in GAP. If C consists of the longest element in W of type F_4 and E is a one-dimensional representation of W other than $1, \mathrm{sgn}$ then the coefficient of E in $\Phi(C)$ is $-2\mathbf{q}^{10}$; thus, $\Phi_{10}(C)$ is not an actual representation of W. A similar thing happens in type B_3 and C_3.

Acknowledgments

The authors would like to thank Andrew Salmon and the referee for their careful reading of the draft.

References

[1] Walter Borho and Robert MacPherson, *Représentations des groupes de Weyl et homologie d'intersection pour les variétés nilpotentes* (French, with English summary), C. R. Acad. Sci. Paris Sér. I Math. **292** (1981), no. 15, 707–710. MR618892

[2] Bomshik Chang and Rimhak Ree, *The characters of $G_2(q)$*, Symposia Mathematica, Vol. XIII (Convegno di Gruppi Abeliani & Convegno di Gruppi e loro Rappresentazioni, INDAM, Rome, 1972), Academic Press, London-New York, 1974, pp. 395–413. MR364419

[3] C. De Concini, G. Lusztig, and C. Procesi, *Homology of the zero-set of a nilpotent vector field on a flag manifold*, J. Amer. Math. Soc. **1** (1988), no. 1, 15–34, DOI 10.2307/1990965. MR924700

[4] P. Deligne and G. Lusztig, *Representations of reductive groups over finite fields*, Ann. of Math. (2) **103** (1976), no. 1, 103–161, DOI 10.2307/1971021. MR393266

[5] Pierre Deligne and George Lusztig, *Duality for representations of a reductive group over a finite field. II*, J. Algebra **81** (1983), no. 2, 540–545, DOI 10.1016/0021-8693(83)90202-8. MR700298

[6] Meinolf Geck and Götz Pfeiffer, *Characters of finite Coxeter groups and Iwahori-Hecke algebras*, London Mathematical Society Monographs. New Series, vol. 21, The Clarendon Press, Oxford University Press, New York, 2000. MR1778802

[7] D. Kazhdan, *Proof of Springer's hypothesis*, Israel J. Math. **28** (1977), no. 4, 272–286, DOI 10.1007/BF02760635. MR486181

[8] G. Lusztig, *Green polynomials and singularities of unipotent classes*, Adv. in Math. **42** (1981), no. 2, 169–178, DOI 10.1016/0001-8708(81)90038-4. MR641425

[9] George Lusztig, *Characters of reductive groups over a finite field*, Annals of Mathematics Studies, vol. 107, Princeton University Press, Princeton, NJ, 1984, DOI 10.1515/9781400881772. MR742472

[10] George Lusztig, *Character sheaves. V*, Adv. in Math. **61** (1986), no. 2, 103–155, DOI 10.1016/0001-8708(86)90071-X. MR849848

[11] George Lusztig, *Green functions and character sheaves*, Ann. of Math. (2) **131** (1990), no. 2, 355–408, DOI 10.2307/1971496. MR1043271

[12] G. Lusztig, *From conjugacy classes in the Weyl group to unipotent classes*, Represent. Theory **15** (2011), 494–530, DOI 10.1090/S1088-4165-2011-00396-4. MR2833465

[13] Bhama Srinivasan, *The characters of the finite symplectic group* $\mathrm{Sp}(4,q)$, Trans. Amer. Math. Soc. **131** (1968), 488–525, DOI 10.2307/1994960. MR220845

[14] M-T. Trinh, *From the Hecke Category to the Unipotent Locus*, arXiv:2106.07444, 2021.

Department of Mathematics, M.I.T., Cambridge, Massachusetts 02139
Email address: gyuri@math.mit.edu

Department of Mathematics, M.I.T., Cambridge, Massachusetts 02139
Email address: zyun@mit.edu

Proceedings of Symposia in Pure Mathematics
Volume **108**, 2024
https://doi.org/10.1090/pspum/108/01969

Geometric crystal on unipotent variety

Toshiki Nakashima

ABSTRACT. For a simply connected, simple algebraic group G over $\mathbb{C}$ and its Lie algebra $\mathfrak{g}$, let $\mathscr{U}(G)$ be its unipotent variety, the set of all unipotent elements and $\mathscr{N}(\mathfrak{g})$ be the nilpotent cone, the set of all ad-nilpotent elements in $\mathfrak{g}$. We shall show that there exists a geometric crystal structure on $\mathscr{U}(G)$ and then $\mathscr{N}(\mathfrak{g})$. For SL_2-case, we give certain positive structure and decoration function on $\mathscr{U}(SL_2)$ and show that its tropicalized crystal is isomorphic to $B(\infty) \otimes T_\lambda \otimes B(-\infty)$, the crystal of λ-component in the modified quantum algebra $\widetilde{U_q(\mathfrak{sl}_2)}$.

1. Introduction

For a simply connected simple algebraic group G over $\mathbb{C}$ let $\mathscr{U}(G)$ be the set of unipotent elements in G called a "unipotent variety", which is a closed irreducible variety of codimension rank(G) and is stable by Ad(G)-action. For a Lie algebra $\mathfrak{g} = \mathrm{Lie}(G)$, let $\mathscr{N}(\mathfrak{g})$ be a "nilpotent cone" which is defined to be a set of all ad-nilpotent elements in $\mathfrak{g}$ (see [**3, 6, 7, 14**]). Those play significant roles in representation theory, such as, classification of nilpotent orbits [**1**, **2**], Springer's resolution([**7, 21**]), Springer correspondence ([**22**]),etc. In this article, we shall try to define a geometric crystal structure on $\mathscr{U}(G)$.

The theory of geometric crystal is initiated by Berenstein A. and Kazhdan D. ([**4, 5**]), which is a sort of geometric analogue of Kashiwara crystal base theory (see e.g.,[**9, 10**]). A remarkable feature of the theory of geometric crystal is that if there exists a "positive structure" for a geometric crystal, by applying tropicalization functor we obtain certain Kashiwara crystal([**4**]). Furthermore, if there exists a "decoration function" on a positive geometric crystal, then one has a normal crystal, like as a crystal base $B(\lambda)$ of an irreducible highest weight module $V(\lambda)$ ([**5**]). The crystal $B(\infty)$ of the negative half subalgebra of the quantum algebra $U_q^-(\mathfrak{g}) \subset U_q(\mathfrak{g})$ is also obtained by considering a half-decorated positive geometric crystal ([**8**]).

Modified quantum algebra $\widetilde{U_q(\mathfrak{g})}$ has been defined by Lusztig([**17**]) and he also found that it holds a canonical base. Kashiwara developed its crystal structure completely ([**11**]). In this article, we shall introduce certain decorated positive

---

2020 *Mathematics Subject Classification.* Primary 17B37; 17B67; Secondary 81R50; 22E46; 14M15.

Key words and phrases. Crystal base, geometric crystal, unipotent variety, modified quantum algebra.

The author was supported in part by JSPS Grants in Aid for Scientific Research #15K04794.

geometric crystal structure on $\mathscr{U}(\mathrm{SL}_2)$ and show that its tropicalization is isomorphic to the crystal base of λ-component of modified quantum group $\widetilde{U_q(\mathfrak{sl}_2)}$. At the end of the article, we conjecture that there would exist certain positive decorated geometric crystal structure on unipotent variety $\mathscr{U}(G)$ for any simply connected, simple algebraic group over $\mathbb{C}$ such that its tropicalized crystal is isomorphic to the Langlands dual crystal for λ-component of modified quantum algebra $(B(\infty)\otimes T_\lambda\otimes B(-\infty))^\vee\cong B(U_q(\mathfrak{g}^\vee)a_\lambda)$.

2. Preliminaries

Let G be a simply connected simple algebraic group over $\mathbb{C}$, $\mathfrak{g}$ its simple Lie algebra generated by standard Chevalley generators $\{e_i,h_i,f_i\}_{i\in I}$, where $I:=\{1,2,\ldots,n=\mathrm{rank}(\mathfrak{g})\}$. Set $\mathfrak{n}_+:=\langle e_i\rangle_{i\in I}$, $\mathfrak{t}:=\langle h_i\rangle_{i\in I}$ and $\mathfrak{n}_+:=\langle e_i\rangle_{i\in I}$, which gives a triangular decomposition $\mathfrak{g}=\mathfrak{n}_+\oplus\mathfrak{t}\oplus\mathfrak{n}_-$. Let $\Delta=\Delta_+\sqcup\Delta_-$ be the root system of $\mathfrak{g}$ and $\Delta_\pm$ a set of positive/negative roots. We also put $\Pi:=\{\alpha_i\in\mathfrak{t}^*\}_{i\in I}$ the set of simple roots, $\{h_i\in\mathfrak{t}\}_{i\in I}$ the set of simple coroots and $(a_{ij}=\langle h_i,\alpha_j\rangle)_{i,j\in I}$ the associated Cartan matrix . We define an inner product on $\mathfrak{t}^*$ such that $(\alpha_i,\alpha_i)\in 2\mathbb{Z}_{\geq 0}$ and $\langle h_i,\lambda\rangle=2(\alpha_i,\lambda)/(\alpha_i,\alpha_i)$ for $\lambda\in\mathfrak{t}^*$. Let $\{\Lambda_i\in\mathfrak{t}^*\mid i\in I\}$ be a basis of $\mathfrak{t}^*$ such that $\langle h_i,\Lambda_j\rangle=\delta_{ij}$ for $i,j\in I$, which is called a fundamental weight.

Set $Q=\oplus_i\mathbb{Z}\alpha_i$, $Q_+=\oplus_i\mathbb{Z}_{\geq 0}\alpha_i$ and $Q_-=-Q_+$. We call Q a root lattice. Define $P:=\oplus_i\mathbb{Z}\Lambda_i\subset\mathfrak{t}^*$, called a weight lattice of $\mathfrak{t}^*$ satisfying $\mathfrak{t}^*\cong\mathbb{C}\otimes_{\mathbb{Z}}P$, and $P^*=\{h\in\mathfrak{t}|\langle h,P\rangle\subset\mathbb{Z}\}$. Let W be the associated Weyl group. Here we have $(\lambda,\alpha_i)\in\mathbb{Z}$ for $\lambda\in P$ by $(\alpha_i,\alpha_i)\in 2\mathbb{Z}_{\geq 0}$. We set $P_+=\{\lambda\in P|\langle\lambda,h_i\rangle\geq 0$ for any $i\in I\}$. An element of P_+ is called a dominant integral weight.

Let T be a maximal torus of G and $B^\pm$ Borel subgroups of G such that $T=B\cap B^-$. Let $U^\pm$ be the unipotent radical of $B^\pm$ respectively, where $\mathrm{Lie}(U^\pm)=\mathfrak{n}_\pm$. Let $U_\alpha:=\exp\mathfrak{g}_\alpha$ $(\alpha\in\Delta)$ be the one-parameter subgroup of G. For any $i\in I$, there exists a unique group homomorphism $\phi_i\colon SL_2(\mathbb{C})\to G$ such that

$$\phi_i\left(\begin{pmatrix}1&t\\0&1\end{pmatrix}\right)=\exp(te_i),\quad \phi_i\left(\begin{pmatrix}1&0\\t&1\end{pmatrix}\right)=\exp(tf_i)\qquad(t\in\mathbb{C}).$$

Set $\alpha_i^\vee(c)=c^{h_i}:=\phi_i\left(\left(\begin{smallmatrix}c&0\\0&c^{-1}\end{smallmatrix}\right)\right)$, $x_i(t):=\exp(te_i)$, $y_i(t):=\exp(tf_i)$, $G_i:=\phi_i(SL_2(\mathbb{C}))$, $T_i:=\alpha_i^\vee(\mathbb{C}^\times)$ and $N_i:=N_{G_i}(T_i)$, which is the normalizer of T_i in G_i.

We have the isomorphism $\phi:W\xrightarrow{\sim}N_G(T)/T$ defined by $\phi(s_i)=N_iT/T$. An element $\overline{s}_i:=x_i(-1)y_i(1)x_i(-1)$ is in $N_G(T)$, which is a representative of $s_i\in W=N_G(T)/T$.

Define the simple reflections $s_i\in\mathrm{Aut}(\mathfrak{t})$ $(i\in I)$ by $s_i(h):=h-\alpha_i(h)h_i$, which generate the Weyl group W.

3. Crystals

3.1. Definition of $U_q(\mathfrak{g})$. The quantized enveloping algebra $U_q(\mathfrak{g})$ is an associative $\mathbb{Q}(q)$-algebra generated by e_i, $f_i(i\in I)$ and $q^h(h\in P^*)$ satisfying the following relations:

$$q^0 = 1, \quad \text{and} \quad q^h q^{h'} = q^{h+h'}, \tag{3.1}$$

$$q^h e_i q^{-h} = q^{\langle h,\alpha_i\rangle} e_i, \qquad q^h f_i q^{-h} = q^{-\langle h,\alpha_i\rangle} f_i, \tag{3.2}$$

$$[e_i, f_j] = \delta_{i,j}(t_i - t_i^{-1})/(q_i - q_i^{-1}), \tag{3.3}$$

$$\sum_{k=1}^{1-\langle h_i,\alpha_j\rangle} (-1)^k x_i^{(k)} x_j x_i^{(1-\langle h_i,\alpha_j\rangle - k)} = 0, \qquad (i \neq j) \tag{3.4}$$

where $x_i = e_i, f_i$ and we set $q_i = q^{(\alpha_i,\alpha_i)/2}$, $t_i = q_i^{h_i}$, $[n]_i = (q_i^n - q_i^{-n})/(q_i - q_i^{-1})$, $[n]_i! = \prod_{k=1}^n [k]_i$, $e_i^{(n)} = e_i^n/[n]_i!$ and $f_i^{(n)} = f_i^n/[n]_i!$.

It is well-known that $U_q(\mathfrak{g})$ has a Hopf algebra structure with a comultiplication Δ given by

$$\Delta(q^h) = q^h \otimes q^h, \quad \Delta(e_i) = e_i \otimes t_i^{-1} + 1 \otimes e_i, \quad \Delta(f_i) = f_i \otimes 1 + t_i \otimes f_i,$$

for any $i \in I$ and $h \in P^*$. We do not describe an antipode and a counit. By this comultiplication, a tensor product of $U_q(\mathfrak{g})$-modules has a $U_q(\mathfrak{g})$-module structure.

3.2. Definition of crystals. Let us recall the definition of crystals [**10, 11**]. The notion of a crystal is motivated by abstracting some combinatorial properties of crystal bases.

DEFINITION 3.1. A *crystal* B is a set with the following data:

$$wt : B \longrightarrow P, \tag{3.5}$$

$$\varepsilon_i : B \longrightarrow \mathbb{Z} \sqcup \{-\infty\}, \quad \varphi_i : B \longrightarrow \mathbb{Z} \sqcup \{-\infty\}, \quad \text{for} \quad i \in I, \tag{3.6}$$

$$\tilde{e}_i : B \sqcup \{0\} \longrightarrow B \sqcup \{0\}, \quad \tilde{f}_i : B \sqcup \{0\} \longrightarrow B \sqcup \{0\} \quad \text{for} \quad i \in I. \tag{3.7}$$

Here 0 is an ideal element which is not included in B. They are subject to the following axioms: For b, b_1, $b_2 \in B$,

$$\varphi_i(b) = \varepsilon_i(b) + \langle h_i, wt(b)\rangle, \tag{3.8}$$

$$wt(\tilde{e}_i b) = wt(b) + \alpha_i \text{ if } \tilde{e}_i b \in B, \quad wt(\tilde{f}_i b) = wt(b) - \alpha_i \text{ if } \tilde{f}_i b \in B, \tag{3.9}$$

$$\tilde{e}_i b_2 = b_1 \text{ if and only if } \tilde{f}_i b_1 = b_2, \tag{3.10}$$

$$\text{if } \varepsilon_i(b) = -\infty, \text{ then } \tilde{e}_i b = \tilde{f}_i b = 0. \tag{3.11}$$

From (3.10), we can consider the following graph structure on a crystal B.

DEFINITION 3.2. The *crystal graph* of crystal B is an oriented and colored graph given by the rule: $b_1 \xrightarrow{i} b_2$ if and only if $b_2 = \tilde{f}_i b_1$ $(b_1, b_2 \in B)$.

DEFINITION 3.3.

(1) Let B_1 and B_2 be crystals. A morphism of crystals $\psi : B_1 \longrightarrow B_2$ is a map $\psi : B_1 \sqcup \{0\} \longrightarrow B_2 \sqcup \{0\}$ satisfying the following axioms:

$$\psi(0) = 0, \tag{3.12}$$

$$wt(b) = wt(\psi(b)), \quad \varepsilon_i(b) = \varepsilon_i(\psi(b)), \quad \varphi_i(b) = \varphi_i(\psi(b)) \quad \text{if } b \in B_1 \text{ and } \psi(b) \in B_2, \tag{3.13}$$

$$\psi(\tilde{e}_i b) = \tilde{e}_i \psi(b) \text{ if } b \in B_1 \text{ satisfies } \psi(b) \neq 0 \text{ and } \psi(\tilde{e}_i b) \neq 0, \tag{3.14}$$

$$\psi(\tilde{f}_i b) = \tilde{e}_i \psi(b) \text{ if } b \in B_1 \text{ satisfies } \psi(b) \neq 0 \text{ and } \psi(\tilde{f}_i b) \neq 0. \tag{3.15}$$

(2) A morphism of crystals $\psi : B_1 \longrightarrow B_2$ is called *strict* if the associated map from $B_1 \sqcup \{0\} \longrightarrow B_2 \sqcup \{0\}$ commutes with all $\tilde{e}_i$ and $\tilde{f}_i$. If ψ is injective, surjective and strict, ψ is called an *isomorphism*.

(3) A crystal B is a *normal*, if for any subset J of I such that $((\alpha_i, \alpha_j))_{i,j\in J}$ is a positive definite symmetric matrix, B is isomorphic to a crystal base of an integrable $U_q(\mathfrak{g}_J)$-module, where $U_q(\mathfrak{g}_J)$ is the quantum algebra generated by e_j, f_j $(j \in J)$ and q^h $(h \in P^*)$.

For crystals B_1 and B_2, we shall define their tensor product $B_1 \otimes B_2$ as follows:

$$B_1 \otimes B_2 = \{b_1 \otimes b_2 | b_1 \in B_1,\, b_2 \in B_2\}, \tag{3.16}$$

$$wt(b_1 \otimes b_2) = wt(b_1) + wt(b_2), \tag{3.17}$$

$$\varepsilon_i(b_1 \otimes b_2) = \max(\varepsilon_i(b_1), \varepsilon_i(b_2) - \langle h_i, wt(b_1)\rangle), \tag{3.18}$$

$$\varphi_i(b_1 \otimes b_2) = \max(\varphi_i(b_2), \varphi_i(b_1) + \langle h_i, wt(b_2)\rangle), \tag{3.19}$$

$$\tilde{e}_i(b_1 \otimes b_2) = \begin{cases} \tilde{e}_i b_1 \otimes b_2 & \text{if } \varphi_i(b_1) \geq \varepsilon_i(b_2) \\ b_1 \otimes \tilde{e}_i b_2 & \text{if } \varphi_i(b_1) < \varepsilon_i(b_2), \end{cases} \tag{3.20}$$

$$\tilde{f}_i(b_1 \otimes b_2) = \begin{cases} \tilde{f}_i b_1 \otimes b_2 & \text{if } \varphi_i(b_1) > \varepsilon_i(b_2) \\ b_1 \otimes \tilde{f}_i b_2 & \text{if } \varphi_i(b_1) \leq \varepsilon_i(b_2), \end{cases} \tag{3.21}$$

Here we understand that $0 \otimes b = b \otimes 0 = 0$. Let $\mathcal{C}(I, P)$ be the category of crystals determined by the data I and P. Then $\otimes$ is a functor from $\mathcal{C}(I,P) \times \mathcal{C}(I,P)$ to $\mathcal{C}(I,P)$ and satisfies the associative law: $(B_1 \otimes B_2) \otimes B_3 \cong B_1 \otimes (B_2 \otimes B_3)$ by $(b_1 \otimes b_2) \otimes b_3 \leftrightarrow b_1 \otimes (b_2 \otimes b_3)$. Therefore, the category of crystals is endowed with the structure of tensor category.

EXAMPLE 3.4. We give some examples of crystals.

(1) For $\lambda \in P$, we set $T_\lambda = \{t_\lambda\}$ with

$$wt(t_\lambda) = \lambda, \quad \varepsilon_i(t_\lambda) = \varphi_i(t_\lambda) = -\infty, \quad \tilde{e}_i(t_\lambda) = \tilde{f}_i(t_\lambda) = 0.$$

We can see that $T_\lambda \otimes T_\mu \cong T_{\lambda+\mu}$ and $B \otimes T_0 \cong T_0 \otimes B \cong B$ for any crystal B.

(2) For $i \in I$, we set $B_i := \{(n)_i \mid n \in \mathbb{Z}\}$ with

$$\begin{aligned} &wt((n)_i) = n\alpha_i, \quad && \varepsilon_i((n)_i) = -n, \quad && \varphi_i((n)_i) = n, \\ &\varepsilon_j((n)_i) = -\infty, && \varphi_j((n)_j) = -\infty \quad \text{for} \quad j \neq i, \\ &\tilde{e}_i(n)_i = (n+1)_i, && \tilde{f}_i(n)_i = (n-1)_i, \\ &\tilde{e}_j(n)_i = \tilde{f}_j(n)_i = 0 \quad \text{for} \quad j \neq i. \end{aligned}$$

(3) For $\lambda \in P_+$, let $(L(\lambda), B(\lambda))$ be the crystal base of a $U_q(\mathfrak{g})$-integrable highest weight module $V(\lambda)$. $B(\lambda)$ is the crystal associated with $V(\lambda)$. For $b \in B(\lambda)$, $\varepsilon_i(b)$ and $\varphi_i(b)$ are given by

$$\varepsilon_i(b) = \max\{n | \tilde{e}_i^n b \neq 0\}, \qquad \varphi_i(b) = \max\{n | \tilde{f}_i^n b \neq 0\}.$$

(4) Let $(L(\pm\infty), B(\pm\infty))$ be the crystal base of $U_q^{\mp}(\mathfrak{g})$. $B(\pm\infty)$ is a crystal associated with $U_q^{\mp}(\mathfrak{g})$. For any $b \in B(\infty)$ (resp. $b \in B(-\infty)$) and $i \in I$ there exists k such that $\tilde{e}_i^k b = 0$ (resp. $\tilde{f}_i^k b = 0$). Then, $\varepsilon_i(b)$ and $\varphi_i(b)$

are given by

$$\varepsilon_i(b) = \max\{n|\tilde{e}_i^n b \neq 0\} \quad (\text{resp. } \varphi_i(b) - \langle h_i, wt(b)\rangle,$$
$$\varphi_i(b) = \langle h_i, wt(b)\rangle + \varepsilon_i(b) \quad (\text{resp. } \max\{n|\tilde{f}_i^n b \neq 0\}).$$

4. Crystals of modified quantum algebra

This section is devoted to review [**11**],[**16**],[**17**].

4.1. Modified quantum algebra and crystal base. For an integral weight $\lambda \in P$, let $U_q(\mathfrak{g})a_\lambda$ be the left $U_q(\mathfrak{g})$-module given by

$$U_q(\mathfrak{g})a_\lambda := U_q(\mathfrak{g})/\sum_{h\in P^*} U_q(\mathfrak{g})(q^h - q^{\langle h,\lambda\rangle}), \tag{4.1}$$

where a_λ is the image of the unit by the canonical projection. We set

$$\widetilde{U_q(\mathfrak{g})} = \bigoplus_{\lambda\in P} U_q(\mathfrak{g})a_\lambda, \tag{4.2}$$

which is called a *modified quantum algebra* and we call $U_q(\mathfrak{g})a_\lambda$ a λ-*component* of $\widetilde{U_q(\mathfrak{g})}$.

We shall see a crystal base of $\widetilde{U_q(\mathfrak{g})}$. Taking $\lambda \in P$ and choosing $\zeta, \mu \in P_+$ such that $\lambda = \zeta - \mu$, we get the following $U_q(\mathfrak{g})$-linear surjective homomorphism:

$$\begin{aligned}\pi_{\zeta,\mu} : U_q(\mathfrak{g})a_\lambda &\longrightarrow V(\zeta)\otimes V(-\mu),\\ a_\lambda &\mapsto u_\zeta \otimes u_{-\mu}.\end{aligned} \tag{4.3}$$

where $V(\zeta)$ and $V(-\mu)$ are as in Example 3.4 (iii) and u_ζ and $u_{-\mu}$ are their highest weight vector and lowest weight vector respectively.

THEOREM 4.1 (cf. [**17**] [**11**]). *For any $\lambda \in P$, there exists a unique pair $(L(U_q(\mathfrak{g})a_\lambda), B(U_q(\mathfrak{g})a_\lambda))$ which satisfies the following properties.*

(1) *We set $A := \{f(q) \in \mathbb{Q}(q)|f$ has no pole at $q = 0\}$. $L(U_q(\mathfrak{g})a_\lambda)$ is a free A-module such that $U_q(\mathfrak{g})a_\lambda \cong \mathbb{Q}(q)\otimes_A L(U_q(\mathfrak{g})a_\lambda)$ and $B(U_q(\mathfrak{g})a_\lambda)$ is a $\mathbb{Q}$-basis of the $\mathbb{Q}$-vector space $L(U_q(\mathfrak{g})a_\lambda)/qL(U_q(\mathfrak{g})a_\lambda)$.*

(2) *For any $\zeta, \mu \in P_+$ with $\lambda = \zeta - \mu$, we have*

$$\pi_{\zeta,\mu}(L(U_q(\mathfrak{g})a_\lambda)) \subset L(\zeta)\otimes_A L(-\mu),$$

and the induced map $\bar{\pi}_{\zeta,\mu}$:

$$\bar{\pi}_{\zeta,\mu} : L(U_q(\mathfrak{g})a_\lambda)/qL(U_q(\mathfrak{g})a_\lambda) \longrightarrow (L(\zeta)/qL(\zeta))\otimes(L(-\mu)/qL(-\mu)),$$

satisfies $\bar{\pi}_{\zeta,\mu}(B(U_q(\mathfrak{g})a_\lambda)) \subset B(\zeta)\otimes B(-\mu)\sqcup\{0\}$.

(3) *There is a structure of crystal on $B(U_q(\mathfrak{g})a_\lambda)$ such that $\bar{\pi}_{\zeta,\mu}$ gives a strict morphism of crystals for any $\zeta, \mu \in P_+$ with $\lambda = \zeta - \mu$.*

Set

$$(L(\widetilde{U_q(\mathfrak{g})}), B(\widetilde{U_q(\mathfrak{g})})) := \bigoplus_{\lambda\in P}(L(U_q(\mathfrak{g})a_\lambda), B(U_q(\mathfrak{g})a_\lambda)).$$

REMARK. $B(U_q(\mathfrak{g})a_\lambda)$ is a normal crystal and then $B(\widetilde{U_q(\mathfrak{g})})$ is a normal crystal.

Let $B(\infty)$, $B(-\infty)$ and T_λ ($\lambda \in P$) be the crystals given in Example 3.4.

THEOREM 4.2 ([**11**]). $B(U_q(\mathfrak{g})a_\lambda) \cong B(\infty)\otimes T_\lambda\otimes B(-\infty)$ *as a crystal.*

COROLLARY 4.3. $B(\widetilde{U_q(\mathfrak{g})}) \cong \oplus_{\lambda\in P} B(\infty)\otimes T_\lambda \otimes B(-\infty)$ *as a crystal.*

EXAMPLE 4.4. We shall see $\mathfrak{sl}_2$-case. Set $I=\{1\}$. Let $B_1 := \{(n)_1 \mid n\in\mathbb{Z}\}$ be the crystal as in Example 3.4 (2) for $\mathfrak{g}=\mathfrak{sl}_2$ and $B(\pm\infty)$ the crystal of $U_q^{\mp}(\mathfrak{sl}_2)$. Now, set for a,b and $\lambda\in\mathbb{Z}$

$$(a,b)_\lambda := \tilde{f}^a(0)_1\otimes t_\lambda\otimes \tilde{e}^b(0)_1 = (-a)_1\otimes t_\lambda\otimes (b)_1 \in B_1\otimes T_\lambda\otimes B_1,$$

where if $a<0$ (resp. $b<0$), we understand $\tilde{f}^a(0)_1=\tilde{e}^{-a}(0)_1$ (resp. $\tilde{e}^b(0)_1 = \tilde{f}^{-b}(0)_1$). By the formula (3.18)–(3.21), we obtain

$$\mathrm{wt}(a,b)_\lambda = \lambda + 2(b-a), \tag{4.4}$$

$$\varepsilon(a,b)_\lambda = \max(a, 2a-b-\lambda) = -\min(-a,-2a+b+\lambda), \tag{4.5}$$

$$\tilde{e}(a,b)_\lambda = \begin{cases} (a-1,b)_\lambda & \text{if } b+\lambda\geq a,\\ (a,b+1)_\lambda & \text{if } b+\lambda< a,\end{cases} \tag{4.6}$$

$$\tilde{f}(a,b)_\lambda = \begin{cases} (a+1,b)_\lambda & \text{if } b+\lambda> a,\\ (a,b-1)_\lambda & \text{if } b+\lambda\leq a.\end{cases} \tag{4.7}$$

Indeed, $B(\infty)=\{\tilde{f}^a(0)_1 \mid a\geq 0\}$ (resp. $B(-\infty)=\{\tilde{e}^b(0)_1\mid b\geq 0\}$) assuming that if $a<0$ (resp. $b<0$) then $\tilde{f}^a(0)_1=0$ (resp. $\tilde{e}^b(0)_1=0$), then we get

$$B(\infty)\otimes t_\lambda\otimes B(-\infty) = \{(a,b)_\lambda \mid a,b\geq 0\}. \tag{4.8}$$

5. Geometric crystals

The basic reference for this section is [**4, 5**]. In this section, G, $\mathfrak{g}$ and $(a_{ij})_{i,j\in I}$, etc are as in Sect.2.

5.1. Definitions.

DEFINITION 5.1. Let X be an irreducible affine algebraic variety over $\mathbb{C}$, γ_i, ε_i, f $(i\in I)$ rational functions on X, and $e_i:\mathbb{C}^\times\times X\to X$ a rational $\mathbb{C}^\times$-action, where we will denote $e_i^c(x):=e_i(c,x)$.

(1) A quadruple $\chi=(X,\{e_i\}_{i\in I},\{\gamma_i,\}_{i\in I},\{\varepsilon_i\}_{i\in I})$ is a G (or $\mathfrak{g}$)-*geometric crystal* if
   (a) the action of e_i is unital, *i.e.*, $e_i(1,x)=x$ and $(\{1\}\times X)\cap dom(e_i)$ is open dense in $\{1\}\times X$ for any $i\in I$, where $dom(e_i)$ is the domain of definition of $e_i\colon\mathbb{C}^\times\times X\to X$.
   (b) The rational functions $\{\gamma_i\}_{i\in I}$ satisfy $\gamma_j(e_i^c(x))=c^{a_{ij}}\gamma_j(x)$ for any $i,j\in I$, where note that $c^{a_{ij}}=\langle\alpha_i^\vee(c),\alpha_j\rangle$.
   (c) e_i and e_j satisfy the following relations (Verma relations):

$$\begin{aligned}
&e_i^{c_1}e_j^{c_2}=e_j^{c_2}e_i^{c_1} && \text{if } a_{ij}=a_{ji}=0,\\
&e_i^{c_1}e_j^{c_1c_2}e_i^{c_2}=e_j^{c_2}e_i^{c_1c_2}e_j^{c_1} && \text{if } a_{ij}=a_{ji}=-1,\\
&e_i^{c_1}e_j^{c_1^2c_2}e_i^{c_1c_2}e_j^{c_2}=e_j^{c_2}e_i^{c_1c_2}e_j^{c_1^2c_2}e_i^{c_1} && \text{if } a_{ij}=-2,\ a_{ji}=-1,\\
&e_i^{c_1}e_j^{c_1^3c_2}e_i^{c_1^2c_2}e_j^{c_1^3c_2^2}e_i^{c_1c_2}e_j^{c_2}=e_j^{c_2}e_i^{c_1c_2}e_j^{c_1^3c_2^2}e_i^{c_1^2c_2}e_j^{c_1^3c_2}e_i^{c_1} && \text{if } a_{ij}=-3,\ a_{ji}=-1.
\end{aligned}$$

   (d) The rational functions $\{\varepsilon_i\}_{i\in I}$ satisfy $\varepsilon_i(e_i^c(x))=c^{-1}\varepsilon_i(x)$ and $\varepsilon_i(e_j^c(x))=\varepsilon_i(x)$ if $a_{i,j}=a_{j,i}=0$.

(2) For a geometric crystal $\chi = (X, \{e_i\}_{i\in I}, \{\gamma_i,\}_{i\in I}, \{\varepsilon_i\}_{i\in I})$, and a rational function f on X, a pair (χ, f) is a *decorated geometric crystal* if f satisfies

$$f(e_i^c(x)) = f(x) + (c-1)\varphi_i(x) + (c^{-1}-1)\varepsilon_i(x), \tag{5.1}$$

for any $i \in I$ and $x \in X$, where $\varphi_i := \varepsilon_i \cdot \gamma_i$. We call the function f a *decoration* or *potential*.

REMARK. The condition (5.1) of f is different from the ones in [**5**].

5.2. Unipotent crystals. In the sequel, we denote the unipotent subgroup U^+ by U. We shall define unipotent crystals (see [**4**],[**18**]).

DEFINITION 5.2. Let X be an irreducible affine variety over $\mathbb{C}$ and $\alpha : U\times X \to X$ a rational U-action such that α is defined on $\{e\}\times X$. Then, the pair $\mathbf{X} = (X, \alpha)$ is called a *U-variety*. For U-varieties $\mathbf{X} = (X, \alpha_X)$ and $\mathbf{Y} = (Y, \alpha_Y)$, a rational map $f : X \to Y$ is called a *U-morphism* if it commutes with the action of U.

Now, we define a U-variety structure on $B^- = U^-T$. The multiplication map in G induces the open embedding: $B^- \times U \hookrightarrow G$, which is a birational map. Let us denote the inverse birational map by $g : G \longrightarrow B^- \times U$. Then we define the rational maps $\pi^- : G \to B^-$ and $\pi : G \to U$ by $\pi^- := \mathrm{proj}_{B^-} \circ g$ and $\pi := \mathrm{proj}_U \circ g$. Now we define the rational U-action α_{B^-} on B^- by

$$\alpha_{B^-} := \pi^- \circ m : U \times B^- \longrightarrow B^-,$$

where m is the multiplication map in G. Then we get the U-variety $\mathbf{B}^- = (B^-, \alpha_{B^-})$.

DEFINITION 5.3. (1) Let $\mathbf{X} = (X, \alpha)$ be a U-variety and $f : X \to \mathbf{B}^-$ a U-morphism. The pair $(\mathbf{X}, f)$ is called a *unipotent G-crystal* or, for short, a *unipotent crystal*.

(2) Let $(\mathbf{X}, f_X)$ and $(\mathbf{Y}, f_Y)$ be unipotent crystals. A U-morphism $g : \mathbf{X} \to \mathbf{Y}$ is called a *morphism of unipotent crystals* if $f_X = f_Y \circ g$. In particular, if g is a birational map, it is called an *isomorphism of unipotent crystals*.

We define a product of unipotent crystals following [**4**]. For unipotent crystals $(\mathbf{X}, f_X)$, $(\mathbf{Y}, f_Y)$, define the morphism $\alpha_{X\times Y} : U \times X \times Y \to X \times Y$ by

$$\alpha_{X\times Y}(u, x, y) := (\alpha_X(u, x), \alpha_Y(\pi(u \cdot f_X(x)), y)). \tag{5.2}$$

THEOREM 5.4 ([**4**]).

(1) *The morphism $\alpha_{X\times Y}$ defined above is a rational U-action on $X \times Y$.*
(2) *Let $\mathbf{m} : B^- \times B^- \to B^-$ be a multiplication map and $f = f_{X\times Y} : X\times Y \to B^-$ be the rational map defined by*

$$f_{X\times Y} := \mathbf{m} \circ (f_X \times f_Y).$$

Then $f_{X\times Y}$ is a U-morphism and $(\mathbf{X} \times \mathbf{Y}, f_{X\times Y})$ is a unipotent crystal, which we call a product of unipotent crystals $(\mathbf{X}, f_X)$ and $(\mathbf{Y}, f_Y)$.
(3) *Product of unipotent crystals is associative.*

5.3. From unipotent crystals to geometric crystals. For $i \in I$, set $U_i^{\pm} := U^{\pm} \cap \bar{s}_i U^{\mp} \bar{s}_i^{-1}$ and $U_{\pm}^i := U^{\pm} \cap \bar{s}_i U^{\pm} \bar{s}_i^{-1}$. Indeed, $U_i^{\pm} = U_{\pm\alpha_i}$. Set

$$Y_{\pm\alpha_i} := \langle x_{\pm\alpha_i}(t) U_\alpha x_{\pm\alpha_i}(-t) | t \in \mathbb{C}, \quad \alpha \in \Delta_{\pm}^{\mathrm{re}} \setminus \{\pm\alpha_i\} \rangle.$$

We have the unique decomposition; $U^- = U_i^- \cdot Y_{\pm\alpha_i} = U_{-\alpha_i} \cdot U_-^i$. By using this decomposition, we get the canonical projection $\xi_i : U^- \to U_{-\alpha_i}$ and define the function χ_i on U^- by

$$\chi_i := y_i^{-1} \circ \xi_i : U^- \longrightarrow U_{-\alpha_i} \xrightarrow{\sim} \mathbb{C}, \tag{5.3}$$

and extend this to the function on B^- by $\chi_i(u \cdot t) := \chi_i(u)$ for $u \in U^-$ and $t \in T$. For a unipotent G-crystal (X, f_X), we define the function $\varepsilon_i := \varepsilon_i^X : X \to \mathbb{C}$ by

$$\varepsilon_i := \chi_i \circ f_X,$$

and the rational function $\gamma_i : X \to \mathbb{C}$ by

$$\gamma_i := \alpha_i \circ \mathrm{proj}_T \circ f_X : X \to B^- \to T \to \mathbb{C}, \tag{5.4}$$

where proj_T is the canonical projection.

REMARK. Note that the function ε_i is denoted by φ_i in [**4**, **18**].

Suppose that the function ε_i is not identically zero on X. We define the morphism $e_i : \mathbb{C}^\times \times X \to X$ by

$$e_i^c(x) := x_i \left(\frac{c-1}{\varepsilon_i(x)} \right)(x). \tag{5.5}$$

THEOREM 5.5 ([**4**]). *For a unipotent G-crystal $(\mathbf{X}, f_X)$, suppose that the function ε_i is not identically zero for any $i \in I$. Then the rational functions $\gamma_i, \varepsilon_i : X \to \mathbb{C}$ and $e_i : \mathbb{C}^\times \times \mathbf{X} \to \mathbf{X}$ as above define a geometric G-crystal $(\mathbf{X}, \{e_i\}_{i\in I}, \{\gamma_i\}_{i\in I}, \{\varepsilon_i\}_{i\in I})$, which is called the induced geometric G-crystal by the unipotent G-crystal $(\mathbf{X}, f_X)$.*

PROPOSITION 5.6 ([**4**]). *For unipotent G-crystals (X, f_X) and (Y, f_Y), set the product $(Z, f_Z) := (X, f_X) \times (Y, f_Y)$, where $Z = X \times Y$. Let*

$$(Z, \{e_i^Z\}_{i\in I}, \{\gamma_i^Z\}_{i\in I}, \{\varepsilon_i^Z\}_{i\in I})$$

be the induced geometric G-crystal from (Z, f_Z). Then we obtain:

(1) *For each $i \in I$, $(x, y) \in Z$,*

$$\gamma_i^Z(x,y) = \gamma_i^X(x)\gamma_i^Y(y), \qquad \varepsilon_i^Z(x,y) = \varepsilon_i^X(x) + \frac{\varepsilon_i^Y(y)}{\gamma_i^X(x)}. \tag{5.6}$$

(2) *For any $i \in I$, the action $e_i^Z : \mathbb{C}^\times \times Z \to Z$ is given by: $(e_i^Z)^c(x,y) = ((e_i^X)^{c_1}(x), (e_i^Y)^{c_2}(y))$, where*

$$c_1 = \frac{c\gamma_i^X(x)\varepsilon_i^X(x) + \varepsilon_i^Y(y)}{\gamma_i^X(x)\varepsilon_i^X(x) + \varepsilon_i^Y(y)}, \quad c_2 = \frac{c(\gamma_i^X(x)\varepsilon_i^X(x) + \varepsilon_i^Y(y))}{c\gamma_i^X(x)\varepsilon_i^X(x) + \varepsilon_i^Y(y)}. \tag{5.7}$$

Here note that $c_1 c_2 = c$. The formula c_1 and c_2 in [**4**] seem to be different from ours.

5.4. Positive structure and tropicalization. In this subsection, we review the notion positive structure and the tropicalization ([**4, 5**]). Note that in [**18, 19**] we used the term ultra-discretization instead of tropicalization. Indeed they are the same notions.

DEFINITION 5.7. Let T, T' be split algebraic tori over $\mathbb{C}$.

(1) A regular function $f = \sum_{\mu \in X^*(T)} c_\mu \cdot \mu$ on T is *positive* if all coefficients c_μ are non-negative numbers. A rational function on T is said to be *positive* if there exist positive regular functions g, h such that $f = \frac{g}{h}$ $(h \neq 0)$.
(2) Let $f : T \to T'$ be a rational map between T and T'. Then we say that f is *positive* if for any $\xi \in X^*(T')$ we have that $\xi \circ f$ is positive in the above sense.

Note that if f, g are positive rational functions on T, then $f \cdot g$, f/g and $f + g$ are all positive.

DEFINITION 5.8. Let $\chi = (X, \{e_i\}_{i \in I}, \{\gamma_i\}_{i \in I}, \{\varepsilon_i\}_{i \in I}, f)$ be a decorated geometric crystal, T' an algebraic torus and $\theta : T' \to X$ a birational map. The birational map θ is called *positive structure* on χ if it satisfies:

(1) For any $i \in I$ the rational functions $\gamma_i \circ \theta, \varepsilon_i \circ \theta, f \circ \theta : T' \to \mathbb{C}$ are all positive in the above sense.
(2) For any $i \in I$, the rational map $e_{i,\theta} : \mathbb{C}^\times \times T' \to T'$ defined by $e_{i,\theta}(c, t) := \theta^{-1} \circ e_i^c \circ \theta(t)$ is positive.

Let $v : \mathbb{C}(c) \setminus \{0\} \to \mathbb{Z}$ be a map defined by $v(f(c)) := -\deg(f(c^{-1}))$, which is different from that in e.g., [**12, 13, 18, 20**]. Here, we have the formula for positive rational functions f and g:

$$v(f \cdot g) = v(f) + v(g), \quad v(f/g) = v(f) - v(g), \quad v(f + g) = \min(v(f), v(g)). \tag{5.8}$$

Let $f \colon T \to T'$ be a positive rational mapping of algebraic tori T and T'. We define a map $\widehat{f} \colon X_*(T) \to X_*(T')$ by

$$\langle \eta, \widehat{f}(\xi) \rangle = v(\eta \circ f \circ \xi),$$

where $\eta \in X^*(T')$ and $\xi \in X_*(T)$.

Let $\mathcal{T}_+$ be the category whose objects are algebraic tori over $\mathbb{C}$ and whose morphisms are positive rational maps. Then, we obtain the functor called "tropicalization functor",

$$\begin{aligned} \mathcal{TR} : \mathcal{T}_+ &\longrightarrow \mathfrak{Set} \\ T &\mapsto X_*(T) \\ (g : T \to T') &\mapsto (\widehat{g} : X_*(T) \to X_*(T')). \end{aligned}$$

Let $\theta : T' \to X$ be a positive structure on a decorated geometric crystal $\chi = (X, \{e_i\}_{i \in I}, \{\gamma_i\}_{i \in I}, \{\varepsilon_i\}_{i \in I}, f)$. Applying the functor $\mathcal{TR}$ to positive rational morphisms $e_{i,\theta} : \mathbb{C}^\times \times T' \to T'$ and $\gamma_i \circ \theta, \varepsilon_i \circ \theta, f \circ \theta : T' \to \mathbb{C}$ (the notations are as above), we obtain

$$\begin{aligned} \tilde{e}_i &:= \mathcal{TR}(e_{i,\theta}) : \mathbb{Z} \times X_*(T') \to X_*(T') \\ \mathrm{wt}_i &:= \mathcal{TR}(\gamma_i \circ \theta) : X_*(T') \to \mathbb{Z}, \\ \widetilde{\varepsilon}_i &:= \mathcal{TR}(\varepsilon_i \circ \theta) : X_*(T') \to \mathbb{Z}, \\ \widetilde{f} &:= \mathcal{TR}(f \circ \theta) : X_*(T') \to \mathbb{Z}. \end{aligned}$$

Now, for given positive structure $\theta : T' \to X$ on a geometric crystal

$$\chi = (X, \{e_i\}_{i\in I}, \{\gamma_i\}_{i\in I}, \{\varepsilon_i\}_{i\in I}),$$

we associate the quadruple $(X_*(T'), \{\tilde{e}_i\}_{i\in I}, \{\mathrm{wt}_i\}_{i\in I}, \{\tilde{\varepsilon}_i\}_{i\in I})$ with a free pre-crystal structure (see [**4**, 2.2]) and denote it by $\mathcal{TR}_{\theta,T'}(\chi)$. We have the following theorem:

THEOREM 5.9 ([**4,5,18**]). *For any geometric crystal* $\chi = (X, \{e_i\}_{i\in I}, \{\gamma_i\}_{i\in I}, \{\varepsilon_i\}_{i\in I})$ *and positive structure* $\theta : T' \to X$*, the associated pre-crystal*

$$\mathcal{TR}_{\theta,T'}(\chi) = (X_*(T'), \{e_i\}_{i\in I}, \{\mathrm{wt}_i\}_{i\in I}, \{\tilde{\varepsilon}_i\}_{i\in I})$$

is a Langlands dual Kashiwara's crystal.

REMARK. Let us denote the crystal $\mathcal{TR}_{\theta,T'}(\chi)$ by $\mathbb{B}$. If we take the map $\nu' : \mathbb{C}(c) \setminus \{0\} \to \mathbb{Z}$ as

$$\nu'(f(c)) = \deg(f(c)),$$

then we also obtain a Kashiwara's crystal $\mathbb{B}'$. Comparing the crystals $\mathbb{B}$ and $\mathbb{B}'$, we obtain

$$\mathbb{B} \cong \mathbb{B}'^{\,\mathrm{op}},$$

where $\mathbb{B}'^{\,\mathrm{op}}$ is the opposite crystal of $\mathbb{B}'$ as in [**5**].Indeed, e.g., the tensor product rules of opposite crystal are different from the one of the original Kashiwara's crystal.

Now, for a positive decorated geometric crystal

$$\mathcal{X} = ((X, \{e_i\}_{i\in I}, \{\gamma_i\}_{i\in I}, \{\varepsilon_i\}_{i\in I}, f), \theta, T'),$$

set

$$\widetilde{B}_{\widetilde{f}} := \{\widetilde{x} \in X_*(T') | \widetilde{f}(\widetilde{x}) \geq 0\}, \tag{5.9}$$

where $X_*(T')$ is identified with $\mathbb{Z}^{\dim(T')}$. Define

$$B_{f,\theta} := (\widetilde{B}_{\widetilde{f}}, \mathrm{wt}_i|_{\widetilde{B}_{\widetilde{f}}}, \varepsilon_i|_{\widetilde{B}_{\widetilde{f}}}, e_i|_{\widetilde{B}_{\widetilde{f}}})_{i\in I}. \tag{5.10}$$

PROPOSITION 5.10 ([**5**]). *For a positive decorated geometric crystal*

$$\mathcal{X} = ((X, \{e_i\}_{i\in I}, \{\gamma_i\}_{i\in I}, \{\varepsilon_i\}_{i\in I}, f), \theta, T'),$$

the quadruple $B_{f,\theta}$ *in* (5.10) *is a Langlands dual normal crystal.*

6. Unipotent varieties and nilpotent cones

We use the notation as in Sect.2. Let us recall that $x \in \mathfrak{g}$ is called *nilpotent* if $\mathrm{ad}(x) : \mathfrak{g} \to \mathfrak{g}$ is nilpotent. It is well-known that for $\mathfrak{g} = \mathfrak{sl}_{n+1}(\mathbb{C})$, $x \in \mathfrak{g}$ is nilpotent iff x is a nilpotent matrix in usual sense. Define

$$\mathcal{N} = \mathcal{N}(\mathfrak{g}) := \{x \in \mathfrak{g} \mid x \text{ is nilpotent}\}.$$

PROPOSITION 6.1 ([**3**, E-61,Theorem 3.3]). *The set of nilpotent elements* $\mathcal{N}$ *is a closed irreducible subvariety of* $\mathfrak{g}$ *of codimension* n *and it is stable by* $Ad(G)$*-action and* $\mathbb{C}^\times$*-action.* $\mathcal{N}$ *is called a nilpotent cone of* $\mathfrak{g}$*.*

Define the set of unipotent elements of G as

$$\mathcal{U} = \mathcal{U}(G) := \{g \in G \mid g \text{ is a unipotent element}\} \tag{6.1}$$

PROPOSITION 6.2 ([**3**, E-51, Theorem 1.8]). *The set of unipotent elements* $\mathcal{U}$ *is a closed irreducible subvariety of* G *of codimension* n *and it is stable by* $Ad(G)$*-action.* $\mathcal{U}$ *is called a unipotent variety of* G*.*

THEOREM 6.3 ([**3**, E-63, Theorem 3.12],[**6**, 1.15]). *There exists an isomorphism of varieties $F : \mathscr{U} \to \mathscr{N}$ which is compatible with the* $\mathrm{Ad}(G)$*-action.*

EXAMPLE 6.4 ([**3**, Example 3.14]).

(1) For $G = \mathrm{SL}_{n+1}(\mathbb{C})$, one has $F(X) = X - \mathrm{Id}$ in the matrix algebra, where Id is the identity matrix.
(2) For $G = \mathrm{SO}_n(\mathbb{C})$, one has $F(X) = (\mathrm{Id} - X)(\mathrm{Id} + X)^{-1}$.

The following theorem is well-known as "Springer's resolution", see e.g.,[**7**],[**21**].

THEOREM 6.5. *Let $\mathcal{B} := G/B$ be the flag variety and $T^*\mathcal{B}$ its cotangent bundle. There exists the resolution of singularities $\mu : T^*\mathcal{B} \to \mathscr{N}$, called "Springer's resolution".*

We shall denote the centralizer of $x \in \mathfrak{g}$ (resp. $g \in G$) in $\mathfrak{g}$ (resp. G) by $\mathcal{C}_{\mathfrak{g}}(x)$ (resp. $\mathcal{C}_G(g)$).

DEFINITION 6.6. An element $g \in G$ (resp. $x \in \mathfrak{g}$) is called *regular* if $\dim \mathcal{C}_G(g)$ (resp. $\dim \mathcal{C}_{\mathfrak{g}}(x)) = \mathrm{rank}(\mathfrak{g})$.

PROPOSITION 6.7 ([**3**, E-61, 3.5],[**7**, Corollary 3.2,13.]). *Let $\{e_\alpha \mid \alpha \in \Delta_+\}$ be a set of positive root vectors. Then, $e := \sum_{\alpha\in\Delta_+} c_\alpha e_\alpha$ is a regular nilpotent element in $\mathfrak{n}_+$ iff $c_\alpha \neq 0$ for all simple roots $\alpha \in \Pi$.*

We also have the analogous statement for unipotent elements:

PROPOSITION 6.8 ([**3**, E-54, 1.13]). *Let $x_\alpha(c) := \exp(ce_\alpha)$ $(\alpha \in \Delta_+)$ be a generator of one-parameter subgroup U_α. Then, $x := \prod_{\alpha\in\Delta_+} x_\alpha(c_\alpha)$ is a regular unipotent element in U iff $c_\alpha \neq 0$ for all simple roots $\alpha \in \Pi$. Here note that the product of* $\exp(c\alpha)$*'s is made by certain proper total order on Δ_+.*

PROPOSITION 6.9 ([**14**],[**6**, Proposition 5.1.2]). *Set*

$$\mathscr{N}^{\mathrm{reg}} := \{x \in \mathscr{N} \mid x \text{ is regular}\} \quad \text{and} \quad \mathscr{U}^{\mathrm{reg}} := \{g \in \mathscr{U} \mid g \text{ is regular}\}. \tag{6.2}$$

Then, $\mathscr{N}^{\mathrm{reg}}$ (resp.$\mathscr{U}^{\mathrm{reg}}$) is a Zariski open, dense, conjugacy class in $\mathscr{N}$ (reap. $\mathscr{U}$).

The following corollary is almost trivial from Proposition 6.8 and Proposition 6.9.

COROLLARY 6.10. *Let $y_\alpha(c) := \exp(cf_\alpha)$ $(\alpha \in \Delta_+)$ be a generator of one-parameter subgroup $U_{-\alpha}$, where $\{f_\alpha \mid \alpha \in \Delta_+\}$ is the set of negative root vectors. Then, $y := \prod_{\alpha\in\Delta_+} y_\alpha(c_\alpha)$ is a regular unipotent element in U^- iff $c_\alpha \neq 0$ for all simple roots $\alpha \in \Pi$.*

PROOF. Let $\overline{w_0} \in G$ be a representative of the longest Weyl group element w_0. Indeed, for a regular unipotent element $x := \prod_{\alpha\in\Delta_+} x_\alpha(c_\alpha) \in U \cap \mathscr{U}^{\mathrm{reg}}$ one finds that $\overline{w_0}\, x\, \overline{w_0}^{-1} = \prod_{\alpha\in\Delta_+} y_\alpha(c'_\alpha) \in \mathscr{U}^{\mathrm{reg}} \cap U^-$ for some c'_α by Proposition 6.9 and $c_\alpha \neq 0$ iff $c'_\alpha \neq 0$, which shows the desired result. □

THEOREM 6.11 ([**3**, E-63]). *Let $F : \mathscr{U} \to \mathscr{N}$ be as in Theorem 6.3. Set $F^{\mathrm{reg}} := F_{|\mathscr{U}^{\mathrm{reg}}}$. Then F^{reg} gives an isomorphism of varieties $F^{\mathrm{reg}} : \mathscr{U}^{\mathrm{reg}} \xrightarrow{\sim} \mathscr{N}^{\mathrm{reg}}$.*

PROPOSITION 6.12 ([**6**, Proposition 5.1.3]). *For a unipotent element u in G, the following conditions are equivalent.*

(1) *u is regular.*

(2) *u lies in a unique Borel subgroup of G.*

Let us show the following lemma

LEMMA 6.13. *Let $\mathscr{U}=\mathscr{U}(G)$ be the unipotent variety in G. Then the intersection $\mathscr{U}\cap B^-U$ is open dense in $\mathscr{U}$.*

PROOF. Since the subset B^-U is isomorphic to the big cell $B\overline{w_0}B$, it is open dense in G. And the unipotent variety $\mathscr{U}$ includes $U^{\pm}$, the intersection $\mathscr{U}\cap B^-U$ is non-empty open subset of $\mathscr{U}$. Furthermore, since $\mathscr{U}$ is irreducible by Proposition 6.2, its non-empty open subset is dense and then we find that $\mathscr{U}\cap B^-U$ is open dense in $\mathscr{U}$. □

7. Geometric crystal structure on unipotent varieties

In this section, we shall construct a geometric crystal structure on a unipotent variety.

7.1. Unipotent crystal structure on the unipotent variety $\mathscr{U}$. First let us show the following lemma.

LEMMA 7.1. *Let X be a subvariety of G such that X is stable* $\mathrm{Ad}(U)$*-action and there exists a subset $O\subset X\cap B^-U$ such that O is open dense in X. Then X is equipped with a unipotent crystal structure.*

PROOF. Let $\pi^-:G\to B^-$ be the rational map as in 5.2. Since X is a U-variety by the $\mathrm{Ad}(U)$-action, it suffices to show that $\pi^-|_X:X\to B^-$ is a U-morphism. For an arbitrary element $x\in O\subset X\cap B^-U$ there exist $b^-\in B^-$ and $u\in U$ such that $x=b^-u$. For any $g\in U$, one has

$$\pi^-|_X(gxg^{-1})=\pi^-|_X(gb^-ug^{-1})=\pi^-(gb^{-1}).$$

On the other hand, one has

$$\alpha_{B^-}(g,\pi^-|_X(x))=\alpha_{B^-}(g,b^-)=\pi^-(gb^-).$$

Then, we obtain $\pi^-|_X(gxg^{-1})=\alpha_{B^-}(g,\pi^-|_X(x))$, which means that $\pi^-|_X$ is a U-morphism. □

Here, we have an open embedding $\iota:\mathscr{U}\cap B^-U\hookrightarrow\mathscr{U}$ by Lemma 6.13. Thus, there exists the inverse birational map denoted by $\omega:\mathscr{U}\to\mathscr{U}\cap B^-U$. By Lemma 6.13 and Lemma 7.1 we obtain the following theorem;

THEOREM 7.2. *Let us define a morphism $\mathbf{f}_{\mathscr{U}}:\mathscr{U}\xrightarrow{\omega}\mathscr{U}\cap B^-U\xrightarrow{\pi^-}B^-$. Then the pair $(\mathscr{U},\mathbf{f}_{\mathscr{U}})$ is a unipotent crystal.*

7.2. Geometric crystal structure on the unipotent variety $\mathscr{U}$. Applying Lemma 7.1 to the unipotent variety $\mathscr{U}$, we obtain a geometric crystal structure on $\mathscr{U}$.

Let us define the rational function $\varepsilon_i:\mathscr{U}\to\mathbb{C}$ by

$$\varepsilon_i:\mathscr{U}\xrightarrow{\pi^-}B^-\to U^-\xrightarrow{\chi_i}\mathbb{C} \tag{7.1}$$

Now, we shall see the following lemma.

LEMMA 7.3. *The function ε_i on $\mathscr{U}$ is non-trivial for all $i\in I$.*

PROOF. By the argument of the proof of Corollary 6.10 one finds that there exists an unipotent regular element $y \in \mathscr{U} \cap U^-$ of the form

$$y = \prod_{\alpha \in \Delta_+} y_\alpha(c_\alpha),$$

where $c_\alpha \neq 0$ for any $\alpha \in \Pi$, which implies the claims. □

Now, we define the morphism $e_i : \mathbb{C}^\times \times \mathscr{U} \to \mathscr{U}$ and the function γ_i on $\mathscr{U}$ for $i \in I$ by

(7.2)

$$e_i^c(x) := x_i\Big(\frac{c-1}{\varepsilon_i(x)}\Big) \cdot x \cdot x_i\Big(\frac{1-c}{\varepsilon_i(x)}\Big), \quad \gamma_i(x) = \alpha_i(\mathrm{proj}_T(\mathbf{f}_{\mathscr{U}}(x))) \quad \text{for } x \in \mathscr{U}, \quad c \in \mathbb{C}^\times.$$

THEOREM 7.4. *The quadruple* $(\mathscr{U}, \{e_i\}_{i\in I}, \{\varepsilon_i\}_{i\in I}, \{\gamma_i\}_{i\in I})$ *is a G-geometric crystal.*

PROOF. Since by Lemma 7.3, one has that ε_i is not identically zero for any $i \in I$, we can apply Theorem 5.5 to the unipotent crystal $(\mathscr{U}, \mathbf{f}_{\mathscr{U}})$. □

COROLLARY 7.5. *Through the birational map* $F^{-1} \circ \mu : T^*\mathcal{B} \xrightarrow{\mu} \mathscr{N} \xrightarrow{F^{-1}} \mathscr{U}$ *(resp.* $F^{-1} : \mathscr{N} \to \mathscr{U}$*), there exists a geometric crystal structure on* $T^*\mathcal{B}$ *(resp.* $\mathscr{N}$*) induced from the one on* $\mathscr{U}$*.*

7.3. SL_2**-case.** Here let us see the SL_2-case. In this case, we can present a positive structure and a decoration on $\mathscr{U}$.

One has

$$\mathscr{U}(\mathrm{SL}_2) = \left\{ \begin{pmatrix} a & b \\ c & d \end{pmatrix} \in \mathrm{Mat}_2(\mathbb{C}) \;\middle|\; \begin{matrix} a+d=2, \\ ad-bc=1 \end{matrix} \right\}. \tag{7.3}$$

Let $(\mathscr{U}(\mathrm{SL}_2), e, \varepsilon, \gamma)$ be the geometric crystal on $\mathscr{U}(\mathrm{SL}_2)$ defined in the previous subsection. Here the open dense subset $O := \mathscr{U}(\mathrm{SL}_2) \cap B^-U$ is given by

$$O = \left\{ \begin{pmatrix} a & b \\ c & d \end{pmatrix} \in \mathrm{Mat}_2(\mathbb{C}) \;\middle|\; \begin{matrix} a+d=2,\ ad-bc=1 \\ a \neq 0 (\Leftrightarrow d \neq 2) \end{matrix} \right\}. \tag{7.4}$$

An element $X = \begin{pmatrix} a & b \\ c & d \end{pmatrix}$ in O can be decomposed as follows:

$$\begin{pmatrix} a & b \\ c & d \end{pmatrix} = \begin{pmatrix} a & 0 \\ c & \frac{1}{a} \end{pmatrix} \begin{pmatrix} 1 & \frac{b}{a} \\ 0 & 1 \end{pmatrix} \qquad \Big(d = \frac{bc+1}{a}\Big). \tag{7.5}$$

Thus, we have

$$\pi^-(X) = \begin{pmatrix} a & 0 \\ c & \frac{1}{a} \end{pmatrix} = \begin{pmatrix} 1 & 0 \\ \frac{c}{a} & 1 \end{pmatrix} \begin{pmatrix} a & 0 \\ 0 & \frac{1}{a} \end{pmatrix}$$

and then

$$\varepsilon(X) = \frac{c}{a}, \qquad \gamma(X) = a^2. \tag{7.6}$$

The rational e^α-action ($\alpha \in \mathbb{C}^\times$) is given by

$$e^\alpha(X) = \begin{pmatrix} 1 & \frac{(\alpha-1)}{\varepsilon(X)} \\ 0 & 1 \end{pmatrix} \cdot X \cdot \begin{pmatrix} 1 & \frac{(1-\alpha)}{\varepsilon(X)} \\ 0 & 1 \end{pmatrix} = \begin{pmatrix} a\alpha & -\frac{(a\alpha-1)^2}{c} \\ c & 2-a\alpha \end{pmatrix} \quad (c \neq 0). \tag{7.7}$$

Now, let us consider certain positive structure on $\mathscr{U}(\mathrm{SL}_2)$. First, let us fix a parameter $\xi = \lambda^{\frac{1}{2}}$. Define a rational map $\Theta_\lambda : (\mathbb{C}^\times)^2 \to \mathscr{U}(\mathrm{SL}_2)$ by

$$\Theta_\lambda(u,v) = \begin{pmatrix} \frac{\xi v}{u} & -\frac{(\frac{\xi v}{u}-1)^2(u+\xi^2 v)}{\xi uv} \\ \frac{\xi uv}{u+\xi^2 v} & 2-\frac{\xi v}{u} \end{pmatrix}. \tag{7.8}$$

Its inverse is given by

$$\Theta_\lambda^{-1}\begin{pmatrix} a & b \\ c & d \end{pmatrix} = \left(\frac{c(1+\xi a)}{a}, \frac{c(1+\xi a)}{\xi}\right), \tag{7.9}$$

which means that Θ_λ is a birational map. Now, we obtain

$$\varepsilon \circ \Theta_\lambda(u,v) = \frac{u^2}{u+\lambda v}, \qquad \gamma \circ \Theta_\lambda(u,v) = \frac{\lambda v^2}{u^2}, \tag{7.10}$$

$$\Theta_\lambda^{-1} \circ e^\alpha \circ \Theta_\lambda(u,v) = \left(\frac{u(u+\lambda\alpha v)}{\alpha(u+\lambda v)}, \frac{v(u+\lambda\alpha v)}{u+\lambda v}\right). \tag{7.11}$$

We define the function F_λ on $\mathscr{U}(\mathrm{SL}_2)$ by

$$F_\lambda\begin{pmatrix} a & b \\ c & d \end{pmatrix} := \frac{c(a+\xi)(\xi a+1)}{\xi a}, \tag{7.12}$$

It follows from (7.8) and (7.12) that

$$F_\lambda \circ \Theta_\lambda(u,v) = u+v. \tag{7.13}$$

THEOREM 7.6. *The 6-tuple* $\chi = ((\mathscr{U}(\mathrm{SL}_2), e, \gamma, \varepsilon, F_\lambda), \Theta_\lambda, (\mathbb{C}^\times)^2)$ *defines a positive decorated geometric crystal on* $\mathscr{U}(\mathrm{SL}_2)$.

PROOF. It suffices to show that

(1) Θ_λ defines a positive structure on $\mathscr{U}(\mathrm{SL}_2)$.
(2) $\gamma(e^\alpha(X)) = \alpha^2\gamma(X) \qquad (\alpha \in \mathbb{C}^\times,\ X \in \mathscr{U}(\mathrm{SL}_2))$.
(3) $\varepsilon(e^\alpha(X)) = \alpha^{-1}\varepsilon(X) \qquad (\alpha \in \mathbb{C}^\times,\ X \in \mathscr{U}(\mathrm{SL}_2))$.
(4) $F_\lambda(e^\alpha(X)) = F_\lambda(X) + (\alpha-1)\varphi(X) + (\alpha^{-1}-1)\varepsilon(X) \qquad (\alpha \in \mathbb{C}^\times,\ X \in \mathscr{U}(\mathrm{SL}_2))$.

It is trivial that (1) holds by (7.10), (7.11) and (7.13). Next, set $X = \begin{pmatrix} a & b \\ c & d \end{pmatrix}$ with $c \neq 0$. Then, we have

$$\gamma(e^\alpha(X)) = \gamma\begin{pmatrix} a\alpha & -\frac{(a\alpha-1)^2}{c} \\ c & 2-a\alpha \end{pmatrix} = (\alpha a)^2 = \alpha^2\gamma(X),$$

which shows (2). We also get

$$\varepsilon(e^\alpha(X)) = \varepsilon\begin{pmatrix} a\alpha & -\frac{(a\alpha-1)^2}{c} \\ c & 2-a\alpha \end{pmatrix} = \frac{c}{\alpha a} = \alpha^{-1}\varepsilon(X),$$

which shows (3). Finally, we obtain

$$F_\lambda(e^\alpha(X)) = F_\lambda\begin{pmatrix} a\alpha & -\frac{(a\alpha-1)^2}{c} \\ c & 2-a\alpha \end{pmatrix} = \frac{c(\alpha a+\xi)(1+\xi\alpha a)}{\xi\alpha a},$$

Then we have

$$\begin{aligned}F_\lambda(e^\alpha(X)) - F_\lambda(X) &= \frac{c(\alpha a+\xi)(1+\xi\alpha a)}{\xi\alpha a} - \frac{c(a+\xi)(1+\xi a)}{\xi a}\\ &= \frac{c}{\xi\alpha a}((1-\alpha)\xi + \xi\alpha a^2(\alpha-1))\\ &= (\alpha-1)ac + (\alpha^{-1}-1)\frac{c}{a} = (\alpha-1)\varphi(X) + (\alpha^{-1}-1)\varepsilon(X),\end{aligned}$$

which shows that F_λ is a decoration on $\mathscr{U}(\mathrm{SL}_2)$. □

7.4. Tropicalization of the geometric crystal on $\mathscr{U}(\mathrm{SL}_2)$. In the previous subsection, we obtain the decorated positive geometric crystal

$$\mathcal{X} = ((\mathscr{U}(\mathrm{SL}_2), e, \gamma, \varepsilon, F_\lambda), \Theta_\lambda, (\mathbb{C}^\times)^2).$$

Here, let us consider its tropicalization and see its explicit crystal structure.

THEOREM 7.7.

(1) *For a positive geometric crystal $\mathcal{X}_0 := ((\mathscr{U}(\mathrm{SL}_2), e, \gamma, \varepsilon), \Theta_\lambda, (\mathbb{C}^\times)^2)$, let $\chi_0 := \mathcal{T}R_{\Theta_\lambda,(\mathbb{C}^\times)^2}(\mathcal{X}_0) = (\mathbb{Z}^2, \tilde{e}, \mathrm{wt}, \tilde{\varepsilon})$ be the corresponding tropicalized crystal. Then we obtain the following isomorphism of $\mathfrak{sl}_2$-crystals:*

$$\chi_0 \xrightarrow{\sim} B_1 \otimes T_\lambda \otimes B_1, \tag{7.14}$$

$$(a,b) \longmapsto (a,b)_\lambda = \tilde{f}^a(0)_1 \otimes t_\lambda \otimes \tilde{e}^b(0)_1 \tag{7.15}$$

(2) *For a positive decorated geometric crystal*

$$\mathcal{X} := ((\mathscr{U}(\mathrm{SL}_2), e, \gamma, F_\lambda), \Theta_\lambda, (\mathbb{C}^\times)^2),$$

set

$$\widetilde{B}_{\widetilde{F}_\lambda} := \{\widetilde{x} \in \mathbb{Z}^2 \mid \widetilde{F}_\lambda(\widetilde{x}) \geq 0\} \tag{7.16}$$

and define a crystal

$$B_{F_\lambda} = B_{F_\lambda,(\mathbb{C}^\times)^2} := (\widetilde{B}_{\widetilde{F}_\lambda}, \mathrm{wt}|_{\widetilde{B}_{\widetilde{F}_\lambda}}, \tilde{\varepsilon}|_{\widetilde{B}_{\widetilde{F}_\lambda}}, e|_{\widetilde{B}_{\widetilde{F}_\lambda}}). \tag{7.17}$$

Then we obtain the following isomorphism of $\mathfrak{sl}_2$-crystals:

$$B_{F_\lambda} \xrightarrow{\sim} B(\infty) \otimes T_\lambda \otimes B(-\infty) \cong B(U_q(\mathfrak{sl}_2)a_\lambda), \tag{7.18}$$

$$(a,b) \longmapsto (a,b)_\lambda = \tilde{f}^a(0)_1 \otimes t_\lambda \otimes \tilde{e}^b(0)_1 \tag{7.19}$$

PROOF.

(1) Set $a = \nu(u)$, $b = \nu(v)$, $\tilde{\lambda} = \nu(\lambda)$, $m = \nu(\alpha) \in \mathbb{Z}$. Then, it follows from (7.10) and (7.11) that

$$\mathrm{wt}(a,b) = \tilde{\lambda} + 2(b-a), \tag{7.20}$$

$$\tilde{\varepsilon}(a,b) = 2a - \min(a, \tilde{\lambda}+b) = -\min(-a, -2a+b+\tilde{\lambda}), \tag{7.21}$$

$$\begin{aligned}\tilde{e}^m(a,b) = (a + \min(a, b+\tilde{\lambda}+m) - m - \min(a, b+\tilde{\lambda}),\\ b + \min(a, b+\tilde{\lambda}+m) - \min(a, b+\tilde{\lambda})).\end{aligned} \tag{7.22}$$

In (7.22), set $m = 1$. Then, by the formula

$$\min(a, b+\tilde{\lambda}+1) - \min(a, b+\tilde{\lambda}) = \begin{cases} 0 & \text{if } b+\tilde{\lambda} \geq a, \\ 1 & \text{if } b+\tilde{\lambda} < a, \end{cases} \tag{7.23}$$

we obtain

$$\tilde{e}(a,b) = \begin{cases} (a-1,b) & \text{if } b+\tilde{\lambda} \geq a, \\ (a,b+1) & \text{if } b+\tilde{\lambda} < a. \end{cases} \tag{7.24}$$

Thus, comparing (7.20), (7.21) and (7.24) with (4.4), (4.5) and (4.6), we find that the crystal χ_0 is isomorphic to the crystal $B_1 \otimes T_{\tilde{\lambda}} \otimes B_1$.

(2) One has $\widetilde{F_\lambda \circ \Theta_\lambda}(a,b) = \min(a,b)$ and then

$$B_{F_\lambda} = \{(a,b) \in \mathbb{Z}^2 \mid a \geq 0,\ b \geq 0\},$$

which implies that the crystal B_{F_λ} is isomorphic to the crystal $B(\infty)\otimes T_{\tilde{\lambda}} \otimes B(-\infty)$ by (4.8). □

CONJECTURE 7.8. *For any simply connected, simple algebraic group G over $\mathbb{C}$, there exists a positive structure $\Theta_\lambda : T' \to \mathcal{U}$ and a decoration F_λ on $\mathcal{U}$ such that the tropicalized crystal $\widetilde{B}_{\widetilde{F}_\lambda}$ is isomorphic to the Langlands dual crystal for λ-component of modified quantum algebra $(B(\infty)\otimes T_{\tilde{\lambda}} \otimes B(-\infty))^\vee \cong B(U_q(\mathfrak{g}^\vee)a_{\tilde{\lambda}})$, where $\lambda \in (\mathbb{C}^\times)^{\mathrm{rank}(G)}$ and $\dim(T') = \dim G - \mathrm{rank}(G)$ and $\tilde{\lambda} = \mathcal{TR}(\lambda) \in P$.*

Acknowledgment

The author would like to thank Toshiaki Shoji for his helpful comments and suggestions.

References

[1] P. Bala and R. W. Carter, *Classes of unipotent elements in simple algebraic groups. I*, Math. Proc. Cambridge Philos. Soc. **79** (1976), no. 3, 401–425, DOI 10.1017/S0305004100052403. MR417306

[2] P. Bala and R. W. Carter, *Classes of unipotent elements in simple algebraic groups. II*, Math. Proc. Cambridge Philos. Soc. **80** (1976), no. 1, 1–17, DOI 10.1017/S0305004100052610. MR417307

[3] A. Borel, R. Carter, C. W. Curtis, N. Iwahori, T. A. Springer, and R. Steinberg, *Seminar on algebraic groups and related finite groups, Lecture Notes in Mathematics*, **131**, Springer-Verlag (1970).

[4] Arkady Berenstein and David Kazhdan, *Geometric and unipotent crystals*, Geom. Funct. Anal. **Special Volume** (2000), 188–236, DOI 10.1007/978-3-0346-0422-2_8. GAFA 2000 (Tel Aviv, 1999). MR1826254

[5] Arkady Berenstein and David Kazhdan, *Geometric and unipotent crystals. II. From unipotent bicrystals to crystal bases*, Quantum groups, Contemp. Math., vol. 433, Amer. Math. Soc., Providence, RI, 2007, pp. 13–88, DOI 10.1090/conm/433/08321. MR2349617

[6] R. W. Carter, *Finite Groups of Lie type, Conjugacy classes and complex characters*, Wiley classics Library, (1993).

[7] Neil Chriss and Victor Ginzburg, *Representation theory and complex geometry*, Birkhäuser Boston, Inc., Boston, MA, 1997. MR1433132

[8] Yuki Kanakubo and Toshiki Nakashima, *Half potential on geometric crystals and connectedness of cellular crystals*, Transform. Groups **28** (2023), no. 1, 327–373, DOI 10.1007/s00031-022-09788-8. MR4552178

[9] M. Kashiwara, *On crystal bases of the Q-analogue of universal enveloping algebras*, Duke Math. J. **63** (1991), no. 2, 465–516, DOI 10.1215/S0012-7094-91-06321-0. MR1115118

[10] Masaki Kashiwara, *The crystal base and Littelmann's refined Demazure character formula*, Duke Math. J. **71** (1993), no. 3, 839–858, DOI 10.1215/S0012-7094-93-07131-1. MR1240605

[11] Masaki Kashiwara, *Crystal bases of modified quantized enveloping algebra*, Duke Math. J. **73** (1994), no. 2, 383–413, DOI 10.1215/S0012-7094-94-07317-1. MR1262212

[12] Masaki Kashiwara, Toshiki Nakashima, and Masato Okado, *Affine geometric crystals and limit of perfect crystals*, Trans. Amer. Math. Soc. **360** (2008), no. 7, 3645–3686, DOI 10.1090/S0002-9947-08-04341-9. MR2386241

[13] Masaki Kashiwara, Toshiki Nakashima, and Masato Okado, *Tropical R maps and affine geometric crystals*, Represent. Theory **14** (2010), 446–509, DOI 10.1090/S1088-4165-2010-00379-9. MR2661518

[14] Bertram Kostant, *Lie group representations on polynomial rings*, Amer. J. Math. **85** (1963), 327–404, DOI 10.2307/2373130. MR158024

[15] Shrawan Kumar, *Kac-Moody groups, their flag varieties and representation theory*, Progress in Mathematics, vol. 204, Birkhäuser Boston, Inc., Boston, MA, 2002, DOI 10.1007/978-1-4612-0105-2. MR1923198

[16] George Lusztig, *Introduction to quantum groups*, Progress in Mathematics, vol. 110, Birkhäuser Boston, Inc., Boston, MA, 1993. MR1227098

[17] G. Lusztig, *Canonical bases in tensor products*, Proc. Nat. Acad. Sci. U.S.A. **89** (1992), no. 17, 8177–8179, DOI 10.1073/pnas.89.17.8177. MR1180036

[18] Toshiki Nakashima, *Geometric crystals on Schubert varieties*, J. Geom. Phys. **53** (2005), no. 2, 197–225, DOI 10.1016/j.geomphys.2004.06.004. MR2110832

[19] Toshiki Nakashima, *Geometric crystals on unipotent groups and generalized Young tableaux*, J. Algebra **293** (2005), no. 1, 65–88, DOI 10.1016/j.jalgebra.2005.07.025. MR2173966

[20] Toshiki Nakashima, *Affine geometric crystal of type* $G_2^{(1)}$, Lie algebras, vertex operator algebras and their applications, Contemp. Math., vol. 442, Amer. Math. Soc., Providence, RI, 2007, pp. 179–192, DOI 10.1090/conm/442/08525. MR2372562

[21] T. A. Springer, *The unipotent variety of a semisimple groups*, in "Algebraic Geometry" (Internat. Colloq. Tata Inst. Fund. res., Bombay, 1968), Oxford Univ. Press, London, pp.373–391.

[22] T. A. Springer, *Trigonometric sums, Green functions of finite groups and representations of Weyl groups*, Invent. Math. **36** (1976), 173–207, DOI 10.1007/BF01390009. MR442103

DIVISION OF MATHEMATICS, SOPHIA UNIVERSITY, KIOICHO 7-1, CHIYODA-KU, TOKYO 102-8554, JAPAN

Email address: toshiki@sophia.ac.jp

Proceedings of Symposia in Pure Mathematics
Volume **108**, 2024
https://doi.org/10.1090/pspum/108/01970

Difference operators for wreath Macdonald polynomials

Daniel Orr and Mark Shimozono

ABSTRACT. We give explicit q-difference operators acting diagonally on wreath Macdonald P-polynomials in finitely many variables.

1. Introduction

Wreath Macdonald polynomials [**Hai2**] originate from the geometry of Γ-fixed loci in Hilbert schemes of points in the plane, where Γ is a finite cyclic subgroup of $SL(2,\mathbb{C})$. Unlike the ordinary Macdonald polynomials (which are recovered when Γ is trivial), their very *existence* relies on the use of extremely deep geometric methods [**BF**, **BK**].

A recent result of Wen [**Wen**] provides a somewhat more direct characterization of wreath Macdonald polynomials. This result is based on the *quantum toroidal algebra* [**GKV**] of type $\mathfrak{gl}_r$, where $r = |\Gamma|$, together with a remarkable symmetry of this algebra known as Miki's automorphism [**Mi**]. However, Wen's result is still quite far from an explicit characterization of wreath Macdonald polynomials. One encounters serious technical obstacles when trying to use it to extract concrete information about wreath Macdonald polynomials, of which very little is in fact known.

The aim of this note is to introduce a new family of q-difference operators for wreath Macdonald polynomials, which are a *direct* generalization of Macdonald's original q-difference operators for ordinary Macdonald polynomials. In particular, we provide explicit first-order q-difference operators $\{\mathsf{M}^{(i)}\}_{i\in I}$ (where $I = \mathbb{Z}/r\mathbb{Z}$) which act diagonally on wreath Macdonald P-polynomials in finitely many variables with an I-graded version of the ordinary Macdonald operator eigenvalues (see Theorem 4.2). Our operators are sufficient to uniquely determine these polynomials and their lifts to multi-symmetric functions.

We have a complete proof that our operators act diagonally on wreath Macdonald polynomials, but we postpone it to [**OSW**], where we will also construct higher order wreath Macdonald difference operators. Our approach is based on: (i) Wen's result mentioned above, and (ii) Negut's shuffle realization of the quantum toroidal algebra [**Ne**]. Based on the rich applicability of Macdonald's original difference operators, we anticipate a number of applications of the operators $\{\mathsf{M}^{(i)}\}_{i\in I}$, which we will take up in future papers.

The first author was partially supported by a Collaboration Grant for Mathematicians from the Simons Foundation and by the Max Planck Institute for Mathematics (MPIM Bonn).

2. Macdonald polynomials

Let $\mathfrak{S}_N$ be the symmetric group and consider the space $\Lambda_N = \mathbb{K}[x_1, \ldots, x_N]^{\mathfrak{S}_N}$ of symmetric polynomials in variables $x_1, \ldots, x_N$ with coefficients in $\mathbb{K} = \mathbb{Q}(q,t)$. Let $\mathbb{Y}$ be the set of all integer partitions $\lambda = (\lambda_1, \lambda_2, \ldots)$ where $\lambda_1 \geq \lambda_2 \geq \cdots \geq 0$, and let $\mathbb{Y}_N$ be the set of all integer partitions $\lambda = (\lambda_1, \ldots, \lambda_N)$ with at most N parts.

2.1. Origins. The *Macdonald operator* is the q-difference operator

$$\mathsf{M} = \sum_{k=1}^{N} \prod_{\substack{l=1 \\ l \neq k}}^{N} \frac{tx_k - x_l}{x_k - x_l} \cdot \mathsf{T}_{q,x_k} \tag{2.1}$$

where the rational functions appearing in M are understood as multiplication operators and

$$\mathsf{T}_{q,x_k} f(x_1, \ldots, x_N) = f(x_1, \ldots, qx_k, \ldots, x_N). \tag{2.2}$$

The operator M stabilizes Λ_N, i.e., the denominators clear and the result is symmetric. The importance of this operator stems from the fact that it belongs to a quantum integrable system:

THEOREM 2.1 ([**Mac**]).

(1) *There exist explicit, algebraically independent q-difference operators $\mathsf{M} = \mathsf{M}_1, \mathsf{M}_2, \ldots, \mathsf{M}_N$, satisfying $[\mathsf{M}_k, \mathsf{M}_l] = 0$ for all $1 \leq k, l \leq N$.*

(2) *The operators $\mathsf{M}_1, \ldots, \mathsf{M}_N$ have a joint eigenbasis $\{P_\lambda\}$ indexed by $\lambda \in \mathbb{Y}_N$. The symmetric polynomials P_λ are uniquely characterized as their joint eigenfunctions, together with the normalization*

$$P_\lambda \in m_\lambda + \bigoplus_{\mu \ntrianglerighteq \lambda} \mathbb{K} m_\mu$$

where m_λ denotes a monomial symmetric polynomial and $\trianglelefteq$ is the dominance order on partitions (restricted to $\mathbb{Y}_N$).

The symmetric polynomials P_λ are the *Macdonald polynomials* (of type GL_N). We will call $\mathsf{M}_2, \ldots, \mathsf{M}_N$ the *higher Macdonald operators.* The eigenvalues of M are:

$$\mathsf{M} \cdot P_\lambda = e(q,t) P_\lambda, \qquad e(q,t) = \sum_{k=1}^{N} q^{\lambda_k} t^{N-k}. \tag{2.3}$$

The eigenvalues of M_k are given by substituting the summands of $e(q,t)$ into the elementary symmetric function e_k.

The Macdonald polynomials P_λ can be stabilized into elements of the $\mathbb{K}$-algebra Λ of symmetric functions in infinitely many variables [**Mac**]. One also considers the modified Macdonald symmetric functions $\widetilde{H}_\lambda$, which are obtained from P_λ by an explicit transformation (see, e.g., [**Hai2**]).

2.2. Geometry of Hilbert schemes. The modified Macdonald symmetric functions $\widetilde{H}_\lambda$ are significant due to their positivity when expanded in Schur symmetric functions, which was the content of Macdonald's famous *positivity conjecture.* Haiman's proof [**Hai2**] of this conjecture recast it as a statement about the geometry of Hilbert schemes $\mathrm{Hilb}_m = \mathrm{Hilb}_m(\mathbb{C}^2)$ of points in the plane. The deepest part of Haiman's proof concerned the existence and properties of a rank $m!$ vector bundle P_m on Hilb_m known as the Procesi bundle.

It is convenient to work with the union $\mathrm{Hilb} = \bigsqcup_{m\geq 0} \mathrm{Hilb}_m$ of all Hilbert schemes. Let us consider their localized T-equivariant K-groups:

$$K_T(\mathrm{Hilb})_{\mathrm{loc}} = \bigoplus_{m\geq 0} K_T(\mathrm{Hilb}_m)_{\mathrm{loc}}$$

where $T = (\mathbb{C}^\times)^2$ acts naturally on $\mathbb{C}^2$ and hence on Hilb, and loc stands for localization with respect to equivariant scalars $R(T) \cong \mathbb{Z}[q^{\pm 1}, t^{\pm 1}]$, which is achieved by tensoring with $\mathbb{K} \cong \mathrm{Frac}(R(T))$. The points of Hilb can be identified with ideals $I \subset \mathbb{C}[x, y]$ of finite codimension. By the localization theorem, $K_T(\mathrm{Hilb})_{\mathrm{loc}}$ has a $\mathbb{K}$-basis given by the classes of $\{[I_\lambda]\}$ of T-fixed points in Hilb, which are precisely the monomial ideals $I_\lambda = (y^{\lambda_1}, xy^{\lambda_2}, x^2y^{\lambda_3}, \cdots)$ where $\lambda \in \mathbb{Y}$ is any partition.

THEOREM 2.2 ([**Hai1**]). *The Procesi bundles P_m induce an isomorphism of $\mathbb{K}$-vector spaces*

$$K_T(\mathrm{Hilb})_{\mathrm{loc}} \cong \Lambda \tag{2.4}$$

under which a fixed point class $[I_\lambda]$ is sent to the modified Macdonald function symmetric function $\widetilde{H}_\lambda$. This isomorphism realizes $\widetilde{H}_\lambda$ as the bigraded $\mathfrak{S}_m$-character of the fiber $P_m|_{I_\lambda}$, where $|\lambda| = m$, implying the Macdonald positivity conjecture.

3. Wreath Macdonald polynomials

For a fixed integer $r > 1$, we consider the finite cyclic subgroup

$$\Gamma = \left\{ \begin{pmatrix} \zeta & \\ & \zeta^{-1} \end{pmatrix} : \zeta^r = 1 \right\} \subset SL_2(\mathbb{C}).$$

Haiman [**Hai2**] proposed a generalization of the *geometric* theory of Macdonald polynomials, replacing Hilb by the Γ-fixed locus Hilb^Γ consisting of finite-codimensional ideals $I \subset \mathbb{C}[x, y]$ which are Γ-invariant.[1] Let $\chi : \Gamma \to \mathbb{C}^\times$ be the representation of Γ given by

$$\chi : \begin{pmatrix} \zeta & \\ & \zeta^{-1} \end{pmatrix} \mapsto \zeta \tag{3.1}$$

The irreducible representations of Γ are given by powers of χ: $\mathrm{Irr}(\Gamma) = \{\chi^i : i \in I\}$ where $I = \mathbb{Z}/r\mathbb{Z}$. The McKay correspondence identifies I with the vertices of the Dynkin diagram of type $A_{r-1}^{(1)}$. Let $Q_{\mathrm{aff}} = \oplus_{i\in I}\mathbb{Z}\alpha_i$ be the root lattice of the affine Lie algebra $\widehat{\mathfrak{sl}}_r$, $Q_{\mathrm{aff}}^+ = \oplus_{i\in I}\mathbb{Z}_{\geq 0}\alpha_i$ the positive affine root cone, and $\delta = \sum_{i\in I}\alpha_i$ the minimal imaginary root. Then the fixed locus Hilb^Γ is the disjoint union of connected components $\mathrm{Hilb}^\Gamma_\alpha$ indexed by $\alpha = \sum_{i\in I} m_i\alpha_i \in Q^+_{\mathrm{aff}}$, namely:[2]

$$\mathrm{Hilb}^\Gamma_\alpha = \{I \in \mathrm{Hilb}^\Gamma : R/I \text{ has } \Gamma\text{-character } \sum_{i\in I} m_i\chi^i\}.$$

Hilb^Γ and Hilb share the same T-fixed points $\{I_\lambda\}_{\lambda\in\mathbb{Y}}$, since $\Gamma \subset T$.

---

[1]In fact, Haiman's generalization applies conjecturally to arbitrary finite subgroup $\Gamma \subset SL_2(\mathbb{C})$; however, we only consider cyclic Γ, as very little is known outside of this case.

[2]The name "wreath" originates from the fact that $\mathrm{Hilb}^\Gamma_\alpha$ is a crepant resolution of $\mathbb{C}^{2n}/\Gamma_n$ for some n (determined by α), where $\Gamma_n = \Gamma^n \rtimes \mathfrak{S}_n$ is a wreath product group.

3.1. Haiman's conjecture. In [**Hai2**] Haiman conjectured the following generalization of Theorem 2.2: there exists a wreath Procesi bundle on each component of Hilb^Γ and that together these afford an isomorphism

$$K_T(\mathrm{Hilb}^\Gamma)_{\mathrm{loc}} \cong \Lambda^\Gamma := \bigoplus_{\gamma \in Q} \Lambda^{\otimes I} \tag{3.2}$$

with properties generalizing Macdonald positivity, where $Q = Q_{\mathrm{aff}}/\mathbb{Z}\delta$ is the root lattice of $\mathfrak{sl}_r$. Parallel to Theorem 2.2, the images of fixed point classes $\{[I_\lambda]\}$ under this isomorphism are the modified *wreath Macdonald symmetric functions* $\tilde{H}^\Gamma_\lambda$, whose precise definition will be recalled below.

REMARK 3.1. In the isomorphism (3.2), the element $\gamma \in Q$ records the r-core of a partition $\lambda \in \mathbb{Y}$ indexing a fixed-point class on the left-hand side. The factors in the tensor power $\Lambda^{\otimes I}$ of symmetric functions have to do with the r-quotient of $\lambda \in \mathbb{Y}$, under the core-quotient decomposition [**JK**].

Haiman's conjecture was proved by Bezrukavnikov and Finkelberg [**BF**]. Their proof relies on deep, non-explicit results of Bezrukavnikov and Kaledin [**BK**] involving quantization in positive characteristic.

3.2. Definition of wreath Macdonald polynomials. The following definitions of $\tilde{H}^\Gamma_\lambda$ and their monic forms P^Γ_λ are due to Haiman [**Hai2**], with existence guaranteed by [**BF**].

3.2.1. *$\tilde{H}$-version.* For $\lambda \in \mathbb{Y}$ denote by $\mathrm{core}_r(\lambda) \in \mathbb{Y}$ and $\mathrm{quot}_r(\lambda) \in \mathbb{Y}^I$ the classical r-core and r-quotient [**JK**]. We identify $\lambda \in \mathbb{Y}$ with its diagram $D(\lambda) = \{(a,b) \in \mathbb{Z}^2_{\ge 0} \mid 0 \le a < \lambda_{b+1}\}$. We identify the set of all r-core partitions with the root lattice Q of $\mathfrak{sl}_r$ as follows. Let $\kappa : \mathbb{Y} \to Q_{\mathrm{aff}}$ be the map $\kappa(\lambda) = \sum_{(a,b)\in\lambda} \alpha_{b-a}$ with subscripts taken modulo r as usual, and let $\overline{\kappa} = \mathrm{cl} \circ \kappa : \mathbb{Y} \to Q$ be the composition of κ with the projection $\mathrm{cl} : Q_{\mathrm{aff}} \to Q$. It is known that $\overline{\kappa}$ restricts to a bijection from the set of r-core partitions to Q [**Hai2**].

Let $\trianglerighteq^\Gamma$ denote the following restriction of the standard dominance order $\trianglerighteq$ on $\mathbb{Y}$ [**Mac**]: two partitions may only be compared if they have the same r-core and the same number of boxes. The map quot_r gives a bijection $\mathbb{Y}^\gamma \cong \mathbb{Y}^I$ from the set $\mathbb{Y}^\gamma$ of partitions with fixed r-core $\gamma \in Q$, to the set $\mathbb{Y}^I$ of r-tuples of partitions [**G**, §6.2].

For $\lambda \in \mathbb{Y}$ we write $s_\lambda[X^\bullet] \in \Lambda^{\otimes I}$ for the tensor Schur function $s_{\mu^\bullet}[X^\bullet] = \prod_{i\in I} s_{\mu^{(i)}}[X^{(i)}]$ indexed by $\mu^\bullet = \widetilde{\mathrm{quot}_r}(\lambda) \in \mathbb{Y}^I$ where $\widetilde{\mathrm{quot}_r}(\lambda)$ is the reversal of the r-tuple $\mathrm{quot}_r(\lambda)$. (Here $X^{(i)}$ stands for an alphabet of variables $x^{(i)}_1, x^{(i)}_2, \dots$ corresponding to the i-th factor in $\Lambda^{\otimes I}$.)

For $a \in \mathbb{K}$ we denote by $\mathcal{P}_{\mathrm{id}-a\chi^{-1}}$ the $\mathbb{K}$-algebra automorphism of $\Lambda^{\otimes I}$ which maps the power sum $p_d[X^{(i)}]$ of degree d living at vertex $i \in I$ of the McKay graph of Γ as follows:

$$\mathcal{P}_{\mathrm{id}-a\chi^{-1}}(p_d[X^{(i)}]) = p_d[X^{(i)}] - a^d p_d[X^{(i-1)}].$$

Let $\langle \cdot,\cdot \rangle$ be the tensor product of Hall pairings on $\Lambda^{\otimes I}$, with respect to which the tensor Schur functions are orthonormal.

The wreath Macdonald polynomials $\tilde{H}^\Gamma_\lambda \in \Lambda^{\otimes I}$ are uniquely defined by [**Hai2**, **BF**]

(1) $\mathcal{P}_{\mathrm{id}-q\chi^{-1}}(\tilde{H}^\Gamma_\lambda) \in \mathbb{K}^\times s_\lambda[X^\bullet] + \bigoplus_{\mu \ntrianglerighteq^\Gamma \lambda} \mathbb{K} s_\mu[X^\bullet]$.

(2) $\mathcal{P}_{\mathrm{id}-t^{-1}\chi^{-1}}(\tilde{H}^\Gamma_\lambda) \in \mathbb{K}^\times s_\lambda[X^\bullet] + \bigoplus_{\mu \ntriangleleft^\Gamma \lambda} \mathbb{K} s_\mu[X^\bullet]$.

(3) $\langle \tilde{H}^\Gamma_\lambda, s_n[X^{(0)}] \rangle = 1$ where n is the total number of boxes in $\widetilde{\mathrm{quot}_r}(\lambda)$.

REMARK 3.2. For each fixed r-core γ, the $\tilde{H}^\Gamma_\lambda$ for $\lambda \in \mathbb{Y}^\gamma$ form a $\mathbb{K}$-basis of $\Lambda^{\otimes I}$. So do the $s_\lambda[X^\bullet]$ for $\lambda \in \mathbb{Y}^\gamma$; for each γ this is the same tensor Schur basis of $\Lambda^{\otimes I}$ but indexed (and ordered) differently.

3.2.2. *P-version.* Define $J^\Gamma_\lambda \in \Lambda^{\otimes I}$ by

$$J^\Gamma_\lambda = \mathcal{P}_{\mathrm{id}-t^{-1}\chi^{-1}}(\tilde{H}^\Gamma_\lambda). \tag{3.3}$$

Then $P^\Gamma_\lambda \in \Lambda^{\otimes I}$ is by definition the unique $\mathbb{K}^\times$-multiple of J^Γ_λ for which the coefficient of $s_\lambda[X^\bullet]$ is 1. Due to the triangularity implied by the definition of $\tilde{H}^\Gamma_\lambda$, the P^Γ_λ form a $\mathbb{K}$-basis of $\Lambda^{\otimes I}$ as λ runs over $\mathbb{Y}^\gamma$ for any fixed r-core γ.

REMARK 3.3. When $r = 1$, one recovers the ordinary Macdonald P functions as $P^\Gamma_\lambda(q,t) = P_\lambda(q^{-1}, t)$.

4. Wreath Macdonald difference operators

4.1. Symmetric functions and symmetric polynomials. Our difference operators will characterize the images of $\{P^\Gamma_\lambda\}_{\lambda\in\mathbb{Y}}$ under projections to finitely many variables. Let $N_\bullet = (N_i)_{i\in I} \in (\mathbb{Z}_{\geq 0})^I$ be a dimension vector and consider the polynomial ring

$$\mathbf{Pol}_{N_\bullet} = \mathbb{K}[\{x^{(i)}_k : i \in I, 1 \leq k \leq N_i\}]$$

and its subring

$$\Lambda_{N_\bullet} = (\mathbf{Pol}_{N_\bullet})^{\mathfrak{S}_{N_\bullet}} \tag{4.1}$$

consisting of polynomials which are invariant with respect to the product of symmetric groups $\mathfrak{S}_{N_\bullet} = \prod_{i\in I} \mathfrak{S}_{N_i}$, with the factor $\mathfrak{S}_{N_i}$ permuting the variables $\{x^{(i)}_1, \ldots, x^{(i)}_{N_i}\}$, for each $i \in I$. We write $N = \sum_{i\in I} N_i$ for the total number of variables.

We write $\pi_{N_\bullet} : \Lambda^{\otimes I} \to \Lambda_{N_\bullet}$ for the natural projection, which is given on simple tensors $f = \otimes_{i\in I} f_i$ by

$$\pi_{N_\bullet}(f) = \prod_{i\in I} f_i(x^{(i)}_1, \ldots, x^{(i)}_{N_i}). \tag{4.2}$$

For $f \in \Lambda^{\otimes I}$ we use the shorthand $f(X_{N_\bullet})$ to denote $\pi_{N_\bullet}(f)$.

For $\gamma \in Q$ indexing a summand of Λ^Γ, a dimension vector $N_\bullet \in (\mathbb{Z}_{\geq 0})^I$ will be called *γ-compatible* if it satisfies

$$N_i - N_{i-1} = \langle \alpha^\vee_i, -\gamma \rangle, \qquad \text{for each } i \in I. \tag{4.3}$$

where the subscripts in $I = \mathbb{Z}/r\mathbb{Z}$ are understood in a cyclic sense and $\alpha^\vee_0$ means the classical projection of the 0-th affine simple coroot.

In what follows, we have to assume that $N_\bullet$ is γ-compatible. But there is no loss of generality in this assumption, since (4.3) continues to hold if we add a constant to each N_i. Thus we can make the components of $N_\bullet$ as large as needed to fully capture the P^Γ_λ as symmetric functions.

4.2. Difference operators. Now we are ready to introduce our wreath Macdonald difference operators. The Macdonald operator M in (2.1) is a sum over all choices of a variable x_k ($1 \leq k \leq N$). Our operators $\mathsf{M}^{(i)}$ are expressed as sums over all nonempty subsets of the variables $\bigcup_{i\in I}\{x_1^{(i)},\dots,x_{N_i}^{(i)}\}$ with at most one variable selected at each vertex. We record these subsets as pairs $\underline{J} = (J, \mathsf{k})$, where J is a nonempty subset of I and k is an integer-valued function on J such that $1 \leq \mathsf{k}(j) \leq N_j$ for each $j \in J$; the variables chosen to be in our subset are $\{x_{\mathsf{k}(j)}^{(j)}\}_{j\in J}$.

Let $\mathcal{S}$ be the set of all such pairs and consider some $\underline{J} = (J, \mathsf{k}) \in \mathcal{S}$. For any $i \in I$, we define:

$$\mathsf{X}_{\underline{J}}^{(i)} = q^m x_{\mathsf{k}(i+m)}^{(i+m)}, \quad \text{where } m \geq 0 \text{ is minimal such that } i+m \in J.$$

For $j \in J$ we have $\mathsf{X}_{\underline{J}}^{(j)} = x_{\mathsf{k}(j)}^{(j)}$, which is simply the variable at vertex j selected by $\underline{J}$. The remaining $\mathsf{X}_{\underline{J}}^{(i)}$ for $i \in I \setminus J$ are obtained by cyclically propagating the selected variables and multiplying by a factor of q at each step.

EXAMPLE. Let $I = \mathbb{Z}/3\mathbb{Z}$ and $N_\bullet = (2,2,2)$. An element of $\mathbf{Pol}_{N_\bullet}$ is a polynomial

$$f = f(x_1^{(0)}, x_2^{(0)} \mid x_1^{(1)}, x_2^{(1)} \mid x_1^{(2)}, x_2^{(2)});$$

dividers separate the variables at each vertex. Let $\underline{J} = (J, \mathsf{k})$ where $J = \{0,1\}$ and $\mathsf{k}(0) = \mathsf{k}(1) = 1$. That is, the selected subset is $\{x_1^{(0)}, x_1^{(1)}\}$. Then

$$\mathsf{X}_{\underline{J}}^{(0)} = x_1^{(0)}, \quad \mathsf{X}_{\underline{J}}^{(1)} = x_1^{(1)}, \quad \mathsf{X}_{\underline{J}}^{(2)} = qx_1^{(0)},$$

where $x_1^{(0)}$ is cyclically propagated to the left and mutlplied by q to determine $\mathsf{X}_{\underline{J}}^{(2)}$.

DEFINITION 4.1. For any $i \in I$, define the wreath Macdonald difference operator

$$\mathsf{M}^{(i)} = \sum_{\underline{J}\in\mathcal{S}^{(i)}} (-1)^{|J|} \cdot \mathsf{A}_{\underline{J}}^{(i)} \cdot \mathsf{T}_{q,\underline{J}} \tag{4.4}$$

where $\mathcal{S}^{(i)} \subset \mathcal{S}$ is the subset determined by the condition $i-1 \in J$,

$$\begin{aligned}\mathsf{A}_{\underline{J}}^{(i)} = \frac{\mathsf{X}_{\underline{J}}^{(i)}}{\mathsf{X}_{\underline{J}}^{(0)}} \cdot \prod_{j\in I} \frac{\mathsf{X}_{\underline{J}}^{(j-1)}}{\mathsf{X}_{\underline{J}}^{(j-1)} - t\mathsf{X}_{\underline{J}}^{(j)}} \cdot \prod_{j\in J\setminus\{i-1\}} \frac{\mathsf{X}_{\underline{J}}^{(j+1)}}{\mathsf{X}_{\underline{J}}^{(j+1)} - q^{-1}\mathsf{X}_{\underline{J}}^{(j)}} \\ \cdot \prod_{j\in I\setminus J} \frac{\prod_{l=1}^{N_{j-1}} (t\mathsf{X}_{\underline{J}}^{(j)} - x_l^{(j-1)})}{\prod_{l=1}^{N_j} (\mathsf{X}_{\underline{J}}^{(j)} - x_l^{(j)})} \cdot \prod_{j\in J} \frac{\prod_{l=1}^{N_{j-1}} (t\mathsf{X}_{\underline{J}}^{(j)} - x_l^{(j-1)})}{\mathsf{X}_{\underline{J}}^{(j)} \cdot \prod_{\substack{l=1\\ l\neq \mathsf{k}(j)}}^{N_j} (\mathsf{X}_{\underline{J}}^{(j)} - x_l^{(j)})},\end{aligned} \tag{4.5}$$

and $\mathsf{T}_{q,\underline{J}}$ is the operator on $\mathbf{Pol}_{N_\bullet}$ given by:

$$\mathsf{T}_{q,\underline{J}} = \textit{for each } j \in J, \textit{ replace } \mathsf{X}_{\underline{J}}^{(j)} \textit{ with } q\mathsf{X}_{\underline{J}}^{(j+1)}. \tag{4.6}$$

THEOREM 4.2. *For any $\gamma \in Q$, $\lambda \in \mathbb{Y}^\gamma$, and γ-compatible $N_\bullet \in (\mathbb{Z}_{\geq 0})^I$, one has*[3]

$$\mathsf{M}^{(i)} \cdot P^\Gamma_\lambda(X_{N_\bullet}; q^{-1}, t) = e^{(i)}_\lambda(q,t) P^\Gamma_\lambda(X_{N_\bullet}; q^{-1}, t). \tag{4.7}$$

where the eigenvalues $e^{(i)}_\lambda(q,t) \in \mathbb{K}$ are given by (recall that $\chi \in \mathrm{Irr}(\Gamma)$ satisfies $\chi^r = 1$):

$$\sum_{i \in I} e^{(i)}_\lambda(q,t)\chi^i = \sum_{k=1}^{N} q^{\lambda_k} t^{N-k} \chi^{k-\lambda_k} \in R(T \times \Gamma) \cong \mathbb{Z}[q^{\pm 1}, t^{\pm 1}, \chi^{\pm 1}]. \tag{4.8}$$

EXAMPLE. Continuing the example above, we have

$$\begin{aligned}\mathsf{A}^{(1)}_{\underline{J}} = \frac{\mathsf{X}^{(1)}_{\underline{J}}}{\mathsf{X}^{(0)}_{\underline{J}}} \cdot \frac{\mathsf{X}^{(2)}_{\underline{J}}}{\mathsf{X}^{(2)}_{\underline{J}} - t\mathsf{X}^{(0)}_{\underline{J}}} \frac{\mathsf{X}^{(0)}_{\underline{J}}}{\mathsf{X}^{(0)}_{\underline{J}} - t\mathsf{X}^{(1)}_{\underline{J}}} \frac{\mathsf{X}^{(1)}_{\underline{J}}}{\mathsf{X}^{(1)}_{\underline{J}} - t\mathsf{X}^{(2)}_{\underline{J}}} \cdot \frac{\mathsf{X}^{(2)}_{\underline{J}}}{\mathsf{X}^{(2)}_{\underline{J}} - q^{-1}\mathsf{X}^{(1)}_{\underline{J}}} \\ \cdot \frac{t\mathsf{X}^{(2)}_{\underline{J}} - x^{(1)}_1}{\mathsf{X}^{(2)}_{\underline{J}} - x^{(2)}_1} \frac{t\mathsf{X}^{(2)}_{\underline{J}} - x^{(1)}_2}{\mathsf{X}^{(2)}_{\underline{J}} - x^{(2)}_2} \cdot \frac{t\mathsf{X}^{(0)}_{\underline{J}} - x^{(2)}_1}{\mathsf{X}^{(0)}_{\underline{J}}} \frac{t\mathsf{X}^{(0)}_{\underline{J}} - x^{(2)}_2}{\mathsf{X}^{(0)}_{\underline{J}} - x^{(0)}_2} \\ \cdot \frac{t\mathsf{X}^{(1)}_{\underline{J}} - x^{(0)}_1}{\mathsf{X}^{(1)}_{\underline{J}}} \frac{t\mathsf{X}^{(1)}_{\underline{J}} - x^{(0)}_2}{\mathsf{X}^{(1)}_{\underline{J}} - x^{(1)}_2}\end{aligned}$$

and $\mathsf{T}_{q,\underline{J}}(f) = f(qx^{(1)}_1, x^{(0)}_2 \mid q^2 x^{(0)}_1, x^{(1)}_2 \mid x^{(2)}_1, x^{(2)}_2)$.

We hope this example convinces the reader that not even the most clever guessing would lead from the ordinary Macdonald operator M to its wreath generalization $\mathsf{M}^{(i)}$. One really needs the quantum toroidal algebra $\mathbf{U}^{\mathrm{tor}}_r$ to find these operators.

4.3. Recovering ordinary Macdonald operators. Let us take $\Gamma = 1$. Then $I = \{0\}$ and $N_\bullet = (N_0) = (N)$. The only possibly subsets of variables are singletons $\{x^{(0)}_k\}$ where $1 \leq k \leq N$. In terms of $\underline{J} = (J, \mathsf{k})$, we have $J = I$ and $\mathsf{k}(0) = k$. The operator $\mathsf{T}_{q,\underline{J}}$ sends $x^{(0)}_k$ to $qx^{(0)}_k$, which agrees with $\mathsf{T}_{q,x^{(0)}_k}$ in (2.1). Finally, we have

$$\mathsf{A}^{(0)}_{\underline{J}} = \frac{1}{1-t} \cdot \frac{\prod_{l=1}^{N}(tx^{(0)}_k - x^{(0)}_l)}{x^{(0)}_k \cdot \prod_{\substack{l=1 \\ l \neq k}}^{N}(x^{(0)}_k - x^{(0)}_l)} = -\prod_{\substack{l=1 \\ l \neq k}}^{N} \frac{tx^{(0)}_k - x^{(0)}_l}{x^{(0)}_k - x^{(0)}_l}$$

and we see that $\mathsf{M}^{(0)}$ reduces to M from (2.1).

4.4. Relation to Shoji's operators. We also mention an interesting similarity between our operators and some introduced by Shoji [**Sho**]. Let us consider only the terms in (4.4) which have $J = I$, meaning that a variable is selected at

---

[3] $P^\Gamma_\lambda(X_{N_\bullet}; q^{-1}, t)$ means that we project to finitely many variables and send q to q^{-1}; cf. Remark 3.3.

each vertex in I. Then, for $\underline{I} = (I, \mathsf{k})$, we have $\mathsf{X}_{\underline{I}}^{(j)} = x_{\mathsf{k}(j)}^{(j)}$ for all $j \in I$ and

$$\mathsf{A}_{\underline{I}}^{(i)} = (-1)^r \frac{\mathsf{X}_{\underline{I}}^{(i)}}{\mathsf{X}_{\underline{I}}^{(0)}} \prod_{j \in I \setminus \{i-1\}} \frac{\mathsf{X}_{\underline{I}}^{(j+1)}}{\mathsf{X}_{\underline{I}}^{(j+1)} - q^{-1}\mathsf{X}_{\underline{I}}^{(j)}} \cdot \prod_{j \in I} \prod_{\substack{l=1 \\ l \neq \mathsf{k}(j)}}^{N_j} \frac{t\mathsf{X}_{\underline{I}}^{(j+1)} - x_l^{(j)}}{\mathsf{X}_{\underline{I}}^{(j)} - x_l^{(j)}}.$$

Shoji [**Sho**] considered the operator

$$\mathsf{S} = \sum_{\underline{I}=(I,\mathsf{k})} \prod_{j \in I} \prod_{\substack{l=1 \\ l \neq \mathsf{k}(j)}}^{N_j} \frac{t\mathsf{X}_{\underline{I}}^{(j+1)} - x_l^{(j)}}{\mathsf{X}_{\underline{I}}^{(j)} - x_l^{(j)}} \cdot \mathsf{T}_{q,\underline{I}} \tag{4.9}$$

and used it to characterize a different family of Macdonald-like polynomials associated with the wreath product groups Γ_n. One might expect these to relate to wreath Macdonald polynomials. Haiman observed some "accidental" agreement between the two [**Hai2**]. Our operators clarify the genuine difference between Shoji's polynomials and wreath Macdonald polynomials. The operators $\mathsf{M}^{(i)}$ for wreath Macdonald polynomials are substantially more complicated than Shoji's operator S.

4.5. The proof. Let us give an indication of how we discovered Theorem 4.2 and how we will prove it and extend it to higher wreath Macdonald operators in [**OSW**] (see Theorem 5.11 therein). The operators $\{\mathsf{M}^{(i)}\}_{i \in I}$ arise from the *horizontal Heisenberg subalgebra* in the quantum toroidal $\mathfrak{gl}_r$ algebra $\mathbf{U}_r^{\mathrm{tor}}$ [**GKV**] acting on the space Λ^Γ, which affords the vertex representation of [**Sai**]. More precisely, the vertex representation space is $\Lambda^{\otimes I} \otimes \mathbb{K}\{Q\}$, where $\mathbb{K}\{Q\}$ is a skew-version of the group algebra of the $\mathfrak{sl}_r$ root lattice Q spanned by formal exponentials e^γ for $\gamma \in Q$ satisfying $e^\beta e^\gamma = \pm e^{\beta+\gamma}$ with the sign depending on $\beta, \gamma \in Q$. We identify the spaces $\Lambda^\Gamma = \bigoplus_{\gamma \in Q} \Lambda^{\otimes I}$ and $\Lambda^{\otimes I} \otimes \mathbb{K}\{Q\}$ so that the summand indexed by $\gamma \in Q$ corresponds to the subspace $\Lambda^{\otimes I} \otimes \mathbb{K}e^\gamma$.

In its natural presentation, $\mathbf{U}_r^{\mathrm{tor}}$ comes equipped with an obvious Heisenberg subalgebra, the *vertical Heisenberg subalgebra*. In both the vertex representation Λ^Γ and the geometric representation $K_T(\mathrm{Hilb}^\Gamma)_{\mathrm{loc}}$ constructed in [**VV**], the vertical Heisenberg algebra acts in a straightforward way. In contrast, it is difficult to describe the aforementioned *horizontal* Heisenberg subalgebra and its actions on Λ^Γ and $K_T(\mathrm{Hilb})_{\mathrm{loc}}$. However, an automorphism due to Miki [**Mi**] swaps the vertical and horizontal Heisenberg subalgebras of $\mathbf{U}_r^{\mathrm{tor}}$.

Extending a result of [**FT, SV**] to the wreath setting, Wen [**Wen**] proved that the isomorphism (3.2) respects the actions of $\mathbf{U}_r^{\mathrm{tor}}$ on both sides, up to some nontrivial scalars *and* a very crucial twist by the Miki automorphism. In particular, Wen's result implies that the horizontal Heisenberg algebra acts in Λ^Γ like the vertical Heisenberg algebra acts in $K_T(\mathrm{Hilb}^\Gamma)_{\mathrm{loc}}$. By construction, the latter acts diagonally on the fixed point classes, with eigenvalues given by T-characters of fibers of certain tautological vector bundles on Hilb^Γ over the fixed points. Since (3.2) maps fixed point classes to wreath Macdonald polynomials, one immediately deduces that the horizontal Heisenberg algebra acts diagonally on wreath Macdonald polynomials. With some care, this leads to the right-hand side of the eigenoperator equation (4.7) in Theorem 4.2.

Obtaining the left-hand side, i.e., the explicit formula for the wreath Macdonald operators $\mathsf{M}^{(i)}$, requires substantial additional work. Using some known Miki automorphism images from [**T**], we transport degree-one elements in the horizontal Heisenberg algebra to the shuffle algebra [**Ne**] and use this to compute the action on symmetric polynomials by a very careful iterated residue computation and further cancellation. In the passage from symmetric functions in infinitely many variables to symmetric polynomials in finitely many variables, symmetric function skewing operators must be converted to the q-translations $\mathsf{T}_{q,\underline{J}}$ in (4.6). When the dust settles, one obtains the formula for $\mathsf{M}^{(i)}$ in (4.4).

Finally, we note that we have extensively verified Theorem 4.2 using software for wreath Macdonald polynomials developed in Sage by the second named author.

Acknowledgements

We thank Mark Haiman and Joshua Wen for helpful discussions.

References

[BF] Roman Bezrukavnikov and Michael Finkelberg, *Wreath Macdonald polynomials and the categorical McKay correspondence*, Camb. J. Math. **2** (2014), no. 2, 163–190, DOI 10.4310/CJM.2014.v2.n2.a1. With an appendix by Vadim Vologodsky. MR3295916

[BK] R. V. Bezrukavnikov and D. B. Kaledin, *McKay equivalence for symplectic resolutions of quotient singularities* (Russian, with Russian summary), Tr. Mat. Inst. Steklova **246** (2004), no. Algebr. Geom. Metody, Svyazi i Prilozh., 20–42; English transl., Proc. Steklov Inst. Math. **3(246)** (2004), 13–33. MR2101282

[FT] B. L. Feigin and A. I. Tsymbaliuk, *Equivariant K-theory of Hilbert schemes via shuffle algebra*, Kyoto J. Math. **51** (2011), no. 4, 831–854, DOI 10.1215/21562261-1424875. MR2854154

[GKV] Victor Ginzburg, Mikhail Kapranov, and Éric Vasserot, *Langlands reciprocity for algebraic surfaces*, Math. Res. Lett. **2** (1995), no. 2, 147–160, DOI 10.4310/MRL.1995.v2.n2.a4. MR1324698

[G] I. G. Gordon, *Quiver varieties, category $\mathcal{O}$ for rational Cherednik algebras, and Hecke algebras*, Int. Math. Res. Pap. IMRP **3** (2008), Art. ID rpn006, 69. MR2457847

[Hai1] Mark Haiman, *Hilbert schemes, polygraphs and the Macdonald positivity conjecture*, J. Amer. Math. Soc. **14** (2001), no. 4, 941–1006, DOI 10.1090/S0894-0347-01-00373-3. MR1839919

[Hai2] Mark Haiman, *Combinatorics, symmetric functions, and Hilbert schemes*, Current developments in mathematics, 2002, Int. Press, Somerville, MA, 2003, pp. 39–111. MR2051783

[Hai3] M. Haiman, unpublished manuscript, 2013.

[JK] Gordon James and Adalbert Kerber, *The representation theory of the symmetric group*, Encyclopedia of Mathematics and its Applications, vol. 16, Addison-Wesley Publishing Co., Reading, MA, 1981. With a foreword by P. M. Cohn; With an introduction by Gilbert de B. Robinson. MR644144

[Mac] I. G. Macdonald, *Symmetric functions and Hall polynomials*, 2nd ed., Oxford Mathematical Monographs, The Clarendon Press, Oxford University Press, New York, 1995. With contributions by A. Zelevinsky; Oxford Science Publications. MR1354144

[Mi] Kei Miki, *Toroidal braid group action and an automorphism of toroidal algebra $U_q(\mathrm{sl}_{n+1,\mathrm{tor}})$ $(n \geq 2)$*, Lett. Math. Phys. **47** (1999), no. 4, 365–378, DOI 10.1023/A:1007556926350. MR1693755

[Ne] Andrei Neguţ, *Quantum toroidal and shuffle algebras*, Adv. Math. **372** (2020), 107288, 60, DOI 10.1016/j.aim.2020.107288. MR4128572

[OS] Daniel Orr and Mark Shimozono, *Quiver Hall-Littlewood functions and Kostka-Shoji polynomials*, Pacific J. Math. **319** (2022), no. 2, 397–437, DOI 10.2140/pjm.2022.319.397. MR4482722

[OSW] Daniel Orr, Mark Shimozono, and Joshua Jeishing Wen, *Wreath Macdonald operators*, `arXiv:2211.03851`, 2023.

[Sai] Yoshihisa Saito, *Quantum toroidal algebras and their vertex representations*, Publ. Res. Inst. Math. Sci. **34** (1998), no. 2, 155–177, DOI 10.2977/prims/1195144759. MR1617066

[SV] Olivier Schiffmann and Eric Vasserot, *The elliptic Hall algebra and the K-theory of the Hilbert scheme of* $\mathbb{A}^2$, Duke Math. J. **162** (2013), no. 2, 279–366, DOI 10.1215/00127094-1961849. MR3018956

[Sho] Toshiaki Shoji, *Macdonald functions associated to complex reflection groups*, J. Algebra **260** (2003), no. 1, 426–448, DOI 10.1016/S0021-8693(02)00673-7. Special issue celebrating the 80th birthday of Robert Steinberg. MR1976702

[T] Alexander Tsymbaliuk, *Several realizations of Fock modules for toroidal* $\ddot{U}_{q,d}(\mathfrak{sl}_n)$, Algebr. Represent. Theory **22** (2019), no. 1, 177–209, DOI 10.1007/s10468-017-9761-5. MR3908899

[VV] M. Varagnolo and E. Vasserot, *On the K-theory of the cyclic quiver variety*, Internat. Math. Res. Notices **18** (1999), 1005–1028, DOI 10.1155/S1073792899000525. MR1722361

[Wen] J. J. Wen, Wreath Macdonald polynomials as eigenstates, `arXiv:1904.05015`

Department of Mathematics (MC 0123), 460 McBryde Hall, Virginia Tech, 225 Stanger St., Blacksburg, Virginia 24061
Email address: `dorr@math.vt.edu`

Department of Mathematics (MC 0123), 460 McBryde Hall, Virginia Tech, 225 Stanger St., Blacksburg, Virginia 24061
Email address: `mshimo@math.vt.edu`

Proceedings of Symposia in Pure Mathematics
Volume **108**, 2024
https://doi.org/10.1090/pspum/108/01962

Universal central extensions of Krichever-Novikov algebras and orthogonal polynomials

Felipe Albino dos Santos, Vyacheslav Futorny, and Kaiming Zhao

Dedicated to the memory of Ben Lewis Cox

ABSTRACT. We give a survey of the theory of the universal central extensions of superelliptic current and derivation Lie algebras of rings of meromorphic functions on Riemann surfaces. These algebras are examples of Krichever-Novikov algebras. Their universal central extensions have finite dimensional centers which defines certain recurrence relations between its elements. The families of polynomials satisfying such recurrence relations are orthogonal polynomials, classical or nonclassical depending on initial conditions. We survey all known cases for hyperelliptic curves. We also discuss recent results on superelliptic derivation Lie algebras.

Introduction

This is a survey based on several papers co-authored by our friend and colleague Ben Cox. We are deeply indebted to him for many years of fruitful joint work and his dedication.

Throughout this paper we will take the set of natural numbers to be $\mathbb{N} = \{1, 2, \dots\}$, the set of nonnegative integers will be denoted by $\mathbb{Z}_+ = \{0, 1, 2, 3, \dots\}$, and we will assume all vector spaces and algebras are defined over complex numbers $\mathbb{C}$.

Let R be the ring of meromorphic functions on Riemann surface with a fixed finite number of poles. Such rings appear in the study of the tensor module structures for affine Lie algebras. In particular, R can be any fine type commutative algebra over $\mathbb{C}$.

For any finite dimensional simple Lie algebra $\mathfrak{g}$ over $\mathbb{C}$ one constructs an infinite dimensional current Lie algebra $\mathfrak{g} \otimes R$ which is a natural generalization of untwisted affine Kac-Moody Lie algebras. This Lie algebra is an example of a Krichever-Novikov algebra introduced by Krichever and Novikov in the connection to the string theory in Minkowski space [**KN**]. Krichever-Novikov algebras have been studied extensively, in particular the Wess-Zumino-Witten-Novikov theory and analogues of the Knizhnik-Zamolodchikov equations are developed for these algebras (see, for example, [**Sc**] and the references therein). These algebras are also known as n-point algebras. Under this name some examples were studied by Bremner [**B1**] and [**B2**], who presented the generators and commutation relations of the universal central extension for the elliptic curve case in terms the families of ultraspherical and Pollaczek polynomials. Passing to algebraic curves of genus one

one finds the DJKM Lie algebras which were introduced by Date, Jimbo, Kashiwara and Miwa in their study of Landau-Lifshitz equation [**DJKM**]. The commutation relations in the universal central extension of the DJKM algebras in terms of generators and families of polynomials were given in [**CF**]. The DJKM algebra acts on the solutions of the Landau-Lifshitz equation as infinitesimal Bäcklund transformations. The study of the universal central extensions of the DJKM algebras led to a discovery of new families of orthogonal polynomials in [**CFT**].

Another appearance of Krichever-Novikov algebras is as derivation algebras of rings of meromorphic functions on Riemann surfaces. This is also a source of many new simple infinite dimensional Lie algebras.

The hyperelliptic curves correspond to the rings of the form

$$R = \mathbb{C}[t, t^{-1}, u \mid u^2 = p(t)],$$

where $p(t)$ is some polynomial in $\mathbb{C}[t]$ of degree ≥ 4.

The hyperelliptic curves are the simplest case of superelliptic curves $u^m = p(t)$, with $p(t) \in \mathbb{C}[t]$ and $m \geq 2$. The superelliptic derivation Lie algebras have been considered recently by Cox, Guo, Lu and Zhao in [**CGLZ17**], while the case of superelliptic current Lie algebras was discussed in [**A**].

1. Universal central extensions and Kähler differentials

A central extension of a Lie algebra $\mathfrak{g}$ over $\mathbb{C}$ is an extension $\tilde{\mathfrak{g}} = \mathfrak{g} \oplus C$ where C is contained in the center of $\tilde{\mathfrak{g}}$. A central extension $\tilde{\mathfrak{g}}$ is said to be a covering of $\mathfrak{g}$ if $\tilde{\mathfrak{g}}$ is perfect, that is $[\tilde{\mathfrak{g}}, \tilde{\mathfrak{g}}] = \tilde{\mathfrak{g}}$. A covering $\tilde{\mathfrak{g}}$ of $\mathfrak{g}$ is said to be universal if for every central extension $\tilde{\mathfrak{g}}'$ of $\mathfrak{g}$ there is a unique homomorphism from $\tilde{\mathfrak{g}}$ to $\tilde{\mathfrak{g}}'$ extending the identity map on $\mathfrak{g}$.

Let R be a commutative algebra defined over $\mathbb{C}$. Consider the left R-module $R \otimes R$ with action $f(g \otimes h) = fg \otimes h$ for $f, g, h \in R$ and let K be the submodule generated by the elements $1 \otimes fg - f \otimes g - g \otimes f$. Then $\Omega_R^1 = (R \otimes R)/K$ is the module of Kähler differentials. The element $f \otimes g + K$ is traditionally denoted by fdg. The canonical map $d : R \to \Omega_R^1$ is defined by $df = 1 \otimes f + K$. The exact differentials are the elements of the subspace dR. The coset of fdg modulo dR is denoted by $\overline{fdg}$. Then $\overline{fdg} + \overline{gdf} = \overline{d(fg)} = 0$ in the R-module Ω_R/dR. The space Ω_R/dR play a crucial role in the construction of the universal central extensions of Lie algebras.

Suppose that $\mathbb{C}[t, t^{-1}] \subset R$ and R is a finitely generated $\mathbb{C}[t, t^{-1}]$-module. Hence there exists $r_1, \ldots, r_n \in R$ such that

$$R = \mathbb{C}[t, t^{-1}]r_1 + \cdots + \mathbb{C}[t, t^{-1}]r_n$$

and

$$R \otimes R = \sum_{k,s \in \mathbb{Z}} \sum_{p,q=1}^{n} \mathbb{C}t^k r_p \otimes t^s r_q.$$

Moreover,

$$t^k r_p t^l r_q = \sum_{m \in \mathbb{Z}} \sum_{s=1}^{n} a_{p,q,s}^{m} t^{k+l+\alpha} r_s, \tag{1.1}$$

for some $a_{p,q,s}^{m} \in \mathbb{C}$.

Then we have in Ω_R:

$$t^k r_p \otimes t^l r_q + t^l r_q \otimes t^k r_p + K = 1 \otimes t^{k+l} r_p r_q + K$$
$$= \sum_{m\in\mathbb{Z}} \sum_{s=1}^{n} a_{p,q,s}^m 1 \otimes t^{k+l+m} r_s + K.$$

Therefore

$$\overline{t^k r_p d(t^l r_q)} = \overline{l t^{k+l-1} r_p r_q dt} + \overline{t^{k+l} r_p dr_q}$$
$$= \sum_{m\in\mathbb{Z}} \sum_{s=1}^{n} l a_{p,q,s}^m \overline{t^{\alpha+k+l-1} r_s\, dt} + \overline{t^{k+l} r_p dr_q}.$$

In particular, consider the ring $R = \mathbb{C}[t, t^{-1}, u \mid u^m = p(t)]$ with the polynomial

$$p(t) = t^n + a_{n-1} t^{n-1} + \cdots + a_0,$$

where $a_i \in \mathbb{C}$.

In this case R has a basis consisting of t^i, $t^i u$, $t^i u^2 \ldots, t^i u^{m-1}$ for $i \in \mathbb{Z}$. The equation $u^m = p(t)$ defines a superelliptic curve.

2. Krichever-Novikov algebras

Let $\mathfrak{g}$ be a simple finite dimensional complex Lie algebra and $\mathcal{G} = \mathfrak{g} \otimes \mathbb{C}[t, t^{-1}]$ the loop algebra of $\mathfrak{g}$ with commutation relations $[x \otimes f, y \otimes g] = [x, y] \otimes fg$, where $x, y \in \mathfrak{g}$ and $f, g \in \mathbb{C}[t, t^{-1}]$. Let R be a commutative algebra. In the construction of the loop algebra above, we may replace the Laurent polynomial algebra $\mathbb{C}[t, t^{-1}]$ by R and obtain the current algebra $\mathcal{G}_R = \mathfrak{g} \otimes R$ with the commutation relations: $[x \otimes a, y \otimes b] = [x, y] \otimes ab$, where $x, y \in \mathfrak{g}$ and $a, b \in R$. In particular, if $R = \mathcal{O}_X$ is the ring of regular functions of an affine scheme X of finite type then $\mathcal{G}_R = M(X, \mathfrak{g})$ is the map Lie algebra of all regular maps from X to $\mathfrak{g}$ [**CFR**].

The universal central extension of the current algebra $\mathcal{G}_R$ is the vector space $\hat{\mathfrak{g}} = (\mathfrak{g} \otimes R) \oplus \Omega_R^1/dR$ with Lie bracket given by

$$[x \otimes f, Y \otimes g] = [xy] \otimes fg + (x, y)\overline{fdg}, [x \otimes f, \omega] = 0, [\omega, \omega'] = 0,$$

where $x, y \in \mathfrak{g}$, and $\omega, \omega' \in \Omega_R^1/dR$ and (x, y) denotes the Killing form on $\mathfrak{g}$ [**K**].

Lie algebras in the form $\mathcal{G}_R$ provide a family of infinite dimensional Lie algebras based on a simple finite-dimensional Lie algebra $\mathfrak{g}$ and infinite dimensional commutative algebra R. Another source of such algebras is related to the derivations of R:

$$\mathrm{Der}(R) = \{f : R \to R \mid f \text{ is linear and } f(ab) = f(a)b + af(b), \forall a, b \in R\}$$

is a Lie algebra with respect to a product $[f, g] = fg - gf$. In particular, the Lie algebra of derivations of the Laurent polynomials $\mathbb{C}[t, t^{-1}]$

$$\mathrm{Der}(\mathbb{C}[t, t^{-1}]) := \bigoplus_{n\in\mathbb{Z}} \mathbb{C} t^n \frac{d}{dt}$$

is the first Witt algebra isomorphic to a subalgebra of the Lie algebra of complex valued vector fields on the circle. Its universal central extension is the celebrated Virasoro algebra.

Let $X \subset \mathbb{A}_k^n$ be an irreducible affine variety, $R_X = k[x_1, \ldots, x_n]/I$ the algebra of polynomial functions on X, where $I \subset k[x_1, \ldots, x_n]$ is a defining ideal of X.

Then $\mathrm{Der}(R_X)$ is the Lie algebra of polynomial vector fields on X. We have the following criterion of simplicity of $\mathrm{Der}(R_X)$ established in [**J**], [**Si**], [**BF**]:

THEOREM 2.1. *Let k be an algebraically closed field of characteristic* 0. *The Lie algebra* $\mathrm{Der}(R_X)$ *of polynomial vector fields on X is simple if and only if X is a smooth variety.*

Observe that the Laurent polynomial algebra $\mathbb{C}[t, t^{-1}]$ can be viewed as the ring of meromorphic functions on Riemann surface with poles in two points: zero and infinity. Let R the ring of meromorphic functions on a Riemann surface with a finite number of fixed poles. Lie algebras of the form $\mathrm{Der}(R)$ and $\mathfrak{g}\otimes R$ and their central extensions are known as Krichever-Novikov algebras. They were introduced by Krichever and Novikov in [**KN**]. These algebras L are not $\mathbb{Z}$-graded in general, but rather quasi-graded, that is there exists m such that $L = \oplus_{i\in\mathbb{Z}} L_i$ and

$$[L_i, L_j] \subset \sum_{k=i+j-m}^{i+j+m} L_k.$$

3. Universal central extension of superelliptic affine Lie algebras

Consider again the superelliptic case: $R = \mathbb{C}[t, t^{-1}, u|u^m = p(t)]$, with $m \geq 2$. The algebras $\mathcal{G} = \mathfrak{g} \otimes R$ are examples of *superelliptic* Lie algebras. Assume that $p(t) = \sum_{i=0}^{n} a_i t^i$, where $a_d = 1$ and a_0, a_1 are not both 0.

One can easily see that Ω^1_R is spanned by the differentials $t^i u^k dt$ and $t^i u^s du$ with $i \in \mathbb{Z}$, $k \in \{0, 1, \ldots, m-1\}$, and $s \in \{0, 1, \ldots, m-2\}$.

Recall the following result:

THEOREM 3.1 ([**B1**, Theorem 2.1]). *The dimension of Ω^1_R/dR is $2g + n - 1$ where g is the genus and n is the number of poles.*

As the genus of the superelliptic curve is $g = \frac{1}{2}(m(n-1) - n - \gcd(m, n)) + 1$ and the number of poles is $\gcd(m, n) + 1$, if $a_0 \neq 0$ and $\gcd(m, n) + m$ if $a_0 = 0$, we have the following formula for the dimension in the superelliptic case [**A**, Theorem 1.10]:

$$\dim \Omega^1_R/dR = \begin{cases} m(n-1) + 1 \text{ if } a_0 \neq 0, \\ m(n-1) - (m-1) + 1 \text{ if } a_0 = 0. \end{cases}$$

THEOREM 3.2 ([**A**, Theorem 2.4]). *The following elements constitute a basis of Ω^1_R/dR:*

$$\overline{t^{-1}dt}, \overline{t^{-1}u^l dt}, \ldots, \overline{t^{-n}u^l dt},$$

with $l \in \{1, 2, \ldots, m-1\}$, where we omit $\overline{t^{-n}u^l dt}$ if $a_0 = 0$.

In the hyperelliptic case $m = 2$ we obtain

COROLLARY 3.3 ([**CF**],Theorem 3.4). *The set*

$$\{\overline{t^{-1}\,dt}, \overline{t^{-1}u\,dt}, \ldots, \overline{t^{-n}u\,dt}\}$$

forms a basis of Ω^1_R/dR (omitting $\overline{t^{-n}u\ dt}$ if $a_0 = 0$).

Now we describe the cocyles for the superelliptic affine Lie algebra.

Set

$$\omega_0 = \overline{t^{-1}dt} \text{ and } \omega_{i,j} = \overline{t^i u^j dt} \text{ for } j \neq 0.$$

For $m, l \in \mathbb{Z}_+$ and $a_0, a_1, \ldots, a_{n-1} \in \mathbb{C}$ define the sequence of polynomials P_s, $s \geq -n$ defined recursively as follows:

$$(nl + (1+s)m)P_s = \sum_{k=0}^{n-1} \left(-a_k \left(kl + (-n+1+s+k)m\right)\right) P_{-n+s+k}$$

with initial conditions

$$P_{-n} = \overline{t^{-n}u^l dt}, P_{-n+1} = \overline{t^{-n+1}u^l dt}, \ldots, P_{-1} = \overline{t^{-1}u^l dt}.$$

We then check that $P_s = \overline{t^s u^l dt}$ for $s \geq 0$.

Also, for $m, l \in \mathbb{Z}_+$ and $a_0, a_1, \ldots, a_{n-1} \in \mathbb{C}$ with $a_0 \neq 0$ define recursively the sequence of polynomials Q_s for $s \leq -n-1$ as follows:

$$(a_0(1+s)m)Q_s = \sum_{k=1}^{n} \left(-a_k \left(kl + (1+s+k)m\right)\right) Q_{s+k}$$

with initial conditions

$$Q_{-n} = \overline{t^{-n}u^l dt}, Q_{-n+1} = \overline{t^{-n+1}u^l dt}, \ldots, Q_{-1} = \overline{t^{-1}u^l dt}.$$

Then $Q_s = \overline{t^s u^l dt}$ for $s \leq -n-1$.

In the case $a_0 = 0$ we define similarly the sequence of polynomials R_s, $s \leq -n-1$ by

$$a_1(l + (1+s+1)m)R_s = \sum_{k=2}^{n} \left(-a_k \left(kl + (1+s+k)m\right)\right) R_{s+k}$$

with initial condition

$$R_{-n+1} = \overline{t^{-n+1}u^l dt}, R_{-n+2} = \overline{t^{-n+2}u^l dt}, \ldots, R_{-1} = \overline{t^{-1}u^l dt}.$$

Then $R_s = \overline{t^s u^l dt}$, $s \leq -D-1$.

Set

$$\psi_{i,j} = \begin{cases} P_{i+j-1} \text{ if } i+j \geq -n+1; \\ Q_{i+j-1} \text{ if } i+j \leq -n \text{ and } a_0 \neq 0; \\ R_{i+j-1} \text{ if } i+j \leq -n \text{ and } a_0 = 0. \end{cases}$$

Then $\overline{t^i u^l d(t^j)} = j\psi_{i,j}$ [**A**, Proposition 3.4].

We can now give explicit commutation relations for $\hat{\mathcal{G}}$.

THEOREM 3.4 ([**A**, Theorem 3.5]). *The superelliptic affine Lie algebra $\hat{\mathcal{G}}$ has a $\mathbb{Z}_m$-grading $\hat{\mathcal{G}} = \oplus_{l=0}^{m-1} \hat{\mathcal{G}}^l$, in which*

$$\hat{\mathcal{G}}^0 = \mathfrak{g} \otimes \mathbb{C}[t, t^{-1}] \oplus \mathbb{C}\omega_0, \qquad \hat{\mathcal{G}}^l = \mathfrak{g} \otimes \mathbb{C}[t, t^{-1}]u^l \bigoplus_{s=1}^{n} \mathbb{C}\omega_{-s,l}.$$

The subalgebra $\hat{\mathcal{G}}^0$ is an untwisted affine Kac-Moody Lie algebra with commutation relations

$$[x \otimes t^i, y \otimes t^j] = [x, y] \otimes t^{i+j} \oplus \delta_{i+j,0}(x, y) j\omega_0.$$

The subspace $\hat{\mathcal{G}}^l$ is a $\hat{\mathcal{G}}^0$-module:

$$[x \otimes t^i u^s, y \otimes t^j] = [x, y] \otimes (t^{i+j}u^{s+1}) + (x, y) j\psi_{i,j}.$$

The remaining relations in $\hat{\mathcal{G}}$ are:

$$[x \otimes t^i u^{l_1}, y \otimes t^j u^{l_2}] = [x, y] \otimes (t^{i+j}u^{l_1+l_2}) + (x, y) \left(\frac{jl_1 - il_2}{l_1 + l_2}\right) \omega_{i+j-1, l_1+l_2},$$

if $l_1 + l_2 \le m-1$, *and*

$$[x\otimes t^i u^{l_1}, y\otimes t^j u^{l_2}] = [x,y]\otimes\left(\sum_{k=0}^{n} a_k t^{i+j+k}u^{l_1+l_2-m}\right) + (x,y)\left(\frac{jl_1 - il_2}{l_1+l_2}\right)\omega_{i+j-1,l_1+l_2}.$$

if $l_1 + l_2 > m-1$.

4. Examples

Example 1 (The three point affine Lie algebras). Let $R = \mathbb{C}[t, t^{-1}, u | u^2 = t^2 + 4t]$. Then $\hat{\mathcal{G}}_R$ is the three point affine Lie algebra considered in [**CJ**]. The space Ω^1_R/dR has a basis $\{\overline{t^{-1}dt}, \overline{t^{-1}udt}\}$ and the cocycle can be written in terms of Gegenbauer polynomials [**B2**, Theorem 3.6].

Example 2 (The four point affine Lie algebras). Let $a \in \mathbb{C}\setminus\{0,1\}$ and $R = \mathbb{C}[s, s^{-1}, (s-1)^{-1}, (s-a)^{-1}]$. It was shown in [**B2**] that $R \simeq \mathbb{C}[t, t^{-1}, u | u^2 = t^2 - 2bt + 1]$. In this case, $\hat{\mathcal{G}}_R$ is called the four point affine Lie algebra. The space Ω^1_R/dR has a basis $\{\overline{t^{-1}dt}, \overline{t^{-1}udt}, \overline{t^{-2}udt}\}$ [**B2**, Theorem 3.3].

Example 3 (The elliptic affine Lie algebras). Let X be an elliptic curve of genus 1. It can be identified with the quotient of complex plane by the $\mathbb{Z}$-lattice with a basis $0, \lambda$, $\operatorname{Im}\lambda > 0$. Let R the ring of meromorphic functions on X which are holomorphic outside of two points $\{0, \frac{1+\lambda}{2}\}$. Bremner [**B1**] showed that $R \simeq \mathbb{C}[t, t^{-1}, u | u^2 = t^3 - 2bt^2 + t]$, where b is some constant determined by the value of the Weierstrass function

$$\wp(z) := z^{-2} + \sum_{0\neq\omega}\left(\frac{1}{(z-\omega)^2} - \frac{1}{\omega^2}\right).$$

The algebra $\hat{\mathcal{G}}_R$ is the elliptic affine Lie algebra $\hat{\mathcal{G}}_R$, where Ω^1_R/dR has a basis

$$\{\overline{t^{-1}dt}, \quad \overline{t^{-1}udt}, \quad \overline{t^{-2}udt}\}.$$

In this case we have

$$\overline{t^{k-2}u\,dt} = p_k(b)\overline{t^{-1}u\,dt} + q_k(b)\overline{t^{-2}u\,dt},$$

where $p_k(b)$, $q_k(b)$ are Pollaczek polynomials that satisfy the following recurrence relation:

$$(k+\gamma)p_k(b) = 2[(k+\lambda+\alpha+\gamma-1)b+\beta]p_{k-1}(b) - (k+2\lambda+\gamma-2)p_{k-2}(b)$$

for the parameters $\lambda = -1/2$, $\alpha = 0$, $\beta = -1$, $\gamma = 1/2$ together with the initial conditions

$$p_0(b) = 0, \quad p_1(b) = 1.$$

The generating function $P = P(x,b) := \sum_{n=0}^{\infty} p_n(b)x^n$ satisfies the following differential equation:

$$x(x^2 - 2bx - 1)\frac{dP}{dx} + \left[-\frac{1}{2}x^2 + 2x + \frac{1}{2}\right]P - \frac{3}{2}x = 0.$$

Example 4 (The Date-Jimbo-Kashiwara-Miwa algebras). Let

$$R = \mathbb{C}[t, t^{-1}, u | u^2 = (t^2 - b^2)(t^2 - c^2)],$$

where $b \neq \pm c$ are complex constants. The Lie algebra $\hat{\mathcal{G}}_R$ is the DJKM algebra whose Kähler differentials are spanned by

$$\{\overline{t^{-1}dt}, \quad \overline{t^{-1}udt}, \quad \overline{t^{-2}udt}, \quad \overline{t^{-3}udt}, \quad \overline{t^{-4}udt}\}.$$

It was introduced by Date, Jimbo, Kashiwara and Miwa [**DJKM**] in the study of integrable systems arising from Landau-Lifshitz differential equation

$$\mathbf{S}_t = \mathbf{S} \times \mathbf{S}_{xx} + \mathbf{S} \times \mathbf{J}\mathbf{S}$$

where

$$\mathbf{S} = (S_1, S_2, S_3), \quad S_1^2 + S_2^2 + S_3^2 = 1, \quad \mathbf{J} = \begin{pmatrix} J_1 & 0 & 0 \\ 0 & J_2 & 0 \\ 0 & 0 & J_3 \end{pmatrix}, \quad J_i \in \mathbb{C}.$$

The Lie algebra $\hat{\mathcal{G}}_R$ acts on the solutions of the Landau-Lifshitz equation by infinitesimal Bäcklund transformations.

The basis Kähler differentials satisfy the following recursion relation:

$$(6+2k)\overline{t^k u\,dt} = 4kc\overline{t^{k-2}u\,dt} - 2(k-3)\overline{t^{k-4}u\,dt}.$$

EXAMPLE 5 (The hyperelliptic affine Lie algebras). Let R be hyperelliptic affine Lie algebra $\mathbb{C}[t, t^{-1}, u]$ in which $u^2 = t^d + a_{d-1}t^{d-1} + \cdots + a_1 t + a_0$. Here $m = 2$. We have that, $\{\overline{t^{-1}dt}, \overline{t^{-1}udt}, \ldots, \overline{t^{-d}udt}\}$ (where we omit $\overline{t^{-d}udt}$ if $a_0 = 0$) is a finite basis for Ω_R^1/dR, as seen in Theorem 3.4 in [**B1**].

EXAMPLE 6 (The superelliptic affine Lie algebras). Let R be superelliptic affine Lie algebra $\mathbb{C}[t, t^{-1}, u]$ in which

$$u^3 = t^3 + bt^2 + c.$$

We have that, $\{\overline{t^{-1}dt}, \overline{t^{-1}udt}, \overline{t^{-2}udt}, \overline{t^{-3}udt}\}$ is a finite basis for Ω_R^1/dR, as seen in Theorem 3.4 in [**B1**].

The following diagram describes the relations between different classes of Krichever-Novikov Lie algebras

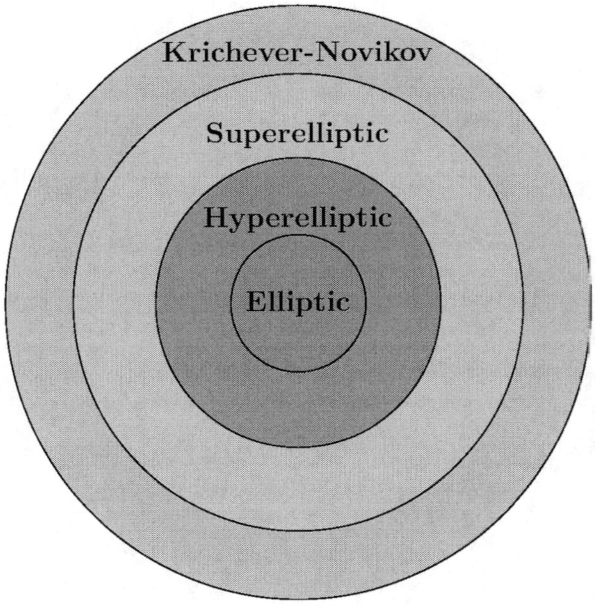

5. Orthogonal polynomials

5.1. The case of the Date-Jimbo-Kashiwara-Miwa algebras. Let $P_k := P_k(c)$, $k \geq 0$ be the polynomial in the indeterminant c satisfying the recursion relation

$$(6+2k)P_k(c) = 4kcP_{k-2}(c) - 2(k-3)P_{k-4}(c).$$

Then the generating function

$$P(c,z) := \sum_{k\geq -4} P_k(c)z^{k+4} = \sum_{k\geq 0} P_{k-4}(c)z^k$$

satisfies the differential equation

$$\frac{d}{dz}P(c,z) - \frac{3z^4 - 4cz^2 + 1}{z^5 - 2cz^3 + z}P(c,z) = \frac{2\left(P_{-1} + cP_{-3}\right)z^3 + P_{-2}z^2 + (4cz^2-1)P_{-4}}{z^5 - 2cz^3 + z}.$$

In order to solve the recursion relation above we need four initial conditions. We will consider the following cases:

1) $P_{-3}(c) = P_{-2}(c) = P_{-1}(c) = 0$ and $P_{-4}(c) = 1$.
2) $P_{-4}(c) = P_{-3}(c) = P_{-1}(c) = 0$ and $P_{-2}(c) = 1$.
3) $P_{-1}(c) = 1$, and $P_{-2}(c) = P_{-3}(c) = P_{-4}(c) = 0$.
4) $P_{-1}(c) = 0 = P_{-2}(c) = P_{-4}(c) = 0$ and $P_{-3}(c) = 1$.

Cases 3) and 4) can be solved in terms of Gegenbauer polynomials, for example

$$P_{-1}(c,z) = \sum_{n\geq 0} P_{-1,n-4}z^n = \frac{1}{c^2-1}\left((c^2-1)z^3 - \sum_{k=2}^{\infty} cQ_n^{(-1/2)}(c)z^{2n+1}\right),$$

where $Q_n^{(-1/2)}(c)$ is the n-th Gegenbauer polynomial. Renumbering the polynomials set $P = P_{-1,2n-3}$, $n \geq 2$. Then P satisfies the second order differential equation:

$$\boxed{(c^4 - c^2)\frac{d^2}{d^2c}P + 2c(c^2+1)\frac{d}{dc}P + (-c^2n(n-1) - 2)P = 0}.$$

The remaining cases correspond to elliptic integrals and the following generating functions:

$$\begin{aligned} P_{-4}(c,z) &:= \sum_{k\geq -4} P_{-4,k}(c)z^{k+4} = \sum_{k\geq 0} P_{-4,k-4}(c)z^k \\ &= z\sqrt{1-2cz^2+z^4}\int \frac{4cz^2-1}{z^2(z^4-2cz^2+1)^{3/2}}\,dz \end{aligned}$$

in case 1), and

$$P_{-2}(c,z) = z\sqrt{1-2cz^2+z^4}\int \frac{1}{(z^4-2cz^2+1)^{3/2}}\,dz$$

in case 2).

We have

$$\begin{aligned} P_{-4}(c,z) &= \sum_{n=0}^{\infty} P_{-4,n}(c)z^n = 1 + z^4 + \frac{4c}{5}z^6 + \frac{1}{35}\left(32c^2 - 5\right)z^8 + \\ &+ \frac{16}{105}c\left(8c^2-3\right)z^{10} - \frac{\left(2048c^4 - 1248c^2 + 75\right)}{1155}z^{12} + O(z^{14}). \end{aligned}$$

Set $P_n(c) = P_{-4,n}(c)$ and $Q_n(c) = P_{-2,n}(c)$. We have

THEOREM 5.1 ([**CFT**]).

(i) *The polynomials $P_n(c)$ satisfy the family of ODE's*

$$\boxed{\begin{aligned}&16(c^2-1)^2P_n^{(iv)}+160c(c^2-1)P_n'''-8(c^2(n^2-4n-46)-n^2+4n+22)P_n''\\&-24c(n^2-4n-6)P_n'+(n-4)^2n^2P_n=0.\end{aligned}}$$

(ii) *The polynomials $Q_n(c)$ satisfy the family of ODE's*

$$\boxed{\begin{aligned}&16(c^2-1)^2Q_n^{(iv)}+160c(c^2-1)Q_n'''-8(c^2(n^2-4n-42)-n^2+4n+18)Q_n''\\&-24c(n^2-4n-2)Q_n'+(n-6)(n-2)^2(n+2)Q_n=0\end{aligned}}$$

(iii) *The polynomials $P_n(c)$ and $Q_n(c)$ are non-classical orthogonal polynomials*

The polynomials P_n and Q_n are the (associated ultraspherical) Jacobi polynomials introduced by Wimp [**W**], Bustoz and Ismail [**BI**].

6. Superelliptic derivation Lie algebras

Let m be a natural number, $p(t)\in\mathbb{C}[t]$. Then we have the algebraic curve $u^m=p(t)$, and its coordinate algebras (the Riemann surfaces) $R_m(p)=\mathbb{C}[t^{\pm1},u\mid u^m=p(t)]$ and $S_m(p)=\mathbb{C}[t,u\mid u^m=p(t)]$ which are commutative algebras over $\mathbb{C}$. The Lie algebras $\mathcal{R}_m(p)=\mathrm{Der}(R_m(p))$ and $\mathcal{S}_m(p)=\mathrm{Der}(S_m(p))$ are called the m-th superelliptic Lie algebras associated to $p(t)$.

If $m=1$, we know that $R_1(p)\simeq\mathbb{C}[t^{\pm1}]$ and $S_1(p)\simeq\mathbb{C}[t]$. The corresponding Lie algebras are the Witt algebra over Laurent polynomial algebra and over polynomial algebra respectively.

If $m>1$ and $p(t)\in\mathbb{C}\setminus\{0\}$, we may assume that $p(t)=1$, and write

$$u^m-1=(u-\omega_1)(u-\omega_2)\cdots(u-\omega_m),$$

where ω_i are all the m-the roots of unity. We see that

$$\mathbb{C}[u]/\langle u^m-1\rangle=\bigoplus_{i=1}^m\left\langle\frac{u^m-1}{(u-\omega_i)\prod_{j\neq i}(\omega_i-\omega_j)}\right\rangle,$$

where

$$\left(\frac{u^m-1}{(u-\omega_i)\prod_{j\neq i}(\omega_i-\omega_j)}\right)^2=\frac{u^m-1}{(u-\omega_i)\prod_{j\neq i}(\omega_i-\omega_j)}\quad\mathrm{mod}\ (u^m-1).$$

As $R_m(p)\simeq\mathbb{C}[t_1^{\pm1}]\oplus\cdots\oplus\mathbb{C}[t_m^{\pm1}]$ and $S_m(p)\simeq\mathbb{C}[t_1]\oplus\cdots\oplus\mathbb{C}[t_m]$, the corresponding Lie algebras are direct sum of Lie ideals, where each Lie ideal is a copy of the Witt algebra over a Laurent polynomial algebra or over a polynomial algebra.

For simplicity of all other such Lie algebras $(n>1)$ we have

THEOREM 6.1 (Theorem 4, [**CGLZ17**]). *Let $m\in\mathbb{N}+1$ and $\deg(p(t))\geq1$.*

(a) *The Lie algebra $\mathcal{S}_m(p)$ is simple if and only if $p(t)$ has no multiple nonzero roots.*

(b) *The Lie algebra $\mathcal{R}_m(p)$ is simple if and only if $p(t)$ has no multiple nonzero roots and $p(t)\neq ct^r$ for any $c\in\mathbb{C}^*,r\in\mathbb{N}$ with $\gcd(m,r)\neq1$.*

When $p = ct^r$ for $c \in \mathbb{C}^*, r \in \mathbb{N}$ with $\gcd(m, r) = 1$, one can easily see that $\mathcal{R}_m(p)$ is isomorphic to the classical centerless Virasoro algebra.

The Lie algebra $\mathcal{R}_2(p)$ is actually a 3- or 4-point Virasoro algebra if $p(t)$ is of degree 1 or 2 respectively. The n-point Virasoro algebras were systematically studied in [**CGLZ14**].

LEMMA 6.2.

(a) *If* $p(t) = t$, *then* $R_2(p) \simeq \mathbb{C}[t^{\pm 1}]$.
(b) *If* $p(t) = t - a^2$ *where* $a \neq 0$, *then* $R_2(p) \simeq \mathbb{C}[t^{\pm 1}, \frac{1}{t+1}]$.
(c) *If* $p(t) = t^2 - 2bt + 1$ *where* $b \neq \pm 1$, *then*

$$R_2(p) \simeq \mathbb{C}[t^{\pm 1}, \frac{1}{t+b+1}, \frac{1}{t+b-1}].$$

(d) *If* $p(t) = t^2 - 2bt$ *where* $b \neq 0$, *then* $R_2(p) \simeq \mathbb{C}[t^{\pm 1}, \frac{1}{t+1}]$.

As for the Lie algebra $\mathcal{S}_2(P)$ with $\deg(p(t)) \leq 2$ we have

LEMMA 6.3. *Suppose that* $p(t) \in \mathbb{C}[t]$ *has no multiple roots.*

(a) *If* $\deg(p(t)) = 1$, *then* $S_m(p) \simeq \mathbb{C}[t]$ *for any* $m \in \mathbb{N}$.
(b) *If* $\deg(p(t)) = 2$, *then* $S_2(p) \simeq \mathbb{C}[t, t^{-1}]$.

As a consequence of Theorem 3.2 in [**S**] we know that all derivations of the simple Lie algebras $\mathcal{S}_m(p)$ and $\mathcal{R}_m(p)$ are inner derivations.

Let $\partial = \sqrt[m]{p(t)^{m-1}}\frac{d}{dt}$. If the Lie algebras $\mathcal{R}_m(p)$ and $\mathcal{S}_m(p)$ are simple, then

$$\mathcal{R}_m(p) = \mathbb{C}[t^{\pm 1}, \sqrt[m]{P}]\partial \text{ and } \mathcal{S}_m(p) = \mathbb{C}[t, \sqrt[m]{P}]\partial.$$

Next we have the universal central extension of the simple Lie algebras $\mathcal{R}_m(p)$.

THEOREM 6.4 (Theorem 9, [**CGLZ17**]). *Suppose that* $R = R_m(p)$ *with* $m \geq 2$ *and* $p = t^l(t - a_1)\dots(t - a_n)$ *for some* $l \in \mathbb{Z}_+$, $n \in \mathbb{N}$ *and pairwise distinct* $a_1, \dots,$ $a_n \in \mathbb{C}^*$. *Then the universal central extension of* $\mathcal{R}_m(p)$ *is* $\mathcal{R}_m(p) \oplus R/\partial(R)$ *with brackets*

$$[f\partial, g\partial] = f\partial(g)\partial - g\partial(f)\partial + \overline{\partial(f)\partial(\partial(g))}, \forall \quad f, g \in R$$

If we can compare this result with Theorem 3.4, we see that the universal central extensions take similar forms but are essentially different.

Now we have the universal central extension of the simple Lie algebras $\mathcal{S}_m(p)$.

THEOREM 6.5 (Theorem 10, [**CGLZ17**]). *Suppose that* $m \geq 2$ *and* $p \in \mathbb{C}[t]$ *is a polynomial of degree* $n \geq 1$ *without multiple roots. Then* $\dim S/\partial(S) = (n-1)(m-1)$ *and the universal central extension of* $\mathcal{S}_m(p)$ *is* $\mathcal{S}_m(p) \oplus S/\partial(S)$ *with brackets*

$$[f\partial, g\partial] = f\partial(g)\partial - g\partial(f)\partial + \overline{\partial(f)\partial(\partial(g))}, \forall \quad f, g \in S.$$

From the above two theorems we need to compute $\overline{f(t)} \in R/\partial R$ for any $f(t) \in R$, and to compute $\overline{f(t)} \in S/\partial S$ for any $f(t) \in S$, or equivalently, to find bases for the spaces $R/\partial R$ and $S/\partial S$ which are easy to use. This is actually very hard in general.

A natural open question is to determine the necessary and sufficient conditions for the isomorphism among the various Lie algebras $\mathcal{R}_m(p)$ (resp. $\mathcal{S}_m(p)$). From [**S**] we need only to consider isomorphisms of the Riemann surfaces $R_m(p)$ (resp. $S_m(p)$). This is a very hard problem (See [**BJ**]). So far we actually know the automorphism groups of compact Riemann surfaces with genus 0 or 1 only, and there are scattered results on higher genus Riemann surfaces, see [**BJ**, **B**] and the references therein. We point out that the Riemann surfaces $R_m(p)$ (resp. $S_m(p)$) are generally not compact which makes the problem even harder.

A related interesting open question is to determine the unit group $R_2^*(p)$ (resp. $S_2^*(p)$) for some $R_2(p)$ (resp. $S_2(p)$), i.e., the multiplicative group consisting of all invertible elements in $R_2(p)$ (resp. $S_2(p)$) for some special polynomial $p \in \mathbb{C}[t]$.

We have partial results for the above two questions.

LEMMA 6.6 (Lemma 12, [**CGLZ17**]). *Suppose that* $\deg(p(t))$ *is odd. Then* $S_2^*(p) = \mathbb{C}^*$. *If furthermore* $p(t)$ *has no multiple roots and* $t|p$, *then*

(i). *The polynomial equation* $f^2 - g^2p = t^k$ *for* $k \geq 0$ *has solutions for* $f, g \in \mathbb{C}[t]$ *with* $\gcd(f, g) = 1$ *and* $g \neq 0$ *if and only if* $\deg(p) = 1$ *and* $k = 1$; *Moreover, the only solutions for the equation* $f^2 - g^2t = t$ *are* $f = 0, g = \pm\sqrt{-1}$.

(ii). $R_2^*(p) = \bigcup_{k\in\mathbb{Z}} \mathbb{C}^*t^k \cong \mathbb{C}x^* \times \mathbb{Z}$ *if* $\deg(p) \geq 3$, *and*

$$R_2^*(p) = \bigcup_{k\in\mathbb{Z}} \left(\mathbb{C}^*t^k \cup \mathbb{C}^*t^{k+\frac{1}{2}}\right) \cong \mathbb{C}^* \times \mathbb{Z} \times (\mathbb{Z}/2\mathbb{Z})$$

if $P = c^2t$ *for some* $c \in \mathbb{C}^*$.

THEOREM 6.7 (Theorem 13, [**CGLZ17**]). *Let* $p(t) = \frac{t^4-2\beta t^2+1}{\beta^2-1}$ *where* $\beta \neq \pm 1$. *Let*

$$\lambda_1 = \frac{t^2+1}{\sqrt{2(\beta+1)}} + \sqrt{\frac{\beta-1}{2}}\sqrt{p},$$
$$\lambda_2 = \frac{t^2-1}{\sqrt{2(\beta-1)}} + \sqrt{\frac{\beta+1}{2}}\sqrt{p}.$$

(a). *As a multiplicative group,* $R_2^*(p)$ *is generated by* $\mathbb{C}^*$, t, λ_1, λ_2.

(b). $R_2^*(p) \simeq \mathbb{C}^* \times \mathbb{Z} \times \mathbb{Z} \times \mathbb{Z}$.

THEOREM 6.8 (Theorem 14, [**CGLZ17**]). *Suppose the polynomials* $p_1(t) = t(t-a_1)\cdots(t-a_{2n})$ *and* $p_2(t) = t(t-b_1)\cdots(t-b_{2m})$ *both have distinct roots where* $a_i, b_j \in \mathbb{C}^*$. *Let* $\phi : R_2(p_1) \to R_2(p_2)$ *be an isomorphism. Then* $m = n$ *and there is some* $c \in \mathbb{C}^*$ *and* $\gamma \in S_{2n}$ *such that*

$$a_{\gamma(i)} = cb_i, \ \forall\, i = 1, 2, \cdots, 2n, \ \text{or}$$
$$a_ib_{\gamma(i)} = c, \ \forall\, i = 1, 2, \cdots, 2n,$$

where S_{2n} *is the symmetry group on* $\{1, 2, \cdots, 2n\}$; *and*

$$\phi(t) = ct, \quad \phi(\sqrt{p_1}) = \pm c^n\sqrt{p_2}, \quad \text{or}$$
$$\phi(t) = ct^{-1}, \qquad \phi(\sqrt{p_1}) = \pm t^{-n-1}\sqrt{ca_1a_2\ldots a_{2n}p_2}.$$

References

[A] Felipe Albino dos Santos, *On the universal central extension of superelliptic affine Lie algebras*, Comm. Algebra **50** (2022), no. 5, 2045–2055, DOI 10.1080/00927872.2021.1998515. MR4401129

[BJ] Mikhail Belolipetsky and Gareth A. Jones, *Automorphism groups of Riemann surfaces of genus* $p + 1$, *where* p *is prime*, Glasg. Math. J. **47** (2005), no. 2, 379–393, DOI 10.1017/S0017089505002612. MR2203507

[BF] Yuly Billig and Vyacheslav Futorny, *Lie algebras of vector fields on smooth affine varieties*, Comm. Algebra **46** (2018), no. 8, 3413–3429, DOI 10.1080/00927872.2017.1412456. MR3789004

[B] Thomas Breuer, *Characters and automorphism groups of compact Riemann surfaces*, London Mathematical Society Lecture Note Series, vol. 280, Cambridge University Press, Cambridge, 2000. MR1796706

[B1] Murray Bremner, *Universal central extensions of elliptic affine Lie algebras*, J. Math. Phys. **35** (1994), no. 12, 6685–6692, DOI 10.1063/1.530700. MR1303073

[B2] Murray Bremner, *Four-point affine Lie algebras*, Proc. Amer. Math. Soc. **123** (1995), no. 7, 1981–1989, DOI 10.2307/2160931. MR1249871

[BI] Joaquin Bustoz and Mourad E. H. Ismail, *The associated ultraspherical polynomials and their q-analogues*, Canadian J. Math. **34** (1982), no. 3, 718–736, DOI 10.4153/CJM-1982-049-6. MR663314

[CFR] Lucas Calixto, Vyacheslav Futorny and Henrique Rocha, *Classification of simple Harish-Chandra modules for basic classical map superalgebras*, arXiv:2104.07517.

[C] Ben Cox, *On the universal central extension of hyperelliptic current algebras*, Proc. Amer. Math. Soc. **144** (2016), no. 7, 2825–2835, DOI 10.1090/proc/13057. MR3487217

[CF] Ben Cox and Vyacheslav Futorny, *DJKM algebras I: their universal central extension*, Proc. Amer. Math. Soc. **139** (2011), no. 10, 3451–3460, DOI 10.1090/S0002-9939-2011-10906-7. MR2813377

[CFT] Ben Cox, Vyacheslav Futorny, and Juan A. Tirao, *DJKM algebras and non-classical orthogonal polynomials*, J. Differential Equations **255** (2013), no. 9, 2846–2870, DOI 10.1016/j.jde.2013.07.020. MR3090080

[CJ] Ben Cox and Elizabeth Jurisich, *Realizations of the three-point Lie algebra* $\mathfrak{sl}(2,\mathcal{R})\oplus(\Omega_{\mathcal{R}}/d\mathcal{R})$, Pacific J. Math. **270** (2014), no. 1, 27–47, DOI 10.2140/pjm.2014.270.27. MR3245847

[CGLZ17] Ben Cox, Xiangqian Guo, Rencai Lu, and Kaiming Zhao, *Simple superelliptic Lie algebras*, Commun. Contemp. Math. **19** (2017), no. 3, 1650032, 22, DOI 10.1142/S0219199716500322. MR3631928

[CGLZ14] Ben Cox, Xiangqian Guo, Rencai Lu, and Kaiming Zhao, *n-point Virasoro algebras and their modules of densities*, Commun. Contemp. Math. **16** (2014), no. 3, 1350047, 27, DOI 10.1142/S0219199713500478. MR3211093

[DJKM] Etsurō Date, Michio Jimbo, Masaki Kashiwara, and Tetsuji Miwa, *Landau-Lifshitz equation: solitons, quasiperiodic solutions and infinite-dimensional Lie algebras*, J. Phys. A **16** (1983), no. 2, 221–236. MR701334

[J] David A. Jordan, *On the simplicity of Lie algebras of derivations of commutative algebras*, J. Algebra **228** (2000), no. 2, 580–585, DOI 10.1006/jabr.2000.8286. MR1764580

[K] Christian Kassel, *Kähler differentials and coverings of complex simple Lie algebras extended over a commutative algebra*, Proceedings of the Luminy conference on algebraic K-theory (Luminy, 1983), J. Pure Appl. Algebra **34** (1984), no. 2-3, 265–275, DOI 10.1016/0022-4049(84)90040-9. MR772062

[KN] Igor Moiseevich Krichever and S. P. Novikov, *Algebras of Virasoro type, Riemann surfaces and strings in Minkowski space* (Russian), Funktsional. Anal. i Prilozhen. **21** (1987), no. 4, 47–61, 96. MR925072

[Sc] Martin Schlichenmaier, *Krichever-Novikov type algebras*, De Gruyter Studies in Mathematics, vol. 53, De Gruyter, Berlin, 2014. Theory and applications, DOI 10.1515/9783110279641. MR3237422

[Si] Thomas Siebert, *Lie algebras of derivations and affine algebraic geometry over fields of characteristic* 0, Math. Ann. **305** (1996), no. 2, 271–286, DOI 10.1007/BF01444221. MR1391215

[S] S. M. Skryabin, *Regular Lie rings of derivations* (Russian), Vestnik Moskov. Univ. Ser. I Mat. Mekh. **3** (1988), 59–62; English transl., Moscow Univ. Math. Bull. **43** (1988), no. 3, 56–58. MR966871

[W] Jet Wimp, *Explicit formulas for the associated Jacobi polynomials and some applications*, Canad. J. Math. **39** (1987), no. 4, 983–1000, DOI 10.4153/CJM-1987-050-4. MR915027

Faculdade de Computação e Informática, Universidade Presbiteriana Mackenzie; and Departamento de Matemática, Universidade de São Paulo, São Paulo, Brasil
Email address: falbinosantos@gmail.com

Shenzhen International Center for Mathematics, Southern University of Science and Technology, People's Republic of China; and Departamento de Matemática, Universidade de São Paulo, São Paulo, Brasil
Email address: vfutorny@gmail.com

Departament of Matematics, Wilfrid Laurier University, Waterloo, Canada
Email address: kzhao@wlu.ca

Proceedings of Symposia in Pure Mathematics
Volume **108**, 2024
https://doi.org/10.1090/pspum/108/01973

Notes on Fock space

Peter Tingley

ABSTRACT. These notes are intended as a fairly self-contained explanation of Fock space and various algebras that act on it, including a Clifford algebra, a Weyl algebra, an infinite rank matrix algebra, and an affine Kac-Moody algebra. We also discuss how the various algebras are related, and in particular describe the celebrated boson-fermion correspondence. We finish by briefly discussing a deformation of Fock space, which is a representation for the quantized universal enveloping algebra $U_q(\widehat{\mathfrak{sl}}_\ell)$.

CONTENTS

1. Introduction
2. Fock space as a vector space
3. Fock space as a representation
Acknowledgments
References

1. Introduction

The term Fock space comes from particle physics, where it is the state space for a system of a variable number of elementary particles. There are two distinct types of elementary particles, bosons and fermions, and their Fock spaces look quite different. Fermionic Fock spaces are naturally representations of a Clifford algebra, where the generators correspond to adding/removing a particle in a given pure energy state. Bosonic Fock space is naturally a representation of a Weyl algebra.

Here we focus on one example of this construction each for Bosons and Fermions. Our fermionic Fock space $\mathbf{F}$ corresponds to a system of fermionic particles with pure energy states indexed by $\mathbb{Z} + \frac{1}{2}$. Our space $\mathbf{B}^{(0)}$ corresponds to a system of bosons with pure energy states indexed by $\mathbb{Z}_{>0}$, and our full bosonic Fock space $\mathbf{B}$ is a superposition of $\mathbb{Z}$ shifted copies of $\mathbf{B}^{(0)}$. There is a natural embedding of the Weyl algebra for $\mathbf{B}^{(0)}$ into a completion of the Clifford algebra for $\mathbf{F}$, which leads to the celebrated boson-fermion correspondence.

We begin by discussing $\mathbf{F}$ and $\mathbf{B}$ as vector spaces. We present various ways of indexing the standard basis of $\mathbf{F}$, and describe the boson-fermion correspondence as a vector space isomorphism between $\mathbf{F}$ and $\mathbf{B}$. Since $\mathbf{F}$ and $\mathbf{B}$ are both vector spaces with countable bases it is immediate that they are isomorphic, the isomorphism we present is only interesting in that it has nice properties with respect to the actions of the Clifford and Weyl algebras. A more intuitive development of the theory (and also how it is presented in [**Kac**, Chapter 14]) is probably to first notice the relationship between the Clifford and Weyl algebras, then derive the corresponding relationship between the vector spaces. However, we find it useful to first have the map explicitly described in an elementary way.

We next discuss the various algebras that naturally act on Fock space, and how these actions are related. These include an infinite rank matrix algebra and various affine Kac-Moody algebras, as well as the Weyl and Clifford algebras. We finish by presenting the Misra-Miwa action of the quantized universal enveloping algebra $U_q(\widehat{\mathfrak{sl}}_\ell)$ on $\mathbf{F} \otimes_{\mathbb{C}} \mathbb{C}[q, q^{-1}]$. Understanding the relationship between the actions of these various algebras on $\mathbf{F}$ has proven very useful, see e.g. [**Kac**, Chapter 14]. It is really this representation theory that we are interested in, not the physics motivations. There will be very little physics beyond this introduction.

These notes are intended to be quite self-contained in the sense that they should be comprehensible independent of other references on Fock space. We do however refer to other sources for some important proofs. Much of our presentation loosely follows [**Kac**, Chapter 14]. Alternate references include [**DJM, KRR**].

2. Fock space as a vector space

Fermionic Fock space $\mathbf{F}$ is an infinite dimensional vector space. It has a standard basis, which can be indexed by a variety of objects. In this section we discuss indexings by Maya diagrams, charged partitions, and normally ordered wedge products. Bosonic Fock space $\mathbf{B}$ is essentially a space of polynomials in infinitely many variables. This has a standard basis consisting of character polynomials, which are related to Schur symmetric functions. Schur symmetric functions are indexed by partitions, which leads to a natural bijection between the standard bases of $\mathbf{F}$ and $\mathbf{B}$, and hence a vector space isomorphism.

2A. Maya diagrams.

DEFINITION 2.1. A *Maya diagram* is a placement of a white or black bead at each position in $\mathbb{Z} + \frac{1}{2}$, subject to the condition that all but finitely many positions $m < 0$ have a black bead and all but finitely many positions $m > 0$ have a white bead. For instance,

··· ○ ○ ○ ○ ○ ○ ○ ○ ○ ● ○ ● | ● ○ ○ ● ● ○ ● ● ● ● ● ···
11 10 9 8 7 6 5 4 3 2 1 0 -1 -2 -3 -4 -5 -6 -7 -8 -9 -10 -11

It is sometimes convenient to think of the black beads as "filled positions" and the white beads as "empty positions" in a "Dirac sea." Note that we label positions from right to left. This is to match the conventions for normally-ordered semi-infinite wedges, where the indices decrease to the right (see §2C and [**Kac**, §14.9]).

2B. Charged partitions.

DEFINITION 2.2. A **charged partition** is a pair (λ, h) consisting of a partition λ and an integer h (the charge).

There is a natural bijection between Maya diagrams and charged partitions: draw a line above the Maya diagram by placing a segment sloping down and to the right over every white bead, and a segment sloping up and to the right over every black bead. See Figure 1. The result is the outer boundary of a partition, where by

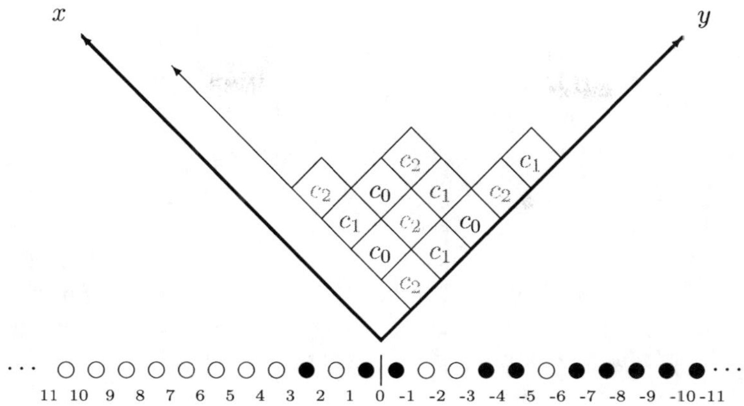

FIGURE 1. The bijection between Maya diagrams and charged partitions from §2B. The black beads in the Maya diagram correspond to the positions where the rim of the partition slopes up and to the right. The parts of λ are the lengths of all the finite "rows" of boxes sloping up at to the left, so here $\lambda = (4, 3, 3, 1, 1)$. The charge is the horizontal position of the vertex of the partition, so here the charge is $h = -1$. Later on in §3E we will choose a "level" ℓ and color the squares of λ with $c_{\bar{i}}$ for residues $\bar{i}$ mod ℓ, where the color of a box b is the position of that box in the horizontal direction mod ℓ.

convention the parts of the partition are the lengths of the "rows" of boxes going up and to the left. The charge is determined by superimposing axes with the origin above position 0 in the Maya diagram, and so that far to the right the axis follows the diagram. The charge is the signed distance between the diagram and the axis far to the left. Equivalently, it is the number under the vertex of the partition.

2C. Semi-infinite wedges.

DEFINITION 2.3. Let $V_{\mathbb{Z}+\frac{1}{2}}$ be the $\mathbb{C}$-vector space with basis $\{e_m\}_{m\in\mathbb{Z}+\frac{1}{2}}$.

DEFINITION 2.4. A semi-infinite wedge is a (formal) infinite wedge $e_{m_1} \wedge e_{m_2} \wedge e_{m_3} \wedge \cdots$. This can be thought of as living inside a formal semi-infinite wedge space $V_{\mathbb{Z}+\frac{1}{2}} \wedge V_{\mathbb{Z}+\frac{1}{2}} \wedge V_{\mathbb{Z}+\frac{1}{2}} \cdots$.

As usual, "wedge" is anti-commutative. So, for example,

$$e_{2.5} \wedge e_{0.5} \wedge e_{-1.5} \wedge e_{-2.5} \cdots = -e_{0.5} \wedge e_{2.5} \wedge e_{-1.5} \wedge e_{-2.5} \cdots$$

DEFINITION 2.5. A semi-infinite wedge $e_{m_1} \wedge e_{m_2} \wedge e_{m_3} \wedge \cdots$ is called *normally ordered* if $m_1 > m_2 > m_3 > \cdots$.

DEFINITION 2.6. A semi-infinite wedge $e_{m_1} \wedge e_{m_2} \wedge e_{m_3} \wedge \cdots$ is called *regular* if $m_{k+1} = m_k - 1$ for all sufficiently large k.

There is a bijection between regular normally ordered semi-infinite wedges and Maya diagrams which takes $e_{m_1} \wedge e_{m_2} \wedge \cdots$ to the Maya diagram with black beads exactly in positions $m_1, m_2, \ldots$. For instance,

$$e_{2.5} \wedge e_{0.5} \wedge e_{-0.5} \wedge e_{-3.5} \wedge e_{-4.5} \wedge e_{-6.5} \wedge e_{-7.5} \wedge e_{-8.5} \cdots \tag{1}$$

corresponds to the Maya diagram shown in Definition 2.1 and Figure 1.

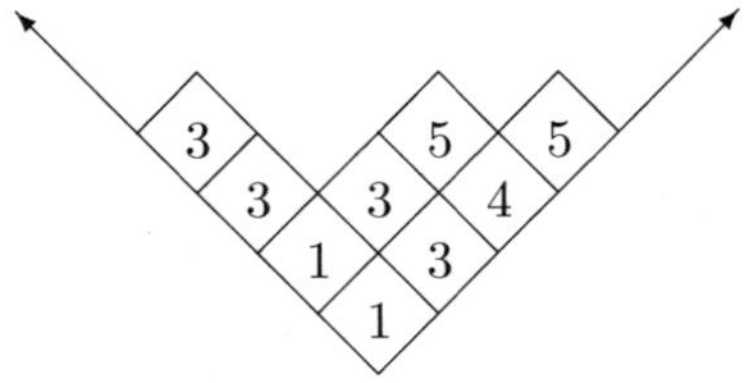

FIGURE 2. A column-strict filling of $\lambda = (4, 2, 2, 1)$. The function t takes b to the integer in box b. Recall that "rows" slope up and to the left, and "columns" slope up and to the right.

2D. Fermionic Fock space F. The fermionic Fock space $\mathbf{F}$ is the $\mathbb{C}$-vector space with basis indexed by any one of the following:

(i) Maya Diagrams.
(ii) Charged partitions.
(iii) Regular normally-ordered semi-infinite wedges.

As in §2B and §2C these three sets are naturally in bijection, and we take the point of view that they all index the same basis of $\mathbf{F}$. We most often refer to a basis element using the corresponding charged partition, using the notation $|\lambda, h\rangle$ to denote the standard basis element of F corresponding to the charged partition (λ, h).

The *charge* h part of $\mathbf{F}$ is

$$\mathbf{F}^{(h)} := span\{\text{charged partitions with charge } h\}.$$

In particular, ordinary partitions index a basis for $\mathbf{F}^{(0)}$. As a vector space, $\mathbf{F} = \oplus_{h\in\mathbb{Z}}\mathbf{F}^{(h)}$.

2E. Bosonic Fock space B. The *bosonic Fock space* $\mathbf{B}$ is

$$\mathbf{B} := \mathbb{C}[x_1, x_2, x_3, \dots; q, q^{-1}].$$

The *charge* h part of $\mathbf{B}$ is

$$\mathbf{B}^{(h)} := q^h\mathbb{C}[x_1, x_2, \dots].$$

As vector spaces, $\mathbf{B} = \oplus_{h\in\mathbb{Z}}\mathbf{B}^{(h)}$.

2F. The boson-fermion correspondence as a map of vector spaces.

DEFINITION 2.7. Fix a partition λ. A *column-strict filling* of λ is a function t from the set of boxes in the diagram of λ to $\mathbb{Z}_{>0}$ which is weakly increasing along rows and strictly increasing along columns. See Figure 2.

DEFINITION 2.8. Let λ be a partition. The *Schur symmetric function* s_λ in infinitely many variables $y_1, y_2, \dots$ corresponding to λ is

$$s_\lambda(\mathbf{y}) := \sum_{\substack{t \text{ a column strict} \\ \text{filling of } \lambda}} \prod_{\text{boxes } b \text{ of } \lambda} y_{t(b)}. \tag{2}$$

COMMENT 2.9. The Schur symmetric functions are symmetric in the sense that they are invariant under permutations of the variables y_i (this can be shown using combinatorics [**Sta**, Theorem 7.10.2], or by appealing to representation theory [**Mac**, §I3)]). They are polynomials in the sense that, if all but finitely many of the variables are set to 0, the result is a polynomial. They have many important properties, most of which arise because (once all but the first n variables are set to 0) they are the characters of irreducible representations of $\mathfrak{gl}_n$.

DEFINITION 2.10. For $k \in \mathbb{Z}_{>0}$, let $p_k = y_1^k + y_2^k + \cdots$ be the kth power symmetric function.

DEFINITION 2.11. The character polynomial χ_λ is the unique polynomial such that

$$\chi_\lambda(p_1, \frac{1}{2}p_2, \frac{1}{3}p_3, \frac{1}{4}p_4 \ldots) = s_\lambda.$$

COMMENT 2.12. To find χ_λ, one may set $y_j = 0$ for all j larger than the longest column of λ. As well, χ_λ cannot depend on p_k for any k bigger than the number of boxes in λ (since s_λ is homogeneous of degree $< k$), so finding any given χ_λ is a finite problem.

PROPOSITION 2.13. *There is an isomorphism of vector spaces $\sigma : \mathbf{F} \to \mathbf{B}$ given by*

$$\sigma(|\lambda, h\rangle) = q^h \chi_\lambda(x_1, x_2, \ldots)$$

for any charged partition (λ, h). This restricts to an isomorphism $\mathbf{F}^{(h)} \to B^{(h)}$ for all $h \in \mathbb{Z}$.

PROOF. As in [**Mac**, §I3], Schur polynomials are a basis for the space of all symmetric functions in infinitely many variables, from which it follows that $\{q^h \chi_\lambda\}$ is a basis for $\mathbf{B}$. By definition, the set of all charged partitions is a basis for $\mathbf{F}$. Thus each restriction $\sigma : \mathbf{F}^{(h)} \to \mathbf{B}^{(h)}$ is a vector space isomorphism, and so is the full map $\sigma : \mathbf{F}^{(h)} \to \mathbf{B}^{(h)}$. □

2G. The inner product. It is often convenient to put an inner product on $\mathbf{F}$, where we declare $\{|\lambda, h\rangle\}$ to be an orthonormal basis.

3. Fock space as a representation

3A. F as a representation of a Clifford algebra. The Clifford algebra is the associative algebra $\mathbf{C}l$ generated by ψ_m, ψ_m^* for $m \in \mathbb{Z} + \frac{1}{2}$ subject to the relations

$$\psi_n \psi_m + \psi_m \psi_n = 0, \tag{3}$$

$$\psi_n^* \psi_m^* + \psi_m^* \psi_n^* = 0, \tag{4}$$

$$\psi_n \psi_m^* + \psi_m^* \psi_n = \delta_{m,n}. \tag{5}$$

One can check that $\mathbf{C}l$ acts on Fock space as follows: Use the description of $\mathbf{F}$ in terms of semi-infinite wedges from §2C. For $m \in \mathbb{Z} + \frac{1}{2}$ and v a regular semi-infinite wedge, define

$$\psi_m \cdot v = e_m \wedge v. \tag{6}$$

The generator ψ_n^* acts on $\mathbf{F}$ as the adjoint of ψ_n with respect to the inner product from §2G. Explicitly, for v a regular semi-infinite wedge,

$$\psi_m^* \cdot v = \begin{cases} 0 & \text{if } e_m \text{ does not appear as a factor in } v, \\ v' & \text{if } v \text{ can be expressed as } e_m \wedge v' \text{ for a wedge } v'. \end{cases} \tag{7}$$

One can obtain any regular semi-infinite wedge v from any other by applying a finite number of operators ψ_m and ψ_m^*. Also, given any linear combination of regular semi-infinite wedges, one can project onto a single term by applying enough $\psi_n\psi_n^*$ and $\psi_n^*\psi_n$. Together this implies that $\mathbf{F}$ is irreducible as a representation of $\mathbf{C}l$.

DEFINITION 3.1. $\widetilde{\mathbf{C}l}$ is the completion of $\mathbf{C}l$ in the topology generated by the open sets $X + \mathbf{C}l\psi_m^*$ and $X + \mathbf{C}l\psi_{-m}$ for all $X \in \mathbf{C}l$, $m > 0$. Explicitly, an element of $\widetilde{\mathbf{C}l}$ is an infinite sum

$$z + \sum_{m<N} X_m \psi_m + \sum_{m>M} Y_m \psi_m^*, \tag{8}$$

where $z \in \mathbb{C}$, $N, M \in \mathbb{Z}$, and all X_m, Y_m are elements of $\mathbf{C}l$.

The completion $\widetilde{\mathbf{C}l}$ acts in a well-defined way on $\mathbf{F}$, since, for each regular semi-infinite wedge v, all but finitely many of ψ_m^*, ψ_{-m} for $m > 0$ act by 0 on v.

COMMENT 3.2. When it seems prudent to distinguish $X \in \widetilde{\mathbf{C}l}$ from its action of $\mathbf{F}$, we use the notation $\pi_{\mathbf{F}}(X)$ to denote the action.

3B. B as a representation of a Weyl algebra.

DEFINITION 3.3. The infinite dimensional Heisenberg algebra $\mathbf{H}$ is generated by α_k for $k \in \mathbb{Z}_{\neq 0}$ and a central element c subject to the relations:

$$\alpha_j \alpha_k - \alpha_k \alpha_j = j\delta_{j,-k} c.$$

DEFINITION 3.4. The Weyl algebra $\mathcal{W}$ is

$$\mathcal{W} := \mathbf{H}/(c-1).$$

PROPOSITION 3.5. *The Weyl algebra $\mathcal{W}$ acts on $\mathbf{B}$ by the follows formulas. Each $B^{(h)}$ is preserved under this action and forms an irreducible representation for $\mathcal{W}$.*

$$\begin{aligned} \alpha_k &\to \frac{\partial}{\partial x_k} && \text{if } k > 0, \\ \alpha_k &\to \text{multiplication by } -kx_{-k} && \text{if } k < 0. \end{aligned}$$

PROOF. A direct calculation shows that, defining the actions of α_k and α_{-k} as above and allowing c to act as multiplication by 1, the defining relations in Definition 3.3 hold. So this gives an action of $\mathcal{W}$ on $\mathbf{B}$. Since none of these operators affect the power of q in any monomial each $B^{(h)}$ is preserved. To see that each $B^{(h)}$ is irreducible as a $\mathcal{W}$ module, notice that

- Any non-zero $p \in B^{(h)}$ can be brought to a non-zero multiple of q^h by applying a sequence of partial derivatives, which are the actions of α_k for $k > 0$.
- Any monomial in $B^{(h)}$ can be constructed from q^h by applying a sequence off multiplications by the generators which, up to scalar, are the actions of α_k for $k < 0$. □

COMMENT 3.6. When it seems prudent to distinguish $Y \in \mathcal{W}$ from its action of $\mathbf{B}$ we use the notation $\pi_{\mathbf{B}}(Y)$ to denote the action.

3C. The boson-fermion correspondence. Proposition 2.13 gives an isomorphism of vector spaces $\sigma : \mathbf{F} \to \mathbf{B}$. This isomorphism was chosen because it reveals an important relationship between the algebras $\mathbf{Cl}$ and $\mathcal{W}$, which is known as the boson-fermion correspondence. We now explain that relationship, referring to [**Kac**, Chapter 14] for rigorous proofs.

DEFINITION 3.7. For $Y \in \mathcal{W}$, let $\pi_{\mathbf{F}}(Y) = \sigma^{-1} \circ \pi_{\mathbf{B}}(Y) \circ \sigma$. That is, $\pi_{\mathbf{F}}(Y)$ is the operator on $\mathbf{F}$ induced from $\pi_{\mathbf{B}}(Y)$ by the vector space isomorphism $\sigma : \mathbf{F} \to \mathbf{B}$.

PROPOSITION 3.8. *(see [**Kac**, §14.10]) For all $k \in \mathbb{Z}_{\neq 0}$,*

$$\pi_{\mathbf{F}}(\alpha_k) = \pi_{\mathbf{F}} \sum_{m \in \mathbb{Z}+\frac{1}{2}} \psi_m \psi^*_{m+k}.$$

COMMENT 3.9. In [**Kac**, §14.10] Kac also defines α_0. Proposition 3.8 can be extended to include α_0, but one must be careful, since $\sum_{m \in \mathbb{Z}+\frac{1}{2}} \psi_m \psi^*_m$ would acts on any $v \in \mathbf{F}$ as multiplication by ∞. Instead, α_0 corresponds to $\sum_{m>0} \psi_m \psi^*_m - \sum_{m<0} \psi^*_m \psi_m$, which acts on $F^{(h)}$ simply as multiplication by h.

EXAMPLE 3.10. The action of α_k on the standard basis of $\mathbf{F}$ has a nice description using Maya diagrams. If v is a Maya diagram, then $\alpha_k v$ is the sum of $(-1)^{j(v,v')} v'$ over all v' obtained from v by moving a single black bead to the right exactly k places, where $j(v, v')$ is the number of black beads that are "jumped" by the bead that moves. If k is negative, this means the beads move to the left instead. For example,

α_2(··· ○ ○ ○ ○ ○ ○ ○ ○ ● ○ ●|● ○ ○ ● ● ○ ● ● ● ● ● ···) =

(9)
−··· ○ ○ ○ ○ ○ ○ ○ ○ ● ○ ○|● ● ○ ● ● ○ ● ● ● ● ● ···
+··· ○ ○ ○ ○ ○ ○ ○ ○ ● ○ ●|○ ○ ● ● ● ○ ● ● ● ● ● ···
−··· ○ ○ ○ ○ ○ ○ ○ ○ ● ○ ●|● ○ ○ ○ ● ● ● ● ● ● ● ···

(10) and

α_{-4}(··· ○ ○ ○ ○ ○ ○ ○ ○ ● ○ ●|● ○ ○ ● ● ○ ● ● ● ● ● ···) =

··· ○ ○ ○ ○ ● ○ ○ ○ ○ ○ ●|● ○ ○ ● ● ○ ● ● ● ● ● ···
−··· ○ ○ ○ ○ ○ ○ ● ○ ● ○ ○|● ○ ○ ● ● ○ ● ● ● ● ● ···
+··· ○ ○ ○ ○ ○ ○ ○ ● ● ○ ●|○ ○ ○ ● ● ○ ● ● ● ● ● ···
+··· ○ ○ ○ ○ ○ ○ ○ ○ ● ○ ●|● ○ ● ● ● ○ ○ ● ● ● ● ···
−··· ○ ○ ○ ○ ○ ○ ○ ○ ● ○ ●|● ○ ○ ● ● ● ● ● ● ○ ● ···

COMMENT 3.11. For $k > 0$, the basis vectors which have non-zero coefficient in $\alpha_k|\lambda, h\rangle$ are exactly those $|\mu, h\rangle$ where μ is obtained from λ by removing a "ribbon" or "rim-hook" of length k. For $k < 0$, they are those $|\mu, h\rangle$ where μ is obtained from λ by adding a ribbon of length k. This is part of the reason ribbons appear in the study of Fock space (see e.g. [**LLT**]).

For $k \neq 0$ we have $\psi_m \psi^*_{m+k} = -\psi^*_{m+k}\psi_m$, so $\pi_{\mathbf{F}}(\alpha_k) \in \widetilde{\mathbf{C}l}$. Thus Proposition 3.8 gives an imbedding of $\mathcal{W}$ into $\widetilde{\mathbf{C}l}$. It is natural to ask if one can go the other way and express $\mathbf{C}l$ in terms of $\mathcal{W}$. All elements of $\mathcal{W}$ preserve the subspaces $\mathbf{F}^{(h)}$ of $\mathbf{F}$, and the generators of $\mathbf{C}l$ do not preserve these subspaces. So we will need to enlarge $\mathcal{W}$. It turns out that it suffices to introduce one more operator.

DEFINITION 3.12. The shift operator $s : \mathbf{F} \to \mathbf{F}$ is defined by $s|\lambda, h\rangle = |\lambda, h+1\rangle$. Note that s corresponds to multiplication by q under the vector space isomorphism $\sigma : \mathbf{F} \to \mathbf{B}$.

Introduce the following power series. In each case all but finitely many terms with negative powers of z act as zero on any fixed $|\lambda, h\rangle$, so acting on any fixed element of $\mathbf{F}$ gives a power series with coefficients in $\mathbf{F}$ and only finitely many negative powers of z.

$$\psi(z) = \sum_{m \in \mathbb{Z}+\frac{1}{2}} z^m \pi_F(\psi_m) \tag{11}$$

$$\psi^*(z) = \sum_{m \in \mathbb{Z}+\frac{1}{2}} z^{-m} \pi_F(\psi^*_m) \tag{12}$$

$$\Gamma_+(z) = \exp \sum_{k \in \mathbb{Z}_{>0}} \frac{z^{-k}}{k} \pi_F(\alpha_k) \tag{13}$$

$$\Gamma_-(z) = \exp \sum_{k \in \mathbb{Z}_{>0}} \frac{z^{k}}{k} \pi_F(\alpha_{-k}). \tag{14}$$

The following is [**Kac**, Theorem 14.10], adjusted slightly to match our conventions.

PROPOSITION 3.13. *As operators on* $\mathbf{F}$,

$$\psi(z) = s z^{\mathrm{ch}+\frac{1}{2}} \Gamma_-(z)\Gamma_+(z)^{-1},$$
$$\psi^*(z) = s^{-1} z^{-\mathrm{ch}+\frac{1}{2}} \Gamma_-(z)^{-1}\Gamma_+(z).$$

Here ch *sends a charged partition to its charge, so* z^{ch} *acts on* $|\lambda, h\rangle$ *as multiplication by* z^h.

Proposition 3.13, in principle, expresses the fermionic operators ψ_m and ψ^*_m in terms of the α_k. Its proof is pretty involved. It proceeds roughly as follows. Kac first shows that the right sides of the equations in Proposition 3.8 generate an algebra of operators on $\mathbf{F}$ which is isomorphic to $\mathcal{W}$. This then implies the existence of some vector space isomorphism from $\mathbf{B}$ to $\mathbf{F}$ which satisfies Proposition 3.8. One then studies commutation relations between the α_k and the generating functions $\psi(z)$ and $\psi^*(z)$ to prove that this isomorphism also satisfies Proposition 3.13. The final step is to show that the isomorphism is given by Proposition 2.13. So really these notes are basically backwards!

EXAMPLE 3.14. Proposition 3.13 predicts, for instance,

$$[z^{3.5}]sz^{\mathrm{ch}+\frac{1}{2}}\Gamma_-(z)\Gamma_+(z)^{-1}(\cdots\circ\circ\bullet\circ\bullet\mid\bullet\bullet\bullet\cdots) \\ =\cdots\circ\bullet\bullet\circ\bullet\mid\bullet\bullet\bullet\cdots. \tag{15}$$

We will check this by explicitly calculating the left side. The only terms in $\Gamma_+(z)^{-1}$ that act without killing this vector are the identity and $-z^{-1}\alpha_1$. So the left side is equal to

$$[z^{3.5}]sz^{\mathrm{ch}+\frac{1}{2}}\Gamma_-(z)(\cdots\circ\bullet\circ\bullet\mid\bullet\bullet\bullet\cdots\ -z^{-1}\cdots\circ\circ\bullet\bullet\mid\bullet\bullet\bullet\cdots).$$

$\Gamma_-(z)$ commutes with $sz^{\mathrm{ch}+\frac{1}{2}}$, since each α_k does, and $sz^{\mathrm{ch}+\frac{1}{2}}$ multiplies these elements by $z^{2.5}$ and then shifts all beads one step to the left. This leaves

$$[z^{3.5}](z^{2.5}\Gamma_-(z)(\cdots\circ\bullet\circ\bullet\bullet\mid\bullet\bullet\bullet\cdots) \\ -z^{1.5}\Gamma_-(z)(\cdots\circ\bullet\bullet\bullet\mid\bullet\bullet\bullet\cdots)).$$

The degree 1 part of $\Gamma_-(z)$ is α_{-1} and the degree 2 parts is $\frac{\alpha_{-1}^2}{2}+\frac{\alpha_{-2}}{2}$, so this is

$$\alpha_{-1}(\cdots\circ\bullet\circ\bullet\bullet\mid\bullet\bullet\bullet\cdots)-(\frac{\alpha_{-1}^2}{2}+\frac{\alpha_{-2}}{2})(\cdots\circ\bullet\bullet\bullet\mid\bullet\bullet\bullet\cdots).$$

Calculating,

$$\alpha_{-1}(\cdots\circ\bullet\circ\bullet\bullet\mid\bullet\bullet\bullet\cdots)=\cdots\bullet\circ\circ\bullet\bullet\mid\bullet\bullet\bullet\cdots \\ +\cdots\circ\bullet\bullet\circ\bullet\mid\bullet\bullet\bullet\cdots$$

$$(\frac{\alpha_{-1}^2}{2}+\frac{\alpha_{-2}}{2})(\cdots\circ\circ\bullet\bullet\bullet\mid\bullet\bullet\bullet\cdots)=\frac{1}{2}\cdots\bullet\circ\circ\bullet\bullet\mid\bullet\bullet\bullet\cdots \\ +\frac{1}{2}\cdots\circ\bullet\bullet\circ\bullet\mid\bullet\bullet\bullet\cdots \\ +\frac{1}{2}\cdots\bullet\circ\circ\bullet\bullet\mid\bullet\bullet\bullet\cdots \\ -\frac{1}{2}\cdots\circ\bullet\bullet\circ\bullet\mid\bullet\bullet\bullet\cdots$$

Subtracting leaves exactly the right side of (15).

COMMENT 3.15. It may seem unfortunate that we need to introduce the power series (11)–(14), but these are in fact very useful, in part because they satisfy nice commutation relations. Each acts on any $v\in\mathbf{F}$ to give a power series in z, z^{-1} with coefficients in $\mathbf{F}$, so each of the products below is well-defined as a map from $\mathbf{F}$ to Laurent power series in two variables with coefficients in $\mathbf{F}$. The following commutation relations follow from results in [**Kac**, Chapter 14] (see also [**OR1**, Appendix B2] for this exact statement, but with slightly different conventions).

$$\Gamma_-(x)\Gamma_+(y)=(1-xy^{-1})\Gamma_+(y)\Gamma_-(x) \tag{16}$$

$$\Gamma_-(x)\psi(z)=(1-z^{-1}x)^{-1}\psi(z)\Gamma_-(x) \tag{17}$$

$$\Gamma_+(x)\psi(z)=(1-x^{-1}z)^{-1}\psi(z)\Gamma_+(x) \tag{18}$$

$$\Gamma_-(x)\psi^*(z)=(1-z^{-1}x)\psi^*(z)\Gamma_-(x) \tag{19}$$

$$\Gamma_+(x)\psi^*(z)=(1-x^{-1}z)\psi^*(z)\Gamma_+(x). \tag{20}$$

Here $(1-a)^{-1}$ is always expanded as $1+a+a^2+\cdots$. Equality is interpreted as saying that, once each side is applied to any fixed $v\in\mathbf{F}$, all coefficients of the resulting power series agree. These facts have been put to great use. For example, in [**OR1**] Okounkov and Reshetikhin interpret the $\Gamma_\pm(z)$ as transition functions and use the above commutation relations to find limit shapes and correlation functions for systems of random plane partitions.

3D. F as a representation of $\mathfrak{gl}_{\mathbb{Z}+\frac{1}{2}}$ and a_∞.

DEFINITION 3.16. A Lie-associative map from a Lie algebra $\mathfrak{g}$ to an associative algebra A is a map σ such that, for all $X, Y \in \mathfrak{g}$, $\sigma([X,Y]) = \sigma(X)\sigma(Y) - \sigma(Y)\sigma(X)$.

An action of a Lie algebra L on a space V is a map $L \to \mathrm{End}(V)$ which preserves the Lie bracket, which is the same thing as a Lie-associative map from L to $\mathrm{End}(V)$. More generally, if an associative algebra A acts on V, then any Lie-associative map $L \to A$ induces an acton of L on V. In the following we use this framework to construct actions of various Lie algebras on **F** by finding Lie-associative maps from these algebras to $\mathbf{C}l$ or the completion $\widetilde{\mathbf{C}l}$.

DEFINITION 3.17. $M_{\mathbb{Z}+\frac{1}{2}}$ is the algebra of matrices with rows and columns indexed by $\mathbb{Z}+\frac{1}{2}$, in which all but finitely many entries are 0. Let $E_{m,n}$ denote the matrix with a single 1 in position (m,n) and zeros everywhere else.

DEFINITION 3.18. $\mathfrak{gl}_{\mathbb{Z}+\frac{1}{2}}$ is the Lie algebra corresponding to $M_{\mathbb{Z}+\frac{1}{2}}$. That is, $\mathfrak{gl}_{\mathbb{Z}+\frac{1}{2}}$ is equal to $M_{\mathbb{Z}+\frac{1}{2}}$ as a vector space, and the Lie bracket is defined by

$$[X,Y] = XY - YX.$$

PROPOSITION 3.19. *(see [**Kac**, §14.9]) There is a Lie-associative embedding of $\mathfrak{gl}_{\mathbb{Z}+\frac{1}{2}}$ into $\mathbf{C}l$ given by $E_{m,n} \to \psi_m\psi_n^*$, and hence an action of $\mathfrak{gl}_{\mathbb{Z}+\frac{1}{2}}$ on* **F**.

EXAMPLE 3.20. Both $\mathfrak{gl}_{\mathbb{Z}+\frac{1}{2}}$ and $\mathbf{C}l$ are associative algebras, so one might hope that the map in Proposition 3.19 was a map of associative algebra. However,

$$E_{12}E_{31} = 0 \quad \text{but} \quad \psi_1\psi_2^*\psi_3\psi_1^* = \psi_1\psi_3\psi_1^*\psi_2^* \neq 0.$$

Working with these as Lie algebras though,

$$E_{12}E_{31} - E_{31}E_{12} = -E_{32}$$

and

$$\begin{aligned}\psi_1\psi_2^*\psi_3\psi_1^* - \psi_3\psi_1^*\psi_1\psi_2^* &= \psi_2^*\psi_3\psi_1\psi_1^* + \psi_2^*\psi_3\psi_1^*\psi_1 \\ &= \psi_2^*\psi_3(\psi_1\psi_1^* + \psi_1^*\psi_1) = \psi_2^*\psi_3 = -\psi_3\psi_2^*,\end{aligned}$$

and the map respects the bracket. One can prove Proposition 3.19 by checking relations like this in more generality.

We would like to extend the action of $\mathfrak{gl}_{\mathbb{Z}+\frac{1}{2}}$ on **F** from Proposition 3.19 to a larger algebra that contains some matrices with infinitely many non-zero entries. To do this, we will need to first take a central extension. The reason for this extension is explained in Comment 3.24 once we've defined a few things.

DEFINITION 3.21. (see [**Kac**, §7.12]) Let $\mathfrak{gl}^c_{\mathbb{Z}+\frac{1}{2}}$ be the central extension of $\mathfrak{gl}_{\mathbb{Z}+\frac{1}{2}}$ by a central element c with Lie bracket defined as follows. We use the notation $\overline{X}$

to mean the matrix X thought of as an element of $\mathfrak{gl}^c_{\mathbb{Z}+\frac{1}{2}}$.

$$[\overline{E}_{m,n},\overline{E}_{p,q}]=\begin{cases}\overline{E_{m,n}E_{p,q}-E_{p,q}E_{m,n}}+\delta_{m,q}\delta_{n,p}c & \text{if } m<0 \text{ and } n>0,\\ \overline{E_{m,n}E_{p,q}-E_{p,q}E_{m,n}}-\delta_{m,q}\delta_{n,p}c & \text{if } m>0 \text{ and } n<0,\\ \overline{E_{m,n}E_{p,q}-E_{p,q}E_{m,n}} & \text{if } m \text{ and } n \text{ have the same sign.}\end{cases}$$

One can verify by direct calculation that the bracket on $\mathfrak{gl}^c_{\mathbb{Z}+\frac{1}{2}}$ satisfies the Jacobi identify, so is a Lie algebra, and the quotient by $c=0$ is $\mathfrak{gl}_{\mathbb{Z}+\frac{1}{2}}$ so it is a central extension.

COMMENT 3.22. This central extension is trivial since there is an isomorphism of Lie algebras $\mathfrak{gl}^c_{\mathbb{Z}+\frac{1}{2}}\to\mathfrak{gl}_{\mathbb{Z}+\frac{1}{2}}\oplus\mathbb{C}c$ given by

$$\overline{E}_{m,n}\mapsto\begin{cases}E_{m,n} & \text{if } m\neq n \text{ or } m>0\\ E_{m,n}-c & \text{if } m=n \text{ and } m<0.\end{cases}$$

PROPOSITION 3.23. *There is a Lie-associative imbedding of* $\mathfrak{gl}^c_{\mathbb{Z}+\frac{1}{2}}$ *into* **Cl** *defined by*

$$\overline{E}_{m,n}\mapsto\psi_m\psi_n^* \quad \text{if } m\neq n \text{ or } m>0, \tag{21}$$

$$\overline{E}_{n,n}\mapsto-\psi_n^*\psi_n \quad \text{if } n<0, \tag{22}$$

$$c\to 1. \tag{23}$$

In particular, the action of **Cl** *on* **F** *introduces an action of* $\mathfrak{gl}^c_{\mathbb{Z}+\frac{1}{2}}$ *on* **F**.

PROOF. Consider the action of $\mathfrak{gl}_{\mathbb{Z}+\frac{1}{2}}\oplus\mathbb{C}c$ on **F** by allowing $\mathfrak{gl}_{\mathbb{Z}+\frac{1}{2}}$ to act via the embedding from Proposition 3.19, with c acting as 1. Then use the isomorphism from Comment 3.22. For $m\neq n$ or $m>0$, the action of $\overline{E}_{m,n}$ is immediate. For n negative, $\overline{E}_{n,n}$ is sent to $E_{n,n}-c$, which acts as $\psi_n\psi_n^*-1$. By the commutation relation (5) this is equal to $-\psi_n^*\psi_n$. □

DEFINITION 3.24. The Lie algebra $\widetilde{\mathfrak{gl}}_{\mathbb{Z}+\frac{1}{2}}$ consists of all infinite matrices $(a_{i,j})_{i,j\in\mathbb{Z}+\frac{1}{2}}$ such that, for some k, $a_{i,j}=0$ whenever $|i-j|>k$. The Lie bracket is the standard bracket for matrices, $[X,Y]=XY-YX$. One can check that matrix multiplication is well-defined on the set of such matrices and hence so is the Lie bracket.

DEFINITION 3.25. a_∞ is the central extension of $\widetilde{\mathfrak{gl}}_{\mathbb{Z}+\frac{1}{2}}$ defined as in Definition 3.21 by

$$[\overline{E}_{m,n},\overline{E}_{p,q}]=\begin{cases}\overline{E_{m,n}E_{p,q}-E_{p,q}E_{m,n}}+\delta_{m,q}\delta_{n,p}c & \text{if } m<0 \text{ and } n>0,\\ \overline{E_{m,n}E_{p,q}-E_{p,q}E_{m,n}}-\delta_{m,q}\delta_{n,p}c & \text{if } m>0 \text{ and } n<0,\\ \overline{E_{m,n}E_{p,q}-E_{p,q}E_{m,n}} & \text{if } m \text{ and } n \text{ have the same sign.}\end{cases}$$

Any $(a_{mn})\in a_\infty$ has only finitely many non-zero entries with $m<0$ and $n>0$, so the coefficient of c that appears in any bracket is finite and the central extension is well-defined.

PROPOSITION 3.26. *The imbedding of* $\mathfrak{gl}^c_{\mathbb{Z}+\frac{1}{2}}$ *into* **Cl** *extends to an embedding of* a_∞ *into* $\widetilde{\mathbf{Cl}}$. *In particular,* **F** *is a representation of* a_∞.

PROOF. This follows from the definition of $\widetilde{\mathbf{Cl}}$, since the image of any element in a_∞ is of the correct form to lie in the completion $\widetilde{\mathbf{Cl}}$. □

COMMENT 3.27. Essentially, the central extension is needed because, using the action in Proposition 3.19, the action of a sum of the form $\sum_{m<0} x_m E_{m,m}$ on $v \in \mathbf{F}$ typically diverges. One fixes this by making a change of basis in $\mathfrak{gl}^c_{\mathbb{Z}+\frac{1}{2}}$: Instead of working with the elements $E_{m,m}$, work with the elements

$$\overline{E}_{m,m} = \begin{cases} E_{m,n} & m \neq n \text{ or } m > 0, \\ E_{m,m} - c & m = n < 0. \end{cases} \tag{24}$$

Then, for any fixed $v \in \mathbf{F}$ and $k > 0$, there are only finitely many $\overline{E}_{m,n}$ with $|m - n| < k$ that act on v to give a non-zero term. One can think of a_∞ as infinite linear combinations of these $\overline{E}_{m,n}$ (along with c) subject to the condition that, for some k, the coefficient of $\overline{E}_{m,n}$ is 0 whenever $|i - j| > k$. Then the action of $\mathfrak{gl}^c_{\mathbb{Z}+\frac{1}{2}}$ extends naturally to a_∞. The bracket in Definitions 3.21 and 3.25 can be derived by taking the normal bracket of the elements $\overline{E}_{m,n}$ as in (24). The formulas in Proposition 3.23 come from the fact that $\psi_m \psi^*_m - 1 = -\psi^*_m \psi_m$.

This discussion also implies that a_∞ is a non-trivial central extension of $\widetilde{\mathfrak{gl}}_{\mathbb{Z}+\frac{1}{2}}$, since the action only works with the central extension.

COMMENT 3.28. The images of the α_m in $\widetilde{\mathbf{C}l}$ from Proposition 3.8 are naturally contained in the image of a_∞; they are matrices with all 1 on one diagonal (other than the main diagonal), and all 0 everywhere else. One can directly calculate that these matrices have the same commutation relations as the generators of $\mathcal{W}$, which in fact proves that the map from Proposition 3.8 gives an action of $\mathcal{W}$ on $\mathbf{F}$. The element α_0 mentioned in Comment 3.9 is the image of $\sum_{m \in \mathbb{Z}+\frac{1}{2}} \overline{E}_{m,m}$.

3E. F as a representation of $\widehat{\mathfrak{sl}}_\ell$ and $\widehat{\mathfrak{gl}}_\ell$. It turns out that each $\mathbf{F}^{(h)}$ carries an action of $\widehat{\mathfrak{sl}}_\ell$. $\mathbf{F}^{(h)}$ is not irreducible under this action, but all the irreducible components are very similar; they are isomorphic as representations of the derived algebra $\widehat{\mathfrak{sl}}'_\ell$. This is explained by the fact that there is a larger affine algebra $\widehat{\mathfrak{gl}}_\ell$ which acts on $\mathbf{F}^{(h)}$ and $\mathbf{F}^{(h)}$ is irreducible under the action of $\widehat{\mathfrak{gl}}_\ell$.

DEFINITION 3.29. Let $\mathfrak{sl}_\ell$ be the Lie algebra of $\ell \times \ell$ matrices with trace 0 and $\mathfrak{gl}_\ell$ the Lie algebra of all $\ell \times \ell$ matrices. We use the notation $X_{i,j}$ to denote the matrix with a 1 in position (i, j) and zeros everywhere else (this is to distinguish $X_{i,j} \in \mathfrak{gl}_\ell$ from $E_{m,n} \in \mathfrak{gl}_{\mathbb{Z}+\frac{1}{2}}$).

DEFINITION 3.30. The affine Lie algebra $\widehat{\mathfrak{sl}}_\ell$ is, as a vector space

$$(\mathfrak{sl}_\ell \otimes_{\mathbb{C}} \mathbb{C}[t, t^{-1}]) \oplus \mathbb{C}c \oplus \mathbb{C}d.$$

The Lie bracket is defined by

$$[X \otimes t^n, Y \otimes t^m] = (XY - YX) \otimes t^{m+n} + n\delta_{n,-m}\mathrm{tr}(XY)c, \tag{25}$$

$$[d, X \otimes t^n] = nX \otimes t^n, \tag{26}$$

$$c \text{ is central.} \tag{27}$$

Here $\mathrm{tr}(XY)$ is the trace of XY (which is also a renormalization of the Killing form).

DEFINITION 3.31. $\widehat{\mathfrak{sl}}'_\ell$ is the derived algebra of $\widehat{\mathfrak{sl}}_\ell$. As a vector space, $\widehat{\mathfrak{sl}}'_\ell = (\mathfrak{sl}_\ell \otimes \mathbb{C}[t, t^{-1}]) \oplus \mathbb{C}c$.

PROPOSITION 3.32. *(see* [**Kac**, Chapter 7.4 and Exercise 7.11]*)* $\widehat{\mathfrak{sl}}_\ell$ *is isomorphic to the affine Kac-Moody algebra with the* ℓ*-node Dynkin diagram*

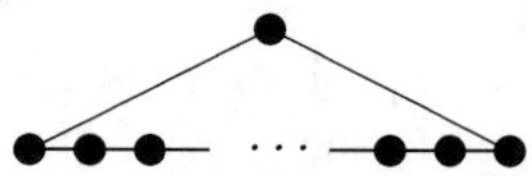

In particular, $\widehat{\mathfrak{sl}}'_\ell$ *is generated by Chevalley generators* $\tilde{E}_{\bar{i}}, \tilde{F}_{\bar{i}}$ *for* $i \in \mathbb{Z}/\ell\mathbb{Z}$*. Furthermore, the isomorphism can be chosen such that the following hold:*

$$\tilde{E}_{\bar{i}} \to X_{i,i+1} \otimes t^0 \quad i = 1, \dots \ell - 1 \tag{28}$$

$$\tilde{F}_{\bar{i}} \to X_{i+1,i} \otimes t^0 \quad i = 1, \dots \ell - 1 \tag{29}$$

$$\tilde{E}_{\bar{0}} \to X_{\ell,1} \otimes t^1 \tag{30}$$

$$\tilde{F}_{\bar{0}} \to X_{1,\ell} \otimes t^{-1}. \tag{31}$$

PROPOSITION 3.33. *(see* [**Kac**, Chapter 14]*)*

(i) *There is a Lie-associative embedding of* $\widehat{\mathfrak{sl}}'_\ell/(c-1)$ *into* $\mathfrak{a}_\infty$ *given by, for any* $X_{i,j} \in sl_\ell$ *and* $m \in \mathbb{Z}$*,*

$$X_{i,j} \otimes t^m \mapsto \sum_{k \in \mathbb{Z}} \overline{E}_{i-\frac{1}{2}+k\ell, j-\frac{1}{2}+k\ell+m\ell}.$$

Thus, using the embedding $\mathfrak{a}_\infty \subset \widetilde{Cl}$*,* $\mathbf{F}$ *carries an action of* $\widehat{\mathfrak{sl}}'_\ell$*.*

(ii) *Each* $\mathbf{F}^{(h)}$ *is preserved by the action of* $\widehat{\mathfrak{sl}}'_\ell$*.*

(iii) *Every irreducible component of* $\mathbf{F}^{(h)}$ *under this action is isomorphic to the highest weight representation* $V_{\Lambda_{\bar{h}}}$ *as a representation of* $\widehat{\mathfrak{sl}}'_\ell$*.*

PROOF. Part (i) can be proven by directly checking that the relations in Definition 3.30 hold. Part (ii) follows from the fact that the image of any E_{ij} in the map from a_∞ to $\widetilde{Cl}$ preserves each $F^{(h)}$. We delay the proof of part (iii) since it follows from the discussion of the $\widehat{\mathfrak{gl}}_\ell$ case below. □

The action of the Chevalley generators of $\widehat{\mathfrak{sl}}'_\ell$ on $\mathbf{F}$ can be described combinatorially:

PROPOSITION 3.34. *Let* (λ, m) *be a charged partition and fix* $\bar{i} \in \mathbb{Z}/\ell\mathbb{Z}$*. Then*

$$\tilde{E}_{\bar{i}}|\lambda, h\rangle = \sum_{\substack{\lambda \backslash \mu \text{ is an} \\ \bar{i} \text{ colored box}}} |\mu, h\rangle \quad \text{and} \quad \tilde{F}_{\bar{i}}|\lambda, h\rangle = \sum_{\substack{\mu \backslash \lambda \text{ is an} \\ \bar{i} \text{ colored box}}} |\mu, h\rangle. \tag{32}$$

Here the boxes are colored as in Figure 1.

PROOF. We will explain the case of $\tilde{E}_{\bar{i}}$, the case of $\tilde{F}_{\bar{i}}$ being similar. Combining formulas (28) and (30) with Proposition 3.33,

$$\tilde{E}_{\bar{i}}|\lambda, h\rangle = \sum_{k \in \mathbb{Z}} \overline{E}_{i-\frac{1}{2}+k\ell, i+\frac{1}{2}+k\ell}|\lambda, h\rangle. \tag{33}$$

Using Proposition 3.23, this is

$$\sum_{k \in \mathbb{Z}} \psi_{i-\frac{1}{2}+k\ell}\psi^*_{i+\frac{1}{2}+k\ell}|\lambda, h\rangle \tag{34}$$

Using the Maya diagram model of Fock space and the bijection in Figure 1,

$$\psi_{i-\frac{1}{2}+k\ell}\psi^*_{i+\frac{1}{2}+k\ell}$$

acts an $|\lambda, h\rangle$ by removing a box in position $i + k\ell$, if possible, and it sends $|\lambda, h\rangle$ to zero otherwise. So, using (34), $\tilde{E}_{\bar{i}}$ is the sum of all ways to remove $\bar{i}$ colored boxes, as stated. □

PROPOSITION 3.35. *The action of $\widehat{\mathfrak{sl}}'_\ell$ on* $\mathbf{F}$ *can be extended to an action of $\widehat{\mathfrak{sl}}_\ell$ on* $\mathbf{F}$ *by letting d act on $|\lambda, h\rangle$ as multiplication by*

$$\mathrm{Num}_0(|\lambda, h\rangle) := \#\ \textit{squares of}\ \lambda\ \textit{colored}\ c_0,$$

where the coloring is as in Figure 1.

PROOF. Using the expression for the Chevalley generators from Proposition 3.32, it suffices to prove that, as operators on $\mathbf{F}$, d commutes with $E_{\bar{i}}$, $F_{\bar{i}}$ for $\bar{i} \neq \bar{0}$, that $E_{\bar{0}}d = (d+1)E_{\bar{0}}$, and that $F_{\bar{0}}d = (d-1)F_{\bar{0}}$. The relations involving $E_{\bar{i}}, F_{\bar{i}}$ for $\bar{i} \neq \bar{0}$ hold because these operators do not affect the number of $c_{\bar{0}}$ colored boxed. The last relations can be seen from the fact that $E_{\bar{0}}$ removes a $\bar{0}$ colored box from λ and $F_{\bar{0}}$ adds such a box. □

The action of $\widehat{\mathfrak{sl}}_\ell$ on $\mathbf{F}^{(h)}$ can in fact be extended to a larger affine algebra $\widehat{\mathfrak{gl}}_\ell$.

DEFINITION 3.36. The affine Lie algebra $\widehat{\mathfrak{gl}}_\ell$ is, as a vector space

$$(\mathfrak{gl}_\ell \otimes_{\mathbb{C}} \mathbb{C}[t, t^{-1}]) \oplus \mathbb{C}c \oplus \mathbb{C}d.$$

The Lie bracket is defined by

$$[X \otimes t^n, Y \otimes t^m] = (XY - YX) \otimes t^{m+n} + n\delta_{n,-m}\mathrm{tr}(XY)c \tag{35}$$

$$[d, X \otimes t^n] = nX \otimes t^n \tag{36}$$

$$c \text{ is central.} \tag{37}$$

Let $\widehat{\mathfrak{gl}}'_\ell$ denote the Lie-subalgebra $(\mathfrak{gl}_\ell \otimes_{\mathbb{C}} \mathbb{C}[t, t^{-1}]) \oplus \mathbb{C}c$.

PROPOSITION 3.37.

(i) *There is an embedding of $\widehat{\mathfrak{gl}}'_\ell/(c-1)$ into a_∞ (and from there into $\widetilde{Cl}$) given by, for $X_{i,j} \in \widehat{\mathfrak{gl}}_\ell$,*

$$X_{i,j} \otimes t^m \mapsto \sum_{k \in \mathbb{Z}} \overline{E}_{i-\frac{1}{2}+k\ell, j-\frac{1}{2}+k\ell+m\ell}.$$

Thus $\mathbf{F}$ *carries an action of $\widehat{\mathfrak{gl}}'_\ell$ which agrees with the action from Proposition 3.33 on the subalgebra $\widehat{\mathfrak{sl}}'_\ell$.*

(ii) *The action can be extended to all of $\widehat{\mathfrak{gl}}_\ell$ by letting d act on $|\lambda, h\rangle$ as multiplication by the number of $c_{\bar{0}}$ colored boxes in λ.*

(iii) *Each $\mathbf{F}^{(h)}$ is preserved by this action, and is irreducible as a representation of $\widehat{\mathfrak{gl}}_\ell$.*

PROOF. Parts (i) and (ii) follow exactly as in the $\widehat{\mathfrak{sl}}_\ell$ case. Each $\mathbf{F}^{(h)}$ is preserved under the action of $\widehat{\mathfrak{gl}}_\ell$, so it remains to show that each $\mathbf{F}^{(h)}$ is irreducible. For this, we use the fact that $\mathbf{F}^{(h)}$ is irreducible as a representation of $\mathcal{W}$. Thus it suffices to show that for all k, $\pi_{\mathbf{F}}\alpha_k$ is in the algebra of operators generated by $\widehat{\mathfrak{gl}}_\ell$. For each residue $\bar{j}$ mod ℓ, let

$$C_{\bar{j}} = \sum_{a=1}^{\ell} \begin{cases} X_{a,a+j} \text{ if } a+j \leq \ell \\ tX_{a,a+j-\ell} \text{ if } a+j > \ell. \end{cases} \tag{38}$$

Here j is the representative of $\bar{j}$ in $\{0, 1, \dots \ell - 1\}$. One then directly checks that, for each $k \neq 0$, the actions of $C_{\bar{k}} \otimes t^{\lfloor k/\ell \rfloor}$ and α_k on $\mathbf{F}$ agree. □

EXAMPLE 3.38. If $\ell = 3$, the elements $C_{\bar{j}}$ as in the proof of Proposition 3.37 are

$$C_{\bar{0}} = \begin{pmatrix} 1 & 0 & 0 \\ 0 & 1 & 0 \\ 0 & 0 & 1 \end{pmatrix}, \quad C_{\bar{1}} = \begin{pmatrix} 0 & 1 & 0 \\ 0 & 0 & 1 \\ t & 0 & 0 \end{pmatrix}, \quad C_{\bar{2}} = \begin{pmatrix} 0 & 0 & 1 \\ t & 0 & 0 \\ 0 & t & 0 \end{pmatrix}. \tag{39}$$

As in the proof of Proposition 3.37, e.g. α_7 should correspond to $C_{\bar{1}} \otimes t^2$. By definition,

$$C_{\bar{1}} \otimes t^2 = \begin{pmatrix} 0 & t^2 & 0 \\ 0 & 0 & t^2 \\ t^3 & 0 & 0 \end{pmatrix} = X_{31}t^3 + X_{12}t^2 + X_{21}t^2. \tag{40}$$

Under the map defined in Proposition 3.37, this is sent to

$$\sum_{k \in \mathbb{Z}} \overline{E}_{3-\frac{1}{2}+3k, 1-\frac{1}{2}+3k+9} + \sum_{k \in \mathbb{Z}} \overline{E}_{1-\frac{1}{2}+3k, 2-\frac{1}{2}+3k+6} + \sum_{k \in \mathbb{Z}} \overline{E}_{2-\frac{1}{2}+3k, 3-\frac{1}{2}+3k+6}.$$

Combining terms and reparametrizing the sums, this is

$$\sum_{k' \in \mathbb{Z}} \overline{E}_{-\frac{1}{2}+k', -\frac{1}{2}+k'+7},$$

which by Proposition 3.23 acts on $\mathbf{F}$ as

$$\sum_{k' \in \mathbb{Z}} \psi_{-\frac{1}{2}+k'} \psi^*_{-\frac{1}{2}+k'+7}.$$

By Proposition 3.8 this is exactly the action of α_7.

One can directly check that the elements $I \otimes t^k \in \widehat{\mathfrak{gl}}'_\ell$ commute with all the generators of $\widehat{\mathfrak{sl}}'_\ell$. Furthermore, these elements for $k \neq 0$ generate a copy of $\mathbf{H}$, where the central element $c_{\mathbf{H}}$ in $\mathbf{H}$ is identified with ℓc. This is because, by direct calculation,

$$[I \otimes t^k, I \otimes t^{-k}] = k\ell c. \tag{41}$$

Together with I itself, these generate all of $\widehat{\mathfrak{gl}}'_\ell$, so $\widehat{\mathfrak{gl}}'_\ell = \widehat{\mathfrak{sl}}'_\ell \oplus \mathbf{H} \oplus \mathbb{C}\mathbf{I}/(\ell c_{\widehat{\mathfrak{sl}}_\ell} = c_{\mathbf{H}})$.

To complete the proof of Proposition 3.33(iii) use the fact that, by Proposition 3.37, each $\mathbf{F}^{(h)}$ is an irreducible representation of $\widehat{\mathfrak{gl}}'_\ell$, so is a tensor product $V \otimes W$ of an irreducible representation V of $\widehat{\mathfrak{sl}}'_\ell$ and an irreducible representation W of $\mathbf{H}$.

3F. The q-deformed Fock space $\mathbf{F}_q$ as a representation of $U_q(\widehat{\mathfrak{sl}}_\ell)$. We now describe the Misra-Miwa Fock space for $U_q(\widehat{\mathfrak{sl}}_\ell)$. This is a representation of $U_q(\widehat{\mathfrak{sl}}_\ell)$ originally developed by Misra and Miwa [**MM**] using work of Hayashi [**Hay**] (see also [**Ari**, Chapter 10]). It can be thought of as a q-deformation of the action of $\widehat{\mathfrak{sl}}_\ell$ on $\mathbf{F}$ described in §3E. See e.g. [**CP**] for the definition of $U_q(\widehat{\mathfrak{sl}}_\ell)$. Note that we use $\tilde{E}_{\bar{i}}$ and $\tilde{F}_{\bar{i}}$ to denote the Chevalley generators, which, in [**CP**], are denoted by X_i^+ and X_i^-.

COMMENT 3.39. Caution: q here is unrelated to the q in the definition of bosonic Fock space in §2E. It is however the normal notation so we will accept this conflict.

DEFINITION 3.40. Let $\mathbf{F}_q := \mathbf{F} \otimes_{\mathbb{C}} \mathbb{C}(q)$.

DEFINITION 3.41. Let (λ, m) and (μ, m) be charged partitions such that λ is contained in μ, and $\mu \backslash \lambda$ is a single box. Set

(i) $A_i|\lambda, h\rangle := \{c_i \text{ colored boxes } n|\ \lambda \cup n \text{ is a partition}\}$.
(ii) $R_i|\lambda, h\rangle := \{c_i \text{ colored boxes } n|\ \lambda \backslash n \text{ is a partition}\}$.
(iii) $N_i^a(\mu\backslash\lambda) := |\{n \in A_i(\lambda)| \text{n is to the left of } \mu\backslash\lambda\}| - |\{n \in R_i(\lambda)| \text{ n is to the left of } \mu\backslash\lambda\}|$.
(iv) $N_i^r(\mu\backslash\lambda) := |\{n \in A_i(\lambda)| \text{n is to the right of } \mu\backslash\lambda\}| - |\{n \in R_i(\lambda)| \text{ n is to the right of } \mu\backslash\lambda\}|$.

THEOREM 3.42 (See [**Ari**, Theorem 10.6]). *There is an action of $U_q(\widehat{\mathfrak{sl}}_\ell)$ on $\mathbf{F}_q$ defined by*

$$\tilde{E}_{\bar{i}}|\lambda, h\rangle := \sum_{\substack{\lambda\backslash\mu \text{ is an} \\ i \text{ colored box}}} q^{-N_i^r(\lambda\backslash\mu)}|\mu, h\rangle, \tag{42}$$

$$\tilde{F}_{\bar{i}}|\lambda, h\rangle := \sum_{\substack{\mu\backslash\lambda \text{ is an} \\ i \text{ colored box}}} q^{N_i^a(\mu\backslash\lambda)}|\mu, h\rangle.$$

By comparing (42) with (32), when q is set to 1 the $U_q(\widehat{\mathfrak{sl}}_\ell)$ action on $\mathbf{F}_q$ becomes the $\widehat{\mathfrak{sl}}_\ell$ action on $\mathbf{F}$ from Proposition 3.34. As in the undeformed case, each $F_q^{(h)}$ decomposes as a sum of infinitely many copies of the irreducible representation of highest weight $\Lambda_{\bar{h}}$.

Acknowledgments

These notes were one consequence of a long series of discussions Arun Ram, so I would like to thank Arun for being so generous with his time. I would also like to thank Antoine Labelle, Tony Giaquinto, Gus Schrader and A.J. Tolland for helpful comments. Finally, I would like to thank my master's advisor Yuly Billig for having me read [**Kac**] and all his help as I tried to understand it.

References

[Ari] Susumu Ariki, *Representations of quantum algebras and combinatorics of Young tableaux*, University Lecture Series, vol. 26, American Mathematical Society, Providence, RI, 2002. Translated from the 2000 Japanese edition and revised by the author, DOI 10.1090/ulect/026. MR1911030

[CP] Vyjayanthi Chari and Andrew Pressley, *A guide to quantum groups*, Cambridge University Press, Cambridge, 1995. Corrected reprint of the 1994 original. MR1358358

[DJM] T. Miwa, M. Jimbo, and E. Date, *Solitons*, Cambridge Tracts in Mathematics, vol. 135, Cambridge University Press, Cambridge, 2000. Differential equations, symmetries and infinite-dimensional algebras; Translated from the 1993 Japanese original by Miles Reid. MR1736222

[Hay] Takahiro Hayashi, *q-analogues of Clifford and Weyl algebras—spinor and oscillator representations of quantum enveloping algebras*, Comm. Math. Phys. **127** (1990), no. 1, 129–144. MR1036118

[Kac] Victor G. Kac, *Infinite-dimensional Lie algebras*, 3rd ed., Cambridge University Press, Cambridge, 1990, DOI 10.1017/CBO9780511626234. MR1104219

[KRR] Victor G. Kac, Ashok K. Raina, and Natasha Rozhkovskaya, *Bombay lectures on highest weight representations of infinite dimensional Lie algebras*, 2nd ed., Advanced Series in Mathematical Physics, vol. 29, World Scientific Publishing Co. Pte. Ltd., Hackensack, NJ, 2013, DOI 10.1142/8882. MR3185361

[LLT] Alain Lascoux, Bernard Leclerc, and Jean-Yves Thibon, *Ribbon tableaux, Hall-Littlewood functions, quantum affine algebras, and unipotent varieties*, J. Math. Phys. **38** (1997), no. 2, 1041–1068, DOI 10.1063/1.531807. MR1434225

[Mac] I. G. Macdonald, *Symmetric functions and Hall polynomials*, 2nd ed., Oxford Mathematical Monographs, The Clarendon Press, Oxford University Press, New York, 1995. With contributions by A. Zelevinsky; Oxford Science Publications. MR1354144

[MM] Kailash Misra and Tetsuji Miwa, *Crystal base for the basic representation of* $U_q(\mathfrak{sl}(n))$, Comm. Math. Phys. **134** (1990), no. 1, 79–88. MR1079801

[OR1] Andrei Okounkov and Nicolai Reshetikhin, *Random skew plane partitions and the Pearcey process*, Comm. Math. Phys. **269** (2007), no. 3, 571–609, DOI 10.1007/s00220-006-0128-8. MR2276355

[OR2] Andrei Okounkov and Nikolai Reshetikhin, *Correlation function of Schur process with application to local geometry of a random 3-dimensional Young diagram*, J. Amer. Math. Soc. **16** (2003), no. 3, 581–603, DOI 10.1090/S0894-0347-03-00425-9. MR1969205

[Sta] Richard P. Stanley, *Enumerative combinatorics. Vol. 2*, Cambridge Studies in Advanced Mathematics, vol. 62, Cambridge University Press, Cambridge, 1999. With a foreword by Gian-Carlo Rota and appendix 1 by Sergey Fomin, DOI 10.1017/CBO9780511609589. MR1676282

Department of Mathematics and Statistics, Loyola University, Chicago, Illinois 60660

Email address: ptingley@luc.edu

SELECTED PUBLISHED TITLES IN THIS SERIES

108 **Pramod N. Achar, Kailash C. Misra, and Daniel K. Nakano, Editors,** Categorical, Combinatorial and Geometric Representation Theory and Related Topics, 2024

107 **Ron Donagi, Adrian Langer, Piotr Sułkowski, and Katrin Wendland, Editors,** String-Math 2022, 2024

106 **Gilles Godefroy, Mohammad Sal Moslehian, and Juan Benigno Seoane-Sepúlveda, Editors,** Geometry of Banach Spaces and Related Fields, 2024

105 **A. Connes, C. Consani, B. I. Dundas, M. Khalkhali, and H. Moscovici, Editors,** Cyclic Cohomology at 40: Achievements and Future Prospects, 2023

104 **A. Kechris, N. Makarov, D. Ramakrishnan, and X. Zhu, Editors,** Nine Mathematical Challenges, 2021

103 **Sergey Novikov, Igor Krichever, Oleg Ogievetsky, and Senya Shlosman, Editors,** Integrability, Quantization, and Geometry, 2021

102 **David T. Gay and Weiwei Wu, Editors,** Breadth in Contemporary Topology, 2019

101 **Avraham Aizenbud, Dmitry Gourevitch, David Kazhdan, and Erez M. Lapid, Editors,** Representations of Reductive Groups, 2019

100 **Chiu-Chu Melissa Liu and Motohico Mulase, Editors,** Topological Recursion and its Influence in Analysis, Geometry, and Topology, 2018

99 **Vicente Muñoz, Ivan Smith, and Richard P. Thomas, Editors,** Modern Geometry, 2018

98 **Amir-Kian Kashani-Poor, Ruben Minasian, Nikita Nekrasov, and Boris Pioline, Editors,** String-Math 2016, 2018

97 **Tommaso de Fernex, Brendan Hassett, Mircea Mustaţă, Martin Olsson, Mihnea Popa, and Richard Thomas, Editors,** Algebraic Geometry: Salt Lake City 2015 (Parts 1 and 2), 2018

96 **Si Li, Bong H. Lian, Wei Song, and Shing-Tung Yau, Editors,** String-Math 2015, 2017

95 **Izzet Coskun, Tommaso de Fernex, and Angela Gibney, Editors,** Surveys on Recent Developments in Algebraic Geometry, 2017

94 **Mahir Bilen Can, Editor,** Algebraic Groups: Structure and Actions, 2017

93 **Vincent Bouchard, Charles Doran, Stefan Méndez-Diez, and Callum Quigley, Editors,** String-Math 2014, 2016

92 **Kailash C. Misra, Daniel K. Nakano, and Brian J. Parshall, Editors,** Lie Algebras, Lie Superalgebras, Vertex Algebras and Related Topics, 2016

91 **V. Sidoravicius and S. Smirnov, Editors,** Probability and Statistical Physics in St. Petersburg, 2016

90 **Ron Donagi, Sheldon Katz, Albrecht Klemm, and David R. Morrison, Editors,** String-Math 2012, 2015

89 **D. Dolgopyat, Y. Pesin, M. Pollicott, and L. Stoyanov, Editors,** Hyperbolic Dynamics, Fluctuations and Large Deviations, 2015

88 **Ron Donagi, Michael R. Douglas, Ljudmila Kamenova, and Martin Rocek, Editors,** String-Math 2013, 2014

87 **Helge Holden, Barry Simon, and Gerald Teschl, Editors,** Spectral Analysis, Differential Equations and Mathematical Physics: A Festschrift in Honor of Fritz Gesztesy's 60th Birthday, 2013

86 **Kailash C. Misra, Daniel K. Nakano, and Brian J. Parshall, Editors,** Recent Developments in Lie Algebras, Groups and Representation Theory, 2012

85 **Jonathan Block, Jacques Distler, Ron Donagi, and Eric Sharpe, Editors,** String-Math 2011, 2012

For a complete list of titles in this series, visit the
AMS Bookstore at **www.ams.org/bookstore/pspumseries/**.